Voices from the Forest

Integrating Indigenous Knowledge into Sustainable Upland Farming

Malcolm Cairns
Editor

RESOURCES FOR THE FUTURE
New York • London

An RFF Press book
Published by Resources for the Future
711 Third Avenue, New York, NY 10017, USA
2 Park Square, Milton Park, Abingdon, Oxon, OX14 4RN

Library of Congress Cataloging-in-Publication Data

Voices from the forest: integrating indigenous knowledge into sustainable upland farming / Malcolm Cairns, editor.
p. cm.
"An RFF Press book"—T.p. verso.
Includes bibliographical references and indexes.
ISBN 978-1-891853-91-3 (cloth : alk. paper) — ISBN 978-1-891853-92-0 (pbk. : alk. paper)
1. Traditional farming. 2. Shifting cultivation. 3. Sustainable agriculture. 4. Fallow lands. 5. Indigenous peoples—Ecology. 6. Forest degradation. 7. Forest conservation. I. Cairns, Malcolm.
GN407.4.V65 2004 306.3'49--dc22

ISBN 978-1-891853-91-3 (cloth) ISBN 978-1-891853-92-0 (paper)

About *Resources for the Future* and *RFF Press*

Resources for the Future (RFF) improves environmental and natural resource policymaking worldwide through independent social science research of the highest caliber. Founded in 1952, RFF pioneered the application of economics as a tool for developing more effective policy about the use and conservation of natural resources. Its scholars continue to employ social science methods to analyze critical issues concerning pollution control, energy policy, land and water use, hazardous waste, climate change, biodiversity, and the environmental challenges of developing countries.

RFF Press supports the mission of RFF by publishing book-length works that present a broad range of approaches to the study of natural resources and the environment. Its authors and editors include RFF staff, researchers from the larger academic and policy communities, and journalists. Audiences for publications by RFF Press include all of the participants in the policymaking process—scholars, the media, advocacy groups, NGOs, professionals in business and government, and the public.

To my parents, William and Helen Cairns of Prince Edward Island, Canada, who taught me the honor and integrity of farmers.

Contents

PART IX: Themes: Property Rights, Markets and Institutions

PART X: Conclusions

Foreword

Shifting cultivation, swidden, or slash-and-burn agriculture, has a bad reputation. It is frequently viewed as a major contributor to deforestation, land degradation, and recently, to widespread smog in Southeast Asia. This reputation is largely undeserved, for the majority of traditional swidden systems are sustainable and feature a high labor productivity at low population densities. However, there are enough cases to the contrary to keep the negative image alive. These usually arise from destabilization of previously sustainable systems as a result of such factors as rapidly increasing population pressure, the encroachment of commercial logging, forced migrations, changing production incentives as a result of market incorporation, or other significant changes in the institutional and policy environments within which swidden farmers work.

These cultivation systems refer to a multiplicity of different fallow and rotational arrangements, associated with a tremendous cultural diversity. It is not surprising, therefore, that the responses to these pressures and opportunities have also been highly variable and on occasions, quite ingenious. For example, it has become widely recognized that several highly productive and sustainable agroforestry systems have their origins in local responses to the need to reduce fallow cycles.

In discussing the intensification of these systems with other researchers, it became apparent that there were many such successful systems of indigenous intensification, and the realization that they had never been systematically reviewed provided the stimulus for *Voices from the Forest*. It was felt that a description and analysis of the multitude of these indigenous strategies would provide useful insights and directions for researchers and development practitioners alike, working on either avoiding or repairing the environmental, social, and economic problems resulting from the destabilization of shifting cultivation.

The book itself illustrates the enormous diversity of shifting cultivation systems and provides a striking testimony to human ingenuity. It sets out six different fallow management typologies and presents case studies of each. The chapters show the richness of farmer experimentation and adaptation, and the frequency of complex or multiple systems within the same agroecosystem.

In order to progress beyond the description of cases, authors have structured their discussions around a number of key questions, in order to further the analysis and draw out the lessons learned.

- What are the key factors that lead to successful indigenous fallow management systems and how can these be transferred to other areas where collapsing swidden systems are endemic?
- What are the elements of a strategic agenda for continued research and promotion of the most promising IFM technologies?
- What are the components needed to make it happen?

The International Development Research Centre (IDRC), as a significant partner in promoting this type of research, has a strong interest in the findings of this book. The work fits squarely within a long-standing agenda dedicated to supporting and disseminating research on farmer innovation, first with a farming systems paradigm, and more recently within a community-based natural resource management (CBNRM) approach. The shift is significant in that CBNRM places greater emphasis on the institutional and contextual factors within which farmer innovation takes place. IDRC's goal is to find ways in which the best of science and farmer experimentation can most successfully be brought together.

Voices from the Forest captures well the problem solving, multi-disciplinary approach which is essential for working with farmers to jointly solve their complex and multi-faceted resource management problems. It is significant, above all, for seeing farmers as innovators and hence as partners in the research process.

Joachim Voss

Joachim Voss, Director General, Centro Internacional de Agricultura Tropical (CIAT), Apartado Aéreo 6713, Cali, Colombia.

Preface

Shifting cultivation is probably as old as agriculture itself, practiced by our ancestors some 10 to 12 thousand years ago when they took their first tentative steps toward intentional husbandry of useful plants. Because it has proven to be a very successful adaptation to the difficult environmental conditions in the tropics, it has continued in widespread use ever since and today is the main source of food for many millions of farmers in Asia, Africa, and Latin America. Yet, despite this success, it is almost always viewed by governments as primitive, inefficient, and a leading cause of deforestation. Practitioners are commonly regarded as ignorant and incapable of adopting better farming methods. The corollary of this rationale is that shifting cultivation is "unscientific," static, and unchanging.

The very terms applied to the practice are symptomatic of its bad reputation. "Slash-and-burn" focuses on the fire phase—only one technique of many used in rotational forest fallows. "Shifting cultivation" emphasizes the movement of cropped fields, while "swidden" refers rather vaguely to a burnt clearing. All fail to see it as a system, let alone recognize its values. *Rotational shifting cultivation* is essentially a system that applies natural vegetative processes as a means of replenishing soil fertility in lieu of chemical fertilizers or alternatively, the very intensive techniques needed to manipulate organic matter in chemical-free systems. Both these latter alternatives are knowledge- and energy-intensive and human-driven—as opposed to the natural processes harnessed by rotational fallowing. Nature is effectively cut out of the game. "Rotational" is an important qualifier to keep in mind when describing shifting cultivation, in that rotational distinguishes such a form of cultivation from the more destructive "pioneering" type that continues cropping until an area is completely exhausted.

The official imperative is to replace shifting cultivation with "scientific" farming methods, but research and development efforts have a disappointing track record in providing alternatives appropriate to the conditions under which shifting cultivators work. Experience has shown that many of the "scientific" agricultural solutions imposed from outside can actually be far more damaging to the environment. Moreover, these external solutions often fail to recognize the extent to which an agricultural system supports a way of life along with a society's food needs.

Voices from the Forest aspires to fundamentally change the way that shifting cultivation is viewed—but it would be a mistake to regard it as offering any "magic bullets" to age-old problems. Rather, it provides a rich menu of farmer-tested innovations that we believe need to be shared with the wider community of shifting cultivators still searching for ways to cope with rising land-use pressures and market economies. In doing so, it resolutely shatters any illusions that shifting cultivators are static or incapable of change. They are clearly adapting, but the magnitude and rate of change are stressing that adaptation process. And there is little or no research to help them. Knowledge needs to be accumulated more rapidly because of the acceleration of change—so that adaptation can keep pace. This book tries to push forward that agenda. Taken collectively, its array of case studies represents an impressive body of work that shows that these fallow management systems do not only occur in one place, but that variations can be found across the Asia-Pacific region—independently discovered by farmers. They are not "curiosities" or "oddities," but their underlying concepts obviously have wide applicability—and therefore deserve scientific attention. The accompanying photos provide a visual sense of that variety.

This volume brings together the best of science and farmer experimentation and vividly illustrates the power of human ingenuity. It is probably unprecedented in scope, with more than 100 scholars from 22 countries—including agronomists, agricultural economists, ecologists, and anthropologists—collaborating in the analyses of different fallow management technologies. This dedicated community of researchers has, in turn, worked closely with a cast of thousands of indigenous farmers of different cultures in a broad range of climate, crops, and soil conditions. By sharing this knowledge—and combining it with new scientific and technical advances—the authors hope to make indigenous practices and experience more widely accessible and better understood, not only by researchers and development practitioners, but by other communities of farmers around the world.

It is important to acknowledge here that much of this book's contents represent the indigenous knowledge, and therefore the intellectual property, of Asia-Pacific's shifting cultivators. It should be used for their benefit. They are intended as the ultimate clients of this work.

Much remains to be done, particularly in assessing the research agenda and policy environment as it impacts on shifting cultivation. Scientific interest in fallow management is relatively recent, and therefore some of the case studies are unavoidably preliminary and descriptive in nature. As further in-depth research continues to focus on "best bet" systems, a sequel to this book will be needed in a few years to re-examine what we have learned about fallow management in light of this next generation of research findings.

This book grew from a regional workshop entitled "Indigenous Strategies for Intensification of Shifting Cultivation in Southeast Asia," organized by The World Agroforestry Centre (ICRAF) in Bogor, Indonesia, in 1997. This work on indigenous fallow management would not have been possible without the generous support of the International Development Research Centre (IDRC), through its Community-Based Natural Resource Management (CBNRM) program. The entire research thrust was in fact the brainchild of Joachim Voss, then Research Manager at IDRC Ottawa and now Director General of the Centro Internacional de Agricultura Tropical (CIAT). John Graham, then of IDRC's Singapore office, has been a valuable colleague and mentor throughout this work. We also extend our sincere thanks to other sponsoring agencies that joined

IDRC in supporting the original Bogor workshop and contributed to making it a success. Among other things, their support enabled national partners from a number of Asian–Pacific countries to attend the workshop. The valuable contributions of the following organizations are thus gratefully acknowledged:

- ASEAN-Canada Fund;
- Australian Centre for International Agricultural Research (ACIAR);
- Cornell International Institute for Food, Agriculture, and Development (CIIFAD);
- The Ford Foundation, Vietnam, P.R. China, and Thailand Programs (FF); and
- The Alternatives to Slash-and-Burn (ASB) Project.

As the book moved closer to press, both IDRC and ACIAR provided important additional grants to help support publication.

Special thanks are owed to Terry Rambo, then of Kyoto University, who took time from his own hectic schedule to review the manuscript and tie everything neatly together in a concluding chapter. Mike Bourke of the Australian National University kindly provided valuable feedback on an earlier draft of the manuscript, and J.F. Maxwell of Chiang Mai University reviewed the botanical index. The charcoal drawings appearing in the opening page of each section reflect the considerable talents of Paradorn Threemake. Most other artwork in the volume was completely redone through the capable efforts of Jenny Sheehan (Cartographic Services, RSPAS, ANU), Kittima (Jaisini) Nungern and Tossaporn Kurupunya. Bob Hill is owed particular gratitude for all his meticulous work in copyediting and formatting the manuscript. A team of outside readers—consisting of Lynley Capon, Dan Powell, David Freyer, and John Cairns helped immensely by reviewing the book one final time in search of any remaining editorial problems. Perhaps none will be happier to see this book completed than Don Reisman and his staff at RFF Press, who have worked with us tirelessly in shepherding it toward publication. We thank them for their patience and professionalism. But the biggest debt is owed to Tossaporn Kurupunya, whose quiet encouragement gave me the stamina to persevere and bring this book project to completion.

Malcolm Cairns
Department of Anthropology
Australian National University

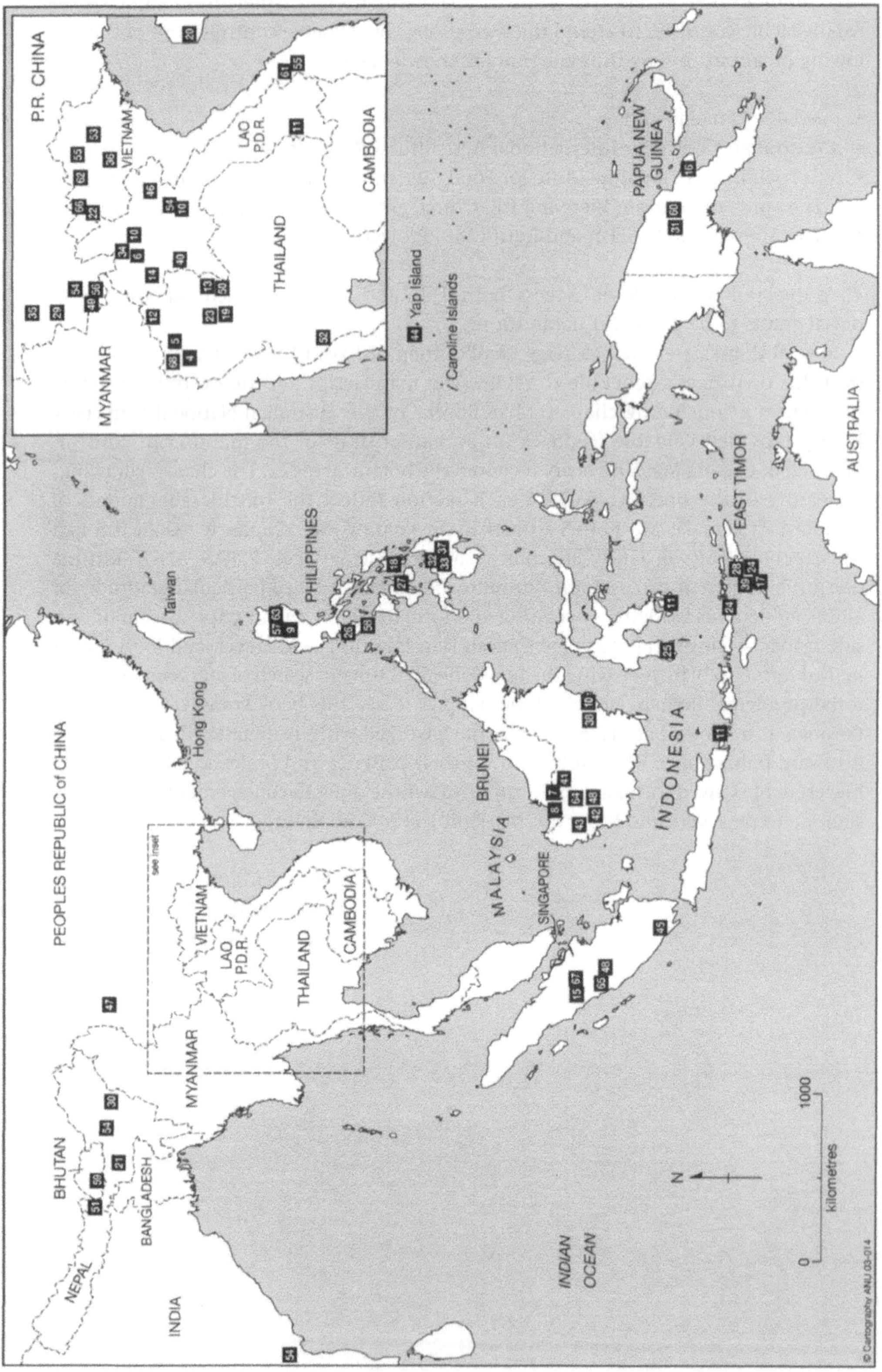

Asia-Pacific Map with References to Case Studies (see Table of Contents)

PART I

Introduction

A Matigsalug villager in Mindanao, the Philippines.

Chapter 1

Challenges for Research and Development on Improving Shifting Cultivation Systems

Dennis P. Garrity

The many contributors to this volume seek to explore and interpret, perhaps in a more determined manner than ever before, the most successful strategies developed by farmers and communities throughout Asia and the Pacific to sustain threatened shifting cultivation systems. From that exploration, we aim to develop a set of methods to put that knowledge to work for the benefit of the many communities facing a crisis of livelihood and declining natural resource quality.

In a region of dynamic economic growth, it is inequitable that upland farming people are not benefiting adequately. Those who make their living through shifting cultivation are resourceful and hard working, and they usually husband their resources carefully. But due to circumstances beyond their control, their efforts are being rewarded with a declining resource base that traps them in poverty. In many cases, they are becoming poorer. Outsiders usually misunderstand them and their farming systems, particularly the very people in their own governments who are charged with helping them to overcome their livelihood challenges with dignity. Consequently, solutions proposed and imposed from the outside usually do not address their needs constructively, and often make matters worse.

It is time to examine the ways of the more successful upland farmers and to understand the systems they have developed to enable them to farm sustainably and intensify their fallow management systems (Sanchez 1999). Many ingenious practices exist, developed through closeness to the land and an awareness of local ecology. What if these practices, these solutions, were really understood, refined for wider dissemination, combined with new scientific knowledge and technical advances, and spread to a vastly larger population of farmers and communities? And how much good might come for the upland peoples, through this sharing of their knowledge?

This volume reviews, classifies, and characterizes outstanding solutions—developed by indigenous upland people—to the challenges of improving fallow management. Our objective must be to collaborate and maintain networks to ensure that, when all is said and done, we will see the impact of these practices upon a much broader population of farm families and communities in Asia.

My own institution, the World Agroforestry Centre (ICRAF), is deeply committed to this mission. Along with many countries and organizations, it is a partner in the Alternatives to Slash-and-Burn (ASB) program. We are trying to identify and highlight best bet alternative land-use systems in tropical countries around the world and to pinpoint the policies that have enabled them to flourish (Buresh and Cooper 1999). We are part of a concerted scientific effort to develop more sustainable and intensive systems of fallow rotation, or shifting cultivation.

Dennis P. Garrity, Director General, World Agroforestry Centre (ICRAF), United Nations Avenue, P.O. Box 30677, Nairobi, Kenya.

Researching and Developing Indigenous Fallow Systems

My concern for volumes like this is that a hundred points of light will shine, but, in the end, they will have no focus. I fear that we could conclude with an outstanding collection of studies that are merely anthropological curiosities. You will note, however, that the vast majority of the exemplary systems documented herein are practiced only on a very limited area. This could mean that they have no extrapolation potential—that there are constraints to their spread that may be subtle and unrecognized. But it could also mean that modest efforts to extend these techniques to wider areas might have great impact. We must not be satisfied to simply report on them and leave them as curiosities. We must ensure that knowledge of the really useful practices is shared widely.

There are many aspects to consider when conducting research on indigenous fallow innovations with the aim of benefiting a wider population of upland people. Special care must be taken to develop research methods specifically for the unique conditions of shifting cultivation. Let me outline some of these issues, with special attention to some of the pitfalls into which we may fall.

The research and development process for improved fallow management systems is a continuum of tasks. These stages are illustrated in Figure 1-1. The process begins with the identification of a promising system or practice. Limited observation indicates that the system has elements that may be of real value elsewhere and the returns to investment in research look positive. This should lead to a characterization of the system and a more thorough description and analysis based on rapid or participatory appraisal methods, perhaps complemented by more in-depth surveys. An indicative analysis of the pros and cons of the system and the nature of its contribution to sustainability should follow.

If, at this point, the system still appears to have development and extrapolation potential, it is time to validate this assumption by sampling soils and fallow vegetation more thoroughly and studying crop performance. This might be done by comparing fields in which the practice is employed with fields in which it is not. However, it may be difficult to achieve valid comparisons using this approach because of site factors that may confound the interpretation. It will often be necessary to conduct new field trials, particularly to test additional management variations. If the innovation still shows promise of wider application, then a dissemination process is next.

To extrapolate the innovation to other communities, one must select new locations where the agroecological and social factors are not too dissimilar. The degree to which the innovation's success may be affected by specific biophysical conditions such as soils, rainfall, and elevation, as well as culture and land tenure, must be kept in mind. When new locations are selected, it is tempting to barge ahead with an extension program in the hope of seeing rapid impact. But it is best to verify the practice with a few key farmers before embarking on wholesale dissemination. This provides the chance to adapt the innovation to the realities of its new environment and possibly avoid a serious failure. As the success of the key farmers becomes evident, it is time to develop an effective extension program that expands adoption widely. The key farmers become the foundation for the diffusion process. Let us now examine some of these issues in more depth.

Phase	Step	Activities
Research Phase	1. Identify promising systems.	Observe, interactions, systems, or practices with communities.
Research Phase	2. Characterize the systems.	Describe and analyze. Estimate sustainability.
Research Phase	3. Validate the systems' utility.	Take samples of soils, fallow vegetation, crops. Conduct field trials. Describe sustainability in more detail.
Research Phase / Development Phase	4. Extrapolate to other communities.	Characterize the agroecological conditions. Compare with other locations. Define recommendations, domains. Select new locations for application.
Development Phase	5. Verify utility in new area.	Assess local conditions carefully. Assess indicators of success or failure.
Development Phase	6. Extend widely in new areas.	Develop appropriate extension methods. Evolve effective extension message. Expand from base of key farmer adapters/adopters.

Figure 1-1. The Process of Research and Development of Indigenous Fallow Management Systems
Source: Cairns and Garrity 1999.

The Critical Assessment

Assessing the utility of an improved fallow innovation is a complex business. There are many snags that researchers and extension workers may find along the way:

- Analyses may show that the innovation has better returns per hectare but ignore the labor requirements. Shifting cultivation is, by definition, a system where labor is a dominant constraint. Increasing the returns to labor is usually much more important than merely increasing the yield. Thus, realistically estimating labor requirements and comparing them against other options is crucial.
- There may be a failure to examine the benefits and costs over the entire fallow rotation cycle. The effects of an innovation must be considered over the whole cycle, and not just for one or two crop seasons. Actual observations should continue even in cases where the entire rotation may extend for several years or even decades. Even if the benefits and costs have to be estimated hypothetically, it is critical to consider the entire cycle.
- Invalid or inconclusive sampling of soils and crop performance can be deceptive. It is difficult to detect unambiguous improvements in soil fertility during a period of a few years' fallow. Many soil scientists believe that conventional soil analyses are simply not precise enough or do not measure the really important parameters. Also, the results of samples analyzed in successive years may be badly confounded by variations caused by changes at the laboratory itself. This can muffle or negate the modest changes expected in the bulk soil properties. The fertility benefit of an innovation may derive from the litter, rather than from changes in the actual fertility of the soil during the fallow period. Alternatively, the nutrients accumulated in the biomass of the fallow vegetation may provide the dominant effect.

- Attempts to compare the performance of an innovation by sampling fields where it is practiced, and comparing the results with nearby fields where it is not practiced, are often fraught with problems. Soils, slopes, cropping history, and many other farm-to-farm management differences confound such comparisons and may easily overwhelm the effects of the innovation. Or they may falsely suggest that the innovation is better than it is. Comparisons based on such sampling methods must be designed very carefully. Even in the best of cases, such results are only indicative and are tricky to interpret. This is why it is often necessary to install new trials that are designed specifically to make valid comparisons. The simplest approach is to conduct paired-plot trials. These compare the innovation with the conventional fallow system side by side on either half of a field. Replication is done across a number of farms.

The above advice directs attention to the serious challenges of conducting research in indigenous innovations. Collecting valid results is the fundamental first step. Subjecting them to a critical analysis that takes into account the shifting cultivator's decision framework is the second step. This sustainability analysis itself must be validated and enriched with local opinions.

At this point, let us assume that we have solid evidence that our innovation is widely useful and deserves to be disseminated widely.

Facing the Constraints to Extension

There are perhaps four major constraints in conducting extension among shifting communities:

- They are usually remote from roads and market infrastructure. This means that they are constrained in participating in the market economy and may be limited in their livelihood options. It also means that extension agencies have little presence in these areas.
- There may be problems with extension agency jurisdiction. Shifting cultivation communities often live on land that is classified as forest and claimed by the state. Agricultural extension agencies are often not permitted to work with farmers on state forest land. As well, these agencies are usually understaffed.
- Land tenure uncertainty plays an important role in household land-use decisions. There is often a conflict between the claims of the state and the realities of the local land tenure system. These may be exacerbated by land conflicts within the community. Adoption of fallow management innovations will be very sensitive to these realities.
- Land use in shifting cultivation communities is often transitional. Land-use intensification is an almost universal, historical process. It typically proceeds from long-cycle fallows to continuous annual farming. Any particular improved fallow management system may be relevant to a farm or community at only one time period in this evolutionary process, but not at others. Successfully introducing innovations in fallow management is, therefore, shooting at a moving target.

Conclusions

There are many unique challenges for research and development on improved shifting cultivation systems. If this volume provides a springboard for a vigorous new international initiative on indigenous strategies for their improvement, then we can take pride that this work has truly come of age and will make a real difference in the lives of upland communities.

References

Buresh, R.J., and P.J.M. Cooper. 1999. The Science and Practice of Short-Term Improved Fallows. Selected papers from an International Symposium, Lilongwe, 1997. *Agroforestry Systems* 47, 1–356.

Cairns, M.F., and D.P. Garrity. 1999. Improving Shifting Cultivation in Southeast Asia by Building on Indigenous Fallow Management Strategies. In: The Science and Practice of Short-Term Improved Fallows. Selected papers from an International Symposium, Lilongwe, 1997, edited by R.J. Buresh and P.J.M. Cooper. *Agroforestry Systems* 47.

Sanchez, P.A. 1999. Improved Fallows Come of Age in the Tropics. In: The Science and Practice of Short-Term Improved Fallows. Selected papers from an International Symposium, Lilongwe, 1997, edited by R.J. Buresh and P.J.M. Cooper. *Agroforestry Systems* 47(1/3), 3–12.

Chapter 2

Working with and for Plants

Indigenous Fallow Management in Perspective

Harold Brookfield

Chapter 3, which follows this one, elaborates on the background and purposes of this volume. It also shows as major and immediate issues the critical situation in the uplands of the Asia-Pacific region and the pressures on shifting cultivators and their land. This chapter therefore sets out to examine broader historical aspects and, in doing so, briefly alludes to what we know of fallow management in some other tropical regions, from which lessons can be learned for Asia-Pacific. I draw only lightly on the United Nations University international comparative project on People, Land Management, and Ecosystem Conservation (PLEC), of which I was scientific coordinator from 1993 to 2002. Some of the arguments of this chapter are illustrated in more detail in Brookfield (2001), a book that drew on the same symposium as that which spawned this volume, and in the two books that came out of the PLEC project (Brookfield et al. 2002; Brookfield et al. 2003).

While writing about indigenous fallow management is a recent phenomenon, the changes reported in this volume did not spring up yesterday, under modern economic, demographic, and political pressures on swidden or shifting cultivators. Indigenous fallow management is not new. The major pre-industrial agricultural revolution in Britain and the Low Countries of Europe in the 16th to 18th centuries was largely about improvement and elimination of the fallow. Farmers, no less "indigenous" in their time than those of modern Asia-Pacific, first innovated by themselves. Then the agronomists followed, reported, and proposed improvements.

In the Asia-Pacific region, there is little written evidence of pre-modern practices. The general assumption is that fallow management is relatively new and a response to very modern forces. Nowadays, there is intense pressure on swidden farmers to eliminate their "wasteful" fallow-based systems and adopt "permanent" cultivation systems in their place. Many believe swidden farmers should be forcibly resettled, though there is now less of this thinking than a few years ago. For this reason, it is important to show that the use and improvement of what is called the "fallow" have long histories. The historical ecology of fallow management is not sufficiently studied in the Asia-Pacific region. Only one or two cases have been examined in any detail.

The practice of fallow management must go back a long way into the history of upland, or dryland, cultivation in this region, as in others. In a high proportion of

Harold Brookfield, Department of Anthropology, Division of Society and Environment, Research School of Pacific and Asian Studies (RSPAS), H.C. Coombs Building, Australian National University, Canberra, ACT 2000, Australia.

contemporary systems described in the literature, when forest is cleared for swidden cultivation some useful trees and other plants are preserved. Others are planted. They are therefore selectively advantaged in the subsequent succession, and rights are maintained to valued trees planted or first found generations earlier (Sather 1990, Peluso 1996). There is no reason to suppose this is new. From Kalimantan, Peters (1996) reports that old and mature *Shorea* species, long ago planted for illipe nuts around villages and in swidden fallows, form dense groves within what later generations often perceive as undisturbed forest. Durian (*Durio zibethinus*) is found almost throughout Borneo, but it may be an introduction of unknown age from mainland Asia-Pacific. Everywhere it occurs, it represents the site of a former settlement (Peluso 1996). In areas such as interior Borneo, where populations suffered severe loss in the late 19th century, forest may appear to be primary, but actually much is secondary (Brookfield et al. 1995, *28–29*). Johns (1990) has similarly remarked on the continuous dynamism of the tropical forests and their species composition. And as Spencer (1966, *39*) put it nearly 40 years ago, "it is possible that old forests are not secondary forests or even tertiary forests, but forests of some number well above three."

Certain species not only thrive in sites disturbed for agriculture, but also survive well in the regrowth forest. These include bamboo, conserved for its utility, though at times and in places becoming an arrested succession (Ramakrishnan 1992). It is possible that the wide range of fruit trees in the forests of Asia-Pacific exists because the trees have been either conserved or planted by forest people over thousands of years. The forests have a history in which people have played a major role. Tree-sized palms flourish in disturbed environments and, because of their height, can retain canopy status in the secondary forest. Sugar palms and coconut palms are regional examples, but the best studied examples are outside Asia-Pacific.

The Uses of the Fallow

The fallow, plus its nature and uses, needs to be closely examined. It is much more than just the resting and recovery period between clearance and cropping episodes. Almost all shifting cultivation systems include foraging elements, and some of the forest products that have entered trade for centuries have come from the fallow. People continue to use wild sources for food, wood, fibers, medicines, and some cash-earning products, finding them both in the more remote wild and in the fallow. In some shifting cultivation systems in this region, wild sources, including fish, provide more than half the diet (Chin 1985). Apart from the fact that some longer-lived crops are still used while new growth is coming up around them, the fallow contains a great deal that is of value to the farmers. Some products obtained from the fallow occur only at a particular stage in the succession (Lian 1987, Colfer and Soedjito 1996). When clearing a new garden site, most farmers will collect everything useful from the forest before cutting, stacking, and burning. Nowadays, some tree cash crops are planted in the fallow. In many parts of Asia-Pacific, rubber is the outstanding example. A Dayak rubber grove in Sarawak, with rubber trees scattered among other secondary growth, bears little resemblance to the carefully lined groves of estates or villages on the Malay Peninsula.

Most of the species used in Asia-Pacific fallows grow spontaneously. But some are planted during the farming phase specifically to yield during the fallow period. In addition to rubber, these include rattan as well as fruit-bearing trees and trees grown for medicinal purposes or for their wood. Only a few of these have an agronomic role in soil enrichment. The number of wild and fallow species described as being used by villagers ranges from a few score to several hundred (Conklin 1957, *125–26*; Kunstadter 1978, *99*). The amount of specific management varies, and is only spottily described in the Asia-Pacific region. Conklin (1957) described in detail the manner in which Hanunóo farmers on Mindoro progressively turn their rice swiddens into root crop swiddens and then into tree crop gardens, which may endure for many years, or become permanent. This is comparable with Amazonian fallows that are managed by slash weeding for a decade or more and then continue yielding useful products for a generation (Denevan and Padoch 1987 a, b; de Jong 1996). The evidence from Borneo led Peluso and Padoch (1996, *133*) to suggest, that "swidden-fallows and other woodlands are far more explicitly managed than is commonly understood." The life of the farm does not end with the harvesting of field crops. Nor does the farm end where the field meets the wood. The "idle" land has many uses.

It needs also to be stressed that fallow improvement is only one element in a set of changes to the plant environment that includes management of the cultivated soil. Other modifications arise from crop, weed-plant, and animal introductions; soil erosion; species extinctions and partial extinctions; wildfire; and both cyclic and secular climatic change. I want briefly to focus attention on deliberate change in management of the cultivated soil, which has consequences for fallow. But first we must discuss two of the surrounding terms and concepts used in this volume: "shifting" or "swidden cultivation," and "intensification."

Shifting or Swidden Cultivation

"Shifting cultivation" is a term used to generalize about a huge range of farming systems. Common to them all is the use of a fallow period of significant length. From the pioneer work of Conklin (1957) to the general text of Ruthenberg (1980), attempts have been made to distinguish shifting cultivation from other systems by the length of the fallow in relation to the period under cultivation. These have enjoyed only limited adoption. Within systems that are commonly described as shifting cultivation, appropriately or otherwise, the fallow period may range from a few months to longer than a human generation. It may be much longer than the cultivation period, about the same length, or shorter. There can be few less satisfactory terms for a large and diverse set of farming systems than shifting cultivation. But the undiscriminating manner in which the term has gained everyday use has led to serious political consequences (Brookfield et al. 1995, *132–140*).

A second distinguishing characteristic of shifting cultivation is often said to be the use of fire. This, however, ranges from burning for total clearance to patch burning and to the complete absence of fire in some fallow-based systems. Hence the derogatory term "slash-and-burn" obscures even more. Though the term "swidden" has the historical meaning of a burned clearing, this is often forgotten, so that "swidden cultivation" is perhaps a more neutral descriptor. So I use it in this chapter. There are other complications. Many agricultural systems, in the Asia-Pacific region especially, are mixed rather than simple or "integral" in the sense proposed by Conklin (1957). They include permanent field—as well as shifting field—arable areas. A considerable number of swidden systems incorporate a range of agrotechnical practices for management of the land, its biota, and water. These include methods of slope management, drainage, and irrigation and a variety of ways of managing soil fertility in the cropping and fallow periods.

Then there is the dynamic aspect. Some systems seem to exhibit little change over long periods. Others are so visibly dynamic that they undergo significant change in only a few years. Between these are many in which farmers proceed more slowly, in an incremental manner, cultivating while progressively investing in new production and management methods over years, decades, or even generations. Some of the investments may perish; others persist. Viewed at a single point in time, such farming systems may appear static. Yet farmers are making new types of fields, digging ditches, planting trees, experimenting with new crops, interplanting systems and rotations, and progressively transforming their production procedures. Farmers everywhere are experimenters, trying out new planting materials, listening to new ideas, responding to price changes and new opportunities, observing the successes or failures of their neighbors, and trying out ideas of their own. This volume distinctively focuses attention on farmers' own experiments and their use of inherited information, acquired information, observation, and experience.

Any one change in a farming or community system requires others. The adoption of a new cash crop may lead to changes in the selection of land for other crops, to changes in the genderization of work and reward, and even to a wholly new distribution of inputs over the face of a farm or village. Improvement or deterioration in the conditions of access to land, relations with neighbors, and personal security all have consequences for settlement patterns and working practices. Threatened resettlement or eviction, seizure of land designated as forest for timber extraction or transmigration settlement, or declaration of land as a nature reserve can have drastic consequences for farming conditions. Decline in the importance of village authorities, patron–client relationships, or the authority of older males in a family may lead to radical changes in how people organize their farms. Access to money can lead to the breakup of communal working arrangements and their replacement by single household operations, leading to other management changes.

The modern acceleration of population growth and of social, economic, and political change everywhere in the Asia-Pacific region has brought many new farmers into the uplands, often with damaging consequences. It has also ensured that long-established upland farming systems in this region have become much more dynamic. But in some areas land degradation has forced its own set of changes in order that agriculture can survive. Most of the farming systems we see today have been radically changed within the past century. Where relative stasis persists, it is often in areas where emigration, tenancy, or sharecropping has reduced the pressure to innovate. But it would be wrong to assume that all change is due to modern forces. Experimentation in agriculture—and in fallow management—is as old as agriculture itself. Even if its pace differs among regions and has varied through time, there has always been the potential for new adoptions and innovations. Today's farmed landscapes and forests have a long human history during which they have undergone many changes.

Intensification and Improvement

We are looking at indigenous fallow management as a strategy for the intensification that everyone demands. But "intensification," also, is not a simple term. For agriculture, intensification commonly refers to means of increasing production from a constant area of land or obtaining the same production from less land. The well-known Boserup (1965) thesis is that such intensification, using more and more labor-demanding systems of agrotechnology, is driven primarily by population growth pressing against declining yields and resource degradation. This certainly has relevance in the Asia-Pacific region. But population growth is only one of a number of reasons for making changes. My question is what actually constitutes intensification in agriculture? Is farmer improvement of the fallow properly intensification in parallel with such measures as more closely interplanting crops, using rotations, and sustainably extending the productive life of the field? Can fallow improvement be separated, for meaningful analysis, from other changes in farming and land management?

The case studies presented in the chapters of this volume describe practices that seem to have either—or both—of two main purposes. The first sets out to generate more effective fallows, leading to improvement of the soil and shortening the intervals between cultivation periods by enriching the soil during them. The second aims to create more productive fallows. This means introduction or encouragement of what are, in effect, additional crops. In some cases, they are already present in the cultivation period and extend into the fallow. In others, the fallow is partly replaced by a long-term crop. We are also concerned mainly with farmers' use or management of what is naturally there, rather than with the deliberate introduction of exotics. Yet some papers are concerned with the benefits to be had from a set of "naturally occurring" plants that are themselves exotics, in that they are introduced *Compositae* weeds. They are manageable, but present without any deliberate farmer selection.

Many farmers who innovate to improve the fallow are "hitching a ride on a multiplicity of processes observed in nature," to use a happy expression attributed to Paul Richards (unpublished, cited in Fairhead and Leach 1996, *207*). This means that farmers are harnessing and directing ecological processes that occur naturally, rather than attempting to innovate over these or override them. There is nothing to be said against hitching rides with nature—quite the contrary. However, to view most farmer improvements in the fallow in this way does help distinguish what is basically intelligent management, which most of these fallow improvements and their use constitute, from the more laborious business in the working fields that goes along with them. Selective use of plants may create "landesque capital" of long life, but it involves little of the labor-demanding investment required to create pond fields, terraces, or patiently improved soil. The content of the landscape is modified, productivity is enhanced, and—from all human points of view—the working environment is improved. Selective use of plants calls for knowledge and use of skills, and the outcome is more productivity on constant land. If this is intensification, it seems a very inexpensive way of doing it. "Innovation" might be a more appropriate term (Brookfield 1984).

Work in the fields can also involve shortening the fallow, transforming it, or even replacing it, by quite different means. One example, from outside Asia-Pacific, comes from the Republic of Guinea in West Africa, where Fairhead and Leach (1996) have convincingly shown that the extent of forest and woodland has enlarged at the expense of savanna during the present century, rather than diminishing, as has long been believed. Moreover, this reforestation is due overwhelmingly to human activity, in nurturing islands of forest around villages, enlarging swamps for rice, and—most significantly from our point of view—by mounding soil and incorporating organic matter into the mounds. This latter method creates conditions that improve soil moisture and texture, and restrict fire. In the presence of a suitable seed bank, this method permits the establishment of pioneer woodland species, followed by forest species, and a lightly disturbed savanna fallow is replaced by a forest fallow. Old settlement sites and termitaria sites are selected first, but mounding extends well beyond these limits, with quite dramatic effects over periods as short as one human lifetime. Planting of chosen trees is an element, but principally the vegetation complex is enriched through edaphic transformation brought about by cultivation. Included in the successional vegetation on such sites are certain nitrogen-fixing and other tree species known elsewhere in the region to be beneficial to crops grown in their vicinity and preserved by farmers wishing to reverse the degradation of their land (Amanor 1994, 1997). The means of transformation involves the whole farming system.

This aspect is worth some emphasis. Asia-Pacific swidden cultivators in the forest usually dibble and only sometimes make substantial mounds. Farmers do till in grasslands and in drier upland areas of the region, with and without the aid of livestock. The people of Roti in eastern Indonesia collect leaves of the lontar palm (*Borassus sundaicus*, pile them, and burn them to fertilize annual gardens. They also collect and spread animal manure (Fox 1977, *30*). Tillage is not always beneficial, but where deep or prolonged tillage incorporates organic matter into the soil, it does transform soils in an enduring manner, and there are major changes in the plant environment. In the Asia-Pacific region, tillage is not new on upland soils and some of it does incorporate organic matter. While much of it is quite modern, there are upland areas of the region where dry land tillage has an antiquity of hundreds of years. Indigenous land management made more labor intensive through tillage does far more than manage the fallow, but this is one of its consequences. Intensified land management can therefore also contribute to indigenous fallow improvement.

Complex Multistoried Agroforests and Some Concluding Questions

Diversion from arable land use to long-enduring or permanent agroforestry is a particular feature of the Asia-Pacific region. It takes us an important step away from simple fallow management. The diverse home gardens, mixed gardens, and forest gardens of Java are the best known examples (Wiersum 1982, Soemarwoto and Soemarwoto 1984, Karyono 1990). The mixed gardens may be created from forest gardens, but also succeed from dry cultivation. The outcome of planting is not a fallow of limited term, but a long-lived managed forest composed principally of useful species. In New Guinea, Clarke (1971) distinguishes between the short-lived mixed crop gardens of the Bomagai-Angoiang and their orchards, which endure for a human lifetime. The orchards may be preceded by short-lived plantings of sugar cane and *Saccharum edule*, but most swiddens go directly back to secondary forest so that the orchards are a separate and deliberate creation, albeit now threatened by change (Clarke and Thaman 1997). The scattered mixed tree crop gardens of the Philippine Hanunóo are always preceded by swidden cultivation (Conklin 1957, *125–126*).

The agroforests of Sumatra, under study by the ICRAF ASB project, are discussed in this volume, as are the even more remarkable agroforests of West Kalimantan. In two Dayak communities studied by Padoch and Peters (1993) and Peluso and Padoch (1996), quite big areas with no substantial surviving high forest and with relatively high population density were devoted to distinctive mixed forest gardens known, from the term for an old settlement site, as *tembawang*. These forest gardens, typical of many others in the same region, originated as fruit gardens surrounding villages, but the villages were later relocated. They now contain up to 74 different species of fruit trees. One sample transect yielded 44 tree species of which 30 produce edible fruits or shoots (Padoch and Peters 1993, *171–172*). Mature tembawang may be several centuries old. They have been developed and are sustained by a combination of deliberate planting, casual planting, and volunteer growth spared in successive weeding around the durian, rambutan, mangosteen, rubber, illipe nut, sugar palm and construction-wood trees, to mention only the principal members. Several species of each genus—and their wild relatives—are consciously conserved and planted. Tembawang grow in a landscape that also includes protected, but used, modified forest, or *tanah adat*. There are also short-term forests of marketable species, especially rubber, which are cyclic in that they both follow and precede swidden. These are called *tanah usaha*. However, not much swidden is now made in this region of Borneo, and its role is increasingly taken over by progressively extended wet rice fields, another form of transformation. The agroforestry systems maintain genetic diversity, conserve soil and water, and meet both domestic and commercial needs. They appear to be highly sustainable. With the extending wet rice, they support population densities in the range of 50 to 100 per square kilometer.

The existence in the Asia-Pacific region of these diverse and productive alternatives to swidden, rather than simple improvements of swidden, raises a final group of questions. If we take full account of the dynamism of many farming systems and of a history of transformation that extends back at least several hundred years, we can better evaluate these indigenous fallow improvements. We need to ask where we are coming from in "improving fallows." Are we coming from modern swiddens under demographic, economic, and political pressure, or from old swiddens under no such pressure? Where are we going? Is fallow improvement likely to perish in order to obtain the clear arable spaces required for commercial production, as is happening in parts of the island Pacific? (Clarke and Thaman 1997). It is perhaps better to speak of "farmer-guided ecological change" than of fallow improvement, since the real direction may be toward forms of agroforestry or permanent arable land uses that replace swidden. The objective may be diversity of livelihood—its quality as much as its quantity—and not just soil improvement or the combating of degradation.

Modern changes may be only an acceleration of long-established trends. If one conclusion is that what farmers do makes sense when studied in its full context, then we shall have made no new discovery. But we will have helped the farmers in their struggle against an enormous body of pressure and ignorance. By exhibiting a willingness to learn from farmers, we are doing something very important. We are joining other initiatives in placing farmers' own practices first, such as the comparative United Nations University project. One hopes that we will together add importantly to growth in real understanding of why the world's farmers do what they do and how much can be learned from them.

References

Amanor, K.S. 1994. *The New Frontier: Farmer Responses to Land Degradation*. Geneva and Atlantic Highlands: UNRISD and Zed.

———. 1997. Interacting with the Environment: Adaptation and Regeneration on Degraded Land in Upper Manya Krobo. In: *Environment, Biodiversity and Agricultural Change in West Africa*, edited by E.A.Gyasi and J.I. Uitto. Tokyo: United Nations University Press, 98–111.

Boserup, E. 1965. *The Conditions of Agricultural Growth: The Economics of Agrarian Change Under Population Pressure*. Chicago: Aldine.

Brookfield, H. 1984. Intensification Revisited. *Pacific Viewpoint* 25, 15–44.

———. 2001. *Exploring Agrodiversity*. New York: Columbia University Press.

———. C. Padoch, H. Parsons, and M. Stocking (eds). 2002. *Cultivating Biodiversity: Understanding, Analysing and Using Agricultural Diversity*. London, ITDG Publishing.

———. H. Parsons, and M. Brookfield (eds). 2003. *Agrodiversity: Learning from Farmers Across the World*. Tokyo: United Nations University Press.

———. L. Potter, and Y.Byron. 1995. *In Place of the Forest: Environmental and Socio-Economic Transformation in Borneo and the Eastern Malay Peninsula*. Tokyo: United Nations University Press.

Chin, S.C. 1985. Agriculture and Resource Utilization in a Lowland Rainforest Kenyah Community. *The Sarawak Museum Journal* 35 (New Series 56), Special Monograph No. 4.

Clarke, W.C. 1971. *Place and People: An Ecology of a New Guinean Community*. Canberra, Australia, and Berkeley, CA: The Australian National University Press and the University of California Press.

———. and R.R. Thaman. 1997. Incremental Agroforestry: Enriching Pacific landscapes. *The Contemporary Pacific: A Journal of Island Affairs* 9, 121–148.

Colfer, C.J.P., and H. Soedjito. 1996. Food, Forests, and Fields in a Bornean Rain Forest: Towards Appropriate Agroforestry Development. In: *Borneo in Transition: People, Forests, Conservation and Development*, edited by C. Padoch and N.L. Peluso. Kuala Lumpur: Oxford University Press, 162–186.

Conklin, H.C. 1957. Hanunóo Agriculture in the Philippines. In: *FAO Forestry Development Paper* No. 12. Rome: U.N. Food and Agriculture Organization.

Denevan, W.M., and C. Padoch. 1987a. Introduction: The Bora Agroforestry Project. In: Swidden-Fallow Agroforestry in the Peruvian Amazon, edited by W.M. Denevan and C. Padoch. *Advances in Economic Botany* 5. New York: The New York Botanical Garden, 1–7.

——— (eds.). 1987b. Swidden-Fallow Agroforestry in the Peruvian Amazon. *Advances in Economic Botany* 5. New York: The New York Botanical Garden.

de Jong, W. 1996. Swidden-Fallow Agroforestry in Amazonia: Diversity at Close Distance. *Agroforestry Systems* 34, 277–290.

Fairhead, J., and M. Leach, with the research collaboration of Dominique Millimouno and Marie Kamano. 1996. *Misreading the African Landscape: Society and Ecology in a Forest-Savanna Mosaic*. Cambridge, U.K.: Cambridge University Press.

Fox, J.J. 1977. *Harvest of the Palm: Ecological Change in Eastern Indonesia*. Cambridge, MA: Harvard University Press.

Johns, R.J. 1990. The Illusionary Concept of the Climax. In: *The Plant Diversity of Malesia*, edited by P. Baas, K. Kalkman and R. Geesink. Dordrecht: Kluwer Academic, 13–146.

Karyono. 1990, Home Gardens in Java: Their Structure and Function. In: *Tropical Home Gardens*, edited by K. Landauer and M. Brazil. Tokyo: United Nations University Press, 138–146.

Kunstadter, P. 1978. Subsistence Agricultural Economies of Lua and Karen Hill Farmers, Mae Sariang District, Northwestern Thailand. In: *Farmers in the Forest: Economic Development and Marginal Agriculture in Northern Thailand*, edited by P. Kunstadter, E.C. Chapman, and Sanga Sabhasri. Honolulu: University Press of Hawaii, 74–133.

Lian, F.J. 1987. Farmers' Perceptions and Economic Change: The Case of Kenyah Swidden Farmers in Sarawak. Ph.D. Thesis, Australian National University, Canberra.

Padoch, C., and C. Peters. 1993. Managed Forest Gardens in West Kalimantan, Indonesia. In: *Perspectives on Biodiversity: Case Studies of Genetic Resource Conservation and Development*, edited by C.S. Potter, J.I. Cohen, and D. Janczewski. Washington, DC: American Association for the Advancement of Science Press, 167–176.

Peluso, N.L. 1996. Fruit Trees and Family Trees in an Anthropogenic Forest: Ethics of Access, Property Zones, and Environmental Change in Indonesia. *Comparative Studies in Society and History* 38, 510–548.

Peluso, N.L. and C. Padoch. 1996. Changing Resource Rights in Managed Forests of West Kalimantan. In: *Borneo in Transition: People, Forests, Conservation and Development*, edited by C. Padoch and N.L. Peluso. Kuala Lumpur: Oxford University Press, 121–136.

Peters, C.M. 1996. Illipe Nuts (*Shorea* spp.) in West Kalimantan: Use, Ecology and Management Potential of an Important Forest Resource. In: *Borneo in Transition: People, Forests, Conservation and Development*, edited by C. Padoch and N.L. Peluso. Kuala Lumpur: Oxford University Press, 230–244.

Ramakrishnan, P.S. 1992. *Shifting Agriculture and Sustainable Development: An Interdisciplinary Study from Northeastern India*, Carnforth, U.K. and Paris: Parthenon/UNESCO.

Ruthenberg, H. 1980. *Farming Systems in the Tropics*. Oxford, U.K.: Oxford University Press.

Sather, C. 1990. Trees and Tree Tenure in Paku Iban Society: The Management of Secondary Forest Resources in a Long-Established Iban Community. *Borneo Review* 1, 16–40.

Soemarwoto, O., and I. Soemarwoto. 1984. The Javanese Rural Ecosystem. In: *An Introduction to Human Ecology Research Systems on Agricultural Systems in Southeast Asia*, edited by A.T. Rambo and P.E. Sajise. Los Baños, Laguna, Philippines: University of the Philippines, 254–287.

Spencer, J.E. 1966. *Shifting Cultivation in Southeastern Asia*. Berkeley and Los Angeles, CA: University of California Press.

Wiersum, K.F. 1982. Tree Gardening and Taungya on Java: Examples of Agroforestry Techniques in the Tropics. *Agroforestry Systems* 1: 53–70.

Chapter 3

Conceptualizing Indigenous Approaches to Fallow Management

A Road Map to this Volume

Malcolm Cairns

Collapsing shifting cultivation systems and attendant environmental damage are pan-regional problems across Asia-Pacific. It is a causal factor in some of the most serious challenges facing the uplands, including deforestation, loss of biodiversity, soil erosion, and deepening impoverishment of swidden-dependent communities. While the dynamics of 'swidden-degradation syndrome' are generally understood, research has been less successful in identifying solutions widely adopted by resource-poor farmers. Most efforts have focused on the cropping phase as the point of intervention for developing improved agronomic technologies. The fallow has been widely viewed as unproductive, unmanaged, and interesting only from the perspective of how it can be shortened. The case studies assembled in this volume soundly debunk this myth and demonstrate the potential of Indigenous Fallow Management (IFM) for contributing solutions to upland problems in the face of enormous land-use pressures and economic change.

Emphasized repeatedly throughout this book, almost like a mantra, is the progressive loss of sustainability of traditional swidden systems across the Asia-Pacific region as land-to-people ratios have declined, the last remaining forest frontiers have been gazetted as protected areas, state policies have continued to discourage or even forbid farming systems that include elements of fire and fallow, national forestry departments have gained increased capacity to enforce these sanctions, and expanding road networks have brought competing demands on land and other upland resources. From this common starting point, the authors diverge to describe an impressive array of farmer-generated fallow management practices that have, in many cases, permitted a sustainable intensification of shifting cultivation. The combined weight of their evidence suggests that there has been a selective blindness to farmer management during the fallow, and that this may have been costly in terms of overlooked opportunities to build on these practices in attempting to stabilize stressed swidden systems.

Malcolm Cairns, Department of Anthropology, Research School of Pacific and Asian Studies (RSPAS), Australian National University, Canberra, ACT 0200, Australia.

This work found its inspiration in promising findings from research undertaken by the World Agroforestry Centre (ICRAF), with the support of the International Development Research Center (IDRC), on several shrub and tree-based fallow management systems (Cairns, Chapter 15, Chapter 30). A more systematic survey revealed that swidden farmers throughout the region's uplands, pushed by increasing land-use pressures, had innovated a compelling array of successful IFM practices that drew on their intimate knowledge of local environments. During this search, we also became aware of other colleagues who, like ICRAF researchers, were working with various IFM practices in relative isolation. The regional workshop, Indigenous Strategies for Intensification of Shifting Cultivation in Southeast Asia, was held at Bogor, in Indonesia, in June 1997 as a means of bringing together many of the individuals and institutions working on different pieces of the same puzzle. The spirit that emerged from that workshop urged the widest possible consultation in order to synthesize current knowledge of farmer-developed and tested technologies for improved fallow husbandry. This introductory chapter briefly outlines a continuum of typologies that provides an organized way of conceptualizing IFM practices and lends structure to this volume.

Conceptualizing Indigenous Approaches to Fallow Management

Clearly there is a wide menu of components from which shifting cultivators may choose to intensify land use (Figure 3-1), but this volume focuses sharply on indigenous innovations to manage fallow land in more productive ways. As illustrated in Figure 3-2, farmer approaches to fallow management may generally be classified as innovations to achieve the following:

- More effective fallows, where the biological efficiency of fallow functions is improved and the same or greater benefits can be achieved in a shorter time frame;
- More productive fallows, in which fallow lengths stay the same or actually lengthen as the farmer adds value to the fallow by introducing economic perennial species; and
- Combinations of the two, where both biophysical and economic benefits may be accrued.

The implications to land use of these major pathways toward swidden intensification are obviously profound. More effective or accelerated fallows often provide an intermediate step in a transition to permanent cultivation of annual crops. Alternatively, in more productive fallows, the phase of reopening and cultivation of annuals may eventually be forgone altogether as the farmer chooses to protect valuable perennial vegetation, allowing it to develop into semi-permanent or permanent agroforests. Although the spectrum proposed in Figure 3-3 provides a useful framework for conceptualizing IFM strategies, it does not suggest that farmers will necessarily move to a linear direction along this continuum. They may or may not- but the dynamism of factors that shape farmer land use decisions defy such easy prediction.

It should also be stressed that our operational definition of managed fallows is very wide and covers the entire spectrum, from growing viny legumes as dry season fallows lasting a few months, to incremental inclusion of more economic perennials into the fallow until it develops into a long-term complex agroforest. The salient point is that we are trying to understand the array of farmer-generated solutions that have successfully permitted an intensification of shifting cultivation in the face of increasing land-use pressures.

As a consequence, the case studies in this volume provide a sweeping cross-section of IFM typologies that have evolved across the Asia-Pacific region. Figure 3-3 attempts to categorize indigenous strategies for fallow management along a continuum of typologies. The map in Figure 3-4 portrays roughly where we know them to be practiced in the region.

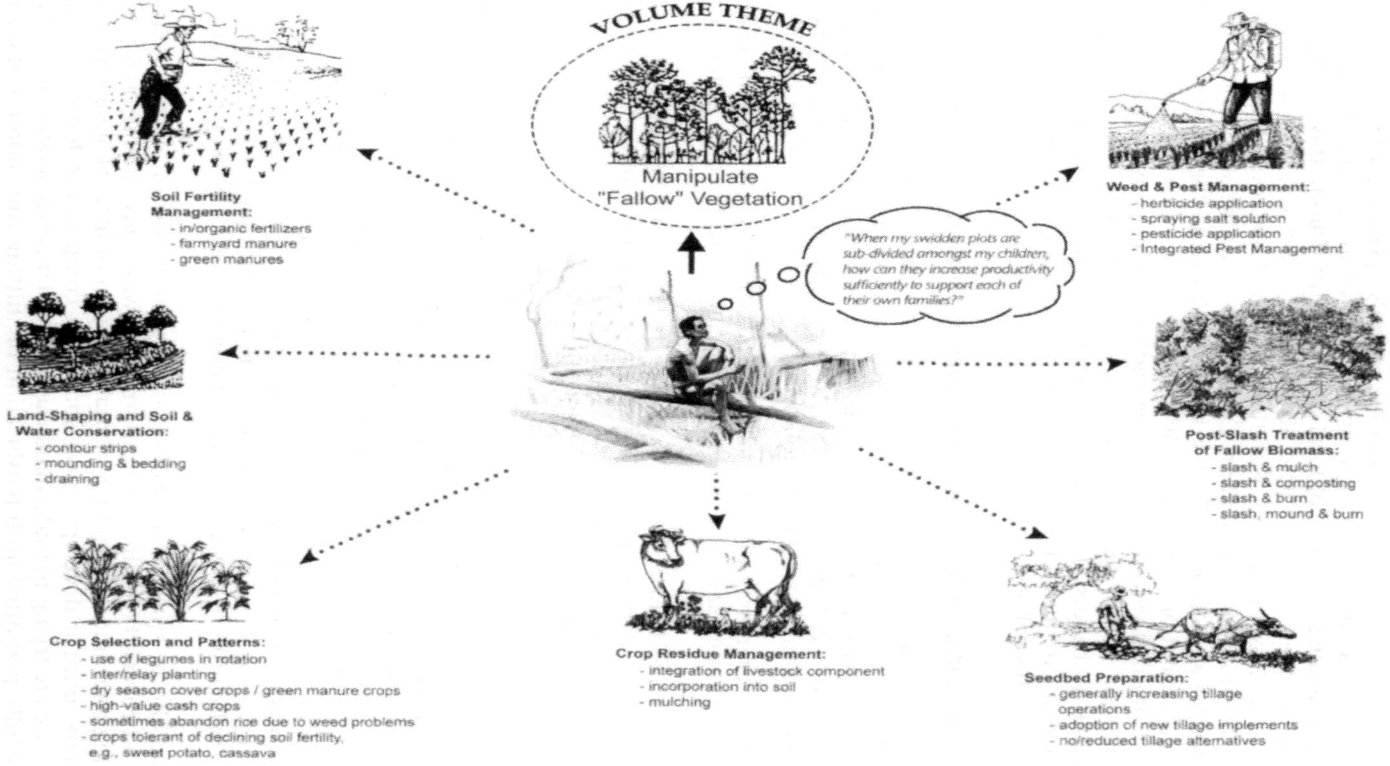

Figure 3-1. Possible Components of Farmer Strategies for Intensification of Shifting Cultivation

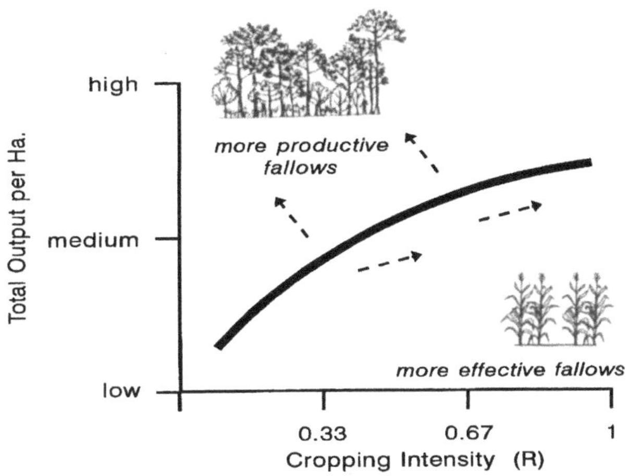

Figure 3-2. Evolution of Intensifying Swidden Systems
Source: van Noordwijk 1996.

It is useful to briefly summarize the distinguishing features of each of the IFM typologies depicted in Figure 3-3. These typologies form the basis for the following sections of the book. This review draws heavily from the case studies compiled in this volume, as indicated by undated citations.

Retention or Promotion of Preferred Succession Species

The most fundamental and widely practiced approach to fallow management is simply the opportunistic use of elements of the fallow succession that provide useful products or ecological services. The division between cropping and fallow is often blurred as, even after the main swidden crops have been harvested, shifting cultivators continue to revisit their fields. Initially, they search for crop vestigials but, later, for the wide variety of natural succession species that make important contributions to household economies through provision of food (see color plates 8 through 13), fiber (see color plates 4 through 7), fodder (see color plate 3), fuel, medicinal herbs, and other useful products (Scoones et al. 1992; Tayanin, Chapter 6; Mertz, Chapter 7; Burgers, Chapter 8; Tangan, Chapter 9; Potter and Lee, Chapter 11). Other practices are conservation-oriented and aimed at encouraging rapid regeneration of the forest during the subsequent fallow (see color plates 1 and 2) (Daniels 1995; Kanjunt, Chapter 5). When clearing fallows, relict emergents of preferred species are retained to disburse propagules and anchor soils on erosion-prone slopes (Schmidt-Vogt, Chapter 4; Tayanin, Chapter 6). This selective retention of desirable species, through successive swidden cycles, gradually alters the composition of forest fallows in favor of those most useful to shifting cultivators (Raintree and Warner 1986). Coppices that sprout from the stumps of felled trees during the cropping period are often protected, allowing the forest to recover quickly after cultivation (Ty, Chapter 55). Complementary practices of avoiding tillage, limiting cultivation to a single year, and controlling fires (Durno et al., Chapter 12; Maneeratana and Hoare, Chapter 13) all have the intended effect of minimizing disturbance of tree seedlings germinating from the soil seed bank.

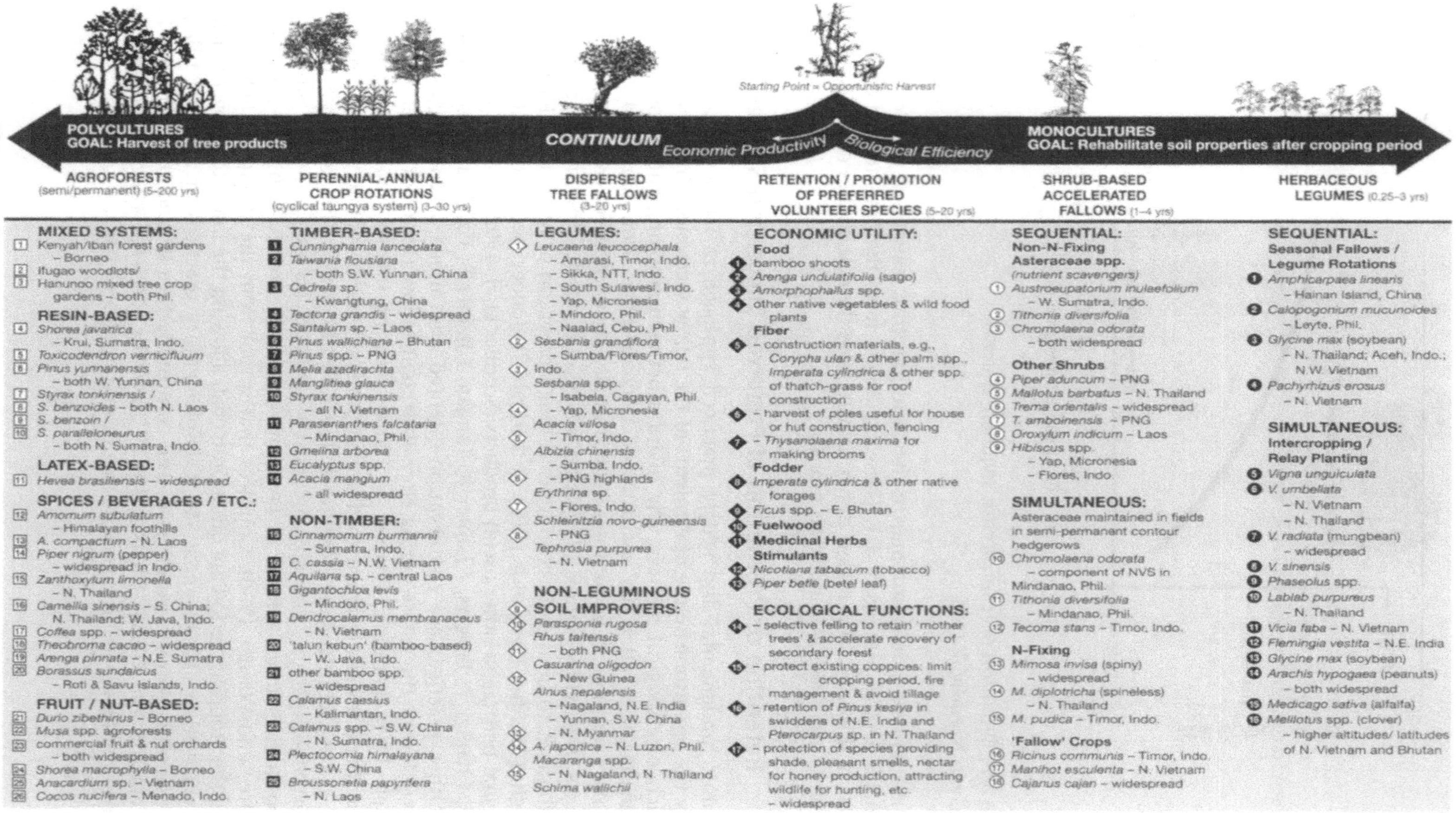

Figure 3-3. Spectrum of Indigenous Approaches to Modify "Fallow" Vegetation in Asia-Pacific

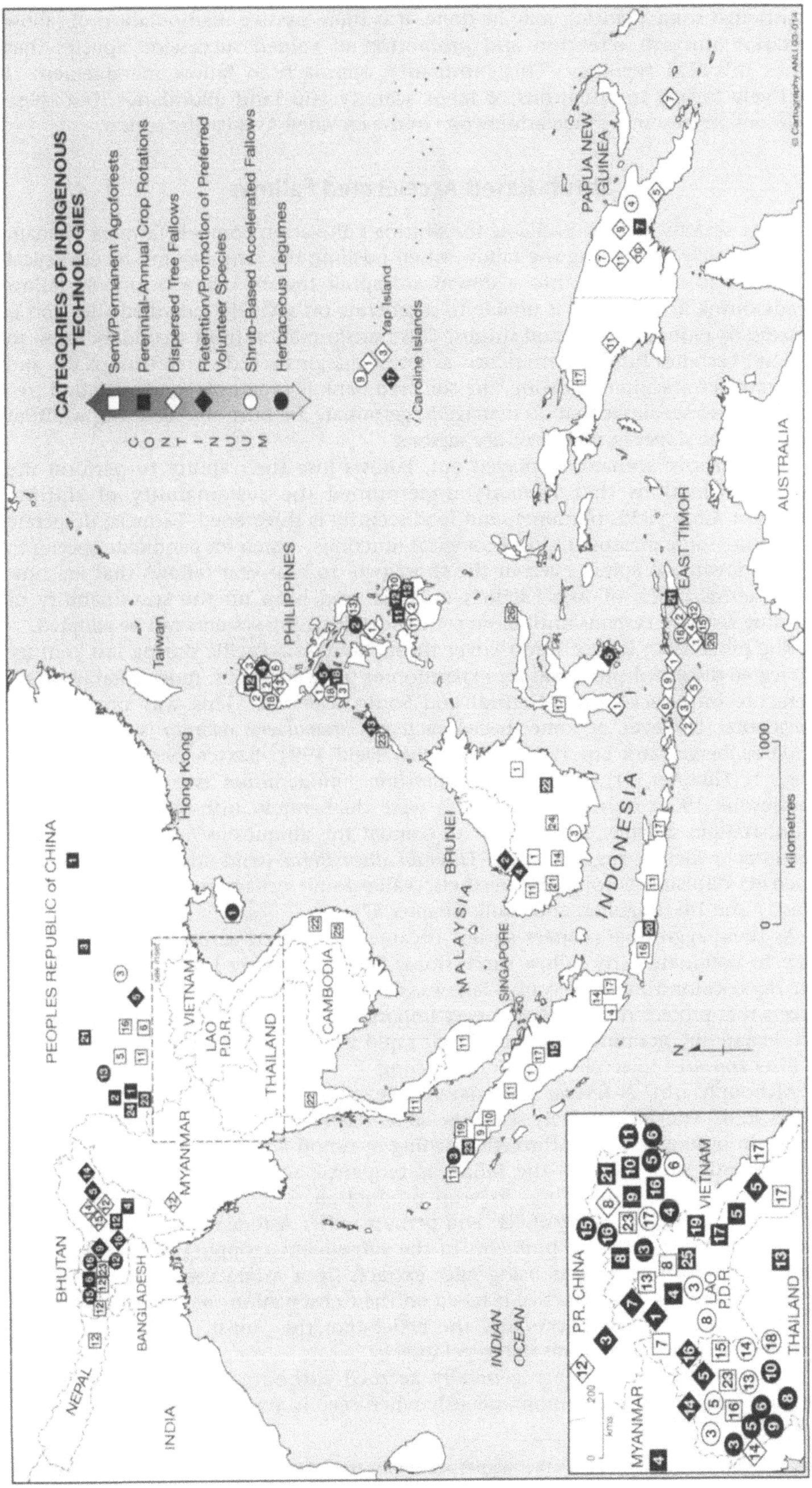

Figure 3-4. Spatial Analysis of IFM Variations (see Figure 3-3 for corresponding legend)

Although some planting may be done, it is more passive manipulation of fallow vegetation through retention and promotion of valued succession species that defines this IFM typology. This minimalist approach to fallow management is intuitively logical in situations of labor scarcity and land abundance, but these conditions are becoming increasingly rare in the crowded Asia-Pacific region.

Shrub-Based Accelerated Fallows

Mounting scarcity of land available for shifting cultivation compels farmers to adapt by progressively shortening the fallow, often pushing the land beyond its ecological resilience and sending it into a downward spiral toward degradation. As fallow periods shrink, forest cover is unable to regenerate on severely disturbed sites and is displaced by pioneer grasses and shrubs. This transformation from secondary forest to grassland becomes further entrenched as forest margins recede, tree stumps die and are uprooted to facilitate plowing, the soil seed bank is progressively depleted of tree seeds, and tree seedlings that do manage to germinate are burnt by recurring wildfires that sweep the slopes during most dry seasons.

As the above scenario is played out, fallows lose their ability to perform the ecological functions that formerly underpinned the sustainability of shifting cultivation. Crop yields plummet, and food security is threatened. Farmers, desperate to recapture some measure of lost ecological functions, search for candidate species to halt the downward spiral, even in the short two- to four-year fallows that are now typical across much of Asia-Pacific's uplands, and prop up the sustainability of degrading swidden systems until more permanent cropping systems can be adopted.

The progressive loss of forest cover throughout Asia-Pacific during last century has created disturbed sites ideal for expansion of pioneer shrubs, most notably exotic Asteraceae introduced from Central and South America.[1] This was not entirely coincidental, however, as some species such as *Chromolaena odorata* (see color plate 15) (Litzenberger and Lip 1961; Dove 1986; Field 1991; Baxter 1995; Roder et al. 1995a, b, Chapter 14) and *Austroeupatorium inulaefolium* (see color plate 16) (Stoutjesdijk 1935; Cairns, Chapter 15) were deliberately introduced by colonial administrations as green manures or to combat the ubiquitous *Imperata cylindrica.* Anecdotal evidence suggests that *Tithonia diversifolia* (wild sunflower) probably earned its diffusion through its aesthetic value as an ornamental plant (see color plates 17 and 18) (Daguitan and Tauli, Chapter 57).

As these aggressive pioneer shrubs became naturalized across the region, they began to dominate early fallow successions. Farmers quickly learned to appreciate their rapid colonization of young fallows, developing dense, almost monospecific thickets that protected the soil and, very importantly, shaded out *Imperata* and other light-demanding gramineous weeds. Their rapid generation of biomass and copious leaf litter appeared to accelerate nutrient cycling.

Although not N-fixers, the Asteraceae play a critical role in nutrient conservation. They aggressively scavenge labile nutrients from the soil nutrient pool that might otherwise be lost through leaching or runoff and hold them in the fallow biomass until such time as the fallow is reopened and they can be channeled, through burning or mulching, to crop production. There is also evidence that attributes *C. odorata, T. diversifolia,* and perhaps other Asteraceae with nematocidal properties, reducing disease problems in the subsequent cropping phase. There are widespread reports of farmers using juice extracts from Asteraceae as ingredients in botanical pesticides. This practice is based on their observation that exotic Asteraceae are seldom attacked by insects and the belief that they must, therefore, contain defensive chemical compounds that repel insects.

Finally, these benefits are generally accrued without additional labor costs; Asteraceae fallows establish spontaneously when seed is available, require no special

---

[1] This invasion of Asteraceae and other aggressive shrubs from Central and South America has been at the expense of displaced native flora and a probable loss of biodiversity.

management, and are easily cleared. Hence, the major ecological functions expected from fallows—soil rejuvenation, suppression of hard-to-control weeds, and disruption of pest and disease cycles—are effectively continued by these shrubs after trees are lost from the system (Oyen 1995; Pandeya 1995).

Other nonleguminous shrubs are also valued by farmers as effective fallow species. There are frequent references in older literature to *Lantana camara* as an effective fallow shrub enabling shortened fallow periods (Ormeling 1955, for example). However, it appears to have been increasingly displaced by *Chromolaena odorata* and other invasive exotics during the latter part of the last century. There are sporadic references to *Trema* spp. as deliberately managed by farmers to improve fallows (Littooy 1989, *67*; Bourke, Chapter 31; Daguitan and Tauli, Chapter 57). *Mallotus barbatus* (Rerkasem 1996) and *Oroxylum indicum* (Fahrney et al., Chapter 40) are preferred fallow species in northern Thailand and northern Lao P.D.R., respectively. *Piper aduncum* is yet another exotic, native to Central America, that has become widely established across parts of the Asia-Pacific region and that is reported to form almost monospecific stands in fallows in parts of Papua New Guinea (see color plate 20) (Hartemink, Chapter 16). All these species share characteristics of being fast-growing pioneer shrubs or small trees that propagate profusely and rapidly establish in swidden clearings. They appear to perform a similar biological role to Asteraceae in accelerating fallow functions within a shortened time frame. In the semiarid ecoregion of eastern Indonesia, shifting cultivators in West Timor have learned to value *Tecoma stans* for its ability to prosper and generate significant biomass in their harsh, dry climate and rocky, coralline soils (see color plate 19) (Djogo et al., Chapter 17). It was introduced to the area from Latin America as an ornamental shrub.

Although usually labeled as noxious weeds, *Mimosa* spp. have been harnessed as effective fallows in a number of noteworthy cases. *Mimosa* spp. share many of the attributes of other shrub fallows but have the added virtue of fixing nitrogen. Interestingly, Thai and Karen farmers in the Wang Chin District of northern Thailand adopted the spineless *Mimosa diplotricha* var. *inermis* for its greater ease of handling (Prinz and Ongprasert, Chapter 19), while farmers in Punta, Leyte, the Philippines, value spiny *Mimosa invisa* Martius ex Colla specifically for its ability to discourage intrusive livestock (see color plate 23) (Balbarino et al., Chapter 18). This highlights the need for a menu of effective fallow species from which candidates can be chosen to best fit local circumstances.

The shrub-based accelerated fallows outlined in this typology are distinguished by farmer manipulation of fallow successions to ensure domination of preferred shrubs or small trees that are efficient in performing the ecological functions needed from fallows in a shortened time frame. They have often escaped scientific attention because management is more subtle and managed species are native or, more often, naturalized exotics. They are consequently mistaken by casual observers as simply "weedy fallows." More careful studies, such as those described in Section Three of this volume, reveal that shifting cultivators are clearly managing selected shrubs as green manures or cover crops in short bush fallow cycles. This provides grounds for arguing that these systems no longer fall under the domain of "shifting cultivation" but should more properly be considered as permanent cultivation with a food crop–green manure or cover crop rotation. The importance of this distinction extends far beyond semantics, since it resonates positively with state policies in the Asia-Pacific region to replace shifting cultivation with more permanent land uses.

Although it does not fit comfortably within this typology, it is relevant to note that farmers attribute some taller crops with the ability, when integrated into cropping patterns, to partially substitute for fallows by performing beneficial ecological functions. Examples include farmers in the Cordilleras, Philippines, who intercrop *Cajanus cajan* (pigeon pea) into swiddens and then, after harvesting, maintain it as an improved leguminous fallow (Daguitan and Tauli, Chapter 57); the use of *Manihot esculenta* (cassava) as a fallow crop in northern Vietnam, where it is alleged to rejuvenate exhausted soils sufficiently to enable another cycle of upland rice (Ty 1997); and the reported use of *Ricinus communis* (castor) as a fallow in Timor, eastern Indonesia (Field 1996).

Herbaceous Legume Fallows

The most direct extension of the previous typology is simply the substitution of passively managed volunteer shrubs for actively planted herbaceous legumes (see color plate 28) to provide both ecological fallow functions and a harvestable bean crop. The legumes continue to be grown in a sequential pattern with other food crops, often as dry season fallows. There is, however, a direct conflict between the harvest index of the herbaceous legume fallow and its capacity to contribute to soil improvement. Advances in one are usually at the expense of the other.

In some cases, these legume fallows receive little care and, although bean yields are low, they make important contributions to household diets and incomes from very little investment. The strategy continues to emphasize "hitching a ride on a multiplicity of processes observed in nature" (as cited by Brookfield, Chapter 2), and farmers opportunistically harvest whatever their fallows can yield under minimal input conditions. Examples of such low input–low output herbaceous legume fallows include the use of *Amphicarpaea linearis* as a seasonal fallow on Hainan Island, China (Lin et al., Chapter 20); *Calopogonium mucunoides* as a spontaneous viny legume fallow in Leyte, the Philippines (Balbarino et al., Chapter 18); and the integration of *Flemingia vestita* as a leguminous fallow crop in Meghalaya and other parts of northeast India (Ramakrishnan, Chapter 21).

Distinct from these systems are those herbaceous legumes that have become valuable cash crops in their own right and hence are more intensively managed. Upland farmers with market access have increasingly adopted *Glycine max* (soybean), *Arachis hypogaea* (peanuts) (see color plate 24) (Hien, Chapter 62), and *Vigna radiata* (mungbean) into their crop rotations. This represents a distinct intensification over the previously outlined low input–low output systems, in terms of management inputs, commercial value, harvest index, and their declining contribution to soil improvement. It becomes hard to justify classifying them as fallows. Rather, these cropping patterns have crossed the threshold and should be considered legume crop rotations grown under permanent cropping systems. It is useful to this discussion, however, to acknowledge them as "falling off" the extreme right pole of the IFM continuum (Figure 3-3). This illustrates the continued intensification of land use until fallows completely disappear from the system. The distinction is not always clear cut, and some legumes, such as *Pachyrhizus* spp. (yam bean) (see color plate 29), may either grow wild as part of the natural fallow succession or be carefully cultivated as a food crop.

The sequential herbaceous legume fallows discussed up to this point produce beans or tubers for human consumption as their main economic products. But the leaves and vines of many also provide livestock fodder as a byproduct. The uplands, with their cooler climate, fewer diseases, and availability of grasslands for grazing, enjoy a comparative advantage in producing livestock for lowland markets. There is now much interest in intensifying animal husbandry in tandem with improved fallow management (Chapman et al. 1998; Horne, Chapter 10). The push of land-use pressures and pull of market opportunities are converging to persuade some farmers, particularly in higher-altitude and higher-latitude zones such as Bhutan (Dukpa et al., Chapter 59) and northern Vietnam (Foerster 1997), to convert fallows into fodder banks by planting clovers (*Melilotus* spp.), alfalfa (*Medicago sativa*), and other temperate forage legumes. This ley farming strategy combines benefits of soil improvement, livestock weight gains, and animal draft power and appears poised to gain wider adoption as upland farmers respond to growing demands for livestock products from the swelling middle class of the Asia-Pacific region.

In a further blurring between fallow and crop, some upland farmers intercrop, or relay plant, many of these same herbaceous legumes into swidden fields to create the effect of simultaneous fallows. This is most commonly done in association with taller crops, such as maize or cassava, which are less vulnerable to damage by climbing vines. Maize is commonly intercropped with *Glycine max* (soybean) by Han Chinese in Yunnan (see color plate 27); relay-planted with *Vigna unguiculata* (cowpea), *V. umbellata* (rice bean), and *Lablab purpureus* (lablab bean) by Lisu in northern Thailand (see color plate 25) (Ongprasert and Prinz, Chapter 23); relay-planted with *Vigna umbellata* (syn. *Phaseolus calcaratus*) by Muòng, Thai and H'mong in northern Vietnam (see color plate 26) (Littooy 1989, *91*; Hao et al., Chapter 22; Hien, Chapter 62); and relay planted with *Vicia faba* at higher altitudes in northern Vietnam (Bunch 1997). After the maize cobs are harvested, the remaining stalks are used as trellises by the viny legumes as they rapidly develop a thick protective canopy over the soil. Cassava fields are similarly relay-planted with *Mucuna pruriens* var. *utilis* in Lampung, Indonesia (Hairiah 1997) (see color plate 31 for an example of spontaneous *Mucuna* fallows in Myanmar) and unidentified beans in northern Vietnam (Littooy 1989, *46, 51, 61*). The *Mucuna*-inspired green manure and cover crop revolution that has swept across West Africa and Latin America in recent years (Bunch 1995; Buckles and Perales 1995; Buckles and Erenstein 1996; Buckles et al. 1998a, b) is conspicuously absent from the farmscapes of Asia-Pacific, raising interesting research questions.

The discussion thus far has taken us from the center of the IFM continuum (Figure 3-3), through three successive typologies, to the extreme right pole. This progressive intensification of fallow management emphasizes food crop production and is marked by accelerating swidden cycles and the loss of tree cover as more of the landscape is brought under arable cultivation.

The alternative direction that fallow management may take is the deliberate integration of more trees, selected for their ecological services or economic products or, more commonly, for a combination of the two. The following sections consider each of the three remaining tree-based typologies, now moving from the center of the IFM continuum toward the left pole.

Dispersed or Interstitial Tree-Based Fallows

This typology is based predominantly on nitrogen-fixing trees and, like the shrub-based fallows, has its origins in farmer observations of their superior ability to perform fallow functions. Most of these systems probably began with farmers selectively retaining or promoting preferred species in fallows, and over continued swidden cycles, they gradually developed into almost pure stands. Agroforestry systems are broadly divided into sequential and simultaneous categories, depending on whether the trees' relationship with the arable crop is temporal or spatial. In the case of dispersed tree fallows, however, this distinction is often unclear. Trees that, on the surface, appear to be completely cleared when fallows are reopened (sequential) may in fact persist through their underground organs and coppice together with the germinating crop (simultaneous) (Olofson 1983). Similarly, tree seedlings that develop from the soil seed bank during the cropping phase and are thereafter protected do overlap temporally with the swidden crop and, technically, would have to be considered as simultaneous systems. The fallow management systems falling under this typology thus appear to be simultaneous, based on the likelihood that there is some degree of temporal overlap in all of them. But many of them remain poorly described in the literature, and this assumption needs to be verified.

Early agroforestry research optimistically dubbed *Leucaena leucocephala* a "miracle tree," so it is perhaps not surprising that *Leucaena*-based fallow systems have been most widely documented. Although *Leucaena* is widespread across Asia-Pacific, its adoption as an improved fallow species has been limited to isolated pockets within the region (see Figure 3-4). This suggests independent discovery. Broadly similar *Leucaena* fallow systems have been innovated by farmers in Amarasi, East Nusa Tenggara (see color plates 32 to 34) (Metzner 1981, 1983; Jones 1983; Piggin and Parera 1985; Field and Yasin 1991; Surata 1993; Yuksel 1998; Piggin, Chapter 24), and Lilirilau subdistrict in South Sulawesi (Agus, Chapter 25), both in Indonesia and in

Naalad village in the municipality of Naga, Cebu (see color plate 35) (Eslava 1984; Subere et al. 1985; Kung'u 1993; Lasco and Suson 1997; Lasco, Chapter 27), and Sitio Sto. Tomas, Barangay Wawa, in Occidental Mindoro (see color plate 36) (MacDicken 1990a, b, 1991, Chapter 26), both in the Philippines. Also some Yapese women in the Western Caroline Islands have recently begun broadcasting *Leucaena* seed into their gardens (Falanruw and Ruegorong, Chapter 44). Comparative research is needed between these sites to identify commonalties and to make clear the conditions under which *Leucaena* can play a useful role in managed fallows. This will help to delineate the region in which farmers could benefit from adoption of these systems.

Shifting cultivators in East Nusa Tenggara, Indonesia, have been particularly prolific in developing other indigenous tree-based systems. This innovativeness may have been spurred by the greatly reduced rate of fallow regrowth in the region's harsh, semiarid climate and the need to bolster fallow functions. These practices include the use of *Acacia villosa* in the Camplong area of West Timor (see color plate 37); *Albizia chinensis* in Sumba (Fisher 1996); *Erythrina* sp. in Flores (Fisher 1996); and *Sesbania grandiflora* in areas across Sumba, Flores, and Timor (Fisher 1996; Kieft, Chapter 28). Further east, growing population pressures have prompted the planting of *Schleinitzia novo-guineensis* (see color plate 38) and *Rhus taishanensis* in fallows on Iwa Island, in Milne Bay Province of Papua New Guinea (Bourke, Chapter 31).

There are undetailed reports of managed fallows based on *Sesbania* spp. in Isabela, Cagayan, the Philippines (Pasicolan 1997), and on Yap Island, Micronesia (Falanruw and Ruegorong, Chapter 44) and *Tephrosia purpurea* in northern Vietnam (Viên 1998).

Most of these leguminous fallow species are unsuited to the acidic soils of higher mountain environments. However, two native pioneer trees, both non-leguminous nitrogen-fixers, play important roles in fallow management in these higher altitude swidden systems. *Casuarina oligodon* is widely managed in the New Guinea highlands both as an improved fallow and in a wide array of other agroforestry patterns (see color plates 41 and 42) (Thiagalingam and Famy 1981; Askin et al. 1990; Bourke, Chapter 31; Bino, Chapter 60). Across the eastern Himalayan foothills, *Alnus nepalensis* also has a long history of use in managed fallows in Nagaland, northeastern India (see color plate 39) (Kevichusa et al. n.d.; Gokhale et al. 1985; Dhyani 1998; Cairns et al., Chapter 30), in northern Myanmar (Troup 1921, *912*; Wint 1996, *39*), and in Yunnan, southwestern China (see color plate 40) (Hong et al. 1960; Zeng 1984; Fu 2003; Guo et al., Chapter 29). A system based on *Alnus japonica* is reported to have recently emerged among Ikalahan shifting cultivators in northern Luzon, the Philippines (Vergara 1995). Although there are many parallels between the agroforestry applications of the *Casuarina* and *Alnus* genera, *Alnus* claims the distinct advantages of prolific coppicing and longevity.

The dispersed tree typology emphasizes the integration and management of recognized soil-improving trees in swidden fallows, many of which also provide fodder, wood for construction and fuel, and other useful products. Such dual benefits are most attractive to farmers trying to increase cash income without sacrificing the sustainability of their shifting cultivation systems (Grist et al., Chapter 32).

Perennial-Annual Crop Rotations

This typology also involves the introduction of selected perennials into swiddens but is distinguished from the previous typology by the choice of species primarily for their harvestable products and not for their efficiency in performing fallow functions. Therefore, the emphasis between provision of ecological services and economic products reverses, and rehabilitation of the field through soil rejuvenation, weed suppression, and other ecological services normally associated with fallowing now become incidental benefits rather than the intended objectives. This economic orientation clearly justifies the consideration of these perennials as crops and, by extension, the shifting cultivation system has arguably been transformed into permanent cultivation under a perennial-annual crop rotation. This again has profound implications for how these land-use systems are viewed by policy makers.

This IFM approach is essentially a *taungya* system, in which trees are relay-planted into swidden fields during the cropping phase. The benefits of this are well known. Little extra labor is needed, and the trees get off to a strong start as they benefit from weeding, fertilizing, and other routine maintenance operations applied to the food crops. By the time the field is ready to be "fallowed," usually in one to three years, the trees are well enough established to compete with weedy regrowth. Weeds are usually slashed around the base of each tree for a further three to four years to control aggressive climbers such as *Mikania* and wild *Mucuna* spp. The canopy then closes, shading out light-demanding weeds, and farmers turn their attention to pruning, thinning, and other silvicultural treatments.

Most of these systems are timber–based. Some date back many hundreds of years and are detailed by historical records, such as the planting of *Cunningamia lanceolata* (Chinese fir) in their swiddens by ethnic minorities in southern provinces of China (Menzies 1988; Menzies and Tapp, Chapter 35). Most examples are more recent and have been triggered by market opportunities and policy reforms. Rising timber prices have fueled wide conversion of upland fields, through the intermediary of the swidden cycle, into *Cunninghamia lanceolata* (see color plate 44) and *Taiwania flousiana* plantations in southwestern Yunnan (author's field notes) and into *Cedrela* sp. (cedar) in Kwangtung, both in China. They have fueled conversation into *Tectona grandis* (teak) (see color plate 43) (Roder et al. 1995c; Hansen et al., Chapter 34) and *Santalum* sp. (sandalwood) (Evenson 1998) in central Lao P.D.R.

Many of these higher-quality timbers are longer term and may require a minimum of 20 to 30 years before harvest. Such a delayed return on investment is likely to be problematic for resource-poor shifting cultivators. With increasing scarcity of agricultural land, many of them cannot afford to lock their land into long-term tree crops without jeopardizing food security. Alternatively, fast-growing trees that reach a harvestable size within 8 to 12 years and can fit into the rhythm of the swidden cycle without disruption are probably more appropriate to smallholder conditions (see, for example, Magcale-Macandog et al., Chapter 37). Nurturing valued fast-growing trees in swidden fields is a traditional practice, as exemplified by management of *Melia azedarach* (neem) by Muóng, Tày, Thái, Cao Lan and Dzao in northern Vietnam (Vien, Chapter 36). Most examples in the region today, however, are recent innovations spurred by new market opportunities for pulpwood and small diameter poles, for which upland farmers enjoy a comparative advantage (Garrity and Mercado 1993). Construction of pulp mills in northern Vietnam has led to expanding hectarage of *Manglietia glauca* and *Styrax tonkinensis* in swidden fields (see color plate 47) (Sam 1994, *48*); Tala-andig swiddenists in the buffer zone of the Mt. Kitanglad National Park in Bukidnon, the Philippines, plant fallows with *Paraserianthes falcataria* to provide logs for a paper mill in Cagayan de Oro (see color plate 46) (author's field notes); and smallholder plantations of *Gmelina arborea* (see color plate 45), *Acacia mangium,* and *Eucalyptus* spp. have sprung up on swidden landscapes across the region. As argued by Pasicolan in his case study in Isabela Province, the Philippines (Chapter 63), the proper combination of property rights, markets, and institutions can lead to underutilized grasslands being transformed into tree-based systems.

Timber-based fallows have the serious constraint that access to nearby road infrastructure is essential for log extraction and transport. More isolated swidden communities need, instead, to think in terms of nontimber tree products that have high value, low volume, and low perishability. Cinnamon bark, for example, fits these criteria. Lucrative world prices in recent years have led to rapid expansion in swidden environments of *Cinnamomum burmannii* in West Sumatra, Indonesia (see color plates 49 and 50) (Suyanto et al., Chapter 65; Werner, Chapter 67), and *C. cassia* in Yen Bai Province of northern Vietnam (see color plate 51) (Hien, Chapter 62). In central Laos, farmer management of *Aquilaria* spp. in swiddens for harvest of the valuable resinous heartwood of fungus-infected trees (agar, aloeswood, gaharu) is another system with potential for wider application, particularly as recent work on fungal inoculation techniques appears promising. Rattan has historically been transported long distances to outside markets. Declining natural stocks led to its propagation in fallows in Yunnan, southwestern China (Chen et al. 1993a, b; Xu, Chapter 56), and Kalimantan, Indonesia (see color plate 52) (Weinstock 1983, 1984, 1985; Godoy and Feaw 1989; Godoy 1990; Peluso 1992; Sasaki, Chapter 38; Belcher, Chapter 64). With herbal treatments gaining increasing currency in the medical world, there may be similar opportunities for domestication of medicinal plants and their cultivation in fallow lands.

A second major constraint to the fallow systems discussed under this typology is that swidden cycles over much of Asia-Pacific have already accelerated to such a degree that the fallow period is now too short for even the fastest-growing trees to reach a harvestable stage. In such cases, bamboo comes into its element, providing short-term fallow species that are both effective and productive (see color plates 53 through 56) (Soemarwoto 1984; Soemarwoto and Soemarwoto 1984; Christanty et al. 1996; Bamualim et al., Chapter 39; Ty, Chapter 55). Propagation of *Broussonetia papyrifera* as a fallow crop in northern Laos (Fahrney et al., Chapter 40) is noteworthy here in that it shows that even short two- to three-year fallows can yield valuable tree and shrub products (see color plates 57 and 58).

The concept of taungya-planting valued perennials into swidden fields clearly has wide application. The income generated by these systems provides farmers with strong economic incentives to resist pressures to continually shorten fallow periods, thus shoring up the sustainability of their swidden systems. However, there is a need to assess the degree to which the fallow's ecological functions are compromised by these systems, since the harvest of many involve exporting significant quantities of biomass and contained nutrients from the swiddens.

Agroforests

If farmers do not restrict themselves to a single economic perennial but continue enrichment planting of other useful species into their fallows, this brings us to the final typology, agroforests. This division is defined primarily by the degree of floristic complexity. Classic examples of such agroforests can be found in descriptions of the Kenyah and Iban forest gardens, or *tembawang*, in Kalimantan, Indonesia (see color plates 59 and 60) (Brookfield, Chapter 2; Wadley, Chapter 41; de Jong, Chapter 43); the Ifugao woodlots, or *pinugo*, in northern Luzon, the Philippines (Conklin 1980); and the mixed tree crop gardens of the Hanunóo in Oriental Mindoro, also in the Philippines (Conklin 1957). Casual observation easily mistakes these systems for natural forests. Their structure mimics that of natural forests and the degree of biodiversity is often comparable. Such complex agroforests are believed to perform most of the same ecological functions as natural forest ecosystems, while providing a wide range of products for household consumption. These characteristics make them well suited for promotion in ecologically sensitive areas, such as the headwaters of important river systems or buffer zones around protected wildlands. They appear to strike an admirable balance between the production needs of local communities and the conservation agenda of wider society (Michon and de Foresta 1990; Michon and Widjayanto 1992; de Foresta and Michon 1994; Michon 1995).

Exposure to markets inevitably leads to specialization and, to varying degrees, agroforests become dominated by a single species whose products command attractive prices and that grows and produces well under local conditions. Hence, there is often a transition from mixed forest gardens to more species-based agroforests. This is well illustrated by some of the resin-producing agroforests that historically evolved across the region in response to world markets. Lucrative prices inspired development of agroforests based on *Shorea javanica* in West Lampung, Indonesia (see color plate 63) (Torquebiau 1984; Michon 1993; Michon et al. 1999, reprinted in this volume as Chapter 45); *Styrax benzoin* and *S. paralleloneuron*, both in northern Sumatra, Indonesia (Pelzer 1978, *278–279*; Watanabe et al. 1996); *S.*

tonkinensis (see color plate 64) (Pinyopusarerk 1994; Fischer et al., Chapter 46) and *S. benzoin*, both in northern Laos; and *Toxicodendron vernicifluum* (Long, Chapter 47) and *Pinus yunnanensis*, both in western Yunnan, China. All these systems are now under economic stress, however, as development of synthetic alternatives has undermined prices of natural tree resins.

Not all agroforest fallows are based on indigenous species. The introduction of *Hevea brasiliensis*, first to Singapore and then to other Southeast Asian countries late in the 19th century, triggered an explosion of fallow enrichment with rubber that had profound effects on the region's agricultural landscape (Thomas 1965; Barlow and Muharminto 1988; de Foresta 1992; Dove 1993; Gouyon et al. 1993; Lawrence et al., Chapter 42; Guangxia and Lianmin, Chapter 49; Werner, Chapter 67). "Jungle rubber," as it is now widely known, was established by inter-planting rubber seedlings into swidden fields during the cropping phase (see color plate 65) so that it dominated the subsequent fallow succession. Fruits, timbers, and other useful trees were also planted or protected, again creating a forest-like environment. Investment in the early systems was low, so farmers could afford to simply stop tapping when latex prices were low and, instead, harvest other useful products from the fallow. Tapping would resume whenever prices improved. The clonal rubber monocultures that now produce most of the world's rubber from plantations in southern Thailand and Malaysia also passed through a jungle rubber stage before government extension programs introduced superior germplasm and management techniques (Penot, Chapter 48). This serves as a stark example of how specialization in, and intensification of, a single element in a diverse agroforest can lead to a dramatic reduction in its diversity of species.

Spices were the lure that first attracted many western trading ships to Asian-Pacific waters. Many of them were grown in the uplands, and this early trade was remarkable for plugging even remote shifting cultivators into European markets centuries before free trade and globalization became buzzwords. Trade was further encouraged by the establishment of trading posts, and later, most of the region was colonized by European powers. In an era when roads were rudimentary or nonexistent, spices fitted the criteria of high value, low volume, and low perishability, and swidden fields provided ideal growing environments. Upland fields were transformed. *Amomum* (cardamom) agroforests were established across the Himalayan foothills (Singh et al. 1989; Sharma, Chapter 51; Dukpa et al., Chapter 59), *Piper nigrum* (pepper) plantations thrived in Sumatra (Pelzer 1978, *276–277*), and *Syzygium aromaticum* (cloves) spread across areas of northern Sulawesi, central Java, Lampung, and West Sumatra, all in Indonesia. These historical trends are now being echoed in northern Thailand, where recent market demands for the seeds of *Zanthoxylum limonella,* a spice used in Thai cooking, has stimulated farmers to begin planting this native tree in their swidden fields (Hoare et al., Chapter 50). Coffee (*Coffea* spp.) and tea (*Camellia sinensis*) were also high on the shopping lists of the colonial powers. Tea tended to be grown in large colonial-administered plantations (see color plate 66), except for the *miang* tea that was planted for local consumption in swidden fields or under natural forest canopy in Yunnan, southwestern China, and northern Thailand (Keen 1978). Coffee[2] and, later, cocoa (*Theobroma cacao*) were routinely established in swidden fields across the region through taungya plantings.

Palm taxa also warrant mention as economic fallow species in low-altitude swidden environments. Notable examples from Indonesia include expansion of *Cocos nucifera* (coconut palm) into shifting cultivation fields in Menado, Sulawesi (Pelzer 1978, *280–282*); plantings of *Arenga pinnata* (sugar palm) on fallow lands in northeastern Sumatra; and intensive management of the multipurpose *Borassus sundaicus* (lontar palm) in shifting cultivation systems on the islands of Roti and Savu in East Nusa Tenggara (Fox 1977). While nut trees are prevalent on South Pacific islands (Bourke 1998), they play a comparatively minor role in fallow enrichment on

---

2 For a description of the introduction of coffee into shifting cultivation areas of Papua New Guinea, see Allen, B.J. (1985).

mainland Asia. Exceptions are the conversion of swidden lands, usually through taungya planting, into cashew nut (*Anacardium* spp.) plantations in lower-altitude tropical zones, and walnut (*Juglans* spp.) plantations (see color plate 67) in the more temperate eastern Himalayas. Illipe nuts (*Shorea macrophylla*) are also an important element of the earlier-mentioned tembawang tree gardens in Borneo (Peters 1996).

Continued penetration of roads and marketing channels into the uplands has steadily broadened the variety of cash crops available to shifting cultivators, to include those that are bulky and highly perishable but which earn high prices from lowland markets. Upland farmers have been able to exploit their cooler climate to grow subtemperate vegetables, flowers and, important to this discussion, a wide range of subtropical to temperate fruits (see color plate 68) (Eder 1981). New market access has, for example, sparked the rapid expansion of apple (*Malus* spp.) and orange (*Citrus* spp.) production in Bhutan, destined for the vast Indian market (Dukpa et al., Chapter 59); litchi (*Nephelium litchi*) orchards, first in northern Thailand and later expanding on fallow lands in Bacgiang Province of northern Vietnam (Long, Chapter 53; Hiên, Chapter 62), as well as passion fruit production (*Passiflora* spp.) in the highlands of Sumatra, for shipment to large markets in Java. These horticultural activities may, in some cases, be new introductions. More often, they represent expansion and intensification of an existing element of the farming system that has become more remunerative. Salafsky (1993, *58*), for example, describes how in the early 1970s the arrival of roads and markets to the township of Benawai Agung, in West Kalimantan, elevated durian (*Durio zibethinus*) from a fruit grown for home use in diverse forest gardens to a valuable commodity (see color plate 62). This sparked a durian boom, as farmers expanded existing orchards and planted new ones. It is likely that the now-commercial banana-based agroforests, or *sagui gru,* of the Karen in western Thailand (see color plate 61) (Srithong, Chapter 52) had their origins in small-scale cultivation for household needs. These examples again illustrate how economic signals may inspire intensification and expansion of a single profitable component of diverse forest gardens, pushing them toward more specialized orchards.

This brings us to the extreme left pole of the IFM continuum and, literally, to the forest margins. Progression from the starting point through these tree-based typologies is generally marked by extended fallows, increasing tree cover on the landscape, and declining importance of arable cultivation. To continue further would mean the cropping phase would disappear completely and we would drop off the end of the continuum into the realm of pure forestry. While it is clearly stretching definitions to consider some of these latter systems as "managed fallows," their inclusion is intended to display the full and logical progression of continued fallow enrichment with trees. Even the complex agroforests, unless regenerated naturally or through gap-planting, will, at the end of their productive life, be subjected to another phase of slash-and-burn and renewed through replanting of seedlings between arable crops. Jungle rubber, for example, declines in latex production after about 30 years and requires renewal by such a process.[3] Hence, the remnants of the crop-fallow rotation that is integral to shifting cultivation continue to be discernable, but the farmers' emphasis has shifted to the fallow vegetation. The cropping phase is now a means to rejuvenate the fallow vegetation, rather than vice versa. Market access is more critical to these economic fallows since farmers increasingly rely on the sale of tree products to finance purchases of food staples.

Although some will argue that the economic productivity of these tree-based systems disqualifies them from being considered as fallows, such systems also perform critical ecological services at both field and watershed levels. Indeed, some of the case studies in Section Two demonstrate that even conventional fallows produce

[3] However, recent surveys in the Muara Bungo area of Jambi Province, Sumatra, suggest that 15% to 50% of jungle rubber producers there interplant rubber seedlings within older stands in a practice known as *sisipan* (Laxman 1999)—an alternative to the more conventional slash-and-burn method to rejuvenate old rubber stands.

a wide range of economic products, in addition to performing their vital ecological functions. It is difficult, therefore, to fix absolute classifications for many of these systems, so that they fit tidily into self-contained boxes. Rather, their classification becomes a question of degree.

Concluding Comments

This thumbnail sketch is intended only to introduce the IFM continuum that provides the structure of this book. Sections Two through Seven examine case studies representative of each typology. Section Eight presents chapters that discuss multiple IFM systems cutting across several typologies. Thematic chapters are presented in Section Nine, dealing with the broad issues of property rights, markets, and institutions as they affect improved management of fallow lands. Finally, Section Ten brings the volume to closure by drawing out important lessons from the case studies. Readers are assisted in locating information by geographic area in an opening map at the front of the book and by plant species, ethnic groups, and general subjects in separate indexes at the back.

This chapter does not suggest redefining terms that are fundamental to this discussion, such as "shifting cultivation," "fallow," and "fallow management." Indeed, wide interpretation of these terms makes it debatable where vertical lines should be drawn through each side of the continuum to indicate the points at which a "managed fallow" becomes a crop and, by extension, when "shifting cultivation" has intensified into permanent land use. That debate is left to another forum. The innovations described in this volume are all farmer-developed and farmer-tested strategies that assist in the sustainable intensification of shifting cultivation systems, regardless of what they are called.

This volume contains overwhelming evidence that the popular perception of fallows as unmanaged, abandoned, and unproductive, is seriously mistaken. Precisely this misinformed stereotype underlies the bad reputation of shifting cultivation as a wasteful and primitive land-use system. At the dawn of a new millennium, an important step in the right direction will be to adopt a more humble attitude, listen to the subaltern voices from the forest, and learn from farmers.

Acknowledgments

This work was made possible by generous support from the International Development Research Centre (IDRC) of Canada. The focus of this research was the brainchild of Joachim Voss, former Senior Research Manager of IDRC, Ottawa, who is now Director General of the Centro Internacional de Agricultura Tropical (CIAT), in Cali, Colombia. Special thanks are also owed to John Graham and Daniel Buckles at IDRC and Meine van Noordwijk at ICRAF for their invaluable mentoring and encouragement in exploring fallow management in the Asia-Pacific region. Most of all, though, the insights in this chapter are owed to the generosity of the many shifting cultivators across the region who contributed their time and wisdom. Final thanks are reserved for Tossaporn Kurupunya, who has always shown more patience than any man has a right to expect.

References

Agus, F. 2006. Use of *Leucaena leucocephala* to Intensify Indigenous Fallow Rotations in Sulawesi, Indonesia. Chapter 25.

Allen, B.J. 1985. Dynamics of Fallow Successions and Introduction of Robusta Coffee in Shifting Cultivation Areas in the Lowlands of Papua New Guinea. *Agroforestry Systems*, 3(3), 227–238.

Askin, D.C., D.J. Boland, and K. Pinyopusarerk. 1990. Use of *Casuarina oligodon* ssp. *abbreviata* in Agroforestry in the North Baliem Valley, Irian Jaya, Indonesia. In: *Advances in Casuarina Research and Utilization. Proceedings of the Second International Casuarina Workshop*, edited by M.H. El-Lakany, J.W. Turnbull, and J.L. Brewbaker. Cairo, Egypt: Desert Development Centre, 213–219.

Balbarino, E.A., D.M. Bates, Z.M. de la Rosa, and J. Itumay. 2006. Improved Fallows using a Spiny Legume, *Mimosa invisa* Martius ex Colla, in Western Leyte, Philippines. Chapter 18.
Bamualim, A., J. Triastono, E. Hosang, T. Basuki, and S.P. Field. 2006. Bamboo as a Fallow Crop on Timor Island, Nusa Tenggara Timur, Indonesia. Chapter 39.
Barlow, C., and Muharminto. 1988. *Smallholder Rubber in South Sumatra: Towards Economic Improvement*. Bogor, Indonesia, and Canberra: Balai Penelitian Perkebunan, Bogor, and the Australian National University.
Baxter, J. 1995. *Chromolaena odorata*: Weed for the Killing or Shrub for the Tilling? *Agroforestry Today*, 7(2), 6–8.
Belcher, B.M. 2006. The Feasibility of Rattan Cultivation within Shifting Cultivation Systems: The Role of Policy and Market Institutions. Chapter 64.
Bino, B. 2006. Swidden Agriculture in the Highlands of Papua New Guinea. Chapter 60.
Bourke, R.M. 2006. Managing the Species Composition of Fallows in Papua New Guinea by Planting Trees. Chapter 31.
———. 1998. Personal communication with the author.
Brookfield, H. 2006. Working with Plants, and for Them: Indigenous Fallow Management in Perspective. Chapter 2.
Buckles, D., and O. Erenstein. 1996. *Intensifying Maize-Based Farming Systems in the Sierra de Santa Marta, Veracruz*. Mexico, D.F., Mexico: Centro Internacional de Mejoramiento de Maíz y Trigo (CIMMYT).
———, and H. Perales. 1995. *Farmer-Based Experimentation with Velvet Bean: Innovation within Tradition*. Mexico, D.F., Mexico: Centro Internacional de Mejoramiento de Maíz y Trigo (CIMMYT).
———, B. Triomphe, and G. Sain. 1998a. *Cover Crops in Hillside Agriculture: Farmer Innovation with* Mucuna. Ottawa and Mexico City: International Development Research Centre and Centro Internacional de Mejoramiento de Maíz y Trigo (CIMMYT).
———, A. Eteka, O. Osiname, M. Galiba, and G. Galiano. 1998b. *Green Manure Cover Crops Contributing to Sustainable Agriculture in West Africa*. Canada: International Development Research Centre.
Bunch, R. 1995. *The Use of Green Manures by Village Farmers: What We Have Learned to Date*. Technical Report No. 3, 2nd ed. Tegucigalpa MDC, Honduras: CIDICCO.
———. 1997. Personal communication with the author.
Burgers, P. 2006. Commercialization of Fallow Species by Bidayuh Shifting Cultivators in Sarawak, Malaysia. Chapter 8.
Cairns, M.F. 2006. Management of Fallows Based on *Austroeupatorium inulaefolium* by Minangkabau Farmers in Sumatra, Indonesia. Chapter 15.
———, S. Keitzar, and T.A. Yaden. 2006. Shifting Forests in Northeast India: Management of *Alnus nepalensis* as an Improved Fallow in Nagaland. Chapter 30.
Chapman, E.C., B. Bouahom, and P.K. Hansen (eds.). 1998. *Upland Farming Systems in the Lao P.D.R.: Problems and Opportunities for Livestock*. ACIAR Proceedings No. 87. Canberra: Australian Council for International Agricultural Research (ACIAR).
Chen, S.Y., S.J. Pei, and J.C. Xu. 1993a. Indigenous Management of the Rattan Resources in the Forest Lands of Mountain Environments: The Hani Practice in the Mengsong Area of Yunnan, China. *Ethnobotany* 5, 93–99.
———. 1993b. Ethnobotany of Rattan in Xishuangbanna. In: *Collected Papers on Studies of Tropical Plants*, Vol. 2. Kunming, China: Yunnan University Press, 75–85.
Christanty, L., D. Mailly, and J.P. Kimmins. 1996. Without Bamboo, the Land Dies: Biomass, Litterfall, and Soil Organic Matter Dynamics of a Javanese Bamboo Talun-Kebun System. *Forest Ecology and Management* 87, 75–88.
Conklin, H.C. 1957. *Hanunóo Agriculture: A Report on an Integral System of Shifting Cultivation in the Philippines*. FAO Forestry Development Paper No. 12. Rome: United Nations Food and Agriculture Organization.
———. 1980. *Ethnographic Atlas of Ifugao: A Study of Environment, Culture, and Society in Northern Luzon*. New Haven, CT, and London: Yale University Press.
Daguitan, F.M., and M. Tauli. 2006. Indigenous Fallow Management Systems in Slected Areas of Cordillera, Philippines. Chapter 57.
Daniels, C. 1995. Environmental Degradation, Forest Protection and Ethno-History in Yunnan: Traditional Practices of Non-Han Swidden Cultivators for the Protection of Forests. *China Environmental History Newsletter* 2(1), 4–6.
de Foresta, H. 1992. Botany Contribution to the Understanding of Smallholder Rubber Plantations in Indonesia: An Example from South Sumatra. Paper presented to BIOTROP Symposium *Sumatra Lingkungan dan Pembangunan*, Bogor, Indonesia.
———, and G. Michon. 1994. Agroforests in Sumatra. *Agroforestry Today* Oct.–Dec. 1994.
de Jong, W. 2006. Forest Management and Classification of Fallows by Bidayuh Farmers in West Kalimantan. Chapter 43.
Dhyani, S.K. 1998. Tribal Alders: An Agroforestry System in the Northeastern Hills of India. *Agroforestry Today* 10(4). Nairobi, Kenya: International Centre for Research in Agroforestry.
Djogo, A.P.Y., M. Juhan, A. Aoetpah, and E. McCallie. 2006. Adoption and Management of *Tecoma stans* Fallows in Semiarid East Nusa Tenggara. Chapter 17.
Dove, M.R. 1986. The Practical Reason of Weeds in Indonesia: Peasant vs. State View of *Imperata* and *Chromolaena*. *Human Ecology* 14, 163–190.

———. 1993. Smallholder Rubber and Swidden Agriculture in Borneo: A Sustainable Adaptation to the Ecology and Economy of Tropical Forest. *Economic Botany* 47, 136–147.
Dukpa, T., P. Wangchuk, Rinchen, K. Wangdi, and W. Roder. 2006. Changes and Innovations in the Management of Shifting Cultivation Land in Bhutan. Chapter 59.
Durno, J.L., T. Deetes, and J. Rajchaprasit. 2006. Natural Forest Regeneration from an *Imperata* Fallow: The Case of Pakhasukjai. Chapter 12.
Eder, J.F. 1981. From Grain Crops to Tree Crops in the Cuyunon Swidden System. In: *Adaptive Strategies and Change in Philippine Swidden-Based Societies*, edited by H. Olofson. Laguna, Philippines: Forest Research Institute College, 91–104.
Eslava, F.M. 1984. The Naalad Style of Upland Farming in Naga, Cebu, Philippines: A Case of an Indigenous Agroforestry Scheme. Country Report presented to a course on Agroforestry. October 1–20, 1984, Universiti Pertanian, Malaysia.
Evenson, J.P. 1998. Personal communication with the author.
Fahrney, K., O. Boonnaphol, B. Keoboulapha, and S. Maniphone. 2006. Indigenous Management of Paper Mulberry in Swidden Rice Fields and Fallows in Northern Lao P.D.R. Chapter 40.
Falanruw, M.V.C., and F. Ruegorong. 2006. Indigenous Fallow Management on Yap Island. Chapter 44.
Field, S.P. 1991. *Chromolaena odorata*: Friend or Foe for Resource Poor Farmers. *Chromolaena* Newsletter, May 1991.
———. 1996. Personal communication with the author.
——— and H.G. Yasin. 1991. The Use of Tree Legumes as Fallow Crops to Control Weeds and Provide Forage as a Basis for a Sustainable Agricultural System. Paper presented to the 13th Asian-Pacific Weed Science Society Conference, Taipei, Taiwan.
Fischer, M., S. Savathvong, and K. Pinyopusarerk. 2006. Upland Fallow Management with *Styrax tonkinensis* for Benzoin Production in Northern Lao P.D.R. Chapter 46.
Fisher, L. 1996. Personal communication with the author.
Foerster, E. 1997. Personal communication with the author.
Fox, J.J. 1977. *Harvest of the Palm: Ecological Change in Eastern Indonesia*. Cambridge, MA: Harvard University Press.
Fu, H. 2003. Study on Alder-based Shifting Cultivation in Yunnan, China. Unpublished MSc thesis, Xishuangbanna Tropical Botanical Garden, Chinese Academy of Science.
Garrity, D.P., and A. Mercado. 1993. Reforestation through Agroforestry: Market Driven Smallholder Timber Production on the Frontier. In: *Marketing of Multipurpose Tree Products in Asia*. Proceedings of an international workshop. December 6–9, 1993, Baguio City, Philippines, 265–268.
Godoy, R.A. 1990. The Economics of Traditional Rattan Cultivation. *Agroforestry Systems* 12, 163–172.
———, and T.C. Feaw. 1989. The Profitability of Smallholder Rattan Cultivation in Southern Borneo, Indonesia. *Human Ecology* 17, 347–363.
Gokhale, A.M., D.K. Zeliang, R. Kevichusa, and T. Angami. 1985. *Nagaland: The Use of Alder Trees*. Kohima, Nagaland: State Council of Educational Research and Training, Education Department.
Gouyon, A., H. de Foresta, and P. Levang. 1993. Does "Jungle Rubber" Deserve Its Name? An Analysis of Rubber Agroforestry Systems in Southeast Sumatra. *Agroforestry Systems* 22(3), 181–206.
Grist, P., K. Menz, and R. Nelson. 2006. Multipurpose Trees as an Improved Fallow: An Economic Assessment. Chapter 32.
Guangxia, C., and Z. Lianmin. 2006. Preliminary Study of Rubber Plantations as an Alternative to Shifting Cultivation in Yunnan Province, China. Chapter 49.
Guo, H., Y. Xia, and C. Padoch. 2006. *Alnus nepalensis*-based Agroforestry Systems in Yunnan, Southwest China. Chapter 29.
Hairiah, K. 1997. Personal communication with the author.
Hansen, P.K., H. Sodarak, and S. Savathvong. 2006. Teak Production by Shifting Cultivators in Northern Lao P.D.R. Chapter 34.
Hao, N.T., H.V. Huy, H.D. Nhan, and N.T.T. Thuy. 2006. Benefits of Nho Nhe bean (*Phaseolus calcaratus* Roxb., syn. *Vigna umbellata*) in Upland Farming in Northern Vietnam. Chapter 22.
Hartemink, A.E. 2006. *Piper aduncum* Fallows in the Lowlands of Papua New Guinea. Chapter 16.
Hien, T.Q. 2006. Some Indigenous Experiences in Intensification of Shifting Cultivation in Vietnam. Chapter 62.
Hoare, P., B. Maneeratana, and W. Songwadhana. 2006. *Ma Kwaen* (*Zanthoxylum limonella*): A Jungle Spice Used in Swidden Intensification in Northern Thailand. Chapter 50.
Hong, J., J. Wang, and W. Feng. 1960. Social Economic Investigation Report on Dulong People in the 4th District, Gongshan County. In: *China's Minority Group Social Economic Investigation Material*, Kunming, China: Yunnan Ethnological Press.
Horne, P. 2006. Farmer-Developed Forage Management Strategies for Stabilization of Shifting Cultivation Systems. Chapter 10.
Jones, P.H. 1983. *Leucaena* and the Amarasi Model from Timor. *Bulletin of Indonesian Economic Studies* 19(3), 106–112.

Kanjunt, C. 2006. Successional Forest Development in Swidden Fallows of Different Ethnic Groups in Northern Thailand. Chapter 5.

Keen, F.G.B. 1978. The Fermented Tea (*Miang*) Economy of Northern Thailand. In: *Farmers in the Forest: Economic Development and Marginal Agriculture in Northern Thailand,* edited by P. Kunstadter, E.C. Chapman, and S. Sabhasri. Honolulu, HI: The University Press of Hawaii, 271–286.

Kevichusa, R., V. Lieze, and V. Nakhro. n.d. Alnus nepalensis: *Alder*. Kohima, Nagaland: Government of Nagaland and the World Bank.

Kieft, J.A.M. 2006. Farmers' Use of *Sesbania grandiflora* to Intensify Swidden Agriculture in North Central Timor. Chapter 28.

Kung'u, J.B. 1993. Biomass Production and Some Soil Properties under a *Leucaena leucocephala* Fallow System in Cebu, Philippines. MSc thesis, University of the Philippines at Los Baños, Laguna, Philippines.

Lasco, R.D. 2006. The Naalad Improved Fallow System in the Philippines and Its Implications for Global Warming. Chapter 27.

———, and P.D. Suson. 1997. A *Leucaena leucocephala*-based Improved Fallow System in Central Philippines: The Naalad System. Paper presented to an International Conference on Short-Term Improved Fallow Systems, March 11–15, 1997, Lilongwe, Malawi.

Lawrence, D., D. Astiani, M. Syhazaman-Karwur, and I. Fiorentino. 2006. Does Tree Diversity Affect Soil Fertility? Initial Findings from Fallow Systems in West Kalimantan. Chapter 42.

Laxman, J. 1999. Field Report: Acquisition of Ecological Knowledge from Jambi Farmers.

Lin W., J. Jiang, W. Li, G. Xie, and Y. Wan. 2006. Growing Ya Zhou Hyacinth Beans in the Dry Season on Hainan Island, China. Chapter 20.

Littooy, S. (ed.). 1989. *Local Farming Technologies Related to Soil Conservation and Tree Planting in Selected Districts of Vinh Phu, Ha Tuyen, and Hoang Lien Son.* Vietnam: Plantation and Soil Conservation Project.

Litzenberger, S.C., and H.T. Lip. 1961. Utilizing *Eupatorium odoratum* L. to Improve Crop Yields in Cambodia. *Agronomy Journal* 53, 321–324.

Long, C. 2006. The Lemo System of Lacquer Agroforestry in Yunnan, Southwestern China. Chapter 47.

Long, T. 2006. Sandiu Farmers' Improvement of Fallows on Barren Hills in Northern Vietnam. Chapter 53.

MacDicken, K.G. 1990a. Agroforestry Management in the Humid Tropics. In: *Agroforestry Classification and Management,* edited by K.G. MacDicken and N.T. Vergara. New York: John Wiley & Sons, 98–149.

———. 1990b. *Leucaena leucocephala* as a Fallow Improvement Crop in Shifting Cultivation on the Island of Mindoro, Philippines. Paper presented to a conference on Research on Multipurpose Trees in Asia, November 19–23, 1990, Los Baños, Philippines.

———. 1991. Impacts of *Leucaena leucocephala* as a Fallow Improvement Crop in Shifting Cultivation on the Island of Mindoro, Philippines. *Forest Ecology and Management* 45, 185–192.

———. 2006. Upland Rice Response to *Leucaena leucocephala* Fallows on Mindoro, Philippines. Chapter 26.

Magcale-Macandog, D.B., and P.M. Rocamora. 2006. Cost-Benefit Analysis of a *Gmelina* Hedgerow Improved Fallow System in Northern Mindanao, Philippines. Chapter 37.

Maneeratana, B., and P. Hoare. 2006. When Shifting Cultivators Migrate to the Cities, How Can the Forest Be Rehabilitated? Chapter 13.

Menzies, N. 1988. *Three Hundred Years of Taungya: A Sustainable System of Forestry in South China.* Plenum Publishing Corporation, 361–375.

———, and N. Tapp. 2006. Fallow Management Strategies in the Borderlands of Southwest China: The Case of *Cunninghamia lanceolata*. Chapter 35.

Mertz, O. 2006. The Potential of Wild Vegetables as Permanent Crops or to Improve Fallows in Sarawak, Malaysia. Chapter 7.

Metzner, J.K. 1981. Old in the New: Autochthonous Approach towards Stabilizing an Agroecosystem: The Case from Amarasi, Timor. *Applied Geography and Development* 17, 1–17.

———. 1983. Innovations in Agriculture Incorporating Traditional Production Methods: The Case of Amarasi, Timor. *Bulletin of Indonesian Economic Studies* 19(3), 94–105.

Michon, G. 1993. The Damar Gardens: Existing Buffer Zones Adjacent to Barisan Selatan National Park. *ITTO Tropical Forest Management Update* 3(3), 7–8.

———. 1995. The Indonesian Agroforest Model: Forest Resources Management and Biodiversity Conservation. In: *Concerning Biodiversity Outside Protected Areas: The Role of Traditional Agroecosystems,* edited by P. Halladay and D. Gilmour. IUCN/AMA.

———, and H. de Foresta. 1990. Complex Agroforestry Systems and Conservation of Biological Diversity. Agroforests in Indonesia: The Link Between Two Worlds. In: *Proceedings of International Conference on the Conservation of Tropical Biodiversity*. Kuala Lumpur: Malayan Nature Society.

———, and N. Widjayanto. 1992. Complex Agroforestry Systems in Sumatra. Paper presented to a workshop on Sumatra, Environment and Development: Its Past, Present and Future, September 16–18, 1992, Bogor, Indonesia. Biotrop Special Publication No. 46.

———, H. de Foresta, A. Kusworo, and P. Levang. 1999. The Damar Agro-Forests of Krui Indonesia: Justice for Forest Farmers. In: *People, Plants and Justice,* edited by C. Zerner. New York: Columbia University Press, Chapter 45.
Olofson, H. 1983. Indigenous Agroforestry Systems. *Philippine Quarterly of Culture & Society* 11, 149–174.
Ongprasert, S., and K. Prinz. 2006. Viny Legumes as Accelerated Seasonal Fallows: Intensifying Shifting Cultivation in Northern Thailand. Chapter 23.
Ormeling, F.J. 1955. *The Timor Problem. A Geographical Interpretation of an Underdeveloped Island.* Groningen and Jakarta: JB Wolters.
Oyen, L. 1995. Aggressive Colonizers Work for the Farmers. *ILEIA Newsletter for Low External Input and Sustainable Agriculture* 11(3), Leusden, Netherlands: Information Centre for Low-External-Input and Sustainable Agriculture, 10–11.
Pandeya, C.N. 1995. We Love and Protect It. *ILEIA Newsletter for Low External Input and Sustainable Agriculture* 11(3). Leusden, Netherlands: Information Centre for Low-External-Input and Sustainable Agriculture, 8.
Pasicolan, P.N. 2006. Productive Management of Swidden Fallows: Market Forces and Institutional Factors in Isabela, Philippines. Chapter 63.
———. 1997. Personal communication with the author.
Peluso, N.L. 1992. The Rattan Trade in East Kalimantan, Indonesia. In: *Non-Timber Products from Tropical Forests: Evaluation of a Conservation and Development Strategy,* edited by D.C. Nepstad and S. Schwartzman. In: *Advances in Economic Botany* 9. New York: The New York Botanical Garden, 115–127.
Pelzer, K.J. 1978. Swidden Cultivation in Southeast Asia: Historical, Ecological, and Economic Perspectives. In: *Farmers in the Forest: Economic Development and Marginal Agriculture in Northern Thailand,* edited by P. Kunstadter, E.C. Chapman, and S. Sabhasri. Honolulu, HI: The University Press of Hawaii, 277–286.
Penot, E. 2006. From Shifting Cultivation to Sustainable Jungle Rubber: A History of Innovations in Indonesia. Chapter 48.
Peters, C.M. 1996. Illipe Nuts (*Shorea* spp.) in West Kalimantan: Use, Ecology and Management Potential of an Important Forest Resource. In: *Borneo in Transition: People, Forests, Conservation and Development,* edited by C. Padoch and N.L. Peluso. Kuala Lumpur, Malaysia: Oxford University Press, 230–244.
Piggin, C.M. 2006. The Role of *Leucaena* in Swidden Cropping and Livestock Production in Nusa Tenggara Timur. Chapter 24.
———, and V. Parera. 1985. The Use of *Leucaena* in Nusa Tenggara Timur. In: *ACIAR Proceedings Series No. 3.* Canberra, Australia: Australian Centre for International Agricultural Research, 19–27.
Pinyopusarerk, K. 1994. *Styrax tonkinensis*: Taxonomy, Ecology, Silviculture and Uses, In: *ACIAR Technical Report 31,* Canberra, Australia: Australian Centre for International Agricultural Research.
Potter, L., and J. Lee. 2006. Selling *Imperata*: Managing Grasslands for Profit in Indonesia and Laos. Chapter 11.
Prinz, K., and S. Ongprasert. 2006. Management of *Mimosa diplotricha* var. *inermis* as a Simultaneous Fallow in Northern Thailand. Chapter 19.
Raintree, J.B., and K. Warner. 1986. Agroforestry Pathways for the Intensification of Shifting Cultivation. *Agroforestry Systems* 4(1). Dordrecht, The Netherlands: Kluwer Academic Publishers, 39–54.
Ramakrishnan, P.S. 2006. Indigenous Fallow Management based on *Flemingia vestita* in Northeast India. Chapter 21.
Rerkasem, K. 1996. Personal communication with the author.
Roder, W., S. Phengchanh, B. Keoboualapha, and S. Maniphone. 1995a. *Chromolaena odorata* in Slash-and-Burn Rice Systems of Northern Laos. *Agroforestry Systems* 31. Dordrecht, The Netherlands: Kluwer Academic Publishers, 79–92.
———, B. Phouaravanh, S. Phengchanh, and B. Keoboualapha. 1995b. Relationships between Soil, Fallow Period, Weeds, and Rice Yield in Slash-and-Burn Systems of Laos. *Plant and Soil* 176, 27–36.
———, B. Keoboualapha, and V. Manivanh. 1995c. Teak (*Tectona grandis*), Fruit Trees and other Perennials used by Hill Farmers of Northern Laos. *Agroforestry Systems* 27. Dordrecht, The Netherlands: Kluwer Academic Publishers, 1–14.
———, S. Maniphone, B. Keoboualapha, and K. Fahrney. 2006. Fallow Improvement with *Chromolaena odorata* in Upland Rice Systems of Northern Laos. Chapter 14.
Salafsky, N. 1993. The Forest Garden Project: An Ecological and Economic Study of a Locally Developed Land Use System in West Kalimantan, Indonesia. Doctoral Dissertation, Department of Environmental Studies, Duke University.
Sam, D.D. 1994. *Shifting Cultivation in Vietnam: Its Social, Economic and Environmental Values Relative to Alternative Land Use.* London: International Institute for Environment and Development.
Sasaki, H. 2006. Innovations in Swidden-Based Rattan Cultivation by Benuaq-Dayak Farmers in East Kalimantan, Indonesia. Chapter 38.
Schmidt-Vogt, D. 2006. Relict Emergents in Swidden Fallows of the Lawa in Northern Thailand: Ecology and Economic Potential. Chapter 4.

Scoones, I., M. Melnyk, and J.N. Pretty. 1992. *The Hidden Harvest: Wild Foods and Agricultural Systems. A Literature Review and Annotated Bibliography*. London: International Institute for Environment and Development.

Sharma, R. 2006. *Alnus*-Cardamom Agroforestry: Its Potential for Stabilizing Shifting Cultivation in the Eastern Himalayas. Chapter 51.

Singh, K.A., R.N. Rai, Patiram, and D.T. Bhutia. 1989. Large Cardamom (*Amomum subulatum* Roxb) Plantation: An Age Old Agroforestry System in the Eastern Himalayas. *Agroforestry Systems* 9, 241–257.

Soemarwoto, O. 1984. The Talun-Kebun System: A Modified Shifting Cultivation in West Java. *The Environmentalist* 4, Suppl. 7, 96–98.

———, and I. Soemarwoto. 1984. The Javanese Rural Ecosystem. In: *An Introduction to Human Ecology Research on Agricultural Systems in Southeast Asia*, edited by A.T. Rambo and P.E. Sajise. Hawaii and Los Baños, Laguna, Philippines: East-West Environment and Policy Institute and University of the Philippines, 254–287.

Srithong, P. 2006. The *Sagui Gru* System: Karen Fallow Management Practices to Intensify Land Use in Western Thailand. Chapter 52.

Stoutjesdijk, J.A.J.H. 1935. *Eupatorium pallescens* D.C. op Sumatra's Westkust. (*Eupatorium pallescens* D.C. on the West Coast of Sumatra.) *Tectona* 28, 919–926.

Subere, V.S., E.B. Alberto, R.V. Dalmacio, F.M.J. Eslava, and M.V. Dalmacio. 1985. The Naalad Style of Upland Farming in Naga, Cebu, Philippines: A Case of an Indigenous Agroforestry Scheme. In: *Report on the Third ICRAF / USAID Agroforestry Course,* October 1–19, 1984, Serdang, Selangor, Malaysia, 71–108.

Surata, K. 1993. Amarasi System: Agroforestry Model in the Savanna of Timor Island, Indonesia. (Paper for National Agroforestry Workshop, Pusat Litbang Hutan dan Konservasi Alam–APAN, Bogor, Indonesia, August 24–25, 1993.) *Savanna* No. 8/93: 15–23.

Suyanto, S., T. Tomich, and K. Otsuka. 2006. The Role of Land Tenure in the Development of Cinnamon Agroforestry in Kerinci, Sumatra. Chapter 65.

Tangan, F.T. 2006. Wild Food Plants as Alternative Fallow Species in the Cordillera Region, Philippines. Chapter 9.

Tayanin, D. 2006. Kammu Fallow Management in Lao P.D.R., with Emphasis on Bamboo Use. Chapter 6.

Thiagalingam, K., and F.N. Famy. 1981. The Role of *Casuarina* under Shifting Cultivation: A Preliminary Study. In: *Nitrogen Cycling in Southeast Asian Wet Monsoonal Ecosystems*, edited by R. Weisilaar, J.R. Simpson, and T. Rosswall. Canberra, Australia: Australian Academy of Sciences, 154–156.

Thomas, K.D. 1965. Shifting Cultivation and Smallholder Rubber Production in a South Sumatran Village. *The Malayan Economic Review* 10, 100–115.

Torquebiau, E. 1984. Man-made Dipterocarp Forest in Sumatra. *Agroforestry Systems* 2(2), 103–128.

Troup, R.S. 1921. *The Silviculture of Indian Trees*, Vol. III, *Lauraceae to Coniferae*. London, U.K.: Clarendon Press.

Ty, H.X. 1997. Personal communication with the author.

———. 2006. Rebuilding Soil Properties during the Fallow: Indigenous Innovations in the Highlands of Vietnam. Chapter 55.

van Noordwijk, M. 1996. Personal communication with the author.

Vergara, N.T. 1995. Technology in the Uplands: Development, Assessment, and Dissemination. Paper presented to the Third National Conference on Research in the Uplands, September 5–9, 1995, Cagayan de Oro City, Philippines, SEARSOLIN.

Viên, T.D. 2006. Indigenous Fallow Management with *Melia azedarach* Linn. in Northern Vietnam. Chapter 36.

———. 1998. Personal communication with the author.

Wadley, R.L. 2006. The Complex Agroforests of The Iban of West Kalimantan and their Possible Role in Fallow Management and Forest Regeneration. Chapter 41.

Watanabe, H., K.I. Abe, K. Kawai, and P. Siburian. 1996. Sustained Use of Highland Forest Stands for Benzoin Production from *Styrax* in North Sumatra, Indonesia. *Wallaceana* 78, 15–19.

Weinstock, J.A. 1983. Rattan: Ecological Balance in a Borneo Rainforest Swidden. *Economic Botany* 37(1), 58–68.

———. 1984. Rattan: A Compliment to Swidden Agriculture? *Unasylva* 36, 16–22.

———. 1985. Alternate Cycle Agroforestry. *Agroforestry Systems* 3, 387–397.

Werner, S. 2006. The Development of Managed Fallow Systems in the Changing Environment of Central Sumatra. Chapter 67.

Wint, S.M. 1996. *Review of Shifting Cultivation in Myanmar*. Yangon, Myanmar: Forest Resource, Environment, Development, and Conservation Association.

Xu, J.C. 2006. Rattan and Tea-Based Intensification of Shifting Cultivation by Hani Farmers in Southwestern China. Chapter 56.

Yuksel, N. 1998. The Amarasi Model: An Example of Indigenous Natural Resource Management. Occasional Paper No. 1, Indigenous Fallow Management. Bogor, Indonesia: ICRAF (World Agroforestry Centre) Southeast Asian program.

Zeng, L. 1984. Alnus nepalensis, *Main Silvicultural Trees of Yunnan*. Kunming: Yunnan Peoples' Press, 119-122.

PART II

Retention or Promotion of Volunteer Species with Economic or Ecological Value

An elder farmer in Mongar, Bhutan.

Chapter 4

Relict Emergents in Swidden Fallows of the Lawa in Northern Thailand

Ecology and Economic Potential

Dietrich Schmidt-Vogt

Since the 1960s, land use in the mountains of northern Thailand has undergone rapid changes, mainly as a result of population growth, improved access, and more effective government control (Schmidt-Vogt 2000). Swidden farming, as the formerly dominant form of land use in this region, is particularly affected by these changes. At elevations above 700 m above sea level (asl), swidden farming is carried out by ethnic groups that are collectively referred to as "hilltribes." Writers on northern Thailand (Credner 1935; Uhlig 1969; Grandstaff 1980; Kunstadter 1980; Hansen 2001) have grouped the many variations of swidden farming practiced by these people into two categories:

- A "sustainable" form of rotational swiddening practiced by long-established highland minorities, such as the Lawa and Karen, who have inhabited the highlands, pursuing a more or less settled way of life, for several hundred years. Their farming system, which consists of short cultivation and long fallow periods, was, in the past, aimed primarily at the cultivation of rice for subsistence. Secondary forests develop rapidly on fallowed swiddens, which are reopened again for cultivation after a period of at least 12 years. The practice does not differ fundamentally from one group to the next.
- More intensive forms of swiddening are pursued by recently established highland minorities such as the Hmong, Akha, Lahu, and Lisu, who migrated to northern Thailand after the middle of the 19th century. These groups have a variety of farming practices, but most are characterized by longer cultivation periods, more intensive cultivation, a strong emphasis on cash crops, the custom of abandoning settlements once the accessible land is exhausted, and a preference for primary forests, as long as these are available, for the establishment of fields.

Changes have been brought about by population growth, which has reduced the amount of available land. This has forced practitioners of rotational swiddening to shorten fallow periods and has limited the ability of those groups with a more intensive form of swiddening to abandon settlements and claim new land. Moreover, the authorities are pressing swidden farmers to abandon or, at least, to limit swiddening for reasons of nature conservation and watershed protection and to hand over swidden land for reforestation. At the same time, swidden farmers are being persuaded by both government and market incentives to convert to permanent farming systems. Projects and advisory institutions are promoting the production of

Dietrich Schmidt-Vogt, Associate Professor, School of Environment Resources and Development, Asian Institute of Technology (AIT), P.O. Box 4, Klong Luang Pathum Thani 12120, Thailand.

temperate crops such as cabbages, tomatoes, potatoes, and soybeans for lowland markets, which have become accessible because of improved transport facilities. These changes are accepted willingly by the more recent immigrants because of their focus on cash crops, but they also have an influence on rotational swidden farmers who are more tied to tradition, as observed by Rerkasem and Rerkasem (1994, *21*):

> Even among those farmers who are practicing "sustainable" rotational shifting agriculture, opportunities for more intensive cropping are often perceived in terms of productivity, against the same amount of effort and resource, as a vast improvement over their traditional practices.

That such a process is a change "from better to worse" has been argued by authors such as the late director of the Tribal Research Center at Chiang Mai, Chantaboon Sutthi (1989), who deplored the replacement of highly diversified cropping systems with monocropping systems which, he pointed out, were both economically vulnerable and ecologically harmful because of their heavy reliance on chemical fertilizers, herbicides and insecticides. At this stage of transition, consideration should be given to alternatives to the present development that are capable of incorporating the more beneficial aspects of the sustainable forms of swidden farming.

The most famous sustainable swiddening system is that practiced by the Lawa. It has been studied thoroughly over the past 30 years (Kunstadter 1974, 1978a, b; Kunstadter et al. 1978; Sabhasri 1978; Schmidt-Vogt 1997, 1999). A significant characteristic of Lawa swiddening is the practice of leaving a rather large number of relict emergents in their swiddens. Relict emergents are mentioned in the literature on swiddening in Thailand but have not yet been studied. Because of this deficit, I decided to collect information on relict emergents, including species, size, number, and selection practices. That information forms the basis of this chapter.

In the course of a comparative research project on the effect of different swidden farming practices on the development of vegetation in fallows, I spent two years, from 1990 through 1992, working in the Lawa village of Ban Tun. It was, at that time, probably the last village practicing the traditional form of Lawa swiddening and thus became the focus of my research on relict emergents.

While my fieldwork was going on, changes to farming practices were afoot. The villagers moved their settlement closer to the main road and built a feeder road to gain access to lowland markets. Their intention was to follow the example of neighboring villages and plant cash crops on at least half of their swiddening area, as a first step in the conversion of swidden farming to permanent agriculture.

One of the most important findings of my research at Ban Tun was the astonishing floristic and structural complexity of secondary forests, as well as the large number of useful species occurring in them. The structural complexity, at least, was partly due to the practice of leaving relict emergents. This observation stimulated ideas for using forests and relict emergents as a basis for transforming swidden systems into agroforestry systems, or into forest-based land use systems. These ideas were discussed with the villagers.

Research Methods

Fieldwork for my project on the investigation of secondary vegetation in swidden fallows consisted of vegetation studies and inquiries concerning knowledge about, and use of, plants occurring in secondary vegetation. Floristic analysis was supplemented by analysis of stand structure, a quantitative method yielding data on the positions and dimensions of trees in a transect. These data were later used for drawing profile diagrams, such as that in Figure 4-6, and frequency histograms of height and diameter size classes, as indicators of age, structure, and development tendency. Relict emergents in the secondary forests were identified by their significantly larger diameter at breast height (dbh). On freshly cleared swiddens, relict emergents were counted, identified, and their height and dbh measured within sampled areas.

Information concerning local names and plant uses was obtained from various informants, but mainly from the Ban Tun village priest, Mr. Um-Kamyan, who was well known as a healer and expert on local flora. For the tree species of secondary forests, the importance value, that is, the sum of the relative abundance, relative frequency, and relative dominance of each species, was calculated according to the method of Curtis and McIntosh (1951). These values were then entered into a table in order from higher to lower importance, together with the uses of the trees, as a means of correlating the ecological and economic significance of each species (see Table 4-1).

The Lawa of Northern Thailand

The Lawa, who are also called Lua, and who call themselves Lavu'a, represent the oldest stratum of settlers in northern Thailand. Their presence in the area predates by several hundreds of years the arrival of the Mons, the Tày, and other minority groups living in the highlands. As members of the Mon-Khmer group in the Austroasiatic language family, they are related to the Khamu and H'tin of northern Thailand and Laos, and to the Wa of Burma and South China. When the Mons expanded into northern Thailand around A.D. 800, the Lawa ruled over kingdoms at the site of present-day Chiang Mai and further to the west. The conflict between the Lawa and the Mons ended in the defeat of the Lawa king, Virangkha, and the withdrawal of some of the Lawa to the hills. However, Lawa kingdoms persisted in the area until they were absorbed with the creation of the Kingdom of Lanna in the 13th century (Condominas 1990).

Today, the Lawa constitute one of the smallest ethnic groups in the region, with a population of 17,346 (Schliesinger 2000). After living in the highlands of Thailand for at least 500 to 600 years (Matzat 1976), the remaining Lawa live mainly in Chiang Mai and Mae Hong Son Provinces, and their population is rather fragmented. The most conspicuous concentration of Lawa settlements can be found in the highlands east of Mae Sariang.

The Lawa Village of Ban Tun

The village of Ban Tun is situated in the Mae La Noi district of Mae Hong Son Province, on the eastern flank of a range forming the watershed between the Mae Nam Ping and the Mae Nam Yuam drainage areas. The slope is dissected into a series of southwest to northeast trending ridges by tributaries of the Huai Mae Ping Noi, the westernmost branch of the Mae Ping drainage. The area used by the villagers of Ban Tun is bounded by two of these tributaries, the Huai Mak Khaeng in the west, north, and northwest, and the Huai Mae Ho in the east and southeast, as well as by the Huai Mae Ping Noi itself in the northeast (Figure 4-1). The boundary in the south and southwest is marked by the watershed at an altitude of 1,300 to 1,500 m asl. Eight hundred meters in altitude below this, the valley bottom of the Huai Mae Ping Noi is situated at 700 m asl, between the confluences of Huai Mak Khaeng and Huai Mae Ho.

When I first chanced upon Ban Tun on July 19, 1990, in my search for a suitable study site, the village was situated on the upper reaches of the Huai Sa Wa Lu at an altitude of 1,080 m asl. There was no road connection then, and my guide and I reached the village by following a foot trail that connected Ban Tun with Ban Chang Mo Noi, on the main road. At the time of this first visit, Ban Tun had been at its Huai Sa Wa Lu location for 27 years. Formerly, the village had been located about two kilometers to the northeast, on the crest of a ridge at an altitude of 1,100 m asl, where its overgrown remnants can still be seen. In 1963, the old Ban Tun, which at that time consisted of 50 to 60 families, was totally destroyed by fire (Kauffmann 1972). After this disaster, the people decided to move away from this inauspicious place to the new site on the banks of the Huai Sa Wa Lu. But only about half of the population moved to the new location and, by 1992, only 177 people were living in the village's 27 households. This was a much smaller population than that of the old village. Natural population increase had to some extent been eroded by migration to the lowlands, especially to Mae Sariang.

It was mainly due to its unique demographic history that Ban Tun had been able to retain a favorable ratio of population to cultivated land and to preserve the traditional Lawa swiddening system, with a fallow period much longer than that found in other villages in the neighborhood. Another important reason for the preservation of traditional subsistence agriculture was the village's relative isolation. It was while staying as a guest in Ban Tun that I compiled the following account of Lawa agricultural traditions.

Agricultural Activities

As in other Lawa villages of the region, land use at Ban Tun is a combination of irrigated and swidden farming. On average, each household farms seven rai, or 1.1 ha, of irrigated land and 13 rai, or 2.1 ha, of swidden land. The amount of swidden land varies from year to year, but the area farmed annually by the Ban Tun community amounts to about 30 ha of irrigated land and 60 ha of swidden fields, or a total area of 90 ha. When fallow land is included, the village's area of usable land amounts to about 800 ha. This is a large area for a population of 177 and Ban Tun is, consequently, not only self-sufficient in rice, but also capable of selling a surplus in Ban Chang Mo.

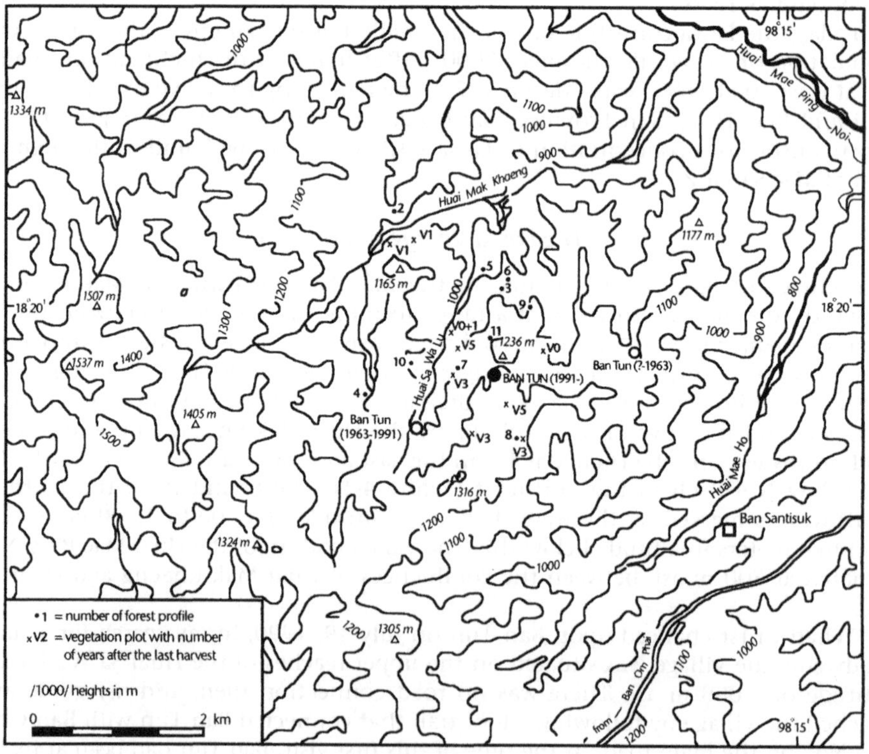

Figure 4-1. The Study Area of Ban Tun in Northern Thailand

Note: Scale 1:50,000.

Source: Royal Thai Survey Department.

Irrigated Farming. Irrigated farming is carried out in the flat sections of the river valleys and on terraces cut into adjacent slopes up to an altitude of 800 m. Several varieties of rice, mostly nonglutinous, are grown in the irrigated fields. There is only one crop per year.

Swidden Farming. Swidden farming is carried out on slopes above 800 m and up to 1,200 m, with gradients varying between 18° and 38°, but mostly around 30°. Forests are felled for swiddening sometimes right up to the crest of hills, but the summit area is never entirely cleared. At least half of it is left under forest.

The annual swiddening cycle begins in January with a ceremony in which the people enter into an agreement with the forest spirits concerning the temporary use of a part of their domain (Kunstadter 1983). The new swiddens are cleared in February. According to Lawa custom, one large and continuous area of about 60 ha is cleared in a communal effort. There is a division of labor according to gender. Women cut the brush and small trees with bush knives (Figure 4-2), leaving stumps up to one meter high. The men climb into the larger trees left standing by the women and trim the crowns in order to prevent shading of the future crop. Sometimes they remove the entire foliage, but more often a few branches are left at the top to ensure the survival of the tree (Figure 4-3). Trees, branches, and foliage are left on the ground to dry in the heat of the pre–monsoon season.

Burning is normally carried out at the beginning or middle of April, just before the onset of the rains. The burning operation is divided into two stages. The first and most spectacular burn is organized as a communal activity. It produces a huge wall of flames, which moves rapidly upslope, consuming the mass of dry foliage and branches, but leaving behind the charred logs. In the second stage, the logs are cut into pieces three to five meters long, gathered into piles along with the unburned slash, and burned in the afternoon hours, when upslope winds fan the flames. After all the plots have been cleared, when field huts have been constructed and fences erected to prevent animals from straying into the fields, the swiddens are finally ready for planting.

Planting is, once again, a communal activity. The young men punch holes in the ground with their planting sticks and the girls follow behind, dropping several grains of rice into each hole. The main crop in the swiddens is rice and, as in the irrigated fields, different varieties are grown. Other crops include sorghum, which is planted in long rows to demarcate field boundaries, maize, several varieties of beans, chilis, tuber plants such as yams and manioc, cucumbers, and cotton.

Secondary succession in the swiddens begins after the first monsoon rains, with the emergence of weeds and the development of woody regrowth from stumps. Weed competition is given as the main reason for cultivating swiddens no longer than one year. As it is, weeding must be carried out about three times in the course of one cropping season. The final weeding is done around September, when the rice begins to ripen. The rice is harvested in October and November. In Ban Tun, the fields are then left fallow for periods of 12 to 17 years, enough time for secondary forest to develop before the swiddens are reopened in the same location.

Secondary Vegetation in Swidden Fallows

Successional Development

At harvest time, about one month after the last weeding, the ground is already covered with a low but quite dense carpet of weeds and resprouting woody plants. One year later, the most successful weed species, which are mainly exotics from tropical America, have achieved dominance and form the upper layers of a two-meter-tall and almost impenetrable jungle. The weed stage persists for about three to

Figure 4-2. A Lawa Woman Helping to Clear a Swidden with a Bush Knife

Figure 4-3. Relict Emergents on a Freshly Cleared Swidden: A Few Branches Are Left at the Top to Ensure Survival of the Tree

four years and is then gradually replaced by a scrub stage, which develops out of the growth of coppice shoots and root suckers from the stumps left in the ground at clearing. The coppice shoots and root suckers begin to show within a few weeks of burning, but they are soon overgrown by the dense tangle of weeds when the field is fallowed. However, they outgrow the weeds after about three years, and by the fourth and fifth years, the resprouting shrubs and trees begin to close their canopy and to suppress the undergrowth by shading.

During the fifth and sixth years, the succession passes from the scrub stage to the secondary forest stage. The trees reach a height of five meters or more, and canopy closure develops to such a degree that only scattered remnants of the former weed cover are left on the forest floor. Seedling establishment, which has been sporadic up to this point, becomes more important. At an age of about 12 years, the secondary forest has reached maturity, in the sense that it is considered ready for swiddening once again. Secondary forests at Ban Tun are cut at an age of 12 to 17 years.

These forests are surprisingly rich in species. The tree layer of sample stands at Ban Tun contained a total of 78 species. Individual stands were made up of a total of about 66 species per 500 m^2, with around 30 species in the tree layer. Trees belonged mainly to the families Theaceae and Fagaceae, followed by Dipterocarpaceae, Leguminosae, Rubiaceae, Myrsinaceae, Anacardiaceae, Lauraceae, Ebenaceae, Styracaceae, Tiliaceae, Sterculiaceae, Burseraceae, Juglandaceae and Dilleniaceae (see Table 4-1). Species richness is partly due to the large number of individuals in the very dense forests. It is also partly due to the location of the study area at an altitude around 1,000 m asl, within a zone of transition, where lower montane forests are being penetrated by floristic elements from forest types of lower elevations, including seasonal rain forest, mixed deciduous forest, and deciduous dipterocarp forest (Santisuk 1988). Of the 78 tree species recorded in the sample plots, 49 were described as useful, and most of them for two or three different purposes. The correlation of ecological significance with usefulness in Table 4-1 shows that those trees that are ecologically important are also useful trees, and that the proportion of useful trees diminishes in relation to their declining importance in fallow forests.

Because of their development history, these forests are also structurally complex, and often consist of up to four different layers, as follows:

- A top layer of relict emergents left standing in the clearing process, with a height of 12 to 14 m or more, and a stem diameter exceeding 20 to 25 cm;
- A main layer of trees that have grown from coppice shoots and root suckers, with a height ranging from 6 to 10 m, and stem diameter from 8 to 10 cm;
- A layer of tall weeds and saplings that have grown from seeds; and
- A ground layer that is generally sparse, except in places where gaps have opened in the canopy, consisting of herbs, climbers, grasses, and usually a sizable number of tree seedlings.

Relict Emergents

The Lawa practice of clearing swiddens by cutting the brush and small trees, but leaving the larger trees standing, creates the parkland scenery of cultivated and fallow swiddens dotted with single trees, which is so characteristic of the Lawa landscape (Figure 4-4). Trees retained in swiddens are, in the literature, sometimes referred to as "seed trees," based on the understanding that villagers consciously preserve these trees in order to assist regrowth of the forest (Mischung 1990; Santasombat 2003). My inquiries at Ban Tun, however, have not yielded sufficient evidence to support such claims. Therefore, I prefer the term "relict emergent," used by Nyerges (1989) in his study of swidden fallows in Sierra Leone. The standard answer to my questions about why some trees were left standing was that they were too thick or their wood too

Figure 4-4. Swidden Fallows with Relict Emergents. View Toward the Southeast, into the Valley of the Huai Mae Ho

tough. This reply was corroborated by measurements of tree and stump diameters on freshly cleared swiddens, summarized in Figure 4-5. According to these measurements, most of the trees felled in the process of clearing swiddens have a diameter of 8 to 12 cm, and virtually no trees are felled with a diameter exceeding 15 to 16 cm. These figures correspond closely with the observation of Nakano (1978, *419*) that on Karen swiddens "a tree with a diameter of more than 15 cm at waist height was not felled, although its branches and twigs were cut off." The impression conveyed by Figure 4-5—that some trees of relatively small diameter are also left standing—can be explained by the frequency of multistemmed trees in a secondary forest and by the custom of leaving the entire cluster untouched, including the smaller stems, even if only one stem is considered too big for cutting.

The average dbh of relict emergents is 18 cm, and the range is 5 to 48 cm. The larger of these relicts have possibly survived more than one cycle of swiddening. Density and species composition of relict emergents have been studied through a record of trees left on swiddens, as well as of those trees in the uppermost tree layer of secondary forests that can be identified as survivors from the previous fallow by having a dbh significantly larger than that of the other trees. The average number of relict emergents is 244 per ha, a density comparable to that of an open forest (Figures 4-6 and 4-7). This makes it seem inappropriate to describe the first stage in the preparation of swiddens as "clearing." The term "thinning" or "selective cutting according to diameter size" comes closer to the truth. Density, however, varies within a range of 66 to 383 trees per ha, probably due to the variable intensity of burning. Relict emergents develop new sprouts after a normal burn, but a fierce fire with high-reaching flames can kill them. I have seen places in three- to four-year-old fallows where all emergents had been killed in this manner. They were mostly located on the upper slopes, and their unscheduled demise was due to fires that, as they move upslope, built up their energy and destructiveness.

The record of relict emergents is listed in Table 4-2. This table, and a comparison with Table 4-1, shows that the frequency of species as relict emergents correlates well with their frequency in the upper tree layers of secondary forests and with their ecological importance in secondary forests in general. This applies especially to *Castanopsis armata, Schima wallichii, Lithocarpus elegans*, and *Castanopsis diversifolia*.

Table 4-1. Ecological Significance and Uses of Forest Fallow Trees

	Ecological Significance A: Relative Abundance (%); B: Relative Frequency (%); C: Relative Dominance (%); D: Importance Value				*Uses*								
Species Name	*A*	*B*	*C*	*D*	*Co*	*Fe*	*Fi*	*T*	*Fo*	*A*	*M*	*Ce*	*D*
Schima wallichii	4.36	2.53	14.59	21.48	x			x	x				
Castanopsis armata	5.54	3.16	10.32	19.02	x		x		x				
Lithocarpus elegans	4.16	3.16	7.61	14.92	x		x						
Shorea obtusa	3.17	1.26	6.89	11.32	x				x				
Aporusa wallichii	5.54	1.90	3.14	10.58					x	x			
Glochidion sphaerogynum	4.55	3.80	1.66	10.01	x	x							
Eurya acumminata	4.75	3.80	1.00	9.55	x	x							
Styrax benzoin	2.77	3.80	1.46	8.03	x	x	x						
Symplocos macrophylla	3.17	2.53	1.87	7.57		x							
Aporusa villosa	3.56	2.53	1.44	7.53	x	x			x				
Castanopsis diversifolia	2.18	1.26	3.79	7.23	x								
Eugenia albiflora	2.57	2.53	1.63	6.74					x		x		
Phyllanthus emblica	2.57	3.16	0.99	6.72					x		x		
Castanopsis tribuloides	1.19	1.90	2.56	5.65	x		x						
Anneslea fragrans	1.39	1.90	2.26	5.55		x	x		x				
Gluta obovata	2.57	1.90	0.88	5.35	x								x
Diospyros glandulosa	1.98	2.53	0.49	5.00		x			x				
Castanopsis sp.	0.99	1.26	2.47	4.72	x				x				
Dalbergia fusca	2.18	1.26	0.86	4.30	x			x					
Wendlandia tinctoria	1.58	1.90	0.65	4.13					x				
Tristania rufescens	1.19	1.26	1.62	4.07	x			x					
Horsfieldia amygdalina	1.19	1.90	0.96	4.05									
Lithocarpus sp.	0.59	1.26	2.16	4.01	x		x						
Elaeocarpus floribundus	1.58	1.26	0.68	3.52	x	x			x	x	x		
Lithocarpus garrettianus	0.59	1.26	1.67	3.52									
Rapanea neriifolia	0.99	1.90	0.51	3.39			x		x				
Phoebe sp.	1.58	1.26	0.45	3.29	x								
Wendlandia sp.	1.39	1.26	0.57	3.22					x				
Olea salicifolia	0.99	1.90	0.29	3.18									
Callicarpa	1.19	1.26	0.64	3.09					x			x	

	Ecological Significance A: Relative Abundance (%); B: Relative Frequency (%); C: Relative Dominance (%); D: Importance Value				*Uses*								
Species Name	*A*	*B*	*C*	*D*	*Co*	*Fe*	*Fi*	*T*	*Fo*	*A*	*M*	*Ce*	*D*
arborea													
Castanopsis indica	0.59	0.63	1.74	2.96									
Turpinia pomifera	0.99	1.26	0.56	2.81									
Camellia oleifera	1.19	1.26	0.35	2.80					x				
Engelhardia spicata	1.19	0.63	0.94	2.76	x	x		x			x		
Dalbergia oliveri	1.39	0.63	0.69	2.71									
Macaranga denticulata	0.99	1.26	0.14	2.39					x				
Helicia nilagirica	0.59	1.26	0.50	2.35									
Quercus vestita	0.20	0.63	1.50	2.33	x		x						
Dillenia parvifolia	0.79	1.26	0.25	2.30									
Eugenia angkae	0.79	1.26	0.19	2.24									
Helicia formosana	0.99	0.63	0.24	1.86	x								
Viburnum inopinatum	0.40	1.26	0.2	1.86							x		
Maesa montana	0.79	0.63	0.39	1.81	x				x				
Archidendron glomeriflorum	0.40	1.26	0.12	1.78									x
Engelhardia serrata	0.40	1.26	0.10	1.76				x	x				
Litsea sp.	0.20	1.26	0.22	1.68	x								
Symplocos racemosa	0.79	0.63	0.24	1.66									
Vaccinium sp.	0.59	0.63	0.41	1.63									
Pyrenaria garretiana	0.59	0.63	0.29	1.51		x					x		
Dalbergia rimosa	0.59	0.63	0.25	1.47							x		
Maesa ramentacea	0.59	0.63	0.23	1.45									
Wendlandia paniculata	0.59	0.63	0.22	1.44									
Beilschmiedia sp.	0.59	0.63	0.20	1.42									

Note: Co = construction; Fe = fences; Fi = firewood; T = tool; Fo = food; A = animal; M = medicinal; Ce = ceremonial; D = decorative.

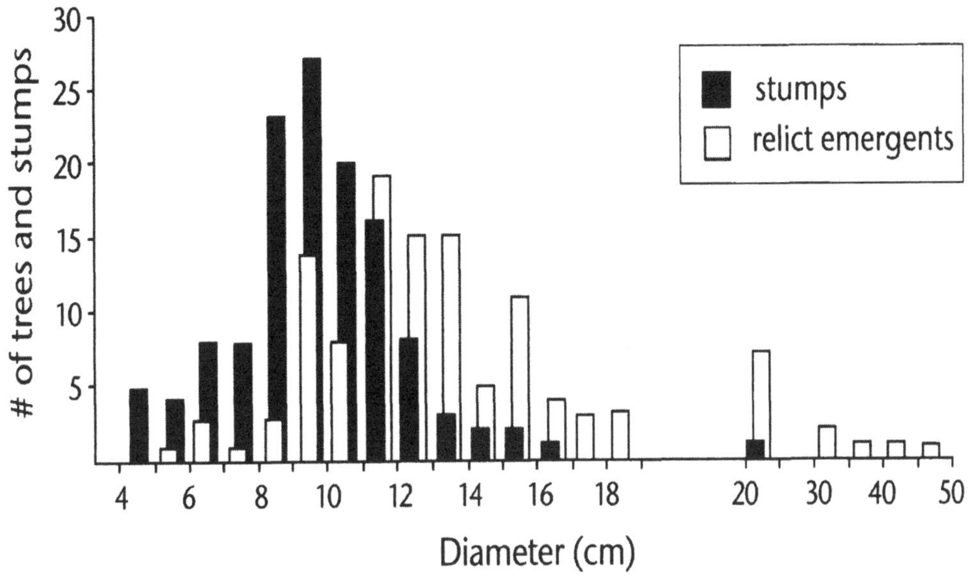

Figure 4-5. Diameter Distributions of Stumps and Relict Emergents on a Newly Cleared Swidden at Ban Tun

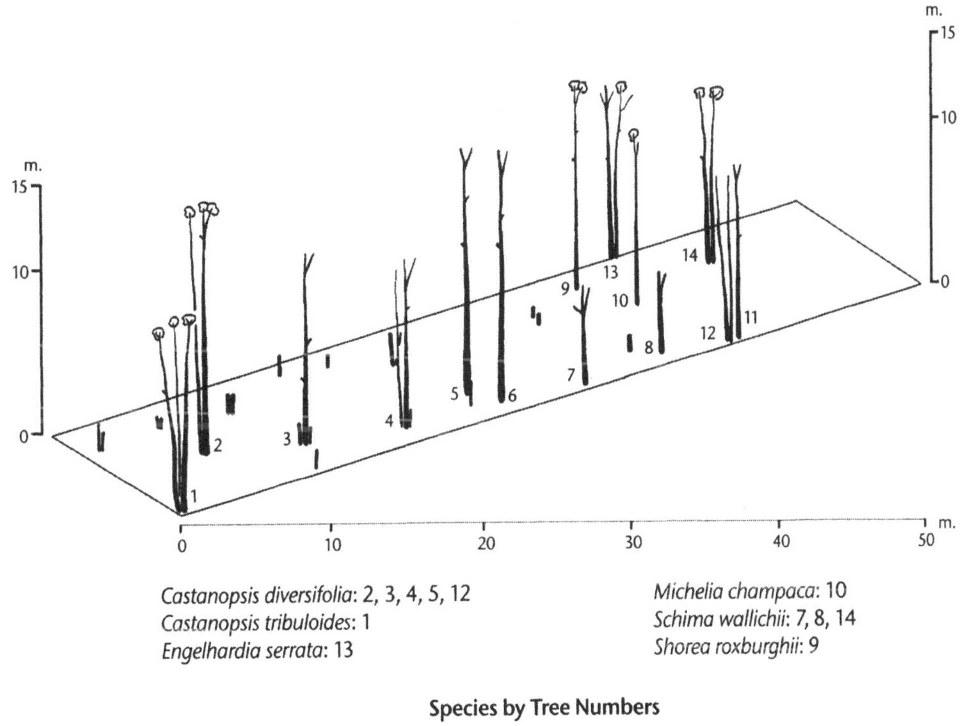

Figure 4-6. Profile Diagram of Relict Emergents on a Swidden Six Months after Clearing and Burning

Table 4-2. Relict Emergents in Swiddens and in the Tree Layer of Secondary Forests

	Number of Relict Emergents by Species	
Species Name	*In Swiddens (area: 4,000 m²)*	*In the Tree Layer (area: 2,400 m²)*
Castanopsis armata	13	9
Schima wallichii	10	10
Tristania rufescens	15	1
Lithocarpus elegans	10	5
Castanopsis diversifolia	8	7
Quercus vestita	8	1
Anneslea fragrans	1	4
Castanopsis indica		4
Dalbergia fusca	4	
Shorea farinosa	4	
Shorea obtusa		4
Engelhardia spicata	2	1
Lithocarpus garrettianus		3
Quercus kerrii	3	
Aporusa villosa	1	1
Callicarpa arborea	2	
Castanopsis tribuloides	1	1
Craibiodendron stellatum	2	
Dillenia parvifolia	2	
Gluta obovata	2	
Shorea roxburghii	2	
Colona floribunda	1	
Dipterocarpus turbinatus	1	
Engelhardia serrata	1	
Eugenia albiflora		1
Glochidion sphaerogynum		1
Gmelina arborea	1	
Horsfieldia amygdalina		1
Michelia champaca	1	
Quercus aliena	1	
Vitex peduncularis	1	

Figure 4-7. The Transect from which the Profile Diagram in Figure 4-6 was Recorded

Land-Use Changes at Ban Tun

During evening conversations in the headman's house, where I stayed as a guest when doing research in Ban Tun, a plan to relocate the village once more became a regular subject for discussion. Finally, in the rice-growing season of 1991, the plan was put into effect and the village moved onto a spur overlooking the Huai Sa Wa Lu at an altitude of 1,100 m asl (see Figure 4-1). At the same time, the villagers built a feeder road to connect their new settlement with a short supply road between a forestry station on the divide and the main road near Ban Chang Mo Luang. The objective was to follow the example of neighboring villages with road access and to plant cash crops on former swidden land for sale in lowland markets.

In the years from 1990 to 1992, when I visited the village regularly, cabbages and soybeans were fetching good prices and were therefore planted extensively. The crops were carried from the fields to the roadside by the farmers and transported to the markets in Mae Sariang or in the Mae Ping valley by middlemen operating a shuttle system with small trucks.

The difference at that time between the domain of Ban Tun and that of its southern neighbor, Ban Santisuk, is shown in Figure 4-8. Whereas the slopes belonging to Ban Tun, on the other side of the Huai Mae Ho, were covered with regrowth in various stages of succession, the land of Ban Santisuk in the foreground had been planted with soybeans.

In those villages already converted to cash cropping, there were many other changes relating to the use of the land. Most importantly, the cultivation period was extended in order to increase the effectiveness of production. Weed infestation, which was previously the main reason for limiting cultivation to only one year, was suppressed with herbicides. This change quickly became evident in the plant succession on fallow land. Coppicing, which was earlier the main agent of vegetational development, was suppressed by the longer cultivation periods and the use of chemicals. Vegetation on the fallow land of Lawa villages with road access was soon composed mainly of grasses such as *Pennisetum pedicellatum*, typical of very intensive swiddening systems such as that of the Hmong.

The villagers of Ban Tun were planning to start out by planting cash crops on only half of their swiddening area. They had been somewhat intimidated by the experience of their neighbors, who had suffered losses in previous years because of low prices. So they were interested in maintaining some degree of diversity by continuing to plant dry rice on the remaining swiddening area. It was at this stage that we discussed other options for intensification of land use on the swidden land, options which were regarded by the farmers as potentially viable.

Alternative Uses of Forested Fallows and the Economic Potential of Relict Emergents

Alternative land uses that incorporated elements of the Lawa swidden system had been envisaged before. Sabhasri (1978) made the interesting suggestion of converting swidden farming to a forest management system, based on the Lawa practice of thinning, for the production of firewood. According to his calculations, the economic returns from forest management would be better, "provided transportation and markets were available" (Sabhasri 1978, *168*).

Transportation and markets are available today, so Sabhasri's scheme deserves reconsideration, along with the development of other options. In the course of my discussions with the farmers of Ban Tun, four alternatives to the traditional form of swidden farming and its replacement with cash crop farming were developed, two involving the use of trees from forested fallows, the other two involving the use of relict emergents. They were as follows:

- Continuation of swidden farming, supplemented by the sale of charcoal;
- Conversion of swidden farming to forest management for the production of charcoal, as well as timber and nontimber products;
- Planting of cash crops supplemented by the sale of nontimber products from relict emergents; and
- Integration of relict emergents into agroforestry systems.

The first option could be implemented more easily than the others. It has been shown that the trees felled in the course of clearing swiddens are not consumed by the first burn, but have to be collected into piles and burned separately. This is a tremendous waste of energy and resources because the ash from the burned logs contributes very little to the fertilization of swidden fields. With almost the same input of time and energy, these logs could be turned into charcoal, for which there is a huge demand in the lowlands. Charcoal can be transported more easily than firewood and brings better economic returns because the majority of trees in forested fallows are hardwoods, which produce a high quality of charcoal.

Transforming swidden farming into a forest management system follows the suggestion of Sabhasri, but with a wider range of forest products. Forests could be thinned at regular intervals by the procedure with which the Lawa are already familiar, and the felled trees turned into charcoal as mentioned above, or used as timber. Income from the sale of charcoal and timber could be supplemented with income from nontimber products, such as edible fruit, resin, or string made from the bark of trees such as *Dalbergia fusca*. These products could be collected from relict emergents in those years when forests were being thinned. This would be an attractive community forestry development project, but it is not feasible at the moment because of the still unresolved issue of community forestry legislation in Thailand. Also, little is known about the market for nontimber products, and more research concerning this aspect would be necessary before this scheme could be seriously considered.

Collecting nontimber products from relict emergents in fields planted with cash crops, as a means of diversifying sources of income as a buffer against price fluctuations for cash crops, would be a compromise that could be put into practice under present conditions. Relict emergents are numerous enough to guarantee a sufficient supply of these products, but the question of marketability remains.

The final alternative—to use relict emergents as a stepping stone for transforming swidden farming into agroforestry—was voiced only tentatively, because farmers had very little experience with agroforestry. One farmer suggested that relict emergents be used as shade trees for the cultivation of coffee, which is widely grown in the highlands.

Figure 4-8. View from the Territory of Ban Santisuk (see Figure 4-1) toward the Territory of Ban Tun, across the Valley of the Huai Mae Ho

Conclusions

The value of fallow vegetation as a resource has not yet been recognized by the authorities in Thailand who, on the contrary, rate fallow vegetation as degraded scrub and aim to remove swidden farming as the chief cause of vegetation change. The example of the Lawa at Ban Tun has shown that these secondary forests can be very complex and rich in useful species, which makes them ecologically and economically valuable. A special role is played in this context by relict emergents, which provide some forest cover on swidden fields and contribute significantly to the complexity of the emerging forest fallows. Both relict emergents and forest fallows could provide the basis for a number of ways to intensify land use by supplementing farming, either swidden or cash crop farming, with income from forest products.

Acknowledgments

This research was carried out with the financial support of the Alexander von Humboldt Foundation in Germany. I have to thank Dr. Thawatchai Santisuk, the Director of the Forest Herbarium, Royal Forest Department, Bangkok, who invited me to conduct research in Thailand and aided my work in many ways, mainly by helping me through the necessary administrative procedures and by placing at my disposal staff members and the infrastructure of the Forest Service. I am indebted to Dr. J.F. Maxwell, who identified all plant specimens collected in the course of my research. I am especially indebted to the headman and the people of Ban Tun for their kind hospitality and support of my work.

References

Brown, S. and A.E. Lugo. 1990. Tropical Secondary Forests. *Journal of Tropical Ecology* 6 (1), 1–32.

Condominas, G. 1990. Notes on Lawa History Concerning a Place Named Lua' (Lawa) in Karen Country. In: *From Lawa to Mon, from SAA' to Thai: Historical Anthropological Aspects of Southeast Asian Social Spaces*, edited by G. Condominas. Canberra: The Australian National University, 5–22.

Credner, W. 1935. *Siam, das Land der Tai*. Stuttgart: J. Engelhorns Nachf.

Curtis, J.T., and R.P. McIntosh. 1951. An Upland Forest Continuum in the Prairie-Forest Border Region of Wisconsin. *Ecology* 32(3), 476–496.

Ewel, J. 1980. Tropical Succession: Manifold Routes to Maturity. *Biotropica*, supplement to 12 (2), 2–7.
Funke, F.W. 1960. Die Stellung der Lawa in der Kulturgeschichte Hinter-Indiens. *Tribus* 9, 138–146.
Grandstaff, T.B. 1980. *Shifting Cultivation in Northern Thailand: Possibilities for Development*. Tokyo: The United Nations University.
Hansen, P.K. 2001. The Forest as a Resource for Agriculture. In: *Forest in Culture—Culture in Forest: Perspectives from Northern Thailand*, edited by E. Poulsen et al. Tjele, Denmark: Research Centre on Forest and People in Thailand, 147–162.
Kauffmann, H.E. 1972. Some Social and Religious Institutions of the Lawa (Northwest Thailand), Part 1. In: *Journal of the Siam Society* 60 (1), 235–306.
Keen, F.G.B. 1972. *Upland Tenure and Land Use in North Thailand*. Bangkok: The SEATO Cultural Program, 1969–1970.
Kellman, M.C. 1960. Some Environmental Components of Shifting Cultivation in Upland Mindanao. *Journal of Tropical Geography* 28, 40–56.
Kunstadter, P. 1967. The Lua' and Skaw Karen of Mae Hong Son Province, Northwestern Thailand. In: *Southeast Asian Tribes, Minorities, and Nations*, Vol.1, edited by P. Kunstadter. Princeton, NJ: Princeton University Press, 639–674.
———. 1974. Usage et Tenure des Terres Chez les Lua' (Thailande). *Études Rurales* 53–56, 449–466.
———. 1978a. Ecological Modification and Adaptation: An Ethnobotanical View of Lua' Swiddeners in Northwestern Thailand. In: *The Nature and Status of Ethnobotany, Anthropological Papers* 67, edited by R.I. Ford, 168–200.
———. 1978b. Subsistence Agricultural Economies of Lua' and Karen Hill Farmers, Mae Sariang District, Northwestern Thailand. In: *Farmers in the Forest*, edited by P. Kunstadter, E.C. Chapman, and S. Sabhasri. Honolulu: East-West Center, 74–133.
———. 1980. Implications of Socio-Economic, Demographic, and Cultural Changes for Regional Development in Northern Thailand. In: *Conservation and Development in Northern Thailand*, edited by J.D. Ives, S. Sabhasri, and P. Voraurai. Tokyo: The United Nations University, 13–27.
———. 1983. Animism, Buddhism, and Christianity: Religion in the Life of the Lua People of Pa Pae, Northwestern Thailand. In: *Highlanders of Thailand*, edited by J. McKinnon and W. Bhruksasri. Kuala Lumpur: Oxford University Press, 135–154.
———, S. Sabhasri, and T. Smitinand. 1978. Flora of a Forest Fallow Environment in Northwestern Thailand. *Journal of the National Research Council of Thailand* 10 (1), 1–45.
———, E.C. Chapman, and S. Sabhasri (eds.). 1978. *Farmers in the Forest*. Honolulu: East-West Center.
Matzat, W. 1976. Genese und Struktur der Dorfsiedlungen des Lawa-Bergstammes. In: *Tagungsbericht und Wissenschaftliche Abhandlungen* 40. Dt. Geographentag Innsbruck 1975. Wiesbaden: Franz Steiner Verlag, 351–358.
Mischung, R. 1990. Geschichte, Gesellschaft und Umwelt: eine kulturökologische Fallstudie über zwei Bergvölker Südostasiens. Habilitation thesis, Frankfurt University.
Nakano, K. 1978. An Ecological Study of Swidden Agriculture at a Village in Northern Thailand. *Southeast Asian Studies* 16 (3), 411–446.
———. 1980. An Ecological View of a Subsistence Economy Based Mainly on the Production of Rice in Swiddens and in Irrigated Fields in a Hilly Region of Northern Thailand. *Southeast Asian Studies* 18 (1), 40–67.
Nyerges, A.E. 1989. Coppice Swidden Fallows in Tropical Deciduous Forest: Biological, Technological, and Sociocultural Determinants of Secondary Forest Successions. *Human Ecology* 17 (4), 379–400.
Rerkasem, K., and B. Rerkasem. 1994. *Shifting Cultivation in Thailand: Its Current Situation and Dynamics in the Context of Highland Development*. International Institute for Environment and Development (IIED) Forestry and Land Use Series No. 4. London: IIED.
Sabhasri, S. 1978. Effects of Forest Fallow Cultivation on Forest Production and Soil. In: *Farmers in the Forest*, edited by P. Kunstadter, E.C. Chapman, and S. Sabhasri. Honolulu: East-West Center, 160–184.
Santasombat, Y. 2003. *Biodiversity, Local Knowledge and Sustainable Development*. Chiang Mai, Thailand: Regional Center for Social Science and Sustainable Development.
Santisuk, T. 1988. An Account of the Vegetation of Northern Thailand. *Geoecological Research* Vol. 5. Stuttgart: Franz Steiner Verlag.
Schliesinger, J. 2000. *Ethnic Groups of Thailand: Non-Thai-Speaking Peoples*. Bangkok: White Lotus.
Schmidt-Vogt, D. 1991. Schwendbau und Pflanzensukzession in Nord-Thailand. *Mitteilungen der Alexander von Humboldt-Stiftung* 58, 21–32.
———. 1995. Swidden Farming and Secondary Vegetation: Two Case Studies from Northern Thailand. In: *Counting the Costs: Economic Growth and Environmental Change in Thailand*, edited by J. Rigg. Singapore: Institute of Southeast Asian Studies.
———. 1997. Forests and Trees in the Cultural Landscape of Lawa Swidden Farmers in Northern Thailand. In: *Nature Is Culture: Indigenous Knowledge and Socio-Cultural Aspects of Trees and Forests in Non-European Cultures*, edited by K. Seeland. London: Intermediate Technology Publications, 44–50.
———. 1998. Defining Degradation: The Impacts of Swidden on Forests in Northern Thailand. *Mountain Research and Development* 18 (2), 135–149.

———. 1999. Swidden Farming and Fallow Vegetation in Northern Thailand. *Geoecological Research* Vol. 8. Stuttgart: Franz Steiner Verlag.

______. 2000. Land Use and Land Cover Change in Montane Regions of Mainland Southeast Asia. *Journal of Geography Education* 43, 52–60.

Sillitoe, P. 1995. Fallow and Fertility under Subsistence Cultivation in the New Guinea Highlands: 1. Fallow Successions. *Singapore Journal of Tropical Geography* 16 (1), 101–115.

Sutthi, C. 1989. Highland Agriculture: From Better to Worse. In: *Hilltribes Today: Problems in Change*, edited by J. McKinnon and B. Vienne. Bangkok: Orstom/White Lotus, 107–142.

Uhlig, H. 1969. Hill Tribes and Rice Farmers in the Himalayas and Southeast Asia. *Transactions and Papers, The Institute of British Geographers* 47, 1–23.

______. 1980. Problems of Landuse and Recent Settlement in Thailand's Highland-Lowland Transition Zone. In: *Conservation and Development in Northern Thailand*, edited by J.D. Ives, S. Sabhasri, and P. Vorauri. Tokyo: The United Nations University, 33–42.

______. 1991. Reaktion von Geoökosystemen auf traditionelle und moderne Landnutzungsformen in Südostasien: Naturpotential und anthropogene Gestaltung in den Tropen. *Nova Acta Leopoldiana NF* 64 (276), 133–164.

Whitmore, T.C. 1983. Secondary Succession from Seed in Tropical Rainforests. *Forestry Abstracts* 44 (12), 767–779

______. 1986. *Tropical Rainforests of the Far East*. Oxford, U.K.: Oxford University Press.

Chapter 5

Successional Forest Development in Swidden Fallows of Different Ethnic Groups in Northern Thailand

Chaleo Kanjunt

For many years, Thailand has experienced severe destruction of its forests. In 1961, the country's forests covered 27.36 million hectares, or about 53% of the land area. This tumbled to 27.9% in less than three decades. In 1989, the government canceled all commercial timber licenses, but this failed to halt the decline. By 1993, the forested area had fallen to 26.6% (Planning Division, RFD 1991, 1993) because of population growth, illegal logging, shifting cultivation, and forest encroachment for agricultural land. Recent estimates put the current rate of forest destruction in the country's north at 32,000 ha per year, mainly because of fires, illegal logging, and swidden cultivation.

Shifting cultivation has long been practiced by both local Thai and "hilltribe" ethnic minority groups living in the mountainous areas of northern Thailand. The minority groups include Lisu (Lisaw), Karen (Kariang, Yang), Hmong (Mèo), Lahu (Mussur), Akha (Kaw), Yao (Mien), Lua (Lawa), H'tin, Khamu and Chinese Haw. The types of shifting cultivation they practice fall into three broad categories, applicable to the lowlands, the foothills, and the high mountains (Kanjunt 1988).

Northern Thai farmers who have migrated into intermediate zones between valleys and hills, at elevations between 300 and 600 m above sea level (asl), practice a short cultivation, short fallow system. Due to high population pressures, these resource-poor farmers have had to clear supplementary land by burning the mixed deciduous or dry dipterocarp forests (Kijkar 1987) in the foothills. Their main crop is rice, both glutinous and non-glutinous, supplemented by cash crops such as cotton, maize, beans, and vegetables. These are usually grown for only one season. Fallow regrowth involves shrubs rather than regeneration of the forest.

The Karen people, who usually live in the mountains at elevations ranging from 500 to 1,000 m asl, practice a short cultivation, long fallow system. The forests at these altitudes are mainly dry evergreen or mixed deciduous types. The Karen cut and burn for cropping and then leave their swiddens fallow, to re-grow, for seven to fifteen years. Their major crop is non-glutinous rice, which is supplemented by a wide variety of other crops, such as maize, sorghum, millet, taro, and beans.

Several ethnic groups, including the Hmong, Lahu, Lisu, Akha, and Yao, practice a system that involves long cultivation followed by a fallow of such length that it regularly amounts to abandonment. It is often criticized, not only because its former main cash crop was opium, but also because of its destruction of forests. It is usually

---

Chaleo Kanjunt, Office of Watershed Development, Huay Kaew Road, Amphoe Muang, Chiang Mai 50000, Thailand.

practiced at elevations between 1,200 and 1,500 m asl. A single forest area is cleared, burned, and cultivated for five or more consecutive years. Then it is usually abandoned after the soil becomes exhausted or it is covered with grass and weeds. The main crops are opium, rice, maize, and potatoes. Under this system, when an exhausted swidden is abandoned, the people look for a new piece of land nearby and, if none is available, they often move to another location.

Sam Mun Highland Development Project

The Sam Mun Highland Development Project was established in 1987. Led by the Royal Forest Department, its main objectives were to improve the quality of life of people living in the area of this study, reduce opium production, and protect highland watershed areas.

An important aspect of the project was its adoption of a philosophy that it was possible for people to live in harmony with the forest. It has made wide use of participatory land-use planning methods to involve ethnic minority communities in the management and protection of the forests and watershed areas in which they live. Goals and guidelines have been established for maintaining permanent and viable agricultural activities, as well as protecting the area's watershed functions. The project has consequently expanded the scope of local land management from field level to community and even watershed levels. Local management is now responsible for the types of land use in different zones within sub-catchment areas, as well as monitoring management practices, such as fallow enrichment, on individual plots.

In this setting, succession and forest stand development on abandoned swidden fields have become relevant to the community's management of the watershed as a whole. Various areas previously used for agriculture have been designated for other land uses, to provide other services for the community, and forest regeneration has taken on a significance beyond the traditional goal of enriching a site for future farming.

The Study Area

Located in Pai district of Mae Hong Son Province, Nam Sa is a small watershed with an area of 158 km^2 (Figure 5-1). It encompasses five microwatersheds. Forest types are mainly mixed deciduous at altitudes between 600 and 1,000 m asl, and evergreen forests above this. As in other mountain areas, forest resources at Nam Sa have been partially destroyed by shifting cultivation. Illegal loggers have also been active in some areas.

In the highland areas of Nam Sa, Hmong and Lisu farmers have practiced a long cultivation, very long fallow or abandonment type of swidden cultivation. This has involved the clearing of steep, erosion-prone ridge tops and exposed upper slopes, the removal of most trees, including their stumps, and burning the entire area. Vast areas are now covered with *Imperata cylindrica* and other grasses and herbs, such as *Eupatorium odoratum, Eupatorium adenophorum, Thysanolaena latifolia,* and *Pteridium aquilinum.*

In midland areas, Karen villagers have practiced both shifting and permanent agriculture, and have combined swiddening with paddy farming. Following their traditional short cultivation, long fallow practices, Karen farmers in the study area only partially cut and burn trees during field preparation, leaving coppice and mother trees in an effort to ensure natural forest regeneration.

Factors Affecting Natural Regeneration

Many areas that have been used for shifting cultivation of agricultural crops and later abandoned are capable of natural regeneration, leading to secondary forest or vegetative recovery. However, forest fire is an important obstacle to the succession process. It destroys seedlings and small trees, as well as depleting soil moisture and fertility.

An exception to this rule is evergreen pine forests, where fire suppression can halt the natural regeneration of *Pinus merkusii* (Koskela et al. 1995). At the beginning of a new succession in these forests, litter consists mainly of needles and leaves and an increasing proportion of dead grass. If there has been no fire for six consecutive years, the dead grass forms a layer thick enough to prevent pine seeds from contacting mineral soil and this suppresses germination. Moreover, if several years have passed since the last fire, then the forest litter may burn so strongly that it will kill all seedlings in the grass stage, as well as small trees, and may even become a crown fire.

Other major factors affecting natural regeneration include logging, firewood collection, wildlife hunting, cattle grazing, and collection of mushrooms and bamboo shoots. The impact of climatic changes on the development of secondary forests is also far from understood.

Methodology

Nine sample plots were established in old swidden areas of three tribal groups living at Nam Sa: Lisu (L), Hmong (M), and Karen (K). The plots covered a range of years since crop cultivation. The oldest of them had been fallow for 18 years. A control plot was also set up in natural forest (NF). The plots were chosen in collaboration with villagers and after talking with village leaders. Each plot measured 20 m by 80 m, equal to one rai, the Thai unit of area (6.25 rai = 1 ha). Fences were erected along the boundaries and grasses and weeds were cut. Then, each plot was divided into four subplots, each 20 m by 20 m, and trees with a girth greater than 30 cm (equal to a diameter at breast height [dbh] greater than 9 cm) were marked with numbers.

Table 5-1 shows the general characteristics of topography, plant cover, and years since abandonment of the sample plots.

Data Collection

The trees in the sample plots were measured once, during June and July 1994, and the following data was recorded:

- Height at first branch;
- Height at top;
- Locations;
- Shape of crown cover;
- Girth at breast height; and
- Local and scientific names.

The plants were identified by both their local and botanical names by the late Dr. Tem Smitinand. Due to time constraints and inaccessibility, plant identification was conducted only once a year.

These data have been analyzed for differences in successional development in each of the old swidden plots, the process of natural regeneration, the forest structure, and the economic aspects of the trees, including timber and non-timber forest products. For the purposes of this chapter, the number of trees, crown projection cover, basal stem area, height of trees, and species composition will be compared directly between the plots.

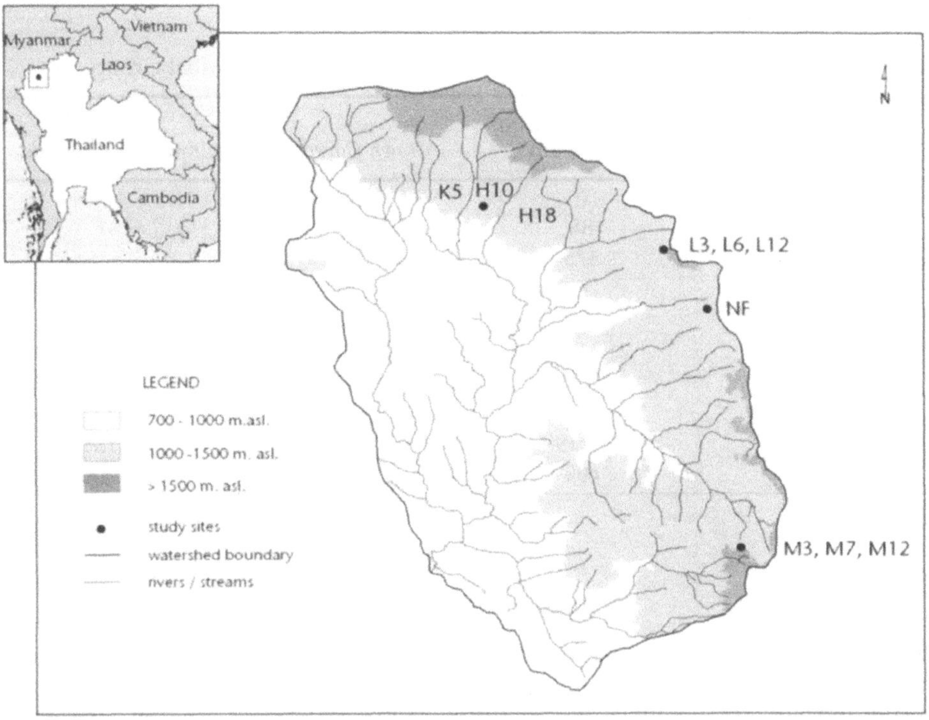

Figure 5-1. Nam Sa Watershed Elevation Map

Table 5-1. Plot Characteristics

Plot Code	*Tribal Group*	*Altitude (m asl)*	*Slope (%)*	*Aspect*	*Soil Character*	*Length of Fallow (years)*	*Vegetative Cover*
L3	Lisu	1,200	25	E	Podzolic	3	weeds
L6	Lisu	1,200	35	SW	Podzolic	6	weeds
L12	Lisu	1,200	50	S	Podzolic	12	saplings, shrubs
M3	Hmong	1,600	60	NW	Podzolic	3	weeds
M7	Hmong	1,600	55	NW	Podzolic	7	saplings, shrubs
M12	Hmong	1,600	30	NW	Podzolic	12	saplings, weeds
K5	Karen	1,000	50	SW	Podzolic	5	saplings, weeds
K10	Karen	900	5	SW	Podzolic	10	saplings
K18	Karen	900	20	SE	Podzolic	18	trees
NF	Control	1,300	65	NE	Podzolic	—	multistoried forest

Note: Plot Code: L = Lisu, M= Hmong, K= Karen. Numbers give the years since cultivation.

Table 5-2. Tree dbh and Absolute and Relative Crown Cover per Tribal Group

Plot Codes	*Trees per ha with dbh >9 cm*	*Crown Cover (%)*	*Crown Cover (m^2/ha)*
L3	0	0	0
L6	88	8	781
L12	375	76	7,575
M3	6	1	69
M7	306	87	8,663
M12	256 (384)	79 (118)	7,919 (11,878)
K5	181	36	3,563
K10	500	103	10,331
K18	1,019	108	10,781
NF	494	158	15,788

Results

Abundance and Location of Trees

The K18 plot had a large number of trees, indicating an even aged forest that had been regenerating largely through coppicing. Trees were also evenly distributed within the plot. These same trends were also obvious in K5 data, because of the usual Karen tradition of leaving stumps in the ground and cultivating each area for only one year, so coppicing trees can survive.

Tree distribution in plot M12 was very irregular. One half was covered with trees, while the other half had very few. Plot M7, on the other hand, had trees scattered all over the plot. This could have been due to firewood cutting or possibly a mistake in establishing the sample plot. Whatever the reason, interpretation of data from the M12 plot was difficult. To get comparable figures, we multiplied the figures from the M12 plot by 1.5, where possible. These are shown in brackets below measured data.

Location of trees in the Lisu plots reflected their agricultural practices. Although the plots were situated not far from an older forest, there were no trees in L3 due to heavy competition from grasses and fires almost every year. Even in L12, there were only a few trees, and many of them had been damaged by fire at the base.

Crown Cover

Crown cover means the vertical projection of a tree's crown to the ground surface. Many methods have been proposed for its measurement. We selected the "crown diameter method" (Mueller-Dombois and Ellenberg 1974). The crown cover projection perimeter was measured at four points perpendicular to each other, with the diameters running parallel to the plot boundaries. We emphasized crown cover as well as basal area because of the great ecological significance of this data. Crown cover pictures were later drawn for all plots. However, these pictures did not allow us to compare cover, distribution of cover among different species, and distribution in cover. Table 5-2 (above) shows data on dbh and area covered by tree crowns.

In the natural forest (NF) plot, the percentage of cover to surface area was as high as 158. This is a good value and represents the multiple stories occurring in this remnant of relatively undisturbed forest. Table 5-2 shows that a crown cover of 100% was reached quite rapidly in the fallowed swiddens of all ethnic groups. The main difference between the plots was that trees with large crowns only occurred in the natural forest. More than one quarter of the crown cover in the natural forest was in ranges where all other plots had no cover. In this regard, all the succession plots were still very different from a natural undisturbed forest. There was no significant difference between the Karen plots K10 and K18. The crowns were small in both plots. In the Hmong M12 plot, the crown cover percentage of 79 (118) was rather high, and might have been even higher than K18 had the whole plot been covered with trees. Table 5-3 compares the percentage of crown cover in the plots and number of individual trees in crown cover ranges. The few individuals in high cover ranges in the natural forest contributed substantially to its high ground cover.

Basal Area

The basal area of tree stands is also a measure of cover (Table 5-4). In this study it is used as a complementary measure of cover, as a figure which can be used for estimation of stock volume and, together with other information, as a tool for addressing various management issues, such as firewood management and household timber production.

As a complementary measure of cover, basal area gives no further insights. As one parameter of volume and increment, however, basal area does give new information. Increment seems to slow down with time and, as the basal area for the natural forest was much higher than all other plots, it was obvious that the point at which the basal area of the succession vegetation equaled that of natural forest was far into the future. Figure 5-2 sets the basal area of the nine study plots against the age of their succession and their tribal background. Significantly, the basal area for the natural forest plot is so far above the others that it does not appear on this graph.

Height

Height analysis deals with the sum of the height of all trees, the average height of all trees, and the average height of the ten tallest trees in one plot, which are supposed to be the most dominant ones. Table 5-4 lists these height measurements for the research plots. However, height information related to just one parameter gives little information about the structure of a forest. The sum height of K18 exceeds the sum height of even the natural forest plot, which is not surprising considering the very high stem number in K18. For the L12, M12, K10, and K18 plots, average height of all trees is very similar, but in the natural forest plot this figure is higher. Looking at the average height of the ten tallest trees, the NF plot is also much higher than all the others.

Table 5-3. Percentage of Crown Cover and Individuals per Crown Cover Range

Crown Cover	*L12*		*M12*		*K10*		*K18*		*NF*	
Range (m^2)	*a*	*b*	*a*	*b*	*a*	*b*	*a*	*b*	*a*	*B*
0–59.9	92	69	88	68	94	70	98	86	89	54
60–119.9	8	31	10	20	5	20	2	14	6	11
120–179.9	0	0	2	12	1	10	0	0	3	12
180–239.9	0	0	0	0	0	0	0	0	0	0
240–299.9	0	0	0	0	0	0	0	0	1	10
300–359.9	0	0	0	0	0	0	0	0	1	13
360–419.9	0	0	0	0	0	0	0	0	0	0

Note: a columns, percentage of individuals per crown cover range; *b* columns, percentage of crown cover.

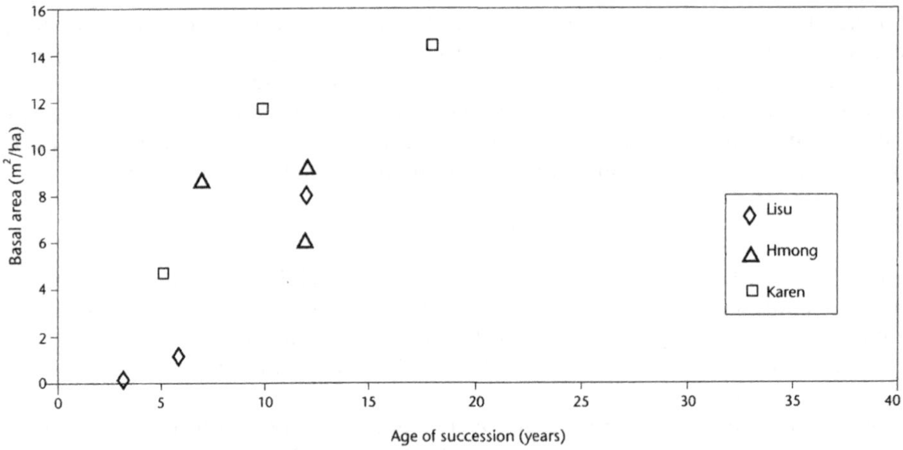

Figure 5-2. Basal Area per Tribal Group and Age of Succession

Table 5-4. Basal Area per Hectare* and Average Heights

Plot Code	*Basal Area (m^2/ha)*	*Sum Height of All Trees (m)*	*Average Height of All Trees (m)*	*Average Height of 10 Tallest Trees (m)*
L3	0	0	0	0
L6	1.25	86.5	6.18	6.2
L12	7.94	586	9.67	15.8
M3	0.06	4	4	0.4
M7	8.69	386	7.88	12.4
M12	6.19 (9.28)	381 (572)	9.29	13.4
K5	4.81	280	9.64	13.2
K10	11.81	822	10.4	16.8
K18	14.50	1692	10.4	15.5
NF	25.94	1111	14.1	27.7

Note: *Trees with a dbh greater than 9 cm.

Species and Species Diversity

Species diversity commonly refers to the number or richness of species and to species "quantity," that is, cover, the number of individuals, distribution, or evenness. In this chapter, I deal only with species richness of vascular plants, which are only one form of life in the ecosystem.

The number of species occurring in each plot was sampled once a year, and the sampling technique allowed direct comparisons between plant communities. Many authors suggest that the best way of comparing species of different communities is to use direct counts. The total number of species per plot is presented in Table 5-5.

Data for the Hmong and Lisu plots are quite similar, whereas Karen plots show a much higher number of species. The most surprising result was the decreasing number of species in the Karen plots in older succession. This might be because K5 showed a high number of annual species that would later be shaded out by forest regrowth.

Some of the species were found in two plots, some in three or even more. Table 5-5 also shows the number of species common to two plots. This is a direct comparison of paired plots and means that at least one plant of the same species occurred in each of the two plots. This is not related to how many individuals of one species were found in one plot. The Karen plots show a much higher number of species common to two plots than the others. As M7 does not include all its species, Hmong plots are impossible to compare. The Lisu plots show low similarity concerning species richness. The low figure of 13 species for K18 versus natural forest may indicate that some species in the lower layers of the Karen forest had been shaded and had not yet been replaced. K10 and K18 had 34 species in common, which may indicate that there is a period of time during which there is very little change in the number of species occurring in Karen forests.

Discussion

In the previous section, successional growth was compared mainly to the natural forest. In comparing successional development, a picture of the end result must always be kept in mind. It will always take a very long time before vegetation reaches the climactic point at which it becomes a secondary forest once again. This section follows succession by tribal group.

Lisu

The Lisu plots showed the lowest number of trees, crown cover, basal area, total height, and average height. This revealed that the process of natural forest regrowth was slowest in these plots. However, the number of species was equal to the Hmong plots. The small increase in the number of species from L3 to L6 showed that the site had suffered a major disturbance, and a very long fallow seemed necessary to restore the soil to its previous condition. After 12 years, the vegetation on L12 still did not resemble a forest. It had taken almost 10 years before one could think of the plot as a regenerating forest rather than grassland with some occasional trees or shrubs. The main problem on these sites was ongoing disturbance, mainly by fire. Each year seedlings could be found, but these died because of fire lack of shade, unsuitable microclimate during the hot season, or fungal infections during the rainy season because of high moisture among the weeds.

Table 5-5. Species Data from Sample Plots

Plots	*Total Species per Sample Plot*	*Number of Species Common to Two Plots*								
L3	31									
L6	44	9								
L12	44	7	10							
M3	38	5	8	7						
M7*	16*	3	9	9	4					
M12	42	3	9	10	13	11				
K5	90	5	10	10	4	3	9			
K10	15	2	5	10	6	4	7	30		
K18	84	0	1	2	4	1	4	28	34	
NF	98	3	4	8	6	4	10	20	20	13
		L3	L6	L12	M3	M7*	M12	K5	K10	K18

Note: * Only tree species.

Hmong

The data is not so clear for the Hmong plots. The partial lack of data from the M7 plot and very uneven tree distribution in M12 caused problems. The total number of species was similar for all ages of succession. The number of trees, crown cover, basal area, and all height parameters seemed to increase steadily with time. The Hmong plots were the only ones that showed slight stories in the canopy. Most trees had low branches, and it would have been hard to find useable wood in these forests. Because these plots were at higher altitudes than those of the other tribal groups, high similarities with Karen and Lisu plots might not be expected. Without ongoing disturbance from fires, overgrazing by cattle and the like, these areas are likely to recover faster than in the Lisu case.

Karen

From a forest regeneration point of view, these were the best sites because they had the highest number of trees and species, crown cover, and basal area and were highest in all height parameters. Successional development had been very fast, and K18 had already been used for firewood collection and small volumes of timber had been harvested for poles. This helped to explain the even basal area and cover distribution. It was a perfect forest for this purpose, and the soils did not seem to be disturbed very much. There was a high number of species, especially in the early years, and the similarity of species among plots was high. Many foresters dream of creating a forest like this in 18 years. The lack of big trees, however, was an issue for consideration because of their importance to the ecosystem. Leaving some "standards" for the next rotation would be beneficial. Any new rotational agricultural use would not harm this system very much. Clearly, the Karen system is not contributing to further deforestation in Thailand.

Diversity

The main focus here was on the total number of different species. Results showed high numbers of species in all plots, although Karen forests had far higher numbers than those of the Hmong and Lisu. The low figures for common species among Hmong and Lisu plots might indicate that the overall number of species in the natural succession was very high.

Fire seems to play an important role in succession. Karen forests were not burned every year, whereas the Lisu plots suffered frequent fires, even in older succession. Fire lowers both the number of trees and occurring species.

Karen forests are very close to their villages. Together with some natural forest areas surrounding the village, the plant diversity is enormous and may be even higher than in a natural forest. Diversity is a value in itself, but it is also a value to the people. It seems as if the Karen people are highly appreciative of this, although they would probably express it differently.

Agricultural practices prior to and during cultivation, as well as interventions after abandonment of crop cultivation, all appeared to have influenced the successional process of forest reestablishment. Traditional Karen management practices aim to rapidly reestablish forest cover for future agricultural use, whereas Lisu and Hmong management practices do not. Therefore, establishment of a secondary forest cover for watershed or community forest purposes appears to be more rapid in areas of Karen management.

Recommendations

To sustain good natural re-growth on former shifting cultivation fields, the following general recommendations should be considered:

- Land-use management should be based on common understanding and cooperation among all parties concerned;
- Post agricultural disturbances such as fire and grazing by cattle should be reduced; and
- Remaining forests or regenerating areas need to have more economic value for local people. Community woodlots and sales by the local people of timber, bamboo, rattan, resins, and herbs, to name just a few forest products, should be allowed on a sustainable scale.

On more seriously disturbed sites such as regenerating Lisu fields, several other measures should be considered to enhance natural regeneration:

- Enrichment planting for grass suppression and to assist natural regeneration;
- Permanent agricultural uses, such as planting fruit trees, on parts of regenerating areas;
- Changes in the management of non-wood forest products that benefit form disturbances, including cattle grazing, harvesting of mushrooms, broom grass, and roof grass; and
- Fire management for large, heavily disturbed areas. For these purposes, fire management does not necessarily mean complete fire suppression, because prescribed burning, spot weeding around seedlings, burning of fire lines, and other measures may be required.

Acknowledgments

This study would not have been possible without the kind support of the Regional Community Forest Training Center in Bangkok and the Rockefeller Foundation. I am very grateful to the late Dr. Tem Smitinand for field identification of plant species. Without the assistance and support of the people from the Lisu village of Ban Lisaw Lum, the Hmong village of Ban Khun Sa Nai, and the Karen village of Ban Mae Muang Luang, I would not have been able to accomplish this research. I also want to thank the personnel of the Watershed Development Unit No. 1, Thung Jaw, for providing facilities during fieldwork. I wish also to thank all of those people I have not personally named, but with whom I had fruitful discussions and who supported this project.

References

Boonkerd S., J. Sadakorn, and T. Sadakorn. 1982. Thai Plant Names. (Thai language).

Kanjunt, C. 1988. Village Settlement: A Solution to Deforestation in Thailand. Masters thesis, Southern Illinois University, Carbondale, IL.

Kijkar S. 1987. *Pines in Thailand.* Bangkok: Silviculture Branch, Royal Forest Department (Thai language).

Koskela, J., J. Kuusipalo, and W. Sirikul. 1995. Natural Regeneration Dynamics of Pinus merkusii in Northern Thailand. *Forest Ecology and Management* 77, 169–179.

Lamprecht, H. 1989. Silviculture in the Tropics.

Mueller-Dombois, D., and H. Ellenberg. 1974. *Aims and Methods of Vegetation Ecology*. New York: John Wiley & Sons.
Planning Division, RFD (Royal Forest Department). 1991 and 1993. *Statistics of Forest in Thailand*. Bangkok: Royal Forest Department.
Zohrer, F. 1980. Forstinventur.

Chapter 6

Kammu Fallow Management in Lao P.D.R., with Emphasis on Bamboo Use

Damrong Tayanin

Shifting cultivation amongst the Kammu in Laos has traditionally followed an 11-year cycle. The land had to be left fallow sufficiently long for the trees to form a canopy and smother out unwanted grasses. High trees overshadow the grass and prevent it from getting any sunshine, so it dies. If grass underlying trees does not get enough rain, it also dies. Swidden fields heavily invaded by ferns and elephant grass need to be left idle for two or three cycles before being reopened. If such places are left fallow for 15 to 20 years, the grass disappears and rice grows well again.

Kammu farmers open new swidden fields every year, rotating around their village territory. People commonly calculate their own age according to the swidden cycle. For example, if someone was born when the villagers opened fields in a certain place, he or she would be 11 years old when they returned to that same location. The swidden cycle follows the Kammu Yuan lunar calendar which, like lunar calendars used elsewhere in Southeast Asia, is structured around two cycles of 10 and 12 years running in parallel. These combine to create a longer 60-year cycle. The same system is also applied to days. This is essentially the same as the better-known Chinese lunar calendar. The villagers cultivate two extensive field areas each year, and the entire village shares the land in both areas. Each family must have at least one field in each place, although some families may have two or more. This is because a big family needs several large fields whereas a small family has neither the use for large fields, nor the labor to manage them properly.

The reason for this segregation of swidden land into two large areas is because two kinds of rice are grown: early and late maturing. The weather also has to be considered, since the amount of rain is unpredictable. One of the field areas is usually located where the soil is somewhat harder and benefits from a lot of rain, and the other is at a place where the soil is softer and needs less rain. If it rains heavily, the rice on the harder soil will grow well, but the rice on the softer soil will not flourish. Conversely, if rain is scarce, the rice on the softer soil will thrive but that in the harder soils will not do well. This is why all families in a village must maintain swidden fields at both sites.

> Editor's Note: This chapter stands apart from others in that its author is himself a member of the Kammu (Lao Theung) ethnic group; he does not report research findings but instead draws on his own experience as a shifting cultivator in Luang Prabang Province of Lao P.D.R.

Damrong Tayanin, Department of Linguistics and Phonetics, Lund University, Helgonabacken 12, S 223 62 Lund, Sweden.

Households may, in some years, run out of rice just before the new crop ripens, and any family that has not sown early-maturing rice will be in trouble. This period before the early crop ripens is the hardest time of the year. Should people run out of rice, it is difficult to buy more from neighboring villages because it rains incessantly, the path is wet and overgrown, there are land leeches and mosquitoes everywhere, dangerous animals move closer to the villages, and the rivers are flooded. It is impossible to travel far during this period, and people stay close to home and work in their fields.

Kammu farmers grow not only rice in their swiddens but also other secondary crops such as cotton, millet, maize, sweet potatoes, taro, peanuts, pumpkins, cucumbers, melons, gourds, pepper, tobacco, vegetables, mung beans, sesame, various cooking ingredients, and some kinds of flowers. Certain flowers, sweet potatoes, and taro are culturally important because they are used in harvest rituals.

One Year of Cropping

Normally, Kammu farmers plant swidden fields with rice and other crops for no more than one year, and their shifting agriculture functions according to strong traditional beliefs. While a family is using a certain area for growing rice and other crops, they are considered its owners and are therefore responsible for it. If something happens in their area—if lightning strikes a tree, for example—that owner is responsible for the situation. They are placed under a taboo until the harvest is finished, and then they have to finish off the old year by driving away the bad luck and the lightning spirit. This is why farmers do not want to retain an area for an extended period or grow anything after the harvest. When a family has completed the old year, they leave the place so that it no longer belongs to them. If something bad happens after they leave, such as a lightning strike on the abandoned field, they are no longer responsible because the land has been given back to the forest.

When the harvest is finished, the entire rice crop is brought home and stored in barns outside the village. However, the women continue to harvest millet, sesame and mung beans. They also dig up all the taro, sweet potatoes, and peanuts and collect the pumpkins and melons. Everything is brought back to the village. Sometimes, when rain falls at this time of year the Kammu refer to it as "rain that destroys the rice stubble." The women then go to dig out rats and collect mushrooms (see color plate 13). It is also the women who cut the tree trunks that have dried out during the cropping period and carry them home for firewood to heat the houses during the cold season and for cooking.

Meanwhile, the men return to the swiddens to tear down the temporary field shelters. They carry the materials back to the village for storage, so they can be used the following year. Then they open the fences around the abandoned fields to let their livestock in. The pigs eat any remaining pumpkins, melons, gourds, and cucumbers, while the buffaloes and cattle graze the rice stubble and young grass.

Another problem that discourages the re-cultivation of fields is that access paths become overgrown. During the single year that a swidden field is cultivated, the path is cleared three times. When a field area is abandoned, the paths are no longer cleared and they quickly become overgrown with grass, weeds, and bushes. There are also many land leeches along unused paths, and once secondary forest begins to re-establish in the fallow, it provides a habitat for wild and dangerous animals that normally live in the very dense forest.

If a farmer grew rice in the same field for two or three consecutive years, the field cycle would also be thrown out of synchrony. There would be a risk that the tree stumps and their roots would die, the trees and bushes would not re-grow, the land would remain naked, and the soil would become hard and unsuitable for cultivation.

The First Year of Fallow: The túh

Many kinds of vegetables, both wild and cultivated, remain in the abandoned swidden fields. These usually include pepper, various varieties of eggplants, and tobacco. Animals do not eat these plants, although they may destroy them by trampling. However, they are usually grown at the lower ends of the fields on slopes too steep for animals, and they may persist for one or two years. No sweet potatoes, taro, mung beans, peanuts, pumpkins, melons or sesame remain because there are eaten by rats, wild boar, and porcupines.

If lightning strikes a tree in a recently fallowed swidden field while it is still yielding crops and is, therefore, still in use, the family is not permitted to return to the abandoned field. All remaining crops in the field are abandoned, even after the family has driven away the evil lightning spirit and its associated bad luck.

During the hot season, women return to the *túh* to collect small bamboo shoots, eggplants, peppers, and tobacco leaves, and may also dig for edible mole-crickets to prepare as food. If there are places on the túh where buffaloes have been kept at night, and dung and urine have accumulated on the ground, farmers often fence off an area of 15 to 20 m^2 to grow tobacco and pepper for an additional year. Beyond that time, grasses and tangled re-growth make walking difficult.

Only a few selected places with exceptionally fertile limestone soils may be planted with rice for a second year. This is known as "clearing the young fallow." When rice is grown for a second year, the farmer can expect big problems with weeds, as well as difficulties in keeping the path cleared and the fence repaired.

The Second Year of Fallow: Still the túh

By the second year of fallow, no cultivated vegetables remain, and grasses, bamboo, and trees have grown to the height of a man. In some places, buffaloes can no longer be seen as they browse under the pioneer trees. In the hot season, women cut thatch grass to make new roofs for their houses. They may also cut some wild vines to weave handbags or make string handles for containers used to fetch water. Buffalo and cattle still graze the thatch grass if the field is not fenced.

The Third Year of Fallow: rèe... saám pií

It is almost impossible to walk through a three-year-old fallow because the vegetation is very dense and trees and grasses are about the height of a man. However, women continue to force their way into the dense new forest to cut thatch grass that continues to grow there.

The Fourth and Fifth Years of Fallow: rèe... sií pií, rèe... haá pií

The vegetation continues to grow very densely and it is impossible to walk through a fallow during the fourth and fifth years. After five years, the trees grow higher and the underlying grass disappears. Wild animals and birds return to live in the fallow area again because some trees are already bearing fruit and berries. People use these areas for hunting and to lay traps for game.

As can be seen, the Kammu have special words for the field at different years during the swidden cycle. When it is used for growing rice, it is called *ré* (field); after the harvest is finished, it becomes *túh* (abandoned field). By the third year, it is known as *rèe... saám pií* (three years fallow). In the fourth and fifth years, it becomes *rèe... sií pií* and *rèe... haá pií,* respectively (fourth and fifth years fallow).

After six or seven years, some trees are big enough to use as house construction materials. Villagers may begin to harvest them for poles. After 8 to 10 years, people refer to the area as *prì* (forest) again, and after 11 years, it is cleared and turned back into a field.

Problems with Landslides and Elephant Grass

There are problems in Kammu swiddens with landslides and invasion by elephant grass. When villagers open a field in a new location that has never been used before, the trees are naturally taller and have larger basal areas. All the trees are cut, the slash burned, and rice is sown in mixtures with other crops. In such new fields, rice may be grown for two years or even more. However, when older trees are cut down, such as those found in deep forest that has not been opened for swiddening for more than 100 years, their roots and stumps usually die. Five or six years later, the area becomes vulnerable to landslides because the roots holding the soil have died. During heavy rains, landslides often sweep from the mountaintops down into the valleys. Such landslide areas cannot be used for agriculture anymore because nothing will grow there except different kinds of elephant grass, and where these grasses grow, the soil is no longer suitable for other plants. Trees generally do not regenerate in landslide areas, with the exception of a tree called *hntaá*. It grows more quickly than other kinds of pioneer trees and its wood is not hard enough to be used for construction.

In contrast to the older trees of new forest locations, young trees often coppice and regenerate some months after cutting because their stumps and roots do not die.

Garden Plots and Fields

In addition to the staple rice crop, Kammu farmers plant and grow many other plants, both in the fields and forests. These include varieties of bamboo, rattan, tea, *Phrynium* plants for constructing house roofs, betel leaf, and screw pine for making raincoats.

Every family in a traditional Kammu village has its own garden near the village, and some families maintain other garden plots in valleys and on riverbanks. These people do not use their fallows for growing other crops or vegetables. Instead, they use riverbank gardens to plant vegetables, tobacco, pepper, and maize during the hot season, when there is no rain. In valley garden plots, they cultivate tea for chewing after meals and for making fermented tea. These tea bushes can reach 50 to 70 years of age before the owner has to dig them up and replant. Gardens surrounding the village are also planted with fruit trees such as pomelo, tamarind, papaya, betel nut, sugarcane, banana, and often also *Strobilanthes*, which is used to make black dye.

Over the past 20 years, some Kammu people have moved their villages from the slopes down to new sites in the valleys. Some families found small flat areas of 40 to 50 m^2 in size, suitable for growing wet rice. However, even after developing small paddy fields, their rice yields were insufficient for household needs and had to be supplemented by continued shifting cultivation on nearby slopes.

Today, in Bo Keo and Luang Namtha Provinces, in the far northwest of Lao P.D.R., shifting cultivation has declined and wet rice cultivation is increasing. Many families have migrated away from their native villages, and this has relieved the pressure on swidden land. Those who remain can now cultivate the areas with highest potential, providing higher yields and fewer weeds.

Kammu villagers eat much more glutinous rice *(kháw niàw)* than ordinary rice *(kháw caáw)*, because glutinous rice is more satisfying, or filling, for a longer time. It is also a very convenient food, and can be eaten with all sorts of stews, soups, and salads, or just with chili sauce. When people eat non-glutinous rice, on the other hand, they usually feel hungry again after a few hours.

Kammu villagers can also find wild vegetables everywhere, both in the forests and valleys. Some grow during the monsoons, while others sprout during the hot season. Taking banana plants as an example, people use banana flowers, young banana leaves, and the core of the stem while it is still white to make stew or soup. Older banana leaves are used to wrap parcels or to cover the roofs of temporary huts when spending the night in the forest. The sap from banana stems is used in making gunpowder.

The Importance of Bamboo

Bamboo—both planted and wild—grows abundantly on mountain slopes and in valleys. It has a multiplicity of uses and people cultivate it outside the village, on riverbanks, and in the valleys. It is used as a building material, in baskets and other handicrafts, and as food. This final section of the chapter describes the uses of bamboo and where it grows, beginning with the largest varieties and working toward the smallest species.

Cultivated Bamboo

Giant Bamboo. This variety reaches heights of 30 to 50 m, so it is planted in high forests. If there are no sheltering trees around the clumps, the stems are likely to break off during storms. Giant bamboo is used to make house floors, to cover the roofs of houses or barns, and to make water containers, rice cookers, bowls, and bamboo clappers for scaring wild animals away from fields. This kind of bamboo is particularly suited for flooring because when the stem is split open and spread, it may be as wide as 50 or 60 cm. The shoots are very tasty when boiled and made into a salad. Giant bamboo grows best in cool, moist places. Its growth will be stunted if its location is too hot and dry.

Second-Largest Bamboo. This bamboo is planted and used in much the same way as described above for giant bamboo.

Oily Bamboo. This type of bamboo is used for making tying strips, which are used when building houses or barns. The young shoots are oily and make a good salad.

Piglet Bamboo (probably *Bambusa vulgaris*). This bamboo species is distinguished by its stem. It is used as a building material, particularly for making floors in field houses. It is not eaten because its shoots are not tasty.

Dark Green Bamboo. This kind of bamboo is bigger than *Dendrocalamus latiflorus* (below) but smaller than piglet bamboo. It is very hard and strong and is therefore good for construction and for making various traps, such as rat snares and deadfall traps. It is preferred for traps because it does not easily lose its spring. This is a critical attribute since the traps sometimes remain set for hours or even days before they are triggered. A softer bamboo would not provide the necessary tension.

Dendrocalamus latiflorus. This bamboo is planted among tree groves surrounding the village. It is a good building material and is also useful for making traps, but its shoots are not particularly edible.

Thin Bamboo. People plant thin bamboo in cool places, such as in valleys near streams. It is especially suitable for making walling material for houses or barns, and many barn floors are made of this bamboo. It is easy to weave and provides thin wall sheets. Rice cookers are also woven from it. The shoots are not very edible.

Small Bamboo. People plant this small species of bamboo in their gardens and use the hard stems to make rods for sucking rice wine from jars. The shoots are not used because they are too small.

Wild Bamboo

There are many species of wild bamboo in the forest, and they are available to anyone who wishes to cut and use them. During the rainy season, the shoots are harvested for food. Types of wild bamboo harvested by Kammu people include the following:

Medium-Sized Bamboo. This bamboo grows among high trees in the valleys and along riverbanks. Its hard stems are useful for many things, including basket weaving, building materials, and making traps. The shoots are valued for preparing many kinds of food. Stem sections are about 78 centimeters long and, therefore, ideal for weaving baskets.

There are two types of wild medium-sized bamboo, one with hairs on the stems and the other without. The variety with hairs is not suitable for basket weaving because its stem sections are too short.

Dendrocalamus hamiltonii. This bamboo also grows in valleys and along riverbanks. People use it to make clappers for scaring animals away from their fields, for building fences around fields, or for making the floors and walls of temporary field huts. The shoots are not particularly good to eat but can be salted, fermented, and then dried for later use. The stem sections are too short for basket weaving.

Thorny Bamboo. This bamboo is not found on mountains and is difficult to handle because it has thorns on both the stems and branches. It is used only for making fences around fields. Its stem sections are short and unsuitable for weaving baskets. However, its shoots are used in the preparation of many kinds of food.

Bambusa tulda. This bamboo species is widely distributed. It grows on mountains, in valleys, or along riverbanks. It may grow close to water as well as in dry places. It has multiple uses—as a material for building houses, traps, strips for tying things, or weaving rope. Because its stem sections can be as long as 100 cm, it is most suitable for weaving any kind of baskets. The shoots are used in preparing many kinds of food. The hard roots are used to make pipes for smoking tobacco.

When farmers open rice swiddens from fallows that have included this bamboo in the succession community, they intentionally choose the burnt bamboo clumps with abundant ash to plant tobacco. Tobacco is known to flourish in soil fertilized with this ash and produce high-quality leaf with a strong flavor. The resulting tobacco leaf has several important uses. It is smoked by people during weeding operations to help repel swarms of mosquitoes, and when someone has a cough or catches a cold while out fishing, he or she chews this strong tobacco leaf in combination with betelleaf, lime and pieces of bark from a tree called *pntrìik*.

Oxytenanthera parvifolia. This bamboo grows mainly in the Puka region, with lesser concentrations in the Yùan region. It is similar to thin bamboo, but the stem walls are thicker and the sections much shorter, so it is not suitable for weaving. It is suitable for making walls for houses and floors for barns. The shoots are edible, but they are not preferred.

Pleioblastus. This bamboo yields several kinds of building materials, but it is not suitable for walls or floors because the stems are not broad enough. The shoots are very bitter, but are used in the preparation of some kinds of food.

Tmaár. This bamboo grows in mixed communities among high trees and its branches intertwine with the trees. The stems are not straight, and have a vine-like appearance. It is preferred for making tying strips, especially those used in building barns, because they can tolerate exposure to rain for some years. Trappers use the strips for making strings for rat snares because of their strength.

Tràal. This bamboo grows along riverbanks. Its stems are not sufficiently hard for use as building materials. However, the roots are used to weave baskets because they are as strong as rattan.

Pléey. *Pléey* bamboo grows both on mountains and in valleys. The stems may be used as building materials, its hard roots are carved into smoking pipes, and the shoots are the very best for eating.

Smaller Bamboo. This smaller kind of bamboo is used for building temporary field houses. The shoots are very tasty and are used in the preparation of several foods.

Smallest Bamboo. This very small type of bamboo grows on mountains. Its stems are used to make mouthpieces for smoking pipes. When a woman gives birth, a piece of this bamboo is used to severe the baby's umbilical cord because it is known to contain no poison. The shoots are prepared in the same way as those of smaller bamboo (above).

The Flowering Phenomenon

Sometimes, bamboo flowers and then dies. Bamboo flowering is inevitably followed by a population explosion among rats and jungle fowl that eat the fallen fruit. To cite examples, *Bambusa tulda* flowered 50 or 60 years ago; *Dendrocalamus hamiltonii* flowered in 1965; and in 1968, some types of *Dendrocalamus latiflorus* bamboo flowered.

The year after bearing flowers, the bamboo dies. The seed is widely disbursed, and many new bamboo seedlings establish over the area. After bamboo flowers, its stems cannot be used for building materials because insects will completely consume them. When farmers notice cultivated bamboo beginning to flower, they quickly cut the affected stems and plaster the stumps with leaves and earth to prevent the flowering from spreading to other clumps.

Conclusions

There are few opportunities for growing wet rice in the mountainous homeland of the Kammu. The river valleys are narrow with little or no flat land adjacent to the riverbanks, and the mountains are high and steep. Fields cultivated on mountain slopes are often so steep that even walking is difficult, and the use of plows is not possible. Even turning the soil with a spade leaves it vulnerable to washing away downslope into streams. Landslides are always a threat, making it critical that tree roots that help to anchor the soil are not damaged or killed while cultivating the fields.

Contrary to conventional wisdom, the Kammu hold a traditional belief that allowing swiddens to revert back to old forest is dangerous because streams will tend to dry up. If this is true, this would spell calamity and, in the long run, would influence the water levels of larger rivers downstream.

Rice is akin to life itself for Kammu farmers, and in their environment, swidden agriculture is perhaps the only practical way to grow it. This highlights the critical need to identify ways to make swidden agriculture more productive and to minimize its damage to fragile upland environments. The sustainability of traditional Kammu practices is proven by the fact that many villages have remained in the same location for centuries. Clearly, this would not be possible if the fieldwork of the Kammu caused regular landslides or devastating forest fires (Roder et al. 1991).

Acknowledgments

I would like to express my deep thanks to Malcolm Cairns for his invitation to take part in the Workshop on Indigenous Strategies for Intensification of Shifting Cultivation. The workshop reminded me of my earlier life. I have now learned many new ways of cultivating rice and other crops from workshop participants who use different techniques. I never realized that people from other countries used special plants to make the soil better. The only plants we use for such a purpose are jack-in-the-bush, *Chromolaena odorata*, which makes the soil soft, and sesame, which kills the *Imperata* grass. I really wish that some of the experts who know about this would come to teach the farmers in my village area.

Related References

Kammu homepages. http://www.ling.lu.se/persons/Damrong/kammu.html

Lindell, K., H. Lundström, J. Svantesson, and D. Tayanin. 1982. *The Kammu Year, Its Lore and Music*, SIAS Studies on Asian Topics No. 4. London: Curzon Press.

Roder, W., W. Leacock, N. Vienvonsith, and B. Phantanousy. 1991. The Relationship between Ethnic Groups and Land Use in Northern Laos. Poster presented at the International Workshop on Evaluation for Sustainable Land Management in the Developing World, September 1991, Chiang Rai, Thailand.

Tayanin, D. 1992. Environmental and Nature Change in Northern Laos. In: *Asian Perceptions of Nature*, Nordic Proceedings in Asian Studies No. 3, edited by Bruun, Ole, and Arne Kalland. Köpenhamn: NIAS.

———. 1994. *Being Kammu: My Village, My Life*. Southeast Asia Program Series, No. 14, Ithaca, NY: Cornell University.

———, and K. Lindell. 1991. *Hunting and Fishing in a Kammu Village*, SIAS Studies on Asian Topics No. 14. London: Curzon Press.

———. 1994. The Kammu Cycles of 60 Days and 60 Years. In: *Mon-Khmer Studies* 23, Nakhon Pathom, Thailand: Mahidol University.

———. 1995. How to Quell Grass? *Indigenous Knowledge and Development Monitor* 3(2), The Hague, The Netherlands.

———. 1996. Kammu Women Suppress Grass Weeds with Sesame, *ILEIA Newsletter* 12(1), Leusden, The Netherlands.

Tayanin, L. 1977a. Kammu Dishes. In: *The Anthropologists' Cookbook*, edited by J. Kuper. London and Henley: Routledge and Kegan Paul.

———. 1977b. Kammu Hunting Rites. *Journal of Indian Folkloristics* 1(2), Mysore.

Chapter 7

The Potential of Wild Vegetables as Permanent Crops or to Improve Fallows in Sarawak, Malaysia

Ole Mertz

The use of wild plants by communities of shifting cultivators in Southeast Asia has recently received increasing research attention, mainly from a conservation point of view, but also because of an increasing demand for forest products from urban populations. In Sarawak, broad ethnobotanical studies have been carried out by Pearce et al. (1987) and Chin (1985), while more specific studies have focused on traditional medicine and plants for decorative uses (Ahmad and Holdsworth 1994; Kedit 1994; Leaman et al. 1996). Most recently, a study on Kelabit and Iban plant uses has resulted in a very comprehensive compilation of indigenous knowledge concerning Bornean rainforest resources (Christensen 2002). A total of 1,144 different species is identified, with an even wider range of associated uses.

These plants, as well as many other rainforest species found in the region, represent an immense economic potential, and their domestication and cultivation in fallowed swiddens or permanent gardens could be an important component of intensification processes in shifting cultivation systems. Most work on domestication has concentrated on tree species, notably leguminous trees for forage and soil improvement, and high-value species producing marketable fruits, latex, resins, or fibers (Okafor and Lamb 1994; Poh 1994; ICRAF 1995). Herbaceous plants used as vegetables have been largely overlooked.

In Sarawak, domestication of wild plants has been on the research agenda of the Department of Agriculture since 1987 (DoA 1987), and observation trials include not only indigenous fruit tree species, but also wild vegetables such as *Gnetum gnemon* L. and various ferns (DoA 1993). The cultivation or "manipulation" potential of these species at farm level has not yet been investigated and this study is therefore a direct part of this initial research.

The objective of this chapter is to test the hypothesis that full or partial domestication of perennial wild vegetables and their cultivation in fallowed swiddens or permanent gardens may offer an important contribution to the intensification of shifting cultivation, in agronomic, ecological, and economic terms. It suggests an alternative to commonly grown exotic vegetable species that tend to be more susceptible to pests and diseases and rely heavily on fertilizer and other chemical inputs (Rahman 1992).

Ole Mertz, Institute of Geography, University of Copenhagen, Øster Volgade 10, 1350 Copenhagen K, Denmark.

On-farm trials designed as a form of "social testing" of interest in cultivating existing wild vegetable species form the core of this chapter, but reference is also made to data from other components of the research project, including fertilizer and shade trials, market surveys, and land-use studies.

The Study Area

The on-farm trials were carried out in the longhouse community of Nanga Sumpa, located in the upper Batang Ai watershed, Lubok Antu district, Sri Aman division, Sarawak (see Figure 7-1). Nanga Sumpa had in 1997 28 households and 180 inhabitants, all ethnic Iban. Situated on the Delok River, the village is accessible only by a boat ride of one-and-a-half hours from the Batang Ai hydroelectric dam, following a bus journey from the towns of Lubok Antu (10 km) or Sri Aman (75 km). Average annual rainfall is 3,450 mm (DID 1993). The area is very hilly, with steep escarpments along rivers and only a few small riverine plains. The soils have not been surveyed, but red-yellow inceptisols and ultisols seem to dominate, as they do elsewhere in the interior of Borneo (DoA 1968). The natural vegetation was once mixed hill dipterocarp rainforest, but today it is dominated by various stages of secondary forest and farmland. Only a few areas of mature forest can be found in the village territory. The main agricultural activities are shifting cultivation of rice and cash cropping of pepper (*Piper nigrum* L.) and para rubber (*Hevea brasiliensis* [Willd. *ex* A. Juss.] Muell.-Arg.). Off-farm income derives mainly from a small tourist lodge at the village, run by a Kuching travel agent, and from migrant work.

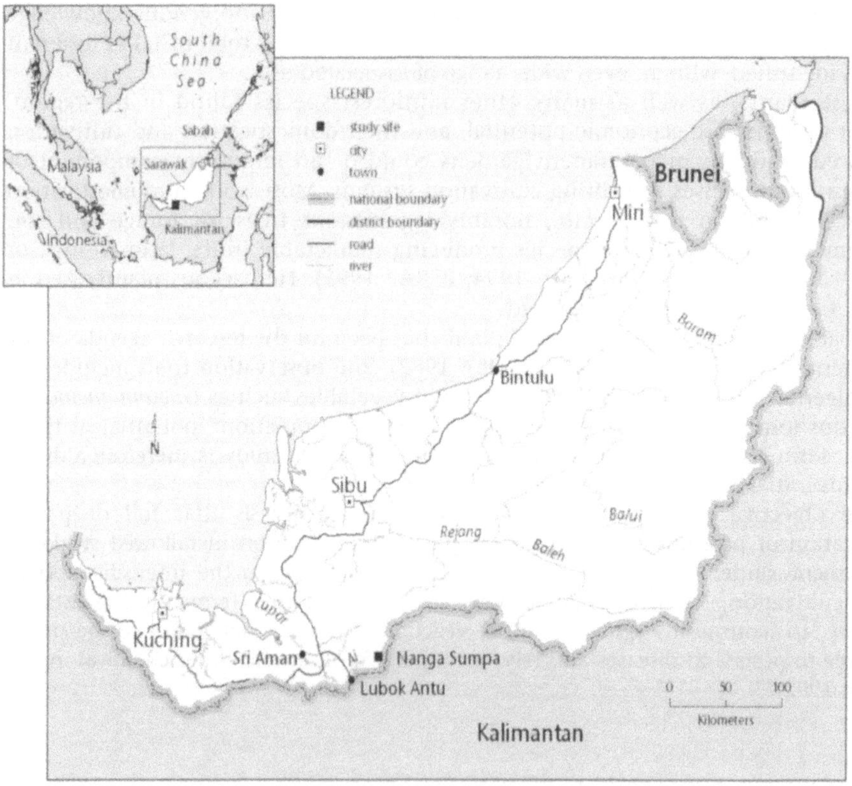

Figure 7-1. Map of Sarawak Showing Nanga Sumpa Study Area

Table 7-1. Wild Vegetables Selected for On-Farm Trials

Family	*Species*	*Iban Name*	*Part Eaten*	*Type and Natural Habitat*
Acanthaceae	*Pseuderanthemum borneense* Hook. f.	Gelabak	Young or older leaves as spinach	Small shrub found in mature secondary forest
Athyriaceae	*Diplazium esculentum* (Retz.) Sw.	Paku Ikan	Young fronds, fried or boiled	Terrestrial fern in partly open secondary forest, swampy and ravine soils
Blechnaceae	*Stenochlaena palustris* (Burm.) Bedd.	Kemiding	Young fronds, fried or boiled	Climbing fern in open areas to old secondary forest, swampy to dry soils
Zingiberaceae	*Etlingera elatior* (Jack) R.M. Smith	Kechala	Heart of young shoots, flower buds, fruits. Condiment or vegetable	Tall herb (3 to 4 m) in open areas or secondary forest
Zingiberaceae	*Etlingera punicea* (Roxb.) R.M. Smith	Tepus	Heart of young shoots, flower buds, fruits. Condiment or Vegetable	Tall herb (3 to 4 m) in open areas or secondary forest

The Plant Material

Of the wild vegetables found in the area, five were identified for use in this study. They are described in Table 7-1. The choice of plants was based on their relative importance in the local diet (Christensen 1997, 2002) and their estimated agronomic and marketing potentials. Of the five, two are ferns that are frequently consumed, notably *Stenochlaena palustris*, which is common around the village, and *Diplazium esculentum*, which is not found in the immediate vicinity. *Pseuderanthemum borneense* is a highly valued but rarely found shrub, and two herbs, *Etlingera elatior* and *Etlingera punicea* are important condiments as well as vegetables. All are perennial and allow for continuous harvesting.

Research Approaches

The research was carried out from February 1995 to July 1997. Initially, interviews were conducted with all households. Local English-speaking interpreters were used, and the interviews were based on questionnaires. These were aimed at determining quantitative elements of the farming system, such as the number of fields, fallow periods, and yields. More specifically, I sought to assess the occurrence, use, and importance of the five vegetables. Local informants assisted with field investigations of the common habitats of the vegetables.

Five households were chosen by the village committee to participate in the on-farm trials. The trials were aimed primarily at determining the local acceptance of cultivating wild vegetables, rather than analyzing specific agronomic parameters. Each household established a small garden of 400 to 600 m^2 in secondary growth or in abandoned rubber or cocoa gardens, near watercourses, and at a distance of no more than 500 m from the village. The gardens were all partly shaded. The soil types consisted of dry red-yellow clay soils and wet organic soils.

The plants were collected from wild groves, sometimes as far away as a one-hour boat ride upriver. All were propagated vegetatively. The households were able to decide which of the species they considered worth planting. Maintenance of the gardens was left to their discretion, with the only requirement during the first seven months being weed control.

The households were offered a payment of one Malaysian ringgit, or about US$0.40 per day, at existing exchange rates, for tending the gardens during the first seven months of the trials, from April to October 1995. There was no compensation paid for the remaining nine months of the trial period, from November 1995 to July 1996. This arrangement aimed to ensure an appropriate establishment of the gardens and, after that, to evaluate continued interest in cultivating the vegetables.

The village was visited about once a month in the period from April 1995 to January 1996, then again in June and July 1996 to evaluate the period without payment, and finally in July 1997 to estimate the viability of the gardens after a long period without project activities.

In July 1996, open-ended interviews were conducted with the households involved to evaluate the success or failure of the gardens and to determine whether they would continue the cultivation. Structured interviews were also carried out with all households in the community to find out whether the idea had caught on.

While the on-farm trials were under way, fertilizer and shade trials were being conducted at Rampangi, a research station of the Sarawak Department of Agriculture, 16 km north of Kuching. These aimed to obtain agronomic data on the growth potential of the five vegetables in a controlled environment. The outlines of these trials are presented in Mertz (1999a,b). The approaches to research into land use and crop diversity are described in Mertz and Christensen (1997).

Results

Conventional Shifting Cultivation and Cash Crop Production

The practices of farming upland rice in Nanga Sumpa are very similar to those in other Bornean shifting cultivation systems, which are well described in the literature (Freeman 1955; Padoch 1982; Chin 1985; Dove 1985; Mertz and Christensen 1997). The cropping season is from August and September to February and March. All Nanga Sumpa households but one plant upland rice in swiddens. The average household has 1.5 fields situated about 1.5 km from the village, often in a cluster with three to five fields belonging to other households. The average area of upland rice per household is 1.2 ha, based on measurements of fields belonging to six households. Wet rice is cultivated in small riverine patches associated with upland swiddens.

Based on data from 1992 to 1996, the average fallow period in Nanga Sumpa seems to be more or less stable at five to seven years. All fields are cultivated for only one year. The main reason given for maintaining at least this fallow period is problems with weeds, notably *lalang*, the local name for *Imperata cylindrica* (L.) Beauv., and the ferns *Blechnum orientale* L., *Pteridium* sp., and *Sticherus truncatus* (Willd.) Nakai. Because of the relatively short fallow periods, there are few large trees in the fallow vegetation. Therefore, there is little need to dry the slashed vegetation for long before burning. Slashing and felling are usually carried out just one month before planting.

Cultivation techniques are traditional, including sowing with dibble sticks and harvesting individual rice panicles. The use of fertilizers is uncommon, although inorganic fertilizers are applied when subsidized. However, weed control with Paraquat is gaining importance, and it is used by about half of the households in the early stages of rice cultivation. Manual weeding remains important in the later stages. Pesticides are almost never used.

Rice yields in Nanga Sumpa averaged 350 kg/ha in 1996, and only two households were able to say that their harvest was sufficient to meet the needs of the coming year. Alternative income sources and production of other crops are therefore important elements of the household economy and, off-farm activities aside, these include the following:

- Cultivation of secondary swidden crops. The most important of these are bananas (*Musa balbisiana* x *paradisiaca* L.), cassava (*Manihot esculenta* Crantz.), cucumbers (*Cucurbita* sp.), Job's tears (*Coix lacryma-jobi* L.), local eggplant (*Solanum* spp.), maize (*Zea mays* L.), pumpkins (*Cucurbita* sp.), and sweet potatoes (*Ipomoea batatas* [L.] Lam.). These are cultivated in traditional patterns and locations in the swiddens, but with very irregular intensity.
- Cultivation of vegetables in small raised beds with seed, fertilizer, and pesticides subsidized by the Department of Agriculture. Mostly exotic species such as various *Brassica* spp. and legumes are cultivated.
- Cash crop production, focused mainly on pepper and para rubber. Cocoa (*Theobroma cacao* L.) was introduced in 1990 and 1991, but it was abandoned in four or five years because of falling prices and high pest and disease pressure. Pepper gardens are intercropped with vegetables such as taro (*Colocasia esculenta* (L.) Schott), chili (*Capsicum annuum* L.), sweet potatoes, and *changkok manis* (*Sauropus androgynus* (L.) Merr.). Subsidized fertilizers and herbicides are used extensively. Rubber is grown in small orchards mixed with fruit trees and other valuable forest trees. Semi-cultivated illipe nut (*Shorea macrophylla* [De Vr.] Ashton) is found along rivers and is an important source of cash income in years of mast fruiting.
- Collection of forest products, notably the wild vegetables, which are often found on fallow land and in riverine wetlands near swiddens. Important forest products other than the vegetables are fruits, firewood, fibers, construction materials, medicinal plants, dye plants, and various game animals and birds (Christensen 2002) (see, for example, color plate 13).

The main advantage of this diversified form of production is that it allows flexibility and reduces risks, factors that are often featured in the literature (Christensen and Mertz 1993; Cramb 1993; Dove 1993). However, the system also has a number of fairly classical weaknesses. The following three are the most important of them:

- Low and unstable productivity of upland rice. This may be caused by the fairly short fallow period, but many other reasons for low yields exist: moisture stress during dry spells, the use of low-yielding plant varieties, inherently low soil fertility, and labor shortages due to off-farm labor and focus on cash crops.
- Limitations on vegetable production. The existing subsistence vegetable production in Nanga Sumpa is very diverse, with 73 cultivated, semicultivated, or naturalized vegetable species and numerous varieties of each (Christensen 2002). However, the swidden vegetables are mostly short seasoned and are mainly consumed during work on upland rice cultivation. Vegetable production in small gardens is very limited. By 1997, the vegetable gardens subsidized by the Department of Agriculture had been abandoned because of substantial labor inputs required for maintaining raised beds, shade construction, and weeding. Fertilizer and pesticide applications, which were also required, had been discontinued because the department stopped supplying them.
- Unstable income levels from cash crops caused by fluctuating prices on unregulated private markets, as well as difficulty in production, mainly related to pest and disease control.

Labor constraints are a cross-cutting issue. These are partly caused by the large numbers of children attending school and young men in temporary or permanent jobs abroad. The problem is intensified when labor-intensive cultures such as pepper and exotic vegetables gain importance. Manual weed control in the later stages of upland rice cultivation, which is very labor consuming (Chin 1985; Dove 1985), coincides with the peak period for migrant work, and this may well be one of the most limiting factors for rice production.

Wild Vegetables and Fallow Management

The relatively low or unstable production of vegetables has not been much of a problem in Nanga Sumpa because wild vegetables have substantially supplemented the diet (see color plate 8). Christensen (2002) reports that ferns alone make up as much as 10% of the vegetables consumed by the population of Nanga Sumpa. A total of 141 wild vegetables are used (Christensen 2002), and *S. palustris* and *D. esculentum* are the most important of them (see color plate 12).

A large number of these species are found in different stages of fallow vegetation, and the collection of *S. palustris*, *E. elatior*, and *E. punicea* often takes place in fallowed swiddens one to four years after cultivation. *D. esculentum* is found in secondary growth in swampy, riverine areas, whereas *P. borneense* (see color plate 9) is most common in old secondary or mature forest. There are two problems involved in the collection of these vegetables:

- Access to fallow vegetables is often hampered by the dense growth of fallow vegetation. Vines, non-vegetable ferns, and other aggressive pioneer plants usually render secondary growth all but impenetrable just five to six months after the rice harvest, and wild vegetable harvesting in these circumstances is often considered not worth the trouble.
- Some vegetables have become fairly rare and their collection requires long walks or upriver travel. The nearest grove with substantial numbers of *D. esculentum* is found 45 minutes upriver from Nanga Sumpa by engine-powered longboat, and *P. borneense* is only found in scattered groves, each with only a small number of individuals.

The only action taken to address the problem of fallow access has been the maintenance of groves of *Etlingera* spp. in certain areas of secondary forest where the plants occur naturally in high numbers. These groves are not actually cultivated and may be exploited by any individual in the vicinity. The land belongs to households either in Nanga Sumpa or the neighboring village Rumah Jambu, but because *Etlingera* spp. are considered wild, there are no limitations on harvesting by outsiders. The land may be cultivated by its owner, but the *Etlingera* spp. would probably dominate the ensuing fallow vegetation as it sprouts from large underground rhizomes that are not easily killed by fire.

It is difficult to establish whether these groves are consciously maintained or exist merely by coincidence. It is a fact that they are known by all individuals in the village and play an important role in supplying these vegetables. Similar "groves" have not been recorded for other wild vegetables, but it is common practice to tend valuable trees when they occur naturally in fallow vegetation, in secondary forest adjacent to swiddens, in cash crop gardens, or along rivers (Christensen 2002).

Table 7-2. Number of Plants in Wild Vegetable Gardens, May 1995 and July 1996

	P. borneense		D. esculentum		S. palustris		E. elatior		E. punicea	
Household	*May '95*	*July '96*	*May '95*	*July '96*	*May '95*	*July '96*	*May '95*	*July '96*	*May '95*	*July '96*
1. Abong	80	70	300	220	20	20	10	10	0	0
2. Andah	60	60	230	230	1	1	50	40	0	0
3. Kuddy	80	80	110	100	10	10	7	5	10	8
4. Ngalih	90	50	140	110	40	40	10	10	10	6
5. Nam	50	30	150	110	0	0	30	30	0	0

Note: Small numbers of other ferns were intercropped with *D. esculentum*, mainly *Paku kelee* (*Pneumatopteris truncata* [Poir.] Holttum, Thelypteridaceae), *Paku lilien* (*Sphaerostephanos polycarpos* [Bl.] Copel., Thelypteridaceae), *Paku manis* (*Helminthostachys zeylanica* [L.] Hook., Ophioglossaceae), and *Paku raba* (*Diplazium asperum* Bl., Athyriaceae).

Table 7-3. Maintenance Log of Wild Vegetable Gardens, April to December 1995

Household	*Apr.*	*May*	*June*	*July*	*Aug.*	*Sept.*	*Oct.*	*Nov.*	*Dec.*
Abong	F^a, W		W		Wx2	W		W	F^a, W
Andah	F^b		F^b,W	W		W		W	
Kuddy	F^c	W						F^d, W	W
Ngalih	F^b, Wx2	F^a,W				W	F^a, W^d	W, P	F^a, W^d
Nam	F^b	W				F^a, W^d		F^a,W, P	

Notes: F = fertilizer application; W = weeding; P = pesticide application. [a]NPK- Mg + TE (12-12-17-2), subsidized for pepper cultivation; [b]Dolomite; [c]Dolomite and urea; [d]Weeding partly with herbicide (Paraquat).

Cultivation of Wild Vegetables

The on-farm trials were designed to address the problem of access to wild vegetables and to test the local interest in their cultivation. The approach of paying the households was carefully considered. Despite the interest in wild vegetables and semi-management of wild groves, there was no previous experience of actual cultivation, and because the establishment of the gardens would require some labor input, financial compensation seemed necessary if the households were to invest time in the project.

This approach differs from most other on-farm research, which normally involves testing existing crops and farming techniques in trial designs that can be treated statistically. Farmer participation is often cited as crucial for the success of on-farm trials (Ashby 1991; Raintree 1994). In this case, with the exception of deciding on the overall concept of wild vegetable cultivation, all aspects of the trials were decided by the farmers.

The sizes and crop composition of the wild vegetable trials are presented in Table 7-2 and the maintenance log, as kept by households, is shown in Table 7-3. The log was not kept after discontinuation of payment. The number of plants indicated is based on plant counts and information given by the households. Based on comments, interviews, and observations, the following key points summarize the outcome of the trials.

Overall Attitude. The concept was generally perceived as good because of the rarity of certain ferns and *P. borneense* and the prospect of selling the produce to tourists staying at the lodge. The latter represents a real potential because the travel agency presently buys wild vegetables in Kuching in order to serve "jungle food" to tourists staying in Nanga Sumpa. The potential for saving labor was not obvious to the farmers because of the initial work involved in establishing the gardens and because collection of forest products is perceived differently than regular farm work.

Choice of Species. The farmer's choice of vegetables planted in the gardens reflects a preference for species based on the difficulty of finding them in the wild, rather than on a dietary preference. According to interviews and food diaries (Christensen 1997), *S. palustris* is the preferred vegetable fern, but because it is found in relative abundance in young secondary growth around the village, there was no incentive to plant it. *E. punicea* is also easily obtained, notably from the above-mentioned groves, whereas *E. elatior* is less common.

Growth and Yields. The plant counts listed in Table 7-2 indicate a fairly stable or slight decline in the number of plants over the 15-month period. They conceal, however, a continuous process of planting and replacement of dead plants, which was necessary because of difficulties in establishing the ferns.

The maintenance log illustrated in Table 7-3 is indicative, as not all operations were recorded, including mainly smaller weeding operations while harvesting or "passing by." Harvesting is excluded from the table as it was recorded in only a few instances. Experiences with the different plants are summarized as follows.

Pseuderanthemum borneense. This species was planted by all households and was particularly successful with households 1, 2, and 3. Good shade and fairly moist soils were common to these households. Household 4 had a more exposed location, and the garden of household 5 was located on a hill, exposed and well-drained. Establishment generally posed no problems, but growth was slow. No significant pests or diseases were recorded, but a number of plants wilted in the gardens of households 4 and 5, probably because of exposure to the sun. Fertilizer trials suggested that *P. borneense* responded to NPK fertilizer but could not survive without shade (Mertz 1997). This was confirmed by the good results of household 1, whose plants were under shade and received NPK on at least two occasions. By July 1996, the successful households reported "many" harvests and evaluated this plant as a good vegetable crop because of its fairly rare natural occurrence and low demands for maintenance in the garden. Many plants were still present in July 1997 and were regularly harvested.

Diplazium esculentum. *D. esculentum* and other terrestrial ferns (see notes below Table 7-2) were the most popular vegetables. Poor survival and slow growth were the main problems, and by July 1996, most gardens had fewer plants than they started with, despite replanting on several occasions. The environmental requirements were similar to *P. borneense,* notably high soil moisture. This may have been a limiting factor in parts of the gardens. More importantly, fertilizer trials showed clearly that *D. esculentum* did not do well without fertilizer. In fact, shoot yields of shaded plants increased linearly with applications of NPK-Mg (12-12-17-2+TE) (Mertz 1999a).

Some problems with pests were encountered, notably a leaf-cutting beetle (*Apicauta ruficeps,* Meloidae) and an unidentified caterpillar. A condition of wilting and crumbling of leaves was observed in the gardens of households 2, 4, and 5 in July 1996. No pathological condition was identified, but as the condition occurred shortly after applications of herbicide, it was assumed that *D. esculentum* was sensitive to Paraquat, even when selective application on the weeds was attempted.

The evaluation of *D. esculentum* was mixed, mainly because the harvest results were poor. Households 3 and 4 all but abandoned the crop, and household 5 was struggling to keep remaining plants alive. The interest, however, was genuine, and by July 1996, experimentation under different environmental conditions had already been initiated by household 5. If fertilizers are not applied regularly, then imitating the natural habitat by planting in fertile soils in small floodplains along streams may be the only solution. It may also be beneficial to reduce weeding, because sprouts from the rhizomes were often damaged or removed in the weeding process.

Stenochlaena palustris. *S. palustris* was planted only by household 4. Households 1 and 3 tended existing plants in the garden by attaching them to small wooden posts and weeding them, whereas households 2 and 5 did not consider them at all. The abundance *of S. palustris* in fallowed swiddens was the main reason for the lack of interest in planting it. However, the few plants in the gardens grew vigorously after a fairly long four-to-five month period of establishment. There were no pest or disease problems. Although *S. palustris* was not seriously considered as a potential crop because it was too common, it seemed obvious that considerable time could be saved during collection if a dense grove was maintained near the village. Because *S. palustris* responded well to fertilizer, it proved it had potential as a cash crop (Mertz 1999a), and it is now being promoted as a profitable commercial vegetable by the Department of Agriculture in Sarawak (Chai 2001).

Etlingera elatior. This species was planted by all households. Establishment was quite easy but growth was very slow, probably because of insufficient fertilization. Shaded conditions were not necessarily an advantage for this plant. Unshaded fertilizer trials produced 3 meter tall plants within one year, and they responded well to NPK-Mg fertilizer (Mertz 1999b). No harvests were obtained by any of the households, but the plants were still growing and were pest free in July 1996. By July 1997 they were the most abundant species in the gardens and had been harvested several times. Very limited tending of the plants was carried out as they were mostly self-contained. Despite the late harvests, the evaluation of *E. elatior* was positive, mainly because of its value as a condiment and its relative inaccessibility in the wild.

Etlingera punicea. This plant was not popular for cultivation, mainly because it can be found as a pioneer plant in fallowed swiddens. Moreover, although the same plant parts are used, it is less popular to eat than *E. elatior*, and the plant is also more difficult to establish (Mertz 1999b).

Generally, all five households evaluated the on-farm trials positively and claimed that the gardens would be continued and further developed. In order to determine whether this positive evaluation was merely a desire to please the researcher, an inventory of interest shown by the remaining 23 unpaid households was taken. This showed that, inspired by the five participating households, ten of the others, representing 43% of the community, had established small wild vegetable gardens for household consumption of *P. borneense*, *D. esculentum*, and the other most popular terrestrial ferns. Other wild vegetables, notably *daun sabong* (*Gnetum gnemon* L.), a small tree with highly valued edible leaves, had also been planted. However, by July 1997, no further gardens had been established. Although existing gardens were still being harvested, households explained that they had been unable to extend the gardens because of increasing work with tourism and migrant labor.

Discussion

Benefits of Wild Vegetable Cultivation

Wild perennial vegetables have the advantage of producing continuous yields and thereby securing a regular supply of vegetables, as well as contributing to a varied diet. Moreover, several of the crops develop a full ground cover, reducing the need for weeding and the risk of erosion. They may ultimately sustain themselves with very little labor and capital inputs, and as garden maintenance may be undertaken at any time of the year, it need not coincide with the peak periods of labor use in upland rice cultivation.

Therefore, establishment of permanent gardens with *P. borneense* and *D. esculentum* seems feasible. The choice of a location with relatively fertile soil and adequate shade is necessary, but swampy patches of fallowed swiddens previously used for wet rice cultivation and planted with bananas provide these conditions. Intercropping with trees such as rubber is another possibility that may increase the value of these plantations as well as provide an incentive for their maintenance, even during periods of low rubber prices.

The potential for development of *S. palustris* and *Etlingera* spp. in fallow vegetation is also promising. The groves with *Etlingera* spp. are one example, and semipermanent "gardens" of *S. palustris* could be established fairly easily by clearing the "weeds" in fallowed swiddens near the village. Intercropping *S. palustris* with pepper is another possibility, which, in addition to providing a regular supply of vegetables, may address the problem of soil erosion in these otherwise clean-weeded gardens.

Household Economics

Being located upriver, the only obvious markets for fresh vegetables produced in Nanga Sumpa are the tourist lodge in the village, which, in the high season, often has a constant occupancy of five or six people or more, and the rapidly developing tourist industry around the Batang Ai lake. The latter includes a large international hotel opened in 1995. So-called jungle vegetables are popular in the "tourist diet," and average market prices in Sri Aman in 1996 were 2.8 ringgits per kg of *D. esculentum* and 2.7 ringgits per kg of *S. palustris*. So daily sales to the tourist lodge alone could be in the order of 10 ringgits, or, at existing exchange rates, about US$4. To the poorer households in Nanga Sumpa, 10 to 20 ringgits per week could represent a valuable addition to the household economy.

For communities located in the more densely populated regions of Sarawak, the sale of wild vegetables in urban markets can be quite profitable (Burgers 1993). Depending on the market price, an annual income of 5,000 to 10,000 ringgits, or between 1997US$2,000 and $4,000, is possible, after deduction of fertilizer costs, from one ha of cultivated *S. palustris*. It would need harvesting twice monthly and fertilizing once monthly with 200 kg/ha of NPK-Mg (12-12-17-2+TE) (Mertz 1999a). The economic benefits of *D. esculentum* are less clear, mainly because the yield potential is lower and prices are more variable. Effective production of *E. elatior* may potentially yield an annual income of 10,000 to 15,000 ringgits per hectare, provided there is sufficient demand for the shoots and flowers (Mertz 1999b). Similar calculations for *P. borneense* are more speculative because it has no current market value.

Ecological Considerations

By increasing the value of tree crop plantations and providing ground cover that reduces erosion, *D. esculentum* and *P. borneense* may contribute to more sustainable use of plantations, possibly avoiding their abandonment and subsequent conversion to swidden farms. This could, in the long term, increase the areas under managed secondary forest and plantations.

The cultivation or manipulation of vegetables emerging in recently fallowed swiddens may prolong the fallow period, if vegetable harvesting proves profitable. In more degraded areas dominated by lalang, *S. palustris* and *Etlingera* spp. may have potential for profitable reclamation of otherwise unproductive land by shading out unwanted monocot weeds, while at the same time providing soil cover and erosion control.

Because wild vegetables are sensitive to pesticides and herbicides, the plants are best grown without resort to these chemicals. Pesticide residues in vegetables have become a public concern in Malaysia, Brunei, and Singapore, and this is partly responsible for the increasing urban interest in wild vegetables. Market development is boosted by consumer belief that these plants are unsprayed.

Possible Application Elsewhere in Southeast Asia

The use of wild vegetables is widespread in the shifting cultivation systems of Southeast Asia, as well as those in the rest of the world. Given the similarity of shifting cultivation practices in the region, the cultivation of these crops may easily be envisaged elsewhere.

D. esculentum is an important vegetable in most areas of South and Southeast Asia and is also used for various medicinal purposes (Zanariah et al. 1986; Bautista et al. 1988; Amoroso 1990; Gaur and Bhatt 1994; Taungbodhitham 1995). Only two reports on cultivation were found, one more than 50 years old from the Philippines (Copeland 1942), and the other mainly focused on development of the plant's ornamental potential in Thailand (Thongtham et al. 1981).

Information on *S. palustris* is scarcer, but it is as widespread as *D. esculentum* (Laderman 1982; Leach 1988; Amoroso 1990; Siemonsma and Piluek 1993). No efforts have been made to cultivate it. *E. elatior, E. punicea,* and *P. borneense* are barely mentioned in the literature, but these or closely related species occur all over Southeast Asia. Notably, the flower of *E. elatior* is widely used as a condiment, vegetable, or as an ornamental under the name "torch ginger" (Kunkel 1984; Smith 1986; Siemonsma and Piluek 1993; Wong et al. 1993).

With rapidly growing economies in Southeast Asia and the emergence of a larger middle class, an increasing awareness of food quality and environmentally sound production methods is expected to develop. As in Malaysia, this may increase the interest in alternative vegetables. Very little information is available on urban markets for wild vegetables, but people in countries such as China, Japan, and Korea have long traditions of consuming wild food products and may represent potential export markets (May 1978). An example is the widespread consumption of pickled shoots of bracken fern (*Pteridium* spp.), a tradition now discouraged because of the strong carcinogenic properties of this plant (Hirono and Yamada 1987; Hirono 1989). If appropriate conservation methods are developed, other nontoxic fern types may gain shares of these markets and increase the profitability of farm-level production in exporting countries.

Key Leverage Points for Development of Wild Vegetable Cultivation

Wild vegetables are a priority of agricultural research in Sarawak, and although a substantial amount of time and money is spent on their development into crop plants, funding represents a major bottleneck. The Fund for Intensification of Research Priority Areas (IRPA) of the Malaysian Ministry of Science and Technology could provide additional funds, not only for research but also for extension work.

Appropriate extension is crucial if shifting cultivation communities are to benefit from research work. National strategies aimed at reducing the extent of shifting cultivation need to incorporate advice on the use of wild plants and indigenous species, rather than focusing on the introduction of exotic cash crops and vegetable species. The genuine interest of indigenous communities in developing the cultivation of wild vegetables can only be further stimulated if they have access to information on appropriate crop husbandry methods and market outlets.

Linkage with research and extension activities in other countries is also essential for a wider adoption of the results obtained in Sarawak, and for the exchange and development of new ideas. The creation of an international wild vegetable network could provide the structure for sharing information. This could be established within an existing plant resource network such as Plant Resources of Southeast Asia (PROSEA) or, with more global implications, under the Non-Wood Forest Products Network of the Food and Agriculture Organization.

Future Research Priorities and Experimental Agenda

Research on wild vegetables and their domestication is very limited and, with respect to the five vegetables in question, almost nonexistent. Given the economic importance and local popularity of *D. esculentum* and *S. palustris* in particular, there is considerable scope for these ferns to play a part in the intensification of shifting cultivation as subsistence or cash crops.

While the trials described in this chapter provide a valuable insight into the interest shown by farmers, they are limited in scope in that they represent only one community, and there is little information on plant performance in various field and garden types. The concept of payment was useful in providing clear "before and after" information within a limited time frame, but true farmer interest must be evaluated on a longer-term basis.

Further trials should involve communities in various locations, and particularly those near markets and with a history of wild vegetable marketing. The time frame should be at least two to three years. In the following list, a number of topics are recommended for new trials. It should be emphasized that these suggested studies should encompass socioeconomic parameters such as labor use and capital investment in order to compare the profitability of cultivation with collection from wild plant populations:

- The performance of groves of *S. palustris* and *Etlingera* spp. in fallow vegetation, through the maintenance of existing plants and new plantings;
- Planting of *D. esculentum* and *P. borneense* under tree crops such as rubber or other tree species planted as fallow improvement crops;
- Maintenance of permanent gardens of *D. esculentum* and *P. borneense* under bananas in swampy riverine soils; and
- Testing of *S. palustris* as an intercrop in pepper gardens, mainly in terms of yield impact and reduction of soil loss.

As mentioned before, the Department of Agriculture, Sarawak, has continued work on *S. palustris* and is promoting the crop to commercial farmers. Research station trials with different training and manuring regimes showed promising results (Chai 2001), and on-farm trials have begun on commercial vegetable farms. Whether this development will reach more remote shifting cultivation communities remains to be seen.

There is also scope for development of fern cultivation in oil palm plantations, where certain ferns, mainly *Nephrolepis* spp., are already treated as beneficial "weeds" and maintained as cover crops for erosion control and conservation of soil organic matter (Hartley 1988).

As a baseline for crop development, further agronomic information is essential and the list of important research topics is very long, including studies on variability, propagation, land preparation, weeding intensity, pruning, and harvesting intervals. Postharvest technology on storage, conservation, and product development should also be investigated, including pickling, canning, and freezing.

The market survey carried out in connection with the present study indicates a fluctuating supply, a fairly stable demand, and variable prices. Further studies on marketing strategies and mapping of urban demand would support efforts to cultivate wild vegetables.

Conclusions

The on-farm trials indicate a genuine interest in wild vegetable cultivation in Nanga Sumpa, and if this applies to other communities in Sarawak, this is a feasible activity with the potential to improve both income levels and the regularity of household vegetable supplies. Whether managed as fallow improvement crops, intercropped with other cash crops, or grown in perennial gardens, the vegetables may contribute to more intensive land use and thereby reduce overall land requirements. Reduced erosion is an additional benefit. As indicated by supportive research, several of the vegetables also have good agronomic potential for cultivation, and the urban demand in Sarawak is substantial. Consequently, there is considerable scope for their development into profitable cash crops.

The use of wild vegetables as fallow improvement crops needs more research, particularly on species such as *S. palustris* and *Etlingera* spp. Systems of intercropping with tree and other cash crops need investigation. Given the interest of the Sarawak Department of Agriculture in research on *S. palustris* and other wild species, these crops should become part of state strategies aimed at intensifying shifting cultivation as soon as appropriate crop husbandry practices have been tested.

Acknowledgments

This project was funded by the Danish Council for Development Research. I would like to thank the Sarawak state government for granting permission to do the field work and the people of Nanga Sumpa for their hospitality and willingness to participate in the project. I am grateful for the support of the staff of the Department of Agriculture, Sarawak, during the field work, and particularly for the assistance of Senior Research Officer Chai Chen Chong. Finally, I wish to express my thanks to Professor Sofus Christiansen and Assistant Professor Søren Kristensen of the Institute of Geography, University of Copenhagen, for their comments and advice.

References

Ahmad, F.B., and D.K. Holdsworth. 1994. Medicinal Plants of Sarawak, Malaysia, Part I. The Kedayans. *International Journal of Pharmacognosy* 32, 384–387.

Amoroso, V.B. 1990. Ten Edible Economic Ferns of Mindanao. *Philippine Journal of Science* 119, 295–313.

Ashby, J.A. 1991. Adopters and Adapters: The Participation of Farmers in On-Farm Research. In: *Planned Change in Farming Systems: Progress in On-Farm Research.* Chichester, UK: John Wiley & Sons, 273–286.

Bautista, O.K., S. Kosiyachinda, A.R. Abd-Shukor, and Soenoeadji. 1988. Traditional Vegetables of ASEAN. *ASEAN Food Journal* 4, 47–58.

Burgers, P.P.M. 1993. Rainforest and Rural Economy in Sarawak. *The Sarawak Museum Journal* 46, 19–44.

Chai, C.C. 2001. *Growing "Miding" as a Crop.* Extension leaflet, Agricultural Research Centre, Semongok. Kuching, Sarawak: State Department of Agriculture.

Chin, S.C. 1985. Agriculture and Resource Utilization in a Lowland Rainforest Kenyah Community. *The Sarawak Museum Journal* 35, 1–322.

Christensen, H. 1997. Uses of Ferns in Two Indigenous Communities in Sarawak, Malaysia. In: *Holttum Memorial Volume.* Kew, UK: Royal Botanic Gardens, 177–192.

———. 2002. *Ethnobotany of the Iban and the Kelabit.* Sarawak, Malaysia: Forest Department; NEPCon, Denmark and University of Aarhus, Denmark.

———, and O. Mertz. 1993. The Risk Avoidance Strategy of Traditional Shifting Cultivation in Borneo. *The Sarawak Museum Journal* 46, 1–18.

Copeland, E.B. 1942. Edible Ferns. *American Fern Journal* 32, 121–126.

Cramb, R.A. 1993. Shifting Cultivation and Sustainable Agriculture in East Malaysia: A Longitudinal Case Study. *Agricultural Systems* 42, 209–226.

DID (Department of Irrigation and Drainage). 1993. *Monthly Rainfall.* Kuching, Sarawak: Hydrology Branch, State Department of Irrigation and Drainage.

DoA (Department of Agriculture). 1968. *Soil Map of Sarawak.* Kuching, Sarawak: Soil Survey Division, State Department of Agriculture.

———. 1987. *Annual Report of the Research Branch.* Kuching, Sarawak: State Department of Agriculture.

———. 1993. *Annual Report, Research Branch.* Kuching, Sarawak: State Department of Agriculture.

Dove, M.R. 1985. Swidden Agriculture in Indonesia. The Subsistence Strategies of the Kalimantan Kantu. In: *New Babylon, Studies in the Social Sciences.* Berlin: Mouton Publishers.

———. 1993. Smallholder Rubber and Swidden Agriculture in Borneo: A Sustainable Adaptation to the Ecology and Economy of the Tropical Forest. *Economic Botany* 47, 136–147.

Freeman, J.D. 1955. Iban Agriculture: A Report on the Shifting Cultivation of Hill Rice by the Iban of Sarawak. In: *Colonial Research Studies.* London,U.K.: Her Majesty's Stationary Office.

Gaur, R.D., and B.P. Bhatt. 1994. Folk Utilization of some *Pteridophytes* of Deoprayag Area in Garhwal Himalaya, India. *Economic Botany* 48, 146–151.

Hartley, C.W.S. 1988. *The Oil Palm* (Elaeis guineensis *Jacq.*). Harlow, UK: Longman Scientific and Technical.

Hirono, I. 1989. Carcinogenicity of Bracken Fern and its Causal Principle. In: *Bracken Biology and Management,* Sydney: Australian Institute of Agricultural Science, 233–240.

———, and K. Yamada. 1987. Bracken Fern. In: *Naturally Occurring Carcinogens of Plant Origin: Toxicology, Pathology, and Biochemistry,* Amsterdam: Elsevier, 87–120.

ICRAF (World Agroforestry Centre). 1995. *Annual Report.* Nairobi, Kenya: ICRAF.

Kedit, P.M. 1994. Use of Plants for Architecture and Decorative Purposes of the Iban in Sarawak. *Sarawak Gazette* 121, 25–31.

Kunkel, G. 1984. *Plants for Human Consumption. An Annotated Checklist of the Edible Phanerogams and Ferns.* Koenigstein: Koeltz Scientific Books.

Laderman, C. 1982. Wild Vegetable Consumption on the East Coast of Peninsular Malaysia. *Malayan Nature Journal* 35, 165–171.

Leach, G.J. 1988. Bush Food Plants of the Blackwater and Karawari Rivers Area, East Sepik Province, Papua New Guinea. *Science in New Guinea* 14, 95–106.

Leaman, D.J., R. Yusuf, H. Sangat-Roemantyo, and J.T. Arnason. 1996. The Contribution of Ethnobotanical Research to Socio-Economic and Conservation Objectives: An Example from the Apo Kayan Kenyah. In: *Borneo in Transition. People, Forests, Conservation, and Development*. Kuala Lumpur: Oxford University Press, 245–255.

May, L.W. 1978. The Economic Uses and Associated Folklore of Ferns and Fern Allies. *The Botanical Review* 44, 491–528.

Mertz, O. 1997. Cultivation Potentials of Wild Vegetables: Their Role as Cash or Subsistence Crops in Farming Systems of Sarawak, Malaysia. Ph.D. thesis. Institute of Geography, University of Copenhagen.

———. 1999a. Cultivation Potential of Two Edible Ferns, *Diplazium esculentum* and *Stenochlaena palustris*. *Tropical Agriculture* 76, 10–16.

———. 1999b. Preliminary Study on the Cultivation Potential of Wild Vegetables *Etlingera elatior, E. punicea* and *Commelina paludosa*, of Sarawak. *Journal of Tropical Agriculture and Food Science* 27, 27–37.

———, and H. Christensen. 1997. Land Use and Crop Diversity in Two Iban Communities, Sarawak, Malaysia. *Danish Journal of Geography* 97, 98–110.

Okafor, J.C., and A. Lamb. 1994. Fruit Trees: Diversity and Conservation Strategies. In: *Tropical Trees: The Potential for Domestication and the Rebuilding of Forest Resources*. London, U.K.: HMSO, 34–41.

Padoch, C. 1982. Migration and its Alternatives among the Iban of Sarawak. In: *Verhandelingen van Het Koninklijk Instituut voor Taal-, Land,- en Volkenkunde* 98. The Hague, Netherlands: Martinus Nijhoff.

Pearce, K.G., V.L. Aman, and S. Jok. 1987. An Ethnobotanical Study of an Iban Community of the Pantu Sub-District, Sri Aman, Division Two, Sarawak. *The Sarawak Museum Journal* 37, 193–270.

Poh, L.Y. 1994. Malaysia. In: *Non-Wood Forest Products in Asia*. Lebanon, NH: Science Publishers, 55–71.

Rahman, S.A. 1992. Management of Pests and Diseases of Vegetable Crops in Malaysia in 2000. In: *Pest Management and the Environment in 2000*. Oxford, U.K.: CAB International, 213–230.

Raintree, J.B. 1994. Farmer Participation in On-Farm Agroforestry Research Prioritization. *Unasylva* 45, 13–20.

Siemonsma, J.S., and K. Piluek. 1993. *Plant Resources of Southeast Asia*. No 8., Vegetables. PROSEA Handbook. Wageningen, The Netherlands: Pudoc Scientific Publishers.

Smith, R.M. 1986. A Review of Bornean Zingiberaceae: II (Alpineae, concluded). *Notes from the Royal Botanic Gardens, Edinburgh* 43, 439–466.

Taungbodhitham, A.K. 1995. Thiamin Content and Activity of Antithiamin Factor in Vegetables of Southern Thailand. *Food Chemistry* 52, 285–288.

Thongtham, C., P. Theeravuthichai, and N. Tumronglaohapunt. 1981. *Final Report: Developmental Research on Economic Ferns and Cash Crops for the Hilltribes of Northern Thailand* [Studies on Morphology, Growth Patterns, Cultivation, Fertilizing, Propagation, Shoot-Tip Culture, Packaging, Production]. Bangkok, Thailand: Kasetsart University.

Wong, K.C., Y.F. Yap, and L.K. Ham. 1993. The Essential Oil of Young Flower Shoots of *Phaeomeria speciosa* Koord. *Journal of Essential Oil Research* 5, 135–138.

Zanariah, J., A.N. Rehan, O. Rosnah, and A. Noor-Rehan. 1986. Protein and Amino Acid Compositions of Malaysian Vegetables. *MARDI Research Bulletin* 14, 140–147.

Chapter 8

Commercialization of Fallow Species by Bidayuh Shifting Cultivators in Sarawak, Malaysia

Paul Burgers

The natural vegetation in the southwest of Sarawak is tropical rainforest. As in most rainforest areas, the soils are poor and infertile. Humus is found only in the topsoil. Permanent use of these chemically poor soils remains an unsolved problem, but shifting cultivation systems have developed with short cropping periods of one to two years and long fallow periods of 15 to 20 years. The fallow period restores soil fertility and its vegetation provides farming households with a variety of products, including food, firewood, and construction materials. Other useful products are harvested from the primary rainforest surrounding villages and swidden fields.

The livelihood of shifting cultivation communities has recently been seriously affected by population growth and large-scale deforestation, resulting from the clearing of forest land for commercial agriculture and unsustainable large-scale logging operations. This has undermined the sustainable use of forest resources. The land-use policies of the Malaysian government favor these developments, and not only can shifting cultivation communities no longer make use of the forests to collect products, they can no longer convert forest areas into cropping land to cope with a growing population.

A case study of these dynamics was carried out in a number of representative communities in the Teng Bukap subdistrict in Sarawak as part of a wider study of processes of agricultural commercialization. This is an area where problems have arisen regarding sustainable production of staple rice crops, as well as difficulties in finding sufficient vital forest products. Scarcity, in addition to a growing demand from urban areas, has increased the commercial value of these products, so swidden communities are attempting to overcome the shortages of supply, as well as taking advantage of the new market demand, by actively propagating them in fallow vegetation. This chapter, therefore, aims to assess the changing management of fallows within the context of agricultural commercialisation and growing connections with urban areas. The following hypotheses were tested:

- Depending on the degree of their market integration, households will search for forest products with potential commercial value.
- The accessibility of markets influences the type of products that are promoted in fallow vegetation.

Paul Burgers, International Development Studies (IDS), Faculty of Geosciences, P.O. Box 80115, 3508 TC Utrecht, The Netherlands.

- Shortages of vital forest products and a lack of alternatives compel farmers to turn to management of fallow vegetation to compensate for the loss of natural stocks.
- More intensive forms of fallow management tend to develop when communities obtain more secure, individual ownership of fallow land and the vegetation it supports.

The Study Area and Methods

The survey was conducted in the Teng Bukap subdistrict, a part of the Kuching division, which consists of the area around Kuching, the capital of Sarawak, extending to the border with Kalimantan, Indonesia (Figure 8-1). In contrast to other parts of Sarawak, river transport is of limited importance in the Kuching division. As well as there being no extensive river system, the terrain consists of lowlands with low hills in the east and a steep mountainous region to the west, along the border with Kalimantan. The main agricultural practice is shifting cultivation of rice.

The Kuching division has the highest percentage of nonagricultural land in Sarawak, as well as having the highest percentage of land under commercial crops. Within the division, the research area is the most commercialized, largely because of its proximity to Kuching. The integration of cash crops into the shifting cultivation system was begun by the Department of Agriculture at the end of the 1960s. Initially, only rubber trees were promoted through subsidy schemes, but to create a diversity of farm products, other crops like pepper and cocoa were promoted soon after. Farming systems have commercialized accordingly and farmland is occupied by these crops on a permanent basis.

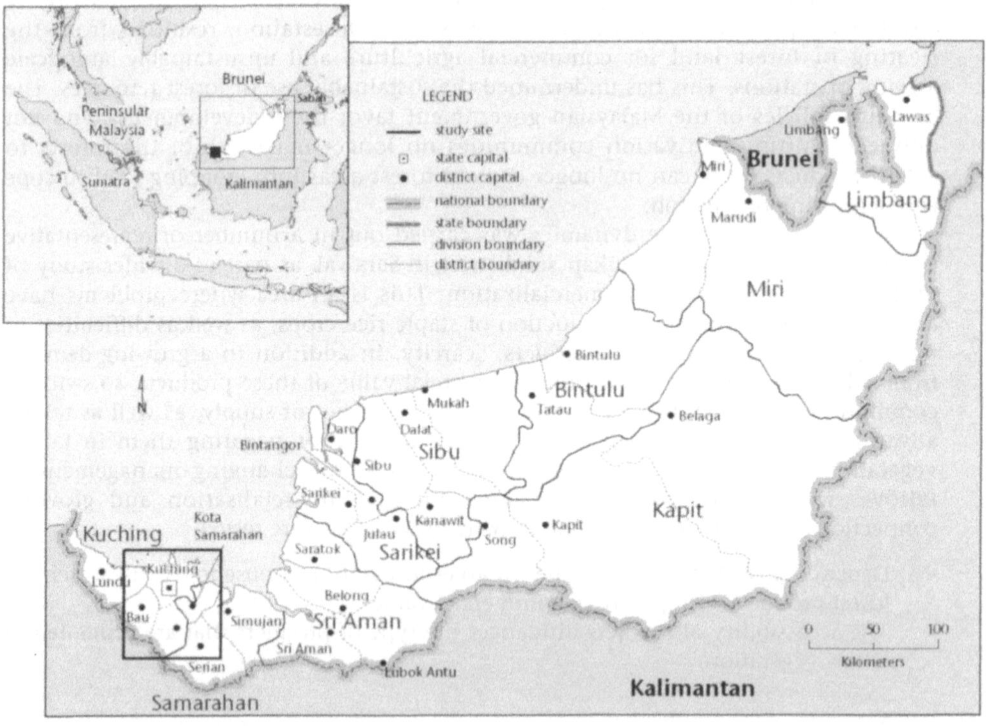

Figure 8-1. Map of the Study Area

Land in the Kuching division is categorized into mixed zone land, covering 15.6% of the total area; native area land, 5.9%; reserved land, 21.4%; and native customary land, 57.1%. Mixed zone land can be owned by anyone, and shifting cultivators meet with severe competition from Chinese farmers and large-scale land development schemes undertaken by Malaysian government organizations. Reserved land consists entirely of forest and is intended to ensure a lasting source of timber. Shifting cultivators are not allowed to use these forests or to collect forest products from reserved land.

The survey was conducted between November 1989 and April 1990. Additional qualitative data were gathered in November 1992. The Teng Bukap subdistrict encompasses 35 villages with a total of 1,470 households. In the first stage of the survey, villages were sampled. Then a random subsample of households was drawn from the sampled villages. The surveyed population was entirely Bidayuh, and the villages were divided into those easily accessible and inaccessible by road. This was done in view of the potential market influence of Kuching, which was only a two to three hours' drive from the research area. It was thought that road access to Kuching could influence farmers' decisions on management of fallow vegetation for a number of farm products. By contrast, the poorly accessible villages required at least a two-hour walk to and from the road.

The survey used semistructured questionnaires and informal interviews. The questionnaire was developed, tested, then slightly revised. The survey proper was undertaken with the help of local enumerators. Discussions were held with key informants, such as district officers and extension workers from the Department of Agriculture, and in the villages with members of village development committees, village heads, and householders. Much information was gathered by living in the study villages, joining in frequent conversations, participating in daily activities, and making joint trips to the Kuching markets.

Farmers in both accessible and inaccessible parts of the study areas had generally tried a combination of different cash crops. At the time of the study, most households grew cash crops. In the accessible area, 97% of respondents had planted at least one cash crop, compared to 86% in the poorly accessible area (Table 8-1).

Family sizes are generally high in Sarawak. The average household size in the study areas was six persons. In the accessible study area, 52% of the respondents were under the age of 20, compared to 42% under 20 in the poorly accessible area. The population growth rate, which had already elevated demands for agricultural land and other natural resources, seemed unlikely to diminish.

Table 8-1. Cash Crops Grown by Surveyed Households in Accessible and Poorly Accessible Areas (%)

Cash Crops	*Accessible*	*Poorly Accessible*
None	3	13
Pepper only	9	10
Rubber only	5	6
Cocoa only	2	11
Pepper and rubber	24	12
Pepper and cocoa	12	19
Rubber and cocoa	2	8
Pepper, cocoa, and rubber	43	20

Note: 129 households surveyed in accessible areas; 104 in poorly accessible areas.

The Shifting Cultivation System without Fallow Management

The season starts in June or July, with the agricultural calendar organized around the shifting cultivation of upland rice. The selection of a plot for a swidden field depends largely on the fallow species found on the land and the fields desired size. From this a farmer is able to judge whether the land has built up sufficient nutrients to grow a single crop of rice. An acceptable plot has usually been fallow for 10 to 15 years. The land is first cleared, and most of the vegetation is slashed. After drying for a few weeks, it is burned so the nutrients accumulated in the vegetation will return to the soil as ash. Large trees or trees that may provide timber in later years are left standing. Fruit trees are also saved. Other trees are cut about 60 cm above the ground, so they will coppice quickly.

This system has several advantages. The roots of retained trees help to stabilize the soil. Coppices usually begin to develop after several weeks, and rapid regrowth of the forest cover after the rice harvest protects the soil from erosion, solar radiation, and soil compaction. A disadvantage of this system is that it only works if fallow periods are long enough to allow the recovery of soil fertility.

Besides rice as the staple crop, vegetables such as cucumber, pepper, okra, and sweet potato are intercropped in the field. Fruit trees, which play a significant role within the system, are interplanted with the rice during the cropping phase. Fruit is later harvested from these "forest gardens" when the trees mature during the fallow. Bananas, mangoes, papaya, and durian are the most common fruit in fallow vegetation.

There is no further management in this system because the land and its vegetation revert to communal property under traditional *adat*, or customary law. Anyone can collect products that grow in fallow vegetation, such as wild vegetables or bamboo, but people are not allowed to gather fruit while it remains on the tree. Only when it falls to the ground does it become communal property and, therefore, collectable by anyone. The collection of durian fruit from forest and fallow vegetation has become one of the largest income earners in the Teng Bukap subdistrict, and during the durian season, many people can be found sitting under the trees waiting for the fruit to fall. At the time of the survey, one durian was worth about US$3 and a strong young man could make about US$180 in one day.

Processes of Change

Increasing population pressures, adverse land policies, the clearing of land for commercial agricultural purposes, and logging have affected traditional shifting cultivation systems both directly and indirectly, and shortened the fallow period (Burgers 1993).

Land Policies and Population Growth

At the time of this study, the opening of new land from virgin forest by local communities was banned. At the same time, the Malaysian government was encouraging population growth. These factors had placed tremendous pressure on existing shifting cultivation systems. The fallow period had been reduced to its ecological minimum, where natural regeneration processes were still capable of restoring soil fertility. In both accessible and poorly accessible areas, the average fallow period was around nine years. However, 40% of the surveyed households said their land holdings were too small to permit ecologically sound shifting cultivation. Of this group, 53% said they could no longer maintain self-sufficiency in rice.

Shifting cultivation was being made difficult by the amount of reserved land within the division, and the threat that farmers could lose their indigenous claims to land if the government needed it for "more productive" purposes. Native area land in at least two of the villages had been converted into plantations, and large-scale agricultural activites initiated by the government, as well as commercial logging, were affecting both the use of forest resources and the capacity of the forest to provide. Remaining "communal" sites were being overexploited, and this was exacerbated by the increasing commercial value of forest products.

Agribusiness and Logging

The integration of cash crops into the shifting cultivation system itself negatively affected the length of the fallow period. In the accessible area, 38% of respondents, compared to 15% in the inaccessible area, agreed that the decline in fallow length was, in part, a result of their expansion into cash crops. This had not occurred to such a marked degree in the inaccessible area because of the more recent introduction there of cash crops.

During the research period, commercial logging was concentrated in the poorly accessible area. It was having one direct result on the area's inhabitants, who were accustomed to using the forest to collect products such as rattan and wild fruit. Their passive management of wild rattan was based on a cyclical collection period intended to avoid overexploitation. When a collection site was "harvested" to a certain level, it was left for about eight years so the canes could recover and grow, and the gatherers moved to other sites. These collection sites were destroyed by the commercial logging, and people were forced to cross the border into Indonesia to gather products from unexploited forests, a trip that could take several days.

Farmer Innovations

The shortfall in forest products, the decrease in land area on which to practice an environmentally sound agriculture, and land claims supported by government agencies forced farmers to seek solutions within their own farming systems. At the same time, a large market for forest products in nearby Kuching offered the opportunity to supply "city-dwellers" who preferred to eat unsprayed "natural" foods untainted by chemical fertilizers and pesticides. For many farmers, the solutions were found in their fallow vegetation.

Farm households in the study area began to intensify their fallow management, firstly by abandoning the old "passive" aproach. Following the tradition in which they enriched their fallow vegetation with fruit trees, they began to incorporate forest products into their fallow vegetation, including food crops such as wild vegetables and fruit, and other products like rattan and bamboo. The aim was to meet both household and cash needs, since all of the products attracted high prices in the markets of Kuching.

*Individual Ownership through Changes in Customary (*Adat*) Law*

In both the accessible and inaccessible areas, ferns, fungi, bamboo, and other valuable products began to thrive in fallow vegetation and were fetching good prices in Kuching's Sunday market. However, a problem soon arose: There was no control over the harvest because fallow vegetation was communal property, and there were many hungry households in the community. The new fallow crops were soon being overexploited and householders were demanding changes to adat law. The system of communal ownership of fallow vegetation had become unsustainable.

In Pesang, one of the research villages in the accessible area, 82% of the residents went to Kuching on a weekly basis to sell forest products and other farm produce. It was the most active research village in selling forest products. Village meetings were organized in Pesang, and it was decided that individual ownership of fallow vegetation would provide for a more sustainable and active system of managing fallow or secondary forest. Individual households were permitted to manage their own fallows, and a fine of 25 Malaysian ringits was set for intruders caught harvesting products from other people's fallows. The fallow was, thus, modified into a more "productive" phase of the shifting cultivation system. There was widespread transplantation of desirable species such as rattan, bamboo, ferns, and fungi from the forest into fallow vegetation, and others were weeded out.

There were simultaneous developments to assist the inaccessible communities because the scope for selling forest products was more limited for these people. A special bus began operating twice weekly at five o'clock in the morning to carry villagers wanting to sell forest products in Kuching. But it was used mainly by people from the accessible area. The journey took too long for people from villages inaccessible by road. They began to turn away from perishable products. However, a local market soon sprang up where the road ended, at Kampong Abang. Middlemen arrived to set up as buyers, and rattan traders from Kuching began visiting Abang several times each month. The research survey revealed that 75% of respondents in the inaccessible areas who sold forest products marketed them in Abang, while the other 25% sold to friends, neighbors, and nearby villagers. Nevertheless, their sales, even those in Abang, were restricted mainly to nonperishable goods such as rattan, hardwood, bamboo, and fruit like durian.

Benefits from Intensified Management

Increase in Cash Income

By selling fallow products several times a month, villagers in the study area have been able to earn a steadier income to meet their daily cash needs. When selling products at the Kuching Sunday market, they have been able to do their shopping at the same time, saving on time and transport.

These important benefits have diversified income opportunities away from conventional cash crop cultivation and provided the villagers with a measure of insurance in case their other cash crops fail. Fallow products that can be sold on a weekly basis, such as ferns, fungi, and sago worms also provide regular income, whereas income from conventional cash crops comes only once or twice a year, at harvest time.

Improved Nutrition

Growing native vegetables and other forest products in their fallows improves the nutritional status of householders because they are getting a more balanced diet. This benefit was confirmed by home economics extension staff of the Department of Agriculture, who were working to eradicate malnutrition in the area at the time of the research. Forest products such as ferns, fungi, wild fruit, and bushmeat were central to their advice about the importance of a diversified and balanced diet in the inaccessible research villages.

Increased Labor Productivity

A good nutritional status is also beneficial to the productivity of labor because the capacity to work increases as physical health improves. In addition, both time and labor are saved by the incorporation of useful plants into fallow vegetation. Farmers previously had to walk long distances to find the forest products they needed. However, firewood, construction materials, and food can now be harvested from nearby fallows, and labor can be deployed elsewhere.

Making handicrafts from rattan and bamboo for the booming tourist industry has become a very important off-farm employment in the area.

Constraints to Adoption of Intensified Fallow Management Systems

Biophysical

Evidence suggests that a serious constraint to the intensified fallow management system has arisen in the Teng Bukap subdistrict. The natural supply of seeds is insufficient to meet both domestic and market needs. Sites where these plants grow naturally are vanishing, and seeds are very difficult to collect. Moreover, the propagation of seedlings is difficult and there is insufficient indigenous knowledge on propagation methods. This is especially true for "woody" nonperishable fallow plants and native fruit trees. This may represent a serious constraint to wider adoption of the system. However, evidence from Sumatra, Indonesia, indicates that hundreds of farmers in one area are intercropping rattan in their rubber gardens without outside technical support (Manurung and Burgers 1999). They followed the advice of one local farmer who had learned to do it by himself. This suggests that, with the correct approach, these constraints may be overcome.

Socioeconomic

The distance to markets and lack of transport are constraints in the poorly accessible areas. These hamper farmers' freedom of choice in deciding what economically valuable products they can grow. Labor might become another constraining factor in the future. Many young people have lost enthusiasm for farming and prefer employment in Kuching. In addition, the labor and resource demands of large-scale operations involved in perennial cash crop cultivation may drain labor and resources from other agricultural practices such as fallow management.

Legal and Policy

According to traditional adat law, vegetation found growing in fallows is owned by the entire village. This has discouraged more productive fallow management because farmers know that crops planted in fallowed fields can be harvested by anyone. A change of tenurial rules toward individual ownership of fallows, such as that at Pesang, is only feasible if an entire village can be mobilized to revise adat law.

Constraints also occur at regional and national levels. The Malaysian government does not give priority to more effective fallow management as an intensification strategy. Socioeconomic development is viewed from a macroeconomic perspective, and the main objective is to convert shifting cultivation into more market oriented agricultural systems based on growing cash crops in pure stands, so that exports will provide the government with foreign exchange income. The central government is of the opinion that fallowing is an extensive form of land use and is, therefore, inefficient. Although the Department of Agriculture in Sarawak is quite positive about intensified fallow management, polices are formulated in Kuala Lumpur, where farming in Sarawak is often misunderstood. Extension programs and seed propagation activities in Malaysia are also undertaken by the central government. Such attitudes have to be resolved before fallow management can be viewed more accurately as a productive system.

Wider Adoption of Improved Fallow Management

On a household level, more intensive management of fallow vegetation seems to hinge upon individual ownership of the fallow. Although consensus of an entire village is needed to make such changes in the tenurial rules, this has already proven to be not only possible, but also successful in the Teng Bukap subdistrict. More intensive fallow management also requires a high demand for forest products that are conducive to domestication and the availability of both extensive fallow lands and labor-saving techniques.

Extension efforts need to concentrate on propagation issues, which were frequently cited by farmers as a major constraint. However, some success with village nurseries of rattan seedlings has already been reported. Depending on market access, extension programs should also concentrate on different crop combinations that improve the productive, as well as the protective, functions of fallows.

Research Priorities

Technical refinements are only one aspect of improving or intensifying shifting cultivation systems. "Full-time" farmers, who depend completely on farming to build a livelihood, need to be distinguished from "part-time" farmers who work off-farm for at least part of the year. Full-time farmers are most likely to adopt and benefit from technological or legal interventions that facilitate improved fallow management. Part-time farmers may not give priority to farming, much less to more intensified forms of fallow management.

Another related research priority should be investigating and promoting opportunities for employment outside the agricultural sector for those farmers not fully engaged in agriculture. Remaining farmers will then have the option of acquiring additional land from those who choose to permanently leave the agricultural sector. The system may thus regain some ecological stability.

Research needs to address these issues on a regional level as part of an integrated approach toward the sustainable development of degrading shifting cultivation systems and more resilient livelihood systems.

Conclusions

Farming systems in the Teng Bukap subdistrict of Kuching division, Sarawak, have undergone major changes from subsistence toward more market-directed production. However, this has not diminished the need for fallowed or forested areas. On the contrary, they have become even more important as a means to improve rural living standards. The main conclusions from this study are as follows:

- The fallow represents a valuable niche for intensified management and generation of income through the sale of commercially important forest products.
- Appropriate "fallow crops" depend for economic success on the degree of their commercialization, access to markets, and their scarcity in natural forest environments.
- A large variety of perishable and nonperishable products can be used to intensify fallow management in areas with access to markets.
- Poorly accessible areas are more limited in their options and should concentrate on nonperishable products.

Supportive Policies

Policies need to be developed that encourage improved fallow management. Importantly, effective extension services and seed distribution systems should be integrated into these policies.

Intensified fallow management offers savings on time and labor because products formerly gathered from distant forests are now available in nearby fallows. This is an attractive aspect, in view of labor constraints.

It is clear that the traditional adat law concept of communal ownership of fallow vegetation is only suitable when population densities are low. In current conditions of high population, this customary tradition may lead to unintended over-exploitation of fallow products and to pressures for change to more individual forms of fallow tenure.

Future policies should also support the creation of increased off-farm employment and other opportunities outside the agricultural sector.

Additional References

Ariff, M., and M. Semudram. 1987. *Trade and Finance Strategies: A Case Study of Malaysia.* Working paper no. 21. Overseas Development Institute.

Beer, J., and M. McDermott 1989. *The Economic Value of Non-Timber Products in South East Asia, with Emphasis on Indonesia, Malaysia, and Thailand.* Amsterdam: Netherlands Committee for IUCN.

Burgers, P.P.M. 1993. Rainforest and Rural Economy. *Sarawak Museum Journal* 44(65), 19–44.

———, M. Nolten, M. Servaas, W. Verhey, and L. van Grunsven. 1991. *Shifting Cultivation in Teng Bukap Subdistrict, Kuching Division, Sarawak: A Socio-Economic Study in Sixteen Communities*. The Netherlands: Department of Geography of Developing Countries, Geographical Institute, University of Utrecht.

Chin, S.C. 1977. Shifting Cultivation: A Need for Greater Understanding. *Sarawak Museum Journal* 25(46).

Chin, Thian Hon. 1989. *Rattan Planting*. Sarawak: Department of Agriculture, 1–14.

Geddes, W.R. 1954. Land Tenure of Dayaks. *Sarwak Museum Journal* 6(4).

Godoy, R. 1990. The Economics of Traditional Rattan Cultivation. *Agroforestry Systems* 12, 163–172.

Hong, E. 1977. Trade, Crops and Land: Impact of Colonisation and Modernisation in Sarawak. *Sarawak Museum Journal* 25(46).

Kessler, J.J., and K.F. Wiersum. 1993. Ecological Sustainability of Agroforestry in the Tropics. Entwicklung and landlicher raum, *Schwerpunkt Agroforstwirtschaft* 5, 8–11.

Manurung, G., and P. Burgers. 1999. Innovative Farmers: Bapak Kanijan. Reviving Rattan in Sumatra is a Booming Business. *Agroforestry Today,* January–June 1999.

Mertz, O., and H. Christensen. 1993. The Risk Avoidance Strategy of Traditional Shifting Cultivation in Borneo. *Sarawak Museum Journal*, 44(65), 1–18.

Noeb, L.M. 1992. A Bidayuh Traditional Romin. *Sarawak Gazette* 119(1520), 4–16.

Raintree, J.B., and K. Warner. 1986. Agroforestry Pathways for the Intensification of Shifting Cultivation. *Agroforestry Systems* 4, 39–45.

Chapter 9

Wild Food Plants as Alternative Fallow Species in the Cordillera Region, the Philippines

Fatima T. Tangan

The Cordillera region of the Philippines is rich in wild food plants that have been used over the centuries by the indigenous Kalanguya and Ibaloi tribes as alternative food resources. The region includes the Mount Pulag National Park, which was closed to occupancy by presidential decree in 1992. The law seeks not only to protect the area but also to strengthen its biodiversity, conservation, and management, so the upland farmers within the national park have taken a lead from the law and have introduced wild food plants into their fallowed swiddens. The innovation has changed the traditional rotational swidden system practiced in the area. The farmers are convinced that the wild food plants are effective cover crops, and that they reduce soil erosion, enhance soil fertility, and contribute to community livelihood.

This study focuses on only two wild food plants, *Rubus niveus* and *Rubus pectinellus* Maxim., although almost 20 species have been identified in the area. Followup research is therefore necessary, not only on the two *Rubus* spp., but also on the others.

Objectives and Methods

The objectives of this study were to describe the conservation practices of indigenous farmers in the Cordillera Region through their management of wild food plants, and to determine how indigenous communities' dependence on forest resources had been affected by the provisions of Republic Act 7586, the law that closed the park to occupancy.

The study was conducted in Barangay Tawangan (Figure 9-1), one of eleven communities occupying Mount Pulag National Park. It was one component of a program implemented by the Department of Environment and Natural Resources (DENR) and funded by the World Wildlife Fund, through the Foundation for the Philippine Environment and Philippine Business for Social Progress. Research methods included direct observation and measurements, interviews and survey questionnaires, use of secondary data, and informal group discussions with upland farmers. Interviews were also conducted with officials from local government units, nongovernment organizations, and DENR field staff. Secondary data were analyzed and cross-checked against information obtained from interviews and field observations.

Fatima T. Tangan, Department of Environment and National Resources (DENR), Cordillera Administrative Region, Loakan Road, Baguio City 2600, the Philippines.

Study Site

Barangay Tawangan covers an area of 2,019 ha. It has 596 people living in 93 households. Most of them are members of the Kalanguya tribe. The community has very limited transportation and communication facilities. It can only be reached by a hike of two hours over foot trails and footbridges. Despite this, local and foreign tourists attracted by two lakes, rainforests, and waterfalls often visit the study site.

Farming Systems

Due to the limited transportation facilities, the people produce practically all their basic food needs. They grow varieties of aromatic rice in paddies, locally called *payew,* and other agricultural crops and vegetables in their swidden fields, or *umas*. They also manage fruit trees and grow limited quantities of semitemperate vegetables. (See Table 9-1 for a typical cropping calendar.)

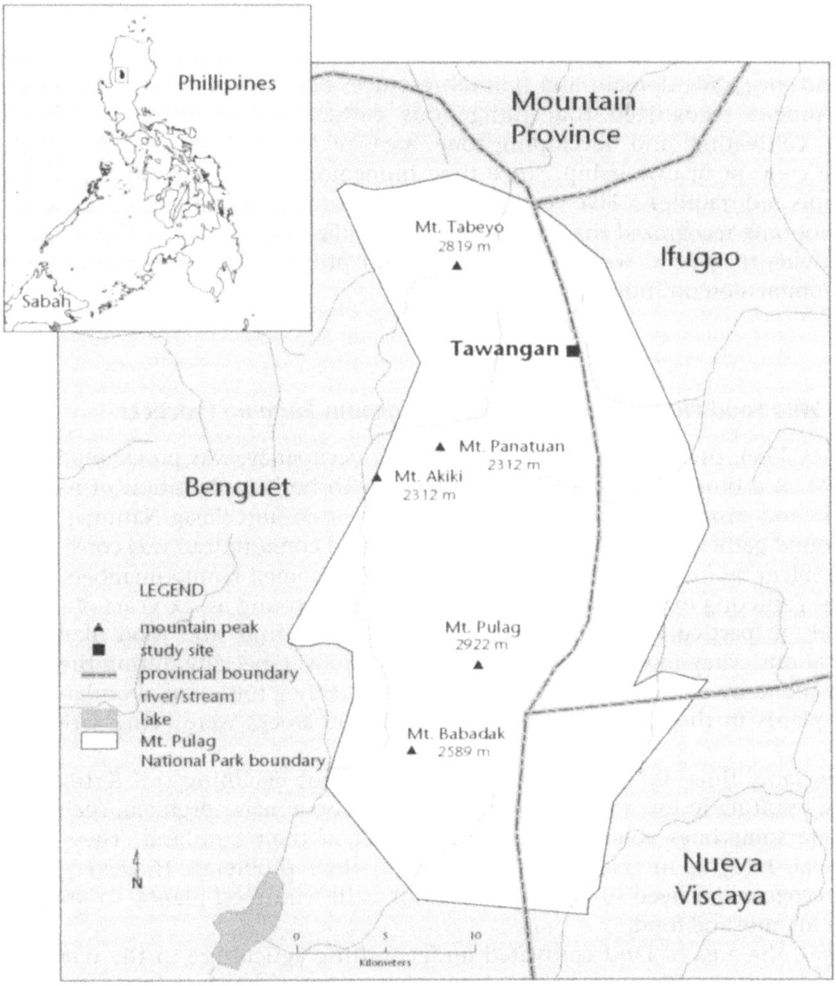

Figure 9-1. Map of Study Site

Table 9-1. Cropping Calendar in Barangay Tawangan

Crops	*Growing Season*
Rice	January – May
Tropical vegetables (sweet potato, peanuts, mung bean)	April – July
Semitemperate vegetables (cabbage, potato, sweet peas)	August – December

Conventional Shifting Cultivation

In the period from the 1960s to the late 1980s, *kaingin,* or shifting cultivation, was rampant in Barangay Tawangan. Umas were cleared by uncontrolled burning, and after the fires, tree stumps and other vegetation not already destroyed was removed. Each family cultivated an average two hectares of land for three consecutive crops before the land was fallowed.

Land Tenure

As a general rule, Philippine legislation on land tenure follows the Regallan Doctrine that all lands of public domain and natural resources belong to the state. However, the government recognized that indigenous cultural communities had been occupying, cultivating, and developing some areas of Mount Pulag National Park, within the concept of ownership, since time immemorial, and had there pursued their customs and traditions. Five years after the original presidential decree, Republic Act 8371 not only recognized the tenurial rights of indigenous cultural communities, but their rights in general were also protected and promoted by the creation of a National Commission on Indigenous Peoples.

Results

Wild Food Plants and Changes to Indigenous Farming Practices

Republic Act 7586, the presidential decree banning occupancy, was proclaimed into law in 1992, at a time when there were problems with both exploitation of natural plant stocks and sustainability of shifting cultivation in Mount Pulag National Park. The year-round gathering of wild food plants by upland communities was considered a major problem by forest rangers. Commonly, all able-bodied family members were involved in gathering expeditions, including children as young as six years old. The women folk, in particular, often hiked for two hours to find wild food plants in distant locations. They found this food gathering tedious, especially during the lean months of the rainy season, from June to October. But they found an abundance of wild food plants in the national park (Table 9-2), and stocks were being seriously depleted.

At the same time, farmers were complaining about declining soil fertility in their umas, resulting in low yields. Water supply was also a major problem, such that farmers were sometimes able to cultivate only half of their farmland. They were unable to fall back upon traditional solutions to their problems. However, they generally recognized a need to enrich their farms with wild food plants, to provide both green manure and food.

In 1993, the area of land cultivated under shifting agriculture in the national park was reduced, the formerly habitual use of inorganic fertilizers and pesticides lessened, and an increased interest in wild food plants began to generate market opportunities. In the rainy season of that year, the Protected Area Management Board, created under the new law, allowed the gathering of wild food plants for planting in swiddens and home gardens. About 30% of households accepted the opportunity and gathered and planted their wild crops. In the following year another 50% of the farmers followed their example.

Although conventional agricultural crops are still grown, the farmers are no longer totally dependent on them. The wild food plants, both fresh or processed, have become a source of livelihood in their own right. Most of them bear fruit during the rainy season, providing a welcome substitute for cereals during the lean months. The plants are also used as animal fodder.

The Kalanguya people prefer the vine and shrub types (*Rubus* spp.) because they bear fruit within one or two years. Wild food from trees requires a much longer wait before harvesting. The farmers also appreciate the fact that, unlike conventional crops, wild food plants require no fertilizers, pesticides, weeding, or tillage, all of which are costly and labor intensive. Most importantly, they claim that soil erosion is reduced in areas where the wild species are grown because there is no longer a need to till the soil. Although these observations are not supported by quantitative data on erosion rates and surface runoff, farmers' perceptions say a lot for the value of cultivating wild food plants.

Table 9-2. Wild Food Plants Growing in Mount Pulag National Park

Local Name	*Botanical Name*	*Family*	*Plant Parts and Their Uses*
Gatili	*Pinanga patula*	Palmae	young shoots eaten
Batbatawang	*Physalis angulata* L.	Solanaceae	fruits eaten raw or cooked
Ladew	*Vaccinium bancanum* Miq.	Ericaceae	leaves boiled and drunk as tea
Nagngay	*Begonia* sp.	Begoniaceae	young stalks are cooked
Pinit	*Rubus pectinellus*	Rosaceae	plants eaten raw or processed
Duting	*Rubus niveus*	Rosaceae	fruits eaten raw or processed
Susuga	*Circium luzonienses* Merr.	Compositae	leaves and roots boiled and drunk as tea; stalks eaten raw
Namey	*Lencasyke capitellata*	Urticaceae	fruits eaten raw or processed
Mangkunetrp	*Rubus chrysophyllus*	Rosaceae	fruits eaten raw or processed
Duting	*Rubus fraxinifoliolus*	Rosaceae	fruits eaten raw or processed
Ayusip	*Vaccinium myrtillcides* Michx.	Ericaceae	fruits eaten raw or processed
Batnak	*Rubus* sp.	Rosaceae	fruits eaten raw or processed
Gatgatang	*Erechtites hieracifolia*	Compositae	young shoots and leaves are cooked
Masap (vine)	*Passiflora edulis* Sims	Passifloraceae	young shoots and leaves are cooked fruits eaten raw or processed
Halmberg	*Begonia merrillii* Merr.	Begoniaceae	young stalks eaten raw or cooked
Oyok	*Sauraria elegans*	Actinidiaceae	fruits eaten raw or processed
Degway	*Sauraria* sp.	Actinidiaceae	fruits eaten raw or processed

Establishing Wild Food Plants in Swidden Fallows

Vines of wild berry (*Rubus niveus* Thunb.) are collected from the forest when they are about 30 to 45 centimeters long. Care is taken not to damage their root systems, and they are carried back to the village in jute sacks or woven baskets.

The roots are soaked in a tub of water or a stream to prevent them drying before transplanting. Alternatively, the sacks containing the vines are sprayed with water and placed in the shade. Then the tips of the roots are trimmed to about 2 cm long prior to planting in the early morning or late afternoon.

The wild berry vines are planted randomly throughout the field with spacing of about 1 to 1.5 m. During the rainy season, new leaves generally appear one month after planting. The plants require no care, except for protection from grass fires.

Adoption of the Technology in Other Communities

In 1993, only 30% of farmers in Barangay Tawangan planted wild food plants in their swidden fallows. Despite encouragement from elders during meetings, younger farmers chose to "wait and see" before adopting the technology themselves in the following planting season. In 1994, the DENR and staff from a state university at Benguet conducted a training course on processing food from wild plants, including lectures and hands-on demonstrations. The participants, mainly women, ranged from 17 to 68 years old. Following this event, three other barangays, Ballay, Lusod, and Bashoy, adopted the planting of wild food plants in their fallow swiddens, and within about three years, 70% of the inhabitants of Mount Pulag National Park were using wild forest plants in their fallows both as live mulch and to provide food and income. (See Table 9-3 for a range of wild plant products.)

Potential Weaknesses of the System

Potential weaknesses arise from what is not yet known about a technology that is still very new. They include the following:

- Planting vines and shrubs in fallows might lead farmers to totally disregard the importance of trees in fallow vegetation.
- Shrub-type wild food plants could be invasive in the long term, and this might discourage a wider adoption of the technology.
- The nutritional content of the biomass of wild food plants remains unknown, and the possibility of allelopathic effects should be considered.
- The nutritional value of wild plant products, whether fresh or processed, is not yet known, and excessive consumption might have adverse health effects.
- Information about the technology, especially on post-harvest handling of products, is nonexistent. Possible pests and diseases have not yet been identified.

Table 9-3. Products Derived from Wild Food Plants

Species		*Products*	*Prices*
Local Name	*Botanical Name*		*(fresh fruit/kg)*
Wild berry	*Rubus niveus*	wine, jam, jelly, candies, and vinegar	₱ 40.00
Sapuan	*Saurauia* sp.	vinegar, fruit preserves	₱ 35.00
Gepas	*Sarcandra* sp	tea	₱ 40.00
Ayusip	*Vaccinium myrtillcides* Michx.	candies, jam, jelly	₱ 50.00
Wild strawberry	*Rubus fraxinifoliolus*	vinegar, wine, jelly	₱ 35.00

Note: 1997US$1 equaled 26.3775 Philippine pesos.

Potential for Disseminating the Technology in Other Areas

It is not unusual for exotic species to be planted in swidden fallows. The use of *Sesbania sesban* in Zambia is just one example. Imported species are also used as live mulch. Wild food plants from the Cordillera region of the Philippines, especially the *Rubus* spp., may similarly be adopted by upland farmers in other places who recognize the benefits of their use.

The people of the Cordillera region claim their wild food plants can conserve the soil and serve as a food substitute for cereals during the lean rainy months, as well as providing a source of livelihood. A wealth of farmer knowledge is revealed by this study. With this technology, problems of forest destruction may be minimized and dependency on forest resources reduced. Such knowledge, even when amassed by indigenous people, should be just as carefully regarded as any technology based on research.

Proposed Research and Development Projects

The rich potential discovered in a few wild food plants opens up a wide domain in which further research is sorely needed. Areas of study should include the following:

- Identification of other wild food plants in the Mount Pulag National Park with potential for domestication;
- The hydrological characteristics of fallows planted to wild food plants, compared with those planted to other species;
- The nutrient uptake, growth performance, biomass production, and allelopathic effects of wild food plants;
- Policies that will institutionalize indigenous fallow management systems;
- Verification that other regions are suitable for extension of this indigenous technology;
- Organization of a wild food plants society, involving farmers, researchers, educators, and horticulturists, to promote these promising but underexploited species;
- Collection of germ plasm of wild food plants for preservation and possible exchange; and
- The ancestral land claim issue, specifically within national parks.

Conclusions

The management of wild food plants in swidden fallows is a new practice in the upland farms of the Cordillera region. Because it has already proven to be effective in

Mount Pulag National Park, it could probably be adopted in nearby areas. The main lessons to be learned from this chapter include the following:

- Farmers took the initiative to solve their problems without government assistance.
- Wild food plants are promising alternatives for introduction into fallow fields under conditions similar to those at Mount Pulag National Park.
- Indigenous knowledge, which often goes unrecognized by outside stakeholders, regularly includes promising technologies, and these warrant research attention.
- Intensive research and development projects involving wild food plants are recommended in existing cropping systems.
- Ancestral land claims of indigenous peoples need to be satisfactorily resolved.
- Technical assistance is needed to help indigenous people develop products from wild food plants.

Additional References

Avila, J.K. 1986. Indigenous Weeds as Food Alternatives. *The Highlands Express* 6(4), 8–9.

DENR-CAR (Department of Environment and Natural Resources, Cordillera Administrative Region). 1988. *Regional Profile*. Manila: DENR-CAR. 8–10.

ERDB (Ecosystems Research and Development Bureau). 1995. *Wildfood Research and Development: An Integrated Sustainable Development Program for CARP-ISF Areas in the Philippines*. Terminal Report on CARP-ISF research and development Program., Vol. 2, 53–74.

Estigoy, D.A. 1990. Wildfood Plants: A Possible Food Alternative. *The Cordillera Gangza* 2(2), April–December 1990.

FAO (Food and Agriculture Organization of the United Nations). 1992. *Forest, Trees and Food*. Rome: FAO.

Fugisaka, S. 1986. Philippine Social Forestry: The Participatory Approach Conceptual Model. In: *Participatory Approaches to Development: Experiences in the Philippines,* edited by T. Osteria and J. Okamura. Manila: De La Salle University.

IIRR (International Institute for Rural Reconstruction). 1996. *Recording and Using Indigenous Knowledge. A Manual*. Cavite, Philippines: IIRR.

Kindtram, J., and K. Kingarukoro. 1991. *Food Forests, Fields and Fallows: Nutritional and Food Security Roles of Gathered Food and Livestock Keeping in Two Villages in Babuti District, Northern Tanzania*, Working Paper 184. Uppsala, Sweden: Swedish University of Agricultural Sciences/IRDC.

Sauwakontha, S.K., J. Chokkanayitak, P. Uttamawatim, and S. Lovikakarorn. 1994. *Dependency on Forest and the Products for Food Security: A Case Study of a Forest Area in Northeast Thailand*, Working Paper 263. Uppsala, Sweden: Swedish University of Agricultural Sciences/IRDC.

Chapter 10

Farmer-Developed Forage Management Strategies for Stabilization of Shifting Cultivation Systems

Peter Horne

The shifting cultivation systems of northern Laos have been the focus of many substantial studies in recent years (see, for example, Gillogly et al. 1990; van Gansberghe and Pals 1993; Chazee 1994; Chapman et al. 1997). The broad dynamics of change in these systems have been reasonably well understood for a long time. Under the combined pressures of increasing population and reduced length of fallow periods, with consequent severe weed problems and lower yields, shifting cultivation has become less sustainable. This has led to continuing efforts by both the Lao government and foreign aid donors to stabilize shifting cultivation.[1] The main objectives of this quest for "stabilization" include the following:

- Alleviation of rural poverty and reduced livelihood risk;
- Reduction of environmental degradation—and the risk of environmental degradation—including that caused by soil erosion, and lessening of the perceived threat from shifting cultivation to old-growth forests; and
- Eradication of opium poppy (*Papaver somniferum*) cultivation.

Approaches to Stabilization of Shifting Cultivation

Several different approaches are being used to stabilize shifting cultivation in Laos. Although, in practice, the distinction between them becomes blurred, they can still be broadly described as follows.

Systems Analysis

In 1993, a meeting on the status of shifting cultivation in Laos accepted a recommendation that a better understanding of the farming systems was needed before alternative practices were tested and demonstrated (van Gansberghe and Pals 1993). This approach suggested that, through systems-wide studies, key components could be targeted for development (ASPAC 1984). The problem with this is that shifting cultivation systems are characterized more by their differences than their

---

Peter Horne, CIAT, P.O. Box 783, Vientiane, LAO P.D.R.

[1] The term "stabilization" is preferred to "intensification" in this chapter because, although stabilization of shifting cultivation may include intensification, other options are also available, as explained herein.

similarities. Immense variability and complexity in land capability, land-use patterns, and population pressures exist, even across short distances and within individual villages. One study in northern Thailand, for example, identified five different household strategies or land-use patterns for dealing with soil erosion problems in the shifting cultivation fields of one village alone (Turkelboom et al. 1996). In addition, most parts of northern Laos, and in particular Luang Prabang Province and parts of Xieng Khouang and Vientiane Provinces, where shifting cultivation occurs, are characterized by unreliable climates and poor soils. The complexity of livelihood systems in shifting cultivation areas has evolved partly to cope with the risk that this variability imposes. As a result, we will never be able to characterize these systems adequately to satisfy the scientific requirement for "certainty before action."

Introduction of New Agricultural Technologies

The approach adopted by many "action-research" groups, including government departments, NGOs, and foreign aid projects, has been to take models that have been successful in other countries and apply them in Laos. Of particular interest has been "sedentarization," or adoption of a settled, permanent style of agriculture in some former shifting cultivation areas of northern Thailand, where the climate, soil, and cultural conditions are not dissimilar to upland areas of northern Laos. This has been achieved largely by the introduction of semicommercial agricultural technologies, including production of fruit, irrigated rice, hybrid vegetable seed, and vegetable crops. This, in turn, has been made possible by the introduction of fertilizers, pesticides, and irrigation, as well as by the expansion of rural roads to allow marketing of the products (Rerkasem 1997). Large, long-term inputs have been required from both government agencies and foreign aid donors (Rerkasem 1994).

Most of these technologies, including production of fruit trees and field crops, and agroforestry systems based on teak, have been demonstrated to work in Laos and there are generally few problems in applying them. However, there are two significant socioeconomic problems:

Lack of Access to Markets. The rural population of northern Laos is sparsely distributed, mostly through rugged, mountainous regions. The rural road network, unlike that in northern Thailand, is generally unable to provide easy access to markets. In 1990, a study found that only 57% of district centers in Laos, excluding the provincial centers, had year-round access by road. Seventeen percent had no access, not even in the dry season (SWECO 1990). For many farmers, the nearest road may be one day's walk away, or more. Even if the road network were to be expanded through the mountains, the sparse distribution of the population would mean that, for the foreseeable future, the number of people gaining easy access to roads and markets would remain small.

Lack of Capital. Even if there were access to markets, a lack of capital at farm level would probably limit the capacity of small farmers to buy into semicommercial technologies. As a result, these introduced technologies would probably remain limited to areas around rural feeder roads, especially where development projects were active and providing access to credit and planting materials. This is a similar situation to that in northern Thailand where, after more than 20 years of intensive effort, "the success of national and foreign-assisted development efforts ... has been only marginal" (Rerkasem 1997).

Strengthening Indigenous Agricultural Technologies

The limited impact of introduced, semicommercial technologies does not mean there are no ways of stabilizing shifting cultivation and reducing livelihood risks in remote areas. Farmers in remote areas frequently demonstrate that substantial improvements are possible simply by introducing their own innovations within existing agricultural practices. The fact that innovations are being made at all indicates the importance of these activities to farmers. Furthermore, farmers who may not have developed innovations usually have a clear idea of the problems they face and know what they would like to do to resolve them.

By working with these farmers, one can strengthen indigenous technologies and innovations through changes or additions developed by the farmers themselves and evaluated in partnership with development organizations, projects, and government agencies. The important point is that there is a role for both indigenous knowledge and introduced raw technologies, but only when they are evaluated on the farm with the full involvement of farmers. This action research approach (described in detail by Horne and Stür 1997) is being used in Lao P.D.R. to develop forage technologies in shifting cultivation areas of northern Laos.

Many indigenous strategies and technologies for stabilization of shifting cultivation are described in this volume. Those on which this chapter focuses are aimed at improved feeding of ruminant livestock in shifting cultivation areas, especially cattle and buffaloes. In some cases, farmers believe their current strategies for improved feeding of livestock, such as storing or reserving rice straw for dry season feeding, are completely adequate. However, there are real opportunities to work with farmers to improve those strategies with technologies such as the cultivation of forage species, either by introducing new plant species, or by suggesting new ways of incorporating forage crops into their farming systems. Many farmers are highly motivated to improve their feeding strategies because of the significant role being taken by livestock in shifting cultivation systems.

The Role of Livestock in Stabilizing Shifting Cultivation

While acknowledging the complexity and diversity of shifting cultivation systems, some generalizations can be made to illustrate the substantial role of ruminant livestock in stabilizing these systems. In the more remote areas of northern Laos, rice shortages are common, either on an annual basis or as a result of frequent climatic disasters. These shortages can last for four to six months or more, and farmers must revert to hunting and gathering in nearby forests, growing less-preferred food crops such as maize or cassava, or buying rice. Traditionally, buying rice has meant selling their labor, opium, some forest products such as medicines and herbs, or livestock, including cattle, buffaloes, pigs, goats, and chickens. These days, the government discourages the production of opium and forest resources are increasingly hard to find because of restricted access, resettlement of villages away from forests, and the fact that more people are gathering them. Farmers are left with a growing reliance on livestock as a source of cash income. However, shifting cultivators find many benefits from raising ruminant livestock (Hansen 1997). They include the following:

- There is an assured market for livestock, with relatively stable prices.
- Livestock can be raised without concern for a lack of transport infrastructure. In one recent example, some Hmong farmers from Xieng Khouang walked 20 bulls to market in the capital, Vientiane, 350 km away.
- Livestock provide a high profit for a relatively low labor input.
- Livestock "store" wealth that can be used at any time.
- Ruminant livestock use natural resources such as grass, rice straw, and tree leaves that would otherwise be wasted.
- They provide a valuable source of manure for maintaining the fertility of irrigated rice fields and home gardens. In some areas, livestock owners sell manure to lowland farmers.

Offsetting these benefits are the almost ubiquitous problems of disease, livestock damage to crops, and limited feed resources. However, the benefits are so substantial that in almost all areas farmers persevere with livestock to reduce their livelihood risks. In shifting cultivation areas, traditional livestock feed resources are becoming scarce or degraded for the following reasons:

- Increased populations of livestock, resulting in overuse of limited feed resources such as grassland, rice straw, and forests;
- Expansion of agriculture into traditional grazing lands;
- Reforestation of grazing land, reducing productivity of native grasses; and
- Limitations on cattle grazing in forests.

These circumstances have been predicted for many years (Remenyi and McWilliam 1986; ADB 1997). However, the significant point now is that farmers themselves are recognizing the problem and, in many cases, are trying to do something about it. The following short case studies illustrate the diversity of livestock-feeding problems and the kinds of innovations farmers are using to alleviate them.

Case Studies

1: Phousy Village, Pek District, Xieng Khouang, Lao P.D.R. Phousy is a village of 36 households of lowland Lao (Lao Loum) people, located in the semiremote forested hills of Xieng Khouang Province. Since 1996, Phousy village has been assisted to plan its future development by the German government–funded Nam Ngum Watershed Management and Conservation Project.

The villagers have traditionally relied on a mixture of shifting cultivation and irrigated rice for their livelihood. During the Vietnam war, the area was heavily bombed, forcing the villagers to flee until the situation became safe. The forest on the hills surrounding the village and the rice paddies were destroyed by the bombing and resulting fires. The stream that irrigated the paddies and provided fish—an important village food source—began flooding in the wet season and drying up in the dry season. The villagers had little choice but to resort to shifting cultivation to survive.

They noticed that, as the forest started to grow back, the stream flooded less frequently in the wet season and flowed more regularly during the dry season. They wanted to allow more forest regeneration and to reduce their dependence on labor intensive and relatively unproductive shifting cultivation, so they began to reclaim their rice fields. As their cattle and buffalo numbers grew, they were able to use the manure to increase the fertility of the lowland soils and to expand the area of paddies. By 1993, all but five households in the village were able to stop shifting cultivation completely.

During the wet season, the cattle had to be sent to grazing lands in the mountains, more than five kilometers away, so they would not damage the rice paddies. This resulted in the loss of much manure, which had become a valuable cash earner. It was also recognized by the villagers as an essential input for paddy rice farming. They wanted to keep their animals penned closer to home, but there was little more than rice straw available near the village to feed the cattle. They also wanted feed at the end of the dry season to condition their buffaloes ready for plowing.

Two farmers heard of a forage trial that was being conducted by the district agriculture office, 40 km distant. One farmer went to the trial and collected a few cuttings of *ruzi (Brachiaria ruziziensis)* to plant on former shifting cultivation fields near his barn. From just one square meter, he has expanded the plot to cover 200 m^2, and three other farmers have joined him. Others in the village want to join in the expansion, but so far, they have been restricted by a lack of planting material. They're expected to join when the four forage farmers once again use vegetative cuttings to expand their areas of ruzi, at which stage there should be enough seed and vegetative planting material for others to participate. This example illustrates the role that new "raw technologies" can play to build on farmer innovation, as ruzi is far from being the best adapted variety for their area but it was all they could access.

2: Nam Awk Hu Village, Xieng Ngeun District, Luang Prabang, Lao P.D.R. Nam Awk Hu is a village of 47 households made up mostly of Hmong people. The Hmong are highlanders, renowned for their livestock-raising abilities and their long involvement in highland shifting cultivation. They settled Nam Awk Hu in 1973 as refugees from the war in Xieng Khouang. The village and its shifting cultivation fields are about three kilometers from a major road, at an altitude of 800 m asl. However, its grazing lands and cash crop fields are at an altitude of 1.200 m asl. Rice yields from the shifting cultivation fields around the village have declined almost threefold (from about 3 tonnes/ha) since the village was settled. Fields in the better soils of the highlands have been overrun by *Imperata cylindrica* and most of them have been abandoned, except for small plots of intensively managed cash crops. Rice shortages are beginning to affect about half of the families in the village.

In order to buy rice, the people of Nam Awk Hu work as laborers, sell the few cash crops they can grow and, most importantly, sell livestock. The 170 cattle owned by 31 of the households in the village are kept permanently fenced in the highland grazing areas, and are managed as a single herd. As a measure of the importance of cattle to these villagers, each household owning cattle has been required by village rule to provide a roll of barbed wire for fencing. Once every three to four days, the cattle must be walked down from the highlands to a river for water. It is a total distance of about 18 km and a drop in altitude of 800 m.

As dependence on the cattle has increased, the herd size has increased, and feed on the grazing land has become inadequate, especially in the dry season. The village cattle raisers' group proposed a requirement that each owner should plant an area of elephant grass (*Pennisetum purpureum*) on abandoned shifting cultivation fields near the grazing land, to be used as cut-and-carry fodder to supplement normal feed during the dry season. This has been successful for more than 15 years. The farmers have not expanded the area of elephant grass beyond locally moist areas with better soils because of its susceptibility to the long dry season. As with the example in Phousy village, *Pennisetum purpureum* was not a species that grew particularly well in their conditions but their strong need for better feed resources motivated them to perservere with it given the lack of any better alternatives.

3: Houay Hia Village, Xieng Ngeun District, Luang Prabang, Lao P.D.R. Houay Hia is a village of 76 households of Lao Theung (middle altitude) people. The village, which had been settled for more than one hundred years, recently relocated to be near a major road. The people rely totally on shifting cultivation for their livelihood. As the population of Houay Hia and neighboring villages has increased, the land area available for shifting cultivation has become limited. Shorter fallow periods have resulted in substantial reductions in rice yields, down to less than 800 kg/ha. As a result, more than 75% of the families in Houay Hia suffer a rice shortage of four to five months each year.

In order to buy rice, the villagers work where possible as laborers and sell livestock. Nearly all of the families have three or four goats and one or two cattle. However, there is no fixed location for grazing, and the animals roam freely, sometimes up to 10 km away. This regularly causes disputes as the animals damage other farmers' upland rice fields, and many are lost through disease, accident, and theft.

Their strong dependence on livestock to provide income to buy rice has led these farmers to try to establish a dedicated grazing area near their village. However, the nearby fallow fields, which used to have many species of palatable plants, are now covered by unpalatable weeds, mainly *Chromolaena*. The villagers had been hoping to find plants they could grow on the surrounding fallow fields as well as near their houses, to supplement the feed of their grazing animals and keep them closer to home, but had not found any suitable varieties.

4: Makroman Village, Samarinda, East Kalimantan, Indonesia. Makroman is a village of transmigrants who moved to the area from Java 20 years ago. When they arrived, the rolling uplands were newly cleared and the moderately fertile soils were ready for cropping. However, the area was large and they were unable to cultivate it all. Gradually, *Imperata cylindrica* spread until, now, the village is located in a "sea" of

Imperata. The farming system of the village is a mixture of irrigated rice production, dryland cropping, and livestock production.

In 1994, some small, informal forage trials were planted with 10 farmers in Makroman. Although many species looked promising to the development workers, at the end of the trials none of the farmers was interested in continuing with the forages as they did not consider the benefits to be great enough when compared with the traditional—and zero-input—grazing resource of *Imperata* all around them.

However, one farmer, with the support of the extension worker, tried oversowing a small 100 m^2 area of corn with the legume *Centrosema pubescens* (CIAT15160), when the corn was two weeks old. He was surprised to find that the crop grew quite well, without needing the chemical fertilizer he would normally have applied. He also did not have to weed the crop, as was his usual practice. At harvest time, the ears of corn were larger than his usual crops and he was able to sell the corn for a substantially increased profit, partly because he did not have to buy any fertilizer. Since then, he has greatly expanded the area, and for six successive crops, he has not had to do any land preparation because the soil is still moist and friable beneath a mulch of *Centrosema*. Thirty neighboring farmers, seeing these benefits, have asked for seed to try the same oversowing practice. This seed is being provided, but at the same time, the innovation is being studied in replicated on-farm trials, in partnership with several farmers, to quantify and better understand its benefits.

5: Pianglouang Village, Pek District, Xieng Khouang, Lao P.D.R. Pianglouang is a mixed Hmong and Lao Loum village located on the treeless Plain of Jars. Fourteen Hmong families were resettled into this village three years ago. They have neither access to forest nor to paddy land, all of which is already used by Lao Loum hamlets. They do not yet have livestock for sale. In short, they have neither their traditional sources of income nor their native food to help them through what is a time of crisis. They rely totally on their upland rice fields and maize crops for survival.

The soils are of moderate to poor fertility, so maize growth is slow. The critical time for weeding the maize is during the first six weeks of growth, but this is also the time that the farmers are busiest in the upland rice fields. So the maize gets minimal weeding, and as a result, yields have been poor. The problem has become so severe that, despite their need for the maize, these farmers will be forced to abandon their maize fields if they cannot find a simpler way to control weeds. Their dire situation makes them farmers who have a real problem that they want to solve in partnership with development workers. They are one group evaluating legume cover crops over sown into maize.

The Potential of Cut-and-Carry

These case studies—and experience from other villages in Southeast Asia—show that farmers in shifting cultivation areas usually have little flexibility to develop strategies to cope with ruminant livestock feeding problems. They are limited to moving livestock between wet and dry season grazing areas, storing or reserving rice straw for dry season feeding, and cultivating grasses on fallow land to provide cut feed for penned animals.

The first two of these strategies are already well developed throughout the region. However, the third strategy, cultivating grasses on fallow land to provide cut feed for penned animals, is rapidly emerging as a practice with significant potential for development in partnership with farmers. Interest at village level also extends to improving grazing areas for the use of communally managed herds of cattle at strategic times.

There is a wide variety of reasons for this interest, ranging in just the first three case studies (above) from wet season supplementation of cattle, to dry season supplementation of buffaloes and cattle, increased manure availability for use on irrigated rice fields, control of animal damage to crops, and lowering of animal losses.

In many cases, the motivation for managing the feed resource is strong but innovation is limited simply by a lack of access to information and planting material.

Successful development of forage technologies does not depend on the quantity of planting material distributed in the first instance, but on the careful selection of farmers who have a real problem that they want to solve in partnership with development workers (see, for example, case study 1). If this is then combined with a broad range of robust technologies, the chances of successful adoption are much higher (Horne and Stür 1997). For example, the upland areas of Bali are now renowned for the widespread use of the shrub *Gliricidia sepium* as a living fence and a source of dry season fuel. The species was introduced to the area as recently as 1970, and with just one hundred cuttings. The key was that the farmers had a real problem that they recognized themselves and the species was robust and easy to manage.

The Role of Introduced Forages in Stabilizing Shifting Cultivation

Through a partnership of farmers and development workers, introduced forage species are currently being developed into technologies that can help stabilize shifting cultivation in northern Laos. This is happening in the following ways:

Using Introduced Species

Regional evaluations of more than 70 forage species at five locations in Laos have identified eight broadly adapted and robust species that are now being assessed by about 100 farmers in three northern provinces for their potential in cut and carry or grazed livestock feeding systems. These species are:

- *Brachiaria brizantha* cv Marandu, and other lines soon to be tested;
- *Brachiaria decumbens* cv Basilisk;
- *Brachiaria humidicola* CIAT6133;
- *Brachiaria ruziziensis* cv Kennedy;
- *Andropogon gayanus* cv Kent;
- *Panicum maximum* T58;
- *Paspalum atratum* BRA9610; and
- *Stylosanthes guianensis* CIAT 184.

At this stage, evaluations are informal and without replication, in order to encourage the participation of more farmers as well as to encourage farmer innovation. Should innovations emerge that have promise, they will be both encouraged by farmer-to-farmer visits and studied in more detail in formal, replicated on-farm trials. This process is similar to that followed in participatory forage evaluations in Makroman village, East Kalimantan, Indonesia.

Incorporating Introduced Forages into Existing Shifting Cultivation Systems

Trials managed by farmers are either in progress or are getting under way in northern Laos to evaluate and, if possible, to adapt the following potentially useful innovations that have been successful elsewhere:

Forage Tree Species for Fence Lines. Livestock damage to crops is a major and constant concern for farmers in the upland areas of northern Laos (Fahrney 1997). A huge amount of effort is spent building solid, semipermanent fences made of wood, wire, and bamboo, particularly in those areas managed by Hmong people. Farmers are already using some living fences, mainly *Jatropha curcas* to either keep their animals fenced in or to fence animals out of their fields. Living fences incorporating *Gliricidia sepium* or *Leucaena leucocephala* on better soils, and *Calliandra calothyrsus* in higher areas, have large potential, from a technical perspective, to provide effective security as well as supplementary feed. However, they need to be evaluated by farmers and development workers together, not only to clarify the technical advantages and limitations, but also the farmers' criteria for accepting or rejecting them.

***Stylosanthes guianensis* (CIAT 184) Oversown into Upland Rice.** Both formal and informal trials with farmers have commenced with *Stylosanthes guianensis* (CIAT 184) being oversown in upland rice fields after the first round of weeding. This species has demonstrated particular potential in other trials because of its rapid establishment, its low impact on rice yields if it is sown late enough, and its ability to grow well on poor soils. Oversowing upland rice with *Stylosanthes guianensis* is not a new innovation (see, for example, Shelton and Humphreys 1972, 1975a,b,c; Madely 1993). As well as providing benefits of reduced weeding and improved fertility for the rice crop, it has the potential of improving subsequent fallows.

The use of forage legume species for fallow fields in shifting cultivation areas has been the subject of much detailed and promising research (see, for example, Roder and Maniphone 1995). The potential benefits over weed fallows include reduced weeding requirements, improved soil fertility, easy establishment after a round of weeding, and reduced risk of erosion. These benefits are well documented (see, for example, Gibson and Waring 1994). However, there has been little adoption by farmers. There are many reasons for this, but the most important is probably that all of the work in Laos so far has been on research stations or in researcher-managed trials, with the expectation that the technologies will then be "extended" to farmers. Informal oversowing trials involving farmers are needed to discover what aspects of oversowing appeal or do not appeal to them, and also to gain insights into what treatments should be investigated in subsequent formal trials. However, sowing fallow fields with forages means that they need to be protected from uncontrolled grazing. Such trials with the farmers of Hoauy Hia village, for instance, would almost certainly fail because of the lack of sturdy fencing. However, in Hmong areas, where individual fallow fields are often sturdily fenced, the potential for success would be much higher.

Conclusions

For the most part, farmers in shifting cultivation areas of northern Laos are strongly dependent on livestock for the security of their livelihoods. Diminishing feed resources for these animals have resulted in some farmer groups taking steps to manage the feed resource, particularly by planting introduced forage species. Others recognize the problems but have had no access to either information or planting materials with which to develop their own forage technologies. Both groups of farmers provide an opportunity for development workers to strengthen local feeding technologies. This can be achieved by both introducing new, robust forage species for comparison with existing species and evaluating new ways of incorporating forages into existing farming systems.

Acknowledgments

Information for this chapter has been gained from field experiences with many dedicated development workers associated with the Forages for Smallholders Project. Special acknowledgment is due to Phonepaseuth Phengsavanh and Viengsavanh Phimphachanhvongsod of the National Agriculture and Forestry Research Institute, the Luang Prabang and Xieng Khouang Provincial Agriculture offices, and Mr. Ibrahim of East Kalimantan Livestock Services, Indonesia, for information included in the case studies.

Since this paper was written, a great deal of progress has been made in developing forage-based livestock technologies with farmers as alternatives to shifting cultivation. For further information, see Stür et al. (2002), Horne et al. (2005), Phimphachanhvongsod et al. (2005) and Phengsvanh et al. (2005).

References

ASPAC (Food and Fertilizer Technology Centre for the Asian and Pacific Region). 1984. *Asian Pastures: Recent Advances in Pasture Research and Development in Southeast Asia*. Taipei, Taiwan: ASPAC.

ADB (Asian Development Bank). 1997. *Technical Assistance to the Lao P.D.R. for the Shifting Cultivation Stabilization Project*. ADB Technical Assistance Report TAR:LAO 29210. Manila, Philippines: ADB.

Chapman, E.C., B. Bouahom, and P.K. Hansen (eds.). 1997. *Upland Farming Systems in Lao P.D.R.: Problems and Opportunities for Livestock*. Proceedings of a workshop, May 19–23, 1997, Vientiane, Lao P.D.R. ACIAR Proceedings Series No. 87. Canberra, Australia: ACIAR (Australian Centre for International Agricultural Research), 156–162.

Chazee, L. 1994. Shifting Cultivation Practices in Laos. Present Systems and their Future. Vientiane, Lao PDR: UNDP.

Fahrney, K. 1997. Livestock in Upland Rice Systems. In: *Upland Farming Systems in the Lao P.D.R.: Problems and Opportunities for Livestock*. Proceedings of a workshop, May 19–23, 1997, Vientiane, Lao P.D.R., edited by E.C. Chapman, B. Bouahom, and P.K. Hansen. Canberra, Australia: ACIAR (Australian Centre for International Agricultural Research).

Gibson, T.A., and S.A. Waring. 1994. The Soil Fertility Effects of Legume Ley Pastures in Northeast Thailand. I. Effects on the Growth of Roselle (*Hibiscus sabdariffa* cv. Altissima) and Cassava (*Manihot esculenta*). *Field Crops Research* 39, 119–127.

Gillogly, K., T. Charoenwatana, K. Fahrney, O. Panya, S. Nanwongs, A.T. Rambo, K. Rerkasem, and S. Smutkupt. 1990. *Two Upland Agroecosystems in Luang Prabang Province, Lao PDR. A preliminary Analysis*. Honolulu, HI: East-West Center.

Hansen, P.K. 1997. Animal Husbandry in Shifting Cultivation Societies of Northern Laos. In: *Upland Farming Systems in the Lao P.D.R.: Problems and Opportunities for Livestock*. Proceedings of a workshop, May 19–23, 1997, Vientiane, Lao P.D.R., edited by E.C. Chapman, B. Bouahom, and P.K. Hansen. Canberra, Australia: ACIAR (Australian Centre for International Agricultural Research).

Horne, P.M, W.W. Stür. 1997. Current and Future Opportunities for Improved Forages in Southeast Asia. *Tropical Grasslands* Special Issue 2, 117–121.

———, W.W. Stür, P. Phengsavanh, F. Gabunada Jr., and R. Roothaert. 2005. New Forages for Smallholder Livestock Systems in Southeast Asia: Recent Developments, Impacts and Opportunities. Chapter in forthcoming book *Grasslands: Developments, Opportunities Perspectives* (FAO, Rome).

Madely, J. 1993. Raising Rice in the Savannas. *New Scientist*, June 19, 1993, 36–39.

Phengsavanh, P., K. Fahrney, V. Phimphachanhvongsod and G. Varney (2005). Livestock Intensification: Forages and Livestock Technologies for Complex Upland Systems. In Bouahom B, Glendinning A, Nillson S and Victor M.(eds.) Poverty Reduction and Shifting Cultivation Stailisation in the Uplands of Lao P.D.R.: Technologies , Approaches and Methods for Improving Upland Livelihoods. Proceedings of a workshop held in Luang Prabang January 27-30 2004. Vientiane, Lao P.D.R. pp. 279-286. (National Agricultural and Forestry Research Institute, Lao P.D.R.)

Phimphachanhvongsod, V., P.M. Horne, P. Phengsavanh and R. Lefroy. (2005) Livestock Intensification: A Pathway Out of Poverty in the Uplands. In Bouahom B, Glendinning A, Nillson S. and Victor M.(eds.). Poverty Reduction and Shifting Cultivation Stailisation in the Uplands of Lao P.D.R.: Technologies, Approaches and Methods for Improving Upland Livelihoods. Proceedings of a workshop held in Luang Prabang January 27-30, 2004. Vientiane, Lao P.D.R. pp. 279-286. (National Agricultural and Forestry Research Institute, Lao P.D.R.)

Remenyi, J.V., and J.R. McWilliam. 1986. Ruminant Production Trends in Southeast Asia and the South Pacific, and the Need for Forages. In: *Forages in Southeast Asian and South Pacific Agriculture,* edited by G.J. Blair, D.A. Ivory, and T.R. Evans. Canberra, Australia: ACIAR (Australian Centre for International Agricultural Research).

Rerkasem, K. (ed.). 1994. *Assessment of Sustainable Highland Agricultural Systems*. Chiang Mai, Thailand: Natural Resources and Environment Program, Thailand Development Resources Institute, Chiang Mai University.

———. 1997. Shifting Cultivation in Thailand: Land Use Changes in the Context of National Development. In: *Upland Farming Systems in the Lao P.D.R.: Problems and Opportunities for Livestock*. Proceedings of a workshop, May 19–23, 1997, Vientiane, Lao P.D.R., edited by E.C. Chapman, B. Bouahom, and P.K. Hansen. Canberra, Australia: ACIAR (Australian Centre for International Agricultural Research).

Roder, W., and S. Maniphone. 1995. Forage Legume Establishment in Rice Slash-and-Burn Systems. *Tropical Grasslands* 29, 81–87.

Shelton, H.M., and L.R. Humphreys. 1972. Pasture Establishment in Upland Rice Crops at Na Pheng, Central Laos. *Tropical Grasslands* 6(3), 223–228.

———. 1975a. Undersowing Rice (*Oryza sativa*) with *Stylosanthes guianensis*. I. Plant Density. *Exp. Ag.* 11, 89–95.

———. 1975b. Undersowing Rice (*Oryza sativa*) with *Stylosanthes guianensis*. II. Delayed Sowing Time and Crop Variety. *Exp. Ag.* 11, 97–101.

———. 1975c. Undersowing Rice (*Oryza sativa*) with *Stylosanthes guianensis*. III. Nitrogen Supply. *Exp. Ag.* 11, 103–111.

Stür, W.W., P.M. Horne, J.B. Hacker and P.C. Kerridge. 2000. Working with Farmers: The Key to Adoption of Forage Technologies. Proceedings of an internal workshop, Cagayan de Oro City, Mindanao, the Phillipines, 12–15 October 1999. ACIAR Proceedings, No. 95

———, P.M. Horne, and P.C. Kerridge. 2002. Forage Options for Smallholder Crop-animal systems in Southeast Asia – Working with Farmers to Find Solutions. Agricultural Systems, 71: 75–98.

SWECO. 1990. *National Transport Study*. Final Report. Vientiane, Lao P.D.R.: Ministry of Communications, Transport, Post and Construction.

Turkelboom, F., G. Trebuil, D. Cools, I. Peersman, and C. Vejpas. 1996. *Land Use Dynamics and Soil Erosion in the Hills of Northern Thailand.* Proceedings of the Ninth International Soil Conservation Organization Conference "Towards Sustainable Land Use: Furthering Cooperation between People and Institutions," August 26–30, 1996, Bonn, Germany. International Soil Conservation Organisation and the German Federal Ministry for the Environment, Nature Conservation, and Nuclear Safety.

van Gansberghe, D., and R. Pals (eds.). 1993. *Shifting Cultivation Systems in Rural Development in the Lao P.D.R.* Proceedings of a workshop July 14–16, 1993, Nabong Agricultural College, Laos. Vientiane: UNDP.

Chapter 11

Managing *Imperata* Grasslands in Indonesia and Laos

Lesley Potter and Justin Lee

The grass *Imperata cylindrica*, which is one of the major volunteer species to emerge after forest clearing in Southeast Asia and a frequent component of swidden fallows, is widely regarded as a troublesome weed and an inefficient land cover. Government policies, which, from colonial times to the present[1] have single-mindedly aimed at eliminating swidden farming, have seized upon the belief that invasive grassy weeds are favored by the opening of the forest canopy and have linked the existence of grasslands with "improper shifting cultivation practices" (Soerjani 1970). During the 1970s and early 1980s there was some reconsideration in Indonesia of the potential usefulness of *Imperata* grasslands for agricultural settlement and grazing activities (Soewardi 1976; Soerjatna and McIntosh 1980; Soewardi and Sastradipradja 1980; Burbridge et al. 1981). It was even suggested as a means of controlling erosion (Soepardi 1980). However, attitudes to *Imperata* have now hardened. While anthropologists and geographers have drawn attention to the deliberate creation, maintenance, and management of grasslands by local people (Seavoy 1975; Sherman 1980; Dove 1981, 1984, 1986), such insights have had little influence on the policies of planners and funding agencies. The push during the 1990s in Indonesia for extensive planting of industrial forests to supply pulp and paper plants redirected energies toward reforesting grasslands and minimizing their value. Local protests were ignored or sidestepped (Brookfield et al. 1995; Potter 1997). There have been similar pressures throughout Southeast Asia, favoring reforestation with plantations of exotics such as *Acacia mangium* or *Eucalyptus camaldulensis*. In addition to the pressures noted above, there is competition from export crops, especially tree crops such as oil palm, rubber, coffee, and cocoa; more intensive agroforestry has been widely promoted, using tree legumes as an alternative to swiddening; and forest conservation has emerged in recent times as a "runaway issue" (Fraser 1989). All have intensified the perception that grasslands are a degraded form of vegetation that must be replaced.

Despite concern about supposedly vast areas of Southeast Asia being occupied by *Imperata* grassland, it is our contention that this particular vegetation type is declining and, in some areas quite rapidly, from competition with other grasses, especially after intensive grazing, and invasion by the weed *Chromolaena odorata* (Compositae), which has been spreading through Southeast Asia for the last 60 years, and is still extending its territory.[2] Because *Imperata* is not valued, and because many

---

Lesley Potter, Associate Professor, Department of Human Geography, Australian National University,Canberra, ACT 0200, Australia. Justin Lee, Department of Foreign Affairs and Trade, Canberra, Australia.

[1] Legal sanctions against swidden cultivation were first instituted in 1874 in both Java and the Philippines (Potter 2003, *40*).

[2] While *Chromolaena odorata* was noted in Myanmar and Laos during the 1930s, it only reached parts of eastern Indonesia, such as Sumba, in the 1970s and still has not penetrated to the heart of

of the most recent vegetation maps are inaccurate and depict a situation that existed perhaps 20 years ago, governments and agencies have ignored the silent retreat of this grass and the possible effects this will have on local people.

Selling *Imperata* Roof Thatch

Preliminary results of research into the origins and uses of *Imperata cylindrica* grasslands suggest that local people may be more dependent upon them than is popularly believed. This chapter describes the management and sale of *Imperata* for roof thatch (Table 11-1) on the islands of Muna and Bali, in Indonesia (Figure 11-1), and in lowland areas of Laos (Figure 11-2).

Table 11-1. Management of *Imperata* for Roof Thatch

Management Technique	*South and East Bali*	*Lowland Areas, Laos*	*Upland Area, Pakse, Laos*	*Muna Island, Sulawesi*
Type of area	Small lowland areas, too dry for wet rice	Small areas near the house in former maize or crop gardens	Open access grasslands and grass emerging in swidden gardens	Open access grasslands and grass emerging in swidden gardens
Planting as a crop	Yes, at Bukit in the far south	Unknown	No	No
Fence, or repair existing fences, around grass in old food crop gardens	No	Yes	No	Probably not
Cut	Annually or biannually	Annually or biannually	Annually or biannually	Occasionally, only where grass is fertile
Burn	Annually	Annually	Annually	Annually
Weed, to create pure *Imperata* stands	Yes	Sometimes	No	No
Deliberately avoid cultivation to avoid disturbing rhizomes	Yes	Yes	No	Yes, even leave fertile open access grassland undisturbed
Timing of harvest	Every 8 months, or 4 months after the wet season	Dry season, shortly before or after rice harvest	Dry season, shortly before or after rice harvest	As needed, normally in the dry season

Kalimantan. It is interesting that some of the largest areas of *Imperata* in West Kalimantan, such as those in the Melawi Basin, are still free of *Chromolaena,* although it has been seen along roadsides within 15 km of Sintang in the middle Kapuas.

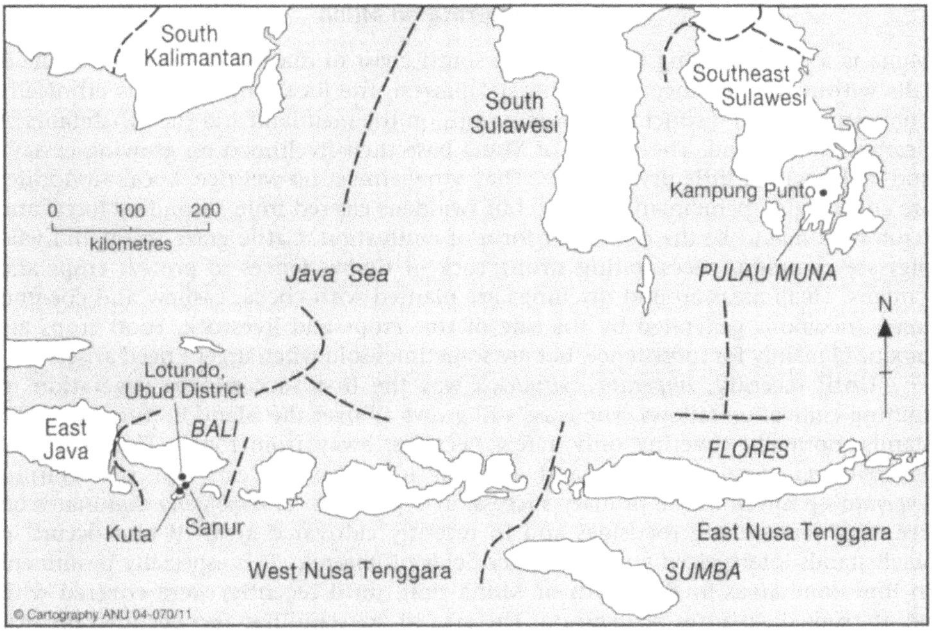

Figure 11-1. Research Sites on the Islands of Muna and Bali, Indonesia

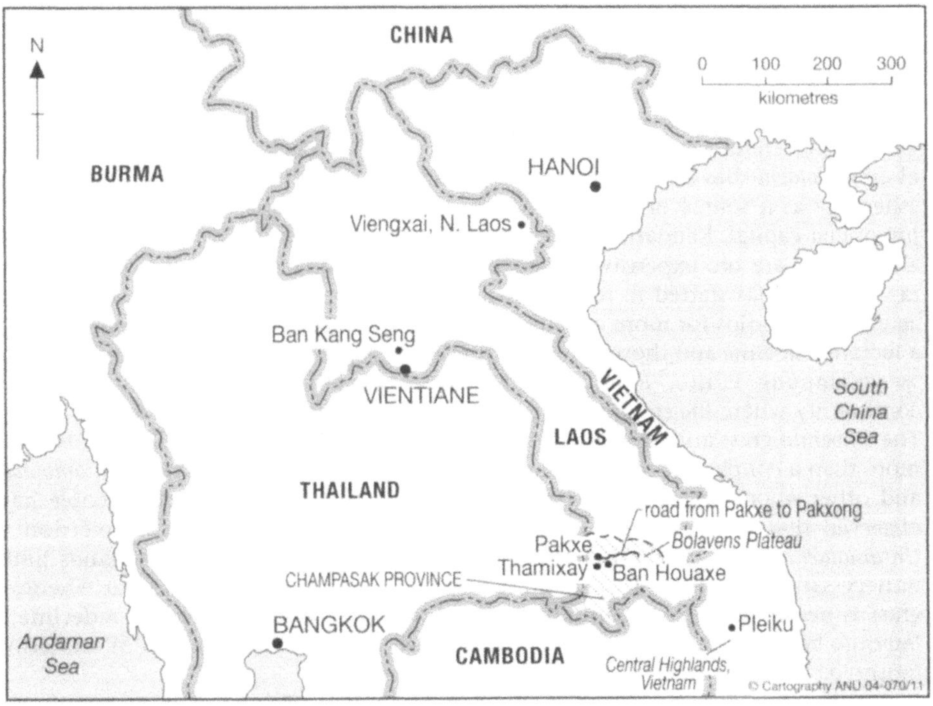

Figure 11-2. Research Site in Champassak Province, Southern Laos

Use of Imperata *on Muna*

Muna is a dry, low-lying island off the south coast of mainland Sulawesi. While it falls within the province of Southeast Sulawesi, the local population is ethnically homogeneous and distinct both from groups on the mainland and the inhabitants of nearby Buton Island. The people of Muna base their livelihood on growing cassava and maize with a little dryland rice. They grow almost no wet rice. Local authorities are encouraging permanent farming, but swiddens cleared from secondary forest and scrub continue to be the dominant form of cultivation. Cattle graze freely and wild pigs are common, necessitating strong rock or timber fences to protect crops and gardens. Small areas around dwellings are planted with cocoa, cashew and coconut trees. Income is generated by the sale of tree crops and livestock. Food crops are produced mainly for subsistence, but are sometimes sold when urgent need arises.

Until recently, *Imperata cylindrica* was the first successional vegetation in shifting cultivation fallows. The grass still grows all over the island in pure but small stands, normally covering only a few hectares, away from roads. The arrival of *Chromolaena odorata* on the island, possibly less than 20 years ago, has usurped *Imperata*'s position as the primary succession vegetation. *Chromolaena* dominates on disturbed lands along roadsides and in recently cultivated areas. It also occurs in small stands interspersed amongst larger fields of *Imperata*. It is especially prominent in limestone areas in the south of Muna that, until recently, were covered with uninterrupted expanses of *Imperata*. The area of grassland has also declined because of the expansion of teak plantations and the relocation of local populations away from waterless areas and onto unused lands.

Observations of the management and use of *Imperata* focused initially on the island's north and east regions, surrounding the capital, Raha. Traditionally, *Imperata* has been extremely important for grazing cattle and hunting deer. The grass is also the best roofing material available locally. Sago and Nipah palms, used for roofing on mainland Southeast Sulawesi, are very rare on Muna. In the past, when *Imperata* was abundant in all locations, local people could obtain all the grass they needed with little effort, and management was limited to annual burning to promote new growth palatable to animals.

With *Imperata*'s retreat, local people have been concerned with managing the grass more carefully in areas where it still occurs. They acknowledge that the arrival of *Chromolaena* has assisted dryland farming, but they still want access to *Imperata*, especially as a source of roofing thatch. Nipah palm thatch is imported from the provincial capital, Kendari, and corrugated iron is increasingly common, but these alternatives are too expensive for many. The increasing difficulty of obtaining the raw material has started to force up the price of *Imperata* thatch. This, in turn, has acted as a stimulus for more careful management, both by those who see selling it as a lucrative sideline and those who do not want to depend on sellers for their supply.

Kampung Punto, a hamlet of Kontunaga village, central Muna, is one community where *Imperata* is being protected for private use and sale as roof thatch. The *Imperata* grasslands are an open-access resource scattered in smallish plots rarely more than a couple of hectares in area. They are often interspersed with *Chromolaena* and other woody weeds and with teak, cashew and other trees. Local people have observed that tilling *Imperata* lands for cultivation hastens their conversion to *Chromolaena* when cultivation ceases, so current management of these lands limits unnecessary cultivation. They continue to be burned every year and cut whenever grass is needed. Farmers maintain that this treatment does not lead to a decline of *Imperata* because the rhizomes are not disturbed and *Chromolaena* cannot establish a foothold.

Punto has become a source of *Imperata* for people throughout its district. Residents of other villages occasionally visit the community to cut *Imperata* for roof thatch from its open access grasslands. There is no payment for cutting rights. However, the majority of visitors seeking thatch do not cut *Imperata* themselves but purchase sheets of thatch already made up by the local people. After being cut and dried, the grass is tied on to long, thin pieces of wood or split bamboo to make a "frond" of thatch about 1.5 m long. As demand has increased, the price of these sheets has risen from 1,000 rupiah (1997US$0.43) for three sheets to 1,000 rupiah for two sheets or 2,000 rupiah (1997US$0.86) for five. The sheets are commonly made by women, usually working at night. One sheet can be made in about 15 minutes and one person can make about 30 in a day, if they work at nothing else. The demand is such that residents of Punto are confident that any thatch they make will be sold. Consequently, they make it whenever they have the time or the inclination, without waiting for orders.

Consequently, the sale of *Imperata* thatch has become a valuable sideline in Punto. One farmer claimed that he had grossed 500,000 rupiah (1997US$215) from it over two years. He praised *Imperata*, saying that it could be cut and sold whenever he needed income, as opposed to food crops, which had a specific harvest time. Considering that grass from an area of about four square meters is needed to make one sheet of thatch, one hectare of *Imperata* can supply the materials for a return of 1,250,000 rupiah (1997US$537.50) per year, although much of this value is generated by the labor needed to transform the grass into sheets of thatch.

Use of Imperata *in Laos*

Known literally as "roof grass" (*nya kha*), *Imperata* is the preferred roofing material for many smallholders in Laos. Observations in lowland agricultural areas of Champassak Province and around the capital, Vientiane, revealed that local people found it difficult to obtain an adequate supply. The shortage not only compelled them to manage the grass carefully when it grew on their land, but it was also the basis of a trade in thatch with people from upland areas where *Imperata* is easily available.

The village of Thamixay, about 30 km southwest of Pakse, in Champassak Province, southern Laos, is typical of many small lowland farming communities. Farmers cultivate rainfed wet rice near their homes on the flat lowlands and commute to small dryland swidden gardens in nearby foothills. They also grow crops and a small number of trees in home gardens. The village is only about 15 years old and has minimal development of tree crops. Home gardens are well fenced to protect crops from cattle and buffaloes, which wander and graze freely in the dry season. These large livestock are relatively numerous but are owned by fewer than half of the households. The community has no electricity or running water and the road, while being improved, is still bad.

Imperata thatch is important in Thamixay for roofing houses, garden shelters, and rice storage huts. However, it is difficult to obtain around paddy fields and along roadways because these areas are heavily grazed. It can be found in swidden fields, but the quantity of grass available is often inadequate or is difficult to transport back to the village. Old people or invalids occasionally pay younger, fitter neighbors to cut and deliver grass for them, which they then make into thatch. The difficulty of obtaining *Imperata* prompts many people in Thamixay to nurture the grass when it grows voluntarily in their home gardens.

With the protection of a fence, a healthy stand of *Imperata* often emerges in home gardens after one season of a dryland crop such as maize. Rather than removing it or allowing livestock to graze the *Imperata*, the farmers protect it as a source of thatch. They permit pure stands to develop, which may then be kept for more than six years, or until the vigor of the grass dwindles and it grows more slowly. They burn the grass every year after cutting it, in an effort to preserve its vigor. As its growth declines, they may cut it for thatch every second year rather than the normal practice of cutting it annually. When the *Imperata* eventually dies back, many of the farmers prefer to turn the garden over to tree crops, especially kapok (*Ceiba pentandra*).

A home garden of *Imperata* can be very lucrative for people in Thamixay. After they have seen to their own roofing needs, they make the remaining grass into sheets of thatch that are sold within the community. The grass is harvested any time from

November to January, around the time of the rice harvest. It is then stored and made into thatch at the householder's convenience, during the dry season. In one example, a household with a garden of pure *Imperata* covering about 0.25 ha cut enough grass to make about 1,000 sheets of thatch in one season. They used half of the sheets themselves and sold the rest for 200 kip (1997US$0.20) each, generating an income of about 1997US$200. This was extremely important to the family. The head of the household was crippled and outside labor was needed to work their paddy fields. As a result, their rice harvest had to be divided. They had no large livestock and had few other income sources apart from remittances sent by a daughter who had migrated to Thailand.

Buyers of *Imperata* thatch tend to be wealthier members of the community. A supply of grass is generally easy to obtain, but making the thatch is a tedious job and more affluent households prefer to buy it. Should there be an insufficient local supply, *Imperata* thatch can also be purchased, sometimes more cheaply, from middlemen. These traders buy it from upland areas and sell it all over the province, especially to people in and near the towns of Pakse and Champassak.

Trading *Imperata* is a profitable business. During the peak roofing period, toward the end of the dry season, at least five families in Pakse and three across the Mekong River in Muang Kao work as thatch traders. Such trading is the sideline of a local public servant's wife in Muang Kao. She fills the family truck with more than 1,500 sheets of thatch, bought for 150 kip apiece (1997US$0.15), and sells them to communities west of the Mekong. With a markup of about 100 kip (1997US$0.10) on each sheet, she earns about 150,000 kip (1997US$150) per trip, minus gasoline cost and wages for a driver. It normally takes her three to four days to buy and sell one truckload of thatch.

On the highway from Pakse to the Bolavens Plateau, where middlemen buy their thatch, many of the families living alongside a 13 km stretch of road sell sheets of *Imperata* thatch, which are displayed outside their houses. Many of them have abundant *Imperata* growing nearby, on the fertile soils of their extensive upland swidden gardens. After two years of cropping dryland rice and pineapples, *Imperata* begins to dominate their fields. This natural fallow growth becomes their raw material, supplemented by grass cut from open access grasslands bordering the forest. Even those without *Imperata* on their own land make money from making and selling thatch. For example, people in Ban Houaxe, 8 kilometers from Pakse, travel 10 kilometers to obtain grass from somebody else's garden. If the garden is fenced and therefore privately owned, they pay 50,000 kip (1997US$50) for one hectare of grass. The combination of a ready supply of *Imperata* and, perhaps more importantly, good road access to traders and markets, means that making thatch is the area's principal dry season activity.

Villagers living along the highways leading into Vientiane take similar advantage of their location. Residents of Ban Kan Seng, 30 km north of Vientiane, hire trucks and travel 50 km to cut *Imperata* from open-access grassland. They make it into sheets and sell it to middlemen plying the highways on the lookout for thatch to be sold on the profitable Vientiane market. These families estimate that they earn up to 400,000 kip (1997US$400) from making thatch every dry season.

Use of Imperata *in Bali*

Resource-poor householders earn valuable incomes from the sale of *Imperata* thatch in Muna and Laos, as they do in many other traditional and semitraditional communities throughout Southeast Asia. However, despite its present value, doubt surrounds the future of cottage industries making *Imperata* thatch. As incomes rise, more people can afford corrugated iron roofs. Iron is regarded as superior because it is more weather resistant, longer lasting, and less flammable. Iron roofs are also more prestigious. It is expected of community leaders that they will reroof with iron, and others will follow when they can afford it. Nevertheless, increasing prosperity does not always reduce the demand for *Imperata* thatch. As reported by Potter, Lee, and

Thorburn (2000), development and diversification of the local economy have not diminished a tradition in Bali of managing, using, and selling *Imperata*. On the contrary, the emergence of a strong tourism and service industry has transformed *Imperata* into a commodity much sought after by the construction industry.

Rifai and Widjaja (1977) wrote that while the grass was always important for medicinal, cultural, and religious purposes, the Balinese also maintained *Imperata* for its economic value, as thatch for houses and temples and as a cover for mud walls between allotments. Arable land was set aside so the grass could grow. Quoting *Teysmannia* (1918), one writer even claimed that in South Bali a field of *Imperata* could be more profitable than rice (Heyne 1950). Recent fieldwork has revealed the persistence of a long tradition of planting *Imperata* on the drier lands of South Bali so it can be sold as thatch. This occurs particularly in the Bukit area (Potter et al. 2000, *1041–2*). While the Balinese once managed and traded *Imperata* in a manner similar to that currently occurring in Muna and Laos, their involvement has now progressed beyond the traditional to a fully fledged commercial activity. Rather than fading into obsolescence, this could be the future direction for *Imperata* in other locations as well.

The tourist industry as well as a desire to maintain Bali's image as a center of Indonesian culture, has generated a massive local demand for *Imperata* thatch. Eighty percent of hotels around Kuta Beach are now purported to have buildings roofed with *Imperata* thatch. It is popular across the island for restaurants, beer gardens, art and handicraft shops, plant nurseries, and temples. It is also used in the largest and most fashionable hotels and shopping complexes, including the Ritz Carlton Bali. In larger buildings, it is often placed beneath an outer shell of tiles as a form of insulation and to create an attractive ceiling. *Imperata's* properties as an insulator are also beginning to attract fresh attention from ordinary householders who earlier abandoned it in favor of earthen tiles or iron. Those with higher incomes are now considering the comfort of their houses, rather than simply their appearance, and are placing *Imperata* thatch beneath the tiles or iron to keep their homes cool in the hot season and warm during the wet.

Dozens of small businesses specializing in making and installing *Imperata* thatch have sprung up across Bali as a result of this new demand. One firm operating a workshop in the Sanur area employs eight full-time staff to make three-meter-long sheets. Each laborer turns out 30 sheets a day, generating a daily output from the factory of about 240 sheets. At the height of the *Imperata* harvest season, three trucks a week bring in 1.5 metric tonnes of grass each, most of which is stockpiled on the premises for later conversion into thatch. Each sheet made by the firm is sold and installed for 2,600 rupiah (1997US$1.10). A permanent roofing crew works as a subcontractor on construction sites, installing the thatch. Depending on current contracts, the whole operation employs between 20 and 30 full-time staff. Supplying *Imperata* thatch to hotels is an extremely lucrative part of the business. One hotel at Kuta needed 15,000 sheets, earning a gross income of about 3.9 million rupiah (nearly 1997US$17,000). The firm also helped supply the Ritz Carlton Bali's requirement of 40,000 sheets. It has recently opened a new avenue of business, exporting *Imperata* thatch on an occasional basis to Batam Island, near Singapore, and to Australia.

Imperata thatching companies buy most of their grass from south central Bali (see color plates 6 and 7). Occasionally, when demand surpasses local supplies, they also import grass from neighboring Lombok, where the rapid rise in tourism has spawned a similar industry (Potter et al. 2000, *1045–7*). Villagers in Ubud district set aside small areas of land specifically to grow *Imperata* for thatch. In Lotundo subvillage (Desa Adat Mawang), farmers allow the grass to grow on plots of less than half a hectare, in areas that are too dry for wet rice. They burn the plots every year to assist fertility and then weed out invading shrubs and herbs to create a pure stand of *Imperata*. Four months after the wet season, the grass is cut and dried, combed to remove imperfect blades, and sold to prearranged buyers. Most fields are privately owned, but landowners usually hire laborers who receive half the return for their help with harvesting. A field of *Imperata* covering 0.4 ha is worth as much as 200,000 rupiah (1997US$90).

Imperata has other characteristics that make it attractive to Balinese smallholders. It needs little labor input, so farmers can devote most of their time to wet rice production or participation in the island's range of off-farm employment and business opportunities. The grass is also a good short-term crop for land that is to be converted to residential or business use. Unlike tree crops, *Imperata* generates an immediate return and is easy to remove with a herbicide when construction is due to begin. There are also tenurial considerations. In Lotundo, for instance, villagers prefer to grow *Imperata* on communal lands because they say that, unlike trees, the grass plants cannot be owned by any individual. When trees are planted on communal land, their owner may eventually be able to make a claim on the land because the trees are perceived as a permanent, individualized crop. Therefore, growing grass protects communal tenure, and when it is cut and sold, half the income is divided among the cutters and the other half goes to the village fund.

Conclusions: The Future of *Imperata* Management and Use

The Balinese situation illustrates that a market for *Imperata cylindrica* as roof thatch can continue even in the face of economic growth and rising incomes. *Imperata* remains an important, managed component of the farming system, making a significant contribution to household earnings. This observation should be considered by those who would intervene to improve farming systems by "rehabilitating" privately owned or open-access lands dominated by *Imperata.* It is often acknowledged that *Imperata* is used or sold in traditional societies, but just as frequently this is dismissed as smallholders making virtue of necessity. It is believed that management and sale of *Imperata* is, at best, a short-term activity linked to poverty and marginality, and at worst, a distraction that diverts farmers from switching to more valuable land uses. The case of Bali illustrates that these interpretations may be erroneous. *Imperata* can persist as a highly valued "crop" not only because of continuing demand and remunerative prices, but also because its distinct characteristics allow farmers to maximize returns from available labor on land types less suited to food or tree crops.

Of course the commercial use of *Imperata* in Bali may be impossible to replicate in other locations. It may require a well-developed tourist industry to drive demand for thatch. However, to say that tourism is the only reason for *Imperata's* rise in Bali does not do justice to the inherent qualities of the grass. It has moved from traditional to commercial use because it is a very good roofing material. It is highly aesthetic—both inside and out—and provides excellent insulation. These are the properties that are attracting both commercial and residential use in Bali. Because Balinese design and architecture are often copied in other parts of Indonesia, it would not be surprising if the very sensible use of *Imperata* as insulation for housing was adopted in other areas as well.

Those aiming at the demise of *Imperata* grasslands should first consider carefully the current uses of those lands and their future potential. They should assess who is using the land, what economic and other advantages are being obtained, how those benefits compare to alternative land uses, and whether removing the grass will really assist smallholders in the long term. It should not be taken for granted that, merely because they dislike *Imperata* in their crop fields, farmers would welcome its total removal from their farming systems. For generations, this plant has occupied a niche either as a permanent cover in specific locations or as a temporary fallow before being replaced by a woody succession. In many cases, farmers have nurtured the grass so that it flourishes in these places. In return, it has made a significant contribution to their livelihoods. This relationship should be respected.

References

Brookfield, H.C., L.M. Potter, and Y. Byron. 1995. *In Place of the Forest: Environmental and Socio-economic Transformation in Borneo and the Eastern Malay Peninsula.* Tokyo: United Nations University Press.

Burbridge, P., J. Dixon, and B. Soewardi. 1981. Forestry and Agriculture: Options for Resource Allocation in Choosing Lands for Transmigration Development. *Applied Geography* 1 (1), 237-258.

Dove, M.R. 1981. Symbiotic Relationships between Human Populations and *Imperata cylindrica*: The Question of Ecosystemic Succession and Preservation in South Kalimantan. In: *Conservation Inputs from Life Sciences,* edited by M. Nordin et al. Bangi, Malaysia: Universiti Kebangsaan, 187–200.

———. 1984. Man, Land and Game in Sumbawa: Some Observations on Agrarian Ecology and Development Policy in Eastern Indonesia. *Singapore Journal of Tropical Geography* 5(2), 112–124.

———. 1986. The Practical Reason of Weeds in Indonesia: Peasant vs State Views of *Imperata* and *Chromolaena. Human Ecology* 14(2), 163–190.

Fraser, N. 1989. *Unruly Practices: Power, Discourse and Gender in Contemporary Social Theory.* Minneapolis, MN: University of Minnesota Press.

Heyne, K. 1950. *De Nuttige Planten van Indonesie,* 3rd edition. Gravenhage, Bandung: N.V. Uitgeverij W. van Hoeve (2 vols.)

Potter, L.M. 1997. The Dynamics of *Imperata*: An Historical Overview and Current Farmer Perspectives, with Special Reference to South Kalimantan, Indonesia. *Agroforestry Systems* 36 (1–3), special issue entitled *Agroforestry Innovations for* Imperata *Grassland Rehabilitation,* edited by D.P. Garrity, 31–51.

———. 2003. Forests versus Agriculture: Colonial Forest Services, Environmental Ideas and the Regulation of Land-Use Change in Southeast Asia. In: *The Political Ecology of Forests in Southeast Asia: Historical Perspectives,* edited by Lye Tuck-Po, Wil de Jong, and Ken-ichi Abe. Kyoto and Melbourne: Kyoto University Press and Trans-Pacific Press, 29–71.

———, J. L. Lee, and K. Thorburn. 2000. Reinventing *Imperata*: Revaluing Alang-Alang Grasslands in Indonesia. *Development and Change* 31, 1037–1053.

Rifai, M.A., and E.A. Widjaja. 1977. An Ethnobotanical Observation on Alang-alang (*Imperata cylindrica* [L.] Beauv.) in Bali, presented at the Sixth Conference of the Asian Pacific Weed Science Society.

Seavoy, R.E. 1975. The Origin of Tropical Grasslands in Kalimantan, Indonesia. *Journal of Tropical Geography* 40, 48–52.

Sherman, G. 1980. What "Green Desert"? The Ecology of Batak Grassland Farming. *Indonesia* 29, 113–149.

Soepardi, G. 1980. Alang-alang and Soil Fertility. In: *Proceedings of Biotrop Workshop on Alang-Alang,* Biotrop special publication No. 5, Bogor, Indonesia: SEAMEO-Biotrop, 57–69.

Soerjani, M. 1980. Symposium on the prevention and rehabilitation of critical land in an area development. In: *Proceedings of the Biotrop Workshop on Alang-alang,* Biotrop special publication No. 5. Bogor, Indonesia: SEAMEO-Biotrop, 9-14.

Soerjatna, E.S., and J.L. McIntosh. 1980. Food Crops Production and Control of *Imperata cylindrica* on Small Farms. In: *Proceedings of the Biotrop Workshop on Alang-alang,* Biotrop special publication No. 5, Bogor, Indonesia: SEAMEO-Biotrop, 135–147.

Soewardi, B. 1976. *Potensi pengembangan wilayah sapi potong di kawasan padang alang-alang di propinsi Kalimantan Selatan dan Kalimantan Timur.* Jakarta: Direktorat Jenderal Peternakan, Dep. Pertanian.

——— and D. Sastradipradja. 1980. Alang-alang and Animal Husbandry, In: *Proceedings of the Biotrop Workshop on Alang-alang,* Biotrop special publication No. 5. Bogor, Indonesia: SEAMEO-Biotrop, 157–178.

Chapter 12

Natural Forest Regeneration from an *Imperata* Fallow

The Case of Pakhasukjai

Janet L. Durno, Tuenjai Deetes, and Juthamas Rajchaprasit

The study at Pakhasukjai was inspired by a conversation with an Akha elder. He said that when the village was founded in 1976, very little forest remained at the site and the main vegetation in the fallow fields was *Imperata cylindrica*. He explained how the Pakhasukjai community helped the forest regenerate by protecting an area around the village from fire. As a result of the discussion, it was decided to carry out a participatory research program to document the process of natural forest regeneration as well as study the management and use of the community forest that grew around the village.

Traditionally, the Akha people have practiced shifting cultivation in the rich forests of China and Southeast Asia. Only recently have land shortages and forest depletion become significant factors in their lives. Within that brief time, their land management systems have been forced to change in response to resource scarcity, and communities have learned to cope not only with swift integration into the cash economy, but also with increasingly complex political and economic interests in their upland environment. Akhas must also live with an uncertain legal status as dwellers in protected watershed areas.

The Pakhasukjai case has several important implications. Firstly, the forest is so vital to the cultures and livelihood of the "hilltribe" people that, if they are forced to settle in a deforested area, they may attempt to create a forest. Secondly, the study demonstrates the ability of hilltribe farmers to adapt their farming systems to changing circumstances. Finally, the study attests to the capacity of the tropical monsoon forest to regenerate, even after it has been cleared and burned. This regenerative capacity is crucial to rotational and shifting cultivators, whose farming cycle relies on the forest to provide fertile soil. When a field is fallowed after one or several years of cultivation, natural forest usually regenerates swiftly because of seeding from the surrounding forest and the coppicing of surviving rootstock in the field. However, if a fallowed field is subjected to frequent fires, *Imperata cylindrica* will likely become the dominant vegetation and forest regeneration will be slowed or even prevented. In these circumstances natural forest regeneration is possible only if viable seeds and rootstock are available, and the area is protected from fire.

Janet L. Durno, Program Manager, Canadian International Development Agency (CIDA). Tuenjai Deetes and Juthamas Rajchaprasit, Hill Area Development Foundation (HADF), 129/1 Mo 4 Pa-Ngiw Road, Soi 4, Tambol Robwiang, Amphur Muang Chiang Rai 57000, Thailand.

Research Methods and the Study Area

The research took place over an 18-month period in 1993 and 1994, in cooperation with the Hill Area Development Foundation (HADF), a Thai nongovernmental organization involved in implementing development programs in the area. Interviews were conducted with individuals and small groups. Men and women were consulted separately to facilitate women's participation, and participatory rural appraisal methodologies were used.

In March 1994, a forest survey was conducted by scientists and students from the biology department of Chiang Mai University, HADF staff, and male and female Akha elders. The aim was to get an overall picture of the ecology and condition of the forest and older fallow fields, as well as to record scientific and Akha names for the different tree species, and the uses of each tree. Five sites were selected for the survey. Two were fields that had been fallowed for forest regeneration. In one field, which had been fallowed five years before the study, supplementary tree planting had taken place. The other field had been left fallow for nine years and no trees had been planted. The remaining three sites were secondary forest areas where the regeneration process had begun 18 years earlier from a fallow of *Imperata cylindrica* with small scattered patches of bamboo and forest.

In the two forest fallows, the underbrush was cleared from an area of about 400 m^2, and all trees and woody climbers were measured for girth at a breast height (gbh) of 1.3 m. The measuring of all trees provided a more detailed comparison of regeneration between the two fallows. However, only trees and woody climbers with a gbh of 10 cm or more were counted when comparing the fallows with the older forest plots. In each of the three secondary forest areas, 20 circular plots, 10 m in diameter, were established in a regular sampling pattern, 30 to 50 m apart, depending on the size of the forested area and the terrain. In each plot all trees and woody climbers with a gbh of 10 cm or more were measured. In all five sites, trees were labeled and identified. Flowering, fruiting, and coppicing were noted. A soil sample was taken from each site for analysis of pH; organic matter content; nitrogen, phosphorus, and potassium availability; texture; and percentage of moisture at field capacity.

Because of time constraints, the survey identified only trees and woody climbers. However, over the course of several visits, a taxonomist from Chiang Mai University identified 260 species of grasses, ferns, herbs, and trees, although this was still only a partial list of species present in the forest.

The study took place in the Akha village of Pakhasukjai, which, at the time, had a population of 420. It is located 15 km from Thailand's border with Myanmar (Burma), at an elevation of 1,120 m above sea level (asl), just below the summit of Pakha Mountain, which rises to 1,164 m asl (Figure 12-1). The mountain is the watershed of two rivers, the Mae Chan and the Mae Chan Noi, both of which flow into the Mekong. A primary evergreen forest once flourished in the area. The 1,000 m elevation of Pakhasukjai marks the transitional zone between evergreen forest, at higher levels, and mixed evergreen and deciduous forest, which is found at 800 to 1,100 m asl and slightly higher in disturbed areas. The primary forest would, therefore, have contained a rich diversity of plants and animals. Common tree species would have included *Castanopsis*, *Pinus*, and *Lithocarpus*.

The village's land ranges from an elevation of 630 m, beside the Mae Chan River, to the summit of Pakha Mountain, with slopes from 20% to 80%. Using FAO classifications, the soils are mainly clay-loam or clayey Regosols and Cambisols on shale, schist, or granite parent materials. In the forest, pockets of stony Leptosols can also be found (Turkelboom 1994).

The average annual rainfall is 1,650 mm, most of it falling from May to October. In the cold season, from November to January, the average minimum temperature is less than 20°C, while in the hot season, temperatures can be well over 30°C.

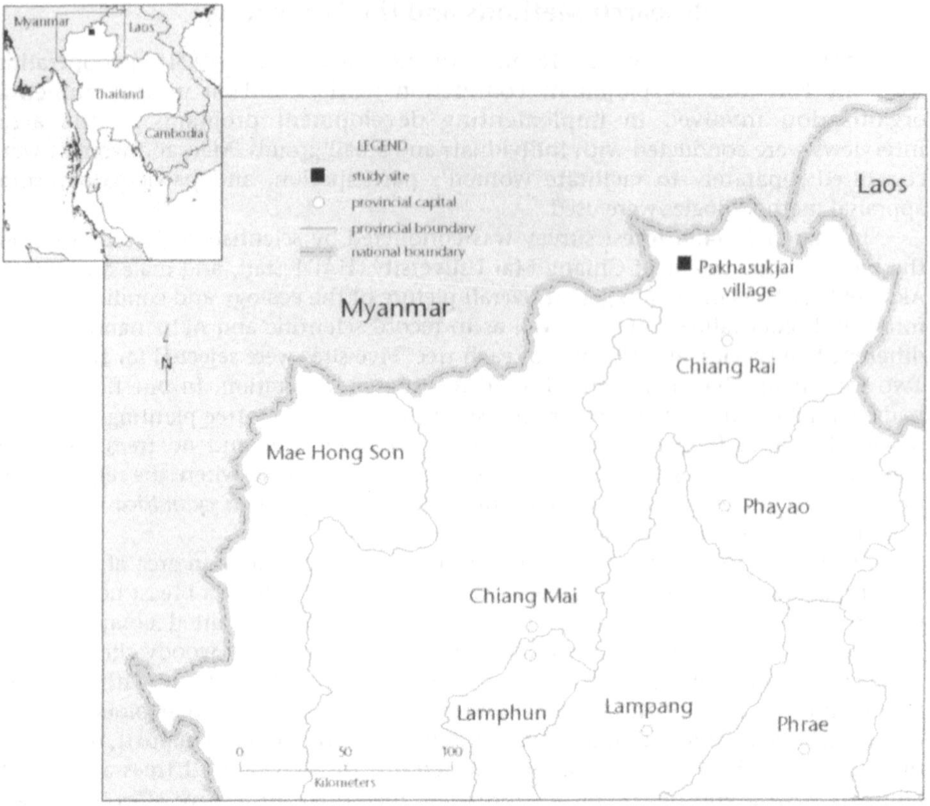

Figure 12-1. Map of the Study Area
Source: Ongprasert et al. 1996.

Since the early 1900s, the Mae Chan watershed has been among the first points of entry for hilltribe peoples migrating southward from Myanmar, and much of the forest has been cleared for agricultural purposes. The Akha at Pakhasukjai settled on land that had earlier been farmed and fallowed by Lahu and Lisu settlers.

Conventional Shifting Cultivation Systems

The Akha, along with the Hmong, Mien, Lisu and Lahu, are classified by researchers as "pioneer" or "shifting" cultivators, as opposed to "rotational" cultivators such as the Karen, whose short cultivation, long fallow cycle allows for permanent settlement of an area (Sutthi 1989). Pakhasukjai villagers say that, in their traditional system, a family would farm a field for two to four years, depending on the soil type and slope, and then leave it fallow for the forest to regenerate. Often, they would return to farm the same field after three to five years, when some soil fertility had been regained. However, after 10 to 14 years, unless the villagers had wet rice fields or were growing opium, which required smaller fields than subsistence crops, some of the households in the village would move on, leaving the fallows behind and opening new fields in the forest.

Like other shifting cultivators, the Akha use the process of natural succession to re-establish the forest on which the shifting cultivation cycle depends. Many of their traditional practices facilitate a rapid regrowth of the forest. For example, tree roots and stumps are left in the ground when the forest is cleared for farming. The fields are not ploughed and are generally cultivated for only a short period, so soil disturbance is minimal and the roots of many tree species survive the farming cycle and begin to coppice. The nearby forest is also a source of seed once the field is fallowed. So are wild fruit trees such as *Castanopsis* spp., *Mangifera caloneura* and *Spondias pinnata,* which are often retained when fields are cleared, protected by a small firebreak. When the primary forest is cleared, conditions also become suitable for a succession of fast-growing pioneer tree species whose seeds require sunlight for germination. Thus, while the species mix is different after the land has been farmed, the forest cover is, nevertheless, restored within a few years unless the area is subjected to frequent burning.

Contextual "Triggers" that Contributed to Farmer Innovations

A broad belt of forest around their village is integral to Akha life and spirituality. It provides timber, fuel, food, water, and ceremonial, medicinal and decorative plants. It acts as a windbreak and moderates the climate of the village. As it lies between the village and the fields, it also acts as a barrier to keep domestic animals out of the fields, as well as providing much of their fodder.

In addition to its environmental importance, the forest is crucial in providing the sites and plants essential to Akha religion. The forest in and around sacred sites, including the shrine to the lords of land and water, the cemetery, and the "pure water source," is traditionally protected under Akha custom. It is forbidden, for example, to cut trees or gather forest products in the cemetery forest. In addition, other small areas of forest, such as the location of a tree struck by lightning, or where someone has been murdered or killed by a wild animal, have a spiritual significance, are avoided by villagers, and are thus protected.

The forest is also the source of the many plants that are essential ingredients of Akha rituals and ceremonies. They are among the most important plants in the Akha world. When male elders were asked which five trees they would take with them to a new land, assuming that bamboo and fruit trees would be present there, the five trees they quickly agreed upon were all essential for important ceremonies. In a separate meeting, women elders listed the same trees when asked which trees were important for ceremonies. The trees are as follows:

- *Callicarpa arborea*: The leaves are used in ceremonies before and after the rice harvest and in an annual ceremony to construct a new village gate.
- *Castanopsis diversifolia*: The leaves are used in new house raising and village gate ceremonies, and in a ceremony when a new baby dies.
- *Eurya acumminata*: The leaves, used in many ceremonies, are placed on the ancestor altar and offered to spirits and the dead.
- *Rhus chinensis*: The leaves are used in the rice ceremony after the harvest, and the wood is used to make the toy weapons used in an annual ceremony to chase spirits out of the village.
- *Schima wallichii*: The leaves are used in ceremonies for new house raising and the sick, and are carried to avert bad luck by a person who has seen mating snakes, crabs, or snails. The wood is used for the pole to which a sacrificial buffalo is tied at the funeral of an elder.

During the forest survey at Pakhasukjai in March 1994, Akha elders identified all the trees in the five research plots and described their uses. In total, 91 indigenous forest trees were listed. Almost all of them had uses, and most had multiple uses (see Table 12-1). It is important to note that the trees were chosen randomly, and were not trees that the Akha themselves chose to describe for the researchers. The data thus provide a clear indication of the depth of Akha knowledge about the tree species in their community forest, as well as the importance to Akha life of virtually every species.

When the new village was founded, the Akha community at Pakhasukjai confronted deforestation for the first time. It was an unprecedented situation calling for an innovative solution. Soon after their arrival, they decided that it was essential to set aside an area where the forest could regenerate. Villagers recall that the need for timber to build houses, as well as their culture and traditions, were motivating factors behind the decision. Although the population was growing and farmland was limited, the elders of Pakhasukjai designated land on the ridges around the village for a community forest. It was just as necessary to them as the agricultural fields.

It is important to note that both forest regeneration and the transition from shifting cultivation to a settled agricultural community began in Pakhasukjai 10 years before government and nongovernment agencies extended their development programs to the area.

Resulting Fallow Management Practices

Based on their knowledge of forest ecology, the elders of Pakhasukjai suggested several simple rules to facilitate the regeneration of forest around their new village. Farming was prohibited in the community forest area, as was the felling of any remaining trees, without prior permission. Only dead wood could be used for fuel, and firewood could not be sold. It was agreed that any villager who cut a tree without permission would not be allowed to keep the tree. In addition, he would be required to provide whisky for the elders to make amends, or be fined a pig if he refused to apologize. However, restrictions on tree cutting were only a beginning. The primary factor in transforming what was an *Imperata* fallow into forest was the need to protect the area from fire (see also Maneeratana and Hoare, Chapter 13).

While fire is used by hilltribe farmers to prepare land for planting and by hunters to flush out game, fires that escape from human control endanger both community forests and the villages sheltered within them. To prevent fire from threatening the village and the regenerating forest, Pakhasukjai villagers cleared a firebreak around the perimeter every year before the dry season. To do the job, they formed community work groups with one laborer from each family. It required five or six days' work each year. The same community groups had the task of fighting fires that threatened to burn toward the village. Any family that failed to contribute labor was fined, with higher fines for refusing to fight fires at night.

Table 12-1. Uses of Trees and the Number of Species Used by the Akha

Uses	*No. of Species Used**
Fuel	78
Timber	64
Edible fruit, nuts, seeds	24
Making tools and utensils	19
Leaves or wood used in ceremonies	18
Medicine for people	11
Edible leaves, shoots, pods	10
Flowers used for decoration	8
Wood used for coffins	8
Bark eaten or chewed with betel	8
Bark used to make rope	5
Medicine for animals	3
Leaves used for wrapping or plates	3
Bark used to coat rice steamer	3
Making toys and games	2
Bark used to make dye	2
Leaves used to make soap	2
Mattress and pillow stuffing	1
Pitch used to light fires	1
Edible flowers	1
Bark used to make fish poison	1

Note: * Out of 91 identified.

When they were no longer subjected to fire, tree stumps and roots began to coppice and seedlings began to grow in the *Imperata* fallow around Pakhasukjai. Grass cutting may also have been another factor contributing to forest regeneration, as large quantities of *Imperata* were used for roof thatching in the new village. Within five years the trees had grown above the grass, which was beginning to die back under their shade.

In more recent years, although community firefighting groups continue to be formed when needed, it has no longer been necessary to construct the annual firebreak. Forest fires have decreased significantly because of the reduced incidence of field burning and improved burning practices. Fields that are planted with contour hedgerows or terraced for wet rice cultivation are no longer burned, and some farmers prefer to use crop residues as mulch rather than burning them. As well, both the HADF and the Thailand government have provided educational programs and have encouraged intervillage agreements on fire prevention and control measures. Farmers who still wish to burn their fields do so earlier in the season, before the vegetation is completely dry, and they avoid burning on windy days. Grass and straw are often cut and piled before burning. An additional factor behind the decrease in forest fires is the disappearance of large game from the area. This has spelled an end to the practice of setting fires to drive wild animals toward the hunters, or to stimulate the growth of new grass to attract deer.

The area of Pakhasukjai's community forest has expanded over time, with several families being asked by the elders to permanently fallow fields located close to the village in order to reduce the risk of fire. More recently, the villagers have become involved in an HADF watershed management program and have developed a land-use plan in which the steepest fields are to be taken out of production and fallowed for forest regeneration. Although it is hard for families to give up fields when food production is already insufficient, the villagers are aware of the importance of the forest in ensuring the environmental sustainability of their land use. In addition, as dwellers in a protected watershed area, they know that their chance of being granted land rights and citizenship depends upon their ability to demonstrate sustainable management and conservation of the watershed forest.

Reforestation programs have also been encouraged by HADF and the Royal Forestry Department. The Pakhasukjai villagers have planted trees in about 9.6 ha of forest fallows. This has involved fast growing multipurpose trees, fruit trees, and ornamentals, depending mainly on the availability of seedlings from government departments. While a few of the tree species were indigenous and adapted to the elevation and climate, most were not. There has been little monitoring of the survival rate of the trees, but it was very low in the forest fallow surveyed in 1994. This field, fallowed in 1989 when its owner returned to Myanmar, was reforested in the same year with seedlings of *Acacia auriculiformis, Acacia mangium, Artocarpus heterophyllus, Cassia siamea, Cassia spectabilis, Diospyros* sp., *Eucalyptus camaldulensis, Eugenia* sp., *Leucaena leucocephala, Mangifera indica, Pinus kesiya, Plumeria acutifolia, Prunus cerasoides,* and *Tamarindus indica*. When an area of 400 m^2 was surveyed five years later, only six trees had survived out of an estimated 100 planted. However, in the same area, 136 indigenous trees had regenerated through coppicing or had grown from naturally dispersed seeds.

The main cause of tree mortality is likely to have been simply that the species were inappropriate for the environment. Other reasons may have included a lack of water in the dry season, browsing and crushing by large animals, and accidental cutting during annual grass clearing carried out to assist seedling establishment. Microenvironmental conditions may also have been crucial to the survival of some species. For instance, *Pinus kesiya,* a species previously indigenous to the area, suffered a high mortality rate in one forest fallow but did well in another. And, although *Tectona grandis* does not occur naturally at the high elevation of Pakhasukjai, a number of teak trees planted by the villagers are growing reasonably well.

Results

Benefits Gained from the Intensified Management of Fallow Land

According to statistics collected by HADF in 1993, the village land area totaled 849.2 ha. Of this total, 578.7 ha, or 68.1% of the village land, was under forest cover, and most of this had regenerated since the Akha arrived in 1976. An additional 103.2 ha, or 12.2% of the total land area, had also been fallowed to increase the area of the community forest. As a result of their land-use plan, the area of farmland cultivated by the villagers had decreased from 285.7 ha, or 33.6% of the total area, in 1987, to 167.3 ha, or 19.7%, in 1993.

Statistics collected in the same year by a researcher from the Soil Fertility Conservation Project showed a significant variation from the HADF data in terms of the relative size of fallows to forest (Figure 12-2). The reason for this difference was probably a matter of definition as, for both sets of statistics, the combined area of fallow and forest, expressed as a percentage of total land area, was almost identical: 80.3% (HADF) and 79.8% (SFC).

The community forest comprises several narrow secondary forest belts located on the upper slopes of the mountain, along ridges, and in small valleys radiating out from the summit. As mentioned earlier, a survey in 1994 covered three sites in different areas of the forest, as well as two fields fallowed for forest regeneration (Figure 12-3). Table 12-2 summarizes the characteristics of each site surveyed.

In 0.55 ha of forest and forest fallow, 910 trees and woody climbers with a girth at breast height of 10 cm or more were recorded, comprising 103 tree species from at least 67 genera and 39 families. Nineteen trees were only tentatively identified due to height or inadequate samples, but it is likely that they represent 14 different species and they have been calculated as such in the data. These results can be compared with a study at Doi Suthep-Pui National Park in the neighboring province of Chiang Mai. This study surveyed tree species in 0.828 ha at an elevation between 670 and 960 m asl, slightly lower than Pakhasukjai, and found 117 tree species from 84 genera and 48 families, making it "the most species-rich dry tropical forest currently known" (Elliott et al. 1989).

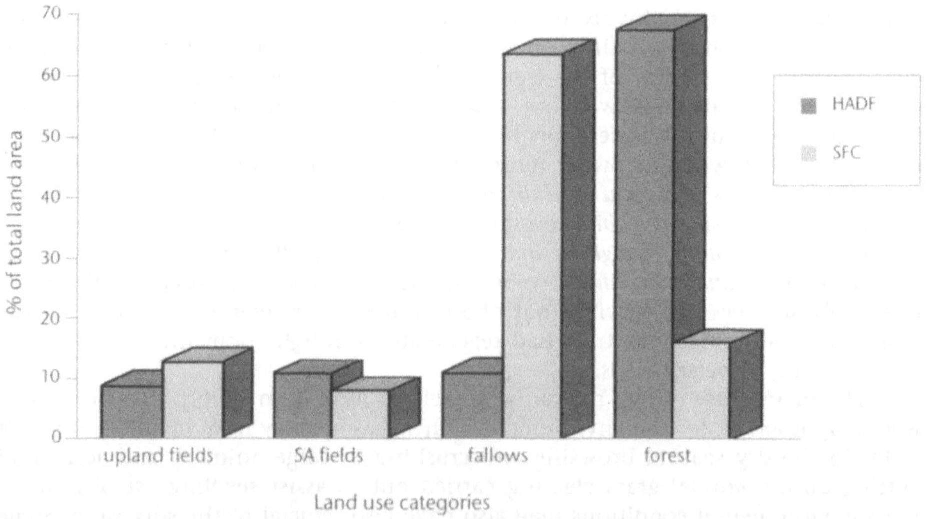

Figure 12-2. Comparison of Pakhasukjai Land-Use Statistics (1993)

Notes: Upland Fields = unmodified fields; SA Fields = "sustainable agriculture" fields planted with contour hedgerows or terraced for wet rice; Fallows = land fallowed for forest regeneration (HADF), future use of the fallow unspecified (SFC); Forest = community forest area.

In the Pakhasukjai forest, the mean number of individuals per tree species was 8.8. Most species were rare, with 55.3% being represented by only one to four individuals and 22.3% by a single individual; 86.4% of species were represented by fewer than 20 individuals (Figure 12-4). The most common tree species were *Castanopsis calathiformis* (80 individuals), *Schima wallichii* (64), *Aporusa wallichii* (51), *Glochidion sphaerogynum* (45), *Wendlandia paniculata* (37), and *Castanopsis diversifolia* (36). Tree species–area curves (Figure 12-5) compare the three secondary forest sites at Pakhasukjai with results of studies on Doi Suthep and Doi Khun Tan, near Chiang Mai. The Doi Suthep curve shows results for primary evergreen forest at 1,100 m asl, while Doi Khun Tan shows evergreen forest at 1,210 m asl that has been disturbed by occasional fires and some cutting, although the damage is considered light (Elliott and Maxwell 1994). The lowest tree species–area curve is found on Doi Suthep, while the highest is found at the Cemetery Forest Hill site at Pakhasukjai. The higher curves on the more disturbed sites can be explained by the high incidence of secondary forest species combined with the many species of primary forest trees that survived the farming cycle. Evidence of coppicing was discovered by the survey in 41 species, although it is likely that more species possess this ability.

Of the tree species identified during the survey, 54.4% were deciduous, 37.8% were evergreen, and the remaining 7.8% were tropophyllous, or an intermediate between deciduous and evergreen. Primary forest species accounted for 51.1% of the trees, and 48.9% were species characteristic of a secondary forest. The mixture of evergreen and deciduous trees can be attributed to the elevation, which places it in a transitional zone between mixed deciduous-evergreen and evergreen forest, as well as to primary forest disturbance, which has created a habitat suitable for a number of deciduous dipterocarp-oak forest species. While this fire-prone forest type occurs from the lowlands up to an elevation of about 850 m asl, some of its species, such as *Anneslea fragrans*, *Aporusa wallichii*, and *Shorea roxburghii*, are found growing at higher elevations following fire or forest degradation (Maxwell 1988).

In the Pakhasukjai forest, the mean girth at breast height for all trees of 10 cm gbh or more was 24.13 cm. This compares with 51.1 cm in the monsoon forest at Doi Suthep (Elliott and Trisonthi 1992). Figure 12-6 shows the frequency histogram of average tree gbh for all five sites. All sites contained large numbers of small trees. Of all the trees surveyed, 56.4% had girths within the range of 10.0 to 19.9 cm. Only 1% had a gbh of more than 100 cm. As expected in a young forest, tree density was high, with an estimated 1,652 trees/ha, a figure which compares with 713 trees/ha in the pristine evergreen forest on Doi Suthep and 795 trees/ha in slightly disturbed evergreen forest at 1,240 m on Doi Khun Tan (Elliott and Maxwell 1994). The volume of wood in the forest, calculated in terms of basal area (m^2 of tree stems per m^2 of ground), is low, averaging .0000947 for all five sites and .0000822 for the three 18-year-old forest sites, compared to .003466 for Doi Khun Tan and .0074155 for Doi Suthep.

Soil samples were taken from the five sites, as well as from a nearby cultivated field (Table 12-3). A block of soil up to 30 cm in depth, with all levels equally represented, was taken from one place in the middle of each site. The resulting data cannot be considered to provide more than a suggestion of soil conditions in the survey sites. The final column in Table 12-3 allows for comparisons with soil conditions in the pristine evergreen forest of Doi Suthep.

During the forest survey it was found that all Akha villagers, from children to elders and both women and men, had an impressive knowledge of the forest and of the many uses of virtually every plant and tree to be found in it. Although easier access to markets and health services has reduced their reliance on certain forest products, the forest is still an essential resource for many of their basic needs, especially those of poorer families.

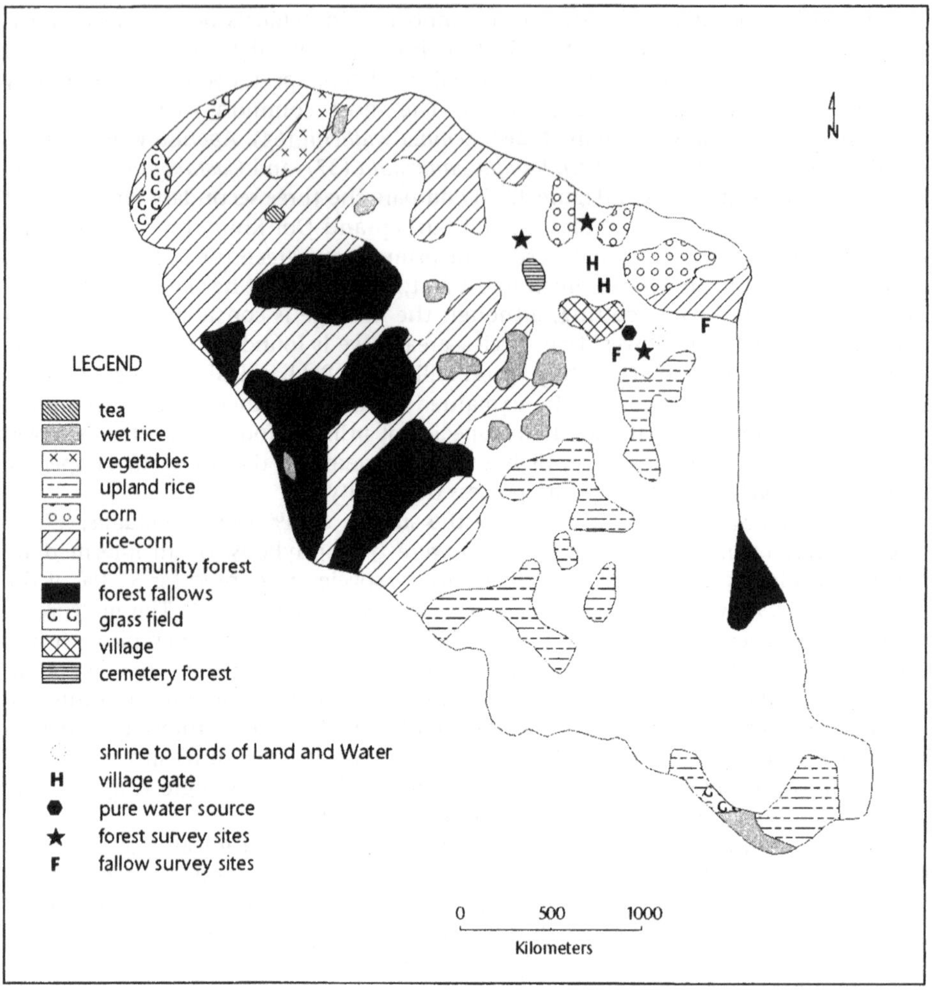

Figure 12-3. Land Use Map of Pakhasukjai Village
Source: Geography Department, Chiang Mai University.

Ecological and Social Sustainability

Natural forest regeneration has considerable potential for rehabilitating degraded upland environments. However, in areas of high population density, there are inevitable tensions between the community's need for a forest and the need to make a living from the land.

At Pakhasukjai, the community forest is heavily used with the exception of protected religious sites, by both villagers and domestic animals. While the villagers generally express satisfaction with the forest regeneration, there is concern that forest resources – particularly firewood – will soon fail to meet demand from the growing population. The forest already shows evidence of degradation, such as trampling by cattle and pigs and tree cutting for fuel and making charcoal, although it is not clear whether the tree cutting was done by Akha villagers or by people from the neighboring village. Most families have planted some fast-growing trees, but not in sufficient numbers to supply future demands for firewood, so villagers have begun to use other materials such as corncobs and bamboo for fuel. In the future, electricity and gas may become viable alternatives to cooking over fire but the cost is presently beyond their reach.

The scarcity of animals and birds, which are important agents of seed dispersal, also threatens the long-term survival of some of the more valuable tree species in the community forest. Currently, only species with wind-dispersed seeds are assured of effective dispersal. *Castanopsis* and *Lithocarpus spp.*, for example, are still relatively abundant in the Pakhasukjai forest, but these trees, which have large seeds, could eventually die out unless the villagers themselves begin to propagate and plant them.

Pakhasukjai's land-use plan requires 60% of families to give up land for forest fallows. Some families agree with the plan, others comply only when pressured to do so by the village committee. Some of the clans reallocate remaining farmland among their member families; others do not. At present, most villagers respect the decision to permanently fallow certain fields. However, as the population grows and land pressure increases, it may become harder for families with insufficient agricultural production to resist the temptation to clear fields from areas of the community forest or from forest fallows. On the other hand, farmers are attempting to intensify agricultural production by adopting soil conservation measures, terracing slopes for wet rice cultivation, planting fruit trees, and growing cash crops (see color plate 24). In an effort to increase their livelihood security, the villagers are marketing vegetables and handicrafts and also working as day laborers for local farmers. A growing number of young people are also migrating in search of employment. These measures, coupled with government requirements for effective watershed management, show some promise of ensuring the preservation of the community forest for future generations.

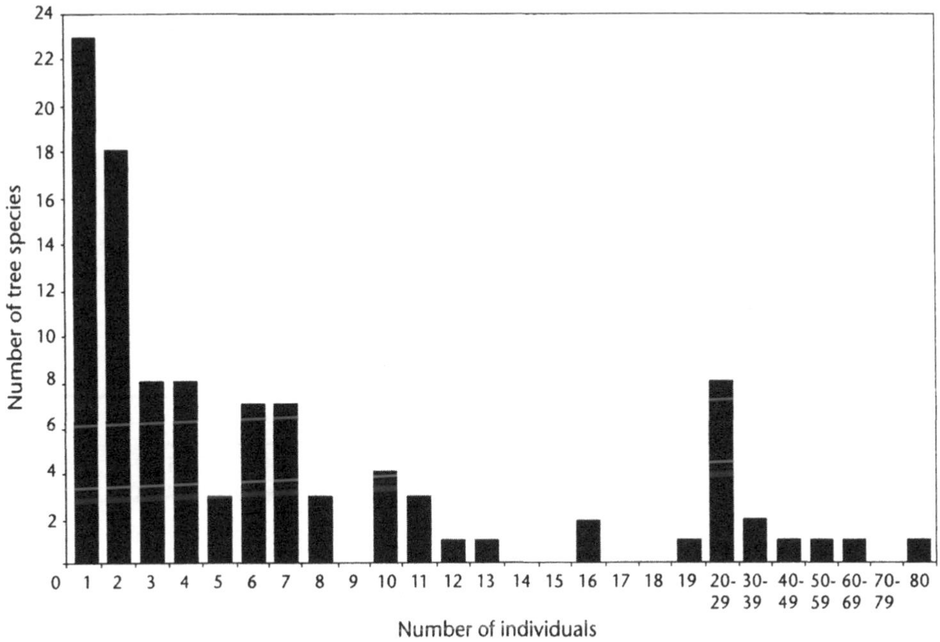

Figure 12-4. Number of Tree Species in the Five Sites by Number of Individuals

Table 12-2. Characteristics of the Survey Sites

Characteristics	*5YF*	*9YF*	*PH*	*CFH*	*FSH*	*Total*
Area (m^2)	400	400	1,570	1,570	1,570	5,510
No. of trees (gbh 10 cm or greater)	38	96	336	269	171	910
Average no. of trees per 100 m^2	9.5	24	21.4	17.1	10.9	16.5
Expected no. of trees/ha	950	2,400	2,140	1,713	1,089	1,652
Average gbh (cm)	17.62	21.63	23.02	23.99	34.40	24.10
Basal area[a]	.0000692	.0002672	.0001026	.0000731	.0000710	.0000947
Total no. of tree species	17	29	56	57	52	103
Average no. of trees per species	2.2	3.3	6.0	4.7	3.3	8.8
Average no. of species per 100 m^2	4.3	7.3	3.6	3.6	3.3	1.9
Species diversity index[b]						
N1:	14.55	18.0	24.62	34.16	33.45	24.96
N2:	18.03	12.56	14.49	26.70	27.68	19.89
Evenness index[c]	1.26	0.68	0.57	0.77	0.82	0.82
Evergreen trees (% of total species)	5.90	13.8	47.2	42.0	37.8	37.8
Tropophyllous trees (% of total)	17.60	20.7	5.6	8.0	11.1	7.8
Deciduous trees (% of total)	76.50	65.5	47.2	50.0	51.1	54.4

Notes: 5YF = 5-year fallow; 9YF = 9-year fallow; PH = Pakha Hill; CFH = Cemetery Forest Hill; FSH = Forest Shrine Hill; [a]. The basal area gives an index of the total volume of wood, which takes into account both tree density and size; [b]. The species diversity index combines in a single value both the total number of species in a community (species richness) and how the species abundances are distributed among the species (evenness). N1 measures the number of abundant species in the sample, while N2 is the number of very abundant species. Both tend toward 1 as one species begins to dominate; [c]. The evenness index approaches 0 as a single species becomes more dominant (Ludwig and Reynolds 1988).

Table 12-3. Forest and Forest Fallow Soils

Parameters	*CF*	*5YF*	*9YF*	*PH*	*CFH*	*FSH*	*DSEF*
pH (in water)	4.55	4.85	4.65	4.65	4.35	4.90	6.22
Organic matter (%)	6.06	6.58	9.53	6.56	6.87	7.64	7.30
Moisture at field capacity (%)	37.67	36.08	46.23	55.85	45.03	47.10	35.35
Macronutrients							
Nitrogen (%)	.282	.293	.341	.319	.336	.334	.370
Phosphorus (ppm. exch)	19.00	41.00	42.50	8.50	8.50	18.00	10.53
Potassium (ppm. exch)	157.50	82.50	87.50	50.00	45.00	195.00	295.67
Texture							
Sand (%)	39.96	50.96	48.36	39.96	39.96	49.96	52.00
Silt (%)	26.08	23.28	23.28	23.28	26.28	23.28	22.00
Clay (%)	33.96	25.76	28.36	36.76	33.76	26.76	26.00

Notes: CF = cultivated field; 5YF = 5-year fallow; 9YF = 9-year fallow; PH = Pakha Hill; CFH = Cemetery Forest Hill; FSH = Forest Shrine Hill; DSEF = Doi Suthep Evergreen Forest.

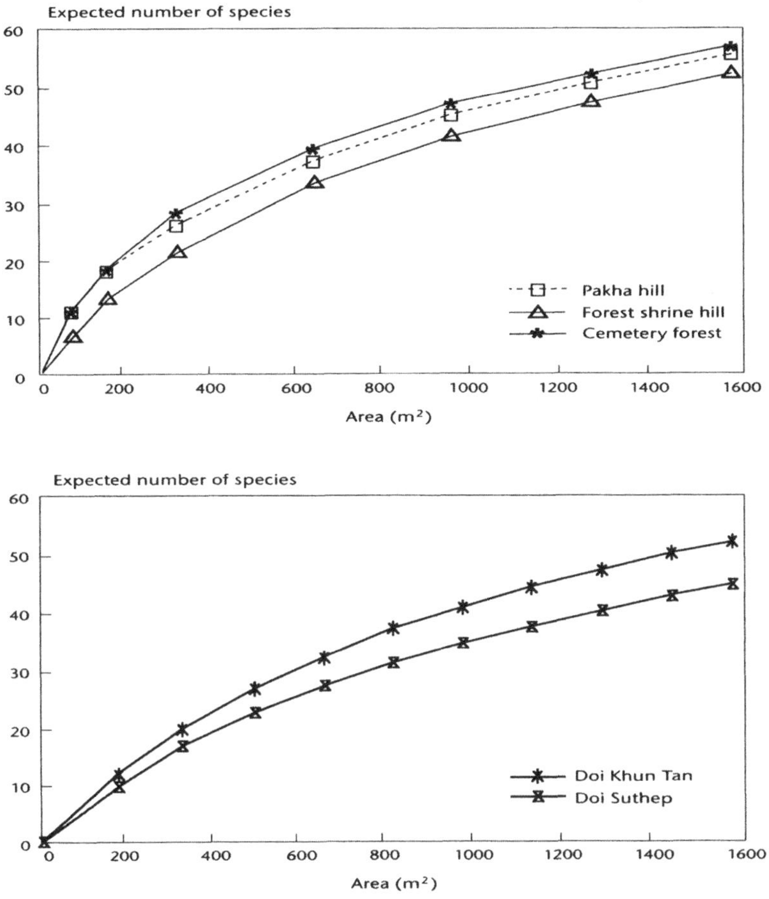

Figure 12-5. Tree Species: Area Curves
Source: Biology Department, Chiang Mai University

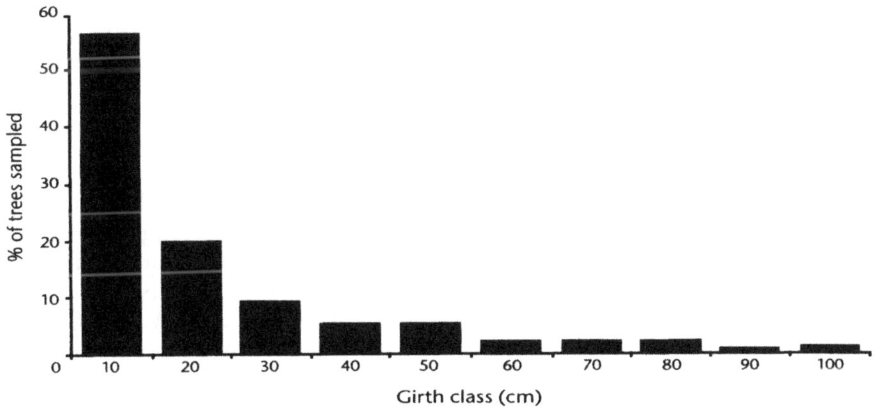

Figure 12-6. Frequency Histogram of Girth at Breast Height, Average of Five Sites

Potential for Application Elsewhere within Southeast Asia's Uplands

The case of Pakhasukjai demonstrates that natural forest regeneration is possible from an *Imperata*-dominated fallow, given protection from fire, as well as surviving root stock and seeds. The villagers of Pakhasukjai have found natural forest regeneration to be more effective than reforestation in restoring forest cover, with the additional advantage that labor requirements are less onerous. Reforestation demands a considerable investment of time for tree propagation, land preparation, planting, and maintenance. Then, many of the seedlings do not survive.

To date, research on natural forest regeneration in Thailand has not been extensive, although several programs have been carried out, or are ongoing. One study at Mae Soi, in Chiang Mai Province, found the following:

> Even in severely deforested sites there is often some natural regeneration. For example, at Mae Soi, the density of naturally regenerating seedlings was 1.1 to $4.3/m^2$, compared with only about $0.1/m^2$ for surviving planted seedlings. Most of the naturally regenerating seedlings were coppicing from surviving root stock. Many were of species previously present in the upper watershed forest, such as *Pinus merkusii* and *Quercus vestita*, but several were deciduous forest species which had been dispersed by wind and birds from lower down. (Elliott et al. 1993)

In an experiment in Lampang Province, an area of degraded teak forest was clear cut and seven species of fast-growing trees were planted. Natural forest regeneration was also allowed to occur. After three years, the planted trees had a very low survival rate, while an average of 344 teak trees per ha had coppiced and grown 9 to 10 m. In the sixth year, the coppiced teak had a diameter of 7 to 18 cm, while the largest of the planted trees had reached only 12.3 cm (Sukwong no date).

Despite these findings, natural forest regeneration and tree planting do not have to be an "either-or" proposition. Selected indigenous species can be planted to enrich a regenerating forest, especially tree species that no longer exist in the area or that cannot easily disperse their seeds due to lack of animals and birds. Despite their preference for natural forest regeneration, the Pakhasukjai villagers continue to plant trees, in part because all surviving trees of whatever species will one day be useful for food, timber, or fuel, and in part because they believe that reforestation will increase their credibility in the eyes of the Royal Forest Department, and thus improve their chances of gaining some measure of legal security.

While tree planting is undoubtedly an important strategy in reforestation and can be effective if species are environmentally appropriate and are properly cared for, the potential of the forest to regenerate itself is one that should not be ignored. Hilltribe farmers can and will assist the process of forest regeneration for a variety of reasons, ranging from economic to spiritual and political. Their efforts could be facilitated by supportive government policies and other forms of assistance, such as training in fire prevention and the propagation of indigenous trees.

Research Priorities and Experimental Agenda

The environmental quality of the forest and its value to villagers could be increased through enrichment planting of indigenous species that can no longer reestablish themselves naturally. There is a need to conduct further research on simple methods of indigenous tree propagation that are appropriate for village conditions.

In Pakhasukjai, tree planting is, in itself, a relatively new activity, and the idea of propagating and planting indigenous forest trees is even more novel, with the exception of a few species such as *Protium serratum* and *Spondias pinnata* which are commonly propagated and planted for their edible bark and fruit. Villagers are unsure of propagation methods for the various trees and doubt that seedlings will survive. Collection of seed from the forest can also be difficult. However, if seedlings of more indigenous tree species were available from government departments and if the staff of nongovernmental organizations and villagers had more information on propagation methods, then planting of indigenous species would probably increase, particularly since the villagers have learned from experience that many exotic or lowland species do not grow well in the highlands.

Conclusions

In a time of smaller populations and larger areas of available land, shifting cultivation was a productive and environmentally appropriate method of farming in the highlands, a system developed by farmers whose lives and livelihood depended upon a harmonious integration with the natural cycles of the environment. When shifting cultivators cleared fields from the forest, they generally left behind them the conditions required for a quick regeneration of forest cover, thus perpetuating the cycle of farming and fallowing for themselves or for others who would follow. The process of natural succession, and the hill farmers' intimate understanding of it, can today be harnessed to regenerate and conserve watershed forests, which are crucial to the environmental health of the entire country.

The case of Pakhasukjai demonstrates that local traditions of forest conservation and management, coupled with the ability of the hill farmers to adapt these traditions as necessary, can make significant contributions to the conservation of Thailand's watershed forests, including regeneration of the forest from *Imperata* fallows.

Acknowledgments

The research on which this paper was based would not have been possible without the assistance of Dr. Stephen Elliott and Dr. J.F. Maxwell of the Biology Department, Chiang Mai University. Many thanks also to Dr. Katherine Warner, then of the Regional Community Forestry Training Centre, Bangkok; to researchers from the Soil Fertility Conservation Program of Mae Jo University, Chiang Mai, and the Catholic University of Leuven, Belgium; to the staff of the Hill Area Development Foundation; and to CUSO and the Canada Fund of the Canadian Embassy, Bangkok, which provided funding. And to the Akha people of Pakhasukjai, we extend our deepest gratitude for sharing so much of their wisdom and experience.

References

Durno, J., and K. Warner. 1993. Community Management and Natural Regeneration Case Study. Paper presented at a seminar on natural forest regeneration, April 1993. Regional Community Forestry Training Centre, Bangkok, Thailand.

———. 1996. From *Imperata* Grass Forest to Community Forest: The Case of Pakhasukjai. *Forest, Trees and People Newsletter* 31, 4–13.

Elliott, S., and J. F. Maxwell. 1994. Personal communication between Dr. Stephen Elliott and Dr. J.F. Maxwell, of the Biology Department, Chiang Mai University, and the authors.

Elliott, S., J.F. Maxwell, and O. Prakobvitayakit Beaver. 1989. Transect Survey of Monsoon Forest in Doi Suthep-Pui National Park. *Nat. Hist. Bull. Siam Soc.* 37(2), 137–171.

Elliott, S., and C. Trisonthi. 1992. *Factors Affecting Distribution, Seed Germination, and Phenology of Trees in Doi Suthep-Pui National Park.* Final report to WCI-Thailand. Chiang Mai, Thailand: Biology Department, Chiang Mai University.

Elliott, S., K. Hardwick, E.G. Tupacz, S. Promkutkaew, and J.F. Maxwell. 1993. Forest Restoration for Wildlife Conservation: Some Research Priorities. Paper presented at the 14th annual wildlife symposium, December 15–17, 1993, Kasetsart University, Bangkok, Thailand.

HADF (Hill Area Development Foundation). 1986–1994. Six-monthly and annual reports. Chiang Rai, Thailand.

Ludwig, J.A., and J.F. Reynolds. 1988. *Statistical Ecology: A Primer in Methods and Computing.* New York: Wiley and Sons, 337.

Maxwell, J.F. 1988. The Vegetation of Doi Suthep-Pui National Park, Chiang Mai Province, Thailand. *Tigerpaper* 15, 6–14.

———. 1992. Lowland Vegetation (450–c.800 m) of Doi Chiang Dao Wildlife Sanctuary, Chiang Mai Province, Thailand. *Tigerpaper* 19(3), 21–25.

Ongprasert, S., F. Turkelboom, and K. van Keer, with other contributors. 1996. Land Management Research for Highland Agriculture in Transition: Research Highlights of the

Soil Fertility Conservation Project (1989–1995). Chiang Mai, Thailand and Leuven, Belgium: Mae Jo University and the Catholic University of Leuven.

Prangkio, C. 1987. *Report on Land Use Study in HADF Project Area*. Chiang Mai, Thailand: Geography Department, Chiang Mai University (Thai language).

Smedts, R. 1994. *Land Use in a Hilltribe Village in Northern Thailand*. Chiang Mai, Thailand and Leuven, Belgium: Mae Jo University and the Catholic University of Leuven.

Sukwong, S., and P. Dhamanitayakul. 1977 (unpubl.). Some Effects of Fire on Dry Dipterocarp Forest Community. Presented to BIOTROP–Kasetsart University Symposium on Management of Forest Production in Southeast Asia, Kasetsart University, Bangkok, as cited in Sukwong, S. (No date) *Potential of Natural Regeneration in Deciduous Forest*. Bangkok: Regional Community Forestry Training Centre.

Sukwong, S. (No date). *Potential of Natural Regeneration in Deciduous Forest*. Bangkok: Regional Community Forestry Training Centre.

Sutthi, C. 1989. Highland Agriculture: From Better to Worse. In: *Hill Tribes Today*, edited by J. McKinnon and B. Vienne. Bangkok: White Lotus-Orstrom.

Turkelboom, F. 1994. *Soil Description of the Study Areas*. Soil Fertility Conservation Project research report, 1993. Chiang Mai, Thailand and Leuven, Belgium: Mae Jo University and the Catholic University of Leuven.

Vlassak, K., S. Ongprasert, A. Tancho, K. van Look, F. Turkelboom, and L. Ooms. 1993. Soil Fertility Conservation research report 1989–1992. Chiang Mai, Thailand: Soil Fertility Conservation Project, Mae Jo University.

Chapter 13

When Shifting Cultivators Migrate to the Cities, How Can the Forest Be Rehabilitated?

Borpit Maneeratana and Peter Hoare

Until the early 1990s, land-use pressures on the uplands of northern Thailand were increasing rapidly and playing a major role in the degradation of forests. However, more recently, the situation in some areas has reversed. Land-use pressures on swiddens in the Upper Nan watershed, in Thailand's far north, for instance, have fallen dramatically and the problem there has become one of land rehabilitation.

The area is the site of the Upper Nan Watershed Management Project, run by Thailand's Royal Forest Department (RFD) and Danish Cooperation for Environment and Development (DANCED). The project area covers 912 km^2 in the upper Nan River basin, one of Thailand's most important watershed areas. Until recent decades, most of the area was covered by dense, moist forests of a type that were not at risk from fire. However, this situation changed drastically because of rapid population growth, the widespread conversion of forest for cultivation, and excessive logging, both legal and illegal. Extensive shifting cultivation led to the expansion of fire-prone grassland, and both the climatic conditions and forest types changed. The dry season became prolonged, with higher temperatures, and a drier type of forest gradually became established. In these conditions the area became vulnerable to forest fires, particularly those lit by people living in or adjacent to the forest (RFD 1996).

Most of the farmers practicing shifting agriculture in the project area are living in a national forest reserve and have no legal land tenure. Even though Nan is one of the most distant provinces from Bangkok, farmers and young adults of working age have been migrating from villages in the area for part or all of the year to work in Bangkok, Chiang Mai, or other regional cities. There they can earn about 1997US$6 per day as unskilled laborers, an income that is hard to match from village swiddens. With decreasing village labor resources, the development problem has become one of land rehabilitation and conversion of fallows dominated by *Imperata cylindrica* to rehabilitated forest. Stakeholder workshops in the villages (RAMBOLL 1996) have shown that older people do not intend to leave, and wish to restore the forest and generate income from forest products, such as *ma kwaen* (RFD–DANCED 1997; Hoare et al. 1997, Chapter 50). However, most of the older people are regarded as "habitual fire lighters" who, in the past, have made no attempt to control the fires they light

Borpit Maneeratana and Peter Hoare, Royal Forest Department, Nan Watershed Management Office (Khao Noi), Tambol Dootai, Amphoe Muang, Nan Province, 55000, Thailand.

on their swiddens, so the Upper Nan Watershed Management Project has focused on fire control to enable natural forest regeneration.

Hypotheses

The aim of this study was to test the following hypotheses:

- When opportunities increase for off-farm employment and the daily wage offered is double the income from family labor in shifting agriculture, then a rapid reduction in land pressure is likely.
- The main development problem then becomes land rehabilitation and, in particular, fire management to enable forest regeneration. The assumption is made that regenerating forest has superior watershed characteristics to *Imperata* grassland, which is subject to annual fires.

The Study Area and the RFD–DANCED Project

Nan Province is in the far north of Thailand, bordering Laos, in one of the country's most important river basins. The Chao Phraya River, which is the "lifeline" for the country's intensively cultivated central plain, has four main tributaries, the Ping, Wung, Yom, and Nan Rivers. The Nan River provides 57% of the annual water discharge available for use in central Thailand.

The RFD–DANCED project area covers 912 km^2 in the upper right catchment of the Nan River. The area is between 350 and 800 m above sea level (asl) in altitude, at about 19 degrees north latitude. The natural forest has been severely reduced by logging concessions, illegal logging, shifting agriculture, and uncontrolled forest fires. The area has 44 villages with about 20,000 people living in 3,843 households. Among them are 28 "hill tribe" villages populated by Khamu, Lue, Hmien (Yao) and Hmong ethnic groups.

The project strategy is to involve these communities in the rehabilitation and protection of forests through participatory land use planning. The project's ideal has been described by Durno (1995) (see also Durno et al., Chapter 12). This study concerned the regeneration of community forest from *Imperata* grassland by an Akha hilltribe community in neighboring Chiang Rai Province. The main emphasis in the Upper Nan watershed is on reforestation and natural forest regeneration in critical areas, changing agricultural practices from shifting agriculture to more environmentally friendly permanent cultivation of paddy rice and horticultural trees, and the development of alternative income sources to shifting agriculture (RFD–DANCED 1995).

One of the project's important tasks in its first dry season was the collection of baseline data on fire management. This involved a survey of the fire management procedures of all the farmers in the 44 villages who practiced shifting agriculture. The data was collected over three months by the 15 community coordinators employed by the project. The community coordinators also assisted RFD personnel in training village fire control volunteers, making firebreaks, appointing fire guards, and encouraging farmers to construct fire breaks around swiddens prior to burning. Fire prevention and control systems alone are expected to increase by 16,000 ha the area of forest cover through natural regeneration within five years.

Research Methods

The research methods included a literature review, the collection of data by the RFD and by field staff of the RFD–DANCED project concerning fire breaks made, fireguards, and farmers practicing shifting agriculture, and the field mapping of areas burnt in the dry season from December 1996 to May 1997.

Table 13-1. Farmers Practicing Shifting Agriculture and Making Firebreaks (1997)

Name of Watershed Management Unit	*Number of Villages*	*Farmers Practicing Shifting Agriculture*	*Area of Fields (ha)*	*Farmers Making Firebreaks*	*Percentage of Farmers Making Firebreaks*
Khun Nam Prik	6	251	200.1	97	39
Nam Yao	6	96	22.6	96	100
Huai Yod	4	95	39.0	95	100
Nam Huai	8	229	329.6	161	70
Nam Haen	11	143	167.8	129	90
Sop Sai	7	138	122.4	82	59
Total	42	952	881.5	660	69

Note: Only 42 of 44 villages included.
Source: RFD–DANCED survey by community coordinators in April and May 1997.

Results

Fire Prevention by Shifting Cultivators

The number of farmers engaged in shifting agriculture, the area of their fields, and the number who made firebreaks before burning in 1997 are shown in Table 13-1. This shows that 952 farmers used fire for land preparation in their swiddens and of these, 660, or 69%, made firebreaks before burning. The average size of the swiddens, mainly planted with hill rice, was 0.92 ha per family. Most farmers informed the village headman on the date of burning and, as an added precaution, took knapsack sprayers and extra labor to the field.

Area Burnt

The area burnt by uncontrolled fires during the 1996–1997 dry season is shown in Table 13-2. The figures apply only to fires that burnt more than 10 ha, and do not include successfully controlled swidden fires. The total area burned was 46 km^2, about 5% of the project area. This data will be checked against satellite images to see if areas burnt in previous years can be traced.

Causes of Fires

According to data collected since 1980 by the Forest Fire Control Office of the RFD, there are no records of fires caused by natural phenomena such as lightning or tree friction. All are lit by people, especially those living in or adjacent to forests. According to the RFD (1996), the following reasons were found for forest fires in 1995:

- Gathering of non-timber forest products. People who traverse the forest during the dry season, usually to collect forest products such as firewood, bamboo, honey, or mushrooms, set fires mainly to clear litter, grass, and undergrowth to make their travel and collection easier (24% of the total).
- Burning of agricultural debris. Farmers traditionally set fires, without any control, to eliminate crop residue and prepare agricultural land after harvesting. This is very common in areas where shifting cultivation is still widely practiced, and the fires often escape into nearby forest (18% of the total).
- Incendiary or grudge fires. These include attempts by rural people to convert forest into cultivation land, or follow conflicts between rural people and forest officials (20%).
- Hunting. In pursuit of small game, rural people set fires to drive animals from concealment (15%).

- Carelessness. These fires mainly originate from campfires and cigarette butts (14%).
- Unidentified. Fires with no apparent cause (9%).

Data collected by staff of the RFD–DANCED project suggest, similarly, that about 40% of fires are lit by hunters and gatherers, about 40% escape from burning swiddens, about 10% are lit by cattle graziers, and about 10% have no known cause.

Discussion

Once established, *Imperata* grasslands can survive almost indefinitely with frequent burning. The process of reversion to secondary forest begins only in the absence of fire. Scrub and trees will successfully establish if fire is kept out of *Imperata* grassland for four or five years. However, the risk of regular fire is substantial. Measurements of *Imperata cylindrica* biomass in northern Thailand, at a similar altitude to the Upper Nan Watershed Management Project, show a peak accumulation of 3,500 kg/ha of dry matter before burning between February and April (TAHAP, 1978).

A large number of farmers still believe that fire does no real harm to the watershed. This is due to the illusion that forest fires in Thailand, which are usually classified as surface fires, appear to be less severe than those that occur in Europe and America. Fire does no apparent damage to mature trees and burns only the undergrowth, which will regenerate in the following rainy season. This misconception about the severity of forest fires diverts attention away from the seriousness of the problem. In order to overcome this misconception, fire prevention programs have been designed around the three components described in Table 13-3.

Village fire codes and regulations have been established in the project area and are being enforced. Fines of 2,000 baht (about 1997US$75) are commonly imposed by village committees where fires lit by farmers cause damage. In addition, the offending farmer is liable for the cost of the damage. In one case in February 1997, where a fire damaged a litchi orchard, the damages were assessed at 22,000 baht (about 1997US$850).

Research Priorities

Ground maps of areas burned, at a scale of 1:50,000, will be compared with satellite images. If it is possible to relate the images, then it may also be possible to examine satellite images from previous years to analyze trends in areas burned.

Table 13-2. Areas Burnt within Upper Nan Watershed Management Project (December 1996 to May 1997)

Watershed Management Unit		*Burned Area*	*Number of Fires*	*Average Area Burned*
Name	*Area (km²)*	*(km²)*	*Over 10 ha*	*per Fire (ha)*
Khun Nam Prik	130	3.19	3	166
Huay Yod	107	6.02	5	120
Nam Yao	95	2.54	4	63
Nam Huai	288	7.26	22	33
Nam Haen	190	5.94	10	60
Sop Sai	102	21.16	18	118
Total	912	46.11	62	74

Table 13-3. Components of a Fire Prevention Program

Component	*Key Elements*	*Strategy*
Engineering	Separate heat sources Village fire fighting equipment	Create and maintain firebreaks by clearing fuel
Education	Create awareness of the need for fire prevention	Public relations meetings with villagers
	Create awareness of the need for controlled burning of swiddens	Training of fire volunteers in fire management
		Topographical models to help define village boundaries for fire control
Enforcement	Village committee enforcement and fines	Investigation and imposition of fines
	Government rules	

Source: Adapted from Ploadpliew 1997.

Conclusions

Survey data from community coordinators show that more than half the population of working age has migrated from some villages in the project area to work in Bangkok and other cities. There, they can more than double the income they might expect from farm laboring, as well as the return from annual crops grown in shifting agriculture. In the latter case, declining soil fertility on steep slopes dominated by *Imperata* has further eroded their local earning capacity.

These circumstances are likely to maintain a rapid migration to urban centers, not only in the north of Thailand, but throughout Southeast Asia. When pressure on land in shifting agriculture is suddenly decreased in this fashion, *Imperata* is left to thrive on abandoned swiddens. The core issue underlying the rehabilitation of this land is the control of fires so the forest can regenerate naturally. Therefore, a participatory fire management program with key elements of engineering, education, and enforcement of village level fines is needed over the long term.

References

Durno, J. 1995. From *Imperata* Grass Forest to Community Forest: The Case of Pakhasukjai. *Forest Trees and People Newsletter* No. 31.

Hoare, P., B. Maneeratana, and W. Songwadhana. 1997. *"Ma Kwaen" — A Jungle Spice Used in Shifting Cultivation Intensification in North Thailand.* Nan, Thailand: RFD–DANCED (Royal Forest Department–Danish Cooperation for Environment and Development).

———. 2006. "Ma Kwaen" (*Zanthoxylum limonella*): A Jungle Spice Used in Swidden Intensification in Northern Thailand. Chapter 50.

Ploadpliew, Apinan. 1997. *Fire Consultant's Report, Upper Nan Watershed Management Project.* Khao Noi, Nan, Thailand: Royal Forest Department.

RFD (Royal Forest Department). 1996. Thailand Country Report on Forest Fire Control, by Supparat Samran and Siri Akaakara. Paper prepared for second meeting of Forest Operation Technical Working Group, June 4, 1996, Chiang Mai, Thailand.

RAMBOLL Consulting Company, Copenhagen, Denmark. 1996. Stakeholder Analysis, RFD–DANCED Upper Nan Watershed Management Project, Nan, Thailand.

RFD–DANCED (Royal Forest Department–Danish Cooperation for Environment and Development). 1995. *Project Document for the Upper Nan Watershed Management Project.* Copenhagen: DANCED.

TAHAP (Thai-Australia Highland Agricultural Development Project). 1978. Annual Report. Chiang Mai, Thailand: Chiang Mai University–Australian Development Assistance Bureau.

PART III

Shrub-based Accelerated Fallows

A Minangkabau mother and child in West Sumatra, Indonesia.

Chapter 14

Fallow Improvement with *Chromolaena odorata* in Upland Rice Systems of Northern Laos

Walter Roder, Soulasith Maniphone, Bounthanh Keoboualapha, and Keith Fahrney

Chromolaena odorata is considered a noxious weed in many parts of the world (Olaoye 1986; Torres and Paller 1989; Waterhouse 1994). It is a major colonizer in slash-and-burn agriculture and is particularly widespread in Africa (de Rouw 1991) and Asia (Nakano 1978; Kushwaha et al. 1981). *C. odorata* was introduced to India in the 1840s (McFadyen 1989) and, by 1920, was regarded as a serious, rapidly spreading weed in Burma (Rao 1920). It reached Thailand in about 1924 and spread to Laos in the late 1920s (Chevalier 1949; Vidal 1960). Although it failed to rate a mention in a 1942 description of upland agricultural systems in Indochina by Gourou, which was based on observations made in the late 1930s (Gourou 1942), Izikowitz (1951) described the invasion of *C. odorata* into what is now Luang Namtha Province during the 1940s.

Finding favorable conditions in northern Laos, *C. odorata* spread rapidly and, by the 1950s, it had already become the most abundant weed in slash-and-burn rice fields and in successive fallows (Vidal 1960). Probably because its appearance coincided with the French presence, it is known in some areas as *nia phalang*, meaning "French weed," or "foreign weed" (Vidal 1960). Interestingly, it is known in the French language as *l'herbe du Laos*, or "the weed from Laos" (Leplaideur and Schmidt-Leplaideur 1985).

Present government policies in Laos give high priority to reducing the area under slash-and-burn agriculture and limiting farmers' access to land. These efforts, combined with rapid population growth, have resulted in shorter fallow periods and, consequently, increased weed problems and soil deterioration (Roder et al. 1995b; Roder 2001). Farmers urgently need technologies that can help them sustain rice production with shorter fallow periods. The improvement of fallow systems has been seen as the most appropriate development in Laos to advance slash-and-burn agriculture toward permanent land use (Fujisaka 1991).

Improved fallow systems are expected to provide similar ecological benefits to those of natural fallows, but over a shorter period (Robison and McKean 1992). *C. odorata* has many of the attributes of an improved fallow species. Although it is not palatable to livestock, it is often considered as a welcome plant, rather than a weed, by slash-and-burn farmers (Ruthenberg 1980; Dove 1986; Keovilayvong et al. 1991).

This chapter examines *C. odorata* in the context of the slash-and-burn agriculture system prevailing in Laos. It uses various data relating to land-use practices, weed

---

W. Roder, c/o Helvetas, P.O. Box 157, Thimphu, Bhutan; S. Maniphone and B. Keoboualapha, Luang Prabang Agricultural and Forestry Service, P.O. Box 600, Luang Prabang, Lao P.D.R; and K. Fahrney, Lao – IRRI Project, P.O. Box 4195, Vientiane, Lao P.D.R.

problems, and soil fertility collected from 1991 to 1996 by the Lao-IRRI (International Rice Research Institute) project. Using pertinent data from these activities and from other references, this chapter pays specific attention to the importance of *C. odorata* during the rice growing and fallow periods and to its potential for fallow improvement.

With only a few exceptions, household surveys and other investigations undertaken in gathering this data (see Table 14-1) were limited to the provinces of Luang Prabang and Oudomxay (Figure 14-1). These two provinces contain more than 35% of the total slash-and-burn rice production area in Laos. All the research activities were carried out in collaboration with agriculture services in Luang Prabang and the Lao-IRRI project. Materials, methods, and findings were published earlier and can be found in the references listed in the notes beneath Table 14-1.

Results

C. odorata *as a Weed in Upland Rice Systems*

During 1991 and 1992, field observations and a survey of households covering a wide area in Luang Prabang and Oudomxay provinces found that labor for weed control was the single most important constraint to upland rice production. When asked to name what they considered were the major constraints, 85% of respondent farmers mentioned weeds. The other major constraints, in order of respondent perception, were rodents, 54%; insufficient rainfall, 47%; availability of land, 41%; insects, 34%; labor, 24%; soil fertility, 21%; and erosion, 15%. Land availability, or the need for shorter fallows, and labor can be directly related to weeding requirements. Weed control in upland rice production currently requires between 140 and 190 labor days/ha, or 40% to 50% of the total labor input (Roder et al. 1997a). What's more, the labor requirement for weeding is increasing with the decline in fallow periods. Average fallow periods have decreased from 38 years in the 1950s, to 20 years in the 1970s, to just five years in 1992. Over the same period, average weeding inputs per cropping season have increased from 1.9 weedings in the 1950s to 3.9 weedings in 1992 (Roder et al. 1997a).

Against this background, *C. odorata* contributes almost 40% of the total weed cover during rice cropping (Table 14-2). Its contribution to the total weed biomass during the 1991 rice growing season was 3% on July 8, 10% on August 28, and 14% on September 27 (Roder et al. 1995a). Although *C. odorata* was introduced to Laos as recently as the 1930s, elderly people cannot recollect which weed species was dominant before *C. odorata* arrived to take over (Roder et al. 1997a). With a coincidental reduction in fallow periods since its introduction, *C. odorata* may have largely replaced tree species coppicing from old plants or growing from seeds. Other important weed species during rice cropping are *Ageratum conyzoides, Commelina* sp., and *Lygodium flexuosum* (Roder et al. 1997a).

Despite the fact that *C. odorata* is the most abundant weed species, farmers generally do not consider it a major weed (Roder et al. 1997a). Usually there are relatively few, but large, *C. odorata* plants, and there is no rooting from aboveground plant parts. Therefore, it is much easier to control by hand weeding than other species, such as *Commelina* sp. or *L. flexuosum* (Roder et al. 1997a). Although its regrowth from rootstocks after burning can make *C. odorata* a serious competitor for young rice plants, farmers find it relatively easy to remove by hand. Plants growing from seeds have a comparatively slow initial growth phase and are less of a problem.

Figure 14-1. Research Focused on Luang Prabang and Oudomxay Provinces in Northern Laos
Source: UNDP 1994.

C. odorata displays a wide range of adaptation. There were no relationships revealed in a correlation analysis between fallow period, selected soil fertility parameters, and the frequency of *C. odorata* (Roder et al. 1995b). Likewise, when ranking in classes for fallow period, cropping period, and soil pH, Roder observed no effect on either the frequency or density of *C. odorata* during the rice crop. Its frequency and contribution to the weed and fallow biomass may decline at elevations above 1,000 m above sea level (asl). However, Nakano (1978) reported *C. odorata* together with *Buddleia asiatica* as major fallow species in the first year after rice harvest in slash-and-burn fields in northern Thailand, at elevations above 1,000 m.

C. odorata as a Component of Fallow Vegetation

Throughout all slash-and-burn areas in northern Laos, *C. odorata* is generally the most important species in the first and second year of the fallow period. In 1991, the average aboveground biomass at four monitoring sites was 1.4 tonnes/hectare (t/ha) at rice harvest. After one year of fallow, by the end of 1992, this had increased to 10 t/ha, and after two years, in 1993, it had reached 15.4 t/ha (Table 14-3). At rice harvest, tree and bamboo species contributed 61% of the nonrice biomass and had frequencies (or presence in 1 m^2 frames) of 32% for bamboo and 95% for tree species

(Roder et al. 1995a). However, their development was too slow to fill the gap left after the rice harvest. After the first year of fallow, tree and bamboo species contributed only 37% of the biomass, and *C. odorata* had risen to account for 48%. The contribution of grass species to the weed and fallow biomass was very minor. Unlike its role in some other Asian slash-and-burn systems, *Imperata cylindrica* rarely threatened to dominate.

Table 14-1. Main Objectives of Studies, Locations, Information Collected, and References

Study and Main Objective	*Reference (see below)*	*Location (province)*	*Information Collected*
1. Household survey (1991–1992) Characterize land-use systems and identify constraints to rice production.	*a, b*	LP, OU	Constraints to rice production Weed cover and frequency Canopy cover of major fallow species Ranking of fallow species
2. Field study—legumes (1992–1994) Evaluate legumes for fallow improvement.	*b, c*	On-station, LP	Biomass produced Litterfall Effect on succeeding rice yield
3. Yield-soil-weed survey (1993) Identify relationships between rice yield, weeds, soil, and nematodes.	*d*	On-farm, LP	Weed cover and frequency Nematode density Soil properties Soil/weed/nematode/yield interactions
4. Field study—fallow (1991–1994) Document changes in soil fertility and fallow vegetation during cropping and fallow period.	*e*	On-farm, LP	Changes in soil properties and vegetation during rice crop and fallow period Nutrients in fallow vegetation and litterfall
5. Field study—rotation/residue (1992–1995) Evaluate effect of residue treatment and cropping intensity on crop yield, weeds, and nematodes.	*f*	On-station and on-farm, LP	Effect of burning on weed biomass and composition Effect of rotation on weed composition Relationships between rice yield, weed density, and nematodes
6. Household survey (1996) Document farmers' assessments of *C. odorata* as a fallow improvement species.	*g*	LP, OU	Ranking of fallow species Perceived properties of *C. odorata* Management of fallow vegetation

Source: a, Roder et al. 1997a; *b,* Roder et al. 1995a; *c,* Lao-IRRI Technical Report 1994 (unpubl.); *d,* Roder et al. 1995b; *e,* Roder et al. 1997b; *f,* Roder et al. 1998c; *g,* Roder et al. 2001. Provinces: LP = Luang Prabang; OU = Oudomxay.

Table 14-2. Cover and Frequency of Major Weeds in Upland Rice Fields of Northern Laos

Weed Species	*Frequency* [a]	*Cover* [b]
Chromolaena odorata	50	5.6
Ageratum conyzoides	31	4.1
Commelina spp.	27	2.1
Lygodium flexuosum	22	1.7
Panicum trichoides	12	0.7
Corchorus sp.	7	0.7
Pueraria thomsoni	6	0.6
Panicum cambogiense	3	0.3
Imperata cylindrica	2	0.1
Total cover (cm/m)		10.5

Note: [a] Frequency (%) in transect segments of 1 m; [b] Cover in cm/m.

C. odorata *for Fallow Improvement*

Among the early proponents of fallow improvement with *C. odorata* were Chevalier (1952) and Poilane (1952). Both recommended its use in Laos to suppress *Imperata cylindrica*. Chevalier even recommended that *C. odorata* be introduced to Africa as a means of suppressing *I. cylindrica* there. In Laos, it is quite likely that *I. cylindrica* would be a greater problem if *C. odorata* was not present.

Species to be used for fallow improvement are expected to establish with ease, provide plant cover after crop harvest, produce large quantities of biomass, suppress weeds, mobilize plant nutrients from lower soil layers, and decompose rapidly (Fujisaka 1991; Rao et al. 1990; Robison and McKean 1992). *C. odorata* excels in most of these attributes.

Ease of Establishment and Provision of Plant Cover after Crop Harvest. The ability of *C. odorata* to expand rapidly and provide a protective cover in the early part of the fallow period is probably its most important attribute making it a good fallow plant for the sloping fields of northern Laos, made possible by *C. odorata's* profuse production of very mobile seeds. Under favorable conditions, *C. odorata* will produce close to half a million seeds per square meter (Kushwaha et al. 1981). The seeds are dispersed by the wind during April and May, and germination starts at about the same time as rice crops are planted.

Table 14-3. Average Aboveground Biomass in Four Slash-and-Burn Fields in Northern Laos (tonnes/hectare)

	Plant Biomass		
Component	*1991 (at rice harvest)*	*1992 (after 1 year fallow)*	*1993 (after 2 years fallow)*
Chromolaena odorata	0.23 ± 0.07 [a]	4.8 ± 0.7	4.5 ± 1.4
Lygodium flexuosum	0.14 ± 0.03	0.6 ± 0.4	0.1 ± 0.05
Other broad-leaved species	0.17 ± 0.03	0.5 ± 0.3	1.3 ± 0.9
Grasses	0.03 ± 0.02	0.1 ± 0.1	0.2 ± 0.1
Bamboo	0.24 ± 0.15	2.1 ± 1.7	4.0 ± 2.0
Tree species	0.51 ± 0.11	1.5 ± 0.9	5.3 ± 1.4
Total	1.4 ± 0.13 [b]	9.8 ± 1.1	15.5 ± 1.9

Note: [a] Mean ± standard error; [b] rice grain harvested and rice stem were 1.1 and 1.2 t/ha.

Table 14-4. Comparison of *C. odorata* with Other Potential Fallow Improvement Plants

Species	*Litterfall (t/ha/year)*	*Biomass after Two Years (Fresh, t/ha)*		
		Species	C. odorata	*Weeds*
Gliricidia sepium	6.4	24.1	1.6	5.6
Calliandra calothyrsus	3.3	7.3	7.1	6.7
Cassia sp.	4.5	3.5	8.1	8.9
Mimosa invisa	2.9	23.6	3.8	1.5
Chromolaena odorata	4.5	18.0	-	1.0
PR > F	0.26	< 0.01	0.10	0.01
CV (%)	32.3	23.8	15.2	30.5

Table 14-5. Effect of Cropping/Mulching Treatment on Residue Load, Weed Biomass, and Nematode Infection

Treatment	*Residue (t/ha)*		Meloidogyne graminicola *(no./mg of root)*	Ageratum conyzoides *(no./m²)*
	C. odorata	*Total*		
Continuous rice burned	1.4 [a]	5.0 [a]	0.74	77 [ab]
Continuous rice mulched [b]	0.9 [a]	3.8 [a]	0.19	131 [a]
Rice-fallow-rice burned	6.5 [b]	9.3 [b]	< 0.01	2 [b]
Rice-cowpea-rice mulched [b]	0.9 [a]	2.7 [a]	< 0.01	26 [b]
Anova (PR>F)	< 0.01	< 0.01	0.11 [c]	0.03
CV (%)	91.1	36.0	34.1	67

Notes: [a] Means with the same letter are not significantly different at 5% level by DMRT; [b] Mulching of weed, rice, or cowpea residue is possible because the quantity is small and less woody compared to fallow residues; [c] ANOVA made with data transformed by (x+0.1)1/2.

Biomass Production. *C. odorata* has produced between four and eight tonnes of biomass per hectare per year in the first years of fallow (Tables 14-3, 14-4, and 14-5). Of various fast-growing leguminous species tested for fallow improvement qualities, *Gliricidia sepium* and *Mimosa invisa* produced higher total biomass than *C. odorata,* but only *G. sepium* produced more litter (Table 14-4).

Weed Suppression. Legumes tested for fallow improvement generally required weeding in the year of establishment. *C. odorata,* however, competed with most of the weeds present in the first year of fallow and did not require weeding. In the study comparing plants for fallow improvement, a *C. odorata*–dominated fallow had the lowest weed biomass (Table 14-4). Weeds are largely suppressed by the vigorous growth of *C. odorata.* However, it also demonstrates some allelopathic effects (Ambika and Jayachandra 1992; Nakamura and Nemoto 1994). In the absence of *C. odorata,* weed species such as *A. conyzoides* and *I. cylindrica* are likely to become more dominant. In a long-term cropping study, the density of *A. conyzoides* was dramatically reduced after a one-year *C. odorata* fallow (Table 14-5).

A. conyzoides and root-knot nematode (*Meloidogyne graminicola)* densities are negatively associated with rice yield (Table 14-6). *A. conyzoides* is a good host for *M. graminicola* (Waterhouse 1994), and the negative association observed between rice yield and *A. conyzoides* density may be due to increased nematode damage. *C. odorata* has been shown to suppress nematodes (Atu 1984; Subramaniyan 1985), and this is a property that may become more important with declining fallow periods.

Mobilization of Plant Nutrients from Lower Soil Layers. Compared to some of the other fallow species, especially bamboo, *C. odorata* contains more nitrogen, phosphorus, and calcium (Tables 14-7 and 14-8) and produces relatively high quantities of easily degradable litter. After one year of fallow, *C. odorata* contributed 62% of the P in the fallow vegetation. Depending on the importance of ash residue contributions to soil fertility and the performance of the rice crop, this higher mineral content may be important. Furthermore, *C. odorata* has a fast decomposition rate and can improve both the quantity and quality of soil organic matter (Obatolu and Agboola 1993). It has also shown good results as a green manure for lowland rice (Litzenberger and Ho Tong Lip 1961).

Farmers' Assessment

Although *C. odorata* is the most abundant weed species, farmers generally appreciate it as a fallow plant. When asked to list "good fallow plants," or plants they like to have in their fallow fields, many farmers favored *C. odorata* (Table 14-9). Some of the plants listed by farmers as bad fallow plants, especially *Cratoxylon prunifolium* and *A. conyzoides,* are generally associated with poor rice yields. In one survey, about 70% of the respondents indicated that they liked *C. odorata* in their fallow vegetation. The main reasons given were better rice yields and the relatively easy control of *C. odorata* as a weed (Table 14-10). Soil fertility and the capacity to enable shorter fallow periods were also mentioned. Farmers interviewed in Savannakhet Province suggested that soil structure was better when *C. odorata* dominated fallow fields, rather than bamboo species (Keovilayvong et al. 1991). Similar preferences for *C. odorata* as a fallow species have been reported in Indonesia and Nigeria (Dove 1986; Ruthenberg 1980). Waterhouse (1994) suggested that *C. odorata* may be beneficial to resource-poor farmers. However, the farmers themselves seem less than convinced that *C. odorata* will suppress weeds (Table 14-11). Forty-six percent of respondents agreed that *C. odorata* suppressed *A. conyzoides,* and 33% believed it suppressed *I. Cylindrica,* but none of them believed it would suppress *Mimosa invisa.*

Farmers' Interventions to Increase *C. odorata* in Fallow Vegetation. While recognizing the positive properties of *C. odorata,* farmers are concerned about labor for weeding. Any intervention that leads to more weeding would not be acceptable, and they are therefore very reluctant to have more *C. odorata* in their fields (Table 14-11). Only a very small number indicated that they were making efforts to increase the *C. odorata* cover. Measures used to enhance *C. odorata* include burning or slashing, selective weeding, and avoiding the last weeding.

Table 14-6. Relationship between Rice Yield and Other Selected Parameters

Parameter	*Range*	*Correlation with Rice Grain Yield*
Straw yield (t/ha)	1.2 to 4.9	0.85[a]
C. odorata weed biomass (fresh, g/m^2)	0 to 448	0.39[b]
A. conyzoides density (no./m^2)	0 to 265	-0.62[a]
Herbaceous weed biomass (fresh, g/m^2)	85 to 819	-0.45[a]
M. graminicola density (no./mg of root)	0 to 3.6	-0.42[b]

Note: [a] Significant at the 1% level respectively; [b] significant at the 0.1% level.

Negative Properties of *C. odorata*

Properties that are considered advantages in a fallow improvement context may easily be serious constraints in others. These include the following:

- High seed production and easy dispersal of seeds by the wind. This results in continuous heavy weed seed influx from fallow land to cultivated areas.
- Plant residues from a *C. odorata* fallow are difficult to manage. Fire may be the only practical field preparation method for areas dominated by this species after a fallow period of more than one year.
- Allelopathic effects on tree growth. *C. odorata* is considered harmful in rubber and teak plantations in Indonesia (Tjitrosoedirdjo et al. 1991), and allelopathic effects on teak have been reported (Ambika and Jayachandra 1992).
- *C. odorata* may impede the development of other preferred perennials. This applies particularly to systems with fallow periods of more than three years.
- *C. odorata* is likely to reduce fodder production on fallow land. It is also liable to become a serious weed in grazing systems.

Table 14-7. Average Nitrogen, Phosphorus, Potassium, and Calcium Content of Major Fallow Species (%)

Plant Species	*N*	*P*	*K*	*Ca*
C. odorata				
Leaf and flowers	2.10	0.130	2.04	0.55
Stems	0.49	0.042	1.02	0.27
Bamboo				
Leaves	2.17	0.042	1.88	0.42
Small branches	0.63	0.022	1.37	0.19
Stems	0.38	0.015	1.19	0.08
Cratoxylon sp.				
Leaves	0.98	0.037	0.91	0.50
Stems	0.56	0.020	0.40	0.32
Lygodium flexuosum				
Whole plant	1.68	0.048	1.84	0.39

Table 14-8. Average Nitrogen, Phosphorus, Potassium, and Calcium in Aboveground Vegetation at Rice Harvest, 1991, and Two Subsequent Years of Fallow, 1992 and 1993 (kg/ha)

Nutrient/Year	C. odorata	*Others*	*Bamboo*	*Trees*	*Rice Straw*	*Total*
N						
1991	0.9	1.5	1.0	2.2	3.5	9.2
1992	32.6	5.5	12.6	9.8		60.5
1993	30.1	7.9	24.0	34.5		97.0
P						
1991	0.4	0.4	0.2	0.8	1.0	2.7
1992	2.4	0.8	0.4	0.3		3.9
1993	2.3	0.6	0.8	1.1		4.6
K						
1991	0.6	0.7	1.0	0.6	2.5	5.4
1992	52.8	4.7	27.1	7.7		92.2
1993	49.5	9.2	51.6	27.0		137.0
Ca						
1991	1.1	1.8	0.7	5.9	4.0	13.5
1992	14.4	4.1	2.9	5.4		26.9
1993	13.5	4.9	5.6	19.1		43.1

Note: Average of four fields.

Table 14-9. Farmers' Perception of Good and Bad Fallow Species

Species	*Good Fallow Plant (% responding "good")*	*Bad Fallow Plant (% responding "bad")*
Good species		
Chromolaena odorata	85	0
Castanopsis hysterix	20	9
Bambusa tulda	15	0
Dendrocalamus brandisii	15	0
Bad species		
Cratoxylon prunifolium	2	55
Symplocos racemosa	0	26
Imperata cylindrica	5	24
Ageratum conyzoides	5	12

Note: Sixty-six respondents were asked which plants they liked to have in their fallow fields (good plants) and which they preferred not to have (bad plants).

Table 14-10. Farmers' Response to the Question "Why is *C. odorata* a Good Fallow Species?"

Reason	*Frequency (% respondents)*
Good rice yield, or rice grows well	50
Easy to control when weeding	18
Good soil or good fertilizer	14
Shorter fallow	9
Easy slashing	9
Burns well	7
Fast growing	2
Many seeds	2
Good soil moisture	2

Note: Forty-four respondents.

Table 14-11. Farmers' Perception of the Potential of *C. odorata* for Fallow Improvement

Questions	*Positive Response (% respondents)*
Is *C. odorata* suppressing weeds?	
Weeds in general	40
Mimosa invisa	0
Imperata cylindrical	33
Ageratum conyzoides	46
Potential for fallow improvement	
Are shorter fallows possible with *C. odorata*?	75
Is it liked in fallow fields?	68
Would you like more *C. odorata*?	32
Are you doing anything to increase it?	11

Summary and Conclusions

C. odorata has various properties that make it a promising fallow improvement species under current conditions in northern Laos. These properties include self-seeding, high biomass production, a wide range of adaptation, and suppression of weeds and nematodes. Although many have suggested that legumes would make better fallow plants (Robison and McKean 1992) and that *C. odorata* has serious adverse effects on agricultural productivity (McFadyen 1992; Waterhouse 1994), we have yet to identify a suitable legume that would satisfy the needs of upland Laos.

Farmers clearly recognize that the presence of *C. odorata* in the fallow has positive effects on a succeeding rice crop. However, they remain anxious about weeding requirements and, given that weeding is the single most serious constraint to upland rice production, their reluctance to encourage *C. odorata* in fallow vegetation is understandable. At this stage, neither farmers nor researchers have sufficient understanding of the effects of *C. odorata* on fallow vegetation, soil properties, and rice crops. Positive effects on rice performance observed by farmers may be mostly due to its suppression of both *A. conyzoides* and nematodes.

It is expected that *C. odorata* will continue to be a preferred fallow species for slash-and-burn systems where no other benefits, such as livestock fodder, are expected. As fallow periods continue to become shorter, *C. odorata's* ability to suppress weeds and nematodes could become more important. However, whether it will be considered a serious weed or a preferred fallow species will depend largely upon evolving land-use systems. There is a danger that the species could become a serious nuisance as slash-and-burn rice production systems are replaced by agricultural practices that include grazed fallow, crop rotation, and fruit or timber production.

It is quite likely that the controversy surrounding this plant will continue. While the debate persists, the potential of *C. odorata* as a fallow plant should not be ignored. Future studies seeking out and comparing improved fallow systems should include *C. odorata*, and particular attention should be given to the following issues:

- Effects on nematodes;
- Allelopathic effects on weeds and crops;
- Efficiency in nutrient mobilization;
- Effects on biological, chemical, and physical soil properties; and
- Integrated weed management systems that reduce weeding labor inputs and optimize *C. odorata* cover after rice harvest.

References

Ambika, S.R., and Jayachandra. 1992. Allelopathic Effects of *C. odorata* (L.) King and Robinson. In: *Proceedings of the First National Symposium on Allelopathy in Agroecosystems*, edited by P. Tauro and S.S. Narwal. Hisar: CCS Haryana Agricultural University and the Indian Society of Allelopathy.

Atu, U.G. 1984. Effect of Cover Plants in Fallow Lands on Root-Knot Nematode Population. *Beitrage zur Tropischen Landwirtschaft und Veterinarmedizin* 22, 275–280.

Chevalier, A. 1949. Sur une Mauvaise Herbe qui Vient D'envahir le S.E. de l' Asie (A Weed which Recently Invaded Southeast Asia) *Rev. Bot. appl.* 29, 536–537.

———. 1952. Deux Composées Permettant de Lutter Contre l' *Imperata* et Empêchant la Dégradation des Sols Tropicaux qu'il Faudrait Introduire Rapidement en Afrique Noire (Two Species of Compositae Controlling *Imperata* and Preventing Degradation of Tropical Soils, which should be Introduced Quickly in Tropical Africa). *Rev. int. Bot. appl.* 32, No. 359–360, 494–496.

de Rouw, A. 1991. Rice, Weeds, and Shifting Cultivation in a Tropical Rainforest. Doctoral thesis, Agricultural University, Wageningen, The Netherlands.

Dove, M.R. 1986. The Practical Reason for Weeds in Indonesia: Peasant vs. State View of *Imperata* and *Chromolaena*. *Human Ecology* 14, 163–190.

Fujisaka, S. 1991. A Diagnostic Survey of Shifting Cultivation in Northern Laos: Targeting Research to Improve Sustainability and Productivity. *Agroforestry Systems* 13, 95–109.

Gourou, P. 1942. *L'utilisation du Sol en Indochine*. Paris: Centre d'études de politique étrangère.

Izikowitz, K.G. 1951. *Lamet Hill Peasants in French Indochina*, Etnologiska studier 17. Goteborg: Etnografiska Museet.

Keovilayvong, K., P. Muangnalad, P. Paterson, P. Phommasay, C. Rambo, R. Rerkasem, D. Thomas, and P. Xenos. 1991. The Agroecosystem of Ban Dong: A Phu Thai (Lao Lum) village. In: *Swidden Agroecosystems in Sepone District, Savannakhet Province, Lao PDR*, Report of the 1991 SUAN-EAPI-MAF Agroecosystem Research Workshop, Savannakhet Province, Lao P.D.R. Khon Kaen, Thailand: SUAN Secretariat, Khon Kaen University, 98–113.

Kushwaha, S.P.S., P.S. Ramakrishnan, and R.S. Tripathi. 1981. Population Dynamics of *Eupatorium odoratum* in Successional Environments following Slash and Burn Agriculture. *Journal of Applied Ecology* 18, 529–535.

Leplaideur, M.A.S., and M.A. Schmidt-Leplaideur. 1985. L'herbe du Laos. *Agricultures Inter Tropiques* 9, 12–13.

Litzenberger, S.C., and Ho Tong Lip. 1961. Utilizing *Eupatorium odoratum* L. to Improve Crop Yields in Cambodia. *Agronomy Journal* 53, 321–324.

McFadyen, R.E.C. 1989. Siam Weed: A New Threat to Australia's North. *Plant Protection Quarterly* 4, 3–7.

———. 1992. A critique of the paper *Chromolaena odorata:* Friend or Foe for Resource Poor Farmers, by S.P. Field (published in *C. odorata Newsletter* 4). *C. odorata Newsletter* 5, 6.

Nakamura, N., and M. Nemoto. 1994. Combined Effects of Allelopathy and Shading in *Eupatorium odoratum* on the Growth of Seedlings of Several Weed Species. *Weed Research Tokyo* 39, 27–33.

Nakano, K. 1978. An Ecological Study of Swidden Agriculture at a Village in Northern Thailand. *South East Asian Studies* 16, 411–446.

Obatolu, C.R., and A.A. Agboola. 1993. The Potential of Siam Weed *(Chromolaena odorata)* as a Source of Organic Matter for Soils in the Humid Tropics. In: *Soil Organic Matter Dynamics and the Sustainability of Tropical Agriculture: Proceedings of an International Symposium*, November 4–6, 1993, Leuven, Belgium, edited by K. Mulongoy and R. Merckx. Chichester, UK: John Wiley & Sons, 89–99.

Olaoye, S.O.A. 1986. *Chromolaena odorata* in the Tropics and its Control in Nigeria. In: *Weed Control in Tropical Crops,* Vol. 2, edited by K. Moody. Los Baños, Philippines: Weed Science Society of the Philippines, 279–293.

Poilane, E. 1952. L'*Eupatorium odoratum* L. et d'autres planted de couverture en Indochine. (*Eupatorium odoratum* and Other Cover Crops in Indo-China). *Rev. int. Bot. appl.* 32 No 359–360, 496–497.

Rao, M.R., C.S. Kamara, F. Kwesiga, and B. Duguma. 1990. Methodological Issues for Research on Improved Fallows. *Agroforestry Today* 2, 8–12.

Rao, R.Y. 1920. *Lantana* Insects in India, Memoirs, Department of Agriculture in India. Calcutta: *Entomol. Series* 5, 239–314.

Robison, D.M., and S.J. McKean. 1992. *Shifting Cultivation and Alternatives: An Annotated Bibliography, 1972–1989.*, Wallingford, UK: CIAT/CAB.

Roder, W. 2001. *Slash and Burn Rice Systems in the Hills of Northern Lao PDR: Description, Challenges, and Opportunities*. Los Baños, Philippines: International Rice Research Institute, 201.

———, S. Phengchanh, B. Keoboualapha, and S. Maniphone. 1995a. *Chromolaena odorata* in Slash and Burn Rice Systems of Northern Laos. *Agroforestry Systems* 31, 79–92.

———, B. Phouaravanh, S. Phengchanh, and B. Keoboualapha. 1995b. Relationships between Soil, Fallow Period, Weeds, and Rice Yield in Slash and Burn Systems of Laos. *Plant and Soil* 176, 27–36.

———, S. Phengchanh, and B. Keoboualapha. 1997a. Weeds in Slash and Burn Rice Fields in Northern Laos. *Weed Research* 37, 111–119.

———, S. Phengchanh, and S. Maniphone. 1997b. Dynamics of Soil and Vegetation during Crop and Fallow Period in Slash and Burn Fields of Northern Laos. *Geoderma* 76, 131–144.

———, B. Keoboualapha, S. Phengchanh, J.C. Prot, and D. Matias. 1998. Effect of Residue Management and Fallow Length on Weeds and Rice Yield. *Weed Research* 38, 167–174.

Ruthenberg, H. 1980. *Farming Systems in the Tropics*, 3rd ed. Oxford, UK: Clarendon.

Subramaniyan, S.T.I. 1985. Effect of *Eupatorium odoratum* Extracts on *Meloidogyne incognita. Indian Journal of Nematology* 15(2), 247.

Tjitrosoedirdjo, S., S.S. Tjitrosoedirdjo, and R.C. Umaly. 1991. The Status of *Chromolaena odorata* (L.) RM King and H. Robinson in Indonesia. In: *Proceedings of the Second International Workshop on Biological Control of* Chromolaena odorata, edited by R. Muniappan and P. Ferrar. Biotrop Special Publication No. 44, 57–66.

Torres, D.O., and E.C. Paller. 1989. The Devil Weed (*Chromolaena odorata* RM King and H Robinson) and its Management. *SEAWIC Weed Leaflets* 4, 1–6.

UNDP (United Nations Development Program). 1994. *Development Cooperation: Lao People's Democratic Republic 1993 Report.* Vientiane, Lao PDR: UNDP.

Vidal, J. 1960. *La vegetation du Laos.* Vol. 2. Toulouse: Douladoure.

Waterhouse, D.F. 1994. *Biological Control of Weeds: Southeast Asian Prospects.* Canberra: ACIAR (Australian Centre for International Agricultural Research).

Chapter 15

Management of Fallows Based on *Austroeupatorium inulaefolium* by Minangkabau Farmers in Sumatra, Indonesia

Malcolm Cairns

Shifting cultivators commonly associate some fallow succession species with accelerated soil rejuvenation and deliberately intervene to promote their dominance. These pioneer species are characterized by prolific seeding, rapid establishment, and aggressive competition for available soil moisture and nutrients. They efficiently scavenge labile nutrients that might otherwise be lost through leaching and runoff during the early fallow period and immobilize them in the vegetation biomass. Later, these nutrients can be applied to crop production when the fallow is reopened. This minimalist approach to fallow management allows fallow functions of soil rehabilitation and weed suppression to be accomplished more quickly, resulting in a shortening of the fallow period and intensification of the swidden cycle.

However, the same properties that are effectively harnessed by swiddenists, who manage them as spontaneous cover crops or green manures, have given some of these species the reputation of noxious weeds in more permanent farming systems. For some years, scientific literature has been debating the seemingly urgent need to control the spread of the "noxious weed" *Chromolaena odorata* (L) R.M. King & H. Robinson (syn. *Eupatorium odoratum* Linn.), which has aggressively colonized increasing expanses of idle land in Asia and Africa (Bennett and Rao 1968; Ivens 1974; Castillo et al. 1981; Cock and Holloway 1982; Garcia 1986; Torres and Paller 1989, to name a few). While no one denies that *Chromolaena odorata* represents a serious weed threat to cattle ranchers and plantation owners, some researchers point out that it has redeeming qualities as a green manure (Mohan Lal 1960; Litzenberger and Ho Tong Lip 1961; Joseph and Kuriakose 1985). Others conclude that it plays a valuable role in the fallow successions of semipermanent cropping patterns (Nemoto et al. 1983; Dove 1986; Agbim 1987; de Foresta and Schwartz 1991; Slaats 1993; Roder et al., Chapter 14), and they caution against embarking on costly eradication programs until the impacts of *Chromolaena* are more completely understood (Field 1991; de Foresta 1993).

Conflicts of interest between ranchers and plantation owners, on one hand, and shifting cultivators on the other, are almost guaranteed because of the nature of their respective enterprises. But problems arise when government policies, programs, and research agenda inequitably address the concerns of affluent and politically

---

Malcolm Cairns, Department of Anthropology, Research School of Pacific and Asian Studies (RSPAS), Australian National University, Canberra, ACT 0200, Australia.

connected plantation owners and ranchers at the expense of typically resource-poor swidden farmers, whose subaltern voice is seldom heard.

This chapter contributes to the debate by presenting the findings of a study of Minangkabau farmers' perceptions of another member of the Asteraceae family, *Austroeupatorium inulaefolium* H.B.K. (see color plate 16). It appears to share similar biological properties and to occupy the same ecological niche as *C. odorata*, but it is restricted to higher altitudes of 200 to 1,800 m above sea level (asl) (Backer and van Den Brink 1965). Because of this, it has not attracted the same attention as *C. odorata*. Preliminary investigations suggest, however, that it may be a promising successional species for high-elevation fallow rotation systems, and it may be capable of rehabilitating critical land colonized by *Imperata cylindrica*.

Objectives

The study set out to achieve the following objectives:

- Explore indigenous knowledge of *A. inulaefolium* accumulated by Minangkabau farmers since its introduction to West Sumatra, and document examples of their innovations and experimentation in its use;
- Provide a voice for resource-poor farmers in the highlands of West Sumatra regarding their valuation of *A. inulaefolium* and its role in their farming systems;
- Elicit farmer perceptions on the succession dynamics between *A. inulaefolium* and less desirable fallow succession species such as *Imperata cylindrica* and Pteridophyta spp. (ferns);
- Compare the aerial biomass production and nutrient content of *A. inulaefolium* fallows at different ages with alternative *I. cylindrica* and Pteridophyta-dominated succession communities; and
- More fully understand the ecological dynamics of fallow succession in semipermanent cropping systems and the role *A. inulaefolium* is playing in the agroecosystem.

Description

Austroeupatorium inulaefolium (Kunth) R.M. King & H. Robinson (syn. *Eupatorium pallescens* DC and *E. javanicum* Boerl.), of the Asteraceae family (previously called Compositae), is an aggressive, fast-growing perennial shrub that has become naturalized in parts of Indonesia following its introduction from South America late in the 19th century (Stoutjesdijk 1935). It first spread into West Java under the local name *kirinjoe* and is now widely distributed in Java, Sumatra, and the Moluccas. It is herbaceous when young, but the stems become woody when mature, often forming dense thickets about three meters high. *A. inulaefolium* is not photoperiod sensitive and forms large inflorescences of small, fragrant white flowers throughout the year (Figure 15-1). Large numbers of seeds with pappi are widely dispersed by the wind. This is the usual means of reproduction, but propagation can also be vegetative, sprouting from root or stem cuttings.

Swiddenists find that *A. inulaefolium*'s prolific reproductive potential and fast growth make it a common weed during the cropping season, but young plants are easily controlled by systematic uprooting. As shown in Figure 15-2, it does not have a taproot but a shallower root system with dense lateral branching.

> Copiously branched, the leaves vary from rhomboid-oblong to rhomboid-lanceolate, with an abruptly contracted, rather short, narrowed base, and a very long-acuminate, acute apex, in the higher part serrate or very rarely entire, on the lower surface light green, shortly pubescent on both surfaces, on the lower finely glandular, 7 to 18 cm. (1/2 to 3 cm. long petiole disregarded) by 2.5 to 8 cm.; upper ones smaller. (Backer and van Den Brink 1965, *379*)

Figure 15-1. Inflorescence and Leaf Structure of *Austroeupatorium inulaefolium* (Kunth)
Source: R.M. King & H. Robinson.

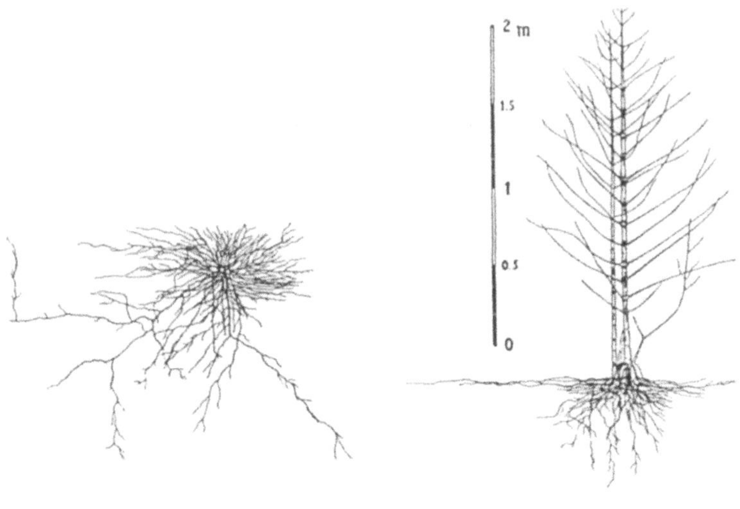

Figure 15-2. Aerial and Root Structure of *Austroeupatorium inulaefolium,*
Source: Coster 1935, *872.*

A. inulaefolium prefers regions that are not overly dry, with an altitude range of 200 to 1,800 m asl. It is often the dominant colonizer in early successional communities on fallowed land and appears to have largely displaced native succession species. After the first year, fallow regrowth is often composed of dense, almost monospecific thickets of *A. inulaefolium* that could prevent the intrusion of pioneer tree species and delay forest regeneration.

In his 1935 report, Stoutjesdijk pointed to *A. inulaefolium's* fast growth and high biomass production, copious shedding of branch and leaf litter, fast fertility regeneration, and reputation for invading and killing *Imperata* stands, and suggested that its promising agronomic properties warranted closer investigation.

Historical Insights

Stoutjesdijk's 1935 paper divulges rich insights into the spread of *A. inulaefolium* following its introduction to West Sumatra around 1890, as a smother crop to combat *Imperata cylindrica.* In the following 20 years, a further three introductions were made to plantations in different parts of the province, and from these centers, *A. inulaefolium* spread rapidly, becoming a dominant shrub in large areas.

In addition to *A. inulaefolium*'s introduction to plantations, Stoutjesdijk described its seeds being spread along the road in the Bukitinggi area because it was expected to benefit the fallow rotation systems of local people. He noted that after kirinjoe became established as a dominant fallow species in *ladang*, or dryland fields, fertility regeneration was sped up so significantly that the fallow period could be cut in half, to three to four years, instead of the six to eight years that had previously been necessary with natural regrowth of grasses, ferns, native *belukar* (secondary forest) shrubs, and coppicing from tree stumps. This had the same effect as doubling the area of fallowed land. There was suddenly no longer any need for farmers to encroach into forest reserves, and pressure on forest margins was relieved for some time (Stoutjesdijk 1935). Of equal significance, in the absence of frequent fires, *A. inulaefolium* gained the reputation of overwhelming grass-fern vegetation, as evidenced by one of its local names, *sialak padang,* or destroyer of *Imperata* fields. When introduced by seed or cuttings into recently abandoned dryland or burnt *Imperata*-fern areas, dense *Austroeupatorium* stands were obtained within a year.

Stoutjesdijk noted that *A. inulaefolium* appeared to spread mainly in an easterly direction before predominant westerly winds. Even at that early stage, farmers widely recognized its value, particularly for growing upland rice, and he thought it likely that villagers were actively introducing cuttings into their own fields from neighboring villages. However, in areas with rubber (*Hevea brasiliensis*) or gambir (*Uncaria gambier* [Hunt.] Roxb.) plantations, the aggressive colonizing habits of kirinjoe were not appreciated, and it was considered a noxious weed requiring much time and labor to control. It posed a similar problem to cattle husbandry. Pastures at the Padang Mengatas cattle breeding station in Pajacombo were invaded, reducing their grazing area. Natural forest regeneration was also thought to be impeded by dense thickets of *A. inulaefolium,* in which few other plants could survive. Interestingly, many of these observations about *Austroeupatorium inulaefolium,* made almost 70 years ago by Stoutjesdijk, echo the current debate on *Chromolaena odorata.*

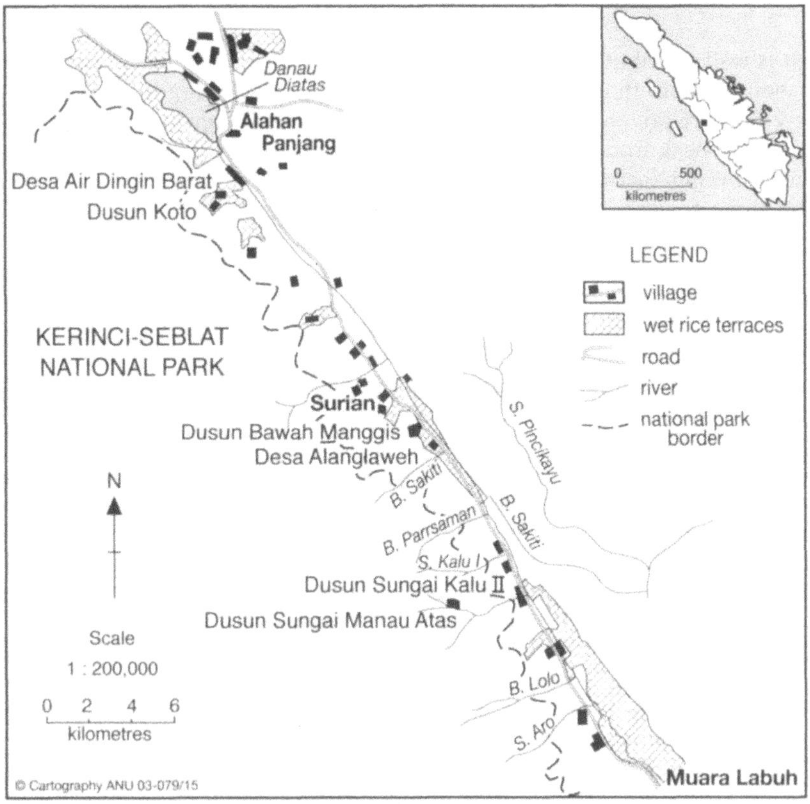

Figure 15-3. Area of Field Study in West Sumatra, Indonesia

The Study Area

The study focused on three research sites situated at intervals within the central rift valley of the Barisan mountain system, which forms the backbone of Sumatra and dominates the topography of West Sumatra Province. Research was conducted at *dusun,* or subvillage, level in Dusun Koto, of Air Dingin Barat village, Dusun Bawah Manggis, of Alang Laweh village, and Dusun Sungai Manau Atas, of Sungai Kalau II village. All three sites are located on or near an asphalt road that stretches between the major towns of Alahan Panjang (01° 05'S, 100° 48'E) and Muara Labuh (01° 28'S, 101° 04'E), in the southern part of the province (Figure 15-3). This entire area is located on the eastern slopes of Kerinci Seblat National Park and lies within the boundaries of Solok district, an area renowned in Sumatra for its superior wet rice production.

The study area comprised the southern fringes of the traditional Minangkabau heartland and included several sites that Stoutjesdijk (1935) mentioned as targets of *A. inulaefolium* introductions. The forest service introduced it from cuttings to the Air Dingin area, the northernmost portion of the study area, in 1926, and sometime later to Muara Labuh, the southernmost portion of the study area. According to Stoutjesdijk's observations, *A. inulaefolium* was well established in the Alahan Panjang area around 1933 or 1934, but it had not yet gained a foothold in the southern portion of the study area. It now appears to thrive at all three sites.[1]

[1] The upper slopes of Air Dingin (1,700 m asl) are nearing the upper limits of *A. inulaefolium*'s altitudinal range. This may be responsible for minor phenotypic differences in leaf structure noticed at this site (i.e., smaller and narrower). This may also have implications for its lack of resilience on the Air Dingin landscape and reduced ability to compete with alternative fern-*Imperata* communities.

Biophysical Environment

Rainfall is well distributed throughout the year, with wet and dry seasons much less clearly defined than in eastern Indonesia. Rainfall peaks during March to May, drops abruptly for a relatively dry season between June and August, and then builds to another rainy peak from September to January. February is usually a dry month and offers a brief respite between the two monsoons.

The three research subvillages receive an average rainfall of about 1,700 mm during 180 rainy days per year. Although this is much lower than the rainfall on the west coast of Sumatra, the total number of rainy days per year is higher. A low-intensity, drizzly rainfall is typical of the study site, posing a lesser threat of large surface runoff and consequent soil erosion, and allowing improved water percolation into the soil (Scholz 1983, *62*).

Temperatures tend to be high and constant, with small yearly amplitudes. However, altitude has a significant influence, particularly at Dusun Koto. If we assume a lapse rate of roughly 0.6°C/100 m, then Dusun Koto, at 1,600 to 1,700 m asl, will have mean daily temperatures considerably lower than those at the other two study sites. Altitude is the main factor in the evolution of very different farming systems at Dusun Koto (1,700 m) than at Dusun Sungai Manau Atas (1,000 to 1,300 m) and Dusun Bawah Manggis (900 to 1,000 m). Due to a lack of short duration rice varieties amenable to higher altitudes and cooler temperatures, Koto is restricted to a single crop of wet rice each year, resulting in large rice deficits. Its altitude is also extremely limiting on the tropical tree crops that can be incorporated into agroforestry systems on its hill slopes. On the positive side, the cooler temperatures are ideal for vegetable crops such as cabbage and potato, and, more recently, passion fruit (*Passiflora quadrangularis*), has proven to flourish there.

Topography and Land-Use Systems

In concert with altitude, topography is the other main determinant of farming systems in the study area. The floor of the central rift valley undulates. In some places it is several kilometers across, but, in others it narrows down to a corridor with steep slopes on either side of the highway. A series of young volcanoes punctuates the length of the Barisan mountain system, and those areas where the valley floor is wide have probably benefited from considerable quantities of volcanic ejecta, leaving pockets of fertile soil. These pockets, with wide valleys suitable for irrigated rice culture, tend to have the highest populations and the oldest settlements. Farming systems in these areas are based on growing irrigated rice, and rainfed cropping on the slopes of adjacent hills is of only peripheral importance. Those areas where the valley narrows, and where there is little land suitable for growing wet rice, have been settled more recently, and farming systems there rely heavily on rainfed annual or perennial crops on the mountain slopes for household food needs and to finance rice purchases.

Soils

Alluvial soils on the floor of the central rift valley are volcanic in origin. They are relatively fertile and suitable for wet rice culture. But the rice paddies are often deep, making them difficult to drain and offering limited scope for other crops.

Sloping lands on adjacent foothills are highly variable. Air Dingin slopes are characterized by black andisols of volcanic origin that are well drained, acid, very low in bases, with high levels of aluminum saturation, high organic matter, and nitrogen. Upland soils in the Alang Laweh and Sungai Kalu II sites are red-yellow ultisols, also acid, but with lower levels of C and N and higher exchangeable Ca, Mg, and K. In most parts of the study area, soils on slopes are unstable in structure, highly exposed to erosion and leaching, and prone to sliding. Soil conservation technologies are seldom practiced, and landslides often scar the landscape after heavy rains.

The Minangkabau

West Sumatra is the traditional heartland of the ancient Minangkabau culture. As early as several centuries ago, the Agam plateau was already densely settled, and irrigated wet rice cultivation was practiced extensively. From this hearth, the expanding Minang population has long been overflowing into surrounding forest frontiers in search of new land. This geographic expansion flowed in a southerly direction down the central rift valley and colonized the study area early in the 19th century, establishing independent and self-supporting villages called *nagari*.

Minangkabau society has probably drawn widest attention as the largest matrilineal culture in the world[2], and for the seeming paradox of its devout following of patriarchal-oriented Islam. The clan is the basic social unit in Minang society. Important property such as land and the large family houses, known as *rumah gadang*, are communally owned within clans and passed down through the women's line of descent. The Minangkabau continue to be governed by an ambiguous fusion of adat regulations, Islamic rules, and conventional civil law. They are noted for keen entrepreneurial qualities and a strong tradition of voluntary out-migration known as *merantau*.

Materials and Methods

Qualitative Data: Farmers' Perceptions

After several windshield surveys and many roadside discussions with farmers, the three subvillages were chosen to provide contrasts in altitude, topography, and age of settlement. The rationale was that villages representing contrasts in these three variables should also exhibit a wider range of land-use practices. It was hoped that this diversity would provide richer insights and a broader range of perceptions of *A. inulaefolium* and that this range may have fostered different kinds of beneficial management practices.

The research team conducted a total of 75 informal and open-ended interviews in the three subvillages over a six-week period in 1994. The team lived with farmers on-site to allow close interaction and to maximize opportunities for observation. Meetings were held with village leaders in the first instance to explain the purpose of the study and seek their cooperation. Respondents were then chosen largely at random but, because *A. inulaefolium* generally grows on idle land, we were biased toward farmers who were practicing fallow rotation on the mountain slopes, and toward older farmers because they could speak from longer experience and could possibly remember when kirinjoe was introduced, and they could talk about adaptive changes farmers had made since then.

Using questionnaires to structure the interviews, we asked farmers about their knowledge and experience in exploiting *A. inulaefolium*'s beneficial properties, the problems it posed to their farming systems, its successional interaction with problem weed species and its role in their bush-fallow system. Data were then tabulated and are presented in this chapter in simple frequencies and means.

Quantitative Data: Soil and Vegetation Analysis

Soil and vegetation samples were collected from a total of 17 fields to characterize upland soils in the study site and to compare aerial biomass accumulation and nutrient content of different-aged fallows dominated by different species. Seven fields dominated by *A. inulaefolium* were located, together with five dominated by *Imperata*

---

[2] Although some refer to Minangkabau as matriarchal, this is inaccurate. Although women do own clan property and enjoy a relatively high status in Minang society, political power and administration of family affairs continue to reside with men.

cylindrica and another five dominated by Pteridophyta spp.[3] The fallow succession communities were of different ages.

Within each field, four plots were selected at random and 2.5 m X 2.5 m quadrants were measured, staked and roped. All aerial biomass in each quadrant was carefully harvested, bagged, and weighed for total fresh weight of the four plots. After weighing, a few representative plants were extracted from each plot and combined as a subsample. Samples were carefully weighed for their fresh weight and then tightly sealed in plastic bags. Soil samples were then collected to a 15 cm depth from each plot where the biomass had been harvested. They were composited and thoroughly mixed, and a 500 gram composite sample was extracted to represent each fallow field.

Litter samples were segregated only in *A. inulaefolium* fallows. After harvesting of the biomass from each 2.5 m X 2.5 m plot, a 50 cm X 50 cm subplot was randomly selected and the litter collected down to the mineral layer. Samples were combined for each of the four plots to obtain a composite sample.

In the laboratory, vegetation samples were chopped and subsamples were extracted and weighed and then oven-dried at 80°C for 48 hours to obtain dry weights. The subsamples were then finely ground and analyzed by the Soils Laboratory of the Faculty of Agriculture, Bogor Agricultural Institute, for analysis of N, P, K, Ca, Mg, and C-organic. Composite litter samples from each fallow field sampled were weighed and chopped into fine pieces. Subsamples were extracted and reweighed. Soil was separated from the litter using a 2 mm mesh. It was weighed and its weight subtracted from the weight of the litter subsample. The remaining vegetative matter was then subjected to the procedure described above for the aerial biomass. Soil samples were air-dried, finely ground, and analyzed for pH, C-organic, total N, available P, Ca, Mg, K, Na, CEC, Al, H+, and texture.

The small number of fields sampled resulted in a limited data set from which definitive conclusions could not be drawn. Rather, the procedure was intended to indicate trends that might enable a degree of quantitative support for qualitative data provided by respondent farmers.

Results

Throughout the study area, *A. inulaefolium* was readily recognized by farmers and known by its vernacular name *rinju*, or close variations, *linju* or *karinyu*. However, to avoid any possibility of confusion, a fresh specimen of *A. inulaefolium* was collected prior to each interview and used as the focal point of the discussion.

Since *A. inulaefolium* was first introduced to West Sumatra in the closing years of the 19th century, it was not surprising to find that most respondents had been familiar with it all their lives and considered it native to the area. The exception was a widely recounted story in Air Dingin that had been handed down from the previous generation. It attributed the origin of *A. inulaefolium* to a Japanese plane that overflew the area during World War Two, "spraying poison across the valley's ladang and broadcasting *A. inulaefolium* seeds."[4] Potato and sweet potato crops were said to have withered, turned black, and died, and to have not grown well in the area

---

3 Although farmers generally refer to ferns, or *pakis*, collectively, thickets were usually populated by *pakis rasan (Gleichenia linearis)* and *pakis gala (Pleocnemia leuzeana)*, both common Pteridophyta species in higher mountain zones.

4 Dove (1986) notes that peasants commonly attribute the origins of new plant species to external political authorities to which they have been subjected, reflecting their belief in the power and purpose of ruling authorities. Broadcasting seed from planes is a common theme in these beliefs. During fieldwork in the Philippines, the author documented similar stories on the origin of *hagonoy (Chromolaena odorata)* by such widely disparate swidden communities as the Tala-andig in Mindanao, Batak in Palawan, and Hanunóo in Mindoro. In Oriental Mindoro, where cattle range land is common and encroachment of *Chromolaena odorata* into pastures is a growing concern, some Hanunóo even suggested a theory that Australia was responsible for broadcasting hagonoy over Mindoro with the intention of destroying Philippine grasslands and reducing competition in global beef markets, a kind of ecological sabotage (author's field notes).

since. Then rinju began to appear on the landscape in ever increasing quantities. The farmers believed that poisoning crops was part of a Japanese strategy to weaken the Dutch food supply in their struggle for control over Sumatra. If *A. inulaefolium* was introduced to the Air Dingin area in 1926 (Stoutjesdijk 1935) and was widespread in the vicinity of Alahan Panjang by 1933 or 1934, then it is conceivable that rinju did, indeed, become an increasingly conspicuous part of Air Dingin's floral landscape during World War Two.

Farmer Valuation of A. inulaefolium

If *A. inulaefolium* began to aggressively colonize the research area during World War Two, then Minangkabau farmers have since accumulated half a century of experience in adapting their farming systems around this pioneer species and experimenting in the benefits of managing it. As illustrated in Figure 15-4, farmer valuation of *A. inulaefolium* in the study area is overwhelmingly positive. The evidence suggests, however, that recent intensification of Minang farming systems, particularly with the introduction of inorganic fertilizers and the growth of permanent cultivation, probably means that *A. inulaefolium*'s role as an improved fallow is becoming redundant.

Utility within Farming Systems

The following is a compilation of insights into the utility of *A. inulaefolium,* gathered from the 75 farmers interviewed during the study. Although agronomic benefits dominate the list, several unexpected household uses were also revealed.

1. Fertility Indicator. Farmers universally recognize the presence of *A. inulaefolium* as a reliable indicator of soil fertility. Soil under rinju thickets will be black and moist, with a soft tilth that enables easy cultivation. If it grows on red, less fertile soils, its leaves will appear yellowish and unhealthy, compared to the dark green vigor of plants on black soils. Hence, fallow communities composed of dense rinju thickets are a farmer's best guarantee of a bountiful crop when the field is reopened and planted. Presence of *A. inulaefolium* and other broad-leafs associated with soil fertility, such as *Crassocephalum crepidioides*, are major criteria used in choosing desirable swidden sites (see Figure 15-5). Field observations by the research team corroborated farmer perceptions of positive physical properties associated with soil in *A. inulaefolium* fallows. The soil was black, with a high humus content, and soft to the extent that seeds could be dibbled without any tillage.

Claims of higher fertility attributed to *Austroeupatorium* soils were verified by laboratory analysis. Table 15-1 presents the mean chemical properties of soils under the sampled fern, *Imperata* and *A. inulaefolium* fallows. A serious weakness in the data is that fallows sampled in the black andisols at the Air Dingin site were only fern communities. Therefore, it is not possible to make direct comparisons between soils found under fern fallows and the very distinct red-yellow ultisols on which *A. inulaefolium* and *Imperata* fallows were sampled at the other two sites. (See Table 15-2 for a comparison of chemical properties of andisols and ultisols in the study area.)

Despite this limitation, several conclusions can be drawn from the data. The clearest trend is that *A. inulaefolium* was observed on less acidic soil (pH 5.4) compared to *Imperata* (pH 4.88) (see also Figure 15-6a). Comparison of *Austroeupatorium* [*A*] and *Imperata* [*I*] soils, all red-yellow ultisols, suggests that soils supporting *Austroeupatorium* communities are markedly higher in available P (1.89 [*A*] and 0.90 [*I*] mg/kg); Ca (9.89 [*A*]: 6.85 [*I*] me/100g); Mg (2.78 [*A*] and 2.01 [*I*] me/100 g); total exchangeable bases (13.50 [*A*] and 9.47 [*I*] me/100 g); and derived base saturation (58.61 [*A*] and 40.26 [*I*] %). Furthermore, the lower pH of soils where *Imperata* dominated suggests both a lower rate of mineralization and nutrient availability to crops. These data clearly support farmer perceptions that fallow vegetation provides an indication of soil chemical properties.

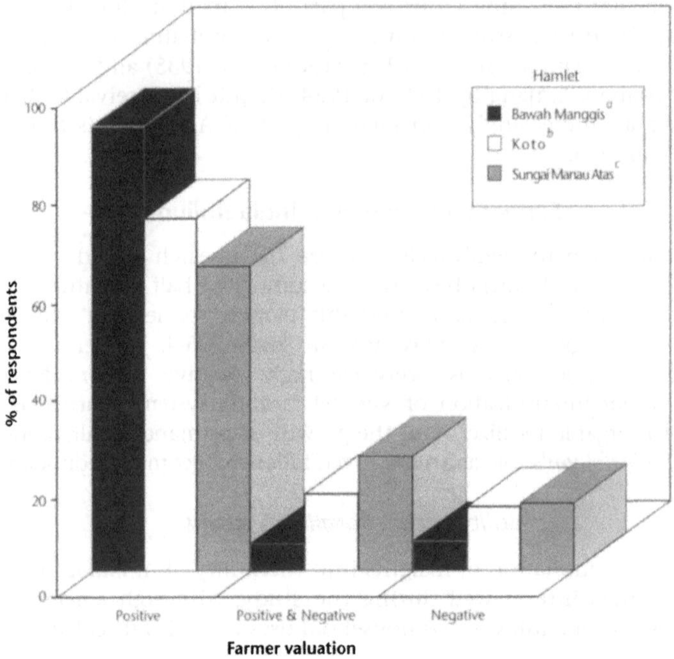

Figure 15-4. Farmer Valuation of *Austroeupatorium inulaefolium*
Note: Total number = 73. [a] Mid-altitude mature (MAM) settlement; [b] High-altitude mature (HAM) settlement; [c] Mid-altitude pioneer (MAP) settlement.

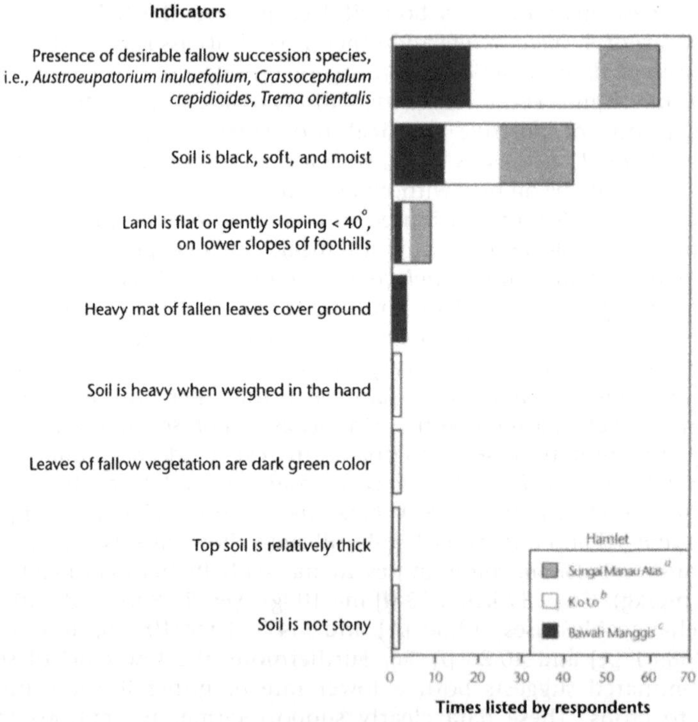

Figure 15-5. Farmer Indicators of a Desirable Site to Reopen a Swidden
Note: Total number = 73. [a] Mid-altitude pioneer (MAP) settlement; [b] High-altitude mature (HAM) settlement; [c] Mid-altitude mature (MAM) settlement.

Table 15-1. Chemical Properties of Soils under Compared Fallow Communities*

Soil Properties	*Pteridophyta spp.* [a]	Imperata cylindrica[b]	Austroeupatorium inulaefolium[b]
pH (H_2O, 1:1)	4.80	4.88	5.48
pH (KCl)	3.80	3.84	4.50
C-organic (%)	11.17	2.85	2.42
N-total (%)	0.65	0.25	0.24
Available P (mg/kg)	0.40	0.90	2.03
Exchangeable bases: (me/100 g)			
Ca	1.04	6.85	10.26
Mg	0.48	2.01	2.96
K	0.22	0.32	0.53
Na	0.31	0.29	0.34
Total	2.06	9.47	14.09
1 *N*-KCl extractable: (me/100 g)			
Al	2.91	2.19	3.87
H	0.29	0.25	0.23
ECEC (me/100 g)	5.26	11.47	14.81
CEC (me/100 g)	47.88	23.72	22.70
Base saturation (%)	4.32	40.26	60.82

Notes: Methods used for soil analysis: C-organic, Walkley & Black; N-total, Kieldhal; Available P, Bray I/Olsen. Soil analysis conducted at Soil Science Department laboratory, Bogor Agricultural University, on June 9, 1994. *Excludes soil from one 15-year *Austroeupatorium inulaefolium* fallow mistakenly sampled and found to test: pH (H_2O, 1:1), 4.8; pH (KCl), 3.8; C, 2.99%; N-total, 0.26%; available P, 0.9 mg/kg; Ca, 6.9 me/100 g; Mg, 1.28 me/100 g; K, 0.33 me/100 g; Na, 0.28 me/100 g; total, 8.79 me/100 g; Al, 2.42 me/100 g; H, 0.26 me/100 g; ECEC, 11.47 me/100 g; CEC, 21.5 me/100 g; base saturation, 40.9%. [a] Mixed communities of *Gleichenia linearis* and *Pleocnemia leuzeana.* Black andisols in Air Dingin Barat village. [b] Red-yellow ultisols in Alang Laweh and Sungai Kalau II villages.

While not comparable to the *Imperata-Austroeupatorium* data (ultisols), data were noteworthy in showing that the andisols colonized by fern thickets in Air Dingin were acidic (4.8);[5] very high in C-organic (11.17%) and total N (0.65%);[6] were impoverished in exchangeable Ca (1.04 me/100 g), Mg (0.48 me/100 g), and total bases (2.06 me/100 g); and contained levels of exchangeable Al (2.91 me/100 g) high enough to be toxic to some crops, such as corn.[7]

2. "Leafy Fertilizer," or *Pupuk Hijau.* Using methods directly related to the above point, farmers have learned to capitalize on the high biomass and nutrient composition of rinju thickets by incorporating it into the soil in several ways:

Slash-and-Burn. When reopening a fallowed field, standard slash-and-burn techniques are used. Sometimes rinju biomass may be gathered from other areas and added to the slash to increase the intensity of the burn and the volume of nutrient-rich ash produced.

Slash-and-Mulch. Annuals: After slashing, the aerial biomass is arranged in rows, and annual crops are planted between them. After the mulch has decomposed for several months, it is gradually pushed to the base of the crop, where it slowly releases nutrients to the soil (see Figure 15-9a). Perennials: Despite the potential for

[5] Pteridophyta spp. are widely associated with acid soils. Potatoes, a popular cash crop at Air Dingin's high altitudes, are relatively tolerant of acid soils. Some farmers in nearby Alahan Panjang are known to taste soil to determine if it needs more calcium.

[6] High N in the andisols reflects high organic matter.

[7] Exchangeable Al/CEC shows that a high percentage of the exchange complex sites is occupied by Al. High total CEC in the andisols reflects their specific mineralogy and high organic matter content.

competition between *A. inulaefolium* and young tree crops, some farmers describe it as a desirable cover crop in plantations. They say it provides shading from direct sunlight, maintains a moist microclimate, and prevents both erosion and incursion by *Imperata cylindrica*, ferns, and other noxious weeds. Although uprooting rinju provides longer-term control, farmers intentionally restrict themselves to periodically slashing it, thus ensuring fast regrowth from the stumps and continued vigor. Rinju slash is left as mulch between tree rows.

Slash-and-Bury. This practice is preferred for younger rinju fallows that still have soft, herbaceous stems. After slashing *A. inulaefolium* and chopping it into pieces, farmers work it into the soil with a hoe or, place young leaves directly into holes with the seed at time of planting, for example, potatoes. Farmers who permanently cultivate their fields without any fallow sometimes collect young rinju from outside areas and incorporate it into raised garden beds as a green manure for vegetable crops.

In addition to actively managing *A. inulaefolium* to build soil fertility, farmers also recognize that by allowing land to lie fallow with rinju-dominated successions, copious quantities of fallen leaves, dead branches, and other organic detritus accelerate the release of mineral nutrients. Fertilizer application guidelines are often cited by farmers to illustrate the superior fertility of *A. inulaefolium* fallows. The following are two typical examples of this:

- If reopening a rinju fallow, application of only 50 kg of a fertilizer mixture will provide good crop yields. If reopening an *Imperata* fallow, 300 kg of the same mix must be applied and the crop will still not perform as well.
- If planting potatoes in a field newly cleared from an *A. inulaefolium* fallow, 1 kg of inorganic fertilizer is needed per 1 kg of potato seed planted. If the fallow community was *alang alang (I. cylindrica)*, the fertilizer application rate has to be doubled to achieve the same results.

To test farmers' contention that *A. inulaefolium* successions not only indicate fertility but also actively enhance soil chemical properties, soils under *Austroeupatorium, Imperata,* and fern fallows were analyzed for significant trends during two-year fallow periods. Results are presented in Figures 15-6a to 15-6g. There is a moderate rise in the C/N ratio in *Austroeupatorium* fallows[8] (Figure 15-6c) and a decline in N in fern fallows (Figure 15-6e). Overall, there is little evidence that *Austroeupatorium* improves soils during a two-year fallow period. This is not unexpected, since two years is quite short to measure significant changes in soil chemical properties. Soil fertility regeneration within shifting cultivation systems is predominantly associated with nutrient accumulation in the aerial biomass that becomes available for crop use only after incorporation into the soil, usually by burning or mulching.

Measurement and analysis of total aboveground biomass, (litter plus aerial biomass), revealed significant trends that strongly support *A. inulaefolium*'s reputation as an effective fallow species (Figures 15-7a to 15-7e, and 15-8). Table 15-3 presents biomass and nutrient accumulation at the end of a two-year fallow period for each of the three succession species and indicates levels of significance. While both *Imperata* and *Austroeupatorium* successions exhibited increases in dry matter biomass during the fallow period, *A. inulaefolium* accumulated 16.9 t/ha by the end of two years, about 2.5 times that of *Imperata* (6.7 t/ha) (Figure 15-7a). This supports farmers' observation that *Austroeupatorium* generates high quantities of biomass in a short time, contributing to soil fertility regeneration, high levels of soil organic matter, and enhanced physical properties.

---

8 All three fallow succession communities examined have globally low C/N, indicating that N should be relatively available to crops.

Table 15-2. Comparison of Properties of Ultisols and Andisols in Study Area

Properties	*Ultisols*	*Andisols*
Research Dusun	Bawah Manggis Sungai Manau Atas	Koto
Soil color	Yellow-red	Black
Fallow vegetation	*Imperata cylindrica* *Austroeupatorium inulaefolium*	Pteridophyta spp.
pH (H_2O, 1:1)	5.14	4.80
pH (KCl)	4.14	3.80
Soil properties:		
C-organic (%)	2.51	11.17
N-total (%)	0.24	0.65
Available P (mg/kg)	1.45	0.40
Exchangeable bases (me/100 g):		
Ca	8.22	1.04
Mg	2.35	0.48
K	0.42	0.22
Na	0.31	0.31
Total	11.29	2.05
1 *N*-KCl Extractable (me/100 g):		
Al	3.12	2.91
H	0.27	0.29
CEC (me/100 g)	22.26	47.88
Base saturation (%)	49.99	4.30
Sand (%)	34.48	19.84
Silt (%)	31.47	36.66
Clay (%)	34.05	43.50

Notes: Methods used for soil analysis: C-organic, Walkley & Black; N-total, Kieldhal; Available P, Bray I/Olsen. Soil analysis conducted at Soil Science Department laboratory, Bogor Agricultural University, June 9, 1994.

Chemical analysis of vegetation revealed similar trends in nutrient accumulation. C accumulation of *Austroeupatorium* (8.2 t/ha) was double that of *Imperata* (3.4 t/ha) (Figure 15-7b). N content of two-year *Austroeupatorium* fallows was 183 kg/ha and statistically significant. In contrast, *Imperata* remained static at 26 kg/ha, showing how little N emanates from *Imperata*-dominated successions (Figure 15-7c). P, which tends to be limiting on red-yellow podzols, was much more abundant in two-year *Austroeupatorium* vegetation (20 kg/ha) than in *Imperata* (6.4 kg/ha) (Figure 15-7d).

Trends in K accumulation were highly significant for both species but, again, yields of *Austroeupatorium* (184 kg/ha) were roughly double those of *Imperata* (95 kg/ha) (Figure 15-7e). Biomass and nutrient accumulation over time in fern vegetation sampled on Air Dingin andisols were consistently low or nonexistent. Finally, chemical analysis of nutrient accumulation in surface litter from fallows dominated by *A. inulaefolium* showed clear increments of C, N, P, and K over the relatively brief two-year period (Figure 15-8).

This data set lends quantitative support to Minang farmers' perception of rinju's ability to substitute for inorganic fertilizer applications and demonstrates the rationality of their practices in managing it as a green manure.

3. Improves Soil Structure. High volumes of litterfall in *A. inulaefolium* fallows are converted by termites, earthworms, beetles, and other invertebrate decomposers into a black topsoil horizon rich in organic matter. This results in reduced soil bulk density and a softer texture requiring minimal tillage before planting.

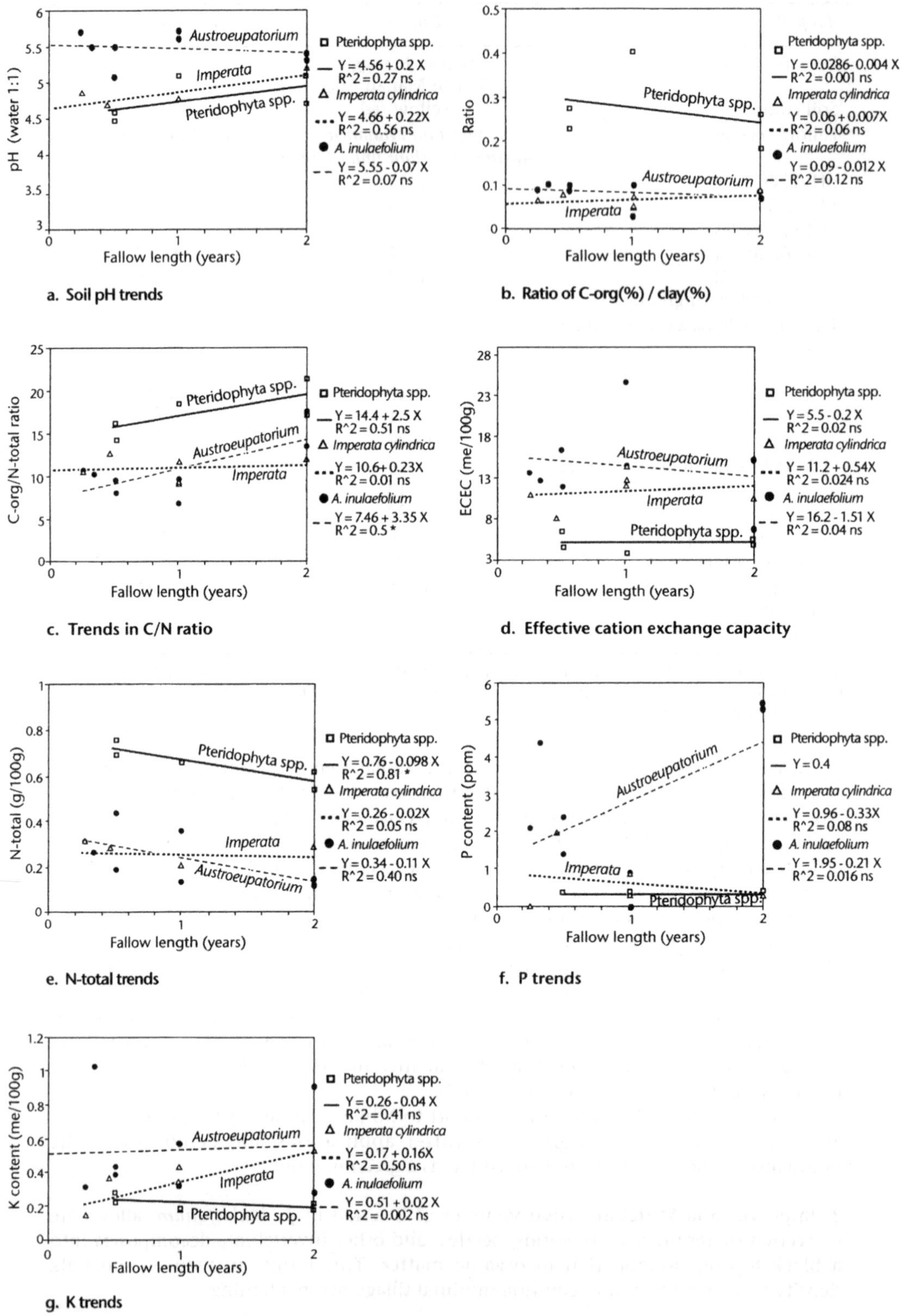

Figure 15-6, a-g. Trends in Soil Chemical Properties during Fallow Period

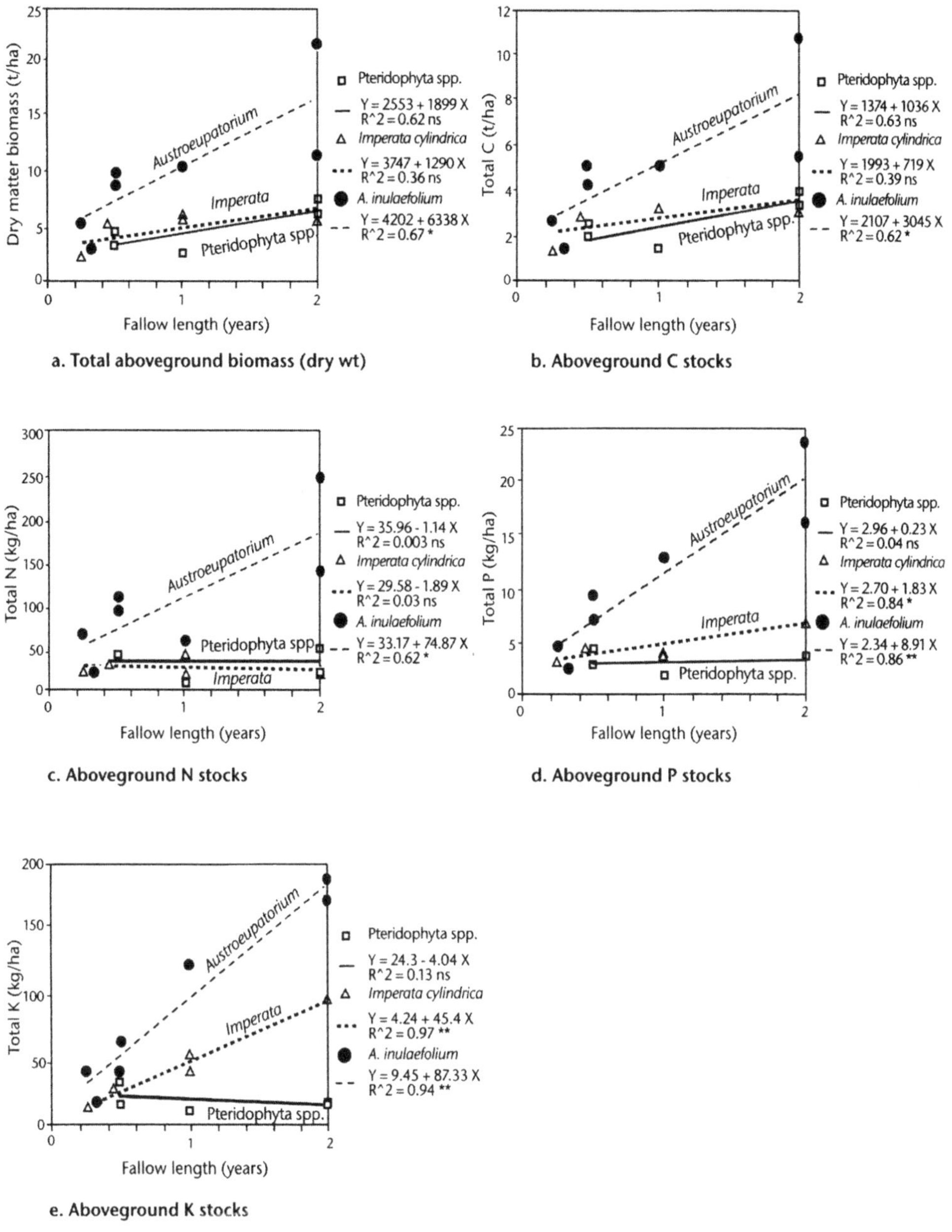

Figure 15-7, a-e. Biomass and Nutrient Accumulation during Fallow Period

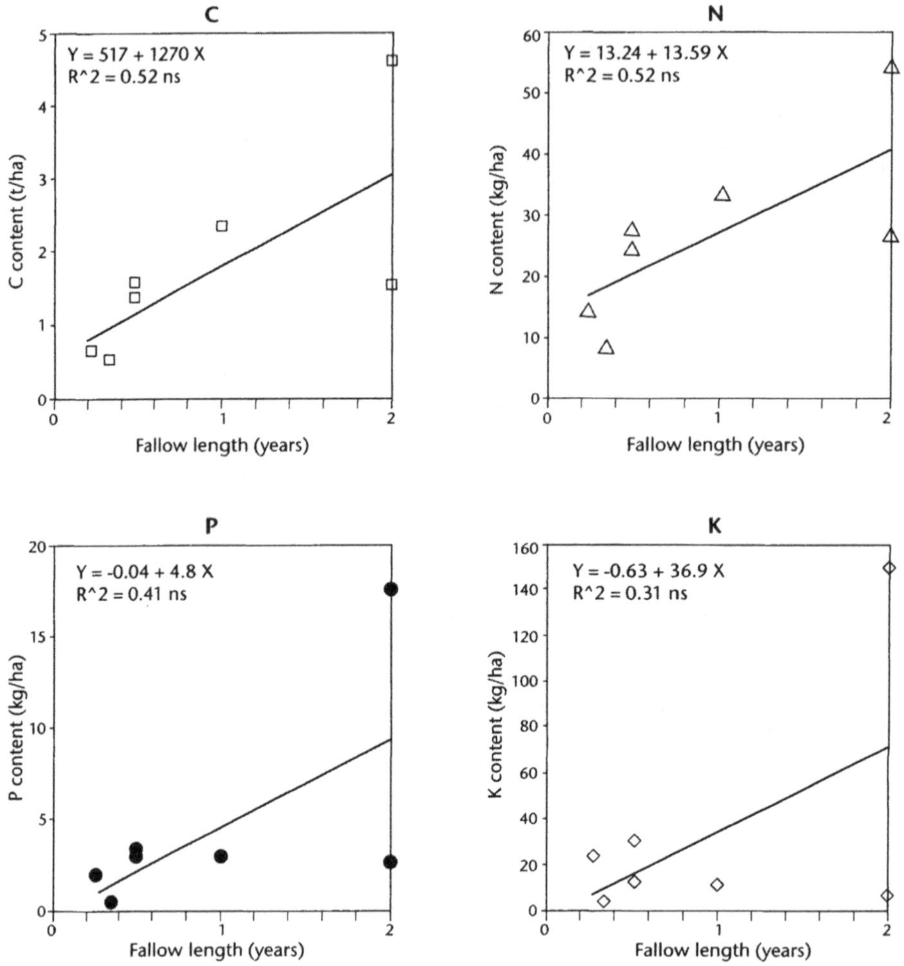

Figure 15-8. Nutrient Accumulation in Surface Litter of *Austroeupatorium inulaefolium*–Dominated Fallows

4. Maintains Soil Moisture. The increase in soil humus probably improves moisture retention, since more rainfall enters the groundwater and less flows away as surface runoff. The humid microclimate is further enhanced by *A. inulaefolium*'s thick canopy, which shields direct solar radiation and reduces water evaporation.[9]

5. Suppresses Noxious Weeds. In enumerating *A. inulaefolium*'s positive attributes, many farmers describe its ability to both shade out and suppress *Imperata* and other problem weeds during the fallow period or to prevent them from becoming established in the first place.

Immediately after field abandonment, the first floral community to colonize fields is often composed of a mixture of fertility-associated herbs such as *Ageratum conyzoides* and *Crassocephalum crepidioides*, and difficult-to-control grasses such as *Imperata cylindrica, Paspalum conjugatum,* and *Panicum palmifolium*. This early succession community plays an important role in stabilizing otherwise vulnerable soil and may provide a moist microclimate conducive to germination of *A.*

[9] The moist microenvironment favors fungal growth, however, and may underlie the increased disease problems in potato and sweet potato crops in Air Dingin following the arrival of *A. inulaefolium,* rather than a Japanese aerial spraying program.

inulaefolium seeds. Three months after crop harvest, the soil is almost completely covered by fallow vegetation. *A. inulaefolium* then emerges as the dominant species, already one to one-and-a-half meters high. By six months, rinju may already be two to three meters high and have a thick canopy. The grasses and herbs will have already disappeared. In subsequent years, the fallow vegetation develops into a dense thicket three to four meters high. Mature *A. inulaefolium* develops woody stems, often several inches in circumference, and becomes heavily branched. When these dense *A. inulaefolium* fallows are reopened for cultivation, most problem weeds will no longer be present. Farmer strategies for actively encouraging establishment of *A. inulaefolium* to avoid encroachment by undesirable species are discussed later.

6. Nurse Crop. Cases where I witnessed *A. inulaefolium* managed as a nurse crop were limited to plantations of young *Cinnamomum burmannii* (cassiavera) seedlings. Farmers placed three poles in the ground, forming a rough perimeter marking the location of each seedling, to prevent accidents while slashing weeds. One of these poles was rinju. It was allowed to root, branch out, and form a light canopy. The fledgling cassiavera seedlings were thus shielded from direct sunlight and afforded a more humid microclimate conducive to early growth. After one year, the cassiavera was considered capable of fending for itself and the rinju trimmed back (see Figure 15-9c). [10]

7. Insecticide. Although not widely practiced, some respondents reported that juice extract from *A. inulaefolium* leaves had insecticidal properties and could be used effectively on chilis and onions. Leaves from *A. inulaefolium*, *Tagetes* sp. (marigold), and tobacco were ground and the resulting extract mixed with water and sprayed. One articulate Javanese farmer described mixing one liter of rinju extract with four liters of water and spraying it on chili and soybeans. He claimed this practice had dual benefits of killing insects and stimulating crop growth.[11] The apparent nonsusceptibility of fast-growing, succulent *A. inulaefolium* to insect pests suggests that it may contain defense compounds that have insect repellant properties.[12] This warrants further investigation for potential as a botanical pesticide. [13]

8. Fencing. Farmers described collecting rinju poles and sticking them into the ground at close intervals to provide protective fencing around seedling nurseries (see Figure 15-9b) or annual crops. Poles would usually take root and become a living fence, thickening and becoming more impenetrable as the "posts" branched and propagated. An alternative method was to simply leave a hedgerow of *A. inulaefolium* around the outside perimeter when reopening a fallow, thus forming a protective

---

10 Faridah Hanum and van der Maesen (1997, *296)* report that *A. inulaefolium* is also used in Indonesia as ground cover in *Pinus merkusii, Cinchona,* and tea plantations.

11 The same farmer also described cutting mature, woody *A. inulaefolium* stems into 30 cm. segments and planting them at intervals in an *Imperata* sward. Within one year, a dense rinju thicket had formed and the *Imperata* was smothered out.

12 Doubless it also reflects *A. inulaefolium's* relatively recent introduction to Indonesia, where insects and large heraivores have not had the opportunity to evolve the strategies to counter whatever toxic or diestioninhibiting chemicals rinju may be using to defend itself. The fact that a search of the Centro Internacional de Agricultura Tropical database in Columbia revealed virtually no research on *A. inulaefolium* in Tropical America suggests that it may not be such a dominating pioneer species in its region of origin. It would be reasonable to postulate that local fauna there have succeeded in evolving a digestive capacity that thwarts *A. inulaefolium's* chemical defense and moderates its expansion.

13 Experiments suggest that *Chromolaena odorata* and other Asteraceae have nematode suppression properties. Farmers in Indonesia often feed *Tithonia diversifolia*, another member of the Asteraceae family, to goats to eliminate worms (Hambali 1994). It is unlikely that *A. inulaefolium* offers any potential as livestock fodder. Not only is it extreamly bitter and unpalatable, but experiments conducted on rats also concluded that it is hepatotoxic, or toxic to the liver (Bahri et al. 1988). Murdiata and Stoltz (1987) confirm that it contains pyrrolizidine, an alkaloid that is probably hepatoxic and is suspected to have caused death of dairy cattle that were imported into Karo district of North Sumatra and the later died with liver damage.

barrier against livestock and wildlife. The impenetrability of the hedgerow could easily be bolstered by pushing vegetative cuttings into the ground to fill any gaps.

9. Firewood. Mature, woody rinju stems appear to be an important source of firewood, particularly for communities distant from forest margins. However, their BTU value is not likely to be high and, where it is accessible, farmers probably prefer other types of firewood. Nonetheless, the role of *A. inulaefolium* in mitigating firewood harvest pressures on protected forests should not be discounted. This view is reinforced by a 1954 report that the high costs of removing *A. inulaefolium* from *Acacia decurrens* plantations were offset by its value as firewood (Hellinga 1954).[14]

10. Poles for Climbing Crops. Mature, woody rinju stems are frequently collected for use as poles for climbing crops such as beans and to support some varieties of chili.

11. Construction. Woody rinju stems are sometimes used in very rough construction projects such as field huts or livestock shelters.

12. Medicine. Respondents widely attributed the juice of crushed *A. inulaefolium* leaves with medicinal properties. It was said to be useful for first aid treatment of wounds, as a blood coagulant, and for itches.[15] Others spoke of its value in treating amoebic dysentery and stomachaches.

Table 15-3. Trends of Biomass and Nutrient Accumulation in Two-Year Fallow Vegetation

	Succession Species								
	Austroeupatorium inulaefolium [a]			Imperata cylindrica [a]			*Pteridophyta spp.* [b]		
Detail	*PV* [c]	*RP* [d]	*Sig.* [e]	*PV*	*RP*	*Sig.*	*PV*	*RP*	*Sig.*
Dry matter biomass (t/ha)	16.9	a = 4,202 b = 6,338	*	6.3	a = 3,747 b = 1,290	-	6.4	*a* = 2,553 *b* = 1,899	-
Soil properties:									
C (t/ha)	8.2	a = 2,107 b = 3,045	*	3.4	a = 1,993 b = 719	-	3.5	*a* = 1,374 *b* = 1,036	-
N (kg/ha)	183	a = 33.17 b = 74.87	*	26	a = 29.58 b = -1.89	-	34	*a* = 35.96 *b* = –1.14	-
P (kg/ha)	20	a = 2.34 b = 8.91	**	6.4	a = 2.70 b = 1.83	*	3.4	*a* = 2.96 *b* = 0.23	-
K (kg/ha)	184	a = 9.45 b = 87.33	**	95	a = 4.24 b = 45.4	**	16	*a* = 24.3 *b* = –4.04	-

Notes: One 15-year *Austroeupatorium inulaefolium* fallow mistakenly sampled contained: 35.12 t/ha dry matter biomass; 14.8 t/ha C; 323 kg/ha N; 36 kg/ha P; and 266 kg/ha K. [a] Red-yellow ultisols in Sungai Kalau II and Alang Laweh villages. [b] Black andisols in Air Dingin Barat village. [c] PV = predicted value at x = 2 years. [d] RP = regression parameters for the model Y = a + b X. [e] Sig. = degree of statistical significance: * Significant at 95% confidence level; ** Significant at 99% confidence level; - Not significant.

[14] In East Java, *Chromolaena* stems are used as firewood for making bricks and burning lime. It provides a hot fire of short duration and is not used for cooking (van Noordwijk 1997).

[15] In a separate study, farmer respondents in the Philippines claimed identical properties for the juice of crushed *Chromolaena odorata* leaves (author's field notes). Ferraro et al. (*1977, 1,618*) report that in Argentina, *A. inulaefolium* is applied externally to cleanse sores and pimples.

13. Weaning Children from Breastfeeding. In what was probably the most novel use of *A. inulaefolium* documented by this study, some women described capitalizing on its bitter taste to wean children from breastfeeding. Juice from crushed leaves was smeared on the mother's breasts often enough to convince even a thirsty and insistent youngster that his meal ticket had suddenly developed a bitter taste and it was time to look elsewhere for sustenance.[16]

a. Slashed Biomass Composted
b. Poles Used to Fence Seedling Nurseries
c. Shade for Young Tree Seedlings
d. Improved Fallow Spieces

Figure 15-9, a–d. Farmer Strategies for Managing *Austroeupatorium inulaefolium*

---

[16] Although the plant used differs from place to place, this strategy is widely employed by Southeast Asian mothers when their toddlers develop teeth and breastfeeding becomes painful. Thai women, for example, commonly use *baurapet (Tinospora* sp.*)* for the same purpose (Kurupunya 1998).

This partial list of Minangkabau strategies for exploiting the useful properties of a single shrub, developed over just 50 years since its widespread appearance, is convincing testimony of the capacity of farmer experimentation to generate innovations and knowledge. It also supports the notion that scientific insights into indigenous knowledge and decision-making processes can help shape the future directions of agricultural research. National agricultural research and extension services working in tandem with indigenous systems, rather than dismissing them, are more likely to generate appropriate solutions that address small farm priorities, use resources readily available to farmers, and are more widely adopted.

Problems Arising from an Aggressive Pioneer Species

Many of the attributes that make *A. inulaefolium* a competitive and valuable species during the fallow period are less welcome after the field has been reopened for cultivation. During the transition from fallow to cropping, farmers stop thinking of *A. inulaefolium* as a valuable, labor-reducing and nutrient-storing cover crop and begin to regard it as a problematic labor and nutrient-consuming weed. This conundrum is reflected in the following inventory of problems that farmers attributed to rinju.

Common Weed. The prolific seed production of *A. inulaefolium* and its ability to re-sprout from slashed stumps guarantees that it will be an invasive weed competing for soil nutrients with annual crops and young tree plantations.

Requires Labor to Eradicate. Its perceived weed status causes farmers to complain of the time and labor they are forced to invest in weeding rinju from cropped areas. Most readily admit, however, that young plants have shallow root systems and are easily controlled by systematic uprooting.

Shades Crops. If allowed to persist on cropped land or even on field perimeters, fast-growing *A. inulaefolium* can quickly shade adjacent crops.

Promotes Bacterial or Fungal Growth on Crops. Rinju's ability to provide shade and a moist microclimate which prompts some farmers to experiment with rinju as a nurse crop, causes problems in other circumstances. This complaint was limited to the Air Dingin study site, representing the northernmost tip of the research area. At 1,500 to 1,700 m asl, Air Dingin is the highest-altitude segment of the study transect, and farmers there have sought to capitalize on their cooler temperatures by specializing in intensive cultivation of semi-temperate vegetable crops. The cloud zone, usually at around 2,000 m, often descends to envelop the valley for days at a time and brings with it a low-intensity drizzle rainfall. At night, heat dissipates quickly and relative humidity increases as temperatures fall, often passing the dew point and leading to water condensation on foliage. These conditions are conducive to fungal and bacterial growth on crops. Air Dingin farmers sometimes lose entire crops to rotting. In these conditions, *A. inulaefolium*'s ability to provide shade and a moist microclimate becomes a handicap. Farmers say it "traps fog." As water-saturated air becomes temporarily caught within its leaves, there is a greater likelihood of water condensing on the rinju and falling to the ground.

Inconvenient during Coffee Harvest. Several farmers mentioned that *A. inulaefolium* caused an inconvenience during coffee harvest. This is unlikely to be a major issue, however, since young coffee plantations are routinely slashed of weeds three times a year and it would be easy to synchronize a slashing operation with harvest. Furthermore, as plantations mature and the canopy closes, light-loving rinju would quickly disappear.

Provides a Habitat for Pests. There is a legitimate concern that fallowed ladang provides an ideal habitat for wild pigs, rats, and insects. During the night, opportunistic wildlife emerges from fallows to feed on crops in adjacent fields. This problem is not specific to *A. inulaefolium,* but its dense, impenetrable thickets provide crop predators with particularly effective cover. One respondent observed that succulent growing tips of rinju were often populated by aphids and speculated that it could act as an intermediate host in their spread to nearby crops.[17]

Despite these concerns, farmers' enumeration of problems posed by *A. inulaefolium* is not nearly as substantive as the benefits they perceive. This explains their overwhelmingly positive valuation of rinju depicted in Figure 15-4.

Ecology of *A. inulaefolium* within Bush-Fallow System

A. inulaefolium's fast growth and high biomass production, its copious shedding of branch and leaf litter, its fast regeneration of soil fertility, and its ability to smother noxious weeds have enabled a sustainable and low-input intensification of Minang bush-fallow systems.

However, a combination of factors is now moving land use in the study area steadily toward permanent cultivation. These include rising demographic pressures on a static land base, improved road infrastructure, access to external markets, pervasion of a cash economy, increased subdivision of land-use rights between clan members, government policy to discourage shifting cultivation, and, perhaps most importantly, the introduction of inorganic fertilizers. The ability to buy fertilizer has meant that farmers no longer have to rely on fallow periods to provide natural nutrient cycling and soil rehabilitation. As a consequence, although farmers speak readily of the beneficial role *A. inulaefolium* has played in their bush-fallow systems, its merits may soon be relegated to mere historical interest.

Not Classical Shifting Cultivators

The Minangkabau have long been a paddy-based society, and shifting cultivation has generally been a peripheral component of their farming systems. Communities were usually founded in wide valleys that offered potential for extensive irrigated terraces and rice security. Shifting cultivation on adjacent foothills provided supplementary food crops and a degree of insurance in the event of failure of the wet rice crop. Hence, Minang obviously do not fit the stereotype of classical shifting cultivators practicing a long fallow rotation. This is evidenced by a complete lack of cultural elaboration, rituals, or taboos within their swidden systems. Unlike integral shifting cultivators, Minang farmers do not view regrowth of secondary forest on fallowed land as desirable or as indicating fertile soil, suppressing weeds, and generating high biomass, which, after slashing and burning, provides the potential for thick layers of fertilizing ash. Instead, they see it in terms of neglect and underuse, requiring a large investment of labor to retrieve the overgrown land for agricultural use. This perception corroborates the assertion of some respondents that even in earlier eras, when land was plentiful and people few, Minang farmers were not in the habit of practicing long fallow rotations within the study area. Stoutjesdijk's (1935) observation of six-to-eight year-fallows in 1935, prior to the introduction of *A. inulaefolium,* is consistent with this hypothesis.

Furthermore, the Minang conceptual distinction between forests and agricultural land encourages short-term fallows and magnifies the importance of *A. inulaefolium* in maintaining an ecological balance within accelerating rotation cycles. As shown in Figure 15-10, the major reasons cited by farmers for maintaining even a short fallow were shortages of capital, labor, and time.[18] If these constraints were removed and Minang farmers continued to follow a trajectory of intensified land use, it would

---

[17] Intari (1975) observed that the incidence of cossid borer attacks on teak plantations was highest (32%) in young stands overgrown with *A. inulaefolium.* He suggested that the borer might be discouraged by weeding the *Austroeupatorium* and creating a drier environment.

[18] Although soil rehabilitation was also commonly mentioned as an obvious rationale for fallowing, farmers viewed this as closely linked with lack of capital. Given financial resources to purchase inorganic fertilizers and hire labor, they would be happy to dispense with a fallow period altogether.

seem logical to predict an end to fallow rotation in the study area within the near future. The following factors illustrate the extent to which this trend is apparent:

- Fallow periods described by respondent farmers have already declined to one to two years (Figure 15-11). At the Sungai Manau Atas study site, there is also a strong tendency to limit fallows to three to six months between annual crops. In this case, leaving land idle is less a strategy for fertility regeneration than it is an attempt to avoid soil compaction and disease buildup. Even within such restricted fallow lengths, they say, *A. inulaefolium* will still form a dense thicket and provide the ecological services already described, that is, biomass production, and to a lesser extent, fertility regeneration, weed suppression, and maintenance of soil moisture and tilth. Table 15-4 outlines farmer rationale in choosing fallow lengths.
- The overwhelming majority of farmers say they are shortening fallow lengths because of increasing population density, limited access to land, increased availability of chemical fertilizers, and a trend toward adoption of permanent cultivation. It is noteworthy, however, that a paradoxical trend exists. There is a significant sector for which ladang cultivation is decreasing in importance, and the land is more often left in indefinite fallows. The reasons for this include the proliferation of off-farm incomes, out-migration of young productive adults, and heavy labor absorption in double-cropping of wet rice. In these cases, ladang are left idle by default rather than design, and the regeneration of soil fertility is an incidental effect, rather than an intended benefit.[19]
- High market prices in recent years have prompted farmers in all three study subvillages to plant cassiavera and, at higher altitude Dusun Koto, passion fruit, on increasing areas of their upland fields. (See Suyanto et al., Chapter 64, and Werner, Chapter 67, for more detailed discussions of this trend.) This rapid expansion of perennial crops, in tandem with increased semipermanent or permanent cultivation of food crops, has been enabled by access to inorganic fertilizers.

Evidence suggests that intensification of land use will continue. Fields judged to have higher agricultural potential by virtue of fertile soils, gentle topography, and close proximity to roads and residential areas will probably be managed under increasingly intensive cultivation of annual crops. As is already happening at the Air Dingin site, farmers will rely increasingly on external inputs not only to maintain fertility levels, but also to control weeds and insects and otherwise manipulate the environment for maximum yields and profitability. Strategies of multiple cropping, intercropping, and relay-planting will become more common where there is limited land and surplus labor. More distant fields with marginal soils or steep slopes will be established to tree crops. At present, farmers commonly plant distant fields to monocultures of cassiavera or coffee as a longer-term investment and use minimal inputs of material and labor. As population density increases and agricultural expansion is further restricted, there is wide scope for enrichment of these plantation monocultures into complex, multistoried agroforestry systems that provide a multiplicity of products, are economically profitable, and, ecologically viable.

---

[19] Minang differentiate between systematic land fallowing intended to rehabilitate soil fertility and land left idle for indefinite periods due to constraints of time, labor, and capital, by referring to the first as *kalapoan* and the latter as *karapuan*.

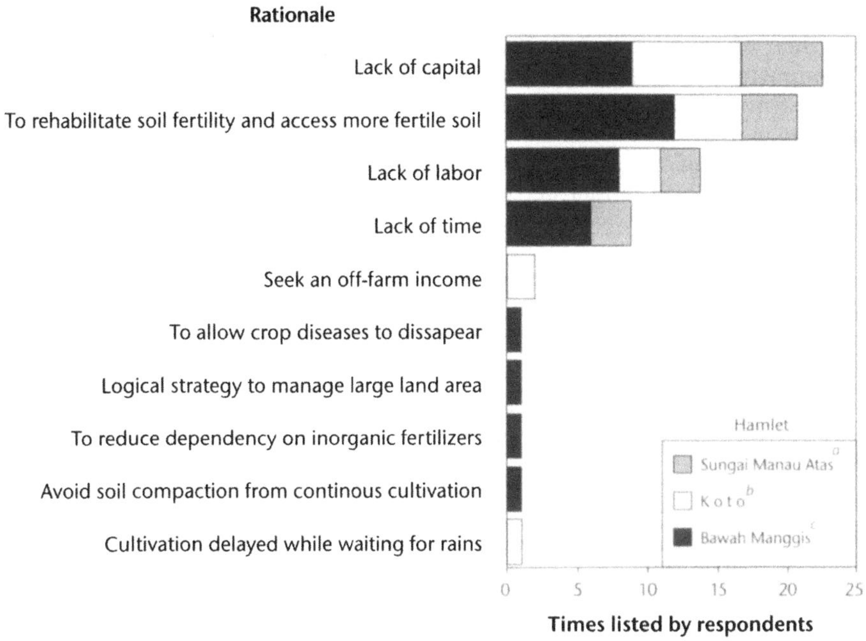

Figure 15-10. Farmer Rationale for Fallowing Ladang
Note: Total number = 53. [a] Mid-altitude pioneer (MAP) settlement; [b] High-altitude mature (HAM) settlement; [c] Mid-altitude mature (MAM) settlement.

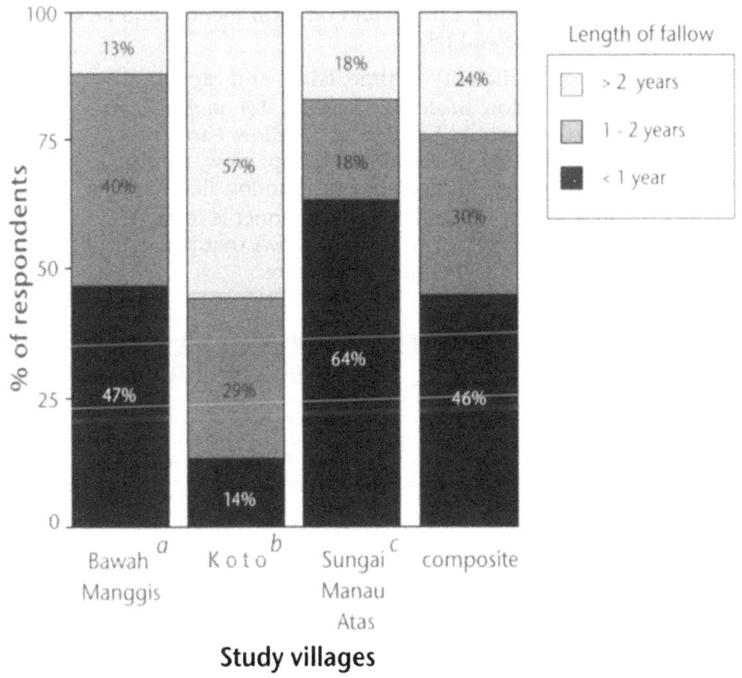

Figure 15-11. Fallow Lengths in Minangkabau Bush-Fallow System
Notes: Total number = 33. An additional 9% of total respondents from Bawah Manggis and 12% from Koto replied that their fallow lengths were uncertain for the reasons outlined in Table 15-4. [a] Mid-altitude mature (MAM) settlement; [b] High-altitude mature (HAM) settlement; [c] Mid-altitude pioneer (MAP) settlement.

This raises the question, if bush fallowing is to disappear from the agricultural landscape, then what role can *A. inulaefolium* usefully play in future farming?[20] As human manipulation of the agroecosystem intensifies, and as linkages between the natural ecosystem and agricultural systems become fewer, will pioneer species such as *A. inulaefolium* be limited to colonizing fence lines, cemeteries, and forest gaps? In this scenario, the utility of *A. inulaefolium* within Minangkabau farming systems may have a limited future, and farmer perceptions of it may reverse, from benign "leafy fertilizer" to malign noxious weed.

Table 15-4. Farmer Rationale Underlying Fallow Lengths

Years	*Rationale*	*Absolute Frequency (n = 71)*
< 1 year	Depends on availability of time, labor, and capital; Short rest period of 2 to 3 months between crops to regain soil fertility and avoid soil compaction; Farmer has accumulated enough money to finance planting another crop; Allows previous crop residues to dry up (disease implications); Wait for *A. inulaefolium* to grow.	38%
1–2 years	Depends on availability of time, labor, and capital; Allows rehabilitation of soil fertility.	25%
> 2 years	Allows *A. inulaefolium* to dominate fallow succession; Cultivation delayed because of lack of capital to buy inputs or because farmer also pursues off-farm income and lacks sufficient time and labor.	20%
Uncertain	Depends on availability of time, labor, and capital; Depends on fallow succession species. If *A. inulaefolium* grows, parcel would be allowed to fallow 1 to 3 years before reopening. If Pteridophyta spp. grow, would probably reopen immediately or abandon field. Land may be left idle indefinitely if farmer is most interested in growing coffee but knows that parcel is not suitable, e.g., may be infertile and dry.	17%

Note: Absolute frequency = $\frac{\text{No. of respondents citing fallow length}}{\text{Total No. of respondents}}$ X 100%

[20] Farmer practices in a major apple-growing area in Malang, East Java, provide an example of how the agronomic properties of *A. inulaefolium* may continue to be exploited in more permanent land-use systems. Orchard growers selectively harvest the aerial biomass of *A. inulaefolium, Tithonia diversifolia, Orthosiphon aristatus* (Bl.) Miq., and *Cestrum nocturnum* L. from along roadsides, fence lines, or disturbed land on the border of a nearby national park. Before planting, 40 cm^3 holes are dug and then filled with a compost consisting of 20 kg of this weed biomass, 20 kg of farmyard manure (FYM), and 1 kg of lime. The compost is covered with soil and allowed to decompose for three months before the apple seedlings are planted. After establishment, these same weedy species continue to be cut annually just prior to the onset of the monsoon season, mixed with FYM, and incorporated into the soil with light hoeing around the base of each tree. When orchards are young, some farmers are reported to plant 30 cm vegetative cuttings of *A. inulaefolium* in rows at 30 cm spacings, on terraces between the rows of apple trees. These fertilizer banks are then trimmed monthly and the loppings applied as a green manure to the young trees. Farmers say that *A. inulaefolium* and *T. diversifolia* were universally used as organic inputs before inorganic fertilizers became available. *T. diversifolia* was generally preferred because of its rapid decomposition, whereas *A. inulaefolium* was considered to be a slower-release fertilizer (author's field notes).

Potential Use of *A. inulaefolium* against Major Noxious Weeds

As fallow periods shorten and soils deteriorate, fertility-associated herbs and shrubs decline and are replaced by noxious weeds that enjoy a competitive advantage on degraded soils. This process is usually accelerated by recurring use of fire in slash-and-burn systems. As noted by Dove (1986, *165*): "Repeated burning favors grasses over woody species in general, and it favors *Imperata* in particular." Throughout the study area, expansion of *Imperata* and fern species, with consequent adverse impacts on soil, represent the largest causes of land abandonment. Like *Imperata*, ferns have deep and extensive root systems and below-ground rhizomes that allow them to survive fire. Thus, both rebound quickly after burning and enjoy a competitive advantage on degraded soils.

Farmers describe the impact of these pernicious weeds in terms of deleterious changes to the soil, hydrology and microclimate; competition with crops for space, sunlight, and nutrients; the large labor investment required to eradicate extensive roots and rhizomes; and their harboring of pests, insects, diseases, and fungi. Given access to forest margins, swidden farmers often prefer to abandon *Imperata*-infested land and clear new swiddens from the forest. It is this scenario of degrading swidden systems and frequent burning that is generally blamed for the 20 to 50 million hectares of *Imperata* grassland across Southeast Asia (Figure 15-12). Rising populations and the urgent need to prevent the conversion of the last remnants of tropical forest to agricultural use has focused attention on how "critical" lands may be best rehabilitated and brought back into productive use.[21]

Intriguing research questions were raised by Stoutjesdijk's observations in 1935, that *A. inulaefolium* had the ability to dominate and smother out *Imperata* grasslands. Within floristic communities colonizing fallowed land, what is the successional relationship between *A. inulaefolium* and noxious weeds such as *Imperata*? What are the ecological determinants that influence which will dominate? And what is the potential for actively introducing *A. inulaefolium* into critical land as a strategy to intervene in the downward spiral of soil degradation, to smother noxious weeds, and to rebuild soil properties?

Detailed, systematic research is needed to understand more fully the ecological processes underlying the transition of fertile soils populated by *A. inulaefolium* into impoverished critical lands, and to identify points of intervention to reverse this trend. Given the resilience of *A. inulaefolium*, its exceptional colonizing ability, and its reputation as a soil-builder, there is strong potential for it to play a central role in rehabilitation efforts.

Farmer Management Strategies to Encourage *A. inulaefolium* Establishment

As in most communities, there is a small subgroup of innovators among the farmers in the study area who have begun experimenting with strategies to assist *A. inulaefolium* succession. Most such activity is in Air Dingin village, where large expanses of the foothills have already been colonized by *Imperata* or ferns and largely abandoned for cultivation. In Sungai Manau Atas, however, where fallow periods are often only three to six months (Figure 15-11), neither *Imperata* nor ferns are a problem, and none of the farmers interviewed was actively encouraging *A. inulaefolium* colonization. Indeed, the question must have seemed nonsensical since, at Sungai Manau Atas, rinju is invariably the dominant fallow species and there is no need for human intervention. Two factors probably contribute to this village's fortunate position. It is a pioneer village first cleared from forest in the late 1960s, so the land is relatively new and fertile. Perhaps more significantly, most farmers are

---

[21] According to the Indonesian Agriculture Research and Development Agency's definition, "critical" land is land that is degraded but still slightly productive for agriculture (Sudihardjo et al. 1992). Rehabilitation requires high inputs, usually by regreening, reforestation, and soil and water conservation measures. The agency divides degraded land into four categories: potentially critical, semicritical, critical, and very critical.

using a mulching system and avoiding burning as an intentional strategy to avoid fern or *Imperata* encroachment.

Elsewhere, a minority of villagers are experimenting with management practices to encourage rinju reestablishment. They are either actively propagating *A. inulaefolium*, manipulating the environment to encourage natural establishment, or maintaining existing stands.

Propagation:

- Stem cuttings of mature, woody plants are cut into 30 to 50 cm segments and planted at intervals in *Imperata* swards. *Austroeupatorium's* higher leaf canopy smothers out *Imperata* within one year, and soil conditions improve by the end of the second year.
- *A. inulaefolium* inflorescences that have gone to seed are distributed around the ladang when the soil is wet. If conditions are dry, seeds will not germinate.

Manipulation of the environment:

- The sod of *Imperata*-fern fallows is turned over, and soil amendments such as lime and manure are added to improve soil fertility and encourage colonization by *A. inulaefolium*.
- Inorganic fertilizers are applied during the cropping phase to ensure a reasonable level of soil fertility before leaving the field idle.
- The field is cleaned of problem weeds and the soil tilled before leaving it fallow.

Stand maintenance:

- Burning is avoided, and a slash-and-mulch system is used instead to ensure that *A. inulaefolium* will persevere as the dominant fallow succession species.[22]
- When periodically weeding cassiavera and coffee plantations, uprooting *A. inulaefolium* is avoided. Instead, it is slashed with a machete to ensure its survival and quick regeneration.
- When a stand of *A. inulaefolium* in a cassiavera plantation is thin and insubstantial, slashing is deliberately avoided and the plants are retained as a seedbank.
- Annuals are intercropped with cassiavera for the first two years, or until the trees are well established and above the weed canopy. *A. inulaefolium* is then allowed to colonize the plantation as a volunteer cover crop. The young cassiavera is not adversely affected by competition from rinju, and weeding is no longer necessary.

---

[22] Since decomposition of large trees requires a long time, farmers often clear new patches from forest several years before intending to cultivate them. In part, this is also a strategy to demonstrate land improvements and make private claim to commons land. Trees and slash are left where they fall. *A. inulaefolium* quickly colonizes these forest gaps and helps ensure a moist microclimate conducive to rotting. In a few years, when the farmer is ready to plant the land, decomposition is well advanced; the soil is covered by a dense thicket of *Austroeupatorium* that has prevented encroachment by noxious weeds and is easily slashed to provide further green manure. When tree crops are to be planted, slash is simply pushed aside to provide small clearings for each seedling. Initially, the farmer has to return to reslash the regenerating *A. inulaefolium* three or four times per year, until tree crops form a closed canopy and light-demanding pioneer species such as *Austroeupatorium* disappear. This more ecologically friendly system of land conversion is a rational alternative to standard slash-and-burn practices for several reasons. Heavy rainfall spread throughout the year makes it almost impossible to dry large-diameter slash and achieve a good burn; upland rice, requiring a much "cleaner" field, is seldom grown in the study area; and farmers widely associate burning with colonization by troublesome ferns.

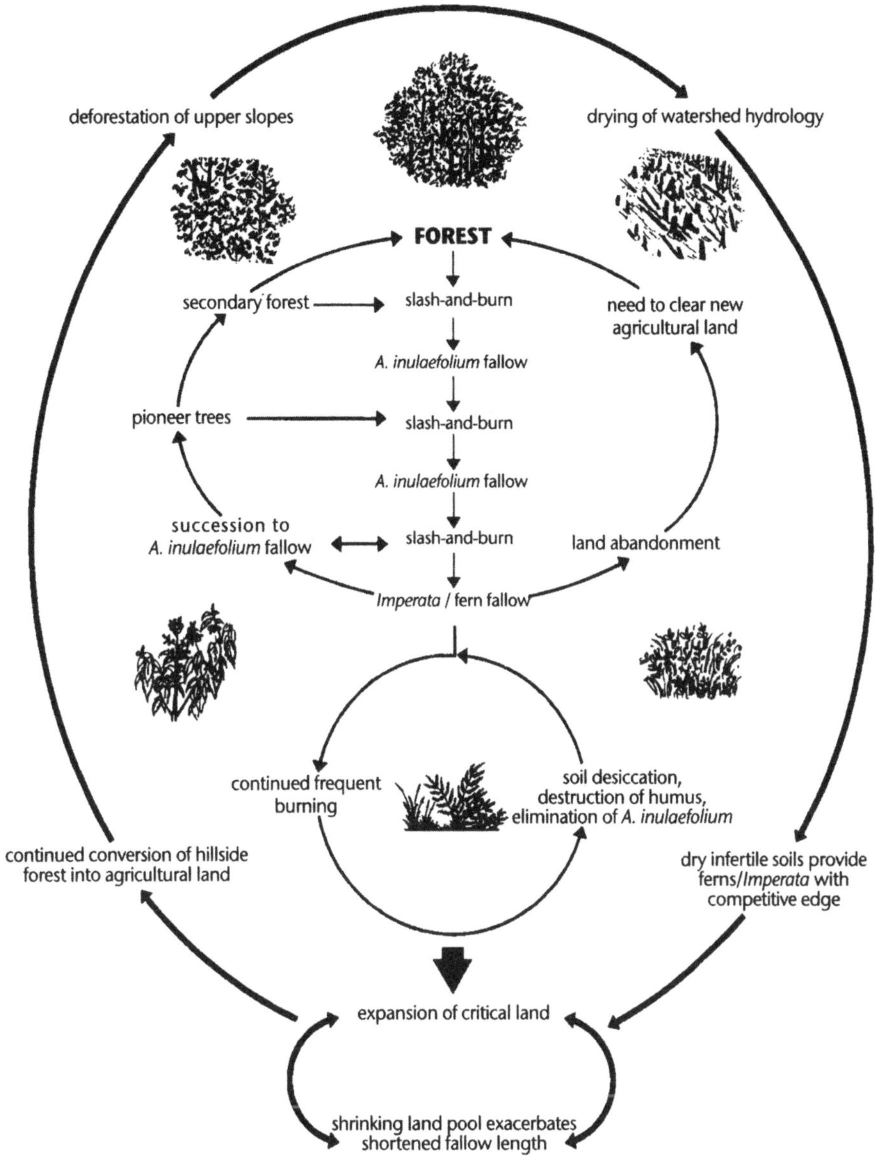

Figure 15-12. Cycle of Land Degradation

If degradation in the study area continues to expand the area of critical land, then farmer experiments and innovations aimed at assisting natural succession of *A. inulaefolium* may become more widespread. Alternately, if current trends toward intensified land use continue, then most land will be permanently cultivated and fallow succession species will not be an issue.

In the interim, even this preliminary list of indigenous innovations aimed at halting expansion of critical land offers a basis for designing farmer-participatory field experiments. Technologies generated in this process could have wide application in the rehabilitation of high-altitude critical lands in the tropics.

The Effect of Succession Communities on Fallow Length

In investigating the ecology of *A. inulaefolium* within Minang bush-fallow systems, a final objective was to understand the effect of succession communities on fallow length. Given *A. inulaefolium's* reputation for fast growth, high biomass production, large litterfall, and efficient nutrient scavenging, we thought it reasonable to postulate that after the cropping period, *A. inulaefolium* fallows could achieve soil rehabilitation at a faster rate than alternative species and, by extension, fallow rotations dominated by *Austroeupatorium* would cycle faster. As shown earlier in this chapter, the first part of this hypothesis has proved valid, but its derived implications were wrong. The error was in misjudging farmers' reactions to fallows dominated by species other than *Austroeupatorium.*

The strong consensus among farmers was that fallows being colonized by *Imperata*, ferns, or other pernicious weeds such as *Mimosa invisa* or *Phragmites karka* should be reopened as soon as possible or at least within the first year.[23] To the farmers, the pertinent issue was not if or when soil fertility had regenerated, but rather, interrupting the process of weed colonization and salvaging the field as quickly as possible. Under fern and *Imperata* successions, farmers were adamant that soil fertility would never recover, regardless of fallow length, and would continue to decline. The logical strategy, then, was to recultivate these fields quickly, before rhizomes and roots became too extensive and laborious to eradicate and before there were deleterious changes in the soil, hydrology, and microclimate. Interestingly, there was an opposing opinion that there was no need to struggle to control *Imperata* because *A. inulaefolium* would, in any case, succeed *Imperata* about five months into the fallow period. This again returns us to the key issue of what ecological conditions determine the continued evolutionary succession to *Austroeupatorium,* on one hand, or an uninterrupted, long-term *Imperata*-fern climax on the other.

Research Issues

A number of research issues emerge from this chapter that require further studies to elucidate processes and practical means of exploiting them:

- This study raises, but leaves unanswered, the critical issue of whether *A. inulaefolium* is actively improving soil conditions or simply colonizing land that is already fertile. The method used in this study, of sampling and comparing fallow successions found across a swidden landscape, does not allow us to filter out confounding variables to enable a clear assessment of the effect of fallow species alone. These variables include site-based issues, such as altitude, aspect, and soil type, and management differences, including the history of previous land use, the number of years it has been cropped, tillage operations, fertilizer applications, and whether fallow regrowth was subjected to periodic burning or grazing by livestock. As a follow-up to this study, replicated agronomic experiments are needed across multiple sites so that these variables can be rigorously controlled and a clear assessment made of the "improver" versus "indicator" roles. The possible mechanisms and limitations of soil improvement need clarification.
- Given the rapid colonizing ability of *A. inulaefolium* and other Asteraceae, a more careful assessment is needed of practices and problems in their control once the field returns from fallow to cultivation.
- In cases such as the Air Dingin study site, where *Austroeupatorium* is receding from the landscape and being displaced by *Imperata* and Pteridophyta spp.,

23 In contemporary farming systems with access to inorganic fertilizers, elimination of the fallowperiod is a viable strategy to keep pernicious weeds at bay. In the past, farmers would not have had the technical means to maintain soil nutrient balances under permanent cultivation and would have had no recourse other than abandonment of fields colonized by problem weeds. Thus, inorganic fertilizers are playing a key role in enabling land rehabilitation rather than further encroachment into forests in search of new land.

research is needed to understand the ecological processes underlying this transition and to identify points of intervention to reverse the trend.

- If bush-fallows are soon to be relegated to the past, as land use in the central rift valley intensifies, how can the beneficial properties of *A. inulaefolium* continue to be exploited in permanent cultivation systems? Since erosion is a serious problem in the study area, is there potential for contour strips of rinju running across ladang slopes to reduce run-off and erosion and promote the formation of natural terraces?
- In view of the severity of farmers' problems with wild pigs, could closely planted *Austroeupatorium* form live, pig-proof fences requiring little maintenance and providing green manure, poles, and firewood, thus helping to reduce farmers' reliance on forest resources?
- Experimental testing of *A. inulaefolium*'s agronomic potential for rehabilitating *Imperata* fields would be of immense interest to ongoing efforts to find practical means of reclaiming critical land. Quantifying the influence of soil moisture, nutrients, and light on the outcome of *Imperata*-Asteraceae competition would provide a clearer understanding of the scope for manipulation of this process.
- *A. inulaefolium* has almost completely replaced native species in fallow regrowth and often grows in almost pure stands. What are the implications of this depauperation of biodiversity on fallowed land? To what extent does *Austroeupatorium*'s competitiveness delay the regeneration of secondary forests?
- If *A. inulaefolium* or other Asteraceae are verified as superior fallow species, what are the implications of promoting the spread of these potentially aggressive species that are desirable for some land-use types but problematic to others? To what extent might native flora be displaced by these exotic species, and how will this affect the wider ecosystem?
- Claims by farmers that *A. inulaefolium* has insecticidal properties require chemical analysis and field trials for verification.
- Many Asteraceae are credited with nematocidal properties. This may be worth investigating to further justify their use as fallow species, particularly in vegetable-growing areas where nematode populations are a problem.
- To further elucidate the role of *A. inulaefolium* in swidden systems, we think it would be worthwhile to repeat a study similar to the one presented in this chapter. However, a more isolated area should be chosen, where bush-fallows are still the rule and application of inorganic fertilizers are still the exception. Focusing on different ethnic groups might also divulge richer insights.[24]
- Research on the above issues should move beyond a narrow species focus on *A. inulaefolium* and consider other Asteraceae spp. that may have similar potential as effective fallow species, for example, *Chromolaena odorata, Tithonia diversifolia, Clibadium surinamense, Eupatorium riparium, Montanoa grandiflora, Mikania cordata,* and *Wedelia triloba.*

Conclusions

The experience of Minangkabau farmers presented in this chapter constitutes an illustrative case study of farmer experimentation in exploiting the beneficial properties of an introduced pioneer shrub. Data from chemical analysis of soils and fallow vegetation provide empirical evidence corroborating the validity of their indigenous knowledge and demonstrate the rationality of their innovations in managing *A. inulaefolium* to their advantage. The Minang experience is particularly

---

[24] The Minangkabau are more widely recognized for their entrepreneurial skills than their knowledge of the natural environment. Other researchers have also commented that, compared to other ethnic groups, Minang seem to have a relatively unimpressive knowledge base about local flora (Werner 1994). Chance encounters with Javanese farmers during the fieldwork suggested that they practiced a more detailed observation of local ecology and had an intricate understanding of how it could be manipulated for their benefit.

cogent because it is an example of successful indigenous intensification of swidden systems with a potentially wide domain of extrapolation.

In 1935, Stoutjesdijk's attention was captured by *A. inulaefolium*'s ability to rehabilitate the soil of fallowed fields in half the time needed by natural forest regeneration. This significantly reduced agricultural pressure on protected forest margins and enabled reclamation of degraded land. Since then, rising population pressures on a static land base have, where fallowing continues at all, forced it to further shorten to about two years. The biomass and nutrient accumulation data documented in this study support farmer claims that even such short fallows can achieve much in terms of soil rehabilitation.

As portrayed in Figure 15-13, the introduction of *A. inulaefolium* into West Sumatra has mitigated the tendency toward ecological decline in intensifying swiddens by playing a valuable bridging role between the relatively long fallow rotations of the past and today's increasing adoption of permanent cultivation. It is precisely this critical stage, when the ecological sustainability of traditional shifting cultivation systems has been lost but appropriate permanent practices not yet adopted, that underlies the serious degradation of swidden environments throughout much of the Asia-Pacific. Minang farmers appear to have found at least a partial solution in *A. inulaefolium*'s spontaneous role as an improved fallow species.

This observation underlines Field's (1991) argument that aggressive pioneer shrubs—regarded as invasive weeds to reforestation projects, large tree crop plantations, and cattle pastures—may provide significant benefits to resource-poor farmers. *A. inulaefolium* not only performs critical ecological services within the farming systems of isolated and marginalized upland communities; its benefits are also specifically targeted at the poorest farmers, who lack the financial resources to purchase inorganic fertilizers to make the transition from fallow rotation to permanent cultivation. The role *A. inulaefolium* has played in enabling intensification of bush-fallow systems and in mitigating pressure on forest margins in the study area suggests that, with skillful management, it could play an even wider role in stabilizing farming systems on sloping highlands.

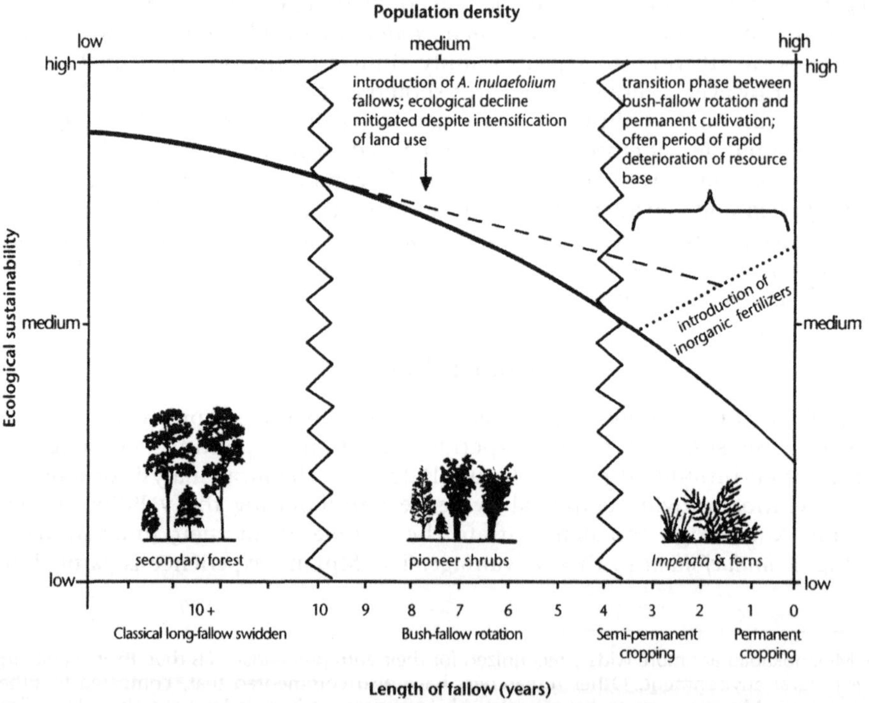

Figure 15-13. Bridging Effect of *A. inulaefolium* in Mitigating Deterioration of Swidden Agroecosystems during Phase of Declining Fallow Length

There are many parallels between the findings of this study and the ongoing controversy surrounding efforts to eradicate *Chromolaena odorata* in Indonesia. The findings strongly suggest that eradication is an inappropriate prescription, favoring some farmers at the expense of others. Impact assessments should be taking a broader view, and considering the ecological role of *C. odorata* in all types of farming systems. Rather than focusing on costly and widespread eradication efforts, the debate should focus on localized control, including stimulation or enhanced use, depending on *C. odorata's* role as friend or foe within local farming systems.

Acknowledgments

This investigation was jointly sponsored by the Southeast Asian program of the World Agroforestry Centre (ICRAF) and the Canadian International Development Agency (CIDA), under its scholarship program. I am grateful to the Forest and Natural Conservation Research and Development Centre for its assistance in facilitating this research, and to Minang farmers in the study area for their hospitality and cooperation. Helpful discussions with Hubert de Foresta are gratefully acknowledged, as well as guidance from Meine van Noordwijk and Fahmuddin Agus regarding the soils aspects of the study. Finally, I wish to thank Dennis Garrity for his invaluable guidance throughout the study and helpful suggestions on an earlier draft of this manuscript.

References

Agbim, N.N. 1987. Carbon Cycling under *Chromolaena odorata* (L.) K. & R. Canopy. *Biological Agriculture and Horticulture* 4, 203–212.

Backer, C.A., and R.C.B. van Den Brink. 1965. *Flora of Java (Spermatophytes Only)*, Vol. II. Groningen, The Netherlands: W. Noordhof.

Bahri, S., D.R. Stoltz, and D. Paramardini. 1988. Hepatotoxic Effect of *Eupatorium* in the Rat. *Penyakit-Hewan* 20(36). Bogor, Indonesia: Balai Penelitian Veteriner, 88–90.

Bennett, F.D., and V.P. Rao. 1968. Distribution of the Introduced Weed *Eupatorium odoratum* Linn. (Compositae) in Asia and Africa and Possibilities of its Biological Control. *PANS* 14(3), 227–281.

Castillo, A.C., E.M. Sena, F.A. Moog, and N.S. Mendoza. 1981. Prevalence of *Chromolaena odorata* (L.) R.M. King and Robinson under Different Grazing Intensities and Methods of Weed Control. In: *Proceedings of Eighth Asian-Pacific Weed Science Society Conference*, edited by B.V.V. Rao, 181–186.

Cock, M.J.W., and J.D. Holloway. 1982. The History of, and Prospects for, the Biological Control of *Chromolaena odorata* (Compositae) by *Pareuchaetes pseudoinsulata* Rego Barros and Allies (Lepidoptera: Arctiidae). *Bull. Ent. Res.*72, 193–205.

Coster, C. 1935. Wortelstudien in de Tropen. V. Gebergtehoutsoorten (Root Studies in the Tropics. V. Tree Species of the Mountain Region). *Tectona* 28, 861–878.

———. 1937. De Verdamping van Verschillende Vegetatievormen op Java (The Transpiration of Different Types of Vegetation in Java). *Tectona* 30, 124.

de Foresta, H. 1993. *Chromolaena odorata*: Calamite ou Chance pour l'Afrique Tropicale? Paper presented at Troisieme Atelier International sur la Lutte Biologique et la Gestion de *Chromolaena odorata*. November 15–19, 1993, Abidjan, Côte d'Ivoire.

———, and D. Schwartz. 1991. *Chromolaena odorata* and Disturbance of Natural Succession after Shifting Cultivation: An Example from Mayombe, Congo, Central Africa. In: *Ecology and Management of* Chromolaena odorata, *BIOTROP Special Publication No. 44*, edited by R. Nuniappau and P. Ferrar, 23–41.

Dove, M.R. 1986. The Practical Reason of Weeds in Indonesia: Peasant vs. State Views of *Imperata* and *Chromolaena*. *Human Ecology* 14(2), 163–190.

Faridah Hanum, I., and L.J.G. van der Maesen (eds.). 1997. *Plant Resources of Southeast Asia: Auxiliary Plants*. Leiden, the Netherlands: Backhuys Publishers.

Ferraro, G.E., V.S. Martino, and J.D. Coussio. 1977. New Flavonoids from *Austroeupatorium inulaefolium*. *Phytochemistry* 16. England: Pergamon Press, 1618–1619.

Field, S.P. 1991. *Chromolaena odorata*: Friend or Foe for Resource Poor Farmers. *Chromolaena odorata* Newsletter 4, (May), 4–7.

Garcia, J.S. 1986. Weedy Spot: A Rancher's Battle against Hagonoy (*Chromolaena odorata*). *NCPC Newsletter* 1(4), 3–4.

Hambali, G.G. 1994. Personal communication with the author.

Hellinga, G. 1954. Problems in Maintaining Soil Fertility in Forest Plantations with a Short Rotation.
Intari, S.E. 1975. Observations on Beehole Borer (*Duomitus ceramicus*) in Teak Plantations in Kendal and Ciamis Forest Districts, Java. In: *Report of Forest Research Institute* 204.
Ivens, G.W. 1974. The Problem of *Eupatorium odoratum* L. in Nigeria. *PANS* 20(1), 76–82.
Joseph, P.A. and T.F. Kuriakose. 1985. An Integrated Nutrient Supply System for Higher Rice Production. *IRRN* 10(2) (April), 22.
Kurupunya T. 1998. Personal communication with the author.
Litzenberger, S.C., and Ho Tong Lip. 1961. Utilizing *Eupatorium odoratum* L. to Improve Crop Yields in Cambodia. *Agronomy Journal* 53(1-6), 321–324.
Michon, G., H. de Foresta, and N. Widjayanto. 1992. Complex Agroforestry Systems in Sumatra. In: *Sumatera, Lingkungan Dan Pembangunan: Yang Lalu, Sekarang Dan Yang Akan Datang* (Sumatra, Environment and Development: Its Past, Present and Future), proceedings of workshop, September 16–18, 1992, Bogor, Indonesia, BIOTROP Special Publication 46, 335–347.
Mohan Lal, K.B. 1960. Eradication of *Lantana, Eupatorium* and other Pests. *Indian Forests* 1960, 86(8), 482–484.
Murdiati, T., and D.R. Stoltz. 1987. Investigation of Suspected Plant Poisoning of North Sumatran Cattle. *Penyakit Hewan* 19(34), 101–105.
Nemoto, M., V. Pongskul, S. Hayashi, and M. Kamanoi. 1983. Dynamics of Weed Communities in an Experimental Shifting Cultivation Site in Northeast Thailand. *Weed Research* (Japan) 28(2), 111–121.
Roder, W., S. Maniphone, B. Keoboualapha, and K. Fahrney. 2006. Fallow Improvement with *Chromolaena odorata* in Upland Rice Systems of Northern Laos. Chapter 14.
Scholz, U. 1983. *The Natural Regions of Sumatra and Their Agricultural Production Pattern: A Regional Analysis*. Bogor, Indonesia: West Java Central Research Institute for Food Crops.
Slaats, J.J.P. 1993. The Use of *Chromolaena odorata* as Fallow in a Semi-permanent Cropping System in Southwest Côte d'Ivoire. Paper presented at Third International Workshop on Biological Control and Management of *Chromolaena odorata*, November 15–19, 1993, Abidjan, Côte d'Ivoire.
Stoutjesdijk, J.A.J.H. 1935. *Eupatorium pallescens* DC op Sumatra's Westkust (*Eupatorium pallescens* DC on the West Coast of Sumatra). *Tectona* 28, 919–926.
Sudihardjo, A.M., U. Affandi, T. Sudharto, Ropik, and Sobari. 1992. Penelitian Identifikasi dan Karakterisasi Lahan Kritis Tingkat Tinjau Daerah Kabupaten Timor Tengah Selatan, Kupang, Timor Tengah Utara dan Belu (Sebagian), Propinsi Nusa Tenggara Timur (Identification and Characterization of Critical Land Research on Surveyed Area of Timor Tengah Selatan, Kupang, Timor Tengah Utara and (Part of) Belu Districts, Nusa Tenggara Timur Province). Agriculture Research and Development Agency, Center for Soil and Agroclimate Research. Suyanto, S., T. Tomich, and K. Otsuka. 2006. The Role of Land Tenure in the Development of Cinnamon Agroforestry in Kerinci, Sumatra. Chapter 65.
Torres, D.O., and E.C. Paller 1989. The Devil Weed (*Chromolaena odorata* R.M. King and H. Robinson) and Its Management. *SEAWIC Weed Leaflet* 4, 1–6.
van Noordwijk, M. 1997. Personal communication with the author.
Werner, S. 1994. Personal communication with the author.
———. 2006. The Development of Managed Fallow Systems in the Changing Environment of Central Sumatra. Chapter 67.

Chapter 16

Piper aduncum Fallows in the Lowlands of Papua New Guinea

Alfred E. Hartemink

Primary forest covers about 75% of Papua New Guinea. Every year about 200,000 ha are cleared for commercial operations, including logging, plantations, and subsistence agriculture. The latter mainly takes the form of shifting cultivation. In many parts of the humid lowlands, secondary fallow vegetation is dominated by the shrub *Piper aduncum* L (see color plate 20). It is not known exactly when and how *P. aduncum* invaded Papua New Guinea from its native Central America, but it was first recorded in the mid 1930s (Hartemink 2001). The invasion has been aggressive and it has spread in a similar fashion to *Chromolaena odorata,* which was introduced to Asia in the late 19th century.

P. aduncum was first described by Linnaeus in 1753. It is common throughout Central America and is also found in Suriname, Cuba, Trinidad and Tobago, southern Florida, and Jamaica. It was introduced in 1860 to the botanical garden of Bogor, in Indonesia, and has naturalized in many parts of Malaysia (Chew 1972). In the Pacific, *P. aduncum* can be found in Fiji and in Hawaii. Australia has listed it as an unwanted weed species (Waterhouse and Mitchell 1998).

P. aduncum is a monoecious shrub or slender tree that grows up to eight meters tall. It has ovate and petioled leaves up to 16 cm long, and its flowers are arranged in a dense spiral (see Figure 16-1). It is commonly found along roadsides and in cleared forest areas on well-drained soils, but is never found in mature vegetation. *P. aduncum* has very small seeds that are dispersed by the wind, birds, and fruit bats. It withstands coppicing, but burning seems to be detrimental. It can be effectively controlled by hand cutting (Henty and Pritchard 1988). Throughout the neotropics, *P. aduncum* extracts are used as folk medicine; and it is mentioned in several ethnopharmalogical databases. It is avoided by livestock (Waterhouse and Mitchell 1998).

It is possible that seeds of *P. aduncum* were deliberately imported to Papua New Guinea, or that it hopped across the border from West Papua (Irian Jaya) (Rogers and Hartemink 2000). Whatever its means of arrival, *P. aduncum* can now be found in many parts of the humid lowlands of Papua New Guinea, whereas 20 or 30 years ago, it was absent (Bourke 1997). It is widespread in the Morobe and Madang Provinces at altitudes up to 600 m above sea level (asl), and it is also found in the highlands up to 2,100 m asl. It often grows in monospecific stands on steep hill slopes (Kidd 1997).

The stems of *P. aduncum* are used for firewood, fence posts, or supporting sticks for yams (*Dioscorea* sp.). In some areas it is even used for building material, but the wood rots quickly. In some coastal villages of Papua New Guinea, the bark or leaves are used to dress fresh knife, axe, or spear wounds, and new leaves are also used as bandages (Woodley 1991). Farmers' perceptions of *P. aduncum* are mixed. Some value its rapid growth and the firewood it provides, while others are convinced that it is

---

Alfred E. Hartemink, ISRIC-World Soil Information, P.O. Box 353, 6700 AJ Wageningen, The Netherlands.

not a good fallow species. Many farmers in the lowlands stress that *P. aduncum* makes the soil dry and loose.

Despite its being widespread in Papua New Guinea, there is no information available on *P. aduncum's* basic growth characteristics nor on its effect on soil. In this chapter I present some results of my research on this rapidly invading fallow species in Papua New Guinea.

The Study Site and Methodology

In October 1996, a trial was set up to investigate *P. aduncum's* biomass and nutrient accumulation compared with that of *Imperata cylindrica* and *Gliricidia sepium*. The location was Hobu (6°34′ S, 147°02′ E), about 20 km northeast of Lae, at the foothills of the Saruwaged Range. Hobu is an area where much of the secondary fallow vegetation is dominated by *P. aduncum*. The altitude is 405 m asl, and the annual rainfall is about 3,000 mm, distributed throughout the year. The mean annual temperature is about 26.7°C. The soils are derived from a mixture of colluvial and alluvial deposits of mostly igneous rocks. They have a high base status and are classified as Typic Eutropepts.

Three plots, each measuring six square meters, were planted with *P. aduncum*, *Gliricidia sepium*, and *Imperata cylindrica* ($n = 4$ each). Planting distances for the *Piper* and *Gliricidia* were 0.75 m by 0.75 m. One year later, the plots were harvested. The plants were slashed at ground level and separated into main stems, branches, leaves, and litter. Each plant part was weighed, oven dried at 65°C for 72 hours, and analyzed for nutrient content at the laboratories of the University of Queensland.

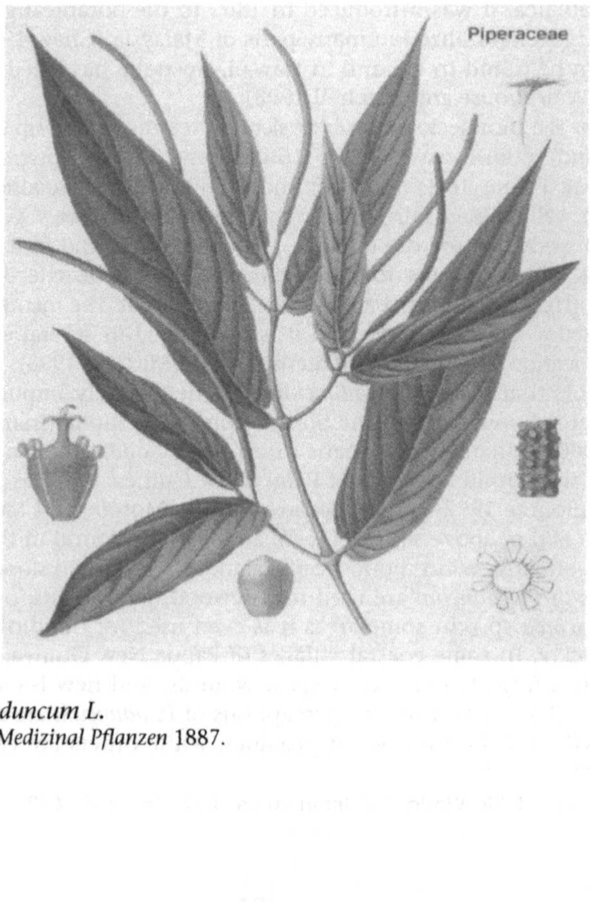

Figure 16-1. *Piper aduncum L.*
Source: H.A. Köhler's *Medizinal Pflanzen* 1887.

Results

Biomass Accumulation

After one year the *P. aduncum* had produced about 13.7 metric tonnes(t)/ha of biomass. Of this, 43% was stems (Table 16-1). About 15% of the total dry matter production, excluding the roots, was found in the litter layer. The *G. sepium* had produced nearly three times more wood and slightly more leaves and litter than the *P. aduncum*. The *I. cylindrica* had also produced slightly more biomass than the *P. aduncum*. After removal of the woody parts, the total biomass returned to the soil was 7.8 t/ha for *P. aduncum* and 8.1 t/ha for *G. sepium*.

In another experiment, *P. aduncum* accumulated about 9 metric tonnes of dry biomass/ha after 11 months. When the trees were nearly two years old, the biomass had increased to 48 t/ha, and the height of the trees was 4.5 m. Growth rates increased with the age of the trees and were mostly linearly related to the amount of rainfall. The highest biomass accumulation rate observed in a two-year period was 134 kg of dry matter/ha/day.

Nutrient Accumulation

The total nutrient content of the fallow vegetation is shown in Table 16-2. *G. sepium* returned the largest amount of N to the soil. *P. aduncum* and *I. cylindrica* returned less than half of this amount of N. The amount of P was similar for all three fallows. The leaves and small branches of *P. aduncum* returned considerable amounts of K to the soil, whereas *G. sepium* returned more than 200 kg Ca/ha. *I. cylindrica* returned relatively few nutrients to the soil.

Table 16-1. Biomass of One-Year-Old *Piper, Gliricidia,* and *Imperata* Fallows (metric tonnes/ha ± 1 SD, dry matter)

Plant Part	Piper aduncum	Gliricidia sepium	Imperata cylindrica
Stems	5.9 ± 1.0	15.2 ± 0.6	
Branches	1.6 ± 0.2		
Leaves	4.2 ± 0.4	5.2 ± 0.3	14.9 ± 2.0
Litter	2.0 ± 0.4	2.9 ± 0.9	
Total	13.7	23.3	14.9

Note: Modified after Hartemink (2003a).

Table 16-2. Nutrients in One-Year-Old Fallow Aboveground Biomass at Hobu (kg/ha)

Fallow Species	*Plant Parts*	*N*	*P*	*K*	*Ca*	*Mg*
Piper aduncum	Total	120	22	299	157	46
	Returned to the soil[a]	97	14	206	147	40
Gliricidia sepium	Total	356	36	248	312	64
	Returned to the soil[a]	192	12	89	222	41
Imperata cylindrica	Total[b]	76	12	89	56	29

Note: Modified afer Hartemink (2003a). [a] *Piper and Gliricidia* main stems were removed from the plots. Totals exclude roots. [b] All biomass returned to the soil.

Table 16-3. Volumetric Soil Moisture Content of Sweet Potato Plots after Different Fallow Vegetation (%)

			Soil Moisture under Sweet Potato after One Year of Fallow with:			
DEP[a]	*Soil Depth (m)*	*Continuous Sweet Potato*	Piper aduncum	Gliricidia sepium	Imperata cylindrica	*SED*[b]
0	0–0.05	34.6	27.4	33.4	31.2	1.43
	0.10–0.15	37.3	29.8	38.8	33.3	3.35
93	0–0.05	41.7	43.4	44.1	46.8	1.04
	0.10–0.15	41.9	42.2	45.8	46.4	2.19
168	0–0.05	39.2	40.2	42.4	42.4	2.61
	0.10–0.15	41.2	39.7	38.3	41.4	3.05

Note: Modified after Hartemink (2004). [a]Days after fallow vegetation was slashed and sweet potato was planted (DEP). [b]Standard error of the difference (SED) in means (9 df).

Fallow Effects on Soil Moisture

Many farmers reported that *P. aduncum* depleted soil water. When the one-year-old fallow vegetation was slashed, volumetric soil moisture was measured (gravimetric content X BD). Soils under *P. aduncum* had significantly lower moisture levels than those under *Gliricidia* and *Imperata* (Table 16-3). Three months after the planting of sweet potato, the plots previously under *P. aduncum* still had significantly lower levels of soil moisture in the 0 to 0.05 m horizon than those previously under *I. cylindrica*. The differences in soil moisture levels, created by the fallow species, disappeared after five months.

Conclusions

P. aduncum's rapid invasion of the humid lowlands of Papua New Guinea can be explained by its dominance in the seedbank and its fast growth (Rogers and Hartemink 2000). In trials at Hobu, *P. aduncum's* total biomass accumulation after one year was lower than that for *Gliricidia sepium* but similar to *Imperata cylindrica*. It returned less than half of the N returned to the soil by *G. sepium* but more than twice the amount of K. It was also confirmed that soils under *P. aduncum* fallows were significantly drier than the other fallows.

Whether the invasion of *P. aduncum* in the Papua New Guinea lowlands is a favorable development from an agricultural point of view remains to be seen. Current research focuses on the effect of different fallows on sweet potato, the main staple crop in Papua New Guinea (Hartemink 2003b). However, from an ecological point of view the invasion is catastrophic because *P. aduncum* prevents the growth of rainforest seedlings. It can therefore be assumed that its dominance will mean a loss of biodiversity, which is frequently regarded as a measure of ecosystem quality (van Groenendael et al. 1998). On the other hand, if *P. aduncum* continues to invade areas currently dominated by *I. cylindrica,* this would have to be regarded as a favorable development (Cairns 1997; Hartemink 2001).

References

Bourke, R.M. 1997. Personal communication between R.M. Bourke, Australian National University, Canberra, and the author.

Cairns, M.F. 1997. Personal communication between Malcolm F. Cairns, Australian National University, Canberra, and the author.

Chew, W.L. 1972. The Genus Piper (*Piperaceae*) in New Guinea, Solomon Islands, and Australia. *Journal of the Arnold Arboretum* 53, 1–25.

Hartemink, A.E. 2001. Biomass and Nutrient Accumulation of *Piper aduncum* and *Imperata cylindrica* Fallows in the Humid Lowlands of Papua New Guinea. *Forest Ecology and Management* 144, 19–32.

———.2003a. Sweet Potato Yield and Nutrient Dynamics after Short-term Fallows in the Humid Lowlands of Papua New Guinea. *Netherlands Journal of Agricultural Sciences* 50: 297–319.

———. 2003b. Integrated Nutrient Management Research with Sweet Potato in Papua New Guinea. *Outlook on Agriculture* 32: 173–182.

———. 2004. Nutrient Stocks of Short-term Fallows on a High Base Status Soil in the Humid Tropics of Papua New Guinea. *Agroforestry Systems 63, 33-43.*

Henty, E.E., and G.H. Pritchard. 1988. *Weeds of New Guinea and their Control.* Lae, Papua New Guinea: Department of Forests.

Kidd, S.B. 1997. A Note on *Piper aduncum* in Morobe Province, Papua New Guinea. *Science in New Guinea* 22, 121–123.

Rogers, H.M., and A.E. Hartemink 2000. Soil Seed Bank and Growth Rates of an Invasive Species, *Piper aduncum,* in the Lowlands of Papua New Guinea. *Journal of Tropical Ecology* 16, 243–251.

van Groenendael, J.M., N.J. Ouborg, and R.J.J. Hendriks 1998. Criteria for the Introduction of Plant Species. *Acta Bot. Neerl.* 47, 3–13.

Waterhouse, B, and A.A. Mitchell. 1998. *Northern Australian Quarantine Strategy: Weeds Target List.* Brisbane: AQIS.

Woodley, E. 1991. *Medicinal Plants of Papua New Guinea.* Part 1: Morobe Province. Wau Ecology Institute Handbook No. 11. Weikersheim: Josef Margraf.

Chapter 17

Management of *Tecoma stans* Fallows in Semi-arid Nusa Tenggara Timur, Indonesia

Tony Djogo, Muhamad Juhan, Aholiah Aoetpah, and Ellen McCallie

Nusa Tenggara Timur (East Nusa Tenggara) is located in the dry zone of eastern Indonesia (see page 273). The province, which includes the western part of the island of Timor, has complex geology, soils, and vegetation, as well as intricate social, political, and economic circumstances. Its agricultural production systems are also very complex. Despite the fact that modernization has brought significant economic development to Indonesia, many farmers must still rely upon marginal natural resources that are significantly degraded. They continue cultivating their small holdings to produce food and cash income against a background of increasing population pressure and a scarcity of agricultural land.

Shifting cultivation is the main traditional farming system in the study area, but many farmers are evolving toward more permanent land-use systems. Better approaches and methods are needed to manage this transition to enable farming systems to become more productive. Improved fallow management is one option that may help.

Traditionally, farmers in West Timor have maintained soil fertility by incorporating both local and introduced trees and shrubs into their cropping patterns. In this way, they have been able to improve or at least maintain productivity. Among the species currently used to improve fallows are *Lantana camara, Acacia villosa* (see color plate 37), *Acacia nilotica, Leucaena leucocephala, Sesbania grandiflora, Bambusa* sp., *Tecoma stans, Ricinus communis, Chromolaena odorata,* and *Zizyphus mauritiana*. Any species, either leguminous or non-leguminous, that produces massive biomass will be used for soil fertility management in a combination of fallow and slash-and-burn techniques. Some species that, elsewhere, may be considered as aggressive weeds will be used in many parts of West Timor as a source of materials or nutrients for soil fertility management.

This chapter describes the use of *Tecoma stans* as a nonleguminous fallow species (see color plate 19) by farmers in three villages in Kabupaten Kupang, West Timor. Its objectives are to characterize the spontaneous adoption of *T. stans* by farmers as an improved fallow and to describe the conditions and production systems of households in the three villages—Bello, Fatukoa, and Tunfeu. The study employed a

---

Tony Djogo, Konphalindo (National Consortium for Nature and Forest Conservation), Jl. Kelapa Hijau No. 99, Jagakarsa, Jakarta Selatan, Indonesia; Muhamad Juhan and Aholiah Aoetpah, Politeknik Pertanian Negeri Kupang (State Agricultural Polytechnic), Kupang, West Timor; Ellen McCallie, Exhibit and Interpretations Coordinator, Missouri Botanical Garden, P.O. Box 299, St. Louis, MO 63110-0299, USA.

simplified agroecosystem analysis approach with rapid rural appraisal and informal interviews, based on a checklist prepared during preliminary visits. It sought to provide insights into future prospects and trends in the livelihood systems of the villages and to explore possibilities for improved management of fallows to enable the villagers to achieve better livelihoods. This chapter also describes the implications for fallow management arising from socioeconomic and biophysical conditions, the state of physical infrastructure, cropping patterns, and alternative income-generating activities.

The Study Region

Biophysical Factors

The study villages are located 5 to 10 km south of Kupang city, the capital of Nusa Tenggara Timur, at approximately 10° south latitude. The area, on the western side of Timor island, is one of the driest in Indonesia, with a prolonged dry season of seven to nine months, followed by three to four months of intense rain. The mean annual rainfall is 1,000 to 1,500 mm (Metzner 1977, 1983). The average and maximum temperatures are 25°C and 33 to 36°C, respectively (McKinnell and Harisetijono 1991).

The island itself is an uplifted coral reef. The topography is rugged and hilly, with craggy plateaus and moderate to steep slopes (Jones 1983). The average elevation is 300 m above sea level (asl), with a range from sea level to about 600 m. In most agricultural areas, rock fragments make–up more than 40% of soil volume in the top 30 cm and the C horizon is less than 50 cm from the surface (Metzner 1977). The soil is primarily Mediterranean, with colors ranging from dust to white and bright, with lots of coral rock. Paddy areas have black sedimented grumosol soils, and several areas have red soil. The latter are primarily used for food crops and vegetables.

Before the land was taken for extensive agriculture, the vegetation on the island was predominantly dense monsoon forest. Nowadays, natural forest remains only at the highest points of Timor. As human and cattle populations increased in the early 20th century, the vegetation became weedy and shrubby, often comprising exotic species transported in by cattle (Metzner 1983). The island is now a mosaic of shifting slash-and-burn agriculture, fallowed land, and occasionally, attempts at high-input plantation cropping by wealthy investors from other islands. While some perennial vegetation can establish in limestone, maize and other annual crops are primarily limited to the minimal soil.

The Study Villages

The study villages of Bello, Fatukoa, and Tunfeu have a coral landscape with Mediterranean-type rocky soils that are silty and yellowish-red in color, with an average solum depth of 20 to 70 cm. The topography is hilly, with 20% to 30% slopes, at altitudes between 300 and 350 m asl. The landscape is dominated by *T. stans* and secondary monsoon forest. Most of the natural trees species observed in the area are *pole* (*Alstonia villosa*), *kom* (*Zizyphus mauritiana*), *matani* (*Pterocarpus indicus*), and *kosambi* (*Schleicera oleosa*). Along roadsides, where most houses have been built, the vegetation has been converted to perennial cash crops: *nangka* (*Artocarpus integra*), *kemiri* (*Aleurites moluccana*), *kelapa* (*Cocos nucifera*), oranges, and ornamental species.

The villages are in transition from a subsistence agricultural economy to a market-integrated system. This is despite the fact that they have limited access to credit, transportation, mechanization, draft power, extension, or inputs. The transition has both direct and indirect influences on the village families. However, this study focuses on aspects of intensification of upland farming systems, and particularly on fallow management.

The populations of the three villages are: Bello, 661; Fatukoa, 1,248; and Tunfeu, 2,018. The major ethnic group is Helong, along with other Timorese. All traditional landholders immigrated to the area at least two or three generations ago.

All three communities were traditionally ruled by *adat*, or customary law. However, under new formal organizations created by the government, a village head represents the village in allocating resources and determines some policy, such as what communal land can be sold, and to whom. The new system consists of a hierarchy of political divisions starting at the subvillage level.

Land tenure is in flux. Most families own more than two hectares of land, unless they are new to the area. Land ownership is patrilineal, or inherited from the father. Some communal land is being sold to outsiders from other villages of Kupang town to grow vegetables and fruit. However, private land seems secure. But, because family fields are not contiguous, crop and cattle protection can be difficult. Rocky coral soils are used for corn and pasture, and soils with vertisol-type characteristics are used for wet rice or vegetable production.

Socioeconomic Factors

The main sources of income for the farming families in the study villages are beef cattle and vegetable production. Collecting limestone rock for urban construction supplements their incomes. Resources are invested in these directions, assuming maize yield goals will be achieved. Cattle are tethered in fallowed cornfields and are not allowed to roam freely (see Figure 17-1).

The farmers do not have access to credit. They may have an elementary school education, but they may not speak, read, or understand the national language. They typically own the land that they cultivate, and with increasing population in the area, they are under pressure to sell underutilized land.

A growing city with market demands is less than 10 km distant, within one hour's travel. Public transportation is irregular and relatively expensive. The only labor that is constant is that of husband and wife, although children from ages 4 to 10 years contribute if they are present. Children often leave home at about 12 years of age to receive further education, so arrangements must be made for extended family and communal labor in times of high labor demand. According to the families, there is little chance of obtaining jobs in the city.

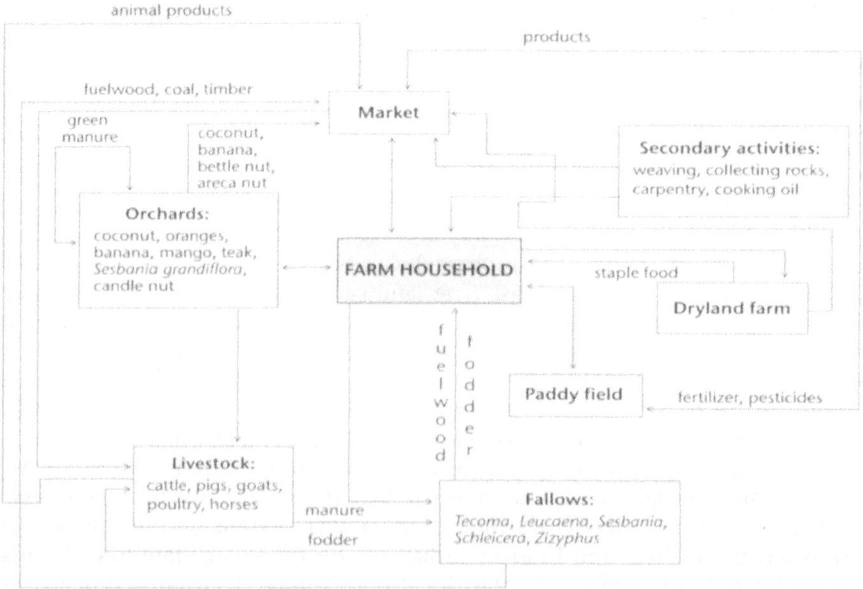

Figure 17-1. Typical Flow of Materials and Products in the Farming System of the Study Villages

Fertilizer inputs for agriculture are limited to farm residues and chemical fertilizers chosen and subsidized by the government. Farm residues are minimal because of the low quantity and quality of biomass produced in the fields. In addition, cooking fuel must also come from farm biomass, usually branches collected from fallowed fields.

The villages have no running water, but each has several wells. The families are eager to raise their standard of living, and particularly to improve their ability to purchase goods. But the transition is not happening as quickly as they would like. Some are resorting to selling their land, and they give the following reasons:

- If the soil is to grow anything, then heavier applications of fertilizers are needed.
- Distant fields are becoming harder to protect, so farmers are focusing on fewer fields with higher inputs. The large number of fields previously cultivated on a rotational basis is no longer considered necessary. Soil restoration through natural processes is no longer regarded as crucial.
- Children are moving to the cities for education and often remain there, so maintaining land to be divided among children is no longer a high priority.

There is now a market for land, and although the price per hectare is low, land sales can return more cash than village farmers have ever received before. However, the harsh lessons of the marketplace have brought realization that money soon runs out and, after all is gone, there is no land left. So more prudent farmers are considering the sale of only their more distant fields, or those with low potential, and keeping those that they think they need.

Outsiders who purchase land are able to invest high capital to establish intensive production systems on the same land where small, resource-poor farmers earlier struggled to survive. Some farmers suggest that the new plantations will perpetuate their poverty because they see no way of competing with them, even if their access to capital improves. However, the newcomers are not producing many of the crops on which the farmers depend, so there is little competition. It should also be noted that farmers have sold only red soil areas that are regarded as offering less than optimum production potential. Black soils capable of producing rice and vegetables have not been sold.

Farming systems in the study area still concentrate mainly on subsistence, but farmers are trying to move toward rice and vegetable production for market sale. Intensification focuses first on producing enough to meet family needs. Remaining resources are then directed to production for market. Subsistence crops such as maize are grown for the family. Maize has little market value because urban people eat rice and, within the farming community, neighbors either do not have the resources to buy it or are bound by cultural relationships that rule out monetary exchange. The barter system is still used frequently.

Crops with market value, on the other hand, are not generally consumed in the home. Opportunities to earn cash are very limited, and cash crops represent an important source of income. Tradition has it that such crops were not grown before fertilizer and seed became available, so they are viewed solely as generators of income, and not as food for the family. Money earned may be saved to pay for electricity or television sets. To generate additional cash income, the men collect rocks from their fields and sell them as building and paving material. Some of the women weave cloth and sell firewood from their homes.

Given conditions where little more than subsistence maize can be grown on the area's rocky, shallow soils, where there's no surplus labor available with which to expand cropping, and where cash cropping is restricted to rare patches of vertisol-like soil, it may be that the production system of the study area has already reached its peak.

Tecoma stans, the Fallow Species

Tecoma stans is called Yellow Bell in the southern United States because of its color and bell-shaped flower. A native of the Andes, it is one of 13 yellow-flowered species composing the *Tecoma* genus (Bignoniaceae), a shrub species with leaves arranged like fingers and winged, white-colored seeds. It has now spread throughout the tropics, particularly in semi-arid regions. In West Timor, it has become common in Kupang Barat and Maulafa, in Kupang District, where it is called *hau suf molo* or *bunga kuning* (yellow flower). It is capable of growing amongst coral rocks, on marginal soils, and in the dry-season. One of its main attributes as a fallow species is its rapid regeneration. It also produces firewood and, in some cases, is used as dry season forage for livestock. This chapter is the first documentation of its use as a fallow species.

In dry areas, the ability of trees and shrubs to regenerate under slash-and-burn systems is considerably reduced. However, *T. stans* begins to resprout from cut trunks at the onset of the rainy season. Most trunks are cut between 30 and 100 cm from the ground and, based on casual observation, those anchored in holes within coral rock appear to survive best. The rock may provide some protection during burning. Resprouting of *T. stans* stumps is preferable to seed regeneration, since canopy closure is quicker and firewood production faster.

The tallest *T. stans* shrubs encountered by the research team were 5 m high, with a diameter at breast height of about 10 cm. Farmer testimony and direct observations suggest that *T. stans* is not prone to pest attack.

About 70% of farmers in the study area use the *T. stans* fallow system. Fallow periods vary from 1 to 10 years, depending on individual management decisions. Farmers observe the growth of *T. stans* as an indicator of soil fertility. When the fallow land has been completely covered by this species, the soil is believed to be rejuvenated. It will be dark brown in color and the topsoil will be covered with litter. When land is left fallow under *T. stans* for an extended period, the litter may accumulate to an average depth of 3 to 4 cm. When fallows are left for five years and there is no fire, the *T. stans* stems may reach 17 cm in diameter at ground level, 9 cm at a height of 30 cm, and 6 cm at 100 cm in height. The total dry matter of the stem will be about 2,790 grams, while the dry weight of leaves will be 230 grams, with an average of two to three branches.

Farmers use *T. stans* as a green manure, a mulch for vegetable crops, or biomass to be slashed and burned to produce fertilizing ash. It is also used as a source of light construction material, for dibble sticks and poles for string beans, and as a means of weed control. Firewood collected from *T. stans* is a valuable source of income. Stems with an average diameter of 3 to 5 cm are cut into 30 to 40 cm lengths, bundled into 10 to 15 pieces, and sold for Rp 100 per bundle. Each three-year-old stand will, on average, produce firewood worth Rp 500,000. A decoction of *T. stans* has reportedly long been used in Mexico to treat diabetes mellitus.

The strengths of this species are its fire and drought resistance and its high germination rate. Farmers have no need to plant *T. stans* but instead rely upon natural reproduction. Its fire resistance is important because, across most of Timor, farmers still practice slash-and-burn cultivation, and they burn range lands to encourage fresh grass regrowth. Therefore, most of the area is susceptible to fire. As far as its drought resistance is concerned, studies showed that although desert willow (*Chilopsis linearis*) had the highest resistance to drought, *T. stans* was not far behind. Both species had a drought resistance more than 1.5 times that of fruitless mulberry.

The main crops cultivated in the study area are maize (*Zea mays*) and groundnuts (*Arachis hypogaea*), either as monocultures or intercropped with cassava (*Manihot esculenta*), pigeon pea (*Cajanus cajan*), or pumpkins (*Cucurbita argyrosperma*). The agricultural production systems of the villages are facing many changes because of proximity to an urban area, the development of physical infrastructure, out migration as settlements expand, increased population pressure, and increasing demand for cash income.

History of the Fallow System

Farming began in the study area around the 1950s, when it was settled and cleared of its vast primary forests for shifting cultivation. After the forest was felled, the first invading vegetation was *Lantana camara*. Although maize yields of that time were not recorded, older farmers say they were higher than present yields. They say *Mimosa pudica* began to dominate fallowed lands between 1961 and 1964. It was not appreciated because its thorns ripped at skin and made management very difficult. Although *Mimosa pudica* still grows in the study area, it is no longer dominant.

T. stans is believed to have been brought to the island of Timor as an ornamental shrub sometime in the 1960s. Some respondents believe *T. stans* first grew around corrals where Chinese traders collected cattle before shipment to other islands. Supposedly because of its beautiful flowers, people took the seedlings and transplanted them as ornamentals in their home gardens. It then escaped and grew spontaneously on fallowed land, spreading either by wind or animal throughout the southern part of Kupang. Later, its merits as a fallow species were recognized and it was intentionally propagated and introduced into other villages. However, the dominance of *T. stans,* like *Lantana camara* and *Mimosa pudica* before it, may be drawing to an end. Around the end of the 1980s and the beginning of the 1990s, the latest weed to invade the territory, *Chromolaena odorata,* first made its appearance. It has since begun to suppress the growth and expansion of *T. stans*. The farmers in the three study villages say they are not yet sure which of the two species will become dominant in their area. They remain unsure about *Chromolaena* as a fallow species, because it is still new. But they didn't recognize the benefits of *T. stans* until it became thoroughly established, some years after it first appeared.

Cropping Practices — Permanent Cultivation

There are two main cropping patterns in the study villages. The first involves permanent cultivation, usually in areas with superior soil and water supply that are close to settlements or hamlets. These lands are continuously cultivated for wetland rice, home gardens, mixed gardens locally known as *mamar*, or traditional agroforestry systems consisting of coconut, banana, and some fruit trees. Vegetables and other horticultural crops are grown in paddy fields when the water level has dropped. Permanent dryland cultivation is restricted to flat land where waterlogging and sedimentation occur during the rainy season. These permanent systems usually require external inputs such as fertilizers and chemicals for pest and disease control.

A mamar is a traditional compound garden or mixed garden planted with different types of crops or naturally occurring species. There are two types of mamar: dryland mamar and wetland mamar. The latter is usually developed around water springs. However, a modified version can be found in dry areas, where most species can be planted except betel nut. Farmers usually grow coconut, betel nut, jackfruit, bananas, and oranges. Beneath the tree canopy they keep cattle and goats. Mamar does not usually require intensive care, and weeds are not a problem because of the dense tree canopy.

A slash-and-burn system practiced in the study area produces one subsistence crop of fast-growing maize with zero to low fertilizer input. After three to four years of bush fallow, farmers slash the vegetation in late August to early November and allow it to dry. Large woody pieces, trunks, and some branches are collected for firewood, and then the plots are burned before planting.

Maize: The Main Food Crop

Planting of maize, without tillage, begins with the onset of rains in about November or December. The harvest is in March. The maize fields occasionally have beans, cassava, pumpkins, or sorghum growing around the borders. *T. stans* is permitted to grow from the third week after planting maize through to the next planting season.

After one to three years of corn, the area is fallowed for 2 to 11 years. *T. stans* resprouts from stumps, particularly those rooted in coraline rocks exposed above ground level.

Land Preparation. The number of plots cultivated by a farmer depends on the availability of both labor and land. Labor comes from the immediate family and some from extended family. Most farmers cultivate from one to three plots.

Planting. This occurs from mid-November to late December, as the rains begin. Most farmers prefer fast-maturing local maize varieties and tend to save seed from one planting season to the next. A hybrid variety, Arjuna, was introduced to the area by a government program (McWilliam 1988), but adoption was low because it was susceptible to pests, and because the ears of *Arjuna* corn were not suitable for the local storage system.

The common planting technique is to punch a shallow hole with a dibble stick and drop in three maize seeds. Planting is done in rows if the soil is extremely rocky and the previous *T. stans* fallow was quite thick. But if the soil has larger but less frequent rocks, then the maize may be planted in circles around lone *T. stans* trunks. Jones (1983), Gunarto et al. (1985), and McWilliam (1988) report that farmers in the area plant in rows, or in a less systematic fashion, with spacing from 25 cm by 75 cm to 2 m by 2 m.

Pest Management. No pesticides are used on maize. A single hand weeding is done approximately three weeks after corn germinates, when the seedlings are about 25 cm tall. This is necessary to prevent maize yields being reduced by weed competition. If fertilizer is used, weeding occurs just prior to its application.

The weeding strategy may be more rational than it first appears. Maize yields are more susceptible to the effects of competition in the early stages of growth, rather than later. The weedy regrowth that occurs during the latter part of cropping helps to form a protective ground cover before the end of the rainy season. This endures throughout the seven or more months of the dry season and is browsed by cattle.

Fertilizer Application. Chemical fertilizers have become available over the past 10 years, so some farmers apply fertilizer to each maize plant as a band application, 5 to 10 cm from the plant and 5 to 10 cm deep, when the plants are about 25 cm tall. Subsidized fertilizer is allocated to the head of the village, who then rations it out to farmers. Farmers have no choice in the fertilizer type, and some of them sell part or all of their ration. However, most of them claim that without these inputs of urea and *complete*, an N-P-K mixture, they would get little to no yield. Fertilizer is not applied at the time of planting.

Table 17-1. Crop Calendar in Bello, Fatukoa, and Tunfeu

Crops	*Calendar*
Maize	November or December to March or April
Rice beans	November or December to April or May
Arbila	November or December to May
Pigeon pea	November or December to May or June
Cassava	November or December for one or three years
Secondary crops	June or July to October or November
Wet rice	November to June
Tree cash crops	Throughout year

Vegetables

String beans are the main vegetable planted in dryland farms. For planting, farmers use a *T. stans* stem as a dibble stick. Other vegetables such as spinach, cabbage, Chinese cabbage, tomato, and chili are only planted in areas with water supply, usually next to mamar, paddy fields, or streams.

Livestock

Livestock are the main source of income. Traditionally, cattle roamed freely around farms or in open lands, but more recently they have been kept tethered and raised in a cut and carry system. Pigs are kept in pens or small barns. Some farmers keep pigs, goats, chickens, and other livestock in a farm plot that is under a *T. stans* fallow.

Crop Calendar

Land preparation is usually done between June and October. Planting usually begins immediately after the first rains, usually in November or December (see Table 17-1). Harvesting varies with the type of crop.

Cropping Practices — Shifting Agriculture

The other major cultivation model is that of nonpermanent agriculture or a shifting, fallow-based system. Traditionally, agriculture relied upon the natural regeneration of fallow vegetation to accumulate biomass and restore soil fertility. This style of shifting cultivation remained very common in the area, even in the 1990s, after farmers gained access to external inputs.

The slash-and-burn operation is usually carried out from July to October, at the end of the dry season. Newly cleared forest areas can usually be planted for three consecutive cropping seasons. When the fertility of the soil declines, the land is fallowed to restore its fertility and to control weeds.

Fallow periods in the study area range from 2 to 10 years. Farmers usually open *T. stans* fallows for cultivation when the canopy completely covers the ground. It also depends on land availability. When a farmer has more than two parcels of land, the fallow period can be extended.

Generally, the fallow vegetation consists of local grasses, trees, and shrubs, including *kosambi* (*Schleicera oleosa*) and *kom* (*Zizyphus mauritiana*). However, some exotic species are also managed as improved fallows. In addition to *T. stans,* these include *lamtoro* (*Leucaena leucocephala*), *gamal* (*Gliricidia sepium*), and *turi* (*Sesbania grandiflora*).

The shallow soils, covered with coral and rocks, are a major limitation to arable cropping. Farmers are often unable to plant for more than two consecutive growing seasons. Corn harvests are generally very low in the third growing season. However, farmers usually have more than one parcel of land, allowing them to cultivate one plot while others are left fallow to rejuvenate.

Strengths and Weaknesses of *T. stans* Systems

Strengths

Most farmers in the study area have more than one plot of arable land, and because the severe limitations of the native soils, *T. stans* is their most favored fallow species.

Soil. Accumulation and decomposition of *T. stans* litter improve the physical, chemical, and biological properties of fallowed soils. The reddish-yellow Mediterranean soils gradually turn dark brown and black as organic matter accumulates. It is assumed that the soil's chemical composition improves as well.

T. stans fallows also reduce erosion by shielding the soil against the direct impact of raindrops and runoff.

Weed Control. The soil under *Tecoma* is usually free from weeds. Weeds are a major constraint in maize production and, when soil fertility declines, weed populations become difficult to control.

Other Benefits. Farmers use *T. stans* as a green manure, as a mulch for vegetable crops, as a source of light construction material, for dibble sticks and poles for string beans, and as a source of valuable firewood. It can also be used as fodder for animals, although it is not regarded as a preferred fodder.

Opportunities

Given the increasing population pressures on a limited land base, demands for income-generating activities, and the general trend toward permanent agriculture, the *T. stans* system has the potential to incorporate other soil-building species to make the fallow more efficient and provide a stronger bridge to more permanent systems. Species listed in Table 17-2 have proven potential to improve fallows. As a combination that may avoid additional labor or management, *T. stans* and *Chromolaena odorata* may be the best option. This would create a situation similar to that when *Lantana camara* still dominated the area and *T. stans* was demonstrating its benefits.

Weaknesses

The major weaknesses of the *Tecoma*-based fallow system in the study villages are declining farm size and increasing population pressure, both of which force a shortening of the fallow period, and which in turn challenges the capacity of the system to restore soil fertility. The biophysical conditions in the study area are also a serious limitation.

Threats

The major threat to this system is the expansion of *Chromolaena odorata*, which is encroaching on areas previously covered with *T. stans*. The other main threat is the expansion of settlements and towns toward the study villages. With better access to markets and a growing demand for cash, farmers are placing increasing emphasis on cash crops. Therefore, farming systems are intensifying toward permanent cultivation, and fallowing is becoming less important.

Recommendations and Discussion

Fallow management helps to stabilize shifting cultivation and acts as a bridge to more permanent land uses. Fallow products provide additional income and thereby move farming practices from subsistence to more market-oriented systems. Fallow management also marks land boundaries and strengthens ownership or tenurial status.

The proximity of the three study villages to Kupang town is persuading farmers away from traditional subsistence farming systems. As generators of income, these are looking less and less attractive. Better access to markets and information, and opportunities for new income-generating activities, are all discouraging farmers from persevering with a fallow system.

Given these conditions, farmers will probably either move away from *T. stans* and adopt other species capable of providing more valuable products and more effective ecological services, or move to permanent cultivation systems. Species listed in Table 17-2 may improve the fallow system by providing better products and services. However, there is one reservation: the invasion of *Chromolaena odorata* may be considered a curse by practitioners of intensive agriculture, while being regarded as a cure by those persevering with fallows under traditional subsistence farming systems.

Changes to the fallow system will not be determined solely by the qualities of alternative species; they also will be determined by the comparative advantages of growing cash crops for market instead. Therefore, the *T. stans* fallow system may fade from use in the future as pressures continue to demand more permanent land–use systems.

Shortening the fallow period, therefore, may offer an intermediate step toward permanent cultivation. There are two major options to accomplish this. The first involves maintaining and improving the use of *T. stans* by the following means:

- Assist its spread;
- Encourage farmers to collect and spread its seeds;
- Integrate complementary terracing or soil conservation techniques into the present system;
- Support cropping with water management systems such as water harvesting or catchment traps;
- Promote in-row tillage; and
- Reduce reliance on slash-and-burn forms of cultivation.

Table 17-2. Comparison of Potential Alternative Species for Fallow Management

Important Parameters	Tecoma stans	Gliricidia sepium	Leucaena leucocephala	Sesbania grandiflora	Chromolaena odorata
Drought resistance	++++	++	++	++	++++
Fire resistance	++++	++	++	+	+++
Fencing material	–	++++	++++	++	–
Fodder	+/–	+++	++++	++++	–
Nitrogen supply	–	+++	+++	+++	–
Firewood	++++	++	+++	+	+++
Resistance to pests and diseases	++++	+	+	+	+++
Biomass production	+++	++++	++++	++	+++
Resistance and response to pruning	+++	++++	++++	+	+++
Adaptation to harsh environments, especially rocky soils	++++	+++	+++	++	++++

Notes: ++++ = very strong, +++ = strong, ++ = fair, + = weak – = no use.

The second option involves a shift to alternative fallow species and pursuit of the following:

- Strengthen soil and water management;
- Reduce reliance on the slash-and-burn technique;
- Promote in-row tillage;
- Replace *T. stans* with more effective or more productive species such as *Gliricidia sepium*, *Leucaena leucocephala*, *Sesbania grandiflora*, *Acacia villosa*, *Acacia angustissima*, or others;
- Clarify farmers' land tenure status;
- Incorporate long-term cash crops; and
- Develop alternative income-generating activities.

The alternatives suggested above need to emphasize biomass management. Cattle, field crops, and weeds should all be considered as potential components of managed fallows.

Risk management is also a crucial issue in considering future farming systems. Informal communications with nongovernmental organizations indicate that, at present, crops completely fail in 6 out of every 10 years. External support is also needed for off-farm activities that generate income, such as rock collecting, vegetable production, and handicrafts.

Green Manure

The use of green manure does not seem to be the best option for future systems, because cattle are the main source of income in the study villages, vegetation is sparse, and moisture to assist decomposition is limited throughout most of the year. Cattle fodder is not only in short supply, even during the cropping season, but it also is often short of nitrogen and therefore protein. A more efficient use of resources might be to encourage the growing of patches of legume trees. Fodder harvested from these trees, as well as maize stover (fodder), could be channeled toward fattening cattle in a cut-and-carry system.

Farm labor is usually limited to one or two family members working full time, so the incorporation of mulches just before and during the cropping season, when it would be most useful, is not very practical. These are times when labor is most needed for preparing fields, planting, and weeding. In addition, the incorporation of green manure into the soil, which minimizes the loss of nutrients due to volatilization, leaching, and runoff, is almost impossible because of the abundance of large surface rocks. While mulches would help protect the soil from the heavy rains, the quantity available is insufficient, unless fallow vegetation is slashed but not burnt.

Fallow Species

Under conditions of short fallow periods, dry climate, and shallow soil, biomass accumulation is severely limited. Fast-growing, drought-resistant fallow species are needed, and *T. stans* stands out as a suitable fallow species in the study area in that it rapidly reestablishes from coppicing trunks. Observation suggests that it can also establish from the profuse seed bank if the area is not burnt for one to two years. As is usually the case in slash-and-burn agriculture, the ash layer created by burning the biomass provides the coming crops with available nutrients. But even with fast-growing species, total biomass accumulation in the dry-season conditions of the study site is low, and burning releases only a small quantity of fertilizing ash. Moreover, since burning often occurs a month or more before the onset of the rains, and gusty winds are not unusual, most of the ash may have blown away by planting time. Remaining ash may later be washed away when the initial rains hit the unprotected ash and soil.

Given this, and the fact that some fertilizer is available, an experiment should be undertaken by splitting a field and comparing yields on burnt versus mulched treatments over several years. In the long term, mulching is probably better in terms of soil protection and organic matter accumulation. As most fallow species are largely nonwoody, more than 40% decomposition of organic matter should occur in the first cropping season, adding some nutrients. Building organic matter should, in time, allow for higher cation holding and exchange capacity.

Weeds and Stover

Weeding labor may be significantly reduced if the slashed vegetation is used as a mulch. The stover, until it is well browsed by cattle, provides additional ground cover. Since soil protection from wind and water erosion is important, crop residues should be left in the field to dry naturally.

It is advisable to allow cattle to graze the stover, since there are few means of composting it or collecting it as fodder and because, in such a dry climate, organic matter breaks down extremely slowly. However, cattle should not be allowed to graze fields down to unprotected soil.

Releasing cattle on land to be cropped the following year might also provide a relatively low labor approach to incorporating nitrogen into the soil. Incorporating the manure into the soil with a stick might slightly improve soil nitrogen levels. Farmers retrieve their cattle every night to prevent theft, so they are already visiting the area where the cattle have been tethered all day. It is difficult to estimate the addition of nitrogen from two to six cows on four to six hectares of land, given this manure management strategy and the low nitrogen diet of the cattle.

In the experiments proposed above, plant densities of fewer than 45,000 plants per hectare should be used. However, different planting densities should be tried, since heavy rain, fertilizer quantity, and soil rooting volumes are difficult to predict. A range from 26,000 to 45,000 plants per hectare could be tested.

Conclusions

Given the marginal, shallow soils dominated by coral and boulders, the farmers of the three study villages have limited opportunities for improving their farming systems. Their own limited technical skills and knowledge of workable alternatives are also an important reason for their continued reliance on bush fallow techniques to restore and maintain soil fertility. However, they now have limited access to fertilizers and their farming systems appear relatively stable.

The present *T. stans* fallow system was developed following its unintentional introduction to the area. Farmers manage *T. stans* because it can survive minor fires and can coppice rapidly. It quickly forms a canopy, improves soils, produces firewood, and, to a lesser extent, provides vegetable poles and animal forage. Maize production following a two- to four-year *T. stans* fallow is marginal, but still possible, without fertilizer. Fallows of longer than four years are recommended if no fertilizer is to be applied.

The *T. stans* fallow system is failing to meet the aspirations of its practitioners but it has some important attributes and seems to provide a viable step in the transition toward something better.

Outstanding questions of interest concerning the system include the following:

- Is the fallow augmenting the capacity of the production system?
- What access do farmers have to markets, including both transportation and services?
- What would be revealed by a social and cultural analysis?
- What are the specific biological and physical aspects of the system?

The maize-*T. stans* system, as practiced in Timor, is a low-productivity system for marginal areas. Since *T. stans* is now found throughout the semi-arid tropics and is self-propagating, the *T. stans*-based fallow system may be transferable to locations where land tenure is secure enough to allow at least four years of fallow, land pressure is not extremely high, families have two or more hectares of land that can be rotated in a bush-fallow system, there are few inputs or little mechanization, soil is shallow and rocky, labor is in short supply, and firewood is needed.

Acknowledgments

The authors gratefully acknowledge funding for this study from the Cornell Agroforestry Working Group of the Cornell Institute for International Food, Agriculture, and Development (CIIFAD), and infrastructure and logistical support from the State Agricultural Polytechnic Kupang, West Timor. This case study would not have been possible without the energy, enthusiasm, dedication, and creativity of our research assistants, Fedi, Mada, Mintje, and Johan.

References

Bouldin, D.R., W.S. Reid, and N. Herendeen. 1968. Methods of Application of Phosphorous and Potassium Fertilizers for Corn in New York: A Summary of Recent Research. *Agronomy Mimeo* 68(7).

———, W.S. Reid, and D.J. Lathwell. 1971. Fertilizer Practices which Minimize Nutrient Loss. In: *Agricultural Wastes: Principles and Guidelines for Practical Solutions*. Proceedings of Cornell University conference on agricultural waste management, February 10–12, 1971, Syracuse, NY, 25–35.

Bryant, R. 1996. Discussions and Class (SCAS 368) on Soil Genesis and Subsequent Management for Crops.

Gunarto, L. 1992. Response of Maize to N,P,K and Trace Element Fertilizers on a Regosol at Maliana, East Timor. (Tanggapan Tanaman Jagung Terhadap Pemberian N, P, K dan Unsur Mikro di Regosol-Maliana, Timor Timur). *Agrivita* 15(2), 65–69.

———, M. Yahya, H. Supadmo, and A. Buntan. 1985. Response of Corn to N,P,K Fertilization Grown in a Latosol in South Sulawesi, Indonesia. *Communications in Soil Science and Plant Analysis* 16, 1179–1188.

Jones, P.H. 1983. Lamtoro and the Amarasi Model from Timor. *Bulletin of Indonesian Economic Studies* 19(3), 106–112.

Landon, J.R. 1991. *Booker Tropical Soil Manual*. Hong Kong: Booker Tate.

McKinnell, F.H., and Harisetijono. 1991. Testing *Acacia* Species on Alkaline Soils in West Timor. In: *Advances in Tropical* Acacia *Research*. Proceedings of a workshop, February 11–15, 1991, Bangkok, Thailand, edited by J. Turnbull. ACIAR Proceedings No. 35, 183–188.

McWilliam, A. 1988. Strategies for Subsistence in West Timor. In: *Contemporary Issues in Development*, edited by D. Wade-Marshall and P. Loveday. Northern Australia: *Progress and Prospects* Vol. 1, 280–290.

Metzner, J.K. 1977. Man and Environment in Eastern Timor: A Geoecological Analysis of the Baucau-Viqueque Area as a Possible Basis for Regional Planning. *Development Studies Center Monograph* No. 8. Canberra: Australia National University, 380.

———, 1983. Innovations in Agriculture Incorporating Traditional Production Methods: The Case of Amarasi (Timor). *Bulletin of Indonesian Economic Studies* 19(3), 94–105.

Soil Survey Staff, 1994. *Keys to Soil Taxonomy*. USDA Soil Conservation Service.

Chapter 18

Improved Fallows Using a Spiny Legume, *Mimosa invisa* Martius ex Colla, in Western Leyte, the Philippines

Edwin A. Balbarino, David M. Bates, and Zosimo M. de la Rosa

The uplands of the Philippines, which are dominated by rugged, hilly topography, have experienced widespread conversion of forest to permanent agriculture based on rice, maize, root crop cultivation, and animal production. Traditionally, upland farmers have relied on fallowing to maintain soil fertility, a practice that is successful as long as the length of the fallow period is sufficient to allow for regeneration of the natural vegetation. However, an increasing human population in the uplands coupled with finite land resources has resulted in a dramatic reduction in the length of the fallow period and its capacity to restore soil fertility. The consequence of continuous cropping, lack of soil conservation measures, and heavy grazing by ruminants has been severe soil erosion and fertility depletion.

Management practices proposed by agricultural research establishments to improve the productivity and sustainability of short-term fallow systems in the uplands have not been widely adopted because of excessive labor requirements, unavailability of planting materials, and destruction of vegetation by uncontrolled fires or communal grazing. Some farmers, recognizing the nature of their problems, have developed local approaches to fallow management that improve its biological efficiency and produce the same or greater crop productivity over a shorter period.

In the uplands of western Leyte, as in other parts of the Philippines, the ability of fallow systems to restore soil fertility is hampered by the cultural practice of permitting fallowed farmlands to be used as communal grazing areas for ruminants. The animals not only compact the soil, they also consume most of the fallow vegetation, which otherwise would hold and enrich the soil. Farmers in western Leyte have found that incorporation of the spiny legume known locally as *benet* (*Mimosa invisa* C. Martius ex Colla) into their crop and fallow system serves to exclude ruminants from grazing the fallow, thereby protecting it and lessening soil compaction to the degree that only two plowings are needed, rather than the usual three, before planting corn. (See color plate 23 for an example in Mindanao.) Furthermore, they say that corn grows more vigorously following a benet-based fallow than one dominated by the weedy legume known locally as *nipay-nipay* (*Calopogonium mucunoides* Desv.).

Edwin A. Balbarino and Zosimo M. de la Rosa, Farm and Resource Management Institute, Leyte State University (LSU), Baybay, Leyte, the Philippines; David M. Bates, L.H. Bailey Hortorium, 462 Mann Library Bldg., Cornell University, Ithaca, NY 14853-4301, USA.

In this chapter, we provide a preliminary overview of the values of benet-based crop and fallow systems in Punta, western Leyte. During the course of this study we have identified questions concerning this system that warrant further investigation. However, we are confident that this first assessment captures its essential elements, and we believe that incorporation of benet would be an effective means of sustaining and intensifying agriculture in other upland areas.

Methods

Staff members of the Farm and Resource Management Institute (FARMI), at Leyte State University (LSU) in the municipality of Baybay, on the western shores of the island of Leyte, were the first to realize that local farmers had devised a crop and fallow system based on the use of benet. They documented the attributes of the system and later used participatory rural appraisal (PRA) techniques in the course of this study with the upland farmers of Punta, a *barangay* (village) in the municipality of Baybay. The PRA approaches involved the following:

- Preparation of transect maps to determine the approximate local distribution of benet in relation to the elevation and character of the topography, the kinds of crops grown, and the nature of the surrounding vegetation;
- Development of a seasonal calendar to describe the benet life cycle in the context of the controlling factors of the crop and fallow system, including crops, labor, land preparation, other plant species, and the periodicity of wet and dry periods; and
- Interviews with individual farmers and groups of farmers to elicit personal experiences with benet.

The results of our analysis were presented to the Punta farm community for its comment and validation. In addition to the on-site PRA activities and the use of data gathered by FARMI staff, photographs were taken to document the life cycle of benet in relation to the annual cropping cycle. Preliminary surveys were also undertaken to determine the distribution of benet in western Leyte and to seek out literature concerning its taxonomic status and reports of its occurrence and behavior in agroecosystems, both as a noxious weed and as a component of fallows.

The Study Site

The farming and fishing community of Punta is located on the west coast of the island of Leyte, about 10 km south of the municipal center of Baybay and about 140 km southwest of Tacloban, the provincial capital of Leyte (Figure 18-1). Punta encompasses four *sitios* (subvillages) with a total of 450 households in a land area of about 600 ha. It has three distinct agroecological zones: coastal, upland farms, and forest. The upland farm zone is stratified vertically into three subzones: the lowest is planted to perennial crops and trees; the middle subzone supports permanent, short-term crop and fallow systems growing corn, sweet potatoes, and annual crops; and the upper subzone remains mostly in traditional *kaingin*, or swiddens (Figure 18-2).

This study focuses on the middle subzone. Here, most farmers cultivate one to three plots of land, with an average total area of about one hectare. Much of the land constituting the middle subzone is in the hands of a single owner. However, the land reform program of the Department of Agrarian Reform proposes to transfer ownership to the present tenants.

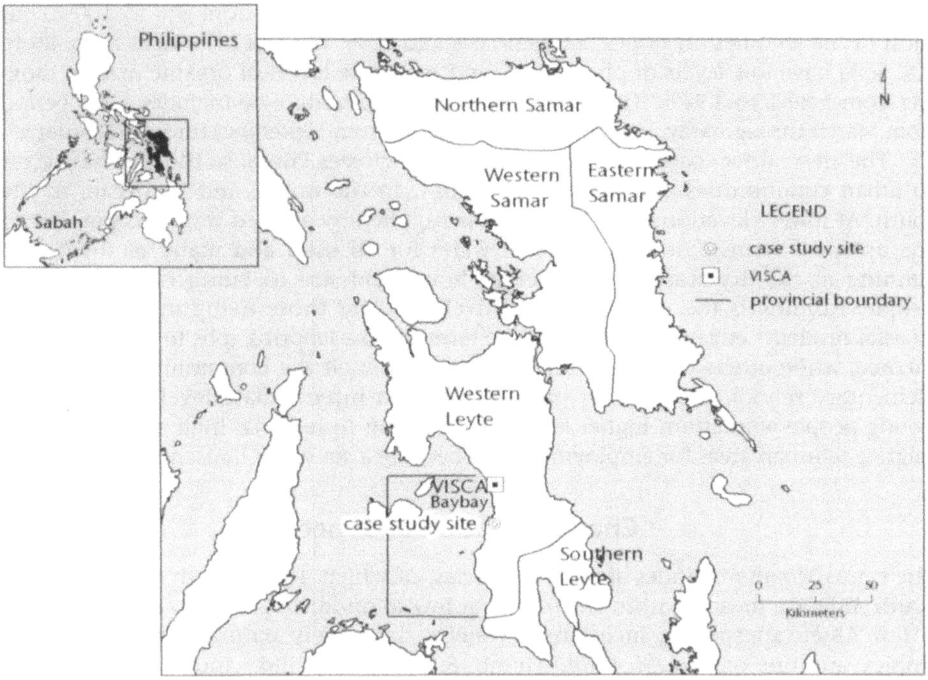

Figure 18-1. Map of Eastern Visayas, Philippines, showing the Study Site in Western Leyte

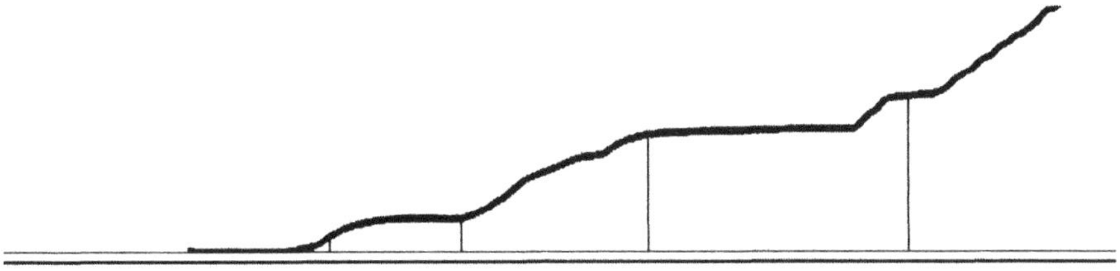

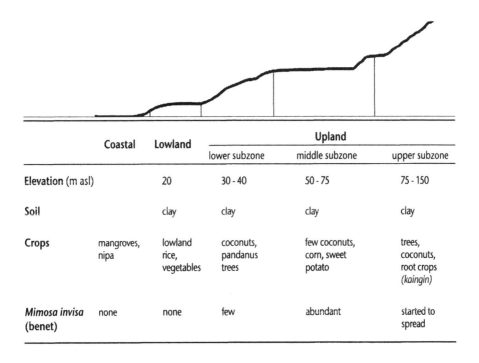

	Coastal	Lowland	Upland		
			lower subzone	middle subzone	upper subzone
Elevation (m asl)		20	30 - 40	50 - 75	75 - 150
Soil		clay	clay	clay	clay
Crops	mangroves, nipa	lowland rice, vegetables	coconuts, pandanus trees	few coconuts, corn, sweet potato	trees, coconuts, root crops *(kaingin)*
***Mimosa invisa* (benet)**	none	none	few	abundant	started to spread

Figure 18-2. Transect Map of the Punta Study Site Showing the Presence of *Mimosa invisa* in the Upland Zone

The topography of Punta is varied, with slopes ranging from 5% to 100%, but most of the land lies on slopes between 20% and 60%. The soil pH ranges from 4.9 to 7.4. Soils have low levels of phosphorus and moderate levels of organic matter, ranging from 1.89% to 3.94%. The rainfall pattern in western Leyte includes a dry period from March through May, with the heaviest rainfall from September through February.

The all-weather coastal road of western Leyte crosses Punta, so the area has access to urban communities and markets at Baybay, in the north, and Inopacan, in the south. At lower elevations drinking water is supplied by a piped water system, but in the uplands, springs are the source of water for all uses, and many of them stop running in the dry season. The average household size in Punta is five to seven people. Farming is the main means of livelihood for those living in upland areas. Coastal residents engage in fishing. Some farmers take laboring jobs to augment their income, while others raise livestock for cash. Although the community has only an elementary school, most adults have attained an intermediate level of schooling. Young people who attain higher levels of education in regional high schools usually migrate to urban areas for employment and become a source of household income.

Characterization of Benet

The genus *Mimosa* includes about 480 species, of which 460 are native to North and South America, mostly within the tropics at low to middle elevations (Barneby 1991). A few American species, including *M. invisa*, are widely naturalized in the Asian tropics, and one or two, *M. pellita* Humb. & Bonp. ex Willd., and *M. pudica* L., are perhaps naturally circumtropical. *Mimosa invisa* is taxonomically complex. Barneby recognizes two subspecies: *invisa* and *spiciflora* (Karsten) Barneby. Subspecies *invisa* includes var. *invisa* and var. *macrostachya* (Bentham) Barneby, both of Brazil and Paraguay. Subspecies *spiciflora* includes var. *spiciflora* of northern South America and var. *tovarensis* (Bentham) Barneby of Venezuela. In addition to the spiny varieties of *M. invisa* that were dealt with by Barneby, a spineless taxon known as *M. invisa* var. *inermis* Adelb. is present in the Asian tropics, apparently having arisen *de novo* in Indonesia and New Guinea (Parsons and Cuthbertson 1992) and perhaps elsewhere. It may occur in western Leyte, for one of us (Balbarino) has noted a spineless *Mimosa* in Matalom, a municipality south of Punta.

Some confusion exists in the interpretation of *M. invisa*. Barneby (1987) has shown that *M. invisa* C. Martius is not the same species as *M. invisa* C. Martius ex Colla, as has been thought, but rather a synonym of *M. diplotricha* C. Wright ex Sauv. Verdcourt (1988) subsequently transferred the variety *inermis* to *M. diplotricha*, as var. *inermis* (Adelb) Verdc. This situation has been reviewed by Maxwell (1988) as it applies to Thailand. These name changes indicate that the literature dealing with *M. invisa* in Asia must be interpreted with caution, for characteristics attributed to that species may actually belong to *M. diplotricha*. We have not yet verified the species or the varietal identity of the plants known as *M. invisa* as they occur in Punta and throughout western Leyte, but for purposes of this chapter and pending further study, we assume them to be *M. invisa* in its typical form.

The most striking characteristics of benet are its long, prostrate to clambering, angular stems, which reach 2 m or more in length and, in mass, form a tangled cover on the ground and over other vegetation. The mass is held in place by recurved prickles 3 to 6 cm long that line the angles of the stem. The leaves, which fold on touch and at night, are from 10 to 20 cm long and are decompound, being divided into four to eight principal segments, each of which bears up to 30 pairs of opposite, sessile, more or less lanceolate leaflets up to about 10 mm long. The flowers form globular, pinkish heads about 12 mm across, which are terminal on short axillary stalks. The fruits, which are borne in clusters, are oblong, segmented, spinescent pods 10 to 30 mm long. Each has three or four seeds. The seeds, 2 to 3 mm long and enclosed in the fruit segment walls, are glossy brown and flattened, with a horseshoe-shaped ring on each side. The root, as described by Parsons and Cuthbertson (1992), is a deep, branching taproot, which bears rhizobial nodules on the root hairs.

Despite the presence of the taproot, but perhaps as the result of plowing, benet is not persistent and behaves more as an annual in Punta. However, Parsons and Cuthbertson (1992) report that in Queensland, Australia, it may persist as a short-lived perennial, and Bolton (1989) observed regeneration of stems from the crown after disturbance. In Punta, benet may reproduce vegetatively when cut by weeding or plowing, but the beginning of each growing season is initiated by germination of its seeds, rather than by new growth from established roots.

The life history aspects of *M. invisa* in Queensland are described by Parsons and Cuthbertson (1992) and in Thailand by Sikunnarak and Doungsa-ard (1985). (Note the taxonomic caution above.) Their observations parallel those made at Punta, which are described below in terms of the crop and fallow cycle at the study site. Baki and Prakash (1994) reported some aspects of floral biology, having studied fruit abortion in *M. invisa* from northern peninsular Malaysia. They attributed this to a failure of sporogenous tissue to develop in anthers. Bolton (1989) and Parsons and Cuthbertson (1992) point out that each seed, held within a single spiny segment of the fruit wall, may be dispersed by adhering to human clothing, animal fur, and perhaps bird feathers, as well as by floating on water, by agricultural implements, or as impurities in crop seeds. Parsons and Cuthbertson (1992) also note that each plant produces up to 10,000 seeds, and they may remain in the soil for 50 years or more before germinating.

Although *M. invisa* was apparently introduced into Asia as a cover crop, its aggressive growth and persistence in soil seed banks have made it a serious weed in tropical plantings, especially sugarcane and pastures. The weed is reported in upland rice fields (Budiarto 1980) and coconut plantations (Prawiradiputra and Siregar 1980) in Indonesia. Control with herbicides can be effective (Parsons and Cuthbertson 1992) but is too expensive for most farmers. Seedlings can be controlled by plowing, and shading will reduce the vigor of older plants. There is interest in biological control of *M. invisa* (Edrolin et al. 1993; Muniappan and Viraktamath 1993). *Heteropsylla spinulosa* (Homoptera: Psyllidae) has been introduced into Australia and Western Samoa to control *M. invisa* and has shown some success, as it has in Papua New Guinea (Wilson and Garcia 1992; Parsons and Cuthbertson 1992). A collaborative project between Australia and Timor that also seeks methods of biologically controlling *M. invisa* has been funded by the Australian Centre for International Agricultural Research (ACIAR).

As a cover crop, *M. invisa* produces thick vegetation. Once decomposed, its residues add organic matter to the soil, increase the level of nitrogen and other soil nutrients, including potassium, calcium, and magnesium, and improve soil tilth (Batoctoy 1982). In Western Samoa, Kaufusi and Asghar (1990) and Tiraa and Asghar (1990) conducted experiments using *M. invisa* and other species, supplemented by phosphorus and potassium, to fertilize corn. They found that all treatments, except those involving *Flemingia macrophylla* and the lowest dosage of *Cajanus cajan* (L.) Huth, increased the growth of corn over that of the controls, with growth generally increasing in relation to the amount of plant material incorporated. Similar results were obtained by Gibson and Waring (1994) from the incorporation of *M. invisa* and other legumes into pastures. As a cover crop in lychee orchards in Thailand, *M. invisa* and *Calopogonium mucunoides* have been moderately effective in covering the ground and in improving soil bulk density and soil organic matter over a three-year period.

An observation reported by Parsons and Cuthbertson (1992) concerning grazing is relevant to the question of palatability of *Mimosa* species and the identity of plants bearing the name *M. invisa*. The spineless variety, now *M. diplotricha* var. *inermis*, is palatable and used as forage in Indonesia and Papua New Guinea, although feeding tests show it to be toxic to sheep in Queensland and to pigs on the Indonesian island of Flores. However, there were no adverse effects when the spiny *M. invisa* was included in the daily rations of a ram. This suggests differences in the chemical constituents of *M. invisa* and *M. diplotricha* that could be useful in resolving taxonomic questions and in selecting chemical variants of these species for agricultural purposes.

The Benet-Based Crop and Fallow System

Conventional Fallow System

To illustrate farmer innovation in developing benet-based crop and fallow practices, we provide a summary of conventional fallowing practices in the Punta region. Fallowing is a traditional practice among Filipino upland farmers, who understand the necessity of resting the soil under vegetation to restore its fertility. Traditionally, fields in the mid-elevation zones of the study site at Punta are cropped for two to three years and then fallowed for three to seven years. Corn is the principal crop, and the dominant fallow species is nipay-nipay *(Calopogonium mucunoides).*

Opening an area for cultivation usually starts by cutting the fallow species and gathering and burning the cut grasses and vines before plowing the field. Farmers plow three times before planting corn. Some farmers also practice green manuring by plowing the nipay-nipay under the soil when hilling up corn. There are no soil and water conservation techniques to protect the soil. During the fallow period, the farms are grazed by *carabao* (water buffalo), cattle, and goats. Upland farmers normally do not plant forage crops for their animals, so this fallow system provides grazing areas for their ruminants, which in turn contribute manure to the soil. Communal grazing, however, causes soil compaction, and the fallow species are not given the chance to produce biomass or, in the case of nipay-nipay, to fix sufficient nitrogen to rejuvenate the soil. The consequence is the need for long periods of fallow.

Motivation for Innovative Fallow Management

Given sufficient land, the conventional fallow system accommodates the needs of both the farming household and the community. However, because Filipino farmers subdivide their farms among their children, farm size is reduced with each succeeding generation, unless new lands can be obtained. The diminishing availability of new land, coupled with an increasing population, are factors that led farmers to intensify production on existing plots. The simplest approach to intensification is to shorten the fallow period while maintaining, if not increasing, the level of crop production. In adopting this approach, the farmers were forced to seek a sustainable fallow system that recovered soil fertility in a considerably shorter period than that needed in the conventional system. In Punta and western Leyte generally, farmers are convinced that leguminous shrubs and vines are good fallow species for restoring soil fertility. This conclusion is based on their comparisons of crop productivity in fields with and without leguminous species. It is not surprising, therefore, that farmers looked to legumes as a means of shortening the fallow period. However, without protection from grazing ruminants, the desired intensification could not be achieved. Fencing of fallowed farms was tried, but it proved to be too expensive and the fences did not last. Furthermore, protection of fallowed farms had broad social implications, because farmers had traditionally pastured their animals on fallowed land whether they cultivated the land or not. This brought the whole need for grazing land into conflict with the need to restore soil fertility.

The Benet Innovation

Benet was introduced to the Punta area in the mid-1960s, presumably by a farmer from Mindanao. It is not common in other areas of western Leyte. The initial reaction of local farmers was hostile, and the plant was regarded as a noxious weed. By the 1980s, however, its value in fallow systems had been recognized. Like other legumes, benet had a positive effect on soil fertility. But, unlike nipay-nipay, and presumably because of its spines, it was not palatable to ruminants, and its presence discouraged their movement onto fallow fields. Soil compaction was reduced, vetiver hedgerows were preserved, and biomass production was increased. Since then, benet has been deliberately managed by Punta farmers as their main fallow species.

In terms of farm management, benet offers the following benefits:

- As an indigenous innovation resulting from local manipulation of the fallow system, benet-based technology works without the "outside" assistance of extension agents, input suppliers, researchers, and other nonfarm parties.
- Farm inputs are not required. Benet seeds profusely and germinates naturally in farm fields. No labor or fertilizers are needed to establish or grow benet, and maintenance is part of regular crop production operations, such as grass cutting, plowing, and weeding.
- The benet-based crop and fallow system restores soil fertility on an annual or biannual cycle. Benet adds a large volume of leguminous biomass to the soil during plowing and provides green manure and mulch during weeding and hilling up.
- Because soil compaction is reduced by excluding animals from fallow fields, only two, rather than three, plowings are required to prepare fields for corn.

While benet is a critical element of the crop and fallow system, it does not operate alone to improve farm productivity. Other management elements also increase the effectiveness of the system, including the following:

- Vetiver contour hedgerows to control soil erosion. Organic matter added to the soil is prevented from being washed away during heavy rains by contour hedgerows of vetiver grass.
- Other fallow legume species. While benet may dominate a fallow, other legumes such as nipay-nipay and other weedy invaders find room to provide additional biomass.
- Slash-and-mulch practice. Farmers cut benet and other herbaceous weeds to open the farm for cultivation. The slashed materials are either used as mulch or plowed under in order to incorporate biomass into the soil.
- Slashed weeds and crop residues are left unburned. After harvesting, crop residues and slash are left to decompose and add fertility to the soil.
- Proper timing of land preparation and weeding. Plowing that is timed to coincide with fruit maturation of benet and other legume species ensures a continued supply of seeds in the soil. Weeding and plowing under benet and other legume species a month after their emergence also provides an excellent green manure to nourish growing corn.
- Planting of sweet potato and other crops. The benet-based crop and fallow system fits in with other cropping patterns. For example, sweet potatoes can be intercropped or grown as a relay crop with corn.

The ultimate goal of the benet-based crop and fallow system is to reduce the fallow period without loss of productivity. This goal has been attained by deterring livestock from grazing on fallowed fields, thereby preventing soil compaction, the breakdown of hedgerows, and the removal of legumes and various weeds and shrubs needed to rejuvenate soil fertility. A study by de la Rosa and Itumay (1996) demonstrated that the crop and fallow system dominated by benet and nipay-nipay, and practiced by Punta farmers, was sustaining crop productivity for three consecutive seasons. In 1993, the average yield of corn in five benet-based crop and fallow farms was 2.67 t/ha. Two seasons later, in 1995, the same farms gave an average of 2.45 t/ha. In PRA interviews, farmers revealed that since the 1980s, when they integrated benet into their management programs, they had not fallowed their fields for more than two years at a time.

Ecology of Benet-Based Crop and Fallow Systems

Benet has now found a significant place in the Punta crop and fallow system. Its life cycle fits the ecological rhythm of the region and interplays harmoniously with other components of the fallow (Table 18-1).

Farmers begin cutting the vegetation of fallow fields in February to early March, when benet pods are fully ripened. The ripe pods are then plowed into the soil, ensuring that the supply of benet seeds is maintained. Benet begins to germinate in late March following the second plowing and about two weeks after the first, more or less coinciding with the onset of the dry season. Corn is planted in May or June with the beginning of the rainy season. Hilling up of the corn is usually done a month after germination. By this time the benet is also growing vigorously, and it is plowed under as green manure. Although farmers weed the corn from its vegetative to early reproductive stages, the benet regenerates rapidly from unpulled plants or from plant parts such as roots or stems, as well as from germinated seeds, to occupy open spaces. Cut benet and other weeds are spread on the soil as mulch or incorporated into the soil as green manure.

Following the corn harvest in August or September, the heavy rains begin, and the benet spreads to cover the stalks and residues of the corn crop. It totally overgrows the field until the time comes for the next cropping cycle in February. If farmers grow sweet potatoes as a relay crop following corn, the field is plowed again after the corn harvest. The sweet potatoes minimize the growth of benet but do not eliminate it. However, the sweet potato crop makes the field out of bounds to grazing animals, so the net effect is the same as if benet were growing vigorously. Benet begins flowering in November and sets pods in January and February.

Table 18-1. Seasonal Calendar of Benet (*Mimosa invisa*)-Based Fallow System

*	*Jan.*	*Feb.*	*Mar.*	*Apr.*	*May*	*June*
1	Rain			Dry months		
2		Vegetation cutting	Plowing	Plowing		
3	Fruiting	Fruit ripening	Seed germ-ination		Vigorous growth	
4						Planting
5	Harvesting					
6	Fruiting	Fruit ripening	Seed germ-ination		Vigorous growth	

*	*July*	*Aug.*	*Sept.*	*Oct.*	*Nov.*	*Dec.*
1	Dry months			Rain		
2						
3		Vigorous growth				Flowering
4	Hilling up	Weeding	Harvesting			
5		Planting	Weeding			
6		Vigorous growth				Flowering

Notes: 1. Seasonal rain and dry; 2. Land cultivation; 3. *Mimosa invisa* (benet); 4. Corn; 5. Sweet potato; 6. *Calopogonium mucunoides* (nipay-nipay)

Benet is compatible with nipay-nipay and other fallow species. Nipay-nipay also produces seeds before the March plowings, and they germinate at about the same time as those of benet. While nipay-nipay is an aggressive species, it cannot outgrow benet because the latter "stands up" and blankets nipay-nipay. There is one clear advantage drawn from having nipay-nipay in the fallow system: it effectively controls the growth of cogon (*Imperata cylindrica* [L.] Raeush).

Discussion

Farmer testimonies concerning productivity from their benet-based crop and fallow systems and results reported by de la Rosa and Itumay (1996) indicate the effectiveness of this intensification approach in maintaining soil fertility over time. The most striking evidence is found in corn production, in which productivity levels have been sustained in an annual or biannual crop and fallow cycle. Farmers say that since adopting the system, they have maintained the same annual production levels for at least five years, except when stricken by natural calamities such as typhoons.

The benet-based crop and fallow system has socioeconomic value. First, it reduces from three to two the number of plowings needed to prepare fields, and one-third of the labor cost of field preparation is eliminated. Second, the organic fertilizer that the system provides to the soil is undoubtedly the equivalent of several thousand pesos' worth of inorganic fertilizer. Third, the value of the crops produced and the assurance of sustained yields each year has lessened concerns about food security among upland families.

In an ecological context, the benet-based system offers an effective means of reducing erosion and maintaining or enhancing soil fertility on sloping lands, even while these lands are in crop production. It has both economic and ecological advantages as a means of rehabilitating degraded upland environments. This aspect gathers significant importance when it is considered that soil is among the most precious of upland resources and it is currently being lost in major quantities every year.

The uniqueness of the benet-based crop and fallow system stems from benet's ability to exclude grazing animals from fallow fields, thereby preserving the fallow vegetation, preventing compaction of the soil, and preserving the integrity of hedgerows. As a consequence, soil loss by erosion is lessened and fertile soils accumulate behind vetiver hedgerows. Soil fertility is improved through the nitrogen-fixing capabilities of benet, nipay-nipay, and other legumes, which, along with other weedy species, are incorporated into the soil as compost and green manure.

The life cycle and behavior of benet have proven to be entirely compatible with the climatic and farming cycles of western Leyte, and although its adoption is limited, its demonstrated value suggests potential for wider adoption in fallow systems throughout the Philippine uplands. Moreover, the benet-based crop and fallow system seems to work most effectively on relatively small farms of less than two hectares, where a farmer and his family manually perform most farm operations. Farms of this character dominate the Philippine uplands.

A major constraint to adoption of the fallow system is the widespread belief that benet is a noxious weed. Of course, under uncontrolled circumstances it may be an appropriate description. A second constraint is the thorny nature of benet, which discourages some farmers from using it. However, we suggest that the testimonies of Punta farmers will convince other farmers to try the technology and sample the benefits it offers. Toward this end, the Punta fallow site will be opened as a learning center for demonstrations of the benet-based crop and fallow system for farmers from other areas of the Philippines, and beyond.

Future Research Priorities and Experimental Agenda

Our studies of the benet-based crop and fallow system are still preliminary. Yet we believe that we understand its basic features and are confident of our assessment of its potential as an effective intensification technique. Simultaneous with the attention from researchers, the farmers of Punta are continuing to refine the system. Without imposing on this process, we identified issues concerning the benet-based crop and fallow system that need to be understood and resolved if the system is to be fully evaluated and adapted to other environments. Therefore, areas for future research include the following:

- A systematic study of *M. invisa* and *M. diplotricha* in order to determine taxon identities, relationships, and distributions. One result of this could be a basis for comparative biological, chemical, and agricultural studies of the taxa involved;
- More complete characterization of the biophysical dynamics of the weed flora and other components of the benet-based crop and fallow system in order to better understand its synergistic interactions;
- Quantitative studies that could effectively compare and assess the impact of a benet-based crop and fallow system on soil characteristics, including fertility, tilth, and retention in a variety of agroecosystems; on socioeconomic conditions of households and communities in total and by gender; and on other fallow intensification strategies and species; and
- Design and evaluation of tools for cutting and handling benet and associated weeds.

Conclusions

Much has been accomplished by attempts to improve upland fallows. However, many innovations have been conceptualized and managed by researchers without their recognition or appreciation of improvements that have already been made to existing crop and fallow systems by the farmers themselves. The benet-based crop and fallow system, which was developed directly by farmers in Punta, Leyte, offers one desirable option in the search for simple, economical, effective, and sustainable approaches to fallow management. For the soil, the system provides organic matter or compost, green manure, and mulch. It protects itself and other fallow species and hedgerows from the depredations of grazing animals and, at the same time, prevents them from compacting the soil. It leads to a reduced fallow period and sustainable levels of crop production. Proper timing of land operations ensures a natural supply of seeds in the soil, making the system self-sustaining. Farmers, by their own observations and experimentation, have proven that the benet-based crop and fallow system is both ecologically sound and of strong potential for the tropical uplands.

Acknowledgments

We are especially grateful to the farmers of Punta who generously and enthusiastically shared their crop and fallow farming experiences with us and actively participated in the analysis of the benet-based crop and fallow system. Sincere thanks to Vivian Balbarino, information officer, ViSCA, for assisting in the preparation of this chapter, and to Malcolm Cairns for some literature citations. Support for this study came from FARMI and the Cornell International Institute for Food, Agriculture, and Development (CIIFAD) through its organization in the Philippines, Conservation Farming in Tropical Uplands (CFTU).

References

Baki, B.B., and N. Prakash. 1994. Studies on the Reproductive Biology of Weeds in Malaysia: Anther Sterility in *Mimosa invisa*. *Wallaceana* 73, 13–16.

Barneby, R.C. 1987. A Note on *Mimosa invisa* C. Martius ex colla and *M. invisa* C. Martius (*Mimosa*ceae). *Brittonia* 39, 49–50.

Batoctoy, G.D. 1982. Optimum Row Spacing of Corn–Soybean Intercropping. Undergraduate Thesis. Visayas State College of Agriculture (ViSCA), Baybay, Leyte, 76.

Bolton, M.P. 1989. The Ecology of Introduced Woody Weeds in Northern Queensland. In: *Noxious Plant Control: Responsibility, Safety and Benefits*. Proceedings of the 5th Biennial Noxious Plants Conference. Vol. 1. Cowra, N.S.W, Australia: Tropical Weeds Research Centre, Queensland Department of Primary Industries, and the New South Wales Department of Agriculture and Fisheries, 136–144.

Budiarto. 1980. Germination of Some Common Weeds (*Ageratum conyzoides, Mimosa invisa, Euphorbia prunifolia* and *Porophyllum ruderale)* found in Upland Rice Fields. MS Thesis, Science Universitas Jenderal Soedirman, Purwokerto, Indonesia, 52.

de la Rosa, Z.M., and J. Itumay. 1996. *Effects of Farmers' Indigenous Crop Fallow Systems in Sustaining Farm Productivity*. Baybay, Leyte, Philippines: Farm and Resource Management Institute (FARMI), Visayas State College of Agriculture (ViSCA).

Edrolin, M., R.L. Miranda, M.O. Mabbayad, C.B. Yandoc, and A.K. Watson. 1993. Biological Control of Some Major Rice Weeds with Fungal Pathogens. In: *Summary of the Annual Scientific Meeting of the Pest Management Council of the Philippines*, Cebu City. Los Baños, Laguna, Philippines: International Rice Research Institute, 86.

Gibson, T.A., and S.A. Waring. 1994. The Soil Fertility Effects of Leguminous Ley Pastures in Northeast Thailand. 1. Effects on the Growth of Roselle (*Hibiscus sabdariffa* cv Altissima) and Cassava (*Manihot esculenta*). *Field Crops Research* 39(2–3), 119–127.

Kaufusi, P., and M. Asghar. 1990. Effects of Incorporating Plant Materials on Corn Growth. University of South Pacific, Alafua Campus, Apia, Western Samoa. *Nitrogen Fixing Tree Research Reports* 8, 81–82.

Maxwell, J.E. 1988. Re-identification of the Weed *Mimosa invisa* Mart. ex Colla (Leguminosae, Mimosoideae) and a New Record and New Combination of the Variant Formerly Known as *Mimosa invisa* var. *inermis* Adelb. in Thailand. *Journal of Science and Technology* (Thailand). (*Warasan Songkhlanakarin*) 10(2), 169–172.

Muniappan, R., and C.A. Viraktamath. 1993. Invasive Alien Weeds in the Western Ghats. *Current Science* 64 (8), 555–558.

Parsons, W.T., and E.G. Cuthbertson. 1992. *Noxious Weeds of Australia*. Melbourne: Inkata Press, 692.

Prawiradiputra, B.R., and M.E. Siregar. 1980. Study on Forage Vegetation of Three Coconut Plantations in North Sulawesi, Indonesia. *Lembaga Penelitian Peternakan Lembaran* (Indonesia) 10(1), 1–5.

Sikunnarak, N., and C. Doungsa-ard. 1985. Biological Studies on the Thorny Sensitive Plant, *Mimosa invisa* Mart. *Journal of Agricultural Research and Extension* (Thailand) 2(4), 189–194.

Tiraa, A.N., and M. Asghar. 1990. Corn Growth as Affected by Nitrogen Fixing Tree and Grass Plant Materials Supplemented by P and K Fertilizer. University of South Pacific, Alafua Campus, Apia, Western Samoa. *Nitrogen Fixing Tree Research Reports* 8, 83–84.

Verdcourt, B. 1988. Two New Combinations in *Leguminosae. Kew Bulletin* 43, 360.

Wilson, B.W., and C.A. Garcia. 1992. Host Specificity and Biology of Heteropsylla spinulosa (Homoptera: Psyllidae) Introduced into Australia and Western Samoa for the Biological Control of Mimosa invisa. Brisbane, Australia: Queensland Department of Lands, Alan Fletcher Research Station 37(2), 293–299.

Chapter 19

Management of *Mimosa diplotricha* var. inermis as a Simultaneous Fallow in Northern Thailand

Klaus Prinz and Somchai Ongprasert

This chapter deals with a group of upland farmers in northern Thailand who faced the problems associated with fundamentally changing their rotational fallow agricultural systems to intensified cash crop production. It sets out to understand and document a remarkable innovation that they began as an experiment, in the absence of outside advice or interference. Over 15 years, it has developed into a successful simultaneous fallow system. It is remarkable not only for its success, but also because policymakers and development or extension agencies frequently neglect or look down upon farmers' perceptions and their wisdom in matters of agricultural production.

The farmers used spineless *Mimosa* (*Mimosa diplotricha* C. Wright ex Sauv var. inermis [Adelb.] Maxw), a nitrogen-fixing plant, to suppress *Imperata cylindrica* in their orange orchards. It not only worked, but it also spread to upland fields, where it delivered the benefits and functions of natural fallow. The farmers have adapted its management for use as a cover crop, live mulch, and green manure. This is despite the recognized drawbacks of the species in other locations.

Methods of dealing with the wild spiny variety of this species (*M. diplotricha* var. diplotricha) are not considered in this chapter. The spiny variety actually occurs on a much wider scale in the uplands of northern Thailand and has a substantial impact on farmers' cropping activities, as well as on public lands. (See Balbarino et al., Chapter 18, for a case study of a spiny *Mimosa* problem being transformed into a livestock solution.) However, the expansion of other species at the study site is discussed, because these may gain greater local importance in the future.

The Study Area and Method

This study involved two villages, Ban Den and Ban Salok, in Wang Chin district of Phrae Province, in northern Thailand. They are among the contact villages where the Agriculture and Development Unit of the McKean Rehabilitation Center seeks to promote understanding of ecological issues through practices of sustainable agriculture.[1] Ban Den and Ban Salok are populated by the Thai and Karen ethnic groups, respectively. Spineless *Mimosa* has been used in the area, in orange orchards

Klaus Prinz, McKean Rehabilitation Center, P.O. Box 53, Chiang Mai 50000, Thailand; Somchai Ongprasert, Department of Soils and Fertilizers, Mae Jo University, Chiang Mai 50290, Thailand.

1 The McKean Rehabilitation Center in Chiang Mai, Thailand, is a nongovernmental organization involved in the physical rehabilitation of disabled persons. As well, its activities include the development and extension of sustainable agricultural production systems.

and upland annual crop cultivation, for more than 20 years. The study was conducted through participatory rural appraisal exercises with farmers. Additional information was obtained through interviews with district officials, village administrators, and research institutions, as well as by field observations.

Ban Den and Ban Salok are situated in the upper Yom River catchment area, on the eastern foothills of the Wiang Kosai mountains. They are 2 to 3 km from the Yom River, at altitudes between 180 and 220 m above sea level (asl). The farmers practice two forms of land use: on rolling upland, with slopes of 5% to 15%, they cultivate annual upland crops and orchards, and on lowland terraces of old alluvium they cultivate wet rice. The parent rocks of soils in the area are shale, slate, and phylite. Soils on the uplands are classified as lithic haplustalfs and lithic paleustults with textures of loam or clay loam over gravelly clay loam, while soils in the lowland paddies are classified as typic paleaquults. The soils are relatively infertile, with a low level of organic matter and pH in the range of 4.5 to 5.5.

The natural vegetation consists of secondary regrowth and open stands of low deciduous forest. The area has a prevailing tropical monsoon climate with a pronounced dry season from November to April. Mean annual rainfall is 1,100 mm (Department of Land Development 1985). Table 19-1 shows some data regarding the two villages.

Results

Farming Systems in the Study Area

Conventional shifting cultivation systems practiced in northern Thailand vary from one ethnic group to another. For the Thai and Karen people, who have a long history of permanent settlement on valley floors and foothills, shifting cultivation is a supplementary farming system, complementing terraced wet rice paddies. Historically, for ethnic Thai farmers it was of the short cultivation, short fallow variety, whereas that of Karen farmers was short cultivation, long fallow (McKinnon 1977). Originally, the farming systems in the study area followed these tribal traditions. However, because of population pressures and the transitional nature of Thailand's economy, both long rotational fallow systems and pioneer shifting cultivation have largely disappeared (Bass and Morrison 1994) and, in Wang Chin district, farmers are now practicing very short cropping and fallow rotations, if they opt for any fallow stage at all. Generally, the present farming systems of the two study villages consist of the following:

- Rainfed and partly irrigated terraced paddies growing wet rice in seasonal rotation with peanuts (*Arachis hypogaea*) and French beans (*Phaseolus vulgaris*);
- Fruit tree orchards, especially oranges; and
- Short cropping shifting cultivation with a short fallow. This supports several annual upland crops, including upland rice, peanuts, cowpeas, and, occasionally, cotton and cassava.

There are now only a few farmers in Ban Den village who continue to fallow in a rotation cycle of more than three years. Others with less land have reduced the rotation period even further, or have given up the cropping and fallow system altogether. As a result, yields of upland rice, peanuts, and cassava have declined to the point where they are unable to meet the needs of an increasing population. Corn cultivation was discontinued 20 years ago.

As they progressively shortened the fallow period, farmers in the area found the quality of their soil was degenerating, the biodiversity of their farming systems was lower, they had increasing weed problems, and they needed more and more inputs to maintain productivity. Improvements in infrastructure led them to look for alternative agricultural systems. Following the lead of other growers in Wang Chin district, and particularly those along the Yom River, farmers in Ban Den village began planting orange trees, which are suited to the local climate, the topography, and the soil conditions. The change coincided with stricter government regulations controlling the conversion of forest land to corn cultivation, and the promotion of agroforestry systems for rehabilitating degraded land (Gypmantasiri and Amaruekachoke 1995). However, the next step in the process of degradation (McKinnon 1977) began to make itself felt in Wang Chin district.

The cover of perennial woody plants, in this case orange trees, was insufficient to keep out the ubiquitous *Imperata cylindrica.* It became an established pest and eventually caused difficulties in the orange orchards. The possibility of fire damage was perceived as the most obvious threat. Other problems were not as readily recognized. But the farmers of Ban Den village regarded available methods of controlling *Imperata* as too costly. So they set out to find a solution that would cost them less in terms of labor input and risk.

Local Introduction of a Cover Crop and Green Manure System

Through their contacts with neighboring villages, as well as by their own observations, farmers' moved their attention to spineless *Mimosa*. It was attractive because of its lack of spines and its capacity for easy establishment. But, most of all, they were impressed with its ability to completely cover and suppress *Imperata* within two years. Importantly, spineless *Mimosa* offered a cover cropping system that required no inputs beyond those that they were able to provide locally. There was little other information available to them, and they were unaware of the plant's serious drawbacks.

Small quantities of spineless *Mimosa* seeds were obtained and planted into the Ban Den orchards. The *Mimosa* flourished, and it successfully suppressed the *Imperata cylindrica.* However, it self-seeded into adjacent cropland and fallowed areas. Without some knowledge of basic management practices, the experiment might have ended there, with the *Imperata* replaced by runaway *Mimosa.* But the Ban Den farmers had worked occasionally as hired labor in other orchards and, with suggestions from extension workers, they not only recognized the additional benefits of the *Mimosa,* but they also adapted management procedures to handle the plants and their residues to suit the requirements of their crops. They learned how to control the growth of the *Mimosa* in upland fields so it did not interfere with crops of upland rice or peanuts. They learned how to apply its biomass as green manure. Over time, the spineless *Mimosa* has become an integral part of a newly evolved agroecosystem. Its development was influenced by the following key factors:

- The growth of infrastructure and better marketing of cash crops;
- Successful use of technology by local farmers; and
- The availability of inputs and low costs of implementation.

Table 19-1. Characteristics of the Study Villages

Factors	*Ban Den*	*Ban Salok*
Ethnic groups	Thai	Karen
Number of households	72	200
Average farm sizes	2 to 5 ha	0.5 to 5 ha
Households with orange orchards	30	50
Size of orange orchards	0.5 to 0.8 ha	0.16 to 0.80 ha
Years of growing oranges	20	20
Households using *Mimosa* in orange orchards*	25	30

Notes: Including 30% of recently planted orchards where spineless *Mimosa* was not deliberately grown, but has established naturally. In upland fields, spineless *Mimosa* occurs in varying densities, depending on the crop and the management. In some places, spineless *Mimosa* already exists in mixed communities with the wild spiny variety.

At the outset, the constraints to the farmers' experimental use of *Mimosa* were not considered, and, fortunately, they had no influence on the outcome. Nevertheless, the constraints could have been summarized as follows:

- Farmers' lack of experince and uncertainty about effects;
- Lack of alternative choices; and
- Lack of advice from extensive agents.

Description of *Mimosa diplotricha* var. inermis

The original classification of *Mimosa invisa* Mart., as separated into spiny and spineless types, has been revised as follows: *Mimosa diplotricha* C. Wright ex Sauv. (Leguminosae, Mimosoidae), var. diplotricha (spiny), var. inermis (Adelb.) Maxw. (spineless), syn. *Mimosa invisa* Mart. ex Colla (The name change is according to Barneby [Brittonia 39] 1987, *49–50*) (Maxwell 1997).

The Thai name for spineless *Mimosa* is *maiyarap rai nam*. Maiyarap is the name of the immortal King of the Underworld in the mythology of the Ramakian. Rai nam means without thorns or spikes (Rerkasem et al. 1992).

In most aspects, *Mimosa diplotricha* var. diplotricha (Md) is similar in appearance to var. inermis: It is a sprawling, straggling to somewhat climbing, vigorous, annual aculeate herb growing up to two or more meters tall. It usually forms dense thickets. Its stems and branches are quadrangular. Its leaves are spirally arranged, double pinnate, well spaced, up to 22 cm long, and tipped with a bristle 3.5 mm long. There are six to nine opposite pairs of pinnae up to 5 cm long, tipped with a bristle up to 2 mm long. There are 24 to 27 opposite pairs of leaflets and the pinnae and leaflets are slowly sensitive, the leaflets folding together dorsally on the dorsal side of the rachis. Flowers are numerous in each head; regular, bisexual, 5-merous, and 7 mm long. Pods are clustered in each infructescence. They are flat and straight to slightly curved near the tip. There are several seeds per pod and one per section. They are nearly orvicular in outline, compressed, glossy light brown, and 3 to 3.5 mm in diameter.

Mimosa diplotricha var. diplotricha (Md) is native to tropical America, but it is now pantropical. It flowers from September to December in northern Thailand and fruits from December to February. The plants have taproots that penetrate as far as 150 cm into the soil, and they can tolerate soils of low pH.

With regard to the plant's biomass production and nitrogen-fixing capacities, a study using spineless *Mimosa* as live mulch for corn reported that 2.4 tonnes of dry matter per hectare had accumulated by the time of corn harvest, and that the N-accumulation was 47 kg/ha. At the end of an entire season, those figures could be expected to double, to 5 t/ha of dry matter and 95 kg/ha of N. Another source indicated that 7.3 t/ha of dry matter was accumulated by spineless *Mimosa*, with nutrient accumulation figures of N, 1.04%; P, 0.04%; and K, 1.03%.

Distribution of Mimosa diplotricha

Spineless *Mimosa* is currently found in only very limited areas of northern Thailand. It was distributed to farmers in the Yom River catchment of Wang Chin district, Phrae (Wang Chin Agriculture District Office 1997), and to upland areas of Chiang Dao district, Chiang Mai (Radanachaless and Maxwell 1994), by the Department of Agricultural Extension. At Ban Den village, conditions were favorable for its adoption into local farming systems. Cross visits were organized between farmers and officials from within Thailand and from Laos and Vietnam to study on-farm experiences, and seed samples were requested for trials in other locations. However, there has been no follow-up to determine whether, or where else, it has been adopted, and only a few farmers are known to be using spineless *Mimosa* as live mulch in corn.

There is controversy concerning any wider use of spineless *Mimosa* because it is known to be suppressed and easily dominated by the spiny variety. The latter has spread in Thailand since about 1935 and is presently widely distributed across most uplands and highlands in the north of the country, at elevations from 300 to 700 m asl. The spiny variety is accepted as a valuable soil improvement plant, but it is extremely difficult to handle without appropriate machinery. On agricultural lands, it is usually observed as live mulch in corn. (See color plate 23 for an example of *M. invisa* succession after corn harvest.)

Since 1965, various stations of the Department of Land Development have produced seeds of spineless *Mimosa* and provided them for use as a green manure crop. However, this program was discontinued after a few years because of the problem mentioned above. There is an additional concern that cross-pollination may occur between the spineless and spiny varieties, creating mutations. However, according to one competent source, and farmers' observations, cross-pollination does not seem to be occurring. There have been no obvious changes to the plants' appearance, and they remain either completely spineless or spiny, with no intermediate genotypes.

General Aspects of Management

The integration of spineless *Mimosa* into agroecosystems at Ban Den and Ban Salok is classified as a simultaneous fallow because soil improvement, or fallow, plants are grown at the same time and in the same fields as cultivated plants or trees (McKinnon 1977). This is done in various ways. For example, spineless *Mimosa* is used as ground cover in orange orchards, live mulch in cassava and corn, and green manure in upland rice and peanuts (see Figure 19-1). In another variation, spineless *Mimosa* is a component of vegetation in fallowed fields, where it drapes over natural shrubs and small trees.

The management of spineless *Mimosa* in orchards and upland fields involves three main phases: establishment, vegetation management, and residue management. Establishment in orange orchards is accomplished by broadcasting seeds, whereas it generally self-seeds into cultivated fields. On a few occasions, seeds have been sown into fields where cassava is to be grown.

Spineless *Mimosa* is a prolific seeder and the seeds scatter sufficiently so that only small quantities are required to achieve optimal plant density within two years. Seed production reportedly amounts to 40,000 seeds per square meter, with 80% to 90% of hard seed in a fresh harvest. It requires only a very low percentage of seeds to germinate and survive in order to maintain a density of 50 plants per square meter (Rerkasem et al. 1992). A complete cover of spineless *Mimosa* is achieved during the second year, and efforts are then needed to prevent it from attaching itself to trees and to manage the developing mass of vegetation and residues that settle on the ground.

When the farmers of Ban Den first used spineless *Mimosa,* the seeds were obtained by collecting small quantities from mature vines in neighbors' orchards and from nearby fallow plots where it had already established. The seeds were then broadcast into *Imperata* grass during the first rains, and the *Imperata* was slashed to provide mulch cover. Seeds are sometimes broadcast into fallows before rain, and the fallow vegetation is then slashed and burned. The heat treatment causes the seeds to germinate more efficiently.

Although farmers may have been able to control *Mimosa*'s natural expansion, its prodigious growth has been tolerated and allowed to proceed. There are several methods of managing the mass of vegetation in orchards. About a year after the young trees are planted, the spineless *Mimosa* vines are flattened and the vegetation cover is suppressed. This normally occurs from June to September. This treatment not only reduces the thickness of the *Mimosa* cover, it prevents the vines from attaching themselves to tree stems and branches and allows the performance of routine orchard maintenance. There are various methods of flattening the vegetation:

- Pressing the vegetation with three-to-four-meter-long bamboo sticks, which may or may not have upright handles;
- Rolling vegetation flat with a 200-liter oil drum;

- Driving over the spineless *Mimosa* with a two wheel tractor, with or without an attached flattening device; and
- Applying a mix of 2,4-D and Roundup to the vegetation.

During a single growing year, two different methods may be necessary, depending on the vigor of the growth. Calculations of the inputs needed to manage the spineless *Mimosa* system are shown in Table 19-2.

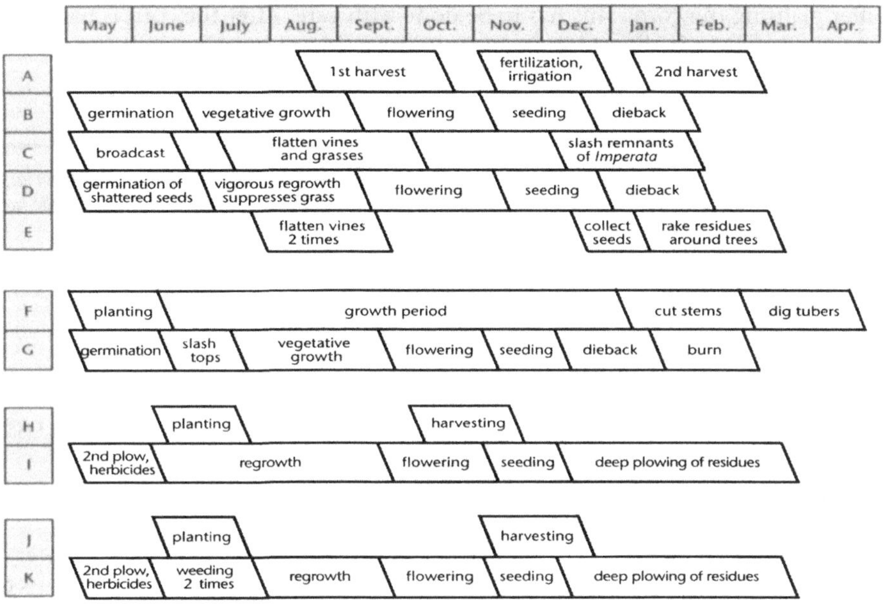

Figure 19-1. Activity Calendar for Management of Spineless *Mimosa* Cover Crops
Notes: A. Oranges. B. First season *Mimosa* in orange orchards. C. *Mimosa* management during the first season. D. Second season *Mimosa* in orange orchards. E. *Mimosa* management during the second season. F. Cassava. G. *Mimosa* as live mulch with cassava. H. Corn. I. *Mimosa* as green manure with corn. J. Upland rice and peanuts. K. *Mimosa* as green manure with upland rice or peanuts. Zero tillage in categories A–G.

Seed Collection and Residue Management

If required for local use, or requested for sale, seeds are collected in January by manually stripping the vines. Residue management depends on the risk of fire. In January or February, the wilted vines are raked around trees as mulch cover. This does not require much labor, and the residues decay quickly.

Live Mulch

Cassava. Farmers prefer, if possible, to plant cassava with no-till management. Fallow vegetation is slashed and burned in a field where spineless *Mimosa* seeds are already in the soil seed bank, and cassava is planted. Both plants grow simultaneously. Thirty days after planting of the cassava, the spineless *Mimosa* is controlled by slashing its tops where they begin to grow over the cassava.

According to one Ban Den farmer who has been growing cassava in combination with spineless *Mimosa* continuously for five years, the vines do not attach aggressively to the cassava stems. When they reach a certain weight, they slip down and remain at a medium level below the cassava canopy.

When cassava is harvested in January or February, its stems are first cut and removed and, prior to digging the tubers, the spineless *Mimosa* residues are burned off. Seeds are scattered on the ground, and the heat treatment enhances germination at the first rains. The roots of the spineless *Mimosa* will have loosened the soil, and digging the cassava tubers is a much easier task than in monocropped fields.

Corn. Because the soil on the Ban Den farms has been improved by spineless *Mimosa,* the farmers have resumed growing corn after having discontinued the practice 20 years earlier. Management of the *Mimosa* in corn is much the same as that for cassava. One weeding or spraying of spineless *Mimosa* regrowth is carried out 30 days after planting the corn. This has to be done with care. After the corn is harvested, the spineless *Mimosa* is allowed to mature and its residues, together with the corn stover, are then incorporated into the soil.

Table 19-2. Inputs Required to Manage *Mimosa*

Crops and Activities	*Labor*[a]	*Number of Times*	*Labor Costs*[b]	*Materials Used/ ha*
Orange orchard				
Flattening spineless *Mimosa* as a cover crop				
By stick	25.00	3	288	
By rolling of drum (gentle slope)	12.50	2	96	
On steeper slopes	25.00	2	192	
By two-wheel tractor	1.25	2	9.6	Fuel: US$5
Raking of residues	6.25	1	25	
Spiny *Mimosa* in orchards	43.75	3	504	
Imperata in orchards				
Manual slashing	50.00	4	770	
Use of grass cutter	9.50	4	146	Fuel: US$8
Herbicide application	9.50	2	146	Herbicide: US$16
Cassava				
Mimosa as live mulch				
Superficial weeding	3.00	3	11.5	
***Mimosa* as a postharvest green manure**				
Deep plowing (March)			72	
Shallow plowing (May)			48	
Pre–emergence spray (UR)	9.5	9.5	73	Herbicide: US$8
Manual weeding	10	20	77	

Notes: [a]Labor: (man-days/ha) each time [b]Labor costs 1997US$/ha/year. Produces less biomass than when managed as live mulch in cassava or corn. UR = upland rice; PN = peanuts.

Green Manure

Spineless *Mimosa* is treated as a green manure crop in fields of upland rice and peanuts. In locations where plowing with a tractor is possible and affordable, plant residues are plowed in March and the regrowth of weeds and *Mimosa* is plowed once more by mid-May, after the first rains. Where fields are not plowed, or get only a single plowing, the regrowth of weeds and spineless *Mimosa* is killed with herbicides. Subsequent weeding is performed manually. However, the first regrowth will die off naturally if it is followed by a lengthy period without rain.

Yields

Using *Mimosa* as a cover crop in orange orchards, the Ban Den farmers recorded annual yields of between 18,750 and 37,500 kg of oranges per ha. With their other crops, they were able to provide comparative yields, with or without *Mimosa.* Cassava, with *Mimosa* as a live mulch, yielded 12.5 t/ha. Without *Mimosa,* the cassava yield was exactly half that figure. Upland rice yielded 3.5 t/ha with *Mimosa* as a green manure. Without it, the yield was only 2.25 t/ha. Peanuts showed no difference, yielding 1.11 t/ha with or without *Mimosa.*

Short Fallow

On some fields, spineless *Mimosa* establishes itself during the cropping period. Then, when the field is left fallow, *Mimosa* spreads over the emerging vegetation. It regrows again in the second year of fallow, but its density may be reduced after the third year, when regrowth is often limited to the fringes of the plot. Meanwhile, scattered seeds of spineless *Mimosa* remain dormant until the fallow vegetation is once again slashed and burned. This system probably cannot be considered a planted successive fallow but may be better described as an improved form of natural, self-seeding fallow.

Elimination of Spineless *Mimosa*

Farmers tend to be very concerned about the elimination of cover crops in much the same way as they are concerned about getting rid of contour hedges, once they are not wanted anymore or are considered a nuisance. This aspect, therefore, needs to be carefully considered.

The need to eliminate spineless *Mimosa* is most likely to occur in situations where it has been used as ground cover or live mulch in orange orchards. In cultivated fields, the methods described below would be difficult to apply without taking the field out of production for a period of time. The reasons farmers may want to eliminate spineless *Mimosa* include the following:

- Spineless *Mimosa* is not necessary anymore because its purpose of suppressing *Imperata* has been achieved. (Farmers in other villages of Wang Chin district establish spineless *Mimosa* in the first year after planting trees and make use of its cover only until the fourth year.)
- Farmers like "clean" fields.
- Mulch cover is inconvenient when erecting support poles during the fruiting period.
- Mulch cover is often unacceptable to hired fruit pickers.
- High labor expenses are incurred for flattening vines in larger orchards.
- The spiny variety of *Mimosa* has moved into the orchard area and has taken over.

In the study area, farmers had several methods of eliminating spineless *Mimosa*:

- Repeated slashing or cutting of regrowth until the soil seed bank is exhausted;
- Repeated herbicide spraying; and
- Overgrazing by cattle, which can apparently consume spineless *Mimosa* without side effects. (It is unknown how much mimosine the plant contains.)

Alternative Cover Crops

Natural Arrivals

After spineless *Mimosa* has been eliminated, care has to be taken to control reemerging *Imperata.* This can be achieved by using herbicides or regular slashing with mechanical equipment. However, some farmers are taking advantage of alternative cover crops that have either spread naturally onto their cultivated land or have been introduced "informally," in much the same manner as spineless *Mimosa* first arrived.

***Vigna* sp.** In Ban Salok and nearby Ban Sop Sai, a *Vigna* species is used that is self-seeding and which is less than normally prone to burning. Its vines reportedly climb tree branches more aggressively, and it requires increased efforts to control. This species, which has the Thai name *tua pee,* may be a wild form of ricebean (Duke 1981).

Centrosema pubescens. *Centrosema* is spreading into many orchards where spineless *Mimosa* has either never been used, has been eradicated, or has been sufficiently reduced for *Centrosema* to invade. This plant, like the *Vigna* sp., is recommended by the Department of Agricultural Extension as a cover crop in orchards. It is an indigenous plant and the source of its expansion into the study area is probably local. For instance, there may have been extension activities in the area by the Department of Land Development, which has been providing *Centrosema* seed to interested farmers (Department of Land Development 1985).

At present, *Centrosema* occurs in a few fields along with spineless *Mimosa,* where the latter has been reduced rather than eliminated. This may be because spineless *Mimosa* has improved formerly exhausted soils, enabling *Centrosema* to gain an ecological foothold (van der Meulen 1985). *Centrosema* has one important advantage over spineless *Mimosa*: it does not dry out completely and provides a continuous green cover. However, its vines climb more aggressively and it is considered difficult to manage in locations where it is now expanding.

The vegetation of *Centrosema* is generally handled in much the same manner as that of spineless *Mimosa.* However, it is also an excellent cattle feed, so farmers may cut and carry its regrowth to feed confined cattle.

Reestablishment of Spineless Mimosa

Among the interviewed farmers, there was one who had previously eliminated spineless *Mimosa* from his orchard but had then been dissatisfied with using herbicide to control the regrowth of *Imperata.* So he sowed spineless *Mimosa* again, just as he had done 15 years earlier.

Discussion

The integration of spineless *Mimosa* into the agricultural production system at our study site is providing benefits related to both cultivation and productivity. Its first purpose was to improve control of *Imperata* with available resources. In this regard, it has been successful. No data are available on the harmful effects of *Imperata* on orange trees. However, the replacement of dense grass by spineless *Mimosa* has improved the soil structure and increased soil porosity, affecting both its aeration and moisture storage capacity (Robert 1982). The soil fertility benefits of *Mimosa* to orange trees have not been investigated, but they should equal the 95 kg/ha of N generated under the corn and *Mimosa* system reported above (Rerkasem et al. 1992).

In addition, farmers at Ban Den are gaining benefits from the use of spineless *Mimosa* as live mulch in cassava, providing biomass and nitrogen. Biomass is produced on the spot, rather than having to be transported into the fields. Cassava yields are stabilized under spineless *Mimosa* mulch, and both soil cultivation and harvest of tubers are easier because of improved soil structure. Having reintroduced corn cultivation, which was earlier abandoned because of soil degradation, the farmers at Ban Den are now interested in raising pigs. They may also develop an interest in alternative multipurpose fallow plants such as lablab or soybean, in the event that spineless *Mimosa* is phased out for one reason or another.

Table 19-3. Strengths and Weaknesses of Spineless *Mimosa* and Its Management

Strengths	*Weaknesses*
Fits criteria for green manure or cover crop use (Bunch 1986): nonwoody stem; grows well in poor soil, low pH; no land preparation necessary; seed sown into grass cover; does not have natural enemies; and good N-fixation.	Does not cover the soil during the dry season.
Taproots eliminate nutrient competition with marcotted fruit trees.	Fire hazard after dieback.
Prolific seed production.	Easily suppressed by spiny var. of *Mimosa diplotricha*.
Acceptable as feed for cattle.	Eradication is difficult.
Absence of spines facilitates handling.	Vines climb onto fruit trees.
Range of control methods available.	Seeds no longer available through official channels.
Farmer-based management and control.	Unsuitable as buffalo feed.
Successful management demonstrated by local farmers.	Mulch cover considered inconvenient for certain orchard maintenance and fruit picking.
Can be used in a mix with other green manures or cover crops.	Seeds do not germinate at the same time.
When used as a green manure or cover crop, reduces greenhouse effect: less burning of biomass, less use of energy and inorganic N-fertilizers.	
On-site production of biomass; no need to carry.	

In the cultivation of upland rice and peanuts, the use of spineless *Mimosa* as a green manure requires additional labor to control regrowth by spraying and weeding. This is in addition to usual weeding operations. However, yields of upland rice have increased by 50%, so the green manuring practice is recommended as long as the household has sufficient labor.

Ecological and Environmental Aspects

Mimosa diplotricha varieties are not indigenous to Thailand. They were introduced to assist soil regeneration and control of *Imperata cylindrica*. At the time of this study, their impact on the environment could not be assessed. However, the spiny variety that has spread throughout the uplands, to as high as 700 m asl, is posing problems for public lands, forest plantations, and agricultural land because of difficulties in its management and dry season fire hazards. The extent to which it displaces native flora or creates other environmental problems has not yet been documented. But environmental concerns are clearly reflected in the decision of the Department of Land Development to discontinue promotion of spineless *Mimosa* because of the danger of mutations into more aggressive spiny types. One environmental impact could then be the effects of harmful chemicals used in its control.

It is perhaps noteworthy that substantial environmental problems have followed the introduction of another species, *Mimosa pigra*, to Thailand. Like *M. diplotricha* var. inermis, it was introduced at about the same time and with much the same purpose, that is, for soil improvement and the prevention of bank erosion. *Mimosa pigra* has infested areas across the entire north of the country and parts of central Thailand, affecting waterways, lakes, reservoirs, and roadsides, and requiring considerable expense for its control (Robert 1982).

However, in the context of agricultural production, *M. diplotricha* var. inermis has successfully contributed to the rehabilitation of marginal uplands. Its tolerance to low pH, deep root system, and high production of biomass have helped it to minimize nutrient losses and preserve moisture. At the same time, it reduces rainy season erosion in field crops like corn and cassava, where it grows as live mulch. In shortened swidden cycles, spineless *Mimosa* also helps to reduce or prevent the growth of *Imperata* and other problem weeds during the fallow phase. The strengths and weaknesses of spineless *Mimosa* and its management are summed up in Table 19-3.

Conclusions

Entirely without outside help or advice, and depending on the experience of their neighbors, farmers in the Wang Chin district of Thailand's Phrae Province adopted cover crop technology and planted *Mimosa diplotricha* var. inermis, or spineless *Mimosa.* It succeeded in suppressing *Imperata* and the practice spread. So did the spineless *Mimosa,* from orchards to nearby upland fields and fallowed areas. But rather than becoming a nuisance, further benefits from its use became apparent, and the enterprise has become a successful simultaneous fallow system. So far, the farmers have managed the tangled vegetation cover by employing their own household resources. However, despite the benefits of growing spineless *Mimosa*, many farmers have rejected it or discontinued its use. They foresee problems with the costs of managing large areas of *Mimosa*, the fire hazard it poses, and the unresolved question of whether it will mutate into a spiny variation whose control will be more difficult.

In general—spineless *Mimosa* is likely to be considered of only temporary importance; as a step toward more intensive cropping patterns where edible and marketable legumes are used in a rotation with grain crops. Therefore, enhancement of soil fertility through the use of spineless *Mimosa* is providing the basis for permanent agricultural systems (Howard 1943).

References

Bass, S., and E. Morrison. 1994. Shifting Cultivation in Thailand, Laos, and Vietnam: Regional Overview and Policy Recommendations. In: *IIED Forestry and Land Use Series* No. 2. London, UK: International Institute for Environment and Development.

Bunch, R. 1986. *What We Have Learned to Date about Green Manure Crops for Small Farmers.* Technical Paper. Tegucigalpa, Honduras: World Neighbors.

Department of Agricultural Extension. 1980. *Cultivation of Orange Orchards.* Bangkok, Thailand: Ministry of Agriculture.

Department of Land Development. 1985. *Thailand: Northern Upland Agriculture.* Thai-Australian-World Bank Land Development Project. Chiang Mai, Thailand: Department of Land Development.

Duke, J.A. 1981. *Handbook of Legumes of World Economic Importance.* Beltsville, MD; New York; and London: USDA and Plenum Press.

Gypmantasiri, P., and S. Amaruekachoke. 1995. Cropping Systems in Sustainable Agriculture in Northern Thailand. In: *Strategies for Sustainable Agriculture and Rural Development*, edited by A. Poungsomlee. Salaya, Thailand: Faculty of Environment and Resource Studies, Mahidol University.

Howard, Sir Albert. 1943. *An Agricultural Testament.* New York and London: Oxford University Press.

Koen, V.K., and C. Vejpas. 1995. Weed Problems in a Transitional Upland Rice-Based Swidden System in Northern Thailand. In: *Highland Farming: Soil and Future?* Edited by F. Turkelboom, V.K. Koen, and K. Van Look. Chiang Mai, Thailand: Soil Fertility Project, Maejo University, 161.

Marten, G.G. 1986. Traditional Agriculture in Southeast Asia. In: *Traditional Agriculture in Southeast Asia: A Human Ecology Perspective*, edited by G.G. Marten. Boulder and London: Westview Press, 6–18.

Maxwell, W.J.F. 1997. Personal communication between W.J.F. Maxwell, Herbarium, Chiang Mai University, and the authors.

McKean Rehabilitation Center. 1996. *Changing Toward Alternative and More Sustainable Agriculture Production.* Study Case of Mr. Samran Manovorn, Ban Den. Chiang Mai, Thailand: MRC.

McKinnon, J. 1977. Who's Afraid of the Big Bad Wolf? Discussion Paper, NADC Seminar: Agriculture in Northern Thailand, Chiang Mai, Thailand.

Nakhaprawes, P., P. Tanyadee, and P. Wasanukul. 1995. *Use of Green Manure Crops to Improve Soils.* Bangkok, Thailand: SWC Division, Department of Land Development (Thai language).

Prinz, D. 1987. Improved Fallow. *ILEIA Newsletter* 13 (No. 1), 4.

Radanachaless, T., and J.F. Maxwell. 1994. *Weeds of Soybean Fields in Thailand.* Chiang Mai, Thailand: Multiple Cropping Center, Faculty of Agriculture, Chiang Mai University.

Rerkasem, B., T. Yonoyama, and K. Rerkasam. 1992. *Spineless* Mimosa (Mimosa invisa): *A Potential Live Mulch for Corn.* Working Paper, Agriculture Systems Program. Chiang Mai, Thailand: Faculty of Agriculture, Chiang Mai University.

Robert, G.L. 1982. *Economic Returns to Investment in Control of* Mimosa pigra *in Thailand.* IPPC Document No. 42-a-82, MCP Agricultural Economics Report No. 15. Corvallis, Oregon: International Plant Protection Center, Oregon State University.

van der Meulen, G.G. 1985. Ecological Soil Regeneration, In: *Agriculture-Man-Ecology*, edited by A.N. Copijn. Groenekan, The Netherlands: AME Foundation.

Wang Chin Agriculture District Office. 1997. *Background of Socio-Cultural and Economic Situation.* Wang Chin, Phrae, Thailand (Thai language).

PART IV

Herbaceous Legume Fallows

A young girl outside Thimphu, Bhutan.

Chapter 20

Growing Ya Zhou Hyacinth Beans in the Dry Season on Hainan Island, China

Lin Weifu, Jiang Jusheng, Li Weiguo, Xie Guishui, and Wang Yuekun

In the coastal tablelands and hilly areas in the southwest of Hainan Island, many farmers grow a legume crop known as Ya Zhou Hyacinth Bean (YZHB) (*Amphicarpaea* sp.) in their upland fields every year, while other crops are discontinued for five to seven months in the winter and spring because of dry weather. YZHB is indigenous to Hainan Island; its bean is a nutritious vegetable, and its vines are used as livestock fodder. Its cultivation is very extensive, and it requires no weeding, fertilizer, or efforts to control diseases or pests. The beans are harvested about 100 days after germination. Although the output of YZHB is only 450 to 2,250 kg/ha, it is still a popular crop because it is so easily managed. Growing Ya Zhou hyacinth beans in the dry season is, therefore, a desirable way of intensifying the use of fallow land.

Despite its local popularity, no record of YZHB can be found in *Flora Republicae Popularis Sinicae* (Editorial Committee of Chinese Academy of Sciences for Flora of China 1998), or in *Flora Hainanica Island* (Chen and Zhang 1965), and there is little available literature about it. A short-term research project by the Chinese Academy of Tropical Agriculture Sciences (CATAS) in 1963 investigated its use as a green manure. YZHB was also mentioned briefly in a report on edible legumes on Hainan Island by Wang et al. in 1992. It has even been the subject of confusion with other species of beans and may continue to be so. Its correct taxonomic identification still requires further study. In any case, YZHB has many superior features that make it an important component of farming systems, and it is deserving of further scientific inquiry.

The Botany of YZHB

YZHB is a small, semiclimbing herbaceous plant. It grows upward at first, and then grows randomly in any direction when its erect stalk reaches 15 to 25 cm tall, and the stalk turns into a flexible trail. The upright stalk branches from the first axil to the seventh axil. It has between five and seven branches when it is grown in low density but does not branch at all when grown in high density. The main branches may branch again after reaching the second to fourth nodes, and these also become flexible trails, which twine about each other or climb on other plants or supports.

Lin Weifu, Jiang Jusheng, Li Weiguo, Xie Guishui, and Wang Yuekun, Rubber Cultivation Research Institute, CATAS, Key Laboratory for Physiology of Tropical Crops of Agriculture Ministry, Dan Zhou, Hainan 571737, China.

The internodes of upright stalks and the several primary internodes of main branches are between 1 and 5 cm long. However, excessive growth may see these extend to as long as 12 cm. Immature stalks and trails are covered with hair, but they become hairless when mature. Some stalks or trails are green when immature and tend to a purplish red color as they mature. The trails are between 50 and 100 cm long but, with excessive growth, may be several meters long and can form adventitious roots when they crawl on moist soil.

Characteristics

The Leaf. The petiole, with pulvinus, is 2 to 4 cm long and trifoliate on its top. The terminal leaflet is oblong and acuminate in shape, 3.2 to 3.9 cm long and 1.3 to 1.9 cm wide, and its petiolule is about 0.2 cm long with an acicular stipel at both sides of its base. The lateral leaflet is slant oval and acuminate in shape, 2.4 to 3.4 cm long and 1.2 to 1.9 cm wide, and its petiolule is about 0.2 cm long with an acicular stipel at its base. With excessive growth, the petiole may become two to three times bigger than normal. The leaflets are chartaceus and are dark green in color. The petioles, the stipels, the main and second veins on leaflets, and the periphery of leaflets are hairy, but the minim vein is rarely hairy.

The Flower and Pod. The flowers are arranged in axillary spikes. These are 2-6, 11-15 mm long flowers in each spike. The persistent calyx is bell-like, with five lobes of unequal length, about 0.6 cm long, and hairy. At the base of the calyx, there are three acicular bracts. The petals vary in color from greenish when in bud to yellowish when in bloom. The standard is obovate, with a sinus at the top, and reflexed when mature. At the center of the standard petals, there is a triangular purple spot with each side measuring about 0.2 cm. The wing petals are on either side of the keel petals. The keel petals are slightly longer than the wing petals, partially connate, and conceal the stamens and pistil. The stamens are nine fused and a free stamen. The style is hairless and forms a beak, and the superior ovary is hairy. The pods are belt-like, slightly curved or sickle shaped, 3.4 to 5.8 cm long (including a hairless beak about 0.6 cm long) and 0.6 to 0.75 cm wide, covered with tomentum and contain three to seven seeds (see Figure 20-1). The seeds are small, flat, and spheroid in shape with a thick gray-yellow testa.

The Roots. The root system of YZHB is highly developed, extending up to 142 cm deep into the earth. The roots bear many nodules, distributed from 3 to 75 cm beneath the surface. The nodules vary from the size of a soybean to the size of millet and are irregular in shape.

Distribution

According to our investigations, YZHB is distributed mainly over Ledong county, Dongfang city, Changjiang county, and Sanya city, at 18°10′ to 19°30′ N, 108°39′ to 109°31′ E, along the southwest coast of Hainan Island. It is grown mainly on the coastal tableland, on coastal terraces, and in hilly areas that are under 400 m above sea level (asl) in altitude. Of the four areas named above, YZHB are most popularly grown in Ledong county and Dongfang city.

Biology and Ecology

The YZHB growing season is from autumn through winter and into the following spring. The area where it is grown has a tropical climate that is generally characterized by high temperatures and drought.

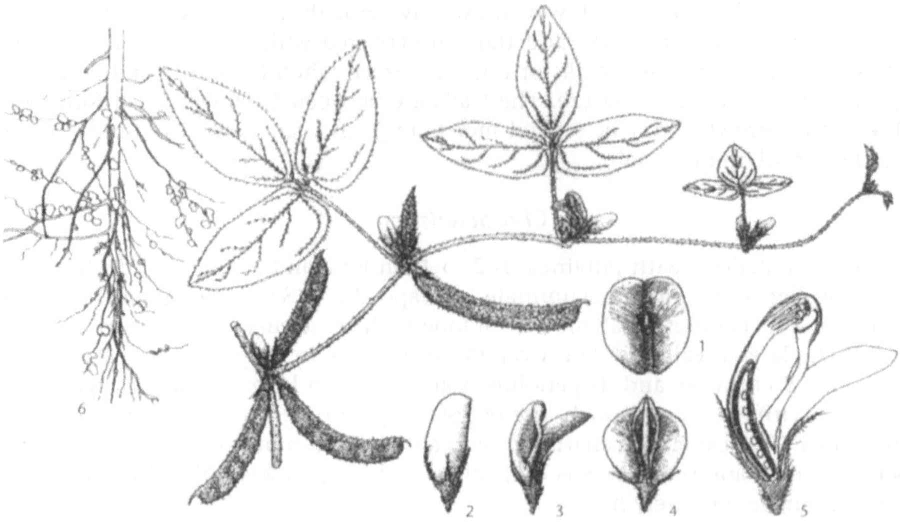

1 - standard (x2) 2,3,4 - flower (x2) 5 - section of flower (x4) 6 - root

Figure 20-1. Ya Zhou Hyacinth Bean
Source: Lin et al. 1999.

High Temperatures and Low Rainfall

In the growing area, and particularly in the YZHB growing season, the duration of sunlight is fairly long (see Table 20-1), so the temperature is relatively high. The monthly average temperature is 23.2 °C, and there are no frosts. In autumn, the temperature is generally above 22 °C and, in winter and spring, it usually remains between 18.5 and 22 °C (Table 20-2). In general, the amount of radiant energy from the sun is high, between 8 and 12 Kcal/cm^2/month (Table 20-3).

The rainfall in the YZHB growing season, from September to March, is between 366 and 575 mm, but there is great variation from month to month. In August and September, more than 250 mm falls per month. But in November, rainfall decreases markedly, and in each of the months of December, January, February, and March, it is generally less than 20 mm (Table 20-4). To a certain extent, the relative humidity follows the rainfall (Table 20-5).

Table 20-1. Hours of Sunlight in YZHB Growing Area

	Month							
Area	*Aug.*	*Sept.*	*Oct.*	*Nov.*	*Dec.*	*Jan.*	*Feb.*	*Mar.*
Sanya	211.0	192.0	206.3	202.8	197.5	195.3	165.6	188.0
Ledong	161.3	166.8	185.3	187.5	175.2	173.9	147.4	166.8
Dongfang	225.6	213.9	228.5	210.0	200.0	195.5	162.4	179.2
Changjiang	197.6	182.2	180.3	187.4	179.8	164.1	153.3	185.8
Mean	198.9	188.7	200.1	194.7	188.1	182.2	157.2	180.0

Area	*Total Sept. to Mar.*	*Annual Average*	*Recorded Period*
Sanya	1347.5	2497.8	59–82
Ledong	1202.9	2151.4	59–82
Dongfang	1389.5	2945.1	62–82
Changjiang	1232.9	2343.1	62–82

Soil Conditions

In the southwest of Hainan Island the land differs markedly in composition and fertility, including dry red soil, brown laterite, and tidal sand. In the brown laterite, the allitization is not marked, the ratio of silica to alumina is 2.0 to 2.5, the cation exchange capacity is 7 to 8 me/100 g, and the ratio of base saturation is 70% to 80%. The epipedon of brown laterite is sandy loam or typical loam. Its structure is grain shaped, the organic matter content is 1.79%, and its pH is 5.5 to 5.8. The allitization of the dry red soil is also not obvious. Its ratio of silica to alumina is 2.6 to 3.3, the cation exchange capacity is 1.5 to 3.5 me/100 g, and the ratio of base saturation is 70% to 90%. The loose epipedon of dry red soil is sand or sandy loam that has a grain shaped or block structure. Its organic matter content is 0.31% to 1.19%, and it has a pH of 6.3 to 6.4. There is iron concretion found widely, about 60 cm beneath the surface. The tidal sand is formed by tidal deposit or shore deposit and river drift. The soil composition consists mainly of sand grains between 0.05 and 1 mm in diameter. Its cation exchange capacity is 0.9 to 2.4 me/100 g. The epipedon of tidal sand is loamy sand, of which the organic matter content is 0.1% to 0.6%, and the soil is neutral or slightly alkaline (Soil and Fertilizer Station 1994).

Table 20-2. Temperatures in YZHB Growing Area (°C)

				Monthly Average				
Area	*Aug.*	*Sept.*	*Oct.*	*Nov.*	*Dec.*	*Jan.*	*Feb.*	*Mar.*
Sanya	27.8	27.3	26.1	24.2	21.9	21.1	22.2	24.2
Ledong	26.5	25.8	24.8	22.2	19.7	19.1	20.2	23.1
Dongfang	28.4	27.3	25.4	22.6	19.7	18.5	19.3	22.1
Changjiang	27.5	26.2	25.0	22.1	19.6	18.9	19.6	22.9
Mean	27.6	26.7	25.3	22.8	20.2	19.4	20.3	23.1

Area	*Mean*	*Annual Average*	*Recorded Period*
Sanya	24.4	25.6	59–90
Ledong	22.7	24.0	59–90
Dongfang	22.9	24.7	62–82
Changjiang	22.7	24.3	66–82
Mean	23.2		

Table 20-3. General Radiant Energy from Sunlight in YZHB Growing Area (Kcal/cm^2)

				Month				
Area	*Aug.*	*Sept.*	*Oct.*	*Nov.*	*Dec.*	*Jan.*	*Feb.*	*Mar.*
Sanya	12.1	11.0	10.8	9.4	8.8	9.0	8.5	10.5
Ledong	10.4	10.0	9.9	8.7	7.9	8.1	7.7	9.6
Changjiang	12.8	11.7	11.4	9.6	8.8	8.8	8.2	10.4
Mean	11.8	10.9	10.7	9.3	8.5	8.9	8.1	10.2

Area	*Total Sept. to Mar.*	*Annual Average*	*Recorded Period*
Sanya	68.0	131.6	59–82
Ledong	61.9	118.8	59–82
Changjiang	68.9	137.1	62–82

Habits and Characteristics of YZHB

The environment described above is arid and barren, but YZHB develops normally in these conditions, demonstrating its high adaptability. It grows normally in bright sunshine and high temperatures, but suffers in overcast and rainy weather with low temperatures. For example, in February 1997, when the daily minimum temperature fell to between 11 and 15°C, and the weather was overcast for 16 days with daily rainfall of between 0.2 and 24 mm, the plants suffered from cold injury, the flowers faded, the tender pods were shed, and the growth of stalks and leaves stopped.

Drought Tolerance

The roots of YZHB can reach as deep as 142 cm into the earth so that the plant can use water stored in deeper soil layers. Moreover, the leaflets move up and down on sunny days because of turgor pressure variations in the pulvini, which are located at the base of the petiolules. The leaflets begin to move up slowly in the early morning and reach their highest position at about 10 a.m. Then they begin to move down at about 5 p.m. and reach their lowest position at about 8 p.m. The terminal leaflets turn up and down within about 100°, and lateral leaflets turn within about 60°. This up and down movement of the leaflets may prevent the upper-layer leaves from overexposure to strong sunlight, while allowing the lower stratum leaves more exposure. It may also result in less evaporation from the leaves. These features contribute to the plant's ability to resist drought. A measure of YZHB's drought-resistance is that it grows well when soil moisture falls to between 3.3% and 8.2% in the top 30 cm. This is lower than the wilting threshold of many other crops (see Table 20-6).

Tolerance of Degraded Land

YZHB's well-developed root system has many nodules in which nitrogen-fixing bacteria provide nitrogen for their host plant. This enables YZHB to grow normally without the application of nitrogenous fertilizer on arid land. Soil parameters such as organic matter content (OMC) and total nitrogen content are not significantly different between fields growing YZHB and those growing other crops, which are regularly fertilized. However, when the soil has a comparatively high capacity to preserve fertility and retain moisture, then OMC and total nitrogen are relatively high in YZHB fields when compared with fields growing other crops. Such soils are found in Potou village (Ledong county) and Gancheng town (Dongfang city).

Table 20-4. Rainfall in YZHB Growing Area (mm)

	Month							
Area	*Aug.*	*Sept.*	*Oct.*	*Nov.*	*Dec.*	*Jan.*	*Feb.*	*Mar.*
Sanya	211.4	257.2	221.7	43.0	7.0	6.2	12.8	20.0
Ledong	326.6	279.0	170.7	39.2	15.6	13.0	11.8	20.9
Dongfang	228.4	175.1	115.5	24.6	8.5	7.7	14.9	19.4
Changjiang	351.8	305.4	185.2	37.2	9.9	8.8	10.8	17.8
Mean	279.6	254.2	173.3	36.0	10.3	8.9	12.6	19.5

Area	*Total Sept. to Mar.*	*Annual Average*	*Recorded Period*
Sanya	376.2	1,383.2	59–90
Ledong	550.2	1,568.2	59–90
Dongfang	365.7	955.9	62–82
Changjiang	575.1	1,642.3	66–82

Table 20-5. Relative Humidity in YZHB Growing Area (%)

	Month							
Area	*Aug.*	*Sept.*	*Oct.*	*Nov.*	*Dec.*	*Jan.*	*Feb.*	*Mar.*
Sanya	85	84	79	74	73	74	76	78
Ledong	87	86	81	78	76	75	77	77
Dongfang	82	83	81	79	79	80	83	83
Changjiang	81	83	80	76	75	77	78	77
Mean	83.5	84.0	80.3	76.8	75.8	76.5	78.5	78.8

Area	*Mean*	*Annual Average*	*Recorded Period*
Sanya	77.9	79	59–82
Ledong	79.6	80	59–82
Dongfang	81.3	80	62–82
Changjiang	79.6	77	66–82
Mean	79.3		

Table 20-6. Soil Moisture in YZHB Fields (%)

	Growth stage of YZHB		
	Tender Pods	*Most Pods Ripe, Stalk Tips Still Growing*	
Soil Layer (cm)	*Clayish Loam*[a]	*Loamy sand*[a]	*Loam*[b]
0–10	3.27	2.20	2.40
10–20	8.05	2.84	2.90
20–30	8.21	4.20	5.19

Notes: Samples were analyzed by the Rubber Cultivation Research Institute, CATAS. [a] Sampling in Yongming, January 24, 1997. [b] Sampling in Potou, January 24, 1997.

However, when it comes to other soil nutrients such as available phosphorus and potassium, there is either much more or much less in YZHB fields than in other crop fields in different areas (see Table 20-7). This may be the consequence of fertilizer applications when other crops were grown in the fields, or differences in the rock that formed the soil.

There is also no marked difference in the soil nutrient content of fields where YZHB has been planted continuously for many years, compared with fields growing other crops on which fertilizer has been frequently applied (see Table 20-8).

Resistance to Diseases and Pests

YZHB is not known to suffer from any serious diseases or attacks by pests. Some insects, such as *Maruca testulalis* Geyer, may damage a few leaves, but this is the limit of its destruction. In practice, chemical inputs have never been used on YZHB to control diseases or pests. The pods are seldom destroyed by rats or birds. The mechanism by which the plant resists diseases and pests remains unknown.

Capacity to Restrain Weeds

Weeds and shrubs are rare in YZHB fields. Its seeds germinate faster than those of weeds, and the seedlings rapidly form a ground cover, which restrains the germination of weed seeds. The trailing branches of YZHB twist around and climb those weeds and shrubs that do manage to grow, making them wither. Also, YZHB's drought tolerance allows it to grow in conditions too dry even for weeds. It is possible that YZHB has other features that allow it to restrain the growth of weeds and shrubs, but these need further research. Farmers call YZHB *Miecao Dou,* meaning "the bean that can eradicate weeds," because it restrains weeds and shrubs when grown in the same field for several years.

Table 20-7. Soil Nutrients in YZHB Fields Compared with Fields under Other Crops

Sampling Area and Depth (cm)	*OMC (%)*		*Total N (%)*		*Available P (ppm)*		*Available K (ppm)*	
	a	b	a	b	a	b	a	b
Banqiao								
0–10	0.520	0.533	0.0265	0.0265	18.19	13.81	85.0	92.5
10–20	0.538	0.449	0.0234	0.0160	14.51	8.39	75.0	46.5
20–30	0.422	0.455	0.0215	0.0297	5.59	8.49	57.5	98.8
Gancheng								
0–10	0.913	0.628	0.0398	0.0271	125.29	274.77	111.5	117.5
10–20	0.754	0.552	0.0371	0.0253	115.78	235.02	81.3	182.5
20–30	0.764	0.591	0.0306	0.0227	76.54	170.65	87.6	207.5
Potou								
0–10	1.480	1.340	0.0773	0.0685	12.02	12.38	185.0	140.0
10–20	0.927	0.954	0.0491	0.0514	4.79	5.36	160.0	116.0
20–30	0.613	0.596	0.0329	0.0325	2.99	5.16	175.0	100.0

Notes: Column a indicates fields with YZHB. Column b indicates fields growing other crops, such as sweet potato in Banqiao and Gancheng towns, and a new planting of sugarcane in Potou village. Where possible, adjacent fields were chosen for sampling to minimize differences between sites. Samples were analyzed by the Rubber Cultivation Research Institute, CATAS.

Fast Maturing

The growth period of YZHB is about 100 days from sowing to harvesting. However, this can be affected by soil fertility and microclimate. If the land is highly fertile and there is ample rainfall, the growth period will be prolonged. If the land is barren and rainfall is light, it will be shortened. The plants begin to bloom after 30 to 50 days of vegetative growth. The raceme can develop from the axils simultaneously with new leaves. YZHB flowers usually bloom in the early morning and fade at dusk, blooming for about 12 hours. The pods ripen 18 to 25 days after flowering. In favorable environmental conditions with some rain, the plants' vegetative and reproductive growth occur at the same time and progress for a relatively long period. Hence, its period of reproductive growth is relatively long compared to its vegetative growth period. Our investigations show that the reproductive growth period of YZHB can be prolonged by three months or more.

Cultivation

Farmers usually plant YZHB from September to October, about one to two months before the end of the rainy season. However, it is planted as early as August in some of the coastal areas where the water retention capacity of the sandy land is poor and the rainfall is comparatively low in September and October. In mountain areas where the water retention ability of the soil is good, YZHB may be planted as late as November.

YZHB is sown one to three days after rain, when the soil is still moist. In coastal areas, farmers usually plow the land first, and then rake and broadcast YZHB seeds evenly on the surface of the soil. Then they rake it again to bury the seeds under a shallow layer of soil. This method is used widely in the rainy season. In mountainous areas, however, farmers sometimes broadcast the seeds evenly on the surface of fallow land, after first clearing it if there are too many shrubs or too much coarse grass. Then they plow the land using cows, or dig it by hand, to bury the seeds fairly deeply beneath the surface. This method is the usual one when the rainy season is ending, or has just ended. The seed is sown at a rate of about 30 to 60 kg/ha, depending on the fertility of the soil. If the land is fertile, the seeding rate is decreased; if it is arid, the seeding density is increased. After sowing, the seeds will germinate in three to five days if there is enough moisture in the soil. If the soil is too dry, the seeds will not germinate immediately but will remain in the soil for a month or more and will germinate when the rains arrive.

All of these details concern the planting of YZHB as a food crop. However, if it is grown only as green manure, the recommendation is that it should be sown at the beginning of the rainy season (Tropical Crops Cultivation Department 1965).

After sowing, YZHB does not need any weeding, fertilizer application, or disease and pest control. It merely requires protection from wandering livestock such as cows and goats.

Although the pods ripen at different times, the harvest can wait until they are all ripe, because the ripened pods do not shatter easily. The harvest period is from December to mid-March. Continuous cropping of YZHB is popular because farmers believe that growing YZHB does not harm the land.

Yields per Unit Area

The yield of YZHB is related to sowing time, rainfall, and soil fertility. The plant's main growing environment is generally very poor, and its structure is small and low, so moderate plant populations per unit area and moderate leaf area per stalk are required to achieve high yields. Although YZHB has high drought tolerance, adequate soil moisture is required for germination and seedling growth. Sowing before the end of the rainy season stimulates seed germination and ensures that YZHB has sufficient time for vegetative growth. But if it is planted too early, the plants will either grow excessively without blossoming or blossom without producing beans. If the sowing time is too late, the germination rate will be low, the plants will not grow well, and they will produce few beans because of a shortage of soil moisture.

In general, it is preferable to sow YZHB one to two months before the end of the rainy season. Hainan farmers plant YZHB at various times, determined mainly by their own experience, because the rainy season ends at different times in different areas. Rainfall during winter and spring can also vary greatly from year to year, so even if planting is timed one to two months before the expected end of the rainy season, the plants may still grow, but they will yield poorly if the dry season arrives earlier than expected. Therefore, the amount of rainfall during the period between sowing and blossoming is important for achieving high yields. Because there are no irrigation systems in the YZHB growing region, seed germination and seedling growth depend on favorable weather.

Table 20-8. Soil Nutrient Content in Fields under Different Planting Regimes

Soil Depth (cm)	*OMC (%)*			*Total N (%)*		
	a	b	c	a	b	c
0–10	0.429	0.736	0.382	0.0144	0.0289	0.0161
11–20	0.279	0.760	0.290	0.0149	0.0193	0.0157
21–30	0.287	0.367	0.367	0.0140	0.0153	0.0152
	Available P (ppm)			*Available K (ppm)*		
	a	b	c	a	b	c
0–10	15.21	48.45	10.89	42.50	71.30	65.00
11–20	8.72	35.42	9.71	42.50	65.00	50.00
21–30	3.86	9.80	6.03	37.50	83.50	43.50

Notes: a. Field continuously cropped with YZHB for five years. b. Field rotated between YZHB and other crops. c. This field rotated with other crops, but not YZHB. The sampling location was in Banqiao town, Dongfang city. Samples were analyzed by the Rubber Cultivation Research Institute, CATAS.

Table 20-9. Main Chemical Composition of YZHB (%)

Sampling Locations	*Protein*	*Fat*	*Starch*	*Carbohydrate*	*Ash*	*Moisture*
Potou[a]	26.07	0.91		4.98	3.68	8.82
Huangliu[a]	23.16	1.12	46.72	5.39	3.85	4.54
Trial Farm[b]	25.9	0.77			2.80	

Notes: [a] Samples were analyzed by the Measurement and Experimental Center of CATAS.
[b] Tropical Crops Cultivation Department of SCUTA, 1965.

Table 20-10. Main Chemical Composition of YZHB Vines (%)

Samples	*Crude Protein*	*Crude Fat*	*Extraction without N*	*Crude Fiber*	*Ash*	*Moisture*
Dry vines at harvest[a]	7.8	2.84	18.89	9.94	5.04	
Fresh vines at blossom[b]	3.46	0.51	15.20	2.70	1.02	77.10

Notes: [a] Samples were analyzed by the Measurement and Experimental Center of CATAS.
[b] Tropical Crops Cultivation Department of SCUTA, 1965.

According to this study, if there are two or three rainfalls, each of 25 mm or more, after sowing in mountainous areas, then a good bean yield can be expected. In coastal areas with sandy soil, five or six similar rainfalls are needed to satisfy the needs of the plants. YZHB can produce high yields from fields with medium soil fertility, whereas it grows vigorously without blossoming or blossoms without producing beans in fields with high soil fertility, and it grows poorly with low yields in fields with poor fertility.

To reinforce that point with real examples, a yield of 1,770 kg/ha was harvested from sloping land with dry red soil and average fertility at Potou village. The land had been cultivated for many years, and its nutrient stocks were OM, 0.613 to 1.480%; total N, 0.0329% to 0.0773%; available P, 3 to 12 ppm; and available K, 60 to 185 ppm. At Yongming village, a medium yield of 1,455 kg/ha was harvested from sandy loam soils with poor fertility. Its nutrient stocks were OM, 1.10%; total N, 0.0476%; available P, 12.89 ppm; and available K, 32.50 ppm. At Banqiao town, a low yield of 675 kg/ha was taken from sandy, low fertility soil with the following nutrient stocks: OM, 0.4% to 0.5%; total N, 0.022% to 0.026%; available P, 5 to 12 ppm; and available K, 57 to 85 ppm.

Harvesting and Storage

The harvesting procedure consists of uprooting the stalks or, alternatively, cutting the upper part of the plants in the morning, exposing them to the sunlight, and then tapping their pods when they are dry. YZHB cannot be harvested at noon or in the afternoon on sunny days because the beans will shatter from the dry pods when they are shaken.

Because the seeds of YZHB are vulnerable to insect damage, it is important that they be protected during storage. Local farmers dry the seeds and put a 5 cm thick layer of ash on the bottom of a jar. Then they place the seeds into the jar and, finally, cover them with another 5 cm layer of ash. The jar is stored in a cool, well-ventilated place.

Chemical Composition

YZHB has a protein content of 23% to 26%. This is equal to, or better than, other beans such as gram beans or red beans. The fat content of YZHB is about 0.9%, and the starch content is relatively high, about 46.7% (see Table 20-9). It is therefore clear that YZHB is not only a healthy food with high protein and low fat, but it is also an organic food because it is produced without pesticides or chemical fertilizers. Also,

YZHB vines make good fodder for livestock, containing crude protein of 7.8%, crude fat of 2.84%, and extraction without nitrogen of 18.89% (Table 20-10).

Methods of Preparation

The main methods of preparing YZHB for human consumption are as follows:

- Growing bean sprouts: In the rainy season, there are few vegetables because of the scorching climate and tropical storms, During this period, local farmers use YZHB to grow bean sprouts as a popular vegetable.
- Making bean sauce: The sauce made form YZHB is a special local condiment. Generally, it is used in cooking fish and meat. The beans are first cooked, and then they are allowed to drip dry. Then some saccharomycetes are added and stirred, the mixture is pounded, salt is added, and finally, the mixture is allowed to ferment for a week. Then it can be eaten as a sauce.
- Eating directly as a protein rich food: In mountain areas where protein sources are scarce, YZHB is boiled and eaten directly.

Conclusions: The Benefits of Growing YZHB

Although its net output is not very high, YZHB can be an economically beneficial crop. Its yield is generally between 1,075 and 1,500 kg/ha, and its market price is 3.6 RMB per kilogram (1997US$1 equaled 8.2916 RMB). Therefore, its output value can reach 4,050 to 5,400 RMB per hectare. The cost of planting YZHB is about 1,200 to 1,350 RMB per hectare, so the net output value per hectare is about 2,700 to 4,050 RMD. This return is quite attractive because of its low input and short growing period.

Planting YZHB enhances land productivity because it uses land resources more fully during the dry season. There is an annual period of drought across large areas of Hainan, during which, monthly rainfall is less than 50 mm for up to seven months in winter and spring. Without irrigation, much sloping land must remain fallow, and land productivity is very low. By planting YZHB, the period of land productivity can be extended by at least one month on sandy soils and two months on soils with a capacity for better moisture retention. Land resources can, therefore, be used more efficiently.

To some extent, planting YZHB also improves the soil. It can be seen from Tables 20-7 and 20-8 that soil organic matter and total N content tend to increase in land with better conservation of water and fertilizer. Even on sandy land with a poor capacity to retain water and fertilizer, continuously planting YZHB results in soils with nutrient contents no lower than those of land planted to other crops that often receive applications of fertilizer. However, if a rotation cropping system is practiced, soil organic matter and other properties are further improved.

Introducing YZHB into New Areas

In 1960, the Tropical Crops Department of CATAS introduced YZHB as a green manure crop and confirmed its beneficial effects (Tropical Crops Cultivation Department 1965). In 1993 and 1995, YZHB was grown under trial at the Rubber Cultivation Research Institute, CATAS, at Danzhou city, north of its original area of distribution. The temperature is lower there, with higher rainfall in winter and spring (see Table 20-11). However, the soil fertility is comparable to the average in its area of origin.

Primary results found that YZHB bloomed normally and bore fruit in its new location. The beans were sown in November 1996 and blossomed and fruited from December 1996 to April 1997. In February 1997, low temperatures with overcast, rainy weather caused a slight cold injury that resulted in the YZHB shedding its flowers. Growth was hindered, but the plants recovered rapidly when the temperature went up again. This shows that if YZHB is introduced to new areas, its growth and yield characteristics may undergo considerable change.

An Agenda for Further Research

Improving Cultivation Techniques

Our investigations have shown that the output of YZHB is relatively low, reflecting its extensive cultivation. However, yields may be increased by improving cultivation techniques. For example, it can be seen in Table 20-8 that soil fertility is improved when YZHB is grown in a rotation system with other crops. Therefore, methods and patterns of rotation cropping with other crops should be further researched and improved across different planting regions, with the aim of changing the present continuous cropping system, improving soil fertility, and raising yield per unit area. YZHB shows great variations in yield when planted in fields with different soil fertility. Therefore, if it is grown in fields with poor soil fertility, the yield should be improved by applying reasonable amounts of fertilizer. Techniques for applying fertilizer to crops of YZHB, especially phosphorus and potash, must be subjected to further research.

There are many differences in growth, yield, and other characteristics between individual YZHB plants. Apart from the influence of soil properties, these differences may be a result of seed quality and the origin of seed stocks. Therefore, the yield per unit area may be further improved by selecting seeds with desirable and consistent traits.

Table 20-11. Temperature and Rainfall Data from YZHB Trial Area

Parameters	*1996*		*1997*		
	November	*December*	*January*	*February*	*March*
Temperature (°C)	22.0	18.1	18.4	17.7	22.4
Rainfall (mm)	162.3	28.4	12.7	96.4	129.6

Note: The trial location was the nursery of the Rubber Cultivation Research Institute, CATAS.

Acknowledgments

We wish to express our appreciation for the helpful comments made on this chapter by Professor Zhou Zhongyu, Professor Hao Bingzhong, and senior engineer Huang Suofen. Thanks also go to Professor Zhong Yi for assisting in identifying the species of YZHB, and Mr. Tao Zhonglian, for providing the meteorological data.

References

Chen, W. and Z. Zhang. 1965. *Flora Hainanica* (vol. 2), Beijing: Academic Publishing House, 236–329.

Editorial Committee of Chinese Academy of Sciences for Flora of China. 1998. *Flora Reipublicae Popularis Sinicae* (vol. 39). Beijing, China: Science Press.

Lin, W., J. Jiang, W. Li and G. Xie. 1999. *The Biological Characteristics of Yazhou Hyacinth Bean.* In: Chinese Journal of Tropical Crops. 20 (1): 59-65.

Soil and Fertilizer Station of Hainan Province. 1994. *Hainan Soils.* Haikou: Sanhuan Publishing House, 73–75, 103–106, 109–110.

Tropical Crops Cultivation Department, SCUTA. 1965. *Ya Zhou Hyacinth Bean.* Internal Report.

Wang, F., Y. Wang, and F. Wang. 1992. *Edible Beans Resources on Hainan Island.* Beijing: Agricultural Publishing House, 244–248.

Chapter 21

Indigenous Fallow Management Based on *Flemingia vestita* in Northeast India

P.S. Ramakrishnan

India's northeast hill areas form a highly complex landscape mosaic. The region is inhabited by more than 100 tribes, each with its own linguistic and cultural characteristics. All the tribes are involved in shifting agriculture, or *jhum*, as it is known locally, and this is a major land-use system in the region (Ramakrishnan 1992). The jhum procedure involves slashing and burning the vegetation on a two-to-three hectare plot at a given stage of forest succession. The land is then cropped, usually for a year. It is seldom cropped for a second or third year, but if this does happen, the crop is usually restricted to perennials, such as bananas. Then the land is allowed to lie fallow, under natural regrowth. Therefore, the landscape is a patchwork of jhum plots at various stages of cropping and fallow.

Supplementing the jhum system is the valley system of wet rice cultivation, and home gardens. The valley system is sustainable year after year because the washout from the hill slopes provides the soil fertility needed for rice cropping without any external inputs. Home gardens, found throughout the region, have economically valuable trees, shrubs, herbs, and vines. They form a compact, multistoried system of fruit crops, vegetables, medicinal plants, and cash crops. There may be 30 or 40 species in a small area of less than a hectare. Linked to these land uses are animal husbandry systems centered traditionally on pigs and poultry. The advantage of those systems is that they are based primarily on recycling food that is unfit for human consumption.

An increasing human population has brought pressure to bear upon the extensive land-use practices of jhum cultivation. Its harmony with the environment is dependent upon the length of the jhum cycle, which must provide a period of fallow long enough to allow the forest, and the soil fertility lost during the cropping phase, to recover. Over the past two or three decades, the period of fallow has fallen drastically from 20 years or more to about 5 years, or even less. However, longer cycles of up to 30 years can still be observed, and in more remote areas, cycle lengths may still be as long as 60 years. But, these days, such cases are rare. Large-scale timber extraction has led to invasion of the landscape by exotic and native weeds. The forest has been replaced, either by an arrested succession of weeds, or by large-scale desertification, resulting in a totally barren landscape. There has also been a sharp reduction in the area of land available for agriculture. Where population densities are high, the burning of slash has been dispensed with, leading to rotational fallow

---

P.S. Ramakrishnan, School of Environmental Sciences, Jawaharlal Nehru University, New Delhi 110067, India.

systems and settled, permanent cultivation. These systems are often below subsistence level, and families attempt to maximize output under conditions of rapidly depleting soil fertility. Inappropriate animal husbandry practices are often introduced, and indiscriminate grazing by goats or cattle leads to rapid site deterioration. The consequent and serious social disruption demands a new and integrated approach to managing the forest-human interface (Ramakrishnan 1992). Under all these circumstances, the following questions arise:

- How do local communities cope with fallow management when jhum cycles are being squeezed down toward fixed, permanent cultivation?
- What are the implications of farmers' concerns about their inability to restore soil fertility through fallow management?
- What role can traditional knowledge and peoples' participation play in the sustainability issues arising from global change?

This chapter sets out to address these concerns, with emphasis on *Flemingia*-based fallow management systems.

Two Contrasting Jhum Systems

The Typical Version

The typical version of jhum is practiced on slopes of 30 to 40°, under a monsoon climate, with an average annual rainfall of about 2,200 mm. Sites for jhum are allotted by the headman in each village community. Then, during the dry winter months of December and January, the undergrowth is slashed and small trees and bamboo are felled. Short tree stumps and large boles are left intact, and the underground organs of different species are not disturbed. This laborious process is often completed by the men from two or three families. A well-knit social organization is one of the essential ingredients for such joint efforts, and they help to promote kinship among members of a village, as does the process of allotment of sites for jhum. This is done by the village headman, who is in overall control of the village community. Such events are typical of many communities, among them the Garos and the Mikirs (Maikhuri and Ramakrishnan 1990).

Toward the end of March or the beginning of April, before the onset of the monsoon, the debris is burned. However, before burning, a fire line is cleared around the field. Any material that survives the first fires is heaped and burned again. A bamboo hut is built at the jhum, and the family takes up temporary residence, among other things, to protect the field from wild animals

Sowing follows the first showers of the monsoon. Seed mixtures used for different jhum cycles vary considerably. Cereal crops are most common under long jhum cycles, whereas perennials and tuber crops are most popular in short jhum cycles, such as those practiced by the Garos in Meghalaya (Toky and Ramakrishnan 1981), and by the Nishis in Arunachal Pradesh (Maikhuri and Ramakrishnan 1991). Table 21-1 lists the species sown at one site in Meghalaya under a 30-year cycle. Between eight and 35 crop species may be grown together. Similarly, up to 35 species may be grown together in Arunachal Pradesh (Maikhuri and Ramakrishnan 1991) or in the Garo and the Naga Hills (Kushwaha and Ramakrishnan 1987). In addition to those in Table 21-1, other crop species could include *Coix lacryma-jobi*, *Eleusine coracana*, *Ipomoea batatas*, and *Dioscorea alata* (Maikhuri and Ramakrishnan 1991).

Seeds of pulses, cucurbits, vegetables, and cereals are mixed with dry soil from the site to ensure their uniform distribution, and broadcast soon after the burn. Maize seeds are dibbled at regular intervals amongst other crops. Similarly, rice is planted into the crop mixture by dibbling with a long stick, after the first rainfall in mid-April. Semi-perennial and perennial crops such as ginger, colocasia, tapioca, banana, and castor are sown intermittently throughout the growing season. *Ricinus communis* is grown for its leaves, which are used for rearing young silkworms. The different crops are harvested as they mature (see Table 21-1), making way for others that are still maturing.

Throughout the cropping period, weeds pose a problem. The most common of them are tree seedlings, grasses, or herbs, and sprouts from roots, rhizomes, and stumps. Under long jhum cycles, the problem is less severe than under short cycles, where many weeds, particularly *Imperata cylindrica*, keep sprouting from underground rhizomes and are difficult to eradicate (Saxena and Ramakrishnan 1984). Others, like *Eupatorium odoratum*, are controlled by frequent slashing. Hand hoeing, a job mainly performed by women, is usually done twice during the cropping season, or up to four times under shorter cycles.

The Modified Version

The jhum system practiced by the Khasis, at altitudes of 1,500 m above sea level (asl) and above in Meghalaya, is a modified version of the typical jhum system outlined above. It is commonly practiced around Shillong (Mishra and Ramakrishnan 1981), where the vegetation consists of sparsely distributed pine trees (*Pinus kesiya*) with some undergrowth of shrubs and herbs. At higher elevations, clear felling and burning of the forest is not feasible because the forest regenerates more slowly in the subtemperate climate (Mishra and Ramakrishnan 1983a). The pine trees are not felled, but the lower branches are slashed in December. The slash is arranged in parallel rows running down the slope and is left to dry. Several months later, in March, soil is placed on top of the slash so as to form ridges alternating with furrows of compacted soil running down the slope (see color plate 14). The slash is then burned, and the fires are slow and controlled. A fire line of cleared vegetation around the plot helps to check its spread.

The preparation of the site into alternate ridges and compacted furrows running down the slope allows the furrows to act as water channels to minimize the loss of nutrients (Mishra and Ramakrishnan 1983c). This is particularly important at these altitudes because soil fertility recovers more slowly (Mishra and Ramakrishnan 1983b, 1983d). The soil under pine forests is also highly acidic, further aggravating the availability of nutrients (Ramakrishnan and Das 1983; Das and Ramakrishnan 1985). Slow burning of the limited slash, after it is stacked in parallel rows and cover with a thin layer of soil to form ridges, is an efficient means of resource management. Crops are grown on the ridges, which are enriched with nutrients. In the current situation of generally reduced soil fertility levels, aggravated by shortened jhum cycles, there has been a shift toward tuber crops, such as potatoes, that can give better economic yields in poor soils (Ramakrishnan 1984). Under a long jhum cycle of 15 years, cropping occupies only one year, and no fertilizer is used. However, under a 10-year cycle, organic manure in the form of pig dung and vegetable matter is applied at a rate of 600 kg/ha/yr (oven dry weight).

The crop mixtures differ from those at low elevations, in that tuber crops such as *Solanum tuberosum*, *Ipomoea batatas*, and *Colocasia antiquorum* are planted on the ridges soon after the burn and before the onset of the monsoon (Mishra and Ramakrishnan 1981). As soon as the rain begins, *Zea mays*, *Phaseolus vulgaris* and a few cucurbits are planted. Along each ridge, potatoes and *Zea mays* are sown together in three distinct rows. *Colocasia antiquorum* is generally confined to the top and bottom part of each ridge and cucurbits are sown at random. *Phaseolus vulgaris* is sown around pine trees, for their support.

After the harvest of the tuber crops in July and August, a winter crop of potatoes is often sown along the ridges. Harvesting of *Zea mays* and the legume *Phaseolus vulgaris* occurs in September and October, after which *Brassica oleracea* seedlings are planted, along with a winter crop of *Solanum tuberosum*. The second potato crop is harvested in November, and the field is left uncultivated between December and March. If there is a second year of cultivation, the same procedures are followed. Otherwise, the land is fallowed for natural regrowth of vegetation.

The mixture of crops varies according to the jhum cycle. Under a five-year cycle, cropping occurs for two to three years continuously after slashing and burning. However, the only crops grown are *Solanum tuberosum*, *Zea mays*, and *Brassica oleracea*. Occasionally a monocrop of potatoes may be raised under shorter jhum cycles (Gangwar and Ramakrishnan 1987). During the first year of cultivation, the first and second crops receive both an organic manure of pig dung and vegetable matter, and inorganic fertilizer consisting of equal quantities of nitrogen, phosphorus, and potassium, at rates of 1,000 kg/ha and 10 kg/ha, respectively. In the second year, the input rates are boosted to 1,850 kg/ha and 20 kg/ha, respectively, for the first and second crops.

A comparative analysis of the modified version of jhum, practiced by the Khasis under varied cycle lengths, shows that the economic returns are very high, although yields decline with shorter jhum cycles. The monetary return, in Indian rupees, under a 10-year jhum cycle (see Table 21-2) is about five times more than that from a similar jhum cycle at lower altitude. The high net monetary return and economic efficiency are despite the high input required for land preparation in the high-altitude system. Potatoes, which have high monetary value, are produced largely for export from the village.

Pressures for Change

Over the past 50 years or more, government agencies have tried in vain to replace shifting agriculture with settled terrace farming, which demands high energy inputs in the form of fertilizers, herbicides, and pesticides (Ramakrishnan 1992). The soil is shallow and infertile, and nutrient losses from the system are very heavy, so more and more fertilizer is often required to sustain such systems, with very low efficiency. The weed problem is also exaggerated in fixed and permanent cultivation, so weed control assumes alarming proportions. For these ecological reasons, as well as for a variety of social and cultural reasons related to land tenure and cultural and religious practices centered on shifting agriculture, farmers rejected settled terrace farming as a permanent solution to the problems of shifting agriculture. At the same time, the highly distorted form of shifting agriculture now being practiced under short cycles of five years or so has become less and less tenable.

During the recent past, there has been a more spontaneous shift to more intensive systems of land use in many parts of the developing world because of increasing population pressure (Boserup 1965; Okafor 1987). Long jhum-based fallow systems have been replaced by shorter bush-fallow systems (FAO/SIDA 1974) and, ultimately, by permanent agriculture. The bush-fallow systems are variously called "semipermanent cultivation" (Nye and Greenland 1960; Allan 1965) or "stationary cultivation with fallowing" (Faucher 1949). Emphasis on crops grown varies considerably (Gangwar and Ramakrishnan 1987).

In the area of this study, city centers such as Shillong in Meghalaya have generated the population pressures under which the modified version of the jhum has often been replaced by fallow or fixed, permanent systems of agriculture (Mishra and Ramakrishnan 1981).

Table 21-1. Sequential Harvesting of Crops Grown by the Garos in Meghalaya on Jhum Plots under a 30-Year Cycle

Species	*Harvesting Time*
Setaria italica	mid-July
Zea mays	mid-July
Oryza sativa	early September
Lagenaria spp.	early September
Cucumis sativus	early September
Zingiber officinale	early October
Sesamum indicum	early October
Phaseolus mungo	early October
Cucurbita spp.	early November
Manihot esculenta	early November
Colocasia antiquorum	early November
Hibiscus sabdariffa	early December
Ricinus communis	(perennial crop)

Note: All seeds were sown in April.
Source: Ramakrishnan 1984.

Table 21-2. Input-Output Analysis of 10-Year Cycle Jhum at Lower and Higher Elevations in Meghalaya, in Indian Rupees

Factors	*Low-Elevation Jhum*	*High-Elevation Jhum*
Input	1,830	3,842
Output	3,354	14,171
Net gain	1,524	10,329
Output/input	1.83	3.9

Note: After Toky and Ramakrishnan 1981; Mishra and Ramakrishnan 1981.

New Systems Arising Out of the Modified Jhum

Fallow Systems

In the fallow system of agriculture (FAO/SIDA 1974), the weed biomass of the fallow phase is slashed in January and organized in parallel rows. Then it is covered by a thin layer of soil and allowed to decompose. The crop is sown on these ridges in March. The rest of the land is compacted into alternating furrows, running down the slope, as in the modified jhum system described above (Mishra and Ramakrishnan 1981).

The land is cropped twice in a year, once between March and June and again between August and November. The first cropping usually involves either a single species or mixtures of two, with emphasis on vegetables. The second crop is always a monoculture of potatoes. *Flemingia vestita* must be raised along with another crop species because, when planted in March, it cannot be harvested until October. This only applies when the system is restricted to one cropping per year. Potato tubers are the only seed stocks purchased outside the farm. Others come from within.

The quantities of fertilizer used vary according to the crop mixture and the location. During every year of cropping, between 312 and 5,400 kg (dry weight) of organic manure is applied per hectare. Inorganic fertilizer with equal quantities of N, P, and K is also applied two or three times during a year's cropping, with a total amounting to between 200 and 1,275 kg/ha.

Economic and energy efficiencies vary considerably, depending upon the cropping pattern (Gangwar and Ramakrishnan 1987). Potatoes are an important crop from a monetary viewpoint. As a second crop along with mustard, potatoes give maximum monetary return to the farmer. However, maximum energy efficiency is achieved when *Flemingia vestita,* a legume crop, is involved. This may be related to the nitrogen accretion in the soil because of the legume's inclusion in the crop mixture (Gangwar and Ramakrishnan 1989). Better performance by tuber crops under reduced soil fertility is understandable because of their more efficient use of nutrients in short fallow systems (Ramakrishnan 1983).

Of all the species cultivated under the fallow system, *Flemingia vestita, Perilla ocymoides,* and *Digitaria cruciata* are three lesser-known food crops of the Khasis (Gangwar and Ramakrishnan 1989). They are of value during leaner months of the year, when traditional crops are in short supply.

Fixed, Permanent Cultivation Systems

The crop mixtures involved in this land use vary considerably, and as a consequence, the economic and energy returns also vary significantly (Gangwar and Ramakrishnan 1987). Vegetable and tuber crops give higher economic returns than seed crops. The energy efficiency is high when *Flemingia vestita* is included in the crop mixture because of the ability of this species to improve the nitrogen fertility of the soil.

Fallow Management

The Value of Nepalese Alder for a Redeveloped Jhum

I will not go into detail regarding this fallow management system, which is discussed elsewhere (Guo et al., Chapter 29; Cairns et al., Chapter 30). I wish, however, to briefly examine the value of this species for land with highly depleted nitrogen budgets under jhum cycles shortened to five years.

We have analyzed in detail the nutrient budgets under different jhum cycles operating in the northeast of India. A comparative analysis of the nitrogen budgets of jhum systems with cycles of 15, 10, and 5 years, practiced by the Khasi tribe at an altitude of 1,500 m asl in Meghalaya, is indicative of the issues involved in nitrogen fertility maintenance under jhum (Mishra and Ramakrishnan 1984).

The net change in the nitrogen pool (see Table 21-3) suggests that shortening the jhum cycle results in lower nutrient capital at the preburn stage, as well as at the end of the cropping period. The loss of nitrogen at the end of one year of cropping under the five-year jhum cycle is higher only if the values are extrapolated onto a 15-year time scale. This then shows a nitrogen loss three times greater than that from a plot under a 15-year cycle and twice that from a plot under a 10-year cycle.

The land-use history of this study site, near Shillong, goes back 20 years. However, if the plots currently under a 5-year jhum cycle had longer fallow cycles before this time, which maintained a nutrient balance (Mishra and Ramakrishnan 1984), then the system seems to have lost about 1,280 kg of nitrogen per hectare over the past 20 years. This is a direct comparison between the preburning nutrient capitals of the jhum with a 15-year cycle and that with a 5-year cycle.

While jhum cycles of 15 or 10 years provide fallow periods long enough to restore the original soil nitrogen status before the next cropping, it seems unlikely that a cycle of merely five years would restore about 800 kg of nitrogen lost per hectare over the course of two croppings in one year (see Table 21-3). One of the disadvantages of a 5-year jhum cycle lies in the reduced nitrogen capital with which the agroecosystem has to operate because of the increased frequency of fire and cropping with too short a fallow phase. Similar conclusions arise from investigations of jhum systems elsewhere in the region (Swamy and Ramakrishnan 1988).

Table 21-3. Change in Soil Nitrogen under Jhum at Shillong in Meghalaya (1,000 kg/ha/yr)

Stage of Cycle	*15-year Fallow Cycle*	*10-year Fallow Cycle*	*5-year Fallow Cycle*	
			1st-year Crop	*2nd-year Crop*
Soil pool before burning	7.68	7.74	6.40	5.98
Soil pool at the end of cropping	7.04	7.15	5.98	5.60
Net difference	0.64	0.59	0.42	0.38

Source: Mishra and Ramakrishnan 1984.

It is in this context that the value of Nepalese alder becomes significant. This early successional tree species grows as part of jhum fallow vegetation in the northeastern hill region at altitudes between 500 and 1,900 m asl. The species has nodulated roots that are colonized by *Frankia,* occurring as an endophyte, and it is effective in biological nitrogen fixation (Sharma and Ambasht 1988). Under short jhum cycles of five to six years, we have already seen that about 800 kg/ha of soil nitrogen is lost during one cropping season, and it takes a minimum of 10 years of natural fallow regrowth to recover this nitrogen back into the system (Mishra and Ramakrishnan 1984). It is in this context of restoring the 800 kg/ha of nitrogen over a shorter fallow period of four to five years that an early successional species such as the Nepalese alder becomes important (Ramakrishnan 1992). In addition to having the capacity for nitrogen fixation, *Alnus nepalensis* also produces litter rich in nitrogen. Mineralization also contributes to the biological buildup of soil fertility. Therefore, the emergence of Nepalese alder during early fallow succession is critical to replenishment of nitrogen in jhum systems with a shortened cycle. The species has a high cultural value and is traditionally conserved by local communities during the slash-and-burn operation in their jhum systems. Because farmers can identify themselves with a value system that they understand and appreciate, Nepalese alder can be used for fallow management with community participation.

Flemingia-*based Fallow Management for the Modified Jhum System*

Flemingia vestita, Perilla ocymoides, and *Digitaria cruciata* are three lesser-known but important food crops grown in rotational fallow or permanent cultivation systems. Of the three, *Flemingia vestita,* when grown along with other traditional crops, improves the ecological efficiency of the system (see Table 21-4).

After a few years of mixed cropping, farmers often raise a pure crop of this legume because of its ability to fix nitrogen through root nodules (Gangwar and Ramakrishnan 1989). It has been shown that *Flemingia vestita* is capable of fixing nitrogen at a rate of 250 kg/ha/yr (see Table 21-5). This compares favorably with many traditional leguminous crop species, which usually have fixation rates between 65 and 224 kg/ha/yr (Nutman 1976), occasionally rising to 500 kg/ha/yr (National Academy of Sciences 1979).

With its matty form of growth and dense crop cover, *Flemingia vestita* also checks nutrient losses during monsoonal water runoff (Mishra and Ramakrishnan 1983c). Therefore, a few years of vegetable or tuber cropping involving traditional species, with or without *Flemingia vestita,* followed by a year of cropping under this legume alone, makes a good rotation practice. With wide variations in patterns of energy and economic output and input over a small area, these systems provide opportunities for farmers to obtain higher returns through simple manipulation of the crop mixture, with minimal inputs of outside technology. However, long-term cropping in high rainfall areas, using either rotational fallow or permanent cultivation systems, often leads to rapid deterioration in soil quality and, eventually, to site desertification

(Ramakrishnan 1985a,b). Traditional technologies, such as the use of *Flemingia vestita* for site quality maintenance, should be combined with appropriate soil conservation measures and organic manure usage if these land-use systems are to be sustained.

Strategy for Sustainable Development

A holistic approach needs to be adopted for sustainable development in the northeast of India, integrating agricultural, animal husbandry, and socioeconomic aspects of village life within the overall context of forest ecosystem function and management.

Over 5 to 10 years, the transfer of traditional jhum technology from one tribe to another should be considered as one of the pathways to sustainable development. There is a high degree of heterogeneity in jhum-related cropping systems, and most differences are based on ecological issues such as altitude, temperature, and rainfall, as well as on cultural diversity. Individual tribal differences are very often highly pronounced because of the insulated social evolution of these societies. We are, in fact, dealing with a wide range of jhum systems, in terms of their cropping patterns, yield patterns, and ecological and economic efficiencies.

Table 21-4. Input and Output Patterns for Two Lesser-Known Crop Species in Northeast India

Production Measure	Digitaria cruciata *var.* *esculenta*	Digitaria cruciata *var.* *esculenta plus Potatoes*	Flemingia vestita	Flemingia vestita *plus Cabbage*
Monetary (Rp/ha):				
Input total	856	6,710	4,891	5,393
Output total	3,890	15,049 (3,685)	11,016	21,404 (10,040)
Output/input ratio	4.5	2.24	2.25	3.97
Energy (MJ/ha):				
Input total	1,114	12,424	984	1,038
Output total	8,973	44,004	21,812	20,643
Output/input ratio	8.04	3.54	22.17	19.89

Note: Values in parentheses are the outputs from lesser-known crop species.
Source: Gangwar and Ramakrishnan 1989.

Table 21-5. Nitrogen Economy under *Flemingia vestita* Cultivated in Pure and Mixed Stands at Shillong in Meghalaya (kg/ha/yr)

Nitrogen Content	*Pure Stand*	*Mixed with Cabbage*
Accumulation in crop biomass:		
Shoots	19 + 0.7	16 + 0.6 (3 + 0.1)
Roots	10 + 0.5	10 + 0.3 (2 + 0.1)
Economic yield removed	31 + 2.5	23 + 1.0 (10 + 0.1)
Crop biomass recycled	29 + 1.7	26 + 1.5
Wood biomass recycled	9 + 0.6	13 + 0.9
Nitrogen fluctuation	207 + 7.2	154 + 5.5
Net gain in soil	245 + 18.3	193 + 15.6

Note: Values for cabbage are in parentheses.
Source: Gangwar and Ramakrishnan 1989.

The agroforestry component of the shifting agriculture system should be strengthened by using local species such as the Nepalese alder (*Alnus nepalensis*), and *Flemingia*-based fallow management should be encouraged where applicable. The valley and home garden components of these farming systems should also be improved by introducing appropriate scientific inputs and linking traditional with modern technologies.

It is interesting to note that Nepalese alder technology is being promoted through village development boards in Nagaland (Gokhale et al. 1985; NEPED and IIRR 1999), and that these same boards are also investigating a whole range of species to strengthen the tree component of the shortened jhum cycle. Community participation in this effort is ensured by the village development boards (VDBs), created on the basis of the value system of individual ethnic groups. With more than 35 ethnic groups involved, all VDBs have the same function: participation in fallow management efforts for a redeveloped jhum. Transfer of technology from one tribe to another could improve management of jhum, valley, and home garden ecosystems. Already, an emphasis on growing potatoes at higher altitudes and rice lower down has led to a manifold increase in economic yields, despite the low fertility of the more acid soils at higher elevations, where tree-based agroecosystems are.

An evaluation of jhum cultivation in the region, using money, energy, soil fertility, biomass productivity, biodiversity, and water quality as currencies, has found that a 10-year cycle is critical for sustainability. A minimum cycle of 10 years should be fixed for jhum, and greater emphasis should be placed on other land use systems, such as traditional valley or home garden cultivation.

Where the length of the jhum cycle cannot be extended beyond five years, it should be redesigned and strengthened as an agroforestry system by incorporating ecological insights into tree architecture. For example, tree canopies should be compatible with the crop species at ground level to permit sufficient light penetration and provide fast recycling of nutrients through leaf turnover rates. Fast-growing native shrubs and trees should be introduced to speed up fallow regeneration after jhum cropping.

The nitrogen economy of jhum cultivation should be improved in both the cropping and fallow phases by introducing nitrogen-fixing species. A species such as the Nepalese alder (*Alnus nepalensis*) would be readily adopted on the basis of traditional knowledge that is being adapted to meet modern needs. In cases where tree-based agroecosystems are not possible because site quality has declined to near desertification, *Flemingia*-based fallow management systems may be appropriate.

One suggestion for a long-term objective for sustainable development in the northeastern region, is a shift toward a plantation economy, with forestry activities based on the home garden concept, and the organization of families into a cooperative production and marketing system. The aim would be to build upon traditional technology and indigenous knowledge by introducing modern scientific inputs and involving family participation in the development process.

Conclusions

Societies living close to natural resources have a rich traditional knowledge arising from centuries of experience in dealing with the environment around them. This is true for their perception of the natural forest ecosystem, with which their agroecosystem functions are very closely linked. Validating this knowledge has to be an important basis for developing improved fallow management practices to replace traditional systems that are rapidly breaking down under the impact of global change (Ramakrishnan 2001). However, building upon this traditional knowledge may require appropriate inputs from the textbook-based formal knowledge of scientists. As well, an important tool in the implementation of any fallow development plan for traditional societies must be the creation of the right type of institutional arrangements, based on a value system that they understand and appreciate, so that they can participate fully in the development process.

References

Allan, W. 1965. *The African Husbandman*. London, UK: Oliver and Boyd, 505.

Boserup, E. 1965. *The Conditions of Agricultural Growth*. Chicago: Aldine, 123.

Das, A.K., and P.S. Ramakrishnan. 1985. Litter Dynamics in Khasi Pine (*Pinus kesiya* Royle ex. Gordon.) of Northeast India. *Forest Ecology and Management* 10, 135–153.

———. 1986. Adaptive Growth Strategy of Khasi Pine (*Pinus kesiya* Royle ex. Gordon). *Proceedings Indian Academy of Sciences* (Plant Sciences) 96, 25–36.

FAO/SIDA (Food and Agriculture Organization of the United Nations/Swedish International Development Agency). 1974. *Report on Regional Seminar on Shifting Cultivation and Soil Conservation in Africa*. Rome: FAO, 248.

Faucher, D. 1949. *Geographie Agraire*. Paris: Genin.

Gangwar, A.K., and P.S. Ramakrishnan. 1987. Cropping and Yield Patterns under Different Land Use Systems of the Khasis at Higher Elevations of Meghalaya in Northeast India. *International Journal of Ecology and Environmental Science* 13, 73–86.

———. 1989. Cultivation and Use of Lesser-Known Plants of Food Value by Tribals of Northeast India. *Agriculture, Ecosystem, and Environment* 25, 253–267.

Gokhale, A.M., D.K. Zeliang, R. Kevichusa, and T. Angami. 1985. *Nagaland: The Use of Alder Trees*. Kohima, Nagaland: Education Department, 26.

Kushwaha, S.P.S., and P.S. Ramakrishnan. 1987. An Analysis of Some Agro-Ecosystem Types of Northeast India. *Proceedings Indian National Science Academy* B53, 161–168.

Maikhuri, R.K., and P.S. Ramakrishnan. 1990. Ecological Analysis of a Cluster of Villages Emphasizing Land Use of Different Tribes in Meghalaya in Northeast India. *Agriculture Ecosystem and Environment* 31, 17–37.

———. 1991. Comparative Analysis of the Village Ecosystem Function of Different Tribes Living in the Same Area in Arunachal Pradesh in Northeast India. *Agricultural Systems* 35, 292–299.

Mishra, B.K., and P.S. Ramakrishnan. 1981. The Economic Yield and Energy Efficiency of Hill Agro-Ecosystems at Higher Elevations of Meghalaya in Northeast India. *Acta Oecologica-Oecologia Applicata* 2, 369–389.

———. 1983a. Secondary Succession Subsequent to Slash and Burn Agriculture at Higher Elevations of Northeast India. I. Species Diversity, Biomass and Litter Production. *Acta Oecologica-Oecologia Applicata* 4, 95–107.

———. 1983b. Secondary Succession Subsequent to Slash and Burn Agriculture at Higher Elevations of Northeast India. II. Nutrient Cycling. *Acta Oecologica-Oecologia Applicata* 4, 237–245.

———. 1983c. Slash and Burn Agriculture at Higher Elevations in Northeast India. I. Sediment, Water, and Nutrient Losses. *Agriculture Ecosystem and Environment* 9, 69–82.

———. 1983d. Slash and Burn Agriculture at Higher Elevations in Northeast India. II. Soil Fertility Changes. *Agriculture Ecosystem and Environment* 9, 83–96.

———. 1984. Nitrogen Budget under Rotational Bush Fallow Agriculture (Jhum) at Higher Elevations of Meghalaya in Northeast India. *Plant and Soil* 81, 37–46.

National Academy of Sciences. 1979. *Tropical Legumes: Resources for the Future*. National Academy of Sciences, Washington, DC. 331 pp.

NEPED and IIRR (Nagaland Environmental Protection and Economic Development and International Institute of Rural Reconstruction). 1999. *Building Upon Traditional Agriculture in Nagaland*. Nagaland, India, and the Philippines.

Nutman, P.S. (Ed.). 1976. Symbiotic Nitrogen Fixation in Plants. Cambridge University Press, Cambridge.

Nye, P.H., and D.J. Greenland. 1960. *The Soil under Shifting Cultivation*. Technical Communication No. 51. Harpenden, England: Commonwealth Bureau of Soils, 156.

Okafor, F.C. 1987. Population Pressure and Land Resource Depletion in Southeastern Nigeria. *Applied Geography* 7, 242–256.

Ramakrishnan, P.S. 1983. Socio-Economic and Cultural Aspects of Jhum in the Northeast and Options for Eco-Development of Tribal Areas. In: *Tribal Techniques, Social Organisation and Development: Disruption and Alternates*, edited by N.P. Chaubey. Allahabad: Indian Academy of Social Sciences, 12–30.

———. 1984. The Science behind Rotational Bush Fallow Agriculture Systems (Jhum). *Proceedings Indian Academy of Sciences (Plant Sciences)* 93, 397–400.

———. 1985a. Humid Tropical Forests. In: *Research on Humid Tropical Forests*. Regional meeting, National MAB Committee of Central and South Asian countries. New Delhi: Man and Biosphere India, Ministry for Environment and Forests, 39.

———. 1985b. Tribal Man in the Humid Tropics of the Northeast. *Man in India* 65, 1–32.

———. 1992. Shifting Agriculture and Sustainable Development: An Interdisciplinary Study from Northeast India. UNESCO-MAB Series, Paris, Carnforth, Lancs, UK: Parthenon Publishers, 424 (Republished, New Delhi 1993: Oxford University Press.)

———. 2001. *Ecology and Sustainable Development*. New Delhi: National Book Trust, India.

Ramakrishnan, P.S., and A.K. Das. 1983. Studies on Pine Ecosystem Function in Meghalaya. *Tropical Plant. Science Research* 1, 15–24.

Saxena, K.G., and P.S. Ramakrishnan. 1984. Growth and Patterns of Resource Allocation in *Eupatorium odoratum L.* in the Secondary Successional Environments following Slash and Burn Agriculture (Jhum). *Weed Research* 24, 127–134.

Sharma, E., and R.S. Ambasht. 1988. Nitrogen Accretion and its Energetic in the Himalayan Alder. *Functional Ecology* 2, 229–235.

Swamy, P.S., and P.S. Ramakrishnan. 1988. Nutrient Budget under Slash and Burn Agriculture (Jhum) with Different Weeding Regimes in Northeast India. *Acta Oecologica-Oecologia Applicata* 9, 85–102.

Toky, O.P., and P.S. Ramakrishnan. 1981. Cropping and Yields in Agricultural Systems of the Northeastern Hill Region of India. *Agro-Ecosystems*, 7, 11–25.

Chapter 22

Benefits of *Phaseolus calcaratus* in Upland Farming in Northern Vietnam

Nguyen Tuan Hao, Ha Van Huy, Huynh Duc Nhan, and Nguyen Thi Thanh Thuy

The degradation of forest land in northern Vietnam is a major environmental problem. It influences the lives of millions of people who live in the mountainous areas of the country, affecting their food and fuel supplies and their access to valuable forest resources. Shifting cultivation is one of the main causes of this forest degradation because pressure from an increasing population has forced the shortening of the fallow period. It is now too short to restore soil fertility for future cropping cycles and on degraded soils, crop productivity tumbles even more quickly.

The government of Vietnam has formulated policies to address these environmental problems. The challenge is to find the means to make agricultural productivity sustainable, so that the environment can be stabilized. As a first step, land is being allocated and land-use certificates issued to households. But the change to fixed, intensive cultivation has brought many problems. In some cases, fields are far from houses, or farmers lack funds to develop new land. Permanent paddy land for growing rice is very scarce, and in some areas there is none at all. Irrigation is difficult and dependent on the natural contours of the land. Much of it is on steep slopes with rocky soils, and food is produced on little more than a subsistence level, with small surpluses sold for the purchase of other goods. Efforts to stabilize farming have not stopped shifting cultivation. The farmers themselves have been particularly innovative in their efforts to conserve the fertility of their soil while intensifying their production. The intercropping of *nho nhe* bean, otherwise known as rice bean (*Phaseolus calcaratus* Roxb., syn. *Vigna umbellata*), with other crops is one indigenous technology aimed at intensifying shifting cultivation.

This study investigates the ecology and growth characteristics of the leguminous nho nhe bean, as well as farmers' appraisals of its use as an intercrop. This information is then considered in the context of land-use intensification, improvement of shifting cultivation, and sustainable land management. This chapter also reviews farmer experience with nho nhe bean under different social and environmental conditions, identifies advantages and constraints, and establishes research priorities for the future.

Nguyen Tuan Hao, Ha Van Huy, and Huynh Duc Nhan, Forest Research Center, Bai Bang, Phu Ninh District, Phu Tho Province, Vietnam; Nguyen Thi Thanh Thuy, World Neighbors, Alley 202B, Doi Can No. 1B, Ba Dinh, Hanoi, Vietnam.

Ultimately, this study seeks to determine whether, by using nho nhe beans intercropped with staple crops, farmers have adequately addressed the problems of erosion control and soil conservation, and are better managing their sloping land to permit intensified but sustainable cultivation.

Research Methods and Study Areas

This research covered different environmental conditions and farming systems. Its main thrust was to investigate the following issues:

- The ecological characteristics and methods of cultivation and management of nho nhe beans;
- Farmers' assessments of the productivity and nutrient values of nho nhe bean, and the labor input required to grow it; and
- Farmers' assessments of the bean's ability to provide soil cover, soil conservation, and soil improvement.

The information in this chapter is, therefore, based on farmer interviews. It develops a framework for future research by analyzing the different technical innovations that have evolved in different farming systems, including the use of cover crops, soil improvement species, and soil fertility management. Surveys were conducted in three districts and provinces, selected on the basis of a combination of factors, including terrain, climate, vegetation, and farming practices.

The main methods used were rapid rural appraisal (RRA) and participatory rural appraisal (PRA). The study team contacted the heads of villages and the staffs of other projects in the study areas to acquire basic information on socioeconomic conditions. Householders interviewed included those with and without experience in growing nho nhe beans. The interviews were structured in combination with a field visit, and the farmers were asked to assess the benefits of nho nhe bean for soil cover, soil improvement, and impact on the cultivation of grain crops. They were also asked to compare the bean's economic value with major crops such as corn, cassava, other beans, and arrowroot. The scoring method was simple: poor, 1; average, 2; and good, 3.

The surveys were carried out at the end of the year, when all crops had been harvested. Only corn stalks, dried vines, and bean leaves were left in the fields. In some places we observed fallen bean seeds that had germinated. In others, farmers were preparing fields. It was a time of year when farmers had free time to discuss the issues of nho nhe bean cultivation. We collected full information on investment, income, and constraints that affected crop productivity, as well as farmers' assessments of these. The data fell into three categories:

- Household economic status;
- Main crops and farming systems; and
- Experiences in cultivation, soil cover, and conservation.

Site Selection

Three sites were chosen for the study: the districts of Da Bac, in Hoa Binh Province; Yen Chau, in Son La Province; and Tua Chua, in Lai Chau Province. These districts lie along the watershed of the Da River in northwest Vietnam (see Figure 22-1). They are populated mainly by three ethnic minority peoples: M-êng, Thái, and Hmong, respectively. The terrain is characterized by elevations between 1,000 and 2,000 m, steep slopes, and poor or rocky soils. Vegetation is predominantly shrubs, bushes, and degraded forests. The climate is cold in winter and hot in summer because of the hot, dry "Lao wind." Rainfall is concentrated from May to July (see Figure 22-2). The three study areas have different farming practices, each of which is suited to the local situation. Table 22-1 gives details of the socioeconomic status and land use at the three sites.

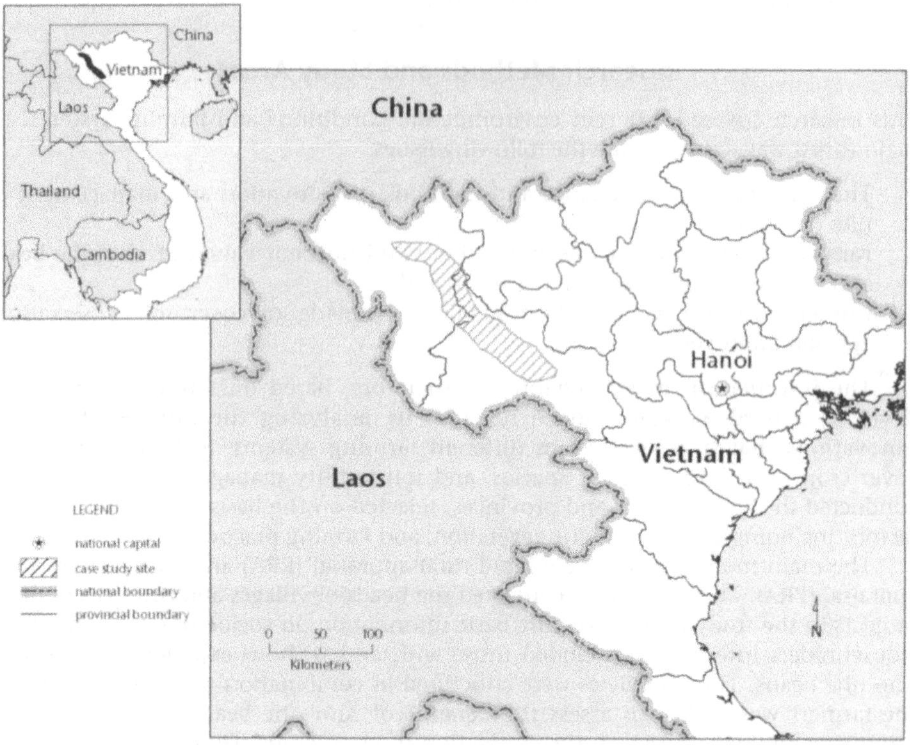

Figure 22-1. The Study Site in Vietnam

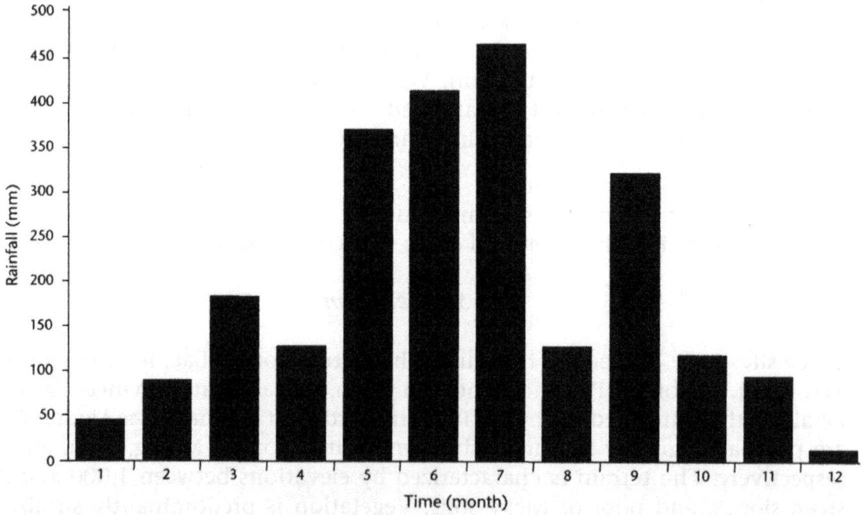

Figure 22-2. Mean Monthly Rainfall in Hoa Binh Province, 1990–1992

Table 22-1. Socioeconomic Status and Land Use in Survey Area

Details	*Doi Village, Da Bac District*	*Chieng Dong Commune, Yen Chau District*	*Sinh Phinh Commune, Tua Chua District*
Village names	Ke, Doi	Luong Me	Sinh Phinh Commune
Ethnic groups	M-êng	Thái	Hmong
Soil types	Sloping land, poor soil	Sloping and paddy land	Rocky, steep slopes
Land types (% of total):			
Paddy	–	1.5	5
Upland field	97	8.8	40.8
Forest land	3	89.7	54.2
Land tenure and management	Land allocated, fixed cultivation	Land allocated, fixed cultivation	Shifting cultivation
Primary crops	Upland rice, cassava	Paddy rice, cassava, corn	Corn
Prevalence of nho nhe bean	Some intercropping	Some intercropping	Commonly intercropped
Households classified as poor (% of total)	37	10	60

Da Bac District in Hoa Binh Province. Da Bac is a mountainous district in Hoa Binh Province that is representative of the midlands of Vietnam. People are mainly of the Muòng ethnic minority group. Acceptance of new technologies is high. The sites for the survey were the villages of Doi and Ke, in the Hien Luong commune. Doi village has 65 households with 268 people and an area of about 500 ha. The major economic activity is cultivation of agricultural crops on sloping lands and animal husbandry. It is accessible by road. Ke village has 61 households with 312 people and an area of about 341 ha. Once again, the major economic activity is cultivation of agricultural crops on sloping lands, but some households are involved in fishing, firewood collection, and animal husbandry. The village is on the same road as Ke.

The people of both villages have been affected by rising water levels in the reservoir of the recently dammed Da River, and have been forced to relocate their communities higher on the surrounding slopes. They have no established paddy fields and are completely dependent on sloping land for agriculture. Their major food crops are cassava, upland rice, corn, and arrowroot and they gather forest products such as *Melia*, bamboo, and *Acacia* spp. The soil type in both villages is yellow feralitic derived from shale. It is infertile and acidic, and provides a very poor basis from which to intensify cropping. Vegetation consists of shrubs and bush.

Yen Chau District in Son La Province. Yen Chau district in Son La Province is located on Highway Number Six. The Luong Me cooperative of Chieng Dong commune was selected for this study. (A commune is the administrative division between a village, or cooperative, and a district). Most people are Thái, with a small percentage of Kinh and Hmong. All households have established paddy fields in addition to upland fields. The per capita land area is 200 m^2 of paddy land and 1,500 m^2 of sloping land. The commune has, altogether, 121 ha of paddy land and 798 ha of sloping land. The main crops are rice, corn, and cassava. As well as providing for household consumption, farmers also produce surpluses for sale. The soil is gray to yellow feralitic, derived from shale. Vegetation cover is degraded forest. The commune population is 5,993 people, and 78%, 12%, and 10% of households are considered rich, middle income, and poor, respectively. Living conditions are rather high, and transportation is developing in the area.

Tua Chua District in Lai Chau Province. In this area, the focus for our study was Sinh Phinh commune. Sinh Phinh has a population of 653 people divided into 93 households, predominantly of the Hmong ethnic group. The area is rocky and mountainous, and most of the land area is devoted to cultivating corn on steep, rocky slopes. Shifting cultivation is a common practice. Corn is commonly intercropped with nho nhe bean or soybean. The soil is reddish-yellow feralitic, derived from limestone, and is quite fertile. The commune has 37 ha of paddy land, 67 ha for growing upland corn, and 130 ha of forest land. Living conditions are poor, and only 37 of the 93 households in the commune are able to produce enough food to meet subsistence needs.

Results

Traditional Practices

Conditions in the study areas are difficult, and living standards are below those in lowland delta regions. Household income depends mainly on upland agriculture and forest products. The primary concern of farmers in these areas is food security. Transportation is difficult, marketing is very unreliable, and agriculture is primarily for subsistence. Swidden agriculture, or shifting cultivation, is the traditional farming practice. When the mountainous areas still had an extensive forest cover, agricultural productivity was high, population density was low, and food was more easily obtainable. Swidden rotations involved about two to three years of cultivation followed by 10 years of fallow. This was sufficient to restore soil fertility. More recently, population pressure and the development of agriculture and forestry enterprises have forced shifting cultivation cycles to become shorter. The current ratio is two to three years of cultivation to only four or five years of fallow. Soil is often prepared by slashing vegetation during the dry season and burning it to prepare the field. On sloping lands this practice contributes to serious erosion, and soil fertility cannot be restored if topsoil is severely eroded.

The government of Vietnam has formulated many policies to provide support for farmers who choose to stabilize swidden agriculture and increase productivity on upland fields. But success has been rare because of an inability, thus far, to solve the issue of food security.

In recent years land has become available for allocation, and land-use certificates have been issued. In these new arrangements, people have become conscious of the needs of the land. Unable to simply move or expand their land area in the style of swidden cultivation, they've been forced to adopt improved management practices to stabilize their production and make it sustainable. In some places, farmers have even begun to use chemical fertilizers.

The Bean Species Used

Low-Growing Nho Nhe Bean. The local name for this plant is *Tau mang*. Its botanical name has not been determined. It is an herbaceous plant that grows to a height of 40 to 60 cm. The leaves have hair on both sides like mung beans (*Vigna aureus* Rock), and the seeds are small and dark green in color. The seeds are used as a vegetable and the leaves as green manure. Sowing is done in June or July for harvest in November or December. The seeds are traditionally used in cakes and soups.

Climbing Nho Nhe Bean. The scientific name is *Vigna umbellata,* syn. *Phaseolus calcaratus* Roxb. This is a climbing, fast-growing viny legume. The root system has a main taproot that grows deep into the soil. The faces of the leaf are hairy. There are different kinds of climbing nho nhe bean, which are differentiated locally by seed colors: white, yellow, black, and violet. People prefer the white-seeded variety, but the yellow variety is more commonly cultivated because it develops more rapidly and is more productive. The variety with violet-colored seeds also has a violet color at the base of the stems. The flowers are yellow.

Climbing nho nhe bean can be intercropped with corn or cassava. Planting is done in February, March, or May, and again in July or August, but to achieve optimum biomass production, the recommended planting time is February or March. Harvesting occurs in November and December. The seeds are used as food and the leaves as green manure or animal fodder. The plants will produce 20 to 25 metric tonnes of biomass per hectare.

Nho nhe bean is affected by water availability. In times of rain, the plant flowers. In times of drought, the crop will fail. The leaves, fruit, and seeds of nho nhe bean may also be damaged by insects Also, plant lice eat the leaves, but damage is not great.

Te Bean. The local name for this plant is *Tau.* Its scientific name is *Glycine max.* It has short bushes like the mung bean. The seeds are green in color. Planting time is June to July, with harvest in November to December. The seeds are considered good to eat.

Soybean. The local name is *Tau pau.* The scientific name has not yet been determined. It is grown on sites where corn has been harvested. Planting time is June to July, with harvesting in September or October. Productivity is high, and the seeds are very good for eating.

Integration into Farming Systems

Farming practices at the study sites can be classified into eight main categories. Five of them involve integration of nho nhe beans. The main systems are as follows:

- Nho nhe beans planted on the perimeter of cassava fields;
- Nho nhe beans planted on the perimeter of maize fields;
- Mixed planting of corn, cassava, and nho nhe beans;
- Nho nhe beans intercropped with cassava, on poor soil;
- Soybeans intercropped with maize;
- Nho nhe beans intercropped with corn;
- Corn intercropped with arrowroot; and
- Upland rice monoculture.

In Da Bac, cultivated land is scarce and therefore land use is more intensified than at the other sites. Farmers often grow two crops per year on good soil, and all fields of corn and cassava have intercropped nho nhe bean. Nho nhe bean may also be planted along the borders of fields. In Tua Chua, farmers grow corn intercropped with nho nhe bean or other bean species, depending on the soil conditions.

Each ethnic group has special farming practices that are suited to local conditions and that illustrate their years of experience in cultivating the nho nhe bean. In this chapter we describe only some of the more common practices (see Figure 22-3).

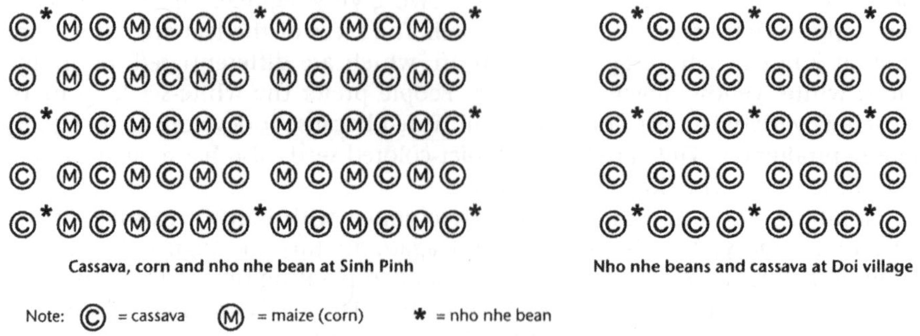

Figure 22-3. Crop Arrangements of Cassava, Corn, and Nho Nhe Bean

Intercropping Nho Nhe Bean with Corn and Cassava

Cassava and corn are sown in February or March. Cassava is planted at 80 x 80 cm spacing, and corn is sown between the cassava plants. When the corn reaches a height of 20 cm, nho nhe beans are sown beside every second or third cassava plant. That is one nho nhe bean about every 2 to 3 meters. As they grow, the nho nhe beans are tended together with the corn and cassava. The first weeding is in February and the second in April. When the nho nhe beans have climbed up the corn and cassava stems to a height of 1 to 1.5 m, the tops are cut to stimulate branching so that the plant develops more fruit. In July, after five months, the corn is harvested, and the stalks are left to support the beans, which spread to cover the ground during the rainy season. The nho nhe beans flower in August and September, and by the end of October or November, they are ready to harvest. The residue from both crops is left on the fields as fertilizer for the following season.

The cassava is harvested in November to December and, after the weeds are cleared, the beans will germinate naturally in the field. By observing their fields, the farmers know when the nho nhe seeds have germinated under the ground. When they sprout, the first set of leaves of the seedling beans remains underground and does not appear above ground like other bean species. The young nho nhe beans are thus protected from birds and animals at their most vulnerable time, making them most suitable for forest environments. According to the estimate of one farmer in Tua Chua, a field of corn and cassava that is intercropped with nho nhe bean can produce 5% to 10% more cassava than a cassava monoculture.

In Ke village in Da Bac, farmers plant nho nhe bean into their cornfields in May, after the first weeding. Beans are sown every 2.5 to 3 m and are harvested in November. During corn harvest, one management option is to bend the corn stalks down for the beans to climb up. Farmers say flowering and fruiting increase with this method. Where corn is intercropped with cassava, without beans, the corn stalks are cut after harvest in order to let the cassava develop.

Intercropping Nho Nhe Bean with Corn on Rocky Land

At Tua Chua, the fields for growing corn are small and situated on rocky mountain slopes. The terrain is high and steep, but the limestone-derived soils are more fertile than those of Da Bac. Corn is planted in February or March and harvested in November. In this area, intercropping with nho nhe beans is the preferred cropping method because farmers say fields are productive over a longer period, with more cropping cycles, and with a slower loss of fertility. The farmers have also recognized that bean productivity will be higher if the seeds are deliberately sown, rather than if farmers depend upon natural regeneration. So the seeds are selected. Before planting and after slashing and burning, the soil is plowed. Normally, this is in January or

February, after the beginning of the rains. The corn is then sown with 80 cm x 80 cm spacing, so that it needs no thinning, and the beans are sown at the same time, with four or five seeds in each hole. In July, the fields of corn and beans are weeded. In September or October, the corn is harvested and the stalks are cut (see Figure 22-4). The beans are harvested in November or December by cutting or uprooting the plants, allowing them to dry, then shelling the seeds in the field. It is interesting to note that only one weeding is needed in a corn and nho nhe bean intercrop, whereas a corn monoculture requires two weedings.

Figure 22-5 shows a cropping calendar covering various farming systems involving nho nhe beans. Experience in Tua Chua shows that if nho nhe beans are grown in very good soil, they will grow very quickly but produce little fruit. Farmers say that newly opened corn or cassava fields in fertile soil must be cultivated for five or six years before the soil fertility declines to the point where nho nhe beans will provide a good harvest. Only when yields of corn or cassava begin to decline are nho nhe beans introduced as an intercrop to stabilize the productivity of the field. Conversely, nho nhe beans will not succeed in soil that is too infertile and has little organic matter.

Intercropping with Cassava

Farmers grow nho nhe bean with cassava on poor soil. This system is followed at Da Bac, where the soil in upland fields is poor. The beans are planted at 2 to 3 meter spacing, or one bean plant for every two or three cassava plants (see Figure 22-3). Alternatively, they may be planted around the field borders to avoid competition with the main crop. When the beans climb up, the shoots are cut to prevent them from covering the cassava. After harvesting, bean stumps, or "coppices," may be protected in the field for up to two years so that they can regenerate to cover the soil, create green manure, and smother weeds.

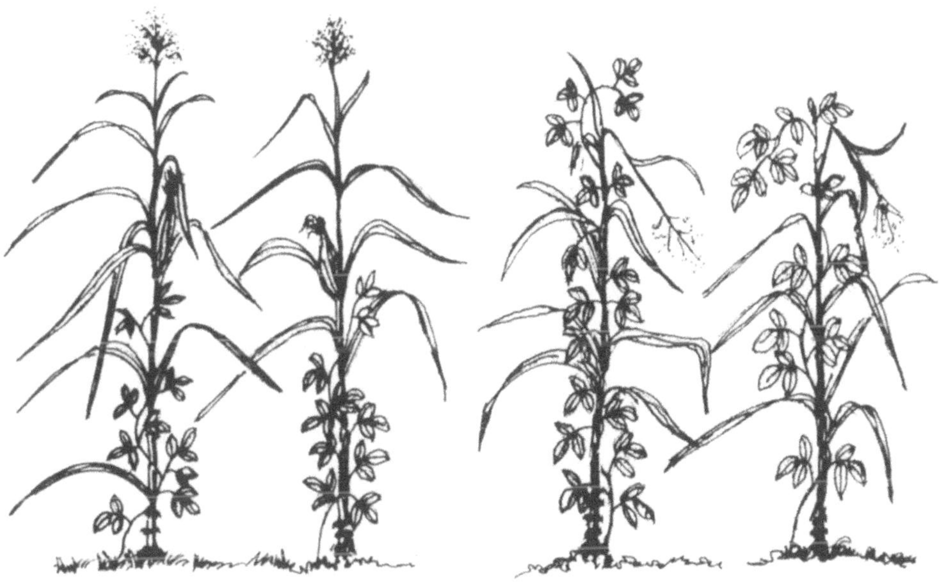

Figure 22-4. Intercropping Nho Nhe Beans with Corn

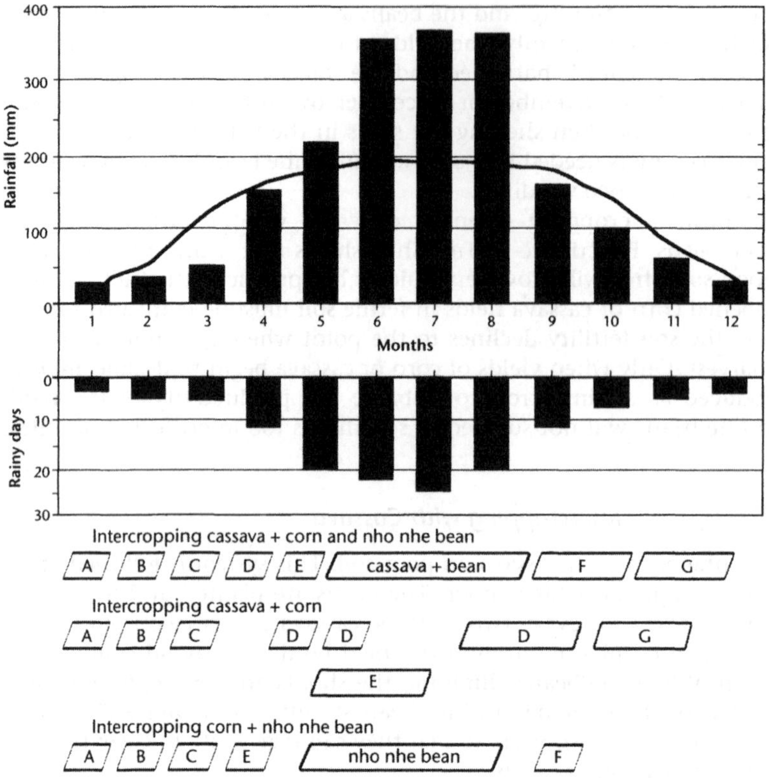

Figure 22-5. Calendar of Nho Nhe Bean Cultivation in Various Farming Systems

Notes: A, Soil preparation; B, Planting; C, Cultivating corn; D, Cultivating cassava; E, Corn harvesting; F, Bean harvesting; G, Cassava harvesting.

Use of Nho Nhe Bean to Eradicate Imperata Cylindrica

Farmers at Da Bac cultivate nho nhe bean to eradicate *Imperata*. This grass species has underground rhizomes that develop very quickly and compete with crops, and fields dominated by *Imperata* are very difficult to rehabilitate. However, *Imperata* requires open space and direct sunlight and is intolerant of shade. Therefore, farmers take advantage of one of the basic characteristics of the nho nhe bean, its ability to grow rapidly and produce a heavy, thick canopy. In the dry season, the *Imperata* is burned and nho nhe bean sown. After two crops of beans, over two years, the *Imperata* is damaged and weakened and can be eradicated with some additional hand weeding. If the grass is too thick and difficult to burn, nho nhe bean is planted into cleared spaces, as well as around the perimeter of the field. The beans climb and cover the weeds. The beans are planted repeatedly in successive seasons, each year gradually reducing the area of *Imperata* grass. By using nho nhe bean as a biological tool to smother out *Imperata*, it thus saves farmers a lot of labor in hand weeding that would otherwise be necessary to rid their fields of this noxious weed.

Discussion

Economic Advantages of Growing Nho Nhe Bean

Nho nhe bean is very widely distributed in northern Vietnam. It is intercropped with both corn and cassava and is planted in home gardens. The plants are easy to grow, they exhibit reliable production on many different soil types, and they are not very susceptible to pest damage. When intercropped on the same plot with staple crops, nho nhe bean diversifies a farmer's production, reduces the need for weeding, and, according to one farmer's estimate, it can increase corn production by as much as 10%. In the course of this study, we saw none of these beans growing in monoculture.

Each vine of nho nhe bean is capable of producing up to 5 kg of seeds, although average yields are 3 to 4 kg. The fruits ripen uniformly, and the seeds have a high germination rate and will sprout and grow very quickly. This makes it a good cover crop that smothers weeds. It also has a strong tendency for natural regeneration.

Farmers believe that if cassava is cropped on its own, yields will decrease by as much as 70% after 10 years due to natural loss of soil fertility. However, if it is intercropped with nho nhe bean and corn, the overall productivity of a field can be sustained (see Table 22-2).

In Tua Chua, farmers have assessed the productivity of corn monoculture and say it is lower than that of intercropped corn and nho nhe beans (see Table 22-3). However, if the beans are not properly managed, corn yields can suffer by up to 30%. This is because local people have the habit of leaving corncobs in the field until they are dry, an inappropriate practice in the midst of a vigorous crop of beans. Nevertheless, farmers are still attracted to intercropping with nho nhe bean because, as well as increasing the output from a field, it reduces the need for labor. This is due to the fact that nho nhe beans smother weeds and that soil preparation for a corn and nho nhe intercrop is less than that for a corn monoculture. Details of labor inputs appear in Table 22-4.

Farmers were asked to assess three commonly grown beans—climbing nho nhe bean, te bean, and soybean—for their productivity, economic value, and labor requirements. The result, shown in Table 22-5, reveals that the economic value of the nho nhe bean is lower than both the soybean and the te bean because the price of its seeds is lower. However, in overall value, nho nhe bean is ranked higher than the other two because the nho nhe bean requires the lowest investment of labor and is more productive. An additional advantage is that it can be grown on different soil types. It not only reduces the labor requirement for planting, but also has the ability to eradicate weeds, both of which are great advantages. Nho nhe bean was also judged the best cover crop.

Environmental Advantages

Over the past decade, as land in northern Vietnam has been allocated to households for their management, the area of land available for rotating fallow systems has become limited. Development efforts have been directed toward the difficult task of establishing permanent cultivation while maintaining soil fertility and stable production. Farmers in our study area recommend the use of nho nhe beans to solve the problems associated with permanent cultivation and soil fertility. They say fields of intercropped cassava, corn, and nho nhe bean can be cultivated for many years, even without fertilizer, and productivity is not reduced.

Table 22-2. Yield Comparison, Cassava Monoculture and Cassava Intercropped with Corn and Nho Nhe Beans

Crop	*Yield (kg/ha/yr)*		
	1994	*1995*	*1996*
Cassava monoculture	700	640	600
Intercrop:			
Cassava	700	600	> 500
Nho nhe bean	200	180	200
Corn	> 200	180	190

Note: Plot size 200 m^2.
Source: Farmer estimates in Tua Chua.

Table 22-3. Inputs and Outputs for Corn Monoculture, Compared with Corn Intercropped with Nho Nhe Beans

Crop	*Seed (kg/ha)*		*Labor (days/ha)*	*Yield (kg/ha)**	
	Corn	*Bean*		*Corn*	*Bean*
Corn monoculture	20		450	180	
Intercropping corn with bean	20	20	350	300	95

Notes: Inputs for one sao (360 m^2) of nho nhe bean and corn in Tua Chua. *Yields were reported by Mr. Sinh A. Giay in Tua Chua from 1 ha of poor soil. Weather was not suitable for the development of nho nhe fruits. Leaf growth was good, but yield was quite low.

Table 22-4. Labor Days Required for Growing Corn Monoculture, Compared with Corn Intercropped with Nho Nhe Beans

Activity	*Corn monoculture*	*Intercropped Corn and Beans*
Soil preparation	3	2
Planting	1	1
Tending	3	2
Harvesting	2	2
Total	9	7

Note: Plot size 200 m^2.

Table 22-5. Comparison of Nho Nhe Bean with Other Bean Species

Parameter	*Nho Nhe Bean*	*Soybean*	*Te Bean*
Productivity	3	2	1
Economic value	1	3	2
Labor inputs	3	1	2
Total score	7	6	5

Note: Scoring method: Poor = 1; Medium = 2; Good = 3.

The characteristics of nho nhe bean, including fast growth, a thick cover, and a widespread root system, all contribute to its ability to cover soil and control erosion. The plant's root system can reach up to two or three meters underground, so that the nho nhe bean has good soil-holding properties (see color plate 26). It is often grown on steeply sloping land because of this ability. It is also probable that, because of the extensive root system, fields with nho nhe bean have improved soil capillary action. After harvest, the whole plant is left in the field. A large amount of biomass is thus returned to the soil, improving soil fertility.

At Tua Chua, farmers report the use of nho nhe beans for land rehabilitation. They say that when nho nhe beans are grown for green manure on poor soil for one to two years, the quality of the soil improves to the point that it can be returned to cultivation. They also claim that if a field has been supporting corn monoculture for many years, its productivity will increase if the corn is intercropped with nho nhe bean and, with continued intercropping, the yield will stabilize. Moreover, crops of upland rice show similar increases in productivity.

Farmers were asked to assess the value of various crops to the environment. They evaluated soil-covering abilities, soil improvement, and production of biomass. Nho nhe bean received a total value equal to that of soybean, at the top of the list. Although the perceived environmental benefits require further rigorous study based on soil tests and the monitoring of soil conditions, the following list summarizes the benefits of nho nhe bean in the eyes of farmers:

- It produces biomass for soil organic matter.
- It has soil rehabilitation capabilities.
- It can eradicate *Imperata cylindrica*.
- It provides soil cover during the rainy season.
- Its roots have a strong capacity to hold soil, even on steeply sloping land.
- Intercropping can delay the need for fallow and prolong the time for which an upland field is productive.
- It possibly increases available nitrogen in the soil because of its nitrogen-fixing ability.

Nho Nhe Beans as Food

Nho nhe beans are eaten in soup, sticky-rice cakes, or processed bean cakes. The leaves are used to feed buffaloes, goats, and horses. At Yen Chau, the Thái people consume the flowers of the nho nhe bean as a vegetable.

Of the four varieties differentiated by seed color—white, yellow, black, and violet—the violet seeds are considered the best quality for food. As mentioned before, yellow seeds are the most commonly grown, because it is believed that they develop more rapidly than the others. In practice, the four types are grown together, and they are very difficult to distinguish in the field because they have the same leaves and yellow flowers, although the violet variety has violet stems.

According to the Hmong people of Sinh Phinh commune, climbing nho nhe beans do not taste as good as soybeans and te beans. Nevertheless, they are commonly grown because of their adaptability and productivity.

Farmers' Acceptance

Farmers in northern Vietnam have been growing nho nhe beans for a long time. In the study areas, the percentage of households currently growing nho nhe beans is Sinh Phinh commune, Tua Chua, 99%; Ke village, Da Bac, 67%; and Doi village, Da Bac, 31%. Overall, the percentage of different bean species planted at the three study sites is nho nhe bean, 5%; black bean, 5%; green bean, 50%; soybean, 40%.

The level of acceptance is not the same in all locations because of differing circumstances and farming practices. In Doi village, for example, the number of households growing nho nhe bean is low because the village is near a road and there is marketing potential for higher-value crops, such as green beans and ginger. In both

Doi and Ke villages at Da Bac, a large number of farmers who plant nho nhe beans only do so along the borders of fields or in their home gardens. At Sinh Phinh commune the number of households growing nho nhe beans is the highest of the survey sites because most of the cultivated land is on steep, rocky mountain slopes, and nho nhe bean is used for soil improvement and maintenance of corn yields.

At Yen Chau, the Thái people do not grow much nho nhe bean because almost all of their fields are a considerable distance from their houses. In addition, cultivated land is readily available to them, so intensification is not considered necessary.

Advantages and Constraints

Advantages

As shown, the nho nhe bean has roles in soil conservation, soil improvement, and environmental sustainability. The plants are easy to cultivate on sloping lands and in poor soil. On good soil, they develop very rapidly and produce heavy masses of leaf matter. Cultivation techniques for nho nhe beans are simple, investment is low, and seed sources are readily available. Nho nhe beans ripen uniformly across a field, so harvesting is rapid. The beans are also easy to store and are less susceptible to weevils than other beans.

Disadvantages

Nho nhe beans are difficult to grow in acid or badly depleted soils. Good harvests are dependent upon the weather, because heavy rainfalls when the plants are flowering will damage the fruit. On the other hand, drought leads to poorly filled pods and reduced yield. Cold weather, such as is common in mountainous areas of northern Vietnam, slows the growth of nho nhe beans, and frost kills them.

Nho nhe beans are also damaged by some insects, although farmers in the study area do not use pesticides because they say pesticides damage the fruit. Fireflies are commonly found on the plants, and they eat the leaves. Another unidentified insect, yellow in color with black spots, appears at the end of July and early August, especially on the flowers, and can lead to crop failure. We hope that the problems of insect damage will be a future research priority.

Although nho nhe beans have many economic advantages, their market price is much lower than other beans, so production is mainly for household consumption. This may restrict large-scale development in the future.

Future Research Priorities

Under the government of Vietnam's policy on land use, the opportunities for shifting cultivation will soon be limited. People will be allocated land and will receive land certificates. As demand for food increases, agricultural intensification will have to be balanced with soil conservation measures if crop productivity is to be maintained. The practices described in this paper suggest that nho nhe beans will become one strategy for sustainable production.

Cultural practices and methods of pest control for nho nhe beans need to be developed. More knowledge is needed, for example, on planting times and the cutting of the shoots.

In order to help farmers to intensify production without negative impacts on the environment, a wide-ranging research program is needed. With respect to the potential of nho nhe beans alone, we need to continue to research the following topics:

- Integration of nho nhe beans into varied upland farming systems with the aim of conserving, rehabilitating, and improving soil fertility on sloping land for sustainable agriculture;
- Growing techniques to develop large-scale production of nho nhe beans;

- The use of nho nhe beans in agroforestry systems, for instance, as soil cover for orchards and plantations;
- The interactions between nho nhe beans and other major crops in different farming systems;
- Techniques for intensive nho nhe bean cultivation; and
- Use of nho nhe beans in fallow management and weed eradication, especially *Imperata cylindrica*, on cultivated land.

Conclusions

Nho nhe bean is a nitrogen-fixing, leguminous crop with wide distribution. It is commonly grown in upland areas of northern Vietnam, where it is used for green manure, soil cover, and soil improvement. It has many environmental and economic advantages, including its ability to eradicate *Imperata*.

Compared with other bean crops, nho nhe beans have distinct advantages in terms of soil cover and soil improvement. This plant needs further research to develop its full uses, especially in the intensification of shifting cultivation and fallow management. We need to consider the important role nho nhe beans can play in crop mulching, erosion control, and soil and water conservation. The results of this initial study should be regarded as a starting point for future research.

Acknowledgments

We gratefully acknowledge the assistance and insights provided by the following people and organizations, without whose help this study would have been difficult to complete: Huynh Duc Nhan, of the Forest Research Center at Phu Ninh; Nguyen Quoc Tho, from the GTZ-sponsored Song Da Social Forestry Project in Tua Chua; and Nguyen Hong Khanh, of Oxfam Belgium, in Da Bac.

Particular thanks are due to the Cornell International Institute of Food, Agriculture and Development (CIIFAD) for providing financial support for the survey through its group for Management of Organic Inputs in Soils of the Tropics (MOIST); ICRAF's Southeast Asia regional research program; and World Neighbors in Vietnam for their support and assistance.

Thanks also to the Cornell University Seminar on Indigenous Strategies for Intensification of Shifting Cultivation in Southeast Asia; workshop coordinator Malcolm Cairns; the area representative of World Neighbors, Karin Eberhardt; and Janet Durno of Oxfam Belgium, all of whom gave up so much of their time to guide us in writing and editing this paper. Despite our debt of gratitude, all errors and opinions remain our own.

References

FAO-RAPA. 1994. *Under-Exploited Legume Crops in Asia* (Review), 348–354.

Forest Research Center. 1989. *Local Farming Technologies*. (Project document). Phu Ninh, Vietnam: Forest Research Center.

Gayfer, J. 1990. *Local Farming Technologies Related to Soil Conservation and Tree Planting in Selected Districts of Vinh Phu, Ha Tuyen, Hoang Lien Son*. (Project document), 46.

Khoi, D.N. 1986. *Research on Fodder in Vietnam* (Part 1). Vietnam: Technical and Science Publisher, 122–126.

Loi, D.T. 1986. *Medicinal Plant Species and Traditional Medicines of Vietnam*. Vietnam: Technical and Science Publisher, 775–778.

Tai, N.D. 1986. *Agroforestry in Northern-Central Vietnam*. Phu Ninh, Vietnam: Forest Research Center, 60.

Thai, P. 1996. Cover Crops and Green Manure Crops on Sloping Land in Vietnam. Paper presented at a meeting of the Agroforestry Working Group, July 29, 1996, Hanoi, 11.

Chapter 23

Viny Legumes as Accelerated Seasonal Fallows

Intensifying Shifting Cultivation in Northern Thailand

Somchai Ongprasert and Klaus Prinz

Land use pressures in northern Thailand have caused the fallow period of traditional shifting cultivation to shorten. As a consequence, productivity has declined and shifting cultivation has become a hazard to the environment. Shifting cultivators have also gained improved access to transportation, communications, and markets, as well as increased attention from extension officers of both government and nongovernment organizations. They have developed both indigenous and exotic alternatives to cope with, and to benefit from, the transition.

This chapter describes an innovative and complex multiple cropping system developed by Lisu villagers at Huai Nam Rin, in Chiang Mai Province. The system consists of first intercropping corn with wax gourd (*Benincasa hispida*) or pumpkin (*Cucurbita moschata*) and then relay planting three viny legumes: cowpea (*Vigna unguiculata*), rice bean (*Vigna umbellata*), and lablab bean (*Lablab purpureus*) (see color plate 25). The system may be considered as a seasonal fallow management technique in an intensified shifting cultivation cycle that both replenishes soil fertility and generates income. It has completely replaced traditional crops of upland rice at Huai Nam Rin and is gradually being adopted by farmers in neighboring villages.

The Study Area and Methods

Huai Nam Rin village is located at a relatively low altitude of 450 to 500 meters above sea level (m asl), on undulating and fertile uplands derived from limestone. It was established in 1978 by Lisu immigrants who formerly grew opium in a national forest reserve. The altitude of the new village was not suited to opium cultivation, so the immigrants were compelled to adopt other cash cropping systems. They had long been shifting cultivators, so intercropping and relay cropping were not regarded as new concepts. The complex, multiple cropping practices they developed, and which are described in this chapter, may have evolved through farmer experience, as well as from external recommendations.

The study involved two participatory rural appraisal (PRA) exercises in Huai Nam Rin and group interviews with farmers in two neighboring villages, Huai Go and Mae

Somchai Ongprasert, Department of Soils and Fertilizers, Mae Jo University, Chiang Mai 50290, Thailand; Klaus Prinz, McKean Rehabilitation Center, P.O. Box 53, Chiang Mai 50000, Thailand.

Pam Norg, which have similar soil types and topography. Soil dynamics resulting from the practice were evaluated by physical and chemical analyses of soil samples taken at the end of the dry season.

The three villages and their cultivated areas are situated on a long narrow foothill between a limestone mountain on one side and a shale/schist mountain on the other. Most of the cultivated areas are located on soils derived from limestone. As a result, the study emphasized these soils. The cultivated lands are relatively flat and lower in altitude than most shifting cultivation areas in northern Thailand (Kunstadter and Chapman 1978). All areas have been declared as national forest reserves.

The natural vegetation in the area is mixed deciduous forest, with a prevailing tropical monsoon climate and a pronounced dry season from November to April. Mean annual rainfall is 1,250 mm. Mae Pam Norg and Huai Go villages are 11 km apart and were established in the 1950s and 1960s by ethnic northern Thai and Karen people, respectively. Later, Akha, Hmong, and Lisu people began to migrate into the villages. Huai Nam Rin was established some 20 years later by the Lisu in a relatively undisturbed forest between Mae Pam Norg and Huai Go when these villages became too crowded. Table 23-1 shows important parameters of the three villages.

Conventional Shifting Cultivation Systems

Conventional shifting cultivation systems in northern Thailand vary among ethnic groups. In general, however, they involve mixed cropping based on rainfed upland rice and corn. Other vegetable crops, such as cucumber (*Cucumis sativus*), wax gourd (*Benincasa hispida*), pumpkin (*Cucurbita moschata*), angled luffa (*Luffa acutangula*), sponge gourd (*Luffa cylindrica*) and chilies are intercropped in and around cornfields. In addition to their shifting systems, Thai, Karen, and Lua groups who have settled permanently in valleys also have rice paddies and orchards.

Table 23-1. Important Parameters of the Three Villages

Parameters	*Huai Nam Rin*	*Huai Go*	*Mae Pam Norg*
Ethnic groups	Lisu	Karen, Akha, Lisu	Thai, Hmong, Lahu
Number of households	70	98	62
Years of establishment	1978	1950s	1960s
Road assessment	unpaved roads difficult access in the rainy season	unpaved roads difficult access in the rainy season	paved roads accessible all year round
Legal land use rights	none	none	some
Distance from Huai Nam Rin	—	7 km	4 km
Main soil types	Clayey Oxic Paleustuls with very good soil structure		
Slopes of cultivated areas	5% to 20%	5% to 20%	5% to 20%
Elevations	460 to 500 m asl	560 to 600 m asl	460 to 500 m asl
Average land holding per household	4 ha	3 ha	2 ha
Off-farm employment	none	yes	yes
Lowland rice fields	none	yes	yes
Upland rice cultivation	none	yes	none
Farmers who practice the system	almost all	more than 50%	about 25%

Being relative newcomers to northern Thailand, Hmong, Lisu, and Akha people had to settle on relatively higher and steeper terrain and became the real shifting cultivators of the area. Traditionally, they had no rice paddies or orchards, grew opium as a cash crop, and moved their entire communities when their cultivated areas were exhausted. Their system was described as "long cultivation, very long fallow," or "pioneer shifting agriculture" (Kunstadter and Chapman 1978).

The altitude of the three study villages was considered too low to grow opium, so the Hmong, Akha, and Lisu migrants had to grow fruit trees as cash crops. Many kinds of fruit trees, mangoes in particular, now grow in the villages and in nearby fields. As a result of a strict reforestation program, planting of mango trees has also spread to distant swidden fields. The farmers believe, in agreement with Pahlman (1992), that planting fruit trees is one way of making their land claims more secure.

Results

Innovations toward Intensified Shifting Cultivation

Before they settled in their present village, the people of Huai Nam Rin knew of lablab bean, but it was not widely grown. They were unfamiliar with rice bean and cowpea. The relay cropping of corn and viny legumes sprang from both their own initiatives and external recommendations. Relay cropping into cornfields was not a new practice for these people. They had previously relay-cropped opium into cornfields about one month before corn harvest (Keen 1978). In 1980, two years after the settlement of Huai Nam Rin, bean seeds were found in unhusked rice bought from another village. They were relay-planted into the cornfields of one woman farmer. Two years later, rice bean was introduced into the village by an immigrant Lisu family from Kampaeng Phet, well to the south. This family had obtained the seeds and planting recommendations from officers of the Department of Land Development (Thirathon 1997). The present bush-type cowpea variety was obtained by the farmers of Huai Nam Rin in 1993, from a business middleman. The seeds were accidentally left lying on the floor of his truck and were freely distributed to some farmers. They were a variety of cowpea grown as a second crop in another village in Phrao district, more than 20 km away. Both rice bean and the bush variety of cowpea had earlier been introduced to northern Thai farmers by the International Institute of Tropical Agriculture (Thirathon 1997).

When asked to rank their major farming problems in order of seriousness, the farmers named marketing, weeds, the reforestation program, declining soil quality, insects and diseases, and drought.

In Huai Nam Rin and Mae Pam Norg, upland rice, which was one of the villagers' ritual crops, was replaced by commercial crops in the 1980s. However, upland rice is still cultivated by about half of the farmers at Huai Go. The farmers of all three villages have adopted a market strategy and appear to be integrated into the mainstream Thai economy.

Relay Cropping with Viny Legumes

The complex system of intercropping corn and wax gourd, with relay cropping of viny legumes, as practiced in Huai Nam Rin and its neighboring villages, is illustrated by the cropping calendar in Figure 23-1. Fields in which the system is used include bare fields and those with small fruit trees. The rotation begins with weeding, piling, and burning of crop and weed residues, usually in March and April, although a late harvest of the previous lablab crop might delay this procedure. Farmers burn crop residues to control weeds, insect pests, and diseases. Corn and wax gourd are planted together in May. Their seeds are mixed in ratios of between 20 and 40 to 1 of corn and wax gourd, respectively. Plant spacing for corn is about 70 cm by 50 cm, with two plants per hill, while the best spacing for wax gourd is 2 m by 2 m. The wax gourd is later thinned if it grows too densely. Some farmers intercrop pumpkin with corn instead of wax gourd. The corn variety used is Suwan 5, an open-pollinating variety.

The three viny legumes—cowpea (*Vigna unguiculata*), rice bean (*Vigna umbellata*), and lablab bean (*Lablab purpureus*)—are planted separately into cornfields about one month before corn harvest. At the time of this study, not all of the corn and wax gourd areas were relay-cropped with legumes because of a shortage of labor. However, the planting time for the legumes can be extended to the end of October if there is enough rain in the second half of that month. Before planting the legumes, the farmers weed the fields. The wax gourd or pumpkin vines are pulled out and piled in circular areas of 1 m to 1.5 m along their hills to open space for the legumes to grow. The farmers say this practice stimulates growth of new shoots and results in higher wax gourd yields. The legumes are planted in the middle of the corn rows, with spacing of 70 cm by 50 cm for both rice bean and lablab bean, and 70 cm by 30 cm for cowpeas.

The use of farm tractors in the villages is prohibited as a measure to control the expansion of cultivated areas, and therefore to protect the forest. But expansion and gradual deforestation still continues. Planting is done using spades, and with no tillage. Weeding is done manually, although herbicide use is increasing because of the growing size of fields. Fertilizers have only recently been used by a few farmers, and small machines have been brought in to plow a few fields.

Harvesting of the crops is spread from October to April; corn in October, cowpea in December and January, rice bean in January and February, wax gourd or pumpkin in February and March, and lablab bean in March and April. Wax gourd and pumpkin fruit lying on the ground are easily gathered after harvesting cowpea and rice bean. On the other hand, harvesting wax gourds in lablab bean fields is more difficult because the bean canopies cover the wax gourd fruit, and the fully-grown green pods of lablab bean contain an oil with an unpleasant smell that causes skin irritation.

Lablab bean is not grown in fields with mango trees because the fields would be covered with undergrowth during the entire dry season, both creating a high fire risk and providing a habitat for rats that could destroy the trees.

Soil Dynamics

For a study of soil dynamics under different years and systems of cultivation, we selected four pairs of cultivated fields, with and without relay cropping of legumes and having comparable years of cultivation, and three patches of disturbed forest with burned undergrowth around the cultivated fields. Field areas were measured by a global positioning system device, and corn yields per unit area were calculated from interviews with the owners. The results are shown in Table 23-2.

The analysis of soil organic matter showed a clear trend of depletion of soil organic matter as well as available phosphorus (Bray II method) according to the length of cultivation. However, three of the four pairs of selected fields showed that relay cropping of corn and viny legumes delayed the depletion of soil organic matter. Depletion of phosphorus was observed in relay-cropped fields over a period of more than 10 years.

Although corn in Huai Nam Rin is grown without chemical fertilizer, its average yield in the selected fields was 3.05 tonnes per hectare (t/ha). This is almost 50% above the national average, indicating the high soil fertility in the area. The average yield of corn with legume relay cropping was 3.63 t/ha, significantly better than the 2.47 t/ha harvested from fields without relay cropping. Higher mineralization of nutrients, especially nitrogen, from a higher content of organic matter was probably a dominant reason for the better performance of corn in the relay-cropping system. However, farmers believed that a difference in weed control could be a partial cause of the yield difference, because those who did not practice relay cropping generally had less household labor for any practice, including weeding.

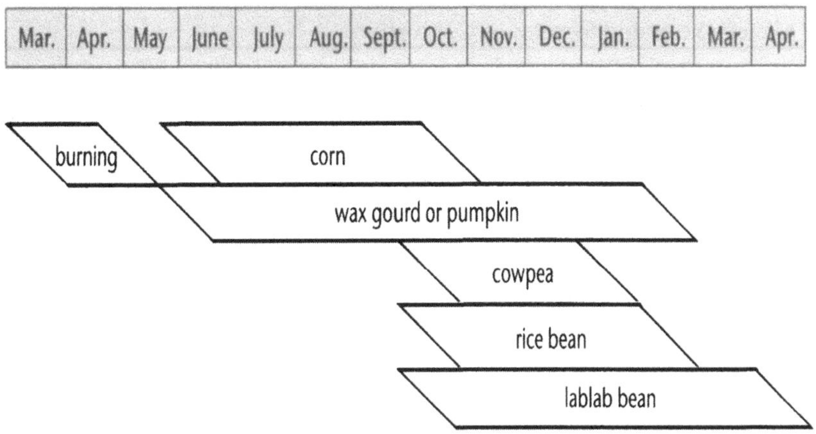

Figure 23-1. Cropping Calendar of the Complex Multiple-Cropping System

Household Economics

The approximate production of Huai Nam Rin village, and therefore the cash income from the complex multiple-cropping system, was calculated from the number and size of trucks that transported the products out of the village. These figures, for the 1996–1997 cropping season, are shown in Table 23-3. In addition to this, some farmers also obtained more cash income from fruit trees. Most of them were satisfied with their income status, which they said was comparable to the cash they would have earned had they been allowed to freely produce opium.

Continuing Transition of the Farming System

In response to questions about their likely future farming systems, farmers in Huai Nam Rin said that within 10 years, most of their fields would be planted with fruit trees, particularly mangoes. They expected to earn more from orchards, with less labor, than the present farming system. The relay-cropping system would gradually decline as the fruit trees got bigger and the acreage of orchards kept expanding. They said there was very little national forest reserve remaining, so they could not significantly expand their land holdings.

Discussion

Contextual Triggers that Contributed to the Innovation

After 1985, opium cultivation was strictly controlled, and all known opium fields were destroyed (Seetisarn 1995). This forced opium growers to adopt new cash cropping systems or find new settlement sites offering possibilities for alternative cash cropping. Many Lisu families, former opium growers, migrated to Huai Nam Rin between 1986 and 1990. Part of the traditional farming system of opium growers was a complex cropping system involving relay cropping of opium into fields intercropped with corn and vegetables (Keen 1978).

Table 23-2. Dynamics of Soil Organic Matter, Available Phosphorus and Corn Yields under Different Years and Systems of Cultivation

Systems	*Years after Forest Clearing*	*pH*	*OM (%)*	*Pavai (g/metric tonne)*	*Corn Yield (metric tonne/ha)*
Disturbed forest		6.4	6.22	100	
Disturbed forest		6.2	6.46	95	
Disturbed forest		6.3	5.61	87	
With relay cropping	3	6.3	5.61	119	3.26
Without	4	6.2	4.42	97	1.57
With relay cropping	5	6.1	5.29	96	4.71
Without	5	6.7	5.43	70	3.55
With relay cropping	10	6.2	4.42	70	3.00
Without	11	6.4	3.23	83	2.12
With relay cropping	17	6.6	3.93	57	3.55
Without	15	6.8	2.92	87	2.62
Average corn yield with relay cropping					3.63
Average corn yield without relay cropping					2.47
Overall average					3.05

Table 23-3. Approximate Production and Income of Huai Nam Rin, 1996–1997

Crops	*Production (metric tonnes)*	*Price (US$/tonne)*	*Income (US$)*
Corn	700	135	94,500
Rice bean	100	290	29,000
Cowpea	80	425	34,000
Lablab bean	80	230	18,400
Wax gourd	220	58	12,760
Pumpkin	15	192	2,880
Average gross income per household			2,736

Availability of Land

Compared to their two neighboring villages, farmers in Huai Nam Rin possessed farmlands big enough to generate an acceptable income from the complex relay-cropping system (Table 23-1). Farmers who adopted the system in Huai Go and Mae Pam Norg were generally those with bigger land holdings. Farmers with smaller holdings in Mae Pam Norg had to generate higher income per unit area by contracting to produce seeds of flowers and vegetables. They were able to do this because their village was located on a paved road, and most farmland was accessible by pickup trucks in the rainy season. Farmers with limited agricultural land who did not adopt the complex relay-cropping system had to earn off-farm income.

The Need for Efficient Weed Control

The prohibition against using farm tractors compelled farmers to try other means of efficient weed control. The fast growth of wax gourd and pumpkins during their early stages provided a solution. They became additional cash crops when some families were able to sell wax gourds and pumpkins to a local military camp. At the time of this study, wax gourds were collected by local businessmen and sent to food factories in Bangkok. Relay cropping of the three pulses was also perceived as an efficient weed control system.

Suitability of Limestone-Based Soils

The selection of Huai Nam Rin as a place for a settlement was deliberate. It had limestone-based soils with big trees in a mixed deciduous forest. In the experience of the villagers, the land was capable of sustaining intensive cultivation over several years. Similar selective judgment on the part of shifting cultivators has been reported elsewhere in Thailand (Keen 1978; Kunstadter and Chapman 1978). High pH and calcium content are essential for efficient nitrogen fixation by rhizobium and, consequently, for the growth of most legumes.

Availability of Transport and Markets

Huai Nam Rin village is just four kilometers from a paved road and is easily accessible by an unpaved road during the dry season. The farmers have no problems marketing their products. They claim that business middlemen normally offer them better prices than those offered to neighboring villages because they are able to produce much larger volumes.

Adoption Constraints

Markets and Transportation

Farmers claimed they would only practice their complex relay-cropping system as long as their products could be sold. If there were no markets, they would only use the system on small areas sufficient for household consumption. This indicates that the farmers' choice of cropping system is based primarily on economic considerations. Therefore, it represents an adoption constraint in areas where markets for the legumes are not available.

This system also requires good transportation, at least in the dry season. In the case of Huai Nam Rin in the 1996–1997 cropping year, large trucks were needed to transport 1,195 metric tonnes of produce from the village center, and pickup trucks were used to collect produce from individual fields and carry it to the village. An obvious conclusion is that this system could not be adopted in remote villages inaccessible by trucks.

Large Land Holdings

Insufficient farm area was a major reason why many farmers did not adopt relay cropping in Mae Pam Norg. Fields were not large enough for the relay-cropping system to produce yields sufficient to generate acceptable cash income. One nonadopter in Mae Pam Norg calculated that the maximum income from the relay-cropping system was US$700 per ha. Therefore, his household would have needed a three- to four-hectare farm to earn an acceptable income. As it was, he was contracted to produce vegetable and flower seeds, and this enabled him to earn a more attractive US$1,900 per ha. Like him, other farmers with only one or two hectares of land preferred to become contracted producers rather than adopt the relay-cropping system.

Soil Fertility

The relay cropping system is a form of permanent intensified farming that, at present, is completely reliant on the mining of soil nutrients. Therefore, the capacity of soils to sustain intensive cropping should be a prerequisite for adoption of this system. The soil analyses in Table 23-2 indicate that soils at Huai Nam Rin originally had a very high phosphorus content and were capable of maintaining a high nutrient availability, even after the 17 years in which the relay-cropping system had been practiced.

Improving the System's Productivity

When farmers in Huai Nam Rin were asked to state their ideas for improving the system's productivity, responses included the use of commercial hybrid corn varieties, tillage using farm tractors, and the use of chemical fertilizers. Almost all households in the village planned to use a hybrid variety of corn in 1997, even though the cost of hybrid seeds was US$5 per kilogram. This followed the experience of some farmers who, in 1996, got 50% higher yields from hybrid corn than from the widely used open-pollinating variety, Suwan 5. At Huai Go and Mae Pam Norg few, if any, farmers planned to use the expensive hybrid corn. The responses demonstrated not only the influence of mainstream agriculture, but also the fact that Huai Nam Rin farmers were integrated into the market system. They were ready to be risktakers.

We proposed the following two measures to farmers as methods to improve the productivity of their system:

- Earlier relay planting of legumes. Research from Chiang Mai indicates that relay establishment of cowpea and lablab bean into corn crops between 60 and 100 days after corn planting does not affect corn yield (Insomphun and Kanachareonpong 1991). Moreover, the legume yields increase with earlier planting. Yields of cowpea and lablab bean planted 60 days after corn were 14 times and 10 times higher, respectively, than those planted 100 days after corn.
- Maximized use of ash. It was observed that almost all ash from the burning of crop and weed residues at Huai Nam Rin was dispersed by the wind. The use of ash should be optimized in this system. In similar farming circumstances farmers have been reported to immediately hoe burned fields to conserve ash (Van Keer 1996).

The farmers of Huai Nam Rin considered the former more interesting than the latter, which they believed was impractical without the use of farm tractors. Some farmers agreed to conduct trials on small patches within their fields, relay planting the legumes about two weeks earlier than currently practiced.

Conclusions

Although nearly all of the 70 households in Huai Nam Rin practice the complex relay-cropping system, its adoption rate in two neighboring villages varies from 25% to 50%. Despite the fact that relay-cropped pulses are excellent cover crops that help control weeds and improve soil fertility, most farmers say that if their products could not be sold, they would restrict their practice of the system to smaller areas, sufficient for household consumption. This indicates that markets and economic returns are important to adoption of the system, as well as transportation and availability of fertile lands. More detailed studies of this innovative system of accelerated seasonal fallow management should be undertaken in the near future.

References

Insomphun, S., and A. Kanachareonpong. 1991. *The Effect of Planting Dates on Growth and Yield of Black Bean (*Vigna unguiculata L.*) and Lablab Bean (*Lablab purpureus L.*) as Relay Crops in Corn under Rainfed Upland Conditions.* Chiang Mai, Thailand: Faculty of Agriculture, Chiang Mai University, 55. (Thai language).

Keen, F.G.B. 1978. Ecological Relationships in a Hmong (Meo) Economy. In: *Farmers in the Forest,* edited by P. Kunstadter, E.C. Chapman, and S. Sabhasri. Honolulu: East-West Center.

Kunstadter, P., and E.C. Chapman. 1978. Problems of Shifting Cultivation and Economic Development in Northern Thailand. In: *Farmers in the Forest,* edited by P. Kunstadter, E.C. Chapman, and S. Sabhasri. Honolulu: East-West Center.

Pahlman, C. 1992. Soil Erosion? That's Not How We See the Problem! In: *Let Farmers Judge*, edited by W. Hiemstra, C. Reijntjes, and E. Van Der Wart. London: Intermediate Technology Publications, 43–48.

Seetisarn, M. 1995. *Shifting Agriculture in Northern Thailand: Present Practices and Problems*. Proceedings of an international symposium on Montane Mainland Southeast Asia in Transition, November 1995, Chiang Mai, Thailand: Chiang Mai University, 17–30.

Thirathon, A. 1997. Personal communication between A. Thirathon of the Department of Agronomy, Mae Jo University, Thailand, and the first author on April 15, 1997.

van Keer, K. 1996. Personal communication between K. Van Keer, of Soil Fertility Conservation Project, Mae Jo University, Thailand, and the first author on October 20, 1996.

PART V

Dispersed Tree-based Fallows

Once known as among the fiercest warriors in northeast India, the Angami Nagas are now recognized for their innovative management of alder trees in swidden fields.

Chapter 24

The Role of *Leucaena* in Swidden Cropping and Livestock Production in Nusa Tenggara Timur, Indonesia

Colin Piggin

Nusa Tenggara Timur (NTT) Province in eastern Indonesia comprises the eastern Lesser Sunda Islands of Flores, Sumba, Roti, Savu, the western half of the island of Timor, and numerous smaller islands. The islands rise from the sea to altitudes of up to 2,500 m above sea level (asl) along central mountain ridges that are rugged and steep. Deep, eroded streams carry water from the mountains to wide, shallow rivers that flow along narrow plains to the sea. Most streams flow intermittently, raging after rains but without water for much of the year. The few big, permanent rivers flow strongly after rains and become trickles as the dry season progresses.

The soils of the outer southern arc of islands, including Timor, Roti, Sabu, and Sumba, have been derived from marine sediments. They are highly calcareous, with a soil pH between 8 and 9, and some soils are sodic (Aldrick and Anda 1987). Poorly adapted crops and forages growing on these soils often exhibit Zn, Fe, and P deficiencies. Chemical properties of soils on the major land units in West Timor have been detailed by Aldrick (1984a, b).

In contrast, the soils of the inner arc of islands, including Flores and Alor, are derived from recent volcanic activity and are generally more fertile than the soils of Timor (Aldrick and Anda 1987). These soils are vulnerable to erosion, slightly acidic, with a pH of 5.5 to 6.0, low in N and P, and low in soil organic matter (SOM) and water-holding capacity (Metzner 1982).

Agriculture is dominated by the semiarid climate, with an extreme dry season that usually extends from April or May until October or November. This is caused by southeast monsoon winds that are dry after blowing over the Australian continent. Northwest monsoons bring rain from November or December to March or April. However, the timing and quantity of rainfall are characterized by extreme variation. Average rainfalls vary from 1,000 to 1,500 mm and generally increase with altitude. Wet season temperatures range from a maximum of 35 to 38°C to a minimum between 22 and 25°C. In the dry season, temperatures range from a maximum between 22 and 35°C and a minimum between 19 and 22°C, with more extreme temperatures in elevated regions. Evaporation rates range from 4 to 9 mm per day with yearly totals of around 2,000 mm. These are extreme conditions for plant

Colin Piggin, Director, Diversification Program, ICARDA, P.O. Box 5466, Aleppo, Syria.

growth, and McWilliam (1986) reported that total crop failure may occur as frequently as one year in every five.

The total land area of NTT is about 50,000 km^2, and the total population about 3 million (see Figure 24-1). Population densities range from 15 to 100 persons per square km. Conditions in NTT have been described by Ormeling (1955), Fox (1977), and Metzner (1982).

Traditionally, Timorese subsistence life was based on hunting and gathering, with some cultivation of ancient crops such as sorghum, Job's tears, rice, millet, mung beans, and cucurbits. According to Fox (1977), slash-and-burn cultivation commenced only after the introduction of maize from the Americas by Dutch and Portuguese colonialists around the 1670s, and extensive grazing by large ruminants only occurred after the introduction of cattle in 1912.

Present-day cropping is based on slash-and-burn cultivation of maize and cassava, with some sorghum, peanuts, mung beans, rice, sweet potatoes, pumpkins, and other vegetables. Most farmers cultivate up to a hectare of land under the shifting slash-and-burn system and also have up to half a hectare of permanent garden around their house. In the flatter areas, many farmers also grow irrigated rice, often quite distant from their homes. Management systems for these crops have been detailed by Pellokila et al. (1991). The importance of different crops in NTT is detailed in Table 24-1.

Tree or horticultural crops are of lesser significance (see Table 24-2). Management systems for fruits such as papaya, bananas, citrus, mangoes, pineapples, soursop, custard apples, and jackfruit and vegetables such as tomatoes, cabbages, beans, eggplant, and garlic have been outlined by Chapman (1986), Baker (1988), and Pellokila et al. (1991).

Many farmers raise cattle, buffaloes, goats, pigs, and chickens. Horses are also common (see Table 24-3). Most livestock feed as scavengers under a free-range system, and management inputs are low. Livestock production systems have been described by Ormeling (1955), Ayre-Smith (1991), and Piggin (1991).

Most crop produce is consumed on the farm. Livestock are a form of wealth and are usually only killed and eaten on traditional or festive occasions and contribute little to the nutrition of rural villagers. In some areas with improved access and management, farmers produce for expanding local, provincial, and national markets.

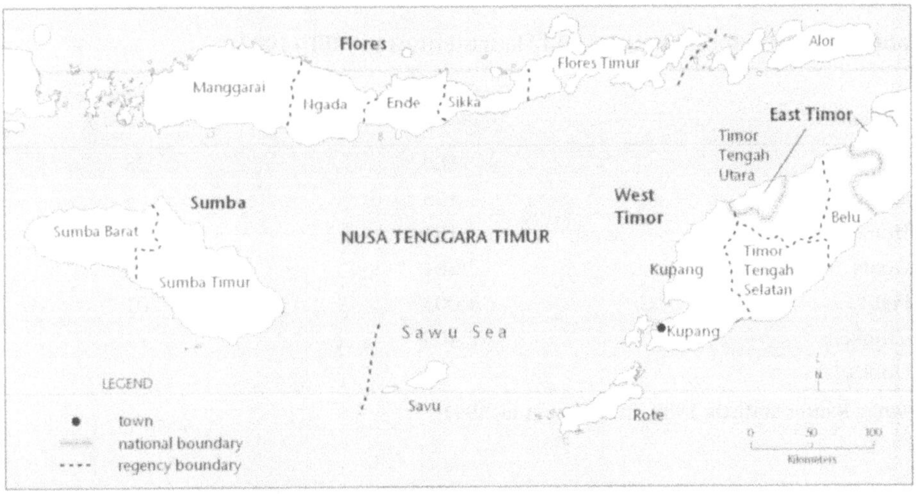

Figure 24-1. Towns, Kabupaten, and Roads of Nusa Tenggara Timur

Note: Timor Barat = West Timor; Timor Timur = East Timor; Nusa Tenggara Timur (NTT) = East Nusa Tenggara; Nusa Tenggara Barat (NTB) = West Nusa Tenggara; the Indonesian province of NTT comprises West Timor, plus the islands of Flores, Sumba, Roti, Savu, and numerous small islands.

Table 24-1. Harvested Area and Production of Major Food Crops in NTT (1987)

Crop	*Harvested Area (thousands of ha)*	*Production (thousands of t)*
Wet rice	59	197
Dry rice	49	77
Maize	188	257
Cassava	72	630
Sweet potatoes	14	105
Peanuts	10	14
Total	391	1,252

Source: Kantor Statistik 1988, in Barlow et al., 1991.

Table 24-2. Planted Area and Production of Major Tree Crops in NTT (1987)

Crop	*Planted Area (thousands of ha)*	*Production (thousands of t)*
Coconut	141	44.7
Coffee	37.2	9.6
Kapok	31.1	4.1
Cashew	30.3	0.6
Candlenut	20.8	3.3
Cloves	4.4	0.1
Cocoa	11.4	1.1
Cotton	1.5	0.3
Total	279	

Source: Kantor Statistik 1988, in Barlow et al. 1991.

Table 24-3. Livestock Numbers and Slaughterings in NTT (1987)

Animals	*Total Number (thousands)*	*Number Slaughtered (thousands)*
Cattle	600	28
Buffalo	175	4
Horses	181	/
Goats	384	108
Pigs	1,003	201
Chickens	2,409	/
Ducks	59	/

Source: Kantor Statistik 1988, in Barlow et al. 1991.

Land Degradation

Various authors have suggested that traditional slash–and–burn cultivation systems can support a maximum of 30 to 50 persons per square kilometer (Fox 1977). Above this limit, there is often progressive degradation of the entire system with shortening of the fallow cycle, a succession from forest to grassland, and severe water imbalance. Over the last century, there has been severe and increasing land degradation throughout much of NTT because of the following factors:

- An increasing human population (see Table 24-4), which has depended largely on slash-and-burn agriculture with progressively longer crop and shorter fallow cycles, and consequent increasing deforestation and reduced forest regeneration. In 1983, 500,000 ha of land was being cropped in NTT by 420,000 small farm households. An estimated 70% of this land was under shifting cultivation (Barlow et al. 1991);
- An increasing cattle population and the introduction and spread of weeds like lantana (*Lantana camara*), which have reduced forest regeneration and placed increasing grazing pressure on grasslands;
- Extensive annual burning of forest and grassland vegetation in the long and extreme dry season, leaving the soil bare and unprotected; and
- High intensity downpours of rain, which are a common occurrence in the short and variable wet season. These cause severe erosion of bare slopes that have been stripped of their vegetation by fires. There is consequent silting of streams and rivers.

Most of the forest in NTT has now been cut, grazed, or burned from mountain areas. Floods and erosion are commonplace, but streams dry quickly because there is little remaining vegetation to catch and hold moisture in catchment areas.

Sustainable Village Systems

There are several parts of NTT where this severe land degradation has been arrested and reversed through the development, largely by local administrators and farmers, of stable swidden systems based on *Leucaena leucocephala*, commonly known as *lamtoro*, or *Leucaena* (see color plates 32 to 34). One is in the Kecamatan district of Amarasi in West Timor, southeast of Kupang and centered on Buraen. The other is in the Kabupaten regency of Sikka on the island of Flores (see Figure 24-1). It is interesting to trace the history of these somewhat contrasting systems and draw conclusions about the reasons for their success. They are examples of successful and robust cropping and livestock systems that have been developed and widely adopted by farmers. Experiences and lessons from these systems are valuable in considering the promotion and adoption of similar systems in other areas.

It was not until 1930 that organized scientific agrarian advice under central direction began in NTT. Significantly, the establishment of an agricultural extension service in Kupang was closely connected to concerns about shifting agriculture (Ormeling 1955). At the time, the Dutch administration was promoting improved systems of food cropping that used various legumes, including *Leucaena* and *Sesbania grandiflora*, for crop rotation and soil stabilization. *Leucaena* had probably been known on the eastern Lesser Sunda Islands for several centuries. It is said to have been used in Java and Sumatra since the early 1800s to provide shade and firewood, improve soil fertility, and reduce erosion (Metzner 1982, 1983). According to Dijkman (1950), it was brought to Indonesia from Central America by early Spanish explorers.

Table 24-4. Area, Population, Cattle Numbers, and *Leucaena* Area in NTT, Amarasi, and Sikka between 1930 and 1987

	NTT *(50,000 km²)*		*Amarasi* *(740 km²)*		*Sikka* *(1,670 km²)*	
Year	*Total*	*No./km²*	*Total*	*No./km²*	*Total*	*No./km²*
Population						
1930			16,800	23	123,000	74
1950			17,600	24	131,500	79
1970	2,260,000	45	25,000	34	188,000	113
1980	2,737,000	55	30,000	41	215,000	129
1987	3,087,000	62				
Cattle Numbers						
1915	234	0.005				
1921	2,700	0.5				
1948–52	108,000	2.2	500	0.7		
1970–76	375,000	7.5	13,000	18	50	0.03
1980–82	414,000	8.3	17,000	23	2,050	1.2
1987	600,000	12			3,400	2
Leucaena (ha)						
1955			440			
1975					8,000	
1982			50,000		20,000 to 43,500	

Source: Piggin and Parera 1985; Barlow et al. 1991.

Kecamatan Amarasi, Kabupaten Kupang, and West Timor

Amarasi occupies a 740 km² strip of land, 10 to 25 km wide and 65 km long, located on the south coast of West Timor. It is undulating land with an average elevation of 300 m asl. The area originally supported dense monsoon rainforests, which were seen as late as 1929 by the naturalist Muller (Metzner 1981). However, because of slash-and-burn cropping, destruction of the forest, development of extensive grasslands, and land degradation became serious problems by the 1930s. Crop yields, in turn, decreased because restoration of soil fertility was slower under grassland than under forest, and famine became an almost seasonal occurrence.

Bali cattle, introduced in 1912 under Dutch encouragement, adapted well to Timor but did little to solve the problems of feeding the population. Livestock, including cattle, buffaloes, and horses, were used mainly for social and ritual purposes and were rarely eaten. Rulers and heads of villages commonly accepted distributed livestock and sold the offspring for slaughter in major centers or for export. Lack of knowledge about livestock and grazing management systems, watering systems, and improved pastures resulted in high mortality and low productivity. Uncontrolled, open range, and indiscriminant grazing became a problem to unfenced crops, as well as for the regeneration of forest areas after cropping (Ormeling 1955; Fox 1977; Metzner 1981).

Land degradation was further exacerbated when lantana (*Lantana camara*), a woody shrub, entered Timor around 1912. It was probably introduced to Kupang as a pot plant, or with cattle, and it spread eastward, probably with the aid of birds, between 1915 and 1935. By 1949, about 80% of Amarasi was covered with lantana

(Ormeling 1955; Fox 1977; Metzner 1981). Livestock owners and cropping farmers had differing opinions on the plant. To the grazier, it was a weed because it dominated grasslands and was not eaten by stock. Metzner (1981) suggested that a decline in numbers of large livestock, including cattle, horses, and buffaloes, from 6,000 in 1916 to 4,000 in 1948, was largely due to lantana. Ormeling (1955) considered that, because of the lantana, livestock numbers in the early 1950s were lower in Amarasi (60 \km^2, or 50 per 1,000 inhabitants) than the Timor average (170 and 450, respectively). Livestock owners were keen to get rid of the plant. To the shifting cultivator, however, lantana was useful because it grew rapidly, quickly provided a soil cover, reduced weed growth, reduced the time taken to prepare land for cropping, maintained good soil structure, and reduced fallow periods from perhaps 15 years to 5 or 6 years (Ormeling 1955).

Increasing land degradation, and the need to find a replacement for lantana that was acceptable to livestock owners and croppers alike, encouraged the search for more useful and sustainable systems. In the 1930s, experimental plantings of *Leucaena* on abandoned fields around the village of Baun were made under the guidance of the Dutch administration (Ormeling 1955; Metzner 1981, 1983). Then, in 1932, the *raja* (ruler) proclaimed an *adat,* or traditional regulation that obliged every farmer in Amarasi to plant contour rows of *Leucaena* on cropping areas before they were abandoned. The contour rows had to be no more than three meters apart. Failure to comply carried the threat of a fine, imprisonment, or both. Planting expanded eastward as the decree was implemented around Oekabiti and Buraen in the early 1940s (Metzner 1981). The adat regulation was reinforced in 1948 when the government introduced the Peraturan Tinkat Lamtoro, or *Leucaena* Increase Regulation. It compelled all shifting cultivators to plant *Leucaena* hedges along contour lines (Ormeling 1955). Over time, the plant spread from the rows to colonize the interrow spaces. An even cover of *Leucaena* was soon formed (Metzner 1981).

In 1938, *Leucaena*-based cropping systems were further promoted with the introduction of land-use zoning regulations. These set aside 10 zones exclusively for cropping. Small livestock, including pigs, goats, and sheep had to be penned, and large livestock such as cattle, buffaloes, and horses had to be tethered. The zones were amalgamated and expanded in 1960 and 1967 to include most of western Amarasi. Any livestock straying onto cropping land could be killed on the spot. Outside the zones, cattle could graze freely but had to be corralled once a week. The successful implementation of land-use zoning eliminated the need to build fences, a pursuit which, according to Ormeling (1955), took up 25% to 30% of the time Timorese farmers spent on cropping.

The increase in land area planted to *Leucaena* (see Table 24-4) underlines the success of the campaign promoting its adoption. In 1948, Kabupaten Kupang had 465 ha of *Leucaena.* All but 28 ha of it was in Kecamatan Amarasi (Ormeling 1955). By 1980, Metzner (1981) estimated that *Leucaena* covered two-thirds, or 500 km^2, of Amarasi. Lantana had been largely eliminated as a weed problem. In the late 1940s, the local ruler, who had earlier decreed that *Leucaena* should be planted by all shifting cultivators, began promoting the cultivation of cassava and fruit trees. By the 1960s, seasonal famine had been eliminated and Amarasi was exporting food.

The widespread and successful adoption of *Leucaena* in Amarasi was only possible because of the supportive regulations introduced and enforced by the adat ruler, or raja, who was later appointed administrative head (*camat*) of Kecamatan Amarasi. He was able to proclaim his regulations because of an adat law stipulating that all land belonged to him. Local farmers were granted the right to cultivate the land by his representatives in each of the 62 Amarasi communities. However, a farmer's right to use the land expired as soon as he ceased to cultivate it. This system was still operating as recently as 1976 because, up to that time, only 100 farmers in Amarasi had decided to have their land surveyed and registered. The majority opted for continuation of the right of usufruct because registration of individual ownership involved surveying costs and payment of a tax (Metzner 1981, 1983).

After 1960, most farmers in eastern Amarasi opted to tether and hand-feed cattle near their homes rather than let them graze freely. Because fence construction was no longer necessary, they had time to look after their cattle (Metzner 1981).

Cattle production was further stimulated in 1971 by the provincial government's introduction of what was called the *paron* cattle-fattening credit scheme. The government bought cattle from central Timor and distributed them to interested farmers for fattening by feeding them with cut-and-carried legume fodder, including the foliage of *Leucaena*, *Sesbania*, *Acacia leucophloea*, and *Tamarindus indica.* After reaching slaughter weight, the animals were sold through traders for export, with 85% of the profit going to the farmer and 15% to the government. More recently, many farmers have bought and sold cattle on their own account. Of all the farmers in NTT, those in Amarasi benefited most from the paron system because theirs was the only district with abundant cut-and-carry fodder. A further adat law obliging each family in Amarasi to fatten between two and seven cattle further increased numbers and evened out the distribution of livestock. In 1949, a total population of 500 cattle was owned by less than 1% of the population. By 1974, 13,000 cattle were owned by 100% of Amarasi families (Metzner 1981, 1983).

The Amarasi farmer of the 1980s was described by Metzner (1981) and Jones (1983b), and many features from this time prevail today. The family is composed of six people and their farm covers two hectares. *Leucaena* grows over the entire farm at a density of 10,000 trees per ha. As previously described, *Leucaena* hedgerows have not been evident in Amarasi for a long time, so cropping and gathering of fodder for livestock require harvesting or cutting the fallow forest of *Leucaena* and associated species. Usually 1 to 1.3 ha is used to provide fodder for tethered or penned livestock and 0.6 to 1 ha is used for crop production. The average farmer raises three head of Bali cattle that he has bought from local markets near the end of the dry season. They are bought at 12 months of age for about Rp 75,000. Fattening takes about 18 months and the cattle are sold at the end of the second wet season for about Rp 200,000. Tethered cattle are each fed 15 to 20 kg of fresh fodder (30% to 40% dry matter) from *Leucaena* and other legumes each morning and evening. This means that more than 100 kg of fresh fodder is required per family per day. This can be gathered from about 1 ha of dense *Leucaena* in the wet season, but supplementary feed is required in the dry season (Piggin et al. 1987). In 1989, Widiyatmike and colleagues (Surata and Komang 1993) reported that farmers raised five to seven head per year. They purchased them at 100 kg and sold them after four to five months at 300 kg body weight, realizing a profit of Rp 200,000 per animal. This excellent weight gain of 1.3 to 1.7 kg/head/day perhaps reflects a high intake of *Leucaena* in the diet. Studies in Australia have shown that steers grazing *Leucaena* pastures during favorable wet season periods can gain 1.03 to 1.26 kg/head/day with a legume intake of around 40% (Wildin 1986; Quirk et al. 1988; Esdale and Middleton 1997; Galgal 2002). Weight gains show seasonal variations, from 1.3 kg/head/day in spring, to 0.76 in summer, 0.56 in autumn, and -0.1 in winter (Quirk et al. 1990).

Water supply for livestock in Amarasi is a problem, given the extended dry season and the very porous soils. The area has many deep wells. However, a convenient system has developed, with increased cultivation of bananas. The tethered cattle are fed banana stems, which contain more than 80% water.

Maize and other crops are grown on one-third to one-half of the farm on a three-year rotation. All *Leucaena* and other vegetation are cut to ground level up to four months before the start of the wet season. The cut vegetation is windrowed at right angles to the contour and is burned, usually one or two weeks before the first rains are expected. Dry mulching has been tried (Metzner 1981) but is not popular because of rodents (Jones 1983b). At the onset of rain, maize is sown on a one-meter grid using a dibble stick and three or four seeds per hole. Cassava, beans, and melons are often sown with the maize. The crop is harvested after 4 to 6 months, after the end of the rainy season. Yields in Amarasi are 10,000 to 20,000 small cobs, or 1,000 to 2,000 kg of grain, per ha. It is not necessary to resow *Leucaena* after cropping because of strong regrowth from cut stems and germination of fallen seed. During cropping, green stem regrowth is broken off to reduce competition and to provide some mulch.

Tree crops, including bananas, papaya, mangoes, and coconuts, are generally grown in *Leucaena* areas, especially around houses once moisture regimes have been restored. Families in Amarasi in 1983 owned between 5 and 100 coconut trees and between 10 and 700 banana trees (Jones 1983b). *Leucaena* systems generated a per capita output of about Rp 100,000 per year from sales of cattle, chickens, maize, bananas, and coconuts, with a similar per capita consumption of *Leucaena* firewood, chickens, maize, bananas, and cassava. Real incomes were estimated to be 20 to 30% higher than the average for West Timor, and this was attributed to the stable farming system based on *Leucaena* (Jones 1983b). The prosperity of Amarasi residents is evidenced by many houses with concrete walls and floors and iron roofs. These have replaced traditional palm leaf houses.

Kabupaten Sikka, Flores

Sikka covers an area of 1,670 km^2. It is 15 to 30 km long and is situated on the eastern end of the island of Flores. The land is undulating, rising from sea level to an elevation of several hundred meters. Sikka has a serious erosion problem, and *Leucaena* was introduced to provide vegetative cover and soil stabilization. Efforts to popularize the plant were first made by the Dutch administration in the 1930s, when it recommended cultivation of *Leucaena* in thickets on nonarable land. Adoption was poor because farmers feared that the thickets would get out of control and spread onto arable land (Metzner 1976).

The need to control soil erosion remained strong, so low bamboo fences were recommended. These were pegged along the contours and often anchored with cassava sticks and covered with grass. Once more, effectiveness and adoption were poor. Traditional terraces were also promoted, and between 1966 and 1973, 750 ha of terraces were built. But enthusiasm was not great because the technique was slow, the work laborious, and the terraces were ineffective without accompanying vegetative stabilization.

Interest in cropping was stimulated, and in 1964, farmers agreed by common consent that horses and small livestock such as pigs and goats should be penned or tethered (Metzner 1982). The need for crop fences was eliminated. However, the need for effective soil stabilization soon became critical. As cropping expanded, the pressure to stabilize about 30,000 ha of erosion-prone land stimulated a search for better erosion control technology.

In 1967, a Catholic priest, Father P. Bollen, was so impressed with the potential of *Leucaena* for land stabilization and rehabilitation that he established a small demonstration garden with contour rows of *Leucaena* near his church at Watublapi, about 30 km southeast of Maumare. The *Leucaena* rows soon became well established and began to collect washed soil and to build up indirect terraces. The demonstration prompted a local farmer, Moa Kukur, to establish a terraced garden using *Leucaena* rows at Wair Muut in 1968. Over three years to 1971, yields from the garden were stable, and there was no need to shift cultivation to a new area (Cunha 1982).

These experiences prompted a farmer group, Ikatan Petani Pancasila (IPP), to trial indirect terracing by establishing a demonstration plot in 1972 at Kloangpopot, 40 km south of Maumere. IPP used contour rows of local *Leucaena* spaced five meters apart, with clove trees between the rows. The demonstration plot was shown regularly to farmers and participants in the group's training courses, and it stimulated great interest in indirect terracing (Metzner 1976; Borgias 1978; Cunha 1982).

In 1973, the district government of Sikka and the Catholic Biro Social Maumere, with the support of IPP, established a program that aimed to stabilize 30,000 ha of land in five years. It was called Program Penanggulangan Erosi Kabupaten Sikka, or the Sikka Erosion Control Program. Farmer training courses were held, water levels were distributed for making contours, seed was purchased and distributed, planting was supervised and evaluated, and prizes were offered in order to encourage farmer cooperation. Within two years, an estimated 8,000 ha of *Leucaena* had been established (Metzner 1976; Borgias 1978; Cunha 1982). *Leucaena* planting was further stimulated by the introduction of giant varieties from Hawaii and the Philippines, and by the local launching in 1974 of the national food crops intensification program (BIMAS), which offered credit for crop inputs if farmers planted *Leucaena* on their farms (Parera 1982a). Parera (1982b) estimated that during the mid-1970s about 20,000 ha of hilly land was terraced with local *Leucaena* and a further 2 million giant *Leucaena* trees were planted. Cunha (1982) concluded that the total area of *Leucaena* at this time was between 27,000 and 43,500 ha.

For indirect terracing, *Leucaena* is sown at a rate of about 70 kg of seed per ha, in furrows or banks formed along the contours of fields with the aid of an A-frame or water level. Early establishment is slow, and the seedlings need protection from weeds and grazing. But with reasonable management, thick hedges form within two years, and these collect soil washed from the upper slopes by the rain and gradually form terraces. These are called indirect terraces because they form naturally and are not constructed. Once established, the hedges are usually cut every four to six weeks during the rainy season and before seeding to a height of 75 to 80 cm. Cut material is thrown on the upper slope to fertilize the soil (Metzner 1976). Unlike in Amarasi, *Leucaena* is maintained in hedgerows in Sikka and cropping takes place between the rows.

The primary aim of the *Leucaena* planting program in Sikka was to control erosion. A measure of its success can be seen in the improvement in water balances. The Batikwair River, which ceased to flow in the dry season in the 1920s, has been flowing continuously since 1979, and Maumare, once a flood-prone town, has not been flooded since 1976 (Parera 1980; Prussner 1981; Metzner 1982).

Other benefits have followed. Established areas are now being cropped more intensively and are more productive. Unterraced fields can be cropped for three to four years, but they need a recovery period of four to nine years because of the loss of soil and fertility. Terraced slopes, on the other hand, can be cropped continuously if *Leucaena* herbage is used as green manure and cereal-legume rotations are used. *Leucaena* also discourages weeds, such as *Imperata cylindrica*, which often appears after the abandonment of unterraced fields. Many terraced fields have been planted with permanent tree crops, such as coconuts, coffee, cocoa, cloves, and pepper. The contour hedges of *Leucaena* provide shade, soil stabilization, increased soil fertility, and improved soil infiltration (Parera 1980, 1982b; Metzner 1982).

Unlike in Timor, cattle have not traditionally played a significant role in the Flores livestock industry, partly because of a lack of both water and extensive grasslands (Metzner 1982). *Leucaena* herbage is, instead, fed mainly to small animals such as pigs, goats, and chickens. There were efforts to encourage cattle farming in 1967, with the introduction of 100 head of Bali cattle under a government credit program. However, according to Cunha (1982), only 50 cattle remained in Sikka in 1970, and they were owned mainly by the Department of Animal Husbandry and the Roman Catholic mission. The cattle industry received a stimulus with the introduction of the giant *Leucaena* varieties K8, K28, K67, and Peru, from the Philippines and Hawaii, in 1978 and 1979. These were planted widely in areas not used for cropping. Further Bali cattle were brought in, and numbers climbed to more than 2,000 in 1982 (Cunha 1982) and to 3,400 in 1987 (Barlow et al. 1991).

Heteropsylla cubana and *Leucaena* Productivity

Before the arrival of the *Leucaena* psyllid (*Heteropsylla cubana*) in NTT, studies in 1982–86 showed that the maximum annual production that could be expected from well-established *Leucaena* was around 6,000 kg of dry matter per ha of leaf and a further 6,000 kg dm/ha of stem. This came from three- to four-monthly cuttings of *Leucaena* with 1.5 m between rows and 10 cm between plants. This level of production is at the bottom of the range of 6 to 18 t/ha of edible dry matter quoted by Horne et al. (1986), no doubt because of the severe moisture limitations in the mid to late dry season. Leaf production rates follow rainfall patterns, falling from 25 to 30 kg/ha/day to just 1 to 2 kg/ha/day in the mid to late dry season (Piggin et al. 1987). Assuming that cattle need 10 kg/day of edible dry matter, it would be possible to feed at least 1.5 cattle/ha/year from a good *Leucaena* stand. This agrees with the frequently mentioned carrying capacity of one or two cattle per ha in the Amarasi district (Piggin and Parera 1985).

The arrival of the psyllid to eastern Indonesia in 1986 initially devastated *Leucaena* plantations. Trees were bared and, in places, died. One study estimated that *Leucaena* productivity was reduced by 25% to 50% (Piggin et al. 1987). For a time, farmers in Amarasi raised fewer livestock and used alternative fodder. This was reflected in an 11% fall in cattle sold in trade markets, from 88,000 head in 1986 to 77,000 in 1987 (Figure 24-2). There were grave fears at the time that the systems of Amarasi and Sikka would be destroyed, with associated long-term hardship and land degradation. However, this has not happened. Over the years, psyllid numbers have declined and productivity of *Leucaena* has gradually recovered. This is perhaps due to a buildup of psyllid predators.

Despite the recovery, the psyllid experience has highlighted the danger of overdependence on a single species and has led to a concerted effort to find alternative shrub legumes. Research has shown that *L. diversifolia, L. collinsii, L. pallida,* and several *Leucaena* hybrids are well adapted and exhibit good resistance or tolerance to psyllids in West Timor (Piggin and Mella 1987a,b; Mella et al. 1989). They also support good animal production (Galgal 2002). Other species like *Sesbania, Acacia villosa, Gliricidia sepium, Calliandra callothyrsus,* and *Desmanthus virgatus* are also well adapted and useful as multipurpose trees. Seed production of these species has been promoted by local NTT departments and NGO groups, and they are being more widely used by farmers.

Reasons for the Success of *Leucaena*

There are many reasons why *Leucaena*-based systems have developed and persisted in Amarasi and Sikka. These can be distilled from the historical and detailed accounts above and include the following:

A Recognized Need for Better Systems

Serious land degradation associated with slash-and-burn cropping, increasing livestock numbers, and the spread of the weed lantana, was causing low farm productivity and poverty. Compounding the problem, population densities had reached 25 to 75 persons per square kilometer, above the limit that might be sustained by traditional slash-and-burn systems (Fox 1977). By the 1930s came the realization that serious efforts had to be made to develop more sustainable farming systems. Such systems, nevertheless, had to permit a continuation of traditional swidden rotation methods of restoring soil fertility, suppressing weeds, and providing timber. As recently as 1983, about 70% of the 500,000 ha cropped in NTT and 85% of the 30,000 ha in Sikka were still under shifting cultivation (Barlow et al. 1991).

Failure of Alternatives

Attempts to control erosion and land degradation with physical structures and traditional terraces in the 1960s and 1970s were not successful because of the labor and cost involved and the general ineffectiveness of the technology. This encouraged the continued development of biological systems using *Leucaena* and other plants.

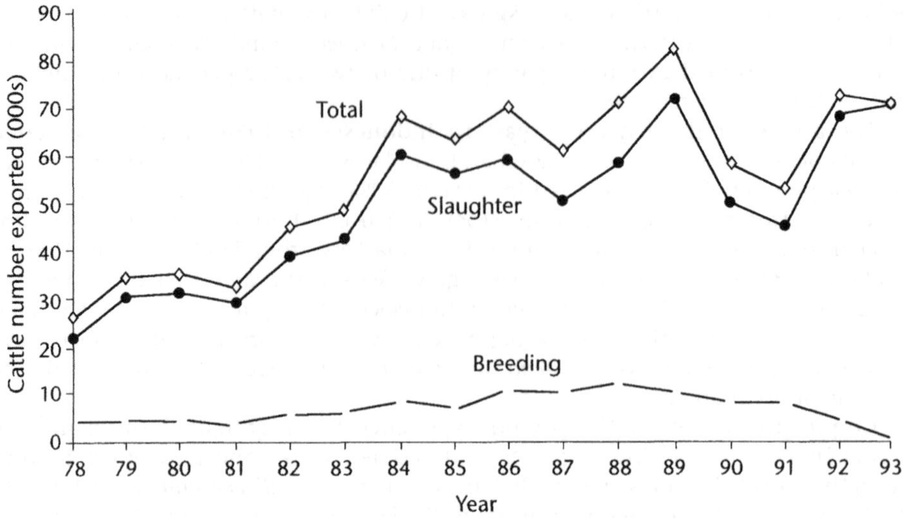

Figure 24-2. Cattle Exported from NTT for Slaughter and Breeding

Adaptation of Leucaena *to the Local Environment*

Leucaena is very well adapted to the semiarid climate and the alkaline or relatively neutral soils of the area. The plant is deep-rooted and drought resistant and, being a legume, is adapted to low N soils. Although its early growth is slow and susceptible to grazing and weed competition, *Leucaena* has proven easy to establish and is very persistent in most situations. Although it was devastated when *Heteropsylla cubana* entered NTT in 1986, *Leucaena* has since recovered much of its productivity and remains a persistent and dominant species in many village areas.

Compatibility of Leucaena *with Local Farming Systems*

Fallow species in NTT must be able to withstand severe treatment. Once established, *Leucaena* is a robust plant able to persist and regenerate despite traditional swidden practices that involve regular and quite severe cutting and burning. It is attractive to cropping farmers because it adds nitrogen to the soil, suppresses weeds, and provides wood for the construction of fences.

Research in NTT has shown that *Leucaena* can be relatively easily established under corn crops without reducing corn yields and can reestablish from cut stumps or seed in subsequent years. With proper management, it competes little with the crop and provides valuable livestock forage and soil N improvement (Field and Yasin 1991; Field 1991a,b,c). Field (1986) has also shown that maize yields can be doubled by including two to three years of *Leucaena* in crop fallow rotations.

Capacity of Leucaena *to Supply Local Needs*

Village life in NTT is harsh and villagers struggle with many constraints. *Leucaena* is a multipurpose plant contributing a multitude of village needs, from firewood and building timber to forage for livestock, mulch for crops, weed suppression, shade for tree crops, and soil stabilization. In Amarasi, it has become like a forest and supports more or less permanent slash-and-burn cropping as a fallow species that improves soil fertility and suppresses weeds. Its capacity to supply nutritious forage has led to the development of a large-scale industry involving the fattening of tethered cattle in Amarasi (see Table 24-4). In Sikka, contour hedgerows of *Leucaena* have been maintained with cropping between the rows. This has led to the buildup of indirect terraces as soil washed downslope by rain is trapped by the hedgerows. Soil loss in the terraced fields of Sikka is consequently much lower than the natural levels in NTT of 200 metric tonnes/ha/year reported by Carson (1979). Demand for fodder has been of little influence on the use of *Leucaena* in Sikka because cattle were not introduced until the 1970s and numbers have remained low. However, cattle numbers are now building up, partly in response to the increased availability of *Leucaena* forage.

Local farming systems have compensated for some weaknesses found in *Leucaena*. Measurements in NTT of the mineral content of *Leucaena* suggest that low Na, P, and Cu may limit its nutritive value and, consequently, its role in animal production. However, farmers in Amarasi already compensate for sodium deficiency in *Leucaena* by the common practice of adding salt to drinking water. Feeding a diverse mix of species to the livestock also helps overcome other deficiencies. For example, *Sesbania*, which is commonly mixed with *Leucaena* in cattle feed, is high in Na.

Commitment of Local Leaders and Groups

Local village heads, NGOs, church groups, and government departments showed great commitment to the need to develop more sustainable systems in Amarasi and Sikka. They were instrumental in recognizing and demonstrating the potential of *Leucaena* to local villagers. Church and farmer cooperative groups were prominent in Sikka, while Dutch administration and local government officials provided the impetus in Amarasi.

Creation of a Favorable Policy Environment

Local administrators recognized the importance of a favorable policy environment to promotion of new technology. They instituted new regulations to encourage not only the planting of *Leucaena,* but the development of more permanent and productive agriculture in general. These measures provided for the following:

- The tethering or confinement of livestock in cropping areas to reduce the need for fences and give farmers more time for crop and livestock husbandry (promulgated in 1938, 1960, and 1967 in Amarasi, and in 1964 in Sikka);
- The availability of cropping credit only to those farmers who were prepared to plant *Leucaena* on sloping land in Sikka;
- The development of erosion prevention programs (established in Sikka in 1973 and 1978);
- The obligatory planting of *Leucaena* in Amarasi, pronounced by the local ruler in 1932 and the government in 1948; and
- The encouragement of cattle husbandry by livestock distribution schemes in Sikka in 1967 and in 1980–82, and in NTT in 1971.

Contribution of Leucaena *to Development of More Commercial Farming Systems*

Leucaena has helped village farmers to move from subsistence to more commercial farming systems. In Amarasi, this has been done through the development of commercial cattle fattening and establishment of orchards of bananas, papaya, mangoes, and coconuts. In Sikka it has followed the development of permanent orchards of tree crops such as mangoes, cloves, pepper, and cocoa. The shade provided by *Leucaena* on previously bare slopes has assisted in the establishment of shade-loving tree crops. In Sikka, increased availability of forage is encouraging the adoption of intensive cattle breeding and fattening. This potential for commercial development has been an important factor in farmer acceptance and enthusiasm for the use of *Leucaena*-based systems.

Modern Development of Legume-Based Systems

The improved fallow systems described above, which are based on *Leucaena,* have been modified and developed in parts of eastern Indonesia to use other species such as *Sesbania, Gliricidia sepium, Calliandra callothyrsus,* and *Acacia villosa.* Such systems, for *A. villosa* at Camplong in West Timor and for *C. callothrysus* in West Flores, have been described by Field (1991b). The use of other species has been partly a response to concerns of dependence on one species, after the *Leucaena* psyllid experience. However, it also recognizes the adaptability and suitability of other species and the need for diversity in farming systems.

Conclusions

It is fascinating that two contrasting systems, both using *Leucaena leucocephala,* should develop and persist in close proximity in eastern Indonesia. Both were prompted by concerns about land degradation, low productivity, and poverty, and both focused on the introduction and promotion of a perennial shrub legume.

The Amarasi system is based on the use of *Leucaena* forests for swidden cropping of corn and feeding of tethered or confined livestock. The Sikka system involves the establishment and maintenance of *Leucaena* hedgerows to support alley cropping of corn, peanuts, and mung beans and permanent tree crop plantations of mangoes, cloves, cacao, and pepper.

The history of these systems suggests that a range of factors has been important in their development, evolution, and persistence. These include a recognized need for better systems, the failure of alternatives, the adaptation of *Leucaena* to the local environment, the compatibility of *Leucaena* with local farming systems, the capacity of *Leucaena* to supply local needs, the commitment of local leaders and groups, the creation of a favorable policy environment, the effectiveness of *Leucaena,* and the contribution of *Leucaena* to the development of more commercial farming systems. Many of these factors can be recognized in descriptions of the processes involved in the successful diffusion of innovations by Rogers (1983).

In a review of slash-and-mulch systems, Thurston (1997) suggested that the jury is still out on the future and potential of alley cropping systems. Sikka and Amarasi provide compelling evidence that, at least in some places, villagers have incorporated shrub legumes into long-established, sustainable farming systems that are supporting a much better quality of life than would otherwise exist.

References

Aldrick, J.M. 1984a. Land Units, Soils and Land Capability of the Middle Mina River Valley, West Timor, November 1984. In: *Indonesia-Australia NTT Livestock Development Project Completion Report,* Kupang, NTT, Indonesia: Indonesia-Australia NTT Livestock Development Project.

———. 1984b. Land Units, Soils and Land Capability of the Besi Pae Area, West Timor. In: *NTT Livestock Development Project Report,* Kupang, NTT, Indonesia: Indonesia-Australia NTT Livestock Development Project.

———, and M. Anda. 1987. *Land Resources of Nusa Tenggara.* Short Term Consultant Report No. 3, Nusa Tenggara Agricultural Support Project. Melbourne: ACIL Australia Pty Ltd.

Ayre-Smith, R. 1991. Livestock Development in NTT. In: *Nusa Tenggara Timur: The Challenges of Development,* edited by C. Barlow, A. Bellis, and K. Andrews. Political and Social Change Monograph 12. Canberra, Australia: Department of Political and Social Change, Research School of Pacific Studies, The Australian National University, 85–104.

Baker, I. 1988. *Fruit Tree Development Program in Timor Tengah Selatan (TTS) and Timu Tengah Utara (TTU).* Project Report. Kupang, NTT, Indonesia: Nusa Tenggara Timur Integrated Area Development Project.

Barlow, C., A. Bellis, and K. Andrews (eds.). 1991. *Nusa Tenggara Timur: The Challenges of Development.* Political and Social Change Monograph 12. Canberra, Australia: Department of Political and Social Change, Research School of Pacific Studies, The Australian National University.

Borgias, F. 1978. *Lamtoronisasi—Usaha Anti Erosi dan Pengawetan Tanah di Kabupaten Dati II Sikka.* Semarang, Jawa, Indonesia: Skripsi Academi Farming.

Carson, B.R. 1979. *Use of the Universal Soil Loss Equation to Predict Erosion of the Timorese Landscape.* Vancouver, Canada: University of British Columbia.

Chapman, K.R. 1986. *Tree Crops for Timor Tengah Selatan (TTS) and Timor Tengah Utara (TTU) in Timor Barat (West Timor)* Project Report. Kupang, NTT, Indonesia: Nusa Tenggara Timur Integrated Area Development Project.

Cunha, I. 1982. *Proses Usaha Lamtorinisasi di Kabupaten Sikka (di P. Flores, Propinsi Nusa Tenggara Timur).* Report from Lembaga Penelitian dan Pembangunan Sosial. Nita, Maumere, Flores, Nusa Tenggara Timur, Indonesia: Institute of Social Research and Development.

Dijkman, M.J. 1950. *Leucaena*—A Promising Soil-Erosion-Control Plant. *Economic Botany* 4, 337–349.

Esdale, C., and C. Middleton. 1997. Top Animal Production from *Leucaena. Leucnet News* 4, 9–10.

Field, S.P. 1986. *Report on the Food Crop Experiment of the NTT Livestock Project.* Kupang, Indonesia: Nusa Tenggara Timur Livestock Development Project.

———. 1991a. Competition Effects of Weeds and *Leucaena leucocephala* on a Maize Crop in a *L. leucocephala* Forest. *Leucaena Research Reports* 12, 55–57.

———. 1991b. *Chromolaena odorata*: Friend or Foe for Resource Poor Farmers? *Chromolaena Newsletter* No. 4, May 1991. Guam: Agricultural Experiment Station, University of Guam.

———. 1991c. The Effects of Undersowing *Leucaena leucocephala* into a Maize Crop. *Leucaena Research Reports* 12, 58–59.

———. and H.G. Yasin. 1991. The Use of Tree Legumes as Fallow Crops to Control Weeds and Provide Forage as a Basis for a Sustainable Agricultural System. In: *Proceedings of the Thirteenth Asian Pacific Weed Science Society Conference*, October 15–18, 1991, Jakarta. Jakarta: Asian-Pacific Weed Science Society, 121–126.

Fox, J.J. 1977. *Harvest of the Palm.* Cambridge, MA: Harvard University Press.

Galgal, K.K. 2002. Forage and Animal Production from Selected New *Leucaena* Accessions. Ph.D. Thesis, University of Queensland, Brisbane, Australia.

Horne, P.M., D.W. Catchpoole, and A. Ella 1986. Cutting Management of Tree and Shrub Legumes. In: *Forages in Southeast Asian and South Pacific Agriculture:* Proceedings of an international workshop, August 19–23, 1985, Cisarua, Indonesia, edited by G.J. Blair, D.A. Ivory, and T.R. Evans. ACIAR Proceedings Series No. 12. Canberra, Australia: Australian Centre for International Agricultural Research, 202

Jones, P.H. 1983a. *Leucaena* and the Amarasi Model from Timor. *Bulletin of Indonesian Economic Studies* 19, 106–112.

———. 1983b. Amarasi Household Survey. Report to Bappeda (Badan Perencanaan Pembungunan Daerah or Regional Planning Development Agency), Kupang, Indonesia: Bappeda.

McWilliam, A. 1986. Profile Survey of Farmers at Besi Pae. Project Report, NTTLDP. Kupang, Indonesia: Nusa Tenggara Timur Livestock Development Project.

Mella, P., M. Zaingo, and M. Janing. 1989. Resistance of *Leucaena* and Some Other Tree Legumes to *Heteropsylla cubana* in West Timor, Indonesia. In: Leucaena Psyllid: *Problems and Management,* edited by B. Nampompeth and K.G. MacDicken. Bangkok: F/FRED (Forestry/Fuelwood Research and Development), 56–61.

Metzner, J.K. 1976. Lamtoronisasi: An Experiment in Soil Conservation. *Bulletin of Indonesian Economic Studies* 12, 103–109.

———. 1981. Old in the New: Autochthonous Approach toward Stabilizing an Agroecosystem: The Case from Amarasi (Timor). *Applied Geography and Development* 17, 1–17.

———. 1982. *Agriculture and Population Pressure in Sikka, Isle of Flores.* Development Series Monograph No. 28. Canberra: The Australian National University.

———. 1983. Innovations in Agriculture Incorporating Traditional Production Methods: The Case of Amarasi. *Bulletin of Indonesian Economic Studies* 19, 94–105.

Nye, P.H., and D.J. Greenland. 1960. *The Soils under Shifting Cultivation.* Tech Comm 51. Harpenden, UK: Commonwealth Bureau of Soils.

Ormeling, F.J. 1955. *The Timor Problem: A Geographical Interpretation of an Underdeveloped Island.* Jakarta: J.B. Wolters.

Parera, V. 1980. Lamtoronisasi in Kabupaten Sikka. *Leucaena Newsletter* 1, 13–14.

———. 1982a. Giant Lamtoro in the Land of the Trees. *Leucaena Newsletter* 3, 44.

———. 1982b. *Leucaena* for Erosion Control and Green Manure in Sikka. Proceedings of a workshop: *Leucaena* Research in the Asian Pacific Region, November 23–26, 1982, Singapore. Ottawa, Canada: IDRC.

Pellokila, C.H., S.P. Field, and E.O. Momuat. 1991. Food Crops Development in NTT, In: *Nusa Tenggara Timur: The Challenges of Development,* edited by C. Barlow, A. Bellis, and K. Andrews. Political and Social Change Monograph 12. Canberra, Australia: Department of Political and Social Change, Research School of Pacific Studies, Australian National University, 121–144.

Piggin, C.M. 1991. New Forage Technologies. In: *Nusa Tenggara Timur: The Challenges of Development,* edited by C. Barlow, A. Bellis, and K. Andrews. Political and Social Change Monograph 12. Canberra, Australia: Department of Political and Social Change, Research School of Pacific Studies, The Australian National University, 105–120.

Piggin, C.M., and V. Parera. 1985. The Use of *Leucaena* in Nusa Tenggara Timur. ACIAR Proceedings Series No. 3. Canberra, Australia: Australian Centre for International Agricultural Research, 19–27.

———. 1987. *Leucaena* and *Heteropsylla* in Nusa Tenggara Timur. Leucaena *Research Reports* 7(2), 70–74.

——— and P. Mella. 1987a. Investigations on the Growth and Resistance to *Heteropsylla cubana* of *Leucaena* and Other Tree Legumes in Timor, Indonesia. Leucaena *Research Reports* 8, 14–18.

———. 1987b, Pengaruh *Heteropsylla cubana* Terhadap Pertumbuhan dan Daya Tahan *Leucaena* dan Jenis Legume Lainnya di Timor, Indonesia. Paper presented to a symposium on psyllid and *Leucaena,* September 24–25, 1987, at the Nusa Cendana University, Kupang, Indonesia.

———, P. Mella, M. Janing, M.S. Aklis, P.C. Kerridge, and M. Zaingo. 1987. *Report on Results from Pasture and Forage Trials, 1985–87.* Kupang, Indonesia: Nusa Tenggara Timur Livestock Development Project.

Prussner, K.A. 1981. *Leucaena leucocephala* Farming Systems for Agroforestry and the Control of Swidden Agriculture. Paper delivered to a Seminar on Agroforestry and the Control of Swidden Agriculture, November 19–21, 1981, Forest Research Institute, Bogor, Indonesia.

Quirk, M.F., J.J. Bushell, R.J. Jones, R.G. Megarrity, and K.L. Butler. 1988. Liveweight Gains on *Leuceana* and Native Grass Pastures after Dosing Cattle with Rumen Bacteria Capable of Degrading DHP, a Ruminal Metabolite from *Leucaena*. *Journal of Agricultural Science (Cambridge)* 111, 165–170.

———, C.J. Paton, and J.J. Bushell. 1990. Increasing the Amount of *Leucaena* on Offer Gives Faster Growth Rates of Grazing Cattle in Southeast Queensland. *Australian Journal of Experimental Agriculture* 30, 51–54.

Rogers, E.M. 1983. *The Diffusion of Innovations*. New York: The Free Press.

Surata, and I. Komang. 1993. Amarasi System: Agroforestry Model in the Savanna of Timor Island, Indonesia. (Paper for National Agroforestry Workshop, Pusat Litbang Hutan dan Konservasi Alam–APAN, Bogor, Indonesia, August 24–25, 1993). *Savanna* No. 8/93 15–23.

Thurston, H.D. 1997. *Slash/Mulch Systems: Sustainable Methods for Tropical Agriculture*. Boulder, Colorado: Westview Press.

Wildin, J.H. 1986. *Leucaena*—Central Queensland Experience. *Tropical Grasslands* 20, 85–87.

Chapter 25

Use of *Leucaena leucocephala* to Intensify Indigenous Fallow Rotations in Sulawesi, Indonesia

Fahmuddin Agus

Efforts to get farmers to adopt technological soil conservation packages often fail because the new technologies are frequently offered without any consideration of farmers' experiences and indigenous practices. Because they have a high-risk livelihood, subsistence farmers are resistant to drastic changes. They are more likely to respond favorably to simple modifications to their traditional practices.

In many parts of Asia Pacific, traditional practices include indigenous agroforestry and soil conservation techniques. For example, West Sumatran farmers value *Austroeupatorium inulaefolium* for its contribution to soil fertility, its provision of firewood, and its mechanical support for viny legumes (Cairns 1994, Chapter 15). In East Nusa Tenggara, the Amarasi fallow rotation system, using *Leucaena leucocephala,* has proven to sustain agricultural production. These systems, however, are only suitable when there is no scarcity of land, and pressure from increasing population has not only reduced the availability of land but has also led to shorter fallow durations and more intensive farming practices, particularly where there is improved access to agricultural inputs.

One modification of traditional crop and fallow rotation that may ease the transition from shifting cultivation to more intensive agricultural practices is the use of contour hedgerows, where shrubs or grasses are planted along the contours of sloping land and annual crops are planted in the alleys between the hedgerows. Instead of fallowing the entire land area, only about 10 to 20% of it is devoted to controlling erosion, producing organic matter, fixing atmospheric nitrogen, and, to a lesser extent, recycling nutrients (Agus et al. 1999).

Such contour hedgerow systems have been tested in several parts of the tropics (Lal 1991). They have proven to prevent downslope soil movement, so that they serve as vegetative "plugs" in areas prone to gully erosion. This chapter, therefore, proposes the modification of existing fallow rotation systems in South Sulawesi by converting them into contour hedgerow systems.

In already intensive farming systems, one disincentive to farmer adoption of a contour hedgerow system is that it takes scarce land out of food crop production. Therefore, in compensation for the sacrificed land, either food crop yields must become significantly higher or the hedgerows must provide byproducts such as nitrogen and firewood (Lal 1991). This problem is not an issue on the outer islands of Indonesia, where traditional shifting cultivation, with a fallow period equal to or

---

Fahmuddin Agus, Indonesian Soil Research Institute, Jl. Ir. H. Juanda 98, Bogor 16123, Indonesia.

longer than the cropping period, is being intensified by simple reduction of the fallow period. Under these circumstances, hedgerows that use only a relatively small portion of a farmer's land would simply replace what is currently in fallow and would not conflict with crop production levels.

Fallow rotation using *Leucaena leucocephala* has long been a traditional farming practice at Lilirilau, in Soppeng district, in the Indonesian province of South Sulawesi. Land is planted with both annual and perennial crops, but annual crops, including onions, corn, and tobacco, occupy more than half of the land. With little or no chemical fertilizers, soil productivity steadily declines and reaches an unacceptable level after three to five years of annual cropping. The land is then fallowed, and *Leucaena* trees sprout from the seeds of previous fallows. The *Leucaena* grows densely and reaches a height of three to eight meters after three to five years. It is then cut. The woody parts are used as firewood, while the green parts are used as mulch.

In this case, a hedgerow system could be established on land currently covered by *Leucaena* simply by cutting the *Leucaena* 30 cm above the surface of the soil to form hedgerows, and at the soil surface in areas to be used as cropping alleys. Farmers cultivating neighboring fields, and who wanted to change to the new system, could make similar adjustments to form continuous contour hedgerows. However, increased agricultural inputs such as fertilizers would be needed to support this relatively more intensive system and there could be unfavorable interactions between the food crops and the hedgerows. These might include competition for light, nutrients, and water; allelopathy; and pests and diseases harbored by the hedgerows (Kang et al. 1990; Lal 1991).

Leucaena leucocephala is not unique in its suitability as a hedgerow species. Several other suitable species have been reported, including *Gliricidia sepium* (Agus et al. 1999; Garrity and Agus 1999) and senna (*Cassia spectabilis*) (Maclean et al. 1992). In addition, several grass species have been used as hedgerow crops and have been received very positively by farmers owning ruminant livestock (Abdurachman and Prawiradiputra 1995). The most appropriate species for a particular area would be the one most adaptable and familiar to farmers. The introduction of exotic species should not be given priority unless they promise significantly greater benefits.

The Study Site

A notable example of a fallow rotation system that would benefit from modification to hedgerow cropping occurs at Tetewatu, a village 27 km east of Watan Soppeng, the capital town of Soppeng district, and about 200 km north-northeast of Ujung Pandang, the capital city of South Sulawesi. The land is undulating to hilly with dominant slopes ranging from 10 to 40%. The village is located 50 to 130 m above sea level (asl). Annual rainfall is 1,540 mm, with three consecutive months having rainfall exceeding 200 mm and two to three months having less than 100 mm (Agus et al. 1995). Mean monthly air temperatures range from 25.4 to 26.6°C.

The soil is classified as very fine, mixed, isohyperthermic, Vertic Ustropept (USDA Soil Survey Staff 1994). It was derived from limestone and has a mixture of 2:1 smectite and illite, and 1:1 kaolinite minerals. This mixture causes the vertic, or shrinking and swelling, property of the soil. Soil cation-exchange capacity is high to very high. However, the concentration of Olsen-P is very low (see Tables 25-1 and 25-2), although 25% HCl-extractable P ranged from 1,300 to 2,100 mg/kg of soil. This indicates that P may be fixed by Ca and Mg, and is therefore unavailable.

Table 25-1. Comparison of Selected Soil Properties

Sample	*Horizon*	*Soil Depth (cm)*	*pH*	*Silt (%)*	*Clay (%)*	*CEC (cmol(+)/kg)*
	A	0–15	7.8	24	75	38
	Bw1	15–40	7.5	23	77	36
L-2	Bw2	40–64	7.7	29	71	39
	Bw3	64–90	7.7	24	75	38
	B/C	90–130	8.0	28	71	38
	A	0–9	7.3	24	75	36
	Bw1	9–32	8.1	25	75	37
Lo	Bw2	32–67	8.0	25	75	36
	Bw3	67–94	7.8	22	77	37
	B/C	94–130	7.9	16	83	37

Notes: L-2 = Fields under annual crop cultivation for two years; Lo = Fields under annual crop cultivation a few months after cutting the *Leucaena* fallow crop, in a 10-year rotation of annual crops and *Leucaena*.
Source: Husen and Agus (1998).

Table 25-2. Selected Soil Properties at Different Years of the Annual Crop–*Leucaena* Rotation

Code and Horizon	*Organic C*	*Total N*	*Olsen P*	*Exchangeable Cation (1 N NH4-OAc, pH 7)* Ca [a]	Mg	K	CEC	*BD*	*Total Pore Volume*	*K-sat*
	(g/kg)		*(mg/kg)*	*(cmol(+)/kg)*				*(g/cm³)*	*(%)*	*(cm/h)*
L-2, A	9.8	1.1	0.9	63	6.9	0.5	38	1.1	59	0.16
L-2, Bw1	3.3	0.6	1.3	56	10.6	0.3	36	1.1	58	1.24
L0, A	7.9	1.1	1.3	65	2.4	0.5	36	1.1	58	2.37
L0, Bw1	3.7	0.6	0.9	73	2.5	0.4	37	1.1	58	0.13
L1, A	7.8	0.8	1.3	65	5.3	0.6	37	1.0	62	1.59
L1, Bw1	3.7	0.6	2.6	66	6.2	0.4	43	1.1	60	2.39
L3, A	12.0	1.0	2.2	64	4.6	0.6	35	1.1	60	2.77
L3, Bw1	5.3	0.7	3.1	64	6.0	0.3	34	1.2	54	0.77
L5, A	14.9	1.6	2.6	71	3.7	1.0	42	1.0	63	3.45
L5, Bw1	6.0	0.8	1.7	70	2.8	0.5	39	1.1	58	1.01

Notes: [a] Soluble cations were not separated from the exchangeable cations; L-2 = two years under corn and onion cultivation; L0 = zero year, after cutting *Leucaena*, and just planted to onion and corn; L1 = one year under *Leucaena* fallow; L3 = three years under *Leucaena* fallow; L5 = five years under *Leucaena* fallow.
Source: Husen and Agus (1998).

Current Farming Systems

Corn, onions, and cacao are the most commonly grown food crops in Tetewatu. Corn is planted either on its own or intercropped with onion or tobacco. At one time, tobacco was the main cash crop, but a drop in its price in 1989 forced farmers to seek a substitute. Cacao then became popular, although its productivity on soils with vertic properties and a sticky B horizon is only about half of that in neighboring subdistricts with Alfisols having good soil structure. The most common farming systems involve mixed cultivation of annual crops with perennial tree crops. Cacao is one of the most popular smallholder plantation crops, and perennial tree crops are also very common in home gardens. Annual crops alone are very common on farms of less than one hectare. On larger farms of more than one hectare, part of the land is devoted to perennial cash crops while the remainder is planted in annual crops (see Figures 25-1 and 25-2) (Agus et al. 1995).

Field observations indicate that the soil is very prone to water erosion, especially rill and gully erosion. At a depth of 10 to 50 cm, there is a very distinctive layer of soil referred to locally as the *bangi* layer. It is pale in color, is sticky when wet and hard when dry. Farmers believe that the bangi layer reduces the productivity of corn, onion, garlic, and cacao. It is very different from the overlying surface layer, which has a crumbly, friable, and soft consistency. This contrast is believed to have resulted from the accumulation of organic matter in the surface layer. Farmers are well aware of the effects of erosion. They believe that the closer the bangi layer is to the surface, the lower the soil productivity will be, and they attribute reduced thickness of the friable surface layer to erosion. In fields covered with five-year-old *Leucaena,* the saturated hydraulic conductivity was found to be 3.5 cm/hr in the A horizon and 1 cm/hr in the B horizon. This drastic change in hydraulic conductivity makes the soil susceptible to erosion.

One-meter-wide gullies with depths up to 1.5 m are commonly found on soils under annual crop production, especially in areas where water flow concentrates. Signs of rill erosion can be seen everywhere, especially on fields planted to annual crops. Farmers have adopted a number of measures to reduce the erosion hazard. They include plant residue management; fallow rotation with *Leucaena leucocephala*; terracing the land; and what is known as the "Tetewatu system," in which perennial crops are planted around the borders of fields.

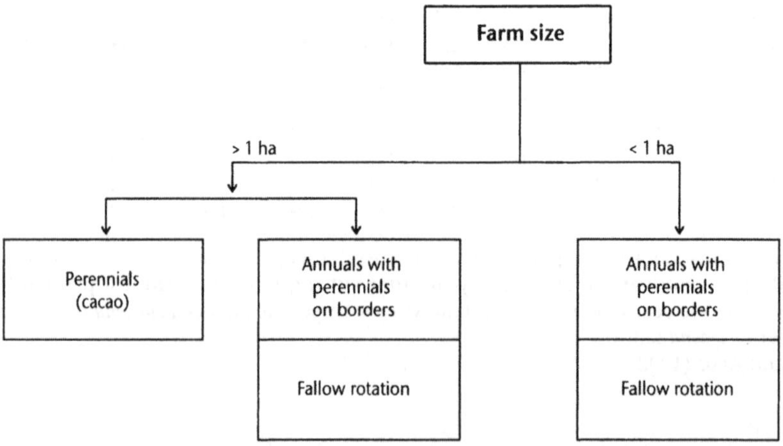

Figure 25-1. Indigenous Decision Tree for Determining Cropping Systems at Tetewatu Village
Source: Agus et al. (1995).

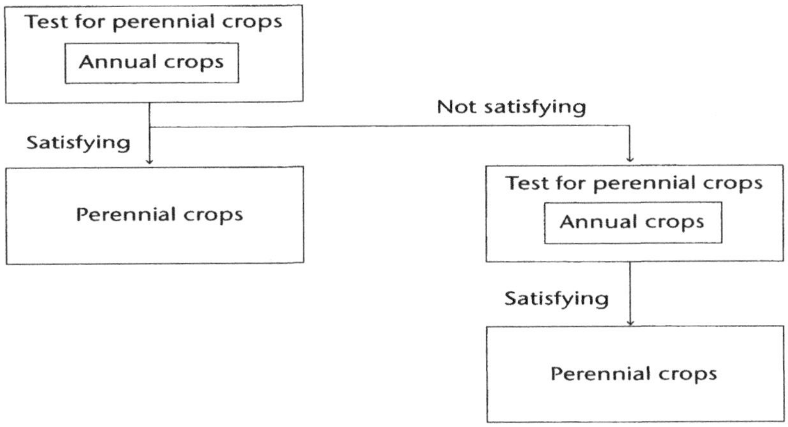

Figure 25-2. Indigenous Approach for Testing New Perennial Crops at Tetewatu Village
Source: Agus et al. (1995).

Plant Residue Management

Plant residues are usually arranged in windrows along the contour lines of annual crop production. Heavier plant residues, such as banana trunks, are usually dumped into gullies. This indigenous technique lessens the erosion problem to some extent, but its effectiveness declines over time.

Fallow Rotation with Leucaena

Annual crops, consisting of corn, onions, tobacco, and peanuts are planted either singly or in multiple cropping arrangements for three to five consecutive years, depending on a farmer's use of fertilizers. After this period, soil productivity becomes uneconomical and the land is left fallow. *Leucaena leucocephala* emerges, appearing to regenerate from the leftover stumps of the most recent fallow or from dormant seeds scattered on the surface soil layer. The former is the more likely because during the annual cropping period *Leucaena* seedlings constantly emerge near tree stumps and are treated like weeds. The fallow period lasts three to five years, after which the *Leucaena* is cut at ground level. The woody parts are harvested as firewood and the green parts are left on the ground to decompose.

This fallow system rejuvenates soil fertility by contributing nitrogen, organic matter, and plant nutrients; increases water penetration and storage in the deeper soil layers after *Leucaena* roots penetrate the sticky bangi soil layer; prevents rill and gully erosion, which disappear after a few years under the *Leucaena* fallow; and allows humus to accumulate. Field observations revealed that the thickness of humus under the one-, three-, and five-year *Leucaena* fallows was two, four, and seven centimeters, respectively (Husen and Agus 1998).

In general, it has been shown that soil properties do not change much between the start of annual cropping and three years into the *Leucaena* fallow. Organic C and total N accumulate after the fallow period reaches five years of age (Table 25-2). There is a significant increase in K, but it may not be important because the soil has a relatively high number of exchangeable cations. Almost no change was observed in the physical properties of the soil.

This study is limited by the fact that the length of fallow and cropping period was confounded by the original soil properties. Soil samples were taken from different fields at different levels of fallow management. Long-term research should seek to better understand the dynamics of soil properties under this fallow system.

Terracing

Bench terracing facilitated by extension workers was found on a few demonstration plots. The demonstration apparently had little or no impact. Farmers claimed that soil productivity declined after the formation of the terraces, especially near the terrace base, where the bangi soil layer was frequently exposed to the surface. Crop growth near the terrace bunds was reportedly fairly good but could not compensate for yield reductions near the terrace base. In addition, terrace risers were vulnerable to sliding in heavy rain, so the construction of bench terraces is not suitable on this soil, with its unstable structure (Agus 2001). To overcome the stability problem, extension agents planted mulberry (*Morus alba*) in hedgerows on the terrace bunds. This was effective on slopes under 15%, but less so on steeper slopes.

The "Tetewatu System"

This system is very specific to Tetewatu village. The land is divided into two parts, a central part that is planted to annual crops such as corn, onion, and tobacco, and a border with widths ranging from two to five meters that is planted to perennial crops such as banana, papaya, cacao, coconut, and jackfruit (Figure 25-1). The spacing of the perennial plants follows the farmers' estimate of canopy diameters. As well as being a means to distinguish land ownership and to generate additional farm products, the border area is used by farmers to trial alternative marketable tree crops. If the crop eventually proves to be either nonadaptive or nonbeneficial, the farmers slash the trees and return to mixes of perennials on the field borders. Most farmers use urea and TSP fertilizers, especially for onions and tobacco. However, the rate and frequency of application is highly variable.

Approach for the Future

This chapter has described a *Leucaena* fallow system that is a well-established indigenous practice. After a five-year fallow period, the leaves and other green parts of the *Leucaena* contribute to each hectare of land as much as 254 kg of N, 12 kg of P, 171 kg of K, 251 kg of Ca, and 38 kg of Mg (Table 25-3). Of these nutrients, N may be the most important. The P recycled by *Leucaena* is a small part of the P required by each annual crop, which is estimated to be 20 to 40 kg/ha. Calcium, Mg, and K are not limiting, so the additional amounts provided by *Leucaena* prunings may not affect crop growth. In addition to N and organic matter, *Leucaena* trees also provide firewood (Table 25-4).

In the existing system, it takes three to five years of *Leucaena* fallow to support three to five years of annual crop production. I propose the modification of the fallow system to make it into a hedgerow system, with a hedgerow width of 0.5 to 1 m, and an alley width of 4.5 to 9 m. The land devoted to *Leucaena* would therefore be about 10 to 20% of the total land area. The hedgerow trees would be cut at 30 cm height about every three to six months, and the supply of organic matter and nitrogen to the soil would be more constant, instead of the present flushes of carbon and nitrogen every five years followed by a serious deficit over the following years. In addition, the hedgerows would serve as gully and rill plugs.

Table 25-3. Nutrient Content of *Leucaena* Leaves and Their Contribution to Fallow Fields

	Concentration in Leaves (g/kg)		*Contribution to Soil (kg/ha)*	
Element	*Mean*	*Standard Deviation*	*Mean*	*Standard Deviation*
N	49	6	254	34
P	2.5	0.7	12.5	4.5
K	41	8	171	35
Ca	49	9	251	47
Mg	8.0	1.0	38.2	4.2

Note: Means were from five samples of four-year-old *Leucaena.*

Source: Husen and Agus (1998).

Table 25-4. Parts of *Leucaena* Trees at the Fifth Year of Fallow

Plant Part	*Sun-Dried Volume (m^3/ha)*	*Value[a] (Rs/m^3)*	*Expected Revenue (Rs/ha)*
Stems for firewood	105	20,000	2,100,000
Branches	23	Not sold, for domestic use	—
Leaves and adjacent green parts	9.7	Not sold, for mulch on the same field	—

Note: Averaged from five fields. [a] US$1.00 = Rs8,500.

Conclusions and Recommendations

The farmers in the study area are not only receptive to new technologies, but they are also using an indigenous method for testing plant adaptability on the borders of their fields. Therefore, any innovation that holds promise of meaningful benefits will be adopted. *Leucaena leucocephala* fallow systems have also been practiced in the area for many years. They have sustained the productivity of annual crops and, to some extent, have reduced erosion. However, significant rill and gully erosion still exists between the fallow fields. The local farming systems would benefit from a continuous hedgerow system planted along contour lines. It is believed this would effectively reduce rill and gully formation, and because it represents only a minor modification to existing management practices, the chances of its adoption are very high.

References

Abdurachman, A., and B.R. Prawiradiputra. 1995. Tinjauan Penelitian dan Penerapan Teknologi Usahatani Konservasi di DAS Jratunseluna dan DAS Brantas: Pengalaman UACP-FSR. In: *Analisis Agroekosistem dan Pengelolaan DAS, Prosiding Lokakarya Pembahasan Hasil Penelitian 1994–1995 dan Rencana Penelitian 1995–1996,* February 15–17, 1995, Bogor, Indonesia: Pusat Penelitian tanah dan Agroklimat, 217–232.

Agus, F. 2001. Selection of Soil Conservation Measures in Indonesian Regreening Program. In *Sustaining the Global Farm: Selected Papers from the 10th International Soil Conservation Organization (ISCO) Meeting, May 24-29, Purdue University,* edited by D.E. Stott, R.H. Mohtar, and G.C. Steinhardt. Purdue, USA: Purdue University Press, 198–202.

______, D.P. Garrity, and D.K. Cassel. 1999. Soil Fertility in Contour Hedgerow Systems on Sloping Oxisols in Mindanao, Philippines. *Soil and Tillage Research* 50:159–167.

———, A. Rachman, and N.L. Nurida. 1995. Analisis Agroekosistem di Daerah Aliran Sungai Billa Walanae: Desa Tetewatu, Kecamatan Lilirilau, Kabupaten Soppeng, Sulawesi Selatan. In *Analisis Agroekosistem dan Pengelolaan DAS, Prosiding Lokakarya Pembahasan Hasil Penelitian 1994–1995 dan Rencana Penelitian 1995–1996*, February 15–17, 1995, Cipayung, edited by A. Abdurachman, D. Santoso, B.R. Prawiradiputra, M.H. Sawit, A.N. Gintings, and F. Agus. Bogor, Indonesia: Pusat Penelitian Tanah dan Agroklimat, 111–134.

Cairns, M. 1994. Stabilization of Upland Agroecosystems as a Strategy for Protection of National Park Buffer Zones: A Case Study of the Co-Evolution of Minangkabau Farming Systems and the Kerinci Seblat National Park. MSc thesis. York University, Ontario, Canada.

Garrity, D.P., and F. Agus. 1999. Natural Resource Management on a Watershed Scale: What Can Agroforestry Contribute? In *Integrated Watershed Management in the Global Ecosystem*, edited by R. Lal. Washington, DC: CRC Press, 165–193.

Husen, E., and F. Agus. 1998. Karakteristik dan Pendekatan Konservasi Tanah di DAS Billa Walanae, Sulawesi Selatan. In *Prosiding Lokakarya Nasional Pembahasan Hasil Penelitian Pengelolaan Daerah Aliran Sungai*, October 27–28, 1998, Bogor, edited by F. Agus, B.R. Prawiradiputra, A. Abdurachman, T. Sukandi, and A. Rachman. Bogor, Indonesia: Pusat Penelitian Tanah dan Agroklimat, 291–304.

Kang, B.T., L. Reynolds, and A.N. Atta-Krah. 1990. Alley Farming. *Advances in Agronomy* 43, 315–359.

Lal, R. 1991. Myths and Scientific Realities of Agroforestry as a Strategy for Sustainable Management for Soils in the Tropics. *Advances in Soil Science* 15, 91–137.

Maclean, R.H., J.A. Litsinger, K. Moody, and A.K. Watson. 1992. Increasing *Gliricidia sepium* and *Cassia spectabilis* Biomass Production. *Agroforestry Systems* 20, 199–212.

USDA Soil Survey Staff. 1994. *Keys to Soil Taxonomy*, 6th ed. Soil Conservation Service, US Department of Agriculture.

Chapter 26

Upland Rice Response to *Leucaena leucocephala* Fallows on Mindoro, the Philippines

Kenneth G. MacDicken

Upland rice is grown on about 19 million ha globally and takes up about 13% of the world's rice-growing area. In some regions it covers more than half of the total rice-growing area (Gupta and O'Toole 1986). It is an important staple crop for shifting cultivators in many parts of the humid tropics. Nevertheless, it is generally grown only as a subsistence crop planted by resource-poor farmers who apply little or no fertilizer, generally resulting in grain yields of less than one metric ton/ha. There are an estimated 250 million shifting cultivators worldwide and about 100 million of them live in Southeast Asia (Christanty 1986). As demand for land increases because of an increasing human population, the area available to shifting cultivators for fallowing becomes smaller and smaller, so to maintain their cropping area, farmers fallow their land for shorter periods. One result of this is lower crop yields (Warner 1991). One strategy for slowing the decline in crop yields is the use of fast-growing, nitrogen-fixing tree species as fallow crops to improve nutrient availability for subsequent crops in the shifting cultivation cycle (Ahn 1979; Unruh 1990).

The leguminous tree species *Leucaena leucocephala* (Lam.) DeWit has been widely used in a variety of indigenous and introduced agroforestry practices and has shown promise as an effective fallow improvement crop (IITA 1980; Field and Yasin 1991; MacDicken 1991a). While substantial research has been conducted on alley cropping with *Leucaena*, there remains a lack of information on the effects of planned or planted *Leucaena* fallows in shifting cultivation systems. Such information should cover effects on crop yields, soil erosion, soil nutrient contributions, adoption by farmers, and sustainability.

Leucaena fallows have sustained the quality and yield of maize and tobacco crops at Naalad, on Cebu, in the central Philippines, for more than 100 years (Lasco 1990, Chapter 27). However, while the potential for *Leucaena* fallows has been described (MacDicken 1981), its use in shifting cultivation has not been thoroughly studied. The experiment described in this chapter was designed to evaluate rice yield differences following natural bush fallows and planned or planted *Leucaena* fallows.

Leucaena fallows were first established in the study area in 1976, using varieties K8 and K28, on lands controlled by members of the Iraya Mangyan tribe. The Iraya are traditional shifting cultivators who, in the past, used secondary forest fallows. In recent years they have used *Leucaena* fallows established through stump cuttings, direct seeding, or assisted natural regeneration (see color plate 36). Fallow periods average just over three years. More than 80% of the households in the study area

---

Kenneth G. MacDicken, Director, Forest Management Services, Winrock International, 85 Avenue A, Suite 301, Turners Falls, MA 01375, USA.

have *Leucaena* in at least one swidden field. However, only 10% of farms sampled in 1990 had at least half of their fields planted to *Leucaena*.

Leucaena fallows in the study area are managed in an identical fashion to natural bush fallows, which are cut between January and March and burned and cleaned in April or May; crops are planted soon after burning. Attempts were made to encourage farmers not to burn, but these failed without exception. An experiment to test the impact of burning on crop yields and regeneration found no significant differences ($p < .05$) between the burned and unburned treatments in grain or straw yield, or in soil or foliar nutrients, although foliar P concentration was lower in the unburned treatment at $p = .06$ (MacDicken and Ballard 1996). There were no detectable differences between soils sampled before burning and those collected 182 days later. Weeding requirements were greater in the unburned treatment, as was stump survival; 41% of stumps survived in the unburned treatment whereas only 4% survived burning.

The use of *Leucaena* fallows with burning makes good agronomic sense. Leaving the slashed biomass unburned does not improve crop yield or the soil nutrient status. Weeding costs are lower when the fields are burned and the fire helps regeneration by scarifying seed. The *Leucaena* seedlings are easy to weed when they are less than about six weeks old and they provide a rich source of nitrogen to the young rice crop.

Methods

The multilocation experiment was conducted on nine sites in Sitio Sto. Tomas, Barangay Wawa, Abra de Ilog, Occidental Mindoro, Philippines (13°29' north latitude and 120°32' east longitude). The watershed where the studies were conducted covers about 800 ha and has a population of about 200. By the time of this experiment, the density of *Leucaena* stands with stems more than five cm in diameter at breast height ranged from 1,860 to 3,725 trees per hectare. Fallow lengths were two to five years.

All nine locations were managed using the Iraya farmers' traditional cultivation methods on sloping lands, including minimum tillage, comparable plant spacings, two manual weedings, and no fertilizer or pesticide inputs. Management was identical for both treatments within each location. Indigenous rice varieties were grown by Iraya farmers based on their preferences and seed availability. However, the same rice variety was used in both treatments at each location. Fallow vegetation was cleared in February and March, it was burned in April, and the rice was direct seeded using a dibble stick. At least two weedings were conducted in each treatment.

The Study Site

The study site is humid monsoonal lowland with distinct wet and dry seasons. Mean annual rainfall is about 2,000 mm, falling mainly from May to November with an isohyperthermic temperature regime. Soils are generally moderately deep, measuring 50 to 100 cm to lithic or paralithic contact, are clay-loam to loam in texture and moderately acidic (pH 5.5 to 6.2). The topography is hilly to mountainous, with most farm fields on slopes of 20 to 60%. Soils are classified as Typic Ustropepts and are freely drained with an ochric epipedon present and high base saturation (greater than 50%). In the study area these soils are commonly used as woodlands or for nonirrigated field crops.

Natural vegetation is predominantly of the monsoonal Molave forest type, characterized by the presence of the deciduous *Vitex parviflora*. However, over the past 50 years it has given way to coarse grasses, including *Imperata cylindrica* and *Saccharum* spp., and degraded secondary forest. Most areas are characterized by fewer than 10 large trees per hectare, scattered or dense shrubs, grasses, climbers, and other herbs. Secondary forest is characterized by species of *Albizia, Intsia, Ficus, Antidesma, Alstonia, Trema, Diospyros, Mitragyna, Terminalia,* and *Macaranga*. *Chromolaena odorata* is the predominant nonwoody perennial in many fallowed fields.

Experimental Design and Measurements

A randomized complete block design was used, with two treatment levels: *Leucaena* and non-*Leucaena* fallows. The experiment was designed and analyzed as a multilocation trial with fields as locations. Three replications were used per location except where total field size was smaller than 2,000 m^2. Two replications were used on these locations. Experiments were established in farm fields where paired treatment plots (*Leucaena* and non-*Leucaena*) were contiguous and on the same slope position, that is, mid-slope or upper-slope, aspect, and slope gradient. Treatment plots were 10 m by 10 m, allocated randomly within blocks.

Rice grain and straw fresh weight, the percentage of filled spikelets, 1,000-grain weight, and the number of tillers/m^2 were measured in eight 50 cm by 50 cm subplots using the methods described by Gomez (1972). Both treatments were harvested together in each field, with the date of harvest set by the farmer-cooperator. Grain was harvested by removing the panicles in the field. Spikelets were removed by hand before weighing. All tillers were counted, effective and ineffective. All counts were made using a multiple tally meter. Subsamples of grain and straw were oven dried to constant weight for use in adjusting fresh weight to oven dry weight.

The analysis of variance and Tukey's HSD were conducted on grain and straw yield, 1,000-grain weight, and tiller number. The analysis was also done using square–root transformed values for the percentage of filled *spikelets* and arc–sine transformed values for grain–to–straw ratio and grain moisture content. Grain moisture contents were adjusted at the plot level to 14 cg H_20 per gram dry matter, using the relationship between grain yield and moisture content described by Nangju and Datta (1970).

Results and Discussion

Grain Yield

Grain yield was significantly higher ($p < 0.01$) in the *Leucaena* fallow treatment compared to the natural fallow (see Table 26-1). Moreover, grain yields were highest in fields that had the longest fallow length (see Figure 26-1). Total soil N was significantly higher ($p < 0.05$) in three- and four-year-old *Leucaena* fallows (see Table 26-2).

The additional nitrogen in the *Leucaena* fallow came from two primary sources: litterfall and *Leucaena* seedlings weeded during the first eight weeks of the rice crop and left as surface mulch. Estimates based on literature values suggest that leaf litter added between 100 and 200 kg of nitrogen per hectare during every year of the fallow period. Nitrogen inputs from weeded *Leucaena* seedlings left as surface mulch during the first eight weeks of crop growth ranged from 32 to 280 kg N/ha.

Farmers in the study area plant their rice so that it germinates with the first reliable rains of the season, usually during May. Soils under *Leucaena* fallows that were more than two years old were higher in total soil nitrogen, nitrate, and ammonium at the time of clearing than soils under non-*Leucaena* fallows. This, and the generally low rate of nitrification that can be expected in dry soils through the dry season, probably provide a flush of nitrate following the onset of rains. Since planting is done to coincide with these rains, the young rice seedlings are able to use this nitrate to speed early development.

Higher yields following the *Leucaena* fallows were not explained by soil nutrients at the time of harvest, or by the foliar P difference in fallow treatments. This is consistent with the findings of investigators working with *Leucaena* prunings in alley cropping systems (Kang et al. 1985; Sanginga et al. 1989). The N:P ratio was higher in the *Leucaena* fallow treatment that produced the highest grain yields, after adjustment for moisture content. This suggested that P alone was not a limiting factor to grain yield, even though P levels in all sites appeared to be low. N:P ratio and grain yield adjusted for moisture content were linearly correlated ($R = 0.78$, $p =$

0.02). Higher N:P ratios in rice straw indicated higher levels of available N for rice after *Leucaena*, which was consistent with the finding of greater available soil N after the *Leucaena* fallow.

These quantities of N, P, and K were significant for rice production. For example, rice grain yield responses to N fertilization are frequently 20 kg or more per kg of N applied to the crop.

Table 26-1. Upland Rice Grain Yield, Yield Component Means, and Analysis of Variance

Treatment		*Grain Yield (t/ha)*	*Straw (t/ha)*	*Mean Grain Moisture Content (cg/g)*	*1000 Grain Weight (g)*	*Percent Filled Spikelets (cg/g)*	*Grain: Straw Ratio*
Natural fallow		2.7	5.4	23.9	25.4	90.9	0.71
Leucaena fallow		3.8	6.2	27.2	24.6	86.5	0.74
Source	*df*	*Grain Yield*	*Straw*	*Grain Moisture Content*	*1000 Grain Weight*	*Grain Fill*	*Grain: Straw Ratio*
Location	8	**	**	**	**	**	**
Reps in Location	11	NS	NS	NS	NS	NS	NS
Fallow	1	**	NS	**	NS	*	NS
Location x fallow	8	NS	NS	NS	NS	NS	NS
Pooled residual	11						

Note: * = F is significant at $p = 0.05$, ** = F is significant at $p = 0.01$, NS = not significant at $p = 0.05$, *df*= degrees of freedom.

Table 26-2. Effects of Fallow Type and Age on Total Soil Nitrogen Concentrations (cg/g)

	Fallow Type	
Age (years)	Leucaena	*Non*-Leucaena
1	0.160a	0.160ab
2	0.195a	0.190a
3	0.180a	0.150b
4	0.145b	0.105c

Notes: Values are the mean of two replications. Mean separations in a column by Tukey's HSD at $p = 0.05$. Column values followed by the same letter are not significantly different at $p = 0.05$. LSD.05 value for comparing fallow ty pe means in a row is 0.022.

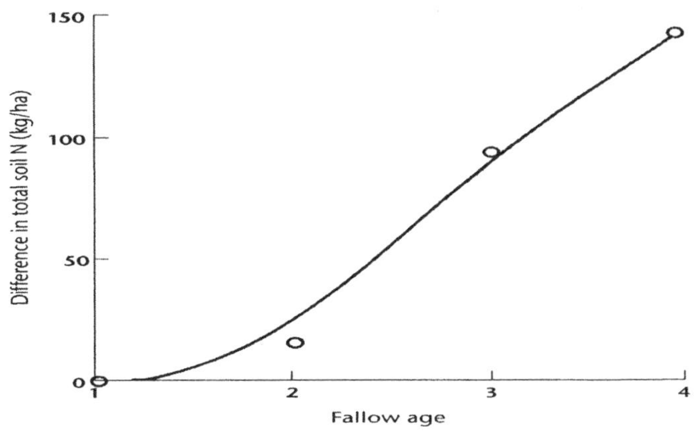

Figure 26-1. Net Difference in Total Soil Nitrogen Levels in *Leucaena* and Non-*Leucaena* Fallows (0 to 30 cm Soil)

Grain Maturation

Consistent and significantly higher ($p < 0.01$) grain moisture content following the *Leucaena* fallows (see Table 26-1) suggested that the development and maturation of grain had been delayed, or the crop developed more slowly. This is further indicated by lower grain filling in the *Leucaena* fallow treatment.

The influence of nitrogen on crop development is dependent on species, soils, and climate. Nangju and Datta (1970) compared grain yields and time of harvest of four rice varieties under three different N fertilizing regimes in a dry season study at the IRRI experimental fields at Los Baños, Laguna, in the Philippines. Each of the four varieties, including one tall indica type, were grown without N, with 60 kg N/ha, and with 120 kg N/ha. The "window" for maximum grain yield in the 120 kg N/ha treatment was about four days later, and that in the 60 kg N/ha treatment about one day later, than that for the rice grown without nitrogen inputs. The same experiment conducted during the rainy season failed to produce these differences in grain maturity.

Conclusions

Grain yield was higher by an average of 42 cg/g in the *Leucaena* fallow treatment, with a range of differences across locations of 19 to 106 cg/g. Soil nutrients generally did not differ significantly between fields at the time of rice harvest, even though there were significant differences between fields in grain and stover yields. Total soil nitrogen under the three- and four-year-old *Leucaena* fallows was higher than in the non-*Leucaena* fallows, suggesting that the potential contributions of nitrogen from *Leucaena* in the field, under the conditions found in Sto. Tomas, take at least three years to be potentially useful to crops. The crop yield data do not conclusively demonstrate a direct fallow age to yield relationship, but yields in the longer fallows were generally higher than those in the shorter fallows. This suggests that the relationship between fallow age and total soil nitrogen detected in this experiment is an important indicator of the amount of time required to accumulate enough soil nitrogen to achieve a positive yield response.

The higher moisture content of grain from the *Leucaena* fallow treatments appears to relate to somewhat slower grain development due to the accumulated total soil nitrogen. Farmers in Sto. Tomas tend to harvest their rice crops while the grain moisture content is higher than the optimum for maximum yield, presumably to reduce the risk of grain loss due to weather or herbivory. Higher grain and straw moisture contents, plus lower grain fill percentages in rice following *Leucaena* fallows, suggest that upland rice grown after *Leucaena* fallows is physiologically less mature than rice grown in non-*Leucaena* fallows.

This experiment was conducted on a site that is well suited to *Leucaena leucocephala,* and on which upland rice yields were exceptionally high in the study year.

Acknowledgments

I wish to thank the farmers of Sto. Tomas, the Bureau of Soils and Water Management, the Philippine-American Education Foundation, the University of British Columbia, and the Winrock International Institute for Agricultural Development for their contributions to this work.

References

Ahn, P.M. 1979. The Optimum Length of Planned Fallows. In: *Soil Research in Agroforestry*, edited by H.O. Mongi and P.A. Huxley. Nairobi, Kenya: ICRAF (World Agroforestry Centre).

Christanty, L. 1986. Shifting Cultivation and Tropical Soils: Patterns, Problems and Possible Improvements. In: *Traditional Agriculture in Southeast Asia*, edited by G.G. Marten. Boulder, Colorado: Westview Press.

Field, S.P., and H.G. Yasin. 1991. The Use of Tree Legumes as Fallow Crops to Control Weeds and Provide Forage as a Basis for a Sustainable Agricultural System. Proceedings of the 13th Asian-Pacific Weed Science Society Conference, 121–126.

Gomez, K.A. 1972. *Techniques for Field Experiments with Rice*. Los Baños, Laguna, Philippines: IRRI (International Rice Research Institute), 48.

Gupta, P.C, and J. C. O'Toole. 1986. *Upland Rice—A Global Perspective*. Los Baños, Laguna, Philippines: IRRI (International Rice Research Institute), 352.

IITA (International Institute of Tropical Agriculture). 1980. *Annual Report for 1979*. Ibadan, Nigeria: IITA.

Kang, B.T., T. Grimme, and T.L. Lawson. 1985. Alley Cropping Sequentially Cropped Maize and Cowpea with *Leucaena* on a Sandy Soil in Southern Nigeria. *Plant and Soil* 85, 267-277.

Lasco, R.D. 1990. Multipurpose Tree Species in Indigenous Agroforestry Systems: The Naalad Case. In: *Research on Multipurpose Tree Species in Asia*, Proceedings of an international workshop, November 19–23, Los Baños, Laguna, Philippines, edited by David Taylor et al. Arlington, VA: Winrock International.

MacDicken, K.G. 1981. *Leucaena* as a Fallow Improvement Crop: A First Approximation. Paper presented at an East-West Center workshop on Environmentally Sustainable Agroforestry and Firewood Production with Fast-Growing, Nitrogen-Fixing, Multi-Purpose Legumes. November 1981, East-West Center, Honolulu.

———. 1991a. Impacts of *Leucaena leucocephala* as a Fallow Improvement Crop in Shifting Cultivation on the Island of Mindoro, Philippines. *Forest Ecology and Management* 45, 185–192.

———. 1991b. *Leucaena leucocephala* as a Fallow Improvement Crop in Shifting Cultivation on the Island of Mindoro, Philippines. In: *Research on Multipurpose Tree Species in Asia,* Proceedings of an international workshop, November 19-23, 1990, Los Baños, Laguna, Philippines, edited by David Taylor et al. Arlington, VA: Winrock International, 34–40.

———. and T.M. Ballard. 1996. Effects of Burning on Upland Rice Crop Yields and Soil Properties Following a *Leucaena leucocephala* Fallow on the Island of Mindoro, Philippines. *Tropical Agriculture (Trinidad)* Vol. 73(1), 10–13.

Nangju, D., and S.K. Datta. 1970. Effect of Time of Harvest and Nitrogen Level on Yield and Grain Breakage in Transplanted Rice. *Agronomy Journal* 62, 468–474.

Sanginga, N., L. Mulongoy, and M.J. Swift. 1989. Contribution of Nitrogen by *Leucaena leucocephala* and *Eucalyptus grandis* to Soils and a Subsequent Maize Crop. In: *Proceedings of a Regional Seminar on Trees for Development in Sub-Saharan Africa,* February 20–25, 1989, Nairobi, Kenya. Stockholm: International Foundation for Science, 253–258.

Unruh, J.D. 1990. Iterative Increase of Economic Tree Species in Managed Swidden-Fallows of the Amazon. *Agroforestry Systems* 11, 175–197.

Warner, K. 1991. Shifting Cultivators: Local Technical Knowledge and Natural Resource Management in the Humid Tropics. In: *Community Forestry Note* 8. Rome: FAO (Food and Agriculture Organization of the United Nations), 80.

Chapter 27

The Naalad Improved Fallow System in the Philippines and its Implications for Global Warming

Rodel D. Lasco

Global warming, or the increase of the earth's atmospheric temperature, is one of today's most pressing issues. Greenhouse gases (GHGs) such as carbon dioxide, methane, nitrous oxides, and chlorofluorocarbons absorb thermal radiation emitted by the earth's surface. A rising concentration of GHGs in the atmosphere could lead to changes in the world's climate and the consequences could be disastrous.

Among the GHGs, carbon dioxide (CO_2) is the most important by weight and is released mainly by the combustion of fossil fuels, burning or decay of vegetation, and by flux with oceans (Moura-Costa 1996). The single most important contribution to climate change originating from the world's forests is the release of CO_2 from deforestation. Of the 7 to 8 billion tonnes of total carbon (C) released into the atmosphere in 1988, deforestation, mainly in tropical countries, accounted for 1.6 billion tonnes (Trexler and Haugen 1995).

However, tropical forests can play an important mitigating role in climate change because they can be both sources and sinks of CO_2. At present, tropical forests are estimated to be a net source of C, primarily because of deforestation, harvesting, and forest degradation. But tropical forests represent 80% of the world's total forests, and they have the biggest long-term potential to sequester C. This can be achieved by protecting forested lands, slowing deforestation, reforestation, and agroforestry (IPCC 1996).

Like forests in general, agroforestry systems can be sources or sinks of GHGs. It is estimated that agrosilvicultural systems in the humid tropics can store between 12 and 228 tonnes C/ha (Dixon 1996). However, there is very little information on the C release and sequestration of specific agroforestry systems.

This chapter, therefore, has a dual purpose. First, it describes the indigenous Naalad improved fallow system. Second, it will attempt to estimate its C-sequestration ability.

Study Site and Methodology

The village of Naalad is located in the municipality of Naga, in the central Philippines province of Cebu. It is about 23 km southeast of Cebu City, at 10°12' north latitude and 123°45' east longitude, with an elevation of up to 300 m above sea

---

Rodel D. Lasco, ICRAF Philippines, 2FCFNR, Admi Bldg., UPLB, College, 4031 Laguna, the Philippines.

level (asl). Annual rainfall in central Cebu ranges from 1,600 mm to 2,000 mm. The area has very mountainous terrain with farms located on slopes of more than 100%.

This chapter uses data from two studies conducted in the area, both to describe the Naalad system and to estimate its C–sequestration ability. The description of the farming system is based on a study conducted from 1990 to 1993 (Lasco and Suson 1997). It documented and evaluated the Naalad system and gathered data on changes in soil properties and crop yields. The second study, by Kung'u (1993), measured biomass accumulation under different fallow ages.

The Naalad Improved Fallow System

Like most traditional fallow practices, the Naalad system has two basic components: the fallow field and the cultivated field. However, there are two vital differences: the use of *Leucaena leucocephala* in the fallow fields and the construction of fascine-like structures to minimize soil erosion in the cultivated fields (see color plate 35). The following discussion is based on Lasco and Suson (1997).

In traditional shifting cultivation, the fallow period is typically much longer than the cropping period. In the Naalad system, farmers discovered more than 100 years ago that by introducing *L. leucocephala* they could shorten the fallow period to just five to six years. When it is time to fallow, *Leucaena* seeds are sown into the fields. In addition, surviving stumps of *Leucaena* are allowed to sprout.

It is worth pointing out that, in economic terms, the fallow field is not entirely unproductive. Farmers gather foliage from the *Leucaena* trees and carry it to their cattle, providing an important source of fodder. It has been suggested that the functions of the fallow, aside from improving soil fertility, are often overlooked (Ohler 1985). Therefore, the functions of the Naalad fallow should be further investigated.

Construction of Fascine-Like Structures in Cultivated Fields

At the end of the fallow, the *Leucaena* trees are slashed. But in contrast to traditional shifting cultivation, they are not burned. Instead, stakes of *Leucaena* about 30 cm long are driven into the ground at regular intervals along the contours. Smaller branches, with a maximum diameter of about two centimeters, are piled against the uphill side of the stakes so that a fascine-like structure is formed. (A fascine was a military defense formed by bundles of sticks). The Cebu farmers call it a *balabag*, or *babag* in their local dialect, which means obstruction. The main function of these structures is to control erosion. In fact, as sediment collects behind the balabags, small terraces are formed after a few years.

The balabags are spaced between one and two meters apart, based on horizontal distance, and crops are planted between them, just like in alley cropping. Farmers recall that in the early years of the system, the space between balabags was much wider. There used to be up to five rows of corn within each alley, but over the years, the number of corn rows has been progressively reduced so that now there are usually only one or two rows of corn between the balabags. The reason for this is that the corn plants nearest to the balabags reportedly grow better than the plants in the middle of the alley. If this is true, then it suggests that the balabags, aside from minimizing soil erosion, also help improve soil properties. Presumably this could be due to nutrient contribution as the balabags decay and to the accumulation of more fertile soil sediments. The favorable microclimate around the balabags could also provide habitat for soil organisms, such as earthworms, resulting in improved physical and chemical properties.

Initially, researchers believed there was a flaw in the system: the decay of the dead *Leucaena* branches. It was feared that with the collapse of the balabags, there would be very high erosion rates, considering the steep slopes of the farms. However, it has been found that farmers use the collapse of the balabags as a key indicator of when a field should be fallowed. Generally, they begin to totally collapse about five to six years after construction. In traditional shifting cultivation, declining yields and weed problems are the main reasons for fallowing land and shifting elsewhere (Sanchez 1976). In Naalad, it is possible that the decay and collapse of the balabags coincides with unacceptably low yields or weed problems.

Cropping and Organic Fertilization

Corn is planted during the wet season from June to October, mainly for the subsistence needs of farmers. After the corn, a dry season crop of tobacco is grown from November to May. The tobacco leaves are dried and sold to meet the farmers' cash needs. During the third year of cultivation, farmers apply a small amount of chicken manure to their fields. This could be an additional reason why they are able to maintain yield levels. Typically no chemical fertilizers are applied.

Length of Cultivation

Just like the fallow phase, cultivation usually lasts for five to six years. Since the length of fallow period is equal to the length of cultivation period, farmers theoretically need only two parcels of land to make the system ecologically sustainable. Since Cebu is one of the most densely populated islands in the Philippines, the Naalad system was most likely developed by farmers to overcome land-use pressures from a rising population.

On some farms, the length of the cultivation or fallow periods may be shorter or longer than the "normal" five to six years. One factor determining the length of the cycle is the availability of labor, because most farm households have members working at nonfarm jobs to provide additional sources of income.

C–Sequestration Capacity of the Naalad System

Data gathered by Kung'u in 1993 showed that the dry weight of *Leucaena's* above ground biomass increased from 4.3 t/ha in the first year of fallow to 63.6 t/ha by the end of a six-year fallow (see Table 27-1). In estimating the C content of the *Leucaena* biomass, the following formula was used: C content = biomass dry weight/ha x 0.5. The assumption was that the C content of the biomass = 50%.

Table 27-1. C–Sequestration Ability of *L. leucocephala* Fallows

Years under Fallow	*Mean Dry Weight of Aboveground Biomass (t/ha)*	*Percentage of Leaves*	*Biomass C (t/ha)*	*Annual Rate of C Accumulation (t/ha)*
1	4.3 d	36.5	2.2	2.2
2	16.1 cd	13.8	8.1	5.9
3	17.6 cd	8.9	8.8	0.7
4	36.4 bc	7.4	18.2	9.4
5	53.8 ab	5.3	26.9	8.7
6	63.6 a	6.1	31.8	4.9
Mean	32		16	5.3

Note: Means in a column with the same letter are not significantly different using DMRT at 0.05.

Therefore, after six years, there were 31.8 tonnes C/ha in aboveground biomass of *Leucaena*. It could be assumed that C storage in the understory, soils, and woody debris was 25% of the aboveground storage (IPCC 1996). Thus, total C storage in the *Leucaena* fallow system amounted to about 40 tonne/ha at the end of the fallow period. On average, there were 16 t C/ha in any given *Leucaena* fallow.

The mean C stored is lower than my initial estimate of 22 t C/ha for Philippine agroforestry systems (Lasco 1997). However, it is within the lower range of 12 to 228 t C/ha stored by agrosilvicultural farms in Southeast Asia, a figure reported by Dixon in 1996. The total C storage of Naalad farms is, therefore, much lower than those of natural tropical forests in the Philippines, which store between 175 and 350 t C/ha, and tree plantations, which store between 29 and 102 t C/ha (Lasco 1997).

On an annual basis, *Leucaena* fallows accumulate 5.3 t C/ha/yr (see Table 27-1). This rate is comparable to tree plantations in the Philippines with annual C-sequestration rates between 3 and 5.4 t/ha (Lasco 1997). As expected, the annual accumulation rate is much higher than the 1 to 2 t/ha estimated for natural old-growth tropical forests (IPCC 1996).

The C stored in the biomass is released when the fields are opened for cultivation. The leaves all go to the soil while the smaller branches become the balabags. However, the larger branches are used for firewood and therefore represent the main loss of C from the system. No quantification of these losses is presently available.

In contrast to traditional shifting cultivators, the Naalad farmers do not burn their fields. This prevents any massive release of C to the atmosphere. The cultivated fields are planted with maize and tobacco. There is no burning of biomass involved in producing maize. However, tobacco leaves are eventually burned, thereby releasing C into the atmosphere. Quantification of the C balance is not possible because data is lacking.

Conclusions

The Naalad improved fallow system has the potential to mitigate global warming through its ability to sequester C in fallow fields. This is primarily because there is no burning after the fallow. This practice should be encouraged in other fallow systems.

The main C loss from the fallowed fields comes from burning *Leucaena* firewood. While this contributes to global warming, there is little that can be done to eliminate it. The alternative, which is to use fossil fuels, would have the same effect.

Tobacco leaf burning is the main C loss from the cultivated fields. Crops other than tobacco should be considered to help reduce C emissions from the system.

References

Dixon, R.K. 1996. Agroforestry Systems and Greenhouse Gases. *Agroforestry Today* 8(1), 11–14.

IPCC (Intergovernmental Panel on Climate Change). 1996. *Climate Change*. Cambridge, UK: Cambridge University Press.

Kung'u, J.B. 1993. Biomass Production and Some Soil Properties under a *Leucaena leucocephala* Fallow System in Cebu, Philippines. Unpublished MS thesis. University of the Philippines, Los Baños, Laguna, Philippines.

Lasco, R.D. 1997. Management of Philippine Tropical Forests: Implications to Global Warming. Paper presented at the 8th Global Warming Conference, May 28, 1997, Columbia University, New York.

———, and P.D. Suson. 1997. A *Leucaena leucocephala*-Based Improved Fallow System in Central Philippines: The Naalad System. Paper presented at an international conference on Short-term Improved Fallow Systems, March 1997, Lilongwe, Malawi.

Moura-Costa, P. 1996. Tropical Forestry Practices for Carbon Sequestration. In: *Dipterocarp Forest Ecosystems: Towards Sustainable Management*, edited by A. Zchulte and D. Schone. Singapore: World Scientific, 308–334.

Ohler, F.M.J. 1985. The Fuelwood Production of Wooded Savanna Fallows in Sudan Zone of Mali. *Agroforestry Systems* 3, 15–23.

Sanchez, P.A. 1976. *Properties and Management of Soils in the Tropics.* New York: John Wiley and Sons, 618.
Trexler, M.C., and C. Haugen. 1995. *Keeping it Green: Tropical Forestry Opportunities for Mitigating Climate Change.* Washington, DC: World Resources Institute, 52.

Chapter 28

Farmers' Use of *Sesbania grandiflora* to Intensify Swidden Agriculture in North Central Timor, Indonesia

J.A.M. Kieft

Swidden agriculture in North Central Timor, Indonesia, has undergone major changes over the past 60 years. It has evolved from subsistence production in a feudal framework toward more market-oriented agricultural systems. Timorese farmers, while undergoing these changes, have also faced increased population pressure and the virtual disappearance of their natural forest vegetation. Much has since been written about the agroforestry systems that have evolved in the search for more intensive land-use strategies in the area (Field et al. 1992; Fischer 1992; Djogo 1995). Most of this literature has focused on the "Amarasi" system, which has been lauded as a successful model of fallow intensification (Field et al. 1992; Fischer 1992). But one system that has been underreported is the use of *Sesbania grandiflora* in North Central Timor district, of the Nusa Tenggara Timur Province (NTT), particularly along the coastal areas near Wini and in the mountainous former kingdoms of Tunbaba and Manamas. Only Rachmawati and Sinaga (1995) mention the use of *S. grandiflora* in the area, its potential as a multipurpose tree, and its common use by farmers. But they did not describe *S. grandiflora's* use by farmers as an improved fallow.

The Timorese situation fits the pattern that is common to most of Southeast Asia. By 1947, the virgin forests were all gone, except for those that had either been set aside by the Netherlands Indies colonial government, or were part of a sacred place (Schulte Nordholt 1971). The disappearing forest and an increasing population forced farmers to reduce the period over which they allowed their land to lie fallow, so after fallowing, there was less biomass available to help maintain soil fertility, and yields plummeted. Ultimately, this led to the so-called swidden degradation complex of declining yields and progressively shorter fallow periods (Raintree 1990).

In 1912, Bali cattle were introduced to Timor. Before then, the farmers had kept only buffaloes. The population of cattle increased progressively, and by 1990, there were about 500,000 head on Timor (Bamualim and Saramony 1995). The study area was part of this intensification of livestock husbandry. By the early 1970s, farmers were starting to keep cattle in a cut-and-carry system, and this demanded fodder of a reasonable quality throughout the year.

Faced with collapsing swidden systems and the need for quality fodder, farmers started to plant *S. grandiflora* on their fallowed lands. Today, it covers the fallow land in most villages in the area. Other legumes are also used, but on a much smaller scale.

Johan Kieft, Program Leader of Agriculture, Disaster Management and Conflict Recovery, CARE International Indonesia, P.O. Box 4743, Kebayoran Baru, Jakarta 12110, Indonesia.

The Study Sites

This research focuses mainly on five sites. Two are located in the mountains, two in the coastal lowlands, and one in the midlands (see Figure 28-1). The study area is characterized by a higher-than-average population density for the district of North Central Timor, and less than average soil fertility. Farmers are subsistence-oriented, although according to Yayasan Timor Membangun (YTM, The Timor Development Foundation), these days many of them work off-farm in Kupang, Attambua, Dili, and other towns. In all five villages, farmers use *S. grandiflora* as an improved fallow, but with different management approaches.

The Villages

The following are brief descriptions of the hamlets studied. The numbers correspond with those in Figure 28-1.

1. Teakas is situated in the mountains at an elevation of about 750 m above sea level (asl). It is an area of steep slopes, varying from 13° up to more than 60°. More than half of the cropped land has slopes steeper than 40°. The upper slopes are part of a forest reserve and therefore remain under forest. The lower parts are either cropped or grazed. Natural water is abundant and is now tapped to supply drinking water to Kefamenanu. Soils are mostly red bobonaro clay with high gravel content, but white soils are also found. The village itself is located on Viqueque soils.
2. Kainbaun is also located in the mountains, at an elevation of about 600 m asl. Part of the land is hilly and is typical of a white bobonaro clay area. Another part is hilly with very steep slopes, and is mostly red bobonaro clay. Farmers use the lower slopes and valley bottoms for agriculture.
3. Nimassi is located in the hills. The soils are basically white Bobonaro with some Viqueque. Slopes are generally not steep, but may be up to 27°.
4. Wini is located on the coast. Soils are alluvial and mostly sandy although clay occurs, especially near the beach. People live by fishing and farming. Ownership of cattle is very limited.
5. Saknati is on the north coast. It is a local transmigration settlement, inhabited by people from mountainous areas. Soils differ from loam to heavy clay. The loamy soils are preferred for gardens, while the clay soils are used for rainfed rice.

Soils

Soils differ between the hamlets. To describe them, farmers' classifications are used and correlated with a classification developed by NTASP (Nusa Tenggara Agricultural Support Project). Soils in the lowlands are, to use the NTASP classification, modern alluvial, but with differing texture. Within the lowland hamlets, farmer respondents did not classify different types of land. In mountainous and hilly areas, farmers distinguish between three different soil types, each with its own characteristics:

1. Black soils *(tanah hitam)*: These soils are most favored and considered most fertile. Villages are often located on this soil type because natural wells can be found here. Black soils are similar to those recognized as the Viqueque formation (Alderick and Anda 1987). Often these soils are lithosols, and are always highly calcareous with a dark A horizon.

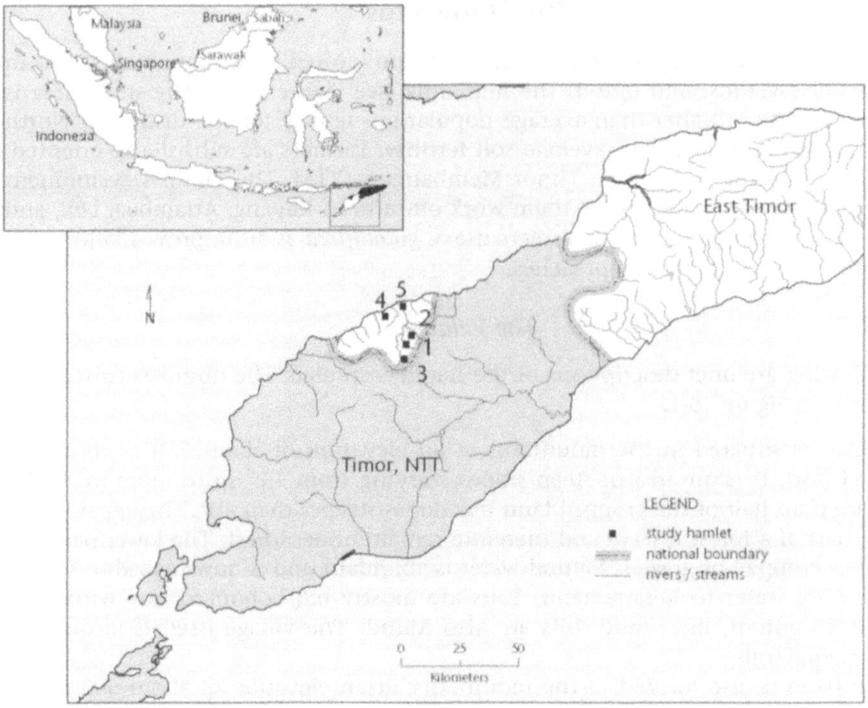

Figure 28-1. Timor Island, Nusa Tengarra Timur (NTT), and the Five Study Hamlets
Note: Numbered according to the text.

2. Red soils *(tanah merah)*: These soils are mostly used for gardens. In one study village, farmers preferred to grow pineapples (*Ananas comosus*) on red soils. According to others, these soils are suited for crop mixtures such as rice, maize, and peanuts. Red soils have high gravel content and are similar to soils categorized as Bobonaro clay category 3 (Alderick and Anda 1987). Red soils are mostly lithosols.
3. White soils *(tanah putih)*: These are the least fertile soils and are used mostly to plant maize and cassava. According to farmers, white soils are unsuited for rice and tree crops, and their use is mostly limited to pasture because of their capacity to shrink, swell, and develop deep cracks. They are equivalent to the Bobonaro clay category 1, and are mostly Grumsols, calcareous with high pH (Alderick and Anda 1987).

Climate

The climate in the study area varies. Coastal areas are warm and have a low annual rainfall of less than 1,000 mm. One of the study sites, Wini, for example, averages 800 mm a year.

Temperatures in the mountains are lower, often falling to 15°C at night. There is a greater difference between day and night temperatures than there is between seasons. Mountainous and hilly areas receive a higher rainfall than the lowlands, varying between 1,400 and 2,500 mm per year, depending upon location. Rainfall is irregular from year to year. On average, coastal areas experience eight dry months, while mountainous areas have about six. Humidity averages between 60% in the dry season and 85% in the wet season (Alderick and Anda 1987).

Methodology

The results presented in this chapter are based on information obtained during group meetings with farmers that were arranged to review YTM's program in the study area. The meetings followed a relatively low rate of adoption by farmers of alley cropping and in-row tillage technologies. During the meetings, participatory rural appraisal (PRA) techniques were used as tools to structure the discussions, including seasonal calendars, participatory cost budget analyses, drawing of gardens, transect walks, mapping, and semistructured interviews. The meetings were held in an informal setting, and, to create an open atmosphere, there were no "outsiders" present.

Field observations were undertaken to verify the results of the meetings. Plant densities, stem diameters, soils, and crop cover were all noted. The aim was to get a clearer picture of how farmers were using *S. grandiflora* in their gardens. Yield data were gathered by field workers at the end of the 1995–1996 cropping year, to assist program evaluation. Data about fodder use in the cut-and-carry system were also gathered, using a seasonal calendar. Farmers were asked to describe the amount and kind of fodder they harvested on an average day during a particular month.

Results

The Farming System

Farmers in the study area usually cultivate more than one garden in a year. The number varies, depending on access to land. On average, most farmers cultivate 3 gardens, or a minimum of 2, while another 5 to 10 gardens are left fallow. These gardens are scattered around the village territory, often on different soil types to spread the risk of crop failure. Land ownership is still based on kinship, or clan membership. This means that every member of a clan has access to land, although no single member owns it. A form of ownership, known as *tanah warisan*, exists for land that is cultivated continuously by the same family. This land is transferable, whereas land owned by the clan is not. The elders of the clan still decide where young adults might start farming, and every male receives land to open his garden after he gets married. Each year, farmers cultivate between 0.5 and 1.5 ha of their land. The remaining 70 to 80% of it is left fallow. It is difficult to get a clear picture of land ownership, but each farmer has access to at least three hectares of land, and this includes fallowed as well as cropped land. In general, two types of gardens can be distinguished. They can be found on both the Bobonaro clays (both red and white):

- Permanent gardens: Although not cultivated with staple crops every year, these gardens are always fenced. They are located near the village and are used intensively. They are planted with perennials such as candlenut (*Aleurites moluccana* [L.] Willd.), bananas (*Musa* spp.), and fodder species such as *Leucaena* or king grass (*Pennisetum purpureum*). The trees are planted along the border, with the fodder species planted along contour lines in an alley cropping arrangement. These gardens are located on Viqueque soils, but should be distinguished from the *mamar*, the area around natural wells. These areas are always planted with tree crops to protect the well. *S. grandiflora* is managed during the fallow.

- Swidden gardens: Use of the garden is not permanent. After the harvest of staple crops, the land is left fallow. Fences are not maintained and cattle can freely enter the garden. When soil fertility has recovered after a three- to four-year fallow period, the garden is cultivated again. During the fallow, *S. grandiflora* is grown. The presence of *Ageratum conyzoides L.*, growing vigorously below the *S. grandiflora* canopy, is often used as an indicator of soil fertility.

Figures 28-2 and 28-3 show a land-use map and corresponding transect for Teakas hamlet. The location of the settlement is typical, with permanent gardens on the Viqueque soils and swidden gardens on the Bobonaro clays. Every farmer cultivates at least one permanent garden and one or more swidden gardens. This differs from year to year and depends upon the time available to open new gardens and their expected yield. Close to their settlements, farmers use *Leucaena, Calliandra calothyrsus, Gliricidia sepium,* and other multipurpose trees as a first step toward cultivating perennials such as candlenut and bananas. After the hedgerows have been established, farmers then plant the perennials as closely as possible to the hedgerows to take advantage of nutrients built up by N-fixation and litterfall. This process of land-use intensification is promoted by both the government and nongovernmental organizations (NGOs), who encourage the planting of perennial crops to improve living standards. The swidden gardens are more distant and scattered throughout the hamlet. Farmers use the swidden gardens according to their perception of soil fertility, and this is related to the color of the soil (see Figures 28-2 and 28-3). After the staple crops are harvested, the land is fallowed with *S. grandiflora.*

In addition to growing staple crops, farmers also raise cattle, which are one of the most important sources of income. In Kainbaun, for instance, half of farmers' cash income comes from cattle sales. Most farmers have at least one or two bulls managed in a cut-and-carry system and one to four cows that are allowed to range freely. Farmers in Wini are the only ones in this study who do not own cattle. Elsewhere, there are three types of cattle management, and the differences affect only the females:

- The cows range freely and are retrieved only once a week to check on them.
- The cows are guarded daily and gathered in a village corral during the night.
- The cows are tethered and moved around according to the availability of fodder.

The choice of system often depends on available time and the number of cows involved.

Bulls are kept in a cut-and-carry system and almost all farmers raise at least one bull. The animal is either their own or is raised on behalf of somebody else. Most of them are simply tied to a tree. Stables are not provided. The manure is left on the ground and its soil nutrients lost to leaching and nitrification. Manure is used only for vegetable production and not in the larger gardens.

Indigenous Strategies for the Use of S. grandiflora

The crop cycle starts in early July when the village elders decide, during a ceremony, who can start cultivating which piece of land. This is called *feknono hawana*. After determining which fields are going to be used, the elders ceremoniously cut wood and ropes to symbolize that the land is going to be cultivated. They also pray to their ancestors and to cosmic forces, asking for good crop yields. Once the ceremony has been performed, the farmers are allowed to begin clearing their gardens. After the fallow vegetation has been slashed, it is burned. Wooden poles are used for fencing. All farmers say that burning increases soil fertility. The ash is seen as a fertilizer.

From a farmer's point of view, as much biomass as possible is required to produce sufficient ash to assure abundant yields. Therefore, biomass production during the fallow period is of crucial importance to adequate food production, and it is seen as one of the main reasons for fallowing land. Around Saknati hamlet, much of the natural vegetation still remains because the area is newly settled. The area was avoided for cultivation in the past because it has heavy alluvial clays with strong shrink-and-swell characteristics. The vegetation is predominantly *gewang* (*Corypha gebanga*) and *lontar* (*Borassus flabellifer*). Although Saknati farmers plant and maintain *S. grandiflora* in their home gardens, it is not on the same scale as the other hamlets. They do not feel the need to grow *S. grandiflora* because they believe their soil is more fertile than that in the mountain villages, and, in any case, they can burn *Corypha gebanga*. They also say the young *S. grandiflora* trees lure cattle into the gardens to eat. This problem was not mentioned in the other study villages.

Farmers in the mountain villages judged *S. grandiflora* and *Leucaena* spp. to be the best trees for burning, although *S. grandiflora* had the advantage of making charcoal that reportedly burned with more heat. These two species were said to produce better quality ash than others. In particular, their ash was reputed to be better than that of *Eucalyptus alba*, which is common on red Bobonaro category 2 soils.

After burning is complete, the land is cooled down by a second ceremony called *si fono nopo*, in which the village elders perform a rite to restore the normal balance in the interplay of cosmic forces. The farmers then need a lot of wood to fence their gardens. In the coastal villages, farmers use the leaf stems of *Corypha gebanga*, but in the mountains they use a mixture of *Eucalyptus alba* and *S. grandiflora*. Farmers prefer not to use *S. grandiflora* for permanent fencing because the wood is not as durable as that of other species and to use it for permanent gardens would risk the need for extensive repairs. For fencing swidden gardens, however, it makes no difference because, by the time of harvest, the *S. grandiflora* that will dominate the fallow is big enough to survive predation by livestock that may break through a decayed fence. Cattle do not browse on its bark, as they do on *Leucaena*, which is often damaged in swidden gardens when cattle intrude through rotting fences to chew the bark of young trees.

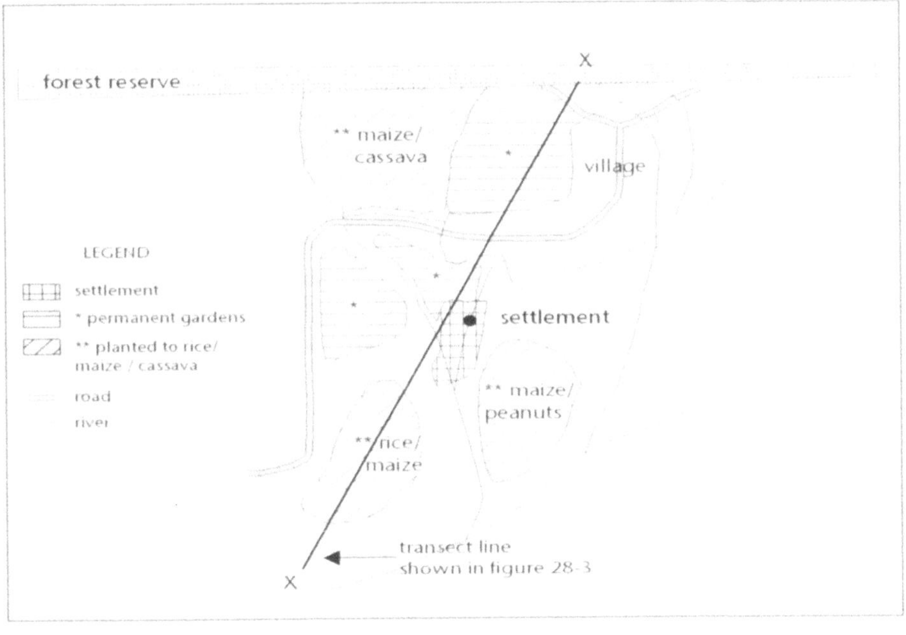

Figure 28-2. Land-Use Map of Teakas Hamlet

Notes: *Used for rice/maize/cassava mixtures. These soils are considered more fertile than white soils. According to farmers, *S. grandiflora* performs better on these soils. **Used for maize/cassava mixed with a local type of *Mucuna*. Because these soils are less fertile, farmers plant maize, rather than rice. *S. grandiflora* grows less vigorously on these soils. The areas where no use is indicated are either savannah or are pasture areas, composed mainly of *Eucalyptus alba, Psidium guajava*, and native grasses.

ıden gardens: rice/maize/peanuts ·w: *Sebania grandiflora*	r	settlement: home gardens	alley cropping *Leucaena*/corn	settlement: home gardens	alley cropping: permanent gardens, vegetables	alley cropping / swidden	forest reserve
onaro cat 2: red soils with high stone tent	i v	alluvial: red/ white soils	Bobonaro 2/ Viqueque	Viqueque	Viqueque	Viqueque	
1. low, severe erosion caused mainly by e and land clearing, low to very low yields	e r	high fertility & O.M.	average	high	good	average	
ion, low soil fertility		crops not well maintained	alleys not maintained	crops not maintained	below optimal use of alley cropping	erosion	uncontrolled use
nsify fallowing, maybe cover crops ı as *Mucuna* spp.		better crop cultivation methods	perennials, e.g., coffee, citrus	better home garden methods	intensify land use by using green manures, intensifying vegetable production	reforestation: social forestry	better use of forest resources (fodder, herbal medicines, etc.

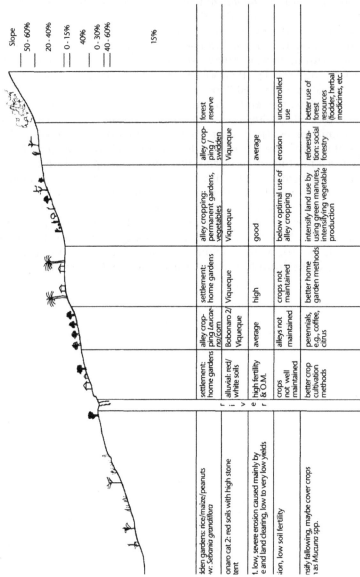

Figure 28-3. Transect of Land Use in Teakas Hamlet
Note: Intersecting line marked "x" is the transect presented in Figure 28-2.

Permanent gardens are planted with a mixture of crops, including *S. grandiflora*. Because these gardens are located close to the hamlet, *S. grandiflora* is always harvested as feed for the tethered bulls. In swidden gardens, *S. grandiflora* is often pruned intensively for cattle fodder and is sometimes shared with relatives who may be suffering a shortage of fodder.

It is interesting to consider why farmers in the study villages prefer *S. grandiflora* to both the common and giant species of *Leucaena*. Their reasons provide insights into why farmers may accept a technology or reject it. They mentioned the following reasons as the most important:

- Farmers would have to guard the trees longer because *Leucaena* has slow initial growth in the study area and free-ranging cattle would destroy the small trees.
- Cattle eat the bark of *Leucaena,* so even when gardens are left fallow, fences must be maintained to protect the trees.
- *Leucaena* is often infested by the psyllid (*Heteropsylla cubana*), and cattle do not like to eat leaves affected by the psyllid.
- *S. grandiflora* feed results in better-quality, "heavier" meat on the bulls.

Of these four reasons, the most important were the first and the fourth.

Management of S. grandiflora *in Swiddens*

Two different methods were used to gain a clearer understanding of indigenous management practices in swidden gardens. The agronomic and forestry aspects were explored, and then management practices were examined from a socioeconomic perspective. Results of the field observations will be presented first, and then they will be cross-checked against what we learned from the group discussions and yield data. Twenty swiddens were selected—10 in Teakas and 10 in Kainbaun—and the following data were recorded:

- The number of trees was estimated by counting all the trees within 5 m x 5 m plots. Five plots were selected in every garden, all located on a diagonal line.
- The crop cover was estimated for all kinds of vegetation.
- The diameter of *S. grandiflora* stems within the plots was measured.
- Management practices, such as cutting and thinning of trees, were observed.
- The height of 10 trees per plot was measured in fallows less than two years old.

In Figure 28-4, the crop cover is correlated to the age of the stand. The choice of a quadratic relation, and not linear, is based upon the following considerations:

- During the first year, the trees grew rapidly. After one year, their height was about 1.5 m.
- After two to three years, trees were disappearing due to pest and disease problems. Farmers said the trees died because of bad cutting practices. Heavy cutting during the early wet season led to increased pest and disease problems.
- After three to four years, the growth rate of the trees slowed and tree density decreased.

Results of the regression analysis are presented in Figure 28-4. The curve shown in the figure is described by the following equation: $Y = -13.5 + 57.7(X) - 13.5 (X)^2$; $R^2 = 0.468$; $P < 0.01$; $N = 20$. Y = crop cover *S. grandiflora*; X = years after establishment. This equation indicates that optimum canopy closure is reached about 2 years 1 month and 19 days after establishment, equating to about one and a half months after the start of the wet season in the third year after planting. By then, the *Sesbania* trees have reached their peak in productivity. By four years, the trees are growing slowly and many are infected with pests and diseases.

The relation between the tree diameter, as a dependent variable, and the number of years after establishment, as the independent variable, is shown in the following equation: $Y = 3.42(X)$; $R^2 = 0.95$; $N = 20$. Y = tree diameter; X = years after establishment. This indicates that the diameter of *S. grandiflora* increases in a linear fashion each year. Although the canopy cover differs, the diameter is not influenced by the differences mentioned above. The tree diameter of *S. grandiflora* at different ages is shown in Figure 28-5. The tree diameter is used as an indication of tree growth, and Figure 28-5 shows that the tree diameter and the age of the stand age are significantly correlated.

Figure 28-6 shows the tree density at different ages. No significant relationship was found between these two variables. The large differences in the first year after establishment are typical, and according to farmers, this depends on the owner and history of a garden. When a garden has previously been sown with *S. grandiflora*, it is able to reestablish itself from the soil seed bank. If there has been a dense crop of *S. grandiflora* before, it will regenerate after burning and after crops of maize, cassava, or upland rice. Differences in soil type also lead to differences in *S. grandiflora* establishment.

Farmers distinguished several different ways of establishing *S. grandiflora*, all of which influenced tree density. The most common method is natural reestablishment, where the seeds germinate after the gardens are burned and the rainy seasons starts. Tree density depends upon the amount of seed in the soil.

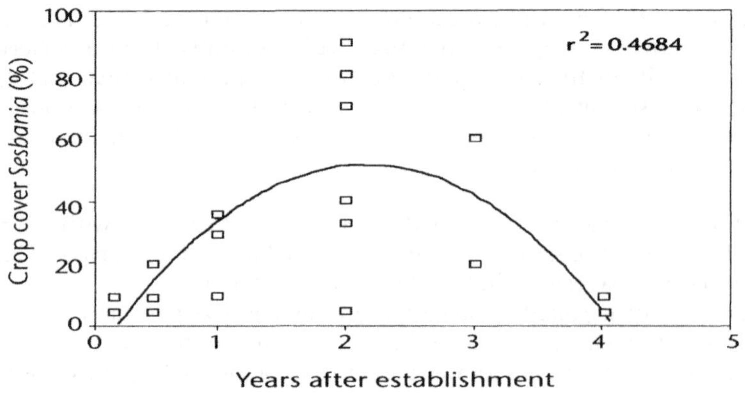

Figure 28-4. Crop Cover for *S. grandiflora* Fallows at Different Ages

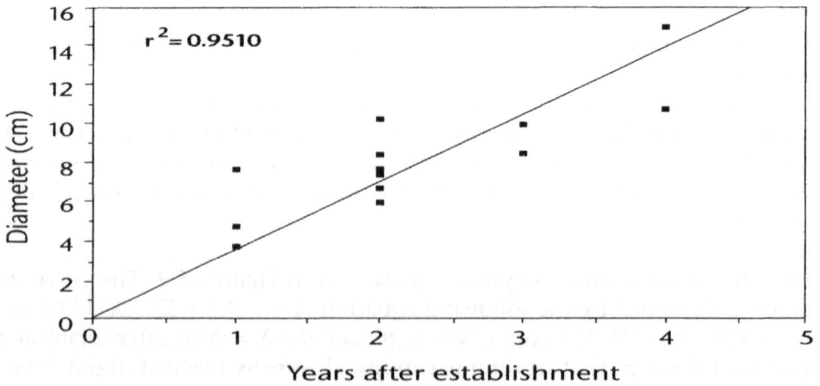

Figure 28-5. Diameter of *S. grandiflora* Trees at Different Ages

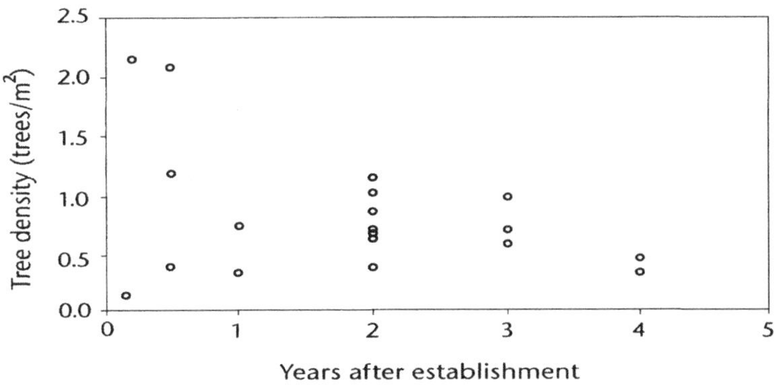

Figure 28-6. Tree Density of *S. grandiflora* Fallows at Different Ages

Some farmers sow *S. grandiflora* into their gardens, and there are two different approaches to sowing. The first is simply broadcasting seeds. The seeding rates depend upon each farmer, but they aim for a complete canopy of *S. grandiflora*. Whether they succeed depends upon land quality. Farmers say *S. grandiflora* grows better on red soils, or heavy-textured Grumsols, than on white soils. Its establishment is reduced on slopes of more than 30%. These findings are based on discussions and interpretation of field observations. The second sowing method involves planting hedgerows of *S. grandiflora*. This practice was found in three cases, two in which *S. grandiflora* was mixed with *Leucaena* and one where it was sown alone. The survey was conducted at the end of the dry season and the trees, having been planted to coincide with the first rain, were about 50 cm high. Farmers said that, in response to government and NGO programs, they were planting *S. grandiflora* in hedgerows as a way to conserve land and soil. Although other nitrogen-fixing trees were recommended, including *Calliandra* spp., *Leucaena* spp., and *Gliricidia sepium*, they preferred to use *S. grandiflora* because of its superior establishment qualities.

Farmers sometimes cut *S. grandiflora* trees from other areas and burn them on plots where *S. grandiflora* has never grown before. They say that *S. grandiflora* can become established by this method.

Finally, there is transplanting. Seedlings are collected from surrounding areas and transplanted into the gardens, usually on the contours. Transplanting is often done in conjunction with the other methods. Seedlings are not raised in a nursery, but are only collected from areas nearby the gardens.

These different approaches to establishing *S. grandiflora* lead to variations in plant density, especially at the start of the fallow. After one year, the farmers thin the *S. grandiflora* and reduce the tree density. Which trees are thinned depends upon their vigor and distribution. Thinning is done in the dry season, with the wood left in the gardens and the leaves harvested for fodder. Farmers thin two or three times, depending upon the density of the *S. grandiflora*. They aim for a tree density of about 1–1.5 trees/m^2 after two to three years. After three years, the tree density continues to decrease, but this is caused by farmers pruning them for fodder during the rainy season. Farmers speculate that this causes disease, but fodder is scarce at the beginning of the rainy season, so they have little choice.

There was no significant linear relationship between tree density and diameter in the plots two years after establishment ($R^2 < 0.01$; $N = 8$). For the same plots, canopy cover and tree density were analyzed for correlation but no significant relationship was found ($R^2 = 0.24$, $N = 8$).

The Uses of *S. grandiflora*

Farmers generally use *S. grandiflora* for the following seven purposes:

- The leaves are used as fodder for bulls in a cut-and-carry system;
- The poles are used for fencing when fallow gardens are opened for the cultivation of staple crops;
- The wood of *S. grandiflora* is burned to produce fertilizing ash;
- *S. grandiflora* maintains soil fertility through litterfall and N-fixation;
- Human nutrition;
- Shading out light-demanding weeds; and
- Erosion control.

Other species have been introduced to serve the same purposes, such as king grass and *Leucaena* for fodder, erosion control, and maintaince of soil fertility. This began in the 1930s when *Leucaena* was introduced to Amarasi (Kab Kupang), 200 km west of the study area. *Leucaena* is grown in the study villages, but only on a very limited basis because most farmers prefer *S. grandiflora.*

Fodder

Figure 28-7 illustrates the importance of different types of fodder. This figure was derived from a seasonal calendar on which farmers indicated, in local units, the amount of fodder they used and its source. Five samples of every type of fodder listed were weighed in kilograms and averaged. The average weights of the sample units were then multiplied by the number of local units.

The result, in Figure 28-7, shows how important *S. grandiflora* is as a protein source. *S. grandiflora* provided 26% of the total fodder input, on a weight basis. Other sources of fodder protein used by farmers include *Leucaena* and *kabesak* (*Acacia leucophloea*). *Leucaena* is mostly planted in hedgerows for erosion control in permanent gardens. Kabesak is found in forests or in savannah-like vegetation. In Figure 28-7 it is not listed separately since it is often fed mixed with *Ficus* sp. and therefore falls under the category of forest leaves.

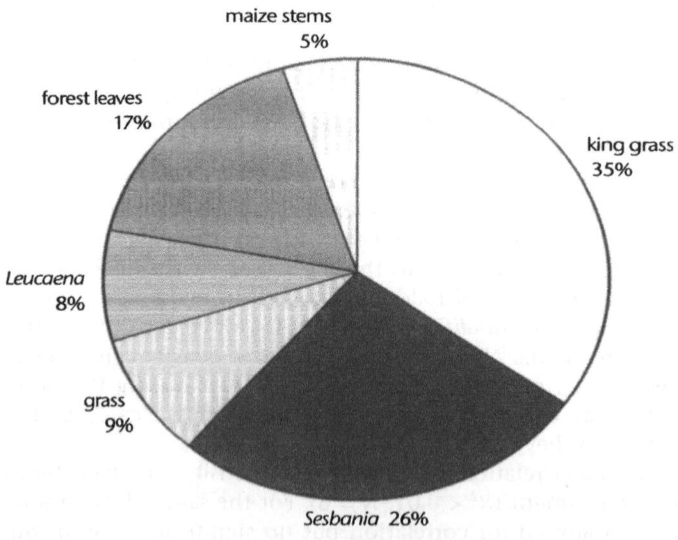

Figure 28-7. Types of Fodder Used in Teakas Hamlet for Bulls Kept in a Cut-and-Carry System

Another major fodder species is king grass. Farmers plant it along contours and near streams. Its production drops after two to three years because it is not fertilized. Farmers then plant new cuttings. They like king grass because of its high production capacity and because, in combination with *S. grandiflora*, it provides better results than other fodder species.

Figure 28-8 shows the use of different types of fodder during the year. The livestock extension service and NGOs recommend that animals should receive fresh fodder at a daily rate of about 10% of their live weight. Teakas farmers have been able to feed their cattle balanced and adequate diets throughout the year. Bali cattle weigh about 150 kg/head when farmers begin to fatten them and they are sold when they reach about 270 to 325 kg/head. On average, farmers achieve a livestock growth rate of about 300 g/head/day with this diet. *S. grandiflora* is most important as a feed resource during the dry season. In the wet season, it is mixed with *Leucaena*, but in the dry season, *Leucaena* tends to be infested with *Heteropsylla cubana*, and farmers are reluctant to use *Leucaena* leaves infested with psyllid because the cattle often reject them and farmers fear that it might cause disease. *S. grandiflora* is most productive during the dry season, but pruning is essential. Farmers note that if the trees are not pruned, *S. grandiflora* leaves will fall during the mid and late dry season, from August through November.

Farmers have developed their own pruning techniques. Figure 28-9 illustrates the different stages of tree development under pruning. The first pruning takes place about 16 months after *S. grandiflora* is planted, around the start of the second dry season. The trees are about 3 to 3.5 m high by then, and are pollarded at 2 m. The lowest two branches are retained. Farmers say that regrowth and further production will be constrained if this procedure is not followed, but they warn that if no branches are left, the trees may die.

Farmers need to prune two trees to gather the 7 kg of leaves that are delivered to each bull on every feed. Based on a productivity of 3.5 kg of fresh leaves/tree/pruning, the annual fodder production can be estimated. One hectare of *Sesbania* yields about 28,000 kg of fresh fodder on each pruning. Hence, farmers harvest approximately 6,000 to 7,000 kg of edible dry matter per hectare from a well-maintained two-year-old stand of *S. grandiflora.*

In Wini, the sale of *S. grandiflora* leaves constitutes an important part of farmers' income. Wini is a collection center for cattle shipped to Java (Surabaya). About 16,000 head are shipped each year, and sometimes more than 400 head can be found waiting for transport. Traders buy *S. grandiflora* leaves from nearby farmers to feed their cattle, and the farmers make about 240,000 rupiah/ha (1997US$1 = 2,400 rupiah). They grow the trees at high densities of at least one tree/m^2. Every four years, the trees are cut and used for fencing and burning. In Wini, trees are established through natural germination. During their first years of growth, the young trees are mixed with staple crops. After the crops are harvested, pure stands of *S. grandiflora* remain for fodder production.

In the hamlets of Kainbaun, Teakas, and Nimassi, free-ranging cattle and their calves are fed *S. grandiflora* leaves during the late dry season. Farmers use gardens more distant from the village that are not frequently pruned. Almost daily, trees are topped and fed to cattle.

Fencing

Fencing is an important activity requiring a lot of wood. Most farmers fence their gardens when they are planted with food crops. Different types of wood are used, depending on the nature of the garden. *S. grandiflora* is used for fences around swidden garden, but for reasons already explained, it is not used for permanent gardens. For these, farmers use *Leucaena* and *Eucalyptus alba*. The latter is particularly favored because of the durability of its wood.

Burning

Burning is very important. For one thing, it allows farmers to clear land quickly and easily. But the remaining ash is also regarded as fertilizer. A key reason that farmers plant *S. grandiflora* is to produce fuel to burn. If they are opening a fallow and feel that there is insufficient wood to produce ash, they may supplement existing biomass by buying wood from outside. Farmers believe there is no difference between the ash of *Leucaena* spp. and *S. grandiflora*, but that of nonlegume trees is believed to be less fertile. As already mentioned, they believe the amount of available ash determines the yield of subsequent crops.

Maintaining Soil Fertility

The effect of *S. grandiflora* on soil fertility is reflected in crop yields. To collect yield data, farmers worked with field staff in undertaking a cost–benefit analysis of their gardens. Only in Nimassi was a different approach used. There, farmers were asked to draw their gardens and provide yield figures in a PRA exercise. The results are presented in Tables 28-1 and 28-2.

Yields varied widely. However, the differences were in the relative amounts of rice and maize in each of the hamlets rather than differences in the total yield of staple crops. These differences depended upon each agroecosystem. Wini and Saknati, for instance, being on the coastal plains, had lower maize yields and higher rice yields. In the high midlands, Teakas, Kainbaun, and Nimassi had lower rice yields and higher maize yields. Nimassi had somewhat higher overall yields because its soils are more fertile.

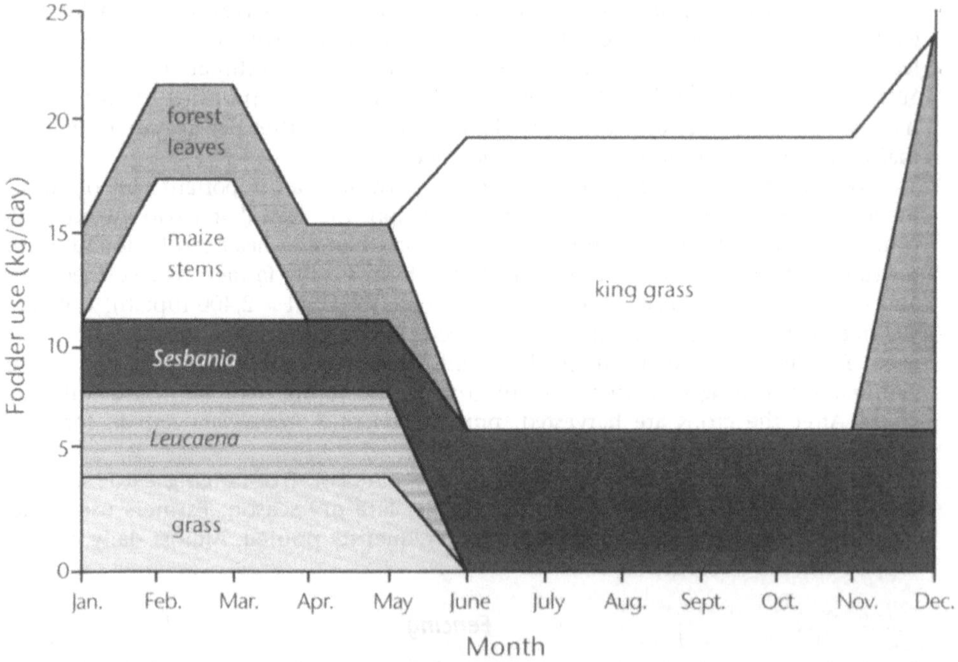

Figure 28-8. Use of Fodder throughout the Year

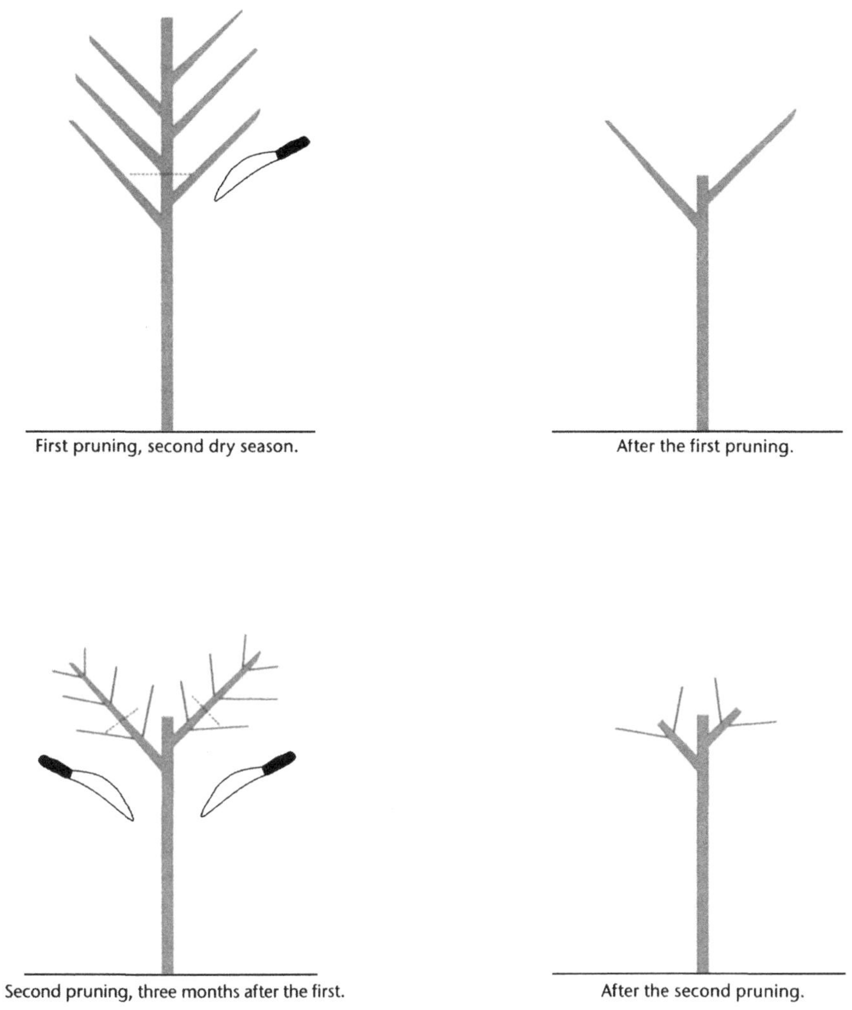

First pruning, second dry season.
After the first pruning.
Second pruning, three months after the first.
After the second pruning.

Figure 28-9. Pruning of *S. grandiflora* for Fodder Production

In general, conditions favor dryland rice cultivation in the lowlands and maize in the mountains. Farmers generally prefer to eat rice rather than maize, and if possible, they plant rice mixed with maize instead of maize alone. But this is not always possible, as reflected by the number of farmers in the mountains who did not plant rice.

A bountiful crop means sufficient food and the ability to share the harvest with relatives. This provides the farmer with status, but only limited economic benefits. It is not often clear why yields differ from farm to farm, but the farmers themselves suggested the following factors:

- Land quality, because white soils are less productive than the red soils;
- Burning, and the quantity of ash produced;
- The time of weeding, because delayed weeding results in reduced yields or crop failure; and
- The time of planting, because any delay at the start of the rainy season may result in crop failure.

Table 28-1. Average Yield, Standard Deviation Yield, and Range of Maize Yields, 1995–1996 (kg/ha)

Hamlet	*Average Yield*	*Standard Deviation Yield*	*Range*	*N*
Teakas/Kainbaun	593.77	372.87	42–1,555.6	48
Wini	367.50	238.34	140–700	4
Nimassi	843.54	832.78	70–3,375	27
Saknati	361.56	253.40	25–729	8

Note: Teakas and Kainbaun are grouped because they are neighboring hamlets with very similar climates and soils.

Table 28-2. Average Yield, Standard Deviation Yield, and Range of Dryland Rice Yields, 1995–1996 (kg/ha)

Hamlet	*Average Yield*	*Standard Deviation Yield*	*Range*	*N*
Teakas/Kainbaun	263.00	177.05	0.00–571.43	10
Wini	766.67	251.66	500–1,000	3
Nimassi	603.85	1045.79	81–5,250	23
Saknati	541.50	588.49	29.17–1,800	8

Note: Teakas and Kainbaun are grouped because they are neighboring hamlets with very similar climates and soils.

Although farmers were familiar with all these factors, they were not always able to avoid their negative impacts on yields.

Of all the villages studied, only in Saknati was *S. grandiflora* not used intensively. However, as Tables 28-1 and 28-2 show, there is no significant difference between the yields in Saknati and those of the other hamlets. This suggests that if there is enough natural vegetation to burn, there is no incentive for farmers to improve their fallows. Farmers say that yields have been stable over the past 25 years. Before that, when only *Eucalyptus alba* was used for burning, crop yields were lower.

To investigate how *S. grandiflora* affects yields, the harvests of permanent gardens, second-year crops in swidden gardens, and newly cleared gardens were compared between the Teakas and Kainbaun hamlets. In Teakas, farmers are sometimes forced to cultivate a swidden garden twice because of land shortage. Permanent gardens are used continuously and are often contour planted with *Leucaena* for erosion control and interplanted with other crops such as candlenut and pineapple. Farmers do not use tree prunings as green manure in permanent gardens because they prefer to use the *Leucaena* for fodder. The yields are shown in Figure 28-10. As expected, farmers realized higher yields from the newly cleared gardens, although these were not significantly higher. Yields from permanent gardens were equal to those from swidden gardens cropped for a second year. This may indicate the effect of *S. grandiflora* on yield, and, if this assumption is accepted as being correct, then fallowing the land with *S. grandiflora* provides a barely significant yield increase of about 75 kg of maize per hectare, on average. According to farmers, however, this difference can be attributed to burning, litterfall, and N-fixation. They were obviously influenced by government and NGO programs promoting the use of N-fixing trees in contour plantings. Burning—and the amount of ash produced by it—is seen as most important.

Farmers say shorter fallow periods are the biggest advantage of using *S. grandiflora*. Before its use as an improved fallow, the gardens needed at least eight years to recover. But with the use of *Sesbania*, this has been reduced to three to four years and yields have reportedly improved. Previously, there was also a severe shortage of firewood and farmers had used mostly bamboo and *Eucalyptus alba* that grew on fallowed land. However, these species have lower growth rates than *S. grandiflora* and do not fix nitrogen.

Human Nutrition

The flowers and fruits of *S. grandiflora* are edible. Although quantitative data are unavailable, the amounts eaten are probably small. But farmers mentioned it as an advantage over other species.

Weed Shading and Erosion Control

Farmers use *S. grandiflora* to shade out weeds, although it requires three years to achieve effective control. *Alang-alang* (*Imperata cylindrica*) is mentioned as one of the most difficult weeds to eradicate. Farmers prefer to use *S. grandiflora* instead of other conventional methods, such as breaking up the soil and allowing the exposed roots to dry in the sun. This latter method is faster, but it requires huge labor inputs of about five man-days/100 m^2. Labor is often unavailable so only limited areas of land are cleared by this method. Farmers place stems of *S. grandiflora* on the contours of sloping land to control water erosion.

Discussion and Conclusions

Since the mid-1980s, farmers have demonstrated their ability to react to changes in their environment. Before this time, a growing population, limited availability of land, and dwindling natural resources had led to the degradation of swidden land, low yields, low cash income, and severe food shortages, culminating in famine in the 1960s. Farmers have since had to develop more intensive ways to use their land. Management of *S. grandiflora* has been a partial solution, enabling them to intensify both land use and livestock production. The cut-and-carry system of livestock production, for instance, would not be possible without the use of fallow land for fodder production.

By means of informal farmer-to-farmer extension, the use of *S. grandiflora* has spread over the study area. Factors that have supported its expansion are its fast initial growth, its easy natural propagation, its good performance as livestock fodder, and the good burning properties of its wood. Farmers have expressed strong reasons for preferring *S. grandiflora* to other leguminous tree species. Although their reasons may be difficult to validate scientifically, they still indicate why farmers make choices and adopt technologies. An interesting case in point is the introduction of *Leucaena* hedgerows for erosion control and green manure production. Although it has been promoted by the government and NGOs for use in staple crop production since the 1980s, its adoption has been limited to permanent gardens only, where *Leucaena* seedlings are able to become better established. One of the main reasons for its rejection, according to extension workers, is that farmers were told they would have to stop burning. However, burning is a powerful part of their culture. Farmers regard it not only as essential but also critical to the success of the crop, through the influence of ancestors and cosmic forces. In a scientific sense, burning has been proven to have no effect whatsoever on yields when compared with mulching (Field 1985). Nevertheless, farmers are unshakeable in their belief that they have to heat the soil by burning to clean it of weeds and diseases, otherwise the harvest will be lost. It explains one of their motives for using *S. grandiflora* in the first place: they needed its wood for burning because there was a widespread shortage of firewood. The opposite is the case in the study village of Saknati, where farmers only use *S. grandiflora* on a limited scale. They still have enough natural vegetation to achieve a good burn. Often NGOs and government workers are unaware of these rationales and cannot explain why farmers choose not to use hedgerows in their swidden gardens. Even

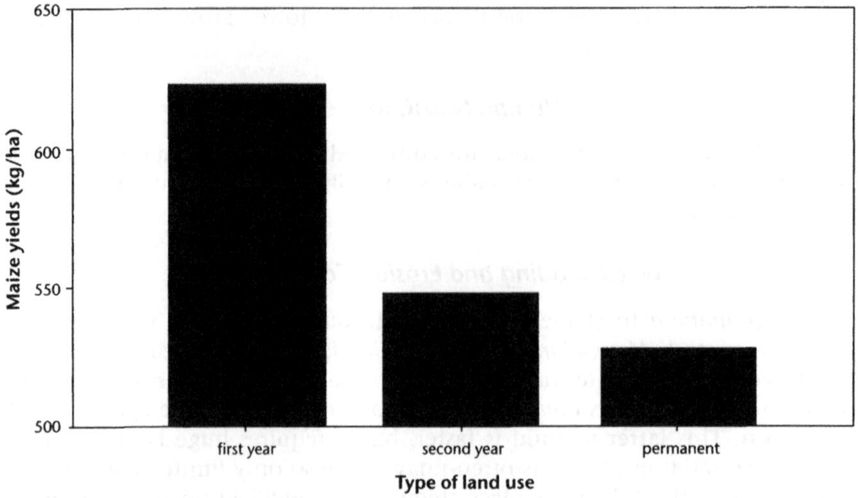

Figure 28-10. Maize Yields across Different Types of Land Use
Note: First year and second year relate to swidden gardens.

worse, it is believed that farmers are incapable of understanding what is good for them. They are not given credit for the ability to develop their own solutions.

The strategies developed by farmers for use of *S. grandiflora* are diverse. This chapter attempts to provide insights into what is happening in the field, but it is not able to account for interfarm variability. This is said to be determined by the rationalities of individual farmers, whose responses to changes in their environments are reflected in the differences from farmer to farmer in crop yield (van der Ploeg 1995) and management of *S. grandiflora*. As demonstrated in Figures 28-4, 28-5, and 28-6, farmers use *S. grandiflora* in different ways, with striking differences in methods of establishment and resulting tree densities. It may be useful to search for relationship patterns between these management variables and yield outputs in both fodder and food staples, as the reasons behind different management practices need to be understood. Many farmers have off-farm employment. It is not clear how this affects their management, but it may be significant. Clarification of the gender aspects of the farming system is also required, including their impact on the use, choice, and management of improved fallows.

Differences in tree density did not affect their growth, and this suggests that most farmers are planting below optimal tree density for biomass production. Research on *Leucaena* has given some insights into the production of biomass (Field 1985). Although measurements in this research were undertaken during the cropping phase, there were significant differences in maize yields according to the number of *Leucaena* stumps left in the field. There were also indications that the amount of fallow biomass before planting was strongly related to the subsequent harvest. Maize yields as high as 3,500 kg/ha were reported. These yield levels may also be possible using *S. grandiflora*. It would be interesting to look for relationships, at the farm level, between the amount of biomass or the number of trees in a garden before burning and subsequent crop yields. This, in conjunction with close discussions with farmers, might help to explain interfarm differences and suggest new directions for agricultural development in the study area.

The most important purpose of the *S. grandiflora* fallows is provision of fodder. Farmers have demonstrated more initiative in developing their fallow land for fodder production than in improving yields of their staple food crops. Technologies used in staple crop production have not changed in the past 50 years (Schulte Nordholt 1971). New technologies, such as in-row tillage, are often strongly rejected, although yield increases of more than 100% have been demonstrated (YTM unpubl.). The reasons for rejecting new technologies are usually related to the need for more inputs for staple food production. Generally, new technologies have higher demands for labor, which is scarce, especially for planting and weeding. The same reason lies behind the rejection of green manures. Even the use of farm manure in production of crop staples is limited.

The reason underlying farmers' fear of losing yield might be the so-called Birch effect (Piggin 1995). When soils are moistened during the first rains, there is an increase in decomposition leading to a flush of N at the start of the season. If farmers plant too late, the N has already been leached from the topsoil.

Farmers have been more innovative in their cattle husbandry and new technologies have been quickly adapted. An example is the use of king grass, which was introduced only five years ago and is already widely accepted as cattle fodder. However, the use of *S. grandiflora* for fodder and the production of food crops are integrated. After a four-year fallow, farmers must clear *S. grandiflora* because it is no longer productive. It needs to be rejuvenated before it can once more produce enough fodder. This is of crucial importance because, after three years, fodder production decreases, the canopy cover shows less vigorous growth, and the tree density decreases as trees begin to die.

The ability of farmers to intensify livestock production by improving their fallow management shows that they are both responsive to market forces and willing to change. However, cattle are a cash commodity, whereas staple crops are produced for subsistence and often fail in the harsh climate. The markets for staple crops are limited, and even if farmers do harvest good yields, they share the bounty with relatives. Cash received from livestock sales is not shared. Innovations such as feeding supplementary fodder from leguminous trees to maintain cattle weight during the late dry season (Bamualim et al. 1991) are readily adopted without any extension effort. Farmers are willing to invest in their cattle because cattle provide their capital and cash income.

Farmers have increased their cash income by intensifying livestock husbandry and improving the management of their fallow lands. This, in turn, has stabilized, although not increased, the production of food staples. Within their cultural framework (*adat*), the farmers have reacted to changing social and economic circumstances, represented by the emerging market for cattle, despite a deteriorating agroecological environment characterized by increasing pressure on land and lower yields. *S. grandiflora* has played a catalyzing role in this process. It enabled farmers to develop the described innovations and react to outside influences. Ultimately, *S. grandiflora* allowed farmers to shift from a feudal, subsistence-oriented society to a more market-oriented one. To support farmers in this process, research and extension on improved fallow management needs to focus on the following:

- Opportunities to increase fodder production from fallows, including improved thinning and pruning techniques, and optimizing tree densities;
- Possibilities of increasing yields of food staples by optimizing the biomass production of the fallow period;
- How improved fallow management technologies can be incorporated into more permanent land-use systems, such as permanent gardens, thereby enabling farmers to intensify land-use and livestock production; and
- Sharing of knowledge between farmers, government agencies, and NGOs, at district and provincial levels, to develop a mutual understanding on improved fallow management.

In the first three issues, it would be useful to develop a decision support model for monitoring nutrient dynamics under agricultural land use, for example, NUTMON crop growth modeling. The model could then be used by researchers, NGOs, and government agencies to forecast the effects of current and alternative land use scenarios, as well as to determine the long-term sustainability of agroecosystems (Smalling and Fresco 1993). It would be interesting to compare different management approaches, both indigenous and nonindigenous, to the use of *S. grandiflora* and other nitrogen-fixing trees for their impact on nutrient balances.

The fourth research issue could be addressed by using Rapid Appraisal of Agricultural Knowledge Systems (RAAKS). RAAKS attempts to develop a mutual understanding between the different stakeholders, including government agencies, NGOs, and farmers involved in the development of swidden systems on Timor. Mutual understanding should lead to more farmer-oriented extension approaches.

In the future, fallows will become more productive but will retain their biological function of rejuvenating soil fertility for the production of food crops. This means that new technologies will have to be developed and new species incorporated into fallows, with the aim of intensifying land use and livestock production. Strategies developed by farmers should be used as a foundation for these technologies, so that they are widely adopted and a mutual understanding is developed between the stakeholders involved.

Acknowledgments

I wish to acknowledge the support of Dr. Marthen Duan, whose help allowed me to attend the Regional Conference on Indigenous Strategies for Intensification of Shifting Cultivation in Southeast Asia. I wish to thank Laurens Foni for his work in gathering data and leading discussions with farmers and, of course, I wish to thank the farmers, without whom this research would not have been possible. Last but not least, I thank my wife, Veby, for her support in finishing this chapter during our home leave.

References

Alderick, J.M., and M. Anda. 1987. The Land Resources of Nusa Tenggara. In: *Technical Report 3 Nusa Tenggara, East Timor and Southeast Maluku.* Kupang, Indonesia: 24.

Bamualim, A., A. Kedang, and S. Field. 1991. The Effect of *Gliricidia sepium* and *Acacia villosa* Supplements on the Growth Rate of Bali Heifers (*Bos sondaicus*). *Nitrogen Fixing Tree Reports* 9, 19–20.

Bamualim, A., and U.P. Saramony. 1995. Livestock Production in Semi-Arid East Nusa Tenggara. Paper presented to an international conference on Agricultural Development in Semi-Arid Areas of East Nusa Tenggara, East Timor, and Southeast Maluku, Kupang, Indonesia, 12.

Djogo, A.P.Y. 1995. Model wanatani Potensial Untuk Pertanian Lahan Kering. In: *Pengembangan wanatani di Kawasan Timur Indonesia. Prosiding lokakarya Wanatani II., Januari 16–18, Ujung Padang,* edited by A. Ng Gintings, R. Fithriadi, S. Riswan, A.P.Y. Djogo, D. Rumpoko, and W. Soeparna. 31–52.

Field, S.P. 1985. *Report on the Food Crop Experimental Program of the NTT Livestock Project for the 1984–85 Wet Season.* NTT Livestock Project, 79.

———, E.O. Momuat, and R. Ayre Smith. 1992. Developing Land Rehabilitation Technologies Compatible with Farmer Needs and Resources. *Journal of the Asian Farming Systems Association* 1, 385–394.

Fischer, L.A. 1992. From Kebe and Lamtorinasi to Farm Planning. Paper presented to an international conference on Alley Farming. Ibadan, Nigeria: International Institute of Tropical Agriculture.

Fox, J. J. 1995. Social History and Government Policy of Agricultural Development in Eastern Indonesia. Paper presented at an international conference on Agricultural Development in Semi-Arid Areas of East Nusa Tenggara, East Timor, and Southeast Maluku, Kupang, Indonesia, 8.

Piggin, C.M. 1995. Agronomic Aspects for Agricultural Development in Semi-Arid Regions. Paper presented at an international conference on Agricultural Development in Semi-Arid Areas of East Nusa Tenggara, East Timor, and Southeast Maluku, Kupang, Indonesia, 24.

Rachmawati, I., and M. Sinaga 1995. Daya dukung pohon serba guna dalam pengembangan agroforestry di Timor (Nusa Tenggara Timur). In: *Pengembangan wanatani di Kawasan Timur Indonesia, Prosiding Lokakarya Wanatani II, Januari 16–18, Ujung Padang,* edited by A. Ng Gintings, R. Fithriadi, S. Riswan, A.P.Y. Djogo, D. Rumpoko, and W. Soeparna, 132–143.

Raintree, J.B. 1990. Theory and Practice of Agroforestry Diagnosis and Design. In: *Agroforestry Classification and Management,* edited by K.G. MacDicken and N.T. Vergara. NY; Chichester; Brisbane; Toronto; and Singapore: John Wiley & Sons, 58–97.

Schulte Nordholt, H.G. 1971. *The Political System of the Antoni of Timor.* Den Haag, the Netherlands: Martinus Nijhof, 511.

Smalling, E.M.A., and L.O. Fresco. 1993. A Decision Support Model for Monitoring Nutrient Balances under Agricultural Land Use (NUTMON). *Geoderma* 60, 235–256.

van der Ploeg, J.D. 1995 (unpubl.). Agricultural Production and Employment: Differential Practices and Perspectives, 19.

YTM (Yayasan Timor Membangun, The Timor Development Foundation). Unpubl. Program Results.

Chapter 29

Alnus nepalensis–Based Agroforestry Systems in Yunnan, Southwest China

Guo Huijun, Xia Yongmei, and Christine Padoch

Swidden cultivation continues to be an important form of land use in Yunnan, where a virtual "swidden belt" encircles the southern, southeastern, and southwestern zones of the province (Ying 1991). This is despite local and central government attempts since the 1960s to eliminate swiddening (Liu and Hu 1990). Efforts to put an end to the cutting of natural forests reached a peak at the beginning of 1990, following a serious flooding disaster in the Yangtze River Basin that was attributed to forest clearing upstream. However, according to the Yunnan Land Administration Bureau (1994), there are still 1 million farmers who rely on swidden cultivation, and they cultivate about 1 million hectares, most of it in lower-lying tropical regions up to an elevation of 2,000 m above sea level (asl) (Guo 1995).

Although it persists on a large scale, swidden cultivation in Yunnan has been changing, as shifts in population, economic systems, and land management policy have all had major impacts on agricultural practices. In some areas, permanent field cropping of annual or perennial cash crops has replaced swiddening. In other areas, swiddens have been displaced by *Eupatorium* grasslands, with *E. cordatum* below 1,000 m asl and *E. coelestinum* above. In cases where traditional swidden systems persist, the fallow period has been shortened (Guo 1995).

Official criticisms of swiddening have recently been tempered. Scientists have recognized and acknowledged that swidden cultivation can be an appropriate form of land management, especially from an ethnobotanical point of view (Pei 1984), and that it can help conserve biodiversity (Guo 1990; Pei 1994). Nevertheless, it remains undeniable that swidden cultivation has had negative impacts in Yunnan. It has been an important motive for conversion and degradation of natural forest, especially community-owned forests. It has been estimated that in the 14 years between 1978 and 1992, 87% of forests owned by communities, amounting to about 15,777,500 ha, has been lost. Swiddening has also contributed to soil erosion in mountainous Yunnan and its destruction of soil microorganisms has also been noted.[1]

Government policy has recently limited swiddening to specific districts. The *siguding* policy, adopted in 1960, included the important *linyeshanding* and *liangshanyidi* regulations, which were enacted in 1982 and 1983. Under these laws,

---

Guo Huijun, Vice Director-General. Forestry Department of Yunnan Province, No. 120 Qingnian Road, Kunming, Yunnan, 650021, China; Xia Yongmei, Associate Professor, Xishuangbanna Tropical Botanical Garden, The Chinese Academy of Sciences, Xuefu Road No. 88, Kunming, China; Christine Padoch, New York Botanical Garden, Bronx, NY 10458, USA.

[1] Although slash-and-burn leads to an increase in soil fertility and microorganisms in the first year, these begin to decline in the second year after burning (Zhang 1955).

local authorities have restricted swidden cultivation to specific geographical regions (Guo 1990, 1995).

Swidden cultivation, therefore, faces further reduction and many people are looking for alternatives that are appropriate from technical, economic, political, and ethnological viewpoints. Local governments have approached the evolving needs of farmers by building upland terraces, promoting the use of hybrid varieties of upland crops such as corn and upland rice to increase yields on limited agricultural land, and encouraging perennial cash cropping on arable uplands.

For their part, the indigenous people of Yunnan have developed many different agroforestry systems that have helped intensify management of land previously used for swiddening. Research in eight prefectures of Yunnan has identified 220 "types" of agroforestry systems that can be categorized into 82 different "forms" of agroforestry. Systems based on the cultivation of alder (*Alnus nepalensis*) trees are some of the most important of these. They may satisfy the need to conserve soils and biodiversity while at the same time providing income and subsistence supplies to local communities (Guo 1995). This chapter describes several alder-based systems and traces their development in a diversity of locations by farmers of several ethnic groups.

Alnus nepalensis in Yunnan

There are about 40 species of the genus *Alnus*, of which eight are found in China, and four in Yunnan. Table 29-1 lists the different uses found for these four species. The most widely distributed and popular species in Yunnan is *Alnus nepalensis*, which can be found from 1,000 to 2,800 m asl, but mostly within a range of 1,500 to 2,400 m asl. Except for the extreme south, southeast, and very high mountain areas in the northwest of the province, *Alnus nepalensis* is found throughout Yunnan. The center of its distribution is the Gaoligong and Nushan mountains along Yunnan's western border (Wang 1985).[2]

Alnus nepalensis is a deciduous broadleaf tree. It occurs in natural forests as pure stands, or mixed with Yunnan pine *(Pinus yunnanensis),* otherwise known as *ketereria*, and other species. *A. nepalensis* is fast growing, with a well-shaped trunk that branches high. It shows a strong ability to regenerate naturally, especially in forest gaps, in swidden fallows, and in wet valleys with enough sunshine. It is an important pioneer species in vegetation succession following both natural and human disturbances not only in Yunnan, but also across the eastern Himalayan region (Sun et al. 1996).

When planted in solid stands, alders grow relatively slowly for the first two years and faster thereafter. The fastest vertical growth rate is reached in year 15, while the greatest diameter increments occur around year 20. Grown in average conditions, an alder will reach 13 m in height and 11 cm in diameter within seven years, and 21.3 m in height and 23 cm in girth by 20 years. Generally, 800 to 1,000 trees can be grown per hectare and 250 to 300 m^3 of timber can be harvested per hectare in 20 years (Zeng 1984).

---

[2] The climatic range of *Alnus nepalensis* is the area with annual average temperatures of 12 to 18°C, active cumulative temperature of 3,000 to 6,500°C, minimum temperature of –13.5°C, and maximum temperature of 34°C, with an annual rainfall of more than 800 mm and relative humidity greater than 70%.

Table 29-1. Usage of Alders in Yunnan

		Usage			
Species	*Elevation*	*Bark*	*Young Branches*	*Young Leaves*	*Trunk*
Alnus cremastogyne Burkill	1,450–2800	Tannin: 5–10%	Medicine: (hot cough, diarrhea, nose bleeds, etc.)	Same as young branches	
A. ferdinandi-coburgii Schneid	1,800–2,800	Medicine: (setting fractures)			Furniture, etc.
A. lanata Duthie ex Bean		Medicine: (diarrhea, leprosy)		Medicine: (snake bite)	
A. nepalensis D. Don	(1,200–) 1,800-3,150	Tannin: 6.82–13.68%; medicine: (diarrhea, dropsy, pneumonia, lacquer disease, broken bones)	Green manure	Green manure	Construction material, packing boxes

Source: Wang 1985; Wu 1984.

The alder is a nonleguminous nitrogen-fixing tree that forms nodules. The main nitrogen-fixing fungus is *Actinomyces alni*. Studies show that one hectare of *Alnus nepalensis* can fix 150 kg of N, equal to 895 kg of ammonium sulphate of nitrogen. Alder leaves are preferred as a green manure by local farmers. Dry leaves contain significant nutrients, including 2.937% N, 0.406% P, and 1.101% K (Zeng 1984). Fresh leaves can be harvested from a four-year-old *Alnus* forest twice to three times per year, yielding 750 to 1,000 kg/mu/year (15 mu = 1 ha). The leaves decompose quickly, improving soil structure and increasing soil fertility.

Alder trees have many features other than N fixation and use as green manure. They produce timber that is used in construction and furniture making, and especially for making tea boxes; their bark is used for tannins; and their leaves, roots, and bark have medicinal qualities. In addition, the alder trunk is particularly suitable for culture of *Auricularia auricula-judae*, a traditionally eaten mushroom famous in Yunnan. These many qualities, together with its ease of regeneration, have made alder a popular component of local agroforestry systems.

The seeds of *A. nepalensis* are small and winged. Germination rates fall from 40% to 10% after one year's storage, so farmers usually sow fresh seeds. The seeds are usually collected in December and planted in upland rice fields in the late spring of the following year, when the rice has reached 4 to 6 cm in height. Professional foresters have also developed nursery and transplanting techniques over recent decades. Alder coppices readily, with alder stumps coppicing vigorously after pruning.

A. nepalensis is not without some problems, however. The main insect pests that attack this tree are *Chrysomela adamsi-ornaticollis* Chen and a scarab beetle. They both feed on the leaves, altering the tree's growth.

Indigenous Management of Alder in Agroforestry Systems

Of the four species of *Alnus* native to Yunnan, only *A. nepalensis* has found a role in agroforestry. Indigenously developed alder-based agroforestry systems in Yunnan can be classified into several subtypes practiced by several ethnic groups. Among the ethnic groups practicing *Alnus*-based agroforestry are the Dulong people in Gongshan county in the northwest of the province; the Lisu and Han in Tengchong county; and the Jingpo of Dehong Prefecture, also in western Yunnan. All of these systems and peoples occur in the Great Gaoligong mountain region, on the far eastern fringe of the Himalayas. This region is also the origin of the genus *Alnus*. Other ethnic groups, including the Wa of the south and the Zhuang in Wenshan Prefecture in southeast Yunnan, also practice *Alnus nepalensis*–based agroforestry. Table 29-2 shows the living conditions of each of these peoples and the type of *Alnus*-based agroforestry practiced, while Figure 29-1 shows the location of the case studies.

Table 29-2. *Alnus nepalensis*–Based Agroforestry Systems in Yunnan

	Counties				
Characteristics	*Tengchong*	*Ximen*	*Gongshan*	*Malipo*	*Fenqing*
Case Village	Hetaoyuan	Amuo	Kongdang	Laodifang	Liguo
Ethnic Group	Han, Lisu	Wa	Dulong	Zhuang, Han	Han, Yi
Natural Resources					
Arable land per person (mu)	2.27	7.83	2.22	2.43	2.27
Percentage of land sloping > 25°	35.01	29.05	91.18	67.49	42.06
Percentage of irrigated paddy fields	47.71	5.71	9.33	13.51	19.09
Avg. forest area per person (mu)	6.98	8.15	84.94	3.78	3.31
Percentage of swidden on upland fields	47.40	89.18	11.68	3.75	40.82
Livelihood					
Avg. grain per person per year (kg)	316	348	276	240	324
Income per person per year (RMB yuan)	782	390	604	721	735
Percentage poor people (< 500 yuan, 300 kg/person/year)	42.91	88.75	39.39	82.24	77.00

Notes: Agroforestry patterns: Tengchong: *a*. Rotational cultivation: Annual cropping 2–3 yrs.; 10 yrs. for firewood forest (leaves used for green manure). *b*. *Alnus* canopy system (Zhaoying village, Shangying township): *Alnus* + tea + corn; Ximen: *a*. Rotational cultivation: 2 yrs. upland rice; 20 yrs. alder forest for timber or 10 yrs. for firewood. *b*. Alley cropping: 1 m for alder hedgerow; 2 m strip for upland rice; Gongshan: *a*. Rotational cultivation (simongmulang): opening fallow; 1 yr. corn; 5 yrs. alder forest; 3 yrs. annual crops (buckwheat, millet, barley); Malipo: *Alnus* canopy system: *a*. Natural alder: clearing shrubs and grass (1st winter); transplanting tsao-ko under alder forest; maintaining tsao-ko for about 10 yrs.; *b*. Introduced alder: 1st yr. transplanting alder; 2nd yr. transplanting tsao-ko; maintain alder for 20 yrs., tsao-ko for about 10 yr; Fenqing: *a*. Alder canopy system:transplanting tea; interplanting alder seedlings. 1997US$1 = 8.29 yuan.

Figure 29-1. Case Studies of *Alnus nepalensis*–Based Agroforestry in Yunnan

The province's great cultural diversity has no doubt been a factor in the development of diverse patterns and techniques of *Alnus* management. Improved transport and communication has resulted in alder-based agroforestry becoming more widespread over the past decade, and it is no longer confined to the agricultural repertoire of only a few ethnic groups. The main variations of alder-based agroforestry include the following:

- Rotational cultivation, with alder harvested after a 10–year fallow for firewood, or after a 20-year fallow for timber. This is followed by two to three years of annual cropping (see color plate 40). This rotational system is used by the Han and Lisu in western Yunnan, and the Wa in the south;
- Alley cropping, with meter-wide strips of *A. nepalensis* alternating with two-meter-wide alleyways for annual crops. This system is employed by the Wa in southern Yunnan;
- Intercropping of alder with upland rice by the Dulong, or with corn by the Lisu in northwest Yunnan; and
- *Alnus* canopy systems over cash crops such as tea (see color plate 66).

Alnus in Tengchong County

Tengchong has long been an important border point and center for the exchange of commodities between China and Myanmar, and even South Asia. Beginning in the Han Dynasty (200 B.C.), the ancient Southern Silk Route passed through Tengchong, connecting interior China with India (Wang and Xu 1995). Tengchong county covers a total of 5,678 km^2, but only 2.54% of it is flat. The rest is mountainous, and ranges in elevation from 930 m to 3,780 m asl.

Most of Tengchong lies between 1,200 and 1,800 m asl. In 1984, 16.2% of the county was used for farming, and less than half of this was paddy for wet rice. Thirty-eight percent was under forest cover.[3] By 1994, 15.02% was farmland, of which more than half was paddy, and the forest cover had increased to 46.1%. Orchards and plantations occupied a little less than 1% of the land area (see, for example, color plates 66, 67, and 68), and most of this was tea. The rest of the county was grassland, settlements, roads, bodies of water, and unused land.

The population of the county increased from 329,617 in 65,881 households in 1964, to 495,181 in 126,000 households in 1995. The main ethnic group is Han Chinese, who have been migrating from interior China, especially southeastern China, since the Han dynasty. The indigenous ethnic groups include the Dai, who lived nearby the county capital during the Han dynasty and gradually moved down to the valleys; the Hui, who began moving into Tengchong during the Yuan dynasty (A.D. 1277); and the Lisu, who migrated from Lushui along the Gaoligong Mountains.

Tengchong county suffers from generally low farmland productivity. Government reports indicate that in 1995, 88.3% of farmland in the county was categorized as having low productivity; that is, producing less than an equivalent of 200 kg of grain per mu. More than 19% of agricultural land has slopes greater than 25°. With population growth and the development of fuel-consuming cash crops such as tobacco and tea, fuel has become scarce in rural Tengchong county. *A. nepalensis* is therefore important as fast-growing and good quality firewood. It is already a popular tree in the county. In the Gaoligong Mountains in the west of the county, for instance, there are 1,651 ha of *A. nepalensis* forest within a nature reserve.

Agroforestry in Xiaoxi Township

According to older residents, alder-based agroforestry systems have existed in Tengchong for more than 200 years. It is said to have begun during the Hongwu period of the Ming dynasty, around A.D. 1368 (Guo 1993). Two villages, Hetaoyuan and Gange, both in Xiaoxi township, were selected for in-depth research during 1991 and 1996. The township has generally steep and unproductive lands, like much of the county.[4] The farmers of Xiaoxi township have included *A. nepalensis* in their agroforests for many generations.

The village of Hetaoyuan has a population of 998 in 228 households. Although the majority of residents are Han Chinese, there are also some members of the Yi, Jingpo, and Dai ethnic groups. The village has a total land area of 2826.8 mu, with 554 mu of paddy (257 of which are rainfed), 672.8 mu of upland fields, 400 mu of *A. nepalensis*-based agroforests, and about 1,000 mu in other vegetation, most of it grassland with sparsely scattered trees. Each family has roughly 0.3 mu of vegetable garden adjacent to their house.

Residents of Hetaoyuan derive most of their cash income from off-farm work (40%) and animal husbandry (20%), as well as sales of tobacco (20%), firewood (10%), and food grains (10%). Paddy fields generally yield 200 to 250 kg/mu. One in every five families fails to harvest enough grain for their own needs.

---

[3] This percentage is considerably higher than the 24% that is average for all Yunnan.

[4] Xiaoxi township has a total of 42,418 mu of farmland; 54.85% is in paddy fields, of which 81.5% is low in productivity. 66.18% of paddy land has productivity lower than 300 kg/mu. Yields of upland fields are also lower than 300 kg/mu, with 87.98% lower than 200 kg/mu.

Historically, the cultivation of *A. nepalensis* has changed considerably. Before 1949, the agroforestry cycle was generally 15 years in length. In the 1960s and 1970s it was reduced to 10 years, then to 8 to 10 years in the 1980s, and 6 to 7 in the 1990s. Nowadays the cycle generally goes as follows:

- Year 1: The land is leveled, grasses are burned, tree trunks and branches are collected for fuel, and upland rice is planted, weeded, and harvested.
- Year 2: Upland rice is planted in May. *A. nepalensis* seeds are sown into fields in June when the rice is 4 to 10 cm high. After the rice is harvested, the main task is keeping cows out of the field.
- Years 3 to 5: Fires must be controlled. *Alnus* leaves are collected for green manure twice a year, leaving only four or five leaves on the top of each tree. Sixty kilograms of fresh leaves can be collected per mu. The green manure is used in paddy fields in the spring.
- Years 6 and 7: Illegal tree cutting must be controlled. Leaf harvest continues.
- Years 8 to 10: Trees are cut, and 10 m^3 of *Alnus* firewood may be harvested.
- Years 10 or 11: Another cycle starts with the planting of upland rice.

Change in Xiaoxi

A. nepalensis–based agroforestry has varied in importance and scale in Xiaoxi. The area planted to alder in the 1970s was substantial, but it has been greatly reduced since 1990. Several reasons for this are apparent, the most important of them being abrupt shifts in social and rural land policy. Before liberation in 1949, large areas were owned by landlords. Not only were decisions concerning land use and agricultural production made almost exclusively by these landlords, but there were also no accurate statistical records kept about farming activities and areas involved.

Between 1950 and 1955, land was allocated to each farming household. But it was taken away again during the cultural revolution in the 1960s and 1970s, when all agricultural and land-use decisions were made by village committees. During this time, *Alnus* firewood was distributed according to "work scores" of families and one or two farmers planted alders for the whole community. This collective agricultural production ended and decision making about farm production was largely returned to individual families by the "responsible system of contracted land" policy in 1978 and the linyeshanding and liangshanyidi policies of 1982 and 1983. Farmlands, forest land, and home garden plots were once more allocated to families and the conditions of alder agroforestry changed again. Land fragmentation was exacerbated by the decision to allocate lands of different quality equally among families. One family we interviewed, for example, had three mu of paddy, but in two locations; six mu of permanent upland, but in three locations; and six mu of *A. nepalensis* fields, in six locations. This fragmentation has resulted in conflicting land uses. For instance, while one family may grow *Alnus* trees, its neighbors may choose to pasture cattle, and when the alder seedlings are young, the livestock destroy them. Where larger areas are in similar production, more people are available to take care of the young seedlings. Xiaoxi had its largest area under alder trees in 1982. After that, the farmers decided to reduce *Alnus* planting.

Alders have also lost popularity because of the growing use of herbicides and chemical fertilizers. The season for sowing *Alnus* seeds is in May, coinciding with the season for weeding upland rice. When herbicides are applied, the young seedlings are usually killed. Increasing use of chemical fertilizer has also reduced the need for green manure, including the leaves of *A. nepalensis*.

Population growth and land conversion to other purposes are also important trends. There were only 58 persons per km^2 in Tengchong county in the 1960s. This grew to 87.6 persons per km^2 by 1995, and the shifting cultivation cycle decreased accordingly. Some land was converted to nonagricultural uses, so additional land for growing food crops was rented from a neighboring village.

Some change has also been due to the local forestry bureau, which has encouraged farmers to grow Chinese fir, an introduced tree species, which is currently mixed with *Alnus*. The price of Chinese fir is higher than that of alder, and the local government provides seedlings to farmers. We believe that between 1992 and 1997, no one actively planted *A. nepalensis* in Hetaoyuan and Gange. However, when we last visited, in January 1997, planting appeared to be increasing again.

The Wa in Ximeng County

The Wa are one of the main swiddening groups of Yunnan Province. Their total population in 1995 was 360,200. There are two autonomous Wa counties in Yunnan: Ximeng and Cangyuan. Ximeng is located in southern Yunnan, bordering the Wa of Shan State in Myanmar. According to land resource surveys, Ximeng has a total land area of 1353.57 km^2, or 2,025,000 mu, of which 725,129 mu are farmland, 594,448 mu are forests, and 6,422 mu are orchards. The surveys list a further 512,494 mu as "waste lands" and 30,235 mu as unusable. More than 78% of Ximeng county lies on slopes greater than 15°. Of the farmland, 32.9% is under permanent upland crops, 62.2% is swidden fields, including fallows, and only 4.7% is paddy. Within the forest land, 62.8% is covered by forest and the rest by shrub and grassland. The county's forest cover is 18.44%. Ximeng's population has grown from 40,000 in 1952 to 80,000 in 1995. The average size of landholdings has obviously decreased during this time, and although yields per unit of land have more than tripled, local people are still short of food grains and lack cash income. Raising the productivity of upland fields is an important issue in the Wa region. The Wa people in Ximeng have developed or adopted many local cultivars of agricultural crops, including 120 varieties of upland rice, 12 varieties of paddy rice, 15 varieties of corn, 15 varieties of wheat, and 50 varieties of beans and other vegetables.

We studied *A. nepalensis*–based agroforestry systems in Amuo village, in Xinchang township, in 1992 and again in 1997. The population of Amuo village in 1992 was 200 persons living in 42 households. The total land area of the village was 558 mu, of which 180 mu were upland fields, 138 mu were paddy terraces, 40 mu were in Chinese fir, and 200 mu were in *A. nepalensis* forest. Each family in Amuo had at least 2 mu of *Alnus* plantations, and some families had more than 5 or 6 mu.

There are no historical records concerning alder agroforestry in Ximeng, but the Wa are known to have long had the custom of preserving *Alnus*-dominated swidden fallows for upland farming since it regenerates easily. In the first year after slashing and burning and growing upland rice, the Wa protect any *A. nepalensis* that regenerates naturally. After 10 to 20 years, the *Alnus* stand is cut for firewood, with the leaves and branches burned and some trees left standing in the new swiddens. Since 1983, when the linyeshanding and liangshanyidi land and forestry policies were adopted by the local government, local swiddenists have been restrained from shifting their fields and have started actively transplanting *A. nepalensis* seedlings into their upland fields. Amuo villagers have developed two ways of establishing alders in agroforestry systems:

- Swidden-fallow agroforestry. After slashing and burning, upland rice is grown for two consecutive years. *Alnus*-dominated fallows are then allowed to take over. Local farmers say that before 1983, the *Alnus* fallow growth came from windborne seeds that naturally germinated and reforested the swidden, and the Wa farmers protected them. However, since 1983, the Amuo villagers have been collecting young seedlings from natural *A. nepalensis* forests for transplantation into upland rice in the second cropping year. The farmers say that if they do not plant *Alnus* trees, *Eupatorium* weeds will invade and occupy all of the uplands. They add that upland fields with *A. nepalensis* trees are more fertile than *Eupatorium*-dominated swidden fallows. After two years of upland rice cultivation, *A. nepalensis* is left for 10 years. Some farmers maintain the trees for 20 years and then harvest them as construction materials. This rotational system is called *geloudanggueang* in the Wa language.

- Alley cropping. This first appeared in Wa villages in the 1980s and is practiced by the same farmers who manage alder-based swidden fallows. Local farmers, professional foresters, and agronomists all agree that alley cropping of alder, interspersed with food crops in the alleyways, was developed locally and is not a technology introduced by outsiders. It was apparently developed as a response to a shortage of arable land.

Methods of Establishing Alnus *Plantations*

There are three methods by which Wa farmers establish *Alnus* plantations:

- Some local people preserve naturally regenerating *Alnus* forest. This is a long-standing traditional method.
- By transplanting. Wa families either collect seedlings from *A. nepalensis* forest or from swidden fields and transplant them into upland fields, or they raise seedlings in nursery gardens. This method was introduced by Ximeng county foresters in the 1980s.
- By broadcasting seeds into upland rice fields during a weeding operation. The forestry department provided farmers with seeds during the 1980s when there were not enough available locally. The seeds had been collected from state forests or from other communities that did not use *Alnus* in their farming systems.

The latter technique is very similar to that of the Dulong people in the Dulongjiang River valley. The Wa have also developed different methods of planting *A. nepalensis* seedlings. It is believed that ethnic groups on the southern slopes of the Himalayas may have developed similar technologies, and, as a consequence, *Alnus*-based agroforestry in western Yunnan is similar to the *taungya* system practiced in Myanmar. This hypothesis remains to be tested through an extensive comparative study across northern Myanmar, northeast India, and southwest China.

Dulong Agroforestry in Gongshan County

A very similar form of *A. nepalensis*–based agroforestry also appears to be an indigenous development by the Dulong who live in the upper reaches of the Irrawaddy River valley. An ethnological survey team in the Dulongjiang region described this system in 1960, when communication between the Dulong community and other parts of China was rare. They found that the Dulong system, using *A. nepalensis,* left the soil more fertile than other farm land, and it was capable of supporting crops for a longer time:

> *Simongmulang* is a type of slash–and–burn with similar management and production processes. *Alnus* is a perennial, fast-growing tree. The soil is fertile after cutting and burning *Alnus*. Dulong have accumulated rich experiences through their long-term agricultural practices, and recognize that alder improves soil fertility. Dulong developed the system from their traditional practice of slashing and burning alder trees in the uplands. Recognizing the benefits of alder, they began to collect alder seedlings to transplant into swidden fields to develop man-made alder forests. The process is as follows: after slashing and burning swidden plots, they grow corn or barley for one year, then plant alder seedlings and allow them to develop into an alder forest fallow for five to six years, and then again slash and burn the alder fallow to grow annual crops. This cycle can support cropping for three years, longer than typical slash-and-burn cultivation, growing buckwheat in the first year, millet in the second, barley in the third, and then starting another cycle. This indicates that the Dulong are in a transitional process from nonfixed agriculture to a semipermanent farming system. Grain yields from this simongmulang system are 30 to 40 times the amount used for seed, compared with a typical swidden where the yield is generally 25 to 30 times the seed sown. Since simongmulang can be

cultivated for three years, Dulong farmers realize its importance, and the area occupied by this system is the second most common use of farming land. (Hong et al. 1960)

The Dulong are one of 25 main ethnic groups in Yunnan, and continue to live in relatively isolated communities. Until 1996, their villages could not be reached by automobile. Communication with outsiders is still very limited during the winter because of heavy snows on the high mountain passes between the Dulong River valley and the county capital. The Dulong community numbered only about 2,501 people in 1953 and about 6,000 in 1995. The great majority of Dulong live in Dulongjiang township, and the rest live in the Nujiang watershed of Gongshan county in northwest Yunnan. Dulongjiang township has four administrative villages, with a total population in 1995 of 4,500 people. The total land area is 1,994 km^2, with a population density for the whole township of only 2.03 persons per km^2. Even that part of the valley under 2,000 m asl has only 24 persons per km^2. Eighty-five percent of Dulong territory has a slope of more than 35°.

Alder-based agroforestry is a popular form of land use in the Dulong community, although it is not evenly distributed throughout the region. Most of it is found in two of the five administrative villages: Kongdang, which is also called the third administrative village, and in Xianjiudang and Longyuan, which together are called the second administrative village. Xianjiudang had a population of 646 people living in 79 households in 1988. It has 1,792 mu of upland fields under cultivation. Of these upland fields, 840 mu (47%) are swiddens, 788 mu (45%) are planted with *Alnus*, and the rest are permanently cultivated for annual crops and gardens. Xianjiudang also has 788 mu of paddy land. The main food crops of the Dulong are corn, rice, wheat, buckwheat, and beans. The main cash crop is *Cannabis sativa*, the bark of which is used for weaving and the seeds are chewed. *Cannabis* is not smoked in China (Zhang 1992).

As reported by Hong et al. in 1960, the Dulong refer to *A. nepalensis*–based agroforestry systems as *simongmulang*. In the Dulong language, *A. nepalensis* is *simong*. The alder-based rotation apparently originated from a simpler swidden system, such as that still practiced on the middle and lower slopes of the Gaoligong and Dandanglika mountains along the Dulongjiang River. A swidden system that does not involve a planted fallow is termed *xiangmulang* in Dulong. Simongmulang, on the other hand, entails transplanting *A. nepalensis* seedlings into the xiangmulang field, that is, intercropping alder seedlings with corn during the first year of cropping. The alder forest that results is usually left standing for about five years, and is then slashed and burned for a new cropping cycle. When alder fallows are reopened, annual crops can be cultivated for three years, longer than in a simple xiangmulang field. During the three years of annual cropping after alder, buckwheat is sown in the first year, millet in the second, and barley in the third. *Alnus nepalensis* seedlings are once more transplanted in the fourth year.

Permanent and semipermanent forms of agriculture, including the simongmulang agroforestry system, have been replacing swiddening in the last several decades. The area is still characterized by low agricultural productivity and the per capita yield of food grains is only 85 kg, meaning that most households run short of their subsistence needs. The average net cash income per person was only 161 yuan in 1990, making the Dulong poor people. Most of their cash comes from the sale of animal products.

A. nepalensis *as a Shade Tree:* Amomum tsao-ko *Cultivation*

Among the most popular agroforestry applications of *Alnus* is its use as a shade canopy over tea or *Amomum tsao-ko*. These agroforestry technologies have played a very important role in the development of cash crops in Yunnan Province since the 1960s.

Amomum tsao-ko, known locally simply as *tsao-ko,* is a perennial herbaceous crop, traditionally used for herbal medicine and as a spice. Tsao-ko is an important product endemic to Yunnan, grown only in some counties in Wenshan, Honghe, and

Baoshan prefectures. Its use is different to that of the more tropical Chinese cardamom (*Amomum villosum)*, as is its climatic preference. Tsao-ko is a highly marketable product much in demand in China. It is rarely exported. Tsao-ko has been traditionally grown in Yunnan for hundreds of years. The province has a total area of 120,000 mu under tsao-ko cultivation, yielding 1,200 tonnes per year (Wu 1989). Most tsao-ko is cultivated under a natural forest canopy at an altitude above 1,000 m/asl, such as in Tengchong and Baoshan counties in the Gaoligong Mountains, where it is grown in or near a large nature reserve. However, *Amomum tsao-ko* is grown under the shade of alder in Minglang township of Yongde county, Lincang Prefecture, in southern Yunnan, as well as in Maguan, Malipo, and Xichou counties in Wenshan Prefecture.

In Malipo county, alders intercropped with tsao-ko are found mostly on former swidden lands where *Alnus* grew naturally as pioneer trees. Only during the past five years have local farmers, in collaboration with foresters, begun to actively plant *Alnus* as shade for tsao-ko (Pi and Guo 1993; Chen 1994). The process normally takes several years. In the first winter and spring season, the shrubs and grasses under a selected stand of *A. nepalensis* are cleared. In the following year, tsao-ko and *Alnus* seedlings are planted together under the existing *Alnus* canopy, at a spacing of 2 m by 2 m. After three years, the *Alnus* canopy closes, and at about the same time, the tsao-ko begins flowering and fruiting. At that time, about 750 kg of tsao-ko can be harvested per hectare. The following year, the crop yield should rise to 1,000 kg/ha, a level that can be maintained for the next five years, after which the tsao-ko yields begin to decrease. When the alder canopy becomes too dense, the farmer can harvest branches for firewood. The alders can also be harvested after twenty years for construction material.

In Laodifang village of Malipo county, we encountered a total forest area of 2,000 mu, with 100 mu, or just 5%, of the forest used for tsao-ko cultivation. The annual cash income from tsao-ko was between 500 and 600 yuan/mu/year, with one mu yielding roughly 270 kg of tsao-ko per year. Cash income from tsao-ko accounted for 10% of the village's total agricultural income.

A. nepalensis *as a Canopy Tree in Tea Plantations*

Tea is one of the four main agricultural cash crops of Yunnan, the others being sugarcane, tobacco, and rubber. The province has a total area of 166,204 ha of tea, with a total yield in 1996 of 64,066 tonnes. It was then ranked sixth among China's 16 tea-producing provinces. Tea planting is believed to have begun during the Donghan dynasty, 2,000 years ago. Most of the tea produced in Yunnan these days is exported to other provinces and overseas. Most of Yunnan's tea plantations are found in the prefectures of Lincang, Baoshan, Simao, and Xishuangbanna.

Traditionally, tea was planted under the canopy of natural forest (Guo et al. 1993), but now only 20,000 mu, or little more than 1,000 ha, continues to be managed in this way. A second approach, intercropping tea with shade trees, including *A. nepalensis*, is widely used in Tengchong and Fenqing counties in western Yunnan. A total area of about 1,000 ha is managed in this fashion (Guo 1993; Dao and Guo 1996), and there has recently been some interest from local foresters in further encouraging and extending these agroforestry systems. Other local forms of tea growing include planting it on paddy bunds or at the margins of upland fields, as well as in home gardens. However, the model most commonly promoted by both scientists and the government is monoculture tea plantations established on upland terraces.

In a case study conducted in Zhaoying Village, Shangyin township, in Tengchong county, we encountered a tea plantation that had been established in 1965. Of its total area of 1,400 mu, 300 mu of the tea was intercropped with *A. nepalensis*. The alders had been introduced into the plantation by the farmers in 1971 for a variety of reasons. Insects such as scarab beetles preferred to eat *A. nepalensis* leaves and left the tea relatively unharmed, making heavy use of insecticides unnecessary; *A. nepalensis* conserved moisture and provided fertilizer for the tea plants so farmers no longer needed to water the tea during the dry season; and trunks and branches of thinned alders could be used as firewood in processing tea. Tea products from this area have become famous and are marketed throughout the county. Corn is also grown among the tea plants.

Fertility and Productivity of *A. nepalensis*-Based Agroforestry Systems

A study by Xia Yongmei et al. in 1989 and 1990 confirmed some of the farmers' perceptions regarding the beneficial aspects of intercropping tea with *Alnus*. Soil samples from four plots on a tea plantation in Fenqing county were collected and analyzed every month to compare soil fertility and tea yields under four kinds of planting patterns: tea with *A. nepalensis,* with and without chemical fertilizers, and tea without *A. nepalensis,* with and without chemical fertilizers. The research results confirmed that the N content of tea plots shaded by *Alnus,* with or without chemical fertilizers; were higher than those without *Alnus*; that tea under an *Alnus* canopy yielded 28.5 kg/mu more than the unshaded plots; and that a tea plantation intercropped with *A. nepalensis* required no applications of chemical N fertilizers (see Tables 29-3 and 29-4).

Table 29-3. Comparison of Soil N (%) in Liguo Tea Garden, Fenqing, Yunnan

Months	*Alder Canopy (no fertilizing)*	*No Alder Canopy (no fertilizing)*	*Alder Canopy (fertilizing)*	*No Alder Canopy (fertilizing)*
January	0.3406	0.3298	0.2295	0.3461
	0.2227	0.2493	0.2948	0.2041
February	0.2761	0.2291	0.3492	0.3795
	0.2987	0.2569	0.3591	0.2438
March	0.2574	0.3198	0.3089	0.2885
	0.3030	0.2741	0.3180	0.2653
April	0.358	0.3529	0.3930	0.2848
	0.3329	0.3064	0.3564	0.2503
May	0.4786	0.4078	0.4525	0.4220
	0.3829	0.3221	0.3818	0.3255
June	0.4960	0.4735	0.4850	0.4140
	0.4850	0.4030	0.4341	0.3565
August	0.3846	0.3403	0.3687	0.2885
	0.3472	0.3451	0.2755	0.2289
September	0.4543	0.3951	0.4553	0.4382
	0.3462	0.4481	0.3799	0.4276
October	0.4413	0.4416	0.3621	0.3526
	0.3959	0.388	0.3360	0.3088
November	0.5035	0.4364	0.4552	0.2091
	0.3864	0.4141	0.3448	0.2571
December	0.3078	0.3224	0.3799	0.1972
	0.3836	0.2228	0.1876	0.2576
Annual Average	0.3915	0.3680	0.3825	0.3291
	0.3527	0.3439	0.3335	0.2845

Table 29-4. Comparison of Tea Yields (kg/month) in Liguo Tea Garden, Fenqing, Yunnan (1989–1990)

Months	*Alder Canopy (no fertilizing)*	*No Alder Canopy (no fertilizing)*	*Alder Canopy (fertilizing)*	*No Alder Canopy (fertilizing)*
June	29	20	33	28
July	44	45.5	50	56.5
August	15	9	17	12
September	31	33	40	43
October	11	8	14	10.5
March	2	2	4	3
April	43	29	42	37
Total annual yield	175	146.5	200	190

Conclusions and Implications

Yunnan Province is characterized by environmental and cultural diversity, but economically it is one of the poorest provinces in China. Much of its territory is mountainous, and agricultural development is severely restricted. The natural forest cover has been decreasing rapidly and soil erosion is a major problem on more than one-third of the province's land area. The human population continues to grow rapidly and the productivity of many upland agricultural systems continues to decrease. The need for chemical fertilizers, merely to maintain yields, is rising. With swiddening still prevalent, a shortage of fuel has become a critical issue in Yunnan because it is needed to process important crops, such as tobacco. How can sustainable development be achieved in the face of these complex challenges?

Indigenous agroforestry practices provide a promising technical approach to these issues, although they are hardly a panacea for all problems. Our research shows that the development of *Alnus nepalensis*-based agroforestry systems by the Dulong in the Dulongjiang River valley, by the Han, Yi, and Lisu in Tengchong county, and the Wa in Ximeng county have all been a successful response to falling soil fertility. Such agroforestry systems have also improved the quality and productivity of tea in Tengchong and Fenqing counties, and *Amomum tsao-ko* in Malipo and Maguan counties. Alder agroforestry has also helped ease fuel shortages in the villages of the Wa, Han, Yi, and Lisu in Tengchong county. The transformation from rotational cultivation to permanent intercropping or alley cropping has also been an appropriate response to increasing population density and scarce agricultural land.

However, when considering the further development and extension of alder-based agroforestry in Yunnan, it is clear that there are many complex factors involved. In reality, because of increasing fragmentation of landholdings and increasing population densities, the traditional *Alnus*-based fallow system is fading. Only in situations with relatively low population densities and coordinated land-use planning by local communities will such rotational systems continue to exist in southern Yunnan.

From a subsistence economy perspective, *Alnus nepalensis* is both a fast-growing tree and good firewood, so it is attractive to local communities and government authorities who should encourage its adoption in areas without natural forests, or where deforestation in nearby state forest farms and nature reserves has been a problem. Often, local communities that depend on tea cultivation are also eager to plant *Alnus* more widely because it can improve the quality and productivity of tea. However, *A. nepalensis* is not a high-quality timber, compared to Chinese fir or *Cinnamomum*, so its price is comparatively low. This makes the production of mature *Alnus* trees relatively unattractive, except for limited uses such as construction of boxes for packing tea.

From an environmental point of view, broader adoption of *A. nepalensis* is very desirable. It is a N-fixer and the leaves have traditionally been used as a green manure. It holds promise to become even more important as a canopy tree over perennial cash crops, such as tea, coffee, and tsao-ko. These cash crops are both popular and economically important in Yunnan, and the role *A. nepalensis* is capable of playing in their wider cultivation fits well with Yunnan's long-term program of "biological resource development."

There is still much work to be done in investigating the potential of alder-based cultivation systems. On-farm experimentation and extension of the many social, economic, and environmental aspects of *Alnus nepalensis*–based agroforestry systems need to be emphasized so that we can provide the necessary scientific data to local communities and policymakers. We also need to further investigate the possible role of alder-based agroforestry in buffer zones of areas set aside for the conservation of biodiversity. Another area requiring further study is the advantages of using alder, a native tree, in silvicultural production instead of introduced trees such as eucalyptus or Chinese fir. Finally, we face the challenge of guiding the development of alder-based systems that might be appropriate for a wide spectrum of communities with different environmental, economic, and cultural conditions.

In order to accomplish these tasks, we hope to gain more knowledge from the experiences of practitioners, scientists, and extension personnel working with other agroforestry systems in other countries. These lessons need to be studied and understood and future communication between policymakers, local communities, and scientists needs to be improved.

Acknowledgments

This chapter is part of the Yunnan Agroforestry System Research Project (YAF), funded by the Ford Foundation and the United Nations University project People, Land Management, and Ecosystem Conservation. The authors would like to thank Malcolm Cairns for his encouragement and sharing of ideas during the process of writing this paper, and Bob Hill for his editing of the final version. Thanks are also owed to Mr. Dao Zhiling for providing recent information about Wa farming systems.

References

Dao Z. and Guo H. 1996. A Traditional Tea Plantation under Natural Forest in Jingmai, Langcang County, Simao. Working Paper for Yunnan Agroforestry System Research Project (YAF-Ford Foundation). (Chinese language)

Guo H. 1990. From Gathering to Cultivation: A Case Study of Agroecosystem Sustainability of Southern Yunnan. In: *Proceedings of the International Symposium on Man Made Community. Integrated Land Use and Biodiversity in the Tropics* (Oct. 26-31, 1990, Jinghong, Xishuangbanna, Yunnan, China). Feng Yaozong and Xu Xiangggyu (ed.). Kunming: Yunnan Science and Technology Press, 237-248.

——— and Z. Dao 1993. *Indigenous Agroforestry Systems in Baoshan, Yunnan, China.* Kunming: Yunnan Science and Technology Press.

———, Z. Dao, and Shen L. 1993. *Indigenous Agroforestry Systems in Baoshan, Yunnan, China.* Kunming: Yunnan University Press. (Chinese language)

———. 1995. The Impacts of Land Use Policy on Forest Resource and Biodiversity in Yunnan, China. In: *Proceedings of the Regional Dialogue on Biodiversity and Natural Resources Management in Montane Mainland Southeast Asian Economies.* Bangkok: Thailand Development Research Institute, 31–43.

———. 1996. Toward Sustainable Development in Yunnan: Current Status of Environment Quality of Yunnan. Paper presented at a School of International Training seminar, October 10, 1996, Kunming, China.

———, and Z. Dao 1995. Land Use Policy and Population Growth on Swidden Cultivation in Southern Yunnan, China. Paper presented at an International Workshop on Alternatives to Slash and Burn Agriculture, March 7–16, 1995, Kunming, China.

———, and C. Padoch. 1995. Patterns and Management of Agroforestry Systems in Yunnan. *Global Environmental Change*, Vol. 5(4), Note 9.

Guan Y., H. Guo, and S. Chen 2000. Cash Crop Cultivation under Natural Forest: Indigenous Practices and Management of Agroforestry Systems in Yunnan. *Acta Botanica Yunnanica* Suppl.XII, 113–122.

Hong J., J. Wang, and W. Feng 1960. Social Economic Investigation Report on Dulong People in the Fourth district, Gongshan County. In: *China's Minority Group Social Economic Investigation Material*. Kunming, China: Yunnan Ethnological Press. (Chinese language)

Liu L., T.Hu 1990. *Investigation Report of Land Resource Economy of Xishuangbanna, Yunnan*. Kunming, China: Yunnan People's Press.

Pei S. 1984. A Preliminary Study of Ethnobotany in Xishuangbannna. In: *Collection of Research Papers on the Tropical Botany*. Yunnan Institute of Tropical Botany (ed.). Kunming: Yunnan Peoples's Press, 16-32. (Chinese language)

Pi W. and H. Guo. 1993. Agroforestry Systems in Wenshan, Southeastern Yunnan. Paper presented at Workshop of Mainland Southeast Asian Cluster Meeting of UNU-PLEC, July, 1993, Chiangmai, Thailand.

Sun H, Z. Zhou, and H. Yu 1996. A Preliminary Probe into the Secondary Succession Series of Tropical Forests in the Big Bend Gorge of Yalu Tsangpo River in Southeast Tibet, Eastern Himalaya. *Acta Botanica Yunnanica* 18(3), 308–316. (Chinese language with English abstract)

Wang W. 1985. *Claves Familiarum Generumque Cormophytorum Sinicum*. Beijing: Sciences Press, 131.

Wang Q. and Y. Xu 1995. *South Silk Route*. Kunming. China: Yunnan People's Press. (Chinese language)

Wu Z. (ed.). 1989. *Development Strategy Study on Biological Resources in Yunnan*. Kunming: Yunnan Science and Technology Press, 186. (Chinese language)

———. 1984. *Index Florae Yunnanensis*. Kunming, China: Yunnan People's Press. (Latin with Chinese descriptions)

Ying S. 1991. *Slash-Burn Cultivation: A Disputed Agroecosystem*. Kunming, China: Yunnan People's Press. (Chinese language)

Yunnan Land Administration Bureau. 1994. *Yunnan's Land Resource*. Kunming, China: Yunnan Science and Technology Press. (Chinese language)

Zeng L. 1984. Alnus nepalensis, *Main Silvicultural Trees of Yunnan*. Kunming: Yunnan People's Press, 119. (Chinese language)

Zhang G. 1992. Utilization and Chemical Composition of *Cannabis* in Yunnan. Master's thesis.

Zhang P. 1995. Ecological Distribution and Biochemical Properties in Secondary Forest in Xishuangbanna. *Chinese Journal of Ecology* 14(1), 21–26.

Chapter 30

Shifting Forests in Northeast India

Management of Alnus nepalensis as an Improved Fallow in Nagaland

Malcolm Cairns, Supong Keitzar, and T. Amenba Yaden

Growing international concerns over global warming, biodiversity loss, and the ongoing reduction of the world's usable arable land have added new urgency to the old problem of replacing shifting cultivation with more permanent land-use systems (Bandy et al. 1993). These fears are based on the premise that swiddening is one of the major causative agents in tropical deforestation, a primitive remnant of the past, and in need of reform. It presupposes that these are forestal lands periodically despoiled by marauding forest dwellers. Even the term *shifting cultivation* suggests, misleadingly, not only nomadism, but also that agriculture is periodically imposed upon permanent forest lands.[1] The logic appears to be that since these lands are intermittently covered with trees, then they should properly fall within the domain of forestry and agriculture should be discouraged.

Careful analysis of the alder fallow management system described in this chapter suggests that this argument should be turned upside down, and, at least in some cases, these areas should more accurately be considered as agricultural lands on which farmers intentionally encourage trees to grow as an integral phase of a cyclical and sustainable farming system. It proposes a revisionist view of "shifting forests on an agricultural landscape," dramatically recasting the role of swidden farmers from forest destroyers to forest planters and managers. This is more consistent with the worldview of tribal peoples such as the Karen of northern Thailand, who describe upland cultivation as *baa muan wiang,* translating literally to "rotating forests."

Indigenous practices of protecting or planting soil-building trees in swidden fallows are neither new nor rare. The extensive pool of swidden-related literature (Robison and McKean 1992) includes documentation of farmer innovations across the Asia-Pacific region to integrate various N-fixing trees into swidden lands, either

Malcolm Cairns, Department of Anthropology, Research School of Pacific and Asian Studies (RSPAS), Australian National University, Canberra, ACT 0200, Australia; Supong Keitzar, Director, Directorate of Agriculture, Kohima 797001, Nagaland, India; T. Amenba Yaden, Deputy Conservator of Forests, Social Forestry Division, Kohima 797001, Nagaland, India.

[1] The pejorative term "slash-and-burn" conjures up even more negative connotations and is deliberately avoided in this chapter, except in cases of direct quotes.

sequentially or simultaneously, to both accelerate fallow functions and provide useful products. The leading species include: *Leucaena leucocephala*[2], *Sesbania grandiflora* (Kieft, Chapter 28), and *Casuarina oligodon* (Thiagalingam and Famy 1981; Ataia 1983; Askin et al. 1990; Bino and Kanua 1996; Bourke, Chapter 31). The work on alley cropping that dominated the early agroforestry agenda was conceptualized as a simultaneous fallow (Kang and Wilson 1987) and mimicked many elements of indigenous variations of fallow management. Within these writings, the potential role of *Alnus nepalensis* (alder) as a superior fallow species has been relatively underreported (Kevichusa et al. n.d.; Hong et al. 1960; Gokhale et al. 1985; Dhyani 1998; Fu 2003; Guo et al., Chapter 29).

Although largely closed to research attention in recent years, historical literature and farmer testimonies indicate that adoption of *A. nepalensis* as an improved fallow species has occurred in a swathe stretching from northeastern India, eastward across northern Myanmar, and into southwestern China (Figure 30-1). The following quotes show both the antiquity and spread of this practice, and provide intriguing glimpses of how farmer observations lead to useful innovations.

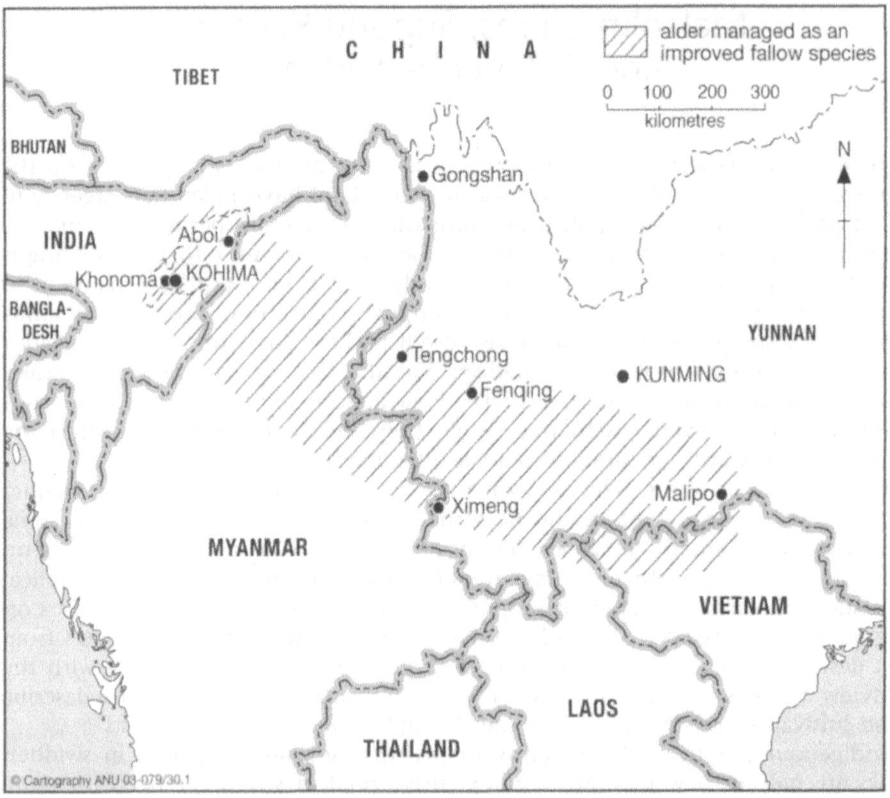

Figure 30-1. Management of *Alnus nepalensis* as an Improved Fallow Species Crosses a Range of Environments and Cultures

[2] For analysis of the "Amarasi system" in Timor, Indonesia, see Metzner 1981, 1983; Jones 1983; Piggin and Parera 1985; Field 1991a, b; Field and Yasin 1991; Surata 1993; Yuksel 1998; and Piggin, Chapter 24. The "Naalad system" in Cebu, Philippines, is discussed by Eslava 1984; Subere et al. 1985; Kung'u 1993; Lasco and Suson 1997; and Lasco, Chapter 27. A comparable practice in Sulawesi is described by Agus in Chapter 25, and relatively recent innovations in Mindoro, Philippines, have been studied by MacDicken 1990a, b, 1991, Chapter 26.

From Northeastern India:

> Indeed, the *jhumming* [swiddening] system of the villages round here is about the finest I have seen. Only millet and Job's tears are grown, but the whole hillside, and very steep it is, is most elaborately laid out in ridges and quasi-terraces with logs cut from the pollarded alders growing all over the slopes and everywhere most carefully preserved. ... Unlike the jhums [swiddens] of other tribes, which are used for at least two successive years, the ground is sown for one year only and then allowed to stay fallow again for 3 or 4 years, instead of the more usual 10, and this rotation is continued with apparently admirable results, showing what really can be done with steep and unpromising land by careful preservation of the alder and precautions against denudation. The Angfang people seem to propagate this alder (*Alnus nepalensis*) from cuttings[3] put in about April, but they told us at Chaoha that they grew it there from seed. Experiments in the Sema county have shown that neither method is at all certain of success when tried by amateurs. (Hutton 1929, *28–29*)
>
> According to oral history, our ancestors have been managing *rupro (A. nepalensis)* in their jhum fields for the past 500 years. Some of the individual trees that you see in the fields are themselves at least 100 years old. They have become an important part of our heritage and we tend them with memories of our distant ancestors. Local history says that when this village was first settled, the land was covered with forest dominated by *Phoebe* sp. and *Michelia* sp. But when our forefathers cleared the forests for jhumming, it was rupro that grew back again. I don't think that *nhalie jhum* [alder-based swiddens] began intentionally, but probably our ancestors noticed that crops growing nearest to rupro performed best. When rupro trees establish naturally in the jhum, they are protected and managed; otherwise they are planted. Rupro is a fertility indicator; the more rupro trees, the better the soil. (Angami village elder in Khonoma Village, Nagaland, in personal communication with the authors, 1996)

From Northern Myanmar:

> An interesting system is in vogue among the *taungya*-cutting [swiddening] tribes in certain parts of the hills of the Bhamo and Myitkyina districts on the northeast frontier of upper Burma. The crops raised in the temporary clearings are maize, cotton, or other hill crops. Alder *[Alnus nepalensis]* seed is sown along with the crop or sown broadcast after it. A dense young crop of alder results. It requires weeding for three years and fire protection for four, after which it requires no further attention. The rotation for shifting cultivation is reduced by this system from twelve years to eight, while the soil is enriched and protected from erosion and desiccation. Experiments have been tried in other parts of Upper Burma in sowing up taungyas in this manner, but it has been found that the alder will not grow below 5,000 ft. (1,524 m asl) elevation, and failure is attributed in some localities to suppression by dense weed growth or to the depredations of deer. (Troup 1921, *912*)
>
> Many of the Kachins are in the habit of reafforesting their *taungyas* [swiddens] by the broadcast sowing of *maibau [A. nepalensis]* and, provided this is insisted upon and the taungya rotation is made sufficiently high to allow for complete reafforestation before cutting again, there is little danger of damage by erosion ... Maibau is favored for afforestation as the Kachins say it improves the soil and produces good ash for cultivation and for this reason it is the most suitable species except in places liable to severe frost.

---

[3] Hutton's understanding of their establishment methods is suspect. Lamichhaney (1995, *13*) contended that there are no records of successful vegetative propagation of *A. nepalensis*.

(Report of Forest Administration in Burma for the Year Ended March 31, 1922, *19–20*)

Moderate supplies [of *A. nepalensis*] may be expected from Burma, where, in the Kachin Hills, the slopes denuded by shifting cultivation are rapidly being stocked by this alder. (Pearson and Brown 1932, *970*)

Broadcasting of seeds of maibau *[A. nepalensis]* in the fallow land for immediate restocking of the land with trees is also practiced in the Chin and Kachin States. (Wint 1996, *39*)

And from Southwestern China:

Simongmulang [an *Alnus nepalensis*–based agroforestry system] is a type of slash-and-burn with similar management and production processes... Dulong developed the system from their traditional practice of slashing and burning alder trees in the uplands. Recognizing the benefits of alder, they began to collect alder seedlings to transplant into swidden fields to develop man-made alder forests. The process is as follows: after slashing and burning swidden plots, they grow corn or barley for one year, then plant alder seedlings and allow them to develop into an alder forest fallow for five to six years, and then again slash and burn the alder fallow to grow annual crops. This cycle can support cropping for three years, longer than typical slash-and-burn cultivation, growing buckwheat in the first year, millet in the second, barley in the third, and then starting another cycle. ... Grain yields from this simongmulang system are 30 to 40 times the amount used for seed, compared with a typical swidden where the yield is generally 25 to 30 times the seed sown. Since simongmulang can be cultivated for three years, Dulong farmers realize its importance, and the area occupied by this system is the second most common use of farming land. (Hong et al. 1960, as cited by Guo et al., Chapter 29)

We generally grew upland rice for one to two years and then left the land to become *qi mu [A. nepalensis]* forest. When the rice was 30 to 50 cm tall, usually around mid-June or earlier (1.5 to 2 months), then we broadcast qi mu seed into the field. By the time of rice harvest in October, the qi mu would already be 20 cm in height. We would return to this field for the next 5 to 10 years to collect qi mu leaves to use as green manure in our irrigated rice fields. After a qi mu forest has been cleared, the land will give much higher yields. I'm not sure how the interplanted qi mu affects rice production because there is no basis for comparison. All our dryland fields are planted with qi mu in this way. (Li Yuexin, a villager from Songshup village, Tengzhong county, Yunnan, in personal communication with the authors, 1998)

There are no historical records concerning alder agroforestry in Ximeng, but the Wa are known to have long had the custom of preserving *Alnus*-dominated swidden fallows for upland farming, since it regenerates easily. ...After slashing and burning, upland rice is grown for two consecutive years. Local farmers say that before 1983, the *Alnus*-dominated fallow that then took over came from windborne *Alnus* seeds that naturally sprouted and reforested the swidden, and the Wa farmers protected them. Since 1983, however, Amuo villagers have been collecting young seedlings from natural *Alnus nepalensis* forest to transplant them into upland rice fields in the second cropping year. The farmers say that if they do not plant *Alnus* trees, *Eupatorium* weeds will invade and occupy all of the upland. And upland fields with *Alnus nepalensis* trees are more fertile than *Eupatorium*-dominated swidden fallows. After two years of upland rice cultivation, *Alnus nepalensis* is left for 10 years. Some farmers maintain the trees for 20 years and then harvest them for construction materials. This rotational system is called *geloudanggueang* in the Wa language. (Guo et al., Chapter 29)

> Farmers broadcast *Alnus nepalensis* seeds into swidden fields and then left them for a two-year fallow period. The young alder seedlings would then be plowed under and crops grown for a further two years before the cycle was again repeated. (Prof. Zhang Jiahe describing management of alder as a green manure crop in Lingxang Prefecture, China, in personal communication with the authors, 1998)

The fact that isolated and disparate swidden communities have independently developed deliberate practices to integrate *A. nepalensis* into fallows suggests that this N-fixing alder offers agronomic or economic benefits, or both, that warrant closer research attention. Furthermore, its wide adoption by a range of tribal shifting cultivators suggests that it is not overly site specific, and has potential for greater dissemination across the Himalayan foothills.

This chapter describes an ancient but little-documented example of farmer manipulation of *A. nepalensis* in Nagaland, northeastern India, which has enabled a significant intensification of the swidden cycle without concomitant ecological decline (see color plate 35). It offers a hypothesis that this intensification was partly prompted by security concerns in an atmosphere of intertribal warfare and headhunting, and gives a brief cultural profile of the main innovators, the Angami Nagas. Much of the historical detail is drawn from early reports by the British colonial government. These provide rich insights into the people of Nagaland and the historical environment that spawned the innovation of alder fallows. The chapter then focuses on a village in Kohima district of Nagaland as a case study, and provides a description of standard jhum cultivation as it is practiced in the region as well as a more detailed diagnosis of the alder fallow innovation. It draws on these findings to elucidate pertinent research issues and to examine the role this system could play in helping to intensify sustainable swidden farming across a broader landscape.

Objectives

This diagnostic survey was intended to be preliminary in nature and to address the following objectives:

- To draw on the literature in elucidating the historical conditions that gave rise to alder management in Nagaland's jhums;
- To characterize traditional farmer management of *A. nepalensis*, its significance as a fallow species, and its apparent role in permitting a dramatic intensification of the swidden cycle;
- To demarcate the area with potential for wider application, and to hypothesize why this apparently successful indigenous system has not been more widely adopted, both within and outside Nagaland;
- To identify an agenda for follow-up research; and
- To generalize lessons of wider use across Asia-Pacific, where degrading swidden systems are endemic.

Methods

The study originates from a collaborative relationship between Nagaland's State Agricultural Research Station (SARS) and the Nagaland Environmental Protection and Economic Development (NEPED) Project, with technical backstopping from the World Agroforestry Centre (ICRAF). Each institution contributed one scientist to assemble a core research team representing the disciplines of agronomy, forestry, and agricultural anthropology. During 1996 to 1997, the team characterized alder-based jhum cultivation and identified corresponding research priorities.

Work began by assembling secondary data such as maps, weather information, and jhum statistics from state documents, archival materials from the British colonial era that commented on alder management in swidden environments, and recent state publications promoting its use (Kevichusa et al. n.d.; Gokhale et al. 1985).

Substantial research was found to have been conducted on *A. nepalensis*, mostly in Nepal and northeastern India, and this work was drawn upon to explain the rationale underlying Naga management of alder.

Khonoma, an Angami village in Kohima district, was selected as the study site because of its development, over many generations, of remarkably sophisticated practices to systematically manage alder as an improved fallow. The team used diagnostic methods (Fujisaka 1991) to elicit farmer practices and knowledge, combined with elements of field observation, ethnographic analysis, and field sampling. Village elders were interviewed to maximize the historical depth of our understanding and to discover how the system may have changed over the past century. Interviews were kept informal, open ended, and structured by a set of guide issues. This avoided the rigidity of formal questionnaires, while maximizing opportunities for the research team to interact with farmer respondents, probe more deeply into issues of particular interest, and gradually build an understanding of alder-based fallows.

Alnus: Applications in Forestry and Agroforestry

The *Alnus* genus (commonly known as alder) is composed of nonleguminous shrubs and trees that fix substantial quantities of nitrogen and are often used effectively in forestry and agroforestry applications. In the western United States, for example, *Alnus rubra* is interplanted into coniferous stands to significantly improve their growth (Hibbs et al. 1994). Although early trials found that pure stands of *Alnus* generated excessive N buildup and leaching, a solution was found in rotating *Alnus* with other timber species (Bauhus 1999). In the Philippines, *A. japonica* is planted to provide a canopy over coffee, is grown as a nurse tree in *Pinus kesiya* plantations, and provides living posts to support a network of wires for chayote (*Sechium edule*) (Faridah Hanum and van der Maesen 1997, *68*). A New World alder, *A. acuminata* (syn. *A. jorullensis* H.B.K.) is traditionally grown as a shade tree over coffee and is also scattered throughout high-elevation pastures in Costa Rica, contributing both shade for livestock and nitrogen for undersown grasses (Russo 1994).

A. acuminata has also been introduced to the highlands of Uganda, where its agronomic properties captured the attention of ICRAF researchers during trials to measure tree crop competition of various multipurpose species. It was found to have a positive influence on crop yields or, more accurately, a "negative competition" effect (ICRAF 1993; Okorio et al. 1994). This led to a hypothesis that it could be managed in a system bearing striking similarities to the indigenous practices developed long ago by Naga swidden cultivators.[4] The concept is for *A. acuminata* to be planted as a rotational woodlot fallow on upper terraces that have become degraded through downward movement of topsoil. It is postulated that the alders will provide both biological benefits of fertility restoration and an economic harvest of poles or firewood at the end of the fallow. The alder stumps will be maintained in the field and the coppices managed to reduce competition while a crop is grown around them. When the fallow effect diminishes and crop yields decline, the woodlot will be allowed to regenerate naturally, either by leaving one coppice sprout for poles or many for firewood production (ICRAF 1996). *A. nepalensis* is also reported to have performed strongly in adaptation trials in the highlands of Rwanda.[5]

Returning the discussion to Asia, Ikalahan swiddenists in northern Luzon, the Philippines, plant *A. japonica* seedlings in fallowed land (Vergara 1995). In other parts of the Himalayan foothills, *Alnus* is known to be managed in swiddens in western Yunnan (Hong et al. 1960; Fu 2003; Guo et al., Chapter 29), northern Myanmar (Wint 1996, *39*), and, although not actively propagated by farmers there, it is an

---

4 This appears to be a rich opportunity for South-to-South technology transfer, in which Nagas could assume an extensionist role in sharing their indigenous knowledge and practices related to alder management.

5 In Rwanda, civil strife forced abandonment of the trials so performance data are not available.

important species in the mid-altitude belt of Bhutan (Roder 1997). This confirms that at least some other alder species may have properties analogous to *A. nepalensis*, as well as the potential to be managed in similar ways.

Alnus nepalensis *in Himalayan Environments*

Across the Himalayan foothills, the *Alnus* genus is represented by both *A. nepalensis* and *A. nitida*. *A. nepalensis* (Figure 30-2) is a common early succession species throughout the Himalayas from as far west as the northern frontier of Pakistan through Nepal, northern India, southern Bhutan, northern Myanmar to southwestern China and Indochina, at altitudes between 1,000 and 2,500 m above sea level (asl) (Faridah Hanum and van der Maesen 1997, *68*).[6] It is widely distributed across Nagaland, with most alder-based swiddens found in the 1,500 m to 2,000 m range. This is consistent with the findings of a West Bengal study that showed alder regeneration, seedling growth, root nodule biomass, and nitrogenase activity were temperature dependent and decreased below 1,500 and above 2,000 m (Sharma, E. 1988). As an early succession species, it thrives under high light intensity and soil moisture (Boojh and Ramakrishnan 1981; Boojh 1982 ab; Storrs and Storrs 1984; Khan and Tripathi 1989 ab) and has useful applications in antierosion work (Basu 1947), stabilizing landslides and plugging gullies (Basu 1951), and as a nurse crop in regenerating clear-felled areas (Ali 1946). Pollination and seed dispersal are by wind. It prefers to establish on slopes with northern aspects where soil moisture tends to be higher. *Alnus* rapidly colonizes land that has been disturbed and is recognized by farmers as a soil fertility indicator. An abundance of vigorous alders with straight boles indicates rich soil with good agricultural potential. Alder will also colonize nutrient-poor areas such as recent landslides or roadsides (Lamichhaney 1995), but growth will be gnarled and branchy.

Taxonomy

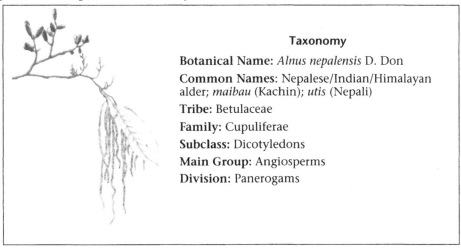

Botanical Name: *Alnus nepalensis* D. Don

Common Names: Nepalese/Indian/Himalayan alder; *maibau* (Kachin); *utis* (Nepali)

Tribe: Betulaceae

Family: Cupuliferae

Subclass: Dicotyledons

Main Group: Angiosperms

Division: Panerogams

Figure 30-2. The Fallow Species

Alder's easy propagation, rapid growth, and prolific coppicing make it an important firewood species (Harrison 1989; Subhash-Nautiyal and Nautiyal 1990; Rimal 1992). *Alnus* roots have heavy concentrations of symbiotic actinomycetes of the genus *Frankia* around their nodules (Sharma, S.K. et al. 1984, 1986; Prat 1989; Brunck et al. 1990) and nitrogen fixation contributes significantly to total N uptake (Sharma, E. and Ambasht 1988; Ambasht and Srivastava 1994). Farmers have learned

---

[6] *A. nepalensis* has been widely introduced as an exotic and plantations now exist across tropical Africa, Costa Rica, and moist mountainous areas of Hawaii. Within Southeast Asia, it has prospered in the Philippines but suffered 95% to 100% mortality in East Java trials. Trial plantings have also been made in Malaysia (Faridah Hanum and van der Maesen 1997).

to exploit its N-fixing capacity by including it in their farming systems in a variety of agroforestry patterns (Neil 1989, 1990; Dhyani and Chauhan 1990; Carter 1992; Amotya and Newman 1993). It is widely used in Sikkim as a shade tree over large cardamom (*Amomum subulatum* Roxb.) (Singh, K.A. et al. 1989, 1991; Zomer and Menke 1993; Sharma, R. et al. 1994, 2002; Sharma, R., Chapter 51) and one study showed that it more than doubled cardamom yields (Sharma, R. et al. 1994). Similar synergistic effects were observed when *A. nepalensis* was evaluated as a shade tree over *Cinchona officinalis* L. plantations. From five shade species evaluated, bark yields were highest in plantations under an *Alnus* canopy (Nandi and Chatterjee 1991). Mountain farmers in Yunnan Province of China cultivate tea under *A. nepalensis* (Shengji 1991) and field trials in northeastern India indicated that it combines well with forage and pineapple cultivation in hilly terrain (Chauhan et al. 1993).

Alnus timber is of medium quality and can be used in construction or making furniture (Visapal et al. 1995). Its pulp is also suitable for newsprint (Guha et al. 1965). *A. nepalensis* has other secondary uses: crushed or chewed leaves can be applied to wounds as a blood coagulant; bark tannins are used for tanning leather; the bark is also used in dying textiles; the leaves are sometimes brewed as a drink; and young leaves can be foraged by cattle.

The Study Area

Nagaland is a mountainous, heavily forested state on the northeast fringe of India. It is bordered by Assam on the west and north, Manipur on the south, Arunachal Pradesh on the northeast, and shares an international border with Myanmar on the east (see Figure 30-3). With a total area of 16,579 km^2, it ranks as one of India's smallest states. Its almost 2 million population (2001 census) is almost entirely tribal, and includes 16 major Naga tribes,[7] each with its own culture and mutually unintelligible dialect of the Sino-Tibetan language family (Elwin 1997; Singh, P. 1972; Horam 1992; Sen 1987; Jacobs et al. 1990; Thong 1997; Stirn and van Ham 2003). A pidgin form of Assamese called *Nagamese* is spoken as the lingua franca between tribes. Their ancestral origins are unclear. Some consider them to be a Mongolian racial group that migrated south from Tibet, while a competing theory claims they were a coastal people from Southeast Asia[8] who migrated west to their present homeland (Naga Institute of Culture 1970; Gundevia 1975, *6*; Shimmi 1988). These tribes, collectively known as Nagas, were renowned as warriors and headhunters and fiercely resisted British penetration into their territory. Work by American Baptist missionaries, beginning in the mid-1800s, contributed to pacification of the Nagas and today the large majority is Christian.[9]

Administratively, Nagaland is divided into eight districts, with Kohima as its capital. More than 70% of the population lives in rural villages, usually perched in defensible positions on mountain spurs and ridges. Agriculture engages most of the working population and is the mainstay of the economy. Shifting cultivation, known locally as *jhumming*, is the dominant land use across much of the state and is widespread throughout the seven states comprising northeastern India[10] (Husain 1988) (Table 30-1). The average jhum area cultivated per household in any given year varies from 0.5 to 2.5 ha, with about 101,400 ha thought to be cultivated annually within the state. Estimates suggest that roughly half of Nagaland's land area is under jhum cultivation, with most of the remainder still under natural forest (Table 30-2).

---

7 There are generally considered to be a total of 34 major Naga tribes in the region; the remainder are distributed between Manipur, Arunachal Pradesh, Assam, and across the international border in northwestern Myanmar. Other non-Naga tribes also inhabit Nagaland, including Garo, Kuki, Kachari, and Mikir.

8 See von Fürer-Haimendorf (1971) for a comparative essay on mountain peoples of Luzon, Philippines and northeast India.

9 See Clark (1978) for an account of early missionary work in the Naga Hills.

10 More than 90% of the estimated 600,000 families of shifting cultivators in India are in the northeastern states, most of them tribal minorities (Keitzar 1998, *51*).

This area of the eastern Himalayas is counted as one of 15 biodiversity hotspots in the world and has a very high rate of endism.

Current trends in Nagaland are placing its forests and biodiversity under serious threat. Rampant logging, combined with the highest population growth rate in India, are degrading its forest resources at unprecedented rates. Nagaland's population grew by 64.41% in the 10 years between the censuses of 1991 and 2001, compared to a 21.34% national growth rate for all of India over the same period. Scarcity of agricultural land has forced both the shortening of fallow periods (Table 30-3) and the conversion of more natural forest into jhum land (Keitzar 1998, *52*). Crop yields are declining and many Naga families, whose average income is already among the lowest in India, are struggling to find enough food. In sharp contrast to India as a whole, in which 90% of forestlands are publicly owned, 88% of Nagaland's forests are privately held, either by individuals or communally by clans or villages. Any proposed interventions, therefore, will have to strengthen the capacity of institutions at village level to manage the vast majority of state lands and forests under their control.

Responding to these challenges, several coordinated research and development projects were launched in 1995. The Nagaland Environmental Protection and Economic Development (NEPED) project encouraged the Nagas to improve their swidden farming by implementing soil and water conservation practices and introducing fast-growing trees into their jhum fields to create an improved fallow that can be harvested and sold at the end of the fallow period. This approach seeks to retain and improve jhum cultivation while introducing trees as a new cash crop. In tandem, the State Agricultural Research Station (SARS) is working with village communities to progressively intensify their land use by drawing on local innovations and genetic resources, as well as introducing useful ideas from outside the state.

Table 30-1. Area Under Swidden Cultivation

States	*Area Available for Jhum (thousands of ha)*	*Planted at Given Point of Time (thousands of ha)*	*% of Area Available*	*Tribal Families Involved (thousands)*	*Area Cultivated per Tribal Family*
Arunachal Pradesh	248.58	92.00	37.00	81	1.13
Assam	498.30	69.60	13.00	58	1.20
Manipur	100.00	60.00	60.00	50	1.20
Meghalaya	416.00	76.00	18.20	68	1.12
Mizoram	604.03	61.61	10.20	45	1.37
Nagaland	608.00	73.54	12.10	80	0.92
Tripura	220.79	22.30	10.10	43	0.51
N.E. Region	2,695.70	455.05	16.60	425	1.07

Source: Keitzar 1998, *52*.

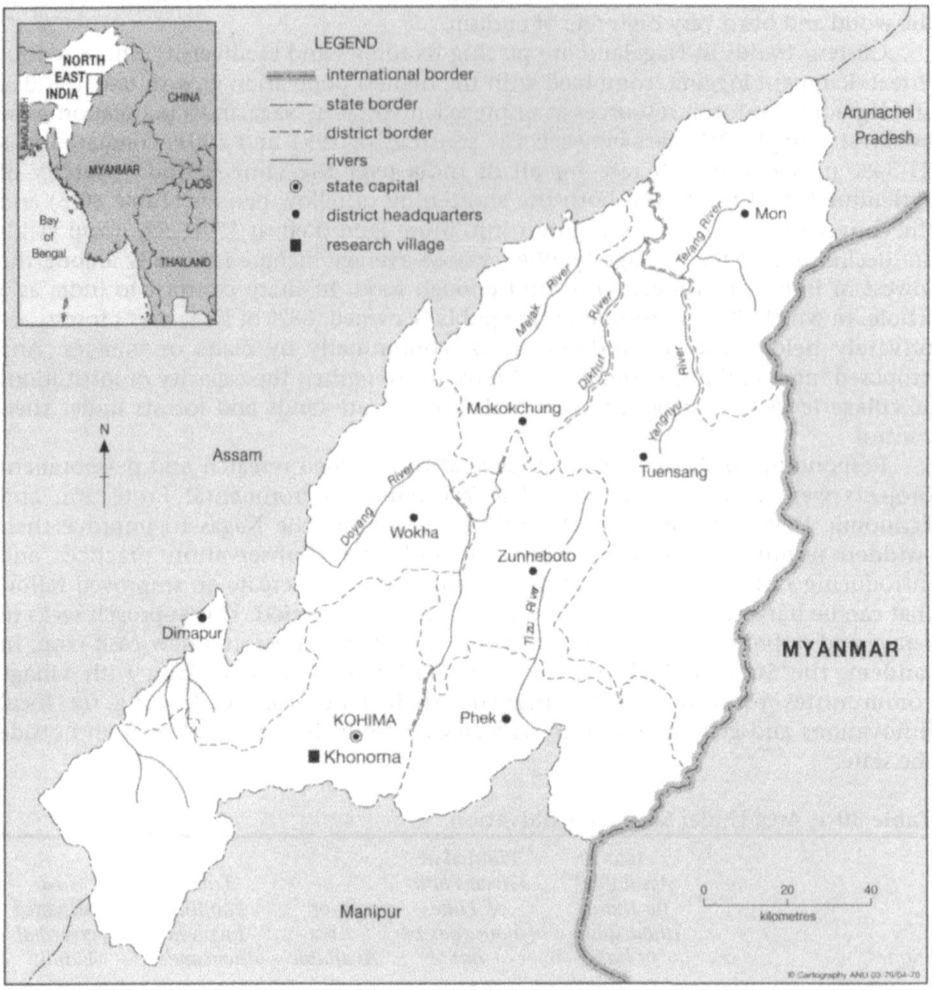

Figure 30-3. The Nagaland Study Site: At the Interface between Northeast India and Mainland Montane Southeast Asia

Table 30-2. Land Use in Nagaland

Land Use	*Area (ha)*	*%*
Forest under government control	100,420	6.06
Forest under control of villages, clans, and families under tribal system	762,112	45.97
Current jhum land	101,400	6.12
Jhum land regrowth	531,600	32.06
Permanent cultivation	50,250	3.03
Area under towns, villages, roads, rivers, etc.	112,118	6.76
Total Area	1,657,900	100.00

Source: Cited in Imnauongdang 1990, *56*.

Table 30-3. Population Growth, Population Density, and Length of Swidden Cycle in Nagaland

District	*Area* (km^2)	*Population* 1981	1991	*Population Increase* (%)	*1991 Population Density per km^2*	*Jhum Cycle (yrs)*
Kohima	4,041	238,747	394,179	65.10	98	5.9
Phek	2,026	70,618	101,823	44.19	50	10.0
Wokha	1,628	58,040	82,394	41.96	51	8.4
Zunheboto	1,255	72,519	97,933	35.04	78	8.1
Mokokchung	1,615	103,736	156,207	50.58	97	6.1
Tuensang	4,228	137,108	232,972	69.92	55	7.2
Mon	1,786	94,162	150,065	59.37	84	7.6
Nagaland	16,579	774,930	1,215,573	56.86	73	7.8

Note: These data predate the creation of Dimapur district in 2001 from land that had formerly been included under Kohima district.
Source: Directorate of Economics and Statistics, Government of Nagaland, 1981–1991 data.

The Practitioners

The study area lies in the territory of the Angami Nagas, one of the largest and, historically, most warlike of the Naga tribes (Figure 30-4). The benchmark anthropological text on the Angami is Hutton's 1921 treatise, reprinted in 1969, which built on the work of Mills (1854), Butler (1875), Davis (1891), Johnstone (1896, reprinted in 1971), and other British predecessors. A short introduction to the Naga Angamis and their circumstances over the past century may shed light on the conditions that triggered Angami innovations in managing alder as an improved fallow.

Warlords of the Naga Hills

When, in 1832 and 1833, the first British military expeditions penetrated the unexplored Naga Hills in search of a direct route between Assam and Manipur, they were forced to fight their way through fierce tribal resistance. It was their first introduction to the powerful Angami Nagas, and marked the start of a protracted struggle to pacify the Naga Hills and bring them under British authority.

The Angami are thought to be of Mongoloid stock and their dialect falls under the Tibeto-Burman subfamily of languages. Their early history is shrouded in uncertainty, for lack of a written script. However, Ghosh (1982, *52*) writes:

> The Angamis have got a tradition that the Karens of Burma, known to the Angamis as Karrennoma, and the Angami themselves, belonged to the same family. They however split into two groups and the Angamis turned westward while the Karens turned eastward.

According to oral tradition, they halted at Khezakenoma, in the south of present-day Nagaland, and from there migrated northwestward of Kohima to their current homeland, possibly in the 13th century A.D. (Imnauongdang 1990, *33*). Angami territory currently extends north from Manipur and occupies the northern part of present-day Kohima district. As heavily armed British patrols pressed into Angami country, they encountered large villages, built in commanding, strongly fortified positions. Their altitude, between 450 and 1,850 m asl, provided a bracing climate: a comfortable zone between the unhealthy lowland heat and the bitter cold of higher peaks. Village outskirts were often marked by huge stone monoliths, often so heavy that several hundred men straining on ropes were needed to drag them into place on wooden sledges. Large mithun (*Bos frontalis*) horns carved out of wood and mounted on the front gable signified Angami dwellings. Early British reports described with admiration the terraced fields carved into hillsides and skillfully irrigated, drawing parallels with the well-known Banaue rice terraces of the Philippines.[11] The less visually striking alder fallows, in contrast, earned comment from only the keenest observers (Hutton 1929, *23, 28–29, 47, 62*; 1969, *76*).

Each village was run akin to a mini-republic, electing its own chief and assembling villagers to make collective decisions. Despite this, unfriendly relations often developed between clans in the same village and disputes were settled by bloody fights. Early reports present a composite picture of Angami as a tall, strongly built people with great powers of endurance and capable of carrying heavy loads over long marches.

Population: 97,408 (1991 census) (8.05% of total Naga population in 1991)

Territory: Southeast of Kohima district

Origins: Indo-Mongoloid stock

Language: Tibeto-Burman subfamily

Village Governance: Elected headmen

Kinship Pattern: Patrilineal

Agricultural Systems: Extensive irrigated rice terraces supplemented by swiddening

Figure 30-4. The Fallow Innovators, the Angami Nagas

[11] Angami skill in constructing irrigated rice terraces is widely known and government schemes have often encouraged their expansion to reduce reliance on shifting cultivation. One innovative program attempted to tap Angami expertise in terrace cultivation by recruiting them as "agricultural demonstrators" who shared their know-how with other Naga tribes that had no tradition of terracing (Sema 1992, *105*).

Prior to the arrival of British authority and Christianity, Naga villages were almost constantly at war. Hostilities focused on headhunting raids, each one often a reprisal for the attack before it. The raids were considered vital to the prosperity of the village. The following series of excerpts illustrates the impetus behind headhunting, the symbolism involved, and its link with farming systems:

> Usually a party of braves would go out either to snatch a head or two from a sleeping village or, more commonly, to try to rush the sentries guarding a party working in the fields, and fall upon the screaming women as they scattered. (Mills, J.P. 1982, *157*)

> The motivation behind a headhunting raid was quite simple, and there are many tribesmen who can still give you disjointed explanations centering round the belief that the soul existed in the head, just under the scalp. ... It was believed that all the strength and force of a person was concentrated in the head and that this was transferable. This soul-force guided the functions of the human body. Further, variation in the sum total of the entire community's soul-force had its impact on the prosperity of the race. If a calamity, an epidemic or a mighty fire broke out or the crops failed, the cause was attributed to a decrease in soul-force. Excessive mortality in population or cattle, or positive evidence of decreasing fertility, could justify demands for a headhunting raid. To compensate for the loss of the force as a result of a raid, counter raids had to be launched to bring back home at least an equal number of heads. Since the loss of heads affected the entire community, it was not the business of the members of the family of the deceased alone to take revenge but of the entire community. ... The Karens tended towards the more physical aspects of the belief and felt that the human head contained a volatile fluid which had a great fertilizing power when the fields were sprinkled with it. This fountain of life-force was supposed to make the soil rich and the increased yield made the animals plump. Both the improved produce of the soil and the fat animals further increased the vitality of the man and, in this way, made him fertile. (Anand 1967, *100–101*)

> The souls of dead men are wanted to fertilize all vegetable and animal life, and to add to the general stock of vital essence in the village. The soul is located in the head above all other parts, and therefore the head at any rate is carried back (with the soul in it) to add to the sum of vital essence in the head-taker's village. ... The Karens have a definitely formulated theory of how this principle actually works out and how the souls of the dead do fertilize crops and animals, including man. The practice, reported from the Maori and from the headhunters of Kafiristan, of warriors cutting off the heads of their own dead to save them from the enemy, a practice which was resorted to by the Chang Nagas who attacked Phomhek and were driven off with loss, but decapitated and took back the heads of some thirty of their own dead, is a case in point. (Mills, J.P. 1973, *225*)

> Exclusive of revenge, however, one of their most barbarous customs is that of cutting off heads, hands and feet of any one they can meet with, without any provocation or pre-existent enmity, merely to stick them up in their fields to assure a good crop of grain. This practice is very common amongst the adjoining tribe of Lotah Nagas, and the Angami Nagas are said to be addicted to it but not so frequently. (Mills, A.J.M. 1854)

> On one occasion they captured an Angami from Tophema, and kept him till the young rice began to sprout, when his fertilizing influence would be most valuable. Then he was killed and his head, hands and feet were hung from the head-tree, as if they were those of an enemy killed in war. ... But the sacrifice did not turn out well. The wretched victim, in his death agony,

> swept the ground with his hand, and this caused the crops to be damaged by wind that year. (Mills, J.P. 1982, *161*)

> In common with the Angamis, in the old days the Kachcha Nagas looked upon no male as worthy of the name of man, unless he had taken at least one head. Any head was sufficient to stamp a warrior; an old woman's or a child's. In all probability, the proud owner waited at the drinking place, or on one of the many paths to the village jhums, and smote some venerable dame toiling home with her basket of sticks. It mattered not, the head was just as valuable in the eyes of the people, and brought him as much in favor with the village belles. (Soppit 1885)

Revenge was considered a sacred duty and each raid called for a series of counterattacks. Blood feuds were handed down from generation to generation. Even by the violent standards that prevailed throughout the Naga Hills in that era, the Angami were considered to carry hostilities to excess.

Early Swidden Systems of a Warring Society

Although historical writings lack detailed descriptions of Naga livelihood systems during the 1800s, fragments can be found, which, together, paint a picture of shifting cultivation and how it was influenced by the prevailing climate of headhunting and warfare. While glossing over significant variations that almost certainly existed between tribes, and even between individual villages of the same tribe, the following generic description was recorded by Allen (1981, *43*):

> A jhum is, as a rule, only cultivated for two seasons in succession and then allowed to fallow for seven or eight years. After the second year, the yield falls off and the weeds spring up and choke the crop. There is a risk, too, that the roots of the scrub jungle may be killed; and the land depends to some extent for its fertility on the regrowth of this jungle, and its subsequent conversion into a bed of ash manure. A village thus requires of culturable land about five times the area actually under cultivation at any given time, and the outlying jhums of the larger communities must, of necessity, be sometimes situated at a considerable distance from the village site.

Around the end of the 19th century, the sustainability of Naga jhumming practices was already an issue of debate among British administrators. Allen's recognition of the need to limit cultivation to avoid damaging tree stumps and root suckers, thus delaying forest regrowth in the subsequent fallow, is noteworthy. The following observations also seem to support the capacity of Nagaland's forests to rebound after cropping:

> It is often held, wrongly I think, that hillsides, if regularly "jhummed," are bound to be worked out in time. The case of the thickly populated Ao country seems to prove the contrary. Provided enough trees are left standing, and the land is "jhummed" at intervals of not less than 10 years, the jungle will grow up strongly on the abandoned "jhums," prevent the soil from being denuded by rain, kill out the useless weeds, and deposit enough mould to keep the soil as rich as ever. (Mills, J.P. 1973, *108*)

Neither of these comments, made early last century, would be out of place in contemporary scientific journals, and they raise the question of just how far the debate over shifting cultivation has advanced in the last hundred years. Both authors correctly stressed sufficient fallow length as the key criterion underpinning jhum sustainability. However, even a century ago there were already examples of system collapse due to excessive population pressures:

> When there is enough land, seven years is usually reckoned the shortest time in which the land can become fit for recultivation, and 10 to 12 years is usually regarded as the normal period for it to lie fallow, while 15 to 20 is regarded as the most desirable time to leave it untouched, though land near a village, being more convenient for cultivation, is rarely if ever left so long as that.
>
> In the Tizu valley, however, and in parts of Kileki valley, where the population has much outgrown the supply of suitable jhumming land, jhums may often be found cleaned after only five years rest and, in some villages, even after three, while loads of earth have to be sometimes actually carried and dumped down in the rocky parts of the field to make sowing possible at all. Of course, under these conditions, the crops are very poor, and the villages live in permanent scarcity. The general introduction of irrigated terraces is a very pressing need, and unless largely carried out in the present generation, it is hard to see how the next can be saved from starvation. (Hutton 1921, *59–60*)

According to the 1901 census, the population of the 7,951 km^2 (3,070 square miles) that then composed the Naga Hills district was 102,402. This makes a population density of fewer than 13 persons/km^2 (Allen 1981, *30*). Hutton explained that land shortages often led to intrusion on neighboring territories and, ultimately, to war.

> Disputes between the villages in the still unadministered country, where there is no superior authority to settle them must, if an amicable agreement is not reached, be subject to the ultimate arbitrament of war. The real causes of war are possibly not more than three in number in the Sema country. First, shortage of land necessitating forcible encroachment on that of neighboring villages; secondly, the protection of trading interests, as an attempt on the part of one village to trade directly with another at some distance has often caused war with an intervening village through which the trade used to pass. The third cause is found in the fits of restlessness that from time to time afflict most Naga villages, the desire of the young men, as yet untried, to prove their manhood and gain the right of wearing the warrior's gauntlets and boar's tusk collar, all culminating in the overwhelming desire to get somebody's head. (Hutton 1921, *167*)

The 1911 Report on the Administration of Assam also confirmed that jhum cultivation, with its hunger for land, was at the heart of many disputes:

> The system of cultivation by jhum which prevails ... demands long periods of rest, during which the land becomes reclothed with forest, and it is often difficult to believe that what seems an uncared-for wilderness is really jealously guarded private property of sib, family, or village. But so it is, and no quarrels have been more enduring or more bitter among these people than those relating to land. (Report on the Administration of Assam 1911)

Clearly the ability of villages to cultivate food crops sufficient to support a sizable population, including many warriors, was critical to avoiding domination. In the absence of land-use intensification techniques, this often meant wresting land from weaker neighbors. The ever-present danger of ambush discouraged Nagas from straying too far from their protective clans. Security concerns provided reason for opening communal swiddens, often covering entire hillsides. It is also probable that mutual distrust and open warfare led to the creation of extensive "no man's lands" between hostile villages, thus further exacerbating land shortages.

> In all Naga villages which do not practice terraced cultivation, it is for many reasons the practice of the village to cultivate together. Patches of jhum

> surrounded by jungle are far more open to the depredations of birds and wild animals, and reciprocal help in cultivation is less easily given. In villages which are liable to headhunting raids, joint cultivation is the only method which offers any safety to the individuals working in the fields. (Hutton 1921, *150*)

A Major Headache for Pax Britannia

The 19th-century British colonial administration in India was reluctant to expand its territory to include the Naga Hills. But in the end, it felt it had no other choice.

> It should first be premised that, for the annexation of their territory, the Nagas are themselves responsible. The cost of administration of the district is out of all proportion to the revenue that is obtained, and we only occupied the hills after a bitter experience extending over many years, which clearly showed that annexation was the only way of preventing raids upon our villages. Had the Angami Nagas consented to respect our frontiers, they might have remained as independent as the tribes inhabiting the hills to the south of Sibsagar and Lakhimpur, but it was impossible for any civilized power to acquiesce in the perpetual harrying of its border folk. (Allen 1981, *9–10*)

Analysts generally divide British relations with the Naga tribes into four phases. The period from 1839 to 1846 was one of attempted control from without. It was characterized by frequent Naga raids to claim heads and slaves from surrounding plains villages and the tea gardens of the East India Company in upper Assam. Determined to protect their subjects, the British responded with punitive military expeditions. During this volatile period, a grouping of six powerful Angami villages west of Kohima, including our Khonoma case study village, were recognized as the main instigators of many of the raids. They became known as the "Khonoma group" and were considered to be Angami par excellence, particularly in matters of warfare. Although the details are beyond the scope of this chapter, it is noteworthy that Khonoma, along with its five other allied villages, led the resistance to British occupation and was singled out for harsh retaliation. It is thus a very historic village in the sense that its incessant predatory raids on British subjects were pivotal in prompting the annexation of the Naga Hills, which drastically altered the history of the area.

When, in 1847, repeated British forays into the hills were failing to achieve the desired results, the administration experimented with establishing outposts in the Naga Hills. But this soon proved untenable, and in 1851, the British troops were withdrawn and the administration adopted a policy of noninterference with the Nagas. However, the Angami took full advantage of the withdrawal and, between 1854 and 1865, launched 19 unanswered raids in which 232 British subjects were killed, wounded, or carried off (Sir Alexander MacKenzie, as cited in Allen 1981, *18*). The situation rapidly grew intolerable for the British. In 1866 the British reversed their policy and began a process of permanent pacification and annexation. But it was not until 13 years later that events came to a head. In October 1879, Deputy Commissioner Damant was ambushed at the gates of Khonoma. He and 35 of his troops were killed and 19 others wounded. Eight days later, the 118-man garrison at Kohima was besieged by an estimated 6,000 Nagas, including contingents from nearly every Angami village. After a six-day standoff, Colonel Johnstone arrived with a strong force of Manipuris and the Nagas disbursed (Johnstone 1971). Concluding that the Nagas needed to be taught a lesson, the British led two Gurkha infantry detachments to assault Khonoma and hostile neighboring villages.

> A detachment of 334 men of all arms, equipped with two three-pounder guns and two four-inch mortars, was dispatched against the fort at Khonoma. So strong, however, were the defenses that, though the guns were finally brought within 75 yards, they did no appreciable damage, and an

> attempt to escalade the fort was foiled by a deep trench. The force accordingly bivouacked before the village for the night and, in the morning, found that the place had been abandoned. The troops then made a demonstration through the hills, and several villages which opposed their progress or declined to furnish them with supplies were burnt. (Allen 1981, *14–15*)

Despite stubborn resistance, one Angami village after another was destroyed until the men of Khonoma, having retreated to a strongly fortified position on the crest of the Barail range, finally submitted in March 1880. Thereafter, British rule was extended throughout the Naga Hills and, by 1905, the Angami had largely given up headhunting and warfare.

Dawn of a New Era

Nagaland's newfound peace had a profound effect on the lifestyle of the people. It is useful, to consider some observations on how peace influenced land-use practices:

> Originally ... each tribe lived almost in an exclusive area with a view to protecting itself from headhunting raids of others. As the administration strengthened and headhunting raids became a matter of the past, the tribes, especially the Semas, spread out into the hills to create new settlements and relieve pressure of the tribal population on their old land. Thus came into existence pockets of some tribes among other tribes. (Singh, K.R. 1987, *20*)
>
> One is glad to think that their material prosperity has greatly increased since our advent to the district. ... Their lives being secure, they can cultivate at greater distances from their villages than formerly. Their cattle, too, are safe from hostile raids, and can be pastured on the best grazing grounds, however far from the village. Judging from the anxiety of tribes outside our border to be brought under our control, our rule appears to be popular among Nagas, and I think the older men enjoy the security of life and property which now obtains. It is, however, inevitable that the younger men should regret that the paths of glory are closed for them. (Report of the Deputy Commissioner in 1901–1902, as cited by Allen 1981, *56*)

The picture that emerges is thus one of expansion of communities, cultivation, and grazing lands as the threat of attack abated. This alleviated land-use pressures and presumably also freed up male labor as warriors no longer had to defend village gates, watch over women as they worked in the fields, or launch their own raids on neighboring villages. Peace, and the gradual improvement of health services, sparked a population explosion that continues today.

In matters of warfare and ritual, the Angami demonstrated the strength of their indigenous institutions and their organizational ability. They could muster, within days, a 6,000-man war party to assault the Kohima stockade; they could launch swift raids, marching under cover of darkness on targets up to 100 miles distant; they could marshal several hundred men to pull a monolith up a mountain slope. Membership of a strong and well-organized clan was essential not only to their prosperity, but also to their very survival.

The security imperative also manifested itself in farming systems. Jhums were made communally to provide safety in numbers, and the need to intensify land use near the relative safety of fortified villages encouraged development of irrigated terraces and alder fallows. These innovations supported higher populations in larger villages, again contributing to security.

Anecdotal evidence and logic both suggest that peace has nurtured growing individualism among Angami youth. They no longer have to rely on the clan for security or look to village elders for advice or traditional wisdom. This may pose a significant threat to traditional innovations such as alder fallows for they may prove economically less attractive in the short term than cash crop monocultures or off-farm employment.

The Case Study Village

The study area is located about 20 km west of Kohima, the capital of Nagaland (Figure 30-3). The village, Khonoma, is perched above a riverine valley (Figure 30-5) that has been carved into an intricate patchwork of terraced rice fields, irrigated from the Dzuza and Khuru rivers and several streams. These are of vital importance to Khonoma's food security. The following is an eloquent description of the agricultural landscape at the beginning of the last century:

> To a stranger suddenly arriving in the Angami country, nothing strikes him with greater surprise and admiration than the beautiful terraced cultivation which meets the eye everywhere, on gentle hill slopes, sides and bottoms of valleys, in fact wherever the land can be utilized in this way. In preparation, upkeep, and irrigation, the greatest care is taken, far in excess of anything seen in the northwest Himalayas. The appearance of the countryside for miles south of Kohima, for instance, is such as to suggest the handiwork and labor of a far higher order of people than these wild Nagas. These terraced fields are often bordered with dwarf alder bushes,[12] are carefully irrigated by an elaborate system of channels bringing water down from mountain streams, and luxuriant crops of rice are grown on them. To pass through the valley where stand the two powerful villages of Khonoma and Mozema [Mezoma] during late October when the crops are ripe is indeed a delight for the eye; a veritable golden valley. (Shakespear 1914, *206–207*)

The "rice granary" function of the valley bottom is complemented by diverse polycultures of secondary food crops cultivated in swiddens on adjacent hillsides.[13] Traditional practices of placing logs or even rock walls at intervals along swidden contours to halt downslope soil movement suggest an early awareness of the need for soil conservation. Internal erosion between these physical barriers has gradually led to natural terrace formation. The swidden fields are generally higher than 1,500 m above sea level. Alder grows in abundance throughout the study area's rugged terrain.

The climate is cool and temperate. Winter and summer temperatures vary from extremes of 5° to 25°C. Frost is widespread during the cool winter months, but snowfall is usually confined to higher mountain peaks. Average annual rainfall is about 2,500 mm, occurring over the six-month period between May and October. Soils in the study area are acidic, rich in organic carbon, but poor in available phosphorus and potash.

Historical Ecology of Land Use

The current agricultural landscape of the study area can be more clearly understood in the context of the historical influences that shaped it. The related pressures of population density and security from attack are most important in terms of explaining what drove land-use intensification.

Scarcity of agricultural land probably has a history almost as long as Naga habitation of the research area. In 1848, the British found Khonoma locked in a land dispute with a neighboring village, Mezoma, and early reports frequently commented

---

[12] Shakespear may have been referring to willows (*Salix* sp.), which are often planted on terrace bunds to stabilize them. If they were indeed alder, presumably their dwarf appearance was the result of frequent pollarding.

[13] Although ideal for wet rice monocultures, irrigated terraces are unable to substitute for the wide range of swidden crops that contibute greatly to the diversity of Naga diets. This probably explains, in part, why Nagas have stubbornly resisted attempts to replace jhumming with alternative agricultural systems.

on land-use pressures. Allen, in particular, pointed to population pressures as having driven the Angami to develop terraces:

> Sometimes the terraces are simply dug out of the earth and are not faced with stone, but near Khonoma they even go so far as to build low walls across their jhum land to prevent the soil from being washed away by the rain. This system of terraced cultivation was probably introduced from the south, and without it the large and populous Angami villages could not exist, as they have not sufficient land in their vicinity to support them by the wasteful system of jhumming. (Allen 1981, *44*)
>
> A country which is entirely composed of hills must obviously include large tracts of land which are quite unfit for cultivation. Most of the remainder is only fit for jhumming, and jhumming postulates a large area of fallowing land. In parts of the district there is no doubt plenty of wasteland on which jhum crops might easily be raised, but this is not the case in Angami territory. Some of the larger villages like Khonoma are positively pressed for land, and the people have carved out into terraced rice fields the most precipitous and unlikely looking slopes. (Allen 1981, *80–81*)

Unfortunately, the significance of the pollarded alders scattered across the sloping hillsides went largely unnoticed by British administrators. However, the ability of Khonoma and other powerful villages to subsidize their existence from weaker neighbors was an issue for comment. It presumably ended with submission to British rule.

> The villages are generally built on the highest and most inaccessible hills north of the great range of mountains separating Assam from Muneepoor and Burmah. Every side is stockaded and a ditch generally encircles the most exposed part of the village that is studded with panjies, alias sharply pointed bits of bamboo stuck into the ground, which form an effective defense against any sudden surprise from an enemy. All the small villages are subject to the large villages of Mozumah [Mezoma], Konomah [Khonoma], Kohemah [Kohima] and Jopshemah [Jotsoma], and they are obliged to secure their own safety by paying them an annual tribute of cloths, fowls, cows, pigs, etc. according to their means, or as much as will satisfy the rapacity of the free-booters. (Mills, A.J.M. 1854)

Security from Attack

As earlier described, to leave the protective walls of the village was not without risk, and armed warriors stood guard over villagers laboring in the swiddens to ward off attack by hostile neighbors. Several minor incidents related by James Johnstone during his tenure as political agent in the Naga Hills, from 1873 to 1886, vividly illustrate Khonoma's feared reputation and hegemony over its neighbors:

> The position of a small village at war with a large one was often deplorable as no one dared to leave the village except under a strong escort. I once knew a case of some Sephema men at feud with Moxuma [Mezoma], hiring two women of the powerful village of Konoma [Khonoma] to escort them along the road as, thus accompanied, no one dared touch them. (Johnstone 1971, *29–30*)
>
> An incident occurred in the month of August which might have proved serious. A native of a Kutcha Naga village within sight of Samagudting came to complain that, while gathering wild tea seed for sale, he had been driven off by a Konoma [Khonoma] Naga. Konoma, though not the most populous village, had long been considered the most powerful and warlike in the hills,

> and a threat from one of its members was almost a sentence of death to a man from a weak village. (Johnstone 1971, *46–47*)

Another example of the fear they commanded occurred in 1880 when, as punishment for instigating a rebellion, the British tried to disband the troublemaking Khonoma men by confiscating their rice terraces and forcing them to vacate the village site.

> The village of Khonoma had its wonderful terraced cultivation confiscated and its clans were dispersed among other villages. The dispossessed villagers found themselves not only deprived of their homes, but by the confiscation of their settled cultivation they were, during a whole year, reduced to the condition of homeless wanderers, dependent to a great extent on the charity of the neighbors and living in temporary huts in the jungles. The result was widespread sickness and mortality. (Elwin 1997, 22)

Despite its weakened state, no other Nagas dared to occupy Khonoma's deserted terraces for fear of possible reprisals. This volatile environment would have discouraged the opening of swiddens on remote slopes where farmers would have been vulnerable to attack and unable to swiftly retreat back to their fortified villages. It made it imperative that villagers synchronize their swiddening to afford safety in numbers, and made swiddening more labor intensive because guards had to accompany laborers. For some centuries, therefore, the security factor had provided a strong incentive for intensification of land use nearby villages, even though more distant slopes were still unclaimed.[14]

Characterization of Alder Management

The Angami villagers of Khonoma appear to have refined the management of alders to a science, through centuries of trial and observation. Other Naga subgroups, including the Chakhesang, Chang, Yimchunger, and Upper Konyak tribes, also maintain alders within their swiddens, but the Angami stand out in terms of intensity of management.[15] Khonoma is thought to be between 600 and 700 years old and, like most Naga villages, is perched on a ridge top where it can be easily defended from invaders. Oral histories recounted by village elders suggest that alders have been managed as an integral part of Khonoma's swiddens for the past 500 years. The fact that some *Alnus* trees currently found in swiddens are themselves thought to be up to 150 years old verifies the system's long history. These trees are akin to family heirlooms that are handed down between generations and tended with fond memories of deceased ancestors. Farmers note that crops growing in close proximity to *Alnus* perform well, and it was probably this observation that led their ancestors to protect *Alnus* seedlings regenerating naturally in their fields. As land pressures mounted through the centuries, Khonoma farmers continued to experiment with alder management. It is likely that passive protection thus intensified into active planting and a very elaborate management regime gradually evolved.

Under high land-use pressures, alder-based swiddens can be managed in cycles as short as four years, with two years of arable cropping between the alder stumps followed by two more years while the soil is rested and the coppices are allowed to grow (Figure 30-6). Much of the landscape surrounding Khonoma is dominated by

---

[14] In contrast, many swidden communities in Asia-Pacific preferred long fallow swidden rotations as an intentional strategy to open more land from the forest commons, and thereby claim it as their own. They thus ensured that the land inheritance of their children would be adequate to meet their needs. This provided a strong disincentive for intensification of land use.

[15] Other areas in Nagaland known to practice similar alder-based jhum systems are Tuli (town) in Mokokchung district; Ruziephema and Intoma (villages), both in the foothills of Kohima district; Jalukie area in the southern region of Kohima district; and in the upper portion of Mon district (Kevichusa 1995).

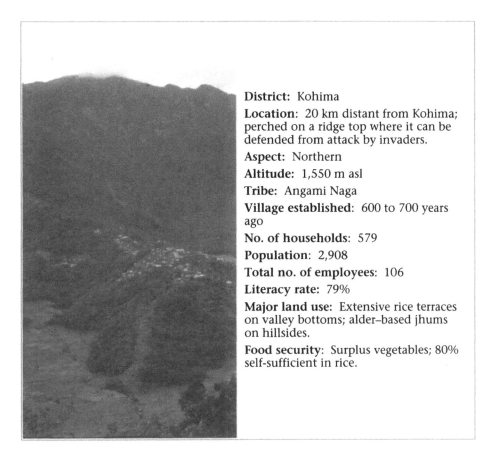

District: Kohima

Location: 20 km distant from Kohima; perched on a ridge top where it can be defended from attack by invaders.

Aspect: Northern

Altitude: 1,550 m asl

Tribe: Angami Naga

Village established: 600 to 700 years ago

No. of households: 579

Population: 2,908

Total no. of employees: 106

Literacy rate: 79%

Major land use: Extensive rice terraces on valley bottoms; alder–based jhums on hillsides.

Food security: Surplus vegetables; 80% self-sufficient in rice.

Figure 30-5. Khonoma: The Study Village That Spawned Alder Fallows
Source: Keitzar (1998).

swiddens at various stages of this cycle. They have been developed over generations, creating a wide variation in individual tree age within alder stands. Respondents felt that the system was continuing to expand, both by protection of naturally regenerating seedlings and intentional planting. Farmers observe that there is no standard tree spacing, although six meters appears to be roughly the norm (Gokhale et al. 1985). Crowded seedlings are easily thinned and must be protected from fire. This slow incremental approach has also been applied to soil conservation. Most alder-based jhum fields are terraced, usually by placing logs across the slopes, or with more permanent rock walls.

First Pollarding of Young Alder Trees

Young trees are pollarded for the first time when the bole circumference reaches 70 to 80 cm and the bark develops rough fissures, usually at around six to ten years. Premature pollarding might leave the alder trunk too small to support the weight of developing coppices and vulnerable to wind breakage.

Pollarding is usually done in the months of December and January. Trees being pollarded for the first time are cut horizontally across the main trunk, leaving a stump height of seven to eight feet. All auxiliary branches are stripped off. It is important that the cutting instrument is kept sharp to keep the cut clean and avoid cracking the stump. Farmer observations that alders pollarded too low result in poor coppicing is supported by Khan and Tripathi (1986, 1989b), who demonstrated a positive correlation between stump height and coppice vigor. Furthermore, low stumps are also more vulnerable to cattle damage, and coppice regrowth would compete with adjacent crops for the same horizontal space. After pollarding, the fresh cut is plastered with mud to prevent drying and cracking, and a stone slab is placed on top. The stone is intended to protect the tree from frost damage and ensures that new coppices will sprout from all sides, resulting in a better horizontal spread of the canopy regrowth. The stone is generally used only for the first pollarding. When coppices are cut back again in subsequent swidden cycles, the wounds are smaller and less vulnerable to frost damage and coppices are spaced by selective pruning.[16]

The Jhum Cycle

Reopening the Fallow

The dynamics of alder management within a four-year jhum cycle are portrayed in Figures 30-6 and 30-9. The cycle begins anew, usually in November, when the farmer returns to his alder fallow and slashes the underbrush. During December and January, he moves from tree to tree, cleanly cutting each coppice flush with the trunk and carefully avoiding splitting. Any side branches are also removed, leaving only the bare stump (Figure 30-7). The felled branches are then stripped of foliage and may sometimes be set aside as small-diameter poles suitable for construction purposes. Otherwise they are cut into firewood lengths,[17] and temporarily stacked in the field for later removal. The pollarding calendar seems several months ahead of findings by the Pakhribas Agricultural Centre in Nepal (1985) that the best coppice regrowth is obtained from pollarding at the start of the monsoons.

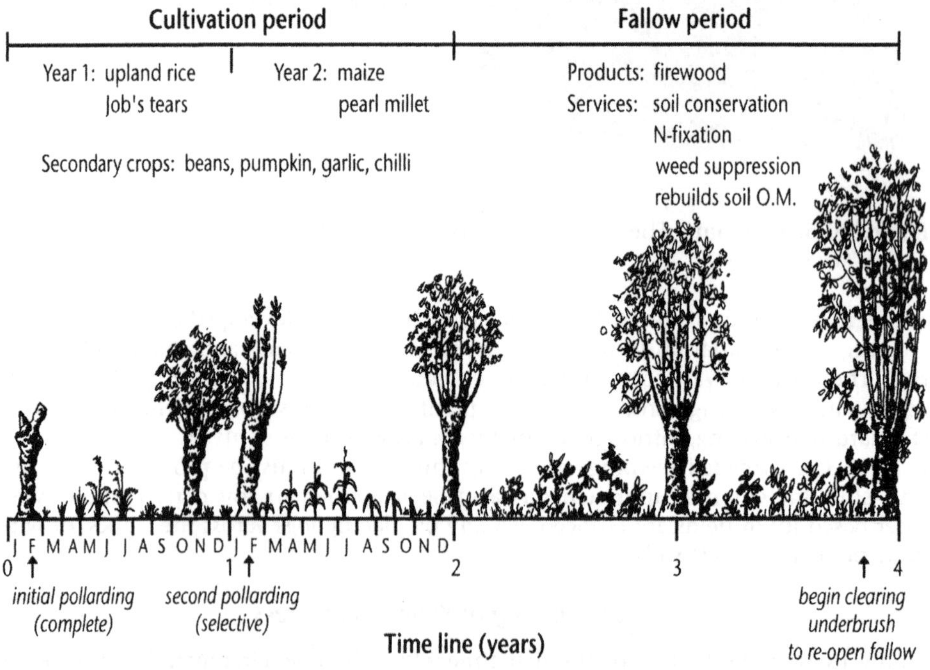

Figure 30-6. Phases of Alder Management throughout the Swidden Cycle

---

[16] Comparable pollarding practices are used on *Quercus* spp. on the west flanks of the Gaoligong Mountains in Yunnan and on *Cassia siamea,* both in northern Lao P.D.R. and further north by Dai farmers in Xishuangbanna. Both species are managed for firewood production.

[17] *A. nepalensis* has a relatively low calorific value of 18,230–20,480 kJ/kg, but dries rapidly and burns easily (Faridah Hanum and van der Maesen 1997). Its wood splits easiest while still green.

Rough field measurements were made to determine firewood yields when alder fallows are pollarded in preparation for arable cropping. A 0.25 ha. sample plot in Khonoma was found to contain 21 mature alders, indicating a sparse density of about 84 trees per ha. After pollarding, the harvested coppices amounted to four stacks of firewood, each a standard size of 2.5 m by 1.5 m by 0.75 m. Total firewood yield from the 21 pollarded stumps was thus 11.25 m^3, breaking down to 0.53 m^3 per stump or aggregating to 45 m^3 per ha. The farmgate price for firewood at the time was Rp 600 per stack,[18] so the economic returns from the "firewood crop" provided a substantial income of Rs114 per tree or Rs9,600 per ha.

Cropping Phase

The main crop during the first year is usually upland rice in warmer areas or Job's tears (*Coix lacryma-jobi*) in cooler regions. Secondary crops such as chilies, cucurbits, garlic, beans, or taro are often sporadically interplanted. A hypothesis that pollarding of alder may cause dieback of root nodules and release a flush of nitrogen at the time of planting is plausible but untested. N-fixing by alders peaks around July in response to increased soil temperature and root nodule moisture (Sharma, E. and Ambasht 1984; Sharma, E. 1988). It may, therefore, be well synchronized with the crops' uptake demands. Accelerated decomposition of remaining litter in the warm rainy season further augments soil fertility (Sharma, E. and Ambasht 1987). Crops are harvested from October to November, after which cattle may be given free range to graze the crop residues.

During the 12 months between the time the alders are completely pollarded and the fallow reopened, the trunks sprout from 50 to 150 new coppices. These are thinned twice, usually first in July and again in December, selectively leaving five to six coppices at the top of the trunk (Figure 30-8). These retained coppices are trimmed of side branches and foliage, leaving only a few leaves at the tips. They are then left to grow, creating a tree canopy during the approaching fallow. Any epiphytes or parasitic growths are removed from the trunks (Gokhale et al. 1985). The pruned coppices are piled to dry and burned in preparation for sowing the second crop, usually foxtail millet. After harvest of the millet in the following August, the field is allowed to lapse back into fallow.

Fallow Phase

During the two-year fallow period, the coppices may reach up to 6 m in length and 15 cm in diameter (Gokhale et al. 1985), eventually forming a full canopy (Figure 30-10. See also color plate 35). One observant farmer pointed to his mature alder fallow and noted that each of the coppices was just as large as a nearby alder of the same age as the coppices, but that had never been pollarded. He pragmatically concluded that nothing was lost by pollarding since stem growth was not diminished, and each pollarded tree yielded five to six poles, compared to a single pole from the unpollarded tree. His observations were supported by a few exploratory measurements of coppice growth rates by the research team. Three coppices retained on one individual alder were noted to have reached diameters of 27.4, 21.1, and 8.6 cm respectively in the four to six years since the stump was last pollarded. As pointed out by the farmer, such rapid growth rates are not achieved by alder saplings planted in afforestation schemes. The superior growth rate of coppices demonstrates the benefits of pollarding as a silvicultural procedure that maintains extensive root systems through successive harvests.

---

[18] Exchange rate at the time: Rs 34 = US$1.00.

Figure 30-7. Newly Opened Alder Jhum, with Khonoma Village in the Background

Figure 30-8. Alder Stumps with Their Coppices Thinned, Ready for the Second Year's Crop

Figure 30-9. Phases of Alder Management through a Swidden Cycle

Cyclical pollarding not only allows crop cultivation two out of every four years, but it is also a strategy to increase the total net productivity of the alder stand. Although it is not well understood, it is possible that pollarding may stimulate other beneficial changes in alder physiology. In comparative studies of different-aged alder stands, Sharma and Ambasht (1988) have shown that, as the trees get older, N-fixation decreases with their decrease in N demand. They also recorded declines in other parameters, including net energy fixation, net primary production rates, production efficiency, energy conversion efficiency, and energy efficiency in N-fixation (1991). No work has yet been done to test if these declines are mitigated by cyclical pollarding. Similar questions apply to the longevity of the alder. Curiously, Lamichhaney (1995, 9) describes *A. nepalensis* as a relatively short-lived species in Nepal that "often dies for no apparent reason before the tree obtains a useful height." This clearly does not reconcile with the Naga experience in managing trees over many generations, raising questions of whether the differences are genotypic, phenotypic, or the result of manipulation.

It is clear that the continued presence of alders in the swidden fields accelerates nutrient cycling (Ramakrishnan 1993). Their extensive roots draw on nutrients from a large soil area, and return them to the surface in high volumes of nutrient-rich litterfall (Puri 1959; Sharma, E. and Ambasht 1987).

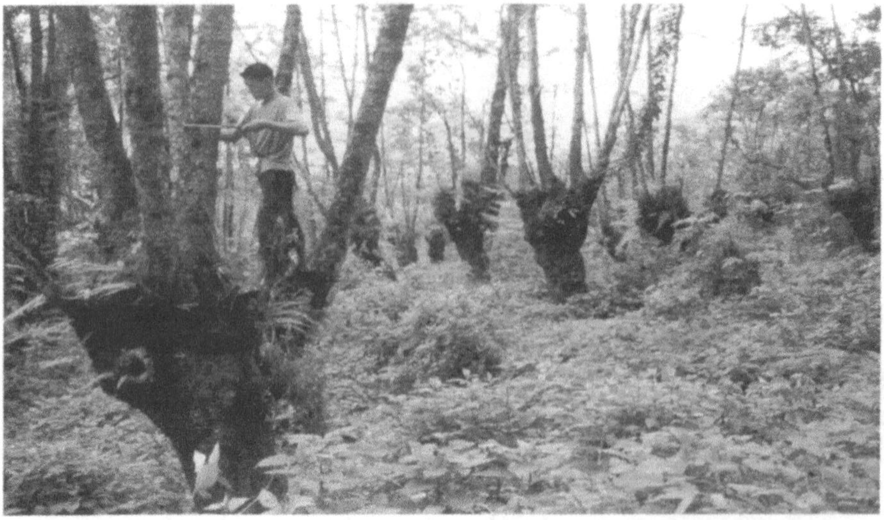

Figure 30-10. When Alder Fallow Periods Are Extended, Coppices Can Grow Sufficiently Large for Timber Purposes

Although alder fallows are sometimes reopened in two years, farmers often choose to allow the coppices to grow bigger and harvest them for poles. Several negative repercussions may arise from this practice. Firstly, if the coppices are allowed to become too large, their combined weight might cause splitting damage to the main trunk in heavy winds. Secondly, farmers observe that soil N accretion is excessive in longer fallows, contributing to a crop lodging problem. Studies conducted in the eastern Himalayas confirm that soils under *A. nepalensis* stands have high levels of organic C and total N that increase with the age of the trees (Sharma, E. et al. 1985). Lastly, since erection of physical fences is prohibitively expensive, farmers instead rely on coordinating their swidden sites so that everyone is cropping in the same area at the same time. It is easier to keep cattle away from the area through community vigilance. But any farmer who alters his swidden cycle will fall out of synchrony with his neighbors and expose his crops to cattle damage.

It is noteworthy that Khonoma and most other villages practicing alder-based swiddening have substantial areas of irrigated terraces for wet rice cultivation (Kevichusa 1995). In contrast, villages with little or no wet rice production generally do not manage alders in their swiddens. This pattern suggests that the degree of rice security provided by irrigated terraces might enable farmers to invest the time and labor needed to integrate alders into their sloping fields as an improved fallow.

The technology apparently does not spread to other communities through out-marriage. The Nagas commonly marry within their own villages, and married couples live with or near the husband's parents. Even in cases of intervillage marriages, a woman would join her husband in his village of birth and assist him in farming his father's fields. A Khonoma woman out-marrying to another community would probably be similarly expected to adopt the land-use system of her husband's village.

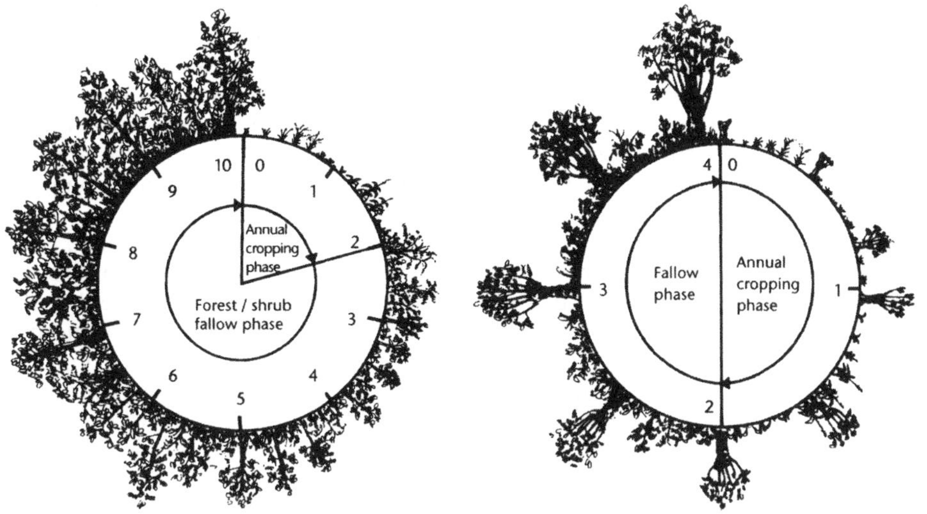

Figure 30-11. Alder's Dramatic Intensification of the Swidden Cycle

Discussion

This brief overview of Nagaland's alder fallow technology is useful for the generation of hypotheses but affords few definitive conclusions. The contextual dynamics that led to its innovation are generally understood. *A. nepalensis* is native to the sub-Himalayan region and thrives in the cool, moist environment of Nagaland's uplands. It is likely that most of the alder systems alluded to in the introductory comments of this chapter began by selective weeding, gradually moving toward a monospecific stand of trees. Increasing population pressures would have prompted farmers to move beyond passive protection to begin more systematic alder planting and management. That the Nagas developed the most intensive alder fallow system may reflect the added pressures associated with constant warfare and headhunting. It was probably no coincidence that Angami Nagas, who were notorious for their ferocity and propensity for warfare, developed extensive rice terraces and alder-based fallow management, both technologies enabling intensified land use nearby the relative safety of their fortified villages.

It is equally clear that this intensified cultivation, through a dramatic reduction in fallow length (Figure 30-11), was not at the cost of system degradation. Yields shown in Table 30-4 are expressed in terms of a given cropping year. If, more significantly, they had been averaged over the compared swidden cycles of 4 years versus 10, then yield benefits would have been magnified by a factor of 2.5. The system's productivity is further boosted by the addition of roughly 45 m^3 per ha of firewood, harvested at the time of pollarding. These data support farmer observations on the inherent value of alder fallows, but they need to be balanced against labor costs to enable a meaningful evaluation.

Alder's success as a fallow species appears to lie largely in the admirable "fit" between its ecology and the rhythm of the jhumming calendar (Figure 30-12). On the most fundamental level, the system's success is due to alder's unique ability to nodulate profusely and fix nitrogen in Nagaland's acidic upland soils,[19] where other N-fixing trees are unsuited (Khurana 1993). Secondly, alder regeneration, seedling growth, root nodule biomass, and nitrogenase activity are temperature–dependent along the gradient of altitude and fall off below 1,500 m and above 2,000 m (Sharma, E. 1988). This is squarely in the belt of Nagaland's high-elevation swiddens.[20] Add to this the probabilities that pollarding may be causing nodule dieback and release of N at the time of planting and that a second flush may occur in July when higher temperatures and moisture stimulate nitrogenase activity (Sharma, E. and Ambasht 1984). Nitrogen accumulation can actually become excessive and farmers complain of crop lodging if alder fallows are extended beyond three to four years.

Alder's coppicing ability is equally vital to the system, and interestingly, Naga pollarding management appears to optimize coppice regrowth. Alder stumps are pollarded in the winter months of December and January, thereby reopening the fallow in preparation for planting at the first rains. It precedes by several months findings by the Pakhribas Agricultural Centre in Nepal (1985) that alder's coppicing capacity is season-dependent and pollarding at the start of the monsoons produces the most vigorous regrowth. Furthermore, stump heights that elevate coppice regrowth above the crop canopy, minimizing shading and spatial competition with adjacent crops and later damage by grazing cattle, have also been shown to maximize coppicing vigor (Khan and Tripathi 1986, 1989b).

Preliminary evidence indicates that pollarding management may also be inducing physiological changes in alder that enhance its role as a fallow species. These include prolonged N-fixation and tree longevity. And as alder trees age, they develop rough fissured bark that protects them from fire damage during the clearing operation. The following is a more comprehensive list of alder's attributes as a superior fallow species. Table 30-5 completes the appraisal by listing the opportunities and constraints of the alder fallow system:

- Alder is widely distributed, from Pakistan in the west through northwest India, Nepal, Bhutan, northeast India, and eastward through northern Myanmar, southwestern China and into Indochina.
- It has wide ecological amplitude in terms of soil types and rainfall zones, and covers the major swidden belt between 1,000 and 2,000 m above sea level.
- It is able to grow on highly degraded, unstable soils.
- It has an extensive lateral root system with heavy concentrations of *Frankia* around nodules that fix atmospheric N.
- Unlike most N-fixing trees, it has the ability to thrive and fix N even in very acidic soils.
- Alder's N-fixation peaks around July in response to increased soil temperature and root nodule moisture and may be synchronized with the crop's uptake demands.
- It has a synergistic, noncompetitive relationship with intercrops.
- It has quick–decomposing leaf litter.
- It maintains favorable C/N ratios in underlying soils.
- It provides poles, firewood, and numerous secondary products such as wood for constructing tea boxes, pulp for newsprint, tannins for tanning leather, and the juice of crushed leaves has medicinal properties.
- Alder is a rapid colonizer of disturbed lands. It grows well in full light and is moderately shade tolerant.

---

[19] Although the association of *A. nepalensis* roots and *Frankia* is most effective in neutral soils, nodulation is still extensive at pH 4.0 (Lamichhaney 1995, *27*).

[20] Nagaland's State Agricultural Research Station (SARS) is currently testing another N-fixer, *Albizia lebbeck*, for its potential to be similarly managed as an improved fallow at lower altitudes.

- It is easily established either naturally or artificially through direct sowing, polybag, or bare root stock. It is an abundant seeder with a high germination rate and is not recalcitrant.
- It has a fast rate of growth, with annual increments of 2.7 m in height and 2.9 cm in diameter recorded in Nepal.
- Alder withstands frequent pollarding and coppices profusely.
- Its coppice regrowth is most profuse when pollarding is done before the start of the monsoon season, dovetailing neatly with the jhumming calendar.
- It develops rough fissured bark that is fire resistant.
- There have been no serious pest and disease problems reported for alder in Nagaland.
- It is long–lived when pollarded. Some individual trees have been managed over the past 100 to 200 years.
- Its leaves are relatively unpalatable, making it less vulnerable to grazing damage by free-ranging livestock.

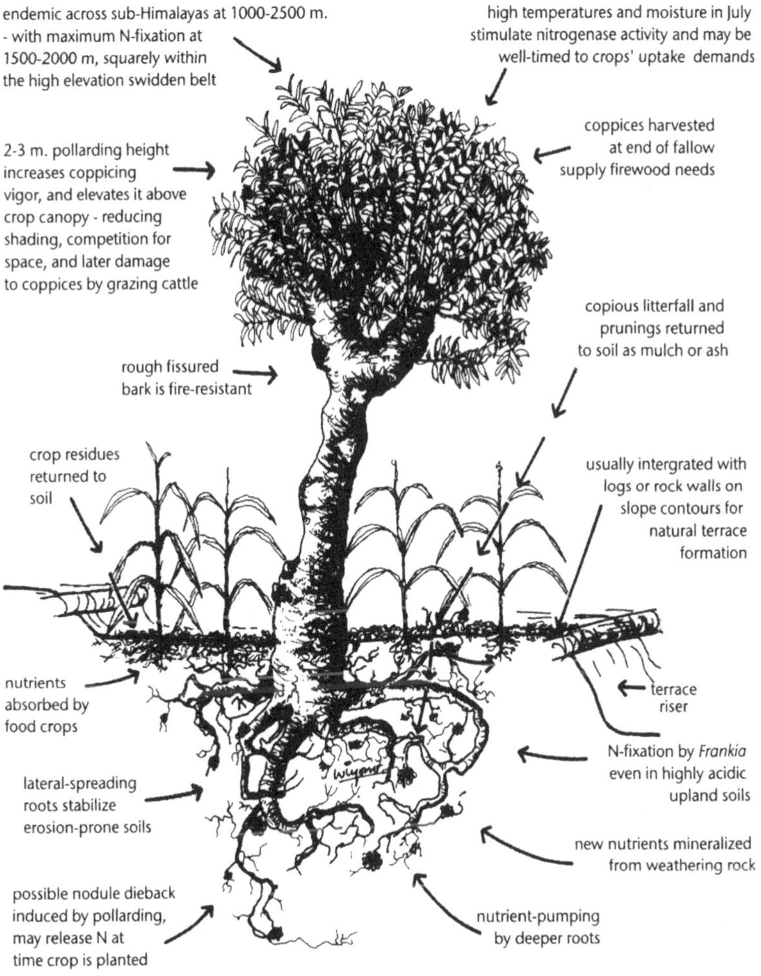

Figure 30-12. Schematic Representation of Alder's Hypothesized Role in Swidden System

Table 30-4. Comparison of Crop Yields between Swiddens with and without Alder Management

	Swidden Fields		
Crops	*With* Alnus *Yield/ha*	*Without* Alnus *Yield/ha*	*Terraced/Irrigated Fields Yield/ha*
Rice	150 tins [a]	—	200 tins
Maize	140 tins	120 tins	—
Job's tears	140 tins	130 tins	—
Potato	200 quintals [b]	120 quintals	250 quintals

Source: Authors' field survey. [a] A *tin* is a local unit of measurement and, for paddy (rice with husks), is equivalent to about 12 kg. [b] A quintal is 100 kg.

Table 30-5. Opportunities and Constraints of Alder Fallow System

Opportunities	*Constraints*
—Possible direct application of technology to other acid upland soils, between 1,000 and 2,000 m asl, with subtropical monsoon climate, where shortened fallows are leading to swidden degradation; —Identification of superior provenances and breeding to further improve alder's performance in agroforestry applications; —Identify superior practices in farmer management of alder fallows for possible refinement and wider dissemination; —Clarify the preconditions that led to adoption of alder fallows in order to help discern how improved fallow husbandry may be promoted on a wider scale; —Scientific principles underlying its management may have application across other species, farming systems, and ecozones; and —It has the potential to evolve into a wide scope of alternative management options (Figure 30-13).	—Labor inputs may be considerable; —Requires a high degree of management, such as timing and technique of pollarding; —Benefits are longer term; —Not likely to be adopted on communal swidden lands; —*A. nepalensis* does not prosper at low altitudes or in drought- or frost-prone areas; —Timber is less valuable and durable than other species that do not have its soil-building properties; —Expanded use may lead to increased insect /disease problems; —Early successional trees are generally more subject to insect herbivory; —Susceptible to wind damage; —Reported to be relatively short lived in Nepal and die for no apparent reason; and —Short seed viability (4 months) under uncontrolled conditions.

It is evident that *A. nepalensis* is admirably suited to its role as a fallow improver or, more accurately, the Naga system has been carefully designed around the ecology of the species and refined through centuries of experimentation.

This positive appraisal brings us back to the vexing question of why development of alder fallow technologies has occurred only in isolated pockets and is not more widespread across the eastern Himalayas, where *Alnus* and shifting cultivation coexist. The answer may lie in a combination of ecology and markets, and the influence of both on farming systems. It is plausible that, historically, alder fallows were more widespread but, in many areas, combinations of population pressure and market opportunities catalyzed a move from shifting cultivation to more settled agroforestry practices in which alder assumed a new role as a canopy species. It is used as a shade tree over large cardamom in Sikkim, southern Bhutan and Nepal; in tea gardens in Yunnan; and in *Cinchona* plantations across the sub-Himalayas. These uses may have their evolutionary origins in alder-based swidden systems.

Alder naturally thrives in uplands between 1,000 and 2000 m asl, with moist but well-drained soils, and a cool subtropical monsoon climate with 800 to 2,500 mm of rainfall. If GIS was used to overlay the ecozones preferred by alder with those areas of the region that continue to be isolated, either geographically, politically, or economically, and that rely heavily on shifting cultivation, then the areas of intersect would probably be those where alder fallow systems can still be found.

The same questions have to be asked within Nagaland to tease out the reasons underlying its sporadic adoption there. The brief details presented in this chapter provide no indication of why the Angami, Chakhesang, Chang, Yimchunger, and Upper Konyak developed such systems while the Semi, inhabiting areas of abundant alder and with similar pressures of population and warfare, did not. In a 1923 trip report, Hutton (Hutton 1929, *47*) made an intriguing offhand comment on diffusion of the technology, stating, "Tuensang received us very well ... Alders are grown here, and the seed is said to have been obtained from Angfang in a raid." The prospect of a warrior pausing in the heat of battle to collect alder seeds from his foes' trees speaks eloquently of the perceived value of the species. It also raises questions of why germplasm of an abundantly available species had to be stolen. Perhaps they mistook Angfang's pollarded trees as a different genotype, rather than the result of management. Clearly, further research is needed to map the natural distribution of *A. nepalensis* within Nagaland; to clarify its ecology and where it performs best; and, within this area, to undertake topical surveys to elicit farmer perceptions of the species and the reasons for its adoption or rejection as an improved fallow species.

Finally, the postulation that alder fallows may have historically been more widespread in the eastern Himalayan foothills but have since evolved into other agroforestry patterns raises the issue of its future prospects in Nagaland. Accelerated development of communications, access to distant markets, and new technologies will place new strains on traditional jhum cultivation. Even if shifting cultivation is eventually displaced by more permanent land uses, examples from across the region suggest that *Alnus nepalensis* will continue to play an important role in Naga farming systems, albeit in modified forestry and agroforestry contexts (Figure 30-13).

Designing a Research Agenda

This chapter overviews Naga innovations to harness the attributes of alder as a superior fallow species. Contrasting models in Yunnan practiced by Dulong, Lisu, Dai, Hui, Wa, and other cultural minority groups have been described by Hong et al. (1960), Fu (2003), and Guo et al. in Chapter 29. Relatively little is known about the status of its use in northern Myanmar (Wint 1996, *39*). However, as emphasized earlier, the fact that culturally distinct swidden communities across the eastern Himalayas have independently developed deliberate practices to integrate *A. nepalensis* into fallows suggests that this N-fixing alder offers agronomic and economic benefits warranting closer research attention. A preliminary set of researchable issues arising from this characterization study includes the following broad objectives, although this broad menu needs to be focused by working closely with farmers to develop a more problem-oriented research agenda that resonates with their priorities and is manageable within available resources:

- Characterization of the wider Naga livelihood system, of which alder-based swiddens are but one component, to understand how alder-based jhums "fit" with other components of the complex whole-farm system, and how farmers are managing their total resources.
- Document the cultural and historical origins of alder management and clarify the underlying social, economic, and ecological conditions that prompted Naga farmers to begin managing alder trees in their jhum fields. What was the underlying problem to which this was a solution?

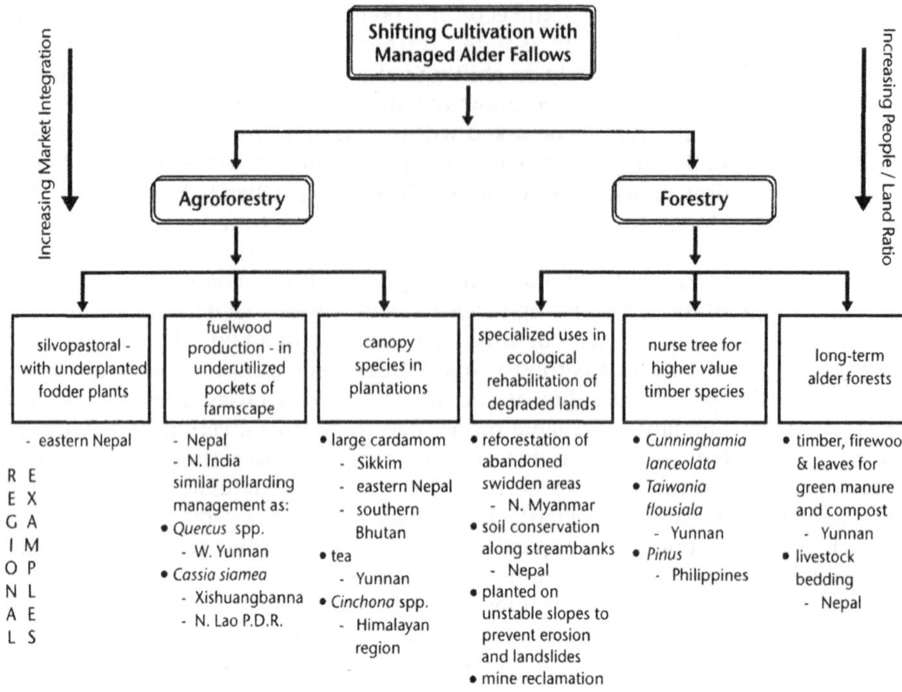

Figure 30-13. Possible Scenarios for Alder Fallows to Evolve into Other Management Patterns

- Explore the role and significance of *Alnus nepalensis* in Naga culture and folklore to understand how this sociocultural context has influenced Naga management of alder as an improved fallow species (for example, possible linkages with ceremonies or rituals);
- Build a rich characterization of traditional management of *A. nepalensis* in jhum fields, its significance as a fallow species, and its apparent role in permitting an intensification of the swidden cycle. Map the current distribution of the system's use, categorize variations in management practices by different tribes, and identify potentially superior innovations that can be tested in controlled experiments. Document if, how, and why the system may be changing;
- Clarify the scientific principles underlying the system. Determine whether these may be transferable to other ecozones where swidden systems are under pressure to intensify. Distill the scientific principles underlying the system that are not limited to the specifics of *Alnus* and Nagaland, but are of wider practical interest to swidden intensification across Asia-Pacific's uplands. In other words, what makes this system work? What are its ecological dynamics?
- Conduct a cost–benefit analysis of alder-based fallows to critically assess their superiority over unmanaged fallows through a careful quantification of the system's inputs (labor) and outputs (improved soil properties and crop yields, firewood, construction poles, etc.);
- Investigate why an apparently successful technology has not expanded more widely. Examine farmers' perceived constraints on adoption, then analyze the system's potential for wider application, both within Nagaland and across the eastern Himalayan ecoregion, where swidden systems are under pressure to

intensify. What are the biophysical and socioeconomic niches where adoption of this system makes sense?

- Examine the ecology of *Alnus nepalensis* and the implications of its management as an improved fallow species. Evaluate the genetics of the species in both wild and domesticated populations and determine the extent to which alder fallows can substitute for natural forest cover in performing ecological functions;
- Discover how Naga resource tenure systems and interactions with markets may be impinging on improved fallow management. What impact might the policy environment have on the adoption of alder fallows?
- What is the future of the Naga system? Within the context of a population doubling every 20 to 30 years and the likelihood that Nagaland will continue to open up and engage more heavily in the cash economy, what role may alders play in future farming systems? What is the likely trajectory of intensification of alder management, and what are the evolutionary pathways by which swiddening may transform into other alder-based forestry and agroforestry systems?
- Develop a comprehensive set of methods to help researchers and extensionists throughout Asia-Pacific work more effectively with indigenous fallow management systems. Using Nagaland's alder technology as an in-depth case study, assist in the development of methodologies to identify "best bet" fallow management systems and map their potential for wider geographic application.

Conclusions

If Asia-Pacific's forest remnants and their contained biodiversity are to be protected and swidden communities are to be afforded a better standard of living, pathways toward stabilizing and enhancing the productivity of stressed swidden systems are urgently needed. One of the most promising approaches to identifying biophysically workable and socially acceptable innovations is to document and understand indigenous adaptations toward improved fallow management.

The alder-based jhum system is a striking example of sustainable land-use intensification evolved through centuries of farmer experimentation, without need of outside technologies, investment capital, or excessive labor inputs. Crop yields are not dwindling, but are as high today as any time within memory. While most farmers in Nagaland cultivate their swidden fields for 2 years within a 10-year cycle (1:4 ratio of cropping to fallow), the traditional *Alnus* system allows crop harvests in 2 out of every 4 years (1:1 ratio of cropping to fallow). Although floristically not that exciting themselves, the increased intensification of land use afforded by alder fallows significantly reduces agricultural pressures on Nagaland's remaining forests and the genetic resources they contain, contributing to their conservation. Its conservation role also extends to Nagaland's diverse crop germplasm. The jhum fields are repositories for a significant diversity of traditional food crops. These precious genetic resources may be rapidly eroded as land use intensifies into permanent cultivation, high-yielding varieties become available, and economies of scale induce farmers to plant monoculture cash crops for lowland markets. By propping up the ecological sustainability of the jhum system, even under dramatically shortened fallows, alder enables the traditional crop germplasm to remain intact, while retaining food security. These global services are in addition to the more localized and quantifiable benefits of N-fixation, reduced erosion, supply of construction poles and firewood, and leaves for mulching the plot. Such cogent case studies need to be carefully documented, understood, analyzed for technological improvements, and evaluated for their potential application in other upland areas.

Given the continued importance of jhum cultivation to Nagaland's agricultural landscape and its contribution to food security, alder's capacity to enable dramatic land-use intensification without concomitant ecological decline is critically important. Although the intricate system of *Alnus* management described in this chapter has evolved in isolated pockets in Nagaland, it is of much wider scientific significance to the entire Himalayan foothills, where *Alnus nepalensis* is a native pioneer species possibly holding potential for stabilizing other stressed swidden systems. Its implications could be even wider. *Alnus* is a large genus spanning tropical and temperate latitudes. Recent anecdotal evidence from northern Luzon, the Philippines, reports that farmers have begun experimenting with *Alnus japonica* as a preferred fallow species, presumably based on observations of tree-crop-soil interactions (Vergara 1995; Rice 2003). The simultaneous existence of *Alnus*-based fallow management technologies on the Asian mainland, highly refined by centuries of farmer experimentation, while other swiddenists on the Philippine archipelago are in the rudimentary stages of testing ways to harness its agronomic benefits, not only reinforces scientific interest in this genus but also points to exciting potential for farmer-to-farmer transfer of indigenous technologies. More fundamentally, research on *Alnus* fallows, as a particularly interesting example of more generic approaches to swidden intensification, promises to draw lessons beyond a narrow species focus that will be of interest to the wider Asia-Pacific region.

Although the historical factors that gave rise to the Naga alder system are somewhat unique, farmer management of shifting forests on agricultural landscapes may be more widespread than generally recognized. The traditional debate surrounding shifting cultivation has widely regarded swidden landscapes as forest lands periodically despoiled by intruding farmers. Such simplified stereotyping glosses over a significant middle ground between farm and forestry interests that could provide a basis for constructive collaboration. Instead, misinformed prejudices are enacted in forestry laws that further marginalize mountain peoples and relegate them to the role of squatters in their ancestral homelands. Documentation and validation of alder management as a superior fallow system helps debunk the common stereotype of shifting cultivators as wanton destroyers of forest ecosystems and more accurately portrays them as forest planters and managers. Such case studies of successful farmer innovations highlight the potential of indigenous technical knowledge to contribute solutions to urgent resource management problems in the uplands and the need for more participatory research. This will open more effective pathways for assisting farmers to stabilize stressed swidden systems.

Acknowledgments

This work was part of a collaborative effort by the India-Canada Environment Facility (ICEF), the International Development Research Centre (IDRC), and the World Agroforestry Centre (ICRAF) to support research on shifting cultivation in Nagaland and its development into improved agroforestry practices. The authors wish to warmly thank the farmers of Khonoma village who were such gracious hosts, always making sure our questions were carefully answered and our stomachs kept full, amidst a steady flow of rice beer (*zu*). Much of the knowledge presented in this chapter is theirs. We are particularly indebted to Mr. A.M. Gokhale, formerly Additional Chief Secretary of Nagaland, and Mr. R. Kevichusa, formerly Team Leader of NEPED's Project Operations Unit, whose assistance made this research possible. Thanks are also owed to Pak Wiyono, of ORSTOM, who contributed his artistic skills in developing several of the accompanying illustrations.

References

Agus, F. 2006. Use of *Leucaena leucocephala* to Intensify Indigenous Fallow Rotations in Sulawesi, Indonesia. Chapter 25.

Ali, A. 1946. Wartime Plantations in Darjeeling Division. *Indian Forestry* 72(8), 350–353.

Allen, B.C. 1981 (Reprint). *Naga Hills and Manipur.* New Delhi: Gian Publications.

Ambasht, R.S. and A.K. Srivastava. 1994. Nitrogen Dynamics of Actinorhizal *Casuarina* Forest Stands and Its Comparison with *Alnus* and *Leucaena* Forests. *Current Science* 66(2), 160–163.

Amotya, S.M. and S.M. Newman. 1993. Agroforestry in Nepal: Research and Practice. *Agroforestry Systems* 21(3), 215–222.

Anand, V.K. 1967. *Nagaland in Transition.* New Delhi: Associated Publishing House.

Askin, D.C., D.J. Boland, and K. Pinyopusarerk. 1990. Use of *Casuarina oligodon* ssp. *abbreviata* in Agroforestry in the North Baliem Valley, Irian Jaya, Indonesia. In: *Advances in* Casuarina *Research and Utilization. Proceedings of the Second International* Casuarina *Workshop*, edited by M.H. El-Lakany, J.W. Turnbull, and J.L. Brewbaker. Cairo, Egypt: Desert Development Centre, 213–219.

Ataia, A. 1983. *Casuarina oligodon* in the Eastern Highlands Province, Papua New Guinea. In: Casuarina *Ecology Management and Utilization: Proceedings of an International Workshop*, edited by S.L. Midgley, J.W. Turnbull, and R.D. Johnston. Canberra, Australia: Commonwealth Scientific and Industrial Research Organization, 80–87.

Bandy, D.E., D.P. Garrity, and P.A. Sanchez. 1993. The Worldwide Problem of Slash-and-Burn Agriculture. *Agroforestry Today,* July–September 1993, 2–6.

Basu, S.K. 1947. Anti-Erosion Work in Kalimpong Forest Division. *Indian Forestry* 73(1), 10–14.

———. 1951. Anti-Erosion Work in Kalimpong Forest Division. Proceedings of 7th Silvicultural Conference, 1946, Dehra Dun, India, 208–212.

Bauhus, J. 1999. Personal communication with the author.

Bino, B., and M.B. Kanua. 1996. Growth Performance, Litter Yield, and Nutrient Turnover of *Casuarina oligodon* in the Highlands of Papua New Guinea. In: *Recent* Casuarina *Research and Development: Proceedings of the Third International* Casuarina *Workshop, March 4–7, 1996, Da Nang, Vietnam,* edited by K. Pinyopusarerk, J.W. Turnbull, and S.J. Midgley. Canberra, Australia: Forestry and Forest Products Division, Commonwealth Scientific and Industrial Research Organization (CSIRO), 167–170.

Boojh, R. 1982a. Growth Strategy of Trees Related to Successional Status. I. Architecture and Extension Growth. *Forest Ecology and Management* 4(4), 359–374.

———. 1982b. Growth Strategy of Trees Related to Successional Status. II Leaf Dynamics. *Forest Ecology and Management* 4(4), 375–386.

———, and P.S. Ramakrishnan. 1981. Germination of Seeds of *Alnus nepalensis* Don. *Science Letters* 4(2). National Academy of Sciences, 53–56.

Bourke, R.M. 2006. Managing the Species Composition of Fallows in Papua New Guinea by Planting Trees, Chapter 31.

Brunck, F., J.P. Colonna, Y. Dommergues, M. Ducousso, A. Galiana, Y. Prin, Y. Roederer, B. Sougoufara, and Y.R. Dommergues. 1990. Control of Inoculation of Trees with Root Symbiotics. A Synthesis of a Selection of Trials in the Tropics. *Bois et Forests des Tropiques* 223, 24–42.

Butler, J. 1875. Rough Notes on the Angamis Nagas and Their Language. *Journal of the Asiatic Society of Bengal* 44, 307–346.

Carter, E.J. 1992. Tree Cultivation on Private Land in the Middle Hills of Nepal: Lessons from some Villagers of Dolakha District. *Mountain Research and Development* 12(3), 241–255.

Chauhan, D.S., S.K. Dhyani, and A.R. Desai. 1993. Productivity Potential of Grasses in Association with *Alnus nepalensis* and Pineapple under Silvi-Horti-Pastoral System on Agroforestry in Meghalaya. *Indian Journal of Dryland Agricultural Research and Development* 8(1), 60–64.

Clark, M.M. 1978 (Reprint). *A Corner of India.* Gauhati, India: Christian Literature Centre.

Davis, A.W. 1891. Naga Tribes. *Assam Census Report, Part* I, 237–251.

Dhyani, S.K. 1998. Tribal Alders: An Agroforestry System in the Northeastern Hills of India. *Agroforestry Today* 10(4). Nairobi, Kenya: ICRAF.

———, and D.S. Chauhan. 1990. Nitrogen Fixing Trees Suitable for Agroforestry in Meghalaya. *Indian Journal of Hill Farming* 3(2), 65–68.

Directorate of Economics and Statistics. 1991. *Statistical Atlas of Nagaland.* Kohima, Nagaland: Government of Nagaland.

Elwin, V. (ed.). 1969. *The Nagas in the Nineteenth Century.* London: Oxford University Press.

———. 1997 (Reprint–First published in 1961). *Nagaland.* Delhi: Spectrum Publications.

Eslava, F.M. 1984. The Naalad Style of Upland Farming in Naga, Cebu, Philippines: A Case of an Indigenous Agroforestry Scheme. Country Report to a Course on Agroforestry, October 1–20, 1984, Universiti Pertanian, Malaysia.

Faridah Hanum, I., and L.J.G. van der Maesen (eds.) 1997. *Plant Resources of Southeast Asia, No. 11: Auxiliary Plants.* Bogor, Indonesia: PROSEA.

Field, S.P. 1991a. Competition Effects of Weeds and *Leucaena leucocephala* on a Maize Crop in a *L. leucocephala* Forest. Leucaena *Research Reports* 12, 55–57.

———. 1991b. The Effects of Undersowing *Leucaena leucocephala* into a Maize Crop. Leucaena *Research Reports* 12, 58–59.
———, and H.G. Yasin. 1991. The Use of Tree Legumes as Fallow Crops to Control Weeds and Provide Forage as a Basis for a Sustainable Agricultural System. Paper presented at 13th Asian-Pacific Weed Science Society Conference in Taipei, Taiwan, 121–126.
Fu, H. 2003. Study on Alder-based Shifting Cultivation in Yunnan, China. Unpublished. MSc thesis, Xihshuangbanna Tropical Botanical Garden, Chinese Academy of Science.
Fujisaka, S. 1991. A Set of Farmer-Based Diagnostic Methods for Setting Post "Green Revolution" Rice Research Priorities. *Agricultural Systems* 36, 191–206.
Ghosh, B.B. 1982. *History of Nagaland.* Delhi: S. Chand & Company Ltd.
Gokhale, A.M., D.K. Zeliang, R. Kevichusa, T. Angami, and Bendangnungsang. 1985. *Nagaland: The Use of Alder Trees.* Kohima, Nagaland: State Council of Educational Research and Training, Education Department, Government of Nagaland.
Guha, S.R.D., M.M. Singh, and K. Kumar. 1965. Newsprint Grade Ground-Wood Pulps from *Alnus nepalensis. Indian Forestry* 91(8), 593–596.
Gundevia, Y.D. 1975. *War and Peace in Nagaland.* Delhi: Palit & Palit Publishers.
Guo, H., Y. Xia and C. Padoch. 2006. *Alnus nepalensis*-Based Agroforestry Systems in Yunnan, Southwest China. Chapter 29.
Harrison, A. 1989. A Yield Table for Firewood Production from Utis (*Alnus nepalensis*). In: *Technical Paper No. 14.* Pokhara, Nepal: Lumle Agricultural Centre.
Hibbs, D.E., D.S. DeBell, and R.F. Tarrant. 1994. *The Biology & Management of Red Alder.* Oregon: Oregon State University Press.
Hong, J., J. Wang, and W. Feng. 1960. Social Economic Investigation Report on Dulong People in the 4th District, Gongshan County. In: *China's Minority Group Social Economic Investigation Material.* Kunming, China: Yunnan Ethnological Press. (Chinese language)
Horam, M. 1992 (Reprint). *Social and Cultural Life of Nagas.* Delhi: Low Price Publications.
Husain, M. 1988. *Nagaland.* Delhi: Rima Publishing House.
Hutton, J.H. 1921. *The Sema Nagas.* London: MacMillan and Co. Ltd.
———. 1929. *Diaries of Two Tours in the Unadministered Area East of the Naga Hills.* Calcutta: The Asiatic Society of Bengal.
———. 1969 (Reprint). *The Angami Nagas, with Some Notes on Neighboring Tribes.* London, U.K.: Oxford University Press.
Imnauongdang. 1990. *Levels of Rural Development in Nagaland: A Spatial Analysis.* A.S. Prakashan.
ICRAF. 1993. *Annual Report* . Nairobi, Kenya: ICRAF.
———. 1996. *Transfer of Agroforestry Technologies: Concept Notes.* Nairobi, Kenya: ICRAF, 5.
Jacobs, J., with S. Harrison, A. Herle, and A. Macfarlane. 1990. *The Nagas, Hill Peoples of Northeast India: Society, Culture and the Colonial Encounter.* London: Thames & Hudson Ltd.
Johnstone, J. 1971 (Reprint). *Manipur and the Naga Hills.* Delhi: Vivek Publishing House.
Jones, P.H. 1983. *Leucaena* and the Amarasi Model from Timor. *Bulletin of Indonesian Economic Studies* 19(3), 106–112.
Kang, B.T., and G.F. Wilson. 1987. The Development of Alley Cropping as a Promising Agroforestry Technology. In: *Agroforestry: A Decade of Development,* edited by H.A. Steppler and P.K.R Nair. Nairobi, Kenya: (ICRAF), 227–243.
Keitzar, S. 1998. Farmer Knowledge of Shifting Cultivation in Nagaland. Project Report.
Kevichusa, R., V. Lieze, and V. Nakhro. n.d. *Alnus nepalensis*: Alder. Kohima, Nagaland: Team of the Agriculture Production Commissioner under the World Bank Aided Extension Project, Government of Nagaland.
Kevichusa, R. 1995. Personal communication with the author.
Khan, M.L., and R.S. Tripathi. 1986. Tree Regeneration in a Disturbed Sub-tropical Wet Hill Forest of Northeast India: Effect of Stump Diameter and Height on Sprouting of Four Tree Species in Burnt and Unburnt Forest Plots. *Forest Ecology and Management* 17(2–3), 199–209.
———. 1989a. Effect of Soil Moisture, Soil Texture, and Light Intensity on Emergence, Survival, and Growth of Seedlings of a Few Sub-tropical Trees. Indian Journal of Forestry 12(3), 196–204.
———. 1989b. Effects of Stump Diameter, Stump Height and Sprout Density on the Sprout Growth of Four Tree Species in Burnt and Unburnt Forest Plots. *Acta Oecologica, Oecologia Applicata* 10(4), 303–316.
Khurana, D.K. and Khosla, P.K. editors. 1993. Agroforestry for Rural Needs: Volume II. Solan, India: Indian Society of Tree Scientists. Proceedings of the Workshop of the IUFRO Project Group "Agroforestry for Rural Needs" held in New Delhi, February 22-26, 1987.
Kieft, J.A.M. 2006. Farmers' Use of Sesbania grandiflora to Intensify Swidden Agriculture in North Central Timor. Chapter 28.
Kung'u, J.B. 1993. Biomass Production and Some Soil Properties under a *Leucaena leucocephala* Fallow System in Cebu, Philippines. Unpublished. MSc thesis, University of the Philippines at Los Baños, Laguna, Philippines.
Lamichhaney, B.P. 1995. *Alnus nepalensis* D. Don: A Detailed Study. In: FORESC Monograph 1 (1995). Kathmandu, Nepal: Forest Research and Survey Centre, Ministry of Forests and Soil Conservation.
Lasco, R.D. 2006. The Naalad Improved Fallow System in the Philippines and its Implications for Global Warming. Chapter 27.

———, and P.D. Suson. 1997. A *Leucaena leucocephala*-based Improved Fallow System in Central Philippines: The Naalad System. Paper presented to an International Conference on Short-Term Improved Fallow Systems, March 11–15, 1997, Lilongwe, Malawi.
MacDicken, K.G. 1990a. Agroforestry Management in the Humid Tropics. In: *Agroforestry: Classification and Management* edited by K.G. MacDicken and N.T. Vergara. New York: John Wiley & Sons, 98–149.
———. 1990b. *Leucaena leucocephala* as a Fallow Improvement Crop in Shifting Cultivation on the Island of Mindoro, Philippines. Proceedings of Conference on Research on Multipurpose Trees in Asia, November 19–23, 1990, Los Baños, Laguna, Philippines.
———. 1991. Impacts of *Leucaena leucocephala* as a Fallow Improvement Crop in Shifting Cultivation on the Island of Mindoro, Philippines. *Forest Ecology and Management* 45, 185–192.
———. 2006. Upland Rice Response to *Leucaena leucocephala* Fallows on Mindoro, Philippines. Chapter 26.
Metzner, J.K. 1981. Old in the New: Autochthonous Approach towards Stabilizing an Agroecosystem. The Case from Amarasi, Timor. *Applied Geography and Development* 17, 1–17.
———. 1983. Innovations in Agriculture Incorporating Traditional Production Methods: The Case of Amarasi, Timor. *Bulletin of Indonesian Economic Studies* 19(3), 94–105.
Mills, A.J.M. 1854. Report No. 309 of 1853 on Relations with Hill Tribes on the Assam Frontier (Jenkins, F.), Report on the Province of Assam, 117–166.
Mills, J.P. 1973 (Reprint). *The Ao Nagas.* London: Oxford University Press.
———. 1982 (Reprint). *The Rengma Nagas.* Kohima, Nagaland: Directorate of Art and Culture, Government of Nagaland.
Naga Institute of Culture. 1970. *A Brief Historical Account of Nagaland.* Kohima, Nagaland.
Nandi, R.P., and S.K. Chatterjee. 1991. Effect of Shade Trees on Growth and Alkaloid Formation in *Cinchona ledgeriana* Grown in Himalayan Hills of Darjeeling. Environmental Assessment and Management: Social Forestry in Tribal Regions. Proceedings of Third Conference of Mendelion Society of India, April 4–7, 1989, Birsa Agriculture University, Ranchi, India.
Neil, P.E. 1989. *Eucalypts*, or other Exotics, or Indigenous Species? *Banko Janakari* 2(2), 109–112.
———. 1990. *Alnus nepalensis*: A Multipurpose Tree for the Tropical Highlands. In: *NFT Highlights 90–06.* Waimanalo, HI.
Okorio, J., S. Byenkya, N. Wajja, and D. Pedon. 1994. Comparative Performance of Seventeen Upperstorey Tree Species Associated with Crops in the Highlands of Uganda. *Agroforestry Systems* 26(3), 185–203.
Pakhribas Agricultural Centre. 1985. *A Review of Forestry and Pasture Trial Work.* Dhankuta, Nepal: Pakhribas Agricultural Centre.
Pearson, R.S., and H.P. Brown. 1932. *Commercial Timbers of India: Their Distribution, Supplies, Anatomical Structure, Physical and Mechanical Properties and Uses.* Calcutta: Central Publication Branch, Government of India.
Piggin, C.M. 2006. The Role of *Leucaena* in Swidden Cropping and Livestock Production in Nusa Tenggara Timur, Indonesia. Chapter 24.
———, and V. Parera 1985. The Use of *Leucaena* in Nusa Tenggara Timur. *ACIAR Proceedings, Series 3*, 19–27.
Prat, D. 1989. Effects of Some Pure and Mixed *Frankia* Strains on Seedling Growth in Different *Alnus* Species. *Plant and Soil* 113(1), 31–38.
Puri, G.S. 1959. Nitrogen Content of Leaves of Some Exotic and Indigenous Forest Tree Species Planted at New Forest (Dehra Dun). *Indian Forestry* 85(7), 426–430.
Ramakrishnan, P.S. 1993. Shifting Agriculture and Sustainable Development: An Interdisciplinary Study from Northeastern India. *Man and the Biosphere Series* 10. UNESCO and Oxford University Press.
Report on the Administration of Assam 1911.
Report of Forest Administration in Burma for the Year Ended March 31, 1922.
Rice, D. 2003. Personal communication with the authors.
Rimal, S. 1992. Preliminary Research Note from Nepal Coppice Reforestation Program, Dhading. *Banko Janakari* 3(2), 40–41.
Robison, D.M., and S.J. McKean. 1992. *Shifting Cultivation and Alternatives: An Annotated Bibliography, 1972–1989.* Wallingford, UK: Centro Internacional de Agricultura Tropical/CAB.
Roder, W. 1997. Personal communication with the authors.
Russo, R.O. 1994. *Alnus acuminata*: Valuable Timber Tree for Tropical Highlands. In: *NFT Highlights 94–03.* Morrilton: Winrock International.
Sema, P. 1992. *British Policy and Administration in Nagaland 1881–1947.* New Delhi: Scholar Publishing House.
Sen, S. 1987. *Tribes of Nagaland.* Delhi: Mittal Publications.
Shakespear, L.W. 1914. *History of Upper Assam, Upper Burma, and the N.E. Frontier.* London: MacMillan & Co.
Sharma, E. 1988. Altitudinal Variation in Nitrogenase Activity of the Himalayan Alder Naturally Regenerating on Landslide-Affected Sites. *New-Phytologist* 108(4), 411–416.
Sharma, E., and R.S. Ambasht. 1984. Seasonal Variation in Nitrogen Fixation by Different Ages of Root Nodules of *Alnus nepalensis* Plantations in the Eastern Himalayas. *Journal of Applied Ecology* 21(1), 265–270.

———. 1987. Litterfall, Decomposition and Nutrient Release in an Age Sequence of *Alnus nepalensis* Plantation Stands in the Eastern Himalayas. *Journal of Ecology* 75(4), 997–1010.
———. 1988. Nitrogen Accretion and its Energetics in the Himalayan alder. *Functional Ecology* 2(2), 229–235.
———. 1991. Biomass, Productivity, and Energetics in Himalayan Alder Plantations. *Annals of Botany* 67(4), 285–293.
Sharma, E., R.S. Ambasht, and M.P. Singh. 1985. Chemical Properties under Five Age Series of *Alnus nepalensis* Plantations in the Eastern Himalayas. *Plant and Soil* 84(1), 105–113.
Sharma, R. 2006. *Alnus*-Cardamom Agroforestry: Its Potential for Stabilizing Shifting Cultivation in the Eastern Himalayas. Chapter 51.
Sharma, R., E. Sharma, and A.N. Purohit. 1994. Dry Matter Production and Nutrient Cycling in Agroforestry Systems of Cardamom Grown under *Alnus* and Natural Forest. *Agroforestry Systems* 27(3), 293–306.
Sharma, R., G. Sharma, and E. Sharma. 2002. Energy Efficiency of Large Cardamom Grown under Himalayan Alder and Natural Forest. *Agroforestry Systems* 56(3), 233–239.
Sharma, S.K., G.D. Sharma, and R.R. Mishra. 1984. Endogonaceous Mycorrhizal Fungi in a Subtropical Evergreen Forest of N.E. India. *Journal of Tree Sciences* 3(1/2), 10–14.
———. 1986. Records of *Acaulospora* spp. from India. *Current Science, India* 55(15), 724–726.
Shengji, P. 1991. Ethnobiology: A Potential Contributor to Understanding Development Processes. *Entwicklung and Landlicher Raum* 25(2), 21–23.
Shimmi, Y.L.R. 1988. *Comparative History of the Nagas: From Ancient Period till 1826*. Delhi: Inter-India Publications.
Singh, K.A., R.N. Rai, Patiram, and D.T. Bhutia. 1989. Large Cardamom (*Amomum subulatum* Roxb.) Plantation: An Age-old Agroforestry System in Eastern Himalayas. *Agroforestry Systems* 9(3), 241–257.
———, R.N. Rai, and I.P. Pradham. 1991. Agroforestry Systems in Sikkim Hills. *Indian Farming* 41(3), 7–10.
Singh, K.R. 1987. *The Nagas of Nagaland: Desperadoes and Heroes of Peace*. New Delhi: Deep & Deep Publications.
Singh, P. 1972. *India—The Land and the People: Nagaland*. Delhi: National Book Trust.
Soppit, C. 1885. *A Short Account of the Kachcha Naga (Empeo) Tribe in the North Cachar Hills, with an Outline of Grammar, Vocabulary, and Illustrative Sentences*. Shillong, India.
Stirn, A. and P. van Ham. 2003. *The Hidden World of the Naga. Living Traditions in Northeast India and Burma*. Munich: Prestel Publishing.
Storrs, A., and J. Storrs. 1984. *Discovering Trees in Nepal and the Himalayas*. Kathmandu, Nepal: Sahayogi Press.
Subere, V.S., E.B. Alberto, R.V. Dalmacio, F.M.J. Eslava, and M.V. Dalmacio. 1985. The Naalad Style of Upland Farming in Naga, Cebu, Philippines: A Case of an Indigenous Agroforestry Scheme. Report on the Third ICRAF/USAID Agroforestry Course, October 1–19, 1984, Serdang, Selangor, Malaysia, 71–108.
Subhash-Nautiyal, and S. Nautiyal. 1990. Some Fast-Growing Short Rotation Fuel-wood Species for Himachal Pradesh. *Indian Journal of Forestry* 13(4) (Publ. 1991), 300–306.
Surata, K. 1993. Amarasi System: Agroforestry Model in the Savanna of Timor Island, Indonesia. *Savanna* 8, 15–23.
Thiagalingam, K., and F.N. Famy. 1981. The Role of *Casuarina* under Shifting Cultivation: A Preliminary Study. In: *Nitrogen Cycling in Southeast Asian Wet Monsoonal Ecosystems,* edited by R. Weisilaar, J.R. Simpson, and T. Rosswall. Canberra, Australia: Australian Academy of Sciences, 154–156.
Thong, J.S. 1997. *Head-Hunters Culture: Historic Culture of Nagas*. Kohima, Nagaland: Khinyi Woch, Chunlikha Tseminyu.
Troup, R.S. 1921. *The Silviculture of Indian Trees, Vol. III, Lauraceae to Coniferae*. London: Clarendon Press.
Vergara, N.T. 1995. Technology in the Uplands: Development, Assessment, and Dissemination. Paper presented at Third National Conference on Research in the Uplands, SEARSOLIN, September 5–9, 1995, Cagayan de Oro City, Mindanao, the Philippines.
Visapal, Z., V. Liezie, and A. Yaden. 1995. *Indigenous and Economic Trees of Nagaland: Project Operations Unit Reference Book*. Kohima, Nagaland: NEPED (Nagaland Environment Protection and Economic Development Project).
von Fürer-Haimendorf, C. 1971. Hill Tribes of the Philippines and Northeast India: A Comparative Essay. In: *Modernization: Its Impact on the Philippines V, Institute of Philippine Culture Paper No. 10,* edited by F. Lynch, and A. de Guzman II. Quezon City, the Philippines: Ateneo de Manila University Press, 169–184.
Wint, S.M. 1996. *Review of Shifting Cultivation in Myanmar*. Yangon, Myanmar: Forest Resource Environment, Development, and Conservation Association.
Yuksel, N. 1998. The Amarasi Model: An Example of Indigenous Natural Resource Management. Occasional Paper No. 1, Indigenous Fallow Management Network. Bogor, Indonesia: ICRAF Southeast Asian Program.
Zhang, J. 1998. Personal communication with the authors, June, 1998.
Zomer, R., and J. Menke. 1993. Site Index and Biomass Productivity Estimates for Himalayan Alder-Large Cardamom Plantations: A Model Agroforestry System of the Middle Hills of Eastern Nepal. *Mountain Research and Development* 13(3), 235–255.

Chapter 31

Managing the Species Composition of Fallows in Papua New Guinea by Planting Trees

R. Michael Bourke

Most Papua New Guineans are rural villagers who grow most of their own food, maintain domestic pig herds, and build their own dwellings. The environmental range in which they live and practice their agriculture is large, covering altitudes between sea level and 2,800 m above sea level (asl), mean annual rainfalls between 1,000 and 8,000 mm, and a wide range of landforms and soil types. The staple foods are sweet potato, sago, bananas, *Colocasia* and *Xanthosoma* taro, yams, and cassava. Soil fertility, in the case of swiddens, is usually maintained by fallowing. However, intensity of land use ranges from very low (r = 5) to semipermanent (r = 33 to 66) and permanent (r = 67 to 100).[1] Population is increasing at a rate of 2.7% per annum, and is doubling about every 30 years. Firewood is the main fuel for cooking and heating. The value of firewood used by all households in Papua New Guinea (PNG) has been estimated at US$105 million by the World Bank Poverty Assessment (Gibson and Rozelle 1998).

Casuarina oligodon, the focus of this chapter, is grown in the central highlands. The highlands and high-altitude zones cover an altitude range from 1,200 m to 2,800 m asl. Population densities in parts of the region are very high, particularly between 1,500 and 2,000 m asl. About 46% of the PNG rural population lives in the highlands, above 1,200 m. Here, sweet potato is the most important staple food. Domestic pigs are common, with about equal numbers of pigs and people. Timber fences are usually built around food gardens to exclude the pigs. Because of the cool temperatures, firewood is needed to heat dwellings. Trees are often scarce in and around the main highland valleys because of the demand for timber.

Throughout Papua New Guinea, there is very little expansion onto previously unused land and land use is becoming increasingly intensive. This is being achieved by replacing certain crops with more productive introduced species, particularly sweet potato, cassava, *Xanthosoma* taro, potato, and maize; growing more productive

---

R. Michael Bourke, Department of Human Geography, Research School of Pacific and Asian Studies (RSPAS), Australian National University, Canberra, ACT 0200, Australia.

1 The intensity of land use can be measured by Ruthenerg's r, where r is the proportion of time that land is under crop out of the total duration of cropping plus fallow, expressed as a percentage (Ruthenberg 1980). For example, a value of 5 indicates that land is cropped for 5 years out of every 100 or, in practice, a 1-year cropping period followed by a 19-year fallow.

cultivars, particularly of sweet potato and banana; lengthening the period of cropping before fallowing; shortening the fallow period; and developing various techniques for maintaining soil fertility.

Techniques for maintaining soil fertility include managing fallow species with tree planting; transferring organic matter into food gardens by composting; rotating a leguminous food crop with a root crop over time; and creating soil retention barriers that reduce soil erosion. Other practices include mulching, the use of animal manure for fertilizer, terrace construction, and transferring organic matter from drainage ditches onto the soil surface. Having been developed by the villagers themselves, these techniques are indigenous technologies.

Casuarina oligodon

The most important tree species planted to form fallows in PNG is *Casuarina oligodon* (see color plates 41 and 42). It is a multipurpose tree species that provides hard timber that is easy to split for fencing, firewood, and house construction. Villagers believe that the species increases soil fertility and reduces soil erosion on steep slopes. It provides shade for the major cash crop in the highlands, Arabica coffee, and also provides "atmosphere" in highland villages, where the sound of wind in the trees is considered favorably. *Casuarina* is used in the following ways:

- It is interplanted with food crops, mainly sweet potato, so that *Casuarina* becomes the dominant species during the fallow phase of a swidden cycle. This chapter concentrates on this usage.
- It is planted in selected sites that are used for "mixed gardens." These gardens contain species other than sweet potato that require a high level of soil fertility. *Casuarina* is used in this way in many parts of the highlands.
- It is planted in and near villages to provide timber and shade. *Casuarina* serves this purpose throughout the highlands, and, in recent decades, this use has spread to highland fringe areas.

Casuarina is the most important long-term shade tree for coffee, having largely displaced *Leucaena* and *Albizia chinensis*, which are sometimes put to this use. A common procedure after fallow vegetation has been cleared is the establishment of "mixed gardens." Seedlings of coffee and *Casuarina* are interplanted with the food crops once the annuals are established. These annuals, and later bananas, provide the initial shade for the coffee, and *Casuarina* and coffee take over after several years (Bourke 1989).

The focus of this chapter is managing fallow species by planting trees, and because *C. oligodon* is the most important species used, it is the main species discussed. However, other tree species are also used for this purpose. Before continuing, these are briefly reviewed.

Minor Tree Species

Parasponia rigida. *P. rigida* (sometimes identified as *Trema orientalis*) forms root nodules and presumably fixes atmospheric nitrogen. It is a species indigenous to the highlands between 1,500 m and 2,000 m asl, although its exact altitudinal range is unknown. In a number of locations, villagers protect self-sown seedlings in sweet potato gardens against competition from grass and other regrowth during the fallow phase. They say the tree enhances soil fertility. Apart from this, very little is known about how villagers manage *P. rigida*. The practice appears to be more common in the Southern Highlands and Enga provinces. Occasionally *P. rigida* dominates the fallow vegetation, for example, in parts of the Sau Valley in Enga Province.[2] When fallow vegetation is cleared, *P. rigida* trees are spared and, rather than being ringbarked and

---

[2] In the nearby Tsak Valley, *Trema cannabina* Lour., possibly the same species, is reported to dominate fallow vegetation in taro gardens (Waddell 1972).

killed, their branches are trimmed. In this way, the trees survive through the cropping phase to grow again during the next fallow. Villagers say that it is not possible to propagate the species by planting seed.

Schleinitzia novo-guineensis. This is a fast-growing indigenous tree related to *Leucaena* (see color plate 38). It is also leguminous. It is common in fallow vegetation on a number of small islands in Milne Bay Province, including Paneati, Munuwata, and Iwa islands (Hide et al. 1994). On Munuwata Island, villagers protect self-sown seedlings as they believe that *S. novo-guineensis* improves soil fertility. On Iwa Island, the population density is about 450 persons/km^2, and there is intense pressure on agricultural land. Here, villagers transplant self-sown seedlings of *S. novo-guineensis*, together with those of *Rhus taitensis*, into their gardens after the first crop of yams is harvested and before the second crop, sweet potato and cassava, is planted. Both species are said to improve soil fertility (Hide et al. 1994). Very little is known about *S. novo-guineensis* and its role in the maintenance of soil fertility. As population pressure imposes increased strain on the agricultural systems of many small islands, it is possible that this species may have a useful role to play. It deserves high research priority.

***Albizia* spp.** *Albizia chinensis* is occasionally used as a shade tree for Arabica coffee. It grows over an altitude range from sea level to 1,900 m asl. An *Albizia* species, possibly *A. chinensis*, is common in fallow vegetation in a number of locations in Morobe Province, including the headwaters of the Erap River, from 500 to 1,300 m asl; the tributaries of the Yakwoi River, from 1,500 to 1,600 m asl; and the upper Watut Valley, from 900 to 1,000 m asl. In these locations, *Albizia* seedlings are reported to be transplanted into sweet potato, banana, and *Xanthosoma* taro gardens so that they survive during the fallow phase (Bourke et al. 1997). Sometimes bamboo, for construction purposes, and *Crotalaria,* for its ability to restore soil fertility, are also planted in the gardens. Very little is known about the use of *Albizia* and there is some doubt as to whether it is planted or self-sown.

Piper aduncum. This species is of South American origin. It is now very common in fallow vegetation on hillsides in Morobe Province. It grows up to about 1,600 m asl and occasionally higher. On the Sogeri Plateau, inland from Port Moresby at altitudes between 600 and 800 m asl, an agricultural system has recently been developed by migrants from Morobe Province using *Piper* sticks. After fallow vegetation has been cleared, *Piper* sticks are cut and hammered into the ground to form small fences. Soil is shoveled against the fences to form terraces and reduce soil erosion. After growing sweet potato in a one-year cropping phase, the *Piper* sticks begin to grow and eventually dominate the fallow vegetation, suppressing the regrowth of grass (Allen et al. 1996). The technique differs from others reported here in that trees are not planted primarily to restore soil fertility. However, this may be the end result of the terrace fences and subsequent *Piper*-dominated fallows.

Casuarina: An Indigenous Tree

C. oligodon is indigenous to the highlands of the island of New Guinea, encompassing PNG and Irian Jaya. Self-sown seedlings and trees are common on sites adjacent to watercourses. Seedlings colonize bare areas, including road cuttings and locations affected by flooding and landslides (Humphreys and Brookfield 1991). Soils in these sites, for example, on the Chim Formation in Chimbu Province, are often low in soil organic matter and nitrogen but rich in other nutrients. Humphreys and Brookfield (1991) documented the colonization by *Casuarina* seedlings of a site devastated by a major landslide adjacent to the Chimbu River. Eleven years after the landslide, the site had a complete cover of large *Casuarina* trees.

Villagers say that transported seedlings grow poorly or not at all on certain sites, because of inadequate soil fertility. Self-sown trees are uncommon in open grassland areas, suggesting that they are vulnerable to fire. There is little information on tree longevity, but some trees in villages and on coffee plantations are more than 30 years old.

Casuarina grows over an altitude range from 700 to 2,600 m asl in PNG, with trees occasionally growing as low as 120 m and as high as 2,820 m asl (Bourke 1988). However, it is most common between 1,400 and 2,100 m asl. The mean diurnal temperature range between 700 and 2,600 m is 9°C to 29°C. Between 1,400 and 2,100 m, it is 11 to 25°C.

Casuarina *as a Planted Fallow*

Throughout most of the PNG highlands, it is common for *Casuarina* seedlings to be planted in food gardens so that they become a component of later fallow vegetation. In Figure 31-1, "minor or insignificant use" of planted tree fallows covers that territory where 10 to 32% of fallow land contains planted *Casuarina* trees. "Significant" means that 33 to 66% of fallow land is dominated by *Casuarina*, and "very significant" covers those areas where more than 66% of fallow land is dominated by *Casuarina* (Bourke et al. 1998). The practice is much more important in four regions of the highlands.

The most important area includes the Sinasina, Chuave, and Gumine districts in the northern part of Chimbu Province, and adjacent locations in the Eastern Highlands Province, including the Watabung, Unggai, and Daulo areas and the edges of the Asaro Valley.

North of Mount Hagen, *Casuarina* dominates fallow vegetation in the middle and upper Kaironk and the upper Asai valleys in the Simbai area of Madang Province. In Enga Province (around and west of Wabag in Figure 31-1), villagers plant extensive stands of *Casuarina* on the edges of the Lai, Ambum, Minyamb, Tsak, Sau, Laigaip, and upper Porgera valleys. The most important technique for maintenance of soil fertility in this region is composting, but this is done on flatter and more fertile land. The fourth region where *Casuarina* planting is significant is the Tekin Basin in the Oksapmin area in the far west of the central highlands (see Figure 31-1).

Overall, about 1.3 million villagers in the PNG highlands, representing 36% of the country's total rural population, plant *Casuarina* trees in food gardens. About one-fifth of these, or 260,000 people, plant *Casuarina* at the significant or very significant level, that is, so that the species dominates more than one-third of fallow vegetation. This is 17% of the rural population living above 1,200 m. The biggest concentration of people using this technique, numbering 165,000 people, lives in northern Chimbu and adjacent area of the Eastern Highlands Province (Bourke et al. 1998).

Descriptions of the physical environment where people plant *Casuarina* fallows at significant and very significant levels was extracted from the databases of Mapping Agricultural Systems of Papua New Guinea and the PNG Resource Information System (see Table 31-1). The most common landform is hills or mountains with weak or no structural control. The most common rock type is sedimentary, including limestone. Field observations suggest that the technique is more common on calcareous mudstone, particularly in north Chimbu and the Tekin Basin.

As expected, the altitudinal range is lower highlands, from 1,200 m asl, to the boundary of the highlands zone and high-altitude zone, at 2,100 m. Planted *Casuarina* fallows are mostly used on steep slopes and are rarely found on gentle slopes or flat land, where the most intensive agricultural land use occurs. The dominant vegetation is grassland, particularly *Miscanthus* cane grass. The practice is more common in the dryer parts of the region, with annual rainfall between 2,000 and 3,000 mm, and less common in wetter locations where rainfall is above 3,000 mm.

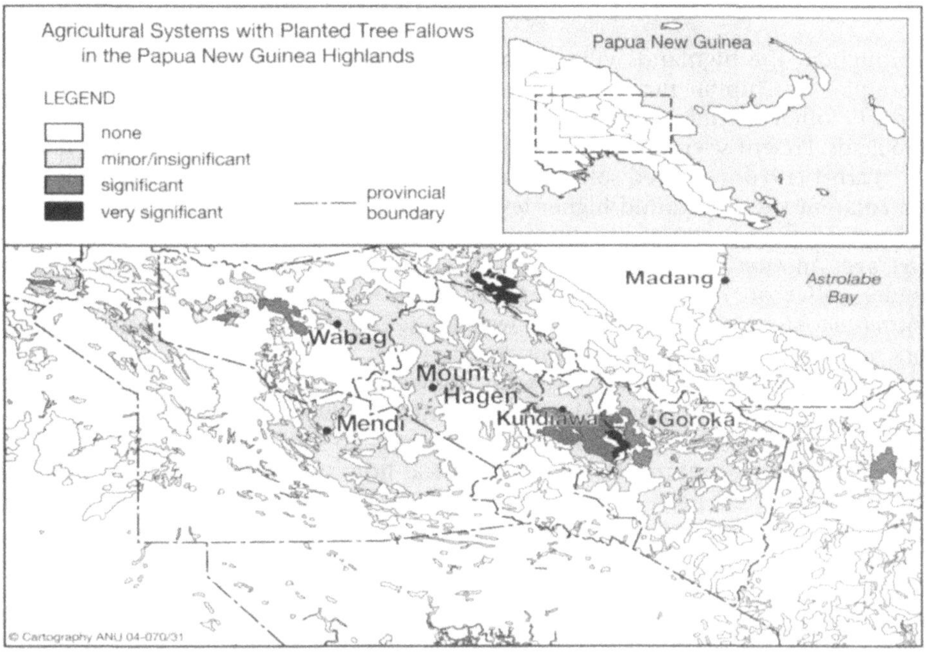

Figure 31-1. Agricultural Systems with Planted Tree Fallows in Five Highland Provinces of PNG
Source: Mapping Agricultural Systems of Papua New Guinea database.

Table 31-1. Environments in Which Planted *Casuarina* Fallows are used in the PNG Highlands

Environmental Factor	*Class* [a]	*Percentage of Planted Fallows in Each Environmental Class, by Area* [b]	*Percentage of Total Area of PNG Highlands in Each Environmental Class* [c]
Landform	Hills/ mountains	92	51
Rock type	Sedimentary	77	62
Altitude	1,200–2,100 m asl	87	54
Slope	Greater than 20°	83	78
Dominant vegetation	Grassland	80	24
Annual rainfall	*2,000–3,000 mm*	*77*	*40*

Notes: [a] Several categories for each environmental factor have been collapsed to form these classes. [b] Planted *Casuarina* fallows are used in agricultural systems that occupy 2,100 km^2 in the PNG highlands. [c] The total area of all land in the PNG highlands, whether it is used for agriculture or not, is 63,800 km^2.
Source: Databases of Mapping Agricultural Systems of PNG and PNG Resource Information System.

Effect of *Casuarina* on Soil Fertility

Throughout the highlands villagers say that *Casuarina* improves soil fertility. For example, in Chimbu Province, discussions about soil erosion and declining soil fertility often include reference to the use of *Casuarina* fallows to reduce these problems. Present scientific evidence, limited though it may be, supports this view.

Parfitt (1976) collected soil samples from under *Casuarina* trees of different ages on adjacent sites. He found higher levels of soil nitrogen under five- or six-year-old *Casuarinas* than under grass fallows or food gardens, and the levels increased with tree age. Another study found that soil N and C were higher under *Casuarina* than under coffee on its own, and that soil N and C increased with the age of the *Casuarina*. Moreover, N and C levels were higher under *Casuarina* than under *Albizia* or *Crotalaria* (Thiagalingam and Famy 1981).

Leaf litter rates of 3.3 tons/ha/year, dry weight, have been recorded in the first three years of *Casuarina* growth (Bino and Kanua 1996). Higher rates of between 7 and 8 tons/ha/year have been recorded under 10- to 15-year-old trees. Estimates of the rate of nutrient returns to the soil from leaf litter were 39 kg of N, 3 kg of P, and 10 kg K per hectare per year for the first three years of growth, and between 84 and 123 kg N per hectare per year for 10- to 15-year-old trees (Bino and Kanua 1996).

Use by Villagers

Only limited information has been published on how villagers select and transplant seedlings, what planting density they follow, survival rates, the length of fallow periods, and utilization of timber. Men appear to be more involved with transplanting seedlings than women, but detailed observations are lacking.

Casuarina seedlings often grow abundantly in sandy sites adjacent to streams. This is where villagers select self-sown seedlings. They are planted in food gardens toward the end of the cropping phase so that the seedlings are about 80 to 120 cm tall by the time the final crop of sweet potato is harvested. At this size, they are able to compete with natural regrowth and invading pigs as the garden reverts to fallow. It is likely that, in locations where *Casuarina* fallows are most dense, many seedlings grow from seed dropped by nearby established trees and transplanting is not necessary. In Chimbu Province, seedlings are transplanted into gardens during the final few plantings of sweet potato (Nilles 1943-44; Brookfield and Brown 1963; Hughes 1966; Sterly 1977). In the Simbai area, seedlings are transplanted to the garden immediately after the food crops have been established (Burnett 1963; Allen et al. 1994). The difference in timing between Chimbu and Simbai reflects the shorter cropping phase, with only two plantings in the Simbai area.

The duration of *Casuarina* fallows is variously reported as being 3 to 8 years for the Simbai area (Burnett 1963); 7 to 9 years for the Chimbu Valley (Sterly 1977); as short as 8 years (Straatmans 1967); and 7 to 10 years in the Baliem Valley of Irian Jaya (Askin et al. 1990). In the Chuave area of Chimbu Province, villagers say that sufficient soil fertility has been restored for another garden phase when *Casuarina* trees reach a diameter of 20 to 25 cm. This probably represents a fallow period of 8 to 12 years. All of the reported periods are estimates only. However, the study of regrowth vegetation adjacent to the Chimbu River on the site of a major landslide indicates that *Casuarina* trees attain a significant size by 10 years (Humphreys and Brookfield 1991).

When villagers judge that a *Casuarina* fallow has restored sufficient soil fertility, the trees are ringbarked, the branches are removed for fencing and firewood, and the trunk is left standing until it is needed for timber. Sometimes trees are left alive, with a small crown to maintain them, until the next fallow phase. Sweet potato and other food crops are planted among the standing *Casuarina* trees.

There are two known problems with *Casuarinas* in the PNG highlands. The first is poor growth caused by a widespread boron deficiency (Bourke 1980), and the second is a problem of unknown cause that results in tree dieback. It is possibly caused by a root pathogen, but this remains speculative. In the case of the boron deficiency, an experiment in the Southern Highlands, where the deficiency is particularly severe, involved the application of three g of boron per tree. Over a 20-month period, the height of treated trees more than doubled, compared with adjacent untreated trees (D'Souza and Bourke 1986).

History of Adoption

The question of how long villagers have been managing the species composition of fallows by planting *Casuarina* is an important one. If it is relatively recent, then the technique has potential for use in other locations in the highlands, on the assumption that its absence means that it has not yet been tried elsewhere and rejected. However, if it proves to be a very ancient practice, then there is less potential for its adoption elsewhere as there has been sufficient time for it to spread throughout the region.

While not conclusive, there is some evidence that the planting of *Casuarinas* in the main food gardens is a relatively new practice. In saying this, I am distinguishing this technique from that of planting *Casuarina* trees in and near villages, or as fallow vegetation for smaller "mixed gardens." The practice of planting *Casuarinas* in the main sweet potato or taro gardens is known to be relatively recent in the Simbai and Oksapmin areas, two of the four regions where the technique is used. In the Simbai area, the systematic planting of *Casuarina* trees developed within the lifetime of the fathers of living men, in other words, since the 1920s (Bulmer 1982). Similarly, in the Tekin Basin in the Oksapmin area, extensive planting of *Casuarinas* is a practice adopted over the past two generations (Cape 1981), once again, since the 1920s (author's fieldwork 1979). It has become more important in this region since the 1960s.

In the Chimbu region, there is evidence that the practice increased in importance during the 20th century, although villagers claim that it has been used for more than four generations. Photographs and observations by Europeans in the 1930s show that *Casuarina* planting for fallows was much less extensive than it became in the late 1950s (Brookfield and Brown 1963). A study of seven villages in the Chuave area of Chimbu Province reported a marked increase in *Casuarina* cover between 1952 and 1961 (Salisbury 1964). In the Sinasina area of the same province, planting of *Casuarinas* was recorded in the mid-1960s as an increasing practice prompted by the loss of forest reserves (Hughes 1966).

In recent decades, *Casuarina* planting has also spread outside of the four main regions shown in Figure 31-1. For example, in the Okapa area of the Eastern Highlands, people started planting *Casuarina* trees in gardens in the mid-1960s, following their introduction as shade for coffee (Sorenson 1976). Writing about the valleys of the northern part of the Eastern Highlands, Ataia (1983) notes that 30 years earlier the area was mainly grassland. At the time of writing, however, trees dotted the landscape, and *Casuarina* was the dominant species. Photographs of the Benabena, Asaro, and upper Ramu valleys taken during the 1930s and 1940s show an almost treeless landscape that contrasts with today's partly wooded aspect. This change is associated with the encouragement of tree planting by Australian administration officials in the 1950s and 1960s to counter deforestation, the use of *Casuarina* for shade for coffee, and movement of hamlets from high ridges to valley floors and lower ridges.

This hypothesis, that the widespread planting of *Casuarina* in food gardens is a relatively recent innovation, differs from the conclusions of Haberle (n.d.). He conducted a comprehensive review of agricultural change in the PNG highlands over a 2,000-year period by examining pollen spores. An increase in *Casuarina* pollen was found at many sites in the highlands from A.D. 500 onwards. Haberle concluded that agriculturists in the Wahgi Valley in PNG and the Baliem Valley in Irian Jaya were the first to adopt silvicultural practices, between A.D. 800 and 1100. Further, he concluded that similar increases in *Casuarina* pollen between A.D. 1000 and 1400 in other highland regions showed that there may have been a diffusion of tree fallowing techniques outward from the main valleys.

I interpret the pollen evidence as indicating that from this period, people were planting *Casuarina* trees in and near villages, but they were not planting extensive areas in the main food gardens as is now common practice in four main regions of the highlands. It is suggested that early tree planting was done to provide timber, because removal of nearby forest had made supplies scarce. The pollen data are from the Wahgi Valley, Tari Basin, Tari Gap, the Baliem Valley, and high-altitude locations in Enga Province. Today, villagers plant *Casuarina* trees in the Wahgi Valley, Tari Basin, and the Baliem Valley, but they do not do so in their main sweet potato gardens.[3] Unless widespread planting of *Casuarina* trees has declined in these locations, and this seems unlikely, then the *Casuarina* trees revealed by the pollen record are likely to represent plantings other than in the main food gardens.

Conclusions

Pollen data showing a rapid increase in *Casuarina* trees after A.D. 500 are interpreted as evidence that trees were planted in and near villages from this period. It is generally accepted that sweet potato was introduced into the PNG highlands about 300 years ago (Yen 1974). This almost certainly resulted in a significant increase in food production and pig numbers, as well as numerous associated changes in land use and possibly social organization and human population. It is hypothesized that, after a long period of expanding production, further increases in sweet potato production were halted in a number of locations by environmental constraints. This was during the 19th and early 20th centuries, and, in order to provide food for more people and pigs, villagers in some regions devised a number of innovative practices, among them composting, soil erosion control, and fallow management with *Casuarina*.

There are significant distances between locations where people use *Casuarina* most intensively. This suggests that each of these groups independently discovered the value of *Casuarina* in maintaining soil fertility. Three factors suggest that there is potential for further adoption of the technique within the highlands. These are (1) the limited number of locations where the technique is now important; (2) the recent expansion of *Casuarina* planting in at least three of the four regions where it is most important; and (3) a continuing demand from an increasing human population for food, stock feed, and cash income.

Many aspects of *Casuarina* trees and their use by villagers remain unknown. It has not yet been established, for instance, that *C. oligodon* fixes atmospheric nitrogen, although there can be little doubt that it does. Growth rates under village conditions are unknown; the cause of the dieback condition has not been investigated; there have been no studies of different provenances; and the role of *Casuarina* in controlling soil erosion, restoring soil fertility, and increasing the yields of following crops has not been documented. Many aspects of human management are also poorly known, including how villagers select and transplant seedlings, planting densities, survival rates, the history of adoption of managed tree fallows, and use of the timber. High priority should be given to researching these aspects of *Casuarina oligodon*.

---

3 In the Pitt River area in the north Baliem, villagers do plant *Casuarina* trees near the end of the cropping phase, but this is the only part of the Baliem area where this is done (Askin et al. 1990).

Acknowledgments

Some of the data reported here were collected together with other members of the land management project at the Australian National University, particularly Bryant Allen and Robin Hide. Robin Grau extracted information from the MASP and PNGRIS databases. Comments on a draft of this paper by Jean Bourke, Harold Brookfield, and Geoff Humphreys are acknowledged with thanks.

References

Allen, B.J., R.L. Hide, R.M. Bourke, D. Fritsch, R. Grau, P. Hobsbawn, M.P. Levett, S. Majnep, V. Mangi, T. Nen, and G. Sem. 1994. *Madang Province: Text Summaries, Maps, Code Lists, and Village Identification.* Canberra: Department of Human Geography, Australian National University.

Allen, B.J., T. Nen, R.M. Bourke, R.L. Hide, D. Fritsch, R. Grau, P. Hobsbawn, and S. Lyon. 1996. *Central Province: Text Summaries, Maps, Code Lists, and Village Identification.* Canberra: Department of Human Geography, Australian National University.

Askin, D.C., D.J. Boland, and K. Pinyopusarerk. 1990. Use of *Casuarina oligodon* subsp. *abbreviata* in Agroforestry in the North Baliem Valley, Irian Jaya, Indonesia. In: *Advances in* Casuarina *Research and Utilization: Proceedings of the Second International* Casuarina *Workshop,* edited by M.H. El-Lakany, J.W. Turnbull, and J.L. Brewbaker. Cairo: Desert Development Centre, 213–219.

Ataia, A. 1983. *Casuarina oligodon* in the Eastern Highlands Province, Papua New Guinea. In: Casuarina *Ecology Management and Utilization: Proceedings of an International Workshop,* edited by S.J. Midgley, J.W. Turnbull, and R.D. Johnston. Canberra: CSIRO (Commonwealth Scientific and Industrial Research Organization), 80–87.

Bino, B., and M.B. Kanua. 1996. Growth performance, Litter Yield and Nutrient Turnover of *Casuarina oligodon* in the Highlands of Papua New Guinea. In: *Recent* Casuarina *Research and Development: Proceedings of the Third International* Casuarina *Workshop,* edited by K. Pinyopusarerk, J.W. Turnbull, and S.J. Midgley. Canberra: Forestry and Forest Products Division, CSIRO (Commonwealth Scientific and Industrial Research Organization), 167–170.

Bourke, R.M. 1980. Boron Deficiency Is Widespread in the Highlands. *Harvest* 6, 175–178

———, 1988. Alitudinal Limits of 230 Economic Crop Species in Papua New Guinea. Human Geography, RSPAS, ANU. Unpublished Paper.

———, 1989. Food, Coffee and *Casuarina*: An Agroforestry System from the Papua New Guinea Highlands. In: *Agroforestry Systems in the Tropics,* edited by P.K.R. Nair. Dordrecht, the Netherlands: Kluwer Academic Publishers, 269–275.

Bourke, R.M., B.J. Allen, R.L. Hide, N. Fereday, D. Fritsch, B. Gaupu, R. Grau, P. Hobsbawn, M.P. Levett, S. Lyon, V. Mangi, and G. Sem. 1997. *Morobe Province: Text Summaries, Maps, Code Lists, and Village Identification.* Canberra: Department of Human Geography, Australian National University.

Bourke, R.M., B.J. Allen, P. Hobsbawn, and J. Conway. 1998. Papua New Guinea: Text Summaries. In: *Agricultural Systems of Papua New Guinea, Working Paper No. 1* (Two volumes). Canberra: Department of Human Geography, Australian National University.

Brookfield, H.C., and P. Brown. 1963. *Struggle for Land: Agriculture and Group Territories Among the Chimbu of the New Guinea Highlands.* Melbourne: Oxford University Press.

Bulmer, R. 1982. Crop Introductions and their Consequences in the Upper Kaironk Valley, Simbai Area, Madang Province. In: *Proceedings of the Second Papua New Guinea Food Crops Conference,* Part Two, edited by R.M. Bourke and V. Kesavan. Port Moresby: Department of Primary Industry, 282–288.

Burnett, R.M. 1963. Some Cultural Practices Observed in the Simbai Administrative Area, Madang District. *Papua and New Guinea Agricultural Journal* 16, 79–84.

Cape, N. 1981. Agriculture. In: *Oksapmin: Development and Change,* edited by S.G. Weeks. Port Moresby: University of Papua New Guinea, 149–190.

D'Souza, E., and R.M. Bourke. 1986. Intensification of Subsistence Agriculture on the Nembi Plateau, Papua New Guinea. 1. General Introduction and Inorganic Fertilizer Trials. *Papua New Guinea Journal of Agriculture, Forestry and Fisheries* 34, 19–28.

Gibson, J., and S. Rozelle. 1998. *Results of the Household Survey Component of a 1996 Poverty Assessment for Papua New Guinea.* Washington, DC: Population and Human Resources Division, World Bank.

Haberle, S. n.d. *Palaeoecological and Archaeological Evidence for Agricultural Change in the Highlands of New Guinea, 0–1930 A.D.* Canberra: Biogeography and Geomorphology Department, Australian National University.

Hide, R.L., R.M. Bourke, B.J. Allen, T. Betitis, D. Fritsch, R. Grau, L. Kurika, E. Lowes, D.K. Mitchell, S.S. Rangai, M. Sakiasi, G. Sem, and B. Suma. 1994. *Milne Bay Province: Text Summaries, Maps, Code Lists, and Village Identification.* Canberra: Department of Human Geography, Australian National University.

Hughes, I. 1966. Availability of Land and Other Factors Determining the Incidence and Scale of Cash Cropping in the Kere Tribe, Sina Sina, Chimbu District, New Guinea, BA (Honors) thesis, University of Sydney, Sydney.

Humphreys, G.S., and H. Brookfield. 1991. The Use of Unstable Steeplands in the Mountains of Papua New Guinea. *Mountain Research and Development* 11, 295–318.

Nilles, J. 1943–44. Natives of the Bismarck Mountains, New Guinea. *Oceania* 14, 104–124.

Parfitt, R.L. 1976. Shifting Cultivation: How It Affects the Soil Environment. *Harvest* 3, 63–66.

Ruthenberg, H. 1980. *Farming Systems in the Tropics.* London: Oxford University Press.

Salisbury, R.F. 1964. Changes in Land Use and Tenure among the Siane of the New Guinea Highlands (1952–61). *Pacific Viewpoint* 5, 1–10.

Sorenson, E.R. 1976. *The Edge of the Forest: Land, Childhood and Change in a New Guinea Protoagricultural Society.* Washington, DC: Smithsonian Institute Press.

Sterly, J. 1977. *Research Work on Traditional Plantlore and Agriculture in the Upper Chimbu Region, Papua New Guinea.* B19. Vienna, Austria: International Committee on Urgent Anthropological and Ethnological Research, Institute für Volkerkunde, 95–105.

Straatmans, W. 1967. Ethnobotany of New Guinea in Its Ecological Perspective. *Journal d'Agriculture Tropicale et de Botanique Appliquee* 14, 1–20.

Thiagalingam, K., and F.N. Famy. 1981. The Role of *Casuarina* under Shifting Cultivation: A Preliminary Study. In: *Nitrogen Cycling in Southeast Asian Wet Monsoonal Ecosystems,* edited by R. Weisilaar, J.R. Simpson, and T. Rosswall. Canberra: Australian Academy of Sciences, 154–156.

Waddell, E. 1972. *The Mound Builders: Agricultural Practices, Environment, and Society in the Central Highlands of New Guinea.* Seattle, Washington: University of Washington.

Yen, D.E. 1974. *The Sweet Potato in Oceania: An Essay in Ethnobotany.* Honolulu: Bishop Museum Press.

Chapter 32

Multipurpose Trees as an Improved Fallow

An Economic Assessment

Peter Grist, Ken Menz, and Rohan Nelson

Traditional shifting cultivation, sometimes called slash-and-burn agriculture, is only sustainable if it involves long periods of fallow. Under pressure from population growth, the fallow periods of tradition are being shortened, year by year, and the fallow lengths now prevailing in most of the uplands of Southeast Asia mean that shifting cultivation is no longer sustainable in either economic or environmental terms (Menz and Grist 1996). Therefore, policymakers and research organizations are encouraging smallholders to adopt alternative forms of agriculture that are more sustainable.

This chapter considers an improved fallow system that is potentially more sustainable. It involves the establishment of a *Gliricidia* plantation in the fallow period and is an adaptation of an improved fallow system conceived by MacDicken (1990) and Garrity (1993). After defining the *Gliricidia* fallow system, we analyze its performance using a computer simulation model. The model consists of two parts: an existing agroforestry model, Soil Changes Under Agroforestry (SCUAF), and a cost-benefit analysis spreadsheet. SCUAF calculates farm outputs, including crop and tree yield, based on climate, site, and soil characteristics. The spreadsheet takes the output from SCUAF plus economic data on input and output levels and prices. We calibrated the model using data from field trials at the Compact research station, Mindanao, the Philippines. The performance of the *Gliricidia* fallow system, calculated in this fashion, is contrasted with that of an *Imperata* fallow system.

The analysis presented in this chapter can be considered within the framework of dynamic technology evaluation and design, outlined by Anderson and Hardaker (1979). Within this framework, the improved *Gliricidia* fallow system can be regarded as being at the initial or "notional" stage of development.

The possibilities of improved fallow systems have been recognized within the scientific community, and there have been isolated reports of indigenous practices, but there has been minimal evaluation and testing. This analysis is intended to move the technology evaluation and design process from a notional to a preliminary, more advanced, stage (Menz and Knipscheer 1981). The computer-modeling approach will not only provide a more formal evaluation of the technology than has been undertaken to date, but will also allow it to be undertaken at minimal cost. If, from

---

Peter Grist, Plantation Taxation Review, Forest Industries Branch, Department of Agriculture, Fisheries and Forestry, G.P.O. Box 858, Canberra, ACT 2601, Australia; Ken Menz, ACIAR, G.P.O. Box 1571, Canberra, ACT 2601, Australia; Rohan Nelson, Resource Economist, CSIRO Sustainable Ecosystems, G.P.O. Box 284, Canberra, ACT 2601, Australia.

this analysis, the *Gliricidia* fallow system appears capable of providing environmental and economic benefits, then a more vigorous program of scientific research is warranted.

Methodology

The Two Fallow Systems

The two systems are an improved fallow system based on a *Gliricidia* plantation and a fallow system commonly used in the more remote upland areas of Southeast Asia, based on the grass species *Imperata cylindrica.* Both systems involve five years of fallow followed by one year of cropping. A summary is given in Figure 32-1.

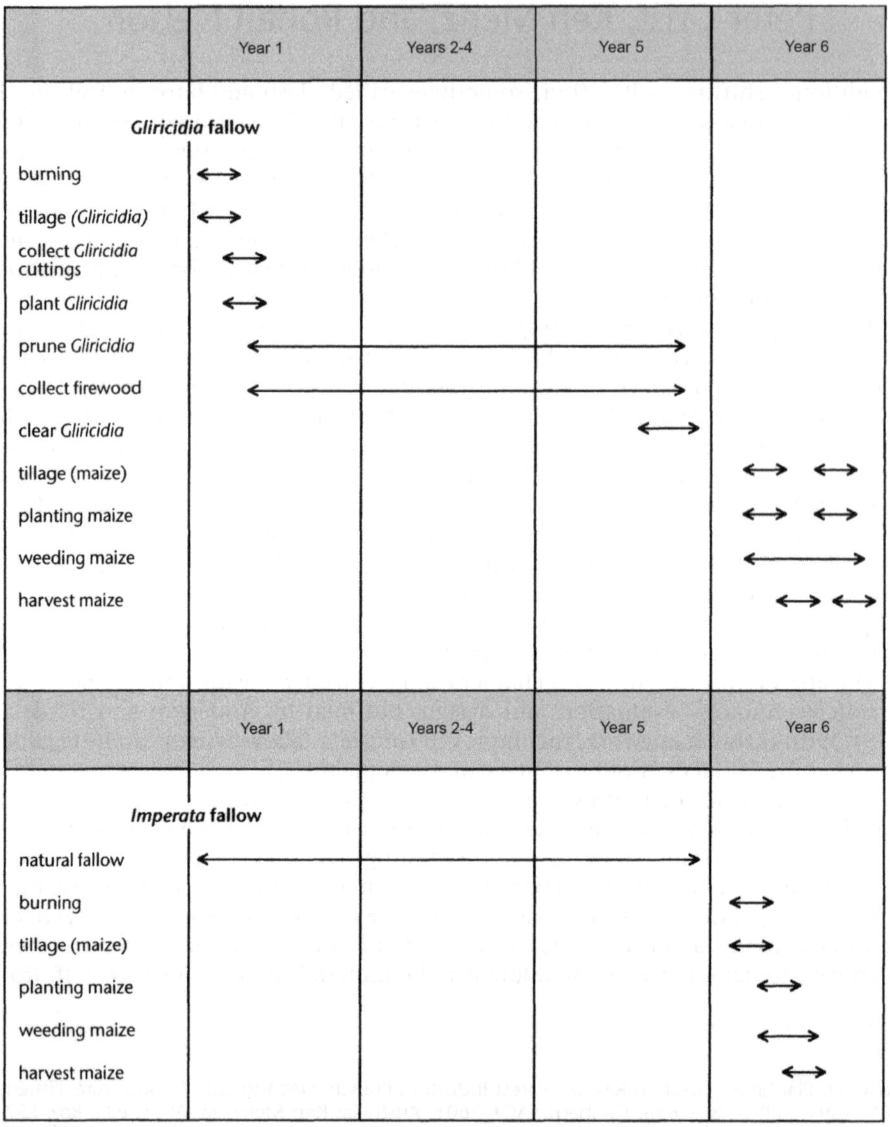

Figure 32-1. Summary of the *Gliricidia* and *Imperata* Fallow Systems

In the proposed *Gliricidia* fallow system, one-sixth of the available land is planted as a *Gliricidia* plantation each year and another sixth is planted as a maize crop. The remaining two-thirds of the farm consists of established *Gliricidia* plantation. In terms of the rotation, *Gliricidia* is planted after the maize crop, and maize is preceded by five years of *Gliricidia* plantation (fallow).

Gliricidia sepium was selected as the tree species for this improved fallow system for the following reasons:

- It is a small, fast-growing tree capable of shading and suppressing *Imperata* and producing firewood and mulch (Ella et al. 1989; Anoka et al. 1991).
- It has a high nutrient value and is capable of cycling nitrogen through mulch or green manure (Simons and Stewart 1994).
- It is able to grow well on acid soils, which are common across Southeast Asia (Lowry et al. 1992).

Preparing land for *Gliricidia* requires burning to remove existing vegetation, including crop residues and *Imperata*. The site is then tilled over, using a hoe, and cuttings are collected. *Gliricidia* is planted in a hedgerow pattern with two rows planted 50 cm apart and with 50 cm between the trees within the rows. A gap of 1.5 m is left between each pair of rows.

Imperata is stimulated by burning and other soil disturbance, such as tillage, so it is important to coordinate burning of the existing vegetation with the end of the dry season, and planting of the *Gliricidia* cuttings with the beginning of the rains. *Gliricidia* is quick to establish and, requires little maintenance. To maximize nutrient recycling, the trees are pruned four times a year and the prunings are used as mulch under the plantation.[1] During years two to five, *Gliricidia* requires little labor other than pruning. At the end of the fifth year, the plantation is cleared for cropping. The *Gliricidia* is cut at ground level and the stumps either hacked or poisoned to prevent coppicing. The foliage is removed from the branches and left on the site as mulch. The branches and trunks are removed and stacked, to be sold as firewood.

In the sixth year, following the clearing of the plantation, the site is hoed in preparation for a maize crop. Maize is planted between the rows of stumps, one row in the 0.5 m space between the paired tree rows, and two rows in the 1.5 m space between the pairs of rows. Although the *Gliricidia* plantation has shaded out *Imperata*, reinfestation from neighboring areas will necessitate some weeding of the maize crop. After the first maize harvest, the site is hoed again in preparation for a second crop. Following the second maize crop, the *Gliricidia* plantation phase begins again.

In the *Imperata* fallow system, the land is abandoned during the five-year fallow. In the sixth year, *Imperata* is burned and the soil hoed in preparation for a maize crop. Maize is then planted in rows 0.75 m apart. Regular weeding of the crop is required because *Imperata* propagules are present. Preliminary analysis with SCUAF advises that the level of soil fertility following a five-year *Imperata* fallow can support only one maize crop per year (Grist and Menz 1996). This is mainly due to the poor replenishment of soil nutrients during an *Imperata* fallow. After the maize crop, the site is again abandoned to *Imperata*.

The Site Used for Model Calibration

The model was calibrated to a specific site. Compact, a research station near Claveria in Mindanao, the Philippines, was chosen for this purpose. The Compact research station was established primarily to trial hedgerow intercropping regimes and is managed by the World Agroforestry Centre (ICRAF). We used biophysical data obtained from field trials at Compact (Agus 1994; Nelson et al. 1996b), and recent economic data obtained from a survey of the surrounding municipality of Claveria (Nelson et al. 1996b, c).

---

[1] Pruning is also an opportune time to collect cuttings for the establishment of *Gliricidia* on other plots.

Economic Data

The data used in the economic analysis originated from three sources: the computer-based biophysical agroforestry model, SCUAF Version 4; a literature review; and an economic survey carried out by Nelson et al. (1996c). The SCUAF modeling will be described later. The other data are discussed here.

In the *Imperata* fallow system, land preparation requires approximately 30 days/ha (Conelly 1992) and planting maize requires approximately 13 days/ha (Nelson et al. 1996c). During the cropping period, there is a significant amount of weed competition because of rhizomes and seed that remain in the soil after clearing (Eussen et al. 1976), so regular weeding requires about 30 days/ha (Conelly 1992).

Labor requirements for the *Gliricidia* plantation are adapted from data obtained from a survey by Nelson et al. (1996c) of *Gliricidia* hedgerow systems. The *Gliricidia* plantation has 20,000 trees/ha, double the figure of the hedgerow system surveyed by Nelson et al. (1996b). The labor required for land preparation in the *Gliricidia* fallow system was calculated to be 10 days. It involved burning, followed by tillage using a hoe. Labor for collection and planting of cuttings was estimated at 70 days/ha and pruning labor at 16 days/ha/yr (see Table 32-1).[2] There are no weeding requirements during the *Gliricidia* plantation phase. To cut and carry the firewood, it was estimated that an extra day would be required at each pruning, making four extra days per year, and another four days would be needed for the major removal of firewood in the year of plantation harvest.

Land preparation for the two maize crops following the *Gliricidia* harvest was estimated to need 30 days labor/ha, and for planting maize, 25 days/ha (Nelson et al. 1996c). Labor requirements for weeding maize after a *Gliricidia* fallow were estimated at 30 days for the two maize crops (Nelson et al. 1996c).

The standard wage for hired agricultural labor in Claveria is ₱60 per day (Nelson et al. 1996b). (1997US$1 = ₱26.38) However, smallholders have limited opportunity for off-farm employment so the hired labor rate overstates the opportunity cost of labor. Consequently, Nelson determined a more realistic opportunity cost of the for off-farm employment so the hired labor rate overstates the opportunity cost of labor.

Table 32-1. Labor Requirements of the Two Fallow Systems (days/ha)

		*Improved Fallow***	
Operation	Imperata *Fallow**	*Crop Cycle*	*Tree Cycle*
Land preparation	30	30	10
Collect/plant cuttings			70
Maize sowing	13	25	
Weeding	30	30	
Gliricidia pruning			16
Maize harvest	9	18	
Postharvest processing	11	22	
Cut-and-carry firewood			4
Additional firewood final year			4
Total crop	93	125	
Tree total (establishment year)			96
Tree total (normal year)			20
Tree total (cut year)			24
Average for whole farm (1/3 ha crop, 5/3 ha fallow)	31		102

Note: * One maize crop during the crop year of the cycle. ** Two maize crops during the crop year of the cycle.

[2] The plot size is only one-third of a hectare; thus, labor requirements for collecting cuttings and planting will be only 23 days per year. Also, as noted earlier, the collection of cuttings can be combined with pruning of the plots under plantation, to reduce labor demand.

Consequently, Nelson determined a more realistic opportunity cost of the farmers' labor, of ₱40 per day, and he called it the "own labor rate." This rate is used as the base price of labor in this analysis. Seed of an improved maize cultivar is bought from

$$NPV_t = \sum_{t=0}^{n} \frac{(B_t - C_t)}{(1+r)^t}$$

local markets at a cost of ₱6.50 per kilogram. There are no other cash costs included in this analysis.

The average farmgate price for the maize crop, in 1994 prices, was ₱4.60/kg in the wet season and ₱5.50/kg in the dry season (Nelson et al. 1996c).[3] The average season price of maize, ₱4.90/kg,[4] is used as the price of maize in the *Gliricidia* fallow system. For the *Imperata* fallow system, the wet season price of maize is used. A price for firewood in Claveria was obtained from a special survey of smallholders near Compact, who had adopted a *Gliricidia* hedgerow system. The average price of firewood was ₱1,000 per metric tonnes.

Twelve percent was chosen as the relevant discount rate. Although not consistent with the market borrowing rate, which is between 16% and 30% (Nelson et al. 1996c), it approximates the social opportunity cost of capital in the Philippine economy (Medella et al. 1990).

Modeling

The model used for this analysis consists of two parts: the biophysical element, which is modeled using SCUAF Version 4 (Young et al. 1996), and the economic element, which uses a Microsoft Excel spreadsheet.

SCUAF is a simple, deterministic model that can be used to predict the effect of interactions between agroforestry systems and soil. Erosion is predicted in SCUAF using a Modified Universal Soil Loss Equation (MUSLE) (FAO 1979), which is based on average-year climate and soil characteristics. Soil carbon, nitrogen, and phosphorus in the soil profile are affected by recycling of plant material, soil erosion, and uptake by plants. By simulating these processes, SCUAF predicts changes in soil fertility, and hence changes in crop yield and biomass production by trees and other vegetation, over time.

Two previous applications of SCUAF give confidence in the model. Vermeulen et al. (1993) used SCUAF to simulate soil nutrient dynamics and plant productivity for the Miombo woodlands and adjacent maize crops in Zimbabwe. SCUAF was judged to provide reasonable predictions for maize and tree growth, although it had no facility to attenuate growth as the woodland approached maturity. The attenuation of growth is not important in this analysis, due to regular harvesting and the short growth cycle. Nelson et al. (1996a) also used SCUAF when researching hedgerow systems in the Philippines. In this case, the results from SCUAF were compared with results from a more complex dynamic process model, APSIM. It was found that SCUAF produced similar estimates of trends in medium-term yield. However, SCUAF is an average-year model that abstracts from seasonal conditions, so short-term yield fluctuations cannot be predicted.

The economic element of the model calculates the net present value (NPV) of the system using a cost-benefit analysis. The cumulative net present value of the system over n years is calculated using the above equation, where B_t and C_t are the total benefits and total costs in year t, and r is the discount rate. Future benefits and costs

---

[3] These prices are calculated on the sale of maize at 18% moisture content. However, SCUAF uses dry matter production figures for maize (0% moisture content). Thus, the dry matter yield of maize was adjusted to its weight at 18% moisture content before returns from maize were calculated.

[4] Given that dry season production is approximately half of the wet season production (i.e., one-third of total production), the average price for wet and dry season crops is ₱4.90 per kilogram (4.6 x 0.667 + 5.5 x 0.333).

are discounted to capture the preference individuals express for present over future consumption. Benefits are calculated by applying the market price to yields of maize and firewood, as simulated with SCUAF.

Returns from the *Gliricidia* fallow system are derived from maize and from firewood. Costs are incurred for tree establishment, and the cutting and carrying of prunings and firewood. With *Imperata* fallow, costs and returns are associated only with maize.

The Calibration of SCUAF

The *Gliricidia* fallow system produces a high level of soil fertility, enabling the planting of two maize crops in the year after the fallow is opened, one in the wet season and another in the dry season. In research station trials of *Gliricidia* hedgerows at Compact, net primary production of maize averaged 7.2 tonnes/ha, including stover, in the wet season and 2.5 tonnes/ha in the dry season. Consequently, the initial wet season maize grain yield in SCUAF was fixed at 2.2 tonnes/ha, and the dry season grain yield at 0.8 tonnes/ha (Nelson et al. 1996b). Biomass production of *Gliricidia* was specified in the model as 11.4 metric tonnes/ha, with leaf production at 7.7 metric tonnes/ha, and wood production at 3.7 metric tonnes/ha. These figures came from measurements of biomass production of *Gliricidia* hedgerows at Compact by Nelson et al. (1996b). Similar biomass production has been reported in *Gliricidia* plantations by Ella et al. 1989; Gunasena and van der Heide (1989); Maclean et al. (1992); and Panjaitan et al. (1993). The figures used in the model are for a stand of 20,000 trees/ha, with four prunings per year or one pruning every 12 weeks. This high pruning intensity increases the leaf-to-wood ratio of *Gliricidia*, providing more foliage for mulch (Ella et al. 1989).

A five-year *Imperata* fallow does not enrich the soil to the same extent as a *Gliricidia* fallow, so the productivity of maize within the *Imperata* fallow system is low. *Imperata* fallowing is not practiced at Compact, so to determine the initial yield of maize following an *Imperata* fallow, preliminary analysis was carried out using SCUAF. Based on the nutrients available in the soil, including carbon, nitrogen, and phosphorus, SCUAF uses the law of the minimum approach to determine plant growth. The preliminary analysis with SCUAF involved reducing the initial net primary production of the crop, which is set as an input parameter, until it was approximately equal to the net primary production in the first year of cropping, which is an output parameter. This ensured that the available soil nutrients were able to support the initial net primary production. The production level that allowed this was about four metric tonnes of biomass per hectare, which implied a maize grain yield of 1.33 tonnes/ha. The low productivity after an *Imperata* fallow also implies that, using this system, the site is only capable of supporting one maize crop. Biomass production of the *Imperata* was modeled as being about four metric tonnes/ha/yr (Castillo and Siapno 1995; Sajise 1980).

To determine nutrient demand by plants, and the fate of soil nutrients, SCUAF equates the nutrient content of the plant parts with their rate of growth. The nitrogen content of *Gliricidia* at Compact was 2.5% for foliage and 0.5% for wood (Agus 1994). The nitrogen content of *Imperata* was measured by Sajise (1980) and found to be 0.94%. For maize, the SCUAF default values of 2.0% nitrogen in crop leaf residues and 3.0% for maize grain were accepted. The phosphorus content of both *Imperata* and maize was calculated by Lowry et al. (1992) to be 0.2%. Maclean et al. (1992) calculated the phosphorus content of the *Gliricidia* leaf to be 0.27%, and Agus (1994) calculated the average phosphorus content of all plant parts to be 0.2%. This implies that the phosphorus content of the wood is 0.065%. Agus (1994) and Nelson et al. (1996b) provide a detailed description of the site conditions at Compact.

SCUAF determines plant growth and soil changes on a per hectare basis. To translate these results into a whole-farm basis relevant to Compact, two hectares was specified as the average available land area (Limbaga 1993; Nelson et al. 1996b).

Results

The *Imperata* fallow system provides a marginal surplus, or profit, with a net present value (NPV) of ₱2,000, over a fifty-year period (Table 32-2). The *Gliricidia* fallow system is far more profitable under the same circumstances, showing an NPV of ₱57,549. This can be attributed to improved soil fertility, and therefore higher crop yields, and to the additional saleable product, firewood. Firewood production also reduces the variability of economic returns over time, through diversification.

The costs associated with the *Gliricidia* fallow system are more than double the costs of the traditional *Imperata* fallow (Table 32-2). However, this is countered by higher maize yield because of improved soil fertility (Figure 32-2), and returns from firewood.

If the profitability of the *Gliricidia* system is calculated upon maize production only, the resulting profit is ₱10,000, which is still five times more than the profit from the *Imperata* fallow system (Table 32-2). Firewood can make almost as much income as maize. Small firewood yields of about 400 kg/ha are provided every year as a byproduct of mulching, because branches are cut in the process of mulching the foliage. A much larger firewood yield, greater than 12 metric tonnes/ha, is provided in the cut year.

Maize yield in the *Imperata* fallow system declines over time, but maize and firewood yields in the *Gliricidia* fallow system increase over time (Figure 32-2). These changes are due to soil fertility, which is depleted over time in the *Imperata* system, but, because of the effect of mulching the *Gliricidia* foliage, improves over time in the *Gliricidia* system. When used as mulch, *Gliricidia* foliage has a half-life of 20 days, which results in a rapid recycling of soil nutrients (Simons and Stewart 1994).

The changes in soil fertility can be observed in the levels of the key nutrients carbon, nitrogen, and phosphorus, which are predicted using SCUAF. For the *Imperata* fallow, all three show strong declining trends (Figure 32-3). This is similar to the trends observed in previous studies of short *Imperata* fallow systems at other locations (Grist and Menz 1996; Menz and Grist 1996). Nutrient levels decline during both maize and *Imperata* phases due to erosion, leaching, and uptake by plants. In sharp contrast, the soil nutrient levels in a *Gliricidia* fallow system steadily increase over time (Figure 32-3).

As well as reducing erosion under the *Gliricidia* fallow system, mulching enhances soil organic matter, hence carbon buildup. Soil nitrogen and phosphorus levels are also improved by mulching the leaves of *Gliricidia*. Phosphorus is thought to be brought up from lower soil levels into the *Gliricidia* leaves and ultimately into the topsoil via mulching.

In the year of maize cropping, there is a decline in soil nutrients. However, overall, the increase in soil nutrients during the *Gliricidia* fallow period is greater than the nutrient demand of the plants in the year of cropping.

Table 32-2. Components of Net Present Value (NPV) for the *Gliricidia* Fallow and *Imperata* Fallow Systems (Philippine Pesos x 1000)

Component	Imperata *Fallow*	Gliricidia *Fallow*
Discounted gross return, maize	21	57
Discounted gross return, firewood		47
Discounted total cost, maize and firewood	19	47
Net present value, maize and firewood	2	57
Net present value, maize only		10

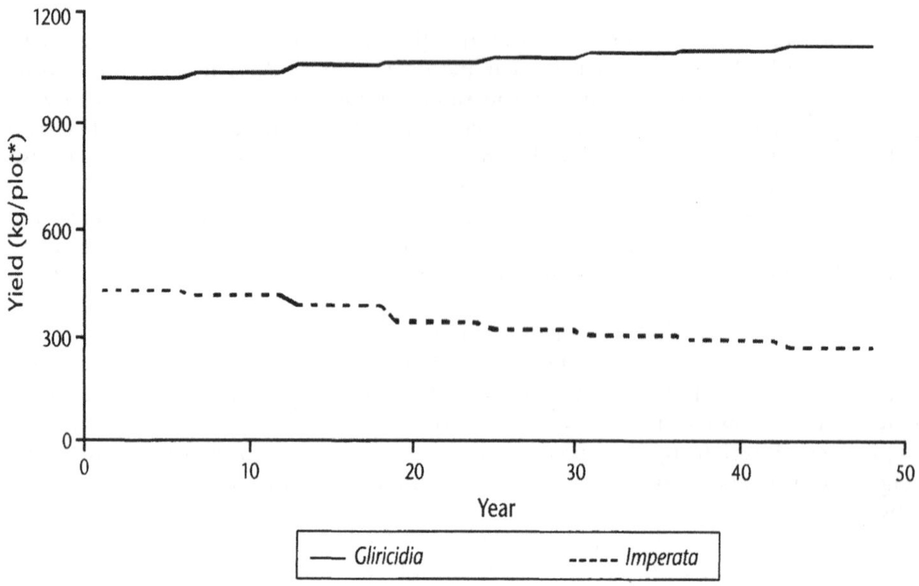

Figure 32-2. Expected Annual Maize Yield for *Gliricidia* and *Imperata* Fallow Systems over an Available Land Area of Two Hectares

Notes: *Annual maize yield is calculated for the whole farm based on a two-hectare plot divided into six equal parcels. Each year, one parcel is cropped and the remaining five are under fallow. Thus, the yield depicted here is kilograms per one-third of a hectare, assuming two maize crops in the year following *Gliricidia* fallow and one maize crop in the year following *Imperata* fallow.

Annual soil erosion for both *Imperata* and *Gliricidia* fallow systems is presented in Figure 32-4. Most soil erosion occurs during the maize cropping years (top two lines in Figure 32-4) rather than during the years of fallow. With the *Imperata* fallow system, annual soil erosion increases over time. Although not depicted here, erosion of soil carbon, nitrogen, and phosphorus is similar to trends for total soil erosion.

Within the *Gliricidia* fallow system, annual soil erosion in maize plots decreases over time. This is a result of the higher levels of soil organic matter associated with mulching. This stabilizes the soil, reducing the impact of rain and surface runoff. Higher soil nutrient levels also increase the production of biomass, which protects and stabilizes the soil. Soil erosion during the fallow periods of both systems is small (bottom two lines of Figure 32-4).

The Transition Period between Imperata *and* Gliricidia *Fallow Systems*

The economic analysis suggests that, in the long run, a *Gliricidia* fallow system is quite profitable. However, for smallholders, the feasibility of the system may depend upon the transition period during which *Imperata* fallow plots are converted to *Gliricidia* fallow plots. Within the system outlined here, one plot would be converted each year, and the complete farm transition would take six years.

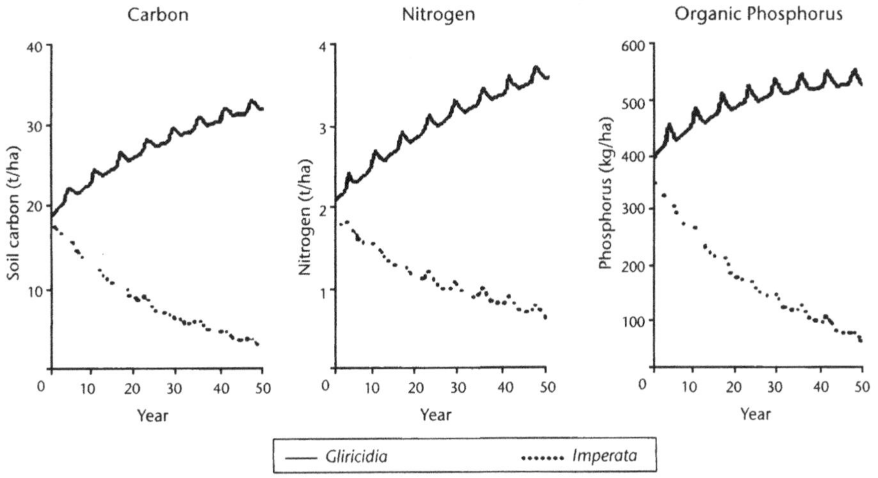

Figure 32-3. Changes in Available Carbon, Nitrogen, and Phosphorus in *Gliricidia* and *Imperata* Fallow Systems

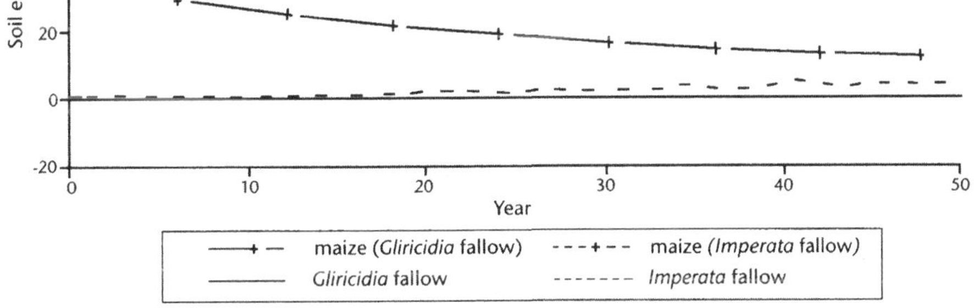

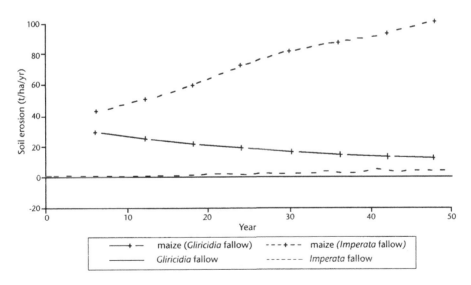

Figure 32-4. Annual Soil Erosion over Time in the *Gliricidia* and *Imperata* Fallow Systems

During the transition period, half of the *Gliricidia* prunings are used as green manure for maize, and the other half are placed as green manure under the *Gliricidia* plantation.[5] This provides for an improvement in maize productivity because, during the transition, the maize plots will not have been preceded by a *Gliricidia* fallow. As the number of *Gliricidia* fallow plots increases, the quantity of prunings available as green manure also increases, as do maize and firewood yields. This situation is slightly different from the full *Gliricidia* fallow system, where all of the prunings are used as mulch under the *Gliricidia* plantation, and the maize does not have green manure applied directly. However, transferring half of the prunings to the maize crop during the transition period immediately increases the economic return from maize because of the improved soil fertility. Without this there would be no increase in maize yield until the seventh year, when the transition to the *Gliricidia* fallow system was complete.

The change in maize yield during the transition period is shown in Figure 32-5. Generally, the maize yield increases steadily over the first three years. By the fourth year, it reaches a plateau, which is approximately equivalent to the yield obtained from the full *Gliricidia* fallow system.

The conversion of one plot each year to *Gliricidia* increases total costs, relative to the costs of maize production only in the former *Imperata* system. It takes time for the investment in *Gliricidia* to be translated into revenue increases, so profitability in the early years of the transition period is lower than that from an *Imperata* fallow system.

The trend observed for maize yield (Figure 32-5) is reflected in the cumulative net present value of the system, represented by the dark, solid line in Figure 32-6. In the first year of transition, the *Gliricidia* fallow system incurs a loss, and in year two it approximately breaks even. By year three, most of the first-year loss is recovered, and after the fourth year, the cumulative net present value of the system has risen above that of the *Imperata* fallow system.

The attractiveness of the *Gliricidia* fallow system is dependent on a smallholder's ability to absorb the loss in the first year, and to accept an income during the first four years of the transition period that is lower than that which might be expected from a traditional *Imperata* fallow system. For subsistence farmers without savings, or with limited capacity to borrow, adoption of the *Gliricidia* fallow system would be difficult, because the ability to survive the first three or four years is critical.

The long-term nature of the *Gliricidia* fallow system, requiring four years to be profitable and six years to complete a fallow crop cycle, also requires that land tenure is secure. Therefore, in order to encourage smallholders to adopt the improved system, government policies providing secure land tenure would be needed, as well as access to credit at reasonable rates. Currently, only those smallholders with secure land tenure and a financial capacity to look beyond immediate time horizons are likely to adopt a *Gliricidia* fallow system.

The availability of labor on a smallholder's farm is relatively high, when compared with the labor required by either the *Gliricidia* or *Imperata* fallow systems. Menz and Grist (1995) estimated labor availability of about 300 person–days per year on a typical smallholder farm. Therefore, although the *Gliricidia* fallow system demands three times the labor of the *Imperata* system—102 person-days compared to 31 respectively, for a two-hectare farm—most smallholders on a farm of that size would be capable of adopting the *Gliricidia* system without placing excessive strain on their labor resources.

---

[5] For example, given 7 metric tonnes of *Gliricidia* prunings per year, in the third year of the transition (and so three plots under *Gliricidia* fallow), each fallow plot would receive 3.5 metric tonnes of prunings and the maize plot would receive 10.5 metric tonnes (3 x 3.5 tonnes) of prunings.

The role of animals in a *Gliricidia* fallow system has been considered in a supplementary preliminary analysis. A forthcoming paper will describe this work in detail, but the results indicate a positive economic role for animals, without significantly compromising the positive environmental effects of the *Gliricidia*.

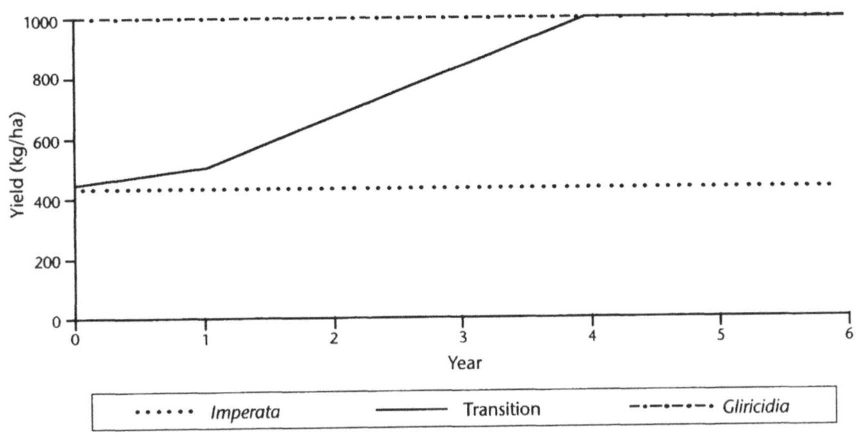

Figure 32-5. Maize Yield during the Six-Year Transition Period to a *Gliricidia* Fallow System

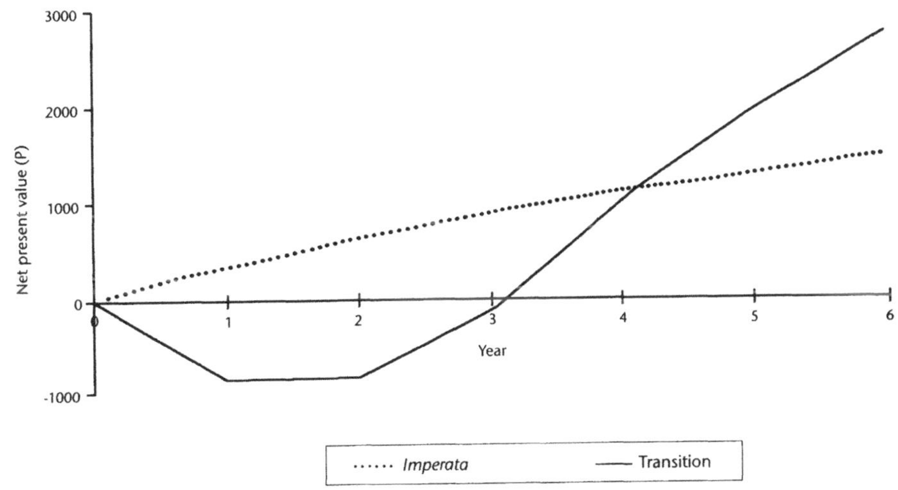

Figure 32-6. Net Present Values during the Transition Period from an *Imperata* Fallow System to a *Gliricidia* Fallow System, Using a Discount Rate of 12%

Sensitivity Analysis

The wage rate and maize price used in this analysis reflect current market values in Claveria. However, because the intention of this analysis was to draw general conclusions on the potential of the *Gliricidia* fallow system on acid soils in Southeast Asia, we have carried out a sensitivity analysis on these prices.

An increase of 50% in the wage rate, to ₱60 per day, or a 50% fall in the maize price, to ₱2.6 per kilogram, would significantly reduce the profit levels of the *Gliricidia* fallow system. However, it would still remain profitable under these substantially less favorable conditions. Given that the *Imperata* fallow system is only marginally profitable at the study levels, any increase in system costs or lowering of revenue would make it unprofitable. Therefore, the choice between the *Gliricidia* and *Imperata* fallow systems is not affected by substantial changes in the price levels of the key inputs and outputs.

Conclusions

The *Gliricidia* fallow system requires no external inputs. It is proposed as an alternative to shifting cultivation—and particularly to *Imperata*-based systems with a short fallow period—for subsistence farmers in upland areas of Southeast Asia. Currently it is not practiced by upland farmers, although there are isolated examples of similar systems. The work reported here should be viewed as technology in the process of development and design, with information about the system still being integrated and evaluated. It was not intended that we should make final recommendations to upland farmers about the immediate relevance of a *Gliricidia* fallow system. The analysis is directed more toward researchers, to indicate that a more intensive investigation is warranted.

However, the analysis does suggest that a *Gliricidia* fallow system has the potential to provide significant improvements to a range of soil biophysical characteristics, compared to an *Imperata* fallow system. The model predicts that the *Gliricidia* fallow system would reduce soil erosion, maintain soil fertility, and enable higher levels of farm outputs. So from environmental and productivity perspectives, a *Gliricidia* fallow system is attractive.

From an economic viewpoint, the *Gliricidia* fallow system also appears attractive. At current costs and prices in Claveria, this analysis indicates that the *Gliricidia* fallow system would have a clear profit advantage over an *Imperata* fallow system. The value of firewood is a major contributor to this result. However, even without a market for firewood, the system is profitable, and more so than an *Imperata* fallow system.

The time taken for smallholders to convert from an *Imperata* system to the new one is important. If they adopt the *Gliricidia* fallow system, it will take smallholders about four years to begin making a profit above that achievable with an *Imperata* fallow system, and they will incur a loss in the first year. Unless smallholders are capable of accepting lower profitability in the first four years, or there is some form of government assistance, they are less likely to adopt the new system. Given the long-term nature of the investment in *Gliricidia*, secure land tenure is also a prerequisite.

This analysis has shown that, in the long term, the *Gliricidia* fallow system has potential for adoption by upland smallholders. Insofar as it has shown the *Gliricidia* system to be potentially attractive from economic, environmental, and productivity perspectives, it has also generated confidence that more substantive research efforts are warranted. The next stage in this process of technology development and design may be the establishment of field trials to refine and validate the prototype system defined here.

The evaluation approach reported in this chapter has merit in being relatively inexpensive, yet powerful. While the model used lacks the sophistication of more process-oriented models, its transportability and ease of use are attractive features. The time frame and cost of this analysis were minute in comparison to what would be involved in field experiments. Furthermore, the modeling approach allows the researcher to exert full "control" over the relevant variables. This factor becomes more important as the number of interesting variables and the time frame of the experiment both increase.

Acknowledgments

This research was undertaken with substantial funding assistance from the Australian Centre for International Agricultural Research (ACIAR) and the Center for International Forestry Research (CIFOR).

References

Agus, F. 1994. Soil Processes and Crop Production under Contour Hedgerow Systems on Sloping Oxisols. Ph.D. dissertation to North Carolina State University, Raleigh, NC.

Anderson, J.H., and J.B. Hardaker. 1979. Economic Analysis in the Design of New Technologies for Small Farmers. In: *Economics and the Design of Small Farmer Technology,* edited by A. Vades, G.M. Scobie, and J.L. Dillon. Ames, IA: Iowa State University Press, 11–29.

Anoka, U.A., I.O. Akobundu, and S.N.C. Okonkwo. 1991. Effects of *Gliricidia sepium* and *Leucaena leucocephala* on Growth and Development of *Imperata cylindrica. Agroforestry Systems* 16, 1–12.

Brady, N.C. 1996. Alternatives to Slash-and-Burn. *Agriculture, Ecosystems and Environment* 58, 3–11.

Castillo, E.T., and F.E. Saipno. 1995. The Current State of Vegetation and Soil Fertility of the Grasslands and Pasturelands of Nueva Ecija, Nueva Vizcaya, and Isabella Provinces. Paper presented to the First National Grassland Congress, Los Baños, Laguna, Philippines.

Conelly, W.T. 1992. Agricultural Intensification in a Philippine Frontier Community: Impact on Labor Efficiency and Farm Diversity. *Human Ecology* 20, 203–223.

Ella, A., C. Jacobsen, W.W. Stur, and G. Blair. 1989. Effect of Plant Density and Cutting Frequency on the Productivity of Four Tree Legumes. *Tropical Grasslands* 23, 28–34.

Eussen, J.H.H., S. Slamet, and D. Soeroto. 1976. Competition between Alang-Alang (*Imperata cylindrica*) and Some Crop Plants. Biotrop Bulletin No. 10. Bogor, Indonesia: BIOTROP (Southeast Asia Ministry of Education Organization Centre for Tropical Biology).

FAO (Food and Agriculture Organization of the United Nations). 1979. *A Provisional Method for Soil Degradation Assessment.* Rome: FAO.

Garrity, D.P. 1993. Sustainable Land Use Systems for Sloping Uplands in Southeast Asia. In: *Technologies for Sustainable Agriculture in the Tropics,* edited by J. Ragland and R. Lal. Madison, WI: American Society of Agronomy.

Grist, P., and K. Menz. 1996. Burning in an *Imperata* Fallow/Upland Rice Farming System. *Imperata* Project Paper 1996/7. Canberra: Centre for Resource and Environmental Studies, Australian National University.

Gunasena, H.P.M., and J. van der Heide. 1989. Integration of Tree Legumes in Cropping Systems in the Intermediate Zone of Sri Lanka. In: *Nutrient Management for Food Crop Production in Tropical Farming Systems,* 327–330.

Limbaga, C.M. 1993. Evaluation of Hilly Lands for Appropriate Conservation-oriented Farming Systems. Ph.D. dissertation to the University of the Philippines at Los Baños, Laguna, Philippines.

Lowry, J.B., J.J. Petheram, and B. Tangendjaja. 1992. Plants Fed to Village Ruminants in Indonesia. ACIAR Technical Report No. 22. Canberra: Australian Centre for International Agricultural Research.

MacDicken, K.G. 1990. Agroforestry Management in the Humid Tropics. In: *Agroforestry: Classification and Management,* edited by K.G. MacDicken and N.T. Vergara. New York: John Wiley and Sons.

Maclean, R.H., J.A. Litsinger, K. Moody, and A.K. Watson. 1992. The Impact of Alley Cropping *Gliricidia sepium* and *Cassia spectabilis* on Upland Rice and Maize Production. *Agroforestry Systems* 20, 213–228.

Medella, E.M., C.M. Del Rosario, V.S. Pineda, R.G. Querbin, and E.S. Tan. 1990. Re-estimation of Shadow Prices for the Philippines. Working Paper 90-16. Philippine Institute of Development Studies.

Menz, K., and P. Grist. 1995. Shading *Imperata* with Rubber. *Imperata* Project Paper 1995/4. Canberra: Centre for Resource and Environmental Studies, Australian National University.

———. 1996. Changing Fallow Length in an *Imperata*/Upland Rice Farming System. *Imperata* Project Paper 1996/6. Canberra: Centre for Resource and Environmental Studies, Australian National University.

Menz, K.M., and H.C. Knipscheer. 1981. The Location Specificity Problem in Farming System Research. *Agricultural Systems* 7, 95–103.

Nelson, R., P. Grist, K. Menz, E. Paningbatan, and M. Mamicpic. 1996a. A Cost-Benefit Analysis of Hedgerow Intercropping in the Philippine Uplands using SCUAF. *Imperata* Project Paper 1996/2. Canberra: Centre for Resource and Environmental Studies, Australian National University.

———, R.A. Cramb, K. Menz, and M. Mamicpic. 1996b. Bioeconomic Modeling of Alternative Forms of Hedgerow Intercropping in the Philippine Uplands using SCUAF. *Imperata* Project Paper 1996/9. Canberra: Centre for Resource and Environmental Studies, Australian National University.

———, and M. Mamicpic. 1996c. Costs and Returns of Hedgerow Intercropping and Open Field Maize Farming in the Philippine Uplands. Working Paper No. 11. Los Baños, Laguna, Philippines: SEARCA-UQ Uplands Research Project.

Panjaitan, M., W.W. Stur, and R. Jessop. 1993. Growth of Forage Tree Legumes at Four Agro-climatic Sites in Indonesia. *Journal of Agricultural Science*, Cambridge 120, 311–317.

Sajise, P.E. 1980. Alang-Alang (*Imperata cylindrica*) and Upland Agriculture. In: BIOTROP Special Publication No. 5, Proceedings of BIOTROP workshop on Alang-alang. Bogor, Indonesia: BIOTROP (Southeast Asia Ministry of Education Organization Centre for Tropical Biology).

Sanchez, P.A. 1995. Science in Agroforestry. *Agroforestry Systems* 30, 5–55.

Simons, A.J., and J.L. Stewart. 1994. *Gliricidia sepium,* A Multipurpose Forage Tree Legume. In: *Forage Tree Legumes in Tropical Agriculture*, edited by R.C. Gutteridge and H.M. Shelton. Wallingford, Oxon: CAB International.

Vermuelen, S.J., P. Womer, B.M. Campbell, W. Kamukondiwa, M.J. Swift, P.G.H. Frost, C. Chivaura, H.K. Murwira, F. Mutambanengwe, and P. Nyathi. 1993. Use of the SCUAF Model to Simulate Natural Miombo Woodland and Maize Monoculture Ecosystems in Zimbabwe. *Agroforestry Systems* 22, 259–271.

Young, A., P. Muraya, and C. Smith. 1996. Soil Changes under Agroforestry (SCUAF). Computer program Version 4. Nairobi, Kenya: ICRAF.

Chapter 33

Pruned-Tree Hedgerow Fallow Systems in Mindanao, the Philippines

Peter D. Suson, Dennis P. Garrity, and Rodel D. Lasco

Alley cropping was a traditional practice of the Ibu tribe of eastern Nigeria (Kang 1997), who developed the concept in response to accelerating pressures on traditional shifting agriculture (Kang et al. 1990). It was described as a stable alternative to shifting cultivation in the sense that farmers no longer needed to shift from one cultivated field to another. Alley cropping combined both the fallow and cultivation phases. The key was the planting of N-fixing woody perennials in rows that followed the contours of the land (see color plate 48). Nutrient recycling was improved, greater quantities of nitrogen were biologically fixed, and nutrients that were once unavailable in deeper soil layers were brought to the surface by a nutrient pumping effect. Nutrient losses were reduced by recapture of leached nutrients or through reduced soil erosion (Kang and Wilson 1987; Raintree and Warner 1986).

However, while alley cropping has proven beneficial under some conditions, more and more farmers who adopted the practice have ceased cropping and have left their hedgerowed fields fallow. This practice has been observed in many parts of the Philippines, particularly in Claveria, Mindanao (Mercado 1997); Aklan, Iloilo; Matalom, Leyte (Fujisaka and Cenas 1993); the Bicol region (Payonga 1997); and Jala-Jala, Rizal (Gomez 1997). Similar observations are reported in Zambia (Franzel unpubl.) and Cameroon (ICRAF 1996). Apparently, there are constraints to the maintenance of contour hedgerow systems that have not yet been adequately understood. As a consequence, two questions arise: Why do farmers fallow their alley farms when alley cropping was supposedly designed to sustain crop production? Are there prospective advantages in doing so?

Our observations led us to hypothesize that farmers were fallowing their hedgerow systems to counteract two constraints commonly faced by shifting cultivators: an inevitable decline in annual crop yields, and an increase in weed pressure. The objectives of this study were to determine the farmers' rationale for fallowing their alley farms; to compare the crop yield performance from fallowed fields with hedgerows versus conventionally fallowed fields without hedgerows, but left fallow for the same period; and to analyze the overall systems productivity and profitability of the two fallow systems.

Peter D. Suson, Research and Planning Officer, Hope for Change, Inc., 9-2 A.F. Celdran St., Corpus Cristi Village, Tubod, Iligan City 9200, Philippines; Dennis P. Garrity, Director General, World Agroforestry Centre (ICRAF), United Nations Avenue, Gigiri, P.O. Box 30677, Nairobi, Kenya; Rodel D. Lasco, Philippine Programme Coordinator, World Agroforestry Centre (ICRAF), 2F CFNR, University of the Philippines, College, Laguna, the Philippines.

The Study Site

The study was conducted at Claveria, Misamis Oriental, in the northern part of the island of Mindanao, in the Philippines. The nearest urban center to Claveria is Cagayan de Oro, about 40 km away. The elevation ranges from 150 to 1,000 m above sea level (asl), and slopes ranging from 8% to 60% dominate the landscape. The rainy season is from April or May to December. On average, rainfall exceeds 200 mm per month for five to six months (Garrity and Agustin 1987). A dry spell usually occurs for about two weeks in August or September. The soils are deep, mainly clay, and strongly acidic, with a pH from 4.3 to 5.2. They are classified as Ultic Haplorthox (Garrity and Agustin 1987).

Methodology

Garrity and Agustin (1987) analyzed changes in land use in the area over the past 100 years. The area was opened from forest during the first half of the 20th century and has since been dominated by *Imperata cylindrica* grasslands. As is typical in large areas of the Philippine uplands, farming in these grasslands has been predominantly based on slash-and-burn systems, with annual crops grown mainly for subsistence. Recently there has been a trend toward higher-value crops, such as coffee and fast-growing timber. However, the dominant cropping pattern is the production of two crops of maize, the first planted in April or May and the second in August or September. Double cropping of maize on the sloping land results in the annual loss of between 50 and 200 tonnes of soil per ha (Garrity 1993). Research and extension activities on contour hedgerow systems began vigorously in the mid-1980s (Fujisaka and Cenas 1993), and scores of farmers established contour hedgerow systems using either leguminous trees or forage grasses. Research trials were superimposed on some of these fields. However, rather than sustaining permanent cropping systems, most of the fields with hedgerow systems using pruned trees had been fallowed within 10 years, and cropping abandoned. These fallowed hedgerow fields provided the basis for this research.

Survey

Five groups of people were interviewed. Group 1 was composed of farmers who had fallowed their alley farms; Group 2 consisted of farmers who had continuously cultivated their alley farms; Group 3 was farmers who had marketed firewood from their hedgerows; Group 4 consisted of people who purchased firewood from these farmers, including bakers and retailers; and Group 5 was composed of farmers who collected wood from their hedgerows as a source of fuel for the family. Interviews with Groups 1 and 2 enabled us to find out why farmers fallowed their hedgerowed fields. Interviews with the other groups helped to determine the feasibility of marketing firewood as an economic product of fallowed hedgerows. Table 33-1 shows the number of those interviewed per grouping. The survey was aided by a prepared questionnaire.

Experiments

Farmers' fields were available in Claveria on which there were large plots with contour hedgerows alongside plots that had been cultivated conventionally as open-field controls. These were fields in which previous experiments had been conducted (Basri et al. 1990; ICRAF 1996). These adjacent plots had a similar history of cropping and fallow length. All of them had been fallowed for at least one year. Four fields were selected. Each contained one replicate of both the treatments (With Hedgerows and No Hedgerows). One site had two replicates of both treatments. Table 33-2 presents background information on the sites. The average period that the With Hedgerow (WH) treatments had been fallowed prior to the start of the experiment was 4.1 years. The average period that the No Hedgerow (NH) treatments had been fallowed prior to the experiment was 4.5 years.

Treatments were assigned using a split-plot design. The NH and WH treatments were main plot treatments. The time of opening of the fallow was the subplot treatment: F0 = opened during 1995, and F1 = opened during 1996. The experiment was still ongoing at the time of writing. For the purposes of this chapter, a combined analysis was used for maize yield that included all cropping seasons and treatments. Yield data were from three crops: first crop, late wet season, 1995; second crop, early wet season, 1996; and third crop, late wet season, 1996. The statistical package used was IRRISTAT 4.0 for Windows.

The cropping pattern was maize-maize using cultivar "Pioneer 3246," a hybrid yellow corn. According to the Pioneer Corn Hybrid Seed Catalogue, this variety was capable of yielding up to 7.7 metric tonnes/ha under favorable conditions. The cultural practices followed were described in Nelson et al. 1996. Several plowing and harrowing operations were conducted for the first and second crops. During the opening of the fallow plots, all of the fallow biomass was returned to the soil. The exception was the woody biomass from the hedgerows in the WH treatment, which was removed from the field with the assumption that it would be used as firewood. No external fertilizer or lime was applied. Tree hedgerow pruning was done immediately after the second plowing. Pruning was usually done once per crop (see Table 33-3), although frequent pruning was not necessary because the trees were old and were no longer coppicing vigorously. The number of trees in the hedgerows had also decreased because of natural attrition over the years.

As hedgerows mature, their production of biomass slowly declines, especially on acid soils (Sanchez 1995). In our case, the average hedgerow age was seven years. According to Nelson et al. (1996), *Gliricidia sepium* hedgerows grow old after three to five years. *Senna spectabilis* also dies back slowly, probably as a result of poor coppicing ability after frequent pruning (Kang 1997), or because of exhaustion of native soil nitrogen (Maclean et al. 1992). Therefore, the hedgerow vegetation at the start of the experiment was a mixture of aged trees and grass strips.

Table 33-1. Composition of the Hedgerow Farmer Survey Groups

Groups	*Number of Respondents Interviewed*	*Remarks*
Fallowed their hedgerows	30	The total number of farmers identified as having fallowed their alley farms was 33.
Continuously cultivated hedgerow fields	6	This group included all the farmers known to be continuously cultivating alley farms in Claveria.
Marketed firewood from hedgerows	30	Respondents came from only one village, Sta. Cruz, in Claveria, which was known for its marketing of firewood.
Purchased firewood from hedgerow farms	6	A total of only 10 suppliers of firewood were known.
Collected firewood from hedgerows for family use	30	Limited to only one subvillage, Lombagohon, in Claveria, which was one of the few places still having difficulty in collecting firewood.

Table 33-2. Details of the Experimental Fields

Cooperator	*Location (village)*	*Treatment*	*Years in Fallow*	*Hedgerow Species*
Codilla	Cabacungan	With Hedgerow	10	*Gliricidia sepium*
	Cabacungan	No Hedgerow	10	
Cuizon	Patrocenio	With Hedgerow	3	*Gliricidia sepium*
	Patrocenio	No Hedgerow	3	
Pabling	Patrocenio	With Hedgerow	3	*Senna spectabilis*
	Patrocenio	No Hedgerow	4	
Pabling	Patrocenio	With Hedgerow	3	*Senna spectabilis*
	Patrocenio	No Hedgerow	3.5	
Rene	Patrocenio	With Hedgerow	1.5	*Senna spectabilis*
	Patrocenio	No Hedgerow	2	

Table 33-3. Pruning Data

Crop	*Prunings per Crop Season*	*n*	*Average (t/ha)*	*Standard Error*
First	Twice	20	4.70	1.33
		15	0.36	0.04
Second	Once	20	1.55	0.81
Third	Once	19	1.50	0.21

Only tree prunings were applied to the alley, as we were specifically interested in the effect of tree prunings. The amount of prunings applied is shown in Table 33-3. Spot prunings were made on a few vigorous hedgerow trees, but these were not applied to the alley because they might have caused variation within the replicates. Instead, these were applied to the base of the hedgerows.

The maize crop was harvested 110 days after planting. Harvest areas were within borders one meter wide on the outside of each treatment plot. All rows across the alleyway were part of the harvest area. Grain yields are reported at 14% moisture content. The grain yield of the WH treatment is reported on field area basis, that is, the field size was reduced by 12.2% to account for the average area occupied by the hedgerows. All maize residues were returned to the soil prior to the next crop.

During the clearing operation, and for the subsequent prunings, the firewood production from the WH plots was determined by sampling a hedgerow length of 4 m. Firewood yield was determined on a dry weight basis. For economic analysis, the volume of firewood was measured by the number of bundles. It was bundled to conform with other firewood sold in Claveria stores, which averaged 29 pieces with circumference of 16 cm and a length of 64 cm. Table 33-4 shows the number of firewood bundles produced per cropping.

An assessment was made of productivity and profitability, and an analysis of sustainability criteria is planned as soon as these data are complete. Productivity in the WH treatment was composed of firewood from tree hedgerows and maize grain yield. For the NH treatment, only maize grain was considered. The fallow biomass was not valued as it was returned to the soil. Costs and returns were computed for both treatments. The amount of labor in all activities, from opening up the fallow to weeding, was timed. A limitation of this method is that extrapolation from "per plot" to "per hectare" tends to overestimate the labor value. Nelson et al. (1996) showed that the amount of labor expended is not correlated with farm size in Claveria. The costs of replanting, pruning for the second and third crops, harvesting, and postproduction were obtained from Nelson et al. (1996), who give production and postproduction estimates from the same locality that fall within the range conducted by other researchers. We monitored labor costs in pruning for the second and third crops. However, our values were low when compared with existing literature (Nelson et al. 1996), and we felt that we needed to verify our values by monitoring pruning labor on a larger hedgerow area. Unit costs were taken from rates prevailing in Claveria in 1996.

Results and Discussion

Most of the hedgerow systems surveyed for this chapter were established between 1987 and 1991, at the time when the classic alley cropping system was being tested in Claveria. Consequently, there were contour hedgerow systems involving four tree species left fallowed in the area. The dominant species was *Gliricidia sepium* (see Table 33-5). After they had fallowed their hedgerow systems, the farmers had left the woody perennials unpruned. When tree hedgerows are well maintained, however, their growth forms a closed, or nearly closed, canopy over the alleyways. Among the farmers interviewed, the average age of the fallow was 4.3 years, with a range of 6 months to 10 years. The majority of the farms had a fallow age falling between one and five years (see Table 33-6). It should be noted, however, that the data show the length of time the fields had been fallow at the time of the survey and not necessarily the total length of the fallow, which may have extended years longer, or indefinitely. Only 23% of the farmers had recultivated their hedgerowed farms after having left them lie fallow. The rest were uncertain as to when they would break the fallow.

Farmers gave many reasons for fallowing their hedgerow farms (see Table 33-7). The main reason was a complex interplay between land, labor, and soil fertility. It was the interaction of these factors, and not necessarily the effect of a single factor, that triggered farmers to fallow their alley farms.

Table 33-4. Volume of Firewood Produced

Season	*Average Number of Bundles*	*n*	*95% Confidence Interval*
First cropping	10,157	20	3,488 to 16,827
Second cropping	1,346	20	743 to 1,346
Third cropping	0	15	n/a

Table 33-5. Species of Hedgerow Fallows on the Surveyed Farms

Hedgerow Species	*Number* [a]	%
Gliricidia sepium	23	55
Flemingia congesta	5	12
Morus alba	5	12
Senna spectabilis	4	10
Gmelina arborea	2	5
Paraserianthes falcataria	1	2
Tithonia diversifolia	1	2
G. sepium and *S. spectabilis*	1	2
Total	42	100

Note: [a] This is not based on the number of farmers interviewed but on the total number of plots with different hedgerow species. Some farmers had more than one hedgerow species.

Table 33-6. Duration of the Fallow among 30 Surveyed Farmers with Fallowed Hedgerows

	Years									
Fallows	*1*	*2*	*3*	*4*	*5*	*6*	*7*	*8*	*9*	*10*
Number	5	8	6	2	6	2	0	0	0	1
Percentage	16.67	26.67	20	6.67	20	6.67	0	0	0	3.33

Table 33-7. Reasons for Fallowing Hedgerowed Fields

Response	*Number* [a]	*%*
Large area–low labor–soil infertility complex	23	52.5
Sickness	4	9.0
Migration	3	7.0
Lack of material inputs	3	7.0
Used tree hedgerows for housing materials	3	7.0
Others		
—Because of trees growing in the alleyways	2	4.5
—For charcoal production	2	4.5
—Mulberry hedgerow contaminated with pesticide due to spray drift	2	4.5
—Minimizes pest infestation	1	2.0
—Good for watermelon production	1	2.0
—Displacement/removal	1	2.0
Total	44	100.0

Notes: [a] Number equals the number of answers and not the number of farmers interviewed. Some farmers had more than one response.

Beginning with farm size as the first factor: among the 30 farmers interviewed, the average farm size was 5.36 ha. Eighty-four percent of them fell within the range of 1.0 to 7.0 ha. This is higher than the national and provincial averages for farm size holdings, which are 3.0 ha and 4.0 ha respectively (Bureau of Agricultural Statistics 1996). The degree of land intensification depends upon the ratio of labor to land. The interview data show that although family size was large, and averaged eight members, the availability of labor for farm work was low. Much of the labor came only from the husband and wife. Hired labor was occasional. Only 21% of the farmers said they got part-time help from their children because they were either at school or were married. Nevertheless, 66% of the children were still dependent on their parents. The farmers said that 25% of their total farm area was left fallowed with every crop. This indicated a relatively low level of available labor for the size of farm.

With regard to the soil fertility factor, farmers said that the tree-based contour hedgerows tended to be on the most degraded portion of their farms. Although they recognized that soil erosion was reduced, they claimed that the level of soil fertility was not substantially improved by the hedgerow system. Such observations were actually validated by ICRAF field trial data (ICRAF 1997), in which it was shown that, after five years, the effect of prunings did not always significantly improve crop yields. Although trees can increase the availability of nutrients by pumping nutrients from the subsoil, the acidic soils of Claveria have only low subsoil nutrient reserves (Garrity 1993). Moreover, even N-fixation can be affected under marginal—and particularly acid—soils because the rhizobial bacteria are dependent upon phosphorus in the production of ATP needed for N-fixation (De la Cruz 1997). Even the quantity of prunings may be affected by the degree of aluminum saturation found in acid soils (Garrity 1994). According to the farmers, under poor soil

Any photographs that are not accredited were taken by the Editor.

1. Shifting cultivators in the Kassam pass area of the PNG highlands (1,200 m asl) prepare forest for cropping *Xanthosoma* taro, bananas, and a few other shade-tolerant food crops by slashing the undergrowth, smaller trees, shrubs, and vines. Less than 25% of larger trees are felled to allow sufficient sunlight to reach the forest floor (Bino, Chapter 60).

2. Valued species are often protected through iterative swidden cycles (Schmidt-Vogt, Chapter 4). In Northeastern India, *Schima wallichii* trees (shown) are often retained in *jhum* fields (swiddens) by selective felling. Farmers highly value their wood for construction and making veneer and firewood; the bark is used for dyeing and processing leather. The trees also reportedly have several medicinal applications.

3. Bhutanese farmers often maintain *Ficus* spp. (mostly *F. roxburghii* and *F. cunia*) and other preferred fodder trees on their *tseri* (swidden) land (Dukpa et al., Chapter 59). Mongar, Central Bhutan.

4. In other cases, such as exemplified by these rattan wildlings in an upland rice swidden in Luang Prabang, Lao PDR, farmers selectively retain desirable species that germinate during the cropping phase. Given market incentives, this has the potential to intensify into more actively planted and managed rattan-based fallows, such as those in Yunnan (Xu, Chapter 56) and Kalimantan (Belcher, Chapter 64; Sasaki, Chapter 38). (See also color plate 52)

5. Broomgrass (*Thysanolaena latifolia* L.) is harvested from swidden fallows, weighed and sold by many upland minorities, such as these Muser (Lao Sung) women in Bokeo, Northern Lao P.D.R. Broomgrass grows best in fields that have been burned, but reportedly causes soil degradation.

6–7. *Imperata cylindrica* is widely viewed as a problem weed. However, Potter and Lee challenge this thinking in Chapter 11 with case studies in which market demand has transformed *Imperata* into a carefully tended fallow "crop", processed for roof thatching (inset). Gianyar, Bali, Indonesia. (Photos: Justin Lee)

8. Harvesting the young shoots of *Pandanus* sp. in a fallow area in Kelabit Highlands, Sarawak. (Photo: Ole Mertz)

9. Picking *Pseuderanthemum borneense* planted in an old rubber garden in Nanga Sumpa Village, Sri Aman Division, Sarawak (Mertz, Chapter 7). (Photo: Ole Mertz)

10–11. Harvesting sago worms (inset) from an old trunk of sago palm previously felled on fallow land in Nanga Sumpa Village, Sri Aman Division, Sarawak. (Photos: Ole Mertz)

12. Tending *Diplazium esculentum* in a wild vegetable garden in Manup Baroh Village, Sri Aman Division, Sarawak (Mertz, Chapter 7). (Photo: Ole Mertz)

13. Swidden fallows are favored locations for hunting and gathering activities. Wildlife often come in search of crop remnants and are later attracted to the protective cover of dense fallow regrowth. Mushrooms, a wide diversity of wild food plants, and other useful products are harvested from fallows, depending on the season and stage of fallow succession (Burgers, Chapter 8; Mertz, Chapter 7; Tangan, Chapter 9; Tayanin, Chapter 6). Many of the items offered at this fresh market in Savannakhet, Lao P.D.R, are fallow products.

14. At higher elevations where forest regeneration is slower, swidden farmers have developed careful ways to maximize the benefits of available biomass. As described by Ramakrishnan in Chapter 21, Khasis farmers near Shillong, in Meghalaya, arrange slash in parallel rows running downslope and allow it to dry. Soil is later pulled up on top of the slash, forming ridges that alternate with furrows of compacted soil (shown). By burning the slash under the soil, the fire is slower, less C is lost through volatilization, and fewer nutrients leached. Crops are then planted on these nutrient-enriched ridges. See Dukpa et al., Chapter 59, for an account of the similar *pangshing* system in central Bhutan. Near Shillong, Meghalaya.

15. *Chromolaena odorata* is the most widely recognized of many exotic Asteraceae that have expanded across Southeast Asia during the last century. Its notoriety as an invasive weed in pasture land and tree plantations has earned it the attention of numerous symposia, a newsletter, and projects aimed at its eradication. However, Roder et al., Chapter 14, and Ty, Chapter 55, add to a growing pool of literature verifying its value as an effective fallow species. Pakhasukjai Village, Chiang Rai, Thailand.

16. *Austroeupatorium inulaefolium* appears to play a similar ecological role in fallows at higher altitudes (200 m to 1,800 m asl). This Minangkabau couple in West Sumatra, Indonesia, is clearing a seven-year-old *A. inulaefolium* fallow. Note that despite the dense forest cover on the upper slope, very few pioneer trees have penetrated through the dense *A. inulaefolium* thicket. This suggests that its aggressive nature may be delaying regeneration of secondary forest (Cairns, Chapter 15).

17. A farmer slashes *Calopogonium mucunoides* undergrowth in preparation to reopen a *Tithonia diversifolia* fallow in Mindanao, the Philippines. This stoloniferous legume/shrub succession is completely spontaneous and may offer a symbiotic combination of N-fixation and nutrient scavenging. It is protective of the soil, mimics the tight nutrient cycling of natural forests, and enables a high photosynthetic efficiency (Daguitan and Tauli, Chapter 57). Lantapan, Bukidnon, the Philippines.

18. Other farmers introduce vegetative cuttings of *Tithonia diversifolia* into *Imperata* grasslands, using it as a biological tool to smother out the *Imperata* and bring the land back into productive cultivation. Farmers say that the *Imperata* is choked out within the first year and, by the end of the second, the land can be reopened and a successful crop grown without fertilizer inputs. Lantapan, Bukidnon, Philippines.

19. Another exotic shrub, *Tecoma stans,* originally native to the Latin American Andes, has proven its ability to thrive and produce significant fallow biomass in the rocky coralline soils and harsh, dry climate of West Timor, Indonesia (Djogo et al., Chapter 17). The understory is clear of weeds and shows a significant buildup of leaf litter. Kecamatan Maulafa, Kotamadya Kupang, East Nusa Tenggara.

20. Yet another example of an aggressive invasion by an exotic shrub from Central America is the case of *Piper aduncum* L. in Papua New Guinea. It has expanded rapidly in the last 20 to 30 years and now dominates many swidden fallows, often in monospecific stands. Its dominance in the soil seedbank and fast growth appear to explain its success (Hartemink, Chapter 16). Nearby Lae, in Morobe Province, PNG.

21–22. In Mon District of northern Nagaland, seeds of *Macaranga indica* are scarified by the burning operation as fallows are reopened for cultivation. The germinating *Macaranga* seedlings are then nurtured in tandem with the rice crop (main photo). They go on to form near monospecific fallows, as seen in this seven-year- old stand (inset). Near Wakching Village, Mon District, Nagaland.

23. *Mimosa invisa* that self-propagates in maize fields may be managed as a green manure intercrop during the cropping phase, and then left to dominate the subsequent fallow succession. As shown in this maize field in Bukidnon, the Philippines, it provides continuous soil cover after harvest, suppresses invasive grasses, and generates large quantities of leguminous biomass. Farmers at Leyte, in the Philippines, further value thorny *M. invisa* for its ability to discourage invasion by free-roaming livestock, thereby solving problems of biomass removal, overgrazing, and soil compaction (Balbarino et al., Chapter 18). However, many farmers remain steadfastly opposed to *Mimosa*.

24. Land use pressures and market opportunities have prompted some swiddenists to integrate herbaceous legumes into their cropping patterns, often as a dry season fallow. This Akha farmer in northern Thailand says integrating peanuts (*Arachis hypogaea*) into his crop rotation has brought soil improvement and additional income.

25. Viny legumes (*Vigna unguiculata, V. umbellata,* and *Lablab purpureus*) are relay-planted into maize by Lisu farmers in northern Thailand, about one month before maize harvest. Ongprasert and Prinz (Chapter 23) describe the system as accelerated seasonal fallow management in intensified shifting cultivation. (Photo: Klaus Prinz)

26. Farmers in upland northern Vietnam also relay-plant *Phaseolus calcaratus* Roxb. (syn. *Vigna umbellata*) into maize (Hao et al., Chapter 22). After harvest, the beans climb up the stalks and form a protective rainy season ground cover. Their extensive roots (shown) also play a valuable role in stabilizing soil on steep slopes. (Photo: Nguyen Tuan Hao)

27. High population densities and a fondness for tofu probably contributed to a traditional practice in China of inter-cropping soybean (*Glycine max*) with maize. The legume-cereal combination boosts total productivity of the land and is suitable for smallholdings without farm machinery. Baoshan Prefecture, Yunnan.

28. In Thailand, a Karen farmer in Kanchanaburi Province harvests beans from his swidden plot. Although lentils are central to South Asian cuisine, beans are not a staple to Southeast Asian diets. This may have contributed to their under-utilization as valuable "fallow crops." (Photo: Payong Srithong)

29. *Pachyrhizus erosus* (Leguminosae), an herbaceous climber often found in disturbed areas, is reported to grow wildly on fallowed lands in northern Vietnam. Its vines have value as a green manure and the fleshy tubers are marketed as a vegetable. The genus, widely known as "yam bean" or "potato bean," originated in the Neotropics but has become naturalized throughout Southeast Asia. Market on the outskirts of Hanoi.

30. Farmer strategies may combine useful elements from several fallow management typologies. This young rubber plantation (agroforest) in Palembang, South Sumatra, is undersown with a cover crop of *Pueraria javanica,* an herbaceous legume. Note the already established rubber plantations in the background. (Photo: Meine van Noordwijk)

31. Although appearing like a carefully managed green manure crop, this stand of *Mucuna utilis* is completely spontaneous, providing benefits of accelerated soil rejuvenation and weed control. The drawbacks of this aggressive fallow species are that it becomes a weed problem when the land is reopened for cultivation, and it suppresses regeneration of secondary forests. Near Myitkyina in Kachin State, Myanmar.

32. The Amarasi subdistrict, West Timor, is dominated by *Leucaena leucocephala* fallows, from which cropping fields are cleared on a rotational basis (Piggin, Chapter 24). Here, a mature *Leucaena* fallow has been cut and burnt and is ready for planting. Binel, South Central Timor, Nusa Tenggara Timor, Indonesia. (Photo: Peter Kerridge)

33. In addition to soil rejuvenation, the *Leucaena* is cut and carried as fodder for tethered or penned cattle, making a critical contribution to farm incomes. The Amarasi system provides a promising model for the intensification of fallow management in tandem with livestock husbandry. Amarasi, Nusa Tenggara Timor, Indonesia. (Photo: Colin Piggin)

34. Another system sees *Leucaena* planted in hedgerows on steeper slopes, with crops occupying th alleys between them. The hedgerows contribute nutrients from *Leucaena's* N-fixation and from mulche prunings, as well as stabilizing sloping soils, preventing erosion, and building up natural terrace Watublapi, Sikka, Flores, Nusa Tenggara Timor, Indonesia. (Photo: Colin Piggin)

35. The sloping uplands of Naalad, in the Philippines, were deforested during the Spanish colonial era According to oral history, farmers began to rehabilitate the anthropogenic grasslands in the mid- 1800 by broadcasting *Leucaena leucocephala* in swidden fallows, smothering out *Imperata* and fixing atmos pheric N. Soil conservation benefits were later added by erecting 30 cm long *Leucaena* stakes at regula intervals along the contours and piling smaller branches against the stakes, resulting in fascine-lik structures (Lasco, Chapter 27). In addition to enhancing fallow functions, the *Leucaena* is harvested fo firewood and is sold in nearby Cebu City.

36. Upland rice cultivation in a swidden field opened from a *Leucaena* fallow in Sto. Tomas, Occidental Mindoro, the Philippines. Rice is cultivated for one season, followed by two to five years of *Leucaena* fallow, after which upland rice is again cultivated. (Photo: Ken MacDicken)

37. Originally brought to West Timor by the Dutch as a ground cover for their teak plantations, *Acacia villosa* has since escaped and now often dominates fallows in the Camplong area. Shortly after the maize harvest, *A. villosa* rapidly regenerates to form a leguminous ground cover.

38. *Schleinitzia novo-guineensis* is a fast-growing leguminous tree commonly found in fallow vegetatior on a number of small islands in Milne Bay Province, Papua New Guinea. Believing that it improves soi fertility, villagers protect self-sown seedlings or actively transplant them into yam gardens (Bourke Chapter 31). (Photo: Michael Bourke)

39. The ancient Naga practice of managing alders within their swidden plots may have wide replicatior potential across the Himalayan foothills where *Alnus nepalensis* is endemic (Cairns et al., Chapter 30) This three-year-old fallow will soon be pollarded in preparation for the cropping phase. Even thougł the trees are widely spaced so as not to interfere with cropping, each alder stump provides an elevatec platform from which five to six coppices are allowed to grow to form a full fallow canopy. Khonom Village, Kohima District, Nagaland.

40. In China's Yunnan Province, many ethnic minority groups also manage *Alnus nepalensis* as an improved fallow (Guo et al., Chapter 29). However, shifting cultivators in Yunnan use a more sequential system in which the trees are completely cleared when reopening the fallow. After cropping, an alder forest is re-established either through natural regeneration or intentional planting. Tengzhong County, Baoshan Prefecture, Yunnan.

41. *Casuarina oligodon* landscapes are a common sight in the highlands of Papua New Guinea, where an estimated 1.3 million people plant *Casuarina* trees in their fallows (Bourke, Chapter 31). Farmers begin by collecting wild seedlings and transplanting them into their sweet potato gardens toward the end of the cropping phase. Once established, the *Casuarina* self-seeds and germinates whenever farmers disturb the site by clearing the fallow in preparation for planting. Near Chuave district of Simbu Province, PNG.

42. When reopening *Casuarina* fallows, several approaches may be taken, depending on how the wood will be used. Trees cut at waist height are used for fencing needs. Alternatively, the trees may be killed by ring barking and eventually harvested for firewood, or, if there is no immediate need for the wood, the side branches are often pruned back heavily (shown) to return biomass to the soil and reduce shading. The trees are then maintained through successive swidden cycles. Simbu Province, PNG highlands.

43. Shifting cultivators in Luang Prabang Province of Lao P.D.R. have responded to rising timber prices by converting increasing dryland areas into teak (*Tectona grandis*) plantations (Hansen et al., Chapter 34). This has raised concerns that the best agricultural land may be tied up in the long term. Most farmers also have difficulty waiting 20 or 30 years before harvest, and are tempted to sell either the land or the harvest rights to urban speculators.

44. Many farmers in China found that after the Green Revolution they were able to meet all their food needs from intensive cropping of valley bottoms. No longer needing their dryland fields to grow food, they turned to growing valuable trees through taungya planting. Although smallholder tree cropping of *Cunninghamia lanceolata* has a long history in China (Menzies and Tapp, Chapter 35), it has expanded in scale in recent years and, in Tengzhong county of Baoshan prefecture, Yunnan Province, *C. lanceolata* (shown) and *Taiwania flousiana* have displaced many of the traditional alder fallows, such as that illustrated in color plate 40.

45. Upland farmers have a comparative advantage in producing tree products along with ruminant livestock, and both provide commodities in high demand. This suggests that silvipastoral systems, such as this Philippine example of grazing cattle under a *Gmelina arborea* fallow crop (Magcale-Macandog and Rocamora, Chapter 37), are a promising approach to managing fallow land more productively. Claveria, Misamis Oriental, Philippines. (Photo: Damasa Magcale-Macandog)

46. Tala-andig swiddenists at Bukidnon, Mindanao, in the Philippines, have developed an improved fallow system that converts degraded grasslands into valuable timber stands. *Paraserianthes falcataria* seeds are broadcast into *Imperata* swards before they are slashed and burned in preparation for cultivation. The fire scarifies the *P. falcataria* seeds, causing them to germinate in tandem with planted food crops. After crop harvest, the *P. falcataria* dominates the fallow succession and can be harvested for timber when 10 to 12 years old. Midway through the fallow, *P. falcataria* begins to contribute seeds to the soil seedbank, so that after tree harvest the subsequent burning of remaining slash sets the cycle in motion again.

47. A market for pulp wood has persuaded many upland farmers in Yenbai Province of northern Vietnam to plant *Styrax tonkinensis* (shown) and *Manglietia glauca* into their swidden fields in a taungya system. The trees achieve marketable size within 10 to 12 years, after which the fallow is reopened for cropping and the logs sold to the pulp mill. In this way, shifting cultivation has been transformed into defacto permanent land use with a food crop–pulp wood rotation. The system has led to increased forest cover in the province.

48. Natural vegetative strips have been widely adopted by farmers at Claveria, Misamis Oriental, Philippines, as a low input alternative to conventional alley cropping. Some of them later relay plant *Gmelina arborea* along the vegetative strips, creating contour rows of marketable trees (Suson et al., Chapter 33). Cropping of annuals inside the alleyways stops when the trees form a canopy and shading becomes intense. The system then reverts from agrosilviculture to silvipastoral (shown). (Photo: Wilbur Sitoy)

49–50. In the Kerinci area of West Sumatra, Indonesia, high cinnamon prices and secure individual land tenure have stimulated many shifting cultivators to interplant *Cinnamomum burmannii* into their fields in a taungya system (Suyanto et al., Chapter 65; Werner, Chapter 67). Intercropping of food crops between the rows of trees is discontinued when shading becomes excessive. Harvest (inset) usually occurs when the trees are 10 to 12 years old and the cycle begins anew. (Photos: Suyanto)

51. Shifting cultivators in Yenbai Province of northern Vietnam are using their local species, *Cinnamomum cassia*, in a similar system (Hien, Chapter 62). Most of the bark is exported, the leaves are processed for oil, and the timber is used to make furniture and packing crates. Such a heavy harvest index leaves little for this fallow system to contribute to soil rejuvenation.

52. A mature 15-year-old rattan garden in Rantau Lajung, Pasir District, East Kalimantan. (Photo: Carmen Garcia, CIFOR)

3–54. This fallow regrowth in Hoa Binh province of northern 'ietnam looks deceivingly like a coniferous forest. It is, in fact, amboo. Many shifting cultivators manage bamboo fallows and arvest the poles when reopening the land for arable cultivation inset). Bamboo is fire resistant due to underground rhizomes; it egenerates rapidly and quickly shades out light-demanding weeds, and it rejuvenates soil through apid accumulation of biomass and nutrients and high rates of litterfall (Hien, Chapter 62; Ty, Chapter 5).

55–56. Sundanese farmers in Ciwidey, West Java, practice an intensive fallow system based primarily on marketable bamboo species. This practice is remarkable for its adaptability to Java's high population densities. The steady supply of bamboo has spawned booming local industries in woven bamboo walling (inset) and furniture construction.

57 and 58. Farmers in northern Lao P.D.R. have long harvested the inner bark of paper mulberr (*Broussonetia papyrifera*), growing wild in bush fallows. Development of a local industry for processin it into a coarse textured parchment has encouraged many farmers to retain *B. papyrifera* growing in ric swiddens and to experiment with propagation (inset) (Fahrney et al., Chapter 40). Paper mulberry is o immense interest because it can produce a harvestable product within the short two to three year fal lows that predominate across much of Southeast Asia's uplands. (Main photo: Keith Fahrney)

59. *Tembawang* forest gardens (above) are man-made forests that have a structure and floristic compo sition that closely resemble natural forest. Outsiders, unaware of their existence, have often mistaker them for natural forests. Jambi, Sumatra. (Photo: Andy Gillison, CIFOR)

60. Tembawang forest gardens supply a diversity of products. Harvested species may be planted, tended, or merely tolerated when they occur spontaneously. This Dayak is collecting fruit that has medicinal properties. Ella Ullu Bukit Baka Raya National Park, West Kalimantan. (Photo: Alain Compost)

61. Under pressure to intensify their land use, Pwo Karen in the buffer zone of Huay Khakhaeng Wildlife Sanctuary, in western Thailand, have begun to develop swidden plots into banana-based agroforests, known locally as *sagui gru* (Srithong, Chapter 52). Bananas are interplanted into upland rice during the cropping phase in standard taungya practice. Once established, the bananas provide protective shade for further enrichment planting with kapok, betel nut, mango, pomelo, jackfruit, coconut, papaya, and numerous useful shrubs and herbs, gradually developing a complex agroforest. (Photo: Payong Srithong)

62. Durian (*Durio zibethinus*) forest gardens in West Kalimantan are thought to have their origins in seeds that were casually jettisoned from houses and field huts in swiddens. Farmers expanded on existing gardens and planted new ones in the early 1970s when access to distant markets suddenly made durian a valuable commodity. This is a similar agroforest in Maninjau, West Sumatra, Indonesia. (Photo: Genevieve Michon)

63. The resin producing (damar) agroforests of Krui, Sumatra, were established by relay planting *Shorea javanica* into swidden fields during the cropping phase (Michon et al., Chapter 45). The system probably evolved from traditional practices of tapping damar resin from *S. javanica* trees that grew naturally on forest lands and intensified into increasingly elaborate management regimes over the past century. Similar resin tapping systems are scattered across Southeast Asia: in northern Sumatra, *Styrax benzoin* and *S. paralleloneuron*; northern Lao P.D.R., *S. tonkinensis* (see color plate 64) and *S. benzoides*; southwestern China, *Pinus yunnanensis* and *Toxicodendron vernicifera*; and the southern Philippines, *Agathis philippinensis*. Krui, West Lampung, Sumatra, Indonesia. (Photo: Genevieve Michon)

64. Shifting cultivators in Nam Bak district of Luang Prabang, Lao P.D.R., manage *Styrax tonkinensis* as a useful fallow species to tap its *benzoin* resin (Fischer et al., Chapter 46). Indigenous to the area, the seeds of *S. tonkinensis* are scarified in the burning operation to clear swidden fields. It germinates together with crops of glutinous rice and, after a single year of cultivation, it is left to dominate the subsequent fallow succession. Tapping (shown) begins when the trees are six years old and can continue until they're 10 to 14 years of age. This traditional system is now under threat because of falling resin prices and shortening fallow periods.

55. The rubber (*Hevea brasiliensis*) agroforests of Sumatra and Kalimantan, Indonesia, known as "jungle ubber," are established by taungya planting rubber seedlings into swidden fields (shown). The subsequent fallow is thus dominated by rubber trees. Tapping may begin when the trees are about eight years old and continue until latex productivity declines between 20 and 30 years of age (Penot, Chapter 48). At this time, the rubber-enriched fallow is slashed and burned in preparation for another cycle. Although the origins of jungle rubber are in shifting cultivation, the farmers' main objective has shifted to the "fallow crop" and the arable cropping phase has become a means to achieve that end (Werner, Chapter 67). Muara Bungo, Jambi Province, Sumatra, Indonesia. (Photo: Hubert de Foresta)

66. This tea plantation in Shangyun township of Tengzhong county, Baoshan district, in China's Yunnan Province, is planted under an *Alnus nepalensis* canopy, because farmers widely recognize that the species has soil building properties (Guo et al., Chapter 29). This practice began as recently as 30 years ago when farmers observed that tea grown under *Alnus* had higher productivity and less insect damage. As a valuable by-product of the system, *Alnus* logs are used as firewood for processing tea, for construction of tea boxes, and as a substrate in culturing several kinds of mushrooms.

67. Chinese walnuts (*Juglans* sp.), such as this grove in Quingshui village, in Tengzhong county Baoshan district, in Yunnan, have also been a major income earner in parts of southern China.

68. In recent years, improved roads and access to markets have allowed many shifting cultivators to pro duce more perishable cash crops, most notably semi-temperate fruits and vegetables. Many highlanc communities have prospered by exploiting their ecological niche and producing such crops for lowlanc markets. This plum (*Prunus* sp.) orchard was taungya planted into a dryland field in Songshupo village also of Tengzhong county in Yunnan province.

conditions, they will optimize their limited labor by selecting for cultivation those parts of the farm that yield the highest output. Fujisaka and Cenas (1993) found, similarly, that farmers with parcels of flat or gently sloping land tended to focus their attention on them, neglecting their steeper hedgerow fields when labor and other inputs were limiting. One farmer commented that, even assuming that parcels of flat and sloping land were equal in terms of soil fertility, he would still choose to work the flat land because it was easier to till and there was less water stress. Therefore, if the ratio between land and labor is low, farmers may have no choice but to do something to improve the productive capacity of their farm. This is the case with farmers in the central Philippines province of Cebu who, with an average farm size of less than 1 ha, invest in labor-intensive activities to make their land more productive. These observations fit the intensification model propounded by Raintree and Warner (1986).

To better understand what prompts farmers to fallow their hedgerowed, alley farming fields, we also interviewed a group of farmers who continuously cultivated their pruned-tree alley farms without fallowing them. Unfortunately, they are now few in number, but they gave two prominent reasons for continuing to farm their hedgerow fields (see Table 33-8). One was that it was necessary to continuously cultivate in order to prevent stray animals from destroying the hedgerows, and the second was that farm size was very limited. The first issue, hedgerow destruction by stray animals, needs further investigation.

The average farm size of those farmers who claimed limited area as their reason for continuously cultivating their alley farm was 1.5 ha. This is one-fourth of the average farm size (5.36 ha) of those farmers who fallowed their alley farms. As well, the proportion of their total farm occupied by hedgerow fields was much higher. An average of 57% of their total farm area had been contoured with tree hedgerows. If these farmers were to abandon their alley fields, the average area left for cultivation would be only 0.65 ha, which is quite small for maize farming. They had much less flexibility to allow any of their farm land to be fallowed. These farmers were asked if they had noticed any decline in crop production. Eighty-three percent said they had, while the rest said they could not tell because they had applied large doses of fertilizer from the beginning. Fortunately, all of these farmers had resources sufficient to surmount the nutrient depletion problem. Implicit in the survey result is that alley farming, without external nutrient application, cannot sustain continuous cropping. But when the farm area is large, fallowing and shifting is perhaps more profitable than maintaining soil fertility by importing nutrients. Szott et al. (1991) are straightforward in their verdict on alley farming in the context of continuous cropping under acid, infertile soils: it is not sustainable without nutrient importation. The main reasons are native soil infertility and insufficient recycling of nutrients from the prunings.

Table 33-8. Reasons That Farmers Continued Cropping Their Alley Farms

Reason	*Number*	*%*
Prevent animals from destroying the hedgerows	4	36
Farm area is small; cannot afford to fallow	3	27
Has the financial resources to continuously crop; finds present area too small for his farming operations	1	9
Hedgerow field is easier to work; the slope is not as steep as the rest of the farm	1	9
Hedgerowed field is near the house	1	9
Prevents the spread of mulberry as a weedy species	1	9

Table 33-9. Productivity of Contour Hedgerowed Fields Compared to Fallowed Open Fields (tonnes/ha)*

Cropping	*Products*	*No Hedgerows*	*With Hedgerows*
First	Maize yield	2.29 b	3.41 a
	Woody biomass	—	29.68
Second	Maize yield	0.69 b	1.40 a
	Woody biomass	—	3.39
Third	Maize yield	0.41 a	0.65 a
	Woody biomass	—	0

Note: * In a row means with different letters are significantly different at LSD 0.05 = 0.59.

Productivity Assessment

The WH treatment produced two economic products: firewood from hedgerow biomass and maize grain. The NH treatment yielded maize alone. Total biomass yield from the WH treatment was therefore much higher than that from the NH treatment (Table 33-9). Maize yield was higher in the WH treatment than in the NH treatment for all three crops, although the difference was statistically significant only for the first two. The higher maize yield in the WH treatment is attributed to several factors. During the fallow period the biomass contribution from litterfall leads to a much higher deposition of organic matter in the WH treatment than in the NH. The shading cast by the trees may also have reduced the rate of oxidation of this organic matter, as a result of lower soil thermal conditions (Main unpubl). This may have allowed a greater accumulation of soil organic matter. The presence of trees in the system may also have improved soil fertility more effectively than the natural grass or *Chromolaena* fallows because of higher rates of N-fixation. Pruning applications during the cropping seasons may also have made significant nutrient contributions through the rapid recycling of the green manure biomass. The trees may also have, to some degree, tapped nutrients deeper in the subsoil during the long fallow period than could be reached by the weedy fallow plants in NH. Analysis of the soils data is awaited to confirm these possibilities. Comparative measurements of soil loss confirm that the tree hedgerows formed very distinct natural terraces that nearly eliminated any evidence of soil erosion. We also noted that the hedgerow vegetation was attractive as natural latrines for people working in the field, possibly causing some gain in nutrients over the years from this source.

The woody biomass yield was a substantial 30 tonnes/ha when the fields with hedgerows were opened up for the first cropping (Table 33-9). This was the accumulated biomass from four years' fallow. The harvest of wood at the beginning of the second cropping was reduced to one-tenth of the initial amount. It was the accumulation from one dry season of only a few months between the first cropping and the second. There was no fallow break between the second and third crops, so virtually no woody biomass was available at the beginning of the third crop.

Profitability Assessment

Table 33-10 compares the costs and returns for maize production in the WH and NH treatments. Maize production costs were slightly higher for the WH treatment, due mainly to higher harvesting and processing costs because of a higher yield. The costs and returns of the firewood production from the hedgerows are not considered in this analysis. The sales of maize from the WH treatment were considerably higher than for the NH in each of the three crops. This resulted in a profit margin for WH that was 68% higher from the first crop. Yields declined in both treatments, but were lower in NH. This resulted in negative returns for NH in both the second crop, minus US$38/ha, and the third crop, minus US$49/ha. Profit margins remained positive in WH, but declined to very modest levels of US$179/ha and US$20/ha for the second and third crops, respectively. Therefore, the hedgerow fallow treatment dramatically increased profitability in the first crop after opening up the land and enabled a modest profitability to continue for one or two more crops. By comparison, four years of natural fallowing under *Chromolaena* and grass was able to ameliorate soil fertility only enough to permit a single profitable maize crop. This suggests that natural fallowing on these soils would require a longer period than just four years to enable cropping to be sustained beyond a single season.

Firewood collection costs, and the value of firewood sales, were omitted from the above analysis. Firewood from the hedgerow species was found to be difficult to market in Claveria. According to retailers, household consumers prefer firewood over coffee or *ulayan,* a local tree. These woods are believed to have a greater heating capacity and ability to make good charcoal than the usual tree species from hedgerow strips. Bakeries, on the other hand, are more concerned with volume than firewood quality. They have recently tapped cheaper sources of firewood from small sawmills that are proliferating along the highways of Misamis Oriental and offering the offcuts from milling *Paraserianthes falcataria.* Sawmill firewood costs 1997US$0.19 for a bundle 143 by 305 cm. A conventional bundle harvested locally in Claveria is 15 by 73 cm and sells for between 1997US$0.04 and $0.05. Therefore, firewood coming from the sawmills is cheaper since their bundles are seven times larger for less than five times the price. In addition, firewood from the sawmills is delivered in bulk with free transport. They are able to sell it cheaply because it is a waste product of their operations.

The economic advantage of growing firewood in farm hedgerows may be minimal when the farm is near the forest or other woodlots. However, we conducted interviews in an area where access to firewood was difficult. We wanted to determine how long it took to collect firewood, the number of bundles collected per collection event, the number of bundles consumed per week, and the time needed to split and pile it. This data was used to compute the volume of consumption and the labor costs. Based on our survey, farmers who do not have their own on-farm source of firewood nevertheless do not buy firewood. Therefore, in Claveria, it is more practical to consider the production of firewood as a source of savings in labor in areas where firewood is not readily available, rather than to consider it as a source of earnings. Savings in the sourcing of firewood contributes to socioeconomic well-being.

Table 33-10. Costs and Returns of Maize Production: Contour Hedgerow Fallow System vs. Open Field Fallow System, after Four Years of Fallow (1997US$/ha)

	First Crop		*Second Crop*		*Third Crop*	
	NH	*WH*	*NH*	*WH*	*NH*	*WH*
Costs						
Maize	277	331	246	242	172	176
Firewood	43	221	43	29	43	-
Total	320	552	280	271	215	176
Sales	687	1022	208	421	123	196
Net profit	367	470	(81)*	150	(92)*	20

Note: Values in parentheses indicate net loss. NH = No Hedgerows; WH = With Hedgerows.

The amount of firewood produced in the first two crops is estimated to provide an assured supply for three years. The amount of cumulative labor saved is substantial, but the cost of processing firewood in newly opened fields is quite high, involving 43 man-days. The high labor investment in cutting and processing the wood obtained in a tree fallow, and the difficulty in marketing it in communities such as Claveria, raises the question of whether there is sufficient incentive for farmers to process and produce firewood. This is time that might be used for other income-generating enterprises. Secondly, can the farmer store large volumes of firewood practicably, for example, 12,000 bundles from a hectare of hedgerows? It may be argued that it is more practicable and useful to burn the bulk of the woody biomass and plow the ash into the soil to improve crop productivity.

Conclusions

Our data indicate that hedgerow fallows have the potential to offer greater system benefits than traditional fallows. However, only one-third of the farmers interviewed in Claveria used hedgerow fallowing as a conscious strategy in managing their farms. Many fallowed their land when yields declined and the returns to their labor in maintaining the hedgerow fields was no longer as remunerative as other uses of their time. The high labor investment required to open fallowed hedgerow fields is a constraint in bringing these fields back into cultivation. Our results showed that the yields and profitability of maize were substantially higher in hedgerow fields after fallowing than in fields following natural fallows. However, in both cases yields declined sharply after the first crop, and returns were negative or near zero by the third season. This indicates that although hedgerowed fields, after four years of fallow, offer distinct advantages over natural fallows of the same duration, the former system does not enable dramatically longer periods of sustained cropping.

This study suggests that farmers are likely to be attracted to this farming system when their farm enterprise has the following characteristics:

- If the farm labor force is relatively low in relation to the total farm area. A low labor-to-land ratio indicates that there is adequate area to practice fallowing. If the farm is too small, fallowing is not practicable. Likewise, if the farm is quite large, the farmer may fallow and open more land each year rather than invest in hedgerow technology to increase returns per hectare.
- If the farm has both relatively flat land and sloping land, the farmer has the option to choose which parcel will yield the highest returns to investment. Sloping fields with hedgerows may be fallowed and farmed less intensively than the flatter fields.
- If the farmer owns the land or the land tenure is otherwise secure. This is a prerequisite for investments such as contour hedgerows.
- If the farmer is unable to buy mineral fertilizers because the farm is remote or he lacks cash. In such cases, the only practical way to restore fertility is through fallowing.
- If the farmer lacks a draft animal. In the absence of animal draft power, *Imperata cylindrica* and other grass weeds are extremely difficult to control. The shading of weeds by the tree hedgerows tends to suppress them during the fallow period, making it easier to open the land at the start of cropping. Farmers practicing manual cultivation also usually have lower returns to labor for land preparation. They would be more likely to find hedgerow fallowing attractive than would farmers with access to a draft animal.
- If the farmer is able to prevent fires, which are an ever-present threat in grassland areas. Fires can damage or destroy tree hedgerow systems. The risk of fire must be low or the heavy labor investment in establishing and maintaining a hedgerow system may not be worthwhile.

Our work indicates that there are yield and profit advantages to be gained from fallowing tree hedgerows on sloping lands. It may therefore be an attractive practice for some upland farmers. However, this recommendation is clearly restricted to farmers whose enterprise meets the conditions listed above. This suggests that the practice is probably not suitable for many upland farmers. Prior to this study, there had been little or no research on the practice of fallowing hedgerowed fields. Further work is needed to validate our findings in other environments, where the practice may be considered suitable for recommendation.

Acknowledgments

The authors are grateful to Mr. Gil Arcenal and Mr. Renato Almedilla for field assistance, and to Mr. Agustin Mercado and the staff of ICRAF Philippines for vital support in the conduct of this study. The work was conducted by the senior author as part of a project of the World Agroforestry Centre (ICRAF) through a grant from the United States Agency for International Development.

References

Basri, I., A. Mercado, and D.P. Garrity. 1990. Upland Rice Cultivation using Leguminous Tree Hedgerows on Strongly Acid Soils. Paper presented to the American Society of Agronomy, October 21–26, 1990, San Antonio, Texas.

De la Cruz, R. 1997. Personal communication with the authors.

Fujisaka, S., and P.A. Cenas. 1993. Contour Hedgerow Technology in the Philippines: Not Yet Sustainable. *Indigenous Knowledge and Development Monitor* Vol.1 (1).

Franzel, S. Unpubl. ICRAF Internal Travel Diary. Nairobi, Kenya.

Garrity, D.P. 1993. Sustainable Land Use Systems for Sloping Uplands in Southeast Asia. In: *Technologies for Sustainable Agriculture in the Tropics*. American Society of Agronomy Special Publication 56. Madison, WI: American Society of Agronomy, 41–66.

———. 1994. Improved Agroforestry Technologies for Conservation Farming: Pathways toward Sustainability. Paper presented at an International Workshop on Conservation Farming for Sloping Uplands in Southeast Asia: Challenges, Opportunities and Prospects, November 20–26, 1994, Manila, Philippines.

Garrity, D.P., and P.C. Agustin. 1987. Historical Land Use Evolution in a Tropical Acid Upland Agroecosystem. *Agriculture, Ecosystems and Environment* 53.

Gomez, A. 1997. Personal communication with the authors.

ICRAF (World Agroforestry Centre). 1996. *Annual Report for 1995*. Nairobi, Kenya: ICRAF.

———. 1997. *Annual Report for 1996*. Nairobi, Kenya: ICRAF.

Kang, B.T. 1997. Personal communication with the authors.

Kang, B.T., L. Reynolds, and A.N. Atta-Krah. 1990. Alley Farming. *Advances in Agronomy* 40, 315–353.

Kang, B.T., and G.F. Wilson. 1987. The Development of Alley Cropping as a Promising Agroforestry Technology. In: *Agroforestry: A Decade of Development*.

Maclean, R.H., J.A. Litsinger, K. Moody, and A.K. Watson. 1992. Increasing *Gliricidia sepium* and *Cassia spectabilis* Biomass Production. *Agroforestry Systems* 20. Dordrecht, the Netherlands: Kluwer Academic Publishers, 199–212.

Main, R.G. Unpubl.

Mercado, A. 1997. Personal communication with the authors.

Nelson, R.A., R.A. Cramb, and M.A. Mamicpic. 1996. Costs and Returns of Hedgerow Intercropping and Open Field Maize Farming in the Philippine Uplands. SEARCA-UQ *Uplands Research Project Working Paper* No. 11.

Payonga, A. 1997. Personal communication with the authors.

Raintree, J.B., and K.W. Warner. 1986. Agroforestry Pathways for Swidden Intensification. *Agroforestry Systems* 4(1), 39–54.

Sanchez, P.A. 1995. *Science in Agroforestry*. Dordrecht, the Netherlands: Kluwer Academic Publishers.

Szott, L.T., C.A. Palm, and P.A. Sanchez. 1991. Agroforestry in Acid Soils of the Humid Tropics. *Advances in Agronomy* 45, 275–301.

PART VI

Perennial-Annual Crop Rotations

A young boy draws on a traditional bamboo pipe in northern Lao P.D.R.

Chapter 34

Teak Production by Shifting Cultivators in Northern Lao P.D.R.

Peter K. Hansen, Houmchitsavath Sodarak, and Sianouvong Savathvong

Shifting cultivation is the dominant cropping system in the uplands and mountains of Lao P.D.R. At least 300,000 families are fully or partially engaged in it, equal to about 1.8 million people or 40% of the population. Assuming that each family plants about 1.5 ha per year, the area used annually for shifting cultivation would be about 450,000 ha. The total area involved in the shifting cultivation cycle, including fallowed land, is difficult to assess, but it may be between 2 million and 2.5 million hectares, equal to about 10% of the area of Laos.

The population of Laos consists of 66 officially recognized ethnic groups, many of which contain several subgroups (Chazee 1995). The ethnic groups are often divided into three main categories: *Lao Lum* (lowlanders), *Lao Theung* (midlanders) and *Lao Sung* (highlanders). The Lao Lum category consists of ethnic Lao and other Thai-speaking groups and accounts for about 60% of the population (NSC 1997b). Although the majority of Lao Lum farmers are engaged in paddy farming, a large number are shifting cultivators. The Lao Theung and Lao Sung groups make up 30% and 10% of the population, respectively. They are relatively more dependent on shifting cultivation than the Lao Lum groups, but their land use is very diverse and ethnic stereotypes often prove misleading (Roder et al. 1991; Hansen 1995).

Most shifting cultivators remain subsistence producers of upland rice, but the commercial production of other crops is expanding in areas with adequate infrastructure and market access. Over the past 20 to 30 years, fallow periods in most places have become critically short. The main causes are an increasing population, government restrictions, and competing land-use objectives, as well as the concentration of people around urban centers and in areas with road and river access. The pressure on land resources has increased problems such as soil degradation, weeds, and pests, and has, therefore, led to lower yields and higher labor requirements.

Shifting cultivation in Laos is, these days, largely based on the cyclical use of young secondary vegetation, although limited swiddening in older forest still takes place in isolated areas. However, over the years, shifting cultivation has considerably reduced the forest area to the detriment of timber resources and natural habitats. Where it is intensely practiced, shifting cultivation has accelerated erosion and

P.K. Hansen, Land Use Policy and Planning Advisor, Policy and Planning Division, Ministry of Agriculture, c/o Box 614, Thimphu, Bhutan; H. Sodarak, Shifting Cultivation Research Project, P.O. Box 487, Luang Prabang, Lao P.D.R.; S. Savathvong, Department of Forestry, Luang Prabang Provincial Forestry Office, P.O. Box 530, Luang Prabang, Lao P.D.R.

brought changes in water discharge capable of impairing water resources for irrigation, hydropower, and domestic use.

Because of these environmental and social problems, stabilization of shifting cultivation has become a major priority of the Lao government. The development strategy (Department of Forestry 1997) includes:

- Land allocation;
- Promotion of permanent cash cropping;
- Expansion of paddy area for wet rice production;
- Expansion of livestock production;
- Tree planting by farmers;
- Infrastructure development; and
- Socioeconomic development work.

Attempts to improve land use in those areas of Laos where shifting cultivation is still widely practiced have proven difficult. Adoption of new technologies is constrained by the mountainous topography, limited infrastructure, low market demand, limited processing facilities, and the poverty of most shifting cultivators (Hansen and Sodarak 1997). Furthermore, alternative land-use practices, especially permanent arable cropping, can be equally or even more damaging than shifting cultivation if carried out in mountainous forest areas. By comparison, shifting cultivation partially restores soil fertility during the fallow periods, limits erosion because of minimal tillage, distributes erosion over a large area, and usually involves little or no use of pesticides.

Because of declining productivity and limited alternatives, shifting cultivators are among the poorest and most disadvantaged groups in Laos. Few farmers would opt for shifting cultivation if alternatives were available, and, when this is the case, farmers have readily adopted new technologies. One such option is teak planting, which has spread rapidly among shifting cultivators in many parts of northern Laos, as described and evaluated below.

Methods and Study Area

This chapter evaluates shifting cultivators' adoption of teak in northern Laos, but much of the analysis is relevant to other areas of Lao P.D.R. The northern region consists of seven provinces: Phongsaly, Luangnamtha, Bokeo, Udomxai, Houaphan, Luang Prabang, and Sayabouly. It accounts for 41% of the area and 33% of the population of Lao P.D.R. It is a predominantly mountainous region with only small areas of basins and river valleys. Permanent upland cropping and paddy farming is therefore secondary to shifting cultivation, and about 65% of the shifting cultivators in Laos live in the north (Souvanthong 1995). Upland rice constitutes 65% of the total rice area in the region, compared to the national average of 31% (NSC 1997a). The concentration of shifting cultivation is partly responsible for the existing forest cover being only about 36% of the land area, compared with 52% and 58% in the central and southern regions, respectively (NOFIP 1992). Improving agricultural production is particularly difficult in the north because of the small potential for paddy cultivation, the hilly topography, and limited infrastructure and market access.

The ministerial and provincial authorities have prioritized tree planting, particularly of teak, in their development strategy for northern Laos, and the authors of this chapter were engaged in extension and promotion of teak planting to shifting cultivators in Luang Prabang Province from 1988 to 1996. In the course of this work, numerous discussions were held with farmers and staff from district agricultural and forestry offices. In addition, a survey was carried out in 14 villages in Nane district of Luang Prabang Province to gauge farmers' perception of teak planting. A survey was also carried out in 16 districts, focusing on the plantation activities carried out by farmers, private investors, and public agencies.

Statistics from the Department of Forestry were also used. In addition, a study of the age and diameter relationship in cut teak logs was carried out to assess the growth of teak and the existing cutting regime.

History of Planting Teak in Laos

Teak occurs naturally in central and southern India, Myanmar, northern Thailand, and in two small areas in Laos close to the western border with northern Thailand. The ecological distribution covers the semiarid to moist lowlands below about 1000 m above sea level (asl).

The teak forests in Laos are believed to be the eastern limit of teak's natural distribution (White 1991). The largest area occurs in Sayabouli Province where mixed deciduous forest with teak inclusions covers between 10,000 and 20,000 ha (Hedegart 1995). Some estimates of the forest area go as high as 50,000 ha (Rao 1993). Small areas also exist in Bokeo Province. These are possibly the only natural stands on the eastern side of the Mekong River. Historically, teak was of little commercial or domestic importance in northern Laos, compared to other species such as *Afzelia xylocarpa* and *Pterocarpus macrocarpus*.

The history of teak planting in Laos may be divided into three periods: the years before the revolution in 1975, the years between 1976 and 1988 when a planned economy was followed, and the years after 1988 when the economy was liberalized. Before 1976, tree planting was very limited, partly because of an ample wood supply from the natural forest. Export opportunities were also limited by the political situation and by the lack of ports and processing facilities.

The first farmer-owned teak plantations in northern Laos were established around 1950 following promotion by the French colonial regime. These plantations were limited to a few areas along main rivers near Luang Prabang town. The river communities were targeted because they were permanent, unlike the predominantly shifting cultivation villages in the uplands. Teak was moreover attractive because of its suitability for boat construction and because the logs could be transported by river. Since most villagers living along the rivers were Lao Lum, teak planting was carried out almost solely by this group.

A plantation program was also started around 1950 by the Department of Forests and Water, mostly in the south of the country, and mostly using teak planted in a *taungya* (swidden) system. The plantations were usually established by shifting cultivators on land belonging to the department, which took over management and ownership of the plantations after farmers ceased interplanting with rice during the first one or two years. These plantations amounted to about 1,500 ha, of which 1,000 ha may have been surviving 20 years later (LARP 1972).

After the revolution in 1975, a centrally planned economy was adopted and the authorities took ownership of all land. Farm collectives and state farms were introduced, although traditional tenure systems generally prevailed in shifting cultivation areas in the uplands and highlands. Responsibility for logging and tree plantations was given to state forest enterprises, and logging became the major earner of foreign currency, along with hydroelectricity generation. During these years the government attempted more ambitious plantation programs in all provinces. Most were carried out by state forest enterprises.

Nevertheless, by 1991, the plantation area reported in official statistics was only 6,250 ha, of which 1,140 ha was established before 1976. Furthermore, sample surveys suggested that only 2,900 ha really existed or had an acceptable stocking rate (LSFCP 1991). As before, teak was the main species, accounting for 47% of the reported plantation area. But other species had also become important, including *Pterocarpus macrocarpus* (20%), *Afzelia xylocarpa* (17%), *Eucalyptus* spp. (6%), and *Alstonia scholaris* (4%).

Since the late 1980s, farmers have greatly expanded teak plantations (see Table 34-1). Luang Prabang Province has been the main center for this expansion because of its better infrastructure, the larger scale of commerce, and the presence of older plantations to supply seed. However, teak planting has expanded around most of the northern provinces.

It is also significant that teak was adopted by shifting cultivators and planted in the uplands where, previously, very little tree planting had taken place. This meant that an increasing number of non-Lao ethnic groups adopted teak, especially people of the Khamu ethnic group.

Factors Facilitating Adoption of Teak

The rapid expansion of teak plantations in the north, particularly in Luang Prabang Province, has been made easier by political and socioeconomic changes. The main factors have been:

- Depletion of natural forest wood supplies and the emergence of a market for relatively young teak timber, only 15 to 20 years old;
- Confidence in the acquisition of secure private land tenure. After periods of insecure ownership and doubts about government land policies, farmers became assured of their rights to land used productively, including management of tree plantations;
- Most villages adopted a permanent settlement pattern. After years of war and unregulated access to land, the government began to encourage permanent settlement and farmers became less inclined to resettle at frequent intervals. Therefore, long-term investments such as teak plantations became more realistic.
- The expansion of the road system, which made plantations possible in new areas. Very few plantations were established off the roads, since farmers did not expect investors to be interested in buying the timber on the land;
- Land allocation schemes that gave additional land for production of perennials. In principle, one additional hectare of land could be allocated to each household specifically for planting timber or fruit trees. Farmers would lose the rights to this land if it were used for any other purpose;
- Promotion by private investors through financial support, the production of stumps for teak propagation, and information dissemination; and
- Promotion and extension by government agencies (see below).

Table 34-1. Teak Plantation Area Established by Farmers in Luang Prabang Province

Period	*Teak Planting (ha)*
1975–79	33
1980–84	42
1985–89	242
1990–94	1,278
1995	1,419
1996	2,039

Source: Luang Prabang Provincial Forestry Office.

The government's strategy for development in areas of shifting cultivation aims at a rapid and considerable expansion of farmers' tree production, including teak, *Eucalyptus* spp., and *Acacia* spp. The plans state that all farmers should plant trees on some of their land, as long as it is suitable and they have the necessary resources, and it is envisaged that about 16% of households currently practicing shifting cultivation will be able to make wood production their main source of income within 5 to 10 years (DoF 1997). The means to achieve these goals include:

- Extension of tree planting to farmers through information, distribution of seedlings or stumps, and through credit schemes;
- Allocation of land to farmers for planting trees;

- Tax exemption on plantations with a density of more than 1,100 trees per hectare;
- Encouraging private companies to establish plantations and processing facilities, in combination with farmer plantations; and
- Plans to finance farmers' basic food requirements while they wait for income from plantations (DoF 1997).

Up to 1994, the authorities, as well as various projects, either gave away or subsidized teak stumps to encourage expansion of teak plantations. Predictably, many stumps were never planted and many plantations were not cared for. The practice also hindered the establishment of private nurseries.

Farmers' Motives for Planting Teak

Farmers say their main reasons for establishing teak plantations include the following:

- To sell timber;
- To use the timber domestically for construction of houses or boats;
- To use the plantations as collateral for obtaining credit;
- To ensure land-use rights; and
- To sell the plantations to investors shortly after establishment.

Farmers usually claim that selling the timber or using the wood domestically is their main reason. This follows official development reasoning, which is imparted in the extension process. However, it is generally believed that the main motivation for upland farmers is the possibility of selling the plantations to investors, such as local businessmen or government staff, when the trees are one to three years old. Farmers' reluctance to state this reason comes from lingering confusion about the laws for transfer of tenure rights. Extension staff estimate that 80 to 90% of upland farmers will sell their plantations if given the chance. Indeed, this has happened over recent years in some villages close to Luang Prabang township. In more distant areas, fewer plantations have been sold off because of lesser interest from investors. In some villages, the establishment of plantations has been financed by external investors, with an understanding that they will take over the land-use rights after intercropping has ceased.

Another important impetus is the possibility of using the teak plantations as collateral for credit. Farmers are able to obtain loans of 40% to 60% of the estimated value of their plantations. At this rate, a three-year-old plantation is able to secure loans of US$1,000 to US$1,500/ha.

In recent years, the authorities have implemented land allocation schemes which grant up to four plots of land per family for cyclical shifting cultivation. In addition, each household can obtain one hectare of land for production of perennials. Many households have taken this opportunity to plant teak. However, the ability or the motivation of many of them to manage teak plantations is doubtful, so the plantations are likely to be sold off. Their own domestic use of teak represents a negligible motivation for farmers to establish plantations, as is indicated by the near absence of teak in villages without road or river access. However, small-scale teak planting for home construction may increase, even in remote areas, when the natural wood supply diminishes, the shifting cultivation communities become more settled, and farmers get easier access to planting material.

Potentials and Constraints

Teak has exceptional properties, which make it one of the most sought after and expensive timber species both locally and internationally. The wood is structurally strong, durable, and very resistant to fungus and termites. The risk of splitting and warping during drying and processing is small and the wood is easily carved and turned. These properties make teakwood particularly suitable for house construction, boat railings, furniture, and carvings.

The international market for plantation teak is likely to expand with increasing consumer demand for certified plantation wood. The high market price of teakwood makes long-distance transport economically feasible, unlike industrial tree species such as eucalyptus and acacias, whose sale depends on nearby processing facilities for pulp or board production.

In Laos, teak is used for a few secondary purposes. A yellow dye used for silk yarn is made from boiling the dry leaves. Occasionally, the leaves are used to thatch roofs and for packaging and young thinnings and branches are used as firewood. These uses, however, are of little importance compared to wood production.

Propagation

Teak is usually established from stumps, which are easy to transport and plant but are sensitive to dry spells in the first two months after planting. Stump production is simple and requires little investment. Since 1993, many private nurseries have been established to produce stumps for family use or sale.

The advantage of genetic improvement of teak is well established. In Thailand, for example, it has been shown that even simple selection of superior trees in plantations may give a 10% to 15% increase in volume production (Hedegart 1995). Nevertheless, hardly any systematic selection of seed trees, stands, or provenances takes place in Laos, and the plantations established in recent years have been propagated from genetically unknown, unselected, and, possibly, inferior sources. Instead, a form of negative selection takes place because the largest and best-formed trees in plantations and natural stands are selectively cut to generate a quick income. This also favors trees that are genetically disposed to early flowering, which is considered a negative trait related to excessive branching and reduced volume increment. This problem increased in recent years following a rise in timber prices and liberalization of the timber trade.

The rapid expansion of teak planting in northern Laos has caused a shortage of seed, so it is likely that unselected seed material will continue to be used for many years. It has also limited teak planting in some areas, especially where teak has not previously been grown.

The scarcity of seed is reflected in the price of dry fruits, which, in Luang Prabang Province, increased from 35 to 800 kip per kilogram (1997 US$0.04 to US$0.87) between 1992 and 1996. However, even at the higher price, the seed costs only 10 to 20% of the price of a teak stump.

Management

Teak is usually established in taungya systems. That is, it is interplanted with agricultural crops during the first one to three years. This facilitates adequate weeding and protection of the teak in the early years. Few farmers would have the labor resources to carry this out if crops and teak were planted in separate areas.

The teak stumps are normally planted in June or July, after completion of the first weeding in the associated crop. Planting after mid-July is not recommended since the small trees would be at risk from weed competition, browsing animals, and fires after the end of the cropping season.

When interplanting ceases after one to three years, little management is required, except for slashing taller weeds and, sometimes, controlled burning early in the dry season. Unfortunately, few farmers prune low branches or forked trees, and forking sometimes occurs in more than half of the plants. Furthermore, thinning is often left until too late, when the trees are 10 to 15 years old. The lack of appropriate management is a major constraint to the growth and quality of the trees. There is, therefore, considerable scope for improving teak production through timely weeding, thinning, pruning, and fire control, as well as improving propagation methods and selecting superior seeds. However, all these things are given low priority if farmers are intent on no more than selling their plantations.

Teak is usually preferred to alternative perennials because it is relatively easy to propagate and manage, grows fast in the early years, and is tolerant of fire. The rapid expansion of teak planting has lead to concerns that teak monocropping may lead to a devastating buildup of pests, especially bee-hole borers and caterpillars. Furthermore, teak offers little soil protection, and sheet, rill, and gully erosion are often seen in older plantations. This aspect warranted special mention by White in 1991 and Hedegart in 1995. Casual observations in 15- to 20-year-old teak plantations also suggested that there was little accumulation of organic matter taking place.

Income Generation

Teak is relatively fast growing compared to most other high-value timber species. Although the rotation in commercial teak plantations is usually 50 to 80 years, teak in farmers' plantations is mostly cut at the age of 15 to 25 years. The average age of trees cut in 1996 was only 21 years. This reflects the fact that very few farmers planted teak before 1970 and also that few farmers are willing to produce timber beyond the minimum diameter required by local timber companies. The limited formation of heartwood in the young trees means a large loss of potential income, but few farmers are able—or willing—to employ longer rotations.

The minimum size of roundwood accepted by the local sawmills is 20 cm diameter at breast height (dbh). In Luang Prabang Province, this size is attained somewhere between 10 and 25 years of age. The average age is about 15 years. On average, four trees of 18 cm dbh make up one cubic meter of roundwood.

Farmers are currently paid about US$25 per tree of 18 to 20 cm diameter, equivalent to about US$100 per cubic meter. The middlemen usually sell the timber for US$130 to $140 per cubic meter to provincial sawmills. Roundwood sold in Vientiane is worth about US$230 per cubic meter, and, on the export market, US$350 to US$600 per cubic meter. Lately, however, farmers seem to be getting a slightly better deal due to improved access to information on prices. Cut wood (planks) in Luang Prabang and Vientiane sell for about US$450 and US$650 per cubic meters, respectively.

Growing teak compares very favorably with upland farming. A well-managed plantation can, after 20 years, produce about 130 cubic meters of wood per hectare, worth about US$13,000 to the farmer, or US$650 for each of the 20 years it took to grow. Upland rice, on the other hand, may produce 1,400 kg of rice per hectare every fourth year, worth US$210. During the 20 years, five crops may be grown adding up to a total value of US$1,050 per hectare. In addition, the labor requirement for growing rice is roughly three times higher than for teak production.

Although these figures are merely indicative, they show the huge economic potential of teak. For farmers, however, the problem is one of waiting 20 to 30 years for their income. Teak is a cash crop for wealthier farmers, businessmen, and government employees: those who can afford to wait. Many poorer farmers have cashed in their plantations to investors after just a few years. Since plantations are often established on land close to roads, farmers also lose many of the best agricultural plots where production of cash crops would be conveniently close to road access. Furthermore, such areas often have relatively gentle slopes, suitable for permanent cultivation.

On the other hand, those farmers who cash in their plantations receive an income that would be very difficult to obtain through other means. Depending on the location—especially proximity to cities—and the age of the trees, plantations sell for between US$700 and US$2,000 per hectare. Normally, the mean annual household income for shifting cultivators, both cash and goods, is about US$500.

While land is still plentiful in most of the country, it is generally scarce in areas where teak planting is most common and feasible, that is, alongside roads near the major cities. The rapid expansion of teak in parts of northern Laos could, therefore, result in poor farmers losing much of their best land to richer people.

Environmental Suitability and Geographical Distribution

Teak is well suited to most of the very diverse environmental conditions in northern Laos. However, depending on local climate, it is limited to altitudes below 700 to 900 m asl. This excludes it from about 30% of the northern region. It is also unable to grow in flood-prone lowland areas or in shallow, gravelly or strongly acidic soils. These conditions effectively exclude teak from another 10% to 15% of the northern region. They also mean that teak plantations do not optimize the use of marginal land.

However, the main restriction on the geographical distribution of teak is the perceived need for transport facilities. Therefore, more than 95% of the plantations are established along roads and rivers, where there is the best chance of selling either plantations or logs, and teak planting is confined to a minority of villages. At present, it is not likely to spread to areas with more difficult access.

Because teak is confined to areas with road or river access, it is mainly planted in the more populous areas where shifting cultivation is under relatively high pressure. In these areas, fallow periods are generally only two to six years long and a majority of farmers are unable to produce sufficient rice for their household requirements. Farmers often give rice deficiency as a reason for selling their plantations.

A larger proportion of ethnic Lao farmers are planting teak compared with the Khamu (Lao Theung) and Hmong (Lao Sung) ethnic groups (Roder et al. 1995; Juville forthcoming). The main reason for this is that Lao people occupy more land close to roads, along rivers, near larger cities, and at the lower elevations. Also, Lao farmers generally have better economic resources, rice security, access to information, and contacts in the cities than the Khamu and Hmong farmers.

These differences are accentuated by the high proportion of Khamu farmers who are selling their plantations and by the many Lao villages that have imposed communal restrictions on the sale of land to outsiders.

Integration with Agriculture

Teak is associated with agriculture through the initial interplanting of teak and agricultural crops and through grazing in the young plantations. The potential for economically improved fallows has also been proposed.

Since teak is usually planted in taungya systems, the plantations occupy agricultural land. However, most of the land is under shifting cultivation and is not necessarily suitable for permanent cultivation. Current extension recommendations promote teak planting on marginal land. While this may be desirable from an overall land-use perspective, it will also limit the growth of the trees and postpone the time of timber sale.

The number of years teak is interplanted depends mainly on whether a reasonable level of production is expected from the agricultural crop. With a 2 by 2 m spacing of teak, usually only one year of intercropping is possible. More commonly, the teak is spaced at 3 by 3 m, and this allows two to three years of interplanting. The choice of crop species also influences the number of years it can be successfully interplanted with teak. Upland rice, for instance, can normally only be grown for one or two years because of weed competition and a rapid yield decline when planted in consecutive years. Other crops such as pineapples, maize, and sugarcane may be planted for two or three years because they are less prone to rapid yield declines, are more competitive with weeds, and are easier to weed. Roder et al. (1995) found that interplanting took place in 86% of teak plantations in year 1, 57% in year 2, 37% in year 3, and just 7% in year 4.

Grazing in the teak plantations is limited because of the risk of damaging the young trees and closure of the canopy after only three to four years. There may be possibilities for cut-and-carry systems or for browsing on leguminous bushes planted for soil improvement and protection, but farmers are unlikely to intensify their livestock production because it has traditionally been a low input–low output practice (Hansen and Sodarak 1997).

The possibility of using teak in improved fallows is limited by the long production cycle of 15 to 25 years, which contrasts with the 2 to 6 years available to most farmers for fallowing their land. Even if long fallow periods were possible, teak plantations, as currently managed, would probably have a limited or negative effect on soil fertility. Furthermore, if farmers were able to keep their teak plantations until logging could start, the high income would make crop production unnecessary. Finally, the high coppicing potential of teak would make conversion of the plantation into swidden fields unrealistic, since farmers would rather opt for a second cycle of teak production.

The possibility of replacing some shifting cultivation with teak planting may be technically and economically feasible. Both the value of the annual increment of teak plantations and the return on labor far exceed those of upland rice cropping. It would, therefore, seem possible to devise schemes that finance farmers' management of plantations until logging can start. The ownership of such plantations could lie with private investors, with the farmers, or with a shared ownership. The forest village projects in Thailand are examples of such mixed ownership (Boonkird et al. 1984). The main constraints, however, are the financing, the organization, and acceptance by farmers.

Given these constraints, the most realistic role for teak is as a supplement to shifting cultivation: planting teak on part of the farm while continuing upland cropping on separate land. Relatively small plantations would be preferable and suitable to most farmers, if they were to apply adequate management and retain the ownership.

Conclusions and Recommendations

The potential income and economic spin-off from teak planting in the uplands of northern Laos is very high, particularly compared with current land use. The large areas of degraded forest in northern Laos and the relatively equitable distribution of land both invite a further expansion of the area planted with teak, and improvements in the road system will open up new teak-growing areas. Until recently, the government's strategy has been to promote as much tree planting as possible. However, various problems and constraints have been identified, and can be summarized as follows:

- The supposedly inferior genetic material currently being planted;
- The inadequate management of farmers' teak plantations;
- Competition with agriculture for the better land; and
- The inability of farmers to hold on to plantations for 20 to 30 years.

Various suggestions for correcting these problems have been made.

Teak Improvement

Following the findings of Hedegart (1995), the following recommendations have been made regarding the genetic improvement of teak in Laos:

- Conservation of the existing natural teak forest;
- Conversion of selected natural stands and plantations into seed stands. This would involve setting up a system for selection, control, and certification;

- Provenance trials using local and foreign material;
- Selection of plus trees; and
- Establishment of progeny trials for selection and multiplication of improved material.

Teak Extension

The current management of farmers' plantations is inadequate. However, there is potential for improving the growth and quality of teak by using relatively simple measures. These include timely weeding, thinning, pruning, and fire control, as well as improved propagation methods and selection of seed sources. Better silviculture can enable quicker sale of logs and thus help farmers retain their teak plantations. The following initiatives should receive priority:

- Recommendations should be formulated for the use of improved technology in propagation and plantation management.
- Better and more accessible information should be produced for farmers and extension workers.
- Sites for demonstrating improved management techniques should be established.
- The establishment of smaller plantations, of perhaps 50 to 200 trees, should be promoted. Farmers could realistically maintain and properly manage these plantations over many years and would be less likely to sell them off.
- Teak planting should be promoted in areas without road or river access to provide timber and seed sources for local usage.

Research Needs

Much research is carried out internationally on teak silviculture, some of it in neighboring countries. Considering the limited economic resources and research capability in Laos, advantage should be taken of foreign research by systematically evaluating it and adapting it to local needs and conditions. Research in Laos should concentrate on solving concrete problems of specific local interest. Implementation of the following recommendations would require considerable staff training and strengthening of national institutions:

- Development of systems for intercropping teak with agricultural crops and other tree species to improve overall production and sustainability;
- Studies of the ecology and structure of natural teak stands to help improve their protection and use;
- Adaptation or development of improved silvicultural practices leading to shorter rotations and higher returns. These should be especially relevant to resource-poor upland farmers;
- Investigations of the suitability of teak for specific environments, especially regarding elevation and temperature, rainfall, and bedrock. This could be linked to provenance trials for specified climatic zones; and
- Surveys of pest and environmental stress problems affecting the performance of teak.

Acknowledgments

We acknowledge suggestions made to an earlier draft of this chapter by Ms. Susan Kirk, Mr. Manfred Fischer, Mr. Bengt Frykman, and Dr. Nicholas Tapp. The Swedish International Development Cooperation Agency (SIDA) has supported our teak research and extension, and this is gratefully acknowledged.

References

Boonkird, S.A., E.C.M. Fernandes, and P.K.R. Nair. 1984. Forest Villages: An Agroforestry Approach to Rehabilitating Forest Land Degraded by Shifting Cultivation in Thailand. *Agroforestry Systems* 2, 87–102.

Chazee, L. 1995. Atlas des Ethnies et des Sous-Ethnies du Laos. Bangkok: private publishing, 220.

Department of Forestry (DoF). 1997. Plan to the Year 2000 for Stabilizing Shifting Cultivation by Providing Permanent Occupation. Paper presented at the Stakeholder Workshop on Shifting Cultivation Stabilization, February 6–7, 1997, Vientiane, Laos. Ministry of Agriculture and Forestry and Asian Development Bank.

Hansen, P.K. 1995. Shifting Cultivation Adaptation and Environment in a Watershed Area of Northern Thailand. Communities in Northern Thailand. Ph.D. dissertation, Department of Crop Husbandry, Royal Veterinary and Agricultural University, Copenhagen, Denmark.

Hansen, P.K., and H. Sodarak. 1997. Potentials and Constraints on Shifting Cultivation Stabilization in Lao P.D.R. Paper presented at the Stakeholder Workshop on Shifting Cultivation Stabilization, February 6–7, 1997, Vientiane, Laos. Ministry of Agriculture and Forestry and Asian Development Bank.

Hedegart, T. 1995. *Teak Improvement Programmes for Myanmar and Laos.* FAO Regional Project Strengthening Reafforestation Programmes in Asia (STRAP), Field Document No. 3. Food and Agriculture Organization of the United Nations, 29.

Juville, M. (forthcoming), *Household Census 1996.* Luang Prabang, Lao P.D.R.: EU Microprojects, Luang Prabang.

LSFCP (Lao-Swedish Forestry Cooperation Programme). 1991. Survey of Forest Plantations in Lao P.D.R. Forest Inventory Report No. 1. Vientiane: Department of Forestry and Environment, Ministry of Agriculture and Forestry, 13–17.

LARP (Laos-Australian Reafforestation Project). 1972. *Manual of Operations.* Vientiane: Forest and Water Department, Royal Laos Government.

NOFIP (National Office of Forest Inventory and Planning). 1992. *Forest Cover and Land Use in Lao P.D.R.*, Final Report on the National Reconnaissance Survey. Vientiane: Department of Forestry, Ministry of Agriculture and Forestry, 71 plus appendices.

NSC (National Statistical Centre). 1997a. *Basic Statistics about Socio-economic Development in the Lao P.D.R.* Vientiane: State Planning Committee, 129.

———, 1997b. *Results from the Population Census 1995.* Vientiane: State Planning Committee, 94.

Rao, Y.S. 1993. Teak: A Plundered World Heritage. Occasional Paper 8. Bangkok: Food and Agricultural Organization of the United Nations, Regional Office for Asia and the Pacific, 11.

Roder, W., W. Leacock, N. Vienvonsith, and B. Phantanousy. 1991. Relationship between Ethnic Group and Land Use in Northern Laos. Poster presentation at the International Workshop on Evaluation for Sustainable Land Management in the Developing World, September 15–21, 1991, Chiang Rai, Thailand. Bangkok: IBSRAM.

Roder, W., B. Keoboulapha, and V. Manivanh. 1995. Teak (*Tectona grandis*), Fruit Trees and other Perennials used by Hill Farmers of Northern Laos.

Souvanthong, P. 1995. Shifting Cultivation in Lao P.D.R.: An Overview of Land Use and Policy Initiatives. In: *IIED Forestry and Land Use Series No. 5.* London: International Institute for Environment and Development, 38.

White, K.J. 1991. *Teak: Some Aspects of Research and Development.* RAPA Publication 1991/17. Bangkok: Food and Agriculture Organization of the United Nations, Regional Office for Asia and the Pacific, 70.

Chapter 35

Fallow Management in the Borderlands of Southwest China

The Case of Cunninghamia lanceolata

Nicholas Menzies and Nicholas Tapp

Upland shifting cultivators such as the Hmong (referred to in Chinese as the *Miao*) and the Yao of southern China and northern Vietnam, Laos, and Thailand have frequently been criticized for their apparent disregard for, and destruction of, their forest environment. Shifting cultivation itself is often treated as a purely agrarian system, with no necessary or beneficial relationship with the forest. Yet since the early pioneering work of Conklin and Geertz in the Philippines and Indonesia, it has been generally realized that in comparison, for example, with wet rice monoculture, in ecological terms, the most distinctive positive characteristic of swidden agriculture is that it is integrated into and maintains the general structure of the pre-existing natural ecosystem into which it is projected, rather than creating and sustaining one organized along novel lines and displaying novel dynamics. (Geertz 1963)

Unprecedented political, social, and economic changes have transformed the upland economy in the region comprising the borderlands of Myanmar, southwest China, Laos, Vietnam, and northern Thailand, a region that is becoming known as Montane Mainland Southeast Asia (Rerkasem 1996). The widespread adoption of cultivated vegetable and fruit crops, often in internationally supported opium substitution programs, has transformed significant areas of shifting cultivation into permanently cultivated areas. These not only require large chemical inputs but also contribute to serious problems of lowland pollution and soil erosion. It is now widely recognized by scholars working in this field that shifting cultivation is seldom the sole cause of deforestation. Rather, the causes can be found in increasing pressure on

Editor's Note: A modified version of this chapter, "Miao Cultural Diversity and Care of the Environment", was published in *Links Between Cultures and Biodiversity*, edited by Xu Jianchu et al. (Yunnan Science and Technology Press, 2000).

Nicholas Menzies, Independent Researcher, 1203 Northwestern Drive, Claremont, CA 91711, USA; Nicholas Tapp, Department of Anthropology, Research School of Pacific and Asian Studies, H.C. Coombs Building, Australian National University, Canberra 0200, Australia.

land, water, and forest resources, and cultivation makes only a small contribution. In fact, in 1981, van der Meer estimated that only a very small proportion of the uplands and highlands was actually planted with swidden crops.

Shifting cultivation requires a low ratio of population to land, and a correspondingly large area of fallow, to be a viable and sustainable system. As Boserup showed in 1965, it is where the population-to-land ratio becomes unbalanced that shifting cultivation itself appears as an unsustainable system. In northern Thailand, for example, the main factor responsible for deforestation in the past has been population increase, particularly the upward movement of landless lowland Thai peasants, combined with large-scale logging and mining operations (Tapp 1986) of the type now threatening the forests of Myanmar and Cambodia. In 1984, Cooper attributed most of the deforestation in northern Thailand to the forestry industry, while there was strong evidence that poor road construction, rather than deforestation, had been the cause of much soil erosion (Talbott 1996).

For about 20 years, field research in northern Thailand has shown that systems of swidden cultivation practiced by a number of ethnic groups not only appear sustainable but have been in existence for several centuries (Kunstadter et al. 1978; Grandstaff 1980). This has led to the distinction between traditional or "integral" swidden cultivation and that practiced by people with less experience or a lesser understanding of the potential impacts of clearing and burning forest land. Field research in northern Thailand in 1981 and 1982 showed that, by contrast with newer, inexperienced swiddenists, the traditional cultivation practices of the Hmong were professional and respectful of the natural environment. For example, they cleared fire breaks before new swiddens were burned off, and left the stumps of large trees in the ground to fix the soil after the site had been cleared (Tapp 1989a). Although Hmong swiddening is invariably considered by governments as being among the principal causes of forest degradation, care for the environment is in fact part of the traditional way of life of the Hmong, and they recognize both the importance of fallow periods and the consequences when these are shortened.

Villages are carefully sited to be in harmony with the surrounding mountains and watercourses. They are surrounded by a belt of forest, and any alteration to the major contours or appearance of the landscape is seen as affecting the prospects of those who inhabit it. The Chinese system of geomancy, practiced by both the Hmong and the Yao, hinges upon the establishment of a basic relationship of equivalence and harmony between the community and its natural environment (Kandre 1991). At the time of this research, an entire Hmong village had to be relocated after a watercourse was built in the valley below it to supply a mining concession. The villagers believed that the "veins of the mountain" (*memtoj*), running from the peak to the village, had been "cut". The Hmong use elaborate physical imagery to describe the features of their landscape, portraying it as a living and dynamic entity. The fate of the living is seen as dependent on the way settlements are inserted into the natural ecosystem (Tapp 1988).

It should be of no surprise, then, to find that for centuries in southern China, where the ancestors of the Hmong and Yao originated, and where their descendants still live, forms of upland dry and shifting cultivation have been successfully combined with the management of plantation forest, both for local use and for commercial sales. Members of ethnic minorities such as the Miao[1] and Yao have played a pivotal role in developing this system since very early times (Menzies 1994). This has particularly been the case in the cultivation of *Cunninghamia lanceolata* (see color plate 40).

Cunninghamia lanceolata

Regularly, but inaccurately, referred to as Chinese fir, *Cunninghamia lanceolata* is a species indigenous to the subtropical region of central and southern China and is an

[1] Tribespeople classified in China as Miao also include people from groups other than the Hmong.

important component of Chinese forest ecosystems. It is virtually unique among conifers in that it has an ability to vigorously regenerate through coppicing and is easily propagated through seed, coppice, or cuttings (Fung 1994). *Cunninghamia* can grow as high as 30 m, with a diameter at breast height of 2.5 to 3 m (Sheng 1996). It thrives across a wide latitudinal range, from 200 m above sea level in the eastern and northern regions, to between 2,000 and 3,000 m in western Sichuan and northeastern Yunnan, in the southwest of China (FAO 1982; Richardson 1990). It prefers a warm, moist environment with a mean annual temperature of 15 to 20°C and a mean annual rainfall between 800 and 2,000 mm. While requiring good drainage, it tolerates acid soils (Richardson 1990). Although it is now grown in large monoculture plantations under state guidance, ecologically it is probably not a monocultural species but one which occurs naturally in mixed deciduous and evergreen broad-leaved forests (Richardson 1966). However, as Chandler (1994a) points out, *Cunninghamia*, or *shamu* as it is known to the Chinese, is such a poor competitor in the early growth stages that it is rare to find natural stands in China today (Huo 1975).

Despite this handicap in natural growth, *Cunninghamia* is now the single most important industrial species planted in China, with the exception of bamboo. It covered at least 4.8 million hectares in 1992, and government policy was to reafforest a further 6 million ha by 2000. It makes up about 24% of the total industrial plantation resource in China and more than 50% in the southern region, as well as 39% of the area planted by 1992 under the World Bank–assisted national afforestation program (Hunter and Evans 1992).

Although *Cunninghamia* is used extensively as a plantation species, Chinese forestry authorities have been concerned for several decades by evidence of severe yield declines in second and third rotations. More recently this problem has been the subject of a research project in which one of the authors of this chapter participated and during which much of the contemporary data was collected.[2]

Social Values of Cunninghamia

Cunninghamia is an enormously valued and valuable local wood, and its usage interacts and intersects with the everyday life of local villagers. Traditionally, a copse of Chinese fir might be planted as a villager approached death, since the wood was the best available for coffins, or on the birth of a baby daughter for the dowry of dressers and chests which would be needed for her wedding. Light, strong, flexible, and durable, it has always been considered the best available material for the building of new houses and repair of old ones, for furniture and cabinet work, for pots and beakers, and for agricultural tools such as ploughs and hoes. Even though it burns badly and gives off little heat, *Cunninghamia* branches and tips are used as firewood in areas where better firewoods are scarce, in a mixture of approximately 30% with deciduous woods, pine, and grasses. Its bark is used for lining roofs, pigsties, and outhouses (Tapp 1996b).

Research in 1994 and 1995 on state forest farms in Jiangxi and Fujian showed that a household of three to five adults needed about two cubic meters of *Cunninghamia* timber every year, although this depended on a number of social variables such as house building or repairs, deaths, and out-marrying daughters. A new house largely built of wood needed about 20 m^3 of timber for four rooms of approximately 10 m^2 per room, although few houses these days are built entirely of wood. Four pieces of furniture were usually needed for the dowry for an out-marrying daughter and this averaged about 2 m^3 of timber. Coffins, on the other hand, required about 0.6 m^3 of timber, and, in just one county, there was an average of 6,000 deaths per year, and about half of the deceased were buried (Tapp 1996a).

---

[2] This project, managed jointly by the Overseas Development Administration and the Chinese Academy of Forestry, with assistance from the Institute of Sociology at Beijing University, was conducted from 1993 to 1996 on state forest farms in Jiangxi and Fujian Provinces. Tapp acted as a sociological consultant, through the Edinburgh Centre of Tropical Forestry at Edinburgh University.

The use of *Cunninghamia*, therefore, is vital to the local economy and to communities in the hilly to mountainous areas of southern China. Of course, carpenters commonly supplement *Cunninghamia* with pine and other woods, but none is so valued as the light, durable, and flexible *Cunninghamia* timber.

Commercial Value of Cunninghamia

Despite this social value to the local people, it is far outstripped by *Cunninghamia's* commercial value. This was very clear at the Jiangxi and Fujian project sites, where the main aspiration voiced by villagers was not to directly avail themselves of more timber, but to secure an increased share of the profits made by the state and collective plantations so they could purchase construction materials, as well as coal for household fuel (cf. Menzies 1991). As Chandler (1994a) points out, the primary motivation for intercropping *Cunninghamia* with edible, medicinal, economic, and oil-producing crops was never the dietary value of the agricultural crops. Rather, it was the opportunity to harvest the cash value of shamu timber. Chandler notes that in the 17th century, Chinese gazetteers reported double to tenfold returns on investments from crops of *Cunninghamia*. It was these commercial values that led farmers to perceive timber as a crop in its own right, rather than as a byproduct of clearing the land or as a secondary crop grown for household needs (Menzies 1988a, 95).

From the 16th century onward, commerce and industry grew rapidly in the Yangtze basin and the southern coastal region. The lowland cities clamored for timber, and the highland farmers increasingly found themselves able to provide it, usually through extensive river transport systems (Menzies 1994). In some areas, intercropping with *Cunninghamia* was used as a means to persuade traditional shifting cultivators to convert to more permanent settlements, and forestry became a viable alternative to establishing terraced ricefields (Menzies 1988a). From the late Ming dynasty, new settlers, who were known as "shed people," were often required, as tenants, to clear land and plant *Cunninghamia* and Masson pine. They could either use the land for grain or cash crops until the trees matured or receive a final share of the profits (Menzies 1996).

Cunninghamia provided cash, rather than a subsistence income, and could be used in times of special need to pay for medical emergencies, weddings, and funerals. This practice is similar to that in other countries where trees serve as a growing asset to be used at times of special need (Chambers and Leach 1987). The commercial importance of *Cunninghamia* can be gauged from the fact that, in the 16th century, a specific system of measurement and valuation was developed just for this species (Menzies 1996).

Historical Aspects of Cunninghamia *Management*

Cunninghamia has been maintained as a plantation species for more than a thousand years in the hilly to mountainous regions of southern China (Menzies 1994). Traditionally, it was regenerated through seedlings, coppice, or cuttings, and different parts of the country specialized in different propagation techniques. Sometimes it was maintained as a pure monoculture species, but more often it was cultivated in mixed plantations with perennial cash crops such as tea oil (*Camellia oleosa*), tung oil (*Aleurites* spp.), some annual staple crops such as wheat or millet, and sometimes with other timber species such as Masson pine (*Pinus massoniana*). Today, large *Cunninghamia* plantations are managed by state-run "forest farms," which are collective organizations attached to local townships and larger administrative villages. Since the reforms of Deng Xiaoping, tracts of these plantation forests have been leased to individual householders or groups of householders. In prerevolutionary China, by contrast, plantations of significant size were owned by rich landlords who hired local or seasonal laborers to work on them, by temples, or by large landholding estates like the lineages common in the southeast of the country. Often there was a division of ownership between a distant owner, perhaps an urban surname association or a trading guild, who retained the subsoil rights, and farmers who leased the topsoil rights and owned the trees, and could mortgage or sell their rights. The farmers usually paid other agricultural workers to plant, harvest, and maintain the plantations and shared the profits with the subsoil owners as rent (Menzies 1988b). While larger plantations tended to be maintained in predominantly Han areas such as Fujian and Jiangxi, smallholder production was more common in the southwest and Hunan, particularly in ethnic minority areas. In more remote regions, poor farmers were also able to plant *Cunninghamia* on unclaimed mountain sites (Menzies 1988a). As early as the year 1173, a Song dynasty official, Fan Chengda, referred to ethnic minorities cultivating *Cunninghamia* in his diary of a journey in southern China:

> The tribespeople do not practice much agriculture. Most of them make a living planting *Cunninghamia. Cunninghamia* is something that grows easily so it may be harvested with little danger of exhausting [the supply]. (Fan Chengda 1173)

Fan gave no indication of the silvicultural system used by the tribespeople, nor is it clear to which of the non-Han people in southern China he was referring. His account was, however, followed over the centuries and references to it can be found in a wide range of publications, including local histories, guidebooks to sacred mountains, briefing manuals for military expeditions, and travel accounts written by western authors from the 19th century onward (Menzies 1996, *574–578* and *626–630*). These accounts offer more detailed descriptions of a system of cultivation practiced by Han and minority ethnic groups, often associating it specifically with the Hmong/Miao, the Yao, the She, and the Dong. While details vary, the system involved land clearance, sometimes followed by burning. Farmers then planted *Cunninghamia,* usually using coppiced sprouts from older trees, and intercropped the plot with a wide variety of annual crops. After a few years, when annual crops could no longer survive under the shade of the trees, shade-tolerant cash crops might be grown until the forest cover prevented any further intercropping. Rotation periods varied over the centuries and, apparently, between ethnic groups, with rotations as short as 15 to 20 years and as long as 60 or even 90 years. After harvesting, the sequence began again with land clearance and intercropping.

The relative intensity with which *Cunninghamia* plantations were managed varied from region to region across China's vast landmass and a range of traditional techniques and systems were employed over the course of a millennium. Nevertheless, traditional management systems had certain common features that we can usefully consider here. Many of these traditional management practices persist in China today in smaller woodlots, in areas where local villagers have some measure of management control over plantations, or where cash investment is unavailable for more intensive methods of cultivation. Of particular interest to us is the system of agroforestry used in the cultivation of *Cunninghamia,* the extent of clear-felling involved, and the role of fire in the management system. These issues are of particular interest because the ethnic minorities who have been prominent in developing *Cunninghamia* cultivation are the same ethnic minorities that practice shifting cultivation. In many of the more traditional but more remote ethnic minority areas, land clearance may take the form of patch cutting rather than large-scale clear-felling, and burning may have been used only at limited points in the sequence of management activities. Clear-felling and extensive site burning are, by contrast, a marked feature of more modern systems of *Cunninghamia* plantation management.

Mixed Cultivation

Very often the practices of ethnic minorities have contrasted favorably, in an ecological sense, with their ethnic Han neighbors, who have settled and colonized southern China largely since the Song dynasty. For example, it was more common in the ethnic minority areas to see *Cunninghamia* in mixed plantations with *wenmu (Cryptomeria fortunei)* (Yu 1983; Wu 1984; Chandler 1994b). Chandler (1994a) contrasts the traditional practices of a She minority lineage (Miao-Yao language group) with two Han Chinese lineages in Fujian Province. One of the Han lineages had been settled in the area for more than 800 years, the other for a mere 225. The longest settled Han Chinese lineage, the Wu, imposed a maximum of three rotations of *Cunninghamia* on a single site before returning it to secondary forest for about 50 years. The more recently settled Han Chinese lineage showed no awareness even of the reasons for the period of fallow, and the minority She lineage practiced mixed cultivation and demonstrated a more ecologically informed understanding of the whole problem. They put it this way:

> The Wu people have that problem [the need to rotate shamu off the growing site] because they do not understand the soil. The soil is like a person. Turnips are good to eat and the Wu like to eat a lot of turnips. But if they eat only turnips, then very soon they become sick. Pork is also good to eat, but if we eat only pork, then very soon we also become sick. If we eat turnips and pork and rice and vegetables together we do not become sick, our health is good, and we are happy. The Wu soil is not happy. The Wu give it only shamu. They do not give it wenmu or *chumu* or *dumu* [both of the latter names refer to several genera of the Fagaceae] or other trees. Only shamu, and the soil becomes sick. Shamu buys a lot of money. The Wu want only money, so the Wu soil becomes sick. (Chandler 1994a)

The She minority lineage quoted here practiced a mixed cultivation of *Cunninghamia* with long rotations and no apparent yield decline. They saw no need for the fallow periods required by the neighboring Han lineage, whose rotations averaged only 35 years.

Integration with Agriculture

The traditional system of *Cunninghamia* management has been described as a remarkable instance of shifting cultivation in forestry (Hunter 1996) and has always been closely integrated with agricultural systems. Even in the state forest farms of Jiangxi and Fujian, villagers have planted medicinal plants, papaya, and gourds under the shade of young *Cunninghamia* trees in remote areas of the plantation (Tapp 1996b). Such intercropping has always been an integral feature of the system.

The 1639 farming manual "Complete Treatise on Agricultural Administration" (*Nong Zheng Quan Shu*), by Xu Guangqi, advised a preliminary plowing of the land before planting *Cunninghamia*, then growing sesame with the young trees for a year or more, after which millet should be planted in the summer and wheat in the winter, as long as the location was suitable for such crops. In this way, it said, weeding and hoeing would not be necessary. It is believed that this system may have been the precursor of the *taungya* swidden system used by the British in Burma (Menzies 1994, 1996). The 17th century advice has since become the general practice. Annual crops planted during the early years of growth before canopy closure include wheat, millet, maize, sweet potato, cassava, and vegetables. Successional systems have also been used in which annual crops are followed by perennials such as tung oil and tea oil (Menzies 1994).[3]

In southwestern China, the related Miao, Yao, and She groups, as well as the Dong, have been particularly associated with the commercial production of timber.

---

[3] Tung oil trees were fully harvested in the fifth and sixth years and became old and feeble only as the canopy closed in the ninth to tenth year (Sheng 1996).

As noted earlier, tribal production of *Cunninghamia* is recorded from 1173, and the local history of Jianghua county in Hunan records commercial production of *Cunninghamia* from the early 17th century. The people of that county, which remains an autonomous Yao county to this day, cultivated *Cunninghamia* and rafted it down the Yangtze River for sale in the cities downstream. Income from the timber trade allowed the people of Jianghua to buy grain and other staple foods that were difficult to cultivate in their steep upland valleys (Menzies 1988a). Fortune (1939) described how the Yao of northwest Guangdong planted trees, conserved them, and cut and carried timber to the riverside for sale. Forestry occupied a crucial place in the agricultural life of the Yao and was a main source of income. In Yao-Ling Pai (Guangdong), which is now a protected forestry area (Tapp 1989b), *Cunninghamia* was traditionally cultivated for sale to wood merchants in Sanjiang (Fortune 1939).

In these montane regions, terraced irrigated ricefields are scarce and most land is devoted to the production of maize or buckwheat, sweet potato, and leafy green vegetables. Animal husbandry forms an important part of the local economy, and *Cunninghamia* has functioned as an important cash crop and source of security, either through laboring on local plantations or through direct revenues from commercial sales of timber to private wood merchants and government licensees. Forest cover in many areas was, therefore, maintained as a fruitful and integral part of the upland agricultural economy.

Intercropping has therefore long been a characteristic of these traditional *Cunninghamia* cultivation systems and has long been associated with minority groups such as the Hmong/Miao and Yao, who are practitioners of shifting and dry upland cultivation. In Guangdong Province, a system of *Cunninghamia* intercropped with cassava is described as having been originally devised by the Miao and subsequently adopted by the Han (Pendleton 1940, in Menzies 1994). Other traditional systems include intercropping for at least two to three years with maize, peanuts, or beans, or *Cunninghamia* grown together with tung oil for three to four years. Sweet potato, soybeans, mushrooms, and tea oil are also commonly intercropped. In parts of Fujian, a complex and lucrative medicinal fungus industry has grown up in the shade of the *Cunninghamia* trees (Tapp 1996b). Chandler (1994b) has shown how, in Fujian, agricultural intercropping with a wide variety of cereal, cash, medicinal, and oil-producing crops has formed a major part of the traditional management system centered around *Cunninghamia*. He argues that such intercropping in the initial stages of growth is an integral part of site preparation and tending to the young trees that contributes importantly to their establishment, growth, and survival.

Commercial production of *Cunninghamia* in mountainous areas, pioneered by upland ethnic minorities, was an integral part of a local economy in which upland dry agriculture, animal husbandry, hunting, and fishing were also important elements. The common practice of intercropping in *Cunninghamia* plantations gives good reason to propose that these farming systems should properly be seen as a form of shifting cultivation. They would appear to fit well with a category of upland farming common in Southeast Asia that Rambo (1996) has described as "composite swiddening agroecosystems."

Shelterwood and Declining Yields

One aspect of what is perceived as the traditional management system for *Cunninghamia* has been widely and intensively replicated on large areas of China's present state forest farms, and the World Bank has advised against it. It is preliminary burning and intensive site preparation before planting. The clear-felling practiced today is also seen as part of the traditional past.

While often described as relics of traditional practice that are difficult to change because of local custom and the lack of available labor, it must be remembered that in traditional systems, *Cunninghamia* was never cultivated in anything like the vast areas common today. In addition, it was intercropped with a variety of medicinal, edible, or economic crops, or cultivated in mixed plantations with other species. In those areas where it was intensely cultivated as a monoculture, it was possible to allow long fallow periods in which the forest reverted to mixed pine, fir, and broadleaf forest. Certainly, both clear-felling and burning before planting were very generally practiced in the past, although on much smaller plots with wider spacing between the trees. As we have already remarked, burning may often have been carried out at limited points in the sequence of activities, and patch cutting was also a common practice, particularly in ethnic minority areas.

The Overseas Development Administration (ODA) research project, from which this chapter partly arises, investigated second rotation declining yields in *Cunninghamia* plantations. The key recommendations from this project singled out intensive site preparation, and particularly burning the site after clear-felling, as the most likely causes of yield decline. The project suggested a change in the silvicultural management of *Cunninghamia*, by leaving the slash unburned and planting seedlings through it, while also experimenting with shelterwood systems as an alternative to complete clear-felling. The shelterwood system would leave a small number of trees from the previous crop standing while seedlings grew in their shade. The elimination of clear-felling would also have made it easier to persuade villagers and forest managers not to burn before planting. In any event, while burning and other trials were conducted, it proved unfeasible to conduct a shelterwood experiment because Chinese silviculturalists, who classify *Cunninghamia* as a yang or sun-loving tree, were convinced that it would not tolerate shade,[4] nor would local villagers take kindly to the additional difficulties of planting among the slash.

The practice of burning the site before planting *Cunninghamia* may have originated with ethnic minorities such as the Miao and Yao. As traditional shifting cultivators, they usually cleared areas of forest in a similar way for their agricultural crops. In a sense, this contributed to the image of traditional *Cunninghamia* management as a system of shifting cultivation adapted for forestry. However, in many areas it seems that, beyond the initial clearing of forest, succeeding rotations were not clear–felled by the ethnic minorities. Here, again, their practices contrast favorably, in an ecological sense, with those of their Han neighbors.[5] In Jianghua county, southern Hunan Province, where traditional management practices are still followed, the Yao leave some trees standing after harvest to continue growing through a second rotation to produce larger sawn timber for house construction. In contrast, their Han neighbors in the same area clear-cut at the end of each rotation (Menzies 1994). Among the Yao, Dong, and Miao minorities on the Guangdong-Guangxi-Hunan border, five to six trees per mu (1/15th ha) are traditionally left to grow through to a second or even a third rotation, and shorter rotations are practiced under these stands. This system closely approximates the shelterwood system recommended by the ODA consultants.[6]

Referring, finally, to our previous example of the She minority lineage system of *Cunninghamia* management in Fujian: it was the She people, not the two Han Chinese lineages, who tended to open the canopy further by not replacing early mortality among the regenerating trees, and only the minority She who avoided clear-felling a site altogether if it had a significant proportion of young natural regeneration of *Cunninghamia*, wenmu (*Cryptomeria*), or useful broadleafs. The rotations of the She, too, were far longer than either of the two Han lineages. To obtain the large sawn timbers needed for house construction, they explained that a complete rotation should last about 95 years (Chandler 1994a).

---

4 For details of the Chinese classification of trees into yin and yang, see Menzies (1996, *610–612*).

5 While the Hmong and Yao have largely remained pioneer shifting cultivators in Southeast Asia until very recently, in many areas of China the transition toward more permanent forms of cultivation was accomplished several centuries ago.

6 Personal observations by Menzies during a field visit to Tongdao Dong Autonomous Prefecture, Hunan, June 1994.

Conclusions

If the soil allelopathy hypotheses preferred by some Chinese silviculturalists (Chou et al. 1987; Cheng 1992; Zhang 1993) prove to be the correct explanation for the declining yields in second or third rotations of monocultural *Cunninghamia* stands, then an initial burning and intensive preparation of the site may well prove to be beneficial. However, if the reasons prove to be an overpreparation of the site before planting, then almost certainly the less intensive practices of land-use management that seem to have characterized traditional ethnic minority *Cunninghamia* cultivation will form part of a solution to the problem.

Extensive intercropping with agricultural and other crops, wider spacing, mixed plantations of *Cunninghamia* with other species, and above all, a far closer integration of industrial forestry with subsistence agrarian systems, are all part of traditional *Cunninghamia* management systems in China. Ethnic minorities, particularly the Yao, Miao/Hmong, Dong, and She, have been prominent in the development of these systems. They demonstrate a possible trajectory of evolution from swidden systems, in which land clearance for subsistence agriculture is the primary objective, to integrated "composite swiddening agroecosystems," in which crops, livestock, and trees all contribute in different ways to rural livelihoods. These management systems, developed by ethnic minority groups over several centuries or more, are important examples of "trees in agricultural landscapes," as agroforestry has recently come to be defined.[7] They also shed a new light on the historical image of the Yao and the Hmong in Southeast Asia as destructive slashers and burners. Clearly, their ancestors in China practiced a careful husbandry of the forest that rationally reduced the pressures of subsistence agriculture on forest resources.

References

Boserup, E. 1965. *The Conditions of Agricultural Growth*. London: George Allen and Unwin.

Chambers, R., and M. Leach. 1987. *Trees to Meet Contingencies: Savings and Security for the Rural Poor*, Social Forestry Network Paper 5a. London: Overseas Development Institute.

Chandler, P. 1994a. Shamu Jianzhong: A Traditionally Derived Understanding of Agroforest Sustainability in China. *Journal of Sustainable Forestry* I, 1–23.

———. 1994b. Adaptive Ecology of Traditionally Derived Agroforestry in China. *Human Ecology* 22: 415–442.

Chen, S., S Pei, and J. Xu. 1993. Indigenous Management of the Rattan Resources in the Forest Lands of Mountain Environment: The Hani Practice in the Mengsong Area of Yunnan, China. *Ethnobotany* 5, 93–99.

Cheng, H.H. 1992. A Conceptual Framework for Assessing Allelochemicals in Soil Environment. In: *Allelopathy: Basic and Applied Aspects*, edited by S. Rizvi and V. Rizvi. New York: Chapman and Hall.

Chou, Chang-hung, Hwang Shih-Ying, and Peng Ching-I. 1987. The Selective Allelopathic Interaction of a Pasture-Forest Intercropping in Taiwan. *Plant and Soil* 98, 31–41.

Cooper, R.G. 1984. *Resource Scarcity and the Hmong Response: Patterns of Settlement and Economy in Transition*. Singapore: Singapore University Press.

Fan, C. 1936 (ed.) *Can Luan Lu* (A Record of Carriages and Phoenixes). Shanghai: Cong Shu Jicheng Series.

Food and Agriculture Organization of the United Nations. 1982. *Forestry in China*. Rome: FAO.

Fortune, R.B. 1939. Introduction to Yao Culture. *Lingnan Science Journal* 18(3), 343–450.

Fung, L.E. 1994. A Literature Review of *Cunninghamia lanceolata*. *Commonwealth Forestry Review* 73(3), 172–192.

Geertz, C. 1963. *Agricultural Involution*. Berkeley, CA: University of California Press.

Grandstaff, T. B. 1980. *Shifting Cultivation in North Thailand*, Resources Systems Theory and Methodology Series. Tokyo: United Nations University.

Hunter, I.R., and J. Evans. 1992. Site Degradation in Chinese Plantations in the Eastern and Southern Monsoon Region of China. Report on a Visit to China for the Overseas Development Administration, November 20, 1992.

[7] This case can usefully be compared with the dongya system practiced by the Hani/Akha on the Myanmar border. See Pei (1985), Wang (1988), Yu (1994), and Chen et al. (1993) for comparative data on the Hani, Dai, Dulong and Nu.

Hunter, I. 1996. Personal communication with the authors.
Huo, Y. 1975. The Effects on Soil Quality and Tree Growth of Burning, Site Preparation, and Intercropping with Shamu. *Guangdong Forest Science and Technology* 4, 7–10 (Chinese language).
Kandre, P. 1991. The Relevance of Ecology and/or Economy for the Study of Yao Religion. In: *The Yao of South China: Recent International Studies*. Paris: Pangu, Editions de l'A.F.E.Y.
Kunstadter, P., G. Chapman, and S. Sabharsi (eds.) 1978. *Farmers in the Forest*. Honolulu, HI: East-West Center, University Press of Hawaii.
Menzies, N. 1988a. Trees, Fields and People: The Forests of China from the Seventeenth to the Nineteenth Centuries Ph.D. thesis. University of California, Berkeley.
———. 1988b. Customary Law and Tree Tenure in China: Whose Trees? In *Proprietary Dimensions of Forestry*, edited by Louise and John Bruce. Boulder: Westview Press.
———. 1991. Rights of Access to Upland Forest Resources in Southwest China. *Journal of World Forest Resources Management* 6, 1–20.
———. 1994. *Forest and Land Management in Imperial China*. London: St. Martins Press.
———. 1996. The History of Forestry in China. In *Science and Civilization in China* Vol 6 Pt.III, edited by Joseph Needham. Cambridge: Cambridge University Press.
Pei S. 1985. Some Effects of the Dai Peoples' Cultural Beliefs and Practices upon the Plant Environment of Xishuangbanna, Yunnan Province, Southwest China. In *Cultural Values and Human Ecology in Southeast Asia*, edited by Karl Hutterer, A. Terry Rambo, and George Lovelace. Ann Arbor: University of Michigan.
Pendleton, R. 1940. Forestry in Kwangtung Province, Southern China. *Lingnan Science Journal* 16(3), 473–495.
Rerkasem, B. (ed.) 1996. *Montane Mainland Southeast Asia in Transition*. International symposium November 12–16, 1995, Chiang Mai, Thailand. Chiang Mai: Chiang Mai University Press.
Rambo, A. T. 1996. The Composite Swiddening Agroecosystem of the Tày Ethnic Minority of the Northwestern Mountains of Vietnam. In *Montane Mainland Southeast Asia in Transition*, edited by Benjavan Rerkasem. Chiang Mai, Thailand: Chiang Mai University Press.
Richardson, S.D. 1966. *Forestry in Communist China*. Baltimore: Johns Hopkins University Press.
———. 1990. *Forests and Forestry in China: Changing Patterns of Resource Development*. Washington, DC: Island Press.
Sheng, W. 1996. A Review on Management of Chinese Fir Plantations. Paper presented to the final workshop of the CAF-ODA Cooperative Project to Correct Yield Declines in Successive Rotations of Chinese Fir, March 25–30, 1996, Chinese Academy of Forestry, Beijing.
Talbott, K. 1996. Roads, People and Natural Resources: Towards a Regional Policy Framework for Transport Infrastructure in Montane Mainland Southeast Asia. In *Montane Mainland Southeast Asia in Transition*, edited by Benjavan Rerkasem. Chiang Mai, Thailand: Chiang Mai University Press.
Tapp, N. 1986. *The Hmong of Thailand*. London: The Anti-Slavery Society and Cultural Survival.
———. 1988. Geomancy and Development: The Case of the White Hmong of North Thailand. *Ethnos* 53, 3–4.
———. 1989a. *Sovereignty and Rebellion: The White Hmong of Northern Thailand*. Singapore: Oxford University Press.
———. 1989b: Reflections on Fieldwork among the Yao. In *Ethnicity and Ethnic Groups in China*, edited by Chien Chiao and Nicholas Tapp. Hong Kong: New Asia Press.
———. 1996a. Social Aspects of Chinese Fir Plantation in China. Paper presented to the final workshop of the CAF-ODA Cooperative Project to Correct Yield Declines of Successive Rotations of Chinese Fir, March 25–30, 1996, Chinese Academy of Forestry, Beijing.
———. 1996b. Social Aspects of Chinese Fir Plantations in China. *Commonwealth Forestry Review* 75(4), 302–308.
van der Meer, C.L. 1981. Rural Development in Northern Thailand: An Interpretation and Analysis. Ph.D. thesis. University of Groningen.
Wang, N. 1988. Preservation and Development of Minority Cultures in Southwest China: A Study on Ethnoecology. Paper presented at the 12th International Congress of Anthropological and Ethnological Sciences, July 24–31, 1988, Zagreb.
Zhang, Q. 1993. Potential Role of Allelopathy in the Soil and Decomposing Roots of Chinese Fir Replant Woodland. *Plant and Soil* 151, 205–210.
Yu, X. 1994. Protected Areas, Traditional Natural Resource Management Systems and Indigenous Women: Case Study in Xishuangbanna, P.R. China. Paper presented at the MacArthur Grantees Meetings/ICIMOD Seminar on Indigenous Knowledge Systems and Biodiversity Management, April 13–15, 1994, Katmandu, Nepal.
———. 1983. *Shamu*. Fuzhou: Fujian Science and Technology Press (Chinese language).
Wu, Z. (ed.). 1984. *Shamu*. Beijing: Forestry Press (Chinese language).

Chapter 36

Indigenous Fallow Management with *Melia azedarach* Linn. in Northern Vietnam

Tran Duc Vien

Upland swidden agriculture constitutes a profound relationship between humans and nature. Countless ethnic groups in Europe, tropical Africa, the Americas, and the Asia-Pacific region have practiced it since the Neolithic era. Today, in tropical and subtropical zones, swidden agriculture still persists under a variety of names. It involves the clearing and burning of primary or secondary forest, dibbling and dropping seeds, and growing crops without fertilizer. Fields are planted with as many as five crops, over several years, before the soil loses its fertility and weed competition becomes too serious. Then the swidden is abandoned, the farmers move to a new plot, and the forest regenerates. The actual area of cultivation takes up only 15% to 20% of the cleared surface area (Cuc 1993).

Swidden agriculture has persisted primarily in tropical zones. It continues to be practiced where a fragile natural environment hinders the development of more intensive types of agriculture. Nearly 74% of Vietnam's national territory, an area covering 24.4 million hectares, can be described as "upland." Of this total, 20 million hectares have slopes greater than 15°. Nevertheless, 2.7 million hectares of this upland are classified as agricultural land, and, according to one estimate, about 1.4 million hectares of it consist of currently cultivated swidden fields (NIAPP 1993). Another estimate says that the total area under shifting cultivation, including fallow areas, is about 3.5 million hectares (Sam 1994).

Many researchers have proven that a minimum standard of living for swidden cultivators can only be maintained when the population density is no more than 5 to 10 people/km^2 of forest. According to Rambo (1997, *27*) traditional systems of swidden agriculture in northern Vietnam are sustainable only if the population density is well under 40 persons/km^2. Only when pressure on the land use is limited in this fashion can the forest still regenerate sufficiently to maintain the cycle.

The uplands of Vietnam are home to 24 million people, including almost all of the country's ethnic minorities. One 1989 estimate says that nearly 2.9 million people, from more than 480,000 households, are engaged in shifting cultivation. What is more, 50 out of Vietnam's 54 ethnic groups practice shifting cultivation, and it is embedded in a diversity of traditions, customs, and cultures. For these reasons,

---

Tran Duc Vien, Center for Agricultural Research and Ecological Studies, Hanoi Agricultural University, Gialam, Hanoi, Vietnam.

upland areas are very prominent in any consideration of economic development and environmental protection.

Swidden cultivation can be both highly destructive to the forest and a cause of serious soil erosion. Policymakers currently view it as the single major cause of deforestation and soil erosion. This claim gains special stature in the face of an estimate by the former Ministry of Forestry that about 100,000 ha of forest area is being lost annually in Vietnam, and about 50% of this is due to shifting cultivation. Swidden cultivation has, consequently, become the focal point of several government programs, most notably the Fixed Cultivation and Sedentarization Program. However, it is rarely recognized that there are many different types of shifting cultivation, with different scales of impact on the natural environment. A distinction can be made between migratory shifting agriculture, where whole villages move from one place to another, and sedentary shifting cultivation, where only the swidden fields are rotated (Sam 1994). Apart from a few ethnic groups, most of the shifting cultivation practiced in northern Vietnam is of the sedentary variety. Swidden cultivation can also be divided into upland-based shifting cultivation and supplementary shifting cultivation (Khiem and van der Poel 1993).

Erosion rates on hilly lands in Vietnam are generally high, although there have been conflicting efforts to quantify soil losses. Some research has shown that under swidden cultivation, topsoil washes away within only two to three years (Toan 1990), whereas a study conducted by the Center for Natural Resources and Environmental Studies (CRES), in the midlands of northern Vietnam, found that soil loss under cassava monoculture amounted to between 147 and 245 tonnes/ha/yr, or between 0.9 and 2.1 cm of topsoil every year. This represented the loss of about 1 ton of humus, 50 kg of N, 50 kg of P, and 500 kg of K. However, the same study found that by intercropping cassava with *Tephrosia candida* or another legume species, soil loss was reduced to between 20 and 30 tonnes/ha/yr (Cuc 1995). Yet another case study, carried out in northwestern Vietnam by the Center for Agricultural Research and Ecological Studies (CARES) showed that the amount of soil lost to erosion was not as great as previously suggested, even in monoculture swiddens with steep slopes of more than 45°. It found that soil loss varied from about 40 to 90 tonnes/ha/yr.

The CARES study also found that the balance of N, P, and K was negative for all crops, even in the first years of cultivation. In almost all cases, and with the exception of K, the nutrients available from recycled materials could not cover the combined nutrient loss from soil erosion and the export of harvested products in the first two years of swidden rice production. This finding confirmed the rationale underlying the usual indigenous practice of cultivating swiddens for no more than three to four years (Vien 1996).[1]

The gradual reduction of soil fertility in swidden fields corresponds to a cyclical transition in land use. This begins with the transition from natural forest to shifting cultivation. The yield of annual crops gradually declines within a swidden cycle. The longer land is under cultivation, the greater the fertility loss, and, under these conditions, swidden cultivators choose drought resistant, traditional crops that produce low but reliable yields. However, when yields fall below a certain minimum threshold, the land is fallowed[2], and, if sufficient land is available, there is potential for forest succession. This is the way swidden farmers intend it, because forest succession is a viable basis for the recovery of lost soil fertility. However, because of population growth, in-migration, and the creation of forest reserves and wildlife habitats, which deny use of this land to swidden farmers, there is a growing scarcity of land and swidden farmers are forced to cultivate largely infertile former forest land on a more permanent basis. Fallow periods are becoming shorter and the forest is not

---

[1] Another negative effect of swidden cultivation is that growing amounts of often acid sediments are deposited on downstream irrigated fields, leading to an increasing workload for those farmers.

[2] In addition to the loss of soil fertility, swidden plots are also abandoned because of increasing weed growth. Weeding requirements eventually exceed available labor.

being given the opportunity to recover. Studies and observations suggest the following main scenarios regarding fallow successions:

- Where a swidden field is cleared from forest, or where good-quality forest remains in surrounding areas, regeneration after cultivation of crops is rapid and soil fertility recovers relatively quickly, usually within seven years.
- On abandoned fields where bamboo dominates the fallow succession, the vegetation cover remains arrested and converts to trees only gradually, over long time periods. However, soil under bamboo forest is relatively fertile.
- In areas where the forest has been heavily degraded and low bushes are dominant, forest rehabilitation may take 15 to 20 years.
- In areas where there is no forest left, fallowed fields are dominated by grasses with low resource requirements. For example, *Imperata cylindrica* is dominant in northwest and central Vietnam. In these areas, forest rehabilitation is very difficult.

In summary, the principal factor contributing to the sustainability of swidden agriculture is the restoration of the forest and thus the recovery of soil fertility for the next cropping rotation.

Fallow Management in Northern Vietnam

In most cases of shifting cultivation in northern Vietnam, the duration of the fallow is now similar to the duration of cultivation. Swiddenists have found many ways of maintaining the productivity of degraded soil in their fields, and approaches to fallow management encompass an astonishing diversity of practices. This section discusses findings from our research on different fallow management practices in northern Vietnam.

First, Figure 36-1 illustrates typical cropping cycles before 1986, compared to those of the present day. Before 1986, rice was grown in swiddens for two to three years, or occasionally four. This was followed by two years of cassava cultivation and then five to ten years of fallow. Currently, rice is grown for one or two years, or occasionally three, followed by one or two years of cassava and, finally, three or four years of fallow.

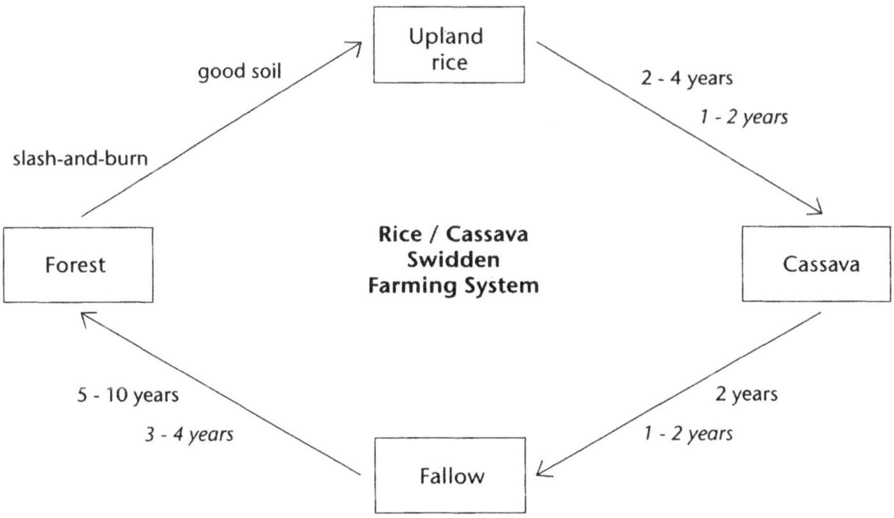

Figure 36-1. The Rice/Cassava Swidden Cycle, before 1986 and Present Day
Note: Data for the present swidden cycle are given in italics.

Among the indigenous innovations aimed at improving the functions of the fallow, Dzao farmers in Moc Chau district, Son La Province, have developed an interesting land classification system to restore the forest after the first crop of shifting cultivation. They have divided their fields according to their location, whether they are nearby the village—or distant—and their soil quality. The characteristics of the fields determine what will be planted, and how.

On fields located not too distant from the hamlet, and regardless of their soil quality, which is usually good, farmers intercrop upland rice with trees. At the same time as they plant their first rice crop, they plant *móc* (*Caryota urens* L.), a tree that provides fiber for making conical palm hats, or *sa nhan* (*Amomum* spp.), a medicinal tree. While the farmers maintain their successive rice crops, they also attend to the interplanted tree seedlings, which need some care in the first three years. After the rice crops, the field is abandoned for five to eight years, at the end of which the trees are harvested and a new rice swidden begins.

On fields with poor soil quality, far away from the hamlet, the Dzao plant fruit trees, including *trâu* (*Aleurites montana*) or *so'* (*Camellia oleifera*), before the field is abandoned for forest regeneration. *Aleurites* and *Camellia* are less popular these days than they were in the past because of unstable markets and their long period of immaturity before providing a harvest. Ethnic minority groups in Yen Bai Province follow a similar practice to that of the Dzao of Moc Chau, but instead plant *Cinnamomum* or ginger as preferred fallow species.

In the Ngoi Lao area, many plantations supply raw material to the Bai Bang Pulp and Paper Mill. Farmers plant *Styrax* and *Manglietia* with their first swidden crops of rice, cassava, or soybeans. The intercropping increases the care provided to the tree seedlings in the first two to three years, through frequent weeding and protection, and improves soil productivity. After the agricultural cropping, the fields are fallowed and the trees dominate. At eight years of age, the height of the trees is about 22 m and they have a diameter at breast height (dbh) of 19 to 20 cm, which is more than the required felling limit. The trees planted in this system can be cut 2 years earlier than conventional plantation trees, which are usually harvested after 10 years. However, in some cases this system reduces the productivity of the agricultural crops by 14 to 20%. The best combination is *Manglietia* trees with soybean in the first year and rice in the second year. The trees are planted at a density of 2,500 per ha and later thinned to a stand density of 1,600 per hectare.

Fallow Management Using *Melia azedarach* Linn.

Swidden cultivators from several ethnic groups in northern Vietnam, including the Muong, Tày, Thái, Cao Lan, and Dzao, incorporate *Melia* into their farming systems as an improved fallow species (see Figure 63-2).

Melia is very popular throughout Vietnam. It is a fast-growing and deep-rooted deciduous tree whose leaves are considered a good source of green manure for use in rice paddies. *Melia* charcoal is also used to make gunpowder for hunting. The timber can be harvested 7 to 10 years after planting, when the dbh of the trunks is about 20 to 30 cm. The trunks are submerged for some time under deep water and, if possible, covered with mud. They are then used as high-quality construction timber. Branches cut during thinning or after harvesting are used as firewood.

There are several ways of introducing *Melia* into swidden fields so that it grows to dominate fallow vegetation after the cropping phase. Seeds can be sown directly, seedlings can be transplanted, or regeneration can be simply left to nature.

Selection and Treatment of Seeds

Melia grows and matures rapidly, and the trees flower and produce fruit after just four years. However, local people select seeds only from those trees that are between five and eight years old, have a good shape, exhibit good growth, and have a dbh of

Figure 36-2. *Melia azedarach* Linn.
Note: 1: leaves; 2: flowers; 3: fruit.
Source: Vu Van Me 1993, *29–30*.

more than 15 cm. When about two-thirds of the fruits have ripened, they are picked to avoid plunder by birds. Usually, they are then sorted, and only those with a diameter of more than 8 mm and a length of more than 12 mm are selected. They are covered for a few days, then the flesh is removed and they are washed and dried in the sun for five to six days. Then they are stored in a dry place. Some farmers adopt a different procedure and delay peeling and washing the fruit until just before sowing or otherwise soak them in water.

Each "nut" contains up to six seeds and has a hard shell, so they must be treated with heat or warm water to stimulate germination. In the case of treatment with heat, people dig a shallow hole, lay the nuts on the bottom, cover them with dry straw or grass, and then burn them. In the case of treatment with hot water, three parts of boiling water are mixed with two parts of cold water and the nuts are soaked in this for 24 hours. During this time, the water is changed twice, and the container is packed in rice husks or straw as insulation to keep the water warm. Then the nuts are placed in a bamboo basket, covered with a jute bag, and turned and watered two to three times per day. After two to three days, depending on the weather, the nut shells begin to crack.

Sowing

After the shells crack, the nuts are promptly sown. Sowing is avoided if the nuts are already showing sprouts because the sprouts can easily be broken and will not withstand dryness. Some farmers, usually those with enough labor, hoe the surface of the soil before sowing.

If sowing is to be done under dry weather conditions or directly into swidden fields, the heat method is usually used; otherwise, the nuts are immersed in warm water and sown without further treatment. They can be expected to germinate within 15 days.

Muong swidden farmers in Hoa Binh and Thanh Hoa Provinces clear the forest and sow *Melia* nuts before burning the site. The fire stimulates germination. (See color plate 42 for a comparable system in the Philippines, based on *Paraserianthes falcataria.*) Later, they sow rice. The rice and *Melia,* thinned to a density of between 1,000 and 1,500 trees per ha, are tended regularly. After three rice crops, the soil productivity declines, but by then the *Melia* is well established and forms a secondary forest, often mixed with bamboo. After 8 to 10 years, the *Melia* and bamboo are harvested, after which the second rotation of food crops starts. During the fallow period, people have been able to harvest bamboo shoots.

In some areas, a similar taungya system is used in which *Dendrocalamus membranaceus* is interplanted with rice and corn. Then, when arable cropping is no longer possible, the *Dendrocalamus* is allowed to dominate the fallow vegetation.

Seedling Cultivation

Some farmers establish nursery gardens to raise *Melia* seedlings before transplanting them into swidden fields. The nursery beds are about 1 to 1.2 m wide and are raised about 15 to 20 cm. The seeds are sown in rows or holes, or they are simply broadcast across the bed. If sown in holes, the distance between holes is about 40 cm, the holes are 2 to 3 cm deep, and one nut is placed into each hole. About 1 kg of nuts cover 10 m^2. After sowing, the nuts are covered with a thin layer of soil and straw and are watered every day. When the seedlings reach a height of 1 to 2 cm, the straw is removed. Because each nut contains several seeds, they germinate to form a clump. When the seedlings reach 10 cm in height, they are thinned and only one or two are retained from each clump. They are weeded about once every two weeks. When the seedlings reach a height of 20 to 30 cm, they are thinned again, this time retaining only the best seedling from each former clump. After five to six months, the seedlings are 40 to 50 cm high and are beginning to develop branches. The branches are pruned to improve their shape and produce a straight trunk. After about one year, when the seedlings are more than 2 m high and have a dbh of 2 to 3 cm, they are ready for transplantation into the field.

Transplanting

The transplanting season is in the spring, when the seedlings are just ready to bud, or just after buds begin to show. The *Melia* seedlings are bare-root transplanted into the swidden fields. If the main root is too long, it can be pruned, usually to about 25 to 30 cm long. The seedlings are transplanted into holes of about 40 by 40 by 40 cm, leaving a distance of about 2 m between holes, which corresponds to a density of about 2,500 to 3,000 trees/ha. If labor is in short supply, farmers may simply open the soil surface to plant the seedlings. If they are not all planted at once, the seedling roots may be placed in mud and kept for up to one week.

Tending and Thinning

In the fourth and fifth lunar months, farmers prune some branches and leaves from the *Melia* trees to prevent them from being blown over in the frequent typhoons and whirlwinds at that time of year. The leaves are used as green manure in paddy fields. Although *Melia* thins itself, farmers still prune branches during the winter and spring for the first two years, usually leaving only three branches on each tree. The branches are cut close and parallel to the main trunk, but care is taken to avoid damaging the bark and risking possible fungal infection. At the same time, poorly performing trees are weeded out, while still maintaining a harvest density of 1,000 to 1,500 trees per hectare.

Intercropping with Yam (Dioscorea esculenta) *or Rattan* (Calamus tetradactylus)

About one year after planting, when the *Melia* trees are 4 to 6 m high, with a dbh of 5 to 6 cm, farmers plant *Dioscorea* next to the *Melia*, at a distance of about 40 to 50 cm. The tubers are cut into pieces, and the pieces are either placed in lime water or their cut surfaces are covered in ash and allowed to dry before planting. Two to three pieces of *Dioscorea* are planted around each *Melia* tree, which then provides support for the climbing *Dioscorea*. However, it is prevented from climbing up into the *Melia* branches. The *Dioscorea* can be harvested 9 to 10 months after planting and yield 2 to 4 kg of tubers per hole.

Dzao farmers in Yen Bai Province grow rattan with *Melia*, so that the *Melia* trunks provide support for the rattan in a similar fashion to their support for *Dioscorea* in the system described above. The price of rattan, when sold in the village, has recently been about 1,200 Vietnamese dong (VND)/ball (a ball is 5 to 6 m long), or 1,800 VND/kg. Because of a lack of labor and unstable prices, farmers often limit their interplanting of rattan and *Melia* to rows around fields, forming barriers to prevent buffalo encroachment. The farmers can earn about 10 million VND (1997US$860) after five to eight years, when the rattan is sold.

Intercropping with *Dioscorea* or rattan depends on the availability of family labor. People generally prefer to grow yams, for the simple reason that it is a food crop. However, most swidden farmers cultivate only *Melia*.

At the harvest density described above, the volume of *Melia* wood reaches 130 to 150 m^3/ha after seven to eight years. The current price of *Melia* wood is 250,000 VND/m^3, so farmers can earn at least 25,000,000 VND/ha (1997US$2,143), equivalent to a rice harvest of 1.9 tonnes/year. If we assume that farmers harvest only half the potential volume, they still obtain an annual income equal to 1 ton of rice from swidden agriculture on the same area, and this does not include the value of the green manure applied to paddy fields or incomes derived from yams or rattan.

Natural Regeneration

In some areas, people neither sow *Melia* nor raise seedlings, but simply rely on natural regeneration after they abandon the swidden field. Because *Melia* seeds can survive in the soil for up to six or seven years, they germinate spontaneously when swidden fields are fallowed. Tày communities in Da Bac district of Hoa Binh Province employ this method to establish tree fallows that are dominated by *Melia*, but which develop into mixed gardens (Rambo 1995).

Conclusions and Recommendations

Swiddening is almost universally denigrated by the Vietnamese government—and by many lowland people—as being wasteful and harmful to forests. Yet, in reality, swidden agriculture includes many complex variations, and generalization is inappropriate. There are many reasons for the persistence of shifting cultivation: some are cultural, because minority groups have a history of swiddening. Others are more practical, such as the very limited areas of paddy fields available to mountain people, their limited knowledge of wet rice cultivation, and their lack of access to higher-yielding rice varieties.

The fundamental requirement of sustainable shifting cultivation is the maintenance of soil fertility. Where land is scarce, the fallow period is reduced and soil fertility declines, leading to soil degradation. Increasing population density is seen as one factor precipitating the slide from sustainability to degradation. However, in a district-level study in Song Da watershed using all available records, no relationship could be traced between population density and the occurrence of shifting cultivation (Khiem and van der Poel 1993). The same researchers also reported another interesting feature: the percentages of forest cover were not directly related to the occurrence of shifting cultivation. These figures confirm that shifting cultivation is not solely responsible for deforestation.

A total ban on shifting cultivation is now official government policy in Vietnam, but, as is obvious, it is not always enforced, and solutions for the sustainable development of agriculture on sloping lands are urgently needed. Therefore, rather than support enforcement of the ban, it might be more useful to consider alternatives. Transforming swidden practices into perennial-based farming systems appears to be a promising solution, but it is not easily achieved in a short time. Some authors have also argued that agricultural development in the uplands needs to take existing swidden systems as a starting point and apply modern agricultural science to improve their productivity.

Since the crucial issue is the shortening of fallow periods, improved management and enrichment of the fallow offers a solution to these problems. This calls for the application of indigenous knowledge in tandem with advanced techniques and sciences. There are many management practices, both indigenous and advanced, that may be suitable. For example, agroforestry models with nitrogen-fixing and multipurpose tree species hold strong promise.

As illustrated by this case study, research is needed to identify suitable ways to improve fallow husbandry using indigenous variations of fallow management. One option to be encouraged is intercropping *Melia* with legume species, providing the dual function of producing wood and improving soil fertility. However, there are some areas where *Melia* does not grow successfully. According to our research, it thrives in areas with fresh weather and humid soils derived from limestone or basaltic stone. If these conditions are not present, *Melia* does not develop as well.

The system of fallow management using *Melia* has been developed by ethnic minority farmers who, over many generations, have developed knowledge of *Melia* husbandry under difficult conditions. Their practices, which are successful in many communities, are a wise strategy in resource management.

We have not yet had the chance to conduct more detailed studies of the many systems discussed in this chapter, but those in which *Melia* is the preferred fallow species warrant clearer understanding. They appear promising for wider application. This observation offers an interesting research focus for the future.

Acknowledgments

I wish to acknowledge the support of Dr. Le Trong Cuc and Dr. A. Terry Rambo, both of whom provided valuable comments and suggestions during the process of this study and preparing this chapter. I would also like to thank my colleagues at CARES who also assisted in the preparation of this chapter. Above all, my most special thanks go to the farmers who provided much additional evidence to reinforce my belief in the value of collaboration between scientists and farmers in undertaking research in rural conditions.

References

Cuc, L. T., K. Gillogly, and A. T. Rambo (eds.). 1990. *Agroecosystems of the Midlands of Northern Vietnam,* Occasional Paper No. 12. Honolulu: East-West Center Environment and Policy Institute.

Cuc, L. T. 1993. *Swidden Agriculture in the Highlands of Northern Vietnam.* Baguio City, Philippines: Cordillera Studies Center.

———. 1995. *Phuc Hoi Dat Suy Thoai Vung Trung Du Mien Bac Viet Nam* (Rehabilitation of Degraded Land in the Midlands of Northern Vietnam). In: *Some Issues of Human Ecology in*

Vietnam. Edited by Le Trong Cuc and A. Terry Rambo. Hanoi: Agricultural Publishing House.

Cuc, L. T., A. T. Rambo, K. Fahrney, T.D. Vien, J. Romn, and D.T. Sy (eds.). 1996. *Redbook, Greenhills: Economic Reform and Restoration Ecology in the Midlands of Northern Vietnam*. EWC Research Report. Honolulu: East-West Center.

Eeuwes, J. 1991. *The Socio-Economic Dynamics of Swidden Fields in Uplands Farming Systems in Dong Nai, Hoang Lien Son, and Son La provinces in Vietnam.*

Ha, N. Q.. 1993. *Doi Moi Trong Chien Luoc Phat Trien Lam Nghiep Den Nam 2000,* Bo Lam Nghiep (Renovation of Strategies for Forestry Development until the Year 2000). Hanoi: Ministry of Forestry.

Khiem, N. D., and P. van der Poel. 1993. *Land Use in the Song Da Watershed*. Hanoi: Vietnam-German Technical Cooperation in Social Forestry Development Project.

Me, V. V. 1993. *Melia azedarach* L. *Scientific, Technological and Economic News of Forestry*, No.2, 29–30.

Ministry of Forestry. 1990. *Bao Cao Tong Ket 22 Nam Ve Phong Trao Dinh Canh Dinh Cu* (Final Report on the 22-year Campaign for Fixed Cultivation and Sedentarization). Hanoi: Ministry of Forestry.

———. 1992. *Tropical Forestry Action Plan: Vietnam*. Hanoi: Ministry of Forestry.

NIAPP (National Institute for Agricultural Planning and Projection). 1993. *Nong Nghiep Trung Du Va Mien Nui. Hien Trang Va Trien Vong*. (Upland Agriculture: Present and Prospective). Hanoi: Agricultural Publishing House.

Rambo, A. T., R.R. Reed, L.T. Cuc, and M.R. DiGregorio (eds.). 1995. *The Challenges of Highland Development in Vietnam*. East-West Center Environment and Policy Institute, Center for Natural Resources and Environmental Studies, and the Center for Southeast Asian Studies, University of California at Berkeley. Honolulu: East-West Center.

Rambo, A. T. 1997. Development Trends in Vietnam's Northern Mountain Region. In: *Development Trends in Vietnam's Northern Mountain Region*. (Vol. 1), edited by D. Donovan, A. Terry Rambo, J. Fox, L.T. Cuc, and T.D. Vien. Hanoi: National Political Publishing House.

Sam, D. D. 1994. *Shifting Cultivation in Vietnam: Its Social, Economic and Environmental Values Relative to Alternative Land Uses*. IIED Forestry and Land Use Series No. 3. London: International Institute for Environment and Development.

Toan, B. Q. 1990. Mot So Van De Ve Dat Nuong Ray o Tay Bac Va Phuong Huong Su Dung Chung. (Some problems of Soil in Slash-and-Burn Cultivation in the Northwestern Region and the Direction of Its Utilization). Ph.D thesis submitted to National Library and Vietnam Agricultural Scientific Institute, Hanoi.

Tiem, L. V. 1995. *Impact of the Transition of Land Use Systems on Soil Fertility Conservation on Sloping Lands in the Northwestern Mountain Region of Vietnam*. Vietnam Agricultural Science Institute (VASI).

Vien, T. D. 1991. The Fabaceae (Soft and Woody Tree) in the Maintenance and Enhancement of the Sustainable Production Capacity of Land. Paper presented to a national seminar on Setting Priorities for Research in the Land Use Continuum in Vietnam, September 9–13, 1991, Hoa Binh, Vietnam. International Institute for Environment and Development, Swedish International Development Cooperation Agency, and Forest Science Institute of Vietnam.

———. 1996. Soil Erosion and Nutrient Balance from Swidden Fields: A Case Study in Dabac District, Northern Vietnam. In: *Agriculture on Sloping lands: Challenges and Potential*, edited by Tran Duc Vien and Pham Chi Thanh. Hanoi: Agricultural Publishing House.

Chapter 37

Cost-Benefit Analysis of a *Gmelina* Hedgerow Improved Fallow System in Northern Mindanao, the Philippines

Damasa B. Magcale-Macandog,
Canesio D. Predo, and Patrick M. Rocamora

Increasing population and dwindling availability of lowland areas for crop production have led to a significant level of migration into the uplands of the Philippines. About 30% of the country's total population lives in upland areas with slopes greater than 18% (Cruz and Zoza-Feranil 1988). Under increased pressure for subsistence food production, a large proportion of the country's upland hilly areas have now been converted to permanent cultivation or fallow rotation (Marchand 1987; Garrity 1993). As one consequence of this pressure on upland areas, there was great interest among researchers and extensionists in soil conservation technology, involving contour hedgerow intercropping, in the mid-1970s (Garrity 1993). Since then, these technologies have been modified by indigenous innovations. Among other things, these local adaptations have made them less demanding of labor.

One of the adaptations tried by farmers was allowing trees and bushes to grow during the fallow period to restore soil productivity after several years of extensive cropping (Fernandez et al. 1992). Fast-growing, multipurpose tree species were introduced to restore vegetation, rehabilitate the soil, stabilize the condition of fields, and ensure higher and more sustainable crop productivity (Gellor and Austral 1996). Furthermore, intercropping of annuals while the multipurpose tree species were in the early stages of growth offered an increase in the productivity and profitability (Miah et al. 1993).

This use of trees in upland farming systems has been taken a step further. In response to a growing timber market, smallholder farmers on the frontiers of infertile grassland soils have begun farming timber trees as a dominant enterprise, using their own capital resources (Garrity and Mercado 1994; Garrity 1994). In particular, they have developed plantations of *Gmelina arborea* (Garrity and Mercado 1994). Their example has been followed by farmers at Claveria, in northern Mindanao, Philippines, who are planting *Gmelina* as an intercrop with their annual crops of maize, upland rice, and cassava.

Damasa B. Magcale-Macandog, Associate Professor, Institute of Biological Science, University of the Philippines Los Baños, College, Laguna 4031, the Philippines. Patrick M. Rocamora, University Research Associate, Agricultural Science Cluster, University of the Philippines Los Baños, College, Laguna 4031, the Philippines. Canesio D. Predo, Assistant Professor, Leyte State University, Leyte, the Philippines.

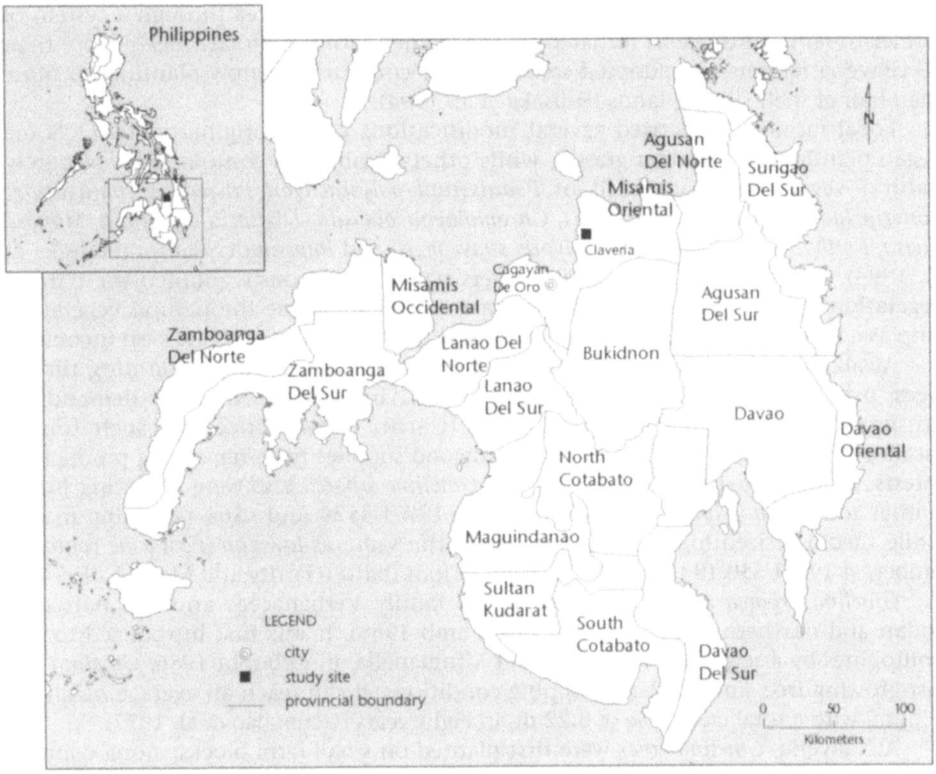

Figure 37-1. Location of *Gmelina* Improved Fallow System, Claveria, Mindanao, Philippines

Based on their results, it is hypothesized that planting *Gmelina* trees to dominate vegetation during the fallow period is a superior system to conventional shifting cultivation, in terms of economic and ecological benefits. This study was undertaken to quantify those benefits.

The Study Area

Claveria is a municipality of Misamis Oriental, on the island of Mindanao. It is 40 km northeast of the city of Cagayan de Oro (see Figure 37-1). It lies on an undulating plateau between a coastal escarpment and a mountainous interior, ranging in elevation from 200 to 1,500m above sea level (asl). Soils in Claveria are classified as acid upland, fine mixed isohyperthermic Ultic Haplorthox, with a depth of more than one meter (Garrity and Agustin 1995). The area has two pronounced seasons, the wet season from May to October and the dry season from November to April. Nearly 80% of upland farms in Claveria are situated on slopes ranging from gentle to as steep as 60%. Erosion is, therefore, a major concern. Aside from upland rice and cassava, the dominant crop is maize, which has adapted well to local conditions and is eaten as a staple food. Maize grain can also be sold as animal feed.

The Evolution of the Gmelina *Improved Fallow System*

Hedgerow intercropping was introduced to Claveria in 1987, after six farmers from the area and two technicians from the International Rice Research Institute visited a World Neighbors project in Cebu, Philippines. They learned how to establish contour lines with an A-frame, construct contour bunds and ditches, and plant hedgerows. The "original" hedgerows were composed of one or two rows of *Gliricidia sepium* and one or two rows of *Pennisetum purpureum* (Stark 1996). Between 1987 and 1991, the technology was passed on to more than 200 farmers in the area through a system of farmer-to-farmer extension initiated by the trained farmers. By late 1992, more than 80 Claveria farmers had adopted some form of contour hedgerow planting on more than half of their sloping lands (Fujisaka et al. 1994).

Local farmers have tried several modifications of the original system.[1] Some began planting more fodder grasses, while others established contour strips of purely natural vegetation, composed of *Pennisetum polystachyon, Paspalum conjugatum, Borreria laevis, Ageratum conyzoides, Chromolaena odorata, Digitaria longiflora, Mimosa invisa, Rottboellia cochinchinensis, Hyptis suaveolens,* and *Imperata cylindrica* (Fujisaka et al. 1994). By 1996, more than 200 farmers had spontaneously adopted the natural vegetation strip or grass strip system (Stark 1996). Some use the natural vegetation strip as a base for establishing lines of fruit and timber trees to generate cash income.

At about the same time, smallholder farmers in Claveria began farming timber trees, using their own capital resources. The driving force was market demand for fast-growing timber like *Gmelina arborea* (Garrity 1994). Prices for such timber escalated following the end of natural hardwood supplies from remaining production forests. By the mid-1990s, small-diameter *Gmelina arborea* logs were attracting prices similar to *falcata* (*Albizia falcata*), between 1997US$24 and $64 per cubic meter, while medium sized logs were priced about the same as *lawa'an* (*Pentacme contorta*) timber, at 1997US$0.19 to US$0.97 per board foot (bdft) (Garrity and Mercado 1994).

Gmelina arborea (Roxb.) belongs to the family Verbenaceae and is a native of Sudan and northern Rhodesia, in Africa (Lamb 1968). It was first introduced to the Philippines by a reforestation program at Minglanella, in Cebu, in 1956. *Gmelina* is a fast-growing tree, and, under Philippine conditions, it can reach an average height of 13.2 m, with a total clear bole of 8.22 m, in eight years (Generalao et al. 1977).

At Claveria, *Gmelina* trees were first planted on small farm blocks, along contour strips, and along farm boundaries. When *Gmelina* seedlings were planted in contour strips, cropping was continued in the alleys between the growing trees for the first two years, but when the tree canopy started to close, the land was left to bush fallow. During the fallow period, in which the *Gmelina* trees continued to grow, associated natural vegetation was grazed by livestock (see color plate 45). By the end of the fallow period, the farmer had both harvestable timber and livestock gains in the form of increased animal weight and draft services.

Research Methods

The Farming Systems

The average farm size in Claveria is 2.5 ha. Two farming systems were compared in this study: the conventional open field maize farming system and the *Gmelina* improved fallow system. The conventional system involves a five-year cycle beginning with three years of continuous maize cropping followed by two years of fallow (Table 37-1). During the fallow period, cattle can graze natural vegetation growing on the land. In the *Gmelina* system, farmers typically devote one hectare of their farm to hedgerow intercropping. Hedgerows are usually six meters apart and *Gmelina* seedlings are planted in the hedgerows at one-meter intervals. During the first two years of *Gmelina* growth, maize is planted in the alley areas. But starting from the third year, the alley areas are left fallow and natural vegetation is allowed to grow. Cattle graze on the natural vegetation until the *Gmelina* trees are harvested at the end of the seventh year. The remaining 1.5 ha of the farm remains devoted to conventional open field maize farming and has a continuous cycle of three years of maize cropping followed by two years of fallow.

---

[1] These modifications entail planting of either *Gliricidia* or napier grass in the hedgerows and annual crops of maize or upland rice in the alley areas.

Table 37-1. Cropping Patterns of Open Field Farming and *Gmelina* Improved Fallow

Year	*Open Field Farming*	Gmelina *Improved Fallow*
1	2.5 ha maize-maize	1.0 ha *Gmelina*-maize-maize (tree establishment) 1.5 ha maize-maize
2	2.5 ha maize-maize	1.0 ha *Gmelina*-maize-maize (tree pruning) 1.5 ha maize-maize
3	2.5 ha maize-maize	1.0 ha *Gmelina*-grasses-animal grazing 1.5 ha maize-maize
4	2.5 ha grasses-animal grazing maize-maize (fallow)	1.0 ha *Gmelina*-grasses-animal grazing 1.5 ha grasses-animal grazing maize-maize (fallow)
5	2.5 ha grasses-animal grazing maize-maize (fallow)	1.0 ha *Gmelina*-grasses-animal grazing 1.5 ha grasses-animal grazing maize-maize (fallow)
6	2.5 ha maize-maize	1.0 ha *Gmelina*-grasses-animal grazing 1.5 ha maize-maize
7	2.5 ha maize-maize	1.0 ha *Gmelina*-grasses-animal grazing (harvest) 1.5 ha maize-maize
8	2.5 ha maize-maize	1.0 ha *Gmelina*-maize-maize (tree establishment) 1.5 ha maize-maize
9	2.5 ha grasses-animal grazing maize-maize (fallow)	1.0 ha *Gmelina*-maize-maize (tree pruning) 1.5 ha grasses-animal grazing maize-maize (fallow)
10	2.5 ha grasses-animal grazing maize-maize (fallow)	1.0 ha *Gmelina*-grasses-animal grazing 1.5 ha grasses-animal grazing maize-maize (fallow)
11	2.5 ha maize-maize	1.0 ha *Gmelina*-grasses-animal grazing 1.5 ha maize-maize
12	2.5 ha maize-maize	1.0 ha *Gmelina*-grasses-animal grazing 1.5 ha maize-maize
13	2.5 ha maize-maize	1.0 ha *Gmelina*-grasses-animal grazing 1.5 ha maize-maize
14	2.5 ha grasses-animal grazing maize-maize (fallow)	1.0 ha *Gmelina*-grasses-animal grazing (harvest) 1.5 ha grasses-animal grazing maize-maize (fallow)

Interviews and Data Collection

Research methodologies included a combination of farmer and sawmill operator interviews, use of secondary data, and simulation of predicted maize yields using SCUAF (Young 1989) (Table 37-2). A survey was conducted in Claveria using the same farmer respondents interviewed in an earlier study (Nelson et al. 1996). SCUAF (version 2.0) is a simple biophysical agroforestry model that can predict crop yield as a function of changes in soil carbon and nitrogen contents (Young 1989). Maize yields were simulated and predicted using SCUAF because of a lack of historical maize yield data and the difficulty of obtaining this type of information because of highly variable climatic conditions and fertilizer inputs. The survey respondents included 17 smallholder timber farmers from the lower Claveria zone (Patrocinio, Tunggol Estate, Ane-i-Lupok, Minsacuba, Hinaplanan, Cabacungan, and Wilcom), six farmers who practiced continuous open maize farming, and five sawmill operators from the nearby towns of Villanueva and Jasaan who were buying and processing *Gmelina* trees for local and overseas markets.

Table 37-2. Information Gathered for the Cost-Benefit Analysis and its Sources

Source	*Information*
Farmer interviews	Farming practices in open field maize farming system
	Farming practices in improved *Gmelina* hedgerow system
	Marketing of *Gmelina* trees
	Ecological benefits of planting perennial trees on the farm
	Animal shelter establishment and maintenance costs, and animal supplements (veterinary drugs, salt, and rope)
	Cost of *Gmelina* planting materials
	Labor requirement for the improved *Gmelina* fallow system
	Benefits from animals (services and animal weight gains)
	Benefits from *Gmelina* trees (firewood and timber)
	Natural vegetation grass yields for the initial year
Secondary literature	
Kearl (1982)	Animal feed requirements (kg/au/day)
Magcale-Macandog et al. (1997)	Total number of productive animal services
	Annual gross benefits obtained from the animal
Nelson et al. (1996)	Cost of farm inputs (fertilizer, seeds, and pesticide) in maize cropping system
	Labor requirements for the open field maize farming system
	Maize grain yield in the initial year
Starck (1996)	Yield of natural grasses in the initial year
Sawmill operators' interviews	Marketing of *Gmelina* trees
SCUAF simulation modeling	Incremental change in maize yields after the initial year
	Incremental change in yields of natural grasses after the initial year

The cost-benefit analysis was based on the inputs and outputs of the two systems (see Table 37-3) over 14 years, corresponding to two cycles of *Gmelina* timber harvesting. Annual net returns and net present value (NPV) were computed in Philippine pesos (₱) to compare the *Gmelina* improved fallow system with the traditional maize farming system.

Choice of Discount Factor

Discounting is required to calculate the present value of a stream of costs and benefits associated with a project or intervention (Hanley and Spash 1993). Future benefits and costs are discounted to account for the cost of forgone investment opportunities when a particular investment option is chosen. This is of critical relevance to upland farmers who have to borrow to establish new farming practices. Generally, however, the choice of a social discount rate is crucial to determining whether the present value is positive or negative. It is a very important factor in the decision of farmers to adopt or reject a new farming technology. Based on the opportunity cost of capital, farmers in Claveria face a real discount rate of 25% (Nelson et al. 1996). Farmers were asked to report the known cost of capital based on recent lending, or to estimate the interest charges for borrowing a nominal amount, based on their knowledge of credit markets. In this study, a lower discount rate of 10% is used to reflect the social opportunity cost of capital. It assumes that government intervention or support, through the likes of farmer cooperatives, reduces the cost of capital to farmers.

Input Requirements

Table 37-4 presents the annual input requirements of the two farming systems for 14 years, equivalent to two cycles of the *Gmelina* improved fallow system. Farmers in

Claveria plant maize twice a year, that is, during the wet season and the dry season. The total annual input requirements and costs for the open maize farming system, valued in Philippine pesos, include the costs of maize seeds, fertilizer, and pesticides during the three-year cropping period. During the fallow periods, in years 4, 5, 9, 10, and 14, cattle can graze on the natural vegetation, so input costs include animal shelter establishment, maintenance, and veterinary supplies. In the *Gmelina* improved fallow system, the alley areas were planted to maize during the first two years of growth of *Gmelina* seedlings. Input requirements for the first two years of maize cropping, that is, years 1, 2, 8, and 9, are similar to those in the open maize farming system (see Table 37-4). Animal shelters had to be re-established every six years and maintained every three years. *Gmelina* seedlings were planted in the first year of the first cycle only. In the second cycle, regrowth from *Gmelina* stumps was pruned and only one coppice allowed to regrow.

Table 37-3. Summary of Assumptions and Vital Information Used in the Computation of Costs and Benefits

Variables/Assumptions	*Information*
Major assumptions	Man-labor wage rate of ₱50/man-day (MD).
	Man-animal labor wage rate is ₱100/man-animal-day (MAD).
	Services provided by the animal are valued at ₱42/animal-day.
	Shelled maize with 18% moisture is sold at ₱4.60 and ₱5.50 per kg in the wet and dry season, respectively.
	Cost of animal shelter establishment and maintenance.
	Discount rate is 10%, reflecting the social rate of preference.
	Time horizon for the analysis is 14 years, corresponding to two cycles of *Gmelina* production.
	Projections were at constant prices. Possible change in prices is examined in the sensitivity analysis.
	Price of *Gmelina* firewood is ₱1,200/metric tonne.
	Price of *Gmelina* lumber is ₱7/bdft.
	Annual income derived from animal weight gain and services is ₱8,551/animal unit.
Variables used in the computation of benefits	Maize yield.
	Harvest from *Gmelina* in terms of lumber and firewood.
	Animal inventory gain and number of days of animal services.
Variables used in the computation of costs	Quantity of inputs used in maize production such as maize seeds, fertilizer, and pesticides.
	Quantity and price of *Gmelina* seedlings.
	Quantity of veterinary drugs and animal feed supplements.
	Value of animal shelter.
	Man labor requirements.
	Animal labor requirements.

Note: At time of study, US$1= ₱40.

Table 37-4. Annual Cost of Inputs for Open Field Farming and *Gmelina* (Philippine Pesos)

Open Field Farming

Year	*Maize Seeds*	*Fertilizers*	*Pesticides*	*Shelter, etc.*	*Animal Drugs, Supplements*	*Total*
1	520	9,337	2,475			12,332
2	520	9,337	2,475			12,332
3	520	9,337	2,475			12,332
4				829	2,479	3,308
5					2,457	2,457
6	520	9,337	2,475			12,332
7	520	9,337	2,475			12,332
8	520	9,337	2,475			12,332
9				745	2,228	2,973
10					2,206	2,206
11	520	9,337	2,475			12,332
12	520	9,337	2,475			12,332
13	520	9,337	2,475			12,332
14				672	2,009	2,681

Gmelina *Improved Fallow*

Year	Gmelina *Seedlings*	*Maize Seeds*	*Fertilizer*	*Pesticide*	*Shelter, etc.*	*Animal Drugs, Supplements*	*Total*
1	15,250	488	8,777	2,340			26,855
2		488	8,777	2,340			11,605
3		312	5,602	1,485	281	841	8,521
4					496	2,315	2,811
5						2,304	2,304
6		312	5,602	1,485	96	808	8,303
7		312	5,602	1,485		808	8,207
8		488	8,777	2,340			11,605
9		176	3,175	855	449	1,343	5,998
10					256	2,096	2,352
11		312	5,602	1,485		753	8,152
12		312	5,602	1,485		743	8,142
13		312	5,602	1,485	88	743	8,230
14					402	1,933	2,335

Notes: Input requirements per hectare and unit costs: maize seeds, 32 kg/ha, ₱6.50/kg; urea, 248 kg/ha, ₱6.50/kg; solophos, 264 kg/ha, ₱5.30/kg; muriate of potash, 134 kg/ha, ₱5.40/kg; pesticide, 2.2 liters/ha, ₱450/liter; *Gmelina* seedlings, 250 seedlings/ha, ₱61/seedling; animal shelter establishment, ₱365/animal unit; animal shelter maintenance, ₱130/animal unit; animal supplements, ₱1,092/animal unit. When the study was conducted, the value of Philippine currency was about US$1 = ₱ 40.

Labor Requirements

Some farm operations require human labor while others require animal labor. The data presented in Table 37-5 do not refer to a specific year in the farming system but simply represent all the operations needed in the whole farming cycle. For example, contour bund construction is done only in year 1, while harvesting of *Gmelina* trees is done in years 7 and 14. The actual total annual labor required and costs of labor for the two farming systems over 14 years, or two *Gmelina* improved fallow cycles, are presented in Table 37-6.

Results and Discussion

Farm sizes of the interviewed farmers ranged from 0.25 ha to 4.0 ha, with an average farm size of 2.5 ha. The average farm size was used in the cost-benefit analysis in this study.

Gross Benefits from the Open Field Maize System

Maize Yield. During the wet season, higher growth rates lead to generally higher yields than those in the dry season. In the first year of maize cropping, yields were 1,920 kg/ha during the wet season and 1,260 kg/ha during the dry season (Nelson et al. 1996). Maize yields during the following two years were predicted by the SCUAF model and reported in previous experiments by Nelson et al. (1996). On this basis, wet season maize yields were estimated to decline to 1,889 and 1,846 kg/ha in the second and third years respectively. The dry season price of maize, ₱5.5/kg, is higher than the wet season price of ₱4.60/kg. Total annual benefits from maize on a 2.5 ha farm were ₱39,405 in the initial year, declining to ₱38,770 and ₱37,893 in the succeeding two years, as predicted by SCUAF (Table 37-7).

Animal Weight Gain. During the fallow period, cattle graze on the natural vegetation that grows in the fields. Wet season yields of natural fallow regrowth in Claveria amount to 1.26 metric tonnes/ha, and 1.06 metric tonnes/ha during the dry season (Stark 1996). This equates to an annual total of 2.32 metric tonnes/ha of animal fodder. An animal with an average weight of 300 kg requires 7 kg feed/day (Junus 1989), amounting to an annual requirement of 2.55 t/ha/animal unit (au). A 2.5 ha plot of natural vegetation can therefore support 2.27 au. Using percentage changes in predicted natural vegetation yields, as reported by Magcale-Macandog et al. (1997), and based on SCUAF calculations when calibrated for Claveria conditions and simulations, natural vegetation yield declined in the succeeding year to 2.30 metric tonnes/ha. Therefore, the number of animal units the farm could support fell to 2.25 au. The average farmgate price for napier grass, as reported by Claveria farmers, was ₱0.35/kg fresh weight. This resulted in a fodder benefit of ₱812/ha for napier grass.

Animal Services. Animal services to the farm include draft, or plowing and harrowing, hauling, and transport. Estimates for the median total number of productive animal services is about 153 days/animal/year, including 88 days for draft and 65 days for hauling and transport (Magcale-Macandog et al. 1997). With an estimate of ₱42 for each animal-day, the value of animal services rendered in one year amounts to ₱6,426. The difference between the animal weight at the start of the year and that at the end of the year is termed the change in animal inventory value. Based on farmer interviews, the mean change in animal inventory value amounted to ₱2,125. The annual gross benefit obtained from animals is a combination of the value of animal services and the change in animal inventory value, amounting to ₱8,551/au (Magcale-Macandog et al. 1997). (See Table 37-7 for total annual benefits from an open field maize farming system.)

Gross Benefits from the Gmelina *Improved Fallow System*

During the first two years of the cycle, when maize is planted in the *Gmelina* improved fallow system, the total maize cropping area is 2.35 ha. Since no data on maize yields in the alley areas were available, it was assumed that competition between maize and *Gmelina* seedlings did not significantly reduce maize yields during the first year and that the initial yields from the alley areas was the same as that from the open areas. However, maize yields in the alley areas were predicted for the succeeding year using SCUAF. Using the same farmgate prices for maize, ₱4.6 during the wet season and ₱5.5 in the dry season, gross benefits obtained from maize were calculated at ₱37,041 in the first year and declining in succeeding years (see Table 37-7).

Table 37-5. Labor Requirements for Each Operation in Open Field Farming and *Gmelina* Improved Fallow

	Open Field Farming		Gmelina *Improved Fallow*	
Operation	*MD 2.5 ha*	*MAD 2.5 ha*	*MD 2.5 ha*	*MAD 2.5 ha*
Hedgerow establishment				
Construct bunds/lay out hedgerows			13	2
Gmelina planting			2.87	
Hedgerow establishment total			15.87	2
Wet season				
Hedgerow weeding (ring)			15.6	
Land preparation		37.5		38.65
Maize sowing and fertilizing at planting	17.5		16.45	
Replanting	5		4.7	
Nitrogen fertilizer	20		17.1	
Interrow weeding		10		9.4
Hand weeding	42.5		35.25	
Gmelina hedgerow pruning			10.94	
Harvesting				
Maize	20		18.8	
Gmelina				
Postharvest processing				
Maize	30		28.2	
Gmelina				
Wet season total	135	47.5	147.04	48.05
Dry season				
Hedgerow weeding (ring)			15.6	
Land preparation		25		23.5
Maize sowing and fertilizing at planting	17.5		16.45	
Replanting	5		4.7	
Nitrogen fertilizer	20		17.1	
Interrow weeding		10		9.4
Hand weeding	37.5		31.85	
Gmelina hedgerow pruning			10.94	
Harvesting				
Maize	15		13.25	
Gmelina			9.38	
Postharvest processing				
Maize	27.5		25.85	
Gmelina			10.94	
Dry season total	122.5	35	156.06	32.9
Annual total	257.5	82.5	318.97	82.95

Notes: Open field farming = maize-maize for 3 years and 2 years fallow; Improved fallow = *Gmelina*-maize-maize; MD = man-days at ₱50/MD; MAD = man + animal-days at ₱100/MAD; Planting of *Gmelina* seedlings = 5.51 mins/tree; Pruning of *Gmelina* seedlings = 21 mins/tree; Ringweeding of *Gmelina* = 5 mins/tree; Harvesting = 18 mins/tree; Postharvesting = 21 mins/tree; Animal feeding (cut and carry) = 1.5 hrs/day. When the study was conducted, the value of Philippine currency was about US$1 = ₱ 40.

Table 37-6. Labor Requirements for Open Field Farming and *Gmelina* Improved Fallow over 14 Years

	Open Field Farming			Gmelina *Improved Fallow*		
	Total Labor			*Total Labor*		
Year	*MD*	*MAD*	*Total Cost*	*MD*	*MAD*	*Total Cost*
1	257.5	82.5	21,125	262.77	82.95	21,434
2	257.5	82.5	21,125	268.78	80.95	21,534
3	257.5	82.5	21,125	228.93	49.50	16,397
4	154.93		7,747	166.57		8,329
5	153.56		7,678	144.01		7,201
6	257.5	82.5	21,125	205.00	49.50	15,200
7	257.5	82.5	21,125	225.32	49.50	16,216
8	257.5	82.5	21,125	246.90	80.95	20,440
9	139.23		6,962	198.23	31.45	13,057
10	137.86		6,893	152.92		7,646
11	257.5	82.5	21,125	223.47	49.50	16,124
12	257.5	82.5	21,125	200.91	49.50	14,996
13	257.5	82.5	21,125	200.91	49.50	14,996
14	125.58		6,279	141.13		7,057

Notes: MD = man-days; MAD = man + animal days; costs in Philippine pesos. When the study was conducted, the value of Philippine currency was about US$1 = ₱ 40.

Table 37-7. Annual Economic Benefits from Open Field Farming and *Gmelina* Improved Fallow (Philippine Pesos)

	Open Field Farming			Gmelina *Improved Fallow*				
Year	*Maize*	*Animal*	*Total*	*Maize*	*Animal*	*Tree*	*Firewood*	*Total*
1	39,405		39,405	37,041				37,041
2	38,770		38,770	36,426			900	37,326
3	37,893		37,893	22,741	6,584		4,008	33,333
4		19,411	19,411		18,128		4,008	22,136
5		19,240	19,240		18,043			18,043
6	39,405		39,405	23,643	6,328			29,971
7	38,770		38,770	23,266	6,328	98,000		127,594
8	37,893		37,893	36,139				36,139
9		17,444	17,444	13,160	10,518		900	24,578
10		17,273	17,273		16,418		4,008	20,426
11	39,405		39,405	23,643	5,900		4,008	33,551
12	38,770		38,770	23,266	5,815			29,081
13	37,893		37,893	22,741	5,815			28,556
14		15,734	15,734		15,135	98,000		113,135

Notes: Maize yield (kg/ha), wet season = 1,920 sold at ₱4.60/kg; dry season = 1,260 sold at ₱5.50/kg; Animal weight gain and services = ₱8551/animal unit; *Gmelina* firewood (tonne/ha) = 1 sold at ₱1,200/metric tonne; *Gmelina* tree (bdft/tree) = 56 sold at ₱7.00/bdft. When the study was conducted, the value of Philippine currency was US$1 = ₱ 40.

In the third year, maize can no longer be planted in the alley areas because of the shading effect from the closing *Gmelina* canopy. Natural vegetation is instead allowed to grow in the alley areas for animal grazing, and benefits include animal services and weight gains (Table 37-7). Years 4 and 5 correspond to the period when the 1.5 ha area used for open maize farming is also fallowed. Natural vegetation is allowed to grow over the entire farm for animal grazing and additional benefits are gained from the animal component during this period.

Gmelina trees are pruned twice each year in years 2, 3, 4, 9, 10, and 11, and the prunings are sold as firewood (Table 37-7). The amount of firewood cut from each *Gmelina* tree was about 1.5 kg/pruning in year 2 and about 6.67 kg/pruning in years 3 and 4. Firewood was priced at ₱1,200/tonne, based on the actual buying rate of bakery owners in the area, so the economic benefits obtained from firewood amounted to ₱900 in year 2 and ₱4,008 in years 3 and 4. Years 7 and 14 were tree harvest years, and, based on farmer interviews, the average amount of timber harvested from the *Gmelina* trees in year 7 was about 56 bdft/tree.

There are two ways of harvesting and marketing *Gmelina* trees in the Claveria area. The first one is bulk contracting directly with a sawmill or a timber trader. In this case, the contractor cuts and hauls the logs or rough-sawn timber to the mill. The second way is direct delivery of logs or sawn lumber to the mill by the farmer. This practice is usually followed only when the quantities are small (Garrity and Mercado 1994). Marketing tends to be more complex as the numbers of intermediaries increases along with the services they provide, so the simplest and most common system involves direct sale by the producer to the consumer (Pabuayon 1989). Despite the fact that timber prices in Claveria fluctuate seasonally, an average value of ₱7 per bdft was used for this analysis. This was based on the direct sale of trees from farmer to sawmillers. The benefit from 250 *Gmelina* trees therefore amounted to about ₱98,000 (Table 37-7).

Net Benefits Obtained from the Two Farming Systems

The net benefits from continuous open field maize farming were greater during years 1, 2, and 3, while maize was being cropped, than the *Gmelina* improved fallow system (Table 37-8, Figure 37-2). The latter had a negative net income in year 1 due to costs incurred in the purchase of planting materials, labor for planting tree seedlings, hedgerow establishment, and reduced maize yields. During the first six years, cumulative benefits were greater from continuous open field maize farming (Figure 37-2) than from the *Gmelina* improved fallow system. However, after the first harvest of *Gmelina* trees in year 7, the cumulative benefits of the *Gmelina* improved fallow system rose beyond those of the conventional system and, for succeeding years, remained well ahead (Figure 37-2). Benefits gained from the *Gmelina* system came on a longer-term basis than those from open field maize farming. Its superior profitability was realized after the returns from harvested timber, which exceeded the value of the crop yield lost as a consequence of growing the trees. By year 14, the cumulative net present value of the *Gmelina* fallow system was double that of open field maize farming.

Ecological Benefits from Gmelina *Trees*

As well as the long-term economic benefits gained from planting *Gmelina* trees, Claveria farmers told of their ecological benefits. Seventy-six percent of interviewed farmers observed cooler air on their farms because of the shading effect of the trees. Half of them spoke of the beneficial effects of leaf litter in building up soil fertility. They said that *Gmelina* leaves had a fast rate of decomposition and were completely decomposed within 48 weeks (Florece 1996). As well as the farmer testimony, hedgerow intercropping has been advocated as a technology to better sustain permanent cropping in the Philippines with minimum or no fertilizer input and as a measure to control soil erosion on sloping lands (Mercado et al. 1996). Areas covered by the survey have slopes ranging up to 30%, and half of the surveyed farmers spoke of the effectiveness of the tree hedgerows and natural vegetation in controlling soil erosion. They also reported that the trees served as very good windbreaks.

Potential for Application Elsewhere

In other parts of Southeast Asia and the world, shifting cultivators deliberately stimulate colonization of fallowed land with tree species to create conditions that accelerate the regeneration of the land between cropping cycles (Garrity 1994). There is much that is unique about the *Gmelina* improved fallow system. It enables shifting cultivators to escape dependence on annual crops for their livelihood and covers large parts of the landscape with perennial species. Natural vegetation growing under the *Gmelina* trees is a source of animal fodder, and integration of the animal component is unique. Economic benefits from the system include cash from the main food crop and timber sold at harvest, cash from off-farm animal services, firewood from pruned lateral branches, and animal weight gains and animal on-farm services. Another benefit obtained from pruning lateral branches is the increased length of clear bole and veneer yield, meaning higher income when the timber is sold (Bhumibhamon et al. 1986). The system also enhances the efficiency of the fallow phase by building the usable soil nutrient reservoir, suppressing weeds, and controlling erosion on steep slopes.

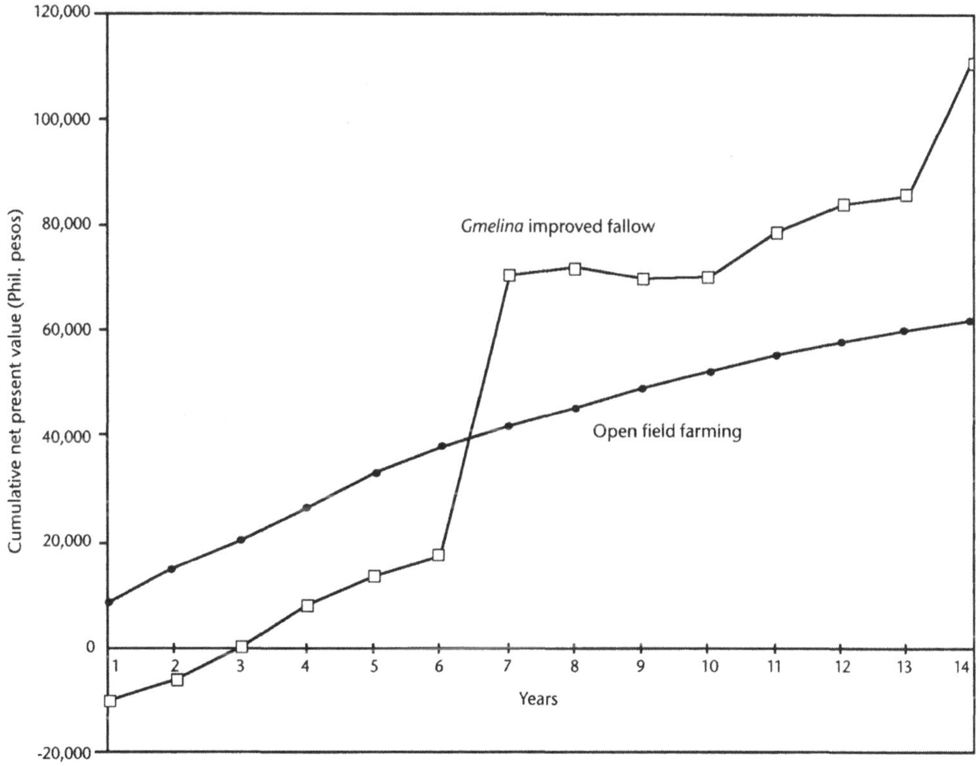

Figure 37-2. Cumulative Net Present Value of Open Field Farming and *Gmelina* Improved Fallow

Table 37-8. Annual Costs and Benefits: Open Field Farming vs. *Gmelina* Improved Fallows

	Open Field Farming			Gmelina *Improved Fallow*		
Year	*Benefit*	*Cost*	*Annual Net Benefit*	*Benefit*	*Cost*	*Annual Net Benefit*
1	39,405	30,882	8,523	37,041	48,288	-11,247
2	38,770	30,882	7,888	37,326	33,138	4,188
3	37,893	30,882	7,011	33,333	24,918	8,415
4	19,411	9,505	9,906	22,136	11,139	10,997
5	19,240	8,599	10,641	18,043	9,504	8,539
6	39,405	30,882	8,523	29,971	23,503	6,468
7	38,770	30,882	7,888	127,594	24,423	103,171
8	37,893	30,882	7,011	36,139	32,044	4,095
9	17,444	8,542	8,902	24,578	19,053	5,525
10	17,273	7,720	9,553	20,426	9,998	10,428
11	39,405	30,882	8,523	33,551	24,276	9,275
12	38,770	30,882	7,888	29,081	23,138	5,943
13	37,893	30,882	7,011	28,556	23,226	5,330
14	15,734	7,704	8,030	113,135	9,390	103,745

Adoption Constraints

Hedgerow intercropping has its limitations. Capital investment is needed during the first year of the *Gmelina* improved fallow cycle for seedlings and contour bund establishment. Because their informal land tenure is not accepted as collateral, upland farmers have difficulty obtaining loans from banks. Credit from informal sources, with an annual interest rate of about 120%, is common but ruinously expensive (Nelson et al. 1996). In some cases, farmers use the *Gmelina* seedling as collateral to borrow money from private entrepreneurs. At the maturity of the loan, usually in five years, the trees are cut and the income from their sale is shared.

Gmelina hedgerow establishment, weeding, and pruning also require significant amounts of labor (see Table 37-5), so availability of labor is another major factor in the adoption of this technology (Duguma et al. 1990; Nelson et al. 1996). Another negative aspect of hedgerow intercropping is competition for light, nutrients, and water between the trees and understory crops planted in the alley areas (Sibanda 1991; Garrity et al. 1995). The ability of the technology to conserve water remains doubtful (Lasco and Carandang 1989). Security of land tenure is also a major consideration because farmers will not plant perennial trees unless their land ownership is secure.

Adoption Niches

With the aim of preserving its remaining natural forests, the Philippines government has enacted several laws, including the illegal logging law, and has declared a total log ban. The resulting scarcity of primary grade wood in the market has made secondary trees more attractive. The price of *Gmelina* trees in the Claveria area has increased continuously under a growing demand for timber on both local and export markets, for construction material and manufacture of furniture. At the same time, there is an intense government campaign to encourage farmers to plant fast-growing timber tree species. This includes dissemination of free seedlings and training for farmers on nursery establishment and transplanting seedlings to the field. These incentives are a major determinant in farmers' adoption of the *Gmelina* improved fallow system, although timber prices still fluctuate seasonally and vary with the distance from farmers' fields to the sawmill. For their part, industrial loggers have turned their attention to timber types like *Gmelina*, although another recent law requires sawmillers and traders to obtain permission from the Department of Environment and Natural Resources before cutting any trees, including *Gmelina*.

Conclusions

The *Gmelina* improved fallow system has considerable potential for enhancing fallow management by increasing economic benefits to farmers and ecological benefits to the land. The system integrates agroforestry and animal pastoral systems, and it has shown the capacity to increase income from harvested timber and firewood, as well as from animal weight gain and services. After 14 years, the cumulative net present value of the *Gmelina* improved fallow system approximately doubled that of a conventional open field maize farming system. Ecological benefits included enhanced soil fertility buildup during the fallow period, soil erosion control, provision of a windbreak, and generally cooler air.

Acknowledgments

This study was jointly funded by the Australian Centre for International Agricultural Research (ACIAR) and the Center for International Forestry Research (CIFOR), in collaboration with the Southeast Asian Ministers of Education Organization, Regional Center for Graduate Study and Research in Agriculture (SEAMEO-SEARCA). Special thanks are due to Mr. Agustin Mercado Jr., Mr. Samuel Nulla, and Ms. Charmaine Pailagao, Senior Researcher and Enumerators, respectively, at the World Agroforestry Centre (ICRAF), for their assistance in the survey at Claveria. Helpful comments and criticisms by Dr. Francisco P. Fellizar Jr. and Dr. Soekartawi (Deputy Directors, SEARCA), Dr. Arturo G. Gomez (Project Leader, SEARCA-ACIAR PN9220), and Dr. Arsenio D. Calub and Roberto F. Rañola Jr. (Advisory Group, SEARCA-ACIAR PN9409) are gratefully acknowledged.

References

Bhumibhamon, S., L. Atipanumpai, and A. Ladplee. 1986. Tree Farming in Thailand: A Case study of Thai Plywood Company. In: *Forest Regeneration in Southeast Asia*, SEAMEO-Biotrop special publication 25, 121–128.

Cruz, M.C., and I. Zoza-Feranil. 1988. Policy Implications of Population Pressure in the Philippine Uplands. Paper presented to World Bank/CIDA Study on Forestry, Fisheries and Agricultural Resource Management, Washington, DC.

Duguma, B., J. Tonye, and D. Depommier. 1990. *Diagnostic Survey of Local Multipurpose Trees and Shrubs, Fallow Systems and Livestock in South Cameroon*. Working paper No. 60. Nairobi, Kenya: World Agroforestry Centre (ICRAF).

Fernandez, E.C.M., D.P. Garrity, L.T. Szott, and C.A. Palm. 1992. Use and Potential of Domesticated Trees for Soil Improvement. In: *Tropical Trees: The Potential for Domestication and the Rebuilding of Forest Resources*. Proceedings of a conference, August 23–28, 1992, Edinburgh, Scotland. Edited by R.R.B. Leakey and A.C. Newton. London: HMSO, 137–147.

Florece, L.M. 1996. Fire Behavior, Fuel Dynamics and the Responses of Trees and Grasses to Fire in Carranglan, Nueva Ecija, Philippines. Ph.D. thesis. University of New Brunswick.

Fujisaka, S., E. Jayson, and A. Dapusala. 1994. Trees, Grasses and Weeds: Species Choices in Farmer-developed Contour Hedgerows. *Agroforestry Systems* 25, 13–22.

Garrity, D.P. 1993. Sustainable Land Use Systems for Sloping Uplands in Southeast Asia. In: *Technologies for Sustainable Agriculture in the Tropics*, ASA Special Publication 56. Madison: American Society of Agronomy, Crop Science Society of America, and Soil Science Society of America.

———. 1994. Improved Agroforestry Technologies for Conservation Farming: Pathways Toward Sustainability. Paper presented to an international workshop on Conservation Farming for Sloping Lands in Southeast Asia: Challenges, Opportunities, and Prospects, November 20–26, 1994, Manila, the Philippines.

———, and A.R. Mercado Jr. 1994. Reforestation Through Agroforestry: Market Driven Smallholder Timber Production on the Frontier. In: *Marketing of Multipurpose Tree Products in Asia*. Proceedings of an international workshop, December 6–9, 1993, Baguio City,

Philippines, edited by J.B. Raintree and H.A. Francisco. Bangkok: Winrock International, 358.

———, A.R. Mercado Jr., and J.C. Solera. 1995. Species Interference and Soil Changes in Contour Hedgerows Planted on Inclines in Acidic Soil in Southeast Asia. In: *Alley Farming Research and Development*, edited by B.T. Kang, A.O. Osiname, and A. Larbi. Ibadan, Nigeria: Iita, 351–365.

———, and P.C. Agustin. 1995. Historical Land Use Evolution in a Tropical Acid Upland Ecosystem. *Agriculture, Ecosystems and Environment* 53 (1995), 83–95.

Gellor, J.M., and T.P. Austral. 1996. Grassland Revegetation and Rehabilitation in Mt. Musuan, Using Timber and Multipurpose Tree Species (MPTS). In: *Strengthening Research and Development for Sustainable Management of Grasslands*. Proceedings of the First National Grassland Congress of the Philippines, September 26–28, 1995, University of the Philippines at Los Baños, Laguna, the Philippines: ERDB, 131–135.

Generalao, M.L, N.A. Andin, M.D. Dimayuga, and F.M. Lauricio. 1977. *Silvical Characteristics and Planting Instructions for High Premium and Fast Growing Species*. Los Baños, Laguna, Philippines: Reforestation Research Center, Forest Research Institute (FORI).

Hanley, N., and C.L. Spash. 1993. *Cost-Benefit Analysis and the Environment*. Cheltenham, UK: Edward Elgar Publishing.

Junus, K. 1989. End Uses of Trees in Village Forests: Two Case Studies from Java. In: *Multipurpose Tree Species Research for Small Farms: Strategies and Methods*, Proceedings of an international conference held November 20–23, 1989, Jakarta, Indonesia, edited by C. Haugen, L. Medema, and C.B. Lantican, 24–31.

Kearl, L.C. 1982. *Nutrient Requirements of Ruminants in Developing Countries*. Logan, Utah: International Feedstuffs Institute, Utah Agricultural Experiment Station, and Utah State University.

Lamb, A.F.A. 1968. *Forest Growing Timber Trees of the Lowland Tropics*, No. 1. *Gmelina arborea*. Commonwealth Forestry Institute, Department of Forestry, University of Oxford, 31.

Lasco, R.D., and W.M. Carandang. 1989. Small Farm Multipurpose Tree Species Research in the Philippines: Method, Issues and Institution. In: *Multipurpose Tree Species Research for Small Farms: Strategies and Methods*, Proceedings of an international conference held November 20–23, 1989, Jakarta, Indonesia, edited by C. Haugen, L. Medema, and C.B. Lantican, 133–139.

Magcale-Macandog, D., C.D. Predo, K. Menz, and A.D. Calub. 1997. Napier Grass Strips and Livestock: A Bioeconomic Modeling, Imperata Project Paper 1997/1. Canberra, Australia: CRES, Australian National University.

Marchand, D. 1987. The Poverty of Slash-and-Burn. *IDRC Reports* 16(1), 7.

Mercado, A.R. Jr, N. Sanchez, and D.P. Garrity. 1996. Crop Productivity using Forage Legumes and Grasses as Contour Hedgerow Species in an Upland Soil. Paper presented to the 12th annual scientific conference of the Federated Crop Science Societies of the Philippines, May 13–18, 1996, Davao City, Mindanao, the Philippines.

Miah, G.M.D., D.P. Garrity, L.T. Szott, and C.A. Palm. 1993. *Comparative Performance of Multipurpose Tree Species Grown Alone and in Association with Annual Crops*. Bogor, Indonesia: World Agroforestry Centre (ICRAF).

Nelson, R., R. Cramb, K. Menz, and M. Mamicpic. 1996. Bioeconomic Modeling of Alternative Forms of Hedgerow Intercropping in the Philippine Uplands using SCUAF. *Imperata Project Paper* 1996/9. Canberra, Australia: CRES, Australian National University.

Pabuayon, I.M. 1989. Marketing Tree Products from Small Farms: Case Studies from the Philippines and Implications for Research. In: *Multipurpose Tree Species Research for Small Farms: Strategies and Methods*, Proceedings of an international conference held November 20–23, 1989, Jakarta, Indonesia, edited by C. Haugen, L. Medema, and C.B. Lantican, 133–139.

Sibanda, H.M. 1991. Indigenous Agroforestry Systems: Livestock-based Agroforestry in Gwenda District, Zimbabwe. In: *Agroforestry Research in the Miombo Ecological Zone of Southern Africa*, summary proceedings of an international workshop, June 16–22, 1991, Lilongwe, Malawi, edited by A.G. Maghembe, H. Prins, and D.A. Brett.

Stark, M. 1996. Natural Vegetation Filter Strips: Farmers' Technology Modification as a Base for Participatory On-farm Research. Paper prepared for Sustainable Agriculture and Natural Resource Management (SANREM) annual workshop, March 18–20, 1996, Malaybalay, Bukidnon, the Philippines.

Young, A. 1989. *Agroforestry for Soil Conservation*. Wallingford, UK: CAB International. Cambridge Press.

Chapter 38

Innovations in Swidden-Based Rattan Cultivation by Benuaq-Dayak Farmers in East Kalimantan, Indonesia

Hideyuki Sasaki

Nontimber forest products have become an indispensable element in any discussion concerning the problems confronting both tropical forests and the people who live in them (Peluso 1983, 1992; Conelly 1984; de Beer and Mcdermott 1989; Panayatou and Ashton 1992). When considering the exhaustion of forest resources, the literature highlights the importance of considering, as a major issue, the economic value of nontimber forest products, along with the economic value of timber. Hand in hand with this consideration comes increasing attention to the significant role of extractive activities in the socioeconomy of forest-dwelling people (Hecht et al. 1988; Godoy and Feaw 1989).

In Borneo, important nontimber forest products include rattan, resins, edible birds' nests, and illipe nuts (Hoffman 1988, *108*). Many of these products are extracted from natural forests by the *Punan*, or forest dwellers, and by swidden cultivators. Some groups of swidden farmers in Borneo have incorporated fruit trees, rattan, rubber, and *tengkawang*, which are *Shorea* species producing illipe nuts, into their swidden fields. They have, thus, created swidden-fallow agroforestry systems.[1] According to the arguments of Dove (1993) and Weinstock (1983), these swidden-fallow systems sustain a balance in respect of both ecology and economy.

Indeed, the rattan cultivation system of the Benuaq-Dayak swidden farmers, which is the main focus of this chapter, provides a similar type of ecologically and economically balanced agroforestry. However, the case of these farmers is different. They were technically innovative and economically motivated in transforming their ways of planting and managing rattan gardens. And, unlike swidden rubber cultivators, for instance, the Benuaq-Dayak invested not only labor but also substantial amounts of money to expand the size of their rattan gardens and to improve cane productivity. This was before a series of state interventions changed the face of Indonesia's rattan industries. The intervening policies were intended to control the exploitation of rattan resources and encourage value-added processing

Hideyuki Sasaki, General Manager, Planning Division, Pacific Consultants International (PCI), 1-7-5 Sekido, Tama-shi, Tokyo 206, Japan.

[1] For swidden rubber, see Thomas (1965), Pelzer (1978), and Dove (1993); for swidden rattan, see Weinstock (1983) and Fried (1995). The cultivation of tengkawang is not integrated into swidden agriculture in the case Chin (1985) describes, but in West Kalimantan, there are many tengkawang plantations created by local people.

industries. But local prices of rattan canes dropped to nearly half of peak price levels, and farmers without means of generating sufficient income turned to wild rattan extraction and informal timber felling within their village territory. Both the swidden-rattan agroforestry system and the livelihood strategies of the Benuaq-Dayak people were destabilized.

Local instability of swidden systems is not new. In most cases, however, the destabilizing factors are found in limited availability of forestlands where shifting cultivation is practiced, and this motivates the enrichment of fallows by tree planting (Angelesen 1995; Mary and Michon 1987). Angelesen (1995) describes how farmers in Sumatra are racing to make rubber gardens in primary forest areas. This is partly a response to market opportunities but also because farmers' tenurial security over traditional forest territory is being threatened by increasing land claims by outsiders.

In contrast to these cases, the destabilization of rattan cultivation in Indonesia is rooted in the political economy of state control over natural resources. A concession system has been imposed on extraction of wild rattan, although Benuaq-Dayak canes are mostly cultivated in gardens, and an export ban has been placed on unprocessed or semiprocessed rattan canes. These efforts to control rattan resources are similar, and linked to, the logic behind Indonesia's timber exploitation policy.[2]

This chapter begins with a general background of the study area and then outlines the traditional Benuaq-Dayak system of rattan cultivation in swidden fields (see color plate 52). It then traces the transformation of rattan cultivation since the last "rattan boom" in the 1980s. Finally, it analyzes the impact of recently implemented government policies on the swidden-rattan economy of the Benuaq-Dayak.

This case study suggests that there are a variety of local implications to the state's action in implementing its timber exploitation policy. It also describes how local farmers have responded to the restrictions imposed on them by state intervention.

The Study Area

Besiq is a Benuaq-Dayak village located in the uppermost headwaters of the Kudang Pahu River, one of the major tributaries to the Mahakam River (see Figure 38-1). The village belongs to Damai subdistrict, and is about 20 km southwest of the subdistrict capital, Damai town. Two to three large longboats operate daily from Samarinda, the provincial capital of East Kalimantan, up to Damai town. The trip takes about 24 hours by large longboat. It takes a further five hours by small outboard motorboat from Damai town to Besiq.

The Benuaq-Dayak farmers in Damai subdistrict mostly practice swidden agriculture in combination with rattan cultivation. Similar practices are prevalent in the neighboring subdistricts of Muara Lawa and Bentian Besar. However, abundant forestland in Besiq allows villagers to practice swidden agriculture in a more extensive manner. Moreover, responding to increased international demands for rattan canes, they have developed commercial rattan cultivation systems based on swidden agriculture.[3] At the same time, Besiq village has been the site of "informal" logging activities, which have waxed and waned since the late 1960s.[4]

According to the national census, the population of Besiq was about 1,126 in 1990. The register of population at Besiq village shows that the village had 253

---

2 In the early 1980s, the Indonesian government began to restrict the export of unprocessed timber and to encourage the development of timber processing industries, such as plywood and sawn timber manufacturing. The aim was to encourage development of local value-added industries.

3 In contrast to Besiq, rattan gardens in Mencimai and other Dayak villages in the Tunjung plateau are small scale, and because they have a shortage of land suitable for rattan cultivation, they are no longer expanding rapidly.

4 Since the early 1970s, local people have lost their access to timber for commercial purposes due to the forest law and regulations.

households in July 1991. The total village area is about 400 km^2, which made the population density in 1990 about 2.8 persons/km^2, one of the lowest in Damai district.

Besiq has two distinct groups of residents. One is a group of indigenous Benuaq-Dayak people who are called either Benuaq-Dayaq or Benuaq-Mumukng. The other group consists of Benuaq-Dayak people who have migrated and settled in Besiq, mainly from the Idatn area, about 20 to 30 km down the river.

The village settlement area of Besiq is located at the junction of two rivers: the Kudang Pahu River and the Peraq River. The village spreads upstream and covers the river basin common to the two rivers. The Benuaq-Idatn tend to live along the Peraq River, while most of the Benuaq-Mumukng live along Kudang Pahu River.

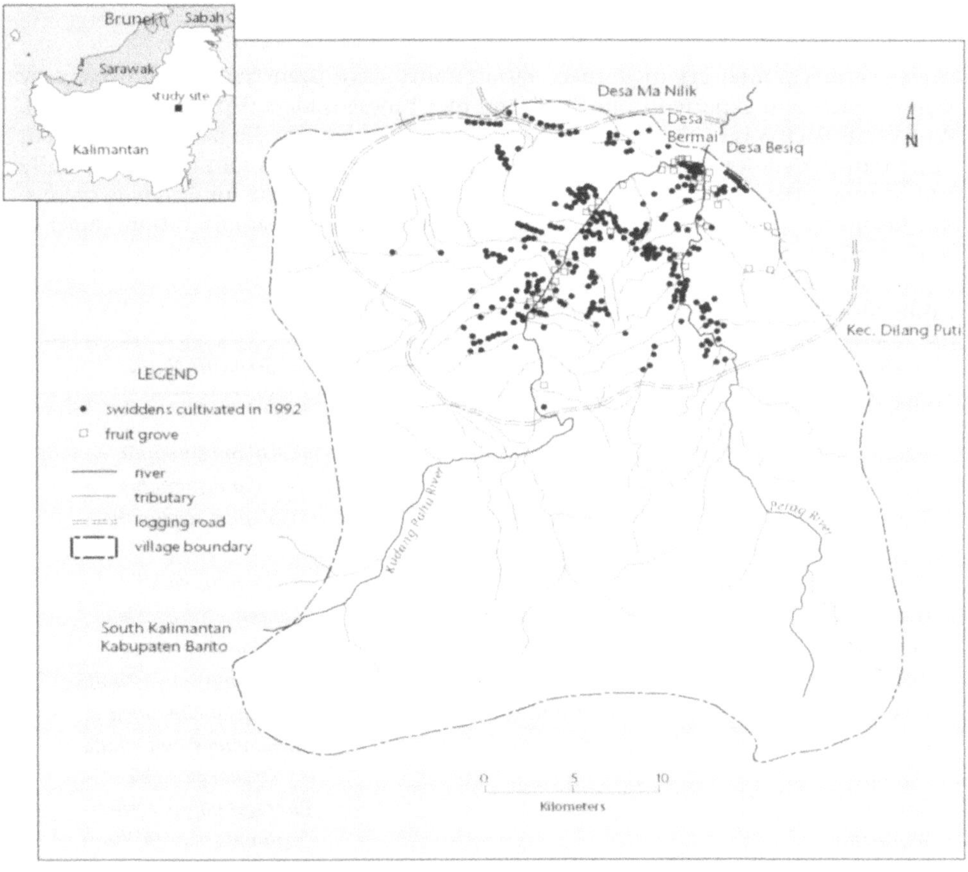

Figure 38-1. The Study Site: Land Use at Besiq Village

Traditional Rattan Cultivation in Benuaq-Dayak Swidden Agriculture

Natural forests in the native territory of the Benuaq-Dayak are rich in rattan species, with more than a dozen species identified (Table 38-1). Traditionally, the people have both extracted rattan from natural forests and cultivated it in swidden fallows, for domestic use and sale. The Benuaq-Dayak use rattan for the following purposes:

- Making woven and nonwoven baskets;
- Making woven and nonwoven mats;
- Making floors of rattan canes;
- Binding materials for house construction;
- Binding the handles of knives, chopping swords, and axes;
- Sewing together sheets of tree bark to construct rice bins;
- Making ropes to climb high trees, for instance, to collect honey; and
- As food, eating the young shoots of some rattan species.

Since the precolonial era in Borneo, rattan canes have been traded extensively by Malay traders and exported from the island by Chinese traders (Hoffman 1988, *108*). When Carl Bock was in the Mahakam in the late 19th century, he noticed that rattan canes transported downstream were taxed by Kutai sultan.[5] It is said that in the early 20th century, the cultivation of the sega species, the local name for *Calamus caesius*, was begun by indigenous people in the Mahakam Basin to supply rattan canes to world markets.

Table 38-1. Rattan Species in Benuaq-Dayak Forests

Benuaq-Dayak Name	*Indonesian Name*	*Scientific Name*
Large diameters (more than 2 cm):		
ngono	manau	*Calamus mannii*
Sidokng		*Calamus ornatus*
Tuu	semambu	*Calamus scipionum*
Medium diameters (between 5 mm and 2 cm):		
Sokaq	sega	*Calamus caesius*
Jahap	irit/jahap	*Calamus trachycoleus*
Boyukng	seltup	*Calamus optimus*
Kotok	kotok	*Daemonorops angustifolia*
uwe meaq	rotan merah	*Korthalsia chinometra*
ia		*Daemonorops hallieriana*
Danan		*Korthalsia rigida*
Bencia		*Plectocomiopsis corneri*
Biyungan		*Calamus polystachys*
siit		*Calamus baratangensis*
Small diameters (less than 5 mm):		
jupukn	pulut merah	*Daemonorops cristata*
pelas	pulut putih	*Calamus penicillatus*
lalutn		*Korthalsia scaphigera*
uwe pakuq		*Calamus exilis*
keheh		

Source: Local Benuaq-Dayak informant in Besiq who is a forestry student at Mulawarman University in Samarinda, East Kalimantan.

[5] Carl Bock, a Norwegian naturalist and explorer, was sent to southeast Kalimantan by the Dutch colonial government (see Bock 1881).

Sega rattan has many uses. It is a primary material in the construction of furniture, baskets, and carpets and is a binding material for furniture. Although the Benuaq-Dayak cultivate several species in their gardens, they favor sega and this chapter concentrates on it solely. Since the 1980s, when Japanese markets for rattan mats began to be successfully exploited, sega species have enjoyed the strongest commercial demand and hence, the highest prices, of all rattan species.

Traditional Techniques of Sega Cultivation

The Benuaq-Dayak traditionally transplant rattan seedlings into young swidden fallows. The seedlings are gathered either from natural forests or from their own rattan gardens. They may be interplanted into swidden rice crops, so the rice provides them with beneficial shade. (See color plate 4 for a similar system in Lao P.D.R.) Transplanting takes place after it has rained, during the rainy season. The seedlings should have not more than two leaves, but at least one new sprout. Locations are carefully selected, with good soil, often near the roots of felled trees. In cases where the soil conditions are not good, sites are often chosen near streams. The seedlings then go on to develop in the fallow vegetation, after the cropping phase has passed.

Rattan canes can be harvested 10 to 12 years after transplanting. Each seedling generates several shoots, developing into canes up to 30 m. When they are cut, at least one meter of cane needs to be left to ensure that the rattan plant can survive to regenerate new canes. Two years after harvesting, the plants will have grown new canes that are long enough to cut again.

In the Benuaq-Dayak system, rattan develops in swidden fallows among other secondary growth. Without the shade provided by the secondary growth, the young rattan plants could not survive. However, without sunlight, the rattan will not produce a satisfactory volume of cane. Therefore, in order to achieve both quantity and quality of canes, rattan gardens need considerable care. Undergrowth around the rattan plants must be slashed to reduce competition for soil nutrients. This process is called *ngowa*. Some tall trees nearby rattan plants must also be killed to allow penetration of sunlight and, once again, to remove competition. The trees are killed by removing bark and surface material from their trunks. In this way, the trees die gradually, resulting in a gradual increase of sunlight into the garden. However, it is also necessary to leave some nearby trees alive so the rattan can climb because rattan that does not climb does not produce strong and flexible canes.

At the same time as they harvest cane from their gardens, the farmers usually clear the undergrowth, cut competing trees, and plant new seedlings. With good care, rattan gardens continue to be productive for several decades. Some Besiq villagers say that well-managed rattan gardens are productive for as long as 70 years.

Many of the people of Besiq were overtaken by a dedication to properly manage rattan gardens during the period when rattan prices were high. They began in difficult circumstances, when rattan was not as plentiful as it has now become, and it was difficult to find seeds or seedlings either in forests or in gardens. A few people had also inherited rattan gardens and they had allowed them to deteriorate to a low density of plants, none of which was producing many canes. However, rather than persevere with the old inherited gardens when the boom time came, the farmers of Besiq began, instead, to establish new gardens in their swiddens.

Integration of Rattan Cultivation with Swidden Agriculture

In terms of allocating both labor and land, the Benuaq-Dayak system of rattan cultivation in Besiq village is closely interwoven within swidden agriculture. It compares with the creation of rubber gardens, where rubber seedlings are planted into swiddens to develop in the fallow phase. Traditionally, planting seedlings of rubber and rattan is similar in that farmers first locate seedlings in gardens or forests and then transplant them into swidden fallows during the rainy season from August to February. This approach does not require much additional labor, and, since the scale of planting is modest, it can usually be done during routine maintenance activities in swiddens.

Because rattan canes are much heavier than rubber latex, rattan gardens are usually located near watercourses for ease of transport. In Besiq, rattan gardens are usually created within 2 km of rivers navigable by small boats. This location preference does not conflict with the need to clear new swidden fields because forest land is abundant along rivers.

Harvesting of rattan cane is not as time-consuming as collecting rubber latex. A strong young man can harvest 100 kg of wet canes in a day, although it depends on the garden's location. Many households with rattan gardens harvest canes only when they need cash to buy goods. Hence, they typically cut only 50 to 60 kg of canes, and labor allocation is much less demanding than, say, for rubber.

Commercial Rattan Cultivation

In the 1980s, Japanese demand for rattan mats increased substantially, and the market price for high-quality sega cane rose along with the demand. The mats sought by the Japanese markets were based on those traditionally made by Dayaks in Borneo, particularly those in southern Kalimantan and the western part of the Mahakam.

It appears that swidden farmers began cultivating rattan because extraction of natural stocks of rattan from forests was neither easy nor rewarding for nonspecialists of extraction. Then after the escalating world demand, they expanded the cultivation of rattan in swidden fallows (Siebert and Belsky 1985; Hoffman 1988; Sellato 1994). Its cultivation was not overly difficult, when its growth in the wild had been carefully observed.

Innovation of Rattan Cultivation Techniques

Progressive Besiq farmers, about 40% of the total number of farmers in the village, began developing methods of intensifying rattan cultivation within their swidden fields. Remaining households were less "progressive," and some did not expand their rattan gardens because their cash needs were modest.

Recently, innovations have been aimed at increasing density and productivity. Farmers are making full use of their swidden fallows for planting rattan. In fact, in order to expand their production, farmers are going further afield, into primary forests upstream from Besiq, to make bigger gardens. They are commuting by outboard motorboat to primary forest areas upriver. To fell the larger trees characteristic of primary forests, they need chain saws. This means that, unlike traditional swiddening or traditional approaches to planting rubber in swiddens, their modern approach to swidden-rattan farming requires capital investment. Benuaq-Dayak farmers are, consequently, under pressure to justify their investments. Failure to plant rattan will mean a loss of invested money; failure to plant rattan throughout an entire swidden means a failure to fully utilize opportunities for higher returns from investment in land. Progressive Benuaq-Dayak farmers refer to such failures with the term *rugi*, which means "to suffer a financial loss." Table 38-2 shows the ways in which recent management techniques and technical innovations have intensified traditional methods of rattan farming at Besiq.

The new drive to increase productivity requires that farmers improve their management techniques. Among the important needs of young seedlings is shade for establishment. But several years after planting, the developing plants need sunlight and nutrition, so farmers slash undergrowth and kill competing trees. Instead of ngowa, they call this practice *bemas*, which means "slash only in the surroundings of rattan plants." Bemas is usually done at the same time as rattan canes are harvested.

Most farmers have found that planting seeds is much easier than transplanting seedlings. With seedlings, the timing is crucial because they should be transplanted when they have new sprouts, but not more than two young leaves, and it should be done in the rainy season, after it has rained. Although the germination rate of planted seeds is much less than the survival rate of properly transplanted seedlings, the farmers have found direct planting of seeds is much more efficient. For one thing, handling, transporting, and planting seeds is easier than seedlings. Sega seeds usually ripen in July and August, at about the same time that rice is planted. Some farmers believe that if they plant rattan seeds before burning their swiddens, the seeds will germinate well. Others advise that planting rattan and rice seeds together saves time. The rattan takes more than one month to germinate, by which time the rice crop is big enough to provide shade for the emerging rattan.

Some farmers have tried new ways to increase the germination rates of rattan seeds. One method is to place the seeds in a cloth sack, in a shaded location, and spray the sack with water until the seeds start germinating. The germinated seeds are then planted into swiddens.

Despite the experiments with direct seeding, many active rattan cultivators continue to transplant seedlings. Most of them have tried new ways of improving their rattan cultivation and have settled on the techniques that appear to be most suited to their situation. This individual selection of suitable practices is called *jawatn* by the Benuaq-Dayak. The different approaches seem to reflect the ecological and economic conditions specific to each farmer.

Social Organization and Strategies for Commercial Rattan Cultivation

The intensification of commercial rattan cultivation by the Benuaq-Dayak required changes to the social organization and working conditions in Besiq village. These included the following:

Swidden Working Groups. These groups are composed of families who have made contiguous swiddens in the same area. This makes it easier for them to work together to guard against crop damage by hungry wildlife. This arrangement is favored because Besiq farmers normally prefer not to work as hired laborers for others and will invest surplus family labor in making larger swiddens to plant more rattan for themselves. The swidden working groups are mostly based on kinship ties.

Table 38-2. Traditional Methods and Innovations in Rattan Cultivation at Besiq

Traditional Methods	*Recent Innovations*
1) Farmers did not always plant rattan in their swiddens.	1) Progressive farmers always plant rattan in their swiddens.
2) Rattan planting in swiddens was sporadic and incomplete.	2) Progressive farmers systematically plant rattan throughout their swiddens.
3) Farmers did not make large swiddens for the purpose of planting a lot of rattan.	3) Progressive farmers prefer to make large swiddens to increase the size of the subsequent rattan garden.
4) Farmers did not always prefer primary forests to secondary forests.	4) In order to make larger swiddens, progressive farmers are choosing locations in primary forests.
5) Farmers maintained planted rattans by slashing undergrowth and killing some trees, but this was done selectively wherever rattan had been planted.	5) To make rattan gardens more productive, progressive farmers slash undergrowth and kill some trees throughout their rattan gardens.
6) Farmers usually transplanted seedlings gathered from forests or gardens.	6) Rather than transplanting seedlings, more progressive farmers plant rattan seeds at the same time as they plant rice.
	7) More progressive farmers are experimenting with more innovative ways of planting.

Extended Family. The Benuaq-Dayak maintain an extended family style, in which one married child lives with the parents, providing each household with increased labor for making larger swiddens and rattan gardens.

Outboard Motors for River Travel. Small boats with outboard motors are used to travel upriver to swiddens and rattan gardens. The first outboard motors appeared in the mid-1970s, and many were purchased in the 1980s, after rattan prices increased. About 60% of households now own outboard motors. These are bought with money saved from the sale of rattan, rice, timber and other forest products.

Chain Saws. Farmers now use chain saws for ease and efficiency in cutting large trees when clearing swiddens from primary forest. Some households own a chain saw; others hire or borrow them from children or relatives. Forty-six percent of surveyed households said they used a chain saw when opening swidden fields; 60% said they used chain saws when clearing primary forest.

Living at Swidden Houses. Swidden fields are often quite distant from the village settlement, sometimes up to three hours' travel by outboard motorboat, so farmers tend to live on the fields, in swidden huts. Younger households may not own a house in the main village settlement, and live permanently on the swidden. They stay overnight at the house of parents or relatives if they have to visit Besiq. Their children go to school in the village and must live apart from their parents.

Labor for Rice Harvesting. Even though households may work together to prepare swidden fields, they usually work individually to harvest the rice crops. This is because the rice on all swiddens ripens at about the same time, and because the swiddens are larger than normal, the labor must be spread more widely to cope with a bigger job. Each household tries to mobilize its own labor resources. Sometimes, relatives from other villages who may have suffered a crop failure are invited to participate. Javanese transmigrants may also be hired to assist. The additional size of the crops means farmers are able to hire labor and still make money. However, in most cases, the rice harvest involves only household members, with some assistance from relatives.

Hiring Labor for Rattan Cultivation. When rattan prices are high enough, more than 500 rupiah/kg, the owners of rattan gardens may hire workers to harvest the canes, particularly from those gardens far from rivers. Wages are about 200 rupiah/kg of harvested rattan weight, although this depends on the distance from the river. The harvesting operation includes both cutting the canes and transporting them back to the village.

Impacts of Government Policies on Benuaq-Dayak Rattan Cultivation

Uncertain Future of Swidden Agriculture

Even before the introduction of the export ban on rattan canes, Besiq's progressive farmers feared that rattan cultivation in swidden fallows probably did not have a bright future. The government was known to be opposed to "shifting agriculture" because of its practice of slashing and burning the forest. Anticipating a total ban on swidden agriculture, the progressive farmers of Besiq decided to expand their rattan gardens as widely as possible because the conversion of primary forests into rattan gardens allowed them to secure tenurial rights over the land. However, the move left them totally reliant upon cash income from rattan for their children's education and for purchasing daily needs.

State Control of Rattan Resources and its Impact

In late 1979, a government regulation, No. 492/KP/VII/79, was introduced prohibiting the export of raw rattan canes. Seven years later, a further government regulation, Keputusan Mentri Perdagangan No. 274/1986, prohibited the export of washed and sulfurized rattan canes. Finally, in January 1989, Keputusan Mentri Perdagangan No. 190/KP/VI/88 banned the export of all kinds of semiprocessed rattan canes. These laws and regulations were intended to encourage the development of domestic rattan processing industries.

In response, the number of rattan-processing companies increased dramatically, along with production capacity. In December 1987, 175 rattan-based manufacturing companies were registered, with an annual processing capacity of 327,181 metric tonnes of semifinished rattan products and 203,574 metric tonnes of finished rattan products (PT Capricorn Indonesia 1988).

The manufacturing industries began to develop in South Kalimantan in the early 1980s, especially in Banjarmasin, a large capital town, and Amuntai, a small rural town specializing in handicrafts. A Japanese entrepreneur started the manufacture of rattan mats in Amuntai. They were handmade, and, by achieving high quality, Amuntai gradually tapped into the Japanese market.

In the beginning, there were only nine companies in Banjarmasin and Amuntai, and the only exporters were in the Banjarmasin area. Then the boom in rattan mat exports to Japan came in the years before 1988, and the number of companies making rattan products for export rocketed to 147. They invested in machinery to improve quality and increase production. But by the end of 1993, only 20 companies were left. The slump was largely due to a recession in the Japanese economy. The rattan companies of South Kalimantan struggled to reduce costs by contracting out parts of their manufacturing processes, especially the most labor-intensive parts, to lesser firms in Amuntai. Rattan mat producers in Banjarmasin survived only by purchasing semifinished mats from Amuntai and finishing them to a high standard. They were able to export cheap products of good quality.

Big rattan mat manufacturers that had grown up in Jakarta and Surabaya were not so fortunate. Unable to "farm out" work to centers of cheap labor and cottage industries, some went bankrupt and others shifted to manufacturing rattan furniture.

Uneven Impacts of the Export Ban on Semiprocessed Rattan

During the period of unrestricted export, rattan canes from the Middle Mahakam could be exported anywhere in the world. However, after the 1986 government ban on the export of washed and sulfurized canes, most rattan from the Middle Mahakam was shipped to mat manufacturers in Surabaya. This was because the largest buyers of rattan in Samarinda, the capital of East Kalimantan, had close relations with the manufacturers in Surabaya, and also because lower transport costs allowed higher cane prices in Surabaya than those in Banjarmasin.

This was to be a strong determining factor in the fate of cane cultivators in the interior of Kalimantan. Most rattan mat manufacturers in Surabaya went bankrupt in the Japanese recession and stopped buying canes from Samarinda, which sourced its rattan from the Middle Mahakam. However, more than 20 manufacturers at Banjarmasin and Amuntai survived because of their ability to exploit cheap labor sources. They continued purchasing rattan canes from Central Kalimantan, but at half the previous prices. Table 38-3 shows the tumble in rattan prices during the late 1980s and early 1990s.

Table 38-3. Price Declines of Rattan Canes in Besiq and Samarinda

		Prices (Rp/kg)			
Location	*Rattan Type*	*Before*		*After*	
Besiq	wet sega cane	700–800	1987/88	350	Apr. 1992[a]
		600	1991	410	Nov. 1992[b]
				400	Dec. 1992[b]
				387	Jan. 1993[b]
				350	Feb. 1993[c]
Samarinda	dry high class sega cane	2,000	1990/91	1,450	Apr. 1992[d]
				1,400	Mar. 1993[e]
	wet sega cane	1,250	1990	800–950	Feb. 1993[f]

Sources: [a, c,] and [f] Local informants in Besiq; [b] Local rattan trader's note in Besiq; [d] and [e] Besiq rattan traders' receipts of transactions in Samarinda. Other data based on the recollections of local informants in Besiq.

Lost Sources of Cash Income

The dual impacts of the export ban on semiprocessed rattan cane and the Japanese economic recession were huge. Not only did the price of rattan cane fall drastically, but also the demand was largely reduced, especially for sega rattan, the main material in rattan mats.

Most Besiq residents were either farmers or traders who depended on canes harvested from rattan gardens for their income, so the whole village was seriously affected by both the decline in prices and the reduced trading volume. Young people sought temporary jobs with nearby logging companies. The standard wage offered for two and a half months of manual labor was Rp600,000 (1990US$300), including three meals daily. However, competition for these jobs was very high and it was not easy to hold on to continuous work.

The impact was most severe for farmers who had invested heavily in rattan gardens. Others who lost out included people engaged in harvesting rattan on a shared basis and those who earned wages from cleaning rattan cane. The former were mostly men who had little mature rattan of their own. One full day of rattan harvesting earned them Rp20,000 to 30,000 (1990US$10 to US$20). During the boom period, even the owners of mature rattan gardens were attracted by the high cash incomes and worked at harvesting others' rattan. Cleaning the rattan was another story. It involved mainly wives and young girls with little mature rattan of their own. They earned little more than Rp5,000 (1990US$2) for a full day's work.

Faced with stagnation or total loss of their income, farmers sought other cash sources, such as extraction of wild rattan and timber. Gathering wild rattan was rewarding but time-consuming. Logging was less rewarding because the timber was illegal, and, when selling it, they obtained only 10% of the price paid by plywood factories to logging companies. Consequently, after having lost nearly everything else, the villagers of Besiq rapidly lost many valuable timber trees as well.

Conclusions

This case study has described an example of innovative intensification of fallow management by Benuaq-Dayak farmers in East Kalimantan. Their transformation of swidden fallows into rattan gardens became possible under a combination of technical and economic innovations. Significantly, these innovations were based on indigenous knowledge, both acquired from their ancestors and generated through their own experiments in the daily effort to improve their standard of living. Government policymakers and development planners frequently overlook such flexible and innovative developments in indigenous swidden agriculture.

Although the swidden-based rattan cultivation described in this chapter involves the clearing of primary forests, it is economically and environmentally justifiable to allocate certain areas of primary forest to such a sustainable use. This case study also suggests that, in the absence of supportive policy environments, it is difficult for people like the Benuaq-Dayak to sustain their efforts to improve their lifestyles through intensified fallow management. It provides an example of how important it is that people are not discouraged from continuing their indigenous agroforestry practices for rural economic development. Equally, it demonstrates the ease with which governments can ignore this principle when pursuing other objectives.

Acknowledgments

This study is based on part of the field work I conducted over 18 consecutive months during 1992 and 1993. It was funded by the Foundation for Advanced Studies on International Development (FASID), Japan, and the Toyota Foundation. Preliminary one-month field visits were undertaken in 1990 and 1991. These preliminary visits were funded by the Nissan Science Foundation and Pacific Consultants International.

References

Angelesen, A. 1995. Shifting Cultivation and Deforestation: A Study from Indonesia. *World Development* 23(10), 1713–1729.

Bock, C. 1881. *The Head-Hunters of Borneo: A Narrative of Travel up the Mahakam and down the Barito.* Oxford University Press.

Chin, S. C. 1985. Agricultural and Resource Utilization in a Lowland Rainforest Kenyah Community. *The Sarawak Museum Journal* 35(56) (new series), Special Monograph No. 4.

Conelly, W. T. 1984. Copal and Rattan Collecting in the Philippines. *Economic Botany* 39(1), 36–46.

de Beer, J. H., and M. J. McDermott. 1989. *The Economic Value of Non-Timber Forest Products in Southeast Asia.* Amsterdam: Netherlands Committee for IUCN.

Dove, M. R. 1993. Smallholder Rubber and Swidden Agriculture in Borneo: A Sustainable Adaptation to the Ecology and Economy of the Tropical Forest. *Economic Botany* 47(2), 136–147.

Fabienne, M. and G. Michon. 1987. "When Agroforests Drive Back Natural Forests: A Socio-Economic Analysis of a Rice-Agroforest System in Sumatra". *Agroforestry Systems* Vol. 5, 1987 (5-27).

Fried, S. 1995. Writing for their Lives: Bentian Dayak Authors and Indonesian Development Discourse. Ph. D. Dissertation, Cornell University.

Godoy, R. A., and T. C. Feaw. 1989. The Profitability of Smallholder Rattan Cultivation in Southern Borneo, Indonesia. *Human Ecology* 17(3), 347–363.

Hecht, S. B., A.B. Anderson, and P. May. 1988. The Subsidy from Nature: Shifting Cultivation, Successional Palm Forests, and Rural Development. *Human Organization* 47(1), 25–35.

Hoffman, C. L. 1988. The "Wild Punan" of Borneo: A Matter of Economics. In: *The Real and Imagined Role of Culture in Development: Case Studies from Indonesia*, edited by Michael R. Dove. Honolulu: University of Hawaii Press.

Kim, S., and M. Kumazaki. 1993. Tenkan-ki wo Mukaeta Shitujyun-Nettairin no Kaihatu - Minami-Kalimantan Niokeru K-sya no Jigyou-Tennkai wo Tyushinn ni, Kankyo-Kenkyu, No. 91, 98–109. (Development of Tropical Rainforest at Crossroads: The Case Study of K-Company's Logging Operation in South Kalimantan.) *Environmental Study*, No. 91, 98–109 (Japanese language).

Panayatou, T., and P. S. Ashton. 1992. *Not by Timber Alone: Economics and Ecology for Sustaining Tropical Forests.* Washington, DC: Island Press.

Peluso, N. L. 1983. Networking in the Commons: A Tragedy for Rattan? *Indonesia* 35, 95–108.

———. 1992. The Political Ecology of Extraction and Extractive Reserves in East Kalimantan, Indonesia. *Development and Change* 23(4), 49–74.

Pelzer, K. 1978. Swidden Cultivation in Southeast Asia: Historical, Ecological and Economic Perspectives. In: *Farmers in the Forest: Economic Development and Marginal Agriculture in Northern Thailand,* edited by P. Kunstadter, E.C. Chapman, and Sanga Sabhasri. Honolulu: University of Hawaii Press.

PT Capricorn Indonesia. 1988. *A Study of the Prospects on the Rattan Industry and Market*. Jakarta.

Safran, E. B., and R. A. Godoy. 1993. Effects of Government Policies on Smallholder Palm Cultivation: An Example from Borneo. *Human Organization* 52(3), 294–298.

Sellato, B. 1994. *Nomads of the Borneo Rainforest: The Economics, Politics and Ideology of Settling Down*. Honolulu: University of Hawaii Press.

Siebert, S. F., and Jill M. Belsky. 1985. Forest-Product Trade in a Lowland Filipino Village. *Economic Botany* 39(4), 522–523.

Thomas, K. D. 1965. Shifting Cultivation and Smallholder Rubber Production in a South Sumatran Village. *The Malayan Economic Review* 10(1), 100–115.

Weinstock, J. A. 1983. Rattan: Ecological Balance in a Borneo Rainforest Swidden. *Economic Botany* 37(1), 56-68.

Chapter 39

Bamboo as a Fallow Crop on Timor Island, Nusa Tenggara Timur, Indonesia

Abdullah Bamualim, Joko Triastono, Evert Hosang, Tony Basuki, and Simon P. Field

The western part of Timor Island is part of Nusa Tenggara Timur (NTT) Province, located in southeastern Indonesia (Figures 24-1 and 28-1). Most of the region is classified as tropical semiarid, with low rainfall, from 1,000 to 1,500 mm per year, and a long dry season of seven to nine months. The population of the province is about 3.5 million, and almost half of this, 1.5 million, is concentrated in West Timor, creating heavy pressures on limited agricultural land.

Dryland farming is the most widely practiced land use on Timor Island. The crops vary between locations, but the main ones are maize, peanuts, pumpkins, and cassava. They are traditionally planted in polycultures to reduce the chances of crop failure. Traditional shifting cultivation has been practiced in the area, but due to contemporary population pressures, the fallow period has been substantially reduced, sometimes to much less than five years. Consequently, land degradation is common in some areas (KEPAS 1986). Tree crops are mostly planted around houses or in abandoned swidden fields. The most important of them are coconuts and bananas, although mangoes, jackfruit, and citrus can also be found. In recent years, the Indonesian government has promoted expansion of industrial crops such as cashews for dryland areas in Nusa Tenggara, including Timor Island.

A large part of Timor is covered with savanna or native grassland. This is communally owned and is either swidden land in the fallow phase or grazing land. Cattle are the most important livestock, although most farmers also own pigs, chickens, goats, and horses. They are usually raised as scavengers, or by some other system of extensive management.

Acacia leucophloea and some bamboos are found on hilly dryland areas. *Bambu duri* (*Bambusa blumeana*) grows naturally in the area and is the dominant bamboo species in this region. In lowland areas, particularly along the western coast of Timor, *lontar* (*Borassus flabellifer*) and *gewang* (*Corypha gebanga*) are commonly found as native plants. Only limited bamboo is found in the lowland area, particularly in wetter regions. Most of it is planted by farmers on limited areas of their own land.

A. Bamualim, Head, Balai Pengkajian Teknologi Pertanian (BPTP), Sumatera Barat, P.O. Box 34, Padang, West Sumatra 25001, Indonesia; J. Triastono, E. Hosang, and T. Basuki, Balai Pengkajian Teknologi Pertanian (BPTP) Naibonat, Jl. Tim Tim Km. 32, Kupang-NTT, Indonesia; S.P. Field, ACIL Australia Pty. Ltd., 854 Glenferrie Road, Hawthorn, Victoria 3122, Melbourne, Australia.

Although it is not a dominant plant, Timorese farmers have recognized bamboo as useful and have integrated it into their farming systems. However, lack of information on its role in farming systems is a major constraint to developing ways of using it to increase land productivity. Based on limited information, this chapter explores the role of bamboo in farming households on Timor Island.

Study Methods

A rapid rural appraisal (RRA) exercise was conducted in four villages: Mio in Timor Tengah Selatan (TTS) district, and Takari, Oekateta, and Tunfeu in Kupang district. The selection of villages was based solely on the existence of bamboo in the area. Farmer interviews were combined with on-site observations. Data were gathered on existing farming systems, farmers' perceptions of bamboo, and its use.

Results

The main results from this study are synthesized in Table 39-1. The data covers four dominant types of bamboo planted and grown by farmers on Timor: bambu duri, *bambu licin, bambu ende,* and *bambu betung.*

Discussion

Observations at Mio village suggest that bamboo is a traditional part of dryland shifting cultivation. The naturally occurring *bambu duri* is the dominant dryland species, and most farmers interviewed said they preferred to cultivate fields where this bamboo grows. The main reasons are:

- Soils under bambu duri are believed to be more fertile.
- Bamboo is relatively easily cleared when reopening fallows.
- Bamboo can be used for fencing and other purposes.

Maize, peanuts, pumpkins, and cassava are the major food crops. Maize is the staple food in the region, whereas cassava provides food security, especially when the maize crop fails due to drought or, as in the 1996–97 rainy season, water-lodging. Peanuts are grown to generate income. They are preferred over other grain legumes because the underground pods are less susceptible to insect pests (Field 1988).

Table 39-1. Data Pertaining to Bamboo on Timor Island

Dominant Species	*Types and Sizes*	*Distribution*	*Agroecology*	*Farming Systems*	*Uses*
bambu duri	Small, grows naturally, 100 stems/ bunch	+++	Hilly drylands	Swidden system	Conservation, construction
bambu licin	Medium, grows naturally, 65 stems/bunch	++	Undulating drylands	Mixed farming	Construction
bambu ende	Medium, grows naturally and planted, 40 stems/bunch	+	Lowland, spring water	Mixed farming	Construction, income
bambu betung	Large, planted, 25 stems/ bunch	+	Lowland, spring water	Mixed farming	Construction, income

Notes: +++; ++; + = extent of bamboo distribution in the study area.

Food crop yields from newly opened fields are usually relatively high during the first two years of cultivation but then decrease in the third and fourth years because no fertilizer is applied. After four years, farmers usually leave the field fallow and move to other sites.

Most farmers own two or more swidden fields, and at any given time, one of them is fallow. Coconuts and bananas are the main perennials planted in these fields, primarily to indicate ownership of the land during the fallow. The average fallow period is about five years, although it can differ according to soil conditions. When a field is planted with *Leucaena leucocephala* before the fallow, farmers may recultivate the land after only three years. However, the field must be fenced to protect the fallow vegetation from free–grazing animals, and few farmers can afford the labor and capital to make perimeter fences.

Farmers in the dryland area recognize the importance of bamboo to their swidden systems. However, it is not normally planted, but instead grows naturally on fallow land. The reason for not planting bamboo may be its low economic value and abundant availability from unused land.

In contrast to the hilly dryland regions, farmers in lowland areas plant bamboo, particularly around springs. These bamboo plantations earn income from sales and are used for a multiplicity of other purposes. Morphologically, bamboo in the lowland is broader in stem diameter and taller, and these differences may be due to the presence of adequate groundwater. The selling price of bambu betung in the lowlands is Rp7,500 to 12,500 (1997US$3 to US$5) per stem, compared to Rp50 to Rp200 for bambu duri harvested from the drylands, Rp2,000 to Rp3,500 for bambu ende, and Rp1,000 to Rp2,500 per stem for bambu licin. These prices are higher than those on Flores and in inland Timor, as reported by Wiendiyati in 1996. This suggests that the bamboo market in the vicinity of Kupang is much better than those at Flores and in other parts of Timor. We were unable to collect data on overall income received from bamboo, especially in dryland areas. However, Wiendiyati (1996) estimated that 10 to 50 clumps of bambu betung might contribute between Rp550,000 and Rp2,750,000 (1997US$220 to US$1,200) to a farmer's income per year. The same author wrote that, on average, farmers owned 20 clumps of bambu betung, providing an average income of Rp1,045,000 (1997US$400) per year.

Although bamboo is invaluable to farmers, there has been little attention given to its development in the area. It has a wide range of uses, including soil conservation, fencing, housing, construction, equipment for fisheries, and providing shade at social gatherings.

It is important, therefore, that the role, the potential, and the problems of bamboo be identified in more detail so as to assist its wider adoption in the region. Special attention should be given to its wider use in dryland areas, including its management as an effective fallow species. As far as expanding and improving the use of bamboo in the farming systems of Nusa Tenggara is concerned, there is a need to gather baseline data and to conduct a systematic study that will direct future development strategies. Government land rehabilitation and reforestation programs may eventually consider including bamboo in their programs, and farming systems research should evaluate the role of bamboo on agricultural land. Promotion through extension services should also be encouraged.

Conclusions

In addition to its many uses, bamboo plays an important role in indigenous farming technologies, including fallow management in dryland areas of Timor, and it helps to conserve soil. Despite all of this, it does not require special management. It grows naturally in hilly dryland areas of the region and is actively planted by farmers in lowland areas as a significant earner of income.

At present, scant information is available on the biological, economic, and ecological niches for bamboo in Nusa Tenggara. There is a need to understand its present status and to develop ways of improving the contribution of bamboo to local farming systems, soil conservation, and farmers' incomes.

References

Field, S.P. 1988. Preliminary Observations of the FSR Component of Nusa Tenggara Agricultural Support Project (NTASP). Internal P3NT Report.

KEPAS (Kelompok Penelitian Agro-ekosistem). 1986. *Agro-ekosistem Daerah Kering di Nusa Tenggara Timur*. Jakarta: Badan Penelitian dan Pengembangan Pertanian.

Wiendiyati. 1996. Prospek Pengusahaan Bambu di Nusa Tenggara Timur. Survey Potensi dan Analisis Ekonomi. Makalah disampaikan dalam Seminar Pemantauan Hasil-hasil Penelitian Kehutanan, tanggal September 28–29, 1996, Jakarta.

PART VII

Agroforests

A Chimbu villager celebrates a festival in the Chimbu Valley of Papua New Guinea.

Chapter 40

Indigenous Management of Paper Mulberry in Swidden Rice Fields and Fallows in Northern Lao P.D.R.

Keith Fahrney, Onechanh Boonnaphol, Bounthanh Keoboualapha, and Soulasith Maniphone

Upland rice is currently grown on about 179,000 ha in Laos, accounting for about 31% of the country's area under rice cultivation and about 21% of the national rice harvest (National Statistical Center 1995). Most upland rice is grown in the mountainous northern region, where it is the predominant cropping system. It accounts for more than 60% to 80% of the cultivated rice area in several provinces where paddy land is limited to narrow mountain valleys (Lao-IRRI Project 1996).

In northern Laos, upland rice is grown almost exclusively in slash-and-burn shifting cultivation systems. Forests, secondary fallow vegetation, and crop residues are burned to provide nutrients for annual cropping, to control weeds, and for land preparation prior to dibbling seeds into untilled soil. Including fallowed lands, shifting cultivation systems occupy about 2.09 million hectares, or about 8.8% of the national land area of Laos (Chazee 1994). Fields planted to rice in any given year cover roughly 10% of this area (see Figure 40-1).

Limited access to land and increasing population pressures have resulted in shorter fallow periods in much of northern Laos. In some areas the shifting cultivation cycle has declined to as short as three years of fallow following two years of rice cropping. This has resulted in declining soil fertility and, in particular, increasing weed problems (Roder et al. 1995). Farmers consider weeds to be the most problematic constraint in upland rice production. Labor inputs for weed control of 140 to 190 days per hectare are common and constitute more than half of the total annual labor requirement for upland rice production (Roder et al. 1997).

To slow down deforestation and to protect the environment, the Lao government plans to reduce the area under slash-and-burn agriculture and to stabilize shifting cultivation by allocating permanent fields to households and diversifying cropping systems (Keoboualapha et al. 1996). It aims to reduce upland rice cultivation by about 75%, to about 52,000 ha (Schiller et al. 1996).

Keith Fahrney, Agronomist and Project Coordinator, International Center for Tropical Agriculture (CIAT) and International Potato Center (CIP), CIAT in Asia, P.O. Box 783, Vientiane, Lao P.D.R.; Bouthanh Keoboualapha, Deputy Director of Provincial Agriculture and Forestry Office, Luang Prabang, P.O. Box 600, Luang Prabang, Lao P.D.R.; Soulasith Maniphone, Lao-IRRI Project, P.0. Box 600, Luang Prabang, Lao P.D.R.

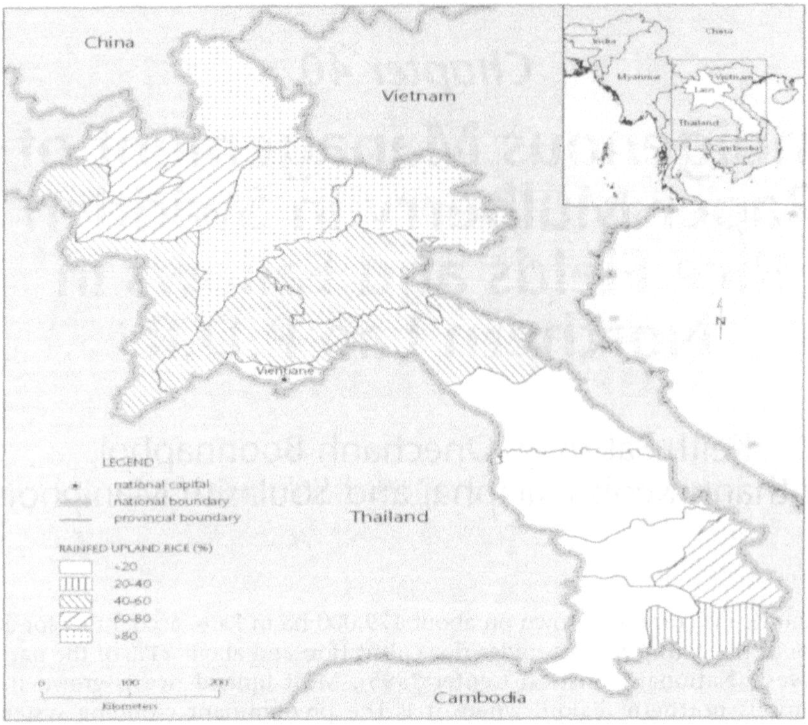

Figure 40-1. Rainfed Upland Rice Production in Lao P.D.R.

The Land Law of 1996 (Ministry of Justice, Lao P.D.R. 1996) provides individual households with long-term tenure of agricultural land. Land area is allocated according to family size, labor capacity, and the productive potential of the land. Taxes are collected from all agricultural lands, including fallow. Importantly, all land that is not productively used for more than three years may be reallocated to other households or designated for community use. Upland farmers will therefore need to make more productive use of smaller land areas, and apply methods that maintain long-term productivity.

Fertilizers and other modern agricultural inputs are generally not available to upland farmers, who are mostly engaged in subsistence production of their staple crop. Sustainable long-term production on fixed fields will require replenishment of soil nutrients and effective weed-control measures. Crop rotation and fallow enrichment are currently seen as possible strategies for achieving this. However, income generation beyond subsistence will also be necessary to provide opportunities for more diversified and intensified production systems and for the purchase of inputs to help provide longer-term sustainability.

Broussonetia papyrifera (L.) Vent., commonly called paper mulberry in English, *paw sa* in Lao, and *sa lae* in Khamu, is a shrubby broad-leaved tree belonging to the family Moraceae (Willis 1973). Probably native to tropical Asia, it is now widely distributed throughout the tropics and subtropics, extending into milder temperate regions of both hemispheres (Haller 1991). In tropical Asia and Oceania, people have long harvested fibers from the inner bark of *B. papyrifera* for processing into paper and clothing (Anderson 1993; Willis 1973).

Paper mulberry grows rapidly, particularly on moist alluvial soils. Although it is most commonly found growing on valley floors and along stream banks (Phetcharaburanin 1981; Saenthaveesouk 1989), it also sprouts spontaneously in swidden fields and in early fallow regrowth on sloping lands at moderate elevations of less than 600 m above sea level (asl) in the mountains of northern Laos.

Encouraged by lucrative prices paid by export merchants in recent years, villagers in upland areas of northern Laos have been harvesting bark from paper mulberry found growing in swidden fallows and selling it to local wholesalers. Extension services in several northern provinces are promoting paper mulberry as a cash crop alternative to upland rice production, and upland rice farmers in many areas are adopting the practice of refraining from weeding paper mulberry that has grown spontaneously in their active rice fields. Some are even beginning to experiment with planting paper mulberry in swidden rice fields on sloping lands in a variety of rotational and interplanting systems (see color plates 57 and 58).

Intensification of paper mulberry on fallowed land may help to reduce areas under shifting cultivation and stabilize upland rice production in two ways:

- As a direct source of income from sales of bark, paper mulberry provides cash for rice purchases, decreasing the land area needed for upland rice production. Since economic harvests persist over a number of years, paper mulberry may also encourage an increased duration of fallow periods; and
- As a fast-growing, invasive pioneer species in natural fallow successions, paper mulberry trees may accelerate mechanisms for weed control and regeneration of soil fertility. Large leaves and aggressive growth result in rapid canopy closure to shade out sun-loving weed species, such as *Ageratum conyzoides* and *Imperata cylindrica*. Canopy closure also shields soil on steep swidden slopes from direct raindrop impact, thus decreasing losses to erosion. The large stature and rapid growth of paper mulberry trees suggests an extensive root system that may help to improve cycling of nutrients that are prone to leaching and runoff losses during annual cropping.

This chapter explores the range of farmer experimentation with intensified paper mulberry production in upland rice swiddens and fallow fields in northern Laos. Knowledge gaps for development of effective management practices are identified, and some potential areas for future research are discussed.

The Botanical Characteristics of *Broussonetia papyrifera*

The genus *Broussonetia* includes 17 species (Willis 1973). The bark of three species is used for making paper. Two of them, *B. kazinoki* and *B. kaempferi*, are small, monoeceous, shrubby trees up to two meters in height that are found in Japan and mild temperate regions of China. *B. papyrifera* is a taller, dioeceous tree that grows up to six meters in height. It is widespread in more tropical climates (Phetcharaburanin and Sirichanthra 1984). In Polynesia, paper made from *B. papyrifera* is called tapa or kapa (Uphof 1968).

The leaves of *B. papyrifera* are large, measuring up to 45 by 25 cm, and are either ovalate or deeply lobed. Male flowers are born in pseudoracemes and have explosive stamens. Female flowers are born in pseudoheads. The fruits are small, about three centimeters in diameter, spherical, and red in color. Seeds are found at the ends of fleshy pericarp protrusions that radiate from the center of the fruit (Srisuriyathada and Bunwathi 1985; Willis 1973).

Paper Production

The technology for making paper from *B. papyrifera* probably originated in China. However, the bark has been collected and processed into paper not only in southern China but in the hilly regions of northern Laos, Thailand, and Myanmar. Traditionally, it is used mainly for making lanterns and umbrellas and for wrapping produce.

Although paper mulberry grows wild throughout the upper Mekhong River valley, Laos is currently the primary producer of the raw material. Previously, Thailand supplied most of the bark, but economic development has created other opportunities for rural households, and villagers are no longer interested in planting or harvesting the trees.

Mulberry bark paper is made at village level by boiling strips of dried inner bark in water. Wood ashes are added and the mixture is stirred and boiled for two to four hours. After the bark has softened, the cooled fiber slurry is poured into wooden frames that support mesh screens of fine cotton netting. Fibers are evenly distributed within the frames, and excess liquid drains through the netting. Frames are then set in the sun until dry. Paper sheets are carefully separated from the netting, removed from the frames, and bundled for sale.

Research on Paper Mulberry

Research on propagation techniques, varietal selection of trees, and productivity of paper mulberry bark was carried out in the early 1980s by the Field Crops Division of the Thai Department of Agriculture at field stations in northern and central Thailand (Phetcharaburanin and Sirichanthra 1984; Phetcharaburanin 1981) and by the Forest Plantation Section of the Saraburi Forest District in central Thailand (Suksomat and Mahannop 1984).

The studies found that the survival of trees propagated by root cuttings was higher (77%) than trees planted from stem cuttings (41%). Older planting material was generally more reliable in both cases. Propagation of horizontally planted stem cuttings was very poor, but planting at an angle of about 45 degrees improved the survival rate (Phetcharaburanin 1981).

After the first bark harvest, locally collected varieties produced from 3 to 27 secondary stems, about 10 of which were considered favorable for bark production. The average diameter of mature plants was around four meters after five months, following the first bark harvest. Researchers believed that a smaller angle between primary and secondary stems was desirable so that the tree density could be increased. Selection of varieties to achieve a decrease in plant diameter to three meters would allow a closer spacing of trees with higher-quality bark (Phetcharaburanin and Sirichanthra 1984).

Table 40-1 shows bark production data from the first and second cuttings of paper mulberry plants grown in central Thailand. The first cutting was at the beginning of the rainy season, one year after propagation. The second cutting was at the end of the rainy season. Yield increased greatly by the second cutting, and the proportion of bark harvested compared to stem weight increased, with a greater number of stems harvested (Phetcharaburanin and Sirichanthra 1984).

Research on paper mulberry has recently begun in northern Laos. The Xiengngeun District Agriculture and Forestry Office, in Luang Prabang Province, has established spacing and productivity trials of monoculture plantings. Research on propagation techniques has been initiated at the Kaeng Ben Teak Improvement Center in Luang Prabang Province.

Methods of this Study

Information about paper mulberry in both natural and managed upland agroecosystems in northern Laos was obtained through interviews with provincial and district agriculture officers in Luang Namtha, Oudomxay, Luang Prabang, and Sayabouli Provinces. In addition, interviews were conducted with "model farmers" identified by agriculture officers and farmer innovators, and household surveys were conducted in eight villages in two northern provinces. Information on market channels was obtained through interviews with provincial agriculture officers and a local wholesaler.

Semistructured interviews were carried out in three villages of Xay district in Oudomxay Province and five villages in the Xiengngeun and Phonxay districts of Luang Prabang Province, in November and December of 1996. Headmen or village elders were interviewed to assess community practices, the local economy, and recent changes related to paper mulberry harvest and cultivation. A total of 44 households, selected somewhat randomly with the assistance of village headmen, were surveyed

Table 40-1. Production of Paper Mulberry Bark, First and Second Cuttings

Harvest	*Bark Yield (kg/ha) (dry weight)*	*Number of Stems Harvested per Plant*	*Ratio of Bark to Stem Weight*
First cutting	274	3	1:6.3
Second cutting	1,156	5	1:4.6

Source: Modified from Phetcharaburanin and Sirichanthra 1984.

to acquire information on household economies, rice yields, fallow management practices, weed control, and farmer knowledge and practices related to paper mulberry harvest and cultivation.

Paper Mulberry Industry and Trade in Northern Laos

All paper mulberry bark collected from Luang Prabang and other northern Lao provinces is exported, except for small quantities of paper made by individual households from bark that they have harvested locally. In central Laos, there is one paper mulberry processing factory located at Kao Lio in Vientiane municipality. It is operated by a South Korean company, and it processes about 50 tonnes of bark per year into paper used for various handicrafts. It processes bark harvested from surrounding districts.

Most exported bark goes to northern Thailand, where there are several large processing factories. The largest, located in Sukhothai Province, processes 1,500 tonnes of bark per year. Another factory near San Kamphaeng, in Chiang Mai Province, processes 80 tonnes of bark per year, and others at Mae Sai and Chiang Khong in Chiang Rai Province process 200 and 100 tonnes per year respectively. The factory at Chiang Khong produces handmade paper for handicrafts.

At the factories, the bark is cleaned and graded. The lowest grade, "C" bark, is sold to small handicraft companies in Thailand where it is processed by hand and made into lanterns, umbrellas, and other handicrafts. Higher-quality "A" and "B" grades of bark are either processed into paper or reduced to loose fibers that are compressed for further export to Germany, Japan, and South Korea. This material is used to make paper flowers, wallpaper, book jackets, articles of clothing, and various handicrafts. The highest-quality, "super A" grade bark is exported to Japan without processing. Its use is unknown. Mulberry fibers are valued for their length and suppleness and their resistance to tearing and creasing. They are reportedly one ingredient used to make paper for printing money.

Although the paper mulberry export trade to Thailand began in Sayabouli Province, Luang Prabang Province is now the largest exporter, sending out 1,000 tonnes of bark per year. It has a well-developed collection and marketing network extending throughout the province. Sayabouli and Oudomxay Provinces each export about 200 tonnes of bark per year. In Sayabouli, paper mulberry collection and trade is limited to the vicinity of the Mekhong River port of Pak Lai and to southern districts along the Thai border. In Oudomxay Province, bark is currently collected from districts adjacent to the Mekhong River port of Pak Baeng.

There are five or six bark collection centers located in the town of Luang Prabang. We interviewed the manager of the largest collection center. He is referred to below as "our wholesaler." He employs about 30 people. His center sends trucks or boats out to collect bark from six subcenters located in five districts of Luang Prabang Province and one district of Houa Phanh Province. They also buy from small traders who bring bark from villages near the collection center. At the time of the interview, our wholesaler's center was handling about 400 tonnes of dried bark per year, about 40% of the paper mulberry harvested in Luang Prabang Province. He also buys smaller amounts, of 30 to 40 tonnes, collected in neighboring Phongsali and Oudomxay Provinces. About half of the bark arrives at the center by truck and about half arrives by boat. The peak buying period is between February and April, when 75% of the annual crop is harvested. Ten tonnes of bark may be collected at our wholesaler's center in a single day during this period. The remaining 25% of the annual crop is harvested between October and December. Bark collection ceases altogether for a month and a half during the cold, dry season in January and early February. Farmers stop harvesting then because the bark becomes difficult to strip.

Our wholesaler's center is the only collection center in Laos that grades the bark prior to export. Buyers in Thailand pay him fixed prices based on his three grading categories that are "A," "B," and "C." Other collection centers receive varying prices according to the general quality of bulk shipments and factory demand at the time of receipt.

Although the distance is similar, our wholesaler prefers to send bark by boat up the Mekhong River to Houayxai district in Bo Keo Province, rather than by boat downriver to Pak Lai district in Sayabouli Province and thence by truck across the land border in Kaen Thao district. Because there are many factories located nearby, there is a better chance of receiving a good price from competing buyers at Houayxai than by dealing with a single buyer at Kaen Thao. Trucks from Thai factories meet his shipments at the border. Laos collects an export tax, and Thailand collects an import tax. In addition to exports to Thailand, our wholesaler recently established a direct relationship with a buyer in South Korea. He exported 20 tonnes of mulberry bark to Korea overland through China, and the buyer now wants to increase purchases to as much as 600 tonnes of bark from Laos every year. This would more than double our wholesaler's turnover.

Our wholesaler believes that 2,000 tonnes of mulberry bark could be exported from Laos to currently known buyers every year without saturating the market and driving prices down. He believes there may be further opportunities to export directly to South Korea. He considers bark quality to be the most important issue in the trade and says he could pay 50% more than his current price for top quality bark.

Quality bark, he says, should be entirely stripped of the darker outer bark, particularly at the buds and branches, be thoroughly dried, and have no discoloration from fungus. He prefers to buy bark from trees that have been planted, rather than harvested from the wild, because farmers have more control over the quality of planted trees. He suggests planting the trees at 1 m by 1 m spacing to encourage upright growth, rather than allowing them to branch, as tertiary branch knots decrease bark quality. He says the bark should be harvested every six to eight months. He believes that bark quality is better from trees growing on flatter land with moister soils because the bark strips more easily. However, the timing of stripping in relation to rainfall could be as important as the trees' microclimate. He does not know of any efforts to select superior varieties for cultivation.

Household Surveys

Characteristics of Natural Fallows

The predominant fallow vegetation at all study sites is *Chromolaena odorata* during the first three to four years. Fallow progression seldom advances to canopy cover by small trees in the surveyed areas of Luang Prabang Province, but in Oudomxay, a fast-growing, small softwood tree species with large leaves, locally known as *thawng khope* (*Oroxylum indicum* [L.] Kurz. [Bignoniaceae]), is found in older fallows. Farmers in all of the surveyed villages say they appreciate the role of *Chromolaena odorata* in their fallow rotations (Roder et al. 1997), but several of the farmers in Oudomxay say they prefer thawng khope because the large leaves shade out annual weed species. In general, fallow rotations, field sizes, and rice yields are higher in Xay district of Oudomxay Province, where there is less population pressure on upland resources (Table 40-2).

Farmer Knowledge of Broussonetia papyrifera

Farmers in northern Laos recognize different naturally occurring varieties of *B. papyrifera*, based on leaf shape and the color of the outer bark. Individual trees with a predominance of round leaves are preferred over those with a greater incidence of deeply lobed leaves. Lighter-colored outer bark is preferred over darker bark. Farmers believe the trees with rounder leaves and lighter-colored outer bark not only grow faster but also have a higher quality inner bark.

It seems that trees with rounder leaves often have lighter-colored outer bark, and trees with more deeply lobed leaves often have darker bark. But since both types of leaves usually occur on the same tree, and even on the same branch, and there is a gradation in bark color with spotted bark patterns, these associations are not altogether clear.

Farmers recognize that male and female flower types occur on separate trees. In the Khamu language, there are different names for male and female trees. Male flowers are picked when green and tender and are either steamed and eaten with chili paste or boiled in soups. The sticky, sweet fruits are eaten raw when ripe.

Some farmers say that fruit develops on trees with darker outer bark and that the long male flowers are more common on trees with lighter outer bark. Because of the gradation and spotting of the bark color and the fact that not all trees flower or fruit in a given year, it is not certain if there is any association between bark color and sexual differentiation. The intermediary bark coloration could indicate hermaphrodism or some other complex sexual differentiation, as has been observed in some dioeceous plants such as papaya (*Carica papaya*).

Table 40-2. Average Field Size, Fallow Duration, and Rice Yields of Interviewed Households

Province	*District*	*Number of Households Interviewed*	*Average Field Size (ha)*	*Maximum Fallow (yrs)*	*Rice Yield (t/ha)*
Oudomxay	Xay	24	1.2	6.6	1.2
Luang Prabang	Xiengngeun	8	1.0	4.1	0.8
	Phonxay	12	0.8	4.6	0.7

Harvesting of Paper Mulberry Bark

Paper mulberry was neither harvested nor sold in 1996 by the farmers we interviewed in the three villages in Xay district of Oudomxay Province. The village headmen in Lak Sip and Na Ngam villages said paper mulberry had never been collected in their villages, and farmers treated it like any other weed in their rice fields and weeded it out. In Khonkaen village, 13 km from the provincial capital on the road to Pak Baeng, the headman reported that villagers collected paper mulberry from fallows in 1996 and sold the bark to a woman trader near the provincial capital. However, in the following year, no one had offered to buy the bark and he doubted that anyone would collect it. None of the villagers in Oudomxay said they had plans to plant or harvest paper mulberry, but most said they would be interested in learning about planting techniques if there were guaranteed buyers for the bark.

More than 90% of upland farmers interviewed in Luang Prabang Province reported harvesting paper mulberry in 1995. Furthermore, most said they had either begun harvesting the 1996–97 crop in November and December, or would harvest it at some time prior to the onset of the wet season in May. Most farmers had been harvesting paper mulberry for only two to three years, although one farmer in Xiengngeun district said he had been harvesting it for four years.

In Phonxay district, about 50% of households reported harvesting from forests or other noncultivated areas and said they walked for up to two hours to reach some trees. The trips were usually made in conjunction with hunting or some other gathering activity. In Xiengngeun district, only about 25% of households reported harvesting from forested areas. The difference probably reflected the relative scarcity of forest areas in Xiengngeun.

Nearly all farmers in Luang Prabang Province reported harvesting paper mulberry from fallow fields of one to four years of age, with two- to three-year fallows the most popular for harvesting because of the greater yields there. Farmers in Luang Prabang said they harvested only from their own fallow fields or from fallow fields belonging to other households, with the owners' permission. There was a different story from Oudomxay Province. Villagers said that, in 1995, when there had been some paper mulberry harvested, it had been gathered from both their own and others' fallow fields, regardless of customary land tenure.

Only two households in Xiengnguen district reported gathering bark from fields where rice had just been harvested, but nearly all farmers said they had avoided weeding paper mulberry out of their active rice fields. Most said they preferred to wait to harvest bark from these trees until just before clearing the field for a second rice crop. One farmer with paper mulberry in his third-year rice crop said that he harvested from first-year trees prior to clearing for the second-year rice crop and then harvested twice in the following year, both at rice harvest and clearing, from coppice regrowth.

Paper mulberry is usually harvested twice a year, at the beginning of the dry season, from late October to early December, and at the end of the dry season, from late February to early June. Farmers said the inner bark was easy to strip at these times, the bark was easily dried in the sun without rain or moisture bringing fungal problems, and these times fit the schedule of their other work. Farmers said they avoided harvesting during the cold season, from late December to mid-February, because during these months the inner bark stuck to the woody part of the stem and was difficult to strip. Some attributed this to the cold temperatures; others claimed it was the lack of rain.

Farmers said that the minimum stem diameter for harvest was between the thickness of a thumb and a knife handle, or about 2 to 4 cm. The optimum size was between knife handle and forearm thickness, about 4 to 8 cm, and maximum diameter was around calf size, between 12 and 15 cm. The quality of the inner bark decreased with the age and the size of the stem, as it became darker in color and less supple.

After cutting suitably sized primary trunks in the first harvest, or secondary stems in later harvests, the tertiary branches are pruned off and the inner and outer bark are stripped off together using a long knife. This is normally done in the field. If the field is not far from the village, farmers will carry the stems home to use as firewood. The inner bark is separated from the outer bark with a knife, either in the field or, more commonly, in the village. The outer bark is discarded. The inner bark is arranged on horizontal wooden poles to dry in the sun, with the outer side facing up. Drying takes about half a day in full sun or longer on cloudy days.

Sales of Paper Mulberry Bark

Householders in Luang Prabang Province said that, in 1995–1996, they harvested and sold an average of 61 kg of paper mulberry bark. In Xiengngeun district the average was 32 kg, and in Phonxay, it was 81 kg. This had been worth an average of US$33 per household at an exchange rate then of 930 kip/US$1. One household in Phonxay district reported selling 200 kg of bark worth, in those days, around US$100. It had been a sizable boost to their household income.

Most households sold the dried bark in 1 kg bundles to traders who came to the village. These collectors then took the bark by small trucks or boats to collection centers located in provincial capitals. In Phonxay district, all of the bark went to Luang Prabang town, where secondary merchants graded it. Most bark collected in

Xiengngeun district also went to collection centers in Luang Prabang town, but some was sent to centers in Sayabouli Province. From Luang Prabang, bark was sent either upriver to Houayxai in Bo Keo Province for export across the border to Chiang Rai Province in Thailand, or by road to Sayabouli Province, were it was exported from Kaen Thao district to Loei Province in Thailand.

The price that village collectors pay for mulberry bark depends largely on the distance to collection centers. Farmers in Phonxay district, 50 to 60 km from Luang Prabang town, received an average of 430 kip for one kilogram of bark in 1996, while farmers in Xiengngeun district, just 30 to 35 km from Luang Prabang, received 504 kip. Some farmers said they preferred to take the bark to collection centers themselves and benefit from the higher prices, but the price paid at collection centers depended on the quality of the fibers, whereas village collectors paid a fixed price regardless of quality.

According to the farmers we interviewed, the average price of mulberry bark in the two districts in Luang Prabang had been steady during 1995 and 1996, at 460 kip/kg, but this represented a substantial increase over 1994, when the price was just 290 kip/kg. Two-thirds of farmers said the price of bark decreased as the harvest season progressed. Collectors, on the other hand, said the buying centers became more discriminating in quality grading as their stocks increased, so they had to pass on the resulting lower prices. Nearly 90% of farmers realized that price depended on quality but said that while collectors told them that they only wanted high-quality fibers, they still paid only one price.

Most farmers said they had seen papermaking, either in their own villages or in neighboring villages, but it had been more common in the past. None of the villages where interviews were carried out were making paper at the time, although several households said they would do it if someone would buy the paper from them for a reasonable price.

Planting of Paper Mulberry

All farmers who planted mulberry trees used root cuttings for propagation because they said it was the most reliable method and trees sprouted more quickly than from stem cuttings. None of them had tried planting seeds. Most farmers said they had developed propagation methods on their own, and some showed remarkable ingenuity in their planting techniques. One farmer demonstrated how he scraped soil away from shallow, horizontal roots so that light and exposure would encourage sprouting. After sprouting, the farmer separated the new shoot with a short section of root from the mother tree and planted it. He said that this method resulted in excellent survival and rapid growth. One farmer said that the village headman taught him how to plant, and another said that a government extension officer advised him.

Farmers said they planted between May and July, with later plantings preferred because of more reliable rainfall. A variety of densities was being tried for various reasons. Dense plantings, at 2 by 2 m and 2.5 by 2.5 m spacing, were preferred for weed control and for conversion of rice swidden fields into paper mulberry plantations. Other farmers chose spacing of 3 by 3 m, and wider, to minimize competition with rice crops.

Labor Requirements and Division of Labor

Labor requirements reported in interviews showed so much variability that data were not considered reliable for meaningful analysis. We asked farmers to recall total time requirements, in person-days, for each of the activities associated with bark production: cutting, first stripping, second stripping, drying, bundling, propagating, and planting. This was a difficult question for most households to answer because each activity was the sum of several trips or repeated events by different family members, not all of whom were present at the interviews.

Both men and women, in nearly equal numbers, cut stems and stripped the outer bark from mulberry trees. Women and children were more likely than men to strip the inner bark and take care of drying and bundling. Both men and women propagated and planted mulberry trees.

Paper Mulberry with Rice: Spontaneous Growth

All but one of the farmers interviewed in Luang Prabang Province said that they allowed paper mulberry to grow in their rice fields. Some had stopped weeding out trees growing spontaneously in their rice fields as early as 1993, and by 1995 or 1996 most had begun protecting the trees. Ninety percent of the farmers interviewed believed that allowing paper mulberry trees to grow with rice decreased their rice yields, and many reported that surrounding rice plants had a higher incidence of unfilled grains, which they thought resulted from shading by the trees. One farmer estimated that rice yields around vigorous mulberry trees were decreased by up to one-third. However, most farmers calculated that the income from the bark was higher than that for rice grown on an equal area. Some nevertheless expressed concern over possible price instability for the bark and felt uncomfortable with the need to buy rice.

In addition to avoiding weeding paper mulberry shoots from their rice swiddens, farmers encourage the mulberries' growth by clearing weeds around them and by removing vines, in particular. They also thin large clumps of shoots that sprout in close proximity and prune secondary stems and branches to encourage faster growth of fewer, thicker stems.

Paper Mulberry with Rice: Planted Trees

Half of the households interviewed in Xiengngeun district reported that they had planted paper mulberry in rice fields in 1996 to boost the density of the trees. Slightly more than half indicated that they intended to interplant in their rice fields in 1997. In Phonxay district, only 8% of households interviewed reported interplanting in rice fields in 1996, but more than one-third said that they intended to interplant in 1997. Some farmers claimed that they would interplant other non–rice crops in the rice and paper mulberry swiddens and would eventually convert them to more permanent mixed garden plantations. Others had no plans beyond increasing the harvests of mulberry bark from their rice swiddens.

Interplanting of Paper Mulberry with Non–Rice Crops

Farmers reported that, when they first began protecting or planting paper mulberry trees, they were nearly always growing upland rice in their swiddens. However, a variety of other annual and perennial crops were sometimes grown in the first year, and these were more common than rice crops in subsequent years. Annual crops included maize, Job's tears, chilies, cucurbits, and sesame. Because paper mulberry trees sprouted randomly and were then used to provide planting materials for further propagation, farmers said they were often obliged to plant annual crops only in those parts of their fields with the lowest density of mulberry trees.

Perennial crops that were either planted or protected if they sprouted spontaneously with paper mulberry included bananas, pineapples, a local tree known as *mahk lin mai*, which is grown for its large edible pods, and teak *(Tectona grandis)*. About one-third of farmers said they had planted, or intended to plant, teak with paper mulberry. The teak was planted at a spacing of at least 3 by 3 m, and most said it was necessary to plant the teak one or two years prior to the paper mulberry to minimize competition. Because they had only recently begun interplanting teak with mulberry, the farmers did not know how long they would be able to harvest paper mulberry among the developing teak trees. Several estimated a harvestable period of two to three years. However, one farmer who had planted teak at 4 by 4 m spacing, along with paper mulberry between the rows in the same year, claimed to have harvested paper mulberry for six years.

Paper Mulberry and Livestock

Farmers say that pigs, cattle, buffaloes, poultry, and fish all benefit from paper mulberry. The leaves are boiled and fed to pigs. Cattle and buffaloes enjoy grazing on the leaves, and farmers report that paper mulberry trees attract large livestock to swiddens. It is, therefore, necessary to fence paper mulberry fallows or plantations wherever livestock graze freely. This has become the prevailing practice in northern Laos. Cattle generally strip lower leaves without destroying the trees, but buffaloes often push the trees down to get at the leafy tops. Two farmers reported feeding leaves to carp and tilapia in fish ponds after cutting branches and stripping the bark. Poultry feed on insect larvae that apparently drop from mulberry leaves. The farmers did not mention goats, but it is quite likely that they also enjoy eating paper mulberry leaves.

Paper Mulberry in Fallow Ecosystems

Farmers said that paper mulberry grew in fallow fields both by sprouting from the roots of existing trees and from seeds that were spread by birds. They regarded the two sources of new growth as nearly equally important.

Eighty-two percent of the farmers interviewed believed that paper mulberry improved their soil, making it cooler and moister. They considered cool soil, or *din yen*, preferable for growing rice and many other crops. All farmers interviewed believed that burning fallow vegetation encouraged the sprouting and growth of paper mulberry trees. However, one farmer was of the opinion that if a fire was too hot, it could decrease the number of subsequent paper mulberry trees. They listed competition from larger trees and vines, shading, overcutting, animal grazing, and time as factors causing a decline of paper mulberry trees in fallow vegetation. Eighty-two percent of farmers considered paper mulberry to be a sun-loving plant that did not tolerate shade.

Discussion

Structure and Function of Paper Mulberry Fallows

It is clear that there is currently a broad spectrum of experimentation with paper mulberry in upland rice systems in northern Laos, with varying degrees of management intensity. The major step in intensification involves propagation from naturally occurring trees and planting to increase population density. The primary motivation for planting mulberry trees is to increase cash income. The function of planted trees in the fallow, and the subsequent length of the fallow, seem to depend upon a variety of factors, including land quality; proximity to roads and villages; and farmer initiative, ingenuity, and exposure to outside ideas or experience. Regardless of the intended duration of a paper mulberry fallow, planting occurs in the last year of rice production in a given field. The importance of the rice crop insures that the developing mulberry fallow will establish well. Rice cropping is not possible for a second year at intensified planting densities of paper mulberry.

Protection of Naturally Occurring Trees

A farmer's first interaction with mulberry trees involves simply refraining from weeding them out of rice fields. Although widespread in Luang Prabang Province, this practice does not yet occur in many areas of northern Laos, including the Xay district of Oudomxay Province. Protection of the trees and a willingness to further intensify management depends upon market opportunities. Where market opportunities exist, farmers believe that the income from bark sales generally offsets decreases in rice yields that may be consequent upon the trees growing in rice fields.

Planting of Paper Mulberry in Short-Term Fallow Rotations

Short-term rotations are systems where paper mulberry is the climax species of the fallow rotation. The field is planted to annual crops, such as upland rice, at the end of the productive period of the paper mulberry. Farmers engaging in this system are generally distant from roads and villages, and this limits their opportunities for interplanting teak or cultivating high-value annual or perennial species that need close supervision. A major limitation to this system is the need to maintain adequate fencing to protect the mulberry trees.

Although farmers were following this system at the time of this study, none of them had actually been through a complete fallow cycle. There was a great deal of uncertainty about the duration of economic productivity of the paper mulberry crop and how the transition back to annual cropping should be managed.

Planting of Paper Mulberry in Long-Term Fallow Rotations

This cropping pattern is generally only found close to roads and villages and often resembles home gardens, with a diverse assortment of annual and perennial species. In this cropping pattern, paper mulberry is an intermediate species that is planted to provide bridging income between annual cropping, usually of upland rice, and the harvest of long-term perennials such as teak. At the common teak spacing of 3 by 3 m, upland rice can be interplanted for only two years. It may be possible to extend the economic productivity of teak plantings, at either 3 by 3 m or a lesser density, by enriching them with paper mulberry. Once more, however, because paper mulberry cultivation is so recent, there is a great deal of uncertainty about how to manage interplanting with longer-term perennials.

Knowledge Gaps and Research Opportunities

The following is a list of researchable questions that should be considered in the development of more intensive and productive paper mulberry fallows.

What Is Optimal Spacing for Paper Mulberry?

Density and spacing of paper mulberry trees will depend on their intended function in the fallow rotation and the productive potential of the land, relative to requirements for optimal paper mulberry growth.

On flat, moist, alluvial soils, growth of paper mulberry trees is extremely rapid. Under such conditions, farmers at Ban Kok Tom in Nan district of Luang Prabang Province say it is not possible to harvest an interplanted rice crop when mulberry trees are planted at 2 by 2 m spacing. However, on steeply sloping swidden fields, under drier soil conditions with generally less fertile soils, mulberry trees grow more slowly. Spacing trials and interplanting studies should be carried out over a wide range of conditions. The growth of mulberry trees should also be correlated with key parameters, including soil moisture, pH, elevation, and slope, to provide results with wide application.

When used as a climax species for fallow regeneration, full canopy closure of paper mulberry is desirable in the first year after harvesting the last rice crop. But since mulberry tree establishment coincides with rice growth and decreases yields, optimal spacing will depend on both agronomic and economic considerations. A trade-off must be found between rice and mulberry bark. Ideal conditions, allowing the best possible harvests from both crops, are likely to depend upon the willingness of individual farm households to tolerate rice yield losses. Experiments should be designed to test time and space interactions, as delayed interplanting of trees within the rice crop cycle could decrease competition and thus minimize rice yield losses or allow for denser tree spacing.

Interplanting paper mulberry with slower-growing perennials will probably require wider spacing. However, it may also be possible to avoid competition from the fast-growing mulberry trees by delaying their planting. Since farmers can generally interplant rice for two years in 3 by 3 m teak spacing, it would be useful to experiment with various spacings of mulberry interplanted with teak after the second rice crop, when the teak is beginning its third year of growth.

What Is the Nutrient Export from Bark Harvests?

Harvesting bark from paper mulberry fallows represents a net export of nutrients from the system. Even though the loss of critical nutrients may not be large, particularly since leaves generally remain in fallow fields, it would be useful to know the actual quantity of nutrients leaving with the harvest. This would allow calculation of nutrient balance in crop and fallow cycles over a long term and may lead to a better understanding of strategies for more sustainable production systems.

Can Interplanting with Tree, Forage, or Cover Crop Legumes Increase Productivity?

Although it is hypothesized that paper mulberry fallows may accelerate nutrient cycling and natural fallow regeneration, paper mulberry is not expected to actually contribute significant amounts of external nutrients to the system, with the possible exception of carbon. Interplanting nitrogen-fixing species into the mulberry fallow could help to increase long-term productivity and shorten the fallow cycle. Therefore, interplanting should be tried with a variety of legume species and planting patterns. Although they will have a space requirement and may compete with paper mulberry for nutrients and sunlight, it may be possible to identify legume species that can be interplanted with paper mulberry and result in greater bark harvests. This could lead to greater crop productivity over the long term by adding of nitrogen to the system and by priming nutrient cycling.

Can Livestock Production Be Effectively Integrated with Paper Mulberry Fallows?

Paper mulberry leaves are attractive fodder to most ruminant livestock, so there may be opportunities to increase the economic productivity of the fallow system by on-site biodigestion of the vegetation. Livestock must be fenced out of paper mulberry fallows to protect the trees. However, after the bark is harvested, it may be beneficial to fence livestock *into* mulberry fallows to both consume a favored feed and to transform the organic matter and its associated nutrients into a stable form available to plants: animal manure. Most livestock production in northern Laos depends on open grazing systems. The major mulberry bark harvest coincides with the dry season period of livestock feed scarcity, so controlled grazing of mulberry fallows could potentially benefit livestock production. Minor damage to mulberry trees by livestock could help to thin stand density and allow longer productive fallow rotations. On the other hand, major damage to mulberry trees may be desirable at the end of the fallow cycle to minimize competition with subsequent annual crops.

What Is the Optimum Duration of a Paper Mulberry Fallow?

The answer depends on the farmer's perspective: optimal for weed control, for long-term rice yields, or for production of paper mulberry bark? As long as the economic yield from mulberry bark production is greater than that from rice production, fallow rotation cycles will likely be determined by production of bark. It is not presently known how long mulberry bark can be economically harvested, what environmental factors increase or decrease the duration of harvest cycles, and what management factors could lengthen the harvest.

What Causes Decline in Production of Mulberry Bark?

Harvesting of mulberry bark cannot go on forever. Is natural decline a genetic function, or is it caused by disease, insects, soil fertility, or other environmental factors?

Are Paper Mulberry Fallows Effective in Controlling Weeds?

It is hypothesized that paper mulberry's large leaves, rapid growth, and canopy closure can decrease weeds faster than a natural fallow composed mainly of *Chromolaena odorata*. This very important possible benefit must be tested. It is quite likely that certain weed species are more susceptible to control by shading while others, such as those with longer-lasting seed viability in the soil, are more resistant to this means of control. The biomass of various weed species should be measured over time in paper mulberry fallows and subsequent rice cropping cycles.

To What Degree Will Paper Mulberry Persist in Subsequent Rice Crops?

After completion of the fallow cycle, which will probably end when bark harvests fall below an economic threshold, farmers will prepare the land to plant rice or other annual crops. They will want their fields to be free of noxious weeds, including paper mulberry. Research is needed to determine the most effective management practices for converting paper mulberry fallows back into fields for rice or other annual crops. Farmers claim that burning probably encourages regrowth of paper mulberry trees, but grazing may help to control it. Timing of these operations may be crucial for effective control. What density of regrowth is considered tolerable or desirable? Will farmers want to plant rice for just one year, or for two years, and to what extent will this be determined by the persistence of mulberry regrowth?

Conclusions

The novelty of growing upland rice followed by a *Broussonetia papyrifera* fallow currently gives rise to more questions than it does answers. The motivation for what is clearly a popular and increasing indigenous technology in northern Laos is clear: it puts money in farmers' pockets from the harvest and sale of the bark. But farmers also believe that growing the paper mulberry trees improves their soil, and it is possible rice yields will increase as a consequence. Until these beliefs are adequately tested, the ultimate question of whether paper mulberry fallows do increase rice yields will remain unanswered.

Acknowledgments

The authors gratefully acknowledge support from the Swiss Agency for Development and Cooperation (SDC), the Ministry of Agriculture and Forestry (MAF) of the Lao P.D.R., and the Agriculture and Forestry Services of Luang Prabang, Oudomxay, and Sayabouli Provinces, which enabled this research. We would also like to express our appreciation to the farmers who shared their knowledge with us.

References

Anderson, E.A. 1993. *Plants and People of the Golden Triangle: Ethnobotany of the Hill Tribes of Northern Thailand.* Chiang Mai, Thailand: Silkworm Books.

Chazee, L. 1994. Shifting Cultivation Practices in Laos: Present Systems and their Future. In: *Shifting Cultivation and Rural Development in the Lao P.D.R.*, Report of the Nabong Technical Meeting, July 14–16, 1993. Vientiane, Lao P.D.R., 67–97.

Haller, J.M. 1991. The Paper Tree: *Broussonetia papyrifera. Pacific Horticulture* 52(1), 50–52.

Keoboualapha, B., J.M. Schiller, and V. Manivong. 1996. Prospects and Priorities for Upland Rice in the Lao P.D.R. Paper presented at an Upland Rice Research Consortium Workshop, January 4–13, 1996, Padang, Indonesia.

Lao-IRRI Project. 1996. *National Rice Research Program.* 1995 Annual Technical Report. Vientiane, Lao P.D.R., 296.

Ministry of Justice, Lao P.D.R. 1996. *Land Law* (draft). Vientiane, Lao P.D.R.

National Statistical Center. 1995. *Basic Statistics about Socio-Economic Development in the Lao P.D.R.* Vientiane, Lao P.D.R. 172 pp.

Phetcharaburanin, C. 1981. Study on Paper Mulberry for Production of Paper Pulp. In: *Experimental Report of Fiber Crops in 1981.* Bangkok, Thailand: Field Crops Division, Department of Agriculture, 41–43 (Thai language).

———, and A. Sirichanthra. 1984. History of Varietal Improvement in Paper Mulberry. In: *Proceedings of a Seminar on the Progress of Plant Breeding Work in the Department of Agriculture.* Bangkok, Thailand: October 4–8 1982, 179–181 (Thai language).

Roder, W., S. Phengchanh, and B. Keoboualapha. 1995. Relationships between Soil, Fallow Period, Weeds, and Rice Yield in Slash-and-Burn Systems in Laos. *Plant and Soil* 176, 27–36.

———. 1997. Weeds in Slash-and-Burn Rice Fields in Northern Laos. *Weed Research* 37.

Saenthaveesouk, S. 1989. *Handbook for Planting Paper Mulberry.* Vientiane, Lao P.D.R.: Office of Paper Mulberry Project in Laos, Indochina Development Company, Ltd. (Lao language).

Schiller, J.M., A. Khamhung, A. Boonnaphol, V. Manivong, and K. Fahrney. 1996. Production Constraints and Research Direction for Sustainable Upland Rice-Based Agriculture in Northern Laos. Paper prepared for the eighth annual meeting of IBSRAM Asialand Network on "The Management of Sloping Lands for Sustainable Agriculture in Asia," September 23–27, 1996. Vientiane, Lao P.D.R.

Srisuriyathada, W., and T. Bunwathi. 1985. Paper Mulberry. *Warasan Kaona* 4(2), 19–26 (Thai language).

Suksomat, P., and N. Mahannop. 1984. Ways to Utilize Paper Mulberry. *Vanasan* 42(4), 235–243 (Thai language).

Uphof, J.C.T.H. 1968. *Dictionary of Economic Plants.* (second edition). Germany: Verlag Von J Cramer.

Willis, J.C. 1973. *A Dictionary of Flowering Plants and Ferns.* (eighth edition, rev. by H.K. Airy Shaw). Cambridge, UK: University Press.

Chapter 41

The Complex Agroforests of the Iban in West Kalimantan and their Possible Role in Fallow Management and Forest Regeneration

Reed L. Wadley

Swidden cultivation with a long fallow period has long been a central component in a complex agroforestry system practiced by the Iban of northwest Borneo. Importantly, the system has also included permanent and semipermanent groves and tracts of forest, and gardens of rubber, fruit, and bamboo. These have not only been preserved and managed for gathering forest products and hunting, but the Iban have also recognized longer-term benefits such as the protection of watersheds and regeneration of the forest after farming. This chapter proposes that the managed forest in the Iban system may have provided not only the seed necessary to reforest fallowed fields, but also the habitat needed by animals that pollinate and disperse seed. These animals may thus have helped sustain both preserved and fallow forests, which, when seen as part of the total agroforestry system, may both have been essential elements in the long fallow cultivation cycle.

Complex agroforests are seen throughout Southeast Asia, with much recent research being focused on the cultivation of forest gardens (Aumeeruddy and Sansonnens 1994; Belsky 1993; Mary and Michon 1987; Michon et al. 1986). Forest gardening in Borneo has also received attention (Lawrence et al. 1995; Padoch 1995; Padoch and Peters 1993; Salafsky 1994). However, in contrast to forest gardens that were apparently developed in response to new markets for specific forest products (Lawrence et al. 1995; Salafsky 1994; Torquebiau 1984; Weinstock 1983), the situation described in this chapter resembles the subsistence-oriented agroforestry of the Kenyah of Sarawak and East Kalimantan (Chin 1985; Colfer, with Peluso and Chin 1997) and the Hanunóo in the Philippines (Conklin 1957). Although mention of it has been made (Michon and de Foresta 1991; Michon et al. 1986), the potentially important role of forest fauna in forest regeneration has received little attention.

This chapter provides a speculative case for further research into these potentially important interactions. However, such has been the pace of change in the study area since the Asian economic crisis of 1997 that many references must be couched in the past tense. These changes have included the expansion of logging, much of it illegal,

Reed Wadley, Department of Anthropology, 107 Swallow Hall, University of Missouri, Columbia, MO 65211 USA.

and commercial black pepper cultivation (Wadley 2001, 2002a; Wadley and Mertz 2005).

Study Site and Methods

The study focused on the Iban longhouse community of Sungai Sedik,[1] located in the remote hills northeast of Danau Sentarum National Park (DSNP) in West Kalimantan (Figure 41-1). Given the unclear boundaries of the national park, the study area may eventually be considered a transition zone, a buffer zone, or even part of the park itself. DSNP is an area of interconnected seasonal lakes and seasonally flooded tropical forests. The water catchment consists of lowland tropical forest in the hills and flooded forest in low-lying areas. The hills are characterized by a patchwork of various forest stages, as a result of both commercial logging and swidden cultivation.

During the study period, the longhouse was a 14-household community. Its population grew from around 60 people in 1979 to 98 people in 1994. The Sungai Sedik territory had a population density of 4.1 persons/km^2 and encompassed about 24 km^2. It was a mosaic of forest succession, agricultural plots, and preserved forest.

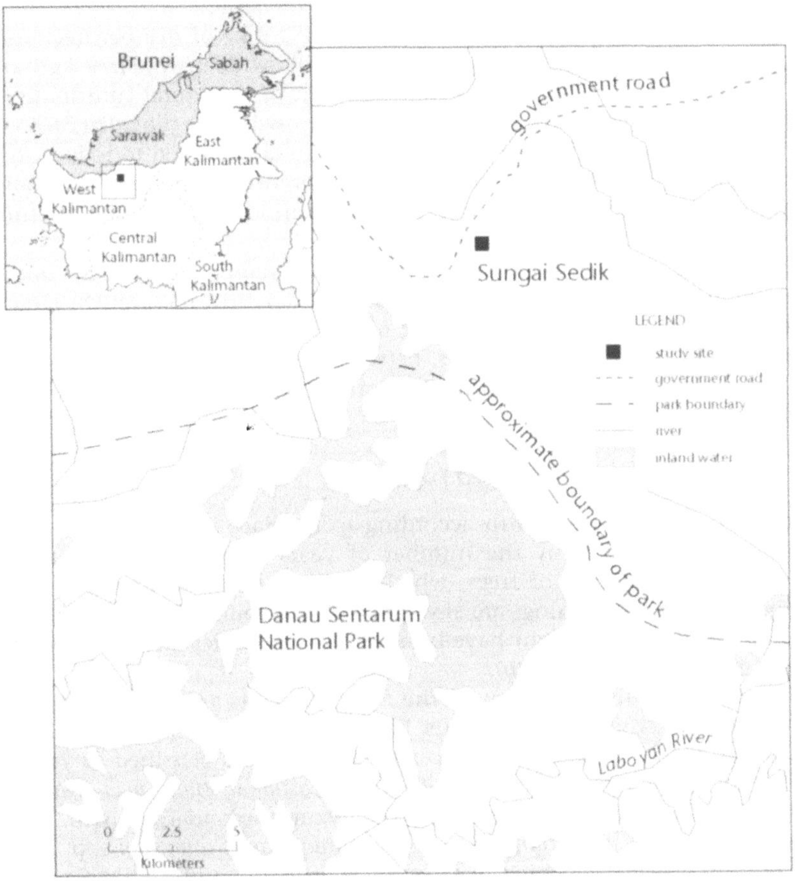

Figure 41-1. Location of Sungai Sedik, West Kalimantan, Indonesia

1 In earlier publications, this community was given the pseudonym "Wong Garai." This is no longer deemed necessary.

The data upon which this chapter is based were gathered during participant-observatory fieldwork at Sungai Sedik from 1992 to 1994. Additional study was done during a brief visit to the area in June 1996. Data on farming were collected using an interview schedule developed by Colfer (Colfer with Dudley 1993) and adapted to the Iban context. People in each household were interviewed concerning a variety of matters related to farming. Data on the role of forest fauna were derived from a study of hunting patterns conducted during 1993 and 1994 (Wadley et al. 1997; Wadley 2002b).

Swiddens and Fallowed Forest

The Iban were generally characterized in the past as pioneering swidden cultivators of upland rice (Freeman 1970, for example). However, most Iban have practiced more established swidden cultivation (Padoch 1982), usually within a delimited territory where forest is fallowed in a long cycle. The Iban of Sungai Sedik were no exception and used an average fallow length of around 20 years for their hill swiddens. Fields were rarely cultivated without an intervening fallow period of at least five years. They also cultivated rice in floodplain and swamp swiddens, which were particularly productive and were fallowed for shorter periods as a consequence: eight to nine years for the former and only about two to three years for the latter.

The swamp swiddens were the most productive (1,553.7 kg of rice per hectare), followed by floodplain swiddens (536.34 kg/ha) and hill swiddens (264.79 kg/ha). However, the hill swiddens were more important in terms of their total contribution to subsistence, making up 59% of the yearly yield. Swamp swiddens provided 36% of the yearly yield. The different ecologies of hill and swamp land helped to reduce the risk of total crop failure in that they were affected differently by drought, rainfall, and pests. Floodplain swiddens were, more or less, periodic substitutes for hill swiddens.

Given the need for long fallows on hill land, and therefore the need for more land, the total amount of land used for hill swiddens was 527.11 ha, a much larger area than that for the other swidden types, 20.62 ha for floodplain land and 21.18 ha for swamp land. Over the course of 15 years, between 1979 and 1994, only 16% of hill land was farmed more than once, while 76% of floodplain and 100% of swamp land was farmed more than once.

Fallowed Forest

The Iban typically classify forest growth according to its place in the fallow cycle. They judge fallowed forest not by the number of years that had elapsed since farming, but rather by the girth of trees, which varies with different regimes of cutting, soil, and slope, so these things are also taken into account. Thus, one area of forest classified as young fallow might have been as old as another area classified as young secondary growth (Wadley 2002b).

The fallow cycle of the Iban begins with the rice field. Plots are usually fallowed after one year of farming, and in the year after harvest, people gather the remaining cassava, bananas, and sugar cane. Forest succession after that is classified as young fallow, characterized by such plant species as *Macaranga costulata*, *Macaranga gigantea*, *Zingiber* spp., and *Glochidion* spp. This is followed by young secondary growth, with such plants as *Mussaendopsis beccariana*, *Nauclea bernardoi*, and *Cratoxylon* spp. Next comes old secondary growth in which was found *Shorea* spp. and *Adinandra* spp.; and, finally, old growth forest that had not been farmed within memory. It includes virgin forest and previously farmed forest that had returned to its primary state, and contains *Shorea* spp., *Eusideroxylon* spp., and *Dryobalanops* spp.

Studies have shown that the burning of vegetation in swidden cultivation may largely destroy the soil seed bank (Riswan and Kartawinata 1991), but those seeds that survive may still contribute to regrowth (Bazzaz 1991). However, forest regeneration appears to rely mainly on dispersed seed, or seed rain, brought by animals or the wind, so mature trees must be nearby to enable rapid regrowth. In the case of Sungai Sedik, fields generally border a mixture of forest vegetation, including both fallowed and preserved forest.

Plant species that pioneer newly fallowed fields are not those usually found in older-growth forest, because they are adapted to high levels of light (Brown and Lugo 1990). Rather, they are species found in various stages of fallowed forest. Therefore, the mosaic of fallowed forest may be necessary in such a system to provide pioneering and secondary plant species to newly harvested fields (Walschburger and von Hildebrand 1991). Later, if the fields are left fallow long enough, preserved forest with older-growth vegetation may contribute seeds and seedlings to the secondary growth.

Studies have also shown that secondary forest is characterized by the rapid accumulation of woody plant species, producing many seeds that become widely distributed. Such forest rapidly accumulates nutrients in its vegetation and shows a fast return in litterfall, with rapid turnover and uptake by roots. However, although restoring some soil fertility, secondary growth appears to be less efficient than older growth in returning nutrients to the forest floor (Brown and Lugo 1990). Notwithstanding this level of efficiency, it has been shown elsewhere that soil fertility adequate for farming is generally restored within a 10-year fallow period (Sabhasri 1978; Zinke et al. 1978; Lawrence and Schlesinger 2001), and the data from Sungai Sedik appear to support this finding (Cramb 1989; Marten and Vityakon 1986).

Productivity of Fallow Land

Between 1979 and 1994, 39% of the total area farmed at Sungai Sedik came from young fallow. Young secondary and old secondary forest accounted for 29% and 32%, respectively. This indicated a strong reliance on long fallowed forest, because the older forest made up 61% of the total.

Fields in old secondary forest averaged 4.94 ha in size and produced an average of 1,187.31 kg of rice. Young secondary fields averaged 3.59 ha and yielded 923.08 kg of rice, while young fallow fields averaged 3.58 ha and produced 1,041.49 kg of rice. Therefore, on average, young fallow provided the higher percentage of crops (43%); while young secondary and old secondary fallows provided 28% and 29% respectively. Again, although young fallow made up 43% of the total rice yield, fields from the older forest produced 57%.

In terms of yield per hectare, young fallow was marginally more productive (290.56 kg/ha) than young secondary (256.61 kg/ha) and old secondary (240.58 kg/ha). These data appear to challenge the notion that fields farmed in younger fallow are necessarily less productive (Mertz 2002). However, what this may reflect is the success of the Iban fallowing regime, with even younger fallow being farmed at an adequate rate.

Preserved Forest

Preserved forest land made up about one-fourth to one-third of the total 24 km^2 territory of the Sungai Sedik longhouse community (Wadley 2002b). This followed the Iban tradition of preserving and managing several types of forest on a permanent and semipermanent basis: old longhouse sites, forest reserves, and forest gardens. These sites were not only protected from farming, but they were also actively managed, although in such an informal way as to have been overlooked as examples of true management (Padoch 1995; Peluso 1995).

Old Longhouse Sites

In the past, the Iban built their longhouses out of nondurable materials and had to rebuild every 5 to 10 years. Usually it was on a different site, although occasionally near or even on the old site (Sather 1990). Farming was forbidden on these sites because of the valuable fruit trees planted around the longhouse while the people still lived there. After a longhouse site was abandoned, the succession of saplings from the original trees was actively promoted by selective clearing of competitors. Forest patches were produced that consisted mainly of fruit trees (Padoch and Peters 1993; Padoch 1995). (See color plate 62 for an illustration of durian forest gardens in West Sumatra that probably evolved from similar beginnings.)

Forest Reserves

Forest reserves at Sungai Sedik covered a number of categories. Some were regarded as sacred. In contrast to old longhouse sites, these patches of forest had been preserved for a variety of reasons, not the least of which was the promotion of certain valuable species. Particular families created nonsacred forest reserves to maintain sites for rattan collection or to preserve timber trees for later cutting; corridors of trees were preserved along the courses of major streams; and strips of forest at varying stages of succession were preserved as boundaries and firebreaks between fields (Sather 1990). Such sites were either permanent or semipermanent. Sacred sites, however, were usually permanently protected. They included forest cemeteries where the ancestors of the longhouse were buried, forests surrounding water and deep pools, and the peaks of mountains.

Forest Gardens

Numerous smallholdings of rubber and other tree crops such as cocoa constituted an important category of semipermanent forest at Sungai Sedik. Rubber groves were felled and farmed after the trees declined in latex productivity. Fruit trees were often interspersed throughout these sites. The floodplain land offered an important mix of managed gardens and fallow. Interspersed throughout fallow forest in this high floodplain were groves of *Arenga pinnata* and bamboo, including *Gigantochloa* spp. and *Dendrocalamus* spp., both planted and deliberately promoted from natural growth.

Ecological Importance of Preserved Forest

The ecological importance of these sites has been explicitly recognized by the Iban. Managed forest, both preserved and fallowed, had long been the source of important products used in daily life: fruit for home consumption or occasional sale, bark for rice bins and field hut walls, fibers for weaving and construction, and wood for lumber were just some. Hanne Christensen has documented an extensive plant classification system used by Iban in Sarawak, with a lexicon of names for about 2,000 plant species. Single longhouses identified and regularly used more than 650 plant species from at least 127 families (Christensen 2002). The Sungai Sedik study supports these findings (Colfer et al. 1993, 1997).

The Iban also had another important reason for preserving areas of forest: the maintenance of habitats for game and, by extension, populations of animals, insects, and plants that were not directly used by humans. By tradition, they also prohibited farming at the peaks of mountains, a ban reinforced by the existence of sacred sites on some mountaintops. However, whether they were sacred or not, mountaintop forests were preserved so there would always be seed to reforest fallowed fields (Jordan 1986). As one Iban man put it: The trees seen in old growth forest, or secondary forest for that matter, didn't grow on their own, but had to be brought to a place by animals or birds or the wind, and areas of preserved forest were necessary for that. The value of both forest and animals in the total agroforestry system was therefore recognized by the Iban, and this suggests that preserved forest sites elsewhere in the tropics may play a similarly important role in forest regeneration (Howe 1984; Terborgh 1995).

Role of Forest Fauna

In the various forest habitats of Sungai Sedik, Iban hunters encountered 41 species of mammals from six orders and 74 species of birds from nine orders (Wadley 2002b). This was by no means exhaustive, so it was unclear how to characterize the level of biodiversity. Superficially, though, Sungai Sedik forests appeared less diverse than the DSNP forests. Much of the difference, however, may have been the result of data collection methods (Wadley 2002b). In this chapter, diversity is considered in relation to important plant-animal interactions, namely pollination and seed dispersal, and their potential relation to the fallow cycle. Other researchers have not explored this relationship extensively (Michon et al. 1986; Michon and de Foresta 1991).

Hornbills, pigeons, and flowerpeckers are some of the frugivorous birds known to be seed dispersers (Lambert and Marshall 1991; Snow 1981; Susilo 1992), while flowerpeckers and sunbirds (Nectariniidae) are also involved in the pollination of some plants (Bawa 1990; Proctor et al. 1996). Bats, civets, squirrels, and langurs are known to be involved in seed dispersal (Emmons 1992; Francis 1994; Lambert 1990; Gautier-Hion 1990), and fruit bats (Pteropodidae) are important pollinators of *Durio* (Bombacaceae), *Parkia* (Fabaceae), and other important fruit trees (Start and Marshal 1976; Proctor et al. 1996). The existence of important invertebrate species that serve as pollinators may be suggested by the presence of some plant species. For example, thrips (Thysanoptera) are regular pollinators of dipterocarps such as *Shorea,* which were abundant in Sungai Sedik forests, and the reproductive cycle of fig wasps (Hymenoptera-Parasitica) is closely tied to that of different species of *Ficus* (Moraceae), which were also abundant at Sungai Sedik (Appanah 1990; Proctor et al. 1996).

The most numerous encounters with both mammals and birds at Sungai Sedik were in forest reserves and old longhouse sites (Wadley 2002b). This was not only because of the hunting effort spent at these sites, but also because animals were attracted to the foods found there (Eisenberg and Harris 1987). Encounters with squirrels were the most numerous, followed by monkeys (mainly macaques), pigs, deer (mainly barking deer), civets, and gibbons. Among avifauna, encounters with bulbuls, flowerpeckers, leafbirds, hornbills, and barbets were most numerous.

Numerous bats (Pteropodidae, Emballonuridae, Hipposideridae, and Vespertilionidae) were sighted (Francis 1994). Monkeys and squirrels were also sighted in large numbers, while the larger game species were less numerous. Of the birds, bulbuls and flowerpeckers were seen in the largest numbers.

Animals were also observed eating a number of forest foods, especially in tree reserves and old longhouse sites. *Lithocarpus* spp. offered the most important fruit among mammals, while *Ficus* spp. provided the most popular fruit among birds. Both wild and domesticated *Nephelium* spp. and *Castanopsis* spp. were also important foods for mammals (Leighton and Leighton 1983; Caldecott 1988).

Although not directly demonstrating seed dispersal and pollination by forest fauna, the presence of animals known to be involved in seed dispersal and pollination certainly suggests this. In forests managed by Kenyah in Sarawak, for example, there are more fruit trees in old secondary forest than in previously unfarmed old growth (Chin and Chua 1984). Not only do light mosaics favor fruit trees, but this also suggests that forest fauna may contribute to seed dispersal as they move through the forest mosaic (Foster et al. 1986). Conversely, the findings here suggest that a mosaic of older preserved forest and secondary growth may be important in promoting faunal diversity.

Forest animals, however, prey on farm and tree crops, and the recognition of their role in forest regeneration must be balanced with the reputation of some as crop pests. Additionally, the hunting of large vertebrates that disperse large seeds may have reduced their population, thereby disrupting regeneration of some old growth species.

This is not necessarily the case for other species. Small vertebrates, which were not intensely hunted, may have played a more important role in dispersing pioneering and secondary plants (Eisenberg and Harris 1987; Maury-Lechon 1991; Salafsky 1993).

Discussion and Evaluation

Iban forest management strategies might be best described as nonintensive or casual (Brown and Lugo 1990; Padoch 1995) and may easily have been overlooked by people expecting to find formal organizational structures. Fallow growth was not managed in the sense that fields were planted with useful trees and shrubs, but rather the fallow cycle was managed by the maintenance of forest as a source of seed and seed dispersal agents, as well as to replenish nutrient stocks. One critical element of their system may have been the long-term view held by the Iban; in managing the forest, they thought not only of present needs but also of those of their future descendants (Colfer et al. 1997; Sather 1990).

This gives rise to a number of questio7ns: Was the long fallow system a result of the total agroforestry system working well, or was it merely a byproduct of the low population density and distance from good markets? If the population was to increase to high levels, or if market access improved, would the system still work, or would the Iban shift to alternative strategies? A shift to more intensive swamp farming and cash cropping would not necessarily mean the total loss of the system, as long as there was adequate land available. When supported by some supplemental cash cropping, long fallow cultivation might still be sustainable with a population density of around 14 persons per square kilometer. However, when densities increase to around 20 persons per square kilometer, the extent of swidden farming might be decreased in order to preserve a long fallow, and more effort might be invested in smallholder cash cropping (Cramb 1993). When populations reach higher levels, such as 88 persons per square kilometer (Padoch 1995), there could be a move away from hill swiddening to more intensive swamp or irrigation farming, with hill land being converted into different types of managed forest.

Merits

The strength of the Iban system possibly lay in its strategy of multiple uses (Weinstock 1985). It provided forest products, habitat for game animals, potential sources of seed and regeneration, and watershed protection. It may also have promoted some degree of floral and faunal diversity. The benefits to Iban households have been quite tangible, and the system may have great potential, not only for sustainability, but, given adequate land, also for rehabilitating degraded areas (Boonkird et al. 1989).

Interestingly, one scheme for promoting forest regeneration and animal habitat has involved long rotational timber harvesting, with old-growth trees being left adjacent to earlier successional species (Harris 1984). This resembles the colonial Belgian corridor system of controlled swidden cultivation, with strips of forest alternating with fallowed fields and newly cut fields (Ruthenberg 1976). Although unintentionally, Iban long fallow cultivation, with its mosaic of fields and secondary growth, may have mimicked these systems on a smaller scale. But it went beyond these schemes by also preserving and managing tracts of forest.

Constraints

Constraints to adoption of this system elsewhere may revolve around issues of land, market involvement, and government policy. First, if there is not enough land available to create such a system, people may need to turn to other alternatives. In areas like Sungai Sedik, with the system in place, there is the issue of how a growing population deals with increased land pressure. Is managed forest land better converted into smallholder cash crop plantations or into more intensive farming? Second, the distance to good markets may influence the extent and nature of cash cropping, thus affecting the area and quality of preserved forest. Although Sungai Sedik has been involved in rubber smallholding and the sale of nontimber forest products, the distance to good markets has been prohibitive to the development of other crops such as pepper and durian. However, as road access improves, the system may change as people respond to market demands. The recent expansion of pepper gardening reinforces this possibility (Wadley and Mertz 2005).

A third constraint may involve government policy toward swidden cultivation, and security of land tenure. Land that is not under permanent cultivation, including fallowed land and forest land, may be converted to logging and plantation concessions, and farmers have been warned that swidden cultivation is technically illegal under national law (Cramb 1989; Dove 1983). However, in light of the system described in this chapter and ample evidence from elsewhere (Colfer with Dudley 1993; Colfer with Peluso and Chin 1997), government policy would do better by promoting such land use practices where they are viable, practical, and necessary. The practitioners must be confident that their efforts over years and even generations will not be wasted, and that managed and fallowed forest will not be felled for development. They must be able to control the use of their forest lands in perpetuity (Colfer et al. 1997).

Another constraint may involve the nature of the Iban management style. Being very casual, unstructured, and informally organized, Iban management can easily be misunderstood by government agents used to a more formal approach. If it is not recognized as management, then the Iban system cannot be adapted for application to other areas. The Iban themselves may even be encouraged to adopt different agricultural practices, and these may actually undermine the traditional system (Weinstock 1985), possibly leading to environmental degradation and loss of important biodiversity.

Application

Niches where this system may be adapted to local conditions would include areas of naturally low soil fertility and relatively abundant land, such as in upland Borneo, where methods of intensification involving terracing or irrigation have not been feasible (Boonkird et al. 1989). Such areas, however, may already have similar systems in place. If not, such a system could be promoted or even created by planting trees and other plants in set-aside areas used by both humans and animals. Planting certain trees that may not be immediately useful to people, but which could be used by animals, and particularly those involved in pollination and seed dispersal, would contribute to habitat improvement and, potentially, to fallow improvement.

Additionally, this system may serve as an efficient buffer zone around nationally protected forest, and its adoption should be encouraged in such areas (Aumeeruddy and Sansonnens 1994; Michon and de Foresta 1991; Salafsky 1993). The sort of mosaic environment created by this type of land use may prove beneficial to fauna within reserved forests by providing a wider range of exploitable habitat. In the case of this study, Danau Sentarum National Park may benefit greatly by being bordered on three sides by Iban-created habitats that may serve as faunal refuges during floods, as well as provide additional sources of food (Wadley 2002b).

All of these possibilities presume the application of indigenous silvicultural knowledge, and this might be highly useful to timber companies. Indigenous knowledge of local plant species and soils could benefit the replanting phases of logging concessions and provide alternatives to the use of exotic seedlings that may not fare well under local conditions (Michon et al. 1986).

Research Needs

There are several gaps in our knowledge about this system, and the relationships outlined in this chapter remain speculative and untested. First, we need more comparative data on Iban areas with higher population densities, better market access, and fewer managed forest sites, to measure the effects of these factors on the fallow cycle and on animal populations subjected to increased hunting and loss of habitat. Second, we need further study of the system itself, to study faunal populations, seed dispersal, soil fertility, rapidity of regeneration, and other variables as they relate directly to the fallow cycle. Third, we need further study of Iban strategies for managing forest. This might include examinations of the decision-making processes in field selection, the strategies involved in managing areas of preserved forest, and knowledge of ecology and animal behavior related to reforestation. Fourth, we need to look more closely at the marketing of forest and garden products to see how access to markets might change the way forests are managed over time.

Conclusions

The indigenous Iban system of agroforestry, involving a long fallow swidden cycle and the preservation and management of forest tracts, reveals a strategy for multiple uses of forest resources, both immediate and long term. The examination of this case has emphasized the potentially important role of forest fauna in the fallow cycle, a function that has been promoted by the Iban in their forest management. Such a system may have great potential for sustaining both long fallow swidden cultivation and floral and faunal diversity. Questions have been raised, however, regarding the effects of human population growth, market involvement, and government policy on this system, and these indicate the need for more comparative studies. As the Iban population grows or gains access to more distant markets, there may be increasing pressures to change the present system, and Iban farmers may need to fall back upon their indigenous ecological knowledge to meet the challenges of the changing environment.

Acknowledgments

Research during the years from 1992 to 1994 was funded by the U.S. National Science Foundation (Grant No. BNS-9114652), the Wenner Gren Foundation for Anthropological Research, Sigma Xi, and Arizona State University, and was sponsored by the Balai Kajian Sejarah dan Nilai Tradisional Pontianak, with permits from the Lembaga Ilmu Pengetahuan Indonesia. Faunal data were collected while I served as a consultant for the Danau Sentarum Wildlife Reserve Conservation Project under the auspices of the Asian Wetland Bureau, which is now Wetlands International-Indonesia Program, the Indonesian Directorate of Forest Protection and Nature Conservation (PHPA), and the Overseas Development Administration (U.K.). Additional study was done in June 1996 during a project for the Center for International Forestry Research (CIFOR) in cooperation with Wetlands International and PHPA. Any conclusions and opinions drawn here are not necessarily those of the funding agencies mentioned above.

References

Appanah, S. 1990. Plant-Pollinator Interactions in Malaysian Rainforests. In: *Reproductive Ecology of Tropical Forest Plants,* edited by K.S. Bawa and M. Hadley. Paris: UNESCO and Parthenon Publishing, 85–101.

Aumeeruddy, Y., and B. Sansonnens. 1994. Shifting from Simple to Complex Agroforestry Systems: An Example for Buffer Zone Management from Kerinci, Sumatra, Indonesia. *Agroforestry Systems* 28, 113–141.

Bawa, K.S. 1990. Plant-Pollinator Interactions in Tropical Rainforests. *Annual Review of Ecology and Systematics* 21, 399–422.

Bazzaz, F.A. 1991. Regeneration of Tropical Forests: Physiological Responses of Pioneer and Secondary species. In: *Rainforest Regeneration and Management,* edited by A. Gomez-Pompa, T.C. Whitmore, and M. Hadley. Paris: UNESCO and Parthenon Publishing, 91–118.

Belsky, J.M. 1993. Household Food Security, Farm Trees, and Agroforestry: A Comparative Study in Indonesia and the Philippines. *Human Organization* 52, 130–141.

Boonkird, S.A., E.C.M. Fernandes, and P.K.R. Nair. 1989. Forest Villages: An Agroforestry Approach to Rehabilitating Forest Land Degraded by Shifting Cultivation in Thailand. In: *Agroforestry Systems in the Tropics*, edited by P.K.R. Nair. Dordrecht, the Netherlands: Kluwer Academic Publishers, 211–227.

Brown, S., and A.E. Lugo. 1990. Tropical Secondary Forests. *Journal of Tropical Ecology* 6, 1–32.

Caldecott, J. 1988. *Hunting and Wildlife Management in Sarawak.* Gland, Switzerland: IUCN.

Chin, S.C. 1985. Agriculture and Resource Utilization in a Lowland Rainforest Kenyah Community. *Sarawak Museum Journal* 35 Special Monograph No. 4.

———, and T.H. Chua. 1984. The Impact of Man on a Southeast Asian Tropical Rainforest. *Malayan Nature Journal* 37, 253–269.

Christensen, H. 2002. Ethnobotany of the Iban and the Kelabit. Kuching, Malaysia: Forest Department Sarawak; Aarhus, Denmark: NEPcon and University of Aarhus.

Colfer, C.J.P., with R.G. Dudley. 1993. Shifting Cultivators of Indonesia: Marauders or Managers of the Forest? Rice Production and Forest Use among the Uma' Jalan of East Kalimantan. In: *Community Forestry Case Study Series,* No. 6. Rome: FAO.

———, R.L. Wadley, B. Suriansyah, and E. Widjanarti. 1993. Use of Forest Products in Three Communities: A Preliminary View. In: *Conservation Sub-project Quarterly Report* (June). Indonesia Tropical Forestry Management Project. Bogor, Indonesia: Asian Wetlands Bureau.

———, N. Peluso, and S.C. Chin. 1997. *Beyond Slash and Burn: Lessons from the Kenyah on Management of Borneo's Rainforests.* New York: New York Botanical Gardens.

———, R.L. Wadley, E. Harwell, and R. Prabhu. 1997. Assessing Intergenerational Access to Resources: Developing Criteria and Indicators. CIFOR Working Paper No. 18. Bogor, Indonesia: Center for International Forestry Research.

Conklin, H.C. 1957. *Hanunóo Agriculture.* Rome: FAO.

Cramb, R.A. 1989. Shifting Cultivation and Resource Degradation in Sarawak: Perceptions and Policies. *Borneo Research Bulletin* 21, 22–49.

———. 1993. Shifting Cultivation and Sustainable Agriculture in East Malaysia: A Longitudinal Study. *Agriculture Systems* 42, 209–226.

Dove, M.R. 1983. Theories of Swidden Agriculture, and the Political Economy of Ignorance. *Agroforestry Systems* 1, 85–99.

Eisenberg, J.F., and L.D. Harris 1987. Agriculture, Forestry, and Wildlife Resources: Perspectives from the Western Hemisphere. In: *Agroforestry: Realities, Possibilities and Potentials,* edited by H.L. Gholz. Dordrecht, the Netherlands: Martinus Nijhoff, 47–57.

Emmons, L.H. 1992. The Roles of Small Mammals in Tropical Rainforest. In: *Proceedings of an International Conference on Forest Biology and Conservation in Borneo,* July 30–August 3, 1990, Kota Kinabalu, Sabah, Malaysia, edited by G. Ismail, M. Mohamed, and S. Omar. Kota Kinabalu, Malaysia: Yayasan Sabah, 512–513.

Foster, R.B., B.J. Arce, and T.S. Wachter. 1986. Dispersal and the Sequential Plant Communities in Amazonian Peru Floodplain. In: *Frugivores and Seed Dispersal,* edited by A. Estrada and T.H. Fleming. Dordrecht, the Netherlands: W. Junk Publishers, 357–370.

Francis, C.M. 1994. Vertical Stratification of Fruit Bats (Pteropodidae) in Lowland Dipterocarp Rainforest in Malaysia. *Journal of Tropical Ecology* 10, 523–530.

Freeman, D. 1970. *Report on the Iban.* London: Athlone Press.

Gautier-Hion, A. 1990. Interactions among Fruit and Vertebrate Fruit-eaters in an African Tropical Rainforest. In: *Reproductive Ecology of Tropical Forest Plants,* edited by K.S. Bawa and M. Hadley. Paris: UNESCO and Parthenon Publishing, 219–230.

Harris, L.D. 1984. *The Fragmented Forest.* Chicago: University of Chicago Press.

Howe, H.F. 1984. Implications of Seed Dispersal by Animals for Tropical Reserve Management. *Biological Conservation* 30, 261–281.

Jordan, C.F. 1986. Local Effects of Tropical Deforestation. In: *Conservation Biology: The Science of Scarcity and Diversity,* edited by M.E. Soulé. Sunderland, MA: Sinauer Associates, 410–426.

Lambert, F.R. 1990. Some Notes on Fig-Eating by Arboreal Mammals in Malaysia. *Primates* 31, 453–458.

———, and A.G. Marshall. 1991. Keystone Characteristics of Bird-Dispersed *Ficus* in a Malaysian Lowland Rainforest. *Journal of Ecology* 79, 793–809.

Lawrence, D.C., M. Leighton, and D.R. Peart. 1995. Availability and Extraction of Forest Products in Managed and Primary Forest around a Dayak village in West Kalimantan, Indonesia. *Conservation Biology* 9, 76–88.

———, and W.H. Schlesinger. 2001. Changes in Soil Phosphorus during 200 Years of Shifting Cultivation in Indonesia. *Ecology* 82, 2769–2780.

Leighton, M., and D.R. Leighton. 1983. Vertebrate Responses to Fruiting Seasonality within a Bornean Rainforest. In: *Tropical Rainforest: Ecology and Management*, edited by S.L. Sutton, T.C. Whitmore, and A.C. Chadwick. Oxford, U.K.: Blackwell, 181–196.

Marten, G.G., and P. Vityakon. 1986. Soil Management in Traditional Agriculture. In: *Traditional Agriculture in Southeast Asia*, edited by G.G. Marten. Boulder, CO: Westview Press, 199–225.

Mary, F., and G. Michon. 1987. When Agroforests Drive Back Natural Forests: A Socioeconomic Analysis of a Rice/Agroforest System in South Sumatra. *Agroforestry Systems* 5, 27–55.

Maury-Lechon, G. 1991. Comparative Dynamics of Tropical Rainforest Regeneration in French Guyana. In: *Rainforest Regeneration and Management*, edited by A. Gomez-Pompa, T.C. Whitmore, and M. Hadley. Paris: UNESCO and Parthenon Publishing, 285–293.

Mertz, O. 2002. The Relationship between Length of Fallow and Crop Yields in Shifting Cultivation: A Rethinking. *Agroforestry Systems* 55, 149–159.

Michon, G., and H. de Foresta. 1991. Complex Agroforestry Systems and the Conservation of Biological Diversity. *Malayan Nature Journal* 45, 457–473.

———, F. Mary, and J. Bompard. 1986. Multistoreyed Agroforestry Garden System in West Sumatra, Indonesia. *Agroforestry Systems* 4, 315–338.

Padoch, C. 1982. *Migration and Its Alternatives among the Iban of Sarawak*. The Hague: Martinus Nijhoff.

———. 1995. Creating the Forest: Dayak Resource Management in Kalimantan. In: *Society and Non-timber Forest Products in Tropical Asia*, edited by J. Fox. Occasional Papers, Environment Series No. 19. Honolulu, Hawaii: East-West Center, 3–12.

———, and C. Peters. 1993. Managed Forest Gardens in West Kalimantan, Indonesia. In: *Perspectives on Biodiversity: Case Studies of Genetic Resource Conservation and Development*, edited by C.S. Potter, J.I. Cohen, and D. Janczewski. Washington, DC: AAAS Press, 167–176.

Peluso, N.L. 1995. Extraction Interactions: Logging Tropical Timbers in West Kalimantan, Indonesia. In: *Society and Non-timber Forest Products in Tropical Asia*, edited by J. Fox. Occasional Papers, Environment Series No. 19. Honolulu, Hawaii: East-West Center, 72–96.

Proctor, M., P. Yeo, and A. Lack. 1996. *The Natural History of Pollination*. Portland, OR: Timber Press.

Riswan, S., and K. Kartawinata. 1991. Regeneration after Disturbance in a Lowland Mixed Dipterocarp Forest in East Kalimantan, Indonesia. In: *Rainforest Regeneration and Management*, edited by A. Gomez-Pompa, T.C. Whitmore, and M. Hadley. Paris: UNESCO and Parthenon Publishing, 295–301.

Ruthenberg, H. 1976. *Farming Systems in the Tropics*. Oxford, UK: Clarendon Press.

Sabhasri, S. 1978. Effects of Forest Fallow Cultivation on Forest Production and Soil. In: *Farmers in the Forest: Economic Development and Marginal Agriculture in Northern Thailand*, edited by P. Kunstadter, E.C. Chapman, and S. Sabhasri. Honolulu, Hawaii: University Press of Hawaii, 160–184.

Salafsky, N. 1993. Mammalian Use of a Buffer Zone. Agroforestry System Bordering Gunung Palung National Park, West Kalimantan, Indonesia. *Conservation Biology* 7, 928–933.

———. 1994. Forest Gardens in the Gunung Palung Region of West Kalimantan, Indonesia: Defining a Locally Developed, Market-oriented Agroforestry System. *Agroforestry Systems* 28, 237–268.

Sather, C. 1990. Trees and Tree Tenure in Paku Iban Society: The Management of Secondary Forest Resources in a Long-established Iban Community. *Borneo Review* 1, 16–40.

Snow, D.W. 1981. Tropical Frugivorous Birds and their Food Plants: A World Survey. *Biotropica* 13, 1–14.

Start, A.N., and A.G. Marshal. 1976. Nectarivorous Bats as Pollinators of Trees in West Malaysia. In: *Tropical Trees: Variation, Breeding, and Conservation*, edited by J. Burley and B.T. Styles. London: Academic Press, 114–150.

Susilo, A. 1992. Seed Dispersal Agents at Mentoko Burned-over Forest, Kutai National Park, East Kalimantan, Indonesia. In: *Proceedings of an International Conference on Forest Biology and Conservation in Borneo*, July 30–August 3, 1990, Kota Kinabalu, Sabah, Malaysia, edited by G. Ismail, M. Mohamed, and S. Omar. Kota Kinabalu, Malaysia: Yayasan Sabah, 516–517.

Terborgh, J. 1995. Wildlife in Managed Tropical Forests: A Neotropical Perspective. In: *Tropical Forests: Management and Ecology*, edited by A.E. Lugo and C. Lowe. New York: Springer-Verlag, 331–342.

Torquebiau, E. 1984. Man-made Dipterocarp Forest in Sumatra. *Agroforestry Systems* 2, 103–127.

Wadley, R.L. 2001. Community Co-operatives, Illegal Logging, and Regional Autonomy: Empowerment and Impoverishment in West Kalimantan, Indonesia. Paper presented at the conference on Resource Tenure, Forest Management, and Conflict Resolution: Perspectives from Borneo and New Guinea, 9–11 April, Australian National University, Canberra, Australia.

———. 2002a. Coping with Crisis—Smoke, Drought, Flood and Currency: Iban Households in West Kalimantan, Indonesia. *Culture & Agriculture* 24, 26–33.

———. 2002b. Iban Forest Management and Wildlife Conservation along the Danau Sentarum Periphery, West Kalimantan, Indonesia. *Malayan Nature Journal* 56, 83–101.

———, C.J.P. Colfer, and I.G. Hood. 1997. Hunting Primates and Managing Forests: The Case of Iban Forest Farmers in West Kalimantan, Indonesia. *Human Ecology* 25, 243–271.

———, and O. Mertz. 2005. Pepper in a Time of Crisis: Smallholder Buffering Strategies in Sarawak, Malaysia and West Kalimantan, Indonesia. *Agricultural Systems* 85, 289-305.

Walschburger, T., and P. von Hildebrand. 1991. The First 26 Years of Forest Regeneration in Natural and Man-made Gaps in the Colombian Amazon. In: *Rainforest Regeneration and Management*, edited by A. Gomez-Pompa, T.C. Whitmore, and M. Hadley. Paris: UNESCO and Parthenon Publishing, 257–263.

Weinstock, J.A. 1983. Rattan: Ecological Balance in a Borneo Rainforest Swidden. *Economic Botany* 37, 58–68.

———. 1985. Alternate Cycle Agroforestry. *Agroforestry Systems* 3, 387–397.

Zinke, P.J., S. Sabhasri, and P. Kunstadter. 1978. Soil Fertility Aspects of the Lua' Forest Fallow System of Shifting Cultivation. In: *Farmers in the Forest: Economic Development and Marginal Agriculture in Northern Thailand*, edited by P. Kunstadter, E.C. Chapman, and S. Sabhasri. Honolulu, HI: University Press of Hawaii, 134–159.

Chapter 42

Does Tree Diversity Affect Soil Fertility?

Findings from Fallow Systems in West Kalimantan

Deborah Lawrence, Dwi Astiani,
Marlina Syhazaman-Karwur, and Isabella Fiorentino

Intensification of management to achieve more efficient and productive fallows in the uplands of Southeast Asia is likely to alter the species composition and structure of the tree community in traditional fallows. Consequent changes in nutrient cycling may compromise the sustainability of these systems, as well as the future productivity of the land. Therefore, understanding the influence of the tree community on nutrient dynamics is critical to formulating long-term solutions to fallow degradation, especially in systems with low potential inputs of capital and labor.

Tree-based alternatives, not only to slash-and-burn, but also to all other forms of deforestation, should facilitate both the preservation of ecosystem function and the conservation of biodiversity, as well as provide economic benefits. The ability of land-use strategies to meet all three goals ultimately depends on the relationship between tree diversity and critical aspects of ecosystem function such as nutrient cycling. This chapter presents initial findings from a study of tree diversity and soil fertility in three indigenous fallow management (IFM) systems in West Kalimantan, Indonesia. It aims to determine the potential of agroforestry systems associated with shifting cultivation to conserve or enhance soil resources.

Our interest grew directly out of discussions about the Alternatives to Slash-and-Burn Project, pursued both in Indonesia and throughout the tropics with the help of the World Agroforestry Centre (ICRAF). It was clear that forest policy in many tropical countries had tended to emphasize short-term economic goals. This is despite the fact that sustained economic benefits for future generations can only be achieved if a long-term view of forest resources is adopted, including adequate attention to environmental effects. Data on the ecological consequences of land-use management systems associated with shifting cultivation are essential to

---

Deborah Lawrence, Environmental Sciences, University of Virginia, P.O. Box 400123, Charlottesville, VA 22902-4123 USA; Dwi Astiani and Marlina Syhazaman-Karwur, Universitas National Tanjungpura, Jln. A. Yani Pontianak, Pontianak, West Kalimantan, Indonesia; Isabella Fiorentino, Laboratory of Ornithology, 159 Sapsucker Weeds Road, Ithaca, NY 14850, USA.

understanding how these systems might be adapted for fallow intensification in traditional agriculture, expansion of agroforestry production, and reclamation of degraded lands.

Although shifting cultivation involves the clearing of forest, some of it primary forest, so do many other widespread forest uses, such as timber extraction, plantation forestry, and industrial agriculture. In each case, changes in the carbon balance between biomass and soil, and between soil and atmosphere, may have serious implications for climate, nutrient cycles, and water balance, on scales from local to global. The loss of forest also represents a loss of suitable habitat for both plants and animals of tropical rainforests. Shifting cultivation involves the rebuilding of forests with higher tree diversity than industrial or estate forests in almost all cases, with the exception of natural forest management for timber. Therefore, shifting cultivation and associated secondary forest management systems seem to represent a better alternative for conserving biodiversity (Michon and de Foresta 1991; Padoch and Peters 1993; Lawrence and Mogea 1996). The effects of these integrated upland rice and tree crop systems on soil nutrient stocks over hundreds of years, and not simply over the course of a single succession, are poorly known. This led us to study the managed and primary forest around Kembera, in West Kalimantan, Indonesia (see Figure 42-1).

Objectives

We had several goals for this study. First, we sought to determine whether traditional tree-based, long-fallow shifting cultivation resulted in significant losses of soil organic matter and essential nutrients from primary forest. Second, we wished to compare the nutrient status of unmanaged fallows (hereafter referred to simply as "fallows") and two current alternatives for more intensive fallow management or expanded agroforestry development: rubber gardens and fruit gardens.

Third, by considering the tree community of each IFM type (these may also be referred to as "forest types," since they are all tree-based systems), we hoped to put the soil nutrient data into a broader context. Specifically, we asked whether species diversity or community structure were related to soil fertility. Fallow intensification often implies a shift in, or loss of, tree species, as useful species replace those deemed useless. The introduction of rubber, for instance, decreased the tree diversity of home gardens in Kerala, India (Jose 1992, cited in Mohan Kumar et al. 1994). Even within rubber gardens, pressure to increase rubber production is likely to lead to a decline in tree diversity (Lawrence 1996). We sought to answer the question: Will such changes in the tree community affect the sustainability of the system, as defined by changes in soil nutrient status?

Methods

Study Site

This research was conducted in the landscape surrounding the Dayak village of Kembera, about 100 km south of the equator in West Kalimantan, north of Gunung Palung National Park. Shifting cultivation and tree crop management have occurred continuously for at least 200 years in this village, on clay-rich ultisols and oxisols. The mosaic landscape is dominated by secondary forest, regenerated after shifting cultivation. Patches of different age, cultivation history, and land use occupy between 3,500 and 4,000 ha of land bordered by a protected primary forest reserve and selectively logged lowland forest. Long-fallow shifting cultivation of upland rice is associated with rainfed wet rice cultivation, smallholder rubber production, fruit crop production, and some extraction of primary forest products (Lawrence et al. 1998).

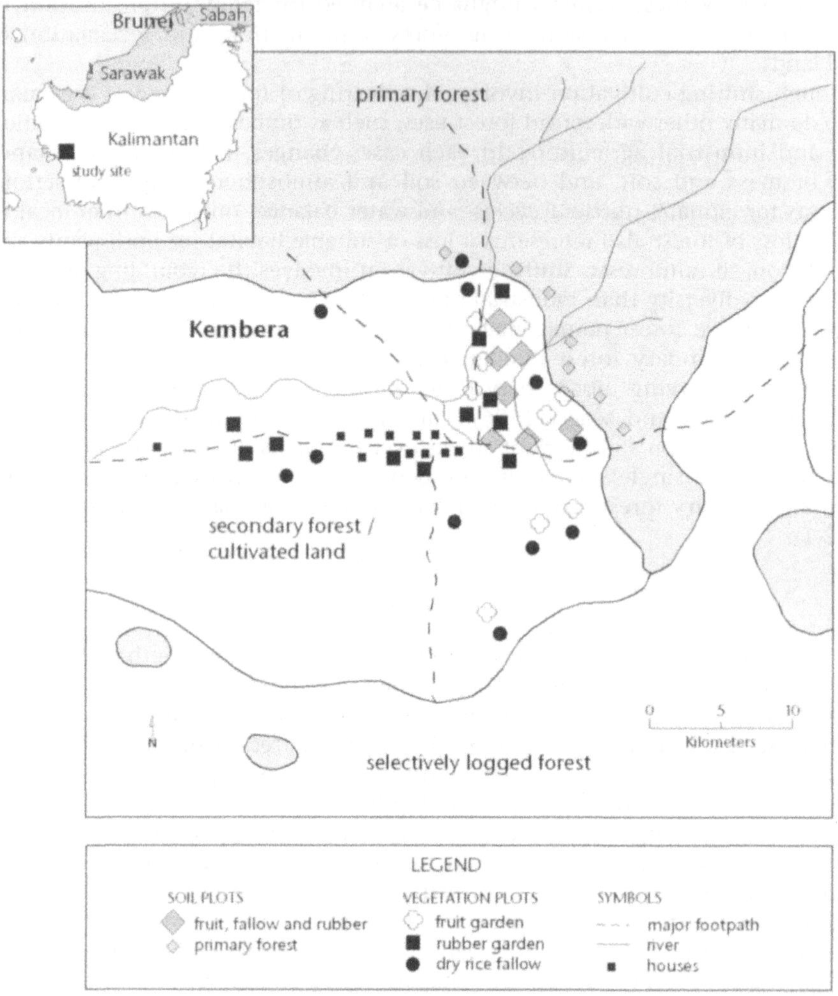

Figure 42-1. Location of the Study Site and Sampling Patches

Approach

We sampled soils and trees in the major upland forest types: upland rice fallows (62% of the landscape, Lawrence et al. 1998), fruit gardens (4%), rubber gardens (17%), and primary forest, which occupies more than 10,000 ha adjacent to the managed lands. In this study, we compared the soils, but not the tree community, between primary forest and the other forest types. Primary forest soils were treated as a control to evaluate the impact of alternative fallow management strategies. Comparative work on the tree communities of primary forest and fallows has been published elsewhere (Lawrence et al 2005).

IFMs Considered

Unmanaged fallows, fruit gardens, and rubber gardens are established following shifting cultivation of upland rice and, therefore, may be considered as alternative indigenous fallow management strategies. A rubber-based fallow involves a longer fallow period than an unmanaged fallow, more labor inputs, lower tree diversity, and significant biomass exports from the site. In Kembera, those exports amount to about 450 kg/ha/yr for about 30 years of productivity. A fruit-based fallow probably involves an even longer fallow period than rubber, moderately more labor than a natural fallow, comparable or increased tree diversity, and variable biomass exports from the site, depending on the products extracted. These alternatives may improve the livelihood of farmers and may help them avoid the ecological and economic risks associated with monospecific plantations or with settling into a system of permanent cultivation. Many factors, including those listed above, should be considered when evaluating the potential costs and benefits of fallow management alternatives. However, in this study, we consider only the effects of the various forest types on soil nutrient levels as a first step toward understanding the dynamics of nutrient cycling in these systems.

Soil Fertility

Our sampling effort was focused on a 4 km^2 area of subwatersheds to the north and east of the village, to control for major differences in topography, parent material, and cultivation history). In August 1995, matched samples for the three managed forest types were taken at seven sites (large diamonds, Figure 42-1). Sampled patches either shared a common border or were no more than 100 to 200 meters apart. Stand age varied from patch to patch, but all the patches were more than 15 years old, and most were more than 20 years old. We controlled for local variation in soils by choosing patches in the same locality rather than controlling for variation in patch age. Our rationale was that differences in average patch age were, in fact, a result of the different management practices associated with each forest type. Evaluating the effect of these practices on soil fertility was our aim. Because of the configuration of the landscape, it was impossible to match primary forest samples in all but one case. We sampled primary forest at seven sites near the boundary of managed lands, between 300 m and 2 km distant from the sites where the alternative fallows were sampled. These sites were further upslope than the IFMs sampled and are not necessarily representative of the primary forest they have replaced. Thus, caution should be taken when interpreting differences between IFMs and primary forest as indicating change through time.

Fruit gardens, which host many long-lived species, tend to be very old (100 to 200 years), as do patches of primary forest. Rubber gardens and upland rice fallows have similar potential life spans of about 40 years. In 1995, the average age of rubber gardens in the Kembera landscape, including both productive and immature gardens, was between 15 and 18 years. Average fallow length varied from 13 to 27 years in the early 1990s, and the six-year average was 19 years (Lawrence et al. 1998). Although differences in age between patches sampled at any one site were likely to contribute to differences in nutrient cycling, we believe that comparisons between forest types across the seven samples will prove robust.

At each of the 28 patches sampled, we collected soil samples at nine points, systematically arrayed in a grid across the patch. The exception was primary forest, where we used a series of short transects. The collection points were 5 to 20 m apart, depending on the size of the patch sampled. We sampled the soil at two depths, 0 to 10 cm and 30 to 60 cm. At several of the sites, soil depth was less than 60 cm, so we sampled to the greatest depth possible. We took one core of 2 cm diameter per point for deep soils and two cores per point for surface samples, compositing each depth separately for the site.

The soils were air dried, passed through a 2 mm sieve to remove roots and stones, and stored in sealed plastic bags prior to analysis. The Soils Laboratory at the Bogor Agricultural Institute (Institut Pertanian Bogor) analyzed the following physical and chemical characteristics of the soils: pH H_20 and pH KCl (both 1:1); cation exchange capacity (CEC) and percent base saturation (from extracts in 1 N NH_4OAc at pH 7.0); and soil texture (percentages of sand, silt, and clay by the hydrometer method). Soil nutrients were analyzed at the biogeochemistry laboratory of Duke University. Samples were analyzed for total carbon and nitrogen (dry combustion on a Perkin Elmer CHNOS analyzer); total phosphorus (Kjeldahl digestion); available phosphorus

(Bray extraction); and available cations (extracted in 1 N NH_4OAc and analyzed with atomic absorption or emission spectroscopy). Bulk density was determined according to an algorithm derived by Rawls (1985) and dependent on soil organic matter and texture. Using these estimates of bulk density, we determined total nutrient stocks for the 0 to 10 cm and 30 to 60 cm layers on a per hectare basis.

Tree Diversity

From April 1991 to April 1992, the tree communities of rubber gardens, fruit gardens, and fallows were sampled in a set of patches distinct from those used in the soil study. Detailed methods can be found in Lawrence et al. (1995). They will be described only briefly here. A randomly located single plot covering 1,000 m^2 was sampled in each of 32 patches scattered throughout the area surrounding the village. These were in 10 fruit gardens, 11 rubber gardens, and 11 fallows. All stems greater than 10 cm in diameter at breast height (dbh) were measured and identified, where possible, by their common names, including trees, lianas, and climbing palms. Unnamed species were given a number. Specimens or leaf samples were matched and vouchers identified by the staff at Herbarium Bogoriense in Bogor, Java. The number of distinct morphospecies, the number of stems, and the total basal area were tallied per plot.

Statistical Analysis

To compare primary forest with the three alternative fallow management strategies, we used one-factor analysis of variance on each of the soil parameters, with forest type as the independent factor. We used the Tukey-Kramer test to determine which comparisons drove a significant forest-type effect. Each depth was analyzed independently because of differences in the variance of most soil parameters with depth. Bonferroni corrections were applied in all ANOVA analyses to restrict the rejection of a null hypothesis of no difference among types, such that critical p-values (p^*) were between 0.002 and 0.007. We used SAS Institute's JMP IN software for all analyses.

We compared the species richness, stem density, and basal area of different forest types using one-factor ANOVA. We used linear regression to determine whether stem density and basal area were correlated with species richness. Analysis of covariance was used to determine whether either stem density or basal area had a significant effect on richness in addition to the effect of forest type.

To examine the relationship between aspects of the tree community and soil fertility among IFMs, we plotted mean species richness, mean stem density, and mean basal area versus the mean value of a given soil parameter. Because there were only three alternatives considered, we did not perform statistical analyses on the means. We inferred trends by visual inspection of the plots.

Results

Soil Fertility

Nutrient Stocks. Soil nutrient stocks tended to be greater in the three alternative fallow types than in primary forest. For the 0 to 10 cm layer, total carbon (ca. 45 to 50 vs. 31 Mg/ha in the top 10 cm), nitrogen (ca. 3.2 vs. 1.8 Mg/ha) and phosphorus (ca. 750 vs. 150 kg/ha) were significantly greater in all three IFMs ($p < 0.003$, Figure 42-2, a-d). Available phosphorus ranged from 6 to 8 kg/ha in the surface soil and did not differ between primary forest and IFMs. The trends were similar for deeper soils, but were statistically significant only for total phosphorus. Total carbon and nitrogen were greater in fallows than primary forest, but the overall effect of soil type in the ANOVA was only marginally significant ($p = 0.004$, where $p^* = 0.003$). Only fruit gardens differed significantly from primary forest in available cations. The surface soils of fruit gardens had more magnesium and calcium and less potassium ($p <$

0.003, Figure 42-2, e-h). There were no significant differences in cations between primary forest and IFMs in the deeper soils. Surprisingly, the only significant differences among IFMs were in the amount of magnesium and calcium in the 0 to 10 cm layer, both of which were higher in fruit gardens than in fallows or rubber gardens.

Physical and Chemical Characteristics. The physical and chemical properties of rubber gardens, fruit gardens, and fallows were generally superior to those of primary forest (see Figure 42-3). The soils were all acid and fell within a narrow range of pH (in H_20 ca. 5.0, in KCl ca. 4.0). The pH of primary forest soils was significantly different from the other land-use types statistically, but probably not biologically. Soil texture was more favorable in all three IFMs. These had higher silt (25% to 35%) and clay (40% to 45%) content and lower sand (20% to 40%) content. The only significant difference between IFMs and primary forest in deep soils was in soil texture, specifically, a higher amount of silt in fallows than in primary forest, and higher clay and lower sand in rubber gardens as well as fallows. Fruit gardens were categorized as clay-loam, rubber gardens and fallows as clays, and primary forest as sandy clay-loam. Perhaps as a consequence of high silt and clay content, CEC at the surface was also significantly higher in all three IFMs (16 to 19 milliequivalents/100 g), as compared with primary forest (8 milliequivalents/100 g). Beyond CEC, base saturation was significantly higher in fruit gardens than the other land-use types. Average base saturation was almost 30% in the top 10 cm of fruit garden soil. It was as low as 5% in rubber gardens and only as high as 15% in primary forest. The difference in base saturation was the only significant difference among IFMs.

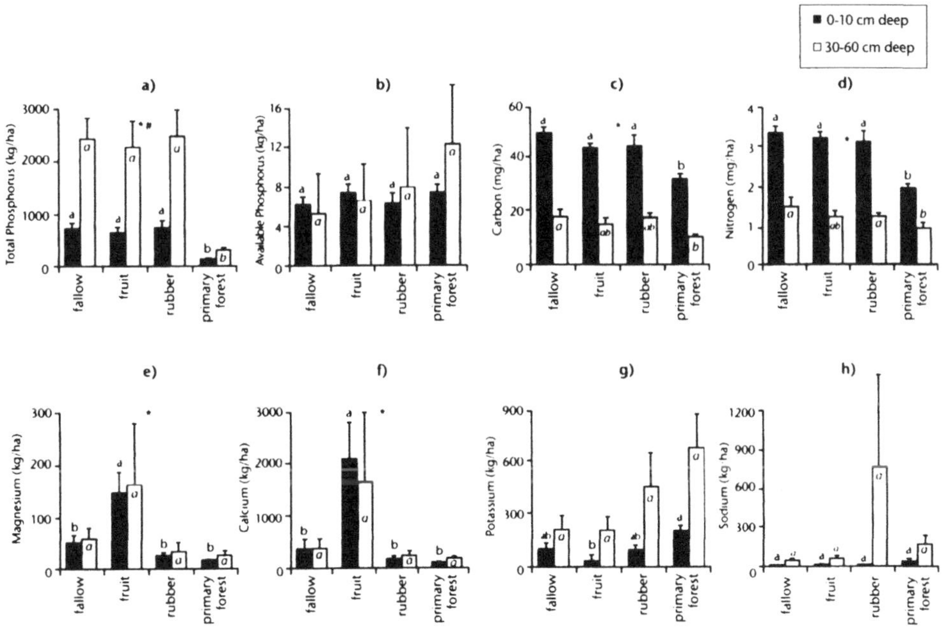

Figure 42-2. Soil Nutrient Stocks[x] in Each of the Three IFM Types and Primary Forest
Notes: x stands for mean ± standard error. Dark bars indicate values for 0 to 10 cm depth; clear bars indicate values for 30 to 60 cm depth. Bars with different letters above are significantly different. Roman letters indicate differences among surface soils. Italic letters indicate differences among deep soils. * indicates significant overall effect of forest type for surface soils. # indicates significant overall effect of forest type for deep soils.

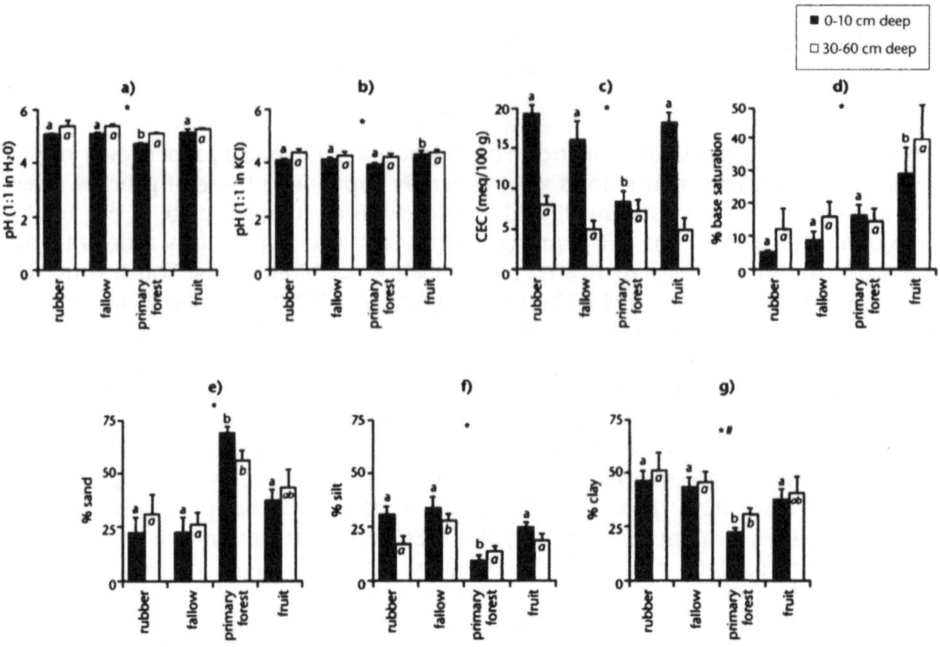

Figure 42-3. Soil Physical and Chemical Characteristics in Each of the Three IFM Types and Primary Forest

Notes: Dark bars indicate values for 0 to 10 cm depth; clear bars indicate values for 30 to 60 cm depth. Bars with different letters above are significantly different. Roman letters indicate differences among surface soils. Italic letters indicate differences among deep soils. * indicates significant overall effect of forest type for surface soils. # indicates significant overall effect of forest type for deep soils.

Diversity and Structure of the Tree Community. Fallows and fruit gardens had significantly higher species richness than rubber gardens ($p < 0.003$, Figure 42-4a), with 21 or 22 morphospecies per 0.1 ha plot, compared with only six. Basal area of trees greater than 10 cm dbh was almost three times greater in fruit gardens (above 60 m^2/ha, $p = 0.0001$). Stem density was ca. 30% to 60% greater in fallows (above 520/ha, $p = 0.006$) than in other IFMs. Species richness was positively but weakly correlated with tree basal area ($p = 0.004$, $r^2 = 0.24$, Figure 42-5a) and stem density ($p = 0.006$, $r^2 = 0.23$, Figure 42-5b). To tease apart the effect of forest type from the effect of basal area or stem density on species richness, we performed an analysis of covariance, considering forest type along with stem density and/or basal area as predictors of species richness. Neither additional factor was significant once the effect of forest type was considered, and the model fits were only marginally better (r^2 of 0.45 increased to 0.53).

Community Structure and Nutrient Stocks. We found no strong trends in the relationship between average species richness and average soil nutrient stocks in this limited sample of IFMs. However, mean stocks of Mg and Ca seemed to increase with increasing species richness (Figure 42-6, m and p). Mean stem density showed no trends with soil nutrient stocks except perhaps for its relationship with N. Total N tended to increase with increasing stem density, but the magnitude of the increase was very small (Figure 42-6e). Strong trends were observed, however, between soil nutrient stocks and mean basal area. Mg and Ca increased as a function of basal area (Figure 42-6, o and r). Available P also seemed to increase with basal area, while total

P tended to decline (Figure 42-6, i and l). Stocks of K may also decline with basal area, but the trend was less clear (Figure 42-6u). Mean C and N stocks were not related to basal area (Figure 42-6, c and f).

Discussion and Conclusions

IFMs vs. Primary Forest

All three IFMs had higher nutrient stocks and more favorable physical and chemical characteristics in the top 60 cm of soil than primary forest had. Therefore, we suggest that rubber gardens and fruit gardens as well as traditional, unmanaged fallows may be sustainable land-use strategies under present socioeconomic and political conditions. These systems have been in place for 40 to 200 years, with rubber introduced latest and unmanaged fallows earliest, and soil fertility apparently has not suffered. However, potassium levels in fruit gardens were lower than those in primary forest and could present a potential constraint on future soil fertility.

Rubber Gardens vs. Fruit Gardens vs. Fallows

Despite the apparent loss of potassium relative to primary forest, soil fertility was enhanced most in fruit gardens. This was indicated by higher levels of available Mg and Ca and higher base saturation. This assessment, however, depends on whether Mg and Ca are considered to be more limiting to the vegetation of future types of land use. If phosphorus is limiting to most upland forest communities on ultisols and oxisols (Sanchez 1976), then the difference in cation stocks among IFMs may be less significant as indicators of current and future soil fertility. This is especially true since soil organic matter and texture, which regulate the availability of water as well as nutrients, were similar among the three. Future work on the nutrient limitations of various fallow management strategies will help illuminate whether fruit gardens are indeed superior to unmanaged fallows from a soil fertility perspective.

Community Structure and Nutrient Stocks

Examining data on base saturation along with data on CEC, it is easier to understand why stocks of Ca and Mg are significantly higher in fruit gardens. The cause of these differences is not clear, since both organic matter (Figure 42-2, a and b) and texture (especially clay and silt content, Figure 42-3, f and g) differ across IFM types. The difference may be related to one of the two factors we hypothesized as relating to soil nutrient stocks, that is, biomass, as indicated by tree basal area, or tree diversity, as indicated by species richness. However, given that species richness is not significantly different between fallows and fruit gardens at the scale of 0.1 ha, we might expect that richness, per se, is not driving the difference in availability of essential cations. Instead, species composition may cause the disparity. On the other hand, a different measure of diversity might better capture the difference in the structure and composition of the tree community, for example, an index that considered the abundance of individual species as well as the total number in a given plot. Tree communities composed of different species, that is, fast-growing secondary forest species vs. mostly rubber vs. a mixture of slower-growing fruit-bearing species, may cycle, release, and retain nutrients at different rates (Glover and Beer 1986; Ewel et al. 1991; Fernandes and Sanford 1995).

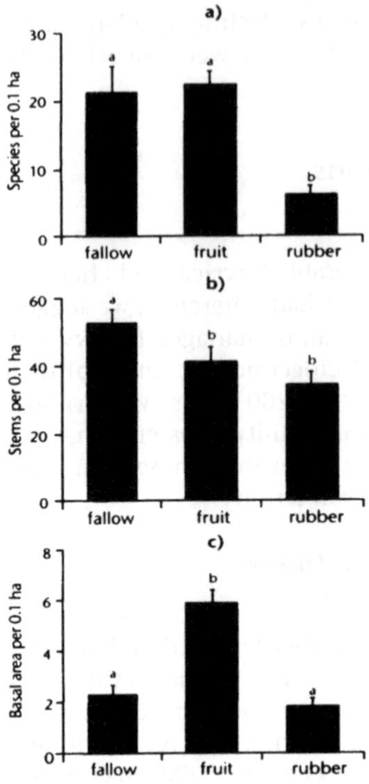

Figure 42-4.

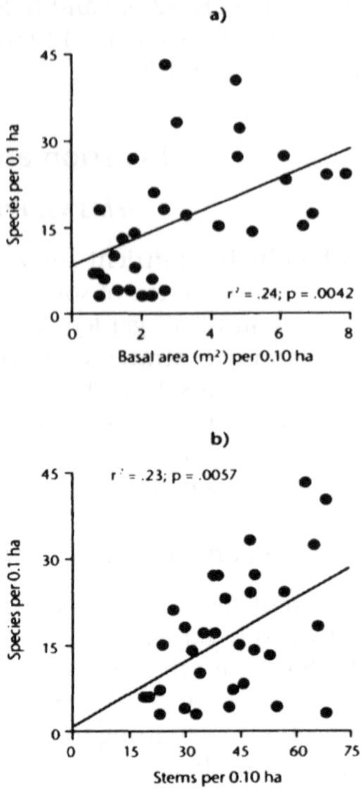

Figure 42-5.

Figure 42-4. Tree Species Density, Stem Density and Basal Area in Fallows, Fruit Gardens, and Rubber

Figure 42-5. Relationship Between Tree Species Density and Forest Structure, in Particular a) Stem Density and b) Basal Area

Total biomass may also drive the difference between fruit gardens and the other two fallow types. A higher biomass above ground may mean a higher density of root biomass below ground. Higher root biomass might indicate a greater flux of nutrients pumped through the vegetation and redeposited on the surface. It may also mean increased deposits directly to the soil through root exudation and root turnover. These are hypotheses that remain to be tested in the systems under study. We can suggest, however, that the apparent loss of K only in fruit gardens is consistent with greater nutrient fluxes there than in rubber gardens and fallows. With each cycling of organic matter through the system, we might expect losses of K due to leaching, since it is the element most easily displaced from the cation exchange complex.

How might the structure of the tree community affect soil nutrient stocks? Decreases in soil P with basal area may mean that much of the P has been taken up into the biomass, especially since P is not likely to be leached from the soil due to the high anion absorption capacity of acid soils. This may be true for K as well; however, K is more likely to be lost from the system by leaching, as discussed above. An increase in soil Mg and Ca is compatible with enhanced nutrient cycling if they are less limiting to the tree communities than P and K. In addition, they are more tightly adsorbed by the cation exchange complex, and even more so at the lower base saturation found in the IFMs. With each cycling of organic matter, some Mg and Ca might be expected to accumulate on soil colloids.

As far as we can tell with this limited sample (both within types and across types), increasing species richness has no negative impact on soil nutrient stocks, and it may contribute to an enhancement of essential cations. Likewise, the density of stems does not appear to deter or augment nutrient loss from the system.

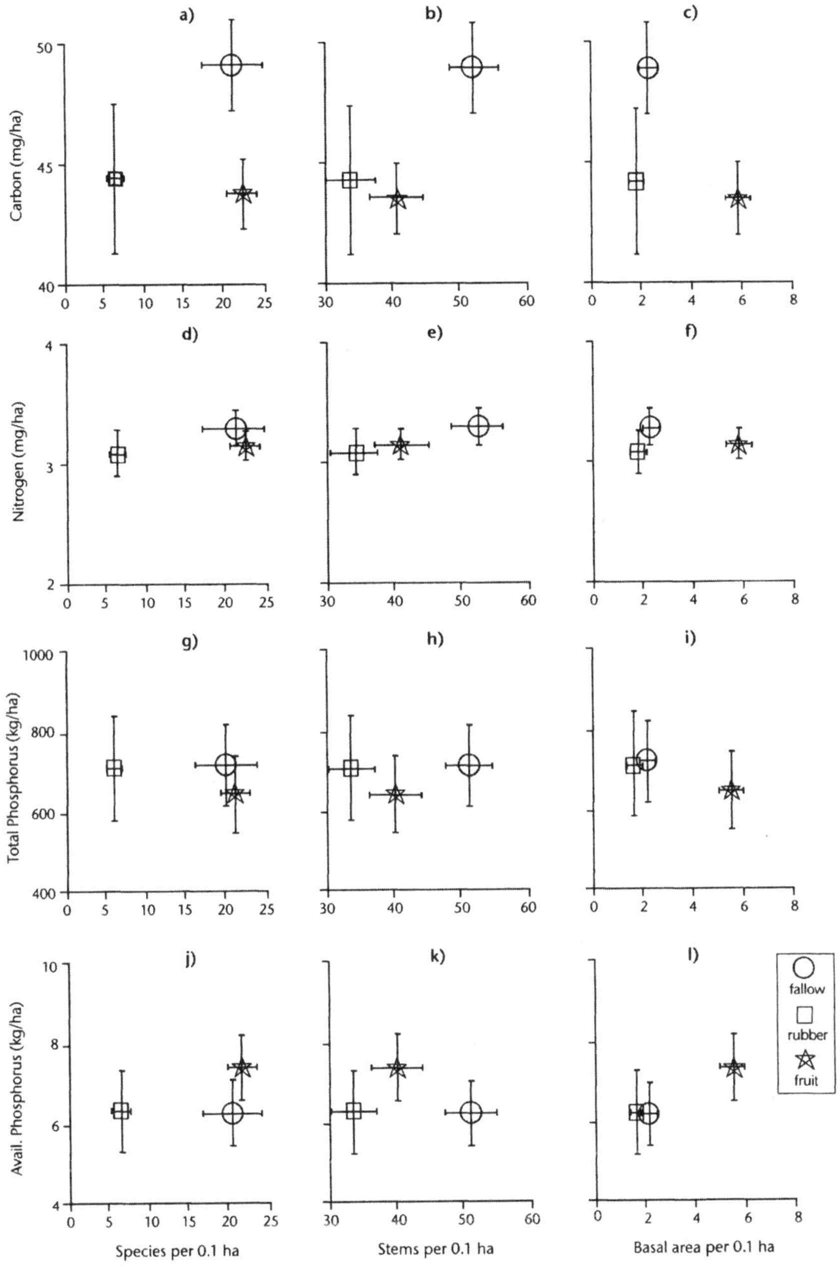

Figure 42-6 (a–l). Relationship between Soil Nutrient Stocks and Mean Species Richness (first column), Stem Density (second column), and Mean Basal Area (third column) of the IFMs

Note: Values are means ± standard error.

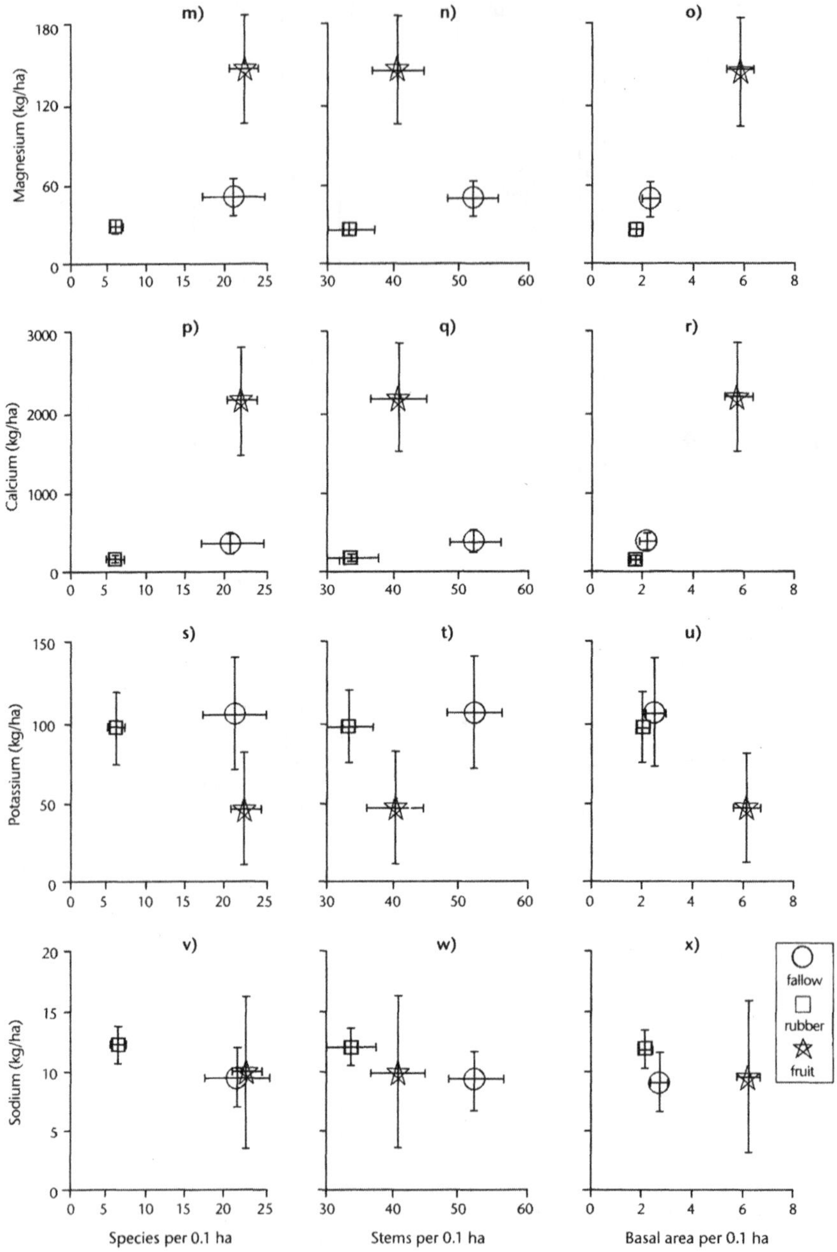

Figure 42-6 (m–x). Relationship between Soil Nutrient Stocks and the Mean Species Richness (first column), Mean Stem Density (second column), and Mean Basal Area (third column) of the IFMs

Note: Values are means ± standard error.

Increasing biomass, however, was related to a potential decline in total P and available K. Does this imply that high biomass systems are somehow unsustainable? Not necessarily. Two key components of the nutrient cycling system were not quantified for this study of nutrient stocks: the standing biomass, both above and below the ground, and the standing litter. There may simply be a shift in the allocation of nutrients among these different pools. As stated above, this is less likely for K and more likely for P. If, in the course of future fallow management or conversion, significant amounts of biomass are removed from sites, those fallows that originally had more biomass may be left with lower soil nutrient capital for P, the most limiting of the nutrients. If, on the other hand, the biomass is left on site, the nutrient budget of those land use types that generated more original biomass would benefit from the augmentation of complementary nutrients, such as Mg and Ca.

Future Research

We have clearly identified two immediate priorities for research. First, the nutrient stocks in biomass and litter must be quantified, as well as those in the soil. Second, understanding the constraints and magnitudes of nutrient fluxes will be essential to evaluating the effects of fallow management types on soil fertility in both the short and the long term. Finally, to better address whether soil fertility and tree diversity are related, we need to sample both soils and the tree community at the same location. Then we would be able to compare diversity and nutrients among sites, instead of simply using means for a given land-use type.

Acknowledgments

We are grateful for the sponsorship of the Indonesian Institute of Sciences (LIPI), the Indonesian Research and Development Center for Biology, and Universitas Tanjungpura. We thank Herbarium Bogoriense for plant identifications and the Institut Pertanian Bogor for soil analyses. For their assistance and friendship, we also thank the people of Kembera. This research was funded through an NSEP Fellowship from the Academy for Educational Development, an NSF Graduate Fellowship, an NSF Dissertation Improvement Grant, and the WWF Garden Club of America Tropical Ecology Award to the senior author.

References

Ewel, J., M.J. Mazzarino, and C.W. Berish. 1991. Tropical Soil Fertility Changes under Monocultures and Successional Communities of Different Structure. *Ecological Applications* 1(3), 289–302.

Fernandes, D.N., and R.L. Sanford. 1995. Effects of Recent Land Use Practices on Soil Nutrients and Succession under Tropical Wet Forest in Costa Rica. *Conservation Biology* 9(4), 915–922.

Glover, N., and J. Beer. 1986. Nutrient Cycling in Two Traditional Central American Agroforestry Systems. *Agroforestry Systems* 4, 77–87.

Lawrence, D.C. 1996. Trade-offs Between Rubber Production and Maintenance of Diversity: The Structure of Rubber Gardens in West Kalimantan, Indonesia. *Agroforestry Systems* 34(1), 83–100.

———, and J.P. Mogea. 1996. A Preliminary Analysis of Tree Diversity in the Landscape under Shifting Cultivation North of Gunung Palung National Park. *Tropical Biodiversity* 3(3), 297–319.

———, M. Leighton, and D.R. Peart. 1995. Availability and Extraction of Forest Products in Managed and Primary Forest around a Dayak Village in West Kalimantan, Indonesia. *Conservation Biology* 9(1), 76–88.

———, D.R. Peart, and M. Leighton. 1998. The Impact of Shifting Cultivation on a Rainforest Landscape in West Kalimantan: Spatial and Temporal Dynamics. *Landscape Ecology* 13, 135–148.

———, V. Suma, and J. P. Mogea. 2005. Systematic Change in Species Composition with Repeated Shifting Cultivation: The Role of Soil Nutrients. *Ecological Applications* 15 (6): 1953-1967.

Mohan Kumar, B., S.J. George, and S. Chinnamani. 1994. Diversity, Structure, and Standing Stock of Wood in the Home Gardens of Kerala in Peninsular India. *Agroforestry Systems* 25, 243–262.

Michon, G., and H. de Foresta. 1991. Complex Agroforestry Systems and the Conservation of Biological Diversity: (I) Agroforests in Indonesia: The Link between Two Worlds. In: *In Harmony with Nature. Proceedings of an International Conference on the Conservation of Tropical Biodiversity, June 12–16, 1990, Kuala Lumpur, Malaysia,* edited by Y.S. Kheong and L.S.Win. Kuala Lumpur: The Malayan Nature Journal. (Golden Jubilee Issue), 457–473.

Padoch, C., and C.M. Peters. 1993. Managed Forest Gardens in West Kalimantan, Indonesia. In: *Perspectives on Biodiversity: Case Studies of Genetic Resource Conservation and Development,* edited by C.S. Potter, J.I. Cohen, and D. Janczewski. Washington, DC: AAAS Press, 167–176.

Rawls W.J. 1985. Estimating Soil Bulk Density From Particle Size Analysis and Organic Matter Content. *Soil Science* 135, 123-125.

Sanchez, P. 1976. *Properties and Management of Soils in the Tropics.* New York: John Wiley and Sons.

Chapter 43

Forest Management and Classification of Fallows by Bidayuh Farmers in West Kalimantan

Wil de Jong

The management of swidden fallow vegetation is an aspect of local natural resource management that is common to all tropical regions. This vegetation follows crop production, which may have lasted any time from one to several years. Its environmental function is to accumulate nutrient stocks and replenish the nutrients lost from the fields during cropping. Often, however, the fallow vegetation is also an important source of useful products and, although most of these are used for family consumption, they are also traded. There are many reports of active management to enhance returns from useful fallow species (for example Denevan and Padoch 1987; de Jong 1996, 1997; Jessup 1981; Smith et al. 1999). Under some circumstances, the useful species in these fallow fields become so valuable that both the fallow and the field's subsequent agricultural function are forgone, and the field is maintained as a forest garden. In some swidden farming communities, this development is anticipated by planting economically important tree species into the fields (de Jong 1996).

Although this conversion of agricultural fields into forest gardens may be the consequence of the high value of useful trees (de Jong 2001), it may also be an important strategy for preserving the value of agricultural land. This is especially the case where there is a threat of *Imperata cylindrica* invasion. Although *I. cylindrica* grasslands are sometimes managed for fodder production or thatching (Dove 1986; Potter 1996), this species is often regarded by swidden agriculturists as one of the main threats to their land. When *I. cylindrica* invades a field, its potential returns are decreased and the need for labor-intensive weeding is increased. *I. cylindrica* is estimated to cover 8.5 million hectares of land in Indonesia (Garrity et al. 1996). In most cases it dominates after forests have been slashed and secondary succession has been inhibited, usually by repeated burning. Although secondary forest will usually return to *Imperata* grassland when no burning occurs, there is an ever-present threat of escaping fire in landscapes scattered with fallowed swiddens or other smallholder land uses.

This chapter presents data on the way Bidayuh farmers in West Kalimantan, Indonesia, classify and manage their fallows and shows how these classifications are used to make choices between the use of land for crops or forestry production. Ultimately, it proposes that fallow classification, and the management practices that

Wil de Jong, Japan Center for Area Studies, National Museum of Ethnology, 10-1 Senri Expo Park, Suita, Osaka, 656-8511, Japan.

relate to this classification, are crucial to the integrated management of agriculture and forestry resources by swidden agriculturists in West Kalimantan. This may also be the case in other parts of the world. It means that these fallow classifications, and their related management practices, should be taken into account whenever proposing alternative agriculture, agroforestry, or forestry options to these farmers.

Considering recent problems with fires in Indonesia, local technologies that reduce the danger of *Imperata* spreading, and thereby reduce the danger of fires, are of wider than local interest.

Bidayuh Swidden Agriculture and Forest Management

This chapter reports a case study of the one Bidayuh group in the subdistrict of Noyan, in the province of West Kalimantan (de Jong 2002), one of four Indonesian provinces on the island of Borneo. With an area of 642,000 km^2, Borneo consists of two East Malaysian states, Sarawak and Sabah, and the Indonesian provinces of East, South, West, and Central Kalimantan. In 2000 the whole island had about 14.7 million people, 10.7 million of whom lived in the Kalimantan provinces. About one-third of the people on Borneo are indigenous Dayak (Brookfield et al. 1995). They live in the more remote interiors and most of them still rely for their daily staples on upland rice grown in swiddens that are cleared every year.

The Bidayuh include several different linguistic subgroups that originated from the Indonesian part of West Kalimantan. The swidden agriculture they practice begins with slashing and burning the vegetation on an area of land, using it for intensive cultivation for anything from one to several years, and then allowing it to lie fallow for a longer period (de Jong 2002). To be able to live in one area for a long time, a Bidayuh farmer must have sufficient land under various stages of fallow. This land holding is built up gradually from the time a newlywed couple becomes an independent farming household, and is large enough only when there are sufficient fields to grow crops and allow adequate periods of fallow.

An important feature of Bidayuh swidden agriculture is the emphasis on tree planting and the conservation of forest resources. As honorable as this may sound, it is often economically motivated. The landscape of a Bidayuh village is usually characterized not only by land under swidden and fallow, but also by many and varying patches of forest.

The Bidayuh have names describing six types of managed forest (de Jong 2002). These include communally held forest, or *hutan tutupan*, which is usually located close to villages and may be as large as 100 ha. Hutan tutupan functions chiefly as a reserve for timber for house construction, rattans, medicinal plants, fishing and hunting sites, and protection of water sources. A smaller version of the same type of forest, between one and three hectares, is *pulau rimba*, or forest islands. Such patches of forest are protected within a household's fields and fallows and, according to local customary law, are the private property of the household. These forests are kept as private reserves and yield the same products as hutan tutupan. In the same category of protected areas of mature natural forest are the *sompuat*, which are patches of forest surrounding honey trees, mostly *Koompassia excelsa* or *K. malaccensis*. The sompuat forests are kept intact to provide an appropriate habitat for the *Apis dorsata* honey bees that build their nests in the central tree (de Jong 2000).

More important for the present discussion are two types of managed forest that develop after land has been used for agriculture: *tembawang*, or fruit forest gardens, and rubber gardens. Individual farmers plant trees, either on sites previously used for agriculture or around the settlement area, to create tembawang (see color plates 59 and 60). Planting may continue for many years until eventually most of the area designated for tree planting is covered. Such tembawang are privately owned and are inherited by descendants of the people who create them. They are usually about one or two hectares in area. Since many farmers in a single village create individual tembawang on adjacent fields, over time these merge into a single stretch of contiguous forest. And because the village is frequently moved a short distance, the tembawang area often trails the path of the settlement as a long band of high closed forest. Tembawang forests have species compositions and structures that compare favorably with those of primary forests (de Jong 2002).

Rubber gardens are usually located adjacent to tembawang. A practice followed by the Bidayuh and other swidden agriculturists on Borneo (Dove 1993) is not to weed rubber until it is fully grown, after about 10 years. Only then are trails established to reach trees that have to be tapped. Very often additional fruit trees are planted in rubber gardens, eventually forming an extension of the tembawang area, rather than creating a spatially segregated area of managed forest. Both tembawang and rubber gardens are sometimes planted directly in natural forest, such as in pulau rimba private forest reserves. Elsewhere, it has been shown that the introduction of rubber has had a positive effect on the expansion of tembawang areas and, therefore, on the entire forest landscape in villages in West Kalimantan (de Jong 2001).

The final category of managed forests is swidden fallow vegetation, which is the main topic of discussion in this chapter.

Bidayuh Fallow Classification and Land Conservation

Both tembawang and mixed rubber gardens, the two most prominent managed forests that are partly planted by farmers, have a clear link with swidden agriculture and, in particular, with its fallow management component. Planting of both tembawang and rubber gardens usually begins in an agricultural field or in the subsequent swidden fallow. As reported elsewhere, even if fields have been planted to rubber or fruit trees, they are still considered to be fallow fields for a considerable time, and in some cases they may still be used for swidden cultivation (de Jong 2001). Final decisions about whether a field should be used as a swidden, a rubber garden, or a tembawang may not be made until several years after trees have been planted (de Jong 2002).

The data collected in Noyan suggest that the choice of which fields are planted to trees, be it rubber or the more common combination of rubber and fruit, depends on the type of fallow. In upland swidden agriculture, fields are commonly used for one year to grow rice and are then fallowed. Farmers recognize that the length of time a field has to be fallowed before it can be used again varies according to the land use history, the topography, and the soil quality. They also understand that the type of vegetation that develops after cropping differs according to the same influences, and that the fallow vegetation indicates the land-use potential of a site.

Bidayuh farmers distinguish three main types of fallow vegetation. When a field has been cleared from mature natural forest, it grows fallow vegetation that the Bidayuh call *jamie rintu*. If needed, these fields can be opened for cropping within only one or two years. *Boruat* is the fallow vegetation that usually develops in a field that has already been used two or three times for rice growing. As a rule, such a field can be used again after three to five years, but if a farmer has enough land, it is preferable to leave it under fallow until older forest develops. *Doda* is the fallow vegetation in a field that has been used to make a swidden four times or more. Dominant species in such a field are low shrubs, ferns and *Imperata cylindrica*. On land of lower agricultural quality, such as mountainous land, doda vegetation may develop after only two or three cycles. Ideally, a doda field should be left fallow long enough to develop vegetation similar to boruat vegetation. This might take 10 years or more, and once a field has been reopened as a new swidden, doda vegetation will soon develop again. Such fields have to be left fallow for extended periods.

Some changes that may occur in fallow vegetation—depending on the number of times a field has been used for agriculture—are indicated in Table 43-1. This shows that the longer a field is used, the further the composition of species shifts to those typical of more disturbed vegetation, like *Vitex pubescens,* which is a common invader of *Imperata cylindrica* grasslands (Utama et al. 1999).

It is common knowledge among Bidayuh farmers that, once a field has reached a stage where doda vegetation develops, there is imminent danger that it will develop a grass vegetation dominated by *Imperata cylindrica,* and such a grass vegetation could become virtually permanent if it is burned every few years. So, rather than continuing to reserve such land for future rice production, farmers reportedly consider it a better option to plant it with trees. In one of the Noyan study villages, between 1984 and 1993, a total of 510 agricultural fields were made into secondary forest. Of these, 130 came from jamie rintu vegetation, 247 from boruat, and 133 from doda vegetation. In that time, 71% of the fields cleared from doda vegetation were planted with trees, while only 13% of fields cleared from jamie rintu and 45% of fields cleared from boruat vegetation were converted to secondary forest. Most of the doda fields, in danger of degrading into *I. cylindrica* grassland, were effectively taken out of the rice cultivation cycle by planting them with trees. Interestingly, these included doda fields that were a large distance from villages, where rubber planting would not normally be expected because of the time needed to travel for future daily rubber tapping expeditions.

These data imply that the land-use choices exercised by the Bidayuh farmers, including choices for crop production or for tree planting, are local methods of preventing the degradation of repeatedly used land into *I. cylindrica* grasslands with little economic value.

Discussion and Conclusions

There are several reasons farmers may want to plant trees in agricultural fields: to create economic opportunities, to establish property rights, or, as this chapter has shown, to allow continued economic use of the fields. In many cases, these reasons arise simultaneously. It is most unlikely, for instance, that farmers will plant trees simply to establish property rights or to avoid losing agricultural fields to *Imperata* grasslands if there is no economic value attached to the trees. However, when the planting has a possible value, then subtle factors that define when and where trees are planted become extremely important, especially when the planting is also of interest to a broader community.

The data indicate that fallow classification and related management practices play a pivotal role in the agriculture and forest complex of Bidayuh swiddenists in West Kalimantan. This chapter shows that, among Bidayuh farmers, tree planting in the fallow is not an ad hoc decision but is carefully related to future land-use options. This adds new weight to the importance of trees as fallow vegetation. They have so far been recognized mainly for their role in the restoration of soil fertility (Smith et al. 1999) and for providing a large number of products for daily consumption or sale (Denevan and Padoch 1987; Padoch 1987).

Table 43-1. Importance Value of the Most Common Species in Fallows, According to the Number of Times the Vegetation Has Been Slashed

	Field Slashed			
Species	*1 Time*	*2 Times*	*3 Times*	*4 Times*
Macaranga gigantean	100	26		
Macaranga sp.	53	70	65	
Coelogyne sp.	53	13	42	
Macaranga beccariana	23	56	27	16
Lygodium sp.	19			
Ilex sp.		46	67	96
Vernonia arborea				31
Vitex pubescens				61
Psychotria sp.				16

Notes: Importance value is the sum of relative density, relative frequency, and relative dominance.

To these should be added a third important function: fallow fields are the initial phase of forestry production, a type of land use that is now widely recognized as common among swidden agriculturists. Besides being the intermediate phase between succeeding cycles of crop production, fallows are also a crucial transition between crop production and forestry production.

The data discussed in this chapter indicate an intrinsic knowledge on the part of Bidayuh farmers about the condition of fallows and the capacity and potential of the underlying soils. This knowledge is translated into a classification of fallow types that appears to determine future land use, whether it will be for crop production or forest production. This has important implications for efforts to improve or intensify production within swidden systems or to increase incomes for swidden agriculturists. It is likely that some of the past failures of social forestry and agroforestry production schemes may be attributed to a lack of awareness of the complex relationships between agricultural fields, fallows, and forest gardens. Options for improved fallow management might best be developed if they are based on these indigenous fallow classifications and their related management practices.

Although it is likely that the case described here is unique to the study location, it is also likely that swidden agriculturists elsewhere have developed similar methods of classifying the land-use potential of various fallow types. This information should be captured first before any outside attempts are made to improve fallow management, forestry production, or even crop production.

Acknowledgments

The results presented here stem from research conducted between 1992 and 1995 on Dayak forest management in the subdistrict of Noyan, West Kalimantan, Indonesia. The research was funded by the New York Botanical Garden, the Tropenbos Foundation, and the Rainforest Alliance, through their Kleinhans Fellowship. The research was sponsored by the Indonesian Academy of Sciences and Tanjungpura University in Pontianak. Carol Colfer kindly reviewed this chapter and made helpful comments.

References

Brookfield, H., L. Potter, and Y. Byron. 1995. *In Place of the Forest: Environmental and Socio-Economic Transformation in Borneo and Eastern Malay Peninsula*. Tokyo: United Nations University Press.

de Jong, W. 1996. Swidden Fallow Agroforestry in Amazonia: Diversity at Close Distance. Agroforestry Systems 34 (3) 277–290.

———. 1997. Developing Swidden Agriculture and the Threat of Biodiversity Loss. *Agriculture, Ecosystems & Environment* 62, 187–197.

———. 2000. Micro-differences in Local Resource Management: The Case of Honey in West Kalimantan, Indonesia. *Human Ecology* 28(4), 631–639.

———. 2001. When New Technologies meet Traditional Forest Management: The Impact of Rubber Cultivation on the Forest Landscape in Borneo. In: *Agricultural Technologies and Tropical Deforestation*, edited by A. Angelsen and D. Kaimowitz. UK: CABI Publishing, 367–381.

———. 2002. *Forest Products and Local Forest Management in West Kalimantan, Indonesia: Implications for Conservation and Development* (Kalimantan Series). Wageningen: Tropenbos.

Denevan, W.M., and C. Padoch. 1987. Swidden-Fallow Agroforestry in the Peruvian Amazon. *Advances in Economic Botany* 5.

Dove, M.R. 1986. The Practical Reason of Weeds in Indonesia: Peasant versus State Views of *Imperata* and *Chromolaena*. *Human Ecology* 14(2), 163–190.

———. 1993. Smallholder Rubber and Swidden Agriculture in Borneo: Sustainable Adaptation to the Ecology and Economy of the Tropical Forest. *Economic Botany* 17(2), 136–147.

Garrity, D. et al. 1996. The *Imperata* Grasslands of Tropical Asia: Area, Distribution, and Typology. *Agroforestry Systems* 36(1–3), 3–29.

Jessup, T. 1981. Why Do Apo Kayan Shifting Cultivators Move? *Borneo Research Council Bulletin* 13(1), 16–32.

King, V. 1993. *The Peoples of Borneo*. Oxford, UK: Blackwell.

Padoch, C. 1987. The Economic Importance and Marketing of Forest and Fallow Products in the Iquitos Region. In: *Swidden-Fallow Agroforestry in the Peruvian Amazon*, edited by W.M. Denevan and C. Padoch. Advances in Economic Botany 5, 74–89.

Potter, L.M. 1996. The Dynamics of Imperata: Historical Overview and Current Farmers' Perspectives, with Special Reference to South Kalimantan, Indonesia. *Agroforestry Systems* 36(1–3), 31–51.

Smith, J., P. van de Kop, K. Reategui, I. Lombardi, C. Sabogal, and A. Diaz. 1999. Dynamics of Secondary Forests in Slash-and-Burn Farming: Interactions among Land Use Types in the Peruvian Amazon. *Agriculture, Ecosystem and Environment* 76(2–3), 85–98.

Utama, R., D. Rantan, W. de Jong, and S. Budhi. 1999. Income Generation through Rehabilitation of *Imperata* Grasslands: Production of Vitex pubescens as a Source of Charcoal. In: *Domestication of Agroforestry Trees in Southern Asia*, edited by J.M. Roshetko and D.O. Evans. Forest, Farm and Community Tree Research Reports, Special Issue, 175–184.

Winzeler, R.L. 1993. The Bidayuh of Sarawak: An Historical Overview with Special Reference to Settlement Patterns. In: *Change and Development in Borneo: Selected Papers from the First Extraordinary Conference of the Borneo Research Council*, edited by Vinson H. Sutlive Jr. Williamsburg, VA: Borneo Research Council, 223–238.

Chapter 44

Indigenous Fallow Management on Yap Island

Marjorie V. Cushing Falanruw and Francis Ruegorong

Islands are useful places to consider agricultural intensification because their limited size makes agricultural production very sensitive to population growth. The people of Yap Island have addressed the need to provide food for a growing population through a diverse set of adaptations, including intensification of shifting agriculture and the development of site-stable tree garden and taro patch agroforestry systems (Falanruw 1985, 1990). Agricultural activities are carried out in a range of wild to domestic habitats. At the wild end of the spectrum, bulbils of *Dioscorea alata* or bits of *D. nummularia* yams are placed in slightly dibbled soil within the forest and left to grow for some years with little attention, then harvested with little disturbance of the forest. At the other end of the spectrum, a wide variety of crops such as *Cyrtosperma chamissonis* taro, breadfruit, bananas, *Inocarpus fagiferus* nuts, betel nut, and *Citrus* spp. are harvested from year to year from almost completely domesticated agroforests where there is no fallow. However, in spite of the efficacy of the agroforestry system, shifting systems are still used as well. This chapter focuses on indigenous strategies for managing the fallow of the shifting systems, where gardens are alternated with wild or semiwild vegetation.

Study Area and Methods

Yap is a close cluster of four high islands with a total area of about 92 km^2, lying within a broad fringing reef at 9°33'N latitude and 138°09'E longitude in the western Caroline Islands (see Figure 44-1). The average rainfall is about 300 cm per year, with irregular periods of drought of one to three months with rainfall below 6 cm per month. Soils are of both metamorphic and volcanic origin, and are relatively infertile. Estimates of the island's population prior to European contact range from 26,240 (Hunter-Anderson 1983) to 50,000 (Hunt et al. 1949), giving a population density of between 273 and 530 people per square kilometer. After contact, the population declined rapidly to a low of about 2,607 in 1947 (Lingenfelter 1975). However, since then it has been growing rapidly, and reached 6,919 by 1994 (Yap State Government 1996).

Descriptions of the agricultural and agroforestry systems used on Yap are based on participant observation over about 25 years. Data on vegetation profiles, species composition, and crop production were gathered from 102 randomly selected agricultural and agroforestry sites. Also, crop and crop production data were gathered

Margie V. Cushing Falanruw, Biologist, U.S. Forest Service & Yap Institute of Natural Science, P.O. Box 215, Yap F.M. 96943; Francis Ruegorong, Forester, Yap Forestry, P.O. Box 463, Yap, F.M. 96943.

from an analysis of records of produce sold at the Yap market between 1973 and 1992 (Falanruw 1995). The species composition of the fallow of an indigenous garden was compared with species found in an introduced gardening system through an inventory of species within transects through the two sites (Ruegorong 1995).

Results

Shifting Gardens

Basically, the development of shifting gardens involves opening the forest canopy and planting a mixed sequence of food crops. Sites used for such gardens are generally secondary forest. The understory is slashed and trees are ringbarked or girdled with fire during the dry season and left standing to serve as yam trellises. Crops are planted as the trees are defoliating, and the ground is generally covered with mulch and crops by the time the heavy rains arrive. A sequence of crops is harvested over a period of one to three years. At the end of the active gardening phase, the site may be converted into a site-stable agroforest or allowed to revert to wild vegetation.

Fallow Enhancement to Develop Site-Stable Agroforestry Systems

If a family settles in the area, the cropping site may be enriched with useful species and developed into a tree garden and taro patch agroforest. This generally involves developing raised areas for house platforms, trails, and tree crops, with soil excavated from low areas, which are developed into taro patches. This landscape enhancement allows the growing of trees in close proximity with root crops, as well as the maintenance of a forest of useful species producing a variety of products from year to year, with no fallow phase. If unwanted trees grow, or the agroforest becomes too shaded, trees are cut or girdled with fire and the area is transformed into an intermittent mixed garden for a time. Breadfruit, *Artocarpus altilis*, provides both fruit and wood and is especially amenable to this type of management. It readily sprouts from roots and, in this manner, migrates into suitable areas and contributes to the reformation of tree gardens.

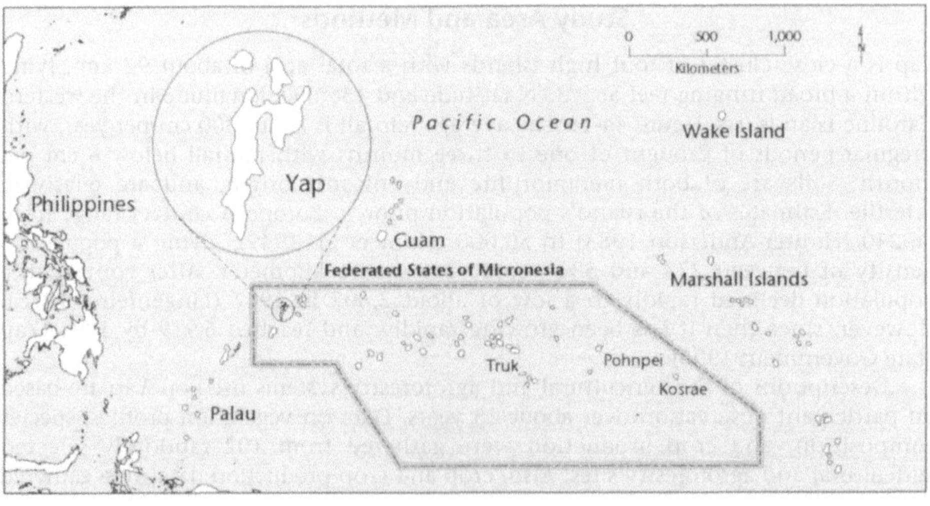

Figure 44-1. The Study Area. Yap is in the Western Pacific, Part of the Federated States of Micronesia

Intensification of Shifting Systems

In upland areas beyond villages, gardens are generally allowed to lie fallow after the cropping phase. However, under conditions of increasing population pressure and decreasing forest resources, a series of methods are adopted to maintain or increase production from these gardens. Where there is an intermittent tree canopy, trees are girdled and left standing to serve as trellises for yam vines. Species whose roots do not compete with crops are retained and managed to provide some shade. Other species deemed to be good for a garden, such as the fragrant giant fern (*Angiopteris evecta*) and useful wild shrubs and trees, may be cut back and allowed to grow again at the end of the active gardening phase, or allowed to grow on the periphery of gardens where they do not shade the crops. Ditches dug around the garden provide drainage and help to prevent the invasion of roots from outside.

A wide variety of crops is planted in suitable microsites. When one crop is harvested, another generally replaces it. For example, when yams are harvested, *Xanthosoma* taro is planted in the hole left by the yam tuber. Pumpkin vines and *Cucurbita* spp. are planted in the early phase of the garden. They serve to protect the soil and produce edible tips and fruit. In the latter phase of the garden, they are replaced by sweet potatoes, which are more tolerant of poor soil and also provide a ground cover, as well as another crop. Bananas also serve to shade out invasive weeds.

Most gardens are small and surrounded by forest, so there is a constant rain of seed carried by fruit bats and birds. Gardens are selectively weeded, and, in the later phases, desirable seedlings are allowed to grow. *Hibiscus tiliaceous*, a shrubby tree believed to contribute to soil fertility, may be planted or allowed to grow in shifting gardens (Zan and Hunter-Anderson 1988). Often, this occurs as a result of planting cuttings around the perimeter of ditched beds to retain piles of stems placed around the perimeter of the garden. The cuttings grow and often dominate the succeeding fallow. *H. tiliaceous* was present in all one- to four-year-old fallows surveyed in this study (Falanruw 1995). The planting of *H. tiliaceous* may have been more systematic in the past, when fibers from special varieties were needed for fish nets, "grass" skirts, and other uses. In recent years, Yapese women have begun to associate *Leucaena leucocephala* with more fertile soil, and some broadcast seed of this species in the garden to enhance the fallow.

Further intensification of shifting gardens includes the segregation of yam beds into more fertile and well-drained parts of the garden. Special attention is also paid to the preparation of yam planting sites, including the development of mounds utilizing fertile soil from ditches around the garden, and heavy mulching with off-site as well as on-site material. Community groups may also create more intensive yam gardens for special occasions, with pyramid-shaped trellises. In such cases, the whole bed is heavily mulched with materials from the site as well as special mulches, soil, and ash brought to the site. Individual yam vines are trained to grow around the pyramid trellis so that the end result is a pyramid of leaves in which shading is minimized.

A more intensive sequence is the conversion of a relatively infertile site into a banana orchard by developing ditched beds that are very heavily mulched with material brought to the site. Fertile soil from surrounding ditches is added periodically. Following one or more harvests of special varieties of bananas, special varieties of yams are grown in the banana mulch and the enriched soil of the ditched beds using pyramid trellises. In this system, the fallow is converted into a banana grove that enriches the site for subsequent cultivation, beginning with yams and continuing with other crops, until the biomass and fertility of the site are exhausted. At this point, with limited materials remaining for trellises and mulch, the ditched garden beds may be used to grow sweet potatoes, which are more tolerant of poor soils.

Among early visitors to Yap, Cantova (1722) mentioned the cultivation of sweet potatoes in gardens surrounded by ditches. In the preparation of sweet potato sites, the on-site vegetation is slashed or merely pressed down. Additional mulch, including sea grass (*Enhalus acoroides),* may be added. Then the layers of mulch are covered with soil from the ditches surrounding the garden bed. This method is similar to that used to grow sweet potatoes in the highlands of Papua New Guinea (Sillitoe 1996), except that the beds are rectangular rather than round.

Extensive Gardening

Yap's population decline was reversed after World War Two. The new American administration provided opportunities for wage employment and the men were less likely to be available for the heavier gardening chores, such as erecting the central poles for pyramid trellises. Women, having more children to care for, turned to more extensive cropping methods, such as the creation of "roll over" gardens. Because the population had dropped so dramatically, they were able to borrow adjacent land and reinvest planting material from a first garden into a new garden next door. In this way, they were able to harvest a wider range of crops, with second-year crops such as bananas harvested from the old garden, and first-year crops coming from the new garden. However, the system resulted in a larger contiguous open area and reduced the opportunity for seedlings of forest trees to germinate and grow. Today's gardeners also "high grade" an area by using a site for one year to produce the most prestigious crop, yams. Then they abandon the site and move to another. Another form of extensification is careless burning, resulting in wildfires that destroy a greater area of forest than is needed for a garden. Such methods more rapidly convert forests into areas of secondary vegetation.

A comparison of vegetation maps based on aerial photographs taken in 1946 (Johnson et al. 1960) and 1976 (Falanruw et al. 1987) shows a 3% increase in secondary vegetation and a matching decrease in upland forest. A more recent timber survey in 1983 (MacLean et al. 1988) indicates more extensive conversion of forest to secondary vegetation. This survey measured the volume of timber in a stratified random sample of plots. The plots were chosen by placing a grid over the 1976 vegetation map and selecting a random sample of plots from grid intersections that fell in areas mapped as upland forest, mangrove, or swamp forest. Field surveys of the 11 chosen plots revealed that 5 fell into small agricultural developments and one fell on a recent burn, leaving only 5 plots with trees big enough to measure. Two of these were in mangroves and three in forest. This suggested that about 66% of the upland forest present on 1976 photos had been converted to agricultural use or burned. Thirty-six percent of the timber volume measured on the plots was composed of fast-growing secondary tree species. It therefore appeared that much of the area demarcated as forest on the 1976 map either contained areas of secondary forest or had been converted to secondary forest by 1983.

In addition to the impact of more extensive methods of shifting cultivation and repeated burning, indigenous gardens are now more likely to be smothered by invasive species such as *Chromolaena odorata* and *Merremia peltata,* or made inaccessible by the noxious spiny *Mimosa invisa*. These species are shaded out in small gardens where the canopy rapidly reforms, but they can dominate sites where clearing has been too extensive.

The island's forest resources are being exhausted, and there is a constant need to intensify agricultural production. Most government and development agencies have invested in introduced systems of open canopy row cropping, but these have not been successful.

The effect of the many selective activities by which traditional gardeners manage both their gardens and their fallows is not as immediately impressive as the development of row crop gardens with bulldozers. But the end result may be more productive and more sustainable. In order to measure the effects of indigenous practices on fallowed land, the species composition of an indigenous garden in its second year of fallow was compared with the species that invaded a government demonstration plot abandoned for two years (Ruegorong 1995). Both gardens were made on the same soil type and were separated by a strip of forest. The demonstration garden was created by bulldozing an area about 10 times larger than most indigenous gardens and then planting rows of bananas and other crops. It was made up of numerous individually owned plots, whose establishment was subsidized by the government, but whose subsequent management was left to the owners. Unfortunately the plots were more open to weeds, and the task of weeding was more intensive than in traditional gardens. The motivation to work harder was limited by the lack of a market for the main crop of bananas,[1] and the yams grown in the demonstration plots were small and unimpressive. Almost all of the 61 demonstration gardens were abandoned within a year after the departure of the consultant who promoted them. The impact of the two systems on forest regeneration can be seen by comparing the species present in the fallows (Table 44-1). The utility of the species occurring in the fallows is compared in Table 44-2.

Table 44-1. Species Composition of a Fallowed Indigenous Garden Compared with an Abandoned Introduced Row Crop Garden

Vegetation	*Species Status*	*Indigenous Garden*	*Introduced Garden*
Trees	Native	14	7
	Introduced	0	1
Shrubs	Native	5	2
	Introduced	1	7
Ferns	Native	5	1
	Introduced	0	0
Grasses	Native	1	4
	Introduced	0	1
	Unidentified	2	1
Herbs	Native	2	2
	Introduced	3	4
	Unidentified	0	2
Vines	Native	3	2
	Introduced	4	1
Total species	Native	30	17
	Introduced	9	15
	Unidentified	2	3
Total		41	35

Source: Ruegorong (1995).

Table 44-2. Summary of Species Composition, by Status and Use, of a Fallowed Indigenous Garden Compared with an Abandoned Introduced Row Crop Garden

Location	*Percentage of Native and Introduced Species*			*Total Percentage of Useful Species*
	Native	*Introduced*	*Unidentified*	
Indigenous garden	73	22	5	66
Introduced garden	48.5	43	8.5	20

Source: Ruegorong (1995).

1 The ability of indigenous systems to produce bananas had earlier been demonstrated when a market opened and, without the aid of a consultant, the production of bananas for the market increased six-fold to meet demand (Falanruw 1995).

The indigenous garden had a greater number of native tree and fern species, while the demonstration garden had more species of grass and weedy shrubs, suggesting that the indigenous garden would develop into a forest fallow, while the demonstration garden was less likely to develop a tree canopy and would instead proliferate weedy species. In the indigenous garden, 73% of the species were native, many of them useful. All but four of the 22% of introduced species were feral crops that would produce small harvests as well as provide planting materials for subsequent gardens. In the demonstration garden, 48.5% of the species were native and the 43% of introduced species were mainly weeds, with only two species of feral crops persisting. The weeds included recently introduced invasive species not yet commonly found in indigenous gardens. In all, there were local uses, such as food, fiber, and medicines, for 66% of the species growing in the indigenous garden, while only 20% of the species in the introduced demonstration garden had any use (see Table 44-2).

Conclusions

The conversion of shifting cultivation sites into tree garden and taro patch agroforestry systems provides the benefits of continuous production from a variety of crops in close proximity to residences. At the same time, these systems provide the ecological services of forests and thereby contribute to ecological stability. These systems are well adapted to providing for household economies with surpluses for sale. This is demonstrated by the fact that most local produce sold between 1973 and 1992 was produced in such systems (Falanruw 1995).

The limitation of the tree garden and taro patch system is that it cannot be rapidly intensified to provide surplus production for special occasions. This function is provided by the shifting cultivation system, and may be a reason why it is maintained even though it is less sustainable. The shifting system allows for extensive management when the ratio of forests to people is high, but it can also be refined and intensified when the ratio of forests to population is lower. The simultaneous maintenance of both site-stable and shifting systems also provides for variety and complementary crop production, as the main crop of the shifting systems, *Dioscorea* yams, is counterpoint to the production of the main tree crop, breadfruit. Maintenance of both systems also provides resiliency resulting from a diversity of crops and production systems in an area of variable rainfall and occasional typhoons. A similar combination of site-stable and shifting systems is found elsewhere, as in the combination of rice paddy and shifting gardens in Indonesia.

While the tree garden and taro patch agroforestry systems appear to be sustainable, there is considerable opportunity for improving current practices of shifting agriculture. Although traditional techniques for intensification of these shifting systems exist, and are still remembered, current practices tend to be more extensive, and forests are being depleted. Traditional technologies need to be re-applied to make the shifting agriculture systems of Yap Island more sustainable. There would also be benefits from an integration of modern scientific technologies and indigenous practices.

Research Priorities

Priority should be given to the following areas for future research:

- Documentation of methods, measurement of production, and evaluation of sustainability of indigenous systems while traditional practitioners are still available;

- Research on the potential contributions of *Hibiscus tiliaceous* to soil fertility. While indigenous gardeners on Yap and other islands in the Pacific regarded this species as enhancing soil fertility, there appears to have been no studies of its efficacy;
- Research on the potential nutrient contributions of microbial enriched silts to garden beds in ditched-bed agricultural systems;
- Management of roots invading gardens by use of ditching technology;
- Suitability of machinery for creating ditches in indigenous gardens; and
- Management of, or substitutes for, species such as *Leucaena leucocephala,* which appears to enhance soil fertility, but which can be invasive.

Acknowledgments

For their assistance, I (M.V.C. Falanruw) wish to thank Dr. Randy Thaman, who encouraged my research on indigenous food production systems; Dr. Bill Clarke and Dr. Harley Manner for helpful literature and discussions; and my many Yapese field associates, especially Elvira Tinag, Teresita Darngun, Anna Gilpin, the late Fathlaan, and the women of Thol. We (both authors) also thank Martin Faimau and Melinda Pinnifen, who worked with us in the field and the laboratory, and Sylvia Stone for reviewing the manuscript and providing helpful suggestions.

References

Cantova, J.A. 1722. Letter to Rev. F.R. William Daubenton of the Same Society, Confessor of his Catholic Majesty. In: *Edifying and Curious Letters Written about the Foreign Missions*, 188–217.

Falanruw, M.V.C. 1985. People Pressure: Management of Limited Resources on Yap. In: *National Parks, Conservation and Development: The Role of Protected Areas in Sustaining Society*, edited by W. McNeely. Washington, DC: Smithsonian Institution Press.

———. 1990. The Food Production System of Yap Islands. In: *Tropical Home Gardens, Selected Papers from an International Workshop held at The Institute of Ecology, Padjadjuran University, Bandung, Indonesia, December 2–9, 1985*, edited by K. Landor and M. Brazil. Tokyo: United Nations University Press.

———. 1995. The Yapese Agriculture System. PhD dissertation, University of the South Pacific, Fiji.

———, C.D. Whitesell, T.C. Cole, C.D. MacLean, and A.H. Ambacher. 1987. *Vegetation Survey of Yap, Federated States of Micronesia*. USDA Forest Service Resource Bulletin PSW-21. Berkeley, CA: USDA.

Hunt, E.E., N. Kidder, and D.M. Schneider. 1949. *The Micronesians of Yap and their Depopulation. Report of the Peabody Museum expedition to the Yap Islands, 1947-48*. Cambridge MA: Peabody Museum.

Hunter-Anderson, R.L. 1983. *Yapese Settlement Patterns: An Ethnoarchaeological Approach*. Monograph 3. Agana, Guam: Pacific Studies Institute.

Johnson, C.G., R.J. Alvis, and R.L. Hetzler. 1960. *Military Geology of Yap Islands, Caroline Islands*. U.S. Army and U.S. Geological Survey report, Washington, DC.

Lingenfelter, S.G. 1975. *Yap: Political Leadership and Culture Change in an Island Society*. Honolulu: University Press of Hawaii.

MacLean, C.D., C.D. Whitesell, T.G. Cole, and E.M. Katharine. 1988. *Timber Resources of Kosrae, Pohnpei, Truk, and Yap, Federated States of Micronesia*. Resource Bulletin PSW-24. Albany, CA: Forest Service, U.S. Department of Agriculture.

Ruegorong, F. 1995. Survey of Soil Characteristics and Vegetation Dynamics of Sites Used in Two Gardening Methods on Yap. Internship report, Micronesian and American Samoa Student Internship Program, Pacific Southwest Research Station, USDA Forest Service, Institute of Pacific Islands Forestry and University of Hawaii at Hilo.

Sillitoe, P. 1996. *A Place against Time: Land and Environment in the Papua New Guinea Highlands*. Amsterdam: Harwood Academic Publishers.

Yap State Government. 1996. *Yap State Census Report 1994*. Yap: Office of Planning and Budget.

Zan, Y. and R.L. Hunter-Anderson. 1988. On the Origins of the Micronesian "Savannahs": An Anthropological Perspective. In: *Proceedings of the Third International Soil Management Workshop on the Management and Utilization of Acid Soils of Oceania, Palau, February 2–6, 1987*. Edited by J.L. Demeterio and B. DeGuzman. Agricultural Experiment Station, University of Guam, 18–27.

Chapter 45

The Damar Agroforests of Krui, Indonesia

Justice for Forest Farmers

Geneviève Michon, Hubert de Foresta, Ahmad Kusworo, and Patrice Levang

From the very beginning of Indonesian history, the forest has been the only available place for demographic, agricultural, economic, and geopolitical expansion. Consequently, the forest has become the means for pursuing wealth and power. In the past, conflicts over forest appropriation or control usually involved more or less equal groups of forest users and warriors. More recently, however, they have involved the state and its political or economic elite on one hand, and local communities on the other. The modern history of forests in Indonesia embodies a continuous process of land and resource appropriation by the state at the expense of indigenous forest people, through a fair amount of ideological imperialism, a convenient use of legal and technical instruments, and a touch of power abuse.

In spite of the important contribution made by wood industries to national development in Indonesia, the ecological, economic, and social damage arising from forest management can no longer be concealed. Cases of resource exhaustion, violation of the basic rights of local people by forestry projects, and local resistance are being publicized more and more. However, even though policymakers at the highest levels see the need for social and environmental justice in forest management, projects aiming to develop and conserve forests are constrained by laws and regulations that deny recognition of local people's practices and rights. The actual benefits of local use and management of forests are seldom analyzed in a nonpartisan way, and the value of customary systems of forest management are

Editor's Note: Conflicts between the state and local communities over forest and land resources threaten traditional property rights and are commonly cited as a major disincentive to more widespread adopition of improved fallow management. This chapter is from *People, Plants, and Justice* edited by Charles Zerner (Columbia University Press, 1999). It is reprinted here, with the permission of the publisher, as a case study that illustrates how even intensive IFM systems with long histories are often denied formal recognition with the state. The reader should bear in mind that this chapter was written during the pre-1998 period before the fall of Suharto.

Geneviève Michon, Senior Scientist, IRD Center in Montpellier, French Research Institute for Development (IRD), B.P. 64501, 34394 Montpellier Cedex 5, France; Hubert de Foresta, Senior Scientist, ENGREF Center in Montpellier, French Research Institute for Development (IRD), B.P. 44494, 34093 Montpellier Cedex 5, France; Ahmad Kusworo, Wildlife Conservation Society, Indonesia Program, Jl. Pangrango No. 8, Bogor 16151, Indonesia; Patrice Levang, Senior Scientist, IRD Center in Montpellier, French Research Institute for Development (IRD), B.P. 64501, 34394 Montpellier Cedex 5, France.

either underestimated or misunderstood. Legal mechanisms for acknowledging the rights of local people over forest lands and resources remain dramatically underdeveloped.[1]

A vital source of support for the formal recognition of local people's rights over forest resources may be found in successful alternative approaches to forest management developed by local communities. Among these approaches, the agroforestry system described in this chapter could be exemplary. In adapting traditional modes of forest extraction, through a logic of agricultural production, farmers have invented new and original agroforestry systems that reshape forest resources and structures. These complex structures, combining forest species and tree crops, can be encountered in many forest farming systems all over the archipelago. They have evolved from shifting cultivation and, in spite of their appearance, they are not natural forests. Rather, they are man-made agroforests, or "forest gardens." As examples of the capacity of local communities to manage the forest, they could allow farmers to affirm, maintain, or regain control over forest lands and resources.

Starting from the broad context of local communities and national power structures with competing ideologies, regulations, and forest management practices, this chapter will focus on the history of an agroforestry landscape in the south of Sumatra. It will describe the originality and efficiency of this particular strategy for controlling forest resources through agricultural development. We will then examine the external pressures that threaten the agroforest, and farmers' reactions to repeated violations of their basic rights, from avoidance of conflict to active resistance. The present ideological and legal conditions in Indonesia obviously do not favor the appropriation of natural resources by local people. Nevertheless, we will attempt to show how conflicts over the use of forest resources could be resolved through an acknowledgment of the benefits to be derived from integrating local forest management into agriculture.

Two Diverging Visions

Local conflicts pitting indigenous communities against the state or its economic elite cannot be analyzed without mentioning the differences in the way each protagonist conceives, appropriates, and manages forests. Everything differentiates these two main categories of actors in forest development, from fundamental concepts of existence through political ideologies and legal and institutional systems, to technical and financial capabilities for exploitation of resources and attitudes toward negotiation and implementation of power.

Forest is the dominant land cover of the Indonesian archipelago. Therefore, for the various ethnic groups presently constituting the Indonesian nation, as well as for the ruling authorities, the forest has been a major element in doctrines defining the origins of the universe and representations of the world. Based on these representations, various management systems have evolved.

Forest: The Center of the World

For many forest communities, most of the myths surrounding the origins of society involve marriages between humans and forest spirits. The forest is the central place for both the spiritual and economic aspects of life. It is considered as a resource with multiple uses.

Forest utilization and management by local communities has developed around two poles. The first includes subsistence resources: game and fish, plant foods and material, and the forest dynamic itself as an essential resource for shifting cultivation.

[1] While being constantly refined, these laws and regulations show only minor differences from those promulgated during the colonial period, at least in terms of underlying concepts (Peluso 1992). With the major changes that have shaped the modern history of forests in Indonesia, present laws and regulations are obviously outdated and are more and more unable to fulfill their overall objectives.

The second refers to extractive resources, harvested for trade: incense, spices, and animal products of the precolonial trade; resins and latexes in the 18th and 19th centuries; and rattan and timber since the beginning of the 20th century. Because of increasing space constraints over the past 50 years,[2] land has also emerged as an essential forest resource for local communities. Interest in land and commercial resources represents the only common ground between indigenous communities, the state, and private companies. It is also the major cause of their conflicts.

Systems devised by indigenous communities for the management of forest lands and resources usually combine "production," commonly through forest clearing, and "harvesting," or *in situ* management of economic resources. However, contrary to what is commonly acknowledged by both policymakers and researchers, production is not limited to agricultural products, and forest resources are not managed through harvesting alone. Indigenous forest management often involves production through active planting of forest crops. Many of these forest production systems deserve more attention, as they represent outstanding examples of sophisticated, multipurpose development of forest resources.

Forests: At the Border of Civilization

From the Javanese kingdoms of the 10th century to the Indonesian Republic, through three centuries of colonial administration, state authorities have used their political and coercive power to try to impose both their representation and methods of forest management on indigenous communities. The perceived need for absolute state control over the forest is dictated by two imperatives: the economic importance of appropriating natural resources and the political imperative of assimilating alien cultures.

The official perception of the forest in modern Indonesia has been shaped by both the Javanese civilization, which valued clearings more than wilderness and permanent rice fields more than shifting agriculture (Dove 1985; Peluso 1992), and by the Western concepts of Dutch colonial forestry. The underlying philosophy that seems to condition all forest policies and management regimes in modern Indonesia, as in any centralized political organization, is that forest is basically a domain that could easily escape state authority and control.

Because its very nature appears based on principles alien to an organized state and because it exists at the periphery of the "civilized" world, the forest—including forest lands, resources, and inhabitants—is perceived by the state as a fundamentally dissident area that must be strictly controlled. While forests contribute to the process of national development through their space and resources, the forest as an entity is not embedded in the philosophy of national development. State-sponsored agents of forest utilization are exclusively state or private companies, and forest dwellers are generally denied any right to participate in forest management. They are considered squatters on state lands and plunderers of state riches.[3]

The economic utilization of forests by the state is characterized by an intense reductionism. For the past 40 years, management has focused exclusively on mining the timber resources. Most areas logged over during the past two decades are no longer covered by forest. This chronic unsustainability of state-sponsored harvesting is another factor that differentiates national and customary systems of forest management.

---

[2] Population pressure is a well-recognized cause behind increased space constraints; other less recognized but major causes are land appropriation by the state (state forest land) and land designation by the state for "development" projects.

[3] For example, the official vision of the underlying causes of deforestation states that shifting cultivators are the main, if not the sole, agent of increased deforestation in the country. Another official dogma is that indigenous people are totally unable to manage forest resources sustainably.

Two Parallel Systems

There is also a juxtaposition of two different definitions and practices of forest appropriation at the policy, juridical, and institutional levels. Although these are not fundamentally incompatible, they are, in fact, mutually exclusive.

National Methods of Control

Forest policies in Indonesia are burdened by several ambiguities: they must simultaneously ensure that the utilization of forest lands and resources serves as a major instrument in national development, and that the same resources are protected for the present and the future. This double task is assigned to a single state body, the Ministry of Forestry, which has to reconcile profit building for the nation with social justice for the nationals. The strategies chosen to reach these political objectives have evolved along three main streams. The first is the separation of forest and agriculture; the second, segregation of forest domains into production, protection, and conversion areas; and the third is delegation of forest utilization and management to concessionaires. The latter releases the state from the practical aspects of forest development and allows it to concentrate on the acquisition of revenue. As a consequence, the first objective, development and profit, has been heavily emphasized, and the second, conservation and justice, has been relegated as an ideal issue for the future.

These policies are implemented through the Basic Forestry Law, issued in 1967.[4] It defines the extent of state forest lands, their functions, and the way they should be used. It is a list that comprises many qualities and land use categories, some of them unexpected,[5] but it denies any right to practice shifting cultivation. The law invests authority and jurisdiction over forest lands with the ministry, which delineates the forest domains, defines their functions, and allocates—or denies—rights over them. Acquisition of revenue occurs through a system of regulations and legal taxation.

The forestry legislation constitutes a rigid body that is not easily changed. Accommodating the needs of local people or developing community forestry agreements, while theoretically desirable and possible, still face many legal impediments. Another chronic disease in the forestry system is a dramatic weakness in law enforcement. Institutions in charge of implementing the forestry regulations are prone not only to political influence, but also to collusion with wealthy partners who do not respect the laws.[6]

Community Methods of Control

Control of and access to forest resources by local communities usually involve a system of rights and regulations, ranging from pure forms of common property to exclusive private rights over land and other resources. The definition of rights commonly varies according to the nature of the resource involved, or the type of ecosystem. It also varies from one community to another, and for any given community it may vary over time. Unlike the national legal system, flexibility, mobility, and adaptability are the main characteristics of customary systems.

Contrary to the common view, "tradition" in appropriation systems is not a rigid concept. Regulations and rights for control of resources, and access to them, constantly evolve to accommodate external or internal changes in resource availability, destination, value, extraction, or production techniques. Another important principle that separates customary and national legal systems is that in most customary systems, land use or property rights are usually accessed through

---

[4] The Forestry Law (1999) does not fundamentally modify this picture.

[5] Watershed protection, forest production, source of livelihood for the community inside and around the forests, flora and fauna protection, transmigration, agriculture, plantation and cattle raising.

[6] However, in relation to indigenous communities, law enforcement is easily achieved with the help of the armed forces.

investment of labor, and not through lobbying or payment. The main weakness of customary systems is their lack of efficiency in controlling the abuses of outsiders.

Management in Practice

Throughout its territory, including that portion designated as state forest land, the Indonesian constitution acknowledges customary rights over land and resources. However, this is most often ignored in the practice of forest management.

Designation, Delineation, and Mapping

The delineation of state forest lands in Indonesia has set aside 144 million hectares, representing 74% of the nation's total lands, for "forest production and protection."[7] Lands outside the forest domain are designated as "appropriated lands." This distinction between forest and appropriated domains implicitly designates forest lands as areas that cannot be legally appropriated by either individuals or groups. More than 95% of the land under customary control is included in state forest lands (Gillis 1988), so indigenous farmers cannot expect to ever receive legal title to their land.

The acknowledgment of customary rights stands "as long as these rights do not interfere with national interests." The wording is sufficiently ambiguous to allow for the widest, as well as the narrowest, interpretation. In the real world of modern Indonesia, customary rights inevitably recede when challenged by conservation projects, mining or logging concessions, transmigration schemes, industrial forest plantations, or any other kind of elite government-sponsored project or activity. The practical interpretation of forest policies not only denies local communities the right to forest lands and resources, but it also denies the very existence of their forest production systems. State forest lands and their boundaries have been mapped for the entire archipelago, and these maps serve as basic documents for development planning. In keeping with state dogma, forest lands are intrinsically uninhabited; indigenous land users do not appear on official documents. Rattan gardens (Weinstock 1983; Fried 1995), fruit forests (Michon and Bompard 1987a; Sardjono 1992; Momberg 1993; Padoch and Peters 1993; de Jong 1994), *damar* gardens (Michon and Bompard 1987b), rubber agroforests (Dove 1993; Gouyon et al. 1993), all the swidden and fallow systems of the outer islands, and sometimes even forest villages, simply do not exist. It is therefore easier for projects to ignore, or even erase, preexisting forest management systems.[8]

Concession Rights and the Criminalization of Indigenous Practices

The practical impact of the concession policy, aimed at controlling resource extraction and production through supervision and taxation, has many perverse effects on local communities. First, it clearly favors private companies at the expense of local communities. It has facilitated the emergence of conglomerates, led by politically connected timber tycoons. The technical, economic, financial, and political power of these tycoons gives them unrivaled advantages in negotiations for land and rights,[9] as well as in the enforcement of these rights in cases of conflicts with local communities. The second consequence of the concession policy is that it clearly criminalizes unlicensed harvesting practices, not only for timber, but also for

---

[7] Current state forest land official figures have recently been revised down to 128 million ha, representing almost 67% of the Indonesian territory (data accessed on December 6, 2004, on the Ministry of Forestry website: http://www.dephut.go.id/INFORMASI/STATISTIK).

[8] Farmers might try to ask for compensation for their lost resources, but in no way will the loss of most of the components of their indigenous agricultural systems, including swidden fields, fallowed fields, rattan gardens, or any kind of agroforestry field, be compensated.

[9] Not more than a dozen timber conglomerates control 56 million hectares of production forest, or about 30% of the total land surface of Indonesia (Gillis 1988).

the most profitable nontimber resources such as rattan, birds' nests, sandalwood, and eaglewood. Communities that have lived upon free extraction of commercial forest products for centuries are presently outlaws, and should purchase temporary rights to harvest these products. But purchasing rights is the privilege of the elite who, by their political or economic influence, dispossess local communities of their most valuable resources.

The Latest Trends in Forest Land Development

A switch from logging to estate development has increased dramatically since the beginning of the 1990s. This mainly involves timber estates, developed on "production forest" lands, and oil palm estates on "conversion forest" lands or empty "appropriated lands."

This carries important implications for local communities. The first is the increased threat of displacement. The traditional land of indigenous people is presently being leased to logging companies, then given to estate firms that not only have legal rights over it, but which also proceed to drastically transform the land by creating plantations. The local people are considered merely as a cheap labor force. The second consequence for local people is an unexpected "diversification" of conflict agents, as well as increased opportunities for making new allies. After dealing exclusively with forest authorities under the forest law, local farmers find themselves having to deal with regional administrators at provincial, district, and subdistrict levels. These administrators have full authority to grant conversion forest lands to private companies. They also have the right to ask for legal revisions of state forest land boundaries[10] and, significantly, they have the authority to defend and support local management systems and to acknowledge their customary foundations.

In this new phase of the land development game, the conflicts are not only between forest authorities and local farmers, but now also between "forest" and "agriculture."[11]

Local communities could benefit from this rebalancing of forces, as there appears to be an opportunity for them to reassert their right of access and control over the lands they have developed. But they are more likely, once again, to be the main losers in the power game that builds up around forest lands. Caught between the various facets of state authority and the prerogatives and greed of estate plantation lobbies, local communities might well lose their last opportunity to have their basic rights acknowledged.

Conflict, and the Farmers of Pesisir

Tracing these conflicts takes us to the densely populated province of Lampung, on the southern tip of Sumatra Island (see Figure 45-1). Westward, beyond the steep slopes of the Barisan Range, the Pesisir subdistricts appear as an estranged appendix to the province. Apparently left behind in the intensive agricultural development that occurred in Sumatra's central and eastern districts, Pesisir still retains large tracts of forest and Pesisir farmers still rely upon forest agriculture. But they are more than

---

10 In most of the outer island provinces, forest land borders are being renegotiated in concert with the regional government and forestry agencies. This negotiation process does not involve any representation from farmer communities.

11 Foresters definitively lose authority and jurisdiction over converted forest lands.

shifting cultivators. They are now truly tree farmers who, over generations, have developed highly original systems of forest cultivation (Figure 45-2).

The situation of Pesisir farmers perfectly epitomizes that of forest farmers in Indonesia. Although having occupied the land for more than five centuries, they still have no legal title to it, because most of the land they have developed and manage is designated as state forest. In spite of this—or, maybe, because of it—they have developed a perfectly balanced forest plantation based on the most valuable species for foresters, Dipterocarps. They have encountered more trouble than effective support from the forest administration at local, district, provincial, and national levels. Only recently have they begun to receive some kind of official recognition of their practices, but this has yet to be translated into formal terms. Their forest gardens cover more than 50,000 ha, yet they do not appear on the maps. After being subjected to repeated, although relatively light, abuses of power by forest authorities, they are presently confronting the insatiable greed of a private sector that is attempting to impose oil palm plantations on lands that are still designated as forest. The conflicts of interest involve various government and nongovernment agents who all impact on local farmers, either through active support, deliberate ignorance, or direct confrontation.

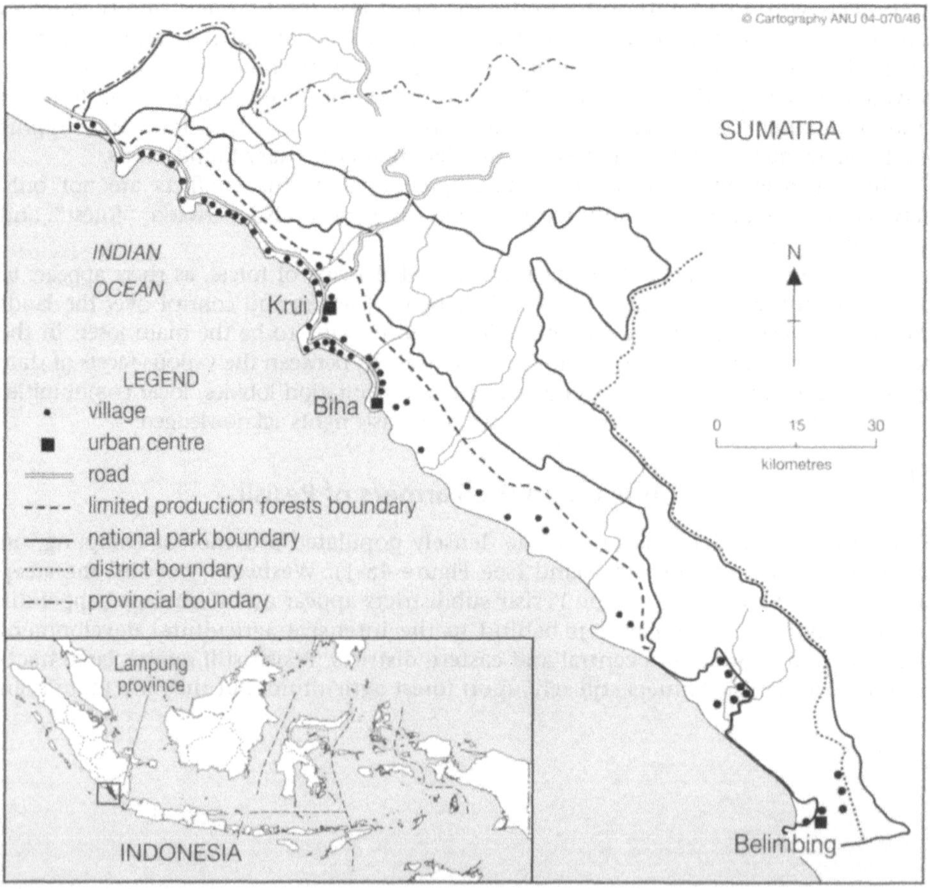

Figure 45-1. Location of the Study Area

Between Forest and Garden

Driving westward from the peneplain, along the Sumatra highway, one first encounters a mosaic of dry fields and pepper plantations, then passes through the Barisan Range, a succession of reddish hills extensively degraded by pioneer coffee growing. Then, one suddenly enters another country, a land of trees that stretches all along the quiet descent to the Indian Ocean. The human mark on this forest landscape is not immediately obvious, although upland paddy nestles in some clearings, and there are a few patches of fallow vegetation. Elsewhere, large trees dominate a venerable jungle. It covers about 100,000 ha, spreading over a long coastal plain that stretches 130 km from the provincial border in the north to Cape Cina, on the Sunda Straits, in the south. Eastward, the dense trees sweep up a steep hilly and mountainous area rising to more than 2,000 m asl. The forest extends into three administrative subdistricts.[12]

Wherever possible, permanent villages, with their associated irrigated rice fields, have been established along the coastal plain, but the topography and the relatively low quality of inland soils have limited opportunities for further permanent agricultural food production. The hills have long been the domain of a classic agroforestry rotation: mosaics of temporary rice fields and coffee plantations with secondary fallow vegetation. Over the past century or so, this pattern of forest conversion to agriculture has evolved into a complex system of forest redevelopment. Planting valuable fruit and resin-producing trees into their swiddens, Pesisir farmers have managed to create a new forest landscape tailored entirely to their needs. This man-made forest, although forming an almost continuous massif, is made up of a succession of individually evolved gardens that the farmers have named damar gardens,[13] after the dominant tree species, *Shorea javanica*. A native to Sumatra, *S. javanica* easily grows to 40 or 45 m in height (see Figure 45-3) (Torquebiau 1984; Michon and Bompard 1987b; Michon and Jafarsidik 1989).

Damar gardens in the Pesisir are totally original examples of sustainable and profitable management of forest resources, entirely conceived and managed by local populations. Their originality lies in the ecological mastery of the main economic resource, the forest tree, not through conventional domestication, which usually involves modification of plant characteristics to achieve adaptation to a cultivated ecosystem, but through an almost total reconstruction of the original forest ecosystem on agricultural lands. Success is due to the proven reproducibility of the system over the long term, as well as to its economic benefits and social basis. Today, more than 80% of the damar resins produced in Indonesia are provided not by natural forests, but by the Pesisir damar gardens. Among the 70 villages scattered along the coast, only 13 do not own damar gardens.

Damar gardens can be categorized as a forest, and indeed, biologically they constitute a forest in their own right—a complex community of plants and animals and a balanced ensemble of biological processes reproducible in the long term through its own dynamics. Casual observers have often mistaken the gardens for natural forest. But they have definitely been established not as a forest, but as an agricultural production unit on an agricultural territory, and are managed mainly as an agricultural enterprise. Occupying the vague interface between the conventional perceptions of agriculture and forest that have been promoted by modern science, the damar gardens fully deserve the name "agroforests" (Michon 1985; de Foresta and Michon 1993).

---

12 Population density ranges from 100 persons/km^2 in the central district, where available space for agriculture has been saturated for more than 30 years, to less than 20 persons/km^2 in the south, where land can still be easily appropriated. The main communication links with regional centers used to be by sea, and several small harbors are scattered along the coast. Recently, the old road to the east, through the central Barisan Range and the national park, was restored, a provincial road to the north was completed, and another is being developed to the south.

13 *Damar* is a generic term used in Indonesia to designate resins produced by trees of the Dipterocarp family.

Figure 45-2. General View of a Damar Garden in the Pesisir Krui

A Tree Plantation with a Rich Forest Ecosystem

While damar trees are clearly dominant in mature gardens, representing about 65% of the tree community and constituting the major canopy ensemble (see Figure 45-4), damar gardens are not simple, homogeneous plantations. They exhibit diversity and heterogeneity typical of any natural forest ecosystem, with a high botanical richness and a multilayered vertical structure[14] as well as specific patterns of forest dynamics (Table 45-1).

Plant inventories in mature damar agroforests have recorded around 40 common tree species, and several more tens of associated species, either large trees, treelets and shrubs, lianas, herbs, or epiphytes. Important economic species commonly associated with damar are mainly fruit trees, which represent 20% to 25% of the tree community. In the canopy, durian and the legume tree *Parkia speciosa* associate with the damar trees.

In the subcanopy ensembles, langsat is the major species and, to a lesser extent, mangosteen, rambutan, jackfruit, and palms, such as the sugar palm (*Arenga pinnata*) or the betel palm (*Areca catechu*). The subcanopy ensembles also include several water apple species (*Eugenia* spp.), as well as trees producing spices and flavorings, such as *Garcinia* spp., the fruits of which are used as acid additives in curries, and *Eugenia polyantha,* the local laurel tree. The final component, comprising 10% to 15% of the tree community, consists of wild trees of different sizes and types that have established naturally and have been protected by farmers, either because they have no adverse effects on the other trees, or because of advantageous end uses. These include bamboos and valuable timber species, such as Apocynaceae and Lauraceae.

---

[14] Tree stands in damar gardens show a mean density of 245 trees per hectare, calculated upon a record of all trees over 20 cm in diameter on eight 4,000 m^2 randomly selected plots. These trees have a mean basal area of 33 m^2/ha. These quite high figures, associated with a well-balanced diameter class distribution, are really close to structural patterns found in natural forests (Michon 1985; Wijayanto 1993).

Nontree species characteristic of a forest ecosystem, including Zingiberaceae, Rubiaceae, Araceae, and Urticaceae, have colonized the undergrowth of gardens, where they contribute to the maintenance of a favorable environment for development of the seedlings of the upper-layer trees.

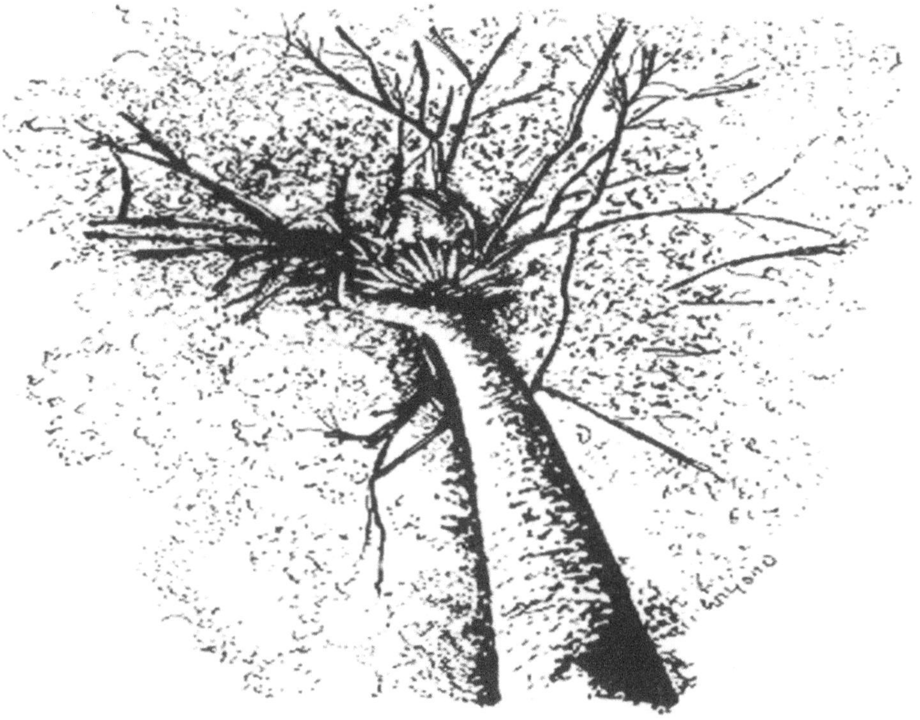

Figure 45-3. The Damar Tree (*Shorea javanica*)

Table 45-1. Structural Characteristics of Damar Agroforest and Primary Forest in Pesisir Krui

Characteristics	*Plot 1*	*Plot 2*	*Plot 3*	*Plot 4*	*Plot 5*
Area (m^2)	600	1,000	400	1,000	2,000
Damar trees over 10 cm dbh (tree/ha)					
Young unproductive trees	200	140	200	150	n.a.
Mature and old productive trees	200	140	250 + 50	190 + 70	n.a.
Total stand density (damar)	400	280	500	410	n.a.
Total trees (all species) over 10 cm dbh (tree/ha)	680	300	650	560	500
Vertical structure					
Number of canopy ensembles	2	3	3	n.a.	4
Distribution of crown coverage					
Emergent trees	0%	0%	0%	n.a.	25%
Upper-canopy trees	130%	88%	114%	n.a.	60%
Lower-canopy trees	34%	5%	8%	n.a.	33%
Treelets		12%	12%	n.a.	13%
Immature trees (trees of the future)	41%	38%	33%	n.a.	45%
Total	205%	133%	167%	n.a.	176%

Notes: Plots 1, 2, and 3: damar gardens in Penengahan, Central Pesisir (Michon 1985); Plot 4: damar garden in Pahmungan, Central Pesisir (Torquebiau 1984); Plot 5: primary forest (Laumonier 1981).

Management of mature gardens is centered on the harvest of resin (see Figure 45-5) and fruits. Labor allocated to routine garden maintenance is mingled with labor devoted to resin harvest, and the tempo of harvests is determined by labor requirements for wet rice cultivation. Work in the damar gardens is postponed during the time of rice harvest or rice field preparation, so that tree gardening never competes for labor with subsistence agriculture. Once established, the damar garden evolves with minimal human inputs. The silvicultural process is not conceived as a mass treatment applied to a homogeneous, even-aged population of trees, as in conventional forest plantations, but aims at maintaining a system which produces and reproduces without disruption either in structural or functional patterns. Natural processes are given the major role in the evolution and shaping of the cultivated ecosystem. Continuity is ensured through a balanced combination of natural dynamic processes prevailing within the tree population[15] and appropriate management of individual trees of economic species. Since the natural decay of planted trees is predictable, farmers anticipate and plan their replacement. The main task of the gardener is to regularly introduce young trees into the garden plot to maintain an uneven-aged pool of replacement trees. In a well-managed garden, the size of the replacement pool ensures the sustainability of the productive stand.

Figure 45-4. Structure of an Old-Growth Damar Agroforest

Notes: 34 = *Shorea javanica* [damar trees]; 13 = *Durio zibethinus* [durian]; 29 = *Parkia speciosa* [petai]; 17 = *Garcinia mangostana* [mangosteen]; 20 = *Lansium domesticum* [langsat]; 26 = *Nephelium lappaceum* [rambutan].

The Economic and Social Values of Damar Gardens

Damar gardens have been established by farmers for commercial production, and their economic management is basically closer to that of an agricultural smallholder plantation than to that of a forest. However, some functions of the gardens still relate to the former harvested forests that complemented rice swiddens in ancient production systems.

15 Pollination, fructification and production, seed dispersion and germination, seedling and sapling development, gap colonization, water and nutrient cycling.

Damar trees provide the main source of cash income for households in the study area (Figure 45-6), and damar collection is far more lucrative than other agricultural activities (Mary 1987; Levang and Wiyono 1993). Resin is harvested on a regular basis; individual trees are usually tapped once or twice a month. A single villager can harvest an average of 20 kg of resin a day. In the central subdistrict villages, average harvests are between 70 and 100 kg per family per month.

In 1993, resin produced by Pesisir farmers was estimated to be worth Rp6.5 billion, or 1993US$3.25 million. When the additional value generated by trade and related wages was added to this, the regional gross value of the damar trade amounted to more than Rp15 billion, or 1993US$7.5 million. Of the 57 regional villages involved in resin production and processing, 11 are "less involved," and damar activity generates 45% of the average household income. In the remaining 46 "damar villages," it generates between 70% and 100% of family incomes. This regular cash income is usually allocated to day-to-day expenses such as purchase of additional foods or weekly costs for children's schooling. Five days of work in damar gardens is usually enough to ensure a month's subsistence for the whole family (Levang and Wiyono 1989, 1993). For those who do not own permanent rice fields, the damar income also allows for the purchase of some rice and, thus, it complements dry rice culture where it still exists. However, the damar income cannot be regarded as a significant amount and it is insufficient to accumulate savings.

Figure 45-5. Damar Agroforest Resin Harvesting

Note: The resin flows at the cambium level and, while collecting it, the tapper stimulates further production by cutting the edges of the wound.

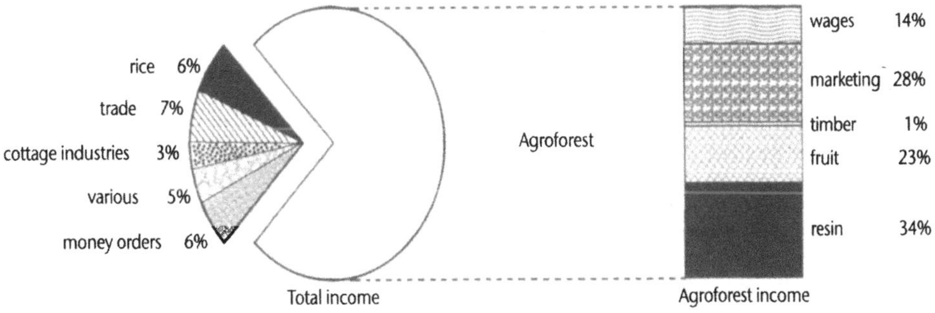

Figure 45-6. Household Cash Income in a Damar-based Village

Source: Levang and Wiyono (1993)

A series of associated activities is generated by the damar gardens. They include harvesting, transportation from the field to the village, stocking, sorting, and transportation to wholesalers in Krui. Harvest, transportation and sorting are carried out either by the grower himself, by members of his family, or by specialized agents who are paid employees. Independent entrepreneurs ensure resin stocking in the village (Table 45-2). These activities raise significant additional income for the village[16] and allow those who do not own a damar garden to participate in the benefits of damar production (Bourgeois 1984; Mary 1987; Levang and Wiyono 1993; Nadapdap et al. 1995).

As in many other parts of Sumatra, the contribution made by fruit to household economies has been increasing in recent years because of the growing importance of urban markets and recent major improvements to road networks. Over recent productive years, marketing of the major commercial fruits, durian and langsat, has doubled the global agroforest income (Levang and Wiyono 1993; Bouamrane 1996). However, due to high irregularities in fruiting seasons,[17] income from fruits cannot be included in daily household budget planning. It is still used mainly for exceptional or "luxury" expenses.[18]

Damar gardens constitute one of the most profitable smallholder production systems in Sumatra (Table 45-3). They ensure reasonable standards of living, including high school attendance by children, which is a top priority in most villages of the area. In addition, the gardens can be managed or used as a safety asset, when the need arises. A garden, or several selected trees, can be pawned through special agreements called *gadai*[19] (Mary 1987; Lubis 1996). These arrangements allow families to overcome difficult periods without resorting to selling trees or land, which is considered as one of the worst things that might happen to a family,[20] for damar gardens also represent a patrimony, an inheritance from male ancestry. The fruits of a farmer's labor in a damar garden are invested for a distant term, mainly to benefit future generations. Therefore, the damar garden constitutes an inalienable lineage property (Mary 1987; Nadapdap et al. 1995). In the very particular social and institutional context of the Pesisir, where families are defined mainly by their land assets, this notion of lineage patrimony defines the agroforest not only as the source of living for a household, but also as the land foundation of a lineage.

Damar Gardens as a Useful Forest

In the economies of forest villages, damar gardens also fulfill a role equivalent to that of natural forests. Wild resources associated with damar trees support a whole range of gathering activities that are more typically linked with natural forest ecosystems, including hunting, fishing, and harvesting of plant products such as noncommercial fruits, vegetables, spices, and firewood, as well as other plant material and timber for housing.[21] These provide important additional resources for households.

[16] The sale value of the resin itself represents less than half (44.5%) of the total income provided by resin production in villages, with related activities accounting for the largest share (Mary 1987; Levang 1992).

[17] Due to adverse climatic conditions, there was no fruit season in the area from 1992 to 1994.

[18] House repair, purchase of furniture, chainsaw, or satellite dish, wedding ceremonies or any festive activity.

[19] Any villager with sufficient funds can become a "pawnbroker," and may provide loans of several hundred thousand rupiah for one garden for an undetermined period (at least one year). Tree production serves as yearly interest for the creditor, who can use the garden for his own convenience during the entire loan period, except for selling or transforming it. The agreement ends as soon as the garden's owner refunds all the money to the creditor or when he claims the profits made by the creditor are sufficient.

[20] Bank credit is still uncommon and unreliable in villages.

[21] Thatching material from palm and *Garcinia* leaves, rattan and other liana, fibers from tree bark, bamboo. In terms of their timber production, damar and fruit trees appear as important as wild species.

As with any natural forest, a damar garden also represents a source of potentially marketable commodities on a larger scale, including timber, rattan, and medicinal and insecticidal plants that can be harvested for sale whenever needed, or if market conditions are considered favorable.[22] As new markets develop, some of the traditional subsistence products have actually emerged as new commodities. Timber, for instance, has become a major "new" commodity that might even revolutionize the management of damar gardens[23] (de Foresta and Michon 1992, 1994a; Michon, de Foresta et al. 1995a; Petit and de Foresta 1996).

Damar gardens have taken over the essential role traditionally played by natural forests in local household economies. They are places open to subsistence gathering and are used to fulfill the family's immediate needs. The forest function also appears in some of the egalitarian social attributes of the local lifestyle, such as product exchanges, sharing and donations,[24] and free harvesting rights.[25] These create important networks of reciprocity that act as a counterpart to trade and commercial activities, and help maintain a social balance between well-endowed people and those without resources.

Table 45-2. Main Characteristics of the Damar Resin Trade Chain Inside Indonesia

	*Relative Profit Margins of Each Agent**		*Activities***					
Agents	*1st Trade Chain*	*2nd Trade Chain*	*Harvest*	*Stocking*	*Drying*	*Sorting*	*Transport*	*Processing*
Damar grower	70%	70%	xxxx	x	x	0	xxxx	0
Village traders	3%	6%	0	xxxx	xx	xx	xx	0
Krui dealers	1%		0	xxxx	xx	xx	xxxx	0
Direct traders		6%	0	xxxx	xx	xxxx	xxxx	0
Krui wholesalers	13%		0	xx	xx	xxxx	xxxx	Xx
Expenses	10%	15%						
Losses	3%	3%						

Notes: *Expressed as a percentage of the resin price in Tanjung Karang or Jakarta. ** xxxx = principal activity; xx = often; x = occasionally; 0 = never.

Source: Bourgeois (1984).

In replacing natural forests by damar agroforests, the villagers' aim has been to amplify commercial strategies linked to the forest ecosystem. This is a dynamic widespread over Indonesia, where slash-and-burn practices are not usually targeted to staple food production alone, but rather to the establishment of income-generating

22 The most valuable, but also the least predictable, extractive commodity in the damar gardens is rattan, the harvest of which is subject to the profit or failure dynamics of local buyers. Rattan fruits once appeared as a valuable product, as well as the canes. This economic unpredictability is the main impediment to the development of rattan harvesting as a more specialized garden activity.

23 As other sources of timber in the area vanish, the economic potential of damar timber is increasing. However, timber harvesting and marketing regulations, taxes, bribes, and police harassment pose major impediments to the development of timber as an integrated product from damar gardens.

24 Poor people and children may harvest resin fallen on the ground, in the latter case to pay their weekly school expenses. They are even allowed to collect resin from the lowest tapping holes. Valuable fruits are traditionally shared by the family, and in season, distant relatives may come and join for a durian party or leave with a basketful of langsat, which is considered as a valuable practice for maintaining family cohesion.

25 Useful garden products such as firewood, sugar palm sap, small fruits, and medicinal plants can be collected in privately owned gardens by whoever needs and asks for them.

agroecosystems (Pelzer 1978; Scholz 1982; Dove 1983; Weinstock 1989). One of the main originalities of the land conversion process in the study area is that, by converting natural forests into commercial plantations, Pesisir farmers have also managed to restore a whole range of economic products and functions originally derived from the forest. Forest conversion did not entail a radical process of biological simplification. Rather, it restored plant and animal diversity through cultivated, preserved, and spontaneously established species. Specialization did not entail economic reductionism. Instead, it restored the whole range of economic choices present in a natural, untransformed ecosystem. From the perspective of an integrated conservation and development program, this preservation of economic diversity appears as important as that of biodiversity.

Species Domestication or Forest Reconstruction?

The damar story in the Pesisir is a highly original example of spontaneous appropriation of a forest resource, the damar tree, by local farmer communities. It was achieved as the wild resource itself was vanishing (Michon, de Foresta et al. 1995a,b, 1996). If human history is rich in examples of the appropriation and cultivation of natural resources to achieve domestication, the originality of the damar example is that, while cultivating this particular forest resource, villagers achieved the total restoration of a forest in the middle of agricultural lands (Michon and de Foresta 1996). Biologists may argue that the damar agroforest is far from a natural, pristine tropical forest and, although close to it, damar gardens cannot totally replace the natural forest ecosystem. But the gardens represent a practically integral resource which is much more significant to local people than a natural forest that increasingly eludes their control. For longstanding and external institutional reasons, the conservation of natural forests is not within their power. Therefore, as well as the technical success involved in establishing and reproducing a vast Dipterocarp plantation over more than a century,[26] it is the appropriation of the total forest resource through an agricultural strategy, and its incorporation into farmers' lands, which are worth analyzing.

[26] This, in itself, is quite remarkable, as foresters have often been unable to achieve success in the industrial growing of Dipterocarps.

Table 45-3. Average Production per Hectare per Year in a Mature Damar Agroforest, Pahmungan Village, Central Pesisir Subdistrict, 1995

Species	*Density Trees/ha > 20 cm dbh*	*Production*	*Traded*	*Family Labor*	*Yearly Income (data: 1995)* Rp	US$
Shorea javanica (resin)	145	1,550 kg	1,500 kg	50	1,500,000	682
Durio zibethinus[a]	25	625 fruits	600 fruits	10	420,000	191
Lansium domesticum[b]	15	600 kg	500 kg	10	250,000	114
Parkia speciosa	8	1,200 pods	1,000 pods	10	100,000	45
Baccaurea racemosa[b]	7	200 kg	50 kg	2	10,000	5
Artocarpus integer – "cempedak"[a]	6	100 fruits	50 fruits	2	50,000	23
Other fruit trees (6 spp.)[b]	10	200 kg	50 kg	3	50,000	23
Timber (all species are used)	250	5 m^3	2.5 m^3	0#	50,000	23
Total labor (man-days)				87		
Average yearly income					2,410,000	1,106
Minimum income (no fruiting season)					1,650,000	750
Maximum income (fruit season)					3,570,000	1,625

Notes: [a]Production every two years; [b]production every three years; # no family labor involved in timber harvesting.
Source: de Foresta and Michon (1997).

The History of Resin Harvesting and Production

Resins, which are sticky plant exudates found in various families of forest trees,[27] are among the oldest traded items from natural forests in Southeast Asia. They were traded over short distances between Southeast Asian islands as far back as 3000 BC, and were probably included in the first long-distance exchanges that developed with China in the 3rd to the 5th centuries (Dunn 1975). The word "damar" appears in the lists of items traded to China from Southeast Asia in the 10th century (Gianno 1981, citing Ma Huan 1451). The first exports to Europe began in 1829 and to America in 1832 (van der Koppel 1932). Locally, damar was used for lighting purposes and for

27 Several types of resin can be harvested from natural forests in Indonesia. Turpentine (pine resin) and copal (*Agathis* resin) used to be known as the major economic resins traded from Indonesia before World War II. Some 115 Dipterocarp species, distributed in 7 of the 10 genera of the family, produce damar (Foxworthy 1922). Due to the dominance of Dipterocarps in lowland forests of the region, damars form the commonest resin type in western Indonesia but are usually considered of lower quality than copal and turpentine. In fact, 2 categories are commonly distinguished which largely differ in quality. *Damar batu* is low-quality damar of dark color, resulting from spontaneous outflows caused by occasional injuries. Large pieces fallen from the bark can be collected by digging the ground around the trees. This usually provides huge quantities from old trees. *Damar mata kucing* is clear to yellow damar of high quality (comparable to copal), obtained by making incisions in the bark. About 40 species, from the genera *Shorea* and *Hopea*, produce damar mata kucing, among which the best are *Shorea javanica* and *Hopea dryobalanoides*.

caulking boats. It was traditionally traded for use in incense, dyes, adhesives, and medicines (Burkill 1935). Around the middle of the 19th century it acquired a new commercial value with the development of industrial varnish and paint factories, and collection intensified for the export trade to Europe, the United States, Japan, and Hong Kong. After 1945, however, exports dropped sharply due to competition from petrochemical resins, which are preferred for most industrial uses.

Indonesia is now the only damar-producing country in the world. Major end users are low-quality paint factories in Indonesia that use the poorest grades. The best quality damar is reserved for export, mainly to Singapore, where it is sorted, processed, and reexported as incense or as a base for paint, ink, or varnish manufacturers in industrial countries. Other users include handmade batik industries and manufacturers of low-quality incense (Bourgeois 1984; Dupain 1994; Anonymous 1995).

In the boom period of intensive harvesting for export, from the beginning of the 20th century until World War Two, the main damar-producing areas were the natural forests of southern and western Sumatra, as well as West Kalimantan (van der Koppel 1932). Today, West Kalimantan and South Sumatra still produce some damar, but the main producing area is Lampung, the southernmost province of Sumatra, which includes the Pesisir.

From at least the 18th century onward, the agricultural economy in the Pesisir involved subsistence farming, with swidden rice production dominating until the end of the 19th century. Then, market-oriented activities took over, with production of copra along the coast, and pepper, coffee, and cloves[28] grown on the hills. As well, there was commercial gathering of forest products, mainly gutta percha, wild rubbers, rattan, birds' nests, and damar. Chinese coastal traders carried all products northward to Bengkulu or southward to Tanjung Karang, Batavia (Jakarta), and Singapore.

Damar production is reported to have been a major activity in the whole area. As early as 1783, the British historian Marsden mentioned a type of resin "yielded by a tree growing in Lampung called Kruyen (one cannot but think about Krui) the wood of which is white and porous... and which differs from the common sort, or dammar batu, in being soft and whitish.... It is much in estimation for [lining] the bottoms of vessels.... To procure it, an incision is made in the tree" (Marsden 1783). Harbor accounts in Teluk Betung from the middle of the 19th century stress that the trade of damar mata kucing raised considerable profit in Lampung, with 285 tonnes being exported in 1843 (Sevin 1989). A map drawn by the Belgian geographer Collet in 1925 mentions damar as one of the three main exports of Krui (Collet 1925). Rappard, a Dutch forester who visited the area in 1936, mentions that damar ranked third in the agricultural exports of Krui, after coffee and copra, but before pepper. Total production in 1936, for the Krui area alone, was more than 200 tonnes (Rappard 1937). In villages, there is still a vivid memory of the importance of wild resins, and people can show old wild damar trees that were protected in the *ladang* (swidden), while the forest itself disappeared.

A Heritage of the Past

When and why did the cultivation of damar trees begin? Farmers say it is a "tradition" inherited from their "ancestors." However, the tradition is certainly no older than a century or so and probably began through a combination of internal factors and external influences. Some villagers trace its origin to the beginning of the 20th century, when two respected Haj visited Singapore. They were convinced of the damar market's bright prospects and returned to establish plantations. Other informants in the ancient Pugung area assert that, about six generations ago, that is, at least 120 years ago, or in the 1870s, villagers came from the central subdistrict to

28 Pepper cultivation probably developed as early as the 17th century, but lost its importance in the 1920s; coffee was developed after the beginning of the 19th century and is still intensively planted; cloves were introduced in 1930, fully developed in the 1970s, but almost totally disappeared due to a serious disease which hit Sumatra after 1975.

ask for damar seedlings, and these were then taken from Batu Bulan forest where natural damar trees were famous (Dupain 1994). Villagers commonly agree that the oldest planted damar trees can be found in the south, where "you can find huge trees that were planted more than 200 years ago." The only written evidence is provided by Rappard, who reported in 1937 that he encountered 70 ha of plantations around Krui, among which several were at least 50 years old. This takes the first plantations back to 1885. Rappard mentioned that 80% of the damar produced in Krui in 1936 was from cultivated trees and that *Shorea javanica* no longer existed in a wild state in the area. He noted that production increased from year to year, with 120 tonnes in 1935, 201 tonnes in 1936, and 358 tonnes expected in 1937 (Rappard 1937).

Among the triggers leading to damar cultivation, the main one was probably the increasing difficulty of collecting it in the wild. This may have parallels with contemporary conflicts over access to other common property resources (Peluso 1983, 1992; Siebert 1989). At the turn of the century, a steep increase in resin prices led to intensive and generalized tapping of trees in natural forests. Overcollection eventually made mother trees rare, and this blocked natural regeneration. At the same time, cultivated territory encroached further and further into the forest itself. Although damar trees were spared in the slash-and-burn process (Figure 45-7) and found it easy to survive in the modified environment of ladang and secondary vegetation, natural regeneration in these conditions was apparently difficult. Serious conflicts broke out between villages, as well as within villages, over access to the remaining damar trees (Levang and Wiyono 1993).

Damar gardens also appeared as a solution to increasing problems in commercial agriculture. Around 1920 a serious disease reportedly wiped out most of the pepper plantations on the western coast of Lampung (Levang 1989). The resulting disturbance of the balance between subsistence and commercial strategies in cropping systems could partly explain the generalized development of damar plantations after 1930. The colonial administration might have played a role in advising local people to continue their process of domestication. It is also most probable that Chinese traders actively encouraged the diffusion of damar cultivation, much as they did for rubber in other parts of Sumatra (Pelzer 1978).[29]

Damar gardens have gradually spread in the Pesisir and, according to an examination by World Agroforestry Centre (ICRAF) and the Department of Forestry of a 1994 satellite image, productive gardens presently cover at least 50,000 ha. The main center of cultivation is located around the city of Krui, where the surrounding hills are almost totally covered with mature damar forest. Yearly production of damar resin was estimated at 8,000 tonnes in 1984 (Bourgeois 1984) and reached 10,000 tonnes in 1994 (Dupain 1994). New gardens are still being established in the northern and southern subdistricts.

How Cultivation Evolved into Forest Appropriation

Reconstruction of the forest by Pesisir villagers was not planned as such. Rather, it appeared as a consequence of a particular cropping system that aimed to minimize labor inputs and maximize use of the natural production and reproduction processes of an artificial ecosystem dominated by trees. In that sense, it was the choice of particular cultivation techniques and patterns, more than the initial selection of a given forest tree, that allowed true forest appropriation.

---

[29] Rappard mentioned that the extension of the damar plantations around Krui was helped by the former governor Helfrish (Rappard 1937). Dupain notes an astonishing coincidence between the oldest centers of damar cultivation and the former centers of residence of *pangeran*, who were the representatives of the village community, or *marga*. Their residences corresponded to territorial units reinforced by the Dutch, and they were an important link between colonial authority and the villages. Dupain formulates the hypothesis that, had the pangeran been more "informed" or "controlled," they could have facilitated the extension of damar cultivation (Dupain 1994). However, these centers of pangeran residency were also the place of residence and activity of Chinese traders (Pelzer 1978).

The main ecological disadvantages of *Shorea javanica* are typical of Dipterocarps: difficult natural regeneration due to irregular and occasional flowering, lack of seed dormancy, and need for mycorrhizae association. But one important advantage should be noted: unlike many Dipterocarp species, *Shorea javanica* appears to be rather light tolerant, which makes it suitable for cultivation in plots already cleared for agriculture.

Villagers solved the regeneration problem through a technology of "assisted storage of seedlings" (Michon and Bompard 1987b; Michon and Jafarsidik 1989). They established small nurseries (see Figure 45-8), where the seedlings could be kept for several years and used whenever planting material was needed. The mycorrhizae problem was avoided through a first phase of direct transplantation of seedlings from the forest to the plantation site.

The ingenious method of "seedling storage" allows the farmers to overcome both fruiting irregularity and lack of seed dormancy. The seed cannot be stored for more than a few days, which poses important problems of supply for planting material. When fruiting occurs, seeds are selected in the gardens, in villages, or in ladang fields. They are then closely planted in the nursery, either 5 by 5 cm or 10 by 10 cm apart, and the seedlings do not develop beyond 20 to 30 cm high. This growth inhibition may be due to high light intensity in the nurseries or high root density among the seedlings, which survive in the nursery conditions with a seemingly low rate of mortality for four to five years, or roughly from one fruiting season to the next. Seedlings are taken whenever they are needed for replacement of old trees or plantation of new land.

Among the other biological constraints is the long renewability rate of damar as a resource: it takes at least 20 to 25 years for a tree to reach a size where its resin can be tapped. The economic consequence is that, for the first 25 years, a pure damar plantation would be of little, if any, use to the planter. This difficulty has been solved through a strategy of crop succession, starting from the ladang and planned over the medium term. The expansion and success of damar cultivation is, therefore, closely related to swidden agricultural practices (Michon and Bompard 1987a; de Foresta and Michon 1994b). It is through the ladang and its traditional crop succession structure that damar trees have been restored to the landscape.

In the former dryland cultivation system, ladang were opened primarily for rice production, but some did not return directly to fallow. Instead, they were further transformed into either coffee or pepper plantations.[30] The first damar trees were introduced into these successional ladang gardens, amidst coffee bushes and pepper vines. After abandonment of the coffee or pepper crops, the damar trees were strong enough to grow along with secondary vegetation and overcome competition from pioneers. The subsequent fallow was a mix of self-established successional vegetation and deliberately planted damar trees, which reached a tappable size about 20 to 25 years after planting, but no more than 10 years after the plot was fallowed (see Figure 45-9). Damar plantations soon became a success story and everyone began to plant seedlings in their ladang gardens. After two decades, traditional fallow land had been transformed into a managed tree garden that included damar trees, as well as other introduced fruit species and self-established trees, bushes, and vines.

This process of establishment is still occurring today as new areas are converted. Ecologically, the development process of these successive crop mixtures imitates natural forest succession,[31] with all its ecological benefits of soil protection and microclimate evolution. Technically, it is similar to a classic agroforestry process of forest plantation establishment, the *taungya* system, in which young seedlings of

---

30 Coffee or pepper, and *Erythrina* as shade trees or living poles, were interplanted with dry rice and vegetables. The productive plantation was maintained for four to six years for coffee, and up to fifteen for pepper, and then returned to fallow.

31 Rainfed rice as the first grassy phase, coffee or pepper as the early pioneer tree phase, subsequent secondary formation with young damar and fruit trees, and damar and fruit trees related to various wild trees, as the mature phase.

economic tree species start to grow in favorable, controlled conditions. Here, maintenance of the coffee and *Erythrina* stand provides good microclimatic conditions of shade and humidity, favoring the survival of transplants, and also provides weed control for the first 4 to 15 years after introduction of seedlings.

Economically, the process of vegetation succession is of tremendous importance as it provides a succession of harvestable commercial products and reduces to about 5 or 10 years the period during which the plantation is unproductive. The costs of labor invested in damar establishment are mingled with those devoted to rice and coffee cultivation in swidden fields, although cultivation of commercial tree crops does not compete for labor with subsistence agriculture. On the contrary, it maximizes returns to labor inherent in the swidden system, including vegetation cutting and field maintenance, through the succession of coffee and trees.

Surprisingly, Pesisir villagers have succeeded in doing what most foresters dream of: establishing, maintaining, and reproducing, at low cost and over huge areas, a healthy Dipterocarp plantation. It remains a unique achievement in the entire world of forestry. The best part of the story is that this success is inextricably linked to shifting cultivation, the agricultural system held in contempt by foresters. The acceptance of the wild tree as a cultivated crop and the subsequent expansion of plantations were achieved only through the particular structure of the swidden production system. Ladang was at the very heart of this success. In achieving the switch from the "natural and sometimes protected" status of the damar tree to its adoption as a new crop in a farming system, the Pesisir villagers have reinvented the common process of appropriating resources through agriculture.

The control of the damar resource based on mimicry of natural forest processes adapts the cultivated ecosystem to the characteristics of the plant. This runs counter to conventional domestication processes, which emphasize modification of biological and ecological characteristics so that the plant will adapt to a cultivated ecosystem (Michon and de Foresta 1996).

Restoring Biodiversity

As in any secondary vegetation dominated by trees, the newly maturing damar plantation provides a suitable environment and convenient niches for the establishment of plant propagules from neighboring forests through natural dispersion. It also offers shelter and food to forest animals. In this natural enrichment process, farmers merely select among the possible options offered by the ecological processes. They favor resources by introducing economical trees and protecting their development, or tolerate nonresource development and reproduction as long as it does not produce "weeds." After several decades of this balance between free functioning and integrated management, the biodiversity levels are fairly high.

Figure 45-7. A Wild Damar Tree Preserved in a Swidden

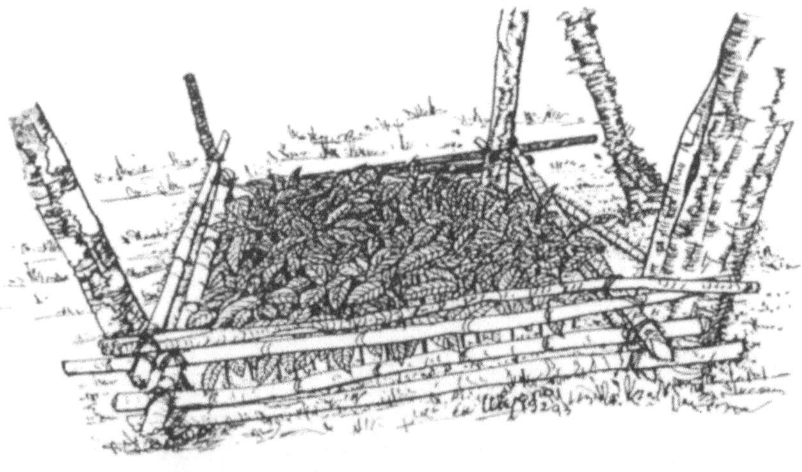

Figure 45-8. A Damar Seedling Nursery

Notes: Fruiting in *Shorea javanica* is highly unpredictable. Seeds are only available once every four to six years and they cannot be stored for more than a few days, so they are collected and planted in these small nurseries. Density is very high, which results in seedlings that do not grow beyond 20 to 30 cm tall and survive for four to five years.

As natural forests below 700 to 800 m asl have almost disappeared in the Pesisir, damar gardens constitute the major habitat for many plant species characteristic of lowland and hill Dipterocarp forests that would otherwise have disappeared (Michon and de Foresta 1992, 1995). The agroforest also shelters many animals, among them some highly endangered species such as the Sumatran rhino and the Sumatran tiger.

In order to assess biodiversity levels, comparative studies have been conducted between agroforests and related primary forests for several fauna and flora groups, including higher plants from ferns to dicotyledons, birds, mammals, and soil mesofauna. Diversity levels of soil mesofauna are quite similar between forest and agroforest. None of the numerically important species of the forest population are absent in the related agroforest. However, because many species in that large group are rare, results do not prove that all forest species exist in the agroforest (Deharveng 1992).

Bird richness in damar agroforests is 30% lower than in primary forest; 96 and 135 species have been recorded in those respective ecosystems. About 57% of the bird species found in the forest have not been encountered in the agroforest, whereas 40% of the agroforest species were not present in the forest surveys (Thiollay 1995). Reduction of bird diversity can be related to biological factors, such as simplification of composition and vertical structure from forest to agroforest, but it is probably mainly due to hunters. Birds are caught for food, but are also often kept in cages in villages or sold to outsiders, as bird keeping is more than a hobby in Indonesia.

As far as mammals are concerned, almost all forest species are present in the agroforest. Densities of the primate population in the agroforest, including macaques, leaf monkeys, gibbons, and siamang, are quite similar to those observed for natural forests. Hoofprints of the rare Sumatran rhino have been recorded in the agroforest, less than 2 km from villages. This is the first record of rhino in this part of Sumatra and it allows us to draw hypotheses on the usefulness of agroforests for the conservation of endangered animals as an important adjunct to protected forests (Sibuea et al. 1993).

Flora richness is reduced to approximately 50% in the agroforest. However, results have to be dissociated by biological groups, as they can be very different from one group to another. The largest loss occurs for trees. Agroforest diversity represents merely 30% of the original diversity levels. This is quite understandable, as economic intensification, and therefore selection, operates mainly on trees. Epiphyte and liana richness in the agroforest is at least 50% of forest richness, whereas for undergrowth plants, our samples show the variety to be twice as high in agroforests as that in natural forest. This is probably related to the common abundance of this group in secondary forests as compared to primary forests.

Seen from the planter's point of view, while the introduction of economic species into the damar agroforest is intentional, reestablishment of its overall biodiversity is "accidental." But it is precisely this accidental establishment of diversified flora and fauna that reconstitutes the forest character of the agroforest.

Both intentional and accidental processes are essential for several reasons. They restore resources that otherwise would not be deliberately conserved because they do not appear economically important. They also permit the restoration of biological and ecological processes that determine the functioning and reproduction of the agroforest as a forest ecosystem. In this sense, even those components that are not obviously economic resources play a valuable role. For instance, nonedible fruit trees in the agroforest help support populations of fruit-eating birds, squirrels, and bats that are essential natural pollinators and dispersers of economic fruit species. These "functional" resources that have no value as commodities, but are nevertheless essential, should never be forgotten, because restoring diversity, either economic or biological, is meaningless if ecological processes are not maintained.

A Process of Conceptual and Cultural Change

Appropriation of the forest resource through damar plantations has led to changes in traditional perceptions of forest resources, as well as the establishment of modified systems of social, legal, and institutional access to resources.

Ancient perceptions and representations of natural forests and forest resources are presently quite obliterated, at least in villages that have reached the limits of their territorial expansion. Having disappeared from the immediate environment of villages, the natural forest has lost its mystical stature in the farmers' imagination, and it no longer dominates their system of beliefs.[32] References to the ancient myths, or to forest spirits and magic, are nowadays very rare. The forest of the past, which was a source of spiritual as well as material wealth, represented an imaginary world as well a major source of life. Today's forest is neither mythical nor mystical; it is no more than the domain of forest administrators and an area of trouble for those who dare to enter it too conspicuously. The agroforest has replaced the natural forest in both the landscape and the local economy. Providing a lifestyle that coexists with the forest culture from which it directly evolved, it represents the last witness of an ancient alliance between Pesisir communities and forest resources. All the interactions between people and forest resources presently transpire through the agroforest. But the agroforest has not replaced the natural forest in local representation systems. It remains first and foremost an agricultural unit[33] (Lubis 1996; Michon, de Foresta et al. 1996), and the determining factor is that it results from a plantation process, even though an important part of its components have regenerated spontaneously.

[32] Among the explanations put forward: The religious evolution toward a "mature" Islam comes at the right time. Islam has shown the errors of former beliefs in which spirits and magic held a determining role.

[33] It is commonly called *darak,* equivalent to the Indonesian ladang, a generic term that defines any field opened in the forest, or *repong,* which formerly designated privately owned fruit gardens surrounding villages. Pesisir farmers also often describe it with the Indonesian word *kebun,* which means "garden" or "plantation."

Figure 45-9. Stages of Development from Swidden Fields to a Damar Agroforest

Notes: A. Year 1: Slashing and burning of secondary vegetation or primary forest and planting of rainfed rice, along with vegetables and fruits such as papaya and banana. B. Year 2: Introduction of coffee seedlings along with the second rainfed rice crop. Damar seedlings are ready to be transplanted. C. Year 3: No more rice crop. Introduction of damar and fruit seedlings in between the coffee rows. D. Years 4 to 8 (but sometimes up to Year 15): Coffee begins to produce a significant amount and is usually harvested for three to five years. However, it is sometimes managed in such a way that it still produces at year 15. E. Years 15 to 20 or 25: The field has been temporarily abandoned after the last economic coffee harvest. A spontaneous component is now developing, including trees, lianas, shrubs, forest herbs, and epiphytes, along with the planted damar trees and fruit trees. F. Damar trees now begin to be tapped. The damar garden continues to develop. Farmers' management ensures that the garden produces and reproduces without ever having to return to a slash-and-burn phase.

However, if damar is actually considered as a crop, the distinction farmers make between cultivated and managed or even wild plants remains somewhat vague and highly subjective. But the agroforest components are never perceived as "forest plants," even though farmers recognize that most agroforest plants can be encountered in the forest as well.

The agroforest itself will never be mistaken for a forest, or *pulan*, except in some very specific activities, usually linked to natural forests. People ordinarily go deer hunting or collect rattans "in the forest." In this case, "forest" refers not to a given garden or a given part of the agroforest, but more generally to the former communal forest space stretching between the reserved forest and the villages. This is the area that the agroforest has replaced, but which is also composed of ladang, successional vegetation, and remnant old-growth forest.

This well-established distinction between forest and agroforest is a logical one. Agroforests result from important initial work and represent long-term investment and years of a process, comparable with capitalization. Identifying agroforests with natural forest would mean denying this work, time investment, and long-term planning of the ancestors for their heirs. It would also deny the whole resource appropriation process achieved through destruction of the forest and plantation of trees. Confusing agroforest with the forest is, in these senses, heresy.

Access Systems

Institutionally, appropriation of the forest resource has entailed a total reorganization of the traditional tenure system for forest lands. It occurred simultaneously with the increasing importance of land as property, and privatization of this property (Michon, de Foresta et al. 1995b).

According to the ancient customary tenure system, forest lands and resources were managed as common property by the village community, or *marga,* and designated as *hutan marga*. Irrigated lands for rice production, on the other hand, were privately owned. Claims over economic resources in the hutan marga were acknowledged for certain species and through certain technical processes. Thus, a wild damar tree could be appropriated by the villager who first began tapping it, and collecting resin from that tree was thereafter considered an exclusive right. However, nobody could claim rights over a piece of unmanaged, pristine forest. Access to land for subsistence and cash cropping was usually gained through clearing a piece of land from the communal forest and cultivating it. Access rights were distributed among the different families of the marga by means of long-term individual usufruct rights, but the land itself remained the property of the marga. The individual usufruct rights were tacitly maintained long after the crops were harvested, and the same family could recultivate the land after a fallow period without asking permission from the marga. However, customary rights strictly forbade the planting of perennials on these communal forest lands, except for short-lived perennials like coffee or pepper. Tree plantation was considered a major investment for land development and was likened to labor invested in irrigation works for rice fields. Because this investment was acknowledged for rice fields, such investment in community land could have led to private appropriation of the land itself.[34]

As more people developed an interest in damar cultivation, the assembly of *pasirah*, responsible for the customary law, formally accepted the removal of the prohibition against planting perennials in the marga lands. This boosted the spread of the plantation movement and led to drastic land appropriation by individuals in the former communal forest domain (Levang and Wiyono 1993). However, land property could only be claimed through tree plantation and the old tenure system of communal ownership of the land and usufruct rights granted to individual families prevailed for nonplanted plots. These were still regarded as hutan marga.

---

[34] Such systems still exist in other parts of Sumatra (Levang 1989).

Privatization, of a Kind

The plantation process was conceived in the context of a relative failure of common property systems, so its success required assurance that the planter's children would enjoy the right to harvest from the trees, implying the need to not only acknowledge and enforce property rights, but also to secure transmission rights.[35] The consequence was that land properties, once created, never returned to the community, and the commons gradually disappeared. However, the privatization process remained original and unpredictable, as it did not entail promotion of individual control or fragmentation of the agroforestry domain.

The customary law makes a clear distinction between *hak milik penuh* (full property rights), a right more or less similar to Western ownership and one that covers newly created, or newly bought land, and *hak waris* (inherited rights), which concern properties inherited by a lineage patrimony. In Pesisir, the owner of a piece of inherited land is bound by traditional restrictions, regarding both the transfer of land and the right to use it. Even though *hak waris* is acknowledged as a right to individual ownership, the owner cannot transfer the land or the trees, nor can he cut productive trees without obtaining permission from the whole extended family.[36]

This rights restriction is more a moral obligation than an enforced regulation. It operates through a strong social control system in which the customary legal authority over land and resources is transferred from the village community to the lineage community. Transmitting an intact family patrimony to one's eldest son is as important as receiving it, and the property rights system cannot be dissociated from a social system in which many community traditions are still alive. The domestic group largely exceeds the limits of the nuclear family,[37] and the lineage maintains overall authority over the family patrimony. The head of the domestic group, who actually holds the *hak waris*, although being the legitimate and real owner of the garden, is socially more the depositary of a patrimony, the continuity of which is under the control of the whole lineage. The local saying: "*Hak waris bukan hak milik saya*" (My heritage is not my property.) summarizes the ethic of the property rights systems in the Pesisir, and this ethic still constitutes, more than any formal regulation, a clear safeguard against total privatization and individualization. This has obvious consequences for the sustainability and efficiency of agroforest management. If it were not controlled by a strong social structure, any breakdown of the agroforest block into individually owned plots could evolve into a mosaic of fields with different structures and vocations. This would lead to a drastic fragmentation of the ecosystem that could greatly endanger the overall reproduction of biological and productive structures[38] (Mary 1987; Levang and Wiyono 1993; Michon, de Foresta et al. 1995a).

---

35 Transmission of rights to the damar gardens commonly follows the traditional patrilineal tenure regime formerly devised for irrigated rice fields and fruit gardens. A piece of land, once acknowledged by the community as the private property of the individual who "created" it, remains in the lineage of the "creator" through its inheritance by his eldest son.

36 Sharing landed properties usually occurs after the birth of the first male child of the eldest son. The newly endowed heir becomes the *kepala keluarga*, or head of a family unit that comprises his children and his parents, as well as his unmarried youngest brothers, his unmarried sisters, and sometimes the children of his married brothers. As the exclusive heir of the family properties, the eldest son is in charge of housing and feeding the entire extended family group. This heavy responsibility, according to the heirs, largely compensates for the inequality of the transmission system, but does not seem to negatively affect individual incentives for production and investment.

37 Usually parents, if present, and direct uncles and brothers.

38 This situation of ecological and economic collapse happened in some villages with the introduction of cloves. The collapse of clove gardens and the related extension of bush and grass vegetation has led in some areas to fire problems, which threaten to destroy the remaining patches of damar gardens.

Reestablishing Common Property Rights and Values

Much as creation of the agroforest has reestablished forest resources and structures, common property traditions have been redefined and reinforced in the context of privatization, and a balance has been struck between formal individual rights and moral obligations toward the "community." Minor resources included in the agroforest remain accessible to members of the domestic group, the lineage, or the village community. The degree of any owner's control over his garden's resources actually depends on the nature of the resource. Important economic resources, such as resin and commercial fruits, as well as the land, are effectively individually owned assets, bound by all the traditional restrictions mentioned above. However, on these private agroforest lands, many resources are still considered common property or accessible to all. In fact, the only strictly privatized resource is the damar resin, and taking resin from someone's tree constitutes a real theft. Other resources such as fruits, sap from the sugar palm, bamboos, and special leaves for thatching, all of them gathered from species commonly considered as "planted," remain at the disposal of the community. But which community may harvest which resources, and to what extent, depends upon the resource and varies from the family group to the lineage or the village itself. Usually, the owner's permission should be sought before collecting what could be considered as "significant quantities," and sharing of benefits usually occurs for those products harvested for commercial purposes. However, picking fruits or bamboo for one's own immediate consumption while passing by a garden is considered normal. Resources considered as pure "forest resources," such as rattan, wild vegetables, medicinal plants, and firewood are covered by rules that fluctuate between a very wide sense of common property and open access. Firewood, for example, may be collected in small quantities by anybody from the village community. In most villages, subsistence hunting and gathering of vegetables or medicinal plants is allowed throughout the agroforest area without any restriction concerning the origin of the collector. The rules also allow income-generating activities, such as commercial gathering of rattan.

In the same way that the technical appropriation of the forest resource did not fundamentally change the Pesisir landscape, the institutional reappropriation of the former forest commons through "controlled privatization" did not represent a revolutionary overthrow of old values. Indeed, the maintenance of a communal philosophy is essential to agroforest management, because such long-term management is more a principle of forestry than a concern of subsistence farmers. Establishing access to productive structures and resources which will start producing for one's child and be fully productive for one's grandchild constitutes a new logic, in which short-term individual considerations need to be buffered by a community concern for perpetuation of these structures and resources. In the way that former common property regulations controlled the permanence of the commons, the new property ethics in the Pesisir ensure that trees and land will be integrally transmitted to future generations.

For village communities, the legal framework for private property could secure a better bargaining position with external bodies than common property, which is still negatively perceived or easily denied by most state bodies, as well as by private companies. Private claims over land are more easily acknowledged, and compensated for, by the Indonesian administration. Privatization could therefore be used as a political strategy for local communities to protect their resources.

The Agroforest Strategy as a New Framework for Local Communities

More than it being a simple success for natural resource management, the establishment of damar agroforests in the Pesisir constitutes a true forestry and agricultural revolution. As a forest plantation strategy, the damar model runs counter to the conventional model of timber estates. While favoring a selected resource, as estates do, the agroforest allows the maintenance of numerous other resources that otherwise would not be conserved, as well as restoring species that are not resources at all. Moreover, the establishment process allows the restoration of biological and ecological processes that are crucial to the survival and reproduction of the agroforest as an ecosystem. The damar agroforest system both contradicts economic reductionism and acknowledges the economic potentialities of the natural forest. Through the restoration of biodiversity in the agroforest, farmers have achieved the restitution of a whole range of economic choices for the present and the future. The Pesisir agroforest development also represents a successful strategy for agricultural intensification without disrupting either the availability of food or living standards, while maintaining intact the productive potentialities of the land itself.

Damar gardens are a fascinating example of agroforestry association. But they convey a totally new dimension of association, not between trees and crops as in conventional agroforestry, but between the forest resource and agricultural logic (Michon, de Foresta et al. 1995a). It is, above all, the integration of forest resource management into the farming system that underpins the success and originality of the damar agroforest.

Damar gardens offer new insights into the technical, ecological, socioeconomic, and institutional basis for managing forest resources within farming systems. They also bring new insights to the open debate on natural resource management by local communities. As a development strategy, the establishment of the damar agroforest is an interesting example of forest product management for commercial purposes, which has entailed a total transformation of the original ecosystem, while preserving most of its resources and retaining an important part of its biodiversity. The transfer of forest functions from the natural ecosystem to the agroforest implied not only a transfer of resources, structures, and economic vocations, but also a guarantee of their renewability. In this respect, the agroforest should be considered by foresters as an ecological model for forest reconstruction with huge potential for reforestation and land rehabilitation programs. The present political, institutional, and socioeconomic atmosphere in Indonesia appears quite unfavorable to long-term maintenance of the forest itself. Therefore, the process of damar agroforest establishment and development appears as an extremely original strategy for local populations to restore their forest resources and to once more use them in the true tradition of peasant economies on the forest margins of Indonesia.

After years of conflict with the forest services in the Pesisir, farmers apparently gave up most of their claims over the national park natural forest. Thereafter, the natural forest was considered more as a geographical unit in an administrative landscape than as a resource in the village landscape. For farmers, the forest is an exclusive, reserved, and closed domain of the state, and entering it is often an act of extreme provocation. Agroforest, on the other hand, represents a man-made structure where the forest resources are appropriated and managed in accordance with the farmers' needs, philosophy, and beliefs. Through the agroforest, farmers can claim that they have restored a privileged space in which their forest resource is protected. What's more, it is in the middle of an agricultural territory over which they feel they have a kind of control.

Forest Farmers versus Foresters, Planters, and the State

The isolation of the Pesisir area and the absence of development projects until the last few years of the 1990s have protected damar agroforests and damar farmers from the outside world. However, the acceleration of regional development since the early 1990s has clearly shown the limits and weaknesses of agroforests as a strategy for appropriation of forest resources by local people. Agroforests are neither recognized by the state, nor are they given legal status.

Until very recently, the history of relations between damar farmers, state officials, and state-supported bodies was dominated by misunderstandings and abuses of rights. It may be divided into an initial period, which was one of exclusive conflicts with forest authorities, and a second phase, which saw the emergence of unexpected but powerful stakeholders: the regional authorities and the private sector.

Conflicting Views over Land Status, Use, and Control

No official map, beyond the village level, mentions the existence of damar gardens. The most recent map for the West Lampung district[39] classifies the land occupied by damar agroforests as either "swidden and dry fields," "secondary forest and degraded vegetation," or "plantation area."[40]

According to the 1995 statistics yearbook of the Department of Forestry (Dinas Kehutanan Lampung 1995 statistics yearbook), a total of 17,500 ha is covered by damar gardens[41] and, of this, 10,000 ha lie on state forest lands, classified either as limited production forest or protection forest. However, village statistics and maps still show these same state forest lands as "customary forest," or hutan marga, as they were known before the 1991 publication of the "Forest Land-Use Master Plan by Consensus" (TGHK). Although the hutan marga status was no longer officially acknowledged, villagers believed that their customary lands were still recognized by the state. They were astonished, therefore, when forestry employees began planting poles in their damar gardens and telling them: "Beyond this limit, the land belongs to the government." Depending on the area, this happened between 1992 and 1996. In some villages, the first poles were located just 100 m from outlying houses. Suddenly, the villagers realized the extent and gravity of coming conflicts.

The Department of Forestry says the 7,500 ha of damar agroforests that are not classified as state forest are on "unclassified" land. Since it is not public land, it is sometimes called private land, but this is not as good as it sounds. Private appropriation by local people is not formally recognized because farmers do not hold any official land certificates for either rice fields or damar gardens. The district administration under whose jurisdiction the private lands fall has indicated that its top priority is to promote estate plantation development based on private investment.

Damar farmers are therefore caught between two mutually exclusive administrative mechanisms, neither of them holding positive prospects. In both cases, their legal position is depressingly weak. To forest authorities, they are outlaws. To conduct any agricultural or harvesting activity on forest lands without permission from the Department of Forestry is constitutionally illegal and penalties are provided by law. Under a "private" regime, but without any official land title, damar farmers may be considered as squatters on empty lands that are reserved for regional development. In either case, they are threatened with eviction in order to make way for officially sanctioned projects.

Damar farmers have never been officially involved in the decision-making processes controlling the future development of the lands that they have actively and efficiently managed for centuries. Neither have they been properly informed of decisions made without their participation, and which might obviously have profound implications for their future. After having labored hard, buoyed by the belief that they were developing a secure future for their children and grandchildren, they suddenly learn, by rumor more than specific explanation, that their lands belong to the state,

---

[39] Dated 1995, but only available in early 1997, this map was displayed by some of the authors in one Central Pesisir village and quickly rolled up on the advice of the village chief, who explained that villagers would riot if they knew how the result of their plantation work was treated by officials.

[40] Already mapped as an oil palm estate plantation in South Pesisir.

[41] This "official" figure was a guess that benefited the Department of Forestry, as it minimized the true area under damar garden cover; the official figure dramatically increased to about 50,000 ha in 1997, when the Department of Forestry agreed to carry out a proper estimation based on recent satellite imagery.

and the state has "better" projects in mind for their development. Not surprisingly, this has led to the outbreak of conflicts between farmers and government-sponsored agents.

The Reserved Forest

The forest reserve was established by the Dutch administration in 1937. Its borders were decided after consultation with local people, and were located far from the agricultural territory of villages. The reserve was upgraded to national park status in 1991, and became Bukit Barisan Selatan National Park. As an old constraint on their territorial development, villagers have always been well aware of its existence and of its borders. However, they do not fully agree with the legitimacy of the park's importance for flora and fauna protection. This disagreement, which is more conceptual than factual, has gained importance in villages where land shortage problems are acute. The fundamental grievance of farmers against the ideology and practice of conservation forestry is that it values the forest more than humans and will always give preference to wildlife and plants, whether or not this results in serious problems for local people.[42]

Because of land shortages in several villages, encroachment of ladang and damar gardens into the forest reserve, especially along the Krui-Liwa road, started as early as 1955. In the late 1960s, a tacit agreement was concluded between farmers and the forestry authorities, allowing several dozens of families to open land in the reserve and establish damar gardens (Mary and Michon 1987). However, police and conservation guards continued to regularly visit the farmers to solicit "rewards" for this agreement. This continuous annoyance led many families to leave the area by the end of the 1970s. Today, in the park where no ladangs have been opened for more than 20 years, the canopy has closed and only the expert eye will distinguish the damar islands in the forest.

The Production Forest

In 1981, the Indonesian government granted a logging company, HPH Bina Lestari, concession rights covering 52,000 ha of land between the reserved forest and the Indian Ocean (Kusworo 1997).[43] This company had formal rights to collect timber all over the three Pesisir subdistricts. Damar farmers were unaware that their territory had been given over to logging, because the company only logged timber in the extreme north and south and did not dare harvest timber planted by local farmers in their agroforests. But under the same circumstances in the province of Bengkulu, private foresters deliberately logged damar gardens belonging to local people. Had they done the same in Pesisir, it would not have been considered illegal and the farmers would have had no right to claim compensation.

HPH Bina Lestari left in 1991 and the area was divided, according to the first officially recognized TGHK maps, into conversion forest covering 7,500 ha in the extreme south, small pieces of protection forest distributed all along the western border of the national park,[44] and about 42,000 ha of production forest. Management of the production area was given to a state-run company, Inhutani V, and rumors quickly arose of a plan for "forest rehabilitation," with large-scale *Acacia* planting, to start in 1992. Fortunately, this never happened and it now seems

---

[42] Wildlife, especially elephants, constitute another source of conflict. For several years, since the remaining production forest was logged over and the related opening of roads attracted migrants who cleared large areas in the logged-over forest for coffee and pepper growing, elephants have frequently come out of the national park, destroying crops and attacking farmers in their *ladang*, and even in their villages. Villagers have sought the right to carry guns to protect themselves, but this is not accepted by the conservation guards.

[43] The Pesisir area is known in the Department of Forestry as "Bina Lestari's forest."

[44] These pieces of protection forest have no rationale other than as compensation for other areas in the province where previously designated protection forest has been declassified to make way for public and private development projects.

unlikely it will ever happen. However, it was at this time, between 1992 and 1996, that the Forestry Service measured the state forest borders in accordance with the new maps, and marked them with poles. Damar farmers began to suspect that their lands were being claimed by the state, but they were never directly informed of the legal consequences of their land being classified as production or protection forest. When they asked, the answer was always that nothing was changed, *"at least for the time being."*

The Ambiguities of Forestry Support

Beginning in 1992, foresters began to change their attitude toward the damar enterprise of Pesisir farmers. The changes followed mounting pressure, led by local and international researchers and nongovernment organizations who promoted the "Krui case" as an outstanding example of reforestation and forest management by local communities.[45] Then, the Department of Forestry itself made a politically correct switch toward allocating more support to forest communities, and regional authorities timidly acknowledged that repeated rights violations and abuses of power were leading to potentially serious social problems. This translated into a more serious consideration of the originality and value of the Pesisir system at various levels of forest and regional administration.

However, the new support may be a double-edged sword for damar farmers. For, while many foresters in Jakarta acknowledge the value and validity of the damar garden system, they seem unable to admit that it arose and worked perfectly well for something like a century without them.

Defeat of the Planters

The Pesisir represents the last "wild frontier" in the already overpopulated province of Lampung. Due to its proximity to Jakarta and ongoing road development, it is tempting territory for private speculators, particularly estate developers and agro-industries. Regional authorities regard these potential investors with great interest. Besides being important taxpayers, which farmers are not, their investments would greatly increase the regional development index and the level of industrial activities in the area (Kusworo 1997).

Following the completion of logging operations in the early 1990s, local authorities began allocating "private lands," as well as part of the logged-over forest lands,[46] in the three Pesisir subdistricts to two oil palm companies. PT Karya Canggih Mandiri Utama was allocated 24,500 ha in the south, and began development in 1994. PT Panji Padma Lestari received 17,352 ha in the north and 4,500 ha in the south, and began development in January 1996. Local farmers were not informed of these projects and began asking questions when they encountered field teams measuring land, including their damar gardens and even their rice fields. They were not always given correct answers.

The local authorities stressed that oil palm would be planted only on "empty" land, although farmers could also be invited to join with their own lands if they wished. They began campaigning for support for the project, asking village heads to speak highly of the economic merits of oil palm planting and to ensure farmers' cooperation. But they also specified that no farmer should be compelled to give up his damar land for the company, and that no damar tree should be felled without the consent of the owner. PT Karya Canggih Mandiri Utama soon applied its own interpretation of "inviting" farmers to join. The company issued a formal invitation through the subdistrict head to village authorities, and there was a lack of

---

[45] As a result of this joint effort, the prestigious National Kalpataru Award for the Environment was given by the president of Indonesia to the "customary community of damar farmers in Pesisir, Krui" on June 5, 1997.

[46] By using the process of revising the district scale of the TGHK as a way to declassify the targeted forest land.

enthusiasm from damar farmers. So the company decided to use fake positive agreements signed by farmers instead of the genuine but negative ones, and started clear-felling damar gardens by moonlight. Angry farmers, who had actually woken up to find that all their damar trees had been cut overnight, publicized the blunt violation of their rights to the provincial assembly and to local newspapers.

Farmers in the northern subdistrict, alerted to the practices of the companies, began affirming and publicizing their resistance to the arrival of PT Panji Padma Lestari even before it actually started measuring land.

The protests of the farmers were supported by nongovernment organizations and international research institutions, which asserted that replacing farmers' damar gardens with oil palm estates was neither ecologically defensible nor socially acceptable. Further, they pointed out that the manner in which the replacement was taking place was clearly a classic case of power abuse by economic and political elites. Finally, the voices of opposition succeeded. In December 1996, the Ministry of Forestry asked PT Karya Canggih Mandiri Utama to suspend its activities and solve the conflicts with damar farmers. Three months later, the provincial governor asked PT Panji Padma Lestari to halt its activities.

Conclusion

The case of the Pesisir damar farmers addresses many justice issues. The main one concerns civil justice. The basic rights of local people over lands and resources they have developed, enriched, and managed sustainably over centuries are not fully recognized by the state, despite the existence of constitutional facilities allowing the acknowledgment and legalization of such rights. This issue is not specific to the Pesisir; it constitutes the major arena of confrontation between the state and communities of forest farmers, and represents a significant impediment to the integration of indigenous communities into the Indonesian nation. The practical interpretation and implementation of land rights granted to public companies or private firms unavoidably erases the expression, and the very existence, of local rights. Moreover, the closure of the damar lands by the state would constitute not only a violation of basic rights, but also pure theft. Replacing damar gardens by forest estates or agricultural plantations, or reserving them for conservation projects or production forestry, would obviously constitute a forceful appropriation not only of other people's lands, but also the fruits of other people's labor. Furthermore, if the damar farmers attempt to defend their legitimate rights and properties, they might lose even their basic human rights, given the fact that past decades have revealed the attitude of both private firms and public bodies where it comes to policy enforcement: fake promises, verbal and physical intimidation, and passivity in the face of violent military intervention.

The second issue is one of economic and social justice. Replacing damar gardens with specialized oil palm or *Acacia* plantations might prove, in the short term and with a partial economic valuation, to be an economic gain for the region. However, it is not certain that such an economic gain would be redistributed to the farmers, who would certainly contribute to it through their poorly paid labor. In terms of equity, the economic characteristic of the damar gardens is that the benefits go to local people, including farmers, wage laborers, and local trade entrepreneurs. But taxes on the damar resin represent less than 0.1% of the district budget. Industrial plantation estates provide much higher profits, but to far fewer people, whereas levies raised by the district through the estates and the related industrial processing units are numerous and substantial. Seen from the viewpoint of regional administrators, the choice is obvious.

The last issue concerns environmental justice. The damar garden system developed by Pesisir farmers has proven to be an almost perfect ecological substitute for natural forests. In fact, when it is remembered that this is a diversified production system, it is probably the best possible substitute for natural forest. Destroying damar gardens to make room for specialized oil palm or *Acacia* plantations would obviously constitute an ecological crime. Its immediate consequences would include the destruction of habitat for many lowland plant species; a significant reduction in the feeding and breeding areas of many endangered mammal and bird species, such as Sumatran rhino, tiger, tapir, elephant, siamang, hornbills, and rapaces; a drastic increase in soil erosion with consequent siltation of the Pesisir coast and irrigation works in the lowlands; and an increase in ecological risks to people as well as to existing plantations. An additional consequence is the uncertain ecological sustainability of monocrop plantations over the long-term, compared to the proven sustainability of the damar gardens over the past 150 years. Crimes of this sort do not result in immediate punishment, but their long-term costs, for locals as well as for the nation itself, are potentially immense.

Without doubt, damar agroforests represent a rare and precious example of successful sustainable management of forest resources in the humid tropics. However, the success story is strongly endangered. Without a significant change in government policy, Pesisir farmers face urgent and threatening choices: either they must become laborers on their own land as their damar agroforests are converted to oil palm estates, or they risk seeing their rights strongly restricted by zealous foresters who confound damar agroforest with natural forest and forget that there are no damar agroforests without damar farmers.

Culturally, biologically, economically, and socially, damar farmers have succeeded in reappropriating their forest resources. However, what the last few years have shown is that their reappropriation probably fell short. It went far enough to ensure the long-term sustainability of the ecosystem, but not far enough to protect its short-term survival. To be ensured against forceful conversion, a fifth element is needed that translates into legal terms the formal and official recognition of the damar farmers' contribution to national and regional objectives.[47] But the agroforest situation does not fit any of the existing legal forest categories, and a new legal status needs to be devised to suit the needs of damar farmers and to ensure a future for damar agroforests. There is new hope that this direction will be followed by the Department of Forestry and that, after years of doubt, justice for damar farmers will be respected. If the Indonesian authorities take this first step, it may set a precedent offering justice for all other agroforest farmers in Indonesia facing similar problems.

The "agroforest" framework offers a good opportunity to escape the formal forestry context and to devise new forms of association between farmers, foresters, and regional authorities concerning forest resources. Ecologically, economically, and socially, the agroforest should not be confused with a natural forest. Indeed, as long as this confusion continues, and as long as local practices for management of forest resources in farming systems are ignored, the chances for survival of agroforests as a unique model of integral forest management continue to decrease. Agroforests, once recognized, open a totally new field for negotiation between foresters and local communities; a field favorable to institutional innovations where ancient conflicts might be resolved without either party losing face. In particular, it could facilitate the formation of new alliances between the conventional forestry sector and local communities. It could also offer new options for management and control of land and other resources without destabilizing existing forestry laws. It would be a pity not to take this as an opportunity to rethink, in practical terms, the whole conventional context of forestry and agriculture.

[47] This high level support led to the "KDTI" decree, issued in early 1998. This decree officially and legally recognizes formal access, management, and transmission rights to the damar farmers in the state forest area already converted to damar agroforest, whatever its development stage, thus preventing further outsider's manipulations. This need was recognized by the Minister of Forestry who, after visiting Krui in 1994, has always been quite positive about damar farmers and their contribution.

References

Anonymous. 1995. Strengthening Community-based Damar Agroforest Management as a Natural Buffer-Zone of Bukit Barisan Selatan National Park, Lampung, Indonesia. Research Report, Lembaga Alam Tropika Indonesia (LATIN). Bogor, Indonesia.

Bouamrane, M. 1996. A Season of Gold: Putting a Value on Harvests from Indonesian Agroforests. *Agroforestry Today* 8(1), 8–10.

Bourgeois, R. 1984. Production et Commercialisation de la Résine "Damar" à Sumatra Lampung, Montpellier, France: Master's thesis, E.N.S.A.M.

Burkill, I.H. 1935. *A Dictionary of the Economic Products of the Malaya Peninsula.* London, UK: Crown Agents for the Colonies, Millbank.

Collet, O.J.A. 1925. *Terres et Peuples de Sumatra.* Amsterdam: Elsevier.

de Foresta, H., and G. Michon. 1992. Complex Agroforestry Systems and Conservation of Biological Diversity. (2) For a Larger Use of Traditional Agroforestry Trees as Timber in Indonesia: A Link between Environmental Conservation and Economic Development. In: *In Harmony with Nature. An International Conference on the Conservation of Tropical Biodiversity,* edited by Y.S. Kheong and L.S. Win. Kuala Lumpur, Malaysia: The Malayan Nature Journal (Golden Jubilee Issue), 488–500.

———. 1993. Creation and Management of Rural Agroforests in Indonesia: Potential Applications in Africa. In: *Tropical Forests, People and Food: Biocultural Interactions and Applications to Development,* edited by C. M. Hladik, H. Pagezy, O. F. Linaret et al. Paris, France: UNESCO and the Parthenon Publishing Group, 709–724.

———. 1994a. Agroforests in Indonesia: Where Ecology and Economy Meet. *Agroforestry Today* 6(4), 12–13.

———. 1994b. From Shifting Cultivation to Forest Management through Agroforestry: Smallholder Damar Agroforests in West Lampung (Sumatra). *APAN News* 6/7, 12–13.

———. 1997. The Agroforest Alternative to *Imperata* Grasslands: When Smallholder Agriculture and Forestry Reach Sustainability. *Agroforestry Systems* 36(1–3), 1–16.

Deharveng, L. 1992. Soil Mesofauna in Agroforests and Primary Forests of Sumatra. Field report, ORSTOM/BIOTROP, Bogor, Indonesia.

de Jong, W. 1994. Recreating the Forest: Successful Examples of Ethno-Conservation among Land-Dayaks in Central West Kalimantan. In: *Management of Tropical Forests: Towards an Integrated Perspective,* edited by O. Sandbukt. Oslo, Norway: Centre for Development and Environment, University of Oslo, Norway, 295-304.

Dove, M.R. 1983. Theories of Swidden Agriculture, and the Political Economy of Ignorance. *Agroforestry Systems* 1, 85–99.

———. 1985. *The Agroecological Mythology of the Javanese and the Political Economy of Indonesia.* Honolulu, Hawaii: East-West Environment and Policy Institute.

———. 1993. Smallholder Rubber and Swidden Agriculture in Borneo: A Sustainable Adaptation to the Ecology and Economy of the Tropical Forest. *Economic Botany* 47(2), 136–147.

Dunn, F.L. 1975. Rainforest Collectors and Traders: A Study of Resource Utilization in Modern and Ancient Malaya. Malaysian Branch of the Royal Asiatic Society Monograph. Vol. 5. Kuala Lumpur. Malaysia.

Dupain, D. 1994. Une Région Traditionnellement Agroforestière en Mutation: Le Pesisir (A Traditional Agroforestry area in Mutation: Pesisir). Montpellier, France: Master's thesis, CNEARC (Centre National d'Etudes Agronomiques des Regions Chaudes) and ORSTOM/BIOTROP.

Foxworthy, F.W. 1922. Minor Forest Products of the Malay Peninsula. *Malayan Forest Records* 2, 151–217.

Fried, S.T. 1995. Writing for their Lives: Bentian Dayak Authors and Indonesian Development Discourse. Ithaca, NY: Ph.D. thesis, Cornell University.

Gianno, R. 1981. *The Exploitation of Resinous Products in a Lowland Malayan Forest.* Research Report, Museum Support Center, Smithsonian Institution, Washington DC.

Gillis, M. 1988. *Public Policies and the Misuse of Forest Resources.* Cambridge University Press.

Gouyon A, H. de Foresta, and P. Levang. 1993. Does "Jungle Rubber" Deserve Its Name? An Analysis of Rubber Agroforestry Systems in Southeast Sumatra. *Agroforestry Systems* 22, 181–206.

Kusworo, A. 1997. *Government Policies that Affect the Damar Agroforests in Pesisir Krui, West Lampung, Sumatra.* Research Report, ICRAF (Center for International Forestry Research) Southeast Asia. Bogor, Indonesia: ICRAF

Laumonier, Y. 1981. Ecological and Structural Classification of Southern Sumatra Forest Types. Unpublished report, BIOTROP, Bogor, Indonesia.

Levang, P. 1989. Systèmes de Production et Revenus Familiaux (Farming Systems and Household Incomes): Transmigration et Migration Spontanées en Indonésie

(Transmigration and Spontaneous Migrations in Indonesia). Departemen Transmigrasi–ORSTOM, 193–283.

Levang, P., and Wiyono. 1993. Pahmungan, Penengahan, Balai Kencana. Enquête Agro-Économique dans la Région de Krui (Lampung). Research Report, ORSTOM/BIOTROP, Bogor, Indonesia.

Lubis, Z. 1996. Repong Damar: Kajian Tentang Pengambilan Keputusan Dalam Pengelolaan Lahan Hutan Pada dua Komunitas desa di Daerah Krui, Lampung Barat. Research Report, P3AE-UI and CIFOR (Center for International Forestry Research), Bogor, Indonesia.

Marsden, W. 1783. *The History of Sumatra*. London. Republished in 1986 by Oxford University Press, Singapore.

Mary, F. 1987. Agroforêts et Sociétés. Analyse Socio-Économique de Systèmes Agroforestiers Indonésiens. Document E.N.S.A.M.–INRA, Montpellier, France.

Mary, F., and G. Michon. 1987. When Agroforests Drive Back Natural Forests: A Socio-Economic Analysis of a Rice/Agroforest System in South Sumatra. *Agroforestry Systems* 5, 27–55.

Michon, G. 1985. De L'homme de la Forêt au Paysan de L'arbre: Agroforesteries Indonésiennes. PhD thesis, U.S.T.L., Montpellier, France.

———, and J.M. Bompard. 1987a. Agroforesteries Indonésiennes: Contributions Paysannes à la Conservation des Forêts Naturelles et de Leurs Ressources. *Revue d'Ecologie. (Terre Vie)* 42, 3–37.

———, and J.M. Bompard. 1987b. The Damar Gardens (*Shorea javanica*) in Sumatra. In: *Proceedings of the Third Round-Table Conference on Dipterocarps*, edited by A.G.J.H. Kostermans. Samarinda: UNESCO, 3–17.

———, and D. Jafarsidik. 1989. *Shorea javanica* Cultivation in Sumatra: An Original Example of Peasant Forest Management Strategy. In: *Management of Tropical Rainforests: Utopia or Chance of Survival*, edited by E.F. Bruenig and J. Poker. Baden-Baden: Nomos Verlagsgesellschaft, 59–71.

Michon, G., and H. de Foresta. 1992. Complex Agroforestry Systems and Conservation of Biological Diversity (1) Agroforestry in Indonesia: A Link between Two Worlds. In: *In Harmony with Nature. Proceedings of an International Conference on the Conservation of Tropical Biodiversity*, June 12–16, 1990, Kuala Lumpur, Malaysia, edited by Y.S. Kheong and L.S.Win. Kuala Lumpur, Malaysia: The Malayan Nature Journal. (Golden Jubilee Issue), 457–473.

———. 1995. The Indonesian Agroforest Model. In: *Conserving Biodiversity Outside Protected Areas: The Role of Traditional Ecosystems*, edited by P. Halladay and D.A. Gilmour. Gland, Switzerland and Cambridge, UK: IUCN (International Union for the Conservation of Nature).

———. 1996. Agroforests as an Alternative to Pure Plantations for the Domestication and Commercialization of NTFPs. In: *Domestication and Commercialization of Non-Timber Forest Products in Agroforestry Systems*, edited by R.R.B. Leakey, A.B. Temu, M. Melnyk, and P. Vantomme. (Non-Wood Forest Products, No.9). Rome: FAO, 160–175.

———, and P. Levang. 1995a. Stratégies Agroforestières Paysannes et Développement Durable: Les Agroforêts à Damar de Sumatra. *Natures-Sciences-Sociétés* 3(3), 207–221.

———, and P. Levang. 1995b. A New Face for Ancient Commons in Tropical Forest Areas? The "Agroforest Strategy" of Indonesian Farmers. Communication to the 5th annual meeting of the International Association for the Study of Common Property Resources, May 24-28, 1995, Bodo, Norway.

———, and A. Aliadi. 1996. Damar Resins, from Extraction to Cultivation: An "Agroforest Strategy" for Forest Resource Appropriation in Indonesia. In: *Ethnobiology in Human Welfare*, edited by S. K. Jain. New Delhi: Deep Publications, 454–459.

Momberg, F. 1993. *Indigenous Knowledge Systems. Potentials for Social Forestry Development: Resource Management of Land-Dayaks in West Kalimantan*. Berlin: Technische Universitat Berlin.

Nadapdap, A., and I. Tjitradjaja, et al. 1995. Pengelolaan Hutan Berkelanjutan: Kasus Hutan Damar Rakyat di Krui, Lampung Barat. In: *Ekonesia* (2), 80–112.

Padoch, C., and C. Peters. 1993. Managed Forest Gardens in West Kalimantan, Indonesia. In: *Perspectives on Biodiversity: Case Studies of Genentic Resource Conservation and Development*, edited by C.S. Potter, J.I. Cohen, and D. Janczewski. Washington, DC: AAAS (American Association for the Advancement of Science), 167–176.

Peluso, N.L. 1983. Networking in the Commons: A Tragedy for Rattan? *Indonesia* 35(1), 95–108.

———. 1992a. *Rich Forest, Poor People: Resource Control and Resistance in Java*. University of California Press.

———. 1992b. The Ironwood Problem: (Mis)Management and Development of an Extractive Rainforest Product. *Conservation Biology* 6(2), 210–219.

Pelzer, K.J. 1978. Swidden Cultivation in Southeast Asia: Historical, Ecological, and Economic Perspectives. In: *Farmers in the Forest*, edited by P. Kundstadter, E.C. Chapman, and S. Sabhasri. Honolulu, HI: The University Press of Hawaii, 271–286.

Petit, S., and H. de Foresta. 1996. Precious Woods from the Agroforests of Sumatra, where Timber Provides a Solid Source of Income. *Agroforestry Today* 9(4), 18–20.

Rappard, F.W. 1937. Oorspronkelijke Bijdragen: de Damar van Bengkoelen (The Damar of Bengkulu). *Tectona* D1(30), 897–915.

Sardjono, M.A. 1992. Lembo Culture in East Kalimantan: A Model for the Development of Agroforestry Land Use in the Humid Tropics. *GFG-Report* 21, 45–62.

Scholz, U. 1982. *Decrease and Revival of Shifting Cultivation in the Tropics of Southeast Asia: The Examples of Sumatra and Thailand.* Ministry of Agriculture, Republic of Indonesia, Agency for Agricultural Research and Development, Central Research Institute for Food Crops (CRIFC), Bogor, Indonesia.

Sevin, O. 1989. Histoire et Peuplement (History and Population). Transmigration et Migration Spontanées en Indonésie (Transmigration and Spontaneous Migrations in Indonesia), Departemen Transmigrasi–ORSTOM, 13–123.

Sibuea, T.T.H, and D. Herdimansyah. 1993. The Variety of Mammal Species in the Agroforest Areas of Krui (Lampung), Muara Bungo (Jambi), and Maninjau (West Sumatra). Final research report, ORSTOM / BIOTROP and HIMBIO. Unpublished report. Bogor, Indonesia.

Siebert, S.F. 1989. The Dilemma of Dwindling Resources: Rattan in Kerinci, Sumatra. *Principes* 32(2), 79–97.

Thiollay, J.M. 1995. The Role of Traditional Agroforests in the Conservation of Rain Forest Bird Diversity in Sumatra. *Conservation Biology* 9(2), 335–353.

Torquebiau, E. 1984. Man-Made Dipterocarp Forest in Sumatra. *Agroforestry Systems* 2(2), 103–128.

van der Koppel, C. 1932. *De Economische Beteekenis der Ned. Indische Harsen* (The Economic Significance of Dutch East Indies Resins). Batavia (Jakarta): Kolff.

Weinstock, J.A. 1983. Rattan: Ecological Balance in a Borneo Rainforest Swidden. *Economic Botany* 37(1), 58–68.

———. 1989. Shifting Cultivation and Agroforestry in Indonesia, Some Notes and Data. In: *Proceedings of the Joint Seminar on Watershed Research and Management*, Bogor and Balikpapan, Indonesia.

Wijayanto, N. 1993. Potensi Pohon Kebun Campuran Damar Mata Kucing di Desa Pahmungan, Krui, Lampung. Field report. ORSTOM / BIOTROP. Bogor, Indonesia.

Chapter 46

Upland Fallow Management with *Styrax tonkinensis* for Benzoin Production in Northern Lao P.D.R.

Manfred Fischer, Sianouvong Savathvong, and Khongsak Pinyopusarerk

The *Styrax* genus consists of about 100 species of deciduous and evergreen trees and shrubs. They are found in the warmer, moister parts of Asia and America, and one species is native to the Mediterranean region. Many species of this genus have some form of medicinal use. *Styrax tonkinensis,* a member of the Styracaceae family, is perhaps best known as the source of *benzoin*, a balsamic resin widely used in the fragrance industry. The collection and sale of benzoin resin from *S. tonkinensis* has been an important cottage industry among the highland people of northern Lao P.D.R. for more than a century. There, *S. tonkinensis* is used as a fallow plant in the traditional slash-and-burn cultivation cycle, following the harvest of agricultural crops. Although its wood has no commercial value in Laos, *S. tonkinensis* is recognized as a source of pulpwood in Vietnam.

This chapter describes fallow management with *Styrax tonkinensis* for benzoin production (see color plate 64). Much of the information was gathered in the area of Nam Bak, a district of Luang Prabang Province, supported by field observations in other benzoin-producing areas.

Styrax tonkinensis

Styrax tonkinensis is a semideciduous tree that grows up to 25 m tall, with a diameter at breast height (dbh) of 30 cm. It is found mainly in secondary forests in the northern parts of Laos and Vietnam (see Figure 46-1). In Laos, it is found mainly at high altitudes between 800 and 1,600 m above sea level (asl). However, in Vietnam it is distributed over low to medium altitudes, mostly below 1000 m asl (Pinyopusarerk 1994).

Most stands of *Styrax tonkinensis* occupy sites previously used for slash-and-burn agriculture. Fires deliberately lit during the shifting cultivation cycle accelerate seed germination. It has many pioneer characteristics: a demand for light, regular production of viable seeds, and rapid early growth. It is, therefore, capable of invading

Manfred Fischer, Consultant in Forestry and Rural Development, Grunberger Strasse 13, D-10243 Berlin, Germany; Sianouvong Savathvong, Chief, Provincial Forestry Office, P.O. Box 530, Luang Prabang, Lao P.D.R; Khongsak Pinyopusarerk, CSIRO Forestry and Forest Products, P.O. Box E4008, Kingston, Canberra, ACT 2604, Australia.

gaps in the forest and, under favorable conditions, may occupy many hectares in almost pure stands.

Benzoin Resin

Benzoin is one of the oldest internationally traded products of Laos. It was mentioned in the travel report of a European who visited the region in 1640, and it can be safely assumed that benzoin has been marketed for a considerably longer period than that. It is known commercially as *Siam benzoin*, or *gum benjamin*. It is used in western pharmaceuticals as an inhalant, or externally as a mild antiseptic, as well as in traditional Chinese medicine. It is used for incense and for the production of fragrances, which are then employed in a wide range of end products such as soap, shampoo, and perfumes. It also has potential as a food additive.

The most important markets for Siam benzoin are in Europe. In Asia, the main buyers are Thailand, China, Vietnam, and Indonesia, and the main trading centers are Bangkok and Singapore. The main user countries are France, the United Kingdom, and Germany. Despite its worldwide use, the demand for benzoin resin has declined over recent years, probably because of increased use of synthetic materials. Prices at farm level have dropped considerably as a consequence. Although there is a long tradition of benzoin production in Lao P.D.R., its domestic use and local processing are minimal.

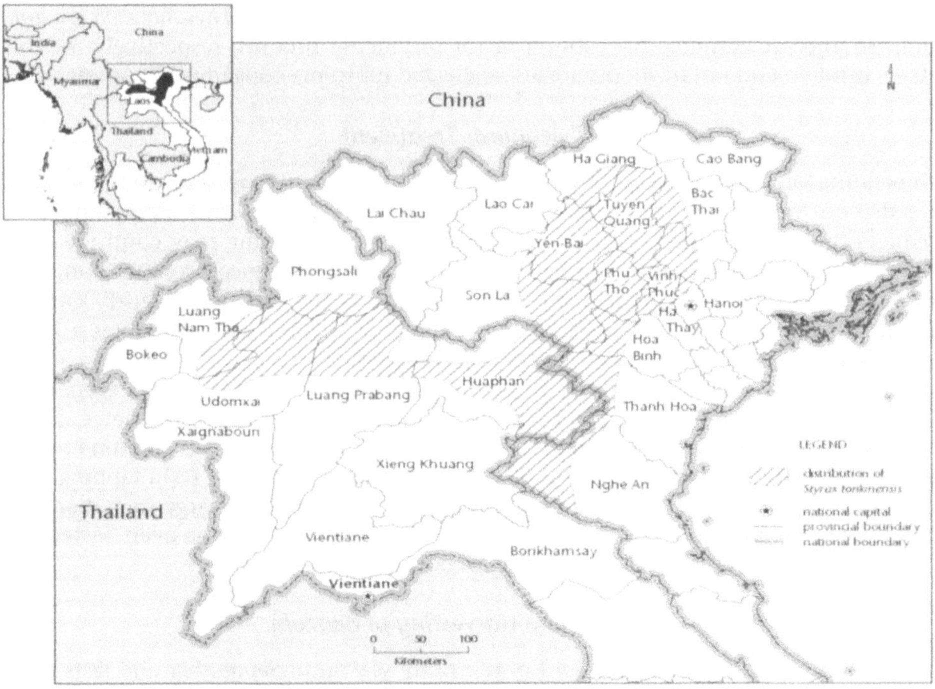

Figure 46-1. Distribution of *Styrax tonkinensis* in Laos and Vietnam
Source: Pinyopusarerk 1994.

Styrax tonkinensis as a Fallow Plant

Site Preparation and Cultivation

Due to natural conditions in northern Laos, shifting cultivators are often forced to slash and burn land on steep slopes at high elevations. In these areas, *Styrax tonkinensis* is indigenous. It regenerates very well in gaps, provided undergrowth is not too heavy, because saplings are sensitive to competition in their early years. Present stands of *S. tonkinensis* in the area are the result of natural regeneration. In October and November, seeds mature and fall to the ground. Those seeds remaining on the trees are knocked down when swiddens are slashed between the end of February and April, and burning of the dry vegetation from March until the beginning of May accelerates germination. Farmers do not plant *Styrax* seeds, nor is it artificially propagated. Therefore, the trees are irregularly spread and growth is usually very variable within each stand.

The main agricultural crop in the swiddens of northern Laos is glutinous rice. Secondary crops include maize, cassava, taro, sweet potato, cucumber, ginger, pumpkin, chili, sesame, and eggplant. The rice and other crops are usually cultivated for one year, after which the area is fallowed. If the soil is sufficiently fertile, the plot may be cultivated for a second year, but this is rare. Every phase of the ensuing fallow has its particular uses for the shifting cultivators, until the area is once again cleared and burned for cultivation. In the first years of the fallow period, the abandoned land is used for grazing cattle, and *S. tonkinensis* is one of the fodder plants. During and after the grazing period, grasses such as *Imperata cylindrica* and *Thysanolaena maxima* flourish and are gathered for roofing or for producing brooms, respectively. At this stage, bamboo and rattan shoots are also collected for home consumption or sale.

Silvicultural Treatment

Depending on the practices of individual families, *S. tonkinensis* seedlings may be weeded out or thinned during the first two years, if their growth is abundant. This reduces their competition with the agricultural crops. Thinning may continue until the fourth year, enabling the young trees to develop with reduced competition from their neighbors and to increase in diameter. The farmers also frequently cut the terminal shoot of the *Styrax* trees. They explain that this not only promotes a wider stem diameter, but also stimulates benzoin production.

After four to six years, the tree canopy in the fallow field closes and a regenerated forest develops. At the age of about six years, the *Styrax* trees' diameter differs considerably. There may be as many as 300 trees per hectare. This is the time when, at the end of the rainy season, about 50 of the trees are selected for resin tapping. The criteria for selection, based on the experience of the tappers, are a large stem diameter and thick, rough bark. The remaining trees in the field are not even tested for tapping. The tappers consider this a waste of both time and labor.

Tapping and Harvesting of Benzoin

The tapping season starts at the end of the rainy season in September and lasts until November. The procedure is similar in all benzoin-producing provinces of Laos. The tools include a traditional Lao forest knife and two or three bamboo sticks 30 to 50 cm in length, each with an attached rope 1.5 to 2 m long.

The tapper begins by cutting the bark about 50 cm above the ground on two opposite sides of the tree. Cutting notches between 5 and 10 cm long, the tapper moves up the tree and cuts similar notches every 30 to 50 cm. The resin accumulates mainly in the space between the bark opening and the stem. The maximum tapping height is 5 to 7 m, depending on the diameter of the tree and the clear height of the bole. The uppermost notches are made with the aid of the bamboo sticks. At a height of about 1.7 m, one stick is fixed to the trunk of the tree with the rope. The tapper then uses the stick like the step of a ladder on which to sit or stand. On average, one person can tap between 30 and 50 trees per day.

The first cool days of winter make the sap hard and brittle, and it is collected in February and March. The harvest uses the same tools as well as two bamboo baskets, one for catching the benzoin at the stem, and a larger one for carrying it home. A countercut is made below each bark pocket. The piece of bark, with the attached resin, is carefully removed. Each tree is capable of producing between 300 grams and 1 kg of benzoin per year. The tapping and harvesting is men's work, while cleaning and sorting is mainly done by the women. Care is necessary during the harvest and transport of the benzoin, because its market value depends on its purity and size.

Tapping can continue for six to seven years in regenerated forests, or until the trees die. According to the tappers, the ability of the trees to produce benzoin begins to decline at the age of 12, and they die naturally at about 15 years old. It is not clear if the tapping is implicated in the death of the trees, since untapped trees reportedly die at a similar age.

Land Tenure and Management

The land on which *S. tonkinensis* grows is usually fallow and is therefore mostly owned and managed by individual families. While exploitation of the *Styrax* is reserved for the owners, access to other nontimber forest products (NTFPs) is open to anybody.

In many cases, land tenure is still based on traditional individual land-use rights. Individual title in shifting cultivation areas of Lao P.D.R. is traditionally acquired by bringing unclaimed land under cultivation. Until recently, land was an unlimited resource, and its use did not require outside control. However, increasing population pressure and competing land-use objectives prompted the government to introduce a variety of regulations regarding the acquisition of land. In Luang Prabang Province, each household receives legal tenure over four plots of land, according to its labor capacity. In practice, between 1.5 and 2 ha are allocated for use per family, or between 6 and 8 ha for a four-year rotation (Hansen and Sodarak 1997). The shorter fallow period enforced by these regulations means there is insufficient time for natural regeneration of soil fertility. More important to the practice under discussion here is the fact that benzoin production is no longer possible because a minimum fallow period of about 10 years is needed.

Economic and Social Aspects of Benzoin Production

The economic importance of benzoin in northern Lao P.D.R. depends on a number of factors:

- The area of *Styrax* forest available per household;
- The male labor force available;
- Access to markets or traders;
- Actual benzoin prices; and
- Real and possible alternatives for generating cash income.

The last two factors, in particular, are crucial for the tappers. Over the past few decades, the farm price for benzoin has fallen continuously to between US$1.50 and US$2.00 per kilogram of mixed-grade resin. In the 1950s, by comparison, the price was between 5 and 6 French silver coins per kilogram, with a current value of more than US$40. As a consequence, many tappers have reduced or even stopped benzoin production, particularly if they have alternative means of generating income.

The main alternatives are other NTFPs and livestock husbandry. Presently, the main NTFPs collected to generate cash income are cardamom (*Amomum* spp.), broomgrass (*Thysanolaena maxima*), *Alpinia malaccensis*, *Zanthoxylum rhetsa*, bamboo and bamboo shoots, paper mulberry (*Broussonetia papyrifera*), rattan, rattan shoots, and rattan seeds. Considering its economic importance to villagers, benzoin ranks behind cardamom. Compared with livestock, paper mulberry, or broomgrass, its importance depends on the village and household situation. Generally, benzoin is far less important than it was around 20 years ago.

Traditionally, people of the Khamu (Lao Theung) ethnic group live at the elevation at which benzoin tapping is practiced.[1] They, together with the lowland Lao (Lao Lum), are the people mainly involved in the cultivation and tapping of *Styrax tonkinensis*. However, as mentioned earlier, low prices for benzoin have led many families to stop their tapping activities. Significantly, more Lao Lum have stopped, whereas the Lao Theung can ill afford to lose this source of income, and must persevere.

Potentials and Constraints of Benzoin Production in Laos

Potentials

- Benzoin is one of the few sources of cash income available to upland people in northern Laos.
- *Styrax tonkinensis* is a multipurpose tree. In Vietnam, it is commonly used for pulpwood production.
- Benzoin production needs a minimum fallow period of six to eight years, providing the added benefit of allowing the soil to recover its fertility.
- During the fallow period, many other products beside benzoin can be gathered on fallow land.
- Marketing channels for benzoin already exist.
- Benzoin has a low volume per kg. This is an advantage for transport in Laos, where the road network is poorly developed.
- Benzoin can be produced on a small scale as a cottage industry.
- No capital investments are necessary to start producing benzoin.
- The tappers possess considerable traditional experience in benzoin production.
- Unlike other cash crops, benzoin is well known in the uplands of northern Laos, and its production has a high acceptance in rural communities there.

Constraints

- Falling prices for raw benzoin have reduced farmer interest in producing it.
- The tendency toward shorter fallow periods in Laos has reduced the area where benzoin production is possible.
- Existing knowledge on silvicultural and processing aspects of benzoin production is still sketchy and unproven.
- The silvicultural treatment of *S. tonkinensis* has been limited. For instance, there are no attempts to genetically improve production stock, such as selection of parent trees.
- No processing of benzoin is done in Laos. The potential for adding value to the product is not exploited.
- Benzoin is only used for export, and its price is dependent on international markets.
- The demand for raw benzoin in Laos comes from only a few traders. They are in a position to dictate prices paid at the village level.
- Benzoin degrades in quality during transport and storage.
- *Styrax* trees do not start producing benzoin until a minimum age of six years.
- The yield per hectare is relatively low, about 15 to 25 kg.
- The traditional tapping technique produces a high percentage of low-grade benzoin.

---

[1] The ethnic groups in Lao P.D.R. are divided into three main categories, according to the elevation of their area: *Lao Lum*, living traditionally in the lowlands; *Lao Theung*, living traditionally on the mountain slopes; and *Lao Sung*, living traditionally on the mountaintops.

Conclusions and Recommendations

The traditional way of producing benzoin, as a part of the shifting cultivation system in northern Laos, is well adapted to local conditions and ensures soil fertility. On the other hand, it shows weaknesses in the silvicultural and processing aspects. This incomplete knowledge should be improved and existing local knowledge verified.

The conditions under which benzoin is produced in Laos have changed considerably, with land becoming scarce and fallow periods shortening. Besides improving the present benzoin production system, agronomists should also assess the possibilities for producing it in plantations or agroforestry systems, so that the minimum time needed for production of benzoin is ensured and other products from the *Styrax* trees can be exploited.

Low prices for benzoin are presently the main constraint to expansion of this product. Without identification of new and stable markets offering attractive prices, farmers' interest in producing it will remain low. Lack of information about production conditions and marketing aspects of benzoin, both within Laos and outside, is another constraint.

The Food and Agriculture Organization of the United Nations is currently providing technical assistance to the government of Lao P.D.R. under a project titled "Improved Benzoin Production" (TCP/LAO/6611). The expected outputs are:

- A silvicultural research and tree improvement program on *Styrax tonkinensis* for benzoin production;
- Agroforestry trials with *Styrax tonkinensis* in shifting cultivation areas;
- Improved processing of benzoin using better tapping techniques, transport, storage, and packaging;
- A market study on benzoin;
- A socioeconomic study in selected benzoin-producing villages;
- Production of extension materials for farmers; and
- Training for technicians and farmers on all aspects of benzoin production.

Acknowledgments

We wish to thank Ms. Silke Stoeber and Mr. Peter Hansen for their helpful comments on an earlier draft version of this chapter.

Additional References

Chazee, L. 1994. Shifting Cultivation Practices in Laos. Present Systems and their Future. Proceedings of Workshop at Nabong Agricultural College. Vientiane, Lao P.D.R: UNDP.

Coppen, J.J.W. 1995. Gums, Resins and Latexes of Plant Origin. FAO (Food and Agriculture Organization of the United Nations) Non-Wood Forest Products series. Vol. 6. Rome: FAO.

———. 1997. *Gum Benzoin: Its Markets and Marketing and the Opportunities and Constraints to Their Improvement in Lao P.D.R.* FAO/TCP Benzoin Project consulting report for FAO Bangkok.

De Beer, J.H. 1993. Benzoin, *Styrax tonkinensis*. In: *Non-Wood Forest Products in Indochina. Focus: Vietnam.* FAO Working Paper FO: Misc/93/5. Rome: FAO

Fischer, M. 1998. Physical and Socio-Economic Conditions of Benzoin Production in Northern Laos. Report on a survey in two villages in Nam Bak district, Luang Prabang Province. Vientiane, Lao P.D.R: FAO/TCP Benzoin Project.

Foppes, J., and S. Ketphanh. 1997. The Use of Non-Timber Forest Products in Lao P.D.R. Paper presented at the Workshop on Sustainable Management of Non-Wood Forest Products, 14–17 October. Selangor, Malaysia.

Hansen, P.K., and H. Sodarak. 1997. Potentials and Constraints of Shifting Cultivation Stabilization in Lao P.D.R. Paper presented to an Asian Development Bank/Department of Forestry Seminar, August 2–4, 1997, Vientiane, Lao P.D.R.

Hoesen, D.S.H. 2000. *Styrax* L. In: *Plant Resources of South-East Asia* (PROSEA) No. 18, Plants Producing Exudates, edited by E. Boer and A.B. Ella. Leiden, Netherlands: Backhuys Publishers, 112–119.

Kashio, M. 2001. Monograph on Benzoin. FAO Publication Series.

Ketphanh, S., and V. Soydara. 1998. The Use of Non-Timber Forest Products in Northern Lao PDR. Paper presented at the Sino-Lao Trans-Boundary Biodiversity Management and Development Workshop, October 26–29, Xishuangbanna, Yunnan, China.

Pinyopusarerk, K. 1994. Styrax tonkinensis: *Taxonomy, Ecology, Silviculture and Uses*. ACIAR Technical Report 31. Canberra, Australia: Australian Centre for International Agricultural Research, 14.

van Gansberghe, Dirk, and Sayamang Vongsack (eds.) 1994. *Shifting Cultivation Systems and Rural Development in the Lao P.D.R.*, Proceedings of a Workshop, July 14–16, 1993, Nabong Agricultural College, Vientiane, Lao P.D.R.

Vidal, J. 1959. Noms Vernaculaires de Plantes (Lao, Meo, Kha) en Usage au Laos. Ecole Francaise D'Extreme-Orient. Paris.

Chapter 47

The Lemo System of Lacquer Agroforestry in Yunnan, China

Long Chun-Lin

Lacquer is a kind of resin widely used in antiseptic preparations and antirust paint. It is an important traditional nontimber forest product in China. The most common lacquer resin is harvested from *Toxicodendron vernicifluum* Barkley, a tree from the family Anacardiaceae. Because *T. vernicifluum* is poisonous, it is usually cultivated on dry terraces and uplands where it is far from food crops. The Lemo are unique as the only ethnic group that propagates lacquer trees in their swidden fields following an agroforestry approach.

The Lemo are a branch of the Bai ethnic group in Lushui county, Nujiang prefecture, in Northwest Yunnan Province, China. They have a population of about 6,000. They live mainly in the Nujiang Gorge or the Thanlwin River valley. Other ethnic groups such as Lisu, Nu, Dulong, Tibet, Yi, and Dai also live in this area. It is very rich in biological resources. The main vegetation types include evergreen broadleaf forest, conifer forest, bamboo forest, scrub, and grassland.

The Lemo use about 300 species of wild plants for different purposes, including timber, food, medicines, fibers, ornaments, weaving, fodder, and resins. Their environment presents challenges to the development of sustainable land-use systems. The altitude of their villages varies from 900 to 3,000 m above sea level (asl) and the land is characterized by steep slopes and poor soil. Despite these harsh conditions, the Lemo have developed an agroforestry system consisting of *Toxicodendron vernicifluum* and *Alnus nepalensis,* together with food crops, to sustain and improve their livelihoods.

Buffered from outside influences by their remoteness, the Lemo retain their traditions in religion, culture, production practices, and lifestyles. They believe in a plurality of gods and worship particular forests, trees, lands, mountains, and rivers. In ancient times, the Lemo used the leaves of plants to communicate. Today, some young people still use plant leaves to express their love (Long and Wang 1994). Their crop production systems include swidden and paddy cultivation, the harvest of timber and nontimber forest products, and animal husbandry. Among these, swiddening is the oldest and most important practice.

Toxicodendron vernicifluum is a traditional resin-producing tree that is both fast growing and widely distributed in southern China. Its resin is a popular lacquer, its seeds are oil-bearing and can be used for industrial purposes, and its timber is yellow and is used in constructing special furniture. Although a rich abundance of agroforestry patterns have been developed in China, the lacquer agroforestry system described in this chapter has not been previously recorded (Guo and Padoch 1995; Zou and Sanford 1990). By comparison, the damar system, a resin-producing agroforestry system in Indonesia that is similar to that of the Lemo, has been studied intensively (de Foresta

Long Chun-Lin, Kunming Institute of Botany, Chinese Academy of Sciences, Kunming 650204, China.

and Michon 1993, 1994a, 1994b; Michon, de Foresta, and Aliadi 1996). (See also Michon et al. Chapter 45.)

Lemo Swidden Cultivation

About 63% of food and cash income in the Lemo community comes from swidden cultivation. They grow corn, buckwheat (*Fagopyrum esculentum*), potato, millet (*Setaria italica*), broomcorn millet (*Panicum miliaceum*), and some vegetables in their swidden fields, in combination with lacquer and alder (*Alnus nepalensis*) trees (Yin 1994).

According to the altitude, characteristics, and quality of the land, the Lemo divide their swidden fields into three categories: *tongkong*, *shenji*, and *kongji*. Tongkong means "the land to be reclaimed," and its altitude is between 800 and 1,700 m asl. Shenji means "the land to be burnt," and its altitude ranges between 1,700 and 1,900 m asl. Kongji means "the land with cultivated lacquer and alder trees," and it occurs at altitudes between 800 and 2,300 m asl.

The tongkong is reclaimed by digging, between January and March. After one month, the dry grasses and shrubs on the fields are collected for burning. Then the tongkong is prepared for crop cultivation. Food crops, mostly corn and some millet, soybean, or pumpkins, are planted in April. If the soil is poor, tongkong will be left fallow after one year of cultivation. However, fields with more fertile soils can be cropped for as long as 15 years. The fallow period is usually three to four years but sometimes more.

The shenji is cleared by cutting trees and scrub in January. The dry trees, shrubs, and branches are burned in April, and the crops are planted two to four days later. Corn, millet, and peas are grown in shenji for one year, and then the land is fallowed, usually for four to five years, but sometimes for as long as 10 years. The process is similar to that of the tongkong.

Tongkong and shenji can become kongji if lacquer or alder trees, or both, are interplanted into them. The kongji is cleared in the same fashion as the tongkong and shenji. In March, seedlings of lacquer or alder, or both, are planted, and food crops like corn, millet, or pumpkins are interplanted around them. After three to four years, when the tree canopy begins to close, the fallow period begins. The fallow duration is seven to nine years when alder trees are planted, or 16 to 20 years if *T. vernicifluum* is grown on its own. Lemo swidden cultivation is summarized in Table 47-1.

Cultivation and Management of Lacquer Trees

Every Lemo household establishes its own nursery to propagate lacquer seedlings. The nursery is usually situated in an area with very rich soil in the highlands, usually above 2,000 m asl. In winter, a field intended for a nursery is cleared by cutting the trees, shrubs, and grasses. After the slash dries, it is burned and the ash scattered across the nursery. The land is then prepared by light tillage and when convenient, application of farmyard manure. The tree seeds are sown in February or March at densities varying between 30 and 60 seeds per square meter. After sowing, the field is lightly irrigated if the land is considered too dry. The lacquer seeds germinate in April with the arrival of the rainy season.

Transplanting Seedlings

By the following March, the lacquer seedlings will usually have grown as high as about 50 cm. They are transplanted into tongkong or shenji, a job that is finished before the rainy season begins. About 600 to 800 seedlings are planted per hectare in fertile fields, and 700 to 900 on poorer fields. Sometimes *A. nepalensis* seedlings are planted together with *T. vernicifluum* on the same field. One or two months later, food crops are interplanted between the tree seedlings.

Managing and Harvesting Lacquer

Lacquer trees are managed by individual households. When Lemo farmers tend to their crops growing in the same fields, the young lacquer trees benefit at the same time. Weeding is done twice yearly when the lacquer trees are young, and excessive branches are pruned back.

Lacquer harvesting cannot begin until the trees are eight years old. Lacquer is tapped near the base of the trunk, using methods similar to those used in collecting pine resins and rubber. Lacquer trees can be economically tapped for 7 to 10 years, after which they are cut and burned for the next rotation of crop cultivation. Fallows cleared from lacquer trees are widely believed to be more fertile than those after a natural succession.

Economic and Environmental Benefits

The lacquer tree is a cash crop unique to Lemo society, and about 20 years ago it was their only source of cash income. They relied on it for buying clothes, salt, iron tools, and other items for daily use. Today, lacquer continues to be the Lemos' main source of income.

Each household has an average of about 1.2 ha planted in lacquer trees and harvests about 250 kg of lacquer every year. During their productive life, the trees yield an average of 210 kg per hectare per year. It is sold to local supply and marketing cooperatives and the price fluctuates according to outside demand. It is divided into three quality classes: good, medium, and bad. In 1995, good-class lacquer was selling for 8 yuan, or less than US$1, per kilogram, and more recently its price has steadily decreased. However, lacquer sales constitute up to 85% of total cash income for Lemo households, so they persevere with traditional planting of lacquer trees into swidden fields and developing lacquer agroforests. Complex agroforests are often similar to natural forests. For example, the damar agroforests of Indonesia and traditional tea gardens in Yunnan, China, mimic forests in structure and function (de Foresta and Michon 1994a; Garrity 1993; Long and Wang 1996; Saint-Pierre 1991). In the case of the *T. vernicifluum* managed by the Lemo in Yunnan, the forest-like landscape is very impressive.

Table 47-1. The Lemo Swidden Cultivation System

Characteristics	*Tongkong*	*Shenji*	*Kongji*
Altitude (m)	800–1,700	1,700–1,900	800–2,300
Land preparation	Digging and burning	Slashing and burning	Slashing and burning or digging
Crops	Corn, soybean, millet, pumpkin	Corn, millet, peas	Corn, millet or pumpkin
Farming duration	1(–15) years	1(2–3) years	3–4 years
Fallow duration	3–4 (–7) years	4–5(–10) years	7–16 (–18–20) years
Tenure	Common property	Common property	Private property
Fallow management	None	None	Growing lacquer or alder trees

The vertical structure of vegetation in kongji fallows consists of two or three layers. The canopy layer, composed mainly of *Toxicodendron vernicifluum* (lacquer) and *Alnus nepalensis* (alder), is 10 to 22 m high. Young alders, *Rhus chinensis*, *Castanopsis* spp., and *Eurya* spp. compose a second layer 3 to 10 m above the forest floor. The third layer, less than 3 m high, is occupied by *Pueraria peduncularis, Desmodium* spp., *Musa acuminata, Conyza canadensis,* and ferns.

The lacquer agroforests also deliver environmental benefits. The uplands in the Lemo area are very steep, with swidden fields in Nujiang Gorge often on slopes as steep as 45°. Sometimes only local people can reach the swidden fields because of the precipitous slopes. Monoculture cropping in this environment can easily compound soil erosion problems, but lacquer agroforests protect both soil and water resources. The Lemo say that without lacquer and alder trees in their swidden fields they would have to grow food crops on rocks. In addition, lacquer agroforests produce substantial quantities of organic matter, thereby improving soil structure and the biophysical environment for growing food crops.

Kongji: Private Property

The Lemo lacquer agroforests are privately owned. The Lemo never plant lacquer trees in common fields because there is open access and anyone can harvest the resin. In former times, an individual Lemo household could claim land from the commons by opening forests and planting lacquer trees on the cleared land. This customary law no longer exists; however, in some villages, lacquer trees that were planted on common fields continue to be recognized as belonging to the original planters.

From the 1960s to 1983, all resources in China became the property of the state or collectives, and at that time, the Lemos' kongji fell under the ownership of production collectives. This change of tenure had a dramatic effect on the traditional lacquer agroforests. Having planted *T. vernicifluum* trees, farmers found they no longer had the right to harvest the lacquer, so they abandoned the practice, and the traditional lacquer agroforests fell into degradation.

In 1983, the Chinese government implemented a new land-use policy, and the Lemo reallocated their swidden fields back to individual households. Every household planted lacquer and alder trees in its kongji, and lacquer agroforests began thriving once again. However, ownership of modern kongji has changed from the traditional tenurial system. Farmers now have rights to use the land for only 30 to 50 years, through contracts between local government and individual households. Therefore, the Lemo have security of tenure for only two or three cycles of the lacquer agroforest system.

Lacquer Planting and Tapping: The Domain of Men

According to tradition, only Lemo men have the right to plant lacquer seedlings and collect the resin. They look after all the processes related to lacquer cultivation, including management, tapping, transportation, and sales. Every man must master the skills of lacquer planting and tapping, beginning as young boys to learn these skills from their fathers. There is a local saying that any Lemo man without skills in managing lacquer will lose his life. Certainly, Lemo girls pay more attention to young men skillful in lacquer cultivation and tapping when considering a suitable marriage partner.

In the years when young lacquer trees are becoming established, Lemo girls and women help to maintain them, often as they attend to the interplanted food crops. But they are generally not permitted to plant lacquer trees or collect the resin. However, there is one exception to that rule: women from households without a son are permitted to tap lacquer trees.

Discussion

Origin of Agroforestry Systems in Lemo Society

The Lemo have lived on and farmed the steeply sloping uplands of the Nujiang Gorge area since ancient times, and they have developed agroforestry systems to protect their soil and water resources and to maintain their agroecosystems. The combination of *Alnus nepalensis* with food crops was probably one of the earliest types of agroforestry developed by the Lemo. *A. nepalensis* is a fast-growing species widely distributed in the area, and it has multiple uses. The modern lacquer agroforests draw on the same agroforestry design as that involving *A. nepalensis* on its own, with food crops.

It is interesting to note that the Lemo do not suffer any reaction to the lacquer, whereas most outsiders who touch it are affected by its poison. This prompts the obvious question: do the Lemo have a secret method of protecting themselves, or have they developed a natural immunity? The question remains unanswered, but it is a curious fact that may provide insights into why the Lemo are the only group known to manage lacquer trees in their swiddens. It may also help to explain the dynamics that led their ancestors to develop the system. The Lemo have both immunity to lacquer poison and a need for the cash income it provides. Therefore, the lacquer agroforestry system and its practitioners are a harmonious match.

Is this a Fallow Improvement System?

The Lemo lacquer agroforests constitute a highly valued swidden agroecosystem, both economically and environmentally, but can they be classified as fallow improvement? In general, leguminous crops are candidates for enriched fallows (Garrity 1993). *Alnus nepalensis* is another promising candidate because of its ability to fix nitrogen. The lacquer agroforests are undoubtedly not as efficient as nitrogen-fixing trees in rejuvenating soils and performing other ecological functions, but their economic value to the Lemo is extremely important. All factors considered, the lacquer agroforests can be regarded as one of the best fallow improvement systems in the Nujiang Gorge area.

Lacquer Price Decline

In recent years, many synthetic products have been developed as effective substitutes for natural resins, including lacquer. Lacquer prices have dropped accordingly. In 1996, the selling price for good-quality lacquer was only 7.4 yuan, or about US$0.90 per kilogram. Although the Lemo are worried about declining prices, it is difficult for them to develop alternative land-use systems because of the biophysical limitations of their environment. The implications of the tumbling prices to the economy, ecology, and culture of the Lemo are not yet known.

Acknowledgments

I would like to thank Malcolm Cairns for sending me useful references and encouraging me to write this chapter. Some of its data were provided by Professor Li Heng, from Kunming Institute of Botany, and by colleagues from the Nujiang prefecture of northwestern Yunnan. This work was supported by the Chinese Academy of Sciences (KSCX2-SW-117), the National Science Research Foundation of China (30170102), the Natural Science Foundation of the United States (DEB-0103795), and the Ministry of Science and Technology of China (2004DKA30430 and 2005DKA21006).

References

de Foresta, H., and G. Michon. 1993. Creation and Management of Rural Agroforests in Indonesia: Potential Applications in Africa. In: *Tropical Forests, People and Food* 13, Man and Biosphere Series, Paris, France: UNESCO, 709–724.

———. 1994a. From Shifting Cultivation to Forest Management through Agroforestry: Smallholder Damar Agroforests in West Lampung (Sumatra). *APA News* 6/7, 12–16.

———. 1994b. Agroforests in West Lampung (Sumatra): Where Ecology Meets Economy. *Agroforestry Systems* 12, 12–13.

Garrity, D.P. 1993. Sustainable Land Use Systems for Sloping Uplands in Southeast Asia: Technologies for Sustainable Agriculture in the Tropics. ASA Special Publication 56.

Guo, H.J., and C. Padoch. 1995. Patterns and Management of Agroforestry Systems in Yunnan: An Approach to Upland Rural Development. *Global Environment Change* 5(4): 273–279.

Long, C.L. and J.R. Wang. 1994. On Social and Cultural Values of Ethnobotany. *Journal of Plant Resources and Environment* 3(2), 45–50. (Chinese language).

———. 1996. Studies of Traditional Tea Gardens of Jinuo Nationality, China. In: *Ethnobiology in Human Welfare*, edited by S.K. Jain. New Delhi: Deep Publications, 339–344.

Michon, G., H. de Foresta, and A. Aliadi. 1996. Damar Resins, from Extraction to Cultivation: An "Agroforest Strategy" for Forest Resource Appropriation in Indonesia. In: *Ethnobiology in Human Welfare,* edited by S.K Jain. New Delhi: Deep Publications, 454–459

Saint-Pierre, C. 1991. Evolution of Agroforestry in the Xishuangbanna Region of Tropical China. *Agroforestry Systems* 13, 159–176.

Yin, S.T. 1994. *A Farming Culture Born out of Forests: Swiddening in Yunnan, China*. Kunming: Yunnan People's Press. (Chinese language).

Zou, X.M., and R.L. Sanford. 1990. Agroforestry Systems in China: A Survey and Classification. *Agroforestry Systems* 11, 85–94.

Chapter 48
From Shifting Cultivation to Sustainable Jungle Rubber

A History of Innovations in Indonesia

Eric Penot

This chapter sets out to show how, on the plains of Sumatra and Kalimantan, fallow management has provided an intermediate stage in evolution from shifting cultivation of upland rice to a more sustainable complex agroforestry system based on rubber. This evolution will be described from two perspectives: farmer-generated innovations, using indigenous knowledge, and their adoption of innovations from outside. These two processes are fundamentally different. They arise from farmers' responses to both market opportunities and the need to adopt more productive and competitive land-use systems. Introduction of rubber to the study area, and the subsequent development of local knowledge about its integration into farming systems, has led to the evolution of sustainable agroforestry practices. A long and important transitionary stage in this evolution has involved gradual improvements to the traditional management of the long, 30-year secondary forest fallow.

The Study Area

The study area encompasses those parts of the central plains of Sumatra and Kalimantan that are regarded as upland, but are below 500 m above sea level (asl). Above this altitude, rubber performance is generally considered marginal. These plains lie between mountainous zones and coastal swampy areas. Soils are leached ferralitic or red-yellow podzolic and, although poor, are suitable for tree crops such as rubber, pulp trees, and oil palm. Average annual rainfall is between 2,000 and 4,000 mm, and the climate is typically equatorial, with a minimum of 100 mm of rainfall per month and a short dry season. Poor soil fertility makes these areas generally unsuitable for intensive cultivation of food crops. They were almost entirely covered by primary forest in 1900, but within 100 years, most of the forest had been displaced by rubber, oil palm, and pulp tree plantations, as well as tree and food crops for government transmigration programs. Slashing and burning of old or young secondary forest or, more frequently, of old jungle rubber, is still a common practice for upland rice production. But in areas where land has become scarce, farmers have converted to permanent tree plantations and no longer grow rice. As farmers have specialized in rubber, they have become increasingly integrated into the cash economy.

Eric Penot, Centre de Coopération Internationale en Recherche Agronomique pour le Développement (CIRAD), B.P. 5035,34032 Montpellier, Cedex 1, France.

Rubber: The Trigger that Transformed Shifting Cultivation into Improved Fallow Management

Rubber (*Hevea brasiliensis*) was brought from Malaysia to North Sumatra, in Indonesia, by the Dutch at the end of the 19th century. It was originally planted in private estates, following the British example in western Malaysia. At the time, Sumatra and Kalimantan were sparsely inhabited, with only one to four people per square kilometer. Shifting cultivation was the predominant agricultural practice, characterized by slashing and burning of primary or old secondary forest, one or two years of upland rice cultivation, and a long fallow period of 30 to 40 years. The peneplains were still largely covered by primary forest, and land was plentiful. There was no particular pressure on farmers to modify their land-use system, which was sustainable as long as the population remained relatively low.

Rubber was first introduced into estates in North Sumatra, and later, between 1910 and 1920, Chinese traders spread it into the south. The first seedlings were introduced to Borneo in 1882 (Treemer 1864, as cited in Dove 1995). The Sarawak government distributed seeds to local people in 1908, and in Kalimantan, Chinese merchants, Catholic missionaries, and a Dutch private company, Nanga Jettah, handed out rubber seeds in 1909 (Uljee 1925, as cited in King 1988). The trees spread quickly into the Kapuas basin, the main river basin in West Kalimantan.

Rubber management in the estates was very intensive, with fertilizer applications and continuous weeding, requiring much labor and capital. However, local farmers rapidly recognized the opportunity offered by rubber production and began to collect seeds from nearby estates to plant in their own fields. They created their own cultivation system to suit their lack of cash and labor limitations. They interplanted rubber trees with upland rice in their swidden fields, using a higher planting density than that of the estates to compensate for tree losses due to competition and the depredations of wildlife. They ended up with between 300 and 500 productive trees per hectare, comparable with the estates (see color plate 61). The rubber was left to grow along with the regenerating forest when the fields reverted to fallow.

Although conventional wisdom at the time viewed rubber as more properly grown in estate monocultures, it proved to be very adaptive to its new forest environment. This was perhaps not surprising because rubber was a forest species in its natural habitat in the Amazon Basin of Brazil, and the management system innovated by the Indonesian farmers mimicked its natural South American environment. Moreover, in Southeast Asia's forests there was none of the crippling leaf blight that prevented its cultivation in pure plantations in its home environment.

The cultivation system of the Indonesian farmers soon became known as *jungle rubber*, because they considered it as basically a swidden fallow enriched with rubber trees. The productive lifespan of rubber, about 35 years, was the same as the traditional fallow period used locally to restore soil fertility and eliminate weeds. The Kantus Dayak, for instance, considered rubber gardens as "managed swidden fallows" (Cramb 1988). A key factor in this transition was a shift in labor demands, from a cyclical seasonal basis in cultivating upland rice, to a permanent daily basis, from 6 a.m. to 11 a.m., for rubber. However, there was no conflict between the two systems because the afternoons were still free for *ladang*, or upland farming, so farmers were easily able to integrate rubber into their traditional swidden systems. This was critical because, under conditions of abundant land and no capital, labor was the main productive input. At the time, rubber was not viewed as an alternative to upland rice, but this changed in some places as rubber management intensified in response to land-use pressures. Writing in 1993, Dove said this about the changed swiddening practice: "The comparative ecology and economy of rubber and upland swidden rice result in minimal competition in the use of land and labor, and even in mutual enhancement, between the two systems." He continued, developing the notion of "composite" systems: "There is little analysis of the relationship between the two systems (rubber and swidden agriculture with rice) and thus little understanding of why this combination historically proved to be so successful" (1993, *137)*.

Until the 1930s, both smallholder farmers and the estates were using the same unselected rubber planting material, and the two systems produced comparable yields of 500 kg/ha/year (Dijkman 1951).[1] Farmer investment in establishing jungle rubber remained minimal. It usually involved about four days of additional work/ha (Levang et al. 1997), and the resulting mixed jungle rubber forest also yielded fruits, nuts, timber for housing, rattan, wild vegetables, and other nontimber forest products (NTFPs). The comparative costs of establishing rubber plantations demonstrate the significant advantages of smallholder methods. These costs, expressed as a ratio comparing estate versus smallholder methods, were estimated at 13 to 1 during the colonial era (Dove 1995) and 6 to 1 in 1982 conditions. Compared with government rubber schemes, the ratio was between 3 to 1 and 11 to 1 in favor of smallholder methods (Barlow and Muharminto 1982). In the case of jungle rubber, the advantages are quite clear: establishment costs are low because unselected seeds are freely available and no fertilizers are used; there is a low investment in labor because the land has already been cleared for upland rice, and only a few days are needed for planting; and no maintenance is required while the rubber is still immature.

Jungle rubber has been well described (Gouyon 1995; de Foresta 1992a) and defined, from a botanical viewpoint, as a "complex agroforestry system" (de Foresta and Michon 1995). The fact mentioned above, that production per hectare from unselected rubber was very similar in both estate and smallholder systems prior to the 1930s, is very important because it demonstrates that rubber can maintain its yield and compete efficiently with a relatively large number of other trees. Whether the same would hold true for clonal rubber trees under the same conditions requires additional verification.

It is clear that rubber has contributed to deforestation (Prasetyo and Kumazaki 1995). But a paradox lies in the fact that jungle rubber has now become an important reservoir of biodiversity (de Foresta and Michon 1995) and is better adapted in this regard than alternative systems such as oil palm, coconut, coffee, cocoa, or pulp trees. Therefore, one could add conservation of biodiversity to the list of benefits, since biodiversity in mature old jungle rubber forests is close to that of primary forest or old secondary forest (de Foresta 1992a, de Foresta and Michon 1995), as are its environmental benefits in soil conservation (Sethuraj 1996) and hydrological functions, due to its forest-like characteristics. The biomass of a 33-year-old rubber plantation is 445 t/ha dry weight. This is similar to that of humid tropical evergreen forests in Brazil, about 473 t/ha (Jose et al. 1986, cited in Wan Abdul Rahaman et al. 1996; Sivanadyan and Norhayati Moris 1992), and in Malaysia, between 475 and 664 t/ha (Kato et al. 1978, cited in Wan Abdul Rahaman et al. 1996).

According to Sethuraj (1996), the potential photosynthetic capacity of rubber leaves is comparable to, and even better than, many other forest species. An area of 10 million ha planted with rubber worldwide would annually fix about 115 million metric tons of carbon. One-third of this would be in Indonesia. Soil fertility is maintained and even improved because rubber increases the nutrient content in the upper soil layer by providing between four and seven metric tons of leaf litter per hectare per year (Sethuraj 1996; Dijkman 1951). The removal of latex exports only a low level of nutrients, estimated at 20 to 30 kg of N, P, K, and Mg ha/year (Tillekeratne 1996; Compagnon 1986). However, rubber wood extraction implies a large nutrient export that should be replaced through heavy fertilization during replanting. Soil moisture is very high under rubber, accelerating the rate of decomposition and nutrient turnover. Mature rubber is a nutritionally self-sustaining

---

1 Rubber yields are always presented in dry rubber content (DRC) of 100%, and not in kg of raw material (rubber sheets or cup lumps) or liters of latex.

ecosystem, unlike, for instance, oil palm. Nutrient cycling is likely to approach that of forest ecosystems (Shorrocks 1995, cited in Tillekeratne 1996). These benefits are not considered by farmers as the main objectives of the system, but as indirect "gifts," comparable to those from a long-term fallow in the original swidden system.

The biodiversity of agroforests has provided the basis for trade between agroforesters and foreign traders, from as long ago as the 5th century A.D. in the case of the Chinese (Wolters 1967), and the Arabs after the 9th century. They sought various nontimber forest products, including resins, aromatics, nuts, and plants. Latex (*gutta percha*) was used for insulating marine telegraph cables as long ago as the 1840s (Dove 1995). However, rubber was the only one of all these NTFPs to trigger the large-scale development of agroforests. By the end of the 20th century, these had grown to cover more than 2.5 million hectares in Indonesia.

It is noteworthy that, originally, farmers were neither forced to adopt rubber, nor were they under pressure to intensify their land use. Rather, they were attracted to rubber by its adaptability to their local environment, its constant market demand, and the opportunity it provided to increase their incomes and improve their living standards. Its success over the past 150 years is now history. The discovery of vulcanization by J. Goodyear in 1839 and the subsequent development of the tire industry paved the way for extensive industrial use of natural rubber. The market has enjoyed a constantly growing demand and by 1997, world consumption of natural rubber had grown to around 6 million metric tonnes/year. In Sumatra, the average population density is now 35 persons per square kilometer, and land is becoming scarce in some provinces. The average area of jungle rubber per family is between 2.7 and 4 hectares (Barlow and Muharminto 1982; Gouyon 1995). In South Sumatra, rubber generates between 55% (Barlow and Muharminto 1982) and 80% (Gouyon 1995) of total farm income.

Constraints inherent in jungle rubber systems are, by comparison with the advantages, relatively minor. They include delayed production—because the trees cannot be tapped until they are 9 to 15 years of age, compared with 5 to 6 years of age in monoculture plantations on estates—and relatively low productivity when compared with the yield of plantation clones (Gouyon 1995).

The Development of Rubber Agroforests

Rubber has now been adopted by more than 1 million farmers in Indonesia. Their holdings total 3 million hectares, of which 2.5 million hectares are devoted to jungle rubber. Four key factors account for this rapid adoption:

- The wide availability of rubber seeds from estates and the ease with which rubber adapted to enrichment planting into regenerating forest growth;
- The availability of land and ease of large-scale expansion, originally through the river system;
- A reservoir of migrants from the overpopulated island of Java, who were favored as laborers by plantation estates. Their numbers were bolstered by spontaneous migration, later followed by official transmigration programs, some of which were centered on rubber; and
- The fact that planting rubber enabled land acquisition, giving the planter both land and tree tenure with security similar to that of full ownership, at least under traditional law.

Historically, the expansion of rubber has followed three stages. The first stage was characterized by the enrichment of fallow land with unselected rubber. At the outset of this "improved fallow" stage, although rubber was recognized as a source of income, emphasis was still placed on rice production in shifting cultivation. However, farmers rapidly intensified efforts to establish rubber agroforestry systems in which rubber latex became the major source of income (Gouyon 1995; SFDP/GTZ 1991).

The second stage was defined by a shift from an improved rubber-based fallow to a genuine rubber-based, complex agroforest. The third stage involved the integration of external innovations into the system to improve the productivity of the rubber agroforestry.

Despite the scale of expansion in rubber agroforestry, only a small percentage of farmers have received assistance from government rubber programs. By the 1980s, only 8% of rubber farmers had received government assistance and, of these, only half developed productive plantations. By 1997, the number who had received government assistance had risen to 13%. By comparison, in this same time frame, 94% of rice farmers with intensive irrigation systems were participating in government programs arising from the "green revolution" (Booth 1988). Therefore, the rubber sector has not received sufficient priority from the government, and the dispersal of techniques, skills, and information on improved rubber has been very limited. There is now an urgent need to disseminate superior innovations to farmers, particularly improved genetic planting material (IGPM) and relevant technical information. This would lead to a wider appreciation of the advantages of rubber agroforests.

During the colonial era, each time a natural resource became the focus of a commercial boom, restrictive measures were introduced to control its exploitation (Dove 1995). Examples include spices in the 18th century, in particular for *jelutung,* from *Dyera* spp., for rubber (the International Rubber Regulation Agreement from 1934 to 1944), and for timber such as teak, in the 1820s. Unitl 1999, farmers did not have the right to exploit, cut, or sell their trees for timber if the land was classified as a forest area. Seventy–four percent of Indonesia's land area is classified as forest area, and is under the control of the Ministry of Forestry. This policy is highly restrictive and provides no incentive for farmers to improve or maximize timber production in agroforests outside the estates sector. (See Table 48-1 for details of historical relations between Indonesian governments and smallholder commodity producers.)

In Kalimantan, the adoption of rubber initiated a political and economic shift for tribal farmers. After an intermediate stage of extensive rubber cropping through the fallow enrichment process, former tribal gatherers became peasant rubber planters. Politically, the situation led to conflicting interests between the state and the peasants, and this is still reflected today in government policies on rubber, wood, timber, and oil palm. In government-sponsored transmigration schemes based on food cropping that continued in West Kalimantan until 1991, tree planting was forbidden.[2] Such policies did not consider either local traditional systems with proven sustainability or their adaptation to economic change. Some traditional systems, such as *tembawang,* which are timber and fruit-based agroforests, and jungle rubber, have proved to be remarkably well adapted to contemporary conditions.

The Importance of Rubber Planting Materials

By 1997, all estates in the study area had adopted improved genetic planting material (IGPM), including rubber clones with the best yields and secondary characteristics. These clones are selected at research stations, the best known of which are Bogor and Medan in Indonesia, Prang Besar and the Rubber Research Institute of Malaysia (RRIM), and RRIC in Sri Lanka. IGPM is propagated by grafting clonal budwood onto unselected rootstock plants grown from seed. This requires a budwood garden, a rootstock nursery, and grafting skills. Without these skills and facilities, farmers are still relying on unselected rubber seedlings for jungle rubber.

---

2 This restriction certainly applied to the first field, or *lahan satu,* provided to transmigrant farmers, and in some cases was also interpreted as applying to the second granted field, or *lahan dua.*

Yields from clonal rubber are 1,400 to 2,000 kg/ha in estates and on the best farms of the SRDP[3] rubber scheme in Indonesia. Other improved rubber-planting materials include clonal and polyclonal seedlings. The former are seeds from plots planted with one clone, and these are often not used because of poor performance. The latter are seeds from isolated gardens planted with several selected clones. In Indonesia, only one estate, BLIG, in North Sumatra,[4] is able to produce proper polyclonal seedlings. Although polyclonal seedlings were widely favored by estates in the 1950s and 1960s, they have generally been abandoned in favor of the more profitable clones, which are more homogeneous, better adapted to high levels of production, and offer good secondary characteristics such as disease resistance. Most of these clones are from the third generation, which has been available since the 1970s, and farmers have yet to benefit from the full effects of the IGPM revolution in terms of boosted productivity. During the 1930s, researchers attempted to compare the "estate" system of rubber production to the "jungle rubber" system. Others evaluated systems with low levels of weeding, such as the *"bikemorse"* system in Malaysia (Sivanadyan and Norhayati Moris 1992) and the "jungle weeding" system in Indonesia (Dijkman 1951). Both were considered failures, leading virtually all research centers, private and national, to consider only monoculture rubber.

Table 48-1. Historical Relations between Indonesian Governments and Smallholder Commodity Producers

Date	*Action*	*Result*
1870	Government passes the Agrarian Act claiming all fallow land as belonging to the state, for granting to European estates, etc.*	Swidden cultivators decide to plant more perennial crops in their fallowed fields.
1910–1943	Government restricts the gathering of forest latexes by smallholders, to protect European concessions.	Smallholders decide to cultivate rubber instead.
1910–1930	Smallholders out-plant estates and increase their market share.	Government decides to protect the estates.
1935–1994	Government imposes punitive export taxes on smallholders, to force a decrease in their production.	Smallholders increase the quantity and quality of their production, to maintain a constant level of revenue.
1951–1983	Smallholders increase their market share from 65% to 84% by expanding the area of cultivation.	Government focuses all capital and technical assistance on the estates, to minimize their loss of market share by increasing yields.
1980–1990	Government promotes nucleus estate schemes, to bring smallholder cultivation under estate control.	Smallholders resist the loss of autonomy implicit in these schemes.
Present	Government supports restrictive markets for cloves, oranges, and coffee.	Smallholders abandon each commodity in turn as prices drop.

Notes: *The Agrarian Act of 1870 classified as state dominion any land not kept under constant cultivation (Holleman 1981). This gave swidden cultivators in disputed areas a strong incentive to plant perennial crops in their swidden fallows (1988).

Source: Dove (1995) and Penot (1997).

[3] The Smallholder Rubber Development Project (SRDP) was a World Bank scheme that ran between 1980 and 1990. It was then replaced by the Tree Crop Smallholder Development Project (TCSDP), which ran from 1990 to 1998.

[4] BLIG = Bah Lias Isolated Garden, London Sumatra, North Sumatra.

Table 48-2. Rubber Planting among Various Projects

Details	*TCSDP*	*SRDP*	*PRPTE*	*GCC/ARP*	*NSSDP WSSDP*	*Total*
Time span	1990–1998	1980–1990	1980–1990	1975–1980	1970s	
Area	69,000	101,149	15,697	112,600	20,019	318,465
Percentage of total area	21.7	31.8	4.9	35.4	6.3	

IGPM Availability to Farmers: The Limitations of Government Programs

In the 1970s, the Indonesian government began to seriously consider support for its smallholder rubber sector. The Thai government was considering similar support, and the Malaysian government had launched a support scheme as far back as the 1950s. By 1990, around 80% of smallholders in Malaysia and 65% in Thailand had been reached by various rubber schemes and had adopted the clonal rubber model of the estate growers. This involved monoculture crops needing high levels of labor and other inputs, and no intercropping while the rubber was still immature, except for cover crops. The objective was to maximize returns to capital and labor investment. The Thais and Malaysians also wanted to develop a simple rubber monocropping system that could be extended across vast areas without major adaptations to local conditions. Adaptations were generally limited to choice of clone and level of fertilization. This model has since proved to be efficient but costly.

To date, fewer than 15% of Indonesian farmers have been reached by such projects, directly or indirectly (see Table 48-2). About two-thirds of them eventually developed productive plantations. Between 1975 and 1980, several projects in Indonesia, such as ARP and GCC,[5] offered partial assistance, while others, such as NSSDP and WSSDP,[6] provided full assistance. The partial assistance consisted of providing farmers with only some specific components of the cropping system, either IGPM, fertilizers, training, or credit. The full assistance provided farmers with all needed components of the technological package, generally under a credit scheme.

In 1979 and 1980, the government launched two new projects. The NES/PIR[7] projects were developed for transmigration areas where migrants were being settled in virgin areas, and the PMU[8] projects, such as SRDP/TCSDP,[9] were designed for established local farmers.

Past projects, as well as schemes similar to the SRDP but funded directly by the Indonesian government, have now been regrouped in the PRPTE.[10] In this new project approach, farmers are provided with credit packages repayable within 15 years. They include the following components:

- Clonal rubber planting materials;
- Fertilizers;
- Pesticides;
- Funds to install terracing (approximately Rp 100,000 per farmer);
- Land certificates; and

5 ARP = Assisted Replanting Project; GCC = Group Coagulating Center.

6 NSSDP = North Sumatra Smallholder Development Project; WSSDP = West Sumatra Smallholder Development Project.

7 NES = Nucleus Estate Project (PIR in Indonesian language), funded either by the World Bank (NES), or directly by the Indonesian government (PIR).

8 PMU = Project Management Unit.

9 SRDP = Smallholder Rubber Development Project, which operated from 1980 to 1990. TCSDP = Tree Crop Smallholder Development Project. This began in 1990 as a continuation of SRDP, and is funded by the World Bank. A similar project, TCSSP, is funded by the Asian Development Bank.

10 PRPTE = Proyek Rehabilitasi Pertanian Tanam Eksport (Rehabilitation Project for Export Crops).

- Wages for the first five years (in NES/PIR only).

Background of the Farming System and Endogenous Innovations

I propose to analyze the development and adoption of innovations through the following four perspectives:

- Endogenous[11] innovations in the jungle rubber system by non–project farmers, that is, smallholders developing their own innovations, which lead to "indigenous" knowledge;
- The transformation from traditional slash-and-burn agriculture to enriched fallows, then later to jungle rubber;
- Endogenous innovations by former project farmers working within the estate-like system of rubber monoculture. After adopting rubber monoculture, along with a set of external innovations generally introduced by development schemes, smallholders go on to develop their own innovations to better adapt the system to their needs and strategies; and
- Rubber agroforestry systems, developed by farmers collaborating with research projects, with a combination of endogenous and exogenous innovations introduced by SRAP/CIRAD/ICRAF.[12]

In this latter case, farmers overcome the usual constraints of jungle rubber and integrate external innovations to move toward an improved rubber agroforestry system capable of competing economically with alternative crops.

Development of the Jungle Rubber System by Non–Project Farmers

I have already described how farmers adapted the estate model into a complex agroforestry system in which secondary forest was permitted to grow in association with rubber. Planting density of rubber was sometimes increased to as many as 2,000 plants per hectare to compensate for expected losses due to competition. This did not entail any additional costs because seeds were collected from old jungle rubber and were free. Five main innovations, introduced by farmers, are observable in the jungle rubber system:

- Stumps of clonal rubber were relatively expensive and were simply unavailable to farmers in many remote rubber-producing areas. So they collected seeds from nearby estates growing clonal rubber and planted these "clonal" seedlings into jungle rubber. They were generally from the widely planted GT1 clone. Production increases were low, but yields reached 700 to 800 kg/ha from pure GT1 seedlings (Gouyon 1995; Dijkman 1951). Because the life span of the jungle rubber system is 30 to 40 years and several generations have now passed, the actual proportion of clonal seedlings within the population of "unselected" rubber seedlings is unknown.
- In the late 1970s and early 1980s, farmers began planting rubber trees in rows within jungle rubber to make tapping easier and to improve returns to labor.
- Since the mid-1980s, farmers have been selectively slashing weeds once a year, conserving timber, fruit trees, and other valued species such as rattan. This compares with 6–12 weedings a year in estate plantations. This limited weeding allows tapping of rubber trees to begin in the sixth or seventh year after planting, instead of waiting 8 to 10 years, as is normal in Sumatra, or even 10 to 15 years as in Kalimantan, where no weeding is done.

---

[11] I understand "endogenous" innovations to be those that came from farmer interventions, not from projects.

[12] SRAP = Smallholder Rubber Agroforestry Project, a research program based on farm experimentation using a participatory approach. It is implemented by the Centre de Coopération Internationale en Recherche Agronomique pour le Development (CIRAD), of France, and the World Agroforestry Centre (ICRAF).

- Some farmers have always intercropped, for a variety of reasons: market opportunities for some products, such as chili and pineapple in Palembang, South Sumatra; the need to intercrop for food if land is scarce, as in the case of transmigration areas; or the natural tendency of some farmers, such as the Minangkabau of East Pasaman, West Sumatra, to continuously cultivate upland food crops in a very intensive way. This innovative practice was restricted in project areas before 1993 by prohibitions imposed by estates and project managers. It was believed intercropping had a negative impact on rubber growth. However, research in several countries has since confirmed that, to the contrary, intercropping actually favors rubber growth.
- Since 1995, many farmers have begun using the herbicide Roundup (glyphosate) to control *Imperata*. This occurs particularly in transmigration areas, but generally in all rubber areas of West Kalimantan, where *Imperata* is a pervasive weed. Its control is usually very time and labor consuming. Rubber growth in a field invaded by *Imperata* will suffer severely from competition, and production will be delayed until the eighth or ninth year. Farmers use Roundup at a rate of two to five liters per hectare, to suppress *Imperata* in upland rice fields. The cost of between Rp40,000 and Rp100,000 per ha is compensated by a reduction of 50 to 70 man-days[13] in labor costs during the rice crop. Labor savings may extend over four to five months if Roundup is used to control weeds between the rubber rows. Costs of the extended chemical control are between Rp25,000 and 50,000 per hectare.

These innovations demonstrate that farmers are gradually adopting those components of the "estate" system that are effective in jungle rubber, such as reducing the period of immaturity through weeding, improving returns to labor by planting in rows, and using herbicides. They have been adopted into the jungle rubber system without external help and, although herbicide is clearly an external innovation, its use has been purely the farmers' decision.

All of these innovations demand very limited costs and limited additional labor. One estimate of extra cost is Rp10 per clonal seedling. The exception is intercropping, which constitutes an important step toward intensification. Intercropping is generally undertaken by farmers who are gradually abandoning shifting cultivation. Although it may not require any cash or material inputs, it needs extra labor, as in the case of intercropping pineapples or chili in South Sumatra. In the absence of inputs, fertilizers in particular, yields may remain low, and intercropping may be relatively risky because of the required labor investment. On the other hand, there is the possibility of slight increases in rubber production as a consequence of intercropping because of the reduction of the immature period, easier tapping, and the improved tree growth.

These innovations have enabled the development of a more complex and sustainable agroforestry system. They have also marked the transition from fallow enrichment to a more intensive cropping system.

Constraints to Adoption of the Monoculture Model

When asked to explain their main reasons for choosing diverse agroforestry systems over monocropped rubber, smallholders provide the following answers:

- Lack of sufficient cash to afford the complete "estate" rubber package;
- The minimized labor requirements of mixed agroforestry systems;
- Savings of time and money in weed control. Farmers point out that only one weeding per year is sufficient in the jungle rubber system;
- Labor returns per farm plot are higher during the immature period of rubber;

[13] Labor costs are generally about Rp3,500 per day, so the comparable weeding cost for 50 to 70 man-days is Rp175,000 to Rp200,000. 1US$ = Rp2200.

- Land was, and in many areas still is available, and this enables a relatively extensive rubber cropping system; and
- Smallholders describe agroforestry systems as efficient in controlling erosion and providing diverse products such as timber and fruits.

The Role of the Marketplace, and the Future of Jungle Rubber

Sustained demand for rubber and a pricing policy generally favorable to farmers have played a significant part in the expansion of jungle rubber and its ability, by 1990, to feed 1.2 million farmers in Indonesia. Market incentives led to both the expansion of areas under cultivation and increases in production. In this respect, Indonesia remains well placed on the world market, offering low labor costs and, if more farmers adopt clonal rubber, a large capacity for increased production. Demand is still sustained and will remain so for the next 20 years, since substitution with synthetic rubber is not possible for at least 25% of total rubber demand. This arises from the heat-and shock-resistance characteristics needed by the tire industry. Natural rubber contributed 35% of total demand in 1997.

Nevertheless, there has recently been a slowdown of innovation adoption, and the jungle rubber system has retreated to remote, pioneer zones. The reasons include:

- The emergence of alternative perennial crops such as oil palm, cinnamon (in Jambi and West Sumatra), and pulp trees;
- Opportunities for off-farm income with the development of industries and trade in expanding cities; and
- The adoption of clonal rubber by leading jungle rubber farmers. As a positive outcome of rubber development schemes, every farmer learned that rubber clones enabled a doubling or tripling of production, and this highlighted the limited productivity of jungle rubber.

Much as fallow enrichment by planting rubber originally triggered a transition to jungle rubber, the jungle rubber system itself has become a transitional phase in the development of improved rubber-based agroforestry systems. These systems boast high productivity but have lower establishment costs than the monoculture estate plantations. They also carry rubber farming far beyond the concept of improved fallow management. The jungle rubber system reached its limits and needed to intensify. The exception is in remote pioneer areas where it is still one of the best alternatives.

The future of the system lies in the possibility of incorporating clonal rubber to boost latex production, while still conserving the agroforestry aspects that have provided farmers with diversified income. This offers a better fit with farmers' limited resources and retains the environmental and biodiversity advantages. The development of rubber-based agroforestry systems represents a third stage of innovation and is a research issue currently being addressed by the World Agroforestry Centre (ICRAF), and the Centre International de Recherche en Agronomie pour le Développement (CIRAD).

Endogenous Innovations in Rubber Monocultures by Former Project Farmers

After having integrated their own rubber growing innovations into traditional swiddening, thereby creating rubber-based fallows, farmers continued to intensify by adopting other external innovations, eventually leading to the development of more productive complex agroforestry systems.

The SRDP/TCSDP and NES/PIR projects, designed to assist smallholder rubber farmers, have been well described in the literature (Gouyon 1995; and unpublished CPIS and Directorate General of Estates project reports). However, one trend in the southern part of North Sumatra and in the Sanggau district of West Kalimantan is particularly notable. It constitutes a major innovation by farmers in SRDP monoculture plantations. This is the planting or selective retention of emerging trees in rubber plots that were originally monoculture plantations. This achieves diversification by planting, or selective retention among natural regrowth, of fruit and timber species between the rows of rubber, leading to tree-to-tree associations. This practice was either forbidden or recognized as undesirable by rubber research and extension services. However, the Rubber Research Institute of Thailand (RRIT) has been developing trials over the past 10 years in which fruit and timber trees are being grown alongside rubber (Sompong 1996; Penot 1997). The Office for Rubber Replanting Fund (ORRAF), which is the rubber growing extension service for rehabilitation, and the RRIT have been actively promoting such systems since 1991.

In Indonesia, farmers have been advised by DISBUN[14] that clonal rubber should be cropped strictly in monocultures, and in projects they are obliged to maintain clean inter-rows, at least during the period before the trees are mature. In West Kalimantan, the SRDP selected GT1 as the main clone for planting. Unfortunately, this clone is very susceptible to a leaf disease, *Colletotrichum*, which causes severe defoliation and reduced production. Defoliation was so severe in some areas that rubber trees lost up to 75% of their foliage for most of the year, and *Imperata cylindrica* and secondary forest regrowth began to invade rubber plots due to increased light penetration. In at least one village, Sanjan, in the Sanggau area of West Kalimantan, some farmers began to select timber and fruit trees from among the emerging vegetation, first to shade the inter-rows and suppress *Imperata* and, second, to obtain production from new "associated trees." These included *meranti* (*Shorea* spp.), teak (*Tectona grandis*), and *nyatoh* (*Pallaqium* spp.) for timber; and durian (*Durio zibethinus*), *pegawai* (*Durio* spp.), *rambutan* (*Nephelium lappaceum*), *duku* (*Lansium domesticum*), *petai* (*Parkia speciosa*), *jengkol* (*Archidendron parviflorum*), jackfruit (*Artocarpus heterophyllus*), and *cempedak*, a wild jackfruit (*Artocarpus integer*), for fruit trees. This same trend has been observed in SRDP plantations in the southern tip of North Sumatra Province. This innovation is remarkable for two reasons:

- Farmers had always believed that it was possible to grow other perennial trees with clonal rubber, just as they had in jungle rubber.
- Farmers were no longer paying attention to prohibitions against associated trees in rubber plantations.

When they decided to experiment, the farmers were unsure to what extent associated trees could be combined with rubber without severely decreasing rubber production. Research had never addressed the issue. However, it is worth pointing out that unselected rubber seedlings had the same yield in both estate plantations and jungle rubber during the 1920s. This suggested that the same might hold true with clonal rubber in a similar density of trees. Among other examples of this innovation, farmers are also combining unselected rubber trees with cinnamon (*Cassiavera*) in the Muara Bungo area, in hilly areas close to the Barisan mountains, and in the immediate vicinity of the Kerinci Seblat National Park.

It is premature to assess the outcomes of this development, but since official agencies are now also experimenting with some combinations, such as rattan (*Calamus* spp.), in North Sumatra and West Kalimantan, the concept of planting fruit and timber trees in association with clonal rubber has the potential to spread rapidly. There are no capital requirements, since the plants are either collected from surrounding jungle rubber or selectively retained from natural regeneration. Cinnamon, however, is more costly. It is usually planted at a density of about 2,500 trees per hectare, each costing Rp 20 to Rp 50, for a total of Rp 50,000 to Rp 125,000 per hectare. Labor is limited to planting trees, or selectively cutting if trees are retained from natural regrowth.

The system provides fruit within 10 years and timber within 40 to 50 years for home use or for sale, contributing to income diversification. Cinnamon can be

---

[14] Dinas Perkebunan (DISBUN) is the extension service for perennial crops.

harvested seven to eight years after planting. Trials in Thailand have also shown interest in fast-growing timber such as *neem* (*Azadirachta indica*), which can be harvested after 10 years as raw material for construction of furniture (Sompong 1996).

Only preliminary observations have been made of this intercropping innovation, and only in selected areas. A more systematic survey is needed to quantify the trend across all rubber projects. In Sanjan village, a rough assessment suggests that at least 20% of farmers are planting or retaining associated trees, out of 50 SRDP farmers. Adoption there is fairly widespread, since Dayak farmers readily draw from traditional agroforestry practices with jungle rubber and with the fruit- and timber-based complex agroforestry system known as *tembawang*.

An important benefit, besides the income diversification and biodiversity aspects, is that it leaves open the possibility that after the normal 35-year life span of rubber, the plot may evolve from a rubber-based agroforest into a diverse fruit and timber agroforest that remains productive for up to 50 years. Rubber wood that is 35 years old can also be marketed, and may provide an important source of income enabling the farmer to make other investments in his farming system. It is clear, however, that in all these systems, latex sales are the main driving force, and rubber wood is only a byproduct. It is technically possible to grow rubber trees specifically for timber, but then tapping would not be possible. Therefore, farmers have to choose between latex and wood production.

Clonal rubber trees have never been produced that are able to simultaneously sustain high productivity of both latex and timber. The economic life span of rubber would be different, depending on whether the objective was latex production or rubber wood, and growing rubber trees for timber only is not economically viable. Apparently the best option for monoculture plantations is to grow clonal rubber primarily for latex at the commonly accepted Indonesian density of 550 trees per hectare, then to extract the timber as a residual product on a 15-year cycle. In rubber-based agroforestry systems, on the other hand, farmers have the choice between cutting and extracting all the timber on the 15th year, as described above, waiting to extract all the timber at the end of rubber's 35-year productive life span, or transforming the field into a fruit and timber agroforest with an extended life span of 45 to 50 years. Agroforestry, therefore, leaves more options open to farmers, who can choose according to their own needs and current market conditions.

Constraints to Adoption

Institutionally, there are virtually no constraints to associating fruit and timber trees with rubber, as long as official projects do not wield authority over farmer plots. However, problems of competition between rubber and associated trees may become apparent after 10 to 15 years for fruit such as rambutan, and after 15 to 20 years for timber such as meranti, or even durian, depending on planting density. Long-term experiments have not yet been conducted to quantify this competition, and little scientific data are available. However, the planting density of associated trees observed in Sanjan is fairly low, at 100 to 200 trees per hectare, compared to the average rubber density of 550 trees per hectare. There is also a limitation on taller trees, such as durian, with a canopy above the rubber trees, and this suggests that farmers are aware of the potential problems of intertree competition.

Land and tree tenure continues to be an important constraint. Under traditional *adat*, or customary law, planting rubber demonstrates land improvement and therefore allows it to be claimed as private property. Under current national laws, farmers are technically not permitted to cut and sell their timber trees, and a tax is levied on rubber wood.

Adoption of the Estate Model in Development Schemes

Some rubber projects, such as SRDP, have success rates of up to 80% of plantations achieving efficient production. This compares with only 24% in PRPTE, and shows that the "estate" model can be suitable to smallholders. These projects have also had a very positive indirect impact on farmers in surrounding areas, in terms of access to information about clonal rubber. This is particularly true around the city of Prabumulih in South Sumatra, and in the Sambas, Anjutan, Sanggau and Sintang areas of West Kalimantan.

However, there are considerable constraints to adoption of the estate model:

- It is costly, requiring 1997US$2,000/ha for SRDP and up to 1997US$4,000/ha for NES. Despite expanded efforts made by the Directorate General of Estates (DGE) since 1980, at this rate it will require 160 years to reach all farmers (Tomich 1992). New low- to medium-cost technologies need to be identified and extended in order to reach more farmers. This will probably be the main research issue in the future, along with environmental concerns.
- The SDRP development approach was successful largely because it provided clonal planting materials and increased security of land ownership to local farmers who already had experience in growing rubber. The NES approach, on the other hand, failed partly because of its complex organization and the heavy pressures it exerted on rubber smallholders, particularly to repay loans. Another contributing factor was that it targeted transmigrant Javanese farmers with no experience in rubber production. Local farmers clearly had a great incentive to intensify their rubber systems by adopting clonal rubber with limited finance and assistance with labor and other input requirements. They had nearly a century of rubber growing experience, so they were more receptive to innovations than the inexperienced transmigrants. However, because jungle rubber is a very extensive system, local farmers may be reluctant to adopt high-input, labor-demanding alternatives such as the estate model, and it is therefore necessary to identify rubber production models that demand less intensification than the conventional estate model.
- Farmers who are not part of development projects are limited in their ability to adopt the estate model because of cost and access to good-quality improved planting materials. Also, there is a lack of cheap phosphate fertilizer. Phosphate applications are a key component of estate management, and rock phosphate is the cheapest and most efficient source. Aside from issues of cost and access, the quality and purity of clonal rubber available from both government projects such as DISBUN and private nurseries are seriously questionable.
- Credit is limited to traditional sources with very high interest rates, and these arrangements are unsuitable for long-term funding of plantations that do not begin production for five or six years after planting.
- Information on the latest management techniques is often not available, particularly from the tree crop extension service for rubber (DISBUN), nor is it regularly updated.

Other Rubber Agroforests in Southeast Asia

Thailand

Thailand has 1.7 million hectares of rubber, mainly in the southern part of the country, although 60,000 ha have recently been planted in the northeast. Farmers no longer tap old jungle rubber and have completely shifted to clonal rubber monoculture. Remnants of old jungle rubber exist as secondary forest plots scattered across the countryside, waiting for conversion to clonal monocultures. Since the early 1990s, the Office for Rubber Replanting Fund (ORRAF) and the Rubber Research

Institute of Thailand (RRIT) have promoted intercropping of fruit and timber trees with rubber, leading to accelerated adoption of agroforestry practices in the south. Both the institutional and the ecological environments are favorable to the development of agroforestry practices, based not only on intercropping food crops during the immature period but also on longer-term fruit, timber, and rubber associations within complex agroforestry systems. Farmers are beginning to grow mangosteen, *Parkia*, longkong, durian, and rambutan in rubber plots that were originally monocultures. Most farmers rely on a single clone, RRIM 600, and this leaves them vulnerable to major disease problems. However, the policy of using clonal rubber on a large scale has been successful. The main trees planted with rubber include the following:

- Timber and rattan: *neem*, or *thiem* (*Azadirachta indica*); *thang* (*Litsea grandis*, a timber tree that regenerates naturally in rubber fields); teak (*Tectona grandis*); mahogany (*Swietenia macrophylla*); *phayom*, or *white meranti* (*Shorea talura*); *tumsao* (*Fagraea fragrans*); and *Acacia mangium*. Of the rattan species, *Calamus caesius* appears to be the most promising.
- Fruit and coffee: coffee (Robusta); *salak* (*Sallaca* spp.); durian (*Durio zibethinus*); *longkong* (*Lansium domesticum*); *petai* or *nita* (*Parkia speciosa*); jackfruit (*Artocarpus heterophyllus*); *cempedak* (*Artocarpus integer*); mangosteen (*Garcinia dulcis*); and bananas.

In Phangnga Province of southern Thailand, jungle rubber more than 40 years old has been enriched with bamboos, rattan, and a variety of multipurpose trees for timber harvesting and consumption of leaves (Pramoth 1997).

Philippines

Rubber is a relatively recent introduction to the Philippines. It has been grown in estate plantations since 1957 and by smallholders on the island of Mindanao since the 1970s (Imbach 1995). Farmers intercrop with bananas, upland rice, maize, peanuts, sweet potatoes, cassava, abaca, and pineapples during the immature period. Maize was intercropped on 57% of sampled farms.

Rubber is also associated with perennial cash crops such as coffee, cocoa, and coconuts; fruit trees such as durian, langsat, and mango; a few timber species such as *Gmelina arborea* and mahogany; and some rattan. Intercropping with rattan is not common and was found on only 7% of surveyed farms. In all cases, the cultivation process is clearly one of establishing a plantation and not improving a fallow. It is a recent introduction to the agrarian system.

Malaysia

Continental Malaysia is covered mainly with rubber monoculture, in a situation similar to that of Thailand. There is still some remaining jungle rubber in northern Borneo, particularly in remote areas of Sarawak. For a clear picture of the part played by Southeast Asian countries in world rubber production, see Table 48-3.

Agroforestry Systems: From Constraints to New Opportunities

ICRAF, CIRAD, and GAPKINDO[15] have collaborated in a development-oriented research program on Rubber Agroforestry Systems in an effort to provide an alternative to both the jungle rubber system, with its low productivity but low cost, and the estate system, with its high productivity and high cost.

The objective of this research is to adopt a participatory approach in experimenting, under farm conditions, with improved rubber agroforestry systems (RAS) as an alternative to both traditional jungle rubber and the classical rubber-based development schemes built around "estate" technology. Various kinds of improved planting materials are being tried at appropriate levels of inputs and labor to determine which grow and produce best in such agroforestry systems, and which are most affordable to smallholders (Barlow 1993; Penot 1994, 1995, 1996).

Table 48-3. World Production of Natural Rubber

Detail	*Thailand*	*Indonesia*	*Malaysia*	*India*	*China*	*Sri Lanka*	*Vietnam*[a]
Production (in thousands of tonnes)							
1910	-	3 (3)[b]	6 (6)	-	-	2 (2)	-
1930	4 (1)	245 (29)	467 (56)	9 (1)	-	77 (9)	11 (1)
1950	114 (6)	707 (37)	761 (40)	16 (1)	-	116 (6)	92 (5)
1970	287 (9)	815 (26)	1,269 (40)	90 (3)	46 (1)	159 (5)	28 (1)
1990	1,271 (25)	1,262 (25)	1,291 (25)	324 (6)	264 (5)	113 (2)	103 (2)
1995	1,786 (31)	1,420 (24)	1,085 (19)	500 (9)	360 (2)	103 (2)	95 (2)
Area in 1990 (in thousands of ha)	1,844 [95][c]	3,155 [83]	1,837 [81]	451 [83]	603 [na][d]	199 [33]	250 [na]
High-yielding trees in 1995 (%)[e]	52	17	95	92	100[f]	75	15
Yield in 1990:							
kg/mature ha	847	677	800	1,057	na	772	na
kg/planted ha	689	400	599	718	438	568	352
Consumption in 1995 (in thousands of tonnes)	150	133	327	516	732	36	30

Notes: See continuation of table on next page.

15 GAPKINDO is the Indonesian rubber association, based in Jakarta.

Table 48-3. World Production of Natural Rubber (cont.)

Detail	*Nigeria*[g]	*Ivory Coast*	*Philippines*	*Cameroon*	*Kampuchea*	*Others*[h]	*World*
Production (in thousands of tonnes)							
1910	14 (14)[b]	g	-	g	-	70 (65)	98
1930	5 (1)	g	-	g	a	20 (2)	838
1950	56 (3)	g	1 (-)	g	a	27 (1)	1,890
1970	65 (2)	11 (-)	20 (1)	12 (-)	3 (-)	172 (5)	3,140
1990	152 (3)	69 (1)	61 (1)	38 (1)	35 (1)	136 (3)	5,120
1995	93 (2)	77 (1)	60 (1)	55 (1)	44 ( )	142 (2)	5,820
Area in 1990 (in thousands of ha)	247 [81][c]	68 [29][a]	88 [75][i]	41 [5][i]	19 (na)	283 [na]	9,085[e]
High yielding trees in 1995 (%)[e]	10	90	30	70	10	na	na
Yield in 1990							
-kg/mature ha	Na	1,712	841	na	na	na	na
-kg/planted ha	615	1,015	693	927	1,842	481	564
Consumption in 1995 (in thousands of tonnes)	19	gj	36	j	10	na	5,920

Notes: Kampuchea and other Southeast Asian production is included under Vietnam up to 1970. [b] Figures in parentheses along production lines indicate percentage shares of each country's natural rubber production within total world production. [c] Figures in brackets along this line are percentages of smallholdings within total planted area in early 1990s. [d] Probably about 30%, with the balance produced by state farms. [e] Estimated by author, using best available information. [f] The smallholder area is poorly managed in China. [g] All African production included under Nigeria up to 1990. [h] Mainly Brazil and Guatemala up to 1970. Subsequently including Myanmar, Liberia, Zaire and several other small producers. [i] Percentage of holdings less than 5 ha. [j] All African consumption included under Ivory Coast. *Sources*: Barlow, Jayasuriya, and Tan (1994); International Rubber Study Group (IRSG), 1946–1996; and Working Papers Trade and Development.

A farming system research program, the Smallholder Rubber Agroforestry Project (SRAP), is being implemented by CIRAD and ICRAF. It is based on farmer experimentation using a participatory approach, and its objective is to improve productivity by optimizing the benefits of labor, minimizing inputs and costs, and preserving the benefits associated with agroforestry practices while keeping very close to current technologies. It is believed this approach will allow farmers to more easily adopt technical innovations.

A network of on-farm experiments has been developed by SRAP in collaboration with 100 farmers in the provinces of Jambi and West Sumatra, in Sumatra, and West Kalimantan, in Borneo. All innovations being tested have been designed in collaboration with farmers to improve the fit of RAS technologies to farmers' resources and priorities. Emphasis is being given to minimizing inputs and labor while conserving the advantages of agroforestry practices, including diversification of income, intercropping to maintain income while the rubber trees are immature, maintenance of biodiversity, and other ecological functions. SRAP is testing three main types of rubber agroforestry systems:

- RAS 1 is comparable to current jungle rubber practices. However, unselected rubber seedlings are replaced by clones selected for their potential ability to adapt to forest-like environments,[16] including competition with natural secondary forest regrowth. Various planting densities and weeding protocols are being tested. This will determine minimum management levels needed for the system, a key factor for labor-short farmers striving to increase their labor productivity. Biodiversity levels are similar to jungle rubber and relatively close to those of primary forest. This system is closest to the concept of fallow enrichment and fits a vast class of farmers.
- RAS 2 is a very intensive, complex agroforestry system that begins with slashing and burning. Then rubber is established at a density of 550 trees per hectare in combination with timber and fruit trees at 92 trees per hectare. Annual crops are interplanted during the first three to four years, with emphasis on improved upland rice varieties, dry season crops such as peanuts, and various levels of fertilizer application. Cash crops such as cinnamon are also being tested in combination with food crops. Several planting densities of rambutan, durian, petai, tengkawang, and other selected species are being tested according to their tree typology. Biodiversity in this system is limited to between 5 and 10 planted species and those that regenerate naturally and are protected by farmers.
- RAS 3 is also a complex agroforestry system featuring rubber and other trees planted in a similar pattern to that of RAS 2. The difference is that it is established on degraded lands covered by *Imperata cylindrica*, or in areas where *Imperata* is a major threat. Its main constraints are the labor or cash needed to control *Imperata* with herbicide. An annual crop, usually rice, is grown in the first year only. Cover crops such as *Mucuna, Flemingia, Crotalaria, Setaria,* or *Chromolaena*; multipurpose plants such as wing-bean or *Gliricidia*; or fast-growing trees such as *Paraserianthes falcataria, Acacia mangium,* or *Gmelina arborea* are then immediately established with the aim of suppressing *Imperata* growth, eliminating the weeding labor, and providing a favorable environment for rubber and the associated trees to grow. Several cover crop combinations are being tested. The biodiversity in this system is expected to be similar to that of RAS 2. RAS 2 and RAS 3 can no longer be regarded as improved fallows. They have intensified into permanent cropping patterns.

Costs and Returns

These variations of RAS have been subjected to cost-benefit analyses (Penot 1996), and results indicate that returns to labor are improved compared to those of the estate model. The contributions of associated annual or perennial crops make the benefits of the RAS systems higher than those of jungle rubber, and similar to or higher than those of rubber monocultures. The scale of the project is currently limited to 100 farmers, each with a rubber plot of between 0.3 and 0.8 ha, or sufficient to reflect actual farm conditions.

It is not known how many farmers are independently experimenting with similar systems. However, although the prototypes are still at an experimental stage, there is already strong demand from surrounding farmers to join the project. Because the prototypes were designed to reflect farmer priorities and include innovations developed by farmers themselves, it is believed that up to 70% of farmers will adopt

---

[16] The selected clones are PB 260, RRIC 100, BPM 1, and RRIM 600.

rubber-based agroforestry systems. However, the experimentation process may yet bias this estimate.

Farmers are aware that, even in an improved agroforestry system, clonal rubber requires significantly more weeding and inputs than the unselected seedlings traditionally planted in jungle rubber. Yet they sometimes underestimate the minimum requirements. While these requirements are still subject to experimental trials, it seems certain that a significant challenge will be for farmers to integrate realistic minimum amounts of inputs and labor into their current practices.

There is also a need to assess trade-offs between the "complete development package," such as currently promoted by development schemes, and a scaled-down approach that supplies only the most essential components. Current research should provide valuable insights into the optimum level of management that is acceptable to farmers.

Economic Comparison of Rubber-Based Farming Systems

Most rubber producers rely heavily on revenues generated by latex sales. In rubber-based agroforests, rubber contributes up to 90% of total income. Non–project farmers generally have between two and five hectares of jungle rubber. Farmers in the SRDP/TCSDP/PRPTE projects generally have one hectare of clonal rubber, and those in NES/PIR projects in transmigration areas, where rubber is the sole source of income, have two hectares.

Improved tree crop systems, including rubber, coffee, and oil palm, deliver far higher benefits than alternative systems such as shifting cultivation of food crops like rice and bananas. The benefits also continue over longer periods, up to 30 years for rubber. When compared to swidden rice, the returns to labor from jungle rubber are higher by a factor of four, and those from rubber monoculture and RAS are higher by a factor of 40. When it is grown in superior conditions, such as those in the West Pasaman area of West Sumatra Province, oil palm gives a return even higher than clonal rubber, and complete reimbursement of credit is possible after the 10th year. But in other regions, such as South Sumatra, oil palm does not yield as much, and benefits and returns to labor are very similar to those from improved rubber systems. It is not clear if rubber agroforestry systems can match the economic performance of oil palm and short-term crops such as pineapples in areas close to cities.

A major concern in farmers' decision-making is maintaining or improving returns to labor. Both rubber monoculture and improved rubber agroforestry systems such as the RAS models developed by ICRAF meet this criterion.

A characteristic feature of rubber is its market flexibility as a nonperishable product. Smallholders seem, in the past, to have sometimes followed an inverse curve to market conditions. When prices were low, such as in the early 1930s, they expanded the area planted to rubber (Boeke 1930, as cited in Dove 1995). When prices were high, as in 1949, they reduced the number of trees to be tapped (Boeke 1953). This is no longer true, and farmers are now acutely sensitive to price signals. They are eager to increase their incomes to meet their expectations of rising living standards and want to maximize returns to their land and labor.

Conclusions

Indonesian rubber farmers have now reached a stage where further increases in productivity can only be achieved by adopting rubber clones and other external innovations. They have reached the stage where complex agroforestry systems can no longer compete with other agricultural systems that may be more risky but are more profitable in the short term.

Improved rubber-based agroforestry systems, therefore, hold promise of a way forward, by achieving an attractive balance between yield gains, continued environmental benefits, and reduced risks.

Farmers have demonstrated their capacity to innovate. Jungle rubber now covers more than 2.5 million hectares in Indonesia. The research challenge is to help them to evaluate new innovations that will achieve higher yields without sacrificing the advantages of their traditional systems.

Barlow (1996) proposed an analytical framework in which the effects of advancing economic growth on rubber plantations are classed into five stages (see Table 48-4). About 15% of smallholders in Indonesia are at stage three, while the remaining 85% are still at stage two. In comparison, Thailand is already at stage four and Malaysia has reached stage five. Historically, development of Indonesia's rubber sector was slowed by political instability up until the 1960s, and then priority was given to achieving rice self-sufficiency. These circumstances did not allow improved rubber-growing technologies to reach farmers on a large scale. So, using the resources available to them, Indonesian farmers integrated rubber into their farming systems with no costs or extra labor inputs during the long immature periods. This was done first by enriching fallows with rubber, and then by moving rapidly toward the development of rubber-based complex agroforestry systems that included innovations to decrease the immature period, facilitate tapping, and otherwise reduce labor.

These rubber-based complex agroforestry systems are still the most widely used in Indonesia. Yet sustained economic growth and new crop opportunities, particularly oil palm, will force farmers to increase the productivity of their rubber systems if they are to compete with alternatives. The development-oriented research described in this chapter is exploring the potential of external innovations to build on indigenous practices without adding to capital or labor inputs.

Indonesia is a vast country with extreme variations in terms of transport infrastructure and economic opportunities. It probably exhibits all of the stages of development shown in Table 48-4, from the shifting agriculture of stage 1 in remote areas of Kalimantan, Sumatra, Maluku, and Irian Jaya, to stage 5 in Java. This diversity presents a vast challenge to provide all types of farmers with access to improved technologies that fit their individual strategies and local resources. However, the variations of complex agroforests developed by farmers across the Indonesian archipelago continue to have a promising future, particularly if environmental benefits are accorded increasing importance. This brings me to two intriguing questions posed by Dove (1995):

- Can the exploitation of nontimber forest products, in this case rubber, be promoted in the absence of a hierarchical political economic structure? The first question raises the issue of "producers' organizations" and their ability to control the evolution of a commodity system. The answer has been positive up to now, but the adoption of external innovations requiring inputs and capital may change that.
- Is it possible to attain goals of both ecological sustainability and socioeconomic equity within a hierarchical structure? And, the answer to the second question is probably *"yes,"* if improved systems such as rubber agroforestry systems are based on existing practices and prove to be easily adopted by farmers.

If the optimism of these responses proves to be unfounded, what then? Policy and technology development need to support rubber farmers' organizations and encourage environmentally friendly systems that may range from semi-intensive rubber-based agroforests to complete monoculture systems. The rubber agroforestry systems described in this chapter are also attractive in that they involve little risk or uncertainty of market demand.

Table 48-4. Analytical Framework for Development of Plantation Tree Crops

Stages	*Characteristics* [a]	*National Positions with Rubber* [b]
1 Backward economy (subsistence agriculture, no plantation crops)	* *Predominantly subsistence agriculture; minuscule services and industry.* * Little trade and fragmented rural markets. * Plentiful and underutilized land, labor with low marginal product, very scarce capital. * Historically established shifting cultivation technology on family smallholdings. * Minimal government. [d]	Thailand to 1920 Indonesia (outer islands) to 1870 [c] Malaysia to 1870 [c] India to 1920 China (Yunnan and Hainan) to 1960 Sri Lanka to 1870 [c] Nigeria (southern area) to 1910 Vietnam to 1920 Philippines (Mindanao) to 1950 Ivory Coast to 1950 Cameroon to 1950 Kampuchea to 1920
2 Early agricultural transformation (simple plantation crops technology)	* *Commercializing agriculture the dominant sector, with estates and smallholdings both involved in plantation crop cultivation. Small but growing services and industry.* * International trade and rural market development commencing. * Land and labor becoming scarce and prices rising, capital becoming more available. * Rapid adoption of simple labor-intensive tree crop technologies, first on estates and then on smallholdings. * Central government begins levying taxes and providing small services.	Thailand, 1920–60 Indonesia, 1900–30 [c] Malaysia, 1900–30 [c] India,1920–40 China, missed out Sri Lanka, 1900–30 [c] Nigeria, 1910–70 Vietnam, 1920–40 Philippines, 1950–70 Ivory Coast, 1950–70 Cameroon, 1950–70 Kampuchea, 1920–80
3 Late agricultural transformation (new plantation crops technology)	* *Agriculture remains one of bigger sectors, but services increasing and manufacturing based on import-substitution passes agriculture toward end of period.* * Rural market development progressing, especially with government interventions, but many imperfections persist. * Land and labor prices rising; capital, management, and transport prices falling. * Generation of new land and labor-saving but more capital- and management-intensive high-yielding tree crop technologies. Adoption first by estates and much later by smallholdings.	Thailand, 1960–85 Indonesia, 1930–present Malaysia, 1930–1970 India, 1940–present China, 1960–present Sri Lanka, 1930–present Nigeria, 1970–present Vietnam, 1970–present Philippines, 1970–present Ivory Coast, 1970–present

		* Central government gradually widening supporting role, and following end of colonialism, providing widespread rural infrastructures and services. Government also promoting import-substituting manufacturing.	Cameroon, 1970–present Kampuchea, 1980–present
4	Early advanced economy (plantation crops becoming less profitable)	* *Manufacturing becomes much larger than agriculture, being increasingly export-oriented and including downstream plantation crop processing into final goods.* * Rural market much better integrated and competitive; pockets of imperfections persist. * Resource price trends of stage 3 continuing, but land and labor price rises accelerating and rural-urban wage differentials widening. Consequent labor migration to towns. * Tree crop technology generation and adoption continuing in directions under stage 3, with concomitant widening of available techniques to assist adjustment to different circumstances. Shifting toward generation and adoption (by plantation crop processors) of quality-improving techniques needed by plantation crop goods subsector. * Government continuing with provisions of rural infrastructures and services under stage 3. Previous regulations in trade being gradually removed.	Thailand, 1985–present Malaysia, 1970–85
5	Late advanced economy (plantation crops not profitable)	* *Manufacturing now predominant, being two or more times bigger than agriculture and including significant plantation crop processing into final goods. Also importation of natural rubber from other countries to supply this goods industry.* * Rural market as for stage 4, albeit superior in integration and competition. * Resource price and exchange rate trends of stage 4 persisting. Traditional plantation crop production is uneconomic, but existing trees still being exploited in "sunset" setting. * Tree crop technology generation and adoption chiefly concentrating on quality-improving techniques for goods subsector. * Government continuing as under stage 4. Remaining plantation crop support measures largely welfare for older generations.	Malaysia, 1985–present

Notes: [a] Divided in each stage between the main characteristic in terms of sectoral balance (in italics) and associated characteristics. [b] Came in after the Backward Economy, stage 1. [c] Coconut, coffee, and cocoa from 1870, rubber first planted in 1900. [d] Traditional government comprising little local kingdoms, which may nonetheless collect tax. "Central government" from stage 2 often entails an invading colonial regime, although by the end of stage 3 it is usually of independent local origin. The modern nation states then involved have varying degrees of devolution at local level. *Source*: Barlow 1996.

As Barlow concluded in 1982: "It is assuredly appropriate to look seriously at policies which basically aim to help people to change, in an evolutionary approach where steady improvements are made from within the beginning framework of traditional agriculture" (1982, *31*). This is an apt summation of what Indonesian farmers have done over the past century with their rubber agroforests. Spurred by fruit and timber markets, neighboring rubber-producing countries such as Thailand and the Philippines are already developing complex rubber-based agroforestry systems, suggesting opportunities for useful links with the experimental RAS work under way in Indonesia.

Acknowledgments

I gratefully acknowledge the invaluable assistance of ICRAF/CIRAD/SRAP field staff at project sites in West Kalimantan, Jambi, and West Sumatra in contributing to the research reported in this chapter.

References

Barlow, C. 1989. *Development in Plantation Agriculture and Smallholder Cash Crop Production.* Conference on Indonesia's New Order. Canberra, Australia: Australian National University.

———. 1993. (Unpubl.) Towards a Planting Material Policy for Indonesia Rubber Smallholdings: Lessons from Past Projects.

———. 1996. Growth, Structural Changes and Plantation Tree Crops: The Case of Rubber. In: *Working Papers in Trade and Development Series,* Paper 94/4. Canberra, Australia: Department of Economics, Australian National University.

———., and Muharminto, 1982. *Smallholder Rubber in South Sumatra: Towards Economic Improvement.* Bogor, Indonesia and Canberra, Australia: Balai Penelitian Perkebunan and Australian National University.

———, S. K. Jayasuriya, and Tan. 1994. *The World Rubber Industry.* London, UK: Routledge.

Boeke, J.H. 1953. *Economics and Economic Policy of Dual Societies, as Exemplified by Indonesia.* New York: Institute of Pacific Relations.

Booth, A. 1988. *Agricultural Development in Indonesia.* Sydney: Allen and Unwin.

Compagnon, P. 1986. *Le Coutchouc Naturel.* Paris: Ed Maisonneuve et Larose.

Cramb, R.A. 1988. The Commercialization of Iban Agriculture. In: *Develoment in Sarawak: Historical and Contemporary Perspectives,* Monash paper on Southeast Asia, No.17, edited by R.A. Cramb and R.H.W. Reece. Melbourne, Australia: Centre of Southeast Asian Studies, Monash University, 105–134.

de Foresta, H. 1990. Personal communication with the author in Bogor, Indonesia.

———. 1992a. Botany Contribution to the Understanding of Smallholder Rubber Plantations in Indonesia: An Example from South Sumatra. Symposium Sumatra Lingkungan dan Pembangunan. Bogor, Indonesia: BIOTROP.

———. 1992b. Complex Agroforestry Systems and Conservation of Biological Diversity for a Larger Use of Traditional Agroforestry Trees as Timber in Indonesia. A Link between Environmental Conservation and Economic Development. In: *Proceedings of an International Conference on the Conservation of Tropical Biodiversity. Malayan Nature Journal* Golden Jubilee issue, 488–500.

———, and Michon G. 1995. The Agroforest Alternative to *Imperata* Grasslands: When Smallholder Agriculture and Forestry Reach Sustainability. *Agroforestry Systems.*

Dijkman, J. 1951. *Thirty Years of Rubber Research.* Miami, FL: University of Miami Press.

Dove, M. 1993. Smallholder Rubber and Swidden Agriculture in Borneo: A Sustainable Adaptation to the Ecology and Economy of Tropical Forest. *Economic Botany,* 47(2).

———. 1995. Political vs. Techno-economic Factors in the Development of Non-timber Forest Products: Lessons from a Comparison of Natural and Cultivated Rubbers in Southeast Asia and South America. *Society and Natural Resources,* Vol. 8.

DGE (Directorate General for Estates). 1995. *Statistik karet.* Jakarta, Indonesia: Ministry of Agriculture.

Gouyon, A. 1995. Paysannerie et hévéaculture: Dans les plaines orientales de Sumatra: Quel avenir poutr les systèmes agroforestiers? Paris, France: Thése INA-PG.

Holleman, J.F. 1981. *Van Vollenhoven on Indonesian Adat Law.* KIT, Translation series No.20. The Hague, Netherlands.

IRRDB (International Rubber Research and Development Board). 1997. Symposium on Farming System Aspects of the Cultivation of Natural Rubber (*Hevea brasiliensis*), November 1996, Beruwala, Sri Lanka. IRRDB.

IRSG (International Rubber Study Group) papers 1946–96. London: IRSG.

Imbach, C. 1995. End-of-study thesis, CNEARC School for Tropical Agronomy, Montpellier, France.

King, V. 1988. Social Rank and Social Change Among the Maloh of West Kalimantan. In: *The Real and Imagined Role of Culture in Development.* Honolulu: University of Hawaii, 219–253.

Levang, P. 1996. De la Jachere Arboree aux Agroforets, des Strategies Paysannes Adaptees a des Milieux de Fertilite Mediocres. Seminaire CIRAD, Montpellier, France, Fertilite du Milieu et Strategies Paysannes sous les les Tropiques Humides.

Levang, P., G. Michon, and H. de Foresta 1997. Agriculture Forestière ou Agroforesterie. *Bois et Foret des Tropiques,* 251(1).

Penot, E. 1994. *The Non-Project Rubber Smallholder Sector in Indonesia: Rubber Agroforestry Systems as a Challenge for the Improvement of Rubber Productivity, Sustainability, Biodiversity and Environment.* ICRAF Working Paper. Nairobi, Kenya: ICRAF.

———. 1995. Taking the Jungle out of Rubber. Improving Rubber in Indonesian Agroforestry Systems. *Agroforestry Today.*

———. 1996. Improving Productivity in Rubber-Based Agroforestry Systems (RAS) in Indonesia: A Financial Analysis of RAS Systems. Paper presented at the GAPKINDO seminar in Sipirok, North Sumatra.

———, and Gede Wibawa 1996. Sustainability through Productivity Improvement of Indonesian Rubber-Based Agroforestry Systems. Paper presented at the 14th International Symposium on Sustainable Farming Systems and Annual Meeting of the International Rubber Research and Development Board, November, 1996, Colombo, Sri Lanka.

———, and A. Gouyon. 1995. Agroforêt et Plantations Clonales: Des choix pour L'avenir. Presented at CIRAD Mission Economie et Sociologie Seminar. Montpellier, France: CIRAD.

———. 1997. Rubber Agroforestry Systems in Thailand: Preparation Mission for INCO proposal. *Mission Report.* Bogor, Indonesia: CIRAD/ICRAF.

Potter, L., and J. Lee. 1998. Tree Planting in Indonesia: Trends, Impacts and Directions. CIFOR Paper. Bogor, Indonesia: CIFOR.

Prasetyo, L.B., and M. Kumazaki. 1995. Land Use Changes and Their Causes in the Tropics: A Case Study in South Sumatra, Indonesia in 1969–1988. *Tropics* 5(1/2).

Pramoth K. 1990. Les jardins a Hevea des contreforts orientaux de bukit Barisan, Sumatra, Indonesia. Diplome d'Etudes Approfondies de Biologie vegetale tropicale. Paris: University de Montpellier.

———. 1997. Personal communication between Pramoth Kheowvongsri and the author.

RRIM (Rubber Research Institute of Malaysia). 1995. *Annual Report 1995.* Kuala Lumpur, Malaysia: RRIM.

Sethuraj, M.R. 1996. Impact of Natural Rubber Plantations on the Environment. Paper presented at an international seminar on Natural Rubber as an Environmentally Friendly Raw Material and a Renewable Resource. Trivandum, India.

Sivanadyan, K., and N. Moris. 1992. Consequences of Transforming Tropical Rainforest to *Hevea* Plantations. Presented at a FADINAP regional seminar on Fertilization and Environment, Chiang Mai, Thailand. RRIM (Rubber Research Institute of Malaysia).

Sompong. 1996. Paper delivered to a Thailand/Malaysia Technical Seminar on Rubber, Phuket, Thailand. In: *Agroforestry Under Rubber Plantations in Thailand,* edited by W. Buranatham and S. Chugamnerd (1997). RRIT (Rubber Research Institute of Thailand).

SFDP/GTZ (Social Forestry Development Project, funded by the German aid agency GTZ). 1991. *Rubber Smallholders Survey in Kabupaten Sanggau, West Kalimantan.* SFDP/GTZ.

Thiollay, J. 1995. The Role of Traditional Agroforests in the Conservation of Rainforest Bird Diversity in Sumatra. *Conservation Biology* 19(2).

Tillekeratne, L.M.K. 1996. Role of *Hevea* Rubber in Protecting the Environment in Sri Lanka. Paper presented at an international seminar on Natural Rubber as an Environmentally Friendly Raw Material and a Renewable Resource. Trivandum, India.

Tomich, T. 1992. Smallholder Rubber Development in Indonesia. In: *Reforming Economic Systems in Developing Countries,* edited by D.H. Perkins and M. Roemer. New York: Cornell University Press.

van Noordwijk, M., and T.P. Tomich. 1995. Agroforestry Technologies for Social Forestry: Tree-Crop Interactions and Forestry-Farmer Conflicts. Social Forestry and Sustainable Management. Jakarta.

Wan Abdul Rahaman, Wan Yacoob, and K. Jones. 1996. Rubber as a Green Commodity. Paper presented at an international seminar on Natural Rubber as an Environmentally Friendly Raw Material and a Renewable Resource. Trivandum, India.

Wolters, A. 1967. Cited in Dove M. 1993. Smallholder Rubber and Swidden Agriculture in Borneo: A Sustainable Adaptation to the Ecology and Economy of Tropical Forest. *Economic Botany* 47.

Working Papers in Trade and Development, series. Canberra, Australia: Department of Economics, Australian National University.

Chapter 49

Rubber Plantations as an Alternative to Shifting Cultivation in Yunnan, China

Guangxia Cao and Lianmin Zhang

China's national campaign for reforestation and conservation of biodiversity has strongly condemned the practice of shifting cultivation, especially since the wasteful burning of "valuable wood biomass" contrasts sharply with serious wood shortages in many parts of the country (Feng Yaozong 1993; Gao Lizhi et al. 1996; Tao Siming 1995; Xu Weishan 1990; Xu Ziafu 1994; Xu Zaifu and Liu Hongmao 1995; Anonymous 1992). As a consequence, both the academic and political communities in China have advocated the development of alternatives to shifting cultivation (Wu Zhaolu and Zhu Hualing 1996; Anonymous 1992). However, wholesale rejection of shifting cultivation ignores the realities of shifting cultivators, for whom practical alternatives are limited. As a result, shifting cultivation continues and is silently accepted by local officials, highlighting the prudence of exploring indigenous solutions (Wu Zhaolu and Zhu Hualing 1996; Xu Youkai et al. 1996; Anonymous 1980; Zeng Juemin 1995).

Across China, expanding rural populations have placed increasing pressure on land resources, resulting in shortened fallows and lengthened cultivation periods, particularly in southern Yunnan. Various strategies have been explored at both national and grassroots levels to tackle this problem, and the national approach generally emphasizes the creation of new land-use paradigms for local communities.

Quite a number of researchers are involved in developing these strategies. For example, researchers have recommended mixed plantations of rubber and tea as one means of increasing the output per unit of land area and improving light use efficiency as well (Wang Huihai 1982; Long Yiming 1993). Others have suggested multilayered agroforestry systems to make use of existing resources in tropical areas (Feng Yaozong 1993; Deng Jiwu 1996). Another national strategy advocates the transfer of technology to assist shifting cultivators in adapting technologies to local conditions (Shi Shan 1993; Sun Hongliang 1996; Li Haifeng 1996). Despite these "top down" strategies, there is increasing recognition of the need for local initiatives to tackle local problems (Xu Zaifu and Yu Pinghua 1987; Long Chunlin and Wang Jieru 1993; Guo Huijun 1994; Liu Dewang 1996; Xu Jianchu et al. 1996).

Local communities have long demonstrated their capacity to develop initiatives to increase their productivity. In southwestern Yunnan, for example, farmers have

---

Guangxia Cao and Lianmin Zhang, Southwest Forestry College, Kunming 650224, China.

planted *Alnus nepalensis* on abandoned fallow land for several centuries, at least (Xue Jiru et al. 1986; Zeng Juemin 1995). *Alnus* aids in soil nutrient enrichment and provides timber and other products. In Xishuangbanna, southern Yunnan Province, Jino farmers have developed sustainable land-use rotations incorporating agroforestry (Long Chunlin and Wang Jieru 1993; Xu Jianchu et al. 1996). Such examples have increasingly focused attention on local or indigenous management systems. However, in discussions of indigenous systems, the question naturally emerges: *"What is an indigenous system or strategy, particularly regarding land use?"*

There are different interpretations of "indigenous systems" or "indigenous strategies." The term indigenous is usually taken as meaning native to a local community, and not shared by most outsiders. Some, however, consider that indigenous systems emphasize traditional values, and from this perspective, outside intervention is necessarily minimal (Long Chunlin and Wang Jieru 1993; Cunningham 1996; Posey 1996; Warner 1991; Xu Jianchu et al. 1996). While this view may be historically correct in China, in the current environment of fast-changing socioeconomic conditions, indigenous systems often reflect external pressures and often have regional dimensions. Therefore, in some circumstances, it is hard to distinguish indigenous knowledge from knowledge originating elsewhere (Agrawal 1995), and to examine indigenous strategies as isolated phenomena, without considering influencing external factors, may misrepresent the real indigenous system.

This is especially relevant to the rubber plantations established by many ethnic groups in Xishuangbanna, southern Yunnan Province. In areas where the altitude is below 900 m above sea level (asl), local farmers have increasingly converted their swidden fallows into rubber plantations as a means of generating family income. However, because rubber is not native to the area and the region is considered by scientists to lie beyond the northern limits of rubber's geographic distribution, the impressive expansion of rubber plantations is not generally viewed as indigenous (Huang Zongdao et al. 1980). The expansion followed a period when rubber was promoted by local government agencies as a major crop in state-owned farms, to meet domestic demand. This was halted because of widespread concern over declining biodiversity in the area. However, as state-owned farms stopped converting tropical forests into rubber plantations, local farmers, nearly all of them individual smallholders, recognized the market opportunity and rushed to convert considerable areas of fallow vegetation into rubber plantations.

Given the spontaneous nature of the conversion, this chapter argues that rubber plantations should be included in the discourse on indigenous systems, in order to reach a more holistic understanding of indigenous strategies developed to deal with intensified stress on land resources. In managing their fallow vegetation, farmers tend to preserve and cultivate those trees that contribute significantly to livelihood needs and improved land use. The intensification of fallow management by incorporating economic perennials is closely related to local socioeconomic conditions. Therefore, this study focuses on patterns of adoption and the willingness of local communities to adopt rubber planting. It is mainly analyzed from an economic perspective, although its ecological impact is also considered.

Research Methods and the Study Area

The study drew on a combination of fieldwork and literature review. The categories of literature consulted included sociological and anthropological works, conservation research, forestry-related publications and reports, and some agricultural journals and publications. The fieldwork was conducted during the periods of July 1995, March 1996, and September to October 1996, in rural areas of central Xishuangbanna. During the investigation, rapid rural appraisal (RRA) techniques were used to collect data on changing patterns of shifting cultivation, household economics, and contextual factors influencing farmers' innovation processes. The study focused on six villages: Manmonei, Chengzizai, Manjianqu, Mannazhuang, Manshaqiao, and

Mannala (see Figure 49-1). More detailed information about rubber plantations, including both qualitative and quantitative information, was collected mainly at Manmonei and Mannala villages.

Yunnan Province is located in southwest China, bordering Vietnam and Laos in the south and Myanmar in the south and southwest, at 21°9' to 29°15' N latitude and 97°39' to 106°12' E longitude. It covers a total area of about 390,000 km^2 and, in 1991, had a population of 37,820,000. Geographically, it may be considered as the northwestern extremity of Southeast Asia, and it is here where the distribution of shifting cultivation reaches its northern limits. The province's many ethnic groups practice variations of shifting cultivation on an estimated one-seventh of Yunnan's land area. While the fieldwork was limited to Xishuangbanna, this study attempts to extrapolate its findings to southern Yunnan generally or, at least, to the more tropical area (*requ* in Chinese) between 21° and 25° N latitude, covering about 55,846 km^2, where the rubber tree, *Hevea brasiliensis* (called *xiangjiao* in Chinese), can be grown commercially (Li Yikun and Huang Zerun 1987; Li Liangsheng 1993).

The climate consists of a dry season between October and May, and a rainy season from June to September. The annual average temperature and rainfall are 19.4°C to 22.6°C and 1,200 to 1,800 mm, respectively. Although some places experience brief strong winds at the beginning of the monsoon season, the annual average wind velocity is one meter per second; the annual total heat above 10°C is 7,000 to 8,200°C; the annual total amount of light is 1,800 to 2,300 hours; and soil pH ranges from 4.5 to 6, with organic matter at 3% to 5% (Wu Zhengyi et al. 1990).

Xishuangbanna, with a total land area of 19,680 km^2, is located between 21°9' and 22°36' N latitude and 99°58' and 101°50' E longitude. Its estimated population of 690,000 includes more than 40 ethnic groups. It is regarded as the ideal area in southern Yunnan for rubber production (Liang Xiaomao 1987; Zhou Yonghua et al. 1994).

Figure 49-1. Yunnan Province in China and Location of the Study Villages
Note: Shaded area of Yunnan is the southern warm area, where rubber can be grown.

In the early 1980s, the usufruct rights to China's agricultural land were changed from communal to individual entitlement, enabling individual rural households to make their own decisions regarding land use. In the study areas of Xishuangbanna, individual households in villages bordering valleys and plains were each assigned 1 *mu* of paddy fields on the plains (15 mu equals 1 ha), and 21 mu of sloping land, used mainly for shifting cultivation. In addition, almost all villages had their own community forest.

Results

Conventional Shifting Cultivation Systems in Southern Yunnan

Shifting cultivation in southern Yunnan has been classified according to the length of cultivation as follows (Ying Shaoting 1994):

- One year's cultivation: After burning, farmers cultivate swiddens for only a single year, before leaving them fallow for the next 7 to 10 years.
- Two years' cultivation: Swiddens are cultivated for two years and crops are rotated. Cotton may be followed by upland rice, or beans followed by upland rice, or upland rice followed by cotton. Then the land is left fallow for 7 to 10 years, or more.
- Three to five years' cultivation: Cultivation continues for three to five years, or possibly as long as 10 to 20 years, with upland rice, corn, cotton, beans, and other crops. Then the land is fallowed for 10 to 20 years.

Despite these categories, our investigations in many parts of Xishuangbanna found that the current pattern of shifting cultivation is typically three to four years of cropping, followed by three to five years of fallow. This takes full advantage of nutrients released by the burned vegetation while providing medium production levels to farming households. The fallow period, even if it is shorter, allows the soil productivity to recover to some extent. It is reported that in Hani (Akha) swidden systems using this cycle, yields of cotton and rice, at 30 to 50 kg/mu and 125 kg/mu respectively, have remained stable for many generations.

Traditionally, shifting cultivation systems were based on a subsistence economy. But local populations began to increase rapidly after 1949 because of improved health conditions, an effective antimalaria campaign, and immigration of people attracted to rubber cultivation. This put more pressure on land resources, and the fallow period decreased. As the fallow shortened, soil fertility declined and chemical fertilizers became essential to achieve reasonable crop yields. However, some villagers believe that the increasing need for fertilizer was a consequence of introducing high-yielding varieties that required both higher nutrient subsidies and extra labor inputs to get a fair output. Perhaps both the shortening of the fallow period and the increasing need to apply external fertilizers made traditional land-use systems less and less attractive. Farmers felt compelled to seek alternatives.

Factors Triggering the Integration of Rubber into Swidden Fallows

Starting in the 1950s, the Chinese government introduced *Hevea brasiliensis* to the study area to meet the national demand for rubber (Huang Zongdao et al. 1980; Li Yikun and Huang Zerun 1987; Pei Shengji et al. 1987; Wu Zhengyi et al. 1990; Li Liangsheng 1993; Luo Jingchun 1994; Zhou Yonghua et al. 1994). At that time, rubber was cultivated mainly by state-owned farms and village community farms (Zhou Yonghua et al. 1994). Cultivation by individual households began only after the 1980s, and in the 1990s, private rubber plantations were expanding rapidly in Xishuangbanna and other areas in southern Yunnan. Private rubber plantations became a dominant trend, while state-owned farms remained stagnant, or even declined. The factors underlying this transformation included the following:

Land Allocation Policy. Beginning in the early 1980s, state agricultural land was allocated to farmers for their private management, providing them with a strong incentive to improve their land management practices. Rubber soon attracted the interest of local farmers as a profitable cash crop.

Food Security. After agricultural land was allocated to individual households, almost all farmers were not only able to feed themselves, but also had surpluses to sell. This supported an improved livelihood that, in turn, allowed farmers to consider more carefully other options to improve the profitability of their farming systems.

Household Income. After local food security was achieved, farmers found that they wanted increased income for other things, such as establishing new settlements and purchasing agricultural machinery to reduce manual labor. Most of them focused on alternative land uses to increase their income. Others resorted to off-farm activities such as operating restaurants, selling their labor on construction sites, brick making, and other small manufacturing industries. Popular agricultural options included growing watermelons and vegetables in paddy fields during the dry season, while some farmers grew *Amomum* and other medicinal plants under a tree canopy (Chen Aiguo et al. 1996; Duan Qiwu et al. 1996; Wu Zhaolu and Zhu Hualing 1996). Many farmers also began to establish rubber plantations.

Market. Both the local market and the wider domestic market for rubber latex in China were very promising. Current prices remain attractive. According to farmer respondents in the surveyed villages, one rubber tree can generate 100 yuan of income per year.

Case Histories

In addition to market opportunities and the demand for increased income, many other complicating factors influenced farmers' decisions to adopt rubber as a component of their farming systems. The following cases provide insights into local initiatives that played an important role in the rapid expansion of private rubber plantations.

- One farmer had relatives working on a state-owned farm. They repeatedly advised him that his fields were suitable for growing rubber, but he hesitated. After years of surplus food production, he finally decided in 1984 to establish a small plot of rubber. When he began tapping the trees six years later and earned a good return, he decided to expand his rubber plantation. He now plans to rent more land from a neighboring village to expand his rubber holdings and increase his production.
- A farmer learned that villagers 30 km distant were earning a good income from rubber. Since there was a state-owned rubber plantation nearby, he concluded that his own fields should also be suitable for rubber and he began planting trees on his fallow land. Having previously been a village leader, he sought the advice of technicians on the state rubber farm. He was later able to establish his own rubber nursery to expand his rubber plantation.
- A farmer began to establish a rubber plantation in his fallow swiddens after observing the success of his neighbors. He noted that most people planting rubber earned good money, and he wanted to be among the beneficiaries.

Other farmers maintain patches of well-established rubber after having been assigned parts of what were once collective rubber plantations. Some even rent former collective rubber farms and, with the profits from that, develop their own plantations. However, not all farmers are interested in cultivating rubber. In one village, for example, several households, instead of planting rubber in their fallows, leased their land to other farmers and concentrated on other businesses such as restaurants, trading, and processing of agricultural products.

It is noteworthy that farmers rarely rely on rubber as their sole source of income. Rather, they grow a variety of crops such as watermelon and vegetables during the dry season to maximize income and spread risk. Although rubber provides an opportunity for a good income, its expansion is limited by biophysical conditions because *Hevea brasiliensis* is a crop requiring tropical temperatures and ample moisture.

Rubber Fallow Management Practices

The geographical area suitable for growing rubber trees is generally considered to be between 15° S and 15° N latitude. Thus, rubber cultivation in Xishuangbanna, between 21° and 22° N latitude, and elsewhere in Yunnan Province, between 21° and 25° N latitude, is far outside the usual limits. Rubber was first introduced to Yunnan in 1907 by Chinese people who had been living in Malaysia. They planted 8,000 trees in Yingjiang county, at 24°50' N latitude. One of the original trees continues to survive today (Wu Zhengyi et al. 1990). In 1948, rubber trees were again planted at Ganlanba, in Xishuangbanna, but with a poor survival rate (Wu Zhengyi et al. 1990). The establishment and expansion of rubber plantations in southern Yunnan in the 1950s was driven by China's desire to meet its domestic need for rubber products (Huang Zongdao et al. 1980). Not surprisingly, the most important research issue in China's rubber industry is the development and selection of cold-resistant varieties with superior yields (Wang Huihai 1982; Xu Wenxia and Pan Yanqing 1992).

Farmers in Xishuangbanna establish their rubber plantations in the following way. After clearing and burning the fallow vegetation on sloping land, crude terraces are developed and rubber seedlings are planted at a density of about 450 per hectare. The seedlings come either from the farmer's own nursery or from state farms, where they can be bought for 3 to 5 yuan each. During the first two years, farmers interplant maize and other annual crops, such as beans and cotton, among the developing trees. Rubber trees can sometimes begin producing latex five years after planting, and build up a good yield after 11 years.

Farmers explain that once the plantation is established, there is no further need to slash and burn the fields every year. They claim the returns were not attractive after the troublesome and labor-intensive slash-and-burn operation. They say rubber provides a higher income with less investment in labor, and the labor thus saved can be channeled into more intensive cropping systems. Farmers also claim that rubber plantations contribute to local government reforestation objectives because they are more effective than traditional swidden systems in conserving soil and water. Despite these claims, this study found that the labor demands of rubber cultivation were no less than those of conventional shifting cultivation. During the tapping period, from late March to early November, farmers must collect latex almost every morning. The trees also need fertilizing. It is unclear if these findings suggest deliberate ignorance on the part of farmers, or a preference for rubber cultivation because of its more attractive returns. It is also possible that farmers simply prefer to work under the shade of the rubber canopy.

More than 200 rubber latex processing factories, under state, community, or private ownership, have been established to process the output of the Yunnan rubber crop. Latex collection stations are established in those villages with sufficient acreage of rubber plantations. Alternatively, middlemen collect latex in the late morning.

Given the benefits that it offers, rubber has become a strong lure to local farmers in southern Yunnan Province. However, it is very vulnerable to market changes, a situation heightened by China's increasing integration into global markets. Previously, limited local production could hardly meet domestic demand. But as China engages with international markets, local rubber producers face competition from neighboring tropical countries, most of which have a comparative production advantage.

The harvest of rubber latex also requires a good tapping technique, and not every rubber grower in southern Yunnan has mastered the job. They should receive some kind of training, either from neighbors or state-owned farms, before beginning to tap their own trees.

Discussion

Household Cost–Benefit Comparison

Based on our survey, a detailed economic comparison has been made between households with rubber plantations and those practicing conventional shifting cultivation (see Table 49-1). Labor savings are not considered in the analysis. It confirms that farmers earn a higher income from rubber plantations, and the net present value (NPV), or the present value of benefits minus the present value of costs, is positive by the sixth year after plantation establishment.

Table 49-1. Costs and Benefits of Rubber Plantations Compared with Conventional Shifting Cultivation

	Rubber Plantations			*Shifting Cultivation*	
Year	*Costs*	*Benefits*	*Net Benefit*	*Costs*	*Benefits*
0	Land preparation, seedlings, corn cultivation	Corn	–3,000	Corn planting, fertilizer	Corn 9,000
1	Fertilizing, corn cultivation	Corn	1,000	Corn planting, fertilizer	Corn 7,500
2	Fertilizing, maintenance			Corn cultivation, fertilizer	Corn 6,000
3	Fertilizing, maintenance			fallow	
4	Fertilizing, maintenance			fallow	
5	Fertilizing, harvesting	Latex	2,000	fallow	
6	Fertilizing, harvesting	Latex	4,000	fallow	
7	Fertilizing, harvesting	Latex	8,000	Corn planting, fertilizer	Corn 9,000
8	Fertilizing, harvesting	latex, seeds	15,000	Corn planting, fertilizer	Corn 7,500
9	Fertilizing, harvesting	latex, seeds	20,000	Corn planting, fertilizer	Corn 6,000
10	Fertilizing, harvesting	latex, seeds	26,000	fallow	
11	Fertilizing, harvesting	latex, seeds	30,000	fallow	
12	Fertilizing, harvesting	latex, seeds	30,000		
		Tractor, new house			None

Note: Indicator NPV > 0 in year 6; discount rate = 15%. Benefits shown in yuan/ha.
Source: Village interviews.

The economic data illustrate that rubber farmers can accumulate more valuable agricultural assets and achieve better living conditions. More importantly, rubber plantations provide a continuous income during the seven- to eight-month latex tapping season, allowing farmers to purchase agricultural inputs for other crops and meet various other urgent cash needs. Moreover, smallholder rubber plantations have proven to be far more profitable than national farms, which are burdened with surplus management personnel and retired workers (Zhou Yonghua et al. 1994). These benefits will continue only if the price of rubber remains stable. In 1988, fluctuating world prices for rubber and illegal rubber imports caused domestic prices in China to tumble, and the industry was in serious trouble. At the time, incentives for local farmers to further invest in rubber plantations were greatly reduced (Li Liangsheng 1993; Zhou Yonghua et al. 1994).

Ecological Sustainability and Potential of Rubber Plantations

The impacts of rubber plantations on soil conservation are summarized in Table 49-2 (Liu Hongmao and Xu Zaifu 1996). The results suggest that rubber plantations are more efficient in conserving soil than shifting cultivation but less so than natural rainforest. Therefore, from an ecological perspective, rubber plantations are a preferable land use when compared to short-cycle swiddening.

However, rubber plantations may reduce the biodiversity of the area. For instance, one study carried out in Xishuangbanna found that various categories of birds had been reduced by more than 50% as a consequence of replacing rainforest with rubber and tea plantations. These categories included:

- Species of oriental birds;
- Species endemic to the southern hilly subregion of Yunnan;
- Species common to the southern hilly subregions of Yunnan and the southwestern mountains; and
- Widely distributed species. None of the species common to the southern hilly subregion of Yunnan and the Hainan Island subregion were found (Liu Hongmao and Xu Zaifu 1996).

These environmental problems, together with the economic vulnerability of rubber production, have attracted criticism from some Chinese authors (Liu Jingyue 1987; Liu Lunhui 1987; Tao Siming 1995; Xu Ziafu 1994; Zhou Yonghua et al. 1994; Xu Zaifu and Liu Hongmao 1995, 1996; Liu Hongmao and Xu Zaifu 1996). It should also be noted that processing rubber latex requires significant energy inputs, and many rural latex processors consume large quantities of firewood (Zhou Yonghua et al. 1994). Some local informants mentioned, however, that a large hydroelectric dam planned for construction in Xishuangbanna would provide cheap and reliable power for processing good-quality rubber.

Table 49-2. Ecological Functions of Rubber Plantations, Shifting Cultivation, and Natural Forest

Type	*Drainage**	*Scouring Volume**
Rainforest	1	1
Rubber plantations	3	43
Slope cultivation (upland rice)	35	78

Note: * Relative values.
Source: Liu Hongmao and Xu Zaifu 1996.

Capacity to Support Increasing Population Densities

Our investigation found that, under current prices, rubber plantations generate improved income for rural households and can create employment opportunities in related processing industries. Rural livelihoods have been improved in the following ways:

- Continuous income during seven to eight months of the year contributes significantly to household budgets.
- Money earned from rubber plantations can be invested in intensive cultivation of cash crops, including the purchase of new varieties and fertilizer.
- Conversion to rubber plantations helps farmers to adopt more intensified agricultural practices, through the purchase of new agricultural tools, and allows them to focus on yield increments, rather than extensification.
- After harvesting rubber latex, farmers turn to cultivating watermelon, vegetables, and other cash crops in their paddy fields, thereby optimizing their labor and complementing the income received from rubber.
- Income from rubber plantations helps finance the purchase of agricultural assets such as tractors, which increase labor productivity and support intensification.

While it can be seen that the expansion of rubber plantations has helped to support increased rural populations, it should be stressed that these points are based on fair rubber prices and an adequate energy supply for processing.

Comparable Systems Elsewhere in Southeast Asia

The literature confirms that other smallholder farmers in Southeast Asia have also adopted rubber-based agroforestry systems to replace traditional swidden practices and that, generally speaking, *Hevea brasiliensis* has played a highly valued role in many smallholder farming systems (see, for example, Chapter 48).

In Indonesia, smallholder farmers, strongly motivated by market opportunities, plant rubber, coffee, and other cash crops into swidden fields. Other farmers focus on intensification of other components of their farming systems, such as wet rice cultivation (Garrity et al. 1993; Scholz 1995).

In Malaysia, farmers regard the rubber tree as an important multipurpose species. Apart from providing latex, it is also used for timber and firewood, contributing significantly to rural household economies (Fui and Chuen 1994).

In Thailand, smallholder plantation owners tap the latex for the productive life of the trees and then harvest the wood for use in construction of furniture (Beldt et al. 1994).

Generalizable Elements

The management of plants in general, and their manipulation in fallow vegetation in particular, reflects the role of those species in the livelihoods of local people. In response to local socioeconomic conditions, farmers will incorporate into their farming systems those trees that best address local market needs. The Indonesian experience illustrates that, as economies become more market-oriented, local farmers will adopt marketable species such as rubber and coffee into their farming systems, and many original species will be displaced (Scholz 1995). Studies in Xishuangbanna have similarly found that many traditional land uses and species have changed in reaction to evolving socioeconomic conditions (Long Chunlin and Wang Jieru 1993; Guo Huijun 1994; Long Chunlin 1994; Wang Jieru 1994; Chen Aiguo et al. 1996; Xu Jianchu et al. 1996). For example, farmers have changed from planting tea under natural forests to planting cardamom in the same environment (Chen Aiguo et al. 1996). Home gardens also contain many economic species that are highly valued (Guo Huijun 1994; Long Chunlin 1994; Wang Jieru 1994). All these examples indicate that indigenous upland farming systems are undergoing dramatic changes, and incorporation of rubber into swidden fallows has been an important strategy on the part of local people to take advantage of market opportunities, intensify their land use, and improve their living standards.

In the study area, farmers have given rubber, an exotic species, a value equal, or perhaps, more than equal to the value that they give to native species. This suggests that these farmers do not have a clear preference for native or exotic species when considering which should be integrated into their farming systems. Rather, their selection is based solely on socioeconomic criteria. This observation contradicts the popular notion that native species may be the result of a long selective process by local communities and that, with exotic species, this learning process must begin anew.

Adoption Concerns

Rubber has been adopted by many smallholders in southern Yunnan Province, often without the direct help of government agencies. A number of factors have contributed to its rapid expansion, and these should be borne in mind if consideration is to be given to the adoption of similar systems elsewhere:

Biophysical Conditions. The case of China illustrates that rubber should be planted in areas having weather no colder than the northern tropics and without severely low temperatures during winter.

Market. The reliability of market prices for rubber latex is essential for the long term prospects of the industry.

Available Technology. Improved technologies should be made available to local farmers to improve the management of their rubber plantations.

Land Tenure. Local farmers should either own their land or have secure rights to it over at least 30 years, as is the case in China.

Economics. Rubber has a clear comparative economic advantage over other cash crops.

Policy. National and local policies in China encourage the establishment of rubber plantations.

Future Research Priorities

The following issues and questions warrant further research attention:

- If integration of rubber into fallows is adopted as a local strategy to intensify shifting cultivation, the first task should be to carefully analyze the indigenous system and discover how it works. This study illustrates that many local farmers of different ethnicity are increasingly converting fallow vegetation into rubber plantations, suggesting that rubber is their preferred fallow species. Any study that considers only native species may not adequately address the content and characteristics of indigenous systems, or the underlying knowledge of farmers.
- Active management of economic species on formerly fallow land has the potential to make important contributions to farmers' daily livelihoods. How do farmers perceive the role of the species? How is it managed? With what criteria do farmers evaluate it? Do farmers' management interventions alter the role of the species and therefore become crucial to its evaluation? How does this dynamic process work and how does it shape management decisions?

- How do farmers view sustainability in their land use, and what is the most important component in complex, whole-farm systems? From our experience in this study, we feel that understanding of this issue is inadequate, and the concept of sustainability may be viewed differently from one group of farmers to the next.
- Past experience in agroforestry extension has shown that while many agroforestry models may be effective in conserving soil and come highly recommended by scientists, they often receive a relatively poor response from farmers. The poor farmer's adoption of alley cropping in the Philippines, in Xishuangbanna, and probably elsewhere, is a good case in point (Watson and Laquihon 1984). Although we may understand many agroforestry models, there is still much work to be done in understanding how these models can best fit into local farming systems. With this in mind, this chapter resists the temptation to recommend rubber as an effective fallow management strategy, but instead tries to better understand the system. Until we have developed a clearer understanding of local farmers and the context of their decision-making, we may not be positioned to suggest improvements. Therefore, we suggest that a research framework be formulated within which we can work toward a sound understanding of farmers and their land-use systems.

Technical Research

Research into the technical aspects of rubber-based fallows might include the following issues:

- Experience has shown that rubber suffers heavy losses during abnormal temperature drops in the dry season (Xu Wenxia and Pan Yanqing 1992; Zhou Yonghue et al. 1994). Therefore, agromists should develop more cold-resistant and productive varieties adapted to Xishuangbanna's biophysical conditions.
- Economic uses of rubber byproducts, such as leaves, seeds, and branches, should be developed. For example, a new technology to process rubber seed into livestock feed could increase the system's profitability.

Conclusions

In southern Yunnan, and particularly in Xishuangbanna below the altitude of 900 m asl and between 21° and 25° N latitude, local farmers have increasingly converted their swidden fallows into rubber plantations to generate income. The conversion generally happens after farmers have achieved a comfortable level of food security.

Rubber plantations are superior to conventional shifting cultivation systems in terms of both soil conservation and profitability. Due to a specific socioeconomic context, rubber has thrived as a profitable crop in China, even outside its usually accepted biophysical environment, and farmers have come to highly value its role in their farming systems.

This study suggests that indigenous knowledge systems are dynamic and do not exclude outside knowledge or germplasm. This analysis also supports the original hypothesis that farmers tend to actively manage those species that hold promise of contributing significantly to their family livelihood.

Therefore, the study suggests the need for a more comprehensive understanding of trees and other perennial crops that are recognized as having the potential to make important contributions to farming systems. Although trends of changes in land use are generally understood, it may still be too early to generalize based on current knowledge.

As farmers are inevitably forced to intensify their swidden systems, their fallow management strategies are shaped in the context of surrounding socioeconomic conditions, particularly market opportunities. Their continual selection of useful species for fallow enrichment, whether exotic or native, will favor those species that most closely respond to farmers' needs within their specific socioeconomic environment. Therefore, it is most meaningful to discuss individual fallow management strategies within a specific local context. Rubber is therefore of high scientific interest in the Xishuangbanna context.

Within the restricted biophysical niche where *Hevea brasiliensis* can grow in China, its expansion has primarily been driven by favorable local, national, and international markets and good extension services. It is interesting to note that the current support system for rubber is built on the remnant network of past government extension efforts. This has now been voluntarily adopted by local farmers for their own benefit.

Acknowledgments

The FTPP (Forests, Trees and People Program) at RECOFTC supported the major part of the fieldwork in September and October 1996. This is gratefully acknowledged. The authors are also indebted to the World Wildlife Fund for Nature and the Ford Foundation who supported the 1995 and early 1996 fieldwork, respectively. Mr. Cor Veer is owed particular thanks for his assistance in formulating the research guidelines for the fieldwork, which was critical to this chapter. Finally, we thank Mr. Yang Bilun, Mr. Long Chunlin, Dr. Sejal Worah, Mr. Dai Chong, Ms. Liu Ping, and Ms. Wang Jieru for their invaluable assistance.

References

Anonymous. 1980. Yunnansheng Renminghengfu Guanyu Baohu Senlin Ziyuan Fazhan Linye Shengchan de Guiding. *Yunnan Linye* 3–6.

———. 1992. Zhonggong Zhongyang Zhengzhiju Changwei Sunping dui Wosheng Linye Gongzuo de Zishi. *Yunnan Linye* 2.

Agrawal, A. 1995. Indigenous and Scientific Knowledge: Some Critical Comments. *Indigenous Knowledge and Development Monitor* 3(3), 3–6.

Beldt, R.V.D., et al. 1994. Farm Forestry Comes of Age: The Emerging Trend in Asia. *Farm Forestry News* 6(2), 1–6.

Chen Aiguo et al. 1996. Transformation of Cultivation under Tropical Forests from Tea into Chinese Cardamom in Jinuo Mountain, Xishuangbanna, Yunnan. In: *Collected Research Papers on Tropical Botany* (IV), edited by Zhongguo Xishuangbanna Redai Zhiwuyuan. Kunming, China: Yunnan University Press, 77–81.

Cunningham, A.B. 1996. Whose Knowledge and Whose Resources? Ethnobotanists and Brokers between Two Worlds. In: *The Challenges of Ethnobiology in the 21st Century*, edited by Pei Shengji et al. Kunming, China: Yunnan Science and Technology Press, 1–6.

Deng Jiwu. 1996. Xishuangbanna Jizhong Rengong Qunluo Moshi Shiyan Yanjiu de Jinzhan. *Development of Research Network for Natural Resources, Environment, and Ecology* 7(1), 39–41.

Duan Qiwu et al. 1996. The Eco-economic Benefits of Amomum Tsao-ko Cultivation under Forest in Tropical Yunnan. *Ecological Economy* 5, 38–40.

Feng Yaozong. 1993. Zhongguo redai senlin shengtai xitong jiegou, gongneng he tigao shengchanli tujing yanjiu: 92nian keti gongzuo zongjie ji 93nian yanjiu jihua. *Development of Research Network for Natural Resources, Environment, and Ecology* 4(4) 38–39.

Fui, L.H., and W.W. Chuen. 1994. Marketing of Agroforestry Products: Some Malaysian Experiences. In: *Marketing of Multipurpose Tree Products in Asia,* Proceedings of an international workshop, December 6–9, 1993, Baguio City, Philippines, edited by J.B. Raintree and H.A. Francisco. Bangkok: Winrock International, 3–14.

Gao Lizhi et al. 1996. A Survey of the Current Status of Wild Rice in China. *Chinese Biodiversity* 4(3), 160–166.

Garrity, D.P., et al. 1993. *Sustainable Land Use Systems and Agroforestry Research for the Humid Tropics of Asia.* Bogor, Indonesia: ICRAF Southeast Asia and Asia-Pacific Agroforestry Network, 99–109.

Guo Huijun. 1994. Transformation from Natural Forest to Home Garden. In: *The Peasants' Home-Garden Economy in China,* edited by Feng Yaozong and Cai Chuantao. Beijing: Science Press, 243–250.

Huang Zongdao, et al. 1980. Assessment of Rubber Planting Areas in Tropical and Subtropical Zones of China: Rational Exploitation of the Natural Resources in the Tropical and South Subtropical Zones to Speed Up the Establishment of Production Bases of Tropical Crops with Rubber Trees In Chief. *Tropical Crop Journal* 1(1), 1–6.

Li Haifeng. 1996. Patterns and Programming for Eco-agricultural Divisions in Jiangsu Province, *Eco-Agriculture Research* 4(3), 82–84.

Li Liangsheng. 1993. *Lun Dianxinan Requ Ziyuan Kaifa*. Kunming, China: Yunnan Renmin Chubanshe.

Li Yikun and Huang Zerun. 1987. *Hevea brasiliensis*: An Important Biological Resource in Yunnan's Tropical Area. In: *Selected Papers of Rational Exploitation and Utilization of Yunnan's Biological Resources*, edited by Xiao Junqin, et al. Kunming, China: Yunnan Science and Technology Press, 81–85.

Liang Xiaomao. 1987. Xishuangbanna shehui lishi gaikuang. In: *Xishuangbanna Ziran Baohuqu Zonghe Kaocha Baogaoji*, edited by Xu Yongchun, et al. Kunming, China: Yunnan Keji Chubanshe, 425–525.

Liu Dewang. 1996. Yingyong zongjiao shi baohu hao yesheng dongzhiwu de youxiao tujing. *Chinese Biodiversity* 4(2), 123–124.

Liu Hongmao and Xu Zaifu. 1996. A Preliminary Study of the Deterioration Mechanism and Recovery Methods of the Tropical Rainforest in Yunnan. In: *Collected Research Papers on Tropical Botany* (IV), edited by Zhongguo Xishuangbanna Redai Zhiwuyuan. Kunming, China: Yunnan University Press, 31–35.

Liu Jingyue. 1987. The Function of Forest Ecosystems in Mountain Area Production. In: *Selected Papers of Rational Exploitation and Utilization of Yunnan's Biological Resources*, edited by Xiao Junqin, et al. Kunming, China: Yunnan Science and Technology Press, 308–312.

Liu Lunhui. 1987. On the Exploitation and Utilization of Land Resources with Regard to the Characteristics of Southern Yunnan's Forest. In: *Selected Papers of Rational Exploitation and Utilization of Yunnan's Biological Resources*, edited by Xiao Junqin, et al. Kunming, China: Yunnan Science and Technology Press, 302–307.

Long Chunlin. 1994. Structure of Home Gardens in Xishuangbanna. In: *The Peasants' Home-Garden Economy in China*, edited by Feng Yaozong and Cai Chuantao. Beijing: Science Press, 143–150.

———, and Wang Jieru. 1993. Forest Management and Biodiversity of Jinuo Nationality. In: *Symposium of Seminar on Biodiversity in Yunnan*, edited by Wu Zhengyi. Kunming, China: Yunnan Science and Technology Press, 189–194.

Long Yiming. 1993. Evaluation of Ecological and Economical Benefits of the Man-made Rubber Tea Community. *Chinese Journal of Ecology* 10(3), 37–40.

Luo Jingchun. 1994. Yunnan Ziran Ziyuan Kaifa Zhanlue Yanjiu. In: *Yunnansheng Fazhan Zhanlue Yanjiu*, edited by Zhao Juncheng. Kunming, China: Yunnan Renmin Chubanshe, 319–336.

Pei Shengji, et al. 1987. Opinions on the Exploitation of Tropical Crop Resources in Yunnan Tropical Area. In: *Selected Papers of Rational Exploitation and Utilization of Yunnan's Biological Resources*, edited by Xiao Junqin, et al. Kunming, China: Yunnan Science and Technology Press, 129–139.

Posey, D.A. 1996. Ethnobiology and Ethnodevelopment: Importance of Traditional Knowledge and Traditional Peoples. In: *The Challenges of Ethnobiology in the 21st Century*, edited by Pei Shengji, et al. Kunming, China: Yunnan Science and Technology Press, 7–13.

Scholz, U. 1995. Change from Slash-and-Burn Agriculture to Alternative Farming Systems in Sumatra. In: *Abstracts of International Symposium on Alternatives to Slash-and-Burn Agriculture*. Kunming, China: 26.

Shi Shan. 1993. Zhongtan shengtai nongyexian jianshe: Jianshe woguo xiandai nongye he xiandai nongcun de yitiao xinlu. *Eco-Agriculture Research* 1(1), 11–13.

Sun Hongliang. 1996. The Main Patterns of Ecological Agriculture in China and the Ecology Principles of their Sustainable Development. *Eco-Agriculture Research* 4(1), 15–22.

Tao Siming. 1995. Ecological Problems in Forestry and Strategies for its Development in China. *Eco-Agriculture Research* 3(1), 8–12.

Wang Huihai. 1982. Discussion on the Design of Cold-Resistant Structure of the Rubber-Tea Community for Fully Utilizing Sunlight Energy of the Winter Season. In: *Collected Research Papers on Tropical Botany*. Edited by the Yunnan Institute of Tropical Botany and Academia Sinica. Kunming, China: Yunnan People's Publication House.

Wang Jieru. 1994. Studies on Home Gardens of Jinuo Nationality. In: *The Peasants' Home-Garden Economy in China*, edited by Feng Yaozong and Cai Chuantao. Beijing: Science Press, 194–203.

Warner, K. 1991. *Shifting Cultivators: Local Technical Knowledge and Natural Resource Management in the Humid Tropics*. Rome: Food and Agriculture Organization of the United Nations.

Watson, H.R., and W.A. Laquihon. 1984. Sloping Agricultural Land Technology (SALT): A Social Forestry Model in the Philippines. In: *Community Forestry: Lessons from Case Studies in Asia and the Pacific Region*, edited by Y.S. Rao, et al. Bangkok: Food and Agriculture Organization of United Nations, 21–44.

Wu Zhaolu and Zhu Hualing. 1996. The Innovation from Slash-and-Burn to Intensive Agriculture in Tiaobahe Region of Xishuangbanna. *Ecological Economy* 4, 31–34.

Wu Zhengyi, et al. 1990. *Yunnan Shengwu Ziyuan Kaifa Zhanlue Yanjiu*. Kunming, China: Yunnan Keji Chubanshe, 139–151.

Xu Jianchu et al. 1996. Diannan ji Dianxinanqu senlin ziyuan de guanli. In: *Zhongguo Yunnan Cunshe Linye Guanli Xianzhuang*, edited by Lu Xing, et al. Kunming, China: Yunnan Daxue Chubanshe, 45–70.

Xu Weishan. 1990. Jiaqiang dui Xishuangbanna Dongzhiwu Ziyuan de Baohu Gongzuo. *Yunnan Linye* 13.

Xu Wenxia and Pan Yanqing. 1992. Progress of Study on Physiology with Cold Resistance of *Hevea brasiliensis* in China. *Tropical Crop Journal* 13 (1), 1–6.

Xu Youkai et al. 1996. Woody Plants Used as Vegetables in Xishuangbanna. In: *Collected Research Papers on Tropical Botany (IV)*, edited by Zhongguo Xishuangbanna Redai Zhiwuyuan. Kunming, China: Yunnan University Press, 110–114.

Xu Ziafu. 1994. Rongshu—diannan redai yulin shengtaixitong zhong de yilei guanjian zhiwu. *Chinese Biodiversity* 2(1), 21–23.

Xu Ziafu and Liu Hongmao. 1995. Palm Leaf Buddhism Sutra Culture of the Dai of Xishuangbanna and Plant Diversity Conservation. *Chinese Biodiversity* 3(3), 174–179.

———, 1996. The Loss and Restoration of Biodiversity in Degraded Tropical Rainforest Ecosystems. *Acta Botanica Yunnanica* 18(4), 433–438.

Xu Zaifu and Yu Pinghua. 1987. The State and Potency of Cultivated Tropical Plant Resources in Southern Yunnan. In: *Selected Papers of Rational Exploitation and Utilization of Yunnan's Biological Resources*, edited by Xiao Junqin, et al. Kunming, China: Yunnan Science and Technology Press, 28–33.

Xue Jiru et al. (eds.). 1986. *Yunnan Senlin*. Kunming, China: Yunnan Keji Chubanshe, 230–234.

Ying Shaoting. 1994. A Farming Culture Born out of Forests: Swiddening in Yunnan, China. Kunming, China: Yunnan Renming Chubanshe.

Zeng Juemin. 1995. Linnong lunzuozhi dui shengtai nongye de tuidong. *Agroforestry Today* 3(1), 5–11.

Zhou Yonghua et al. 1994. Yunnan yi xiangjiao weizhude redai jingji zuowu kaifa yu baohu shengwuhuanjing de maodun fenxi ji jiejue tujing. In: *Yunnan Jingji Kaifa yu Shengtaihuanjing Xiaotiao Fazhan Yanjiu*, edited by Yao Neizhe and Zhou Yonghua. Kunming, China: Yunnan Renming Chubanshe, 117–144.

Chapter 50
Ma Kwaen
A Jungle Spice Used in Swidden Intensification in Northern Thailand

Peter Hoare, Borpit Maneeratana, and Wichai Songwadhana

Ma kwaen, also called *ma kaen* in Nan Province, is the northern Thai name for a small forest tree found at altitudes between 350 and 1,000 m above sea level (asl) in northern Thailand and Laos. When mature, it is 5 to 10 meters tall and has a diameter of about 20 cm. It is found mainly in upland forest reserves and highland areas where the farmers have no legal title to land. When it is three to five years old, ma kwaen flowers in racemes. The seed pods have been harvested for generations by the northern Thai people in the cool, dry season in November and December. The pods are harvested just before the seeds mature and are then dried and sold in local markets. The black seeds are not eaten, but the seed coats contain an essential oil used for flavoring food. They are pounded by hand mortar and pestle and mixed with meat dishes and soups. They give a sweet lemon flavor to the food.

Hypotheses

Ma kwaen (*Zanthoxylum limonella*) is well suited to intensifying shifting cultivation in priority watershed areas where the government retains land tenure rights but allows the harvest and sale of minor forest products.

Furthermore, ma kwaen has potential in northern Thailand to accelerate the move from shifting agriculture based on monocultures of annual crops to polycultures with higher returns per unit area.

The Study Area

Nan Province, in the far north of Thailand and bordering Laos, is one of Thailand's most important river basins. The Nan River provides more than half the annual water discharge for the Chao Phrya River, the lifeline of the large population on the country's central plains. The study area for this chapter is in the upper right

---

Peter Hoare, Project Coordinator, Royal Forest Department—Danish Cooperation for Environment and Development (RFD-DANCED) Upper Nan Watershed Management Project, P.O. Box 31, Nan 55000, Thailand; Borpit Maneeratana and Wichai Songwadhana, Royal Forestry Department, Nan Watershed Management Office (Khaki Noi), Tambon Dootai, Amphoe Muang, Nan Province 55000, Thailand.

catchment of the Nan River, an area of 912 km^2 covered by the Royal Forest Department-Danish Cooperation for Environment and Development (RFD-DANCED) upper Nan watershed management project. The altitude is between 350 and 800 m asl. The natural forest has been severely reduced over recent years due to legal and illegal logging, shifting agriculture, and uncontrolled forest fires.

In part, this is a legacy of the years up to the mid-1980s, when Thailand's economy was still driven by its agricultural sector and before the country's massive industrialization. Thailand had been able to remain the largest food and agricultural exporter in Southeast Asia because, unlike most other countries in the region, it had an abundance of surplus land until the 1980s. This surplus was used up in the race to significantly boost agricultural production without any increase in agricultural investment or changes in technology levels (Australian Department of Foreign Affairs and Trade 1994).

Maize was the first major crop in the expansion and diversification drive, and it became Thailand's largest crop after rice in harvested area and production. The growing of maize in the upper Nan watershed in the 1970s and 1980s contributed to the destruction of large areas of forest. Cotton also became an important cash crop on upland fields and, along with the production of upland rice for subsistence food production, played its part in forest destruction.

The study area has a total of 44 villages with about 20,000 people living in 3,843 households. These include 28 so-called hilltribe villages, populated by Khamu, Lue, Hmien (Yao), and Hmong ethnic groups. Many Hmien and Hmong people, who formerly lived on higher slopes, have moved to lower altitudes in recent years.

The strategy of the RFD-DANCED project is to involve these local communities in rehabilitation and protection of forests in the area through participatory land-use planning. The main emphasis is on reforestation and natural forest regeneration of critical areas; changes from shifting cultivation to more permanent and sustainable forms of agriculture, including horticulture and paddy rice; and the development of alternative income sources (RFD-DANCED 1995).

Research Methods

Research methods for the study detailed in this chapter included:

- A literature review;
- Participation in a Thailand Department of Agricultural Extension field day on ma kwaen in January 1997; and
- Interviews with farmers and field staff of both the Royal Forest Department and the RFD-DANCED project with experience in growing ma kwaen in Nan and Chiang Mai Provinces in northern Thailand.

Results

The traditional cultivation system in the study area was based on rice subsistence farming until the 1960s. Glutinous rice was grown in paddy fields in narrow mountain valleys until population growth demanded the shifting cultivation of rice on upland fields.

People of the Hmien (Yao) and Hmong ethnic groups migrated from Laos to highland parts of the project area, and their production of nonglutinous upland rice, maize, and other crops contributed to widespread deforestation. Uncontrolled fires and shifting cultivation by successive migrant groups created large areas of grasslands dominated by *Imperata cylindrica*.

The introduction of cash cropping after the 1960s has led to the development of monocultures and, more recently, diversified annual cropping and polycultures. These cropping systems include:

- Lowland paddy: rainy season rice followed by dry season irrigated rice; rainy season rice followed by dry season soybean; rainy season rice followed by barley under brewery contract.

- Floodplain: maize (flooding July-August) followed by peanuts; maize (flooding July-August) followed by maize or vegetables.
- Upland fields: upland rice for two to three years followed by fallow; upland rice for two to three years, during which litchi is established, followed by cotton for two years then fallow; or continuous cropping.

Hmong and Hmien villages are now developing litchi gardens at altitudes of around 600 m asl. At this attitude, the litchis are harvested earlier than the main lowland crops grown in Chiang Rai and Chiang Mai Provinces and gain consequent market advantages. The harvesting of the fruit of forest palms also provides considerable income in some hilltribe villages.

Contextual Triggers Leading to Farmer Innovations in Fallow Management

The "triggers" that appear to have been responsible for the change from gathering the seed of ma kwaen in the forest to its more intensive propagation and transplanting by farmers include:

- A steady increase in the market price of fresh ma kwaen, from $US 0.1/kg to $US 2/kg; and
- A declining market for traditional northern Thai fermented tea (*miang*), forcing producers of miang in the Chiang Mai area to diversify.

Management Practices

The areas planted to ma kwaen in Nan Province were initially upland fields bordering paddy areas in the villages of local northern Thai people and Lue ethnic groups. There are now 500 ha planted in five villages, at an average density of 84 trees per hectare. However, more recent plantings are at higher densities of up to 600 trees per hectare.

In the past five years, Hmien and Khamu villagers have started to plant ma kwaen as a cash crop. Also, there are now middlemen buying and repacking ma kwaen purchased from producers in Laos at bimonthly border markets.

The indigenous practice is to gather ma kwaen seedlings beneath parent trees in the forest. This is similar to the practice followed by the northern Thai in developing gardens for the production of miang, or fermented tea.

Attempts to improve the system include burning straw under mature trees to scarify the hard seed coat and accelerate germination. The seed takes about two months to germinate. However, losses of young seedlings are often as high as 60% after heavy rains.

A common method is used for establishment of ma kwaen gardens containing large numbers of trees. When sufficient parent trees are established at the top of slopes, dissemination of seeds is aided by the downslope movement of soil and transport by birds. Fields below parent trees are slashed and burned, and many seedlings establish naturally. Some supplementary planting is done. Simple nursery propagation techniques have been developed by a community coordinator working with the RFD-DANCED project.

Discussion

The main benefits from intensified management of fallow land using ma kwaen are increased farmer income as the market price increases, and improved fire management in watershed areas because ma kwaen is very susceptible to damage by fire.

At a field day for ma kwaen cultivation in January 1997, at Pa Luk village in the Song Kwae district of Nan Province, an extension leaflet in the Thai language was handed out. It gave the following information.

A total of 155 families in five villages in the study area have planted 2,164 *rai*, or 346 ha, with 29,129 ma kwaen trees, and these are now fruiting. Sales from the trees in the 1996–1997 dry season were estimated at $US120,000 to $US160,000. The leaflet said there were another 978 rai, or 156 ha, with 12,800 young trees not yet fruiting. All these trees were growing at altitudes of 400 to 600 m asl.

These figures indicate an overall density of only 84 trees per hectare. However, many more recent plantings are at about 4 by 4 m spacing, providing around 600 trees per hectare. Yield data for trees of different ages is presented in Table 50-1.

In the 1997 harvest, each 20-year-old tree provided an income of about $US104 without any cash inputs for agricultural chemicals. Family labor was required between November and January for harvesting the fruit and making firebreaks.

Yield data are not available from Pa Pae district of Chiang Mai Province, the other major production area. Pa Pae has a higher altitude, of between 600 and 800 m asl. This area also produces fermented tea, and ma kwaen was introduced into low-density tea gardens to diversify income as the market for fermented tea declined. The ma kwaen from this area is reported to have smaller seed pods that are of higher quality than those grown in Nan Province. They earn a price premium of about 10%.

Sustainability and Potential to Rehabilitate Degraded Upland Environments

Ma kwaen scores well on ecological sustainability and the potential to rehabilitate degraded upland environments. No soil tillage or agricultural chemicals are necessary for its establishment or maintenance. In villages where farmers plant ma kwaen, fire management is also greatly improved, as the trees are very susceptible to fire damage. Radiant heat from a fire just a few meters away will kill them. As a consequence, many villages where ma kwaen is planted impose heavy fines on farmers who light fires that damage trees. In Ban Yot village, for example, the fine is $US4,000 for every 13 trees killed by fire.

Capacity to Support Increasing Population Densities

One rai (0.16 ha) of 6- to 10-year-old ma kwaen trees at the higher planting density of 4 by 4 m spacing would have produced an annual income of $US2,000 in 1996. This was more than the income generated by one hectare of traditional upland crops of maize and cotton. Ma kwaen can therefore support a population density at least five times higher than that supported by traditional cash crops.

Key Leverage Points to Encourage Adoption

The main leverage point with which to encourage expansion of ma kwaen cultivation is to convince farmers that prices will not collapse in an oversupplied market. Such fears were well illustrated at a seminar held in conjunction with the January 1997 field day at Pa Luk village, mentioned above. Farmer representatives attended a discussion on the topic: "How to increase the area planted to ma kwaen."

Table 50-1. Yield Data for Ma Kwaen Trees of Different Ages

Age of Tree (years)	*Average Yield per Tree (kg fresh weight)*	*Average Return per Tree at $US 2/kg*
3–5	2	4
6–10	10	20
11–15	30	60
21–25	50	100

Note: $US 2/kg fresh weight was the average farm gate price in 1997.
Source: Thailand Department of Agricultural Extension (1997).

It soon became apparent that existing producers were afraid that their market price would decline if many additional villages started planting the jungle spice. They gave conflicting advice on methods of propagation and maintenance of the trees, showed a lack of interest in assisting a project to develop village nurseries to sell seedlings at the local price of 15 baht each, and even claimed that a middleman from a neighboring province had attempted to light fires to destroy their trees in an apparent attempt to monopolize the market.

Research Priorities and Experimental Agenda

Market Potential

A study of the market potential for ma kwaen is needed during the harvest period between November and January, and the marketing chain in Thailand needs to be defined. This also applies to the ma kwaen imported into Thailand from Laos. What are the destinations after the dry jungle spice is sorted, trimmed, and packed into bamboo baskets of about 12 kg weight and sent to markets in the lower north and central Thailand?

Traditionally, ma kwaen has been consumed only by northern Thai people and the Lue ethnic group. However, other Thais are starting to use the spice, and this suggests that the domestic market may expand greatly. There is therefore a need to estimate the potential size of the market for ma kwaen, both within Thailand and elsewhere. The ability of ma kwaen farming to become established in other parts of Southeast Asia would depend upon the existence of a local market returning adequate prices and the development of export markets.

There could be a potential export market for a new "environmentally friendly" jungle spice. After all, the European trade in tropical spices has been an important factor in the history of many Asian countries.

Improvement of Cultural Practices

Many development workers consider that ma kwaen seed is difficult to germinate. Tests with different treatments, including hot water and heating the seed, were reported by Gijka (1986) but were inconclusive. Farmers attending the seminar in January 1997 considered that seed contained in bird droppings provided the best germination.

The main problem, according to Somsak Sangngam (1997), is that immature seed is often planted. Ma kwaen is harvested when the seed coat is still green and the seed immature, and many development workers have tried to germinate the immature seed, with poor results. Khun Somsak uses small plastic trays measuring 20 by 30 cm containing several hundred seeds mixed with subsoil, and reports the germination of up to 500 seedlings per tray after two months. The young seedlings are transplanted at the two-leaf stage, about a week after germination, using a spoon. They are sensitive to root damage and this sometimes results in high losses.

Farmers report losses of up to 60% of seedlings transplanted during heavy rains in July and August. Experience suggests that these losses can be reduced by starting the nursery as early as February and having the seedlings well established before the heavy rains begin. Alternatively, the seedlings should be transplanted into the field at the end of the wet season and watered during the dry season from November to March.

Genetic Selection

There is considerable genetic variation in ma kwaen. This has shown up in laboratory tests at the Multiple Cropping Center at Chiang Mai University. No research has yet been undertaken to measure the yields of the different types at various altitudes. It has been suggested that ma kwaen provenances should be collected from various sites for planting trials at a range of altitudes in northern Thailand.

Small-Scale Processing of Spices

The possibility of adding value to ma kwaen by small-scale village processing should be investigated. In Malaysia, for example, more than 150 small-scale processors of spices each produce on average 100 to 200 kg of finished product daily. Ground pepper and curry powder are common spices, but each processor produces more than one product line. Most is for local markets and the technology is very simple: coarse grinding, followed by fine grinding (Senik 1995).

Conclusions

Ma kwaen is a nontimber tree and a local spice that has received little attention in the literature covering forestry and shifting cultivation. Its economic importance has gradually increased, and if an export market can be developed, it could become of considerable economic importance to several provinces in Thailand and Laos.

Ma kwaen can easily be established in swidden fields once the population of fruiting mother trees increases around villages. After slashing and burning, the germination of seed transported by downslope soil movement and bird droppings is often enough to give a good establishment of ma kwaen, although some additional enrichment planting may be needed.

Ma kwaen is suitable for intercropping with upland rice and cotton. It can also be integrated into agroforestry systems with fruit trees and other tree crops. Although traditionally used as a food flavoring by northern Thai people, it is now being planted by neighboring Hmien (Yao) and Khamu hilltribe farmers.

The Royal Forest Department–Danish Cooperation for Environment and Development (RFD–DANCED) upper Nan watershed management project considers itself fortunate to have such a tree crop, developed by indigenous technology, with which to intensify shifting agriculture. The future rate of its adoption will depend largely on the market potential for this jungle spice. However, ma kwaen is expected to play an important role in the project's objectives of naturally regenerating 16,000 ha of forest, reducing forest fires by 50%, and reducing the area of shifting agriculture by 35%.

References

Australian Department of Foreign Affairs and Trade. 1994. *Subsistence to Supermarket: Food and Agricultural Transformation in Southeast Asia*. Canberra: East Asia Analytical Unit.

Gijka, Somyos. 1986. *"Ma Kwaen," a New Interesting Spice*. Saraburi, Thailand: Seed Center, RFD.

RFD–DANCED. 1995. Project document for upper Nan watershed management project. Copenhagen: DANCED.

Senik, Ghani. 1995. *Small Scale Food Processing Enterprises in Malaysia*, Ext. Bulletin 409. Taipei, Taiwan: FFTC.

Somsak Sangngam. 1997. Personal communication between Sangngam, a community coordinator with the RFD–DANCED project, and the author in January 1997.

Thailand Department of Agricultural Extension. 1997. *Extension Bulletin on "Ma Kwaen,"* Amphoe Song Kwae, Nan Province, Thailand: Department of Agricultural Extension.

Chapter 51

Alnus-Cardamom Agroforestry

Its Potential for Stabilizing Shifting Cultivation in the Eastern Himalayas

Rita Sharma

Shifting agriculture, or *jhum*, is the major economic activity in northeastern India. This highly organized agroecosystem is based on empirical knowledge accumulated through centuries, and as long as the jhum cycle is long enough to allow the forest to recover the soil fertility lost during the cropping phase, it is in harmony with the environment. However, increased population pressure and declining land area resulting from extensive deforestation for timber has resulted in a shortening of the jhum cycle (Ramakrishnan 1992). Shifting cultivation is becoming not only less productive, but also unsustainable. Areas allocated to shifting cultivators have been decreased substantially. In these circumstances, the adoption of stabilized systems such as agroforestry based on large cardamom could be a promising means of halting the slide into widespread land degradation.

Large cardamom (*Amomum subulatum*) is the most important perennial cash crop of the Sikkim Himalayan region. It is grown in traditional agroforests presently covering about 26,000 ha. Production from this area amounts to four million kilograms of capsules annually[1], worth about 1997 US$8 million. Large cardamom is a high-value, low-volume, nonperishable cash crop that is particularly suitable for the inaccessible mountain areas of northeast India. The agroforestry system is well adapted to the mountain niche at altitudes between 600 and 2,000 m above sea level (asl). The crop does not need much management, except occasional weeding and annual harvesting. Large cardamom is a shade-loving plant requiring high moisture, and it is usually grown in areas where mean annual rainfall varies between 1,500 and 3,500 mm. It is cultivated on steep hill slopes under tree cover, either in natural forests or in plantations. Large cardamom starts yielding three to four years after planting and continues producing profitably until it is 15 to 20 years old (Zomer and Menke 1993).

Alnus nepalensis and *Schima wallichii* are highly preferred shade trees in the cardamom agroforestry system, as well as in the jhum of shifting cultivators. *Alnus nepalensis*, an actinorhizal nitrogen-fixing tree, is used as a shade or nurse tree in

---

Rita Sharma, G.B. Pant Institute of Himalayan Environment and Development, Sikkim Unit, P.O. Tadong, Gangtok, Sikkim 737102, India, or Water, Hazards and Environmental Management, International Centre for Integrated Mountain Development, G.P.O. Box 3226, Khumaltar, Lalitpur, Kathmandu, Nepal.

1 "Capsule" refers to the cardamom fruit, which develops at the base of the tillers and contains the seeds. The whole capsule is used as a spice, or condiment. After being harvested the capsules are dried in traditional kilns before being sold.

cardamom agroforestry and as a main fallow species in the jhum system. *Alnus* also produces efficiently up to 20 years of age, and by then it has achieved timber quality (Sharma and Ambasht 1991). Therefore, in a jhum cycle of about 15 years, an *Alnus*-cardamom agroforestry model could be adapted to provide high annual economic returns during the period when the land is normally left fallow. Some shifting cultivators in northeast India have already begun trials using the *Alnus*-cardamom system.

This chapter discusses findings on yield, soil fertility, and stability of cardamom agroforestry. The role of the nitrogen-fixing *Alnus* has proven to be beneficial for a number of reasons, so the potential of the *Alnus*-cardamom partnership is examined from several perspectives. First, can the *Alnus*-cardamom system be adopted by shifting cultivators for fallow management if they move to longer jhum cycles? Second, is it possible to reunite fragmented family *jhumias*, if only during the fallow period, by planting fallow crops of perennial large cardamom with the incentive of high economic returns? And third, can the *Alnus*-cardamom system provide farmers with a stable and permanent alternative to shifting cultivation?

Alnus-Cardamom Agroforestry System

The influence of nitrogen (N_2)-fixing *Alnus nepalensis* in cardamom agroforestry was studied by selecting and comparing a stand with *Alnus* and a control stand without it. That with *Alnus nepalensis* was referred to as the "*Alnus*-cardamom stand," and the control stand with mixed, non-N_2-fixing trees as the "forest-cardamom stand." The sites were located in the Mamlay watershed in the south district of Sikkim, India. Density and basal area of trees were 517 trees per hectare and 5.6 m^2/ha in the *Alnus*-cardamom stand, and 850 trees per hectare and 6.3 m^2/ha in the forest-cardamom stand, respectively.

Estimations of nitrogenase enzyme activity in the root nodules of *A. nepalensis* in the *Alnus*-cardamom stand were made throughout the growing season, from April to October, 1994. The N_2-fixation rate was 55 μmol N/g nodule dry weight/day. Nodule biomass of *Alnus* during the peak growing period was 388 g/tree, or 201 kg/ha. Nitrogen accretion through atmospheric fixation was 126 g/tree/year, amounting to 65.34 kg/ha/year (see Table 51-1).

The total stand biomass, tiller number, basal area, and biomass of cardamom were much higher under the influence of N_2-fixing *Alnus* than in the forest-cardamom stand. Despite their lower stand density, the annual net primary productivity of *Alnus* trees was slightly higher than that of the mixed tree species. The productivity of cardamom, however, when grown under *Alnus*, was more than double that in the forest-cardamom stand. The agronomic yield of cardamom was also 2.2 times higher under the *Alnus* canopy (Table 51-1). Under the influence of the N_2-fixing species, the greater biomass accumulation, net primary productivity, yield, and higher litter production were a direct expression of better performance consequent upon higher soil fertility.

Total annual production of litter plus cardamom residue was 1.6 times higher in the *Alnus*-cardamom stand than in the forest-cardamom stand. This difference was mainly attributed to higher cardamom residue production under the influence of *Alnus*. The floor litter was 1.3 times higher in the *Alnus*-cardamom stand, and the ratio of litter plus residue production to floor litter was also higher. Nitrogen and phosphorus contributions in litter plus crop residue production and floor litter were conspicuously higher under the influence of *Alnus* (Sharma et al. 1997a, b).

In detailed studies of the cardamom agroforestry systems, total soluble polyphenolics in decomposing litter fractions showed an exponential relationship with time, having 77% to 93% losses during the first three months. The initial concentration of total soluble polyphenolics was higher in the leaves of *Alnus* trees than in those of mixed species trees. Most of the total soluble polyphenolics were leached out exponentially during the initial phase of decomposition, with levels falling to less than 1% of the original level within three months.

Table 51-1. Tree Density, Productivity, and Nutrient Dynamics in Cardamom-Based Agroforestry Systems

Parameter	*Agroforestry Systems* Alnus-*Cardamom*	*Forest-Cardamom*
Stand tree density (trees/ha)		
N_2-fixing trees	517	—
Mixed tree species	—	850
Total	517	850
Stand biomass (kg/ha)	28,422	22,237
Productivity (kg/ha/year)	10,843	7,501
Agronomic yield of cardamom (kg/ha/year)	454	205
Nitrogen accretion through atmospheric fixation (kg/ha/year)	65.34	—
Nutrient dynamics		
Soil status (kg/ha up to 30 cm depth)		
Total nitrogen	5,880	7,590
Total phosphorus	1,278	1,149
Total uptake (kg/ha/year)		
Nitrogen	143.83	80.56
Phosphorus	13.18	6.52
Total release to soil through decomposition (kg/ha/year)		
Nitrogen	83.67	29.23
Phosphorus	6.15	2.35
Nutrient exit from agronomic yield (kg/ha/year)		
Nitrogen	4.04	1.78
Phosphorus	0.70	0.33

Sources: Sharma et al. 1994; Sharma 1995.

A decrease to below 1% level was critical for rapid decomposition and it was presumed that, at higher levels of polyphenolics, some of the early succession microbes were inhibited from colonizing the litter. Cumulative ash-free mass loss from tree leaf litter and crop residue fractions was negatively dependent on the polyphenol/nitrogen ratio of the material (Sharma et al. 1997a).

Both the turnover time and the time required for the loss of half the initial ash-free mass, nitrogen, and phosphorus from litter fractions were least in N_2-fixing leaf litter. The litter from *Alnus* trees generally decomposed faster than that of non-N_2-fixing species, and the addition of *Alnus* litter might also have accelerated the decomposition of other litter types. The turnover times for ash-free mass and nitrogen were nearly the same for twigs of *Alnus* and non-N_2-fixing species. But phosphorus turnover in *Alnus* twigs was found to be nearly three times faster than in mixed species twigs.

Sampling interval and cumulative loss of both nitrogen and phosphorus from decomposition bags were highest in *Alnus* leaf and twig combined, compared to mixed tree species leaf, twig, and cardamom residue. Losses of nitrogen and phosphorus were lowest in the forest-cardamom stand. These losses were high in N_2-fixers, mainly because of higher concentrations of nitrogen and phosphorus in their tissues. The mobility of phosphorus was higher than that of nitrogen from all the litter fractions (Sharma et al. 1997a). Most of the litter fractions were nutrient rich, showing no net immobilization of either nitrogen or phosphorus, except for a brief period in the twigs of tree litter. However, the nitrogen and phosphorus release from a unit area of floor was much higher in the *Alnus*-cardamom stand, where a higher contribution was recorded from all litter fractions, especially from the leaf litter of *Alnus* compared to that of non-N_2-fixing species. Therefore, the role of *Alnus* in cardamom agroforestry was found to be the acceleration of nitrogen and phosphorus cycling—through aboveground litter plus residue production—and a greater release of these nutrients to the soil.

Soil organic carbon levels are good indicators of soil fertility status. The soil organic carbon levels in the *Alnus*-cardamom stand were lower than those in the forest-cardamom stand. Productivity is limited by the availability of nitrogen, and in many agroforestry systems, the nitrogen supply rate is closely linked to N-limitation. Total nitrogen in the soil was higher in the forest-cardamom stand than it was in the *Alnus*-cardamom stand. The levels of soil total-N showed annual fluctuations that seemed very small relative to the total pool. Therefore, to see more of seasonal dynamics, the inorganic-N was estimated and measured as ammonium-N and nitrate-N. Ammonium-N was highly seasonal in its concentration in the soil, showing very low values in the rainy season in both the agroforestry stands. The ratio of seasonal maximum and minimum for ammonium-N showed up to a sixfold variation. Nitrate-N also showed seasonality, with up to a sixfold difference between minimum and maximum. The ratio of total-C to total-N in soil is sometimes regarded as a better index of N-availability than total-N alone. In the cardamom agroforestry stands, a higher C/N ratio indicates lower N-availability. The *Alnus*-cardamom stand showed a relatively lower C/N ratio, indicating higher N-availability than the forest-cardamom stand (Sharma et al. 1997b).

Nitrogen mineralization is highly dependent on levels of soil organic matter and moisture, and these factors, along with the temperature, regulate mineralization in the field. Organic matter and moisture levels were both high at the study sites. The presence of the N_2-fixing species, while changing land use from a natural forest tree mixture to sparse tree-based agroforestry with traditional crops, helped to both maintain levels of soil organic matter and achieve a higher rate of nitrogen mineralization. This shows its potential to enhance the sustainability of cardamom agroforestry (Sharma et al. 1997b).

In most soils, the amount of organic phosphorus exceeds the level of inorganic phosphorus, and turnover of organic-P pools provides a large portion of phosphorus taken up by plants (Sharma 1995). In both the agroforestry stands of this study, most of the total phosphorus in the soil was in organic form. Availability of phosphorus is also highly pH dependent, a factor that was quite evident in this study because soil pH values were low in all the seasons, and as a result, the levels of available phosphorus were also low (Sharma et al. 1997b).

The general concept of an inverse relationship between nutrient availability and conservation is well illustrated by this study. Nitrogen and phosphorus concentrations in different tissues of *Alnus* were higher than those of mixed tree species, but nitrogen and phosphorus back-translocation from leaf to branch before abscission were lower. This was because the availability and uptake of these elements were higher in the *Alnus*-cardamom stand than in the control stand. Since more of these elements were available to *Alnus* than were available to the mixed tree species of the forest-cardamom stand, the *Alnus* trees recorded lower back-translocation, pointing out their poor conservation strategy (Sharma et al. 1994).

Annual uptake and return of nitrogen to the soil in the *Alnus*-cardamom stand was higher than that in the forest-cardamom stand, and this was attributable to N_2-fixation by *Alnus*. The rates of phosphorus uptake and return through litterfall and decomposition were also higher in the *Alnus*-cardamom stand, and this probably resulted from an increase in the rate of phosphorus supply attributable to geochemical and biological factors influenced by *Alnus* (Ho 1979; Malcolm et al. 1985; Gillespie and Pope 1989; Ae et al. 1990). Nitrogen and phosphorus cycling appeared very flexible under the influence of *Alnus*.

The nutrient-use efficiencies in the cardamom-based agroforestry systems were generally consistent with the hypothesis that nutrient-use efficiency falls as utilization of that nutrient increases, because availability of some other resource limits production (Melillo and Gosz 1983; Binkley et al. 1992). Nutrient use efficiencies in the study areas decreased as an influence of N_2-fixing *Alnus*.

However, the *Alnus*-cardamom agroforestry system was more productive and had faster rates of nutrient cycling. The poor nutrient conservation and low nutrient-use efficiency of *Alnus*, and the flexibility of nutrient cycling under its influence, made it an excellent associate in the cardamom-based agroforestry systems being studied because it promoted both higher availability and faster cycling of nutrients.

Potential of *Alnus* for Stabilizing Shifting Cultivation

At the study sites, 517 *Alnus* trees per hectare seemed quite a reasonable density to maintain the N and P balance in large cardamom-based plantations. However, an average density of 1,300 *Alnus* trees per hectare has been reported in *Alnus*-cardamom plantations surveyed at 66 sites in the mid-hills of eastern Nepal (Zomer and Menke 1993). The production, energy conservation and N_2-fixing efficiencies of *Alnus* trees have been reported to decrease with age (Sharma and Ambasht 1991). *Alnus* has high nitrogen demands and is a poor competitor for soil nitrogen uptake, depending heavily on N_2-fixation. Beyond 20 years of age, it becomes costly in terms of energy and low N_2-fixation efficiency, and productivity reduces substantially (Sharma and Ambasht 1988, 1991). Therefore, to maintain nitrogen and phosphorus balance for a rotational cycle of 20 years under an *Alnus*-cardamom system, a density of at least 1,000 trees per hectare should be planted in the beginning. Using the data from the present study, we estimate that such an increase in *Alnus* density would increase by 1.9 times the annual quantity of nitrogen fixed per hectare by the trees. This would also enhance the production and nutrient cycling rates in the associated cardamom plants.

After five years, *Alnus* trees should be thinned every year, and this would meet firewood requirements for cardamom curing and domestic use. With an initial plantation density of 1,000 trees per hectare, thinning of *Alnus* trees at an average of 30 trees per hectare per year for up to 20 years would ensure about 550 standing trees per hectare by the end of the rotational cycle. The remaining *Alnus* trees would provide about 120 m^3/ha of timber wood and about 40 metric tons/ha of branch wood, which would generate a high cash income.

Cardamom production also reduces substantially after the plants reach 20 years of age (Singh et al. 1989, Zomer and Menke 1993). Therefore, the *Alnus*-cardamom agroforestry system should be reestablished after 20 years and the rotational cycle planned in compartmentalized plots. Each landholding should be divided into four equal plots, with plantations established at a rate of one plot per year over four years. This will ensure cardamom yield and *Alnus* wood returns every year without interruption, even during the replanting period after 20 years. This form of plantation management would make *Alnus*-cardamom agroforestry a sustainable system in terms of soil nitrogen and phosphorus balance as well as cardamom yield. It would be self-sufficient in terms of firewood production for cardamom curing and domestic consumption, reducing pressures on adjoining forests.

The management of fallow species in between periods of arable cropping has become a vital issue for sustaining the traditional practice of shifting cultivation in northeast India. *Alnus* is an important swidden species, and this chapter has discussed its role in the sustainable management of cardamom-based agroforestry. *Alnus* can also play a key role in the management of the fallow period in jhum, or shifting cultivation. Stabilization of the jhum cycle can be achieved by adopting *Alnus*-cardamom plantations during a long fallow period of 15 to 20 years. This would ensure high economic returns from cardamom sales after an establishment period of three to four years. The *Alnus*-cardamom system is not labor intensive and requires one visit for weeding and another for cardamom harvest. It maintains soil fertility and would allow swiddens to recover for the cultivation of traditional crops according to the usual cycle.

Conclusions

The overall benefits from large cardamom agroforestry are:

- High economic returns;
- Less need for labor;
- A nonperishable crop with an established market;
- A low-volume crop, resulting in reduced nutrient export;
- Supplies of basic resources such as firewood, timber, and fodder; and
- Minimal soil and nutrient losses.

The following are benefits of planting *Alnus nepalensis* as a shade or nurse tree in large-cardamom agroforestry systems:

- Stand biomass, productivity, and agronomic yield of large cardamom are increased by 1.3, 1.5, and 2.2 times, respectively, compared with cardamom grown under non-N_2-fixing forest trees.
- Nitrogen accretion by *Alnus* through biological fixation is 65 kg/ha/year, at a mean N_2-fixation rate of 55 μmol N/gram of nodule dry wt/day.
- Nitrogen and phosphorus cycling are accelerated under the influence of *Alnus*.
- *Alnus* effectively helps to maintain soil fertility on fragile slopes.
- The poor nutrient conservation and low nutrient-use efficiency of *Alnus*, together with the flexibility of nutrient cycling under its influence, make it an excellent associate, promoting higher availability and faster cycling of nutrients.
- Annual income from large-cardamom yields may more than double under the influence of *Alnus*, compared with cardamom grown under mixed forest species.

Acknowledgments

I am thankful to the director, GB Pant Institute of Himalayan Environment and Development, for facilities, and to the Council of Scientific and Industrial Research for financial support in the form of a research associateship.

References

Ae, N., J. Arihara, K. Okada, T. Yoshihara, and C. Johasen. 1990. Phosphorus Uptake by Pigeon Pea and its Role in Cropping Systems of the Indian Subcontinent. *Science* 248, 477–480.

Binkley, D., P. Sollins, R. Bell, D. Sachs, and D. Myrold. 1992. Biogeochemistry of Adjacent Conifer and Alder/Conifer Stands. *Ecology* 73, 2022–2033.

Gillespie, A.R., and P.E. Pope. 1989. Alfalfa N2-fixation Enhances the Phosphorus Uptake of Walnut in Interplanting. *Plant and Soil* 113, 291–293.

Ho, I. 1979. Acid Phosphatase Activity in Forest Soil. *Forest Science* 25, 567–568.

Malcolm, D.C., J.E. Hooker, and C.T. Wheeler. 1985. *Frankia* Symbiosis as a Source of Nitrogen in Forestry: A Case Study of Symbiotic Nitrogen Fixation in a Mixed *Alnus-Picea* Plantation in Scotland. In: *Proceedings of the Royal Society of Edinburgh* 85B, 263–282.

Melillo. J.M., and J. Gosz. 1983. Interactions of Biogeochemical Cycles in Forest Ecosystems. In: *The Major Biogeochemical Cycles and their Interactions*, edited by B. Bolin and R. Cook. New York: John Wiley & Sons, 177–221.

Ramakrishnan, P.S. 1992. *Shifting Agriculture and Sustainable Development: An Interdisciplinary Study from Northeastern India*. Carnforth, Lancs, UK: MAB, UNESCO Paris, and Parthenon Publishing Group.

Sharma, E., and R.S. Ambasht. 1988. Nitrogen Accretion and its Energetics in the Himalayan Alder. *Functional Ecology* 2, 229–235.

———. 1991. Biomass, Productivity and Energetics in Himalayan Alder Plantations. *Annals of Botany* 67, 285–293.

Sharma, R. 1995. Symbiotic Nitrogen Fixation in Maintenance of Soil Fertility in the Sikkim Himalayas. Ph.D. thesis, H.N.B. Gargwal University, India.

Sharma, R., E. Sharma, and A.N. Purohit. 1994. Dry Matter Production and Nutrient Cycling in Agroforestry Systems of Cardamom Grown under *Alnus* and Natural Forest. *Agroforestry Systems* 27(3), 293–306.

———. 1997a. Cardamom, Mandarin, and Nitrogen-Fixing Trees in Agroforestry Systems in India's Himalayan Region. I. Litterfall and Decomposition. *Agroforestry Systems* 35, 239–253.

———. 1997b. Cardamom, Mandarin, and Nitrogen-Fixing Trees in Agroforestry Systems in India's Himalayan Region. II. Soil Nutrient Dynamics. *Agroforestry Systems* 35, 255–268.

Singh, K.A., R.N. Rai, Patiram, and D.T. Bhutia. 1989. Large Cardamom (*Amomum subulatum* Roxb) Plantation: An Age-old Agroforestry System in the Eastern Himalayas. *Agroforestry Systems* 9, 241–257.

Zomer, R., and J. Menke. 1993. Site Index and Biomass Productivity Estimates for Himalayan Alder-Large Cardamom Plantations: A Model Agroforestry System of the Middle Hills of Eastern Nepal. *Mountain Research and Development* 13(3), 235–255.

Chapter 52
The Sagui Gru System

Karen Fallow Management Practices to Intensify Land Use in Western Thailand

Payong Srithong

The Karen are the largest of the 12 ethnic minority groups, or "hilltribes," inhabiting Thailand. They make up 60% of the country's total tribal population of about 600,000, and are thought to have been the first of the tribal groups to settle in Thailand. Most Karen communities are located in forest lands in the north and west of the country, near Thailand's border with Myanmar. They are swidden cultivators who usually clear fields from secondary forest, or fallow, to cultivate upland rice and other food crops for subsistence. They cultivate the land for two to three years, and then, when weeds proliferate and soil fertility declines, they move on to reopen other fallows.

By tradition, the Karen are also hunters and gatherers of minor forest products, including bamboo shoots, mushrooms, and edible wild plants. However, over the 1980s and 1990s, a number of factors contributed to the destruction of primary forests in Thailand, including those near Karen settlements. Those factors include the following:

- Unsustainable commercial logging;
- Forest encroachment by lowland farmers in search of land for cultivating maize and cassava;
- Construction of large hydroelectric dams; and
- Illegal trafficking in logs by locally influential people.

As a consequence, forest degradation has contributed significantly to prolonged drought and other microclimatic changes. Streams and watersheds have dried up, and wild food plants and other nontimber forest products have become increasingly scarce.

The Study Site and Methods

This chapter reports on a study conducted by the Project for Agroecology Development and Plant Genetic Resources Conservation (AGRECO/PGEC) in five villages of Pwo Karen, located in the protected forests of three national parks and in

Payong Srithong, Coordinator, Project for Agroecology Development and Plant Genetic Resources Conservation (AGRECO/PGRC), P.O. Box 15, Danchang, Suphanburi 72180, Thailand.

the buffer zone of the Huay Khakhaeng Wildlife Sanctuary, all in western Thailand (see Table 52-1). Eighty-eight Karen households out of a total of 120 were selected for the study and were divided into two groups. The first group, involving 58 households, represented normal swidden cultivators. The second group consisted of 30 households that owned permanent *sagui gru*, or banana gardens (see color plate 61).

The Karen communities were found to be facing the following constraints to their traditional farming systems:

- Degradation of forests and other natural resources;
- Loss of agricultural land due to encroachment by lowland farmers;
- State regulations restricting them to cultivation within demarcated areas and prohibiting the use of minor forest products;
- Changes in cultural, social, and economic environments as a result of state policies encouraging modernization and social integration; and
- Penetration of a market economy that imports commodities in exchange for siphoning off local resources.

Results

Upland rice (*Oryza sativa*) is the staple food of the Karen, and they normally try to grow enough rice for year-round consumption. However, because of the loss of forest cover and associated changes in microclimate, only 52% of interviewed households succeeded in harvesting enough rice for their needs over the four years prior to 1997. In a survey involving all households in the study, families were asked in how many of those four years they had enjoyed sufficient rice. Fifteen percent said that at no time had they had enough. Responses are detailed in Table 52-2.

Karen Livelihood Strategies

To supplement their subsistence in the absence of sufficient rice, or to provide cash income for various household needs, the Karen pursue the following:

- Gathering nontimber forest products (NTFPs) such as bamboo shoots, honey, mushrooms, edible wild plants and medicinal plants;
- Taking off-farm employment, mainly laboring for wages or working for government agencies;
- Cultivating cash crops, particularly corn;
- Marketing surplus yields of traditional crops planted in swiddens;
- Operating small businesses;
- Selling surplus products from home gardens; and
- Making traditional handicrafts.

Table 52-1. Details of the Surveyed Karen Communities

Communities	*Households*	*Total Population*	*Forest Types*	*Location and Forest Status*
Chaowat	17	67	DE/MD	Huay Khakhaeng WS/BZ
Huay Hindam	46	252	MD	Phutei NP
Buengchakho	38	176	DE/MD	Srinakarin NP
Namphu	109	506	DD/MD	Srinakarin NP
Khaolek	61	278	DE/MD	Chalermratanakosin NP
Total	251	1277		

Notes: Figures compiled from various sources in 1996. DE = dry evergreen forest; DD = dry dipterocarp forest; MD = mixed deciduous forest; WS = wildlife sanctuary; NP = national park; BZ = buffer zone.

Table 52-2. Rice Self-Sufficiency Status of Surveyed Karen Households

Number of Years Families Have Had Enough Rice	*Households in Each Village*						
	CWT	*HHD*	*NMP*	*KLK*	*BCH*	*Total*	*%*
4 years	1	2	15	5	7	30	52
3 years	0	0	1	6	4	11	19
2 years	0	0	1	1	1	3	5
1 year	0	2	0	3	0	5	9
0 years	1	5	0	3	0	9	15
Total	2	9	17	18	12	58	100

Gathering of Nontimber Forest Products

NTFPs have always been collected for home consumption, with limited barter trade with outside communities. However, with new opportunities for sale to outside markets, NTFPs have been transformed into saleable commodities for generation of cash. The survey found that all Karen households earned some income from NTFPs, and NTFPs constituted 39.4% of total income.

Intensification of Karen Swidden Cultivation

Karen farmers in the study households have adopted two different strategies for intensifying their use of limited land. The first is simply growing more of their traditional crops, including bird chilies, pumpkins, wild eggplants, melons, and cucumbers, in answer to market demand. The other means of intensification, innovated by the Karen themselves, is the establishment of the sagui gru system. "Sagui" means banana in the Karen dialect. "Sagui gru" means, simply, banana garden.

These complex fruit tree–based agroforests are an introduced agricultural system that builds on Karen traditional agricultural practices. They employ traditional principles of both upland rice cultivation and home garden management. Karen swidden fields are planted predominantly to several varieties of upland rice, as well as a large number of secondary food crops, herbs, and medicinal plants. Bananas are interplanted into the upland rice during the cropping phase in standard *taungya* practice. As the name implies, bananas are the dominant species in the system. However, the farmers interplant many other species, including betel nut, mango, pomelo, jackfruit, coconut, numerous herbs and shrubs, and other useful native species (see Table 52-3). The general practice is that other species are planted only after the bananas have become firmly established, so they can provide protective shade for young seedlings. But there was one case where a farmer started his garden with kapok trees, and another planted bananas and kapok at the same time.

Ninety percent of the income generated from the sagui gru system comes from banana sales, with most of the remaining 10% coming from mangoes and kapok. When compared with the incomes of those households practicing traditional swidden and home garden cultivation, households with sagui gru achieved substantially higher percentages of income from their land (see Table 52-4). At the time of the study, sagui gru ranged in size from 1.3 to 12 rai, with an average of 3.4 rai.[1] The oldest garden was seven years old, but the average age was only four years.

Plant genetic diversity is a dominant characteristic in Karen swidden cultivation. The study found that most swidden cultivators who developed sagui gru gave priority to growing food crops for household consumption. There was therefore a wide variety of plant species but only a small plot of each. However, all communities in

[1] 1 rai equals 1,600 m^2.

the study had been exposed to market forces, and many of them were expanding their area under fruit trees, while still attempting to maintain the plant diversity of their farms. They envisaged that, in the near future, sagui gru would provide an alternative source of cash income.

Development of the Sagui Gru *System*

The study showed that the Karen households that established sagui gru were mainly those that had no home gardens. They were usually recent immigrants to the village or young families recently separated from their parents' households. However, there were broader, perhaps more important, factors contributing to the development of sagui gru in Karen communities. They included:

- Penetration of the market economy into remote areas through road construction, providing access to "middlemen" traders;
- Scarcity of land for shifting cultivation, necessitating more intensified use of available land for producing food and cash; and
- Pressures from the Royal Forest Department and the military, enforcing prohibitions on the use of minor forest products.

Table 52-3. Ten Major Fruit Species Planted in the Sagui Gru Fallows of Surveyed Farms

Farm Number	*Banana*	*Betel*	*Mango*	*Jackfruit*	*Longan*	*Coconut*	*Bamboo*	*Kapok*	*Pomelo*	*Lime*
1	350	0	0	0	0	0	0	0	0	0
2	300	0	0	0	0	0	0	0	0	0
3	140	0	10	0	0	0	0	0	0	0
4	350	170	2	0	4	0	0	0	0	1
5	57	31	7	5	7	0	0	0	10	2
6	80	0	0	0	0	0	0	0	0	0
7	200	0	15	0	0	0	0	0	0	0
8	350	0	0	5	0	0	0	0	0	0
9	100	0	10	0	0	10	0	0	0	0
10	30	450	15	20	15	9	15	0	12	6
11	84	0	6	5	0	0	0	0	0	3
12	1000	40	15	20	20	7	0	0	5	3
13	105	60	50	30	27	16	15	30	15	30
14	121	270	0	40	2	14	0	15	13	10
15	360	20	0	0	19	16	0	0	30	8
16	57	120	5	1	30	1	14	6	4	0
17	402	180	20	8	9	10	47	0	8	1
18	1000	0	20	100	25	10	0	0	15	0
19	266	0	13	0	25	0	50	3	0	0
20	0	0	40	25	0	0	0	0	2	3
21	300	0	6	6	0	1	0	0	0	0
22	130	0	10	0	2	0	0	0	0	0
23	60	0	15	8	0	0	0	20	0	1
24	20	0	20	10	0	0	0	10	0	0
25	0	0	40	8	0	30	0	0	0	0
26	110	0	4	8	0	0	0	0	0	0
27	100	0	20	20	0	8	0	0	0	20
28	40	0	20	5	0	0	0	50	0	0
29	100	0	0	0	0	0	0	0	0	0
30	0	2	10	0	0	20	0	6	0	20

Table 52-4. Sources of Income (%), Sagui Gru Adopters versus Nonadopters

Income Source	*Nonadopters*	*Adopters*
Surplus from swidden	6.9	9.3
Home gardens	1.5	0
Sagui gru	0	11.9
Cash crops	15.1	29.3
NTFPs	39.4	25.1
Handicrafts	1.0	0.9
Livestock	6.3	2.8
Off-farm	26.7	17.0
Others	3.1	3.6
Total	100	100

Conclusions

Although the development of complex fruit tree–based agroforests by Karen swidden farmers is an innovative example of permanent cultivation developed as a solution to limited agricultural land, the majority of households in the studied communities continue to rely on traditional shifting cultivation, which needs sufficient area to permit both regular rotation and adequate fallow duration. If these communities are to overcome their current constraints, Karen farmers, with the help of development agencies, will have to refine their fruit tree–based agroforestry systems so that they can meet their needs and limitations. This may require the identification of agricultural technologies and practices to enable them to manage their fallow lands more productively.

Chapter 53

Sandiu Farmers' Improvement of Fallows on Barren Hills in Northern Vietnam

Ta Long

It is estimated that barren, underused hills occupy about 5 million hectares in Vietnam, or about 15% of the country's total land area. They stretch from the mountainous north, through the central region, and down into the eastern part of southern Vietnam (Institute of Ecological Economy 1993). There has consequently been a determined research effort in Vietnam to find agricultural systems capable of providing a sustainable use of sloping land. Sloping agricultural land technology has been extended, and various other models have been developed (Vien and Thanh 1996). As a result, there has been a transition from traditional nomadic and extensive farming systems to more settled and intensive agriculture (Long, T. 1997).

This chapter focuses on the agricultural systems of Sandiu people, an ethnic minority group, living in two communes, Namduong and Quyson, in Lucngan district. It is in the far north of the country, in Bacgiang Province, about 100 km northeast of Hanoi. Prior to 1989, the hills of the communes were fallowed and barren; used only for grazing the farmers' buffaloes. But the development of markets has seen the barren hills reclaimed for fruit trees, mainly litchi. The district has become green, and the livelihoods of the Sandiu families have been transformed. Their more intensive use of fallow lands has improved the farmers' labor efficiency and is providing a basis for widespread changes to sustainable development of other rural communities.

Transformation of Sandiu Farming Systems

The Sandiu people living in the study region account for about 18% of the ethnic group's total population (Vietnam Population Investigation 1991). They refer to themselves as *Sanyaonhin* (meaning "Dao living in mountainous regions"), but most people know them as *Man* (another name for *Yao*) or *Manruong* ("Yao who work in the wet field") (Bang 1983). In Lucngan district, they are called *Trairuong* ("people living in farmhouses and doing wet field work").

Historically, some Sandiu families neither owned land nor were able to rent it, simply because they did not own buffaloes. While the land around them, on the barren hills, remained fallow, they had to search for arable land in other localities. The barren hills accounted for 47% of the total land area in Quyson commune and 64% in Namduong commune.

Ta Long, Academy of Social Sciences of Vietnam, Institute of Anthropology, 27 Tran Xuan Street, Hanoi, Vietnam.

Conditions began to change in 1986, when the People's Committee of Lucngan district introduced a policy of allocating fallow land to individual households for planting eucalyptus trees to regreen the barren hills. This plan was widely rejected by farmers because they preferred to use the grasslands for grazing their buffaloes. However, in 1989, a system called "10 contract" was introduced, under which sloping lands were contracted out for use by households, tax free. At first, families who accepted the deal grew eucalyptus and pineapples, but in the early 1990s, demand for litchis surged. The first farmers to grasp the opportunity were mostly migrant Kinh people, who planted litchis in their household gardens and on barren hills. The trend caught on, and litchi orchards spread to transform the formerly barren hills (Long, C.A. et al. 1996). Local livelihoods have improved, and the agricultural and economic structure of Lucngan district has changed dramatically. (See also Chapter 62.) The data presented in Table 53-1 reflect these changes.

Land Improvements and Uses

Attempts to improve agricultural land use usually rely on the selection of suitable species to provide superior ecological and economic benefits. The Sandiu are skilled farmers and understand the use of N-fixing plants to build soil fertility. However, land use strategies depend not only on farmers' knowledge, but also on the availability of labor, the structure of households, and market demands. Generally speaking, too little attention has been paid to the human ecology aspects of farming systems research. Nevertheless, the following discussion is limited to aspects of soil improvement.

Because of the sloping topography of the barren hills, the soil remains dry and is easily eroded during heavy monsoon rains. Interventions are needed to control erosion and to improve soil fertility and moisture.

From Barren Hills to Terraced Fields

The barren hills are covered with *Melastoma*, *Mimosa*, and other pioneer weeds. These are cleared away, and terraced fields are constructed on the foothills. Pineapples are planted in contour rows to prevent soil erosion and to maintain soil moisture. Channels are dug to guide excess rainwater off the fields. There are no water wells or ponds on the hills. On hills with steeper gradients, eucalyptus trees are planted on the summits and upper slopes and fruit trees such as litchis, custard-apples, and oranges are planted on the middle and bottom slopes. Litchis account for 75% of the area under fruit trees. Subsidiary crops are scattered throughout. Hills with more moderate slopes are typically planted to litchis, maize, cassava, sweet potatoes, and white beans, in mixed arrangements.

Table 53-1. Changing Use of Arable Land in Quyson Commune, 1980–1996 (ha)

Crop	*1980*	*1990*	*1993*	*1996*
Paddy rice	41.9	59.4	46.9	28.8
Maize	1.78	0.92	2.61	1.16
Cassava	20.1	12.7	9.72	6.12
Sweet potatoes	3.34	2.13	3.16	4.50
Assorted beans	1.42	0.79	1.77	1.75
Soybeans	6.13	2.13	2.17	0.50
Peanuts	1.67	1.42	1.20	0.53
Sesame	1.39	0.79	0.49	0.12
Pineapples	11.1	5.67	3.43	1.38
Litchis	0.00	14.2	28.6	54.7

Source: Quyson Commune People's Committee reports.

On hills with steeper gradients, eucalyptus trees are planted on the summits and upper slopes and fruit trees such as litchis, custard-apples, and oranges are planted on the middle and bottom slopes. Litchis account for 75% of the area under fruit trees. Subsidiary crops are scattered throughout. Hills with more moderate slopes are typically planted to litchis, maize, cassava, sweet potatoes, and white beans, in mixed arrangements.

Litchi Varieties

There are four indigenous and seven imported litchi varieties grown in different parts of the study area, depending on slope gradients. However, the main varieties are as follows:

- Sour litchi, which is indigenous to Lucngan district, is grown widely on steeper slopes. This variety grows rapidly and has some cold resistance. The fruit is sour and yields are low. But because it ripens earlier than other litchi species, it is able to maintain its share of the market. This variety accounts for about 20% of the total litchi acreage in the study area.
- Phuho litchi is also indigenous and is commonly cultivated on moderate slopes. This variety also grows rapidly. It has a stable yield of roughly 60 kg per tree, usually harvested in mid-June. It accounts for 25% of the total litchi acreage.
- Thieu litchi was introduced from Haiduong Province and is cultivated on gentle slopes and the foothills. It thrives in the soil and ecology of Lucngan district. Its average yield is 50 kg per tree. Harvesting time is generally June or July. It accounts for 50% of the total litchi acreage.
- Some other litchi varieties have been imported from China and other countries, but they compose only 5% of litchi acreage. They grow well and mature early, but yields on flat land are lower than those from Thieu litchis (Long, C.A. et al. 1996).

Litchi seedlings are planted 60 cm deep. The roots are covered with soil, and small dykes are built up around the seedlings to retain water. Recommended planting distances are 5 by 5 m or 7 by 7 m, providing a density of 200 to 400 trees per hectare.

After planting one *sao* (a unit of area measurement in Northern Vietnam equal to 360 m^2) of seedlings, farmers water the small trees. A second watering follows three days later, and for the first four months the seedlings are watered once every five or six days. When the trees have become established, watering is no longer necessary. They are weeded two or three times a year. When the trees reach two years of age, 20 kg of farmyard manure (FYM) and 3 to 4 kg of phosphate fertilizer (P_20_5) are applied to the base of each tree. Thereafter, fertilizers are applied once a year.

Intercropping in Young Litchi Orchards

Soil accumulated at the foot of slopes is used to cultivate a variety of crops suitable for dry conditions. They include maize, cassava, sweet potatoes, taro, yams, sesame, peanuts, soybeans, green beans, white beans, and aquatic taro. Their yields are very low.

At the time of writing, intercropping of arable crops between the young litchis and other trees had been practiced for only five years, because the reclamation of fallowed land on the barren hills had begun only five years earlier than that. Therefore, intercropping was still at an experimental stage. But by five years of age, litchi trees have already developed a well-distributed root system, and there is a danger that continued intercropping might damage the litchi roots.

According to Sandiu informants, the pattern of intercropping during the initial five years was as follows:

- Year 1: Two crops of soybeans were interplanted between the small litchi trees, a winter-spring crop and a summer-autumn crop.
- Year 2: One crop of peanuts was followed by a crop of white beans.
- Year 3: Maize was planted in spring and soybeans were planted in summer. Sweet potatoes were also planted in the autumn-winter season from October to December.
- Year 4: Part of the acreage was planted with cassava from March to November, followed by a crop of peanuts from December to February. Other parts were planted to a single yam or taro crop from March to August, followed by one peanut crop.
- Year 5: A crop of green beans from February to April was followed by a crop of white beans from June to August.

Cultivation Methods

Soybeans. Land preparation begins with removing weeds and cultivating the soil. Planting beds about 30 cm high and 15 cm wide are fashioned along the contours. The recommended fertilizer application rate is 60 to 70 kg FYM, 1.5 kg of nitrogenous fertilizer (N), and 10 kg of phosphate fertilizer (P_2O_5) per sao. After incorporating the fertilizer into the raised seed beds, the farmers plant soybeans at 0.8 kg per sao, and cover them with a layer of soil.

About one month after sowing, the field is weeded, 5 kg of nitrogenous fertilizer is applied per sao, soil is piled up around the roots of the soybeans, and pesticide spray is applied. No further care is required prior to harvest. The soybeans mature in three months and are expected to yield 640 kg/ha.

Green Beans. The management of green beans is essentially the same as described above for soybeans. They will mature after 100 days and yield about 1,400 kg/ha.

White Beans. Preparation of land and seedbeds is the same as described for soybeans. Fertilizer is applied at a rate of 30 to 40 kg FYM and 2 kg of phosphate fertilizer per sao. The recommended seeding rate is 1.2 kg per sao. When the beans flower, soil is hoed up around their roots and pesticides are applied. Herbicides are unnecessary. White beans mature after three months.

Peanuts. Land preparation and sowing is the same as described above for beans, but more care is required. Soil amendments should be applied at 500 kg FYM, 150 to 200 kg ash manure (see "Soil Amendments Used by Sandiu Farmers," below), 30 kg phosphate fertilizer (P_2O_5), 2 kg potassium fertilizer (K_2O), and 10 kg of lime per sao. Plants require close care, and soil should be hoed up around their roots at the two-leaf stage. Pesticides and herbicides should be sprayed at the time of flowering. Peanuts mature in 100 days.

Maize. Land preparation is the same as described for beans. Recommended fertilizer application rates are 600 to 700 kg FYM, 15 kg nitrogen fertilizer, and 10 kg phosphate fertilizer per sao. When maize starts to flower, about one month after sowing, an additional 5 kg of nitrogenous fertilizer per sao should be applied and the soil hoed up around the roots. Maize matures in 100 days.

Cassava. After plowing and harrowing the soil, farmers hoed it into planting beds 30 cm high, 90 cm long, and 50 cm wide. Fertilizer is typically applied in two treatments. The first, consisting of 300 kg FYM per sao, is applied during planting, and the second, 30 kg of nitrogenous fertilizer and 1 kg of potassium fertilizer (K_2O) per sao, is applied when the cassava is 30 cm tall. At one month of age, the cassava is weeded and additional soil hoed up around its roots. Cassava takes nine months to mature. The yield per hectare should be 25,000 kg of fresh cassava or 8,300 kg dried.

Sweet Potatoes. The land is prepared by developing raised planting beds and applying 2 kg of nitrogen fertilizer per sao at the time of planting. The soil later needs to be hoed up around the roots. Sweet potatoes mature in 90 to 100 days.

Yams. After a single plowing and harrowing operation, raised planting beds are formed, similar to those used for cassava. Tubers are planted in the beds at about 30 kg per sao. Fertilizer may be applied in two portions. The first, at the time of planting, should be 300 to 400 kg FYM per sao. The second application should consist of 4 kg of nitrogenous fertilizer applied when the first leaves emerge. The soil should then be hoed up around the roots. Yams mature in six months, and their yield per hectare should be about 15,300 kg, fresh weight.

Impacts of Reclamation of Barren Hillsides

For the purposes of this chapter, the impacts of the expanding litchi orchards on the Lucngan district are defined by comparison with alternative crops. Table 53-2 shows the land area planted in litchis and pineapples in Quyson commune during the period 1980 through 1996.

Economic Effects

The economic performance of alternative crops to litchis are summarized in Table 53-3. Sweet potatoes are a low-input crop that demands the least capital and labor investment but offers a return on invested capital second only to litchis. However, the total profit from sweet potatoes is lower than that from both rice and green beans. In fact, returns from sweet potatoes are only half those from rice paddy, the staple food of the Sandiu. Cassava ranks about equal to sweet potatoes in all indexes, but it requires more care (labor inputs) than all other crops monitored. Low production costs and high returns on investment make both sweet potatoes and cassava attractive to resource-poor farmers. However, farmers with more resources at their disposal may prefer crops promising higher total profits and rates of return.

The economic performance of legume crops, including green beans, soybeans, and peanuts, is generally poor except in the case of green beans, which parallel rice. Because legumes tend to be few and sparsely scattered, their beneficial impact on soil properties is also expected to be extremely limited.

In all respects, litchis are an economically rewarding crop. Their rapid expansion in Lucngan district, and the demand by Sandiu farmers for FYM as a production input, has made the formerly barren hillsides a highly profitable enterprise. Litchi yields from Lucngan are very high. Yields compare with those from Thanhha district, in Haiduong Province, which is famous in Vietnam for its litchi growing, and are 3.47 times higher than yields from Sonla Province (Tuc et al. 1996). A strong and stable market demand for litchis in Lucngan district has obviously played a pivotal role in this transformation.

Table 53-2. Area of Hillside Fallows Reclaimed for Fruit Trees and Pineapples in Quyson Commune

Crop	*1980*	*1990*	*1993*	*1996*
Pineapples (ha)	200	80	60	40
Litchis (ha)	0	200	578	1,590
Total yield from fruit trees compared to 1980 (%)	100	140	319	815

Source: Quyson Commune People's Committee reports.

Positive Effects on Soil Properties

Control of Soil Erosion. Construction of physical barriers such as contour lines, gutters, small dykes, and raised planting beds, combined with the protective canopy of fruit trees, significantly reduces soil erosion, maintains soil moisture, and reduces aluminum toxicity.

Rebuilding of Soil Fertility. Soil fertility gradually improves because of the heavy applications of FYM and inorganic fertilizers and the inclusion of legumes in the cropping mix. Plowing and harrowing create conditions favorable for biological processes and beneficial soil fauna (Long, C.A. et al. 1996).

Negative Effects on Soil

Insufficient Use of Legumes. Legume crops are commonly planted on land with high agricultural quality, including fields and alluvial plains. They are not a dominant feature in the newly reclaimed hillside fields. However, Canh hamlet in Namduong commune appears to be something of an exception, with some households planting 20% of their cultivated land to legumes.

In fact, farmers are probably planting more legumes on newly opened fields, with the intention of rebuilding the soil fertility, than the figures would suggest. In many cases, this land is being used without having been "officially" contracted out to households, and the farmers should be paying tax on earnings from it. But they simply "neglect to enter it into the official records" of the People's Committee in the commune. However, if everything was declared and the percentage of land planted to legumes was as much as 30%, it would still be insufficient to contribute significantly to the rehabilitation of soils and regreening of the barren hillsides.

Table 53-3. Costs and Returns per Hectare in Canh Hamlet, Namduong Commune, 1996

		Farm Products					
Crops	*Total Labor Units*	*Returns (000 VND)*	*Costs (000 VND)*	*Return on Investment (000 VND)*	*Ratio of Profit to Costs*	*Returns on Labor (000 VND)*	*Returns on Other Crops, Compared with Rice %*
Paddy rice	222	9,473	3,124	6,349	2.03	28.0	100.0
Maize	222	4,000	2,250	1,750	0.78	7.9	27.6
Cassava	1,056	3,889	745	3,144	4.22	3.0	49.5
Sweet potatoes	167	4,194	573	3,621	6.32	22.0	57.0
Yams	389	5,000	1,652	3,348	2.03	8.6	52.7
Green beans	222	9,028	2,223	6,805	3.06	31.0	107.0
Soybeans	194	2,417	2,000	417	0.21	2.2	6.57
Peanuts	500	3,417	2,983	434	0.15	1.2	6.84
Litchi	400	1,666	745	165,000	222.00	413.0	2,599.00

Note: Initial invested capital: 2,778,000 VND/ha on average.
Source: Nam Duong Commune People's Committee report.

Table 53-4. Proportion of Farm Land Planted to Food Crops (%)

	1980	*1990*	*1993*	*1996*
Area in Food Crops	75.6	75.0	62.4	40.8

Note: Data drawn from Table 53-1.
Source: Quyson Commune People's Committee reports.

Insufficient Livestock Manure. The reduction in grazing land consequent upon the expansion of litchi orchards, combined with the increased use of small farm machinery, has led to reduced numbers of buffaloes, which are forced to graze on forest land. There is now a severe shortage of livestock manure in the area to maintain the topdressing regime on the hillsides. Adding considerably to the problem is the fact that the acreage of land planted to food crops in the study area is decreasing (see Table 53-4). This means there are fewer crop residues available as livestock fodder. Areas remaining for grazing livestock amount to about 23% of the land area in Namduong commune and about 35% in Quyson commune. Livestock populations in the area are declining, as is the availability of FYM.

The shortage of FYM is compensated for to some extent by the use of green manure, ash, silt, and pond mud as fertilizer. However, farmers' ability to gather vegetative biomass from the hillsides for use as green manure, or for reduction to ash, has also been reduced as more sloping lands have been brought under cultivation.

Soil Amendments Used by Sandiu Farmers

Ash Manure. Plant biomass gathered from the hillsides and from cultivated land is dried and burned to produce fertilizing ash.

Green Manure. Herbaceous stems and leaves are gathered from wild plants such as *Sesbania javanica* or *Lantana camara*, or from the residues of harvested legumes such as peanuts, for use as green manures on cultivated fields.

Animal Manure (FYM). Pig and buffalo manure are mixed with rice husks and the stems and leaves of peanuts or *Lantana camara*. Before field application, the livestock manure is further mixed with ash and pond mud or silt at a ratio of 35:25:40, respectively.

Conclusions

Under the previous collective economy and its associated bureaucracy, the hills were barren and fallow. At best they supported a few crops of pineapples. Sandiu and other ethnic groups in Lucngan district lived in difficult conditions. Many Sandiu were forced to move away in search of a better life, and Kinh settlers had to return to their home villages.

Under the newly adopted market economy, litchi orchards have expanded rapidly to spearhead the reclamation of the district's barren hills. The higher ratio of profit to costs in litchi farming, as well as its returns on labor, has seen a reduction in cultivation of other food crops. Nevertheless, the transformation of the agricultural landscape and the improved livelihood of the Sandiu farmers have been remarkable.

This case study is a compelling example of the potential to improve the livelihoods of rural people through market opportunities and by rehabilitating degraded agricultural lands and bringing them back into productive use.

References

Bang, Ma Khanh. 1983. *Sandiu People in Vietnam*. Hanoi.

Cuc, Le Trong, and A.T. Rambo (eds.). 2001. *Mountain Regions of Northern Vietnam: Issues of Environment and Socio-Economy*. Hanoi: Agricultural Publishing House.

Cuc, Le Trong, and Chu Huu Quy (eds.). 2002. *Sustainable Development of Mountainous Vietnam: Ten-Year Review and Problems That Have Arisen. A Workshop Proceedings*. Hanoi: Agricultural Publishing House.

Institute of Ecological Economy. 1993. *The Problem of Vietnam's Ecology*. Hanoi, Vietnam.

Long, Cao Anh, et al. 1996. Collecting Genetic Resources of Fruit Trees Cultivated on Mountainous Ecological Zones in Northern Vietnam. In: *Agriculture on Sloping Lands: Opportunities and Challenges. Selected Research Results during 1991–1996*, edited by Tran Duc Vien and Pham Chi Thanh. Hanoi, Vietnam: Agricultural Publishing House.

Long, Ta. 1997. Swidden Culture and Nomadic Life: A Comparative Study of Ethnic People's Traditions in Vietnam. *Economic Studies*.

Long, Ta, and Ngo Thi Chinh. 2003. *Environmental Changes Under Impact of Humanity Systems at Dienbien District, Laichau Province, from 1945 to the Present Day*. Hanoi: Publishing House of Social Sciences.

Quang, Trinh Hong, et al. 1996. Initial Research on Peanuts in the Crop Structure during Winter on Infertile Soils in Socson District. In: *Agriculture on Sloping Lands: Opportunities and Challenges. Selected Research Results during 1991–1996,* edited by Tran Duc Vien and Pham Chi Thanh. Hanoi: Agricultural Publishing House, 347.

Quyson Commune People's Committee. 1980, 1990, 1993, and 1996. *Reports on Agricultural Production*.

Tuc, Tran The, et al. 1996. Ability to Develop Fruit Trees along Route Six in Sonla Province. *Agriculture on Sloping Lands: Opportunities and Challenges. Selected Research Results during 1991–1996,* edited by Tran Duc Vien and Pham Chi Thanh. Hanoi: Agriculture Publishing House, 309.

Vien, Tran Duc. 2001. *Indigenous Management in Vietnam. A Workshop Proceedings*. Edited by Tran Duc Vien. Hanoi: Agricultural Publishing House.

Vien, Tran Duc, and Pham Chi Thanh (eds.). 1996. *Agriculture on Sloping Lands: Opportunities and Challenges. Selected Research Results during 1991–1996*. Hanoi: Agricultural Publishing House.

Vietnam Population Investigation. 1991. *Comprehensive Research Results, Vol. 1*. Hanoi.

PART VIII

Across Systems and Typologies

Women typically perform much of the work in Asia-Pacific's farming systems. This woman tends a swidden crop of upland rice in northern Thailand.

Chapter 54

Strategies of Asian Shifting Cultivators in the Intensification Process

Dev Nathan, P.V. Ramesh, and Phrang Roy

Shifting cultivation, also known as "swidden" or slash-and-burn cultivation, is arguably the original form of agriculture. It is an intrinsically extensive form of production and consequently is highly susceptible to ecological constraints and population pressures. As a result, much of the world's agrarian history has pivoted on man's efforts to increase the returns per head from agriculture, either through intensification of production or reduction of the population dependent upon the land, or usually a combination of both.

This chapter draws attention to some of the ways in which present-day shifting cultivators are attempting to ease the ecological limits to their subsistence. It is based on the experience of the United Nations Office of Project Services (UNOPS) during supervision and related work for projects supported by the International Fund for Agricultural Development (IFAD) in shifting cultivation areas. Most of its illustrative field experience has been drawn from "tribal" areas in Andhra Pradesh, Orissa, and the Himalayan foothills of northeast India. However, other examples in support of the analysis and its arguments come from Nonghet district, Xien Khouang Province, in the Lao P.D.R., and Manglun county, Yunnan Province, in China.

These are regions of low hills, with shifting cultivation as the predominant agricultural system. Recently, they have been under stress due to the shortening of the fallow period. This has been caused by a number of factors. One of them is local population growth, with limited land available for expansion of shifting cultivation systems to maintain subsistence. Another factor, and probably more important in some areas, is migration from the plains. Landless peasants from these areas, historically denied land by feudal landlords, have been moving up into the hills. These people are largely from the former untouchable, or *dalit,* castes, and others are those displaced by hydroelectricity projects.

The tribal people of the hills have their own social systems, distinct from those of the plains. But they have remained in regular contact with plains people,

Dev Nathan, Senior Visiting Fellow, Institute for Human Development, IIPA Campus, IP Estate, New Delhi, 110002, India. P.V. Ramesh, Senior Advisor (Institutional Development), UNOPS, UNOCA Compound, Jalalabad Rd., Kabul, Afghanistan. Phrang Roy, Assistant President, External Affairs Department, International Fund for Agricultural Development (IFAD), Via del Serafico 107, 00142 Rome, Italy.

participating in trade and, in turn, being exploited by them. At the same time, various governments over the past 50 years have tried, most often unsuccessfully, to bring about changes in the agricultural production systems of the tribes with the principal objective of eliminating, or at least controlling, shifting cultivation.

Consequently, visible changes in fallow management that are intended to intensify production are the result of a number of factors. One such factor is what we usually call the "top-down" approach, where not only the idea, but also the initiative for adopting it, comes from outside the social system. This top-down approach is very likely to fail, and it is this probable failure that, in part, provides justification for looking at indigenous innovations in the intensification process. However, to understand changes in the fallow systems of shifting cultivators, it is also relevant to analyze not only farmers' ideas, but also initiatives arising from within their communities.

Some Examples of Intensification

Terracing

The development of terraces on slopes formerly used for shifting cultivation is not an infrequent method of intensifying agricultural systems. Sloping land that would have remained fallow under the swidden system is converted into terraces in the period between one harvest and land preparation for the next.

In Andhra Pradesh there is a long history of terracing, as a result of both independent local initiatives and formal development assistance. In Yunnan Province of China, a Hani village called Manmo provides another example. In the mid-1960s, this village, learning from its Dai neighbors, initiated a construction program for terraces and related water-harvesting systems.

There are two major constraints, however, to the effective use of terracing as an economically viable form of intensification. First is the high cost of physical development, which is particularly onerous in the kind of economies where shifting cultivation is practiced. Secondly, there needs to be an assured water supply. In all the areas of India covered by the field experience cited in this chapter, farmers consistently reported that dry terraces were not worth the effort.

In Andhra Pradesh, terracing carried out on an individual's initiative usually requires mobilization of labor from within the community. This has to be paid for in cash or otherwise compensated by a feast, and costs run at around Rp25,000, or about US$700, per hectare. Financial constraints in the Hani example meant that the village could not afford concrete structures and had to make do with stone and brush alternatives.

The overall impact of terracing and related water harvesting on local agriculture appears to be a function of the prevailing political economy and absolute availability of water. In the Indian cases, benefits from terracing and water-harvesting systems seem to have traditionally been limited to the political and economic elite, since only clan or tribal elders have been able to mobilize and pay for the necessary labor. This has become more egalitarian in recent times.

In the Chinese instance, greater emphasis was placed on collective action, so that all those families with land to be improved pooled their labor for initial construction and later maintenance. However, the extent to which irrigated terraces can be usefully constructed in any area and the number of people who may benefit are limited not only by the availability of water, but also on local controls on water use and relative costs within the local economy.

Experience suggests that terracing and water-harvesting systems will never be cheap enough—or simple enough to build and maintain—to replace shifting cultivation in any given locality. However, they can reduce pressure on upland areas and open windows of opportunity for their transformation. For example, they can offer an alternative source of subsistence or income generation so that upland communities can undertake the sometimes lengthy transformation of their land from swiddening to sustainable agroforestry or other plantation alternatives.

Vegetable Cultivation

Vegetable cultivation offers an avenue for intensification if there are both sources of capital for production and local markets to provide demand and to solve the problem of perishability. Eight villages in the Araku valley of Andhra Pradesh are an example of how vegetable production has intensified local agriculture and reduced dependence on shifting cultivation.

To begin with, the villages had the benefit of some paddy land in the valley bottoms, and the local production system combined lowland rice with upland swidden crops. The wet rice production was sufficient to generate a regular surplus, and this was used to buy inputs for vegetable production. About 10 years ago, the farmers began experimenting with vegetables such as cauliflower, cabbage, and potatoes. The government supplied some seed, and after about five years, vegetable production was well established in the area. Now, virtually every family has a vegetable plot. Production is concentrated in "home gardens" on the slightly sloping lands just above the rice paddies. The location has an assured water supply and the climate enables year-round production.

Shifting agriculture in the area has been considerably reduced. Families found that labor requirements for vegetable cultivation and those for slash-and-burn conflicted, and the returns to labor from vegetables were higher. Being unable to work in both places at once, one of the villages agreed with the suggestion of a local nongovernmental organization that one of its hillsides should be left uncultivated so that the forest could regenerate to meet local timber requirements. Others have since followed its example.

Vegetable growing in home gardens is a now an expanding activity in many parts of Andhra Pradesh, Orissa, and northeastern India, where swiddening once played a relatively major part in subsistence.

Horticulture

Horticulture plantations are perhaps the most frequently attempted intensification strategy promoted by governments and other formal development agencies. Interestingly, the following case study is of a village that initiated horticultural expansion into its cropping systems without intervention from the outside world.

Muthyalu village in Seethampeta Mandal, Andhra Pradesh, initially grew cashews only for home consumption. Until about 20 years ago, the hillsides around the village were given over to shifting cultivation. However, as the village population grew, the people faced a land shortage because surrounding areas belonged to others and there was no "reserved forest" into which encroachment was possible. The local farmers turned to their experience in cashew production and expanded their output to make it a commercial operation. Later, they added pineapples to their garden production, having learned how to do so from a nearby roadside village. They also introduced turmeric and ginger as intercrops.

In an interesting reversal of the case described in the preceding section, the Muthyalu villagers used surplus income from horticulture to buy paddy land, thereby reinforcing their food security and reducing risks associated with a relatively unfamiliar fruit-producing and marketing system.[1] In contrast with the proceeding case, it appears that a combination of paddy cultivation and horticulture is less demanding of labor than paddy cultivation and vegetable production.

---

[1] See Scherr (1995) for a discussion on how farmers reduce their risks by gradually adopting agroforestry systems.

High-Value Crops

The examples so far have involved meeting the challenges of rising populations and land-use pressures by reorienting crop mixes and displacing subsistence production with cash crops. There are, however, some instances where market opportunities are the dominating factor, so much so that a separate category of "high-value crops" is justified here as an observed alternative to subsistence swiddening.

For example, the comparative advantage of the uplands in the production of pulses and oilseeds has led farmers in some parts of India to focus on these crops to such an extent that they buy in, rather than produce their own, food grains. In Nonghet district of the Lao P.D.R., Hmong farmers trade maize and pigs to Vietnam in exchange for rice. In Yunnan, there is a strong trend to increase areas under tea, rubber, spices, and medicinal plants in response to a growing market demand for these commodities. In northeastern India also there is a trend toward crop production for sale in the marketplace, and a study of six districts in the region points out:

> Within the food crops, pulses have appeared at the cost of millet, and vegetable products, mainly cabbage, cauliflower, tomato, etc., are replacing the old pumpkins and brinjals. Thus the trend is toward high-value quality products for the market instead of common and coarse food crops for self-consumption. (NIRD 1996, 8)

A controversial example of high-value cropping as a substitute or complement to upland slash-and-burn cultivation is, of course, growing opium poppies. Raw opium is of high value, it is portable, easily marketable, and it does not deteriorate. Poppies are also suited to the ecology. In the absence of attractive alternative income sources, poppies are widely grown in the Nonghet district, mentioned above.

Tree Products

It has always been a universal feature of shifting cultivation that any trees with a perceived subsistence or commercial value are spared when clearing the land (see color plate 2). Farmers have long recognized the value of nontimber forest products (NTFPs) to their livelihoods, and it is an area where new products are being developed and new market opportunities are being created by modern communications and processing. However, the role played by NTFPs in the livelihoods of shifting cultivators would seem to be sensitive to the policy environment and the distribution of value addition in the production system.

In Andhra Pradesh, mango and tamarind have traditionally been left standing on cleared land, the former as a source of food and the latter as a valued commodity for sale in local markets. At one time there was also significant production of gum *karaya*, a resin used in the pharmaceutical industry. However, the market for the gum collapsed in the 1970s and the gum trees were felled and left to rot in the fields because they were not suitable as firewood. Later research revived the gum karaya market and increased its value. Farmers once again began saving and protecting the gum-producing trees.

Another such development also underscored the importance of NTFPs to shifting cultivators. The fruit from a local tree was named "the cleaning nut." It was used locally for cleaning water and also in some food preparations. A project came up with a commercial-scale use of the product for cleaning municipal water supplies. It was so successful that the new market for cleaning nuts led to overexploitation of the trees. In one case, farmers in Chintamanuguda were reportedly so eager to maximize the short-term returns that they broke the branches off their trees and later faced a considerable wait for their regrowth.

In contrast to the important role played by NTFPs in the income composition of forest dwellers in Andhra Pradesh, it appears, at least to some, that shifting cultivators in northeastern India have little interest in NTFPs. The NIRD commented in 1996: "Surprisingly, nonwood forest products do not feature prominently in the production system of the shifting cultivators. It seems shifting cultivation and growth of nonwood products is inversely correlated."

Closer observation suggests that this conclusion may be misleading. The general and simple conclusion to be drawn from field experience is that the defining factor for the role of tree products—and the part they might play in replacing swiddening with more intensified cultivation—is their marketability and the returns that can be expected by the collector or producer. The insignificant overall role of NTFPs in the northeast is largely due to restrictive policies. While cultivators may use the products of any tree to meet their own consumption or other local requirements, any trade in these products or their movement outside the village is the sole privilege of monopolists who have acquired the right to trade in the product. This reduces village gate prices and the income received by village collectors or producers. The fact that these traders are also nonlocal, even nonregional, further inhibits the development of local processing industries. For instance, the hill areas continue to export broomsticks and miss the value addition that would accrue to them were they to export manufactured brooms.

Despite the apparently minor importance of NTFPs in northeastern India, those shifting cultivators with fewer irrigated rice terraces and nonswidden sources of income, other than wage labor, still rely to a great extent on collection and sale of NTFPs. It would seem, therefore, that there is a direct, rather than inverse, correlation, even in the northeast, between the extent of swiddening and reliance on NTFPs.

Changes in Inputs

The changing management of the fallow and the introduction of other kinds of production, such as wet rice and vegetable growing into what were once shifting cultivation systems, usually requires increased and varied inputs. For example, more draft power may be needed, and fertilizer must be applied in some form—but not as ash, after burning, as in conventional swidden systems. These all represent new and increased material inputs needed to make the most of infrastructure and cropping innovations.

In northeastern India, the NIRD study mentioned earlier found that communities did not apply manure to their land, even in permanent terraced cultivation. However, in contrast, it was observed in this study that manuring of paddy land was carefully and consistently carried out following the construction of terraces in Andhra Pradesh and Orissa. Furthermore, in villages in Orissa where green manure (*dhanicha*) was applied, farmers with terraced paddy fields were very careful to save the seed for planting in the following year.

At Demudavalasa village in Andhra Pradesh and Silagam village in Orissa, vegetable farmers not only used household waste as organic manure, a practice they followed even in their original home garden cultivation, but they also used compost. Every vegetable-growing family in both villages had a compost pit.

In the project areas of this study, inorganic fertilizers appeared not to be extensively used. But where they were applied, they were limited to the cheaper nitrogenous fertilizers. The more expensive phosphate fertilizers were not being used at all. By comparison, use of pesticides was common in vegetable growing areas.

Land-Use Management

Land Tenure

A consistently important aspect of the intensification process in the cases mentioned in this chapter has been security of tenure over land. Land used for shifting cultivation is traditionally accessed through various types of communal tenure, under which exclusive usufruct or ownership rights do not rest with individuals or particular farm families. As such, customary tenure systems for swidden land usually keep it in the public, rather than the private, domain. There can be little doubt that this is a disincentive to intensification and stabilization of land use. In the absence of secure tenure, whether it is full ownership or simply secure use rights, shifting cultivators will not undertake the investments of labor and other resources needed to intensify cultivation.

In the example of the Hani village in Yunnan, neither the increased production due to the irrigated rice terraces nor the new labor demand for plowing, transplanting, and caring for the rice crops changed the fallow system on the hill slopes, although it did relieve pressure on the hills. It was only in the 1980s, after redistribution of land to households in common with other Hani villages (Henin 1996), that Manmo began to fundamentally change hillside use by introducing tea plants into its fallows. The tea was grown along with other regular swidden crops such as maize, tapioca, dry rice, and thatch grass. Later, a variety of other new crops were introduced, including bananas, pineapples, and peanuts.

It would seem that while surplus harvests from terracing were important to finance potential changes to fallow management, the actual changes were triggered by security of tenure, since this made the lengthy wait between investment and return from tea acceptable to the Manmo community.

A similar example of the importance of security of tenure comes from Orissa's Kashipur block. Here, the government recently decided to give usufruct titles to hill lands for horticulture and agroforestry plantations and to undertake the plantation development itself. It was only after the plots had been designated and the process of issuing titles was under way that farmers showed any interest in contributing voluntary labor for weeding or any other kind of care. And, yet again, it is worth pointing out that the horticultural intensification in Andhra Pradesh, reviewed earlier, took place only on land for which farmers held secure titles.

Conversely, the absence of secure titles appears to perpetuate shifting cultivation even in circumstances where other factors such as population pressure, markets, and labor requirements might otherwise induce intensification and stabilization. For example, in Andhra Pradsh, India, shifting cultivators find themselves in competition with immigrants for land suitable for long-term cultivation. It seems that uncertainty about the legal and administrative status of the land dissuades them from switching to more stable forms of production, for in the event of eviction by the state, the losses of invested labor and material would be very great.

These examples imply that security of tenure is a necessary condition for intensification but not, in itself, sufficient justification for changes in the fallow system. In the absence of land and labor constraints, but without financial inducements and means to change cropping patterns, shifting cultivators will tend to simply extend their area of production under customary usage regardless of tenurial considerations. Nevertheless, without security of tenure, it is unlikely that shifting cultivators will invest the labor and materials needed to intensify their production systems, regardless of the *prima facie* benefits.

Control of Grazing

A customary practice in areas of shifting cultivation is to allow livestock to forage freely on fallow swiddens in the period between one harvest and sowing the next crop. However, it seems that farmers are likely to institute more rigorous grazing regulations with the advent of intensification in its various forms.

In both the Andhra Pradesh–Orissa region and in northeastern India, uncontrolled grazing once remained the normal use of fallow, after the swidden harvest. But in cases where cultivation has been extended beyond one season, or where horticulture has been developed, farmers quickly realized that controlling grazing was a vital requirement. For example, the Andhra Pradesh vegetable growers quickly appreciated that livestock roaming around in their gardens would seriously threaten their year-round production. The village of Demudavalasa solved the problem by recruiting a family to work as herders for the village livestock. A fining system was introduced to punish stock owners who failed to obey grazing regulations.

Adequate control of grazing is only possible when an entire village agrees to implement it. The need for some form of collective decision to control grazing is the main reason why it is difficult for one or two, or even more, families to extend their cropping season unilaterally or to introduce permanent crops. But when all or the great majority of farmers and their families are involved, then effective grazing control becomes possible.

There is some evidence that grazing controls, enforced primarily to protect investments in relatively high-value crops, have had a more general impact upon forest regeneration on former swidden hills. In Andhra Pradesh, grazing control and tree regeneration in former swiddens were closely connected to a local shortage of building timber. The same was the case for Silagam village, Kashipur, in Orissa. Villagers planted fallowed swiddens with timber species supplied by the Forest Department. A ban on grazing ensured that surviving rootstock in the ground was also able to regenerate.

In all the observed cases except one, farmers did not seem concerned about intensification of fodder production and firewood supplies. In Andhra Pradesh and Orissa, there was a tendency for these requirements to be met by encroachment into forest reserves rather than through controlled grazing and replanting measures on villagers' own land. The apparently exceptional case was that of Naga villages in Manipur, northeastern India, where the young people conducted a campaign for stall feeding of pigs. Although the campaign entailed intensification of fodder production, this seemed incidental. Health was the main reason for stall feeding the pigs, as well as keeping them out of home gardens.

The Human Dimension

Economic Rationality

It is possibly useful in the interests of prediction—and the design of policies and interventions—to consider whether the activities of shifting cultivators reviewed here could be interpreted as consistent with an economic rationale.

It is often difficult to offer a conventional economic account of peasant decision making because of the need to deal with a partial, rather than full, articulation of peasant production and consumption interactions with the market. Problems arise, for example, with the definition and valuation of household or family labor as a cost and hence, whether or not producers can be said to be maximizing their returns to labor. It is perhaps appropriate to start from Chayanov's (1966) view that the farm family "...as a result of its year's labor receives a single labor income and weighs its efforts against the material results obtained."

This approach recognizes that there is an objective value in principal to be attached to peasant labor, defined as the value of the gross farm product minus nonlabor input costs. It also acknowledges the subjective consideration of whether the material returns justify the expenditure of energy involved. Chayanov goes on to consider the rationale behind peasants' decisions to invest in farm development. He argues that since development costs must be met from products that would otherwise be consumed, farm families must be able to perceive a higher value from investing the amount, rather than consuming it (Chayanov 1966).

The production activities in the villages mentioned in this chapter suggest that while farmers do set out to maximize their returns to land, capital, and labor, the value of their returns is qualified by the nature of the political economy in which they operate. A wide and complex range of cultural and political issues, differing with every location, complicates the process of economic decisionmaking. Management of resources is disparately influenced by politics, legal processes and customary institutions, access to information, and gender definition.

The historical idea that there was an age when shifting cultivators produced only enough to satisfy their basic needs and thereafter led a leisurely life in which they were indifferent to maximizing their returns is probably rather romantic. Certainly, in the present day, when there is increasing contact between shifting cultivators and the rest of the world, consumption patterns are changing and expectations are rising. Shifting cultivators these days are more likely to want to send their children to school and to enjoy the health benefits of modern medicine. Even access to television appears to be having an impact on farmers' aspirations. There is no doubt that they try to maximize their returns, including returns from labor, in an attempt to access modern goods and services. However, the ways in which they set about it reflect the varying complexities mentioned above.

Whether cultivators move into, out of, or continue swidden production depends upon the relative returns of the alternatives in the given political and economic context. Some form of intensification has been taking place in each of the case studies in this chapter and the process has been economically rational in the sense of farmers attempting to maximize their returns in the given context. The initial impetus for intensification appears to be land pressure. This has led people to convert to alternative forms of production on nonswidden lands, such as wet rice in valley bottoms, vegetables in home gardens, and plantation cash cropping and agroforestry on former swidden fields. The nature of these conversions has also depended upon other factors such as local ecology, markets, input availability, security of tenure, and the ability to raise investment costs.

In some cases, the intensification has been successful enough to both relieve pressure on swidden lands and to generate surpluses that producers have been able to recycle into the development of more sustainable cropping patterns, or to allow tree regeneration.

By contrast, it is possible to cite instances of people moving for the same reasons of land pressure and lack of returns in the opposite direction: out of intensified settled agriculture and into shifting cultivation. For example, migrants from plains areas have moved into shifting cultivation in hilly areas of Andhra Pradesh and in the Upper Pokhara valley of Nepal (Thapa and Weber 1991).

An interesting and important aspect in these processes has been the pattern of labor deployment. It seems that so long as land and labor are still available after intensification, then the higher incomes accruing from intensification do not, in themselves, lead to abandonment of swiddening. It was repeatedly found, for example, that families who had constructed terraces still continued shifting cultivation, provided that land and labor were available. The crucial periods for labor in shifting cultivation in Andhra Pradesh and Orissa are in March and April for clearing land, which is largely men's work, and then from August through November for weeding, which is entirely women's work. For terraced rice, labor is needed in June for plowing and in July for transplanting. Therefore, the combination of terraced rice and swidden cultivation does not entail competition for labor.

Conversely, swiddening will be abandoned if its labor demands compete with those from crops yielding a higher return. For example, the terraced rice and vegetable-growing combination of the Araku valley villages in Andhra Pradesh requires labor through most of the year. There are crops in both the rainy season and the dry season and production is very labor intensive, entailing frequent land preparation, weeding, harvesting, and sale. Therefore, the labor requirements for commercial-scale vegetable growing compete with those of shifting cultivation. If shifting cultivators in this area have the opportunity to move into vegetable growing, they will tend to move out of swiddening entirely as a result of needing to redeploy all their labor on the production system yielding the higher income.

There was also rational economic decision making over labor deployment in Nonghet, Lao P.D.R., where women were diverted from poppy production to laboring on road construction because the returns were higher. Interestingly, the substantial cash wages earned by the women were used to increase livestock holdings, mainly pigs.

In concluding this section, it seems reasonable to assert on the basis of the case studies that production decision making on the part of shifting cultivators, both with respect to different product combinations and deployment of labor, is economically rational. They will go, if possible, to where the returns are highest. However, while rational economic self-interest may be the guiding principle, its scope and impact may be subject to noneconomic factors, some of which are discussed in the following sections.

Politics

Politics might usefully be defined as a nonmarket means of resource management. The political dimension is important in the swidden experiences reviewed in this chapter because of the role of surplus in allowing intensification to occur. It would seem that the political elite are more likely to be able to generate this surplus—and more likely to secure tenure and access inputs—than the rest of the community. The field experience suggests that, in many cases, exclusive reliance on swiddening for a living indicates both economic and political marginalization. It is the poor and the powerless who are most likely to be caught between exclusion from intensification, on one hand, and administrative and police action to prevent swidden expansion into ecologically valuable forests, on the other. The poor man's constraints can only be dealt with by development interventions that offer alternative income sources without provoking reaction from the politically dominant.

Collaboration

Another way of getting around a poor family's difficulties in earning the surplus income needed for intensification of production would be a collective mobilization of labor and the pooling of small, individual surpluses. However, collaboration seems to be rather sensitive culturally. The Hani, as noted earlier, did collaborate and the benefits of their water-harvesting systems were relatively widespread throughout the community. It is notable, also, that Naga villages in northeastern India have built magnificent churches with collective labor. Each family raising a pig for sale meets cash requirements for construction and regular church expenses. Such collective labor could be, but has never been, mobilized for production purposes. It would require strong and accepted village leadership, and such leadership does exist in many upland villages. Additionally, however, collective mobilization of labor could only be achieved if all families were assured of a share, if not an equal share, of returns. If, for example, water-harvesting systems and terraces were to be built with collective labor, redistribution mechanisms would be necessary within clans and families, and these would have to take account of loan-financed purchases.

Gender

The economic "rationality" of intensification and labor deployment by shifting cultivators tends to be distorted by male self-interest being considered ahead of the interests of women. The case studies indicate that while upland social systems tend to be more gender-equal than those on the plains, most of them nevertheless exhibit varying degrees of male domination. Types of tasks, time spent working, and patterns of expenditure for investment and consumption all indicate that male work and male leisure are valued more highly than that of women. Investment in women, or with a view to improving returns from women's work, holds lowest priority in the making of economic decisions. There is a spurious *prima facie* rationale to this prioritization: women already do work for the least returns, so therefore the costs of failing to grasp an opportunity to invest in it are less than those of failing to invest in men's work. This "rationality" breaks down, however, when it is apparent that men's leisure is, in effect, valued more highly than women's work. For example, hunting is no longer a meaningful economic activity in the study communities, but many men still go out to "hunt." It has become a pastime rather than a production activity. Even in the evening, when men and women return from the fields or from "hunting," it is noticeable in a Hani village in Yunnan that men relax on a sofa listening to music while the women fetch water and cook the evening meal. Individual families spend money on a cassette player, even borrowing for the purpose, but do not invest in either improved stoves or piped water. The social and economic marginalization of women means that their customary work of fodder production, firewood collection, and other household chores is the last to be lightened by investment in improvements, even if it would release them for more profitable work. Moreover, women's awareness of the world outside their immediate community, and their scope for trade and travel to grasp economic opportunities, is precluded and subordinated to male interests.

Sources of Change

So far, this review has looked at why shifting cultivators change their production patterns, how they change them, the rationale for particular patterns of change, and who the main beneficiaries are as a result of change. It is perhaps useful, finally, to examine in the context of the case studies the sources of change. One of the reasons why this is arguably important is that over the past 50 years, governments worldwide, including administrations in India, have tried to bring about various changes in shifting cultivation with the ultimate objective of eliminating or at least "controlling" it. Mostly these efforts have been unsuccessful.

According to Rogers (1969), changes in the use of the fallow in shifting cultivation can be categorized on the basis of the initiative for a change and the source of the idea. Indigenous change or innovation involves a change from within the social system being considered. On the other hand, "contact" change means the new idea comes from outside the social system. If, however, the source of the idea comes from outside but the initiative for adopting it comes from within, it is a case of "selective contact" change. Finally, in what is currently known as the "top-down approach," not only the idea but also the initiative for adopting it comes from outside the social system.

As government attempts to control shifting cultivation have shown, the top-down approach is very likely to fail. And, as mentioned earlier, it is the failure of this approach that, in part, provides the justification for looking at the role of indigenous change in the intensification process. It is also necessary to look at whether selective contact change is relevant to analysis of the intensification process in shifting agriculture. For instance, the conversion of fallow land into terraces stands out as an important step in the intensification process. But the construction of terraces and the related system of wet rice cultivation were both probably learned by upland farmers from those who had been settled in the valley bottoms before the need for intensification was felt. The Hani in Yunnan, for instance, were very clear that they had learned terracing and wet rice cultivation from their lowland neighbors, the Dai. In fact, even the architectural character of their houses has changed under the influence of contact with the Dai.

At the same time, apparently contradictory reports have come from northeastern India on the adoption of terraces by upland shifting cultivators. Ramakrishnan's

(1993) study of northeastern India concludes: "terracing of land, promoted through the government agencies, has not found acceptance from the local communities."

However, field observations in Ukhrul district of Manipur and the West Garo Hills of Meghalaya dispute this finding, and a study of six districts of northeastern India by the National Institute of Rural Development commented:

> With the increase in population and exposure to the outside world, terraced cultivation emerged and increased in all six districts, but the extent of terraced cultivation varied according to the varying topography and varying permissive or restrictive customary practices followed by the communities (NIRD 1996).

It seems clear that to succeed, changes in shifting cultivation require an indigenous element. Attempts to oblige farmers to change their practices through external legislative and policing measures are likely to fail. The case studies indicate that cultivators are aware of alternative agricultural systems in their neighborhoods and can—and do—adopt them if they are perceived to bring an improved income under the prevailing circumstances. The problem would appear to be largely one of practical possibility within a given political economy, rather than obstinacy on the part of shifting cultivators. Field experience suggests that the bulk of continuing slash-and-burn producers are poor people who have no alternative means of livelihood.

Conclusions

The experiences cited in this chapter suggest some conclusions that may be inferred about factors affecting intensification and stabilization of shifting cultivation systems:

- The farm family must be able to produce a surplus from its original swidden system. If they cannot produce beyond their subsistence requirements, they cannot raise the investment capital needed for intensification.
- Intensification is dependent upon changes in prevailing local property rights, especially land tenure. For investments to be made in intensification, investors require at least security of tenure and, in practice, this means land must be taken out of communal, and into private, ownership.
- When adopting intensification measures, shifting cultivators can—and do—learn from neighboring settled cultivators. They are also responsive to market opportunities and to development agency inputs, provided it is clear that these inputs are in their interests.
- The markets rationally determine the role of nontimber forest products as part of overall subsistence and income generation in swidden areas.
- Intensification is usually induced by land and labor constraints, rather than by higher income. Investment of surpluses in intensification appears only to occur when land and labor are perceived to be in increasingly short supply.
- There appears to be a gender dimension in the relationship between labor supply and intensification. Labor shortages involving activities designated as women's work are less likely to trigger intensification than shortages in the male domain. There is a self-fulfilling economic "rationality" to this phenomenon in that men have historically captured those activities that bring the highest returns, and therefore the opportunity costs of accommodating labor shortages in these areas are greater than those applying to the low-return drudgeries of women.

In summary, economically attractive and sustainable alternatives need to be offered to shifting cultivators. Moreover, if conservation objectives are to be achieved, these alternatives need to be attractive before, and not after, the onset of ecological breakdown. The case studies suggest that development policies and programs in areas currently given over to shifting cultivation will need to have the following characteristics:

- Since shifting agriculture tends to be practiced in remote, physically arduous, and ecologically fragile areas, the development of viable alternatives is likely to require a multifacet approach, rather than one or two simple interventions. These alternatives are likely to require aspects of education and training, input supply, credit, marketing, crop and livestock development, agroprocessing, adaptive research, new product development, off-farm income generation, provision of production support services, and rural infrastructure development. Most of these features appear in the examples of change given in this chapter, although not as a coherent package and usually informally. Overall, agriculture needs to be intensified and a greater proportion of the population moved out of primary agricultural production into services, processing, and nonfarm forms of production.
- The examples in this chapter indicate that interventions are more likely to succeed if they are poverty and equity oriented. Most shifting cultivators are poor rather than rich.
- It seems clear that moving out of swidden production requires either access to capital as an independent agricultural producer or to attractive paid employment. Either route implies the need for distributive mechanisms within local socioeconomies. Instances have been given in this chapter of the ways in which slash-and-burn is perpetuated by the absence of such mechanisms and by the restriction of opportunities imposed by the political and economic elite.
- In view of the substantial role of women in shifting cultivation, it would seem that poverty and equity-oriented interventions will, functionally, amount to women's development programs. Therefore the cultural issues surrounding gender will need to be considered in the design and implementation of development initiatives.

References

Chayanov, A.V. 1966. *The Theory of Peasant Economy.* Homewood, IL: Richard D. Irwin Inc.

Henin, B. 1996. Ethnic Minority Integration in China: Transformation of Akha Society. *Journal of Contemporary Asia*, 26(2).

NIRD (National Institute of Rural Development). 1996. *Socio-Economic Production Study of Tribal Communities in Northeast India.* Shillong, India: NIRD.

Ramakrishnan, P.S. 1993. *Shifting Cultivation and Sustainable Development.* Paris and Delhi: UNESCO/Oxford University Press.

Rogers, E.M. 1969. Motivations, Values and Attitudes of Subsistence Farmers: Toward a Subculture of Peasantry. In: *Subsistence Agriculture and Economic Development,* edited by Clifton R. Wharton. Chicago: Aldine Publishing.

Scherr, S.J. 1995. Economic Factors in the Adoption of Agroforestry: Patterns Observed in Western Kenya. *World Development*, 23(5).

Thapa, Gopal B., and K.E. Weber. 1991. Deforestation in the Upper Pokhara Valley, Nepal. *Singapore Journal of Tropical Geography*, 12(1).

Chapter 55

Rebuilding Soil Properties during the Fallow

Indigenous Innovations in the Highlands of Vietnam

Hoang Xuan Ty

The government of Vietnam has made sustained efforts to halt the spread of shifting cultivation and its impacts on mountain areas. However, despite these efforts, it does not seem to have decreased in importance. More than 2 million people, mostly ethnic minority groups, still depend upon swiddening for their livelihood (Toan 1991; Huyen 1991; Sam 1996) (see Figure 55-1 and Table 55-1).

Under pressure from an increasing population over the past two to three decades, shifting cultivators have been forced to shorten the length of their fallow periods from the traditional 15 to 20 years down to about 5 to 8 years. In some areas of particularly dense population, the fallow is now restricted to as little as 4 to 5 years (Binh 1991; Thong 1994). The consequences have been catastrophic, with degrading soils, dwindling productivity, and swidden farmers forced to move more frequently to find new areas of fertile land in mountain forests (Thang 1995). The magnitude of the problem has been countered, to some degree, by the ingenuity of local shifting cultivators in developing innovations for rapid and effective restoration of soil fertility. Many of these methods have potential for replication in other upland areas where shifting cultivation systems are under stress (Warner 1991; Thang 1995).

This chapter presents the results of preliminary research into three farmer-generated innovations that aim to improve fallow management. They are:

- Protection and maintenance of small-sized stumps at a height of 1 to 1.5 m above the ground during clearing of vegetation, in order to promote early restoration of forest cover and soil productivity during the subsequent fallow;
- Promoting regeneration and growth of *Chromolaena odorata* to suppress the spread of *Imperata cylindrica* and improve soil conditions during fallows; and
- Following traditional practices to ensure conservation of bamboo in swidden fields.

The objective of the study was to use scientific methods and tools to assess the practical values of these innovations and, at the same time, to suggest further improvements and applications.

Hoang Xuan Ty, Research Center for Forest Ecology and Environment, Forest Science Institute of Vietnam, Chem, Tu Liem, Hanoi, Vietnam.

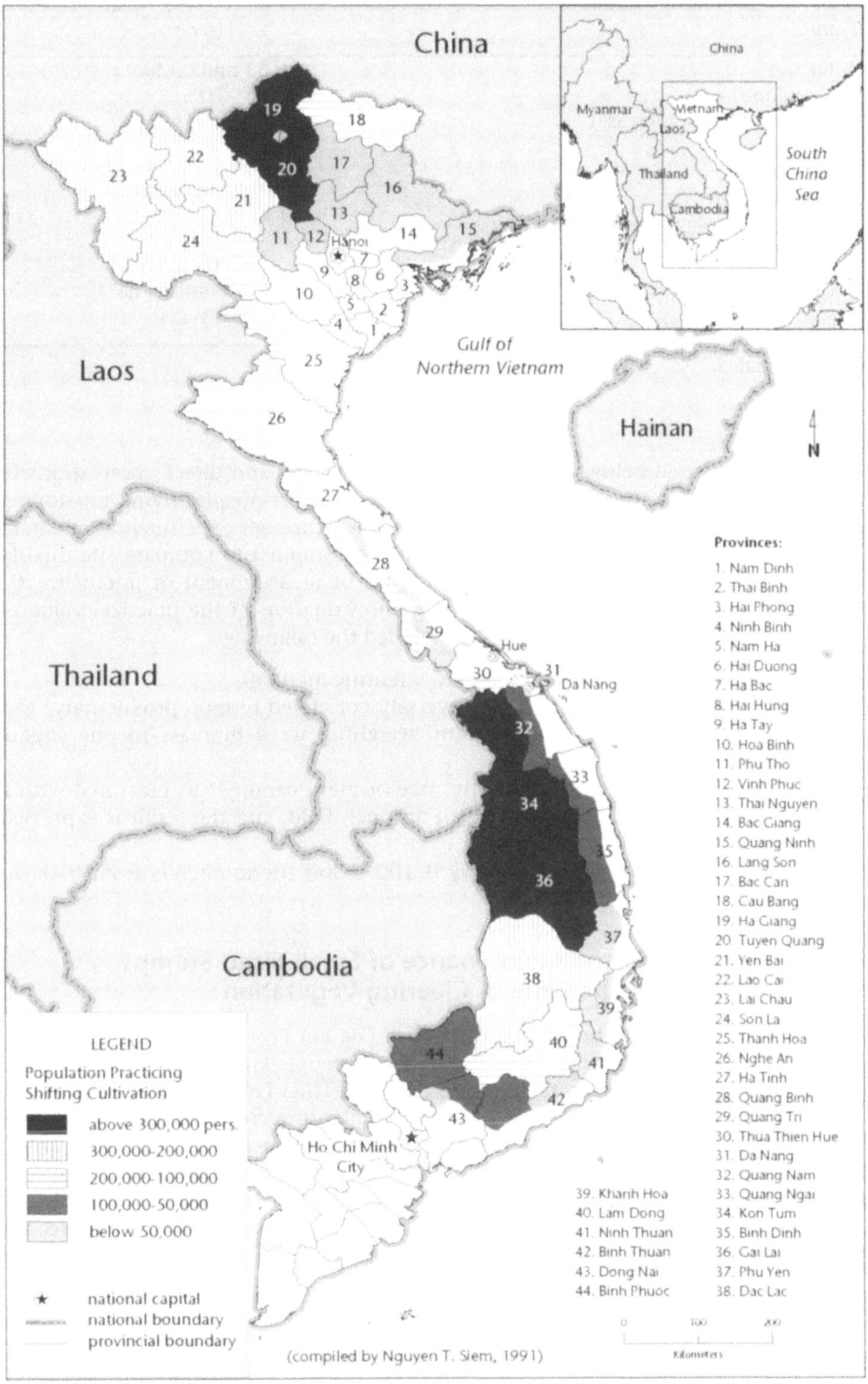

Figure 55-1. Distribution of Shifting Cultivators in Vietnam (1991)

Table 55-1. General Features of Shifting Cultivation in Vietnam

Feature	*Figures*
Total area	33 million ha
Mountainous area	70%
Population (rate of increase)	76 million (2.1%)
Land per capita:	
Agriculture	0.10 ha per person
Forested land	0.15 ha per person
Number of people involved in shifting cultivation*	3 million
Deforestation rate*	120,000 ha/yr
Total area under shifting cultivation*	2.5–3.0 million ha
Fallow rotation (years)	5–8 years

Note: * Estimated.

Methodology

The results presented below are drawn from field surveys and direct observations to document the indigenous knowledge developed by forest peoples living on sloping lands; they are not systematic conclusions obtained through experimental research. However, a wide range of methods and tools were applied to compare site quality through, for example, analyses of weed density or measurement of microclimatic conditions on different plots, thus allowing an evaluation of the practical value of the innovations. The main methods used included the following:

- Soil sample analyses following standard scientific methods.
- Weed biomass, which is generally inversely correlated to crop productivity, was assessed by counting individuals and weighing weed biomass in one square meter plots.
- The relative light intensity under the tree or plant canopy was measured with a luxmeter, then compared with that in an open field, and the result is expressed as a percentage.
- Soil temperatures were measured by 0–100°C soil thermometers from 0700 to 1700 hrs.

Protection and Maintenance of Small-Sized Stumps at the Time of Clearing Vegetation

This study was made in the K'bang district of Gia Lai Province, on swidden fields cultivated by Gia-Rai tribesmen, and also by forest people known as the Catu, or the K'ho, in the high mountain areas of Quang Nam and Thua Thien Provinces.

When clearing vegetation in preparation for planting crops, all small-sized trees of 5 to 15 cm diameter at breast height (dbh) are cut at a height of 1 to 1.5 m above ground. The number of stumps varies from 800 to 2,000 per ha. After burning, between 40% and 60% of them can still coppice, while the remainder die in the fire.

At the beginning of the rainy season, from April to June, when maize and rice are beginning to germinate and grow, the shading provided by the coppicing stumps is beneficial to the crops. During weeding, the coppices are managed to create the most favorable shading. Under conditions of no drought, milder air temperatures, and controlled sunlight, all the young coppices may be pruned and the prunings used to mulch the swidden field. By manipulating the stump coppices in this fashion to control shading, the yields of food crops can be stabilized under a variety of conditions.

After two to three years of cultivation, the swidden field is left fallow to restore soil fertility. The conservation of the tree stumps accelerates the restoration process. Forest cover is said to be restored after only one year of fallow, by which time the stump coppices have created a canopy over 100% of the field. All light-demanding weeds that would otherwise invade the site, such as *Imperata cylindrica, Thysanolaena maxima,* and *Saccharum arundinaceum*, are suppressed by the dense vegetative cover. At the same time, the deep root systems of the retained stumps pump nutrients from deeper soil layers, returning them to the surface through litterfall and thus accelerating soil rejuvenation (see Figure 55-2).

Changes in Soil Properties

Working in two districts, K'bang in Gia Lai and Tra My in Quang Nam Da Nang, data were collected from fallows with stumps protected and maintained as described above and also from control fallows where no such treatments were carried out. The data are summarized in Table 55-2.

Reliable information on the yields of food crops in the sampled fields is not available to enable final conclusions, but on-site differences between the treatment and control fields allow some preliminary observations. The treatment swiddens with retained 1 to 1.5 m stumps rapidly developed a dense tree canopy during the postcropping fallow that, compared to the control fields, exhibited lower weed populations, improved soil nutrient status, and moderated soil temperatures. All of these conditions are conducive to improved crop performance.

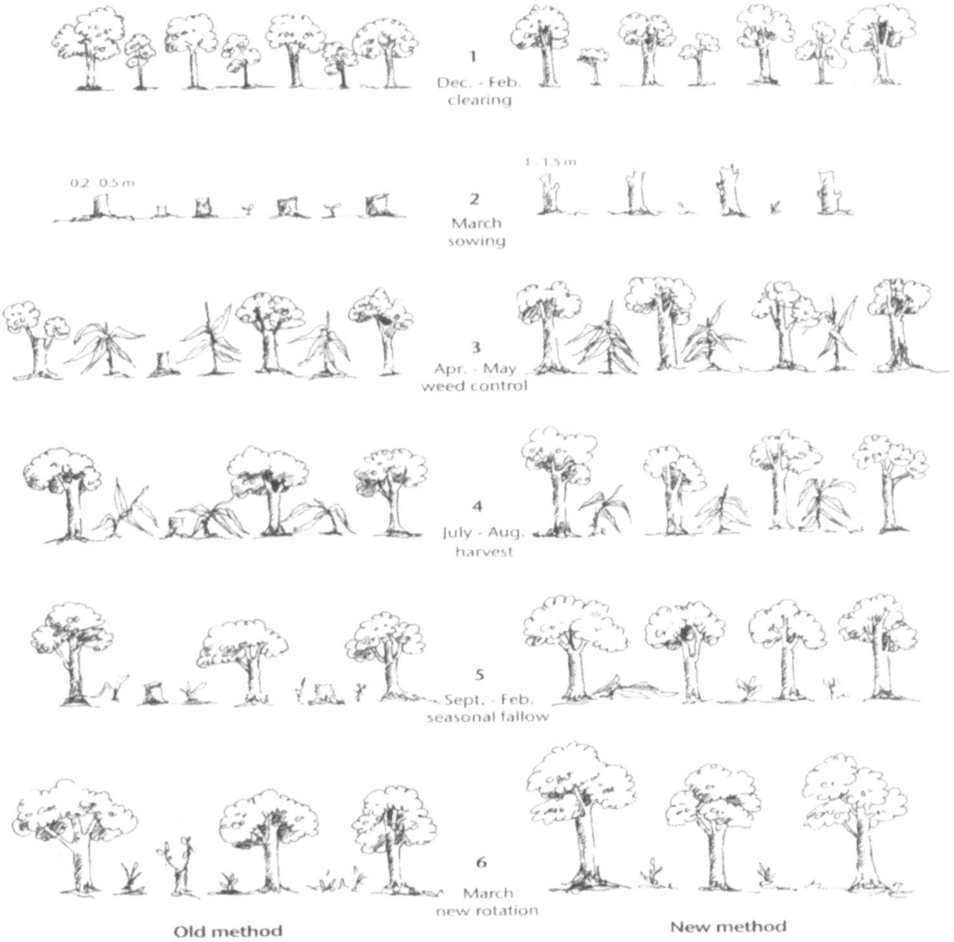

Figure 55-2. Stump-Based Fallow Management System

Table 55-2. Comparison of Treatment and Control Swiddens after a Three-Year Fallow Period

	District of K'bang		*District of Tra My*	
Description	*Stumps 1 to 1.5 m High*	*Standard Stumps 0.1 – 0.2 m High*	*Stumps 1 to 1.5 m High*	*Standard Stumps*
Coppicing stumps				
Living stumps per ha	1,230	308	860	106
Stump height (m)	1.42	0.35	1.27	0.55
Forest canopy (%)	78	41	76	35
Weeds				
Clumps per m^2	3.5	25	4.6	47
Fresh biomass (g/m^2)	48	321	62	508
Relative light intensity under canopy (%)	12	45	10	49
Soil properties (0 to 10 cm)				
Humus (%)	4.31	4.10	4.48	4.02
Total N (%)	0.20	0.18	0.22	0.20
pH (KCl)	4.8	4.8	4.5	4.6
Available P_2O_5 (mg/100g)	3.1	3.3	2.6	2.5
Available K_2O (mg/100g)	8.6	7.2	9.7	6.3
Soil temperature at noon on a hot July day (°C)	35.0	41.5	34.1	39.2

These positive outcomes are directly attributable to the larger number of living stumps retained in a field which, in turn, accelerate forest recovery. The canopy of the forest developing from the high stumps was almost two times higher than that from stumps of conventional size. Treatment fields had populations of 860 to 1,230 living stumps per hectare, and this was three to seven times more than the control fields where trees were cut at standard heights. There are two reasons for this:

- The standard stumps, cut at 20 to 30 cm above the ground, are severely damaged by the high temperatures of fires when the fields are burned.
- Coppices sprouting from standard stumps, just 10 to 30 cm above the ground, risk hampering the growth of nearby food crops. As a consequence, the coppices and the stumps producing them would probably be destroyed by the cultivators themselves.

Weed Populations

Effective weed control is a fallow function of critical importance. *Imperata cylindrica*, one of the most pervasive grassy invaders, was almost completely suppressed in the fields with high stumps. The fresh biomass of weeds was only 48 to 62 grams per m^2 in the treatment fields, whereas that with standard stumps was 321 to 508 grams per m^2.

The soil nutrient status, although superior in fields with high stumps, did not show a big difference. It was clear, however, that soil temperatures in fields with high stumps were lower by 5 to 6°C because of their denser vegetative cover.

The farmers contend that if 800 to 1,200 living stumps of small-sized trees, between 8 and 12 cm dbh, can be protected and maintained to produce profuse coppices, then the fallow period can be reduced to only four to six years. If, on the other hand, tree stumps are generally destroyed, fallows need to be at least 10 to 12 years long, or more, to achieve similar results.

General Assessment

The practice of protecting the stumps of small-sized trees, at heights of 1 to 1.5 m, can easily be adopted by shifting cultivators elsewhere. No additional management is needed, less labor is required, and household food security can be maintained. The practice encourages sustained forest protection by better managing plant resources to enhance soil conservation and food production. The retained stumps also stabilize erosion-prone slopes. They reduce the rate of soil loss while, at the same time, increasing the uptake of nutrients from deeper soil layers for use by food crops.

Despite its apparent advantages, there are limitations to this fallow management approach that need to be highlighted. Most tropical tree species are not resistant to fire and not many stumps will survive burning. Even if they do survive, tropical species do not coppice profusely. The practice seems best suited to older, 15- to 20-year-old fallow vegetation that is dominated by heat- and drought-resistant trees.

Priorities for Research

The results of these field surveys, in search of farmer-generated innovations, appear promising. Therefore, these activities should continue and all findings should be integrated into a series of experimental trials that aim to achieve the following:

- To identify practical field methods of improving the survival rate of trees in slash-and-burn systems. The research might focus on the optimum season for clearing and burning fallow vegetation, on burning practices, and the cutting height of trees.
- To select tree species most suitable for timber plantations in areas where timber resources are depleted. This should give priority to drought-resistant legumes able to be multiplied through vegetative cuttings. In Vietnam, *Cassia siamea, Indigofera teysmannii,* and *Dalbergia tonkinensis* appear promising (Eklof 1994; Neave and Quang 1994; Ty 1996).

Promoting *Chromolaena odorata* as a Fallow Species to Limit the Spread of *Imperata* and Improve Soil Conditions

Chromolaena odorata (syn. *Eupatorium odoratum*) is a shrub belonging to the Asteraceae family (see Roder et al., Chapter 14, and color plate 15). It grows between 2 and 2.5 m high and is heavily branched. During every dry season, from December to February, the aboveground part dies off but the stumps coppice again at the beginning of the next rainy season, in March or April. *C. odorata*'s small seeds have pappi and they are widely disbursed by wind, so it spreads rapidly. It has a number of beneficial agronomic properties that qualify it as a soil improver:

- It grows rapidly, is a prolific seeder, and its wind-borne seed is widely disbursed in a very short time during the early rainy season.
- Its stem density is high, from 5 to 15 clusters per m^2, enabling it to control soil erosion and prevent invasion by *Imperata*.
- On good sites, its dry biomass can reach 10 to 25 tonnes/ha after eight months.
- The nutrient content of its biomass is high, especially in potassium.
- After the fallow period, only modest amounts of labor are required to reopen the field for cropping. *C. odorata* competes very weakly with agricultural crops and does not, itself, represent a serious weed problem.

At present, farmers are collecting and planting *C. odorata* mainly for its value as a green manure. Its value as a soil improver is very clear, but the challenge is how to encourage this plant's ability to dominate grasses that otherwise demand heavy labor investments to control by hand weeding.

Promoting C. odorata *during Fallows*

Hmong and Thái ethnic minorities in the province of Son La, in northwest Vietnam, have developed methods to encourage *C. odorata* as the dominant succession species in young fallows.

After three to four years of cultivation, and when the last crop has been harvested, all the crop residues are collected and burned in the early dry season, around October and November, just before *C. odorata's* seeds ripen and spread in the wind. This usually happens between September and December. When the seeds fall to the burned ground they are covered by a thin layer of ash, and they store well during the winter months (see Figure 55-3).

In the following March or April, when the first rains come, the *C. odorata* seeds germinate. After two months, a dense carpet of *C. odorata* appears and is able to smother most kinds of grasses. In the spring of the following year, *C. odorata* coppices from its dry season stumps and grows more rapidly than in the first year. It is not yet known for how many years *C. odorata* can continue coppicing from its stumps in this manner in Vietnamese conditions. However, casual observation confirms that four-year-old *C. odorata* stands continue to coppice vigorously in the spring and grow well.

C. odorata *as an Effective Fallow Species*

C. odorata's value as a superior fallow species comes from two important points:

- Its high biomass accumulation, which contains nutrients that, when the fallow is reopened and the slash is burned, are available to planted crops.
- The ability of its dense canopy to smother out *Imperata* and other troublesome weeds.

Table 55-3 presents data on *C. odorata* fallows collected from two different locations in northwest Vietnam. The first location was in Ky Son district, Hoa Binh Province, at an altitude of 300 m above sea level (asl); the second in Moc Chau district, Son La Province, at 900 m asl. From both locations, data on biomass and associated environmental parameters under *Chromolaena,* or treatment, fallows are compared with nearby alternative, or control, fallow successions. The data show three important consequences of *C. odorata* fallow domination:

- Annual biomass increments in *C. odorata* ranged from 1,042 to 1,671 grams/m^2, or 10.4 to 16.7 tonnes/ha. *Imperata* almost completely disappeared from the succession. Control plots without *C. odorata* were dominated by *Imperata* and *Miscanthus* grasses.
- Relative light intensity in *C. odorata* plots was four to seven times less than that in the grass-dominated plots.
- Soil properties in the *C. odorata* plots were superior, especially in humus and potassium content in the top layer.

Laboratory analyses of *C. odorata* biomass revealed a high nutrient content, particularly of potassium (see Table 55-4).

C. odorata offers good prospects for wider application across many bioclimatic zones. However, while farmers have learned to value *C. odorata* as an effective fallow species, it has several noteworthy limitations:

- *C. odorata's* shallow root system, which grows no deeper than 30 cm, cannot pump leached nutrients from deeper soil layers.
- It is not suitable as fodder for livestock.

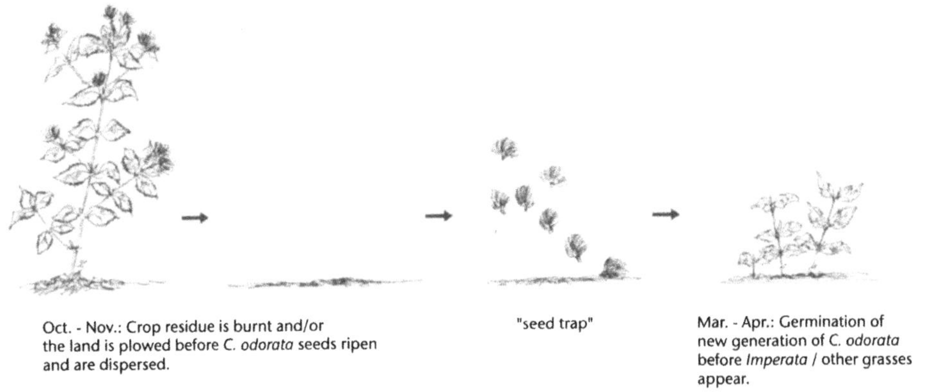
Oct. - Nov.: Crop residue is burnt and/or the land is plowed before C. odorata seeds ripen and are dispersed.
"seed trap"
Mar. - Apr.: Germination of new generation of C. odorata before Imperata / other grasses appear.

Figure 55-3. "Recruiting" *C. odorata* as a Spontaneous Cover Crop in Swidden Fallows

Table 55-3. Comparison of Fallow Plots Dominated by *C. odorata* vs. Alternative Succession Communities

	Ky Son District (300 m asl)		*Moc Chau District (900 m asl)*	
Environmental Factors	*Plot Dominated by C.* odorata	*Plot Dominated by Grass*	*Plot Dominated by C.* odorata	*Plot Dominated by Grass*
Most recent crop	dry rice	dry rice	corn	corn
Length of fallow (years)	3	3	3	3
Plants presently dominating fallow succession	*C. odorata, Panicum montanum*	*Imperata, Miscanthus japonicus*	*C. odorata*	*Imperata, Miscanthus japonicus*
Dry biomass of *I. cylindrica* and *Miscanthus japonicus* (g/m^2)	15	650	6	1,040
Dry biomass of *C. odorata* (g/m^2)	1,042	19	1,671	45
Solar radiation under canopy, at 50 cm above ground. (% of direct sunlight)	16	70	12	85
Soil properties: (0–10 cm)				
humus (%)	3.85	3.11	4.61	4.21
total nitrogen (%)	0.18	0.17	0.21	0.18
pH (KCL)	4.6	4.8	4.5	4.6
available P_2O_5 (mg/100g)	2.1	2.4	3.5	3.1
available K_2O (mg/100g)	8.3	6.7	9.1	7.5
exchanged CaO (meq/100g)	1.75	1.82	1.35	1.42
exchanged MgO (meq/100g)	0.80	0.67	0.61	0.54

Table 55-4. Nutrient Content of *C. odorata* Biomass

Elements	*Content in Variation*	*% (DW) Average*	*Annual Accumulation per ha (kg/ha/yr DW)*
N	0.4–0.7	0.52	54.2–86.8
P	0.08–0.10	0.09	9.3–15
K	0.80–1.16	0.97	101–162

Note: DW = dry weight.

Research Priorities

- Identify simple methods of encouraging *C. odorata* domination of fallow successions. This could focus on propagation techniques using both seeds and stumps.
- Study the feasibility of collecting *C. odorata* varieties and distributing superior strains to farmers practicing bush-fallow rotations.

Protection of Bamboo in Slash-and-Burn Systems

In the mountainous region of northern Vietnam, bamboo is a preferred fallow species in slash-and-burn systems because it is effective in rehabilitating both soil and ecological conditions in swidden fallows (Trung 1970; Ty 1987) (see Bamualim et al., Chapter 39, and color plates 53 and 54). Compared, for example, with shrub successions, bamboo has properties that allow farmers to shorten fallow periods (see Figure 55-4). This raises the need for practical techniques to ensure the easy regeneration of bamboo after cropping. Interviews with shifting cultivators in several provinces of northern Vietnam have provided useful indigenous knowledge on this issue.

The Significance of Bamboo

Bamboo is widespread in forest and mountain ecosystems across Vietnam, as it is in many countries of the humid tropics. According to the Ministry of Forestry, the country had nearly 1 million ha of bamboo in 1993, of which 580,000 ha were pure bamboo forests and 326,000 ha were mixed with broadleaf trees. Most of these bamboo forests originated from forests dominated by large timber trees. The trees fell victim to the combined effects of overlogging and shifting cultivation, and the forests gradually became pure stands of bamboo.

The species of most significance to shifting cultivation systems in Vietnam are *Neohouzeaua dulloa*, *Dendrocalamus patellaris*, *Arundinaria racemosa*, *Arundinaria sat Balan*, *Bambusa procera*, and *Phyllostachys pubescens*. The first three cover the largest areas where slash–and–burn is endemic.

Bamboo has a number of important attributes that contribute to its role as an effective fallow species. It regenerates and grows quickly, needing only 100 days to reach the maximum height of mature plants. Because of its dense rhizomes, bamboo can regenerate to form a forest within the first year of fallow, in this way improving the environment more rapidly than alternative shrub and grass succession communities.

Bamboo is highly adapted to burning and does not compete strongly with agricultural crops. Its rhizome system, often lying 10 to 20 cm under the soil, can easily survive the fires of swidden clearing. Its aboveground biomass is converted into high volumes of ash, supplying large quantities of nutrients to the soil. The remaining clumps do not represent a serious obstacle to weeding or other management operations. Trimming back regenerating bamboo shoots during the cropping phase requires little labor compared to shrubby trees or perennial grasses.

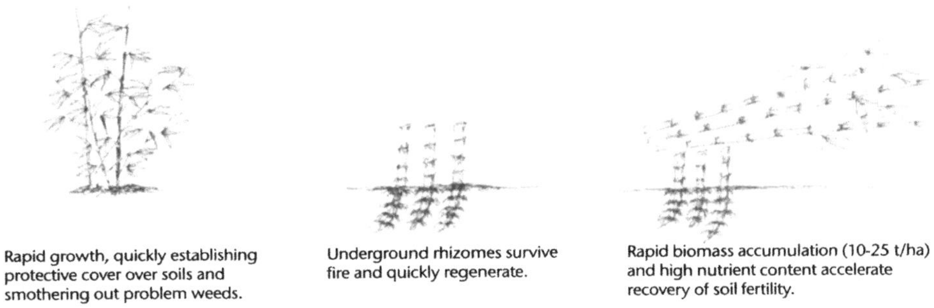

Rapid growth, quickly establishing protective cover over soils and smothering out problem weeds.
Underground rhizomes survive fire and quickly regenerate.
Rapid biomass accumulation (10-25 t/ha) and high nutrient content accelerate recovery of soil fertility.

Figure 55-4. Attributes of Bamboo as a Desirable Fallow Species

Bamboo has an outstanding ability to improve the soil and the environment. The density of bamboo stands is typically between 400 and 2,000 clumps per ha. If the rhizomes are well protected, a young bamboo forest will establish after only one year, reaching heights of 5 to 8 m. Annual litterfall in a six- to seven-year-old bamboo forest reaches 3.8 to 9.5 tonnes/ha, dry weight. This contains 240 to 350 kg of macroelements, such as N, P, K, Ca, and Mg, and a high silica content (Ty 1986).

In addition, the dense root systems of bamboo are concentrated in the topsoil layer, improving soil porosity. Soil moisture in *Neohouzeaua dulloa* forests, for example, is maintained at high levels and has a lesser variation between the seasons. Therefore, biological processes within the soil occur more rapidly than in shrub savannas (Ty 1986).

The canopy of a bamboo forest is very dense, so it effectively smothers out light-demanding weeds such as *Imperata*. If swidden soil is covered by bamboo for six or seven successive years, there will be no troublesome grasses during the first two years of the following cropping phase.

Bamboo is also an important minor forest product. After fallow periods of five or six years, between 5,000 and 20,000 bamboo poles, with diameters of two to four centimeters, can be harvested per hectare. These provide an important source of short-term income for swidden communities.

Farmers also recognize bamboo as a good "indicator" plant. Its presence and condition suggest soil moisture and fertility. When the diameter of *Neohouzeaua dulloa* and *Arundinaria racemosa* stems surpasses 2.5 to 3.0 cm, this tells farmers that the soil is sufficiently fertile to plant another crop of rice or maize (Ty 1986).

Drawbacks to Bamboo-Based Fallows

As part of their life cycle, many bamboo species flower and then die, often affecting thousands of ha at a time. The life cycle varies between 20 and 40 years. Young bamboo regenerating from seed needs four to six years to become a forest with stem diameters of about two centimeters. It is critical to protect young bamboo forests during this time to ensure their survival.

Protection of the Rhizome System

The ability of a bamboo forest to regenerate depends upon the health of its rhizome system. Field surveys and interviews with various ethnic farming communities in the provinces of Phu Tho and Tuyen Quang have provided a number of methods of protecting bamboo during the clearing and burning operations of shifting cultivation.

Following standard swidden practices, the aboveground biomass of bamboo is completely cut down at the end of the dry season. After two to three months, sometime between February and June, the dry bamboo slash is burned and the field prepared for planting various kinds of rice, maize, manioc, and other subsidiary crops. This same method of shifting cultivation within bamboo forests has been practiced for centuries. But while bamboo regenerates vigorously in some fallows, it is completely wiped out in others. What are the differences in treatment that might explain this variation?

Many old and experienced farmers contend that the rhizome survival rate does not depend on the volume of burned biomass or the heat generated by the fire. Rather, they suggest that it depends on the weather during the fire and up to three hours afterward, when the temperature of the top five centimeters of soil can reach 100 to 200°C. If there is a sudden heavy rain during the fire or while the bamboo clumps and topsoil are still smoldering, then the rhizomes will almost completely die out, destroying any chance for the bamboo to regenerate. Experienced farmers are therefore skilled in forecasting the weather three to five days in advance and can judge the most appropriate time to burn the bamboo slash to avoid damaging the rhizomes.

Regeneration of Bamboo from Seed

The Tày and Dao ethnic groups have developed methods for maintaining bamboo forests around their villages, for the collection of forest products. Use of these methods would be unlikely in the context of swidden fallows.

The two bamboo species, *Neohouzeaua dulloa* and *Arundinacea,* flower on a 20- to 40-year rotation in north and central Vietnam. Flowering can continue for up to two years and involves both old and young plants. Then the entire bamboo forest dies, en masse. In the early rainy season of the following year, bamboo seedlings begin to sprout at a density of between 20 and 150 per m^2, looking much like rice plants. The villagers have adopted the following methods of helping the young bamboo forests become established:

- The seedling forests are protected from humans and wild animals. Slash-and-burn is prohibited in the area. After three or four years, the new bamboo forest will be well established, and when, after five or six years, stem diameters reach 2.5 to 3.0 cm, the land can be reopened for cropping.
- For plots where the bamboo forest does not regenerate naturally, seedlings can be transplanted during the rainy season when they are just one centimeter in height, at densities of 500 to 1,000 clumps per hectare. This practice is rarely followed because of the large expenditure of labor involved. But it remains an option for maintaining stands of bamboo when the plants reach the end of their life cycle and the seedlings are usually abundantly available.

Research Priorities

- Further documentation and validation of villagers' traditional techniques for conserving bamboo in slash-and-burn systems.
- Building on and improving these techniques for important bamboo species across a variety of different environments.

References

Binh, T. T. 1991. Effects of Cultivation Types of the Minority People on Sustainable Land Use in the Uplands, Vietnam. In: *Proceedings of a Workshop on Land Use in Vietnam.* Hoa Binh, Vietnam: Forest Science Institute/IIED, 82–88.

Eklof, G.. 1994. Sustainable Agroforestry: Issues and Strategies in Da Bac District, Hoa Binh Province, Vietnam. In: *Proceedings of a Workshop on Sustainability in Uplands Agroforestry Programs.* Hanoi, Vietnam: Agricultural Publishing House, 18–19.

FAO (Food and Agriculture Organization of the UN). 1990. *Farmer Systems Development: Guidelines for the Conduct of a Training Course in Farming Systems Development.* Rome, Italy: FAO.

Garrity, D., L. Fisher, and M. Cairns. 1996. Farmer-Generated Innovations towards Improved Fallows in Southeast Asia. Concept paper.

Gradwohl, J., and R. Greenberg. 1990. Long-term Cultivation of Swidden Fallows by Bora Indians, Peru. In: *Saving the Tropical Forests.* 127–129.

Huyen, T. G. 1991. The Status of Land Use in Vietnam and Policies. In: *Proceedings of a Workshop on Land Use in Vietnam.* Hoa Binh, Vietnam and London: Forest Science Institute of Vietnam/IIED.

Neave, I., and B. N. Quang. 1994. Lessons Learned from a Community Forestry Project in the North of Vietnam. Paper No. 5. (funded by International Care).

NFTA (Nitrogen Fixing Tree Association). 1985. Leucaena, *Wood Production and Use.* Waimanalo, Hawaii: NFTA.

Older, Carine, et al. 1993. Linking with Farmers: Network for Low External Input and Sustainable Agriculture. In: *ILEIA Reading in Sustainable Agriculture.* Intermediate Technology Publications.

Rerkasem, K., and B. Rerkasem. 1994. *Shifting Cultivation in Thailand: Its Current Situation and Dynamics in the Context of Highland Development.* Forestry and Land Use Series No 4. London: IIED.

Sam, D. D. 1994. Shifting Cultivation in Vietnam: Its Social, Economic and Environmental Values Relative to Alternative Land Use. Forestry and Land Use Series No. 3. London: IIED.

Thang, N. V. 1995. The Hmong and Dao Peoples in Vietnam: Impacts of Traditional Socio-economic and Cultural Factors on the Protection and Development of Forest Resources. In: *Proceedings of a Workshop on the Challenges of Highland Development in Vietnam.* Honolulu, Hawaii: Program of Environment, East-West Center, 101–120.

Thong, N. 1994. The Fixed Cultivation and Sedentarization Campaign in Son La Province. In: *Proceedings of a Workshop on Sustainability in Uplands Agroforestry Programs.* Hanoi, Vietnam: Agricultural Publishing House, 59–62.

Toan, B. Q. 1991. Planning for Sustainable Land Use in the Hill and Mountain Areas. In: *Proceedings of a Workshop on Land Use in Vietnam.* Vietnam and London: Forest Science Institute of Vietnam/IIED, 39–48.

Trung, T. V. 1970. The Forest Vegetation Types in Vietnam. Hanoi, Vietnam: Scientific-Technical Publication House.

Ty, Hoang Xuan. 1986. Soil Conditions for *Styrax tonkinensis* Planting in the North of Vietnam: A Case Study. In: *Selected reports of the Forest Science Institute of Vietnam.*

———. 1987. Soil Conditions for Planting *Styrax tonkinensis* for pulp wood in the North of Vietnam and the Effects of *Styrax pure* Stands on the Soil Properties. In: *Selected Works of the Forest Science Institute of Vietnam.* Hanoi, Vietnam: Agricultural Publishing House.

———. 1994. The Role of Leguminous Trees in Sustainable Land Use in the Northwest Region, Vietnam. In: *Proceedings of a Workshop on Sustainability in Uplands Agroforestry Programs.* Hanoi, Vietnam: Agricultural Publishing House, 25–27.

———. 1996. Matching Trials of Leguminous Tree Species for Soil Improvement in Wastelands of Vietnam. Results of a Case Study.

Warner, K. 1991. *Shifting Cultivators: Local Technical Knowledge and Natural Resource Management in Humid Tropics.* Community Forestry Note 8. Rome: FAO.

Chapter 56

Rattan and Tea-Based Intensification of Shifting Cultivation by Hani Farmers in Southwestern China

Xu Jianchu

The forest cover in the Xishuangbanna Prefecture of China's Yunnan Province has decreased significantly, from 62.5% in 1950 to only 33.9% in 1985. According to the government census, the population has increased from 200,000 to nearly 800,000 over roughly the same time. Indigenous people, particularly swidden cultivators, are often blamed for destroying the forests of Xishuangbanna.

The concept of carrying capacity is often applied in measuring population densities in agroecosystems. However, the potential for indigenous technological innovation and institutional evolution to contribute to increased carrying capacity is often ignored. For instance, swidden farmers have often accumulated comprehensive indigenous technical knowledge about crop selection, cropping patterns, and crop and land rotations. They have also learned how to use different microenvironments and niches within swidden agroecosystems in response to changes in biophysical or socioeconomic conditions. In addition, they have developed practices for enhancing forest regeneration. They protect useful tree species through successive swidden cycles, combine annual crops with perennial tree crops, selectively weed their fields to preserve forest tree seedlings, and plant favored trees and plants for both economic and ecological benefits. As stated by Warner (1991):

> The swidden cultivator's goal is not to destroy, but through clearing and then managing the regeneration of the forest, to obtain a continuous harvest of cultigens on the way to a new forest of rich diversity, containing stands of trees that are highly valued.

The Hani, who are also known as Akha, have cultivated rattan (*Calamus* spp.) in fallow fields for about one hundred years in Mengsong, and they have grown tea plantations in both natural forest and swidden fallows for about eight hundred years in Nannuoshan. That they began planting rattan in fallow fields in the first place was due to a scarcity of rattan in the wild, illustrating the fact that indigenous innovations in swidden cultivation are often triggered by resource scarcity. However, these Hani practices, now that they are mature and proven, can be extended into other degrading swidden systems.

Xu Jianchu, Department of Ethnobotany, Kunming Institute of Botany, The Chinese Academy of Sciences, Heilongtan, Kunming, Yunnan 650204, China.

Research Methods and the Study Area

This study involved the following methods:

- Key informant interviews. The traditional village chief and government officials were interviewed in group meetings in order to understand customary and statutory institutions, practices, and management of agricultural and forest lands.
- Household interviews. These were used to reveal the rationale underlying farmers' decisions on land use and management and to document indigenous technical innovations in swidden cultivation.
- Genealogical survey. The Hani have no written language, but a recitation of ancestral lineage can provide detailed information about a tribe's history. The history of the village, particularly that pertaining to conservation and cultivation of rattan, was related by older men and women during group discussions.
- Household sampling. A household questionnaire was used to gather data on cash income from nontimber forest products, animal husbandry, and cash crops.
- Participatory mapping. A group of local villagers was invited into the fields to assist with mapping local zones for different land uses and distribution of important resources, such as rattan and tea gardens.

Two Hani administrative villages, Mengsong, bordering Myanmar in Jinghong county, and Nannuoshan, in Menghai county, were chosen as case studies because of their cultural identity and because they were representative of local biophysical conditions. Both are in Xishuangbanna Dai Autonomous Prefecture.

Mengsong is composed of 10 natural villages, including nine Hani villages and one Lahu village. In 1992, they had a total population of 2,698 people, living in 540 households, and a total area of 100 km^2. This made the population density 27 persons per km^2.

Nannuoshan is composed of 12 natural villages, which had a total population of 3,726 in 710 households in 1992, and a total area of 88 km^2. This made the population density 42 persons per km^2. Nannuoshan's territory is covered mainly by evergreen broadleaf forest, but pine forest is interspersed through both study villages. Elevation ranges between 800 and 1800 m above sea level (asl).

Social Structure

The Hani probably originated from central Yunnan about one thousand years ago (Ma 1983) but were forced by wars and land scarcity to migrate south, finally reaching the uplands of Northern Thailand and Myanmar by the early 20th century. According to the 1990 census, there was then a total population of about 1.2 million Hani, living mainly in the uplands of Yunnan. Most of them live in montane subtropical areas and practice subsistence farming on terraced paddies. The composite swidden system typically practiced by the Hani includes traditional tea gardens in forests, intensive paddy cultivation, home gardens, shifting cultivation, and grazing livestock.

The Hani are patrilineal and tend to live together in clans. The position of the village chief, or *zoema*, is either hereditary or filled through selection by the most knowledgeable and prosperous villagers. Although the village chief is often highly respected, that does not necessarily mean that he acts as a ruler. The chief makes decisions through consultation with members of the male clan, called the *pamou*, and heads of households. Thus, a communal committee consisting of the village chief, the headman or *palu*, and the heads of clans consults to deal with the daily affairs of the community, such as delineation of sanctuary forest boundaries or site selection for swiddening within the forest (Xu 1991; Xu et al. 1995).

The Hani are generally animistic in their beliefs, with a strong emphasis on the cult of ancestors, as evidenced by their strict protection of cemetery forests and attribution of supernatural powers to familiar objects such as trees or animals (Lebar et al. 1964). They believe that any disturbance or violence inflicted on the supernatural, including ancestral spirits, will cause illness.

Customary and Statutory Institutions on Forest Management

Customary institutions play a very important role in forest resource management in Hani society. The Hani often classify all forest lands into swidden fields, or *dongya,* water source forests, village aesthetic forests or *puchang,* road protection forests, economic forests, cemetery forests, and holy forests. They have detailed regulations for management of their forest lands, including designation of swidden areas, fire protection, and punishment for illegal cutting. The regulations have a rather complex system of penalties, paid either in money or in goods such as pigs, according to the type of crime committed.

Results

Conventional Shifting Cultivation Systems, or Dongya

The literal translation of dongya in the Hani language is "nonirrigated uplands." The Hani practice a very diverse cultivation system in which they produce a mix of subsistence and cash crops under complex patterns of intercropping, crop rotations, and fallow management. The location of swiddens and the cropping patterns adopted depend on soil fertility, relief, distance from the houses, and even village arrangement. The Hani like to live at high elevations with pleasant lower temperatures. Swidden sites are opened at lower elevations, often within two hours' walking distance, because the more fertile soils and warmer temperatures there provide better crop yields. The Hani prefer a 13-year fallow period, but this depends on the availability of land. Some useful trees are often carefully preserved in swidden fields to serve as shade trees during farming, as mother trees for seed dispersal, and to provide fodder, fruit, or timber.

Rattan at Mengsong: From Open Access to Strict Protection

According to ethnographic evidence, the Hani were among the earliest dwellers in Mengsong. They are thought to have lived there for 12 generations, or about 240 years. Before their arrival, the area was clothed in dense primary forest rich in rattan.

According to customary laws, all forested land and the products from it belonged to Dai headmen, who lived in the lowlands near Mengsong. The Hani, as later arrivals, did not own the land and its forests, but only had rights to use them. In return, they paid a tax to the Dai headmen in the form of wild game, rattan, and other valued products. Soon after their arrival at Mengsong, the Hani often gathered young rattan tips as a vegetable to exchange for rice with the Dai in the lowlands. This was a very important source of livelihood, particularly for the poorer sector of Hani communities. Usually, one basket of rattan tips was exchanged for about three kilograms of rice. Such was its value that overcollection eventually led to depletion of natural rattan stocks in surrounding forests.

As rattan resources declined under the onslaught of overcollection, a Dai headman in the lowlands, drawing on his power as ruler over minorities living in the uplands, designated forest areas with high rattan stocks as a sanctuary in which rattan collection was banned. The declaration of the sanctuary forest was made about 150 years ago and it is still respected, despite the fact the forest covers more than 200 ha, with an annual potential of 10 tons of cane production.

The customary regulations governing use of the rattan sanctuary forest, or *sangpabawa,* are as follows:

- Cutting any trees is strictly forbidden.
- Harvest of rattan canes or young shoots is prohibited. However, rattan seeds may be collected for cultivation, as well as medicinal plants and timber for coffins.
- The local community is allowed to collect small amounts of rattan to repair farm tools before the planting season.
- The Hani are also permitted to collect rattan for constructing swings during their *yeku* swing festival in July.
- Villagers may collect rattan as a binding material when building new houses, but they must first obtain permission from the community committee.
- Anyone violating the rules will be fined one pig and wine for a village feast.

Rattan Cultivation in Swidden Fallows

The Hani swidden cultivators in Mengsong have interplanted rattan into their swiddens for about a hundred years, as a response to the scarcity of natural stocks. Their indigenous rice-rattan swidden agroforestry is called the *qaiya-aneya* system. When farmers open new swidden fields for upland rice, they interplant rattan seeds, particularly near remaining stumps. After several rice harvests, the land is left fallow and rattan cane can be harvested after 7 to 10 years. *Qaiya* refers to the grain crop or upland rice stage. This is often intercropped with tubers such as yams and taro and Cucurbitaceae such as *Cucumis hystrix* and *Benincasa hispida*. The main function of the qaiya stage, and the shifting cultivation system of which it is a part, is production of food crops.

Although the Hani have reportedly used more than 100 varieties of upland rice in the past, this number appears to have been reduced to about 25 at present. This degree of agrodiversity allows careful selection of different rice varieties to best fit different agricultural microenvironments, with varying soil fertility, moisture, slope, and altitude. The qaiya stage continues for two to three years, and when yields decrease, the food crops are rotated with rattan, bamboo, or even fruit trees. The Hani also plant perennial crops at the perimeters of their upland rice fields during the first cropping year. With these well established, the qaiya gradually evolves into *aneya*, a rattan and bamboo-based fallow (Weinstock 1983). About seven years later, the farmer can enjoy the cash income from harvesting both the rattan and the bamboo, while once more preparing to plant annual crops.

Tea Plantations in Swidden Fields in Nannuoshan

Tea gardens are the main land use resulting from fallow enrichment in Xishuangbanna. The original distribution of tea is believed to have centered on southwest China, northern Myanmar and northeast India. One of the oldest known tea trees is found in Nannuoshan. It is thought to be about 1,500 years old and has a diameter at breast height of 1.5 m. Oral history suggests that it was planted by ancestors of the Bulang, another ethnic minority in Xishuangbanna, before the arrival of the Hani.

The tea plantations in Nannuoshan can be classified into two main types: traditional and improved (see Table 56-1).

Table 56-1. Land Use in Nannuoshan

Land Use	*Area (ha)*	*% of Total Area*
Traditional tea gardens	320	3.6
Improved tea gardens	360	4.1
Active swidden fields	500	5.7
Paddy fields	302	3.4
Total cultivated area	1,482	17.0

Source: Nannuoshan administrative village, 1995.

Traditional Tea Gardens

Traditional tea gardens, in both natural forests and fallow fields, have been managed in Xishuangbanna Prefecture for hundreds of years by many ethnic groups, including the Bulang, Jinuo, and Hani. Both the Hani and the Bulang make room for tea planting by thinning out some shrubs in natural forests. Alternatively, they plant tea directly into swidden fallows at about 2 by 2.5 m spacings, or about 2,000 plants per ha. Useful tree species are often preserved as an upper story, including *Cinnamomum glanduliferum,* for food seasoning; *Docynia indica*, an edible fruit; *Schima wallichii,* for its timber and fire resistance; *Bauhinia variegata,* for its edible flowers; and *Castanopsis* spp., for their timber and nuts. In the middle layer of the forest, tea bushes, mixed with rattan, thrive in the shade of the upper canopy. Useful shade-tolerant plants, such as *Baphicacanthus cusia,* for its dye, and *Pandanus tonkinensis,* a source of fiber, can often be found in the understory. No chemical fertilizers or pesticides are used in traditional tea gardens. Management is minimal, usually amounting to weeding once a year and some pruning, and production is low. The cultivator always has secure rights to access and harvests all cash crops and trees that he manages in swidden fields or natural forests. The tea gardens are inherited by male members of the family.

Improved Tea Gardens

In the 1980s, with assistance from research technicians, local farmers began to plant tea in very compact rows along the contours of fallowed swidden fields. The shade trees recommended for interplanting were *Cinnamomum spp., Poularia* sp., *Melia* sp., and walnut, at a density of 120 trees/ha. Driven by increasing land scarcity, this model of improved tea garden has replaced most traditional tea gardens in Nannuoshan because of its superior production, although it depends on high inputs of chemicals and labor. The area covered by traditional tea gardens has decreased significantly, from 700 ha to about 320 ha. Because of the many "mother trees" growing in traditional tea gardens, they were easily converted into forests for firewood and timber production. The area of forest land has therefore increased as traditional tea gardens have been abandoned, and, because many swidden fields have been taken out of arable cropping and converted into improved tea gardens, Hani farmers have channeled some of their increased cash income into constructing more paddy fields and reestablishing their food security.

Discussion

Benefits from Intensified Fallow Management

The Hani in Xishuangbanna Prefecture continue to depend for their livelihood and cash income on livestock and the cultivation and gathering of nontimber forest products. Table 56-2 shows that nontimber forest products such as tea, rattan, bamboo, mushrooms, and fruit account for half of the total cash income for the Hani in Mengsong. Pig husbandry is the biggest income earner, providing 42.4% of income. The actively cropped and fallowed swidden fields provide large quantities of forage, such as maize and banana stems, suitable for pigs. There is also a large demand for pig meat from workers who have settled in Mengsong but cross the border to work in mines inside Myanmar.

Table 56-3 shows that the household variation in rattan cultivation is greater than tea planting in the swidden fields. Cash income from tea is more stable and equitable than that from rattan.

Although the Hani in Mengsong have a long tradition of cultivating rattan in their swidden fields, there are only a few craftsmen in each village with the necessary skills to process it into handicrafts. Those families who lack the skills to make rattan stools and other handicrafts hand over their rattan canes to craftsmen for weaving and then receive half of the finished products back. Pieces of rattan furniture, and particularly rattan stools, are very popular gifts and property among the Hani. Rattan stools and tables are commonly given by parents as wedding gifts to young couples. The stools are often buried with the dead as personal property. Therefore rattan has important cultural and symbolic value in Hani society. Due to constantly increasing market demand for rattan furniture, some farmers have started to plant rattan for commercial purposes. One farmer in the case study area harvested a total of 3,000 kg of rattan in 1988 and 1989, earning about US$900.

Table 56-4 shows that rice yields from paddy fields are higher than those from swidden fields. When swiddens are opened from forest, their production is marginally higher than those opened from grasslands. The difference is more dramatic when measured in terms of working-day productivity. Rice paddies provide much higher returns to labor than upland swidden fields, particularly those opened from grasslands. The net income per working day ranks highest for tea picking when compared with that from either irrigated paddy or swidden fields.

Table 56-2. Sources of Cash Income for Hani in Mengsong, 1990

Income Source	*Units*	*Production*	*Price/Unit (US$)*	*Income (US$)*	*%*
Tea	kg	39,900	0.6	23,940	15.0
Fruit	kg	412,500	0.06	24,750	15.5
Rattan cane	kg	39,000	0.3	11,700	7.3
Bamboo	piece	23,180	0.4	9,272	5.8
Dry bamboo shoots	kg	2,240	0.4	96	0.5
Mushroom collection	kg	2,656	1.1	2,942	1.8
Rattan and bamboo processing	piece	5,496	1.4	4,894	3.1
Pigs	head	1,129	60	67,740	42.4
Cattle	head	171	80	13,680	8.6
Total income				159,814	100.0
Income per capita				US$61 per capita	

Table 56-3. Household Sampling on Cultivation of Rattan and Tea in Mengsong Swidden Fields*, 1990

Parameters	*Total*	*Average*	*CV (%)*
Population	181	5.84	32
Labor force	100	3.23	46
Productive rattan (clumps)	883	28.48	129
First harvesting year	320	11 years	36
Self-consumption of rattan (kg/year)	808	26.06	128
Productive tea	4,721	168.61	84

Note: *Number of households = 31.

Table 56-4. Productivity of Different Activities, 1990

Activities	*Labor Input*	*Yield (kg/ha)*	*Productivity (kg/working day)*	*Income (US$/day)*
Rice paddy	315 days/ha	2,985	9.5	1.34
Rice in grassland swidden	660 days/ha	2,235	3.4	0.48
Rice in forest swidden	375 days/ha	2,385	6.4	0.90
Picking tea (dry weights)	1 day/load	250	3.0	1.80

Ecological Sustainability

In their farming practices, indigenous people often imitate natural ecosystems. The Hani practices of rattan and tea cultivation in swidden fallows are good examples. Both are well adapted to the local ecosystem. As well as swidden fallow tea cultivation, farmers in Xishuangbanna Prefecture have been imitating the natural ecosystem by planting tea in forests for more than 1,000 years. Within their traditional tea gardens, they have preserved shade trees with economic and cultural value. The forest structure of traditional tea gardens can be divided into three layers: the shade canopy, from 15 to 35 m high; the tea layer, from 2 to 8 m; and a ground layer that includes herbaceous plants and tree seedlings. The canopy provides conditions of ideal sunlight and humidity for growing tea. No chemicals are needed in traditional tea gardens because tree litter recycles nutrients and natural predators control insects.

As for rattan, Hani farmers in Mengsong report that it needs a minimum of 11 years to mature. Therefore, its cultivation in fallow fields is well matched to their traditional practice of a 13-year swidden cycle. Mature fallows, rich in rattan, provide both economic returns and good soil fertility when opened for the next cropping phase.

Capacity to Support an Increasing Population

Changes in indigenous technical knowledge are inevitable, particularly when driven by population growth and increasing demands for cash. In Xishuangbanna, the extensive traditional tea garden is being replaced by a more productive and intensified version that relies on chemicals for both replenishment of soil fertility and pest control. Production from an improved tea garden can be six times higher than its traditional forerunner (see Table 56-5). As a consequence of their relative inefficiency, many traditional tea gardens are being converted back into forest for timber and firewood production.

At one time, tea ranked as Yunnan's foremost agricultural product. It was exported, both domestically to other provinces in China and internationally to other countries in Southeast Asia. However, a recent slump in the international market for tea has seen a reduction in exports and tea has slipped down to fourth position on the table of Yunnan's agricultural products.

Extension, Conservation, and Knowledge Transfer

The indigenous practice of rattan cultivation in swidden fallows, developed by Hani communities in Mengsong, has been successfully transferred to Mingzhishan administrative village of Simao Prefecture and other Hani communities through cross-farmer visits and training. After negotiating with the Bureau of Forest, permission was also granted for a trial replication of the technology on degraded state forest lands near a Bulang community in Menghai county, Xishuangbanna.

Table 56-5. Comparison of Traditional and Improved Tea Gardens in Nannuoshan

Parameter	*Traditional Tea Garden*	*Improved Tea Garden*
Planting method	random, scattered	contour planting
Management	extensive	intensive
Chemical fertilizers	none	1,500 kg/ha
Yield (dry)	250 kg/ha	1,500 kg/ha
Income	1,500 yuan/ha	9,000 yuan/ha

Note: Based on field interviews in 1993.

Wild rattan resources have been overharvested in both protected and unprotected areas of Yunnan, and it is now difficult to collect mature rattan seeds for propagation purposes. In response to this depletion, the Kunming Institute of Botany has developed tissue-culture technology for both field propagation of rattan and germplasm conservation.

Adoption: Ecological Niches

Both rattan and tea cultivation are well adapted for integration into swidden agroecosystems in tropical mountainous areas. "Invisible" traditional tea gardens under the natural forest canopy in Xishuangbanna have contributed significant ecological and socioeconomic benefits, including control of soil erosion, biodiversity conservation, aesthetics, subsistence use, and cash generation.

Rattan, one of the most valuable nontimber forest products in Southeast Asia and southwestern China, is adaptive to a range of ecological niches in tropical areas because of the diversity of species and their respective ecologies. There are eight species recommended for planting at different elevations in Xishuangbanna (Chen et al. 1993). Five species recommended for altitudes lower than 1,000 m asl are *Calamus gracilis, C. aff. multinervis, C. wailong, C. rhabdocladus,* and *C. platyacanthus,* and those for higher elevations are *C. yunnanensis, C. nambariensis* var. xishuang-bannanensis, and *Plectocomiopsis himalayana.*

Adoption Constraints

Indigenous and state resource management systems differ and clash in many areas. Indigenous systems of classifying forestlands are more resource based or function oriented, such as the rattan reserve forests, *sangpabawa*, or the aesthetic forests, *puchang*. The government, on the other hand, is more focused on controlling use of forest lands rich in timber. Many forest lands have been declared protected areas or state forests, meaning that, by law, local people are no longer allowed to collect or plant nontimber forest products in these areas. At the same time, planting these long term cash crops is no longer possible in swiddens because fallow periods have, in many cases, been reduced to the point where they are now too short. Scarcity of rattan seeds and seedlings is also a big constraint to the extension of this technology into new areas, and farmers without a tradition of cultivating rattan need to learn about its propagation, transplanting, and general management before committing their time to its cultivation. Because rattan is a relatively long-term crop, secure land tenure is also essential if farmers are to be encouraged to plant it in their fallows and forest lands.

Research Priorities and Experimental Agenda

Future research should aim to answer the following questions:

- What are the inputs and outputs of rattan and tea-based fallows, in comparison to other on- and off-farm activities?
- What are the socioeconomic conditions conducive to adopting such fallow management systems?
- How do the products flow? Who receives benefits, and at what percentage, through the processing and marketing chain, and which marketing systems are most beneficial to smallholders?
- What kinds of technical and credit assistance are needed by local farmers?
- Does government policy on land tenure and marketing encourage or discourage the planting of rattan and tea in swidden fallows?

Conclusions

Population growth, including in-migration, often leads to resource depletion. As reported in this chapter, the Hani have responded to these pressures by controlling access to certain areas and imposing supportive sanctions. At the same time, they have intensified their management of swidden fallows. The integration of rattan and tea into fallows provides both environmental and economic benefits to local communities and holds promise for replication in other upland areas of Southeast Asia. However, these technologies have been developed under, and are dependent upon, local institutional arrangements for property rights and marketing. Therefore, as well as the benefits mentioned above, the integration of rattan and tea into fallows is also an intentional strategy to strengthen tenure security. Government policy impinging on these key factors may undermine the socioeconomic stability of rattan and tea production systems. When policies are being crafted with a view to encouraging a more sustainable use of upland resources, policymakers should first seek insights from indigenous institutions regulating resource management.

Acknowledgments

The author gratefully acknowledges the generous support of the Ford Foundation, Beijing, for the project "Biodiversity in Swidden Agroecosystems in Xishuangbanna, China," under which some of the field data presented in this chapter were collected from 1993 to 1995.

References

Chen, S.Y., S.J. Pei, and J.C. Xu. 1993. Indigenous Management of the Rattan Resources in the Forest Lands of Mountain Environment: The Hani Practice in the Mengsong Area of Yunnan, China. *Ethnobotany* 5, 93–99.

———. 1993. Ethnobotany of Rattan in Xishuangbanna. In: *Collected Papers on Studies of Tropical Plants*, Vol. II. Kunming, Yunnan, China: Yunnan University Press, 75–85.

Lebar, Frank M., Gerald C. Hickey, and John K. Musgrave. 1964. *Ethnic Groups of Mainland Southeast Asia*. New Haven, CT: Human Relations Area Files Press.

Ma, Y. 1983. *A Brief Introduction of Yunnan History*. Kunming, China: Yunnan People's Press (Chinese language).

Pei, Shengji. 1982. Some Effects of the Dai People's Cultural Beliefs and Practices upon the Human Ecology in Southeast Asia. In: *Cultural Values and Human Ecology in Southeast Asia*, edited by K.L. Hutterer et al. East-West Environment and Policy Institute.

Warner, K. 1991. Shifting Cultivators: Local Technical Knowledge and Natural Resource Management in the Humid Tropics. FAO, Rome, 20-47.

Weinstock, J.A. 1983. Rattan: Ecological Balance in a Borneo Rainforest Swidden. *Economic Botany* 37(1).

Xu, J.C. 1991. Study on Indigenous Agroecosystems in a Hani community. Unpublished. MSc thesis presented to the Chinese Academy of Sciences.

———, S.J. Pei, and S.Y. Chen. 1995. From Subsistence to a Market-Oriented System. In: *Regional Study on Biodiversity: Concepts, Frameworks and Methods*, edited by S. Pei and P. Sajise. Kunming, China: Yunnan University Press.

Chapter 57

Indigenous Fallow Management Systems in Selected Areas of the Cordillera, the Philippines

Florence M. Daguitan and Matthew A. Tauli

Research has confirmed that traditional shifting cultivation practices employed by peasants of the Philippines Cordillera are not necessarily destructive. On the contrary, many of the systems used in the cultivation of their swiddens, locally called *uma,* have a strong conservation orientation (Angelo and de los Santos 1987), and, as a consequence, uma cultivation has been sustainable over a long period. This is attributed to the wisdom of the farmers, their careful use of resources, and their practice of controlled burning. As Olofson described it in 1981:

> Burning requires a good deal of skill and precise evaluation of the microenvironment and the general climatic context to make certain that a thorough and even fertilizer layer is achieved and that adjacent forest and dwellings are not accidentally damaged. Burning among traditional shifting cultivators is therefore controlled.

After several harvests, the traditional shifting cultivator of the Cordillera lets the uma lie fallow for 7 to 20 years to allow the soil to regenerate its fertility. However, in recent times, farmers have had to shorten the fallow period because they have fewer and fewer swidden sites to cultivate. This has been due to population increase; the incursion of extractive industries into agricultural areas, particularly commercial logging and mining; and the conversion of many traditional farming areas to large-scale, cash crop production. The shorter fallow periods have resulted in a significant decrease in crop yields. Nevertheless, uma harvests continue to make a major contribution to the food needs of the local population.

Given this context, shifting cultivation is often accused of being the main culprit in forest degradation and many development planners consider it inappropriate to contemporary conditions. They have campaigned indiscriminately for its replacement with intensive cash crop cultivation, and this usually requires heavy use of external inputs.

The Montañosa Research and Development Center (MRDC) advocates an alternative approach. It believes that agricultural development should build on traditional practices that have sustained rural communities for centuries. Therefore,

---

Florence M. Daguitan and Matthew Tauli, Montañosa Research and Development Center, Inc. (MRDC), Makamkamlis, Sagada 2619, Mountain Province, the Philippines.

it sees the need to develop a detailed knowledge of traditional farming practices in selected areas and to determine their underlying principles, so that better informed recommendations can be made for the development of shifting cultivation.

This Study, its Limitations, and the Study Area

The methodology of this study was limited to key informant interviews and actual observation. The data have been processed by the research team, but the resulting analyses have not yet benefited from group discussions in the communities where the research was undertaken.

The Northern Luzon Cordillera

The Cordillera Administrative Region of the northern Philippines is located within the coordinates 16°15' to 18°30' N latitude and 120°30' to 121°45' E longitude, on the island of Luzon (Figure 57-1). The topography is an assortment of rugged mountain peaks, narrow river valleys, sloping foothills, and flatlands. It constitutes a critical watershed for six major river systems vital to the economy of northern Luzon. The Cordillera has two distinct seasons: wet and dry. It is the coldest region in the country, with monthly mean temperatures ranging from 16.1°C in January to 18.8°C in May.

Research was conducted in six areas in the three provinces of the region. In addition, information from interviews conducted in Bayyo, and Samoki, Bontoc, both within the Mountain Province, was incorporated in this study.

Almost all of the areas listed in Table 57-1 are *barangays*, the smallest administrative unit in the Philippines. The exception is Mangali, which is a tribal unit composed of three barangays. Information from the barangay of Pidlisan is true for the whole area occupied by the Pidlisan tribe, which is composed of four barangays.

Agricultural systems in the study areas are primarily subsistence oriented. Farming methods are characterized by low levels of external inputs and are heavily influenced by cultural beliefs and practices. An exception is Ankileng, where most of the farmers have adopted commercial vegetable production as a second crop in their ricefields, with high inputs of inorganic fertilizers and pesticides. Despite this development, traditional farming practices continue.

Table 57-1. Study Areas and Their General Features

Name of Area	*Location*	*Population*	*Elevation (m asl)*
Pidlisan, Sagada, Mountain Province	17°08' N latitude 120°54' E longitude	369	1,000–2,000
Ankileng, Sagada, Mountain Province	17°04' N latitude 150°54' E longitude	763	1,000–1,800
Maligcong, Bontoc, Mountain Province	17°07' N latitude 121°59' E longitude	713*	1,000–1,500
Bangbang, Hungduan, Ifugao	16°50' N latitude 120°03' E longitude	210	800–1,800
Ngibat, Tinglayan, Kalinga	17°15' N latitude 121°05' E longitude	167	900–1,000
Mangali, Tanudan, Kalinga	17°10' N latitude 121°17' E longitude	2,390	600–1,000

Note: *Year 2000 population, the rest of the table is year 1995 population.

Figure 57-1. The Cordillera Administrative Region, Showing Research Areas

Fallow Management Practices

Use of Tithonia diversifolia

Across the Mountain Province, wild sunflower *(Tithonia diversifolia)* is widely used as an organic fertilizer. It is used in three different ways:

- During preparation of both irrigated and nonirrigated land, slashed sunflower plants are directly incorporated into the soil as green manure;
- *Tithonia diversifolia* often serves as a main component of compost, being mixed directly into compost piles on farms; and
- *Tithonia* is often thrown into pigpens to be incorporated into pig manure.

In order that it is readily available for these purposes, *Tithonia* is planted as hedgerows, near stonewalls, and at the edges of rice paddies, home gardens, and *camote* (sweet potato) fields. The Bontok people also plant *Tithonia* next to streams and rivers. The aerial biomass is cut at designated times and immersed in the water to accelerate decomposition, and the released nutrients are carried downstream and into canals irrigating rice paddies.

Farmers at Maligcong, Ankileng, and Pidlisan make intensive use of *Tithonia diversifolia* as a green manure crop to rebuild soil fertility in fallowed fields. When opening swiddens, part of the final phase of land preparation is the planting of wild sunflower cuttings. The cuttings, about 0.5 m in length, are planted in an "X" formation or in parallel strips on the steepest portion of the uma. After the first crop is harvested, in four to nine months, the sunflowers are pruned and their biomass is incorporated into the soil in the more gently sloped parts of the field. Alternatively, it may simply be laid on the ground and left to decompose. Management varies according to the crop grown. When pruning the *T. diversifolia*, at least 0.5 m of stem is left to ensure rapid regrowth. Systematic pruning of the sunflowers in this manner, each time leaving the biomass in the field, is a routine part of swidden management.

During the fallow period, the *Tithonia* is left to multiply and choke out *cogon* (*Imperata cylindrica*) and other weeds. It is a common practice at Bayyo for farmers to plant additional sunflower cuttings within the swidden to increase the density of the *Tithonia* stand before leaving the field fallow. This ensures that *Tithonia* dominates the fallow succession (see color plates 17 and 18 for examples in Mindanao).

T. diversifolia has the following characteristics that recommend it as an effective fallow species:

- It produces large amounts of biomass in a short time;
- Its dense canopy protects the soil from direct sunlight, thus providing a more favorable environment for soil organisms and, at the same time, preventing the growth of *Imperata*; and
- Its widely distributed roots trap debris and inhibit downslope soil movement.

However, as a fallow species, *T. diversifolia* is limited by its shallow root system, which is unable to recycle leached nutrients from lower soil depths. Despite this limitation, field trials involving a two-year *T. diversifolia* fallow showed a marked increase in yield from a subsequent crop of *bush sitao (Vigna sinensis),* one of the most widely grown legume vegetables in the Cordillera region (Montañosa Research and Development Center 2002). When planted half a meter apart over the experimental plots, the *T. diversifolia* fallow resulted in a virtual doubling of yield from the subsequent sitao crop. Wider spacing of the *Tithonia* resulted in a lesser increase in crop yield. Researchers reported that the trial plots also showed a marked increase in soil organic matter after the two-year *Tithonia* fallow.

Management of Trema orientalis *(Anablon) as a Fallow Crop*

Trema orientalis, like *Tithonia diversifolia*, is fast growing and accumulates biomass rapidly. At Bangbang, Hungduan, in Ifugao Province, shifting cultivators collect *T. orientalis* seeds in August and sow them in swidden fields at the start of the fallow period.

The Use of Pigeon Peas as Both a Main and a Fallow Crop

At Ngibat, Kalinga, farmers who plant pigeon pea (*Cajanus cajan*) as a food crop in the cropping phase extend its use into the fallow as a green manure crop. *C. cajan* is planted throughout the swidden at intervals of roughly one meter, with no fixed alignment. Short-term crops such as legumes, corn, or root crops are interplanted with the pigeon pea. After the second year's harvest, the *C. cajan* is pruned to a height of about one foot above the ground, and the field is planted to other crops for the last time. Management of the lopped biomass differs. Some farmers distribute it over the ground as mulch, while others throw it to the edge of the field or even discard it. By the time the field is left fallow, the *C. cajan* has once more formed a canopy sufficient to protect the soil. Farmers at Ngibat claim that by integrating pigeon pea as an improved fallow, their land regains its fertility after two to three years, allowing an intensified use of scarce land.

The Apa System of the Mangali Tribe

The apa system originates from the belief that any misfortune befalling a farmer or *kaingero*, or any member of his family, is caused by spirits dwelling within or near the swidden, or *kaingin*. If, for instance, there is a death in the family, a portion of swidden land may be declared off-limits to appease the spirits, and cultivation, cutting firewood, or any other use of the area banned in order to give the spirits peace and freedom from disturbance for an extended period (Daguitan and Maduli 1992). Land areas set aside under the apa system range from 150 to 350 ha, and complete abstinence from any economic activity is believed to be crucial.

One 67-year-old informant recalled three apa declarations in her lifetime. The first apa began when she was a teenager and the land was off-limits for more than 20 years. The second lasted for 10 years, and the third began in 1979 and was still in force at the time of this study.

Of the tree apas mentioned, the second was considered unsuccessful for not having reached the desired 20 years. It involved an area of more than 150 ha near the village. Because of its easy access, there was pressure form neighbors who wanted to cut firewood before the apa period had elapsed. The declarent was forced to bow to their demands, and, despite the threat of ill fortune, he undertook the elaborate rituals to seek the permission of the spirits and reopened the area prematurely. Aside from the supernatural aspects of the case, there was marked increase in the first harvest following the reopening, and this reinforced the Mangali people's belief in the apa system.

Casual observation seems to support claims that Mangali swidden areas have been better conserved than those of neighboring tribes, where significant areas of swidden land have degraded into grassland and farmers have encroached upon critical watershed areas, threatening water supplies.

As well as the apa system, other forms of indigenous resource management have helped the Mangali tribe's ability to maintain a relative balance in their ecosystem. The use of some uma land is intensified by Mangali farmers who gradually enrich the plant diversity by planting root crops, legumes, leafy vegetables, fruits, and other useful species. The result is the creation of semipermanent umas called *amak*, which, in part, compensate for the productivity lost by setting aside apa land. Table 57-2 details the surprising plant diversity found in one such amak.

Table 57-2. Partial Inventory of Plants in a 5,000 m^2 Amak

Local Name	*Common Name*	*Scientific Name*
Plants used as hedgerows		
atolba		*Callicarpa* sp.
bubog		*Erythrina* sp.
daun		*Hibiscus tiliaceous* (Linn.)
kakaute	madre de cacao	*Gliricidia sepium* (Jacq.)
pulawol	wild sunflower	*Tithonia diversifolia*
isit		*Ficus* sp.
bayabat	guava	*Psidium guajava*
bugas	prickly narra	*Pterocarpus echinatus*
kawayan	bamboo	*Bambusa* sp.
buwa	betel palm	*Areca catechu*
Wild vegetables		
burburtak		*Bidens pilosa*
allagiya		*Nasturtium indicum*
amti*		*Solanum nigrum*
Mushrooms		
kudilap		
bulung		*Perotus* sp.
kesop		*Schizophyllum commune*
kulat		*Lentinus sajor-caju*

Note: *Amti* is considered semi-wild. Although it grows naturally, the berries are gathered, crushed, and scattered over fields to ensure a continuous supply.

Conclusions

The study revealed four systems of fallow management. The first involves planting biomass-producing plants such as *Trema orientalis* and *Tithonia diversifolia*, and the second, planting, and extending into the fallow period, leguminous crops such as *Cajanus cajan*. The *C. cajan* canopy protects the soil from direct sunlight and rain impact, while its root nodules fix atmospheric N. There is wide scope for refining both of these fallow management approaches, and there may be potential for integrating components of both systems, that is, planting both leguminous and biomass-producing plants.

The third is the apa system, a type of forced long fallow. Through its belief in the spirit dwellers, it demonstrates respect for the environment. It also shows community solidarity, supporting the apa declarant's relationship with the rest of the village and the preeminence of long-term sustainability over short term gain. It is a noteworthy example for policymakers, development planners, and farmers alike.

The fourth is the amak system, which is an intensified use of uma land. The land is planted to a variety of food crops, and over time, fruit-bearing and forest tree species are added, along with biomass-producing shrubs.

The study of indigenous farming systems offers many lessons in maintaining soil fertility, managing fallows more effectively, and maintaining an ecological balance within a land-use system. In the present context of limited agricultural lands, research should be focused on workable pathways to intensify shifting cultivation, and on alternative agroforestry options. This work should begin with a detailed assessment of indigenous farming systems and be tailored to site-specific conditions.

References

Angelo, J., and Aloma M. de los Santos (eds.). 1987. *Igorot: A People Who Daily Touch the Sky*. Comtemporary Life and Issues Vol. 3. Baguio City, Philippines: Cordillera Schools Group.

Daguitan, E., and P. Maduli. 1992. Upland Rice Cultivation in Mangali (Kalinga-Apayao),

Montañosa Research and Development Center. 2002. Study on the Effectiveness of Sunflower as Follow Crop in Uma Cultivation. Research report submitted to the International Development Research Centre (IDRC), Ottawa.

Montañosa Research and Development Center occasional paper, August 1992. MRDC.

Olofson, H. (ed.) 1986. *Adaptive Strategies and Changes in Philippine Swidden-Based Societies*. Laguna, the Philippines: Philippine Forest Research Institute.

Chapter 58

Management Systems in Occidental Mindoro, the Philippines

Michael Robotham

The spread of upland cultivation in the Philippines and the concurrent trend toward more intensive land use have led to a number of warnings about negative consequences (Sajise 1986, Garrity et al. 1993). Similar pressures on upland areas are being felt elsewhere in Southeast Asia, and the concerns are no less disquieting (Garrity 1993). In the Philippines, the upland population is approximately 30 million and is growing at nearly 3% per year (Garrity et al. 1993). Migrants from lowland areas make up the fastest-growing segment of this population, a reality seemingly diminished by the fact that most studies of upland farming systems have focused on ethnic minority groups.

Migration from lowland to upland areas in the Philippines has occurred more or less continuously since the early 20th century (Cruz 1986). As a result, there are members of lowland ethnic groups who have lived in the uplands for many years, and at least some of them have developed successful, productive, and potentially sustainable land management strategies. This chapter, therefore, describes and analyzes the local land management systems used by lowland migrants now living in two small upland communities in the municipality of San Jose, Occidental Mindoro.

Methods

This analysis uses data collected from informal interviews conducted during visits to the area in 1996, as well as information from a formal survey of all 63 resident households in November of the same year. Additional information includes a general soil fertility assessment, based on analysis of surface soil samples collected from various land uses and landscape positions. This was supplemented by my observations and discussions with residents during 1996 and my experience in the area as a Peace Corps volunteer from 1988 to 1990. Climate and land-use information was also provided by Philippine government agencies. The data was collected and compiled as part of a larger study on Philippine upland management systems that formed the basis for my doctoral dissertation (Robotham 1998).

Michael Robotham, Department of Natural Resources and Environmental Management, University of Hawaii at Manoa, 1910 East-West Road, Honolulu, HI 96822, USA.

The Study Site

This case study focuses on land management systems used in two small upland communities, Imbarasan and Himamara, located in the municipality of San Jose, Occidental Mindoro. Mindoro is one of the major Philippine islands and is located south of the central part of the island of Luzon. The province of Occidental Mindoro occupies the western half of the island. The municipality of San Jose is located in the southwestern corner of the province (Figure 58-1). Imbarasan and Himamara are located adjacent to each other in the foothills, about 10 km east-northeast of San Jose town.

The communities have about 300 ha of land, consisting of small valleys bisecting relatively low but steep hills, ranging in elevation from 20 to 200 m above sea level (asl). About 65 households are resident in the area for at least part of the year. They are all members of lowland ethnic groups. Most of the older residents came to the area from Panay Island between 1960 and 1980, and many of the younger residents are the children of these original migrants. They face significant constraints on their land management systems.

Occidental Mindoro is characterized by a strong monsoonal climate. Although the average annual rainfall in the San Jose area is nearly 2,400 mm (PAGASA 1996), it is strongly seasonal (see Figure 58-2). The long dry season, from January through April, severely limits cropping options. Fifty-six of the communities' 63 farmers have small parcels of valley land, averaging 1.1 ha in size, which can be used for rainy season rice production. However, all but two of them have upland parcels averaging 2.0 ha in size that they rely on for a significant percentage of their annual production.

Most upland parcels are not well suited for annual crop cultivation. In an area where 95% of cultivated uplands are said to have slopes exceeding 30%, they are steep and the soil has low fertility. Although acidity is not a problem, the soils are generally deficient in potassium and phosphorus.

Access to much of the area is difficult, especially during the rainy season. An all-weather road runs along the southern boundary of the area and is serviced by public transportation to and from San Jose at least twice daily. However, the distance from individual land holdings to the road must be traversed on foot or by buffalo and may take several hours. As a consequence, transport of farm products to markets is often extremely difficult and costly.

The earliest settlers say the area was forested until the start of commercial logging in the early 1950s. Logging continued through the mid-1970s. Now the only remaining high forest is in scattered, inaccessible patches further into the mountains. Imbarasan and Himamara were completely logged out by the early 1960s. However, the land remains classified by the Philippine government as forest land and is therefore under the jurisdiction of the Department of Environment and Natural Resources (DENR). Imbarasan and Himamara are an Integrated Social Forestry Program (ISFP) site. Through participation in this program, a majority of the residents have gained some tenure security by obtaining renewable 25-year leases on their current holdings.

Results

Most farmers in the two communities use a combination of land management practices. These include the traditional practice of long fallow shifting cultivation of annual crops, as well as the following alternatives:

- Retention of beneficial species throughout the annual cropping period;
- Planting of leguminous perennials at the end of the annual cropping period; and
- Planting fruit trees, forest trees, or other perennials such as bamboo during the annual cropping period.

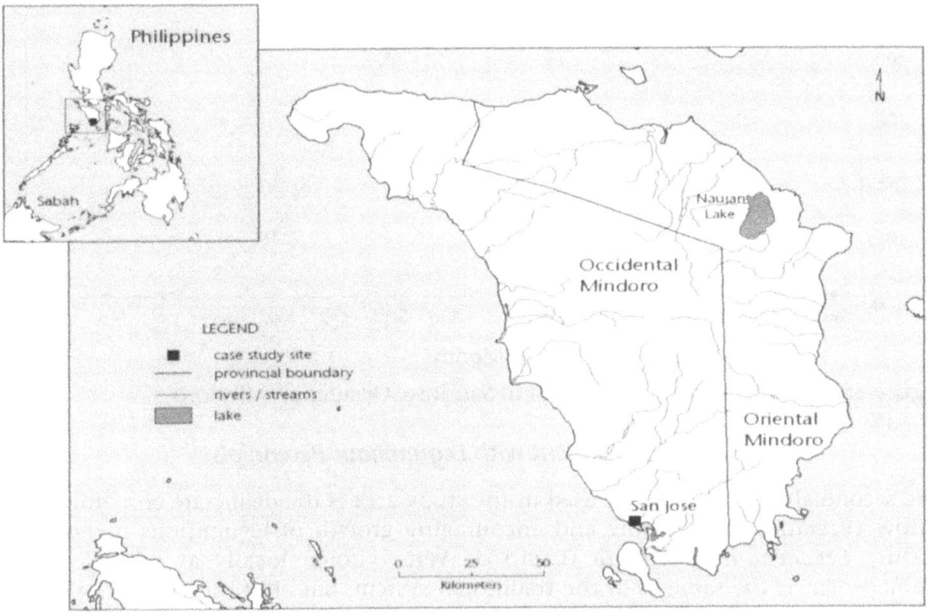

Figure 58-1. Case Study Location: San Jose, Occidental Mindoro, Philippines

Traditional Shifting Cultivation

The traditional practices are similar to the shifting cultivation systems described for other areas of the Philippines (Conklin 1957; Olofson 1981; Lopez-Gonzaga 1983) and elsewhere in the humid tropics (Warner 1991). Typical shifting cultivation plots are about 0.5 ha in size. Existing vegetation, usually secondary forest, is slashed and burned on the field during the dry season. Crops are planted after the first rains in late May. Farmers generally grow one crop per season, although, if the rains are early, some manage to grow a second crop of corn. The land is left fallow over the dry season, at the end of which the vegetation is slashed and sometimes burned before the cultivation of a second year of annual crops. Depending on crop yields, this process may be repeated for a third year, but two years is more common.

With sufficient land available, farmers prefer to leave the plot fallow for at least seven or eight years before returning to plant annual crops. In this system, fallow vegetation is subject to low levels of management. Farmers retain control over the area throughout the fallow cycle, and often harvest useful forest products such as fruit and wood for charcoal production or cooking fuel.

Retention of Useful Species

The least intensive alternative management practice now being used is the retention of useful species, primarily *buri* palm (*Corypha ulan* Lam.), throughout the entire shifting cultivation cycle. Buri palm fronds are used for roofing and weaving material, the stem fibers are used for rope, and the sap is used to make vinegar and an alcoholic beverage. When a patch of land is slashed to prepare it for annual cultivation, the scattered buri palms are pruned. They are sometimes as dense as 400 trees per ha but are usually more sparse. Since buri is fire resistant, the palms are not killed when the slash is burned. The buri is cut back again before the second annual crop is planted. Buri trunks are sometimes used during the annual cropping cycle as supports for climbing plants such as squashes and beans.

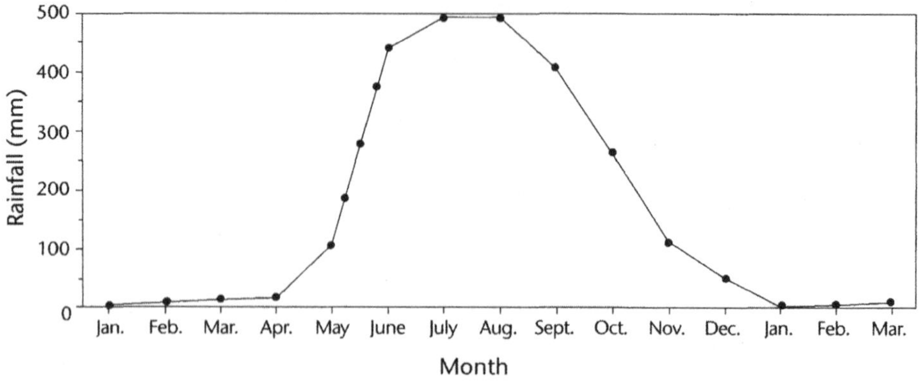

Figure 58-2. Average Monthly Rainfall in San Jose, Occidental Mindoro

Fallow Enrichment with Leguminous Perennials

The second alternative practice used in the study area is the deliberate enrichment of fallow vegetation by planting and encouraging growth of leguminous perennials, mainly *Leucaena leucocephala* (Lam.) de Wit., known locally as *ipil-ipil*. Land management is the same as in the traditional system, but after the cropping period, farmers encourage the establishment of ipil-ipil as the dominant fallow species, usually by broadcasting locally collected seed.

Fruit Trees, Forest Trees, and Bamboo

The third alternative practice is the incorporation of fruit trees, forest trees, or bamboo—or a combination of all three—into the management system. Tree or bamboo seedlings are interplanted with the annual crops during the first or second year of cultivation. They are then managed more or less intensively throughout the fallow period. Traditionally, farmers have planted small numbers of trees on the borders of fields and near house lots and have managed bamboo clumps. While these practices continue, a small number of farmers has recently decided to plant larger numbers of seedlings with the goal of converting a shifting cultivation plot into a managed orchard, bamboo plantation, or woodlot. Three fruit species, banana (*Musa* spp.), mango (*Mangifera indica* Linn.), and cashew (*Anacardium occidentale* Linn.), are most common. However, survey respondents reported planting more than 20 different fruit tree species. Typical forest species include *melina* (*Gmelina arborea* Roxb.), a fast-growing timber and pulpwood tree that has been promoted by the Philippine Government, and mahogany (*Swietenia macrophylla* King). The main bamboo species being planted on a large scale is *Gigantochloa levis* (Blanco) Merr., known locally as *patong*.

Frequencies of Use and Mix of Practices

The communities of Imbarasan and Himamara appear to be in an early stage of transition from the traditional long-fallow shifting cultivation of predominately annual crops to a more diverse set of alternative management practices. The majority of farmers is using one or more of these alternatives on at least part of their land, with the most common being the retention and management of buri palm (84%), the cultivation of fruit trees (83%), and the cultivation and management of bamboo (83%).

Most residents are managing a relatively small number of palms, assorted fruit trees, and bamboo. Twenty-one survey respondents (33% of the total) are managing 50 or more buri palms, 13 (21%) are managing more than 50 banana plants, 12 (19%) are managing 10 or more mango trees, 13 (21%) are managing 10 or more cashew trees, and 9 (14%) are managing at least 20 clumps of bamboo. Over recent years, residents have sold buri, bananas, mangos, cashews, bamboo, and small amounts of other fruits. Although these sales can potentially produce significant income, only one family has converted to a land-use system based exclusively on perennials, in this case, bananas.

Other alternative practices are also being incorporated into the existing management systems. Nearly half of the resident families manage ipil-ipil and two-thirds of these sell charcoal or firewood harvested from their trees. Twenty-five percent of households manage some species of forest tree, and four households have followed local extension recommendations and planted melina. One of these four households has made a major shift toward forest tree cultivation by planting 1,000 melina and 1,000 ipil-ipil trees on its upland holdings.

Discussion

Traditional Management Practices

As long as sufficient land is available, traditional management practices appear to be ecologically sustainable and capable of providing long-term subsistence production. They closely mimic the management strategies that have been used by the ethnic peoples of Mindoro for centuries (Conklin 1957; Lopez-Gonzaga 1983). Based on soil sampling from fallowed and secondary forest areas, the present fallow period of seven to eight years appears to result in a significant increase in available nutrients in surface soils. Although the productivity of these plots is low, the effort and monetary resources required for their management are also low. Most farmers continue to grow native or seminative crop varieties and use few, if any, outside inputs.

During interviews and informal discussions, residents identified three major weaknesses in the traditional fallow rotation system. First, the annual crop productivity on upland plots was low, even in the initial year of cultivation. Average maize yields were reported to be about one metric ton per hectare and, for upland rice, about 600 kg/ha. Second, there was low overall production from upland holdings because the fallowed plots provided few usable products beyond fruit and firewood. Third, the traditional system provided insufficient cash income to meet household needs such as school expenses.

Residents also said that, even if they possessed both lowland and upland parcels, the upland plots provided the only venue for experimentation and innovation in their farming practices. Cultivation on most valley parcels was constrained by a lack of water in the dry season and was limited to paddy rice in the wet season because of too much water.

Overall Productivity of Land Holdings

A common reason given by farmers for switching to more intensive management practices is the need for increased productivity, defined here as production of usable biomass, such as grain, fruits, or forest products, from a given area of land. This increased productivity is usually demanded to support an increasing population (Boserup 1965). The population of Imbarasan and Himamara is increasing and residents acknowledge rising land-use pressures. Although ample secondary forest land is available for clearing and cultivation further up into the mountains, its use is technically illegal. These laws, however, are seldom enforced. More important reasons why new land is not cleared include poor access, the presence of malaria, and the belief that antigovernment rebel forces of the New Peoples Army (NPA) still occupy the area. At the time of this study the presence of the NPA was a big concern. However, rather than fearing for their personal safety, most local residents were concerned about harvests being collected as "taxes" and of being caught in the middle of encounters between NPA fighters and government soldiers.

All of the alternative systems discussed in this chapter provide equal or higher levels of productivity when compared with the traditional system, at least in the medium term, and would conceivably allow the area to support a higher population. However, all of the alternative practices involve a shift away from subsistence production, and any larger population in these upland areas would be left dependent on the outside world for basic foodstuffs.

Financial Returns

All three types of systems described here provide tangible economic benefits to farming households, in excess of the benefits from the traditional shifting cultivation system. The annual returns and cumulative net present value (NPV) of production from an average upland farm of two hectares were estimated over a 20-year time frame for five land management systems: traditional; traditional plus buri palm retained during cropping (buri); traditional with *Leucaena*-enriched fallow (*Leucaena*); banana orchard (banana); and bamboo plantation (bamboo). Three discount rates, 13%, 27%, and 80%, were used. These approximate the interest rate on savings certificates, the bank lending rate for large borrowers, and the lending rate from local credit sources, respectively. For the sake of this financial assessment, the total land holding was assumed to be divided into five identical 0.4 ha parcels. The yearly management sequence used on the first of the parcels is shown in Table 58-1. The start of this cycle is shifted two years into the future for each of land parcels 2 through 5. Years 11 to 20 are a repeat of years 1 to 10 in the original, buri, and *Leucaena* systems, and of years 9 and 10 in the bamboo and banana systems.

The following assumptions were made in calculating the annual returns and cumulative net present value for the five land use systems. All monetary calculations in Table 58-2 are in Philippine pesos. At the time of the study, US$1 was worth about ₱ 27.

Traditional System. Shifting cultivation of maize on 0.4 ha for two years, eight years of fallow.

Benefits. Maize yield of one ton per hectare in the first year. Yield in the second year is equal to the first year's yield multiplied by the yield decline factor.

Costs. Ten days' labor for clearing fallow every other year, 25 days' labor for cropping maintenance and harvest in all years.

Table 58-1. Land-Use Sequences for Land Parcel 1 Under Various Management Systems

	Year									
System	*1*	*2*	*3*	*4*	*5*	*6*	*7*	*8*	*9*	*10*
Original	M	M	F	F	F	F	F	F	F	F
Buri	M/Bu	M/Bu	Bu	Bu	Bu	Bu	Bu	Bu	Bu	Bu
Leucaena	M	M	L	L	C	C	C	C	C	C
Banana	M	M/Bn	Fr	Fr	Fr	Fr	Fr	Fr	Fr	Fr
Bamboo	M/Bm	M/Bm	M/Bm	P	P	P	P	P	P	P

Notes: M = Maize; F = Unmanaged fallow; Bu = Buri harvested; L = *Leucaena*-dominated fallow; C = Charcoal from *Leucaena*; Bn = Banana; Fr = Banana fruit; Bm = Bamboo; P = Bamboo poles.

Table 58-2. Prices, Costs, and Factors Used in Calculating NPV

Selling prices	
Maize	₱ 10.00 per kg
Buri	₱ 2.50 per frond
Charcoal	₱ 50.00 per 20 kg sack
Banana	₱ 0.50 per piece
Bamboo	₱ 50.00 per pole
Wage rate	₱ 70.00 per person-day
Transport costs	
Charcoal	₱ 5.00 per sack
Banana	₱ 5.00 per 250 pieces
Bamboo	₱ 5.00 per pole
Other factors:	
Yield decline factor in shifting cultivation	0.75
Competition factor for buri	0.75
Competition factor for trees, year 1	0.90
Competition factor for trees, year 2	0.75
Leucaena fallow enrichment factor	1.1

Buri System. Shifting cultivation as in traditional practice, but buri retained at 200 palms/ha.

Benefits. One buri frond harvested/palm/year. The maize harvest is reduced by the buri competition factor.

Costs. Equivalent labor to the traditional swidden system. No transport cost.

***Leucaena* System.** Shifting cultivation as in the traditional system, but the fallow is enriched with *Leucaena*. A yearly charcoal harvest starts in the third year of fallow.

Benefits. Shifting cultivation as before, but the land parcel is planted to *Leucaena* after two years. One hundred sacks of charcoal are harvested/ha/year, beginning in the third year after the *Leucaena* is planted. The yield of maize is multiplied in years 11 to 20 (the second cycle) by the *Leucaena* enrichment factor.

Costs. Shifting cultivation costs as before. No cost for establishment. During charcoal production, labor costs of 10 person-days/plot/year are incurred. Transportation costs are as above.

Banana System. Shifting cultivation crops are planted first on each parcel of land, then bananas are planted in the second year at a density of 600 plants per hectare. The first bananas are harvested in the third year.

Benefits. Shifting cultivation yields in first year as in the traditional system. Second-year yields are depressed by both the yield reduction and tree competition factors. Bananas are harvested at a rate of 100 pieces per plant, beginning in year 3 on land parcel 1.

Costs. Shifting cultivation labor as in the traditional system for the first 10 years. In years 2, 4, 6, 8, and 10, there are 30 extra person-days of banana planting labor. In years 3 through 20, there is maintenance labor of 30 person-days/ha/year and harvest labor of 50 plants per person-day. Transport costs are as above.

Bamboo System. Shifting cultivation crops and bamboo are planted in the first year on each parcel. Bamboo is planted at a density of 400 clumps/ha. It is first harvested in the fourth year.

Benefits. The same as the traditional system, but with tree reduction factors in first 10 years. Bamboo is harvested at a rate of one pole/clump/year, starting in year 4.

Costs. Shifting cultivation labor as in the traditional system for the first 10 years. In years 1 through 9, 45 person-days/year of labor is provided for nursery and clump establishment. Harvest labor is provided at a rate of 25 poles/person-day in years 4 through 20. Transport costs are as above.

The banana-based system produced the highest cumulative NPV over the 20-year scenario, under all three discount regimes. The original and buri-based systems produced very similar returns. These were slightly less than the returns to *Leucaena* cultivation at lower discount rates (13% and 27%) and virtually equivalent at the high rate (80%). Because of high estimated establishment costs, the bamboo-based system produced high cumulative NPV at the lower discount rates, but it actually produced a negative cumulative NPV at the high rate (see Figure 58-3).

Although cumulative NPV calculations provide a good overall picture of the system, they may obscure important characteristics of the distribution of benefits and costs. A graphical representation of the annual returns to each management system (Figure 58-4) illustrates these characteristics. One important observation is that the banana and bamboo systems actually require an investment of capital, and they have negative returns in the initial stages of cultivation. This is likely to be a severe and possibly insurmountable problem for cash-poor farmers.

Residents of the two communities cited two other systems as potential land management options: the cultivation of mango trees for fruit production and the cultivation of melina for pulp and timber production. Rough calculations of relative returns suggest that these systems may be very lucrative. Based on estimated growth rates and farmer-reported prices, gross returns from a melina plantation harvested on a seven-year cycle could be as high as 1 million pesos/ha/harvest. For a 10-year-old mango plantation, gross returns, again based on local yield and price estimates, could be as high as 1.5 million pesos/ha/year. However, the data necessary to develop even an estimated production budget for either of these enterprises was lacking, particularly data related to harvest, transportation, and processing costs. As a consequence, these systems were not considered in this analysis.

Diversity

Another important criterion mentioned by residents is increased diversity of food and income sources. In the traditional system based on annual cropping, maize and root crops grown on upland plots provide a valuable source of food when rice is not available. They are also regularly used as food for livestock and can be sold on a limited basis. Because they augment, but do not eliminate, the traditional practice of annual crop cultivation, buri and *Leucaena,* when included in the management system, appear to increase income source diversity with minimal effects on food source diversity. However, even though fruits can be consumed by the household, the practices that lead to a complete shift to perennial crop-based systems may increase the diversity of income sources but decrease the diversity of basic food sources.

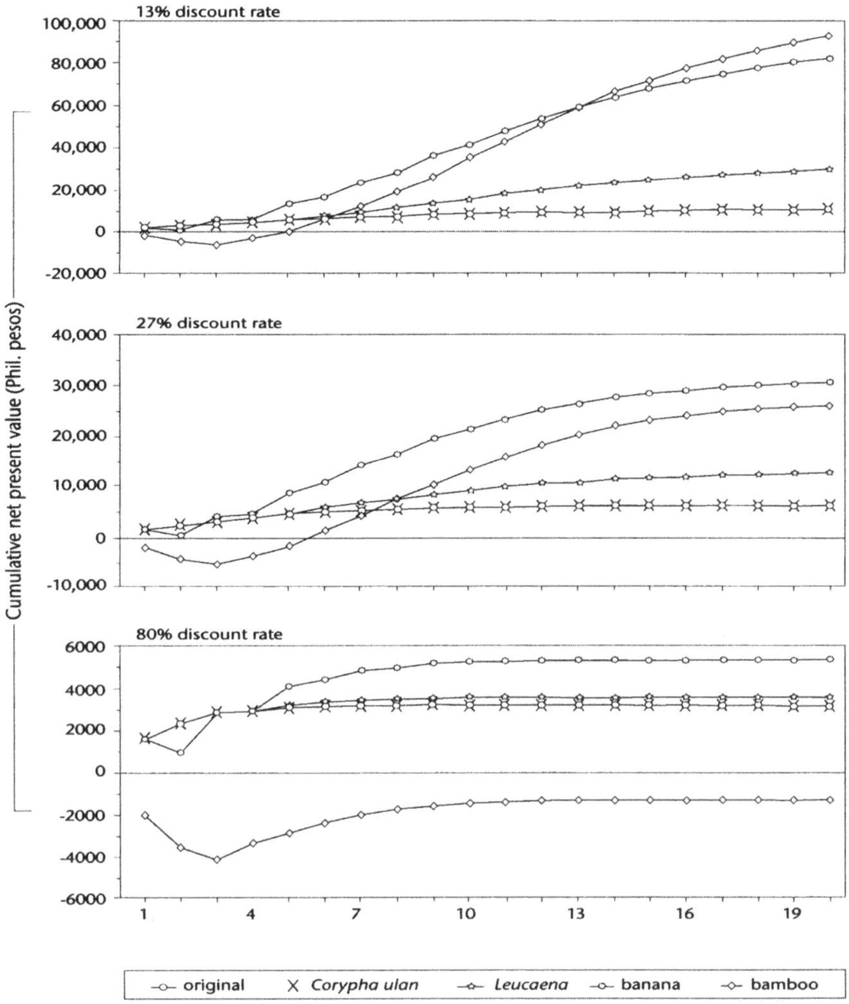

Figure 58-3. Cumulative NPV of Compared Management Systems

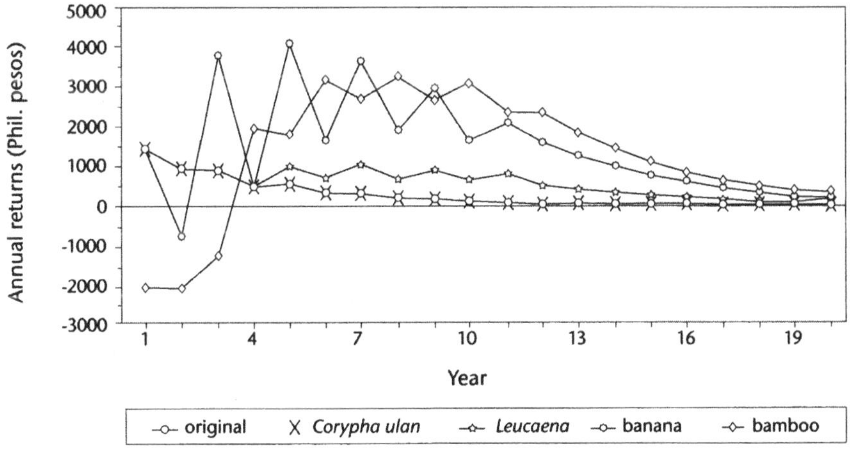

Figure 58-4. Annual Returns from Management Systems at 27% Discount Rate

Flexibility

Another issue related to diversity is flexibility. It will be divided into three subissues: land management flexibility on an annual or seasonal basis; flexibility in the use of other productive resources, particularly labor; and flexibility of harvest. With regard to land management, the traditional practices are very flexible. They allow residents to change strategies year-to-year in response to weather conditions, household needs and preferences, and other factors. The buri and *Leucaena* systems are also very flexible. However, management practices based on fruit trees, timber trees, or bamboo tie up land in one specific use for a long time.

The degree of flexibility in the use of other productive resources, especially labor, also differs between management practices. Buri harvest and charcoal making are generally dry season activities and use labor when it is more available. If perennial species such as fruit trees are grown, labor will be required during the annual crop cultivation season. However, tree planting is the only activity that will require significant extra labor since land clearing and weeding will be necessary whether trees are present or not.

Both cashew and mango trees produce fruit in the dry season when harvest labor is most likely to be available. Bananas, however, fruit year-round and so harvest needs may conflict with the labor needs of annual crops. Although the labor market is not well developed in the study area, laboring jobs are available on a semiregular basis throughout the dry season on larger, irrigated farms closer to San Jose town. If residents are to widely adopt alternative land-use practices, they must perceive the performance of additional on-farm labor as more productive than working off-farm for wages.

An additional component of flexibility is the level of flexibility in harvest time. The traditional practice has relatively low flexibility since crops are usually harvested only once each year, near the end of the rainy season. Buri palm management is the most flexible practice since it can be harvested at any time. Because of heavy and continual rain during the wet season, charcoal production using traditional methods is mainly a dry season activity, although firewood may be harvested year-round. Fruits need to be harvested when they are ripe. For bananas, this occurs throughout the year. Mangos and cashews, on the other hand, have well-defined fruiting periods near the end of the dry season. Timber species offer some flexibility in harvest, because after the trees are of marketable size, any number of them may be harvested at any time. However, access and transportation difficulties are likely to confine timber harvests to the dry season.

Other Factors

There are several other factors that may influence residents' adoption of various alternative management practices. These include their vision of the future, initial investment capital, infrastructure, information availability, and management skills. Since several of the alternative practices that appear to be the most lucrative only produce significant benefits several years into the future, residents' vision and initial investment capital may be important considerations. The family that has planted more than 2,000 tree seedlings repeatedly discussed their vision of the trees providing them with income in their retirement and a legacy for their children. Others who were planting tree seedlings expressed similar sentiments. However, the families also said that the initial capital needed to start tree orchards or plantations was a significant constraint. In addition, the lag time between tree planting and harvest was also a major concern. The couple that is converting their land into a tree plantation has found off-farm employment to support themselves while the trees mature. A more common strategy has been to plant trees on small plots of land or in marginal areas that have produced very poor annual crop yields.

Infrastructure and the availability of information present other potential difficulties for residents of the two communities. The adoption of some alternative practices brings a need to transport products out of the area for sale and, possibly, the need to transport management inputs in. This would be very difficult without improvements in the current infrastructure, including roads, bridges, markets, and secondary processing facilities. A successful switch to a more market-dependent system also requires access to timely and accurate information on markets, prices, potential management problems, and a variety of other issues. At present, this access does not exist.

In addition, most residents of Imbarasan and Himamara do not possess the skills necessary to successfully and profitably manage systems based on fruit and timber trees. Skills that are currently in short supply include the identification, prevention, and treatment of pest and disease problems and better silvicultural techniques, including grafting of fruit tree seedlings and pruning of both fruit and forest species.

Summary of Reasons for Adoption of Alternative Systems

Land management in Imbarasan and Himamara appears to be at a transition point between traditional shifting cultivation based primarily on annual crops and a more diverse set of management systems based on perennial species. The primary factors driving this transformation are concerns about the continued low productivity of land holdings and the need for additional cash income to meet household expenses. However, the present mix of practices is unlikely to produce either the highest productivity or the optimal cash income. This, and interviews with residents, suggest that issues such as the complementarity of labor demands and the level of initial risk and investment play a pivotal role in decision making. The alternative practices adopted by nearly all residents, management of buri and of limited clumps of bamboo and fruit trees, generally have low risk and initial investment and require little additional labor.

Several factors seem likely to retard more extensive adoption of some of the more productive alternative systems. First, the local environment severely limits options to two fruit tree species, mango and cashew, except for specific microenvironments that are suitable for bananas. Second, the infrastructure, information, and management skills necessary for successful and profitable management of fruit and forest tree-based systems are in short supply.

Unique and Generalizable Elements

The management systems being used in Imbarasan and Himamara are not unique in either the Philippines or Southeast Asia. The retention of valuable perennial species is common to many traditional shifting cultivation systems throughout the humid tropics (Warner 1991). Fallow enrichment with leguminous species and their potential use for firewood or charcoal has been documented in other locations (Grist et al. 1997, MacDicken 1990, 1991). Upland dwellers throughout Southeast Asia have traditionally planted fruit trees in fallow areas (Garrity 1993, Warner 1991), and there are other examples of transition from a shifting cultivation system to one based on fruit trees (Fujisaka and Wollenberg 1991). However, this case study provides an opportunity to examine a community that appears poised to intensify a shifting cultivation system based on annual crops and long fallow rotation. It also illustrates that absolute land scarcity is not the only driving force for management intensification. In Imbarasan and Himamara, the potential shift to perennial crop-based systems is being driven by residents' cash income needs and by the generally low productivity of their existing system. Although there is a level of perceived land scarcity due to nonpopulation factors, including malaria and insurgency, residents did not cite lack of land as a primary reason for adopting alternative land management practices.

This example illustrates what may be the first stages in a gradual change in management strategies. As such, it does not appear to be applicable to situations where the existing management systems are already under severe stress. Additionally, the management alternatives discussed here have been developed in the context of a moderate level of basic resources. They may not be well suited to the rehabilitation of a severely degraded resource base.

Key Leverage Points to Encourage Wider Adoption

There are several key leverage points that appear likely to have positive impacts on the adoption of alternative fallow management practices. On the policy side, the provision of adequate infrastructure is certainly the direct responsibility of the local and national governments. Government can also formulate policies that encourage the development of market networks and economic infrastructure, as well as secondary processing facilities for fruits and forest products. Increasing tenure security for local residents may also be beneficial (Riddell 1987), although this was not a major concern for residents in this study.

As far as extension is concerned, the provision of quality seedlings is a necessary prerequisite to encouraging expansion of fruit tree production. Residents in the study area expressed a willingness to use scarce cash to purchase high-quality seedlings if they were available. Extension activities would also be useful to help farmers identify pests and diseases in perennial species, as well as the most appropriate methods of prevention and control. This is especially important for fruit trees. Lastly, the alternative management systems now present incorporate only a small number of fruit and forest species. Extension efforts could be focused on providing both the knowledge and the planting materials for farmers to diversify these systems.

Extension may also be very effective in educating farmers in efficient charcoal production. Given that traditional pit methods achieve only about 25% recovery (PCARRD 1985), there is wide scope for improvement and any increase in efficiency would directly benefit the farmers.

Future Research Priorities and Experimental Agenda

There are a number of areas where additional research could provide the information with which to better understand and improve local land management systems. First, there is a need to better quantify the costs and benefits of alternative practices and of systems integrating several different practices. Implicit in this is the need to better characterize the yields of various alternative products under the conditions and management regimes found in upland areas, as well as positive and negative interactions between various practices. Yields and expenses associated with fruit crop cultivation in other parts of the Philippines do not necessarily reflect the situation of upland smallholders in Mindoro. In the case of buri management, studies have simply not been conducted.

Another potentially valuable area of research is the development of low-cost pest and disease control systems for fruit crops. One constraint to mango cultivation mentioned frequently by local residents was the perceived need for high-cost chemical pesticides to ensure a marketable harvest. The identification of alternative crops and trees and the development of simple methods of propagation to ensure quality planting materials could also be extremely useful. There is ample room for research on the development of integrated farming systems for upland areas that would include livestock and possibly fish, as well as annual and perennial crops.

Conclusions

Imbarasan and Himamara provide a valuable opportunity to describe and analyze an area where a traditional long-fallow shifting cultivation system is starting to change. Residents are now incorporating at least three alternative practices into their management systems: retention of useful species, fallow enrichment with leguminous perennials, and fruit and forest tree cultivation.

Unlike other upland situations, population pressure does not appear to be a major driving force behind these land-use changes. Instead, farmers cite the need for additional sources of cash income and a desire to improve overall productivity of land holdings as their primary motivations. However, the lack of transportation and a market infrastructure, the initial costs of shifting to perennials, and the added risk associated with not producing basic foodstuffs on the farm have influenced all but a small number of residents to make gradual, rather than extensive, changes to their management systems.

As a group, the residents of Imbarasan and Himamara appear to have taken the first steps on an ongoing journey of experimentation, adaptation, and change that will hopefully lead to new land management systems better able to meet household needs and goals within the constraints and opportunities provided by their upland environment. Only time will tell if they will succeed, but they appear to be headed in the right direction.

Acknowledgments

Field research support for this study was provided by the Philippine-American Educational Foundation (Fulbright–Hays Fellowship). Support for the data analysis was provided by the East-West Center.

References

Boserup, E. 1965. *The Conditions of Agricultural Growth.* London: Earthscan Publications.

Conklin, H.C. 1957. Hanunóo Agriculture. In: *FAO* (Food and Agriculture Organization of the United Nations) *Forestry Development Paper No. 12.* Rome, Italy: FAO.

Cruz, M.C.J. 1986. Population Pressure and Migration in Philippine Upland Communities. In: *Man, Agriculture and the Tropical Forest.* Morrilton, AR: Winrock International.

Fujisaka, S., and E. Wollenberg. 1991. From Forest to Agroforest and Logger to Agroforester: A Case Study. *Agroforestry Systems* 14, 113–130.

Garrity, D.P. 1993. Sustainable Land Use Systems for Sloping Uplands of Southeast Asia. In: *Technologies for Sustainable Agriculture in the Tropics.* ASA Special Publication 56. Madison, WI: American Society of Agronomy.

———, D.M. Kummer, and E.S. Guiang. 1993. The Philippines. In: *Sustainable Agriculture and the Environment in the Humid Tropics.* Washington, DC: National Academy Press.

Grist, P., K. Menz, and R. Nelson. 1997. Multipurpose Trees as Improved Fallows: An Economic Assessment. In: *Imperata Project Paper* No. 97-2. Canberra, Australia: Center for Resource and Environmental Studies, Australian National University.

Lopez-Gonzaga, V. 1983. *Peasants in the Hills.* Diliman, Quezon City, Philippines: University of the Philippines Press, 226.

MacDicken, K.G. 1990. Agroforestry Management in the Humid Tropics. In: *Agroforestry: Classification and Management,* edited by K.G. MacDicken and N.T. Vergara. New York: John Wiley & Sons.

———. 1991. Impacts of *Leucaena leucocephala* as a Fallow Improvement Crop in Shifting Cultivation on the Island of Mindoro, Philippines. *Forest Ecology and Management* 45, 185–192.

Olofson, H. (ed.) 1993. *Adaptive Strategies and Change in Philippine Swidden-based Societies.* Laguna, Philippines: Forest Research Institute, 181.

PAGASA, 1996. Climatological Averages (photocopy). Quezon City, Philippines.

PCARRD (Philippine Council for Agricultural and Resources Research and Development), 1985. The Philippines Recommendations for Fuelwood and Charcoal Utilization. *Technical Bulletin Series* No. 56. Los Baños, Laguna, Philippines: PCARRD, 95.

Riddell, J.C. 1987. Land Tenure and Agroforestry: A Regional Overview. In: *Land, Trees, and Tenure. Proceedings of an International Workshop on Tenure Issues in Agroforestry,* edited by J.B. Raintree. Nairobi, Kenya: International Center for Agroforestry (ICRAF).

Robotham, M.P. 1998. Staying Alive: Sustainability in Philippine Upland Management Systems. Doctoral Dissertation. Department of Agronomy and Soil Science, University of Hawaii, Honolulu, HI.

Sajise, P. 1986. The Changing Upland Landscape. In: *Man, Agriculture, and the Tropical Forest,* edited by S. Fujisaka, P. Sajise, and R. del Castillo. Morrilton, AR: Winrock International.

Warner, K. 1991. Shifting Cultivators: Local Technical Knowledge and Natural Resource Management in the Humid Tropics. In: *Forest, Trees and People Program Community Forestry Note 8.* Rome, Italy: FAO, 80.

Chapter 59

Changes and Innovations in the Management of Shifting Cultivation Land in Bhutan

T. Dukpa, P. Wangchuk, Rinchen, K. Wangdi, and W. Roder

The Himalayan kingdom of Bhutan has an area of about 46,500 km^2 and a population of about 0.6 million. Within this small area, there is a relatively wide diversity of climates, vegetation, and topography resulting from a dramatic range in altitude and rainfall. About 85% of the rural population is engaged in agriculture, which is the mainstay of the country's economy.

Shifting cultivation is still used in many parts of Bhutan (see Figure 59-1 and Table 59-1), with two distinct production systems. A grass-fallow system, known locally as *pangshing,* is practiced at elevations ranging from 2,500 to 4,000 m above sea level (asl). A bush-fallow system, known as *tseri,* is practiced in the subtropical regions of the country. Both systems are well adapted to prevailing conditions and have been self-sustaining for generations. However, increasing population pressure; new rules and regulations limiting access to land; higher economic expectations; rising labor costs; and a gradual change from a closed, subsistence farming system to a market-oriented system are combining to force changes.

The Royal Government of Bhutan is determined to identify land-use options that are suitable to the country's socioeconomic and physical conditions, so poor farming communities can enjoy better living standards. Given this context, the government decided that tseri cultivation should be phased out by the end of 1997, the final year in Bhutan's seventh five-year plan.

The objectives of this chapter are, first, to describe shifting cultivation systems practiced in Bhutan, then to analyze recent changes in these systems and, finally, to investigate causes underlying the changes.

T. Dukpa, Rinchen, K. Wangdi, and W. Roder, Renewable Natural Resources Research Centre Jakar, P.O. Jakar, Bhutan; P. Wangchuk, National Project Manager, Land Use and Natural Resources Planning, Policy and Planning Division, Ministry of Agriculture, Thimphu, Bhutan.

Figure 59-1. Bhutan: Districts Where Tseri and Pangshing Are Concentrated

Table 59-1. Area of Tseri and Pangshing in Bhutan (hectares)

District	*Tseri*	*Pangshing*	*Total*
Samdrupjonkha	9,528	308	9,836
Chhuka	5,609	43	5,652
Samtse	4,964	60	5,024
Trashigang	3,427	852	4,279
Bumthang	458	3,765	4,223
Zhemgang	3,147	47	3,194
Pemagatshel	2,076	50	2,126
Monggar	1,435	300	1,735
Lhuentse	1,346	232	1,578
Trongsa	972	536	1,508
Trashiyangtse	635	758	1,393
Sarpang	1,185	179	1,364
Dagana	1,061	30	1,091
Wangduephodrang	153	635	788
Ha	631	110	741
Punakha	52	502	554
Thimphu	84	430	514
Paro	246	231	477
Tsirang	361	45	406
Gasa	2	8	10
Total	36,763	8,837	45,600

Source: Land documents for each district.

The Study Areas

Investigations for this study were carried out in two regions of Bhutan with high concentrations of tseri and pangshing cultivation. The climate in both areas is characterized by wet summers and dry winters.

Tseri Cultivation Region

Tseri cultivation is concentrated along the foothills of Samdrupjonkha, Chhuka, Samtse, Pemagatshel, and Zhemgang districts and in some parts of Trashigang, Monggar, Lhuentse, and Trashiyangtse. The climate in these areas is wet to dry subtropical with annual rainfall ranging from 4,000 mm in Samdrupjonkha to less than 1,000 mm in rain-shadowed areas of Trashigang and Monggar districts. Many of the tseri cultivation areas are very steep and remote. The nearest road is often a several days' walk away.

Pangshing Cultivation Region

This type of shifting cultivation is practiced at higher elevations, from 2,500 m to 4,000 m asl, in areas with a temperate climate. Bumthang, located in central Bhutan, is the single most important pangshing district, accounting for about 42% of the total pangshing area registered in the country. Soils are generally derived from coarse-grained granite-gneiss and are poor in phosphate but rich in potassium. Road access reached the area in 1974.

Methods

The data in this chapter were generated through the following investigations:

Tseri System

- Previous investigations by the authors during the 1980s (Roder et al. 1992);
- A tseri survey carried out by the Ministry of Agriculture, in which information on the current use of tseri areas and potential future alternatives was collected by visiting every household cultivating tseri in the districts of Samdrupjonkha, Pemagatshel, Trashigang, Trashiyangtse, Lhuentse, Monggar, and Zhemgang; and
- Formal and informal interviews with cardamom growers in Zhemgang district.

Pangshing System

- Earlier research by the authors during the 1980s (Roder et al. 1992, 1993);
- A survey conducted in 1996 by the Renewable Natural Resources Research Centre, Jakar, covering 206 households. A formal questionnaire was used to document changes in soil fertility management in buckwheat cultivation systems (RNR-RC Jakar 1996);
- Investigations carried out by the forestry research section of the Integrated Forestry Development Project involving blue pine increments in a silvipastoral trial (Rosset et al. 1997); and
- Additional information obtained from RNR-RC Jakar and the National Fodder Seed Production Center (Rosset et al. 1997).

Results and Discussion

The Tseri System

This system resembles the widely practiced slash-and-burn systems found throughout many tropical and subtropical regions of Asia. Its characteristics are listed in Table 59-2. The vegetation, consisting of trees, shrubs, and other plants, is cut during the dry season, allowed to dry, and burned shortly before planting. Seeds are either dibbled or broadcast. The main crops include maize, millet, rice, and buckwheat. Because of the short cropping period, the absence of tillage, and the stability provided by roots and residues of the fallow vegetation, crops, and weeds, erosion hazards are minimized in spite of the often steep gradient of the land. The most important labor requirement is for weeding (Upadhyay 1988).

The Pangshing System

The fallow vegetation that develops under grazing consists of short grasses and forbs, interspersed with blue pine (*Pinus wallichiana*). Prior to cultivation, the topsoil layer is dug with a hoe to a depth of five to seven centimeters and is allowed to dry for several months (Table 59-2). Dry topsoil is then collected in conical mounds about 2 to 3 m apart, making between 1,200 and 2,500 mounds per hectare. Small quantities of fuel, consisting of dry manure or blue pine needles collected on site or imported from nearby forest, are added to each mound. After the fuel is ignited, the organic material in the mounds of dry soil continues to burn slowly and temperatures often reach 500°C or higher. Burning lasts for several hours, resulting in high losses of organic matter and nitrogen. Because almost all the vegetation is killed, the soil is fully exposed to erosive forces. Buckwheat seeds broadcast in April may be covered by using either a traditional bullock-drawn plow or a hand rake. In the pangshing system, up to 85% of labor input is required for land preparation. No weeding is necessary.

Recent Changes and Land-Use Options

Road access to markets within Bhutan and in neighboring India has gradually become available over the past three decades. This has dramatically increased the options available to some shifting cultivators (Table 59-3). With access to imported rice, there is no longer a need for self-sufficiency in food production, and some farmers have begun to exploit their special climatic niche. Crops produced in temperate or cooler subtropical hill regions, such as apples, oranges, potatoes, and vegetables, attract premium prices in markets on the nearby Indian plains. The fastest changes have taken place in the pangshing regions, where potatoes have become a popular and profitable cash crop with immediate returns. In 1987, only 15 years after the road was opened, 50% of all households in Bumthang were cultivating potatoes (Guenat 1991). Other options include perennial fruit trees, timber plantations, and livestock production, but most require substantial investment before returns can be expected and have, therefore, been adopted at a much slower rate. With an expected change in forestry laws that will afford farmers full rights to use and sell timber, the option of timber plantations is soon expected to become very attractive in many areas. (Table 59-4 shows the rate of conversion of pangshing areas in Bumthang district, while Table 59-5 lists the potential economic returns from various land-use options.)

Alternatives for Tseri Systems

Since the completion of the tseri survey, which covered eight districts with wide areas of slash-and-burn agriculture, substantial areas have already been converted into permanent cropping, paddy land, orange orchards, or cardamom plantations. Although they are already converted to other uses, these areas remain recorded as tseri in official government records. Options for future land use include conversion to dryland cultivation of maize, millet, barley, and buckwheat, orange and cardamom plantations, private forestry, pasture, conversion to paddy fields, and bamboo. The feasibility and economic benefits of these options depend largely on access to markets and on location-specific climatic conditions. Areas where slash-and-burn agriculture still predominate are often very far from roads capable of carrying vehicular traffic. Another fact that should be noted while considering alternative land uses is that preliminary analyses suggest that about half of tseri land is on slopes steeper than 50%. More detailed research is needed to identify those options that are not only best suited to the physical conditions but are also socially, economically, and environmentally sound.

Table 59-2. Characteristics of Shifting Cultivation Systems in Bhutan

Characteristic	*Tseri*	*Pangshing*
Altitude range (m asl)	300 to 2,500	2,500 to 4,000
Major crops	maize, millets	buckwheat
Major fallow vegetation	shrubs, trees	grasses, blue pine
Fallow period (years)	2 to 8	6 to 20
Aboveground biomass when burning (t/ha)	> 10 (range 5 to 40)	< 2 (range 0.5 to 15)
Origin of N and C losses by burning	aboveground biomass	soil organic C
Erosion hazard	low to medium	high
Major limitations to longer cropping periods	weed competition, topography	phosphate, nitrogen
Labor input (days/ha)	290	380
Labor productivity (kg/day)	10	4
Major labor requirements	weeding	land preparation
Main tools	knife, sickle	hoe, plow
Benefits during fallow period	firewood, minor forest products	grazing
Most promising land-use options	cardamom, orange plantations, private forestry	private forestry, pasture, potato
Area at national level (ha)	36,700	8,800

Source: Adapted from Roder et al. 1992.

Table 59-3. Tseri Area Converted and Conversion Options Suggested by Farmers

Option	*Converted (%)*	*Option Suggested*
Permanent dryland cultivation	11.0	53
Paddy cultivation	0.7	1.5
Cardamom	0.4	1.2
Oranges	1.3	3.0
Pasture		2.2
Private forestry		3.0
Cash compensation		4.5
Land substitution		10.8
Resettlement		1.0
Others		13.2

Notes: Percentage of the total area in Samdrupjonkha, Trashigang, Pemagatshel, Monggar, Lhuentse, Trashiyangtse, and Zhemgang districts. The numbers given indicate the proportion % of tseri area for which a particular option was recommended.

Table 59-4. Estimated Conversion of Pangshing Land in Bumthang District and Its Potential Future Use

	Percentage of Total Area	
Alternatives	*Existing*	*Potential*
Revert to blue pine forest	20	90
Permanent grassland	30	40
Permanent cultivation	5	10
Private forest	—	60
Apple orchards	< 1	3
Silvipastoral systems	—	50

Note: Indication of what the area could be used for. The total is greater than 100%.

Table 59-5. Potential Economic Returns to Land-Use Options

Option	*Annual return* (BTN x 1000 per ha)*	*Advantages*	*Constraints/Limitations*
Tseri			
Dryland crops (maize, millet, etc.)	< 15	Provides food for farmer	Low economic returns, limited market, erosion
Rice	20	Provides food for farmer	Low economic returns, limited market
Cardamom	30	Low inputs required, high-value product. Not dependent on road access.	Requires high rainfall, limited to elevations below 1,800 m, resistance by Forest Department
Oranges	80	Low inputs required	Five years before first return, requires road access
Private forest	30	Low inputs required, optimal for soil conservation	Long-term investment, first returns after 20–30 years, requires road access
Pasture	35	Low inputs required, road not required	Investment in animals
Pangshing			
Permanent crop cultivation	< 15	Provides food for farmer	Low labor productivity, poor economic returns
Potato	150	Fast returns	Road access, wild animals, high investment for seed, high fertilizer inputs
Apple	175	High returns, low inputs required	Initial investments, skill
Blue pine	50	No inputs required	First return after 30 years
Silvipastoral	60	Medium input, early cash returns compared to private forest	Requires initial investment in livestock

Note: Bhutanese Ngultrums earned under optimal production conditions (1997US$1 = 35.92 BTN).

Alternatives for Pangshing Systems

Conversion to livestock pasture may be the most attractive option for most areas where pangshing is currently practiced. Only those sites that are below 2,900 m asl, near a road, and close to settlements are favorable for apple production. Good potato yields can be expected at altitudes up to almost 4,000 m asl, but lack of road access and damage by wild animals seriously limits this as a viable option.

The fast-growing blue pine (*Pinus wallichiana*) is well adapted to prevailing conditions where grass-fallow shifting cultivation is practiced and can produce annual increments of between 8 and 15 cubic meters of timber per hectare (RNR-RC Jakar 1996). In areas accessible by road, silvipastoral systems combining timber and dairy production can generate annual cash returns of US$1,000 to US$2,000 per hectare. In contrast, the same land area managed under the traditional grass-fallow system would provide a crop of buckwheat every 8 to 12 years with a value of about US$250/ha/year.

Conclusions: What Made the Changes Possible?

Farmers' Initiatives

- Use of locally available sources of plant nutrients. In the traditional farming system, cattle play a key role in the collection, transport, and transformation of plant nutrients. Cattle are often tethered in fields overnight to fertilize the land. Forest litter and ash are other major sources of plant nutrients (Roder 1990).
- A change to permanent dryland farming and inclusion of legumes in the system. Maize, millet, potato, and wheat are often intercropped with *Phaseolus* sp., *Vigna* sp., peas, or soybean.
- Use of a large variety of tree fodder. The most important fodder tree species are *Ficus roxburghii* and *Ficus cunia* (see color plate 3). Tree fodder is used mainly during the dry winter period. The dung adds substantial quantities of nutrients to the cropping systems.
- Introduction and cultivation of oranges and cardamom.

Access to Markets

- Imports of rice allowed farmers at high elevations to reduce their buckwheat production and shift to more remunerative cash crops, such as potatoes and apples, and to livestock production. The livestock production system, because it depends on long term or permanent grass and legume cover with a high content of white clover, can be considered an improved fallow system.
- Maize and millet producers were able to grow oranges for the Indian and Bangladeshi markets. The oranges are often interplanted with annual crops, especially *Vigna* and *Phaseolus* species.

Changes to Rules and Regulations

- Conversion of tseri land to permanent land uses was encouraged by the Land Act.
- Regulations introduced under the Forest and Nature Conservation Act (1995) gave farmers full rights to use and sell timber grown on private forest lands.

Government Development Programs

- Introduction of fertilizers. According to findings from a recent household survey, the most important factor leading to changes in the pangshing system has been the use of nitrogen and phosphorus fertilizers (RNR-RC Jakar 1996). With applications of fertilizer, buckwheat is now cultivated in rotation with wheat, potato, and a short fallow period.

- Introduction of new germplasm, particularly of apples and wart-resistant potatoes, and government assistance in marketing apples, potatoes, oranges, and cardamom.
- Cattle crossbreeding programs.
- Fodder development programs.

References

Guenat, D. 1991. Study of the Transformation of Traditional Farming in Selected Areas of Central Bhutan: The Transition from Subsistence to Semi-Subsistence, Market-Oriented Farming. Dissertation submitted to the Swiss Federal Institute of Technology, Zurich.

RNR-RC (Renewable Natural Resources Research Centre), Jakar. 1996. *Annual Report 1995–1996.* Jakar, Bhutan: RNR-RC Jakar.

Roder, W. 1990. Traditional Use of Nutrient Inputs. *ILEIA Newsletter* 6(3), 3–4.

———. 1993. Effect of Burning on Selected Soil Parameters in a Grass Fallow Shifting Cultivation System in Bhutan. *Plant and Soil* 149, 51–58.

———. O. Calvert, and Y. Dorji. 1992. Shifting Cultivation Systems Practiced in Bhutan. *Agroforestry Systems* 19, 159–158.

Rosset, J., Rinchen, and T. Dukpa. 1997. *Silvipastoral Trial on Abandoned Pangshing Land Colonized by Blue Pine Forest,* Jakar, Bhutan: Renewable Natural Resources Research Centre, Jakar.

Upadhyay, K.P. 1988. Shifting Cultivation in Bhutan: Present Situation and Alternatives. In: FO:TCP/BHU/6653 Field Document 1. Rome, Italy: FAO.

Chapter 60

Swidden Agriculture in the Highlands of Papua New Guinea

Bire Bino

Traditional farming practices in the Papua New Guinea (PNG) highlands were once not only stable but were also capable of sustaining subsistence agriculture. There was only low pressure on natural resources to ensure adequate agricultural productivity, and the most common form of traditional farming was extensive shifting cultivation, both in forested areas and grasslands.

However, when the Europeans arrived they brought with them various forms of development, including better health services and a cash economy. The population more than doubled and with it came mounting pressure on natural resources to sustain agricultural production. People began moving from marginal areas to the more fertile highland valleys. Swamp areas were drained for farming and settlement. The overall consequence was increased population density. Farming practices had to change as a result, and several of the new practices observed in the highlands have been identified as agroforestry.

The PNG government realized the problems associated with agricultural productivity and the need for studies of sustainable farming systems, including agroforestry research. One resulting research program, centered at the research stations of the Agricultural Research Division (see Figure 60-1), involves identification of suitable multipurpose tree species for agroforestry systems, intercropping, and maintenance of soil fertility. The recently established National Agriculture Research Institute (NARI) has taken leadership of the program and will focus its work on the highlands, where problems of agricultural sustainability are most prevalent.

This chapter gives a brief description of traditional highland farming practices and the changing patterns of land use in response to various constraints and opportunities, and how research is trying to address the problems associated with these changes.

The Highlands of Papua New Guinea

The highlands of PNG support 40% of the country's 4.5 million people. The area is broadly delineated on the basis of altitude (Bourke 1981), and ranges from 1,000 to 5,000 m above sea level (asl). Much of the population is concentrated in the highland valleys where, in some places, population densities are as high as 320 persons per

---

Bire Bino, EHDAL Program Advisor, Eastern Highlands Division of Agriculture and Livestock (EHDAL), P.O. Box 766, Goroka, Eastern Highlands Province 441, Papua New Guinea.

km^2. Grasslands dominate the vegetation in most valleys. Agriculture is based on sweet potato and coffee cultivation. Sweet potato gardens are managed using fairly intensive techniques in which land can be cropped for 3 to 5 years before being allowed to revert to either forest or grassland fallow for the following 3 to 15 years. High-altitude valleys commence at 1,800 m and people live and cultivate crops up to as high as 2,850 m, although population densities at these higher altitudes are generally low. Agriculture is based on sweet potato, Irish potato, and vegetables, both introduced and traditional. Fairly intensive cultivation systems are used, including compost mounding for sweet potatoes. Intermediate-altitude areas cover the edges of the highland ranges, with altitudes between 1,000 and 1,200 m. Population densities are also low at these intermediate altitudes because of endemic malaria, and agriculture is of very low intensity.

Traditional Shifting Cultivation

The majority of rural people living in the highlands are essentially shifting cultivators. Their traditional systems introduced the first and oldest forms of agroforestry to the area. They begin by clearing fallow land from the forest by felling the trees and shrubs and burning the dried slash. The soil is usually broken up, and food crops are neatly planted to meet domestic subsistence requirements. New gardens are usually characterized by a large variety of crops and wide species diversity. This is because ash from the recently burned vegetation has provided generous quantities of K, Ca, Mg, and micronutrients. However, subsequent cultivation is often limited to staple crops like sweet potato. After a series of crops on the same site, the land is fallowed for periods varying from 3 to 15 years. The fallow land is often selectively planted to *Casuarina* trees, but it is increasingly common to see nonselective regrowth of shrubs, *pitpit* (*Miscanthus* spp.), and *kunai* (*Imperata*) grassland.

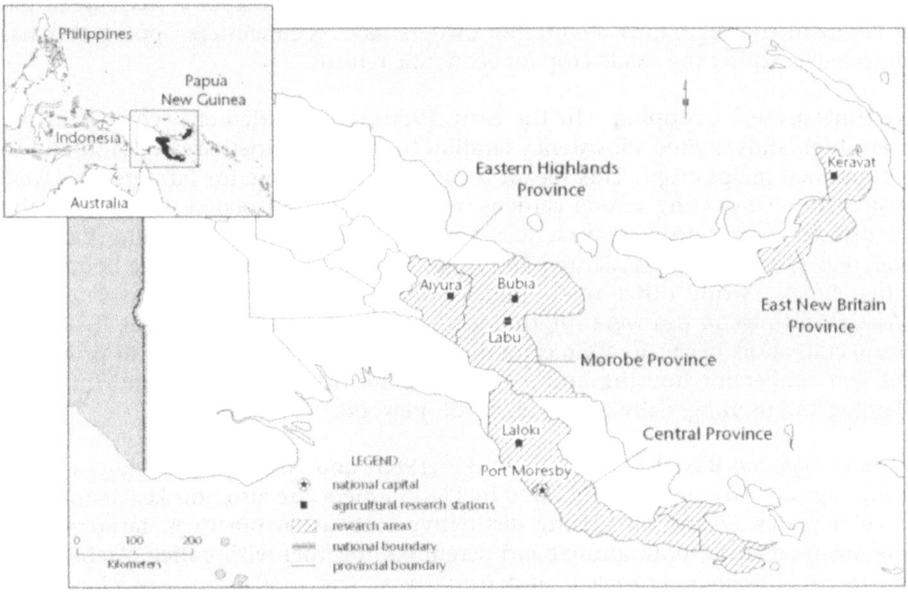

Figure 60-1. Research Stations of the Agricultural Research Division

Areas that are densely populated are already showing signs of declining agricultural productivity and shortages of basic needs, such as food and fuel, because of excessive human pressure on the natural resource base. Farmers have been forced to change both the intensity of cultivation and their cropping patterns in an attempt to sustain subsistence productivity. Among other things, they have adopted planted fallows, crop rotations, and soil conservation techniques. However, native forests on hillsides continue to retreat as more and more land is converted to farming. Because of more intensive and more frequent cultivation, accompanied by regular burning, formerly forested areas are not being given the chance to regenerate tree cover.

Recent Innovations

Highland farmers realize the importance of the forest fallow for maintaining crop productivity in fallow rotation systems, so in densely populated areas, they are now selectively using certain tree species, either as improved fallows or in agroforestry systems. The most commonly managed tree in the highlands is *Casuarina oligodon*. It is planted as a fallow species, as a shade canopy over coffee, as a component of various other agroforestry patterns, and as an ornamental tree around homesteads. Other tree species that warrant mention include *Eucalyptus grandis*, which is planted in swampy areas, *Pinus* spp. in grasslands, and *Parasponia* spp. in cooler high-altitude areas. Farmers have selected these useful tree species to perform both service and productive functions in particular environmental conditions. A few examples of resulting agroforestry practices in the major highland zones are discussed below. These in no way exhaust the full range of agroforestry practices found in PNG.

Plantation Agroforestry. The intercropping of bananas with coffee in the highlands is a recent phenomenon but is being widely adopted. It may even threaten to displace traditional shade tree and coffee mixtures, although this may depend upon the variety of coffee grown. Tree species, such as *Casuarina oligodon* and *Albizia* spp., have been used as permanent shade in coffee plantations ever since coffee cultivation first began in the highlands. Plantation owners have begun intercropping bananas with coffee to exploit the shade crop for economic returns.

Mixed/Integrated Cropping. In the early 1960s, land settlement schemes in the western highlands invited subsistence families to farm a defined area of land and to grow specified major crops. This meant a considerable change for farmers who were accustomed to cultivating several gardens over a number of isolated sites. It gave rise to a distinctive agroforestry practice: an integrated mixed tree and foodcrop production system. Farmers planted *Eucalyptus* and *Casuarina* around the boundaries of their blocks, while other trees such as *Albizia* and shrubs such as *Crotalaria, Tephrosia,* and pigeon pea were intermingled with coffee and tea. For the first time, shifting cultivators began to appreciate the contribution made by trees in providing poles and timber for housing and, because of the need for regular and intensive pollarding and pruning, daily requirements of firewood.

***Casuarina oligodon*-Based Systems.** Bourke (1985) and Allen (1985), describe two dynamic agroforestry practices followed by PNG farmers (see also Bourke, Chapter 31, and color plates 37 and 38). Using distinctive cultivation practices, farmers grow numerous food crops, both annual and perennial, together with coffee (Arabica and Robusta), *C. oligodon, Leucaena,* and fruit trees. The food crops are planted in established stages and are managed to continue production even under mature coffee trees. The coffee earns cash for the farmers; *Leucaena* and *Casuarina* provide shade for the coffee and firewood for the household, as well as enriching the soil through leaf litter and N-fixation; and the food crops supply the daily needs of the household and its domestic livestock.

Neither these practices, nor any other agroforestry system, have ever been evaluated in PNG for their economic benefits and ecological sustainability. However, the farmers involved appear to be reaping substantial benefits.

Forestry and Livestock. Commercial forestry plantations have been used for grazing livestock beneath the canopy. This system is used extensively in *Eucalyptus* plantations in the Whagi valley and in *Araucaria* plantations at Bulolo. Plantation owners reap dual benefits: the cattle keep the weeds down, and the highly palatable grass under the trees fattens the cattle, creating a good cash income.

Cropping under Rainforests. This cropping system is noteworthy, even though it is a very remote and location-specific adaptation by a small group of people living at an altitude of 1,200 m asl in the transitional fringes of the Kassam Pass area. Unlike in conventional shifting cultivation, the primary rainforest is only partially cleared. The undergrowth, shrubs, smaller trees, and less than one-quarter of the larger trees are felled. The decision on how many trees should be felled appears to depend on the amount of sunlight penetrating the canopy. No lopping, ringbarking, or other tree management practices are used. It is essentially a "minimum forest tillage" practice in which large trees are deliberately but nonselectively allowed to remain. A minimum of clearing, burning, and tillage is employed. Generally, fewer than three species of shade-tolerant food crops are grown (see color plate 1). These gardens are abandoned to fallow after fewer than three years of *Xanthosoma* cropping, and the farmers move to a new site in the forest.

Sustainable Farming Systems Research

Research on sustainable farming systems in the highlands has been going on for more than 20 years. It emphasizes the development of low-input practices, and the concept is to build upon traditional knowledge and practices by adding scientific innovations.

Major subprograms include research on soil management on sloping lands, aimed at identifying appropriate soil husbandry practices, and on agroforestry, emphasizing the use of suitable trees in combination with crops or livestock as a means of sustaining subsistence agricultural production. The PNG government originally funded both subprograms, but with irregular budget support. The recently established National Agricultural Research Institute (NARI) will make them core programs of its highland agriculture research agenda.

Extension Work

Having recognized the problem of deforestation in the highlands, several organizations are raising and distributing tree seedlings. The National Forest Authority is developing commercial forest plantations and is assisting local farmers to develop woodlots. Landcare of Australia is also working closely with the North Simbu Rural Development Project to establish mobile nurseries and to raise and distribute *Casuarina oligodon* and fruit tree seedlings to farmers in targeted areas. A smallholder market access and food supply project is also doing similar work. The Agricultural Research Division is also providing seeds and seedlings of promising multipurpose tree species, such as *Casuarina*, *Leucaena* spp., *Calliandra* spp., and *Acacia* spp., to highland farmers with the aim of integrating them into sustainable farming systems.

Conclusions

The PNG government has given a mandate to agricultural researchers to improve and develop low-input, sustainable farming practices. But their work in the highlands is quite new. The only existing studies include the evaluation of multipurpose tree species, studies on the maintenance of soil fertility, and intercropping trials. Irregular funding has impeded the research, but it is hoped that with the recent establishment of NARI, solutions to these problems may be found.

Unless measures can be taken to address the problems of agricultural sustainability, the future for much of the highlands region looks bleak. With increasing population and expansion of cash cropping, resources available for subsistence farming are under stress, and a major research effort is required to identify suitable solutions.

References

Allen, B.J. 1985. Dynamics of Fallow Successions and Introduction of Robusta Coffee in Shifting Cultivation Areas in the Lowlands of Papua New Guinea. *Agroforestry Systems* 3, 227–238.

Bourke, R.M. 1981. Agriculture in Papua New Guinea. *Alafua Agricultural Bulletin* 6(3), 77–82.

———. 1985. Food, Coffee and *Casuarina*: An Agroforestry System from the Papua New Guinea Highlands. *Agroforestry Systems* 2, 273–279.

Chapter 61

The Problems of Shifting Cultivation in the Central Highlands of Vietnam

Phan Quoc Sung and Tran Trung Dung

Vietnam's central highlands cover 5.6 million hectares and range mostly between 450 to 900 m above sea level (asl). The total population is 3.1 million people, 30% of which consists of ethnic minorities representing 37 distinct ethnic groups whose livelihood is essentially based on swidden agriculture. The largest ethnic minority groups are listed in Table 61-1.

According to 1990 statistics, shifting cultivation extended over about 407,000 ha in Vietnam and accounted for 7.5% of total upland rice area in the country. From this national total, 128,000 ha were in the central highlands. Daklak Province had 51,000 ha; Lamdong, 27,000 ha; and Gialai-Kontum, 50,000 ha. Across the central highlands, the consequences of shifting cultivation have become clear: degraded, barren lands that can no longer support agriculture are expanding at a rate of deforestation that has recently accelerated to average 25,000 ha per year. Therefore, studies aimed at solving the problems created by shifting cultivation and mitigating its consequences are needed urgently.

The Current Situation in the Central Highlands

Traditional shifting cultivation in the central highlands is classified as rotational swiddening. This is characterized by the rotation of crops and fallow in space and time. Usually, a 3- to 5-year cultivation period is followed by a fallow period of 5 to 15 years. Durations of cultivation and fallow vary from location to location, depending on soil fertility, vulnerability of the land to erosion and runoff, the structure of vegetational cover, and the climate. These farming systems cover large areas, although fields actively cultivated at any given time occupy only a fraction of the "used" area. Families clear plots around their village and cultivate them for a number of years, until the soil fertility becomes too low or weed growth too vigorous. Then, because every household generally has access to different lands for swidden cultivation, they move to a new site.

Phan Quoc Sung, Director, Western Highland Agro-Forestry Science Research Institute (WASI), Km, 8, National Rd. No. 27, Hoa Thang Commune, Buon Ma Thuot City, Dak Lak Province, Vietnam; Tran Trung Dung, Chief, Forest Ecology and Environment Section, Agro-Forestry Faculty, Central Highland University, 567 Le Duan St., Buon Ma Thuot City, Dak Lak Province, Vietnam.

Table 61-1. Most Populous Ethnic Minority Groups in Vietnam

Tribe	*Population*	*Tribe*	*Population*
1. Gia Rai	239,950	6. Mnong	66,654
2. E De	199,398	7. Ra Glai	59,793
3. Ba Na	138,939	8. M'o	19,347
4. Xe Dang	96,326	9. Gie Trieng	15,918
5. Ko Ho	85,968		

Source: Do Dinh Sam 1996.

These shifting cultivators have rich experience in selecting swidden sites. Being indigenous to the highland areas, they have an intimate knowledge of the natural conditions and particularities of their land. This guides them in selecting the best sites for growing crops, where soils are fertile and successful harvests most likely. They traditionally plant a diverse range of crops and varieties in a single field, in intercropping and crop rotation patterns, suggesting that they have learned to conserve their soil fertility. Given the fast regrowth of fallow vegetation and a fallow period of sufficient length, there is relatively little soil erosion in this type of swidden system. It actually shares some similarity with natural forest gap regeneration, and soil fertility is restored through natural processes.

Some of the ethnic minority communities actively protect their forests, believing that by appeasing the forest spirits in this fashion they will bring prosperity and prevent calamities and "devil-inspired" events. In these cases, forest lands are strictly managed by the community through a board of elders that establishes regulations on land use, forest clearing and protection, and allocates forest land to every member of the village for farming purposes. Farming practices for the uplands and sloping lands have been passed down from generation to generation through legends and stories memorized and recounted by tribal elders.

Traditional farming practices in the highlands also involve a "dibbling holes and depositing seeds" method of land preparation. This minimizes soil disturbance and hence reduces erosion losses on sloping lands during the rainy season. There are many such aspects of traditional shifting cultivation that, in the past, when population densities were low and land resources extensive, made it a sustainable system.

However, that historical situation has changed drastically over the past three decades because of a rapidly escalating population. In the decade of the 1980s, for instance, the population of the central highlands increased by 2.5 million people. A major contribution to this growth was an influx of immigrants from provinces in Vietnam's central coastal area and from the country's north. These population movements were sponsored by the government to redistribute human resources and expand agricultural production. More recently, a wave of illegal migrants flooded into the central highlands from the mountainous north, where they had grown accustomed to forest logging and pioneer shifting cultivation.

In general, shifting cultivation is reported to be sustainable at population densities below 50 to 70 people per square kilometer. The actual threshold also depends on other factors such as the susceptibility of the land to deterioration, previous farming history, and the crops grown, but population density is critical because, once the population grows beyond the threshold level, sustainability may collapse. Shortening of fallow periods in order to intensify production leads to declining soil fertility and productivity. The exhausted land is then abandoned and becomes barren, characterized by grass or shrub cover and low agricultural productivity. Shortened fallow periods have made the forest cover of the central highlands more and more susceptible to degradation by shifting cultivation as the natural environment is pushed beyond its ecological resilience and loses its buffering capacity.

The transfer of land ownership from local communities to the state also brought negative impacts to land management because the government was incapable of extending its management down to the lowest administrative levels in rural areas. Illegal selling of land by tribal cultivators to Viet (lowland) immigrants contributed to the scarcity of agricultural land and accelerated encroachment into forests. At the same time, agroindustry, and its demands for both markets and land, brought a major influence to the livelihoods of minority ethnic groups.

An Economic, Social, and Environmental Assessment

Shifting cultivation is characterized by its minimal inputs, usually only human labor. No complicated tools, draft animals, irrigation, or fertilizers are used. Its low demands for investment make it suitable to resource-poor ethnic minorities. In terms of output, crop yields can only marginally meet the food needs of swidden populations and there is almost no surplus for marketing. Shifting cultivation, therefore, is subsistence oriented and its practitioners often lack the resources to meet other basic needs.

Shifting cultivation not only generates low income, but also appears to be economically ineffective compared to alternative land uses. If, for instance, shifting cultivators converted their efforts to management and harvest of forest products, they would realize a far greater benefit. The growth rate of evergreen forest in the central highlands is very high and can reach volumes up to 300 m^3/ha/year. The potential value of forest products is two to three times that of the products from shifting cultivation.

Because the land available for shifting cultivation is decreasing rapidly and crop yields within these degrading systems are themselves declining, traditional farming methods can no longer meet the food demands of the population, to say nothing of their other basic needs. Most farmers who rely heavily on shifting cultivation live far below the poverty line. According to criteria set by the Ministry of Labor, War Invalids, and Social Affairs, a "starving household" is defined as earning a per capita income equivalent to less than 8 kg of rice per month. A "poor household," on the other hand, earns less than 15 kg of rice per month. In shifting cultivation communities, none of the households has sufficient food. Sixty percent of them fall into the "starving" category and the remaining 40% are regarded as "poor."

Table 61-2. Food Security Indices in Vietnam, Central Highlands and Daklak Province

Indices	*Vietnam*	*Central Highlands*	*Daklak Province*
Annual food production (kg/capita)	346.20	230.60	244.70
Food security (%)	115.40	76.87	81.57
Annual rice production (kg/capita)	214.90	115.20	123.90
Rice security (%)	107.45	57.60	61.95

Source: Easoup multipurpose project, 1995.

Throughout the central highlands, and particularly in Daklak Province where shifting cultivation is common, the food deficit is a matter of great concern. Annual food production per capita can serve as an indicator of food security. Assuming that 200 kg of rice and 300 kg of other foods is a normal rate of consumption, Table 61-2 shows that the central highlands and Daklak Province have food security indexes far below the rest of the country.

Loss of Forest and Contained Biodiversity

Shifting cultivation systems under this kind of stress can cause serious environmental damage, including loss of forest and biodiversity, declining soil fertility, and reduced capacity of forests to regenerate on fallowed lands. As mentioned earlier, the rate of deforestation in the central highlands is about 25,000 ha/year. But the extent of shifting cultivation is much larger, accounting for 128,000 ha. During the period from 1978 to 1992, the average loss of forest in Daklak Province was about 8,000 to 10,000 ha/year. In 1978, the province had a forest area of 1,263,136 ha with a biomass of 104 m^3/ha. By 1992, only 1,148,836 ha remained, with a reduced biomass of 95 m^3/ha.

Apart from direct forest loss, collapsing shifting cultivation systems are causing serious land degradation through overcultivation. Recurring fires are preventing the regeneration of natural forests on barren lands, and species richness is declining. Land degradation is being accelerated by the invasion of noxious weeds. Up to 200 species of weeds can be found on barren lands, particularly *Echinochloa* and *Imperata,* which compete directly with crops for soil nutrients.

The removal of forest cover by shifting cultivation may also lead to severe soil erosion. The average soil loss from one ha of swidden on an 8° slope is about 150 to 200 metric tons. This is equivalent to 6 metric tons of humus, 150 kg of nitrogen, 80 kg of phosphorus, and 200 kg of potassium. Even on gentler slopes, the rate of soil loss is still significant, as shown in Table 61-3.

After the cropping period, the soil is characterized by reduced moisture, low exchangeable potassium, calcium, magnesium, and extractable phosphorus, and increased acidity, as shown in Table 61-4. Soil moisture plays an important role in supporting crop growth. In the dry season in the central highlands, soil moisture from the surface to a depth of 50 cm usually drops below the wilting point and may damage plants, especially young trees, seedlings, and shallow-rooting crops.

Forest Regeneration after Shifting Cultivation

Studies on forest regeneration after shifting cultivation are very few. However, the following points are well established:

- If the size of the cultivated field is small, similar to forest gaps, and there is still forest cover surrounding this land, then secondary forest may regenerate rapidly, in 7 to 10 years. Soil fertility will be maintained longer under these conditions.
- If bamboo communities dominate fallow successions, then it will take longer before a mixed stand of bamboo and timber trees will regenerate.
- If the size of the cultivated field is large and surrounding forest cover is low, then it takes more than 20 years to regenerate secondary forest cover.
- If fallow successions are dominated by weedy shrubs or coarse grasses such as *alang-alang* (*Imperata cylindrica*) that are prone to fire in the dry season, then the regeneration of timber trees will be difficult.

Table 61-3. Soil Loss by Water Erosion in Daklak Province*

Slope (degrees)	*Crops*	*Cover (%)*	*Soil Loss (t/ha)*
0–3	coffee	60	22
0–3	upland rice†	30	50
3–5	coffee	75	32
3–5	grass and bushes	55	40
3–5	upland rice†	30	65

Note: * Based on 48 samples collected over three years of monitoring. † Cultivated in swiddens.
Source: Tran Trung Dung 1987.

Table 61-4. Changes in Soil Physical and Chemical Properties in the Arable Layer of Sampled Swidden Fields

Crops	*Clay (< .001mm)*	*pH (H_2O)*	*Humus (%)*	*Nitrogen (%)*	*Hydrolytic Acidity (mg/100 g)*	*Exch. Ca^{+2} Mg^{+2} (me/100 g)*	*Available P_2O_5 (mg/100 g)*
1st year of cultivation	13.2	4.9	5.5	0.23	12.8	3.7	5.0
2nd year of cultivation	14.0	5.9	4.1	0.22	5.9	8.0	3.1
4th year of cultivation	19.6	6.1	4.2	0.21	4.4	10.5	0.8
3rd year of fallow period	10.8	5.6	3.8	0.17	6.3	4.2	1.2
7th year of fallow period	21.6	4.1	4.9	0.36	10.3	0.8	0.6
14th year of fallow period (secondary forest)	21.6	4.8	6.2	0.24	14.0	2.7	1.5

Note: Collated over more than 30 years of research in northern Vietnam.
Source: Bui Quang Toan 1991.

Table 61-5. Changes in Soil Physical and Chemical Properties after Three Years of Incorporating Green Manure Legumes

Green Manure Legumes	*pH (KCl)*	*Humus (%)*	*P_2O_5 (mg/100 g of soil)*	*Ca^{2+} Mg^{2+} (me/100 g of soil)*	*CEC (me/100 g of soil)*
Tephrosia candida	3.92	4.8	4.4	6.10	31.2
Cassia surattensis Burm.	4.00	4.1	3.2	5.52	31.1
Vigna sinensis	3.98	3.9	3.4	5.51	28.3
Stylosanthes guianensis	3.97	4.7	4.3	5.70	29.5
Control: no green manure use	3.65	3.0	1.8	3.84	23.6

Source: Luong Duc Loan 1996.

Studies on the rehabilitation of basaltic soils degraded by swidden overcultivation have shown the following:

- The incorporation of organic fertilizers such as compost, leafy biomass from natural vegetation, or green manures into the soil in large quantities will provide relatively rapid improvements in physical and chemical properties.
- Green manure legumes have a good growth rate on degraded basaltic soils and recycle large amounts of biomass and nutrients back to the soil.
- After three years of planting green manure legumes and incorporating all the biomass into the soil, soil properties improved markedly (Table 61-5) and upland rice yields were between 130% and 156% of those where green manures were not used.

Conclusions

Shifting cultivation is an important element of the cultural identity of ethnic minority groups living in Vietnam's central highlands. At one time, traditional swidden fields imitated gaps in the forest and were cultivated on a rotational basis, providing almost all the subsistence needs of the farmers and their families. However, the socioeconomic environment in both the central highlands and in Vietnam generally has changed and shifting cultivation is no longer a suitable land use. Stressed swidden systems are unable to meet the subsistence food needs of local communities living by this practice, and they are contributing to the degradation of the natural environment.

Planting green manure legumes and incorporating all biomass into the soil or applying other organic fertilizers is needed to rehabilitate degraded soils. Appropriate alternatives to improve the situation may include judicious use of inorganic fertilizers to "jump start" the production of large quantities of organic matter.

Recommendations

In view of the controversy surrounding shifting cultivation, researchers need to carefully assess it as a dynamic farming system and elucidate its strengths and weaknesses along its trajectory of intensification. Individual studies of shifting cultivation as it is practiced by each of the ethnic groups, under different ecological and economic conditions, are needed as a guide to programs which aim to settle these populations into more permanent forms of agriculture and provide for their economic development.

In exceptional areas, where forest cover is still high and population densities are low, rotational shifting cultivation is still an appropriate land use and should be supported by the government through accelerated allocation of agricultural and forest lands to farmers, with secure tenure. The government should further initiate a credit program to help local communities acquire new technologies and germplasm. These should include:

- New crop varieties with high productivity, to increase rural food security; and
- The application of organic fertilizers and intercropping of multipurpose legumes to improve soils during both cropping and fallow phases.

Programs aimed at reduction of shifting cultivation should prioritize areas around national parks, nature reserves, and protected areas. Such programs should also focus on areas with access to transportation and markets, and where sufficient knowledge exists in the local community to support adoption of alternative land uses such as industrial tree plantations, rice paddies, or livestock husbandry. In these cases, the most attractive options will depend on the biophysical and social conditions of the area, and the customs of the ethnic groups concerned.

In tandem, training and education is needed to improve knowledge of environmental issues and sustainable development. Credit facilities should also be available, together with training on both technical issues and fiscal management. Local communities should, in this way, be supported in choosing and pursuing their own development priorities.

References

Easoup Multipurpose Project. 1995. Initial Report, Dak Lak, Vietnam.
Loan, L.D. 1996. Research on Erosion Control for Coffee Trees and Some Short Term Crops in the Central Highlands of Vietnam. Sustainable Cultivation System in Highlands of Vietnam. Publishing House of Agriculture. Hanoi, Vietnam.
Sam, D.D. 1996. Overview of Shifting Cultivation in Vietnam. Hanoi.
Toan, B.Q. 1991. Some Problems in the Shifting Cultivation in the Northwestern Region of Vietnam. Doctoral Thesis. The Agricultural Research Institute of Vietnam, Hanoi.

Chapter 62

Some Indigenous Experiences in Intensification of Shifting Cultivation in Vietnam

Tu Quang Hien

The northern midlands and highlands of Vietnam consist of 15 provinces covering an area of 182,965 km^2 and populated by 12.9 million people. The highlands are classified as those mountain areas with an average altitude above 600 m above sea level (asl) and the midlands are mountain areas below 600 m. The climate of the area is characterized by yearly average temperatures of 19°C to 23°C, humidity of 78% to 83%, and 1,300 to 2,300 mm of annual rainfall.

The area is inhabited by 20 ethnic groups, of which the largest is the Kinh group, making up about 70% of the population, followed by the Tày, Thái, and Muòng. Generally, the lifestyles fall into three main categories: fixed residence, permanent cultivation; fixed residence, shifting cultivation; and nomadic lifestyle, shifting cultivation.

The ethnic groups that live in high mountain areas, such as the Hmong and the Dao, belong to the third type: nomadic lifestyle, shifting cultivation. Other groups, such as Tày, Thái, and Muòng, live in the low mountain areas and belong to the fixed residence, shifting cultivation category. Finally, the majority Kinh group, along with some other ethnic groups living in the lower mountains and hills, is typified by fixed residence, and permanent cultivation. This chapter reviews some of the practices and innovations of shifting cultivators within each category.

Intercropping Upland Rice with *Melia azedarach*

Muòng shifting cultivators in Thanh Hoa and Hoa Binh Provinces have developed a special technique for integrating *Melia azedarach* into their swiddens as a superior fallow species (see also Tran Duc Vien, Chapter 36). After slashing the vegetation to open a new swidden, *Melia* seeds are broadcast over the field. When the slash is burned, the high temperatures cause the *Melia* seeds to germinate, and the seedlings sprout together with the upland rice crop. If the *Melia* is overly dense, young seedlings may be thinned during weeding operations, leaving a density of about 1,000 plants per hectare. After three years, planting upland rice is discontinued

Tu Quang Hien, Director of Scientific and International Relations, Department of the Vice-Director, Center for Scientific Research and Technology Transfer in the Northeast Region, Thai Nguyen University, Bac Thai Province, Vietnam.

because the tree canopy begins to close. After 10 years, the *Melia* trees are harvested and the next crop of upland rice is planted, beginning a second cycle.

In some cases, a similar system is practiced with bamboo, which is then harvested after an eight-year fallow. However, the bamboo clumps pose greater difficulties to land preparation in subsequent cycles.

Using *Phaseolus calcaratus* to Improve Soil Fertility

Phaseolus calcaratus, known locally as *caobang bean,* has a strong root system, is fast growing, and produces 10 to 15 tonnes of green biomass per ha. The Tày, Nùng, Lang Son, and Hagiang use *P. calcaratus* to improve soil fertility (see also Hao et al., Chapter 22, and color plate 26). After two or three crops of maize, soil fertility decreases and farmers begin to relay plant the bean into maize fields, usually 24 or 25 days after the maize is planted. After maize harvest, the beans are left to climb over the maize stalks and completely cover the field surface. Leaves of *P. calcaratus* drop to the soil and provide large quantities of organic matter, making the soil soft. Its root nodules also fix significant amounts of nitrogen. After harvesting the beans, farmers clear the area and use the bean residues as organic fertilizer for the next maize crop. Yields from the subsequent maize crop can improve by 15% to 20%. Use of *P. calcaratus* allows farmers to extend the length of cropping and provides more stable maize yields.

Cropping Peanuts with Cassava to Enable Prolonged Cultivation

Slash-and-burn farmers in northern Vietnam typically plant upland rice in the first year, and maize or cassava in the second. In years 3 and 4 they plant cassava, after which the land is left fallow. In upland fields with a moderate slope of 10° to 15°, farmers commonly intercrop peanuts or beans with the cassava to extend the cropping cycle. This method was pioneered by farmers in midland provinces such as Bacgiang, Phutho, and Thainguyen.

Research has confirmed that this cropping system is effective in mitigating soil erosion. After peanuts are harvested, the residues, amounting to 7 to 9.5 tonnes per ha, are applied as a green manure to other crops in the same field. A field with a slope of about 10° lost 18.27 tonnes of soil per ha per year to erosion under this system, whereas under a cassava monoculture the loss was about 72.23 tonnes of soil per ha per year. Cassava yields under the monoculture system were 15.6 tonnes per ha, while those of cassava intercropped with peanuts were 15.42 tonnes per ha of cassava plus 730 kg of unshelled peanuts. The cash value of the cassava and peanuts was 50% higher than that from cassava monocultures. This cropping system allows farmers to extend cultivation for a few more years and to earn higher income from the same land area.

Constructing Terraced Fields

Nguyenbinh and Nganson are high mountainous districts in Caobang Province, reaching altitudes of 1,000 m asl, and populated by Dao, Hmong, and many other ethnic groups. Unlike the Hmong, the Dao have adopted wet rice cultivation and have constructed terraced, irrigated fields across the mountain slopes.

The terraces are usually carved into hills and mountains with slopes in the range of 10° to 40°. The mountains chosen for terracing usually have underground water sources, and to maintain these, the farmers strictly protect the forest cover on the summits of the hills. The terraces may be as small as 50 to 70 m^2 near the hilltops, but grow as big as 100 m^2 near the bottom. Width varies from 2 to 10 m, and the length from 20 to 100 m around the slopes.

The terraces are enclosed by earthen dikes stabilized by naturally growing vegetation. To seal the terraces and prevent water from percolating away, farmers chase buffalo around the terrace to compact the soil and create an impervious layer similar to a plow pan. This is repeated in the second year. Water is systematically channeled from fields at the top of the slope to those at the bottom. When water is scarce, the top fields are kept full of water, allowing it to percolate downslope to the lower fields.

Local varieties of rice are planted in the terraces, with good drought tolerance and relatively long growing seasons. Seedlings are planted at high densities in small nurseries near water sources. Young plants of the long duration varieties can wait 10 to 15 days for rain without adverse effects on their later growth and development. When the monsoons arrive and the terraces fill with water, the seedlings are transplanted from the nursery into the main fields.

At present, government and agricultural organizations do not attach much importance to developing terraced fields on mountainsides, considering them small and unproductive. However, they play an important role in assuring the food security of adoption villages and reducing their reliance on slash-and-burn practices. Table 62-1 shows that rice yields from terraced fields are double those from dryland cultivation on sloping lands. This suggests that, in the absence of terraced fields, farmers would have to open swiddens twice the area of their terraced fields, or about 0.3 ha per household, to achieve the same food production.

In comparing the two districts, the terraced fields of households in Nganson district (0.13 ha) are smaller than those in Nguyenbinh district (0.18 ha), while, in contrast, the swidden areas of Nganson (0.70 ha) are bigger than those in Nguyenbinh (0.61 ha). It is clear that if farmers do not make terraced fields, they have to open more areas for shifting cultivation. A terraced field can be cultivated continuously for about four years. If they were not available, then farmers would have to clear at least 0.3 ha of forest per household for cultivating crops during those four years.

Extrapolating these data across Vietnam's northern mountainous area, it suggests that if the area's 600,000 farming households had terraced fields, then within the four-year cultivation period, 180,000 ha of forest would be saved from clearing for shifting cultivation. This highlights the importance of encouraging farmers to invest in terraced fields, thereby enhancing their food security and mitigating agricultural pressure on forest lands.

Intercropping Cinnamon with Upland Rice and Cassava

Yenbai is one of the mountainous provinces in northern Vietnam. It has an area of 6,808 km^2, and a population of 640,000. The average temperature is 20°C, with a relative humidity of 80% and annual rainfall of 2,000 mm. The soil is derived from Macma stones and is quite fertile, with pH ranging from 5.5 to 6.5.

Table 62-1. Agricultural Area and Productivity in Nguyenbinh and Nganson Districts

	Area (ha/household)		*Yield (kg/ha)*		*Total Production (kg/ha)*	
Land Type	*NB**	*NS***	*NB**	*NS***	*NB**	*NS***
Swidden fields	0.61	0.70	1,405[a]	1,350[b]	857	945
Terraced fields	0.18	0.13	2,960	2,430	533	316
Total	0.79	0.83	—	—	1,390	1,261

Notes: * Nguyenbinh district: [a] Upland rice yield: 1,440 kg/ha; Maize yield: 1,370 kg/ha; Average yield: 1,405 kg/ha. ** Nganson district: [b] Upland rice yield: 1,320 kg/ha; Maize yield: 1,380 kg/ha; Average yield: 1,350 kg/ha.

Farmers have found the climate and soil to be well suited to growing cinnamon and have increasingly adopted it as a lucrative cash crop (see color plate 51). They usually interplant it into upland fields with rice and cassava in a *taungya*-style pattern[1], thus improving the survival and growth of cinnamon seedlings and providing a short-term income while the trees become established.

After slashing, burning, and clearing the fields, farmers plant cinnamon seedlings at a density of 3,000 to 4,000 plants per ha. Upland rice is then dibbled into the same field, using standard swidden practices. The rice crop grows quickly and can reach a height of 70 cm after two to five months, providing shade for the cinnamon seedlings. In the second cropping year, when the cinnamon trees are higher, they are intercropped with cassava in the same fashion. The cassava reaches a height of 1.5 m, again providing protective shade for the cinnamon. Sometimes the cassava is harvested in the first year and replanted again, but in most cases it is harvested after two years, allowing the cassava to provide even better shade.

This taungya system improves cinnamon seedling survival rates by 10% to 15%, and height increments are 20% greater than when they are grown under conventional plantation practices. During this early establishment phase, about two tonnes of upland rice per ha is harvested in the first year and 12 to 17 tonnes/ha of cassava during the second and third years. In those cases where the cassava is harvested after two years, yields can reach 28 tonnes/ha.

From years 9 or 10, farmers begin to harvest the cinnamon once a year. The average yield of cinnamon has a cash equivalent of about 12 tonnes of rice. After 14 years, the stand of cinnamon trees has been considerably thinned and farmers resume planting food crops.

This method has been adopted by many farmers in Vanchan, Tranyen, and Vanyen districts of Yenbai Province. Vienson commune, in Tranyen district, is particularly well known for cinnamon production and each household plants up to 3 ha. Cinnamon is therefore playing a critical role in providing farmers with cash incomes and reducing reliance on slash-and-burn.

Growing Litchi

Bacgiang is a mountainous province in northeastern Vietnam. Lucngan district, with a 1996 population of 169,795 and area of 101,224 ha, is located in the northeast of the province. Within the district, forest land occupies 26% (26,279 ha); nonagricultural land, 23% (23,898 ha); barren hilly land, 37.6% (38,099 ha); and agricultural land, 12.7% (12,947 ha). There are 7,038 ha of rice paddies, of which about 40% are irrigated and can support two rice crops per year. The remaining 5,909 ha of agricultural land is mainly on hillsides with slopes of 5 to 20°. The average temperature is 22°C, relative humidity is 80%, and rainfall is lower than other areas, about 1,400 mm per year. During the dry season, from October to April, rainfall is only about 50 mm per month. The low rainfall in February and March is conducive to litchi production.

The district is dominated by hills and mountain ranges. The soil is classified as yellowish-brown Feralit, formed on shale rock of average depth. It is acid, with a pH of 5 to 6, and relatively infertile. Lucngan can be divided into two microclimatic areas, the southeast and the northeast. The southeast is relatively flat and has more than 70% of the district's paddy rice. The northeast region is more hilly and

---

[1] "Taungya" is a reforestation technique in which trees are planted as an intercrop with agricultural crops. The young tree seedlings thus benefit from routine maintenance operations directed at the crops, helping them to become firmly established during the critical early years. The food crops are then discontinued and the plot reverts into a pure tree plantation. This system was popularized in Myanmar by British colonial foresters.

mountainous. It has only a small area of rainfed rice paddies that produce a single crop per year. Wet rice cultivation supplies 60% to 70% of the area's food needs, while the remainder comes from dryland crops such as corn, rice, cassava, and beans.

The population of Lucngan district increased by 64%, from 104,719 to 169,743, over the 16 years from 1980 to 1996. During the same period, agricultural land increased by only 5.8%, from 12,235 ha in 1980 to 12,947 ha in 1996. Forest land was under government management and slash-and-burn was forbidden, so farmers were left with only one option: to increase the productivity of their existing fields. Continuous cropping soon exhausted the land and crop yields declined. In 1980, yields of corn, cassava, and rice were from 1.7 to 2; 18 to 22; and 1.1 to 1.3 tonnes/ha respectively. By 1995, the respective yields had fallen to 0.8 to 1.2; 9 to 14; and 0.6 to 0.8 tonnes/ha. With their traditional farming systems seriously degrading and food security threatened, farmers in Lucngan district eventually identified what appears to be a sustainable solution. They began developing litchi orchards on sloping lands (see also Ta Long, Chapter 53).

Ecology of Litchi

Litchi (*Nephelium*) favors conditions between 19 and 24°C, with medium relative humidity and soil fertility. However, it can grow and fruit heavily even in semiarid areas with poor soils. Litchi trees flower in March and rainfall damages the blossoms, so only areas with low rainfall in spring can produce high and sustainable yields.

Conversion from Annual Crops to Litchi Orchards

Litchi trees begin bearing fruit four to five years after planting, and, if farmers grew only litchi trees instead of arable crops, their food supply would fall short by 30% or 40%. Therefore, the trees are intercropped with food crops.

In the first and second years, maize or cassava is interplanted with the litchi. These taller crops provide shade for the litchi seedlings, increasing their survival rate. Farmers also apply increased fertilizer to sustain production levels because the litchi trees reduce the cropped area.

In the third and fourth years, green beans or soybeans and peanuts are planted with the growing litchi trees. These legumes are chosen for their ability to help remedy decreasing soil fertility.

Some farmers intercrop pineapples with the litchi, and after planting in the first year, the pineapples are harvested over the following four to five years. However, this system is not popular because it quickly drains soil fertility.

Economic Benefits of Growing Litchi

In the mountainous terrain in the northeast of Lucngan district, the average household has six family members and farms 0.25 ha of rice paddy, of which 30% is irrigated and can support two crops of rice per year, with an average yield of 3 tonnes per ha per crop; 0.3 ha on which they plant two dryland crops per year, including corn, which yields about 1.1 tonnes/ha; and 0.3 ha of litchi orchards containing about 60 trees. By the fifth and seventh year after planting, each tree produces about 15 kg of fruit annually, and by the eighth and tenth years, this can reach 25 kg per year. Commodity prices are 2.2 million dong per tonne for rice, 1.8 million dong per tonne for corn, and 12,000 dong per kilogram for litchi. The estimated annual income for a household is shown in Table 62-2. Litchi orchards have increased household annual incomes by almost 660%. At present, about 9,000 households, or nearly 50% of the total, have planted litchi and their standard of living has improved significantly. Deforestation has been reduced, and the tree cover better protects the soil. What is more, there appears to be the opportunity for continued expansion of litchi production without threatening to saturate Vietnam's domestic market. The conditions underpinning the Lucngan farmers' success in growing litchi can be summarized as follows:

- The soil and climatic conditions are suitable for litchi growing. Importantly, there is little rain during the flowering season. Persistent rain at this time will damage the blossoms.
- The farmers use a taungya system providing continued harvests of food crops during the four to five years before the litchi trees begin producing.
- Transportation infrastructure allows access to markets all over the country.
- Loans for agricultural development were available from the government.

Interplanting *Acacia* with Tea

Bac Thai is a mountainous province in northeastern Vietnam, with a population of about 1.2 million. It covers an area of 650,280 ha, of which 173,550 ha (26.7%) is forest land; 268,030 ha (41.2%) is barren land; and 137,209 ha (21.1%) is nonagricultural land, including roads, construction sites, rocky mountains, and streams. Agricultural land covers only 71,416 ha, or about 11% of the total, and this amounts to 0.06 ha per capita.

Bac Thai has a tropical monsoon climate, characterized by high heat and humidity. The average temperature is 21°C, annual rainfall is between 2,100 and 2,300 mm, and the average humidity is 82%. The province is dominated by hills and mountain ranges. The soils include red and yellow Feralit of medium depth, formed on gabro and shale rocks.

The area of tea grown in Bac Thai is 7,940 ha, accounting for 75% of the total area of perennial crops in the province (10,575 ha). Tea grown in Bac Thai is of high quality. It enjoys a stable market, and tea revenues provide between 30% and 60% of household income in the northern region. Tea has traditionally been intercropped with other trees for the purposes of shading, maintaining soil moisture, and providing timber and firewood. *Melia azedarach, Artocarpus integer* (jackfruit), *Aleurites fordii,* and *Manglietia glauca* are the trees most commonly grown as a canopy over tea plantations. However, since 1986, farmers have increasingly converted to *Acacia*. Independent of any extension advice, more than 80% of tea producers in one district, Dai Tu, interplanted *Acacia* with their tea. Dai Tu has 2,600 ha of tea, accounting for 31.7% of the province's total tea area. When asked why they had adopted *Acacia* so rapidly, the farmers stressed two main factors:

- *Acacia* is a fast growing species that provides shade for tea, as well as timber and firewood, in a short time.
- *Acacia* improves soil fertility, but does not compete with the tea crop.

Table 62-2. Composition of Household Income in Lucngan District, Bacgiang Province (million dong)

Income Sources	*Before Planting Litchi*	*5–7 Years after Planting*	*8–10 Years after Planting*
Paddy rice	2.11	2.11	2.11
Other crops	1.18	0.39	0
Litchi	0	10.80	18
Total	3.29	13.30	20.11
Comparison (%)	100	404	611

Note: 1997 US$1 = 11,666 Vietnamese dong.

Characteristics of Tea and Acacia auriculiformis *Cunn.*

Tea (*Camellia sinensis* [L.] Okize) prefers acid soil with a pH of 4.5 to 5.5 and average moisture content. If grown under a canopy providing 40% to 50% shading, tea has superior growth and yields compared to that grown in full sunlight.

A. auriculiformis is a fast-growing tree that can reach 25 m in height and 60 cm diameter at breast height (dbh). It is known to attain 18 m in height and 20 cm dbh within 12 years. *Acacia* prefers plenty of sunshine, temperatures above 20°C, and an annual rainfall above 1,000 mm. It can, however, grow under quite adverse conditions, such as in low-fertility loam, sandy or rocky soils, and climates with five or six dry months. It flowers throughout the year. Its wood is suitable for pulp and firewood.

Acacia is initially planted at a density of 1,000 to 1,600 trees per ha. Farmers begin to prune and thin the stand at four to five years of age and continue to do so every two years. By the tenth year, the *Acacia* density is ideally about 400 to 600 trees per ha in highly fertile sites, 600 to 800 trees per ha in medium fertility sites, and 800 to 1,000 trees per ha in low fertility sites. At these densities, *Acacia* provides less than 50% shading.

Effects of Acacia *Canopy on Tea Quality and Yields*

The benefits provided by the *Acacia* canopy include moderated temperatures, reduced evaporation and wind speed, protection from direct sunlight, and increased air and soil moisture. This managed environment under an *Acacia* canopy provides tea yields 9% to 10.4% higher than when tea is grown in direct sunlight. Tea grown under *Acacia* is also of higher quality. It contains higher nitrogen organic matter, such as caffeine and protein, and less nonnitrogen matter, like tanin and glucid.

Effects on Soil Properties

Our studies found that under densities providing 30% to 50% shade, *Acacia* has a litterfall of four to eight tonnes/ha. This is about 25% of what would be expected from a broadleaf forest in the tropics.

Soils under an *Acacia*–tea combination have a thicker A horizon, reduced erosion, and soil moisture levels 10% to 15% above those found under tea without a protective *Acacia* canopy. The distinction between soil layers is also much clearer than under monocropped plantations. Nitrogen content in the A horizon is 10% to 29% higher in *Acacia*-shaded tea plantations, reflecting *Acacia*'s ability to fix significant quantities of atmospheric N. The A horizon also has P_2O_5 and K_2O levels that are 10% to 15% higher than those in the absence of any canopy.

Firewood as an Important Byproduct from Pruning

The *Acacia* trees are pruned every two years to maintain shading at 50%. This provides a substantial yield of 2.5 to 3 m^3 of firewood from the lopped branches that can then be used when processing the tea, thereby mitigating harvest pressures on natural forests.

Conclusions from Acacia*-Tea Studies*

Tea growers in Bac Thai innovated the system of integrating *Acacia auriculiformis* as a canopy into their plantations and it has since spread rapidly. It offers compelling advantages, primarily in boosted tea yields, N-fixation, and provision of firewood. This field-tested technique warrants further evaluation for its potential application in other tea-growing areas.

PART IX

Themes: Property Rights, Markets, and Institutions

A Han Chinese elder in Tengchong, Yunnan.

Chapter 63

Productive Management of Swidden Fallows

Market Forces and Institutional Factors in Isabela, the Philippines

Paulo N. Pasicolan

The rapid loss of forest in the Philippines, which began in the 1890s, has transformed huge areas, once rich with vegetation, into grassland. According to the Master Plan for Forestry Development devised in 1990, the Philippines had about 16 million hectares of primary forest remaining in the 1950s. However, wanton logging, combined with slash-and-burn cultivation (locally known as *kaingin* farming), reduced the country's area of natural forest to less than 1 million hectares by the 1990s. Removal of protective cover on this scale, over just four decades, degraded the soil over large areas. It is estimated that grasslands now cover more than 6.5 million hectares, or approximately 22% of the country's total land area (Concepcion and Samar 1995). This is believed to be the final product of this denudation of the land (Myers 1993).

A large portion of this grassland is classified as abandoned swidden fallows. Generally, it is marginal land with low productivity per unit area. It is often invaded by *Imperata cylindrica,* and the light-loving nature of this aggressive grass species, assisted by frequent burning, perpetuates its dominance over other plants. The rest of the grassland is classified as "pasture lease" areas.

From the perspective of many governments and international organizations, treeless tropical grasslands are simply "wastelands" (Turvey 1994). Ideally, such areas should be converted to hardwood plantations to meet the demand for tropical hardwoods (Grainger 1988), but regenerating grasslands with trees costs money and much effort, and the track record of such projects is not good. The Philippines' Contract Reforestation Program is one example. With plans to reforest 225,000 ha, the program borrowed a total of US$240 million from the Asian Development Bank and the Overseas Economic Cooperation Fund from 1988 to 1992. In the project's first three years, 24,000 pesos, or US$923, was spent per hectare on seedling establishment and maintenance. Yet only about 10% of the targeted area was successfully regenerated (Pasicolan 1996), and the dismal result fell far short of justifying the huge investment in the project (Korten 1993).

At the farm level, on the other hand, a number of self-motivated farmers have been able to establish agroforestry systems, or woodlots, on swidden fallow lands

---

Paulo N. Pasicolan, Professor and Director, Upland Resource and Development Center, Isabela State University, Cabagan, Isabela 1303, Philippines.

without direct financial support from the government (Pasicolan and Tracey 1996). Despite their success, these cases are not generally considered when public-sector, community-based, reforestation plans are being designed. The stark contrast in performance between government-driven reforestation and spontaneous tree growing at the farm level suggests the presence of critical success factors in the latter that are not evident in the former.

The aim of this study was to establish an empirical basis for how market forces and other institutional incentives could motivate subsistence farmers to grow trees on their swidden fallows, even without direct government financial assistance. The following main hypotheses emerge:

- Subsistence farmers may spontaneously introduce tree crops on their fallowed swiddens, despite their "risk-averse" nature, if they are driven by market forces and encouraged by other institutional incentives.
- Successful conversion of idle grasslands is a matter of forming a partnership between the tree growers, the market sector, credit institutions, and the Forest Management Bureau.

Methodology

Data for this study were gathered in the municipalities of Ilagan and Dinapigue, in Isabela Province, northeastern Philippines (see Figure 63-1). However, most of the secondary information used in this chapter was obtained from the Australian Centre for International Agricultural Research (ACIAR) SEARCA *Imperata* Project Paper 1996(10) (Pasicolan et al. 1996).

I interviewed 25 key respondents (10 from Ilagan and 15 from Dinapigue) on a staggered schedule from June 1995 to December 1996, using an unobtrusive and informal approach. Vayda's (1983) progressive contextualization and de Groot's (1992) actor-in-context analyses provided the main social inquiry techniques. The respondents were chosen on the basis of their initiative to plant *Gmelina arborea* on their idle land without direct financial support from the government.

Factors that triggered the respondents' decision to plant tree seedlings were closely examined. As an exploratory investigation, this did not involve statistical analysis. The respondents' tree-growing perceptions and motivations were all examined in relation to their development plans and options for the use of their fallow land.

The Case Studies

Salingdingan, Ilagan

Salingdingan is a *barangay*, or village, in Ilagan, Isabela, with an area of 173 ha. It is located on the western foothills of the Sierra Madre mountains, in northern Luzon. According to early residents, the area was covered by virgin forest, dominated by Dipterocarp and other premium timber species, in the 1940s. Four logging companies began operating in the area in 1952, and within one year, each was clearing forest at a rate of about 200 ha/year. The 1960s were considered a logging boom. Government incentives allowed the logging companies to extract 100% or more than the annual allowed cut during that period, so clearing was intensified.

As the loggers moved further afield in the 1960s, migration to the area increased. A large portion of the logged-over area was cleared and cultivated. However, population increases could not keep pace with the deforestation rate. Uncultivated logged-over areas slowly turned to grassland savanna, and shrubs, *Imperata,* and *Saccharum* finally dominated the land.

In the late 1960s, more people from nearby provinces moved in. The majority were peasants in search of land to cultivate. There were vast tracts of open public land suitable for farming, and it was grabbed up by both old and new occupants. Land that at one time was dominated by *Imperata* was planted in crops. However, as the soil became exhausted, the kaingins, or swidden fields, were abandoned or, while under long fallow, were used for communal grazing and became pasture lease areas.

As the community grew and became a more functional political unit in the 1970s, basic government infrastructures and social services arrived and people became more organized. Farm zoning and the boundaries of landholdings were strictly followed. Salingdingan became a farming community with a mosaic landscape of varied land-use types. In the late 1990s some lands were under long fallow, some areas were grassland pasture leases, but most of the area was cropland. Yellow corn remained the dominant crop, followed by bananas, vegetables, and tobacco. The area's mean net annual income per household was about 36,600 pesos.

In 1989, the Integrated Social Forestry Program (ISFP) was launched in Salingdingan. Under this program, farmer-participants were given a 25-year stewardship certificate that was a formal guarantee of their right to stay and use the public land they had been occupying for many years. A minimum of five hectares was given to each participating household. This tenurial instrument could be renewed for another term if the farmer developed the area into agroforestry by planting at least 30% of the land area with forest trees. The remainder could be devoted to short-term crops.

The ISFP set out to be a springboard for successful tree growing at farm level. However, soil conservation soon became its main concern. Farmers were forced to construct terraces and adopt sloping agricultural land technology (SALT) in order to comply with the program's prescribed technology. *Leucaena leucocephala* or *Gliricidia sepium* were recommended for hedgerows. Because of the emphasis on SALT adoption, participants were paid for every linear foot of hedgerow constructed on sloping land. At the start, there was a remarkable level of farmer participation. In 1990 the program was responsible for creating the Department of Environment and Natural Resources' first model upland farm in Isabela. But then the payments for hedgerow construction stopped, and so did the hedgerows.

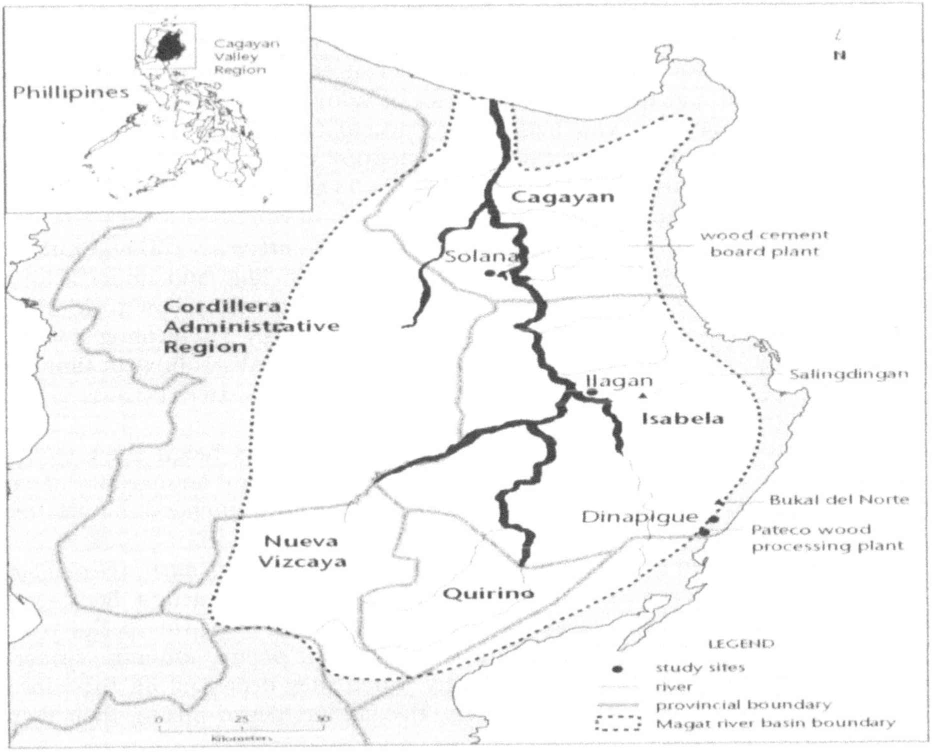

Figure 63-1. Location of Study Sites

Out of 45 original participants, 12 expressed their disgust with the project for failing to make full payment for their labor. Five disgruntled participants uprooted the *Leucaena leucocephala* hedgerows and replaced them with *Gmelina arborea*. Others allowed the hedgerows to grow without trimming, until they began to shade adjacent crops and their roots made contour plowing difficult. So they clear-cut the overgrown hedgerows for firewood and turned the area into open fields for yellow corn.

While the ISFP was designed to spearhead the vegetative regeneration of the area, little attention was paid to farm forestry. Yellow corn production remained the dominant preoccupation of the farmers, who were never encouraged to grow trees because there was no wood market nor financial assistance for plantation establishment and maintenance.

Meanwhile, in Cagayan, about 70 kilometers from Salingdingan, a wood–cement board manufacturing plant was established in 1994, despite the region's scarcity of wood. It hoped to supply growing domestic and international markets for wood board, and it needed a sustained supply of raw material. The plant launched a campaign to recruit smallholders with three to five hectares of land for tree growing. Contract growing agreements were made with farmers' cooperatives, local organizations, and other groups who were interested in tree farming. As an incentive, the plant offered market security as well as free seedlings and technical assistance. It also assured prospective tree growers that they would be able to obtain production loans from the government's Land Bank. Furthermore, harvesting and transport services were provided, albeit at the farmers' expense.

The plant also completed a cost-benefit study to show farmers what they could expect to earn from a three-year-old stand of *G. arborea*. Besides the short cropping cycle of three years, the farmers were impressed by the high rate of return on investment. Whereas they were averaging an annual income of only 36,600 pesos from yellow corn and other farm produce, they could expect a net gain of up to 51,992 pesos per hectare from three-year-old *Gmelina*. Table 63-1 shows the company's cost-benefit analysis in detail.

In June 1995, the ACIAR-SEARCA *Imperata* project's technical advisory team conducted a community meeting in Salingdingan as part of its information-dissemination drive. The meeting highlighted the concept of farm-based tree growing for rural livelihood and detailed the prospect of tying up with the wood–cement board plant. Fifty farmers who had earlier participated in the Integrated Social Forestry Program immediately expressed their intention to venture into tree growing. They intended not only converting a portion of their cornfields, but also re-cultivating their fallow lands for *G. arborea*.

Three months later, 10 of these farmers planted *G. arborea* seedlings, both in farmlots concurrently with intensive cultivation, and in idle land under a long fallow. In the former, *G. arborea* seedlings were interplanted with yellow corn, while in the latter, seedlings were planted in a monoculture with 2 by 2 m seedling spacing. At the same time, about 12 to 15 farmers began raising potted seedlings in time for planting during the rainy season. All of them wanted to grow trees because of the ready local market.

The enthusiasm of the farmers for tree growing was heightened by an educational tour of the wood–cement board manufacturing plant arranged in March 1996 by the ACIAR-SEARCA *Imperata* project study group. The number of farmer–tree growers increased to 30.

More recently, the government imposed a cutting ban on *narra* (*Pterocarpus indicus*), a premium species for furniture making and house construction. Because of *G. arborea's* comparable qualities, local furniture shops began using the species as an alternative raw material. This increased the demand for *G. arborea*, and more farmers began planting the species. Unlike previously, when they depended on the Forest Management Bureau for planting materials, the farmers began raising their own seedlings in their backyards.

Table 63-1. Projected Costs and Benefits of a Three-Year-Old *Gmelina arborea* Plantation, per Hectare (Philippine pesos)

Details, Costs, and Benefits	*Figures*
Given	
Tree spacing	2 m X 2 m
Number of trees	2,500
Gestation period	3 years
Cost per seedling establishment in the first years	
Land cultivation (tractor: 2,100 pesos/ha)	0.84
Seedling	3.00
Labor for planting (12 persons at 65 pesos/person/day)	0.31
Fertilization (organic): (6 bags/ha at 140 pesos/bag)	0.34
Applying fertilizer: (12 persons, 65 pesos/day)	0.31
Water impounding: 100 pesos/m^2 X 100 m^2	4.00
Maintenance: 6 times a year, 12 persons at 65 pesos/day	1.87
10% contingency (10.67)	1.07
Total cost:	11.74
Plantation establishment cost for 3 years	
First year total input	29,348.00
Second year input (30% of the first year input)	8,804.40
Third year input (20% of the first year input)	5,869.60
Total cost:	44,022.00
Assumptions	
Estimated harvestable trees at 20% mortality rate	2,000.00
Estimated weight per tree in kilograms (minimum)	150.00
Total weight per hectare	300,000.00
Projected income	
Gross income at 1 peso per kilogram: buying rate	300,000.00
Less 3 years establishment and maintenance cost	44,022.00
Less production inputs	225,978.00
Less cutting and transport costs	100,000.00
Net profit	155,978.00
Annual net income	51,992.66
Monthly net income	4,332.72

Source: Cagayan Wood Manufacturing Company, as cited in Pasicolan et al. 1996.

This change in land use, which had never before been achieved despite huge government spending on reforestation, can be attributed to the presence of a ready wood market.

Bukal del Norte, Dinapigue

Bukal del Norte is an upland village of Dinapigue. It is situated in the eastern coastal zone of the Sierra Madre mountains, and a large portion of the village is swampy. Paddy rice is grown as the staple crop. Swidden farms occupy adjacent hillsides.

The area was once covered with virgin forest. Commercial logging peaked in the 1970s to early 1980s, and during those years, members of almost all of the village's 200 households worked for the logging companies as forest guards, scalers, haulers, and mill workers. Only two of the five logging companies that began work in the area in the 1960s are still operating. The other three closed down in 1993. The government allowed LUZMATIM and PATECO to continue because of their good record of strict adherence to forest regulations, particularly with regard to reforestation. However, both companies scaled down their logging activities and intensified reforestation. The company closures and changes of operation meant a major setback to the livelihood of the people of Bukal del Norte. Their main source of income was drastically reduced.

However, one of the remaining companies, PATECO, saw the need to establish a community forest plantation to meet its future raw material needs. It initiated a tree-growing program at household level in 1993 as part of a community livelihood outreach project. The program was launched with a massive information campaign conducted in collaboration with the Community Environment and Natural Resources Office (CENRO), the local unit of the Department of Environment and Natural Resources (DENR). As well as outlining tree growers' responsibilities, the campaign offered incentives to prospective participants. PATECO offered starting capital and seedlings, and tree growers contributed labor and land. PATECO also agreed to buy wood from the growers at a fair price.

Then, the CENRO released public land for farm-based tree growing, on the basis that this would save a large amount of government money being spent on reforestation. To safeguard each party's interests, a tripartite agreement was forged between PATECO, the tree growers, and the CENRO.

In the first year, only five farmers entered into agreement with PATECO. They planted *G. arborea* and *Swietenia macrophylla* on areas of land that had been idle for many years. Much of it was abandoned swidden fallow. In the second year, after the CENRO started providing additional tree growing areas, the number of participants increased to 30, then to 105 in the third year. Eventually, almost all of the households in the community began growing *G. arborea* on their farmlots, even without direct government financial support. The size of the individual farmlots ranged from one to five ha.

When I asked the 15 farmer-respondents from Dinapigue why they were easily convinced to grow trees despite there being no direct financial support from the government, they cited the following reasons, in order of highest frequency: presence of a sure wood market, 100%; provision of starting capital, 87%; provision of additional planting area, 80%; subsidized price of seedlings, 60%; and free technical assistance, 33%.

With market security and starting capital for seedling establishment provided by PATECO, tree growers in Dinapigue successfully converted idle lands into tree-based systems.

Factors Leading to the Creation of Tree-Based Systems

On the basis of the findings of this case study, some emerging trends can be identified as conditions for successful conversion of swidden fallows into tree-based systems.

A Ready Wood Market

At both study sites, tree growers were eagerly anticipating a big income from their tree produce because of an assured market for their wood. This encouraged them to invest in tree planting, especially of fast-growing species.

This confirms the findings of Chowdhry (1985) that many farmers in Uttar Pradesh, India, converted part of their agricultural fields to *Eucalyptus* plantations because of a ready market with a promising wood price, along with other government incentives. Likewise, Hyman (1983) found that contract growing for the Paper Industries Corporation of the Philippines paper mill eased the risk and uncertainties of participants to such an extent that it became a contributing factor to the corporation's success.

Therefore, as a reforestation policy, tree growing should be tied to an assured market so that small tree growers may have a long-term stake in protecting and maintaining the trees they plant.

Availability of Credit Assistance

The readiness of the wood–cement board manufacturing plant to pay establishment costs to prospective tree growers at Ilagan reinforced the farmers' desire to plant *G.*

arborea on their farmlots. A similar offer, made by PATECO to tree growers at Dinapigue, provided an added incentive for farmers there to recultivate idle land for tree crops.

A similar experience in the late 1970s and 1980s involved the Paper Industries Corporation of the Philippines. In 1968, during the first year of a contract tree-farming project, only 22 farmers engaged in tree farming on an area totaling 220 hectares. In 1981, a loan for tree growers from the Development Bank of the Philippines caused the number of participants to grow to 3,778, covering 22,600 ha (Hyman 1983).

Similar cases have been reported in other countries. For instance, production loans attracted many poor farmers into tree growing at farm level in Plan Bosque, Ecuador (Gradwohl and Greenberg 1988), and in the Saemeul Udong reforestation project in South Korea (FAO 1982).

Generally, resource-poor farmers lack basic production inputs. Loans or credit are meant to solve cash flow problems or the lack of capital needed to start and sustain a tree-growing project. Resource-poor farmers will benefit from loans if interest rates are low and there is a grace period equaling the time it takes for tree crops to become marketable.

In principle, loans and other types of credit appear to be attractive incentives that could draw both small and large farmers into growing trees. Interest rates should be low and the mode of payment liberalized to help poor farmers who are unable to pay the prevailing interest rates of commercial banks. However, under favorable conditions, small farmers should also be able to pay commercial interest rates, if the banks involved restrict their lending to small amounts and they have good field infrastructure or socialized institutional arrangements. This is an area where the government could provide assistance by providing a security or collateral system for small farmers who are unable to repay their loans to private banks and other lending institutions on time.

Provision of Additional Planting Area with Security of Tenure

When the CENRO offered farmers the use of idle public land with secure tenure, farmers in Dinapigue had an added incentive to begin growing trees. For farmers who are keen to produce trees as an income source under good market conditions, there are various means of enhancing their motivation without necessarily using direct payments for tree planting. The most practical incentive from the government can be the provision of additional planting areas to interested individuals, with security of tenure. Allowing enterprising farmers to encroach upon and rehabilitate abandoned and idle lands, coupled with providing more in-kind incentives such as fencing materials, farm animals, production inputs, and technical support, would likely encourage them to occupy less-developed and inaccessible areas and successfully plant them in trees.

In the Philippines, the government recently launched a program known as the Socialized Industrial Forest Management Agreement (SIFMA). It invites individuals, family units, associations, or cooperatives to venture into tree farming on idle grasslands, brushlands, and open and denuded forest lands (SIFMA Primer 1996). An individual or single-family unit can have 1 to 10 ha, and associations or cooperatives can obtain 10 to 500 ha. The participants are provided with security of tenure for 25 years, renewable for a further 25-year term.

Partnership between Government and Private Sectors

A common and important aspect of successful tree growing at the two study sites was the partnership between tree growers, private industry, and the government. Such interinstitutional collaboration relieves would-be tree growers of very substantial concerns about marketing their future wood produce. The Philippines government spent much money for tree planting in the first three years of its contract reforestation program. However, because no formal market agreement had been made with the wood industry, participants were unable to see any reason to persevere with the project, especially when payment for seedling maintenance ceased. In other words, there was no guarantee that tree growers would profit from their investment, because of discontinuous or nonformalized involvement of the private sector.

Instead of viewing the establishment of forest plantations from the three different perspectives of government, private industry, and local communities, and considering separately the different motivations of each (Turvey 1994), researchers should treat them as one complete package, combining the concerns and inputs of all three interested parties.

A well-organized tree-growing program is one where the community or individual family establishes the forest plantation, the private wood industry guarantees the market, and the government, through the Forest Management Bureau, provides the policy instruments and other institutional incentives. Unless the private wood industry has a stake in any tree-growing program, there will be no guarantee of its success.

Effective Extension Program

By nature, most rural peasants are conservative. They are reluctant to change their traditional patterns of subsistence, particularly if such changes require sacrificing what they currently consider expedient. They prioritize their immediate, direct, and specific needs and are always conscious of the inherent risks of trying something new (Leach and Mearns 1988).

When many mill and field workers in Dinapigue lost their jobs because of the closure of three local logging companies, it should have discouraged them from any further involvement in the forestry industry, particularly something like tree growing, which was perceived as being highly risky. However, PATECO succeeded in convincing them to participate in its smallholder tree-farming project because of its strong information dissemination campaign. Similarly, farmers in Salingdingan were eager to grow *G. arborea* as a result of the consultative community meeting arranged in June 1996 by the ACIAR-SEARCA *Imperata* Project.

These two interventions succeeded in convincing farmers to begin growing tree crops. Two factors were vital to the success of these extension activities: a ready wood market and the availability of credit assistance.

Conclusions

The successful introduction of tree-based farming in swidden fallows in the two municipalities of Isabela stemmed from the interplay of the following factors:

- The presence of wood markets, which became the pivotal motive for farmers to grow trees on their present and idle croplands;
- The availability of credit assistance, which circumvented tree growers' cash flow problems;
- The provision of additional planting areas, which gave tree growers added incentive to develop other idle lands for tree growing;
- The close partnership between the tree growers, the wood industry, credit institutions, and the Forest Management Bureau, which provided security and mutual support; and
- An effective extension program that linked the tree growers to the market, credit institutions, and government support.

Policy Implications

Some of the lessons from these case studies may have policy implications on a wider geographical stage:

- Conversion of idle grasslands to tree-based systems in debt-ridden countries does not necessarily have to depend on huge external funding to propel successful tree-growing programs.
- Spontaneous tree growing at the farm level can be triggered by market forces, provided the other basic institutional infrastructures are also in place, such as the availability of credit support, security of land and tree tenure, and enabling government policies.
- Tree growing at the farm level should go beyond just meeting households' direct needs. It should be market-oriented to bolster the entrepreneurial capacity of small farmers.
- For maximum impact, a tree growing program should consist of a complete package of incentives agreed upon under a formal contract between tree growers, private wood markets, credit institutions, and the Forest Management Bureau.
- A farm forestry for rural livelihood campaign should be packaged by the wood industry, representing the market sector, to boost the confidence of small tree growers. This should be carried out in collaboration with the local government unit, Forest Management Bureau, credit institutions, and the Department of Education, Culture, and Sports.
- The government should provide more enabling policies to enhance the active participation of the wood industry and the community in tree growing at the farm level. These policy measures should include tax exemptions or discounts for private tree growers, access to credit institutions, favorable tenurial arrangements, brokering and networking assistance, and provision of good infrastructure in support of farm forestry programs.
- Isabela's experience in the transformation of idle grasslands into more productive tree-based systems could provide a prototype for larger-scale grassland rehabilitation projects.
- Spontaneous grassroots transformation of swidden fallows into farm-based tree growing areas can be possible if support systems are in place to circumvent farmers' financial, physical, legal, and institutional constraints.
- Failure to create the right blend of institutional incentives from the government, investment options from the private sector, and resource capacity from farmer–tree growers will continue to see national funds for reforestation siphoned off down the drain.

Acknowledgments

Special thanks go to the Australian Centre for International Agricultural Research (ACIAR), Canberra, Australia, and the Southeast Asian Regional Center for Graduate Study and Research in Agriculture (SEARCA), Los Baños, Philippines, for their financial and technical support in conducting the *Imperata* Project titled Improving Smallholder Farming Systems in *Imperata* Areas of Southeast Asia: A Bioeconomic Modeling Approach. Great appreciation is extended to Dr. Ken Menz of the Australian National University for his valuable suggestions and assistance in preparing this chapter.

References

Chowdhry, K. 1985. Social Forestry: Who Benefits? In: *Community Forestry Socio-Economic Aspects*, edited by Y. S. Rao et.al. Bangkok, Thailand: East-West Environment and Policy Institute and the U.N. Food and Agriculture Organization. RAPA.

Concepción, R.N., and E.D. Samar. 1995. Grasslands: Development Attributes, Limitations and Potentials. Paper presented to the Forest National Congress, September 26–28, 1995, Los Baños, Laguna, Philippines. de Groot, W.T. 1992. *Environmental Science Theory: Concepts and Methods in a One-World Problem-Oriented Paradigm*. Amsterdam, London, New York, Tokyo: Elsevier.

DENR (Department of Environment and Natural Resources). 1990. *Master Plan for Forestry Development*. Quezón City, Philippines: DENR.

———. 1996. *SIFMA (Socialized Industrial Forest Management Agreement) Primer*. Quezón City, Philippines: Forestation and Afforestation Section, FRDD, DENR.

FAO (Food and Agriculture Organization of the UN). 1982. Village Forestry Development in the Republic of Korea: A Case Study. GCP/INT347/SWE. Rome: FAO.

Gradwohl, J., and R. Greenberg. 1988. *Saving the Tropical Forest*. London: Earthscan Publications.

Grainger, A. 1988. Estimated Areas of Degraded Tropical Lands Requiring Replenishment of Forest Cover. *International Tree Crop Cover* 5, 31–61.

Hyman, E.L. 1983. Pulpwood Tree Farming in the Philippines from the Viewpoint of the Smallholder: An ex post Evaluation of the PICOP Project. *Agricultural Administration* 14.

Korten, F.F. 1993. The High Cost of Environmental Loans. *Asia Pacific Issues 7*. Honolulu: East-West Center.

Leach, G., and R. Mearns. 1988. *Beyond the Woodfuel Crisis: People, Land and Trees in Africa*. London: Earthscan Publications.

Myers, R.J.K. 1993. Introduction to the Network on the Management of Acid Soils in Asia. *IBSRAM Network Document* 6. Bangkok, Thailand: IBSRAM.

Pasicolan, P.N. 1996. Tree Growing on Different Grounds: An Analysis of Local Participation in Contract Reforestation in the Philippines. Ph.D. dissertation, Centre of Environmental Science, Leiden University, The Netherlands.

———. A. Calub, and P.E. Sajise. 1996. The Dynamics of Grassland Transformation in Salingdingan, Ilagan, Isabela, Philippines. *Imperata* Project Paper 1996/10. Canberra, Australia: ACIAR (Australian Centre for International Agricultural Research).

———. and J. Tracey. 1996. Spontaneous Tree Growing Initiatives by Farmers: An Exploratory Study of Five Cases in Luzon, Philippines. *Imperata* Project Paper 1996/3. Canberra, Australia: ACIAR (Australian Centre for International Agricultural Research).

SIFMA Primer. 1996. Forest Management Bureau, Forestation and Afforestation Section, FRDD, DENR, QC, Philippines.

Turvey, N.D. 1994. Afforestation and Rehabilitation of *Imperata* Grasslands in Southeast Asia: Identification of Priorities for Research, Education, Training and Extension. In: *ACIAR Technical Report* 28. Canberra, Australia: ACIAR (Australia Centre for International Agricultural Research).

Vayda, A. 1983. Progressive Contextualization: Methods for Research in Human Ecology. *Human Ecology* 11(3). Plenum Publishing.

Chapter 64

The Feasibility of Rattan Cultivation within Shifting Cultivation Systems

The Role of Policy and Market Institutions

Brian M. Belcher

One of the options available for increasing the productivity of shifting cultivation systems is to grow high-value products within the forest that develops during the fallow period. This kind of agroforestry system can provide a range of benefits to the farm household, including additional cash and in-kind income, as well as the benefits of income diversification, risk spreading, and evidence of land occupancy. Moreover, this approach may make longer fallow periods economically feasible on a given plot of land, with associated benefits of soil regeneration, weed control, and enhanced ecological functions.

In developing innovations of this kind, it is tempting to focus first and foremost on technical feasibility. However, unless an innovation is socially and economically feasible, its technical merit is only of academic interest. There must be a demand for the product. In the increasingly cash-based economies of Asia, farmers appreciate products that can be sold for cash. While there are numerous forest products that can be grown in forest fallows, the benefit of adopting their cultivation ultimately depends on their value in the marketplace. However, even when there is a potential market, the actual benefit gained by farmers will depend on the institutional framework within which the trade is carried out, that is, the structure of the market and the policy environment that guides it.

The case of rattan very clearly illustrates the sensitivity of forest or agroforest product markets to institutional and policy influences. Rattan is one of the most important of all of the tropical nontimber forest products (NTFPs). This group of plants exhibits many of the important characteristics for which NTFPs have become popular to rural poor people, including their renewability and accessibility. There are large, well-established markets and an existing range of production-to-consumption systems for rattan in Asia and, to a lesser extent, in parts of tropical Africa. Moreover, at least two traditional shifting cultivation systems incorporate rattan cultivation in the fallow. These are useful models, demonstrating that the approach can be both technically and economically feasible.

Brian M. Belcher, Principal Scientist, Forest and Livelihoods Programme, Center for International Forestry Research (CIFOR), P.O. Box 6596 JKPWB, Jakarta, 10065.

However, rattan raw material producers, whether they collect rattan from the wild or cultivate it, often have very weak bargaining positions and have to accept low prices for their product, even under conditions of strong demand at the manufacturing stage. The structure of the market and sweeping policy changes over the past decade have conspired to worsen their position. The rattan cultivators of Kalimantan, one of the two documented examples of traditional rattan cultivation, have been particularly hard hit (see color plate 52). In some places rattan cultivation has been rendered uneconomic, and farmers are abandoning it.

It is instructive to examine this system to understand the role of institutions and policy on the economic feasibility of an improved fallow system. This chapter will begin with a brief overview of the rattan sector in Asia. It will then focus on the Kalimantan rattan production-to-consumption system, with emphasis on the current policy environment and market structure. Under current conditions, rattan cultivation is economically feasible only in relatively accessible areas. This is reflected in a financial cost-benefit analysis and corroborated by reports that farmers are reducing their efforts to cultivate rattan. From this case, a number of lessons can be drawn that may be applicable to the encouragement of intensified fallow systems.

Background

Rattan is a collective term for a large group of climbing palms. Their importance as tropical NTFPs is felt at many different levels of the economy. Rattan provides a source of employment and income to people with very limited opportunities to earn cash. It is a versatile raw material for numerous small- and medium-scale enterprises, and the finished products, as well as the raw materials themselves, find important international demand. Sales of rattan and rattan products generate valuable foreign exchange needed by developing countries in an increasingly global economy.

Rattans are primarily plants of the tropical forest, occurring mainly in Asia, although a few species can also be found in the equatorial rainforests of Africa. They grow as spiny, woody vines, climbing up trees to reach the canopy. They typically require a degree of shade during their establishment; the optimal light penetration is around 50% for the main commercial species. Most also require relatively high humidity. Rattan grows best in a forest environment, with overstory vegetation to provide shade and regulate microclimatic conditions, as well as to provide trees for the rattan to climb. Rattan will grow along the ground, but stems touching the ground are susceptible to damage from infections. For this reason cultivation in a pure stand is not effective for production of stems. Some arrangement of rattan with trees is necessary.

There are two principal growth habits among the rattans: solitary and clustering. Many, though not all, of the larger rattans, which reach more than 20 mm in diameter, grow as a single stem from the root. Once cut, these plants will not regenerate. They are called solitary species. Most of the smaller-diameter rattans, with stems of less than 20 mm in diameter, have a clustering, multistemmed, architecture. It is possible to cut the stems of these species, and the plants will continue to grow and to send out more stems. This characteristic lends itself to incorporation in managed systems.

The furniture industry is the largest consumer of rattan. The larger-diameter rattans are mainly used as structural elements in manufactured articles and especially as frames in furniture. The rattan stem (or cane) may be used with the skin on, especially if it is smooth and unblemished, or it may be peeled, although peeling reduces its strength. Small-diameter rattan stems are usually split or peeled to get thin strips of the tough outer layer for weaving, to make rattan webbing or wickerwork. This material is used for cane chair seats. The inner core is also used, and increasingly so, as raw material becomes more expensive and as coring machines become more widely available to facilitate processing. The core is weaker but still very usable in wickerwork.

As well as being used in weaving, rattan is used in various handicraft applications, as binding, and as cordage. In some areas, such as Lao P.D.R., northeastern Thailand, and the northeastern states of India, young stems of rattan are considered a delicacy and are eaten in soups and stews. The fruits of some species are edible, and the fruits of other species are used to produce dyes. Rattan is also still in demand for use as walking canes and in specialized applications in sporting goods, such as the handles of cricket bats and polo mallets.

The global market for rattan is substantial. For example, export earnings from rattan in Malaysia went from US$1.76 million in 1984 to US$14.6 million in 1988 (IFAR 1991). Exports of finished rattan products, including furniture, woven mats, baskets, and handicrafts, earned US$119 million for Indonesia in 1988, increasing to US$293.6 million in 1992 (Nasendi 1994). China exported more than US$42 million worth of rattan products in 1993 (Zhong et al. 1995). The Philippines' exports of rattan furniture, baskets, and wickerware earned about US$240 in 1994, a performance that seems to have been supported by increased and unsustainable domestic rattan harvesting (Pabuayon et al. 1996).

The bulk of the rattan coming onto the market is harvested from the wild. IFAR (1991) estimates the proportion at 90%. It grows mainly in Southeast Asia, with Indonesia by far the leading producer. Some estimates put the Indonesian contribution at 80% of world supplies. Other historically important producers include the Philippines, Malaysia, and Thailand. All of these countries previously exported unprocessed rattan to other Asian countries, especially Taiwan, Hong Kong, and Singapore, and to European countries, such as Italy and Spain, for processing. However, a combination of overexploitation and the destruction of habitat through widespread deforestation and forest degradation have exhausted rattan supplies in many areas. Harvesting efforts have now shifted to new, relatively unexploited areas, and when the best species can no longer be found, lower-quality species are being used as substitutes (IFAR 1991). In several countries, wild populations of the most important rattan species are nearing commercial extinction, for example, *Calamus manan* in Thailand and Malaysia.

As a response to the rapid overexploitation of rattan resources, and as a means to promote increased domestic processing, all the major producing countries imposed bans on the export of unprocessed cane in the late 1970s to mid-1980s. This enabled other Southeast Asian countries to exploit the market, and unprocessed rattan is now being exported from Vietnam, Myanmar, Papua New Guinea, and Lao P.D.R.

Surprisingly, even in the face of increasing global demand for rattan and decreasing natural supplies, rattan cultivation has not been broadly adopted. There are a number of cultivation models that hold potential for increasing production. For example, there appears to be scope for enrichment planting and intensified management within natural forests. There are some anecdotal reports of rattan harvesters experimenting with enrichment planting, and in some places rattan cutting licenses require harvesters to replant, but, to date, enrichment planting of rattan has been relatively insignificant (Belcher 1997).

Two traditional rattan cultivation systems have been described in the literature. One of these is practiced by the Hani people of Yunnan Province, in southwestern China (Pei and Chen 1990; Chen, Pei, and Xu 1993; Pei et al. 1994; Xu, Chapter 56). The Hani plant rattan in swidden fields, although it is not clear from the articles when the planting takes place, and allow it to mature for 10 to 12 years before harvesting. The second traditional rattan cultivation system is practiced in several areas of Borneo, mainly in the Indonesian states of East Kalimantan and West Kalimantan. This is discussed below.

Much more recently, commercial rattan plantations for stem production have been established within rubber and timber plantations. These plantations, using a variety of large- and small-diameter species, have been undertaken on a very large scale in Malaysia, where they now cover close to 20,000 ha (Lim and Nur Supardi 1994). Experimental efforts have also been undertaken in the Philippines and Indonesia, but these are not on a scale sufficient to make a significant impact on the trade (Pabuayon et al. 1996). Some of the rattan from these plantations should begin to come onto the market in the next few years.

Rattan Cultivation in Long Fallow: Costs and Benefits

Growing rattan in the fallow appears to be a good approach to intensifying a fallow-farming system. The system practiced in Kalimantan has been well described in the literature (Weinstock 1983; Mayer 1989; Setyawati et al. 1989; Godoy 1990; Godoy and Ching Feaw 1991; Fried and Mustofa 1992; Peluso 1992; Boen et al. 1996; see also Chapter 38). This discussion will therefore provide only a general overview of the cultivation system, followed by a more detailed look at the policy and market institutions that influence its economic feasibility.

Agriculture in the rattan garden areas is mainly based on a swidden system in which fields, or *ladang* in Indonesian, are prepared by cutting and burning an area of primary or secondary forest. The fields are typically between 0.7 and 1.5 ha and the main crop is dryland rice.

Descriptions of how the rattan is managed vary somewhat among the various authors and may reflect both actual differences in practice from place to place and modifications to the system over time (Sasaki, Chapter 38). For example, in Weinstock's (1983) account, farmers plant rattan in swidden fields that are ready to be fallowed and return after 7 to 10 years to cut the rattan and reestablish the fields for cropping. With rattan established in this way in several swidden fields, a farmer can count on a rattan harvest each year. Some authors say that planting may be done after the rice is established (Mayer 1989). In the Pasir district of East Kalimantan, farmers plant rattan seeds at the same time as rice is planted in a newly cleared ladang, or after the rice is established. According to participant interviews, these practices can vary even within a community. The rattan seedlings are protected during the one to two years that the field is used for rice cultivation.

In any case, after one or two rice crops are harvested from the ladang, the area is allowed to revert to secondary forest. This secondary forest is really a garden (or *kebun*) containing a relatively high density of rattan plants and numerous other valuable plants that have been protected or planted by the farmer. The garden is maintained as a source of fruits, wild animals, medicinal plants, wood, and rattan (Fried and Mustofa 1992; Michon 1997).

In contrast to the description provided by Weinstock, the rattan gardens are managed on an ongoing basis. The rattans used are multistemmed, small-diameter varieties that lend themselves to multiple harvests. This is a good strategy, as the establishment time for rattan is fairly long, but once the rootstock is established, stems can be harvested every two to four years.

A little extra labor is required for gathering seeds and planting them, and then later for protecting young rattan plants and training them to available trees, but most of this work can be done along with regular farm jobs. Also, rattan takes a long time to mature, so the land must be kept "fallow," as a rattan garden, for at least 8 to 10 years, but more likely for 15 to 20 years or even longer. Once established the rattan needs very little maintenance, and the only other cost involved is the labor to harvest it and haul it to a pickup point.

The most important benefit of the rattan garden system is the cash income generated from the forest fallow. However, the system is also attractive to farmers for its ability to spread their risk. That is, they have a ready source of cash available in case crops or other sources of income fail. The ability to defer harvest during periods of low prices, or to save up for festivals or other events requiring large cash outlays, is also appreciated. According to the farmers, this flexibility is one of the great advantages of the rattan fallow system. The rattan continues to grow in the forest, with low rates of deterioration or overmaturity. As one farmer put it: "A man who has a rattan garden has money in the bank." The ability to generate cash revenue from the "fallow" also makes longer fallows feasible, with many resulting ecological benefits. The longer period under secondary forest permits improved soil regeneration from the buildup of organic matter, reestablishment of microflora and microfauna, erosion protection, more complete weed reduction through shading and competition from trees, and habitat for forest plants and animals. Farmers also husband and collect a number of other products from the fallow. Finally, in areas where farmers typically do not have official land titles, perennial crops "contribute to family wealth by bearing witness to land occupancy" (Gouyon et al. 1993).

Financial Analysis

A simple financial cost-benefit analysis is a useful tool with which to assess the potential profitability of an activity, even though such analyses fail to capture many of the economic benefits enumerated above.

Despite the apparent technical feasibility of a range of rattan cultivation options, very little work has been done to determine the financial feasibility of possible models. Very few growth and yield studies have been undertaken on rattan. Some experimental planting work has provided information on survival and growth in young rattan plants, mainly from the most valuable commercial species. Additionally, some studies have estimated harvest rates from natural stands. However, the wide variety of rattan species used and the different growing conditions encountered make it difficult to apply existing data beyond specific situations. The few financial and economic assessments of rattan cultivation models have been based mainly on extrapolations of studies of the survival and growth of young rattan plants.

Most of the research attention on rattan cultivation to date has been focused on relatively intensive rattan plantations. Work at the Forest Research Institute of Malaysia provides some of the key references. For example, Halim and Aminuddin (undated), Latif and Rahman (1986), Salleh and Aminuddin (1986), and Rauf (1982) all consider the financial feasibility of planting *Calamus manan* within timber and rubber plantations. Similar work has been done in the Philippines on *Calamus merillii* (PCARRD 1991). These studies projected financial returns (internal rate of return, or IRR) ranging from 11% to 20%. Similar studies considered the financial potential of planting small-diameter rattan species in India (Muraleedharan and Seethalakshmi 1992) and in China (Xu et al. 1990).

The rattan gardens of Kalimantan have also attracted attention in this respect. Setyawati, Rachmat, and Ayi (1989) did a financial analysis, and Godoy (1990) conducted both financial and economic analyses. The economic analysis in that case adjusted some of the prices to account for distortions in the exchange rate and in the labor market but disregarded other economic benefits. The former study estimated an IRR of 27%, while the latter came out at 21%.

My analysis uses data collected as part of a study of the Kalimantan rattan production-to-consumption system (PCS) supported by the International Network for Bamboo and Rattan (INBAR). A research team from the Forest Products and Forestry Socio-Economic Research and Development Centre of the Indonesian Ministry of Forestry conducted interviews with farmers, traders, rattan product manufacturers, and laborers, as well as with government officials and a range of key informants, between September and November 1995. I participated as a member of the team in November 1995. The research is reported in Boen et al. (1996). Additional information for this analysis has also been gathered from the secondary sources cited above. Therefore, this analysis represents a composite picture, but generally reflects 1995 conditions in the Pasir District. Prices are given in US$ for ease of comparison, and sensitivity analyses help to take into consideration the impact of changes in important variables.

The financial assessment is done for a one hectare plot over 24 years. It follows the model employed in the Pasir District of East Kalimantan, where rattan seeds are planted in newly cleared swidden fields and grown with field crops and then in the forest fallow. Two or three seeds are planted in suitable locations at a density of between 50 and 100 locations per hectare (250 seeds per hectare). Assuming germination failures and mortality, the rattan garden will have about 80 plants per ha. Technically, a higher density is feasible, but farmers dislike too much rattan, as it interferes with their movement while carrying out other activities in the area.

The cost of establishing a rattan garden includes the cost of land preparation, seeds, and planting. Clearing of a fallow field is necessary as part of the shifting cultivation system used by farmers in the area, and it is difficult to attribute the appropriate cost to the rattan component. The total cost of field preparation is estimated at about US$70 (Boen et al. 1996). This cost must be borne whether or not rattan is planted. However, for the purpose of this analysis and following Boen et al. (1996), it is conservatively assumed that one-half of this cost, or US$35, is attributable to rattan garden establishment. This is somewhat lower than the approximately US$75 assumed by Setyawati et al. (1989) and more than the US$23.46 assumed by Godoy (1990). Seeds are gathered locally. The cost is equivalent to the labor cost of gathering and preparing them, or US$15 (US$0.06 times 250). Planting is estimated to take two days, or US$5. Total establishment cost is therefore estimated at US$55. Maintenance is required for the first four years of establishment. This involves a labor cost of three days, or US$7.50, per year.

The opportunity cost of land is very low because, under the existing shifting cultivation system, land is left fallow anyway. Presumably, if the rattan gardens were not there, the farmers might bring the area into cultivation sooner and perhaps reduce the pressure on other lands. However, other benefits from the long fallow, including bushmeat and other valuable food and medicinal plants, offset the costs of keeping the land out of agricultural production for a longer period, so the land cost is assumed to be zero for the purpose of this analysis.

The first harvest is assumed to be possible after eight years, or one year longer than assumed by Godoy (1990). Subsequent harvests are possible every three years, for a total of eight harvests over 30 years. Each cluster yields about 20 kg wet weight of green rattan in the first year, or 1.6 metric tonnes from one hectare. This increases to a steady 25 kg per cluster per year in subsequent harvests.

Harvesting costs include only the cost of labor for cutting, bundling, and carrying the rattan to the road. On average, a rattan cutter can manage about 50 kg of green cane per day, so harvesting will cost about 32 man-days at US$2.50 per day for a total of US$80 for the first harvest, and 40 days for each subsequent harvest for a total cost of US$100.

The current farm gate price for *rotan sega* in Indonesia is about US$123 per metric ton of green cane, so a conservative estimate of revenue from the first harvest is US$197, and US$246 for subsequent harvests.

These data were used to estimate the costs and returns from a rattan garden over a 30-year period. Future costs and benefits were discounted at 15% to determine their present values. Calculations included the net present value (NPV), which is the present value of benefits minus the present value of costs, and the benefit:cost ratio (BCR), or the present value of benefits divided by present value of costs. The internal rate of return (IRR) of the investment, that is, the discount rate at which the present value of costs and benefits are equal, was also calculated. A sensitivity analysis was done to show the effect of independent increases and decreases in yield, price, and labor costs, and the effect of higher and lower interest rates on the NPV and BCR. All other conditions were assumed to remain stable. The results appear in Table 64-1.

This analysis shows that in Indonesia, rattan cultivation is profitable but returns are low. Raw material prices are still high enough to justify planting and harvesting activities in the relatively accessible areas of Pasir District. However, even in these areas, there are rattan farmers who have reduced or discontinued their harvesting activities because they consider prices to be too low.

In more remote areas, higher transportation costs, in-transit losses, and risk translate into lower raw material prices. The sensitivity analysis demonstrates that the system is very price sensitive. A further decrease in price, or higher labor or discount rates, strongly reduces or reverses profitability. Rattan cultivation quickly becomes untenable, and in places like the Middle Mahakam region, large numbers of farmers who expanded their rattan gardens under conditions of relatively high prices reduce or abandon them (Haury 1997).

The curious thing is that these low prices prevail even in the face of a stable or increasing world demand for rattan raw material. Therefore, we need to look beyond production of the raw material and examine the entire system in which the raw material is produced, processed, and marketed to discover the reasons.

The Production-to-Consumption System

A production-to-consumption system is defined here as the entire set of actors, materials, activities, and institutions involved in growing and harvesting a particular raw material, transforming the raw material into higher-value products, and marketing the final products. The system includes the technologies used to grow and process the material, as well as the social, institutional, and economic environments in which these processes operate. Changes at any point in the PCS can influence demand for a raw material, such as rattan, in terms of quality, quantity, and timing, with related effect on the price (Belcher 1997).

In the Kalimantan rattan PCS, the main rattan species planted in the gardens is known locally as *rotan sega* (*Calamus caesius*) or, in wetter areas, *rotan irit* (*Calamus trachycoleus*), although a number of other species are also used. These two species are the main raw material used in the valuable *lampit* industry in Indonesia. Lampit is a mat made of strips of rattan sewn together with heavy thread. In 1992, the industry was worth US$3 million in exports (Nasendi 1994). These two species also provide much of the small cane used in the Southeast Asian furniture industry.

Market Participants

The main market participants in the PCS are the following:

- Farmers: Farmers may grow rattan as one of several crops. Rice is by far the most important food crop, along with bananas, vegetables, root crops, and rubber. However, rattan provides upward of 75% of some farmers' cash income (Boen et al. 1996). Farmers typically wait for an order from a trader, and an adequate price, before harvesting. The trader may pay in advance or on delivery or collection.
- Village traders: A few of the rattan farmers in producing areas also do some trading. They accept orders from regional traders and purchase rattan from other farmers in their area. They may have some rudimentary storage facilities. In some cases, they deliver the rattan to regional traders or, rarely, to a manufacturer, or the regional trader may pick it up from them. The village traders may also do some sorting. In Kutai, where most of the rattan is shipped to Java, village traders dry, wash, and sulfur treat the rattan, adding to its value.

Table 64-1. Rattan Cultivation in Long Fallow Model: Sensitivity Analysis

Assumptions	*Internal Rate of Return (IRR)*	*Net Present Value (NPV) at 15%*	*Benefit:Cost Ratio*
Base case	19%	US$45.38/ha	1.27
+ 20% yield (price)	22%	US$87.50/ha	1.53
- 20% yield (price)	15%	US$3.27/ha	1.02
+ 20% labor cost	16%	US$12.34/ha	1.06
- 20% labor cost	23%	US$78.43/ha	1.59
Base case at 10%	NA	US$151.39/ha	1.63
Base case at 20%	NA	–US$5.46/ha US$34.08/ha	0.96

- Regional traders: The lampit manufacturers normally use an intermediary to procure raw material. These regional traders tend to have long-term relationships with one or a few buyers. They also maintain relationships with a number of local rattan traders who can arrange for cutting and collection on order. These traders normally arrange for collection and provide transportation. They also frequently provide advance money for the rattan orders, arrange the necessary transport permits, and pay the official and unofficial charges required to transport the rattan.
- Manufacturers: The bulk of the rattan from gardens in Kalimantan is used to make lampit. The main market for more than 90% of the production is Japan, where it is sold as *tatami*, the traditional floor covering. In addition to the lampit, a number of other rattan-based products are manufactured in the area. Most are woven products, including various mats and webbing, made from peel. In 1993, there were 53 lampit enterprises in South Kalimantan, employing more than 12,000 people (NRMP 1996a). They were concentrated in two cities in South Kalimantan, Amuntai, and Banjar Baru. There were also numerous small family-based enterprises, several with between 40 and 70 workers, and one with 145 workers. Some do all their processing and manufacturing in-house, while others subcontract parts of the processing procedure. The subcontracted work is often undertaken by families, in their homes.

  Other important stakeholders in the system are workers, including factory workers and subcontractors, and the Association of Manufacturers and Exporters (ASMINDO).

Vertical Linkages

Trading in the Kalimantan rattan PCS is strongly based on personal relationships and formal and informal contracts. As it is in most of Asia's major rattan PCSs, price competition is less important in this PCS than hierarchical forces (Belcher 1997).

The market functions in a very "top down" way; the manufacturers typically place orders for raw material through regional traders who, in turn, place orders with village traders, who organize local farmers to supply it. The price seems to be determined largely by the manufacturers, with limited bargaining room, and offers are made on a take it or leave it basis.

Sellers in the early stages of the PCS, and especially the producers, may have only one or very few buyers to choose from. Market information is limited and sometimes confusing. In almost all cases, the buyers are the primary source of price and market information. In general, the rattan farmers are price takers, with their buyer-traders controlling the price.

A variety of informal contracts exist in the system, with traders commonly advancing money or supplies to producers. Even at higher levels in the system, long-term personal arrangements between business partners tend to be more important than "spot sales."

Horizontal Linkages

In contrast to the strong vertical linkages described above, the links between various actors at any given stage, such as between rattan farmers or between traders, are relatively weak, especially in the early stages of the PCS. Rattan farmers may have informal contact with other farmers in their village or neighboring villages, and they may share information about prices, technology, or other issues. However, there are no formal growers' organizations or collective bargaining mechanisms. Similarly, there are no formal organizations of village traders or of regional traders. Higher up the PCS though, at the level of the lampit manufacturing and exporting industry, horizontal linkages through the trade organization ASMINDO are very strong. This government-approved exporters' system and joint marketing board is very powerful, as demonstrated by the imposition of quantity- and quality-based export restrictions, discussed below.

Market Institutions

The strong vertical linkages and relatively weak horizontal linkages found in the rattan market of Kalimantan are not unique. Indeed, a series of studies on both bamboo and rattan PCSs throughout Asia have found remarkable similarities in market structure, which leaves raw material producers in relatively weak positions (Belcher 1997). This can be explained, at least in part, by the nature of the material.

Most importantly, quality is inconsistent. Rattan is not homogeneous. The size, density, and other inherent characteristics important to users vary with species, with the particular variety, and with its age and growing conditions. As processing and manufacturing become increasingly mechanized, and as the manufacturers try to supply larger orders, uniformity becomes more and more important. Even color is important. According to rattan producers I interviewed, recent demand favors lighter-colored material than in the past, leaving producers of darker material facing lower prices than expected.

Furthermore, rattan is perishable. It is prone to fungal staining, called *blue stain*, cracking and checking, and damage by various pests. Quality very much depends on proper postharvest treatment, especially drying. There are also numerous reports of unscrupulous sales practices in which sellers, at some point in the PCS, adulterate their product by concealing immature or otherwise poor-quality material inside bundles of better-quality canes. Sellers also sometimes soak rattan to increase its weight.

Under these conditions, it is not possible for buyers at any stage in the PCS to simply purchase their rattan sight unseen. Each and every consignment must be examined to determine its quality, and the price must be negotiated. This is a very costly approach. Alternatively, buyers may enter into longer-term relationships with sellers. In these cases, trust and informal longer-term contractual agreements can reduce the transaction costs of buying and selling.

From the sellers' point of view, longer-term arrangements are also attractive. Rattan production is highly dispersed geographically. Much of it is grown far from the processing centers, and transportation and other communications infrastructure are typically poor. Collectors and growers of rattan have no direct access to the buyers. They have no information, or access to information, on prices, quality requirements, or other market information. As individuals, the costs of accessing this information, making the contacts, arranging for transportation, and negotiating the sales would be too high. The higher prices they might be able to get would be unlikely to compensate for the higher costs of making the sale.

Therefore, the rattan collectors and growers are willing to accept the terms offered by traders, as long as the price is higher than their costs. In many cases, where opportunities are limited, this means that they are actually satisfied with anything better than break-even prices (Satria 1997). In this respect, the rattan cultivators have a significant advantage, compared with those who gather rattan from the wild. Rattan continues to grow if it is left unharvested and, although thefts are not unknown, the ownership of rattan gardens appears to be respected, so farmers can delay harvesting until prices rise. Rattan gatherers, on the other hand, compete for an open-access resource and cannot afford to wait for higher prices lest someone else gets the best or the most easily accessible rattan.

The same imbalance of bargaining power also applies at transformation points higher up the PCS. Local traders do not have the necessary business relationships with, and probably cannot supply the volumes required by, the semiprocessors or manufacturers. Many of these business relationships are based on family ties and ethnicity, so the barriers to entry may, in practice, be quite high (Peluso 1992). In general, there are large inequalities in social, economic, and political power that influence the market as well.

Policy

The natural disadvantages faced by rattan producers in Indonesia have been exacerbated by government policy. The most influential policies affecting the rattan sector generally were the ban on the export of unprocessed, or raw, rattan in October 1986 and a subsequent ban on the export of semifinished rattan in January 1989. These bans were replaced in 1992 with a prohibitive export tax of US$10 to US$15 per kilogram, which had the same effect: raw rattan, and even semifinished rattan, could only be sold to manufacturers within Indonesia. In the early days of the bans there were very few local manufacturers of rattan products, so the market was extremely limited. Part of the rationale behind the policy was the encouragement of domestic rattan processing and, indeed, the bans acted as a subsidy for domestic processors by artificially reducing demand for raw material (NRMP 1996a). In that respect, the policy was successful; the rattan-processing industries grew substantially. However, the bans had a strong depressing effect on raw material prices, at great cost to the people involved in rattan extraction and cultivation.

Up until 1992, rattan webbing[1] was classified as a finished product. This allowed one important outlet for rattan from the gardens. However, in 1992, webbing was reclassified as a semifinished product, effectively shutting that door as well. This reclassification reduced the demand for wild and cultivated *sega* and *irit* as sources for rattan peel used in webbing. It also reduced the competitiveness of these species as a source of core within the domestic rattan-processing industries. Rattan is sold by weight in Indonesia, and small-diameter canes have a higher proportion of their weight in peel than do larger-diameter canes. Therefore, with the peel devalued, rattan processors switched to higher-volume, larger-diameter canes as a source of core.

Official policy regulating rattan processing also restricted investment in these industries. In 1989, foreign and domestic investment in raw rattan processing and semifinished rattan production was forbidden. Foreign investment in the manufacturing of finished products was also banned. Later, these restrictions were relaxed to allow investment in rattan processing outside Java. However, a number of factors, such as higher labor costs, poor infrastructure, and lack of trained workers, made investment outside Java less attractive. The policy was reversed in 1995, but not before depressing Indonesia's rattan-processing capacity for nearly six years.

Other official decisions affecting rattan farmers included Ministry of Trade decrees establishing a joint marketing board, an approved exporters system, and an export quota system for lampit. The measures were taken to prevent "unhealthy competition" among lampit exporters, and also to prevent overexploitation of raw material.

Also, governments in most rattan-producing countries levy a royalty on rattan based on the principle that forest products are a part of the rent accruing from state land. However, no distinction is made between rattan that is extracted from wild sources and rattan that is cultivated and managed. Any investment by an individual to enhance production is therefore a subsidy to the state. In cases such as the rattan gardens of Kalimantan, the rattan is essentially an agricultural crop, and royalty charges do not seem justifiable. The royalties are levied through a system of transit permits. A permit to transport a load of rattan is issued upon payment of the royalty, and the permit must be displayed at checkpoints along the road. This system creates an additional burden on producers and increases the costs of getting the rattan to the manufacturers. In any case, the relatively high levels of graft within the system mean that very little of the royalty, whether justified or not, actually reaches the state coffers.

---

[1] Woven rattan skin used for chair seats and backs and other decorative purposes.

Trends in the System

As a result of these policies, the Indonesian lampit industry has gone through pronounced growth, then recession, all in the 12 years since the raw rattan export ban. To illustrate the point: in 1984, there were just 21 lampit manufacturing enterprises in Amuntai, making 64,000 m^2 of lampit for a value of around US$180,000. By 1987, the industry was at its peak, having swollen to 435 producers making more than 1 million square meters of lampit worth US$3.2 million. The total value of output was slashed to just US$236,000 in 1993, the last year for which statistics are available, and the number of enterprises in Amuntai had dropped to 201. In South Kalimantan as a whole, production of lampit went from 537 tonnes in 1987, to a peak of 19,000 tonnes in 1991, then down to a low of 9,800 tonnes in 1994 before rebounding to 15,400 tonnes in 1995. In the process, the unit price changed, in nominal terms, from US$6.38 down to as low as US$1.22 and back up to US$8.39 (NRMP 1996b). Demand and prices for raw material were also caught up on the roller-coaster ride.

The lampit manufacturing industry is concentrated in South Kalimantan, and most of its raw material originates in the rattan gardens. These events have therefore had a severe impact on rattan farmers. Current raw material prices are almost the same, in nominal terms, as they were in 1987 (Godoy 1990), so, in real terms, they have decreased significantly. Similar, or more pronounced, trends are reported in other rattan-farming areas of Kalimantan. In more remote areas, with higher transport and other costs, there have been no buyers for several years. This is leading farmers to discontinue the practice of planting rattan, and even to abandon existing rattan gardens.

Conclusions and Recommendations

The technical feasibility of rattan cultivation within forest fallows is clearly demonstrated by the traditional systems used in Kalimantan and Yunnan for more than 100 years. If the climatic and other ecological conditions are favorable for growing rattan, and if adequate land is available to permit relatively long rattan garden fallows, then the system can be practiced (see color plate 4 for an embryonic system in Lao P.D.R.). Potential benefits include the generating of income from forest fallows, as well as diversifying income sources, reducing risk, and allowing farmers to "save" by deferring rattan harvesting. The ability to generate cash revenue from the fallow also makes longer fallows feasible, with many consequent ecological benefits. Moreover, there is a large and growing demand for rattan products, and a strong potential demand for rattan raw material grown within shifting cultivation systems.

However, under current conditions, rattan cultivation is a relatively unattractive option except in areas with easy access to the main processing centers. Within the Kalimantan rattan PCS, and in many other NTFP systems, raw material producers tend to have very weak bargaining positions. Vertical linkages are strong and power tends to be centralized with the manufacturers. This can be explained, at least in part, by the nature of the material. Rattan, like most biological materials, is not homogeneous. Its quality varies, it is prone to damage in storage and in transit, and there is the possibility of adulteration. To reduce the transaction costs inherent in buying and selling an inconsistent commodity, larger-scale buyers rely on the services of middlemen, and many parties choose to use long-term contractual arrangements, both formal and informal.

Long-term arrangements are also attractive from the sellers' point of view. For individual rattan producers, the costs of contacting processors or manufacturers, accessing information on prices and quality requirements, then arranging for transportation and negotiating sales are prohibitive. Therefore, the collectors and growers are willing to accept the terms offered by traders, as long as the price is higher than their cultivation costs.

Many of these characteristics are also found in other forest products. However, the rattan sector generally—the Kalimantan rattan PCS in particular—has been strongly influenced by a range of policies that have drastically reduced the size of the market for rattan produced from the rattan gardens, and both demand and price have fallen steeply.

A cost-benefit analysis of the rattan cultivation system shows that financial feasibility is sensitive to price, and that under current conditions it is not very profitable. This reflects the real situation in which rattan cultivators in accessible locations, with good road transport and proximity to the main market, are still able to produce rattan economically. But even in these areas, some farmers are reducing their efforts in rattan cultivation, and in more marginal areas, with higher transportation and other costs, rattan cultivation has been rendered uneconomic.

There are two kinds of lessons from this. First of all, any strategies to intensify production from shifting cultivation systems by increasing production for the rattan market must take the market linkages carefully into consideration. The PCS approach, described briefly here and in more detail elsewhere, provides a framework within which to assess opportunities and constraints. Lessons can be drawn from the existing system that can be applied to new or evolving approaches.

Second, there is an opportunity for policy and project-level interventions that could help to shift the balance of power in favor of raw material producers. This could be done by improving market institutions. Possible interventions include making market information available, along with other agricultural-commodity price broadcasts and publications, and developing and promoting the use of appropriate rattan grading standards or rules. Then perhaps auction markets could be developed for rattan and other NTFPs, where competitive bidding would help to increase raw material prices and reduce other costs. In practice, collectors might still have to rely on traders to take their material to market, but an auction mechanism would help move price and quality criteria into public view.

Horizontal linkages are relatively weak in existing rattan PCSs, including that in Kalimantan. Stronger linkages among actors in the early stages of a PCS would help to increase bargaining power for producers and local traders. This could be done through cooperatives, training in management, and other skills.

Generally speaking, improved horizontal linkages in the industry may be beneficial in terms of increasing capacity and reducing costs. However, the case of ASMINDO in Indonesia provides a warning that too much horizontal control can be detrimental to wider national interests. The statutory marketing and exporting activities of ASMINDO seem to have had a negative effect on the sustainability of Indonesia's production of rattan raw material.

There is also a need to improve and extend appropriate small-scale rattan cultivation technologies. Further research is needed in several areas, including the following:

- Selection or improvement of high-quality, clustering, large-diameter rattan varieties suitable for small-scale cultivation;
- Selection or improvement of rattan species with faster growth;
- Development and extension of improved small-scale cultivation technologies, including planting material; and
- Development and extension of an efficient and safe technology for preserving rattan.

There may also be an opportunity for helping small scale rattan cultivators by assisting small scale processors. Training and technology transfer could help to improve their efficiency and the quality of their products and stimulate a broader based demand for raw material.

Acknowledgments

This work draws lessons from a series of studies on a range of bamboo and rattan production-to-consumption systems undertaken through the International Network for Bamboo and Rattan (INBAR), with support from the World Agroforestry Centre and the International Fund for Agricultural Development (IFAD). I am grateful for having had the opportunity to participate and, in so doing, to benefit from the ideas and energy of the national program scientists involved. In particular, I would like to acknowledge the effort of the authors of the study on the Kalimantan rattan PCS: Boen Purnama, Beni Nasendi, Hendro Prahasto, Satria Astana, Wesman Endon, Setyawadi Hadi, Rachman Effendi, Jasni, and Djarwanto. Thanks are also due to my colleagues and companions during fieldwork in East and South Kalimantan, Rahayu Syupriadi, Manuel Ruiz-Perez, Genevieve Michon, and Dieter Haury. Finally, my understanding of the role of policy in the Indonesian rattan sector has gained a great deal from work published by the Natural Resources Management Program (NRMP) and from discussions with Christopher Bennet.

References

Belcher, B.M. 1997. Commercialization of Forest Products as a Strategy for Sustainable Development: Lessons from the Asian Rattan Sector. Ph.D. thesis, University of Minnesota.

Boen, P.M., P. Hendro, and A. Satria. 1996. Study on the Socio-Economic Aspects of the Rattan Production-to-Consumption System in Indonesia: A Case Study in Kalimantan. Draft report to INBAR (International Network for Bamboo and Rattan).

Chen Sanyang, Pei Shengji, and Xu Jianchu. 1993. Indigenous Management of the Rattan Resources in the Forest Lands of Mountain Environment: The Hani Practice in the Mengsong Area of Yunnan, China. *Ethnobotany* 5, 93–99.

Dransfield, J. 1988. Prospects for Rattan Cultivation. *Advances in Economic Botany* 6, 190–200.

Fried, S.T., and Mustofa Agung Sardjono. 1992. Social and Economic Aspects of Rattan Production in the Middle Mahakam Region: A Preliminary Survey. *GFG (German Forestry Group) Report* 21, 63–72.

Godoy, R.A. 1990. The Economics of Traditional Rattan Cultivation. *Agroforestry Systems* 12, 163–172.

——— and T. Ching Feaw. 1991. Agricultural Diversification among Smallholder Rattan Cultivators in Central Kalimantan, Indonesia. *Agroforestry Systems* 13, 27–40.

Gouyon, A., H. de Foresta, and P. Levang. 1993. Does "Jungle Rubber" Deserve Its Name? An Analysis of Rubber Agroforestry Systems in Southeast Sumatra. *Agroforestry Systems* 22, 181–206.

Halim Bin Haji Ibrahim, and Aminuddin Bin Muhamad. No date. Large Scale Planting of Rattan in Rubber: Kurnia Setia Berhad Experience (mimeograph). Malaysia: Kurnia Setia Berhad.

Haury, Dieter. 1997. Personal communication with the author.

IFAR (International Fund for Agricultural Research). 1991. *Research Needs for Bamboo and Rattan to the Year 2000.* Arlington, VA: Tropical Tree Crops Program, IFAR.

Latif Bin Nordin, and Rahman Bin Fatah. 1986. Rotan Planting under Rubber Plantation in Pahang Darul Makmur (mimeograph).

Lim Hin Fui, and Nur Supardi Md Noor. 1994. Social, Economic, and Cultural Aspects of Rattan in Malaysia. Paper presented to FAO Asia Regional Expert Consultation on Non-Wood Forest Products, November 28–December 2, 1994, Bangkok, Thailand.

Mayer, J. 1989. Rattan Cultivation, Family Economy and Land Use: A Case from Pasir, East Kalimantan. *GFG (German Forestry Group) Report* 13, 39–53.

Michon, G. 1997. Personal communication with the author.

Muraleedahran, P.K., and K.K. Seethalakshmi. 1992. Rattan Plantation and its Profitability. In: *Rattan Management and Utilization,* Proceedings of the Rattan (cane) Seminar, January 29–31, 1992, India, edited by S.C. Basha and K.M. Bhat. Trichur, India: Kerala Forest Research Institute and IDRC (International Development Research Centre), 311–315.

Nasendi, B.D. 1994. Socio-Economic Information on Rattan in Indonesia. In: *Working Paper 2.* INBAR (International Network on Bamboo and Rattan).

NRMP (Natural Resources Management Project). 1996a. Value-added and Resource Management Policies for Indonesian Rattan: Aims, Outcomes and Options for Reform. Draft Summary of Findings presented to FORDA/INBAR Rattan Seminar, November 4-5, 1996, Bogor, Indonesia.

———. 1996b. Raising the Value of Rattan Exports: Policy Challenges to Achieving Improvements in Efficiency, Equity, and Ecosystem Management. Annex 3: Rattan Policy Framework. Jakarta: NRMP.

Pabuayon, I.M., M.N. Rivera, and L.H. Espanto. 1996. The Philippine Rattan Sector: A Case Study of an Extensive Production-to-Consumption System. Final Report to INBAR (International Network for Bamboo and Rattan).

PCARRD (Philippine Council for Agriculture, Forestry and Natural Resources Research and Development). 1991. *The Philippines Recommends for Rattan Production.* Philippines Department of Environment and Natural Resources.

Pei Shengji, and Chen Sanyang. 1990. The Resolution of Rattan Resource Crisis in Yunnan Province of China. Paper presented to SUAN V Symposium on Rural-Urban Ecosystems Interactions in Development, May 21–24, 1990.

Pei Shengji, Chen Sanyang, Wang Kanlin, Xu Jianchu, and Xue Jiru. 1994. Ethnobotany of Indigenous Non-Wood Forest Products in Xishuangbanna of Yunnan in Southwest China. Paper presented to the Fourth International Congress of Ethnobiology, November 17–21, 1994, Lucknow, India.

Peluso, N.L. 1992. The Rattan Trade in East Kalimantan, Indonesia. In: *Non-Timber Products from Tropical Forests: Evaluation of a Conservation and Development Strategy*, edited by D.C. Nepstad and S. Schwartzman. *Advances in Economic Botany* 9. New York: New York Botanical Garden, 115–127.

Rauf Salim. 1982. A Financial Appraisal of Rotan Manua (*Calamus manan*) Cultivation at the Bukit Belata Forest Reserve. *The Malaysian Forester* 45(4), 576–582.

Salleh bin mohd. Nor, and Aminuddin bin Mohamad. 1986. Rattan as a Supplementary Crop in Rubber Plantations. Paper presented at a Rubber Growers' Conference, October 20–23, 1986, Ipoh Perak, Malaysia.

Sasaki, H. 2006. Innovations in Swidden-Based Rattan Cultivation by Benuaq-Dayak Farmers in East Kalimantan, Indonesia. Chapter 38.

Satria, Astana. 1997. Personal communication with the author.

Setyawati Hadi, Rachmat Hidayat, and Ayi Suyatna. 1989. Analisis Finansial Penaman Rotan Di Long Kali Kabupaten Pasir, Kalimantan Timur (A Financial Analysis of the Establishment of Rattan Plantations at Long Kali, Pasir, East Kalimantan). *Rimba Indonesia* 23(1-2).

Weinstock, J.A. 1983. Rattan: Ecological Balance in a Borneo Rainforest Swidden. *Economic Botany* 37(1), 58–68.

Xu Huangcan, Zhou Zaizhi, Ying Guangtian, and Zhang Weiliang. 1990. A Financial Appraisal of Rattan Plantation Development. Paper presented to International Union of Forestry Research Organisations Congress, Montreal.

Xu Jianchu. 2006. Rattan and Tea-Based Intensification of Shifting Cultivation by Hani Farmers in Southwestern China. Chapter 56.

Zhong Maogong, Xie Chen, Fu Maoyi, and Xie Jinnzhonget. 1995. Bamboo and Rattan Socio-Economic Database, P.R. China. Final Report to INBAR (International Network for Bamboo and Rattan).

Chapter 65

The Role of Land Tenure in the Development of Cinnamon Agroforestry in Kerinci, Sumatra

S. Suyanto, Thomas Tomich, and Keijiro Otsuka

When traditional shifting cultivation becomes unsustainable because of population pressure and scarcity of land, the adoption of agroforestry systems can be an alternative. These systems often start with the introduction of trees into cleared swidden land during or after food crops have been harvested. The trees grow to partly replace the natural vegetation of the shifting cultivation system and are then known as an "improved tree fallow." Typically, continuing population pressure forces an evolution from low- to high-intensity management of planted tree fallows. The final stage of this enrichment of shifting cultivation is referred to as an agroforest. It is a major land use in Sumatra (de Foresta and Michon 1991), where important tree crops such as rubber (*Hevea brasiliensis*) and cinnamon (*Cinnamomum burmannii*, known as cassiavera) have begun in this fashion. This transformation of farming systems, driven by population pressure, has been summarized by Tomich (1994) and appears in Table 65-1.

This chapter examines cinnamon agroforestry as an example of an improved tree fallow, in the Kerinci district of Jambi Province in Sumatra (see color plates 49 and 50). Specifically, it focuses on the effect of land and tree tenure on the development of cinnamon agroforestry. Although some cinnamon is grown in agroforest-type systems in the Kerinci valley, a significant portion of the local crop is essentially a monoculture. Our analysis is based on a community-level survey of 19 villages and an intensive follow-up survey of 100 households in two of the 19 villages. The sample villages are located in the zone surrounding Kerinci Seblat National Park (Figure 65-1), the largest area of continuous primary forest in Sumatra, covering 14,847 km^2.

S. Suyanto and Thomas Tomich, World Agroforestry Centre (ICRAF), Southeast Asian Regional Research Programme, P.O. Box 161, Bogor 16001, Indonesia; Keijiro Otsuka, Foundation for Advanced Studies on International Development (GRIPS/FASID Joint Graduate Program), 2-2 Wakamatsu-cho, Shinjuku-ku, Tokyo 162-8677, Japan.

Table 65-1. Farming System Transformations

	Increasing Population Density →				
Rotation (years)	*Sustainable Shifting Cultivation*	*Unsustainable Shifting Cultivation*	*Agroforests*	*Agroforestry*	*Tree Monoculture*
0	Slash–and–burn	Slash–and–burn	Slash–and–burn	Slash–and–burn	Slash–and–burn
1–2	Food	Food	Food	Food	Trees
3–5	Fallow	Food	Trees	Trees	Trees
5–10	Fallow	Weeds	Trees	Trees	Trees
10–19	Fallow	Weeds	Trees	Trees	Trees
20	Slash–and–burn	Weeds	Trees	Trees	Slash–and–burn
21–22	Food			Trees	Trees
23–24	Fallow			Trees	Trees
25	Fallow			Slash–and–burn	Trees
26–27	Fallow			Food	Trees
28–39	Fallow			Trees	Trees
40	Slash–and–burn	↓	↓	Trees	Slash–and–burn
		Infinity (?)	Infinity (?)		

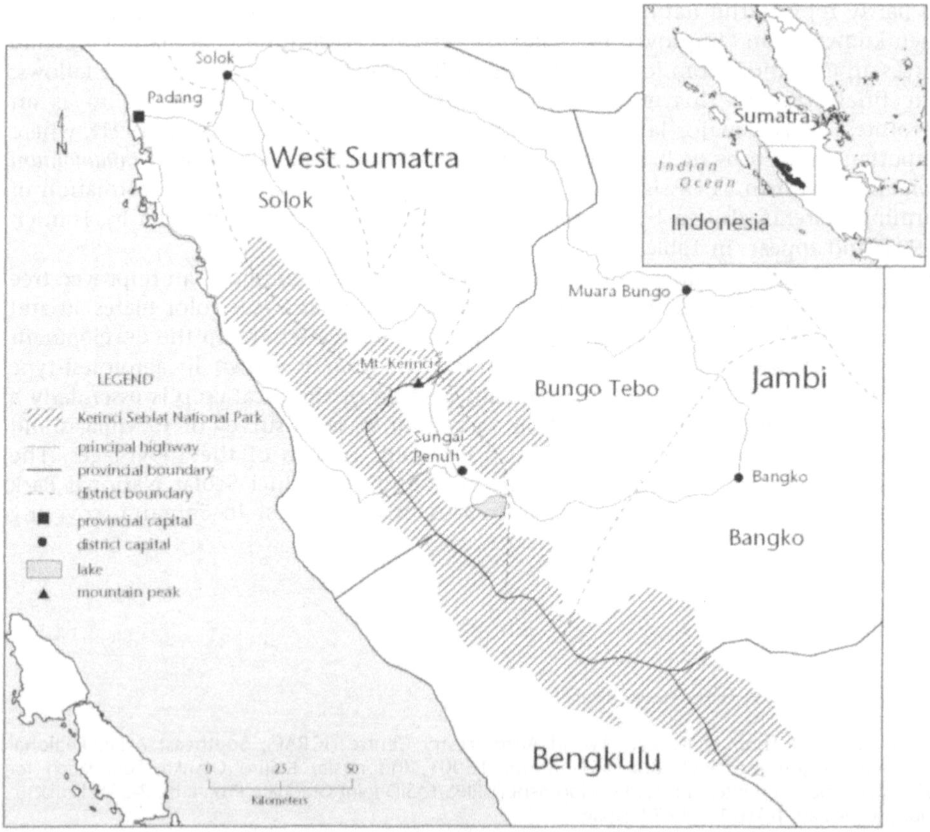

Figure 65-1. Map of Study Site

Results

During the 19th century, the main agricultural practice in Kerinci was shifting cultivation (Marsden 1811). At that time, *Cinnamomum burmannii*, a species native to Kerinci, grew and was harvested in the Bukit Barisan mountain range. The change from traditional shifting cultivation to cinnamon agroforestry began in the 1920s, when the Dutch colonial government began developing roads, the population began to grow, and farmers began planting *Cinnamomum* in their fallowed swiddens. Cinnamon production increased rapidly, and the agricultural landscape in Kerinci became a mosaic of cinnamon trees. Today, it is the major crop in Kerinci, accounting for nearly 70% of Indonesian cinnamon production (Scholz 1983, Rismunandar 1989). Almost all of it is cultivated by smallholders.

Throughout this region, indigenous societies have traditionally followed a matrilineal inheritance system, with joint ownership that limits individual rights to land and others assets. However, along with the change of land use from shifting cultivation to agroforestry systems, land tenure has also been evolving toward individual ownership. In cinnamon agroforestry, individualized tenure institutions are dominant and their incidence is increasing. In fact, land tenure appears as a major determinant in the development of cinnamon agroforestry, and we suggest that secure individual land rights actually stimulate farmers to adopt more efficient and sustainable land uses.

We have two specific hypotheses: that greater individualization, or the evolution of tenure systems toward full private-property rights, will lead to the adoption of more intensive agroforestry systems, and that greater profitability of tree crop production will lead to the greater individualization of land tenure. If, on the other hand, individual land rights do not evolve, we expect there to be less development of agroforestry and a continuation of less-intensive shifting cultivation of annual crops.

It follows from these hypotheses that, with secure land and tree tenure, high profitability of cinnamon production, and diminished access to forest land, farmers will tend to continue converting bush-fallow areas to cinnamon agroforestry, rather than maintaining traditional shifting cultivation of annual food crops.

In this chapter, we present a preliminary analysis of the survey data to characterize land-use patterns, cinnamon agroforestry practices, and the evolution of land tenure in response to population pressure.

Land Use Patterns and Cinnamon Agroforestry Practices

The three main agricultural land uses in Kerinci are wet rice fields located in valleys and flat areas, 18%; perennial crop systems on surrounding hills, dominated by cinnamon agroforestry, 77%; and annual crops, including bush-fallow areas, 6% (Scholz 1983). The total area of cinnamon in Kerinci increased at a rate of 3.64% per year between 1985 and 1995, from 36,766 to 52,564 ha. Annual production over that decade nearly tripled, from 5,737 tonnes in 1985 to 16,357 tonnes in 1995 (Biro Pusat Statistik 1995).

Our survey of 100 farmers in two villages (Lempur Mudik and Lempur Tengah) confirmed Scholz's earlier finding that tree-based systems are the most important land use. The average household owns 6.2 ha of land, to which must be added land that is sharecropped or borrowed, making an average operating unit of 6.9 ha. Of this, the average area of perennial crops per household is 5.3 ha, or 77% (see Table 65-2).

In this analysis, we distinguish between young perennial crops less than four years old and mature perennials older than four years. Within a total sample of 424 plots, 153 plots are young and 271 mature.

Cinnamon accounts for 96% of young perennial plots and all of the mature perennial plots. Most farmers grow cinnamon without other trees. However, 31% of young cinnamon plots are planted in association with other trees, compared to only about 18% for mature cinnamon gardens. This suggests that intercropping cinnamon with other trees is becoming more important.

Coffee is the major tree planted in association with cinnamon. Among the plots of mature cinnamon planted in association with other trees, 63% are planted with coffee, 25% with *surian* (*Toona sinensis*), and 13% with fruits. For young cinnamon plots that are intercropped with other trees, 92% are planted with coffee and only 8% with surian and fruits (see Table 65-3).

When the cinnamon is young, food crops can be grown under the trees until they are four years old. At this stage, the canopy closes out the sunlight. More than 86% of all cinnamon plantations in the survey were either intercropped with food crops or had been intercropped during the immature stage (see Table 65-4). In more than 90% of these cases, chili was, or had been, the most important annual crop. In the case of mature cinnamon plots, most farmers had planted chili in association with tobacco (39%); chili with other food crops such as peanuts, potatoes, and vegetables (26%); or chili only (32%). Most young cinnamon plots were currently planted with chili only (52%), chili with tobacco (23%), and chili with other food crops (21%).

Food crops had been grown mostly in the first to third years after planting the cinnamon. Only a few farmers intercrop annuals into fourth year; the figures were just 6% of currently mature plantations and 4% of young cinnamon (See Table 65-3).

Cinnamon is harvested by felling the trees, and the stumps remain in the field. While the stumps coppice, annual crops are grown in between. Farmers burn the biomass, but the fires are very limited compared to slash–and–burn because intense fires can destroy the young cinnamon coppices. Only light burns are used on more than half of all cinnamon plots (Table 65-5). Therefore, the release of CO_2 to the atmosphere from burning to clear land for agroforestry is reduced compared to traditional practices of shifting cultivation.

Table 65-2. Average Land per Household (ha)

Land-Use Type	*Owned*	*Cultivating Other People's Land*	*Total*	
			ha	*(%)*
Wet rice fields	0.73	0.26	0.99	14
Cinnamon	4.91	0.69	5.31	77
Agroforestry	0.56	0.00	0.56	8
Bush-fallows	6.20	0.95	6.86	100
Total land				

Table 65-3. Intercropping Cinnamon with Other Tree Crops

Characteristics		*Young Perennial*	*Mature Perennial*
Number of plots		153	271
Main trees (%)	Cinnamon	96	100
	Coffee	4	0
Intercropping with other trees (%)	Yes	31	18
	No	66	82
Trees intercropped with cinnamon (%)	Coffee	92	63
	Surian	0	25
	Fruit trees	8	13

Table 65-4. Intercropping Cinnamon with Food Crops

Intercropping Details		*Young Cinnamon*	*Mature Cinnamon*
Number of plots		153	271
Intercropping with food crops (%)	Yes	87	90
	No	13	10
Number of years food crops planted (%)	1	31	36
	2	36	39
	3	22	19
	4	4	6
Type of food crops intercropped with cinnamon (%)	Chili	51	32
	Chili and tobacco	23	39
	Chili and other food crops	21	26
	Others	5	3

Table 65-5. Planting Materials and Land-Clearing Techniques

Details		*Young Perennial*	*Mature Perennial*
Number of plots		147	271
Type of planting material (%)	New seed	52	40
	Stumps	48	60
Use of fire in land-clearing technique (%)	High	33	31
	Low	31	49
	None	19	17

Farmers usually harvest cinnamon after eight or more years. However, there are possibilities for smaller harvests before that. These are called *panen pucuk* and involve the pruning of branches, which yield the lowest-quality cinnamon. Our survey showed that this type of pruning was performed on only 16% of the plots, usually after the third year. On about 46% of plots, farmers selectively felled low-quality trees when they were between four and six years old to reduce competition with higher-quality trees (Table 65-6).

Cinnamon quality is classified according to the part of the tree from which the bark comes and the bark's thickness. Generally, there are three different qualities: Kc comes from branches and is about 1 mm thick; Kb also comes from branches but is between 1.5 and 2.5 mm thick; and Ka comes from the trunk of trees ages 8 to 15 years, and is also between 1.5 and 2.5 mm thick. In addition, there are three premium quality grades: KA1 which comes from the trunk of trees ages 15 to 18 years and is between 2.5 and 3 mm thick; KF, which comes from the trunk of trees ages 18 to 25 years and is between 3 and 5 mm thick; and KM, which comes from the trunk of trees over 25 years of age and is between 5 and 10 mm thick.

Table 65-6. Harvesting of Intercropped Plots

Details		*Mature Perennial*
Number of plots		271
Harvesting of pucuk/pruning (%)	Yes	16
	No	84
Year of harvesting pucuk	Year 3	11
	Year 4	41
	Year 5	25
	Year 6 and above	23
Selective thinning	Yes	47
	No	54
Year of thinning	Year 4 and below	25
	Year 5	32
	Year 6	24
	Year 7 and above	19

Inheritance Rules

Kerinci society basically follows matrilineal inheritance and matrilocal residence patterns. Property is classified into two types: *harta pusako,* or ancestral property, and *harta pencaharian,* or earned property. Ancestral property such as paddy fields and houses are inherited by a daughter and cannot be sold. Earned property may be inherited either by sons or by sons and daughters. This inheritance rule has undergone substantial change over time. In our 19 survey villages in Kerinci we found significant changes, from matrilineal inheritance by daughters, to undifferentiated inheritance by both sons and daughters. Some villages use both systems.

Table 65-7 shows the percentage of villages that follow the various inheritance rules. For wet rice fields, for instance, 35% of the villages follow the matrilineal inheritance rule, 35% follow the undifferentiated inheritance rule and, in 30% of villages, there is a combination of both. The transformation from matrilineal to undifferentiated inheritance seems to have gone faster for dryland than for wet rice fields. About 50% of the villages follow undifferentiated inheritance for dryland fields, 28% follow a combination of both systems, and only 22% stick with the old matrilineal inheritance rule.

Land Tenure Categories

Table 65-8 shows the major land tenure systems in Kerinci and their characteristics, with the tenure systems arranged from the weakest individual ownership rights to the strongest.

Communal Land Tenure. Under this system, land belongs to a group of lineages or jointly to all people in the village. Members of specific lineages or villagers have rights to cultivate this land, but they cannot rent, pawn, or sell it. Moreover, they cannot plant tree crops without the approval of the customary or village chief.

Lineage Land Tenure. Under this system, the land belongs to descendants of a common grandmother. Land is collectively owned by the lineage and can be inherited by daughters, sisters, and nieces of a woman who dies. A lineage head is selected among uncles, that is, male members of the second generation, and he exercises strong authority over land inheritance.

Table 65-7. Percentage of Villages Observing Inheritance Rules, by Field Type

Inheritance Rule	*Wet Rice Fields (%)*	*Dryland Fields (%)*
Matrilineal	35	22
Matrilineal and undifferentiated	30	28
Undifferentiated	35	50

Table 65-8. Land Tenure Categories and Their Major Characteristics

Land Tenure Categories	*Members*	*Inheritance*	*Joint Ownership*
Communal	A group of lineages or all people in the village	Inherited communally	Yes
Lineage	Lineage coming from a common grandmother	Daughters, sisters and nieces	Yes
Joint family	Members of nuclear family		
Among daughters		Daughters	Yes
Daughters and sons		Daughters and sons	Yes
Among sons		Sons	Yes
Single family	Family members		
Among daughters		Daughters	No
Daughters and sons		Daughters and sons	No
Among sons		Sons	No
Private land	Those who have privately acquired land (cleared forest land or purchased land)	Daughters Daughters and sons Sons	No

Source: Otsuka 1996.

Joint Family Tenure. In this category, joint ownership of land involves more than one family and encompasses families of the same generation or two generations. Inheritance may be among daughters, daughters and sons, or only sons, depending on the specific village. The joint family system is an equitable one because rights to use a field are rotated among households.

Single Family Tenure. Single family land tenure seems to have evolved from joint family tenure. Land may be inherited by daughters or sons, or daughters and sons, depending on local inheritance rules. Under this system, a single family has clear rights to use, rent, or mortgage the land, even though the consent of other relatives is often required in order to sell it.

Private Land Tenure. Full private-property rights are acquired by clearing forest land or by purchasing land. Owners of private land are free to rent, mortgage, or sell it.

Distribution of Land by Tenure Type

Table 65-9 divides land into paddy fields, perennial gardens, and bush fallows, then shows the percentage of each land-use type held under the various tenure categories. The area of communal and lineage land is very small, so for this analysis we have combined these two tenure types with joint family tenure. The area of paddy fields under joint family tenure is 74%; single family tenure, 8%; and private ownership, 14%. The traditional tenure system predominates for paddy fields because there is only a very small investment required. The irrigation systems are of a simple, traditional variety that requires a minimum of effort to maintain and repair.

In contrast, cinnamon agroforestry occurs on land with a higher degree of individualization. Around 62% is on land with single family tenure, 33% is on private land, and only 6% is on joint family land. This suggests that secure land tenure with a high degree of individualization may be a prerequisite to investment in cinnamon agroforestry. Its establishment requires significant effort in land preparation, tree planting, weeding, and pruning. Without secure individual ownership rights, those who plant and grow the trees might not be able to reap the benefits.

It is reasonable to hypothesize that population pressure influenced the evolution of land tenure toward a higher degree of individualization. Demand for land has increased as the population has grown, and as a consequence, land has become scarce. Under these conditions, communal land tenure systems cease to work efficiently and may evolve toward more secure individual ownership (Ault and Rutman 1979).

According to the 1930 census, Kerinci district then had a population of 50,248 (Table 65-10). By 1990, this had grown to 280,017 people and, in 1994, Kerinci's population density was 68 people per km^2. This compared with an average 40 people per km^2 for the whole of Jambi Province, of which Kerinci is a part (Table 65-11).

Tables 65-9 to 65-11 suggest that the influence of population pressure on land tenure depends on land-use types. Ownership of wet rice fields is the least individualized of the land uses; ownership of perennial plots is the most individualized. A large investment in clearing forest land and planting trees creates stronger rights than a small investment in establishing a rice field. This confirms Shepherd's argument in 1991 that it is an investment of labor, more than anything else, that creates ownership.

The Change from Shifting Cultivation to Cinnamon Agroforestry

Boserup (1965) hypothesized that farming systems change from shifting cultivation to more intensive systems in response to population pressure. Our survey data show that, in the Kerinci area, upland food crop production always occurs in association with young cinnamon. This suggests a change from shifting cultivation, which Marsden found in 1811 to be the area's dominant agricultural system, to cinnamon agroforestry.

Farmers are clearing forest and bush-fallow land to establish cinnamon agroforestry, rather than to open swiddens for shifting cultivation. Of forest land, 83% is converted to perennials within one year of farmers clearing it, and at present, 84% of forest-land plots are planted with perennials. The same trends apply to bush-fallow land. Of bush-fallow land, 79% is converted to perennial plantations within one year of its acquisition, and at present, 86% of bush-fallow plots are planted with perennials (Table 65-12).

Table 65-9. Percentage of the Area under Different Land Tenure Systems by Land Use Type

Land Use Type	*Joint Family Ownership*	*Single Family Ownership*	*Private Ownership Purchase*	*Private Ownership Clearance*
Paddy fields	74	6	6	8
Cinnamon agroforestry	6	62	14	19
Bush-fallows	30	43	14	13

Table 65-10. Population of Kerinci District between 1930 and 1994

Year	*Population*	*Growth (%)*	*Population Density (people/km²)*
1930	50,248		11.96
1960	155,874	3.85	37.11
1971	186,615	1.82	44.43
1980	241,081	2.59	57.40
1990	280,017	1.51	66.67
1994	285,621		68.00

Source: Biro Pusat Statistik 1994.

Table 65-11. Population in Jambi Province by District in 1994

District	*Area (km²)*	*Population*	*Population Density (people/km²)*
Kerinci	4,200	285,621	68
Bungo Tebo	13,500	383,108	28
Saralangun Bangko	14,200	372,749	26
Batanghari	11,130	358,831	32
Tanjung Jabung	10,200	384,202	38
Kodya Jambi	206	343,489	1,670
Jambi Province	53,436	2,128,000	40

Source: Biro Pusat Statistik 1994.

The area of forest land acquired for cinnamon agroforestry increased until the 1970s and then started to decline. This was a result of increasing enforcement of the conservation status of the forests, which in 1991 became a national park. As a result, forest clearing declined, and the conversion of bush-fallow to perennial areas increased (Table 65-13). Along with that, the tenure of bush-fallow ownership became stronger. Our survey data show that 37% of bush-fallow land is now under single family tenure and 35% is privately owned.

It is not surprising that the incentive to plant commercial trees is very high when it is an important way to obtain and maintain secure land rights. According to customary rules, people who clear communal forest land and plant commercial trees receive relatively strong individual ownership rights. Such rights are given to those who make the land fruitful, or it reverts back to the communal pool. Although these owners may leave the land from time to time, they still retain their ownership because the land is covered by perennial crops that do not require continuous maintenance (Aumeeruddy 1994).

Conclusions

There is significant evidence that customary land tenure institutions in Kerinci district have evolved toward greater security of ownership as a response to a growing population. However, the change to individualized tenure depends on the type of land use. Most of the perennial crops, in this case cinnamon agroforestry, are under a high degree of individualized land tenure, while most paddy fields are under a lesser degree of individual ownership. Secure land tenure with a high degree of individualization may be a prerequisite to investment in cinnamon agroforestry.

Table 65-12. Land-Use Conversion to Cinnamon Agroforestry

Land Use	*Number of Plots*	*Plots Planted to Cinnamon Agroforestry within One Year of Acquisition (%)*	*Plots Planted to Cinnamon Agroforestry at Present (%)*
Cinnamon	215	99.53	99.53
Bush-fallow	163	78.53	86.50
Forest	82	82.93	84.15

Table 65-13. Conversion of Forest and Bush-Fallow to Cinnamon Agroforestry (ha)

		Converted to Cinnamon Agroforestry			*Converted to Cinnamon Agroforesry*	
Year	*Forest Area Acquired*	*One Year after Acquisition*	*1996*	*Bush-Fallow Area Acquired*	*One Year after Acquisition*	*1996*
Pre–1959	9.15	9.15 (100%)	9.15 (100%)	7.58	5.38 (71%)	6.55 (86%)
1960–69	38.44	36.44 (95%)	29.59 (77%)	31.31	31.31 (100%)	31.31 (100%)
1970–79	74.02	50.32 (68%)	69.70 (94%)	31.65	23.88 (75%)	30.42 (96%)
1980–89	36.62	27.14 (74%)	26.28 (72%)	77.07	63.81 (83%)	68.44 (89%)
1990–96	17.23	15.84 (92%)	15.84 (92%)	68.00	50.16 (74%)	48.51 (71%)

With more secure land tenure linked to the profitability of cinnamon, the traditional land use, shifting cultivation, has been largely converted to a tree crop system of cinnamon agroforestry.

People are now clearing forest to establish cinnamon agroforestry rather than for traditional shifting cultivation. Also, when they clear communal forest land, the incentive to plant commercial trees, such as cinnamon, is very high because this is a way to obtain more secure land rights.

Acknowledgments

We used data from a land and tree tenure survey conducted by ICRAF and Jambi University, with financial support from the International Food Policy Research Institute, the government of Japan, and the Overseas Development Administration. We also acknowledge the excellent work of our field assistants, Noviana Khususiyah, David Varianto, Idris Sardi, A Khoiri, and Delfi Andra.

References

Ault, D.E., and G.L. Rutman. 1979. The Development of Individual Rights to Property Rights in Tribal Africa. *Journal of Law and Economics* 22.

Aumeeruddy, Y. 1994. Local Representations and Management of Agroforest on the Periphery of Kerinci Seblat National Park, Sumatra, Indonesia. *People and Plants Working Paper 3*. Paris: UNESCO.

Biro Pusat Statistik. 1994. *Statistik Penduduk Propinsi Jambi Pertengahan Tahun 1994*. Indonesia: Kantor Statistik Propinsi Jambi.

———. 1995. Kerinci Dalam Angka 1995. Kantor Statistik Kabupaten Kerinci.

Boserup, E. 1965. *Conditions of Agricultural Change*. Chicago, IL: Aldine.

de Foresta, H., and G. Michon. 1991. Indonesia Agroforestry System and Approach. In: *Harmony with Nature, Proceedings of an International Conference on Conservation of Tropical Biodiversity*, edited by Y.S. Kheong and Lee Su Win. Kuala Lumpur: Malayan Nature Society.

Marsden, W. 1811. *The History of Sumatra* (reprint of 3rd edition). Singapore: Oxford University Press.

Otsuka, K. 1996. Land Tenure and Forest Resource Management in Sumatra: Analytical Issues and Research Plans for Extensive Survey. Mimeograph, IFPRI (International Food Policy Research Institute).

Rismunandar. 1989. *Kayu Manis*. Jakarta: Penebar Swadaya.

Scholz, U. 1983. The Natural Regions of Sumatra and their Agricultural Production Pattern. In: *Regional Analysis 1A*. Bogor, Padang, Indonesia: CRIFC (Central Institute for Food Crops) and SARIF (Sukarami Research Institute for Food Crops).

Shepherd, G. 1991. The Communal Management of Forest in the Semi-Arid and Sub-Humid Regions of Africa. *Development Policy Review* 1.

Tomich, T.P. 1994. Putting Slash-and-Burn in Context: Socioeconomic and Policy Issues. Paper presented at Methodology Workshop on Participatory Rural Appraisal, November 21–23, 1994, Bogor Indonesia. Slash-and-Burn Project.

Chapter 66

Effects of Land Allocation on Shifting Cultivators in Vietnam

Dinh Van Quang

More than 50 ethnic minorities live in Vietnam's uplands, and from these, nearly three million people still practice shifting cultivation. This has led to serious destruction of the natural environment, and the country's forest assets are shrinking at a rate of 200,000 ha per year. As a consequence, the Government of Vietnam has given priority to encouraging the adoption of more-permanent farming systems. Recognizing the difficulties rural households face in developing settled and permanent farming systems while their land tenure remains insecure, the government reshaped its land allocation policy, allowing forests and forest lands to be allocated to economic entities, including industry. In 1993, the Land Act was passed, and land uses under specific conditions received the sanction of law. Land could be transferred, including to offspring, it could be leased, and it could be used as collateral to secure loans and other transactions. With new security of tenure, farmers were encouraged to develop more intensive cultivation methods, to diversify their agricultural production, and to adopt soil protection technologies.

Since the policy was introduced, about 7 million hectares of forests, denuded hillsides, and barren lands have been allocated to farm households, collectives, and even army units. Of this, nearly 2 million hectares is forest. The rest has been allocated for agriculture, the cultivation of fruit, and tree crops.

This chapter describes changes in land use in shifting cultivation communities in the highlands of Vietnam following the introduction of the Forest and Forest Lands Allocation (FFLA) policy. It uses as a case study an upland commune populated by people of the Hmong ethnic group. Overall, the FFLA policy has led to the adoption of more sustainable farming systems on sloping lands, there has been less pressure on forests, and agricultural land is being managed with a longer-term view.

The Attitude of Farmers

Across the midlands, the FFLA policy was not only welcomed by local farmers, it catalyzed positive changes in Vietnam's national economy. Land values and land-use rights immediately became clearer. Farmers had always regarded secure access to land as critically important to their long-term prospects and to the future of their children. Many had long held visions of how they wanted to develop their land, and were eager not only for legal tenure, but also for the ability to use their land as collateral to secure development loans. Many of them had already planted fruit or other tree crops on their small plots and were waiting for formal recognition of their land rights.

In the uplands, the situation was different. Farmers there were not enthusiastic about receiving extra land because their holdings were already sufficient. What they

---

Dinh Van Quang, Research Centre for Forest Ecology and Environment under FSIV, Chem Tu Liem, Hanoi, Vietnam.

needed most were other resources, such as funds and technical know-how, to help develop their farming systems. This was particularly true for the poorest of them. They feared that the new policy would bring with it a ban not only on slash-and-burn practices, but also on encroachment into new forest areas to find fertile soil. It soon became clear that, to be effective in the uplands, the FFLA needed other projects and incentives introduced in tandem.

Therefore, the government set about creating other economic and technological development projects and programs to support FFLA objectives of developing sustainable land uses. Worth mentioning are Program 327, which provides technical and financial assistance to farm households in the uplands, and the Resettlement and Sedentarization Program, intended to benefit shifting cultivators. Furthermore, the Swedish International Development Cooperation Agency-funded Vietnam-Sweden Mountain Rural Development Program aims to improve the use of sloping lands by local ethnic people and to improve road access to their villages. In most areas of the country, agricultural extension workers have been appointed to guide farmers in developing and diversifying their agricultural practices.

Overall Effects of the FFLA Policy

As a result of FFLA and its support programs, monoculture is declining and product diversification is beginning to develop. In some remote and inaccessible areas, farmers have adopted better farming methods with crop rotations and the use of legumes. Sloping agricultural land technology (SALT) is appearing in the countryside.

Statistics from the Forest Inventory and Planning Institute are shown in Table 66-1. These data suggest that the objectives of better environmental protection and reduced clearing of forested areas for shifting cultivation are both being achieved. When farmers' land-use rights are well defined, there is a tendency for forested areas, and particularly natural forests, to increase, and for forest clearing for swiddening to decrease.

The Economic Development of Sinh Phinh Commune

Sinh Phinh commune is located in the northern part of Tua Chua district in the province of Lai Chau. The commune has a land area of 7,863 ha. At the time of this study it was home to 5,160 people of the Hmong ethnic group, living in 850 households. The area enjoys a cooler upland climate, with an average annual temperature of 19°C, and three months below 15°C. Annual rainfall is 1,800 mm, most of it falling from May to October.

Table 66-1. Forest Inventory, 1990–1995 (x 000 ha)

Description	*1990*	*1991*	*1992*	*1993*	*1995*
Forested area	9,175	9,005	8,886	9,066	9,302
Natural forests	8,430	8,207	8,047	8,270	8,252
Man-made forests	745	798	839	796	1,050

Note: In 1994, no data relating to forests were released by the Vietnamese Government.

At the time of this study, Sinh Phinh had 1,641 ha under shifting cultivation, or an average of 2 ha per household. The area under wet rice was 180 ha, averaging 0.2 ha per household, and there was a large area of denuded hillsides, totaling 5,182 ha, or 6.1 ha per household. The commune also had 860 ha of forest, representing 1 ha per household. The main crops were wet rice, upland rice, and maize. Some of the wet rice fields were irrigated and provided two crops per year for a yield between 3,000 and 3,500 kg/ha. Yields from the single crop in the remaining wet rice fields were between 2,000 and 2,500 kg/ha. Upland rice yielded 800 kg/ha, and maize, grown in swidden fields, yielded between 800 and 900 kg/ha. Every household had two or three head of livestock, either oxen or cows, horses, or pigs.

Constraints to Development

The main constraints to developing and diversifying the farming systems at Sinh Phinh and thereby improving the livelihoods of the people, were as follows:

- The commune's sloping uplands were strongly dissected by an underdeveloped road network. Transportation was poor, as were communication facilities. These difficulties were most acute during the rainy season.
- The population was scattered over a large area and was living on and practicing shifting cultivation in traditional fashion. There was a generally low level of education.
- Household incomes were low and subject to fluctuating yields from traditional swidden crops.
- The area suffered negative impacts from forest destruction, including flash floods, severe soil erosion, and lack of water for farm and domestic uses, especially during the dry season.
- There was a lack of flat areas for developing wet rice fields. Sloping agricultural land technology had not been adopted, and there was free grazing of cattle.

Potential for Development

Despite the constraints, there was a willingness to change, and Sinh Phinh had resources and potential for the following developments:

- Using valley bottoms and flat areas more efficiently for rice, maize, and bean cultivation, and pursuing more settled agricultural practices to improve the commune's food production;
- Promoting livestock husbandry, including buffaloes, cattle, and horses, for draft services, farmyard manure, food, and possibly sale. This would first require an improvement in grazing systems; and
- Improving forest protection, restoration, and establishment, using species suitable for better environmental protection, soil conservation, water supply, and raw materials for cottage industries.

Outlines for Development

It was decided that development at Sinh Phinh should first focus on improved food security, and the main approach would be to develop technologies to support sustained farming on sloping land. This, in turn, would create a foothold for further economic and social development centered on improvements to the resource management system, including both forest management and swiddening. The latter would be combined with cattle grazing and gardening. These would be the first steps toward longer-term improvements in living conditions and educational standards.

A first assessment showed that to achieve 2,500 kg of food and a cash income of 400,000 dong per year, a household of six to seven people would need 3.5 ha for dryland farming, using existing technologies. They would also have to keep cattle and pigs. Therefore, the need to develop cropping systems on sloping land was very important.

Implementation

First, the total area of land within the commune was divided as follows: forest land, 2,704 ha; swidden fields, 4,065 ha; wet rice fields, 180 ha; home gardens, 170 ha; homestead areas, 51 ha; grazing lands, 125 ha; reserved lands, 500 ha; and other lands; 68 ha. Then, the allocation of forest and forest lands to households in the two villages of Sinh Phinh commune, Ta La Cao and Ta La De, proceeded through the following steps:

- Announcement of project objectives and goals;
- Registration of local farmers' priorities in developing their land;
- Discussions and negotiations with local communities;
- Delineation and area measurements of individual household land holdings, in the presence of the land users, followed by plot mapping and area estimation; and
- Issuing of land-use certificates by authorities of the district people's committee.

Due to limited resources, the project could not cover the whole commune. Only those farmers who volunteered to take part were actually involved. The following broad principles were applied to land allocations.

Primary Forest Land. Because the existing forested area was quite small, allocation trials were made for forest maintenance and resource development along springs and watercourses and for restoration of young forests regenerating after swiddening. These forest areas, although small, play a very important role in the lives of local farmers. They are subject to intricate perceptions of the local people, and as a consequence, their allocation to individual households was difficult. Under these circumstances, it was decided the allocation of land-use rights should be made according to the following conditions:

- An allocated plot should be located next to the beneficiary's house or garden, or adjacent to an already allocated forest area, to encourage integrated land use and careful management.
- Wherever the forested area was too small or was located in a vulnerable area, it should be allocated to a neighboring local community, or to a group of households, for conservation. The same procedure should apply to forest lands capable of easy regeneration through human intervention.

Swidden Fields. These fields were the main source of income for local farmers. They were widely distributed over a large area and managed in a piecemeal fashion according to the current customary law of the commune. Therefore, the allocation of swidden fields was based upon the rights of the first occupiers, and compromises were forged between them and other interested farmers during community meetings. At the same time, attention was paid to securing the minimum land area needed for subsistence, which, under existing technologies at Sinh Phinh, was estimated to be 3.5 ha for a household of seven people.

Home Gardens. These were smaller areas and were usually located next to households, so they were allocated without changes in land holdings. After the allocation was complete, the following had been decided:

- 215 ha of forests were allocated to 65 households for protection and restoration on a long-term basis. This was an average area of 3 to 4 ha per household.
- 250 ha of swidden fields were allocated to 65 households, or an average 3.8 ha per household.
- 10 ha of home gardens were allocated to 92 households.

Following settlement of the new landholding system, preparations began for the introduction of new supportive technologies aimed at building up household economies in the Sinh Phinh commune.

Technology Transfer

Forest and Forest Land Management

Among the first efforts was the development of procedures for the protection of natural forests. These had to follow the customary law of the Hmong, as well as the provisions of the Forest Act. Among the activities involving local farmers at Sinh Phinh were the establishment of fire breaks around the forests for protection and regeneration of allocated lands; promotion of *Amomum* cultivation under the forest canopy; and enrichment planting of *lat hoa (Chukrasia tabularis), voi thuoc (Schima superba),* and *keo la tram (Acacia auriculiformis)* in forest gaps, degraded forest, and regenerating areas.

Two tree nurseries were also established for the production of about 70,000 seedlings per year, and soil and site preparation undertaken for 20 ha of man-made forest consisting mainly of *thong ba la (Pinus kesiya), sa moc (Cunninghamia sinensis),* and *Schima superba.*

Improved Swiddening

Alley cropping was introduced into the swiddens, with one to three rows of *cot khi (Tephrosia candida)* planted along slope contours in upland rice and maize fields. There was 20 to 30 m of space between the rows. This was aimed at decreasing erosion by water, reducing soil loss, and helping maintain soil fertility.

Ten households were supported in intensifying their farming methods on 20 ha of swidden fields. In another case, five households on 2 ha experimented with sloping agricultural land technology practices, using rice terraces and alley cropping with tea, tung oil, *Acacia,* and other shade legumes. New seeds and cultivars were introduced to the commune, including new cultivars of maize—VN10 and Q2—and upland rice varieties imported from Thailand.

While this was happening, the food production of 50 households was being constantly monitored to measure the result of applying well-cured farmyard manure and other fertilizers to crops. Food production doubled and sometimes tripled.

Gardening

Gardening practices also improved. Quickset hedges were established for soil protection and to facilitate introduction of new perennials such as *Prunus salicina, Armeniaca vulgaris* from Yunnan, and mango from Yen Chau. Tea and coffee were also planted on 5 ha held by 70 households.

Strong emphasis was placed on stabilizing and improving lands managed under variations of forest swiddening, and expansion of home gardens appeared to be the most suitable technical alternative. New households, recently segregated from their parental homes, were encouraged to establish better systems of gardening over larger areas of between 2,000 and 3,000 square meters per household. They planted fruit and cash crops most suitable to the environment, as well as grew beans and soybeans, either around the gardens or under the canopy of fruit trees. In addition, water ducts and storage tanks were constructed in some settlements. All these activities were supported with technical assistance and funding.

Results

After three years of project activities at Sinh Phinh commune, the household economies of the Hmong residents have improved significantly. In summary, the following results have been achieved:

- Forests and forest lands have been allocated to 100 households in the commune, with more than 600 ha of land being managed for sustained production.

- The 215 ha of forest that was originally allocated to households is not only well protected, it has been enriched with indigenous tree species. Over three years, the forested area has been increased by the addition of 20 ha of plantations.
- Swiddening practices are evolving toward settled and permanent agriculture. Farmers are using higher yielding food crops and improved farmyard manure, and yields from rice and maize have increased from 800 to 2,500 kg/ha. They have also been encouraged to practice alley cropping and to establish rice terraces for soil conservation. Food production is estimated to have increased two- to threefold, and food shortages are now a thing of the past. Previously underutilized labor is now absorbed into gardening and forest management activities.
- Traditional gardens have been improved with the introduction of fruit and cash crops.
- Social conditions are also improving. Two hundred students are attending formal and informal training courses in farming techniques on sloping lands and other new technologies. The former tendency to exploit natural resources is being replaced by a new concept of nature conservation for the benefit of both individual farmers and local communities. Subsistence agriculture has developed toward commercialized agriculture and agrobusiness, with continuing emphasis on conservation. Lands that were formerly extensively exploited are now intensively managed.

Lessons Learned

The commune of Sinh Phinh is typical of the uplands in Vietnam's northwest, where 85% of the farmland is sloping. Farmers there struggle with adverse physical, economic, and social conditions. Forest cover is less than 10%, transportation and communication facilities are almost nonexistent, water supplies are negligible, and both education and living standards are low.

Under these conditions, the development of more settled farming methods is a basic need, combined with well-controlled cattle grazing and intensive forest management. It requires an integrated approach in which forests and swidden fields play important roles in building up household economies. It should be noted, however, that each ethnic minority has its own traditions and customary laws, and the cultural context must be understood if forest and forest land allocation, and its associated transfer of technology, is to be effective. The transfer of new technologies following land allocation should also be carried out with voluntary participation, so that farmers are willing to participate, to learn, and to develop appropriate techniques. Any attempt to force edicts upon farmers is doomed to failure. The learning process is most productive when it is carried out in the field, where farmers feel at home and are better able to absorb the necessary knowledge.

To reach farmers, these projects should begin by working hand-in-hand with local authorities and mass organizations, such as the Farmers' Association and the Women's League. These authorities and organizations, together with individual farmers, then become the promoters and main agents for change.

Priorities for Future Research

It is important to remember that swiddening and dibbling seed with an iron bar are traditional practices of upland farmers. By itself, the traditional system does not cause serious impacts on the environment, particularly when the fallow period is long enough to allow forests and soils to recover.

However, the practice of swiddening on sloping lands has become harmful under the pressures of a growing human population, the adoption of a market economy, and shortening of fallow periods. Therefore, the integration of legumes into appropriate agroforestry patterns, as a strategy to extend the period of cultivation and shorten that of the fallow, should receive much more attention in forthcoming research activities and projects.

Chapter 67

Managed Fallow Systems in the Changing Environment of Central Sumatra, Indonesia

Silvia Werner

Shifting cultivation has a long history in Sumatra and in Indonesia as a whole. It is one of the oldest agricultural systems in the world and was originally focused on subsistence production. However, over the past 150 years it has been changed by the introduction of the monetary, market-oriented economy and subsequent cash crop cultivation. This transformation process is still continuing and, depending on environmental and infrastructure conditions, local land-use systems are evolving in different directions. This chapter describes recent changes in shifting cultivation systems in central Sumatra and evaluates the factors responsible. Furthermore, it attempts to analyze the conditions under which these systems are diverging and proceeding in different directions.

For the people involved, shifting cultivation is part of their culture. However, this does not imply any obligation to strive for rice subsistence. Indeed, as a land-use system, shifting cultivation is able to adapt and transform to meet changing socioeconomic and environmental conditions. This process, involving the integration of new aspects, may represent the development of entirely new land-use patterns.

Research Methods

The results presented in this chapter are part of a larger study carried out within the framework of my doctoral research, from 1993 to 1994 (Werner 2004). The study focuses on traditional resource management in the buffer zone of the Kerinci Seblat National Park and the scope of involvement of the local people in the sustainable management of these areas. Three villages on the park boundary were studied using a human-ecological and ethnoecological approach.

As a framework within which to analyze the relationships between the people and their environment, the local land-use systems and traditional agricultural practices were studied through guided field walks and household interviews. The ecological impact of these systems and practices was assessed through qualitative and quantitative analyses of soils and vegetation from different fallow stages as well as of cassiavera and rubber gardens. Furthermore, to understand the rationale behind local land-use practices, ethnoecological methods were used, involving group interviews and discussions with key informants.

Silvia Werner, Osterburg Str. 32, 31737 Rinteln, Germany.

The Study Area

The study was carried out in the lowland boundary zone of the Kerinci Seblat National Park (KSNP), in central Sumatra. Stretching along the volcanic Barisan mountain chain, the protected area is located between 100°31′8″E to 102°44′1″ E and 1°7′13″ S to 3°26′14″ S and measures about 345 km in total length. The park straddles the boundaries of four provinces: Bengkulu, South Sumatra, West Sumatra, and Jambi. It encompasses ecosystems from lowland primary forest to alpine flora, within altitudes from 300 to 3,804 m above sea level (asl). The climate of Sumatra is humid tropical, categorized as Af when using the Köppen classification. Soils are of limited fertility, being mainly acrisols or red-yellow podsolic soils with low exchange capacity, base saturation, and shallow topsoil susceptible to erosion.

The study villages are located at about 300 to 350 m asl, two in the province of Jambi (Dusun Birun and Pemunyian) and one in West Sumatra (Lubuk Malakko). The local agricultural production systems are based on subsistence rice production and economic tree crop cultivation, including mainly rubber, cassiavera (*Cinnamomum burmannii*), and some coffee, as cash crops.

Results

Shifting cultivation remains the major agricultural production system in the remote villages of the KSNP lowland boundary areas. Typically, these involve upland rice produced for subsistence and rubber produced as a cash crop. There are also some coffee gardens, but these have fallen into neglect because of low world market prices. On the other hand, cassiavera, a tree originating from the Kerinci highlands, is increasingly planted because of favorable world market prices. Rice is usually grown in the upland fields, along with some vegetables and tuber crops, mainly cassava. However, in villages having a natural potential for irrigated rice cultivation, not all families persevere with upland rice crops.

These findings relate to the thesis of Scholz (1988, *19*), that shifting cultivation is not related to any state of cultural development, nor is it caused by some "ecological forcing factor." Nevertheless, hilly or mountainous conditions, where a lack of flat land restricts irrigated rice cultivation, clearly favor the practice of shifting agriculture.

Shifting Cultivation Systems in Central Sumatra

The major agricultural system for rice production in the research area varies according to the physical conditions. As in other areas where shifting agriculture is practiced, irrigated rice is cultivated wherever the terrain permits (Ramakrishnan 1992). In areas like Pemunyian and Dusun Birun, there are only small patches of level land between rolling hills. In other areas, such as the village lands of Lubuk Malakko, level land dominates the landscape, with dispersed hills and ridges. Shifting cultivation has, therefore, been the major traditional means of rice production in Pemunyian and Dusun Birun, since irrigated rice cultivation at these locations, given the availability of abundant water, would have required a substantial investment in terracing. The conditions at West Sumatra's Lubuk Malakko, on the other hand, favor the establishment of paddy rice fields, so shifting cultivation in this village has, for many years, been only supplementary to irrigated rice production. The farmers of Lubuk Malakko, in fact, discontinued the cultivation of upland rice when technical irrigation was introduced in 1991, since this allowed two crops of wet rice a year instead of one.

In the Jambi study villages, upland rice production has different purposes. As might be expected, the percentage of farmers owning rice-based upland fields is highest in those areas where people most depend on upland rice as their staple food crop. Pemunyian is not only a remote village, it has a relatively small population and abundant land resources. Almost every family in the village keeps upland fields, with the primary aim of subsistence rice production and growing of vegetables. Tree crops may be planted into old rice fields to establish gardens, or simply as an expression of land ownership. However, the primary purpose in creating upland fields remains rice production.

Dusun Birun, on the other hand, although it has no irrigated rice production, has good market connections. Like other villages in these circumstances, its upland rice production is always linked to cassiavera cultivation. The primary purpose of the upland field is for establishment of a cassiavera garden, and upland rice loses its priority. It is planted between the cassiavera trees during the first three years.

In regions where rice is generally grown only in irrigated paddies (Lubuk Malakko), the principal reason for opening upland fields is the establishment of new cassiavera, coffee, or rubber gardens. Vegetables are also grown in these fields while the gardens are still young, but rice is no longer planted.

In these last two areas, the land-use system has a food subsistence component only during the first years. It then changes to become predominantly cashcrop producing. The main motive for establishing upland fields has become the generation of a monetary income.

Integration of Economic Tree Crops

Economic, cash-generating tree crops are planted extensively by smallholders as an outgrowth of shifting cultivation in many parts of the humid tropics (Raintree 1986). Formerly, the spread of tree crops was constrained by market imperfections and the fact that products like rattan, resin, and aromatic gum could be harvested for cash income from abundant primary forests. Increasing pressure on primary resources and the resulting scarcity of these naturally occurring products has played a part in turning shifting cultivators to planting their own economic tree crops (Dove 1994).

In Indonesia alone, most shifting cultivators have already integrated some kind of economic tree crop into their shifting cycle. In the lowlands of West Java, Kalimantan, and Sumatra, the main cash crop has been rubber (Barlow and Thomich 1991; Dove 1993). In East Kalimantan, the fallow is economically enriched with rattan (Weinstock 1984), in West Kalimantan with illippe nuts (*Shorea* spp.) (Werner 1993), and in the uplands of Sumatra with coffee, and in some areas, cinnamon (Godoy and Bennett 1989; Scholz 1988).

After *Hevea brasiliensis* was introduced into central Sumatra in the 1920s, rubber production quickly became the main income-generating activity in that area (see Penot, Chapter 48, and color plate 61). Because rubber trees adapt easily to acid upland soils of low fertility, grow without any fertilizer input, and are easy to grow and propagate, rubber cultivation spread rapidly in Sumatra and other moist tropical areas of Indonesia, including West Kalimantan and West Java. It was considered an ideal peasant smallholder crop. In 1906, South and Southeast Asia accounted for only 1% of world rubber production. Just two decades later, the area's production from cultivated plantations amounted to more than 95% of world supply (McHale 1965).

Rubber remains the main economic tree crop cultivated in the peneplains and hill zones of Sumatra, extending over an area almost 1,000 km long and 120 km wide, covering about one-quarter of the island (Scholz 1988). It has been especially popular where infrastructure and the availability of agricultural inputs have been limited. However, village economies have been changing with infrastructure development, including the building of roads and the flow of information and goods.

In the study villages, farmers have been incorporating other economic tree crops into their rubber gardens. The number of farmers interplanting tree crops and the nature of the new crops differ greatly between the three locations. In the most remote area, only about 15% of all rubber gardens are mixed with either cassiavera or coffee, while in tree crop–oriented villages with good market access, more than 77%

are now mixed. Whereas coffee has been planted in the area for several decades, the planting of cassiavera is a relatively new phenomenon. In tree crop–oriented villages with good market access, the percentage of farmers who mix rubber with cassiavera is highest. This process is exemplified by Dusun Birun, where all families own young cassiavera gardens, and rubber is rarely planted in exclusive stands anymore without being mixed with cassiavera.

Cassiavera has been gathered from the forests surrounding the Kerinci valley for many human generations (see Suyanto et al., Chapter 64, and color plates 45 and 46). Cultivation in the area began in the 1920s, and, within 50 years, Kerinci was producing 63% of Sumatra's cassiavera, or 40% of Indonesia's national production. By 1986, the amount of cassiavera coming from Kerinci was nearly 78% of Sumatra's production (Aumeeruddy 1994).

However, cassiavera prices, like those of many agricultural products, are prone to severe fluctuations, and the attractiveness of the crop to farmers waxes and wanes with the price. During the period of this research (1993 to 1994), high prices were attracting many farmers to plant cassiavera. Prices were similarly high during the 1950s and 1960s, but there was a dramatic drop in the early 1970s (Belsky 1993), and it was about 10 years before cassiavera prices recovered sufficiently to once again make the crop an interesting one for farmers (cf. Scholz 1988). Similar price instability is reported for the Sri Lankan cinnamon crop, and plantings there fluctuate accordingly (Smith et al. 1992).

There are several reasons why cassiavera has spread, even into less suited lowland areas, and has begun to be favored over rubber cultivation. While young rubber plantations are often disturbed by wild pigs, cassiavera is not. It needs less care, can be harvested sooner, and provides cash income according to a farmer's needs. Although rubber is also a very flexible crop and can be tapped according to the need for cash, its adaptability to individual circumstances cannot match cassiavera. In times of low demand, only a few branches of cassiavera can be cut and sold, while on occasions of high need, such as paying for school fees or marriages, whole trees can be harvested.

From Subsistence-Based Shifting Cultivation to Economic Tree Gardens

According to Dove (1993, *142*), the imperative of shifting cultivation societies is always to prioritize rice subsistence. This is no longer true for the changing village economies of central Sumatra. When observing that Kerinci farmers seemed to schedule all other economic activities around the cycle of rice cultivation, Belsky (1993, *138*) concluded that tree crop cultivation had a largely supplementary function. According to my observations during this study, and earlier observations among the Land Dayak in West Kalimantan (Werner 1993), this assessment, which has been shared by many authors, must be revised. While rice production is favored and considered superior to the cultivation of other crops on Java, partially due to mythical reasons (Soemarwoto and Soemarwoto 1984; Dove 1988), this is not necessarily the case elsewhere. During times of difficult market access, and in areas where this difficulty persists, rice production for subsistence remains important because local rice prices are generally high due to high transportation costs. However, farmers generally act economically, and rice production may be neglected if the production of other crops is more profitable and rice can be purchased locally at reasonable prices (Chin 1985; Mackie 1986; Mayer 1989).

This changing attitude is obvious in the study villages. As already mentioned, the pattern of rice production varies according to the ecological suitability of the village land. Dependence on rice subsistence also differs according to local access to roads and markets. In all villages, tree crops produce the main commodities for cash generation, and the extent to which local communities focus on tree crop production depends on both road access and ecological potential for irrigated rice cultivation.

In all villages, the average area planted with tree crops is generally close to the size of freshly opened upland fields. Therefore, it appears that farmers tend to plant all their old *ladang*s with tree crops, and not just part of them. This seems reasonable, since the establishment of upland fields demands major efforts, compared to the planting of tree crops in already established fields.

In Dusun Birun, the primary reason for opening an upland field is to plant a cassiavera garden, and all upland fields are planted with economic tree crops. This is not surprising, because these fields are no longer opened primarily for rice production. Rather, they are opened to establish mixed gardens, which during the last 5 to 10 years, have all been dominated by cassiavera trees, whose harvest is used to buy rice.

In remote Pemunyian, rubber is the main means of cash generation, and about half of the respondents have planted their old ladangs with tree crops. However, a major motivation for these plantings may be a demonstration of land ownership, and not all gardens will continue to be managed.[1] A process of land grabbing is reported in both the study villages described above, and farmers are competing with timber concessions, protected areas, commercial plantations, and transmigrants for remaining primary forest resources.

In Lubuk Malakko, the recently low rate of farmers planting tree crops in their upland fields is mainly a reflection of the overall low number of respondents that had upland fields during the time of my interviews. Farmers at Lubuk Malakko do not have to sell tree crop products to purchase rice, so they have a lower need for cash income. Furthermore, the demand for labor in the rice paddies limits the availability of manpower to open and manage tree crop gardens. This differs from traditional shifting cultivation areas like Pemunyian, where tree crop garden establishment is simply a byproduct of subsistence rice cultivation, in terms of overall labor input. However, as mentioned earlier, it is a relatively new process, and large areas of fallow land remain available. As an indication of a clear trend, all recent young fallows have been transformed into intensively managed tree crop gardens.

Each village exhibits a distinct pattern in terms of the nature of tree crops it prefers in its upland fields. In Pemunyian there is a clear preference for rubber, and less attention is paid to cassiavera or even to coffee. In both of the other areas cassiavera is the tree crop most frequently planted. Following from the fact that farmers in Dusun Birun tend to neglect their harvestable rubber gardens, the new planting of rubber in this village is much lower than that of cassiavera. Nevertheless, most existing cassiavera gardens are mixed with some rubber. Because most of the cassiavera is still very young, it might be possible that, after the cassiavera is older and needs less care, farmers will again pay more attention to rubber and their rice fields. Rubber may also be a kind of "insurance crop," in case cassiavera suffers the same fate as cloves did some years earlier: diseases and a sharp drop in prices. In Lubuk Malakko, only 15% of respondents planted tree crops in the final year of the study. The majority was cassiavera, with only a few respondents planting rubber or coffee. Cassiavera has been planted increasingly in this village since 1987, and particularly since 1991.

In Pemunyian, another development process is taking place: the trend toward high-yielding varieties of rubber. Supported by a government program, but also influenced by a nearby transmigration area with an economy based on high-yielding rubber, most farmers planting new rubber gardens are using improved varietal saplings. Given the fact that overall rubber yields are low in Pemunyian, further extension is needed concerning improved management and processing practices.

In addition to environmental potential and road access, the agricultural development of these areas seems to be heavily influenced by their proximity to successful "centers" of tree crop cultivation. This proximity is important for receiving information concerning profitability, management, and marketing of the respective crop, as well as access to planting materials, other agricultural inputs, and marketing

[1] This process was also observed by Dove (1993) for rubber planting in West Kalimantan.

facilities. Villages influenced by the powerful cassiavera economy of the Kerinci enclave, for instance, will try to imitate the success of the Kerinci farmers. Those living close to successful coffee growers will do the same; and those villages witnessing the profitability of high-yielding rubber compared to their traditional varieties will try to adopt this improved germ plasm. The new influences merge with traditional methods, and management practices slowly adapt to local conditions as the farmers gather experience with their new crops.

Benefits from the Economic Use of the Fallow Period

The changes to agricultural systems in central Sumatra are mainly based on intensified use of what otherwise would have been fallow land. Farmers are not only making more productive use of the fallow period; in some areas, the entire purpose of the shifting cultivation cycle now focuses on productivity during the fallow period. Besides providing income, tree crop gardens are also fulfilling the traditional functions of fallow vegetation, by providing building and handicraft materials, firewood, fruit, and medicinal plants, as well as providing recovery of soil fertility, protection from erosion, and conservation of local biodiversity.

This transformation has been catalyzed by increased market access, making the production of cash crops more attractive than subsistence upland rice production. In the case of Lubuk Malakko village, where technical irrigation was introduced, the development away from shifting cultivation moved in two directions: toward an increase in tree crop production and an intensification of paddy rice cultivation.

The ability of Sumatran farmers to adapt to market prices and changes in the economy was demonstrated on several occasions last century. When coffee prices dropped during the world recession in the 1930s and the costs of foodstuffs rose, farmers converted their gardens into upland rice fields (Belsky 1993). In times of booming coffee prices, between 1973 and 1980, many rice paddies in the coffee zone of the Barisan range were fallowed, only to be cultivated again when coffee prices dropped in the early 1980s (Scholz 1988). In the early 1960s, before the green revolution, high rice prices in relation to rubber drove West Sumatran farmers to produce as much rice as possible, to reduce their dependence on purchases from the market (Thomas 1965). Later, between 1972 and 1979, rising rubber prices led to declining rice production and increased rubber tapping. In the same period the cost of living index rose 2.5 times and rice prices only doubled, marking a steady decrease in real prices for rice[2] and a further reason why farmers found rice cultivation unattractive (Deuster 1982).

Farmers were quick to see the advantages of combining rubber with rice production. Rubber provided excellent prospects for a very attractive cash income from relatively small inputs of money, capital, and labor (McHale 1965). Farmers could hedge against the loss of their rice crop by drawing more heavily on their rubber, and when rubber prices were low, at least they had their rice crop for subsistence (Thomas 1965; Dove 1983).

The reasons for the increase in cassiavera production are similar to those which, 80 years ago, prompted the rapid expansion of rubber production. Compared to the integration of rubber into the fallow period, however, the difference with cassiavera cultivation is that, in those villages well connected to markets, it has intensified beyond fallow enrichment and has become the primary purpose for creating upland fields.

Cassiavera promises a good harvest and good income, needs little care, and is not prone to pests and diseases, so its cultivation spread rapidly through the hills and mountains of the Kerinci area. Investment costs are low; in 1994 the price for an 8 to 12-month-old sapling was only 100 rupiahs (US$0.05), but most farmers were able to propagate their own saplings or collect them from the wild. A further advantage of cassiavera is that the harvest stores well and transportation costs are low because the

---

2 Changes in rice prices as compared with changes in prices of other vital goods.

bark weighs very little. During the first three years of cassiavera cultivation, before the canopy closes out the sunlight, various food crops are planted between the trees. Routine weeding and fertilizing of the intercrops simultaneously benefit the young cassiavera. The food crops are generally produced for home consumption, but some may be sold on local markets. When it is planted in fields distant from villages, cassiavera is not often mixed with food crops because of likely damage by wild pigs, monkeys, deer, or other wild animals. After four years, the young trees form a closed canopy and the intercropping ceases.

The major patterns of cassiavera production vary between the research sites, partly because of its differing economic importance in the three areas. However, like the expansion of rubber before it, the rising popularity of cassiavera and consequent neglect of rice production, results from farmers responding to changes in relative rice prices and market access.

Conclusions

In the study area of central Sumatra, farmers with a new set of alternatives are progressing away from shifting cultivation. This process has accelerated since the mid-1960s, when increasing market access resulted in expansion of cash crop production, especially tree crops, and a decreasing emphasis on subsistence rice cultivation.

The most important feature of this process is not whether irrigated rice or economic tree crops expand, at the expense of upland rice. Rather, it is mainly the increasing role that cash plays in village economies, and the greater dependence villagers have on marketing their produce.

In his 1984 review of Conklin's work about the Philippine Ifugao people, Dove states: "[agricultural intensification] is likely to rarely involve the precipitous and complete abandonment of one system of cultivation for another. It is likely that long stages in the evolution of agriculture have been occupied by composite systems, consisting of both more or less intensive sub-systems (55)."

This is also true in central Sumatra, with one major exception: in this case, these "composite systems" do not consist of distinct land-use patterns but take place within one and the same. In the "cassiavera boom" example, people have integrated shifting cultivation of rice and vegetables with the establishment of tree crop gardens. The same principle can be seen in more remote areas, where people plant rubber into their old rice fields. The main change is the primary aim of the agricultural system. It is no longer food subsistence, but now seeks to generate cash instead. In Lubuk Malakko, the system has moved even further from the original shifting cultivation. People no longer plant rice in their upland fields. They still establish upland fields by means of shifting cultivation, but now they skip the rice phase and directly establish gardens. Vegetables are still planted, but now they are mixed with tree crops. From an ecological point of view, a move from shifting agriculture to tree crops ensures permanent ground cover and would be an ideal alternative in hilly terrain (Ramakrishnan 1992).

Originally, the "dual system" of subsistence-oriented shifting cultivation and cash-oriented tree crops began when rubber was introduced to Sumatra at the beginning of the 20th century. Now, at the beginning of the 21st century, tree crop production has replaced subsistence rice production on upland fields.

References

Aumeeruddy, Y. 1994. Local Representation and Management of Agroforests on the Periphery of Kerinci Seblat National Park, Sumatra, Indonesia. People and Plants Working Paper 3. Paris: UNESCO.

Barlow, C., and T. Thomich. 1991. Indonesian Agricultural Development: The Awkward Case of Smallholder Tree Crops. *Bulletin of Indonesian Economic Studies* 27(2), 29–53.

Belsky, J.M. 1993. Household Food Security, Farm Trees, and Agroforestry: A Comparative Study in Indonesia and the Philippines. *Human Organization* 52(2), 130–141.

Chin, S.C. 1985. Agriculture and Resource Utilization in a Lowland Rainforest Kenyah Community. *The Sarawak Museum Journal* 35(56), Special Monograph No. 4. Kuching.
Deuster, P. 1982. The Green Revolution in a Village of West Sumatra. *Bulletin of Indonesian Economic Studies* 18(1), 86–95.
Dove, M. 1983. Theories of Swidden Agriculture, and the Political Economy of Ignorance. *Agroforestry Systems* 1, 85–99.
——. 1984. Review of H.C. Conklin's Ethnographic Atlas of Ifugao and its Implications for Agro-Ecological Studies in Indonesia. *Prisma* 31 (March 1994), 48–56. LP3ES.
——. 1988. Introduction: Traditional Culture and Development in Contemporary Indonesia. In: *The Real and Imagined Role of Culture in Development: Case studies from Indonesia*, edited by M. Dove. Honolulu: University of Hawaii Press, 1–37.
——. 1993. Smallholder Rubber and Swidden Agriculture in Borneo: A Sustainable Adaptation to the Ecology and Economy of the Tropical Forest. *Economic Botany* 47(2), 136–147.
——. 1994. Transition from Native Forest Rubbers to *Hevea brasiliensis (Euphorbiaceae)* among Tribal Smallholders in Borneo. *Economic Botany* 48(4), 382–396.
Godoy, R., and C. Bennett. 1989. Diversification among Coffee Smallholders in the Highlands of South Sumatra, Indonesia. *Human Ecology* 16(4), 397–420.
Indrizal, E., Hazwan, and Musral. 1992. *Pola Pertanian Lahan Miring di Kawasan Kerinci: Analisa Proses Perubahan Ekologi dan Pertemuan Multi-Kepentingan Dalam-Luar*. Padang: Faculty of Sociology and Anthropology, Andalas University and Worldwide Fund for Nature Project 3941 (Kerinci-Seblat National Park Development Program).
Mackie, C. 1986. The Landscape Ecology of Traditional Shifting Cultivation in an Upland Bornean Rainforest. In: *Proceedings Workshop on Impact of Man's Activities on Tropical Upland Forest Ecosystems*. Serdang, Selangor: Faculty of Forestry, University Pertanian Malaysia, 425–464.
Mayer, J. 1989. Rattan Cultivation, Family Economy and Land Use: A Case from Pasir, East Kalimantan. In: *Forestry and Forest Products*, GFG-Report 13, edited by Samarinda. Indonesia: Indonesian-German Forestry Project (IGFP) at Mulawarman University, 39–53.
McHale, T.R. 1965. Rubber Smallholdings in Malaya: Their Changing Nature, Role and Prospects. *The Malayan Economic Review* 10(2), 35–48.
Raintree, J.B. 1986. Agroforestry Pathways: Land Tenure, Shifting Cultivation and Sustainable Agriculture. *Unasylva* 154, 38(4), 2–15.
Ramakrishnan, P.S. 1992. Shifting Agriculture and Sustainable Development. An Interdisciplinary Study from Northeast India. *Man and the Biosphere* 10. Paris, Casterton Hall, Carnforth, and New Jersey: UNESCO and Parthenon.
Scholz, U. 1988. *Agrargeographie von Sumatra: Eine Analyse der Räumlichen Differenzierung der Landwirtschaftlichen Produktion*. Giessener Geographische Schriften Heft 63. Giessen: Geographisches Institute, Justus Liebig Universität.
Smith, N.J.H., J.T. Williams, D.L. Plucknett, and J.P. Talbot. 1992. *Tropical Forests and Their Crops*. Ithaca and London: Comstock Publishing Associates.
Soemarwoto, O., and I. Soemarwoto. 1984. The Javanese Rural Ecosystem. In: *An Introduction to Human Ecological Research on Agricultural Systems in Southeast Asia*, edited by T.A. Rambo and P.E. Sajise. Honolulu, HI: East-West Centre, 254–287.
Thomas, K.D. 1965. Shifting Cultivation and Smallholder Rubber Production in a South Sumatra Village. *The Malayan Economic Review* 10(1), 100–115.
Weinstock, J.A. 1984. Rattan: A Complement to Swidden Agriculture. *Unasylva*, 36(143), 16–22.
Werner, S. 1993. Unpub. Traditional Land Use Systems of Dayaks in West Kalimantan, Indonesia: Ecological Balance or Resource Destruction? A Study of Vegetation Dynamics and Soil Development. M.Sc. thesis, Department of Geography, Free University of Berlin, Germany.
——. 2004. Environmental Knowledge and Resource Management: Sumatra's Kerinci-Seblat National Park. Berliner Beitrge zu Umwelt und Entwicklung, Vol. 22.

Chapter 68

Community-Based Natural Resource Management in Northern Thailand

Thawatchai Rattanasorn and Oliver Puginier

Over the past 15 years, the situation of ethnic minority groups living in the highlands of northern Thailand has changed considerably. What were once very remote areas with purely subsistence shifting agriculture and cash-generating opium cultivation have become communities fully integrated into the Thailand government's highland development schemes. These programs began as long ago as the late 1960s and attracted considerable sums of development assistance. Many of the projects focused either on social welfare and community development matters or on single-sector agricultural development issues. Over the years, the emphasis of government policies has changed, and the approaches used in working with the highlanders have been progressively modified.

Attempts to harmonize sectoral development priorities led to the First (1992-1996) and Second (1997- 2001) Master Plans for Highland Development and Narcotic Crops Control, both focusing on the socioeconomic improvement of hill tribes, their settlement in permanent villages, environmental conservation and community organization (RTG 1997). However, there has also been a recent shift towards decentralized planning through the enactment of the Tambon Council (TC) and Tambon Administrative Organization (TAO) Act, in March 1995. The objective is the propagation of democracy at grass-roots level by organizing villages into Tambons (sub-districts), with elected village leaders having mandates for local government functions (Nelson 2000). In spite of these promising developments there remain a number of contradictions, and the plethora of policies has led to a situation whereby hill tribes are caught between three divergent policies regarding forest settlement and farming:

- The restoration of forest cover to 25% conservation and 15% production forest, enforced by the Royal Forest Department (RFD), using the restrictive watershed classification. This went as far as hill tribe resettlement (Arbhabhirama et al. 1987).
- Village registration by the Department of Local Administration (DOLA) under the Ministry of Interior, classified by population and long-term residence, progressing from satellite village with no official status to key village with recognized village leaders (Aguettant 1996).

Thawatchai Rattanasorn, GTZ Technical Adviser. Programme de Conservation et de Gestion des Ressources Naturelles (ProCGRN), B.P. 322, Natitingou, Benin. Oliver Puginier, GTZ Technical Advisor, Programme de Conservation et de Gestation des Ressources Naturelles (ProCGRN), B.P. 322, Natitingou, Benin.

- The classification of highland communities according to their potential for permanence, assessed in terms of household numbers, permanent settlement and land suitability for permanent agriculture, yet without the inclusion of hill tribe land classifications. The Department of Land Development carries this out (RTG 1997), though without coordination with RFD regarding the forest status.

The livelihood basis of the hill tribes - shifting cultivation - has no place in the policy framework, hence it is important to help them overcome their marginalization and facilitate the transformation of their farming systems to adapt to new realities. Hill tribes have practiced various forms of shifting cultivation for centuries in a sustainable way and are often victimized as land destroyers by the government instead of being accepted as partners in land use planning (Puginier 2003). The highlands of northern Thailand are thus a prime example for a conflict between a centralized government system with divergent priorities of forest preservation and integration of ethnic minorities that extends its control to the remote areas, where traditional shifting cultivation clashes with centralized planning – an ideal case study for land use planning. The issue has thus become one of mediation and conflict resolution in order to overcome the apparent dichotomy between forest protection and agricultural subsistence.

This chapter springs from the work of the Thai-German Highland Development Program (TG-HDP), a long-term rural regional development project funded by the German development-assistance agency GTZ. When the project was launched, considerable international concern was focused on opium poppy cultivation. Later, however, the main concerns became erosion and natural resource protection, and the project's goals settled on improving the quality of life of the highland population, reducing drug abuse problems, and improving the maintenance of the area's ecological balance.

Of particular significance are the working approaches developed in the course of the TG-HDP. Whereas many applied researchers and some government organizations still maintain a demonstration of new technology mentality, those involved in the TG-HDP grew to believe that learning from and with farmers provided the best avenue for developing farming practices. This led to a shift of extension approaches from crop substitution in the first instance to soil and water conservation, and finally to community-based land-use planning and local watershed management. As a consequence, the guidelines for the final phase of the TG-HDP, from October 1994 to September 1998, declared: "The highland population of Mae Hong Son Province is increasingly self-reliant in applying economic, ecological and social practices, which have been developed and successfully tested in the TG-HDP pilot areas" (Anonymous 1998). Futhermore, a progress report in 1996 stated: "As of 1997, community based natural resources management is accepted at policy level as a sustainable and ecologically sound practice" (Dirksen 1996).

This chapter briefly explores the different extension approaches and then concentrates on experiences over the past few years with the two predominant agricultural systems in the project areas. Much greater attention has recently been paid to building on the traditional, or indigenous, practices of the different ethnic groups. Finally, the effects of the different extension approaches will be discussed, and the future of shifting cultivation will be considered.

The TG-HDP has concentrated on three project areas (see Figure 68-1). These are Tambon Wawi in Chiang Rai Province, which was first selected in 1981 and concluded in 1994; Nam Lang in Mae Hong Son Province, begun in 1983; and Huai Poo Ling in Mae Hong Son Province, begun in 1990.

The two main types of shifting cultivation will be discussed in the context of Nam Lang, where pioneer swiddening is being practiced by Lahu and Lisu hilltribes, and Huai Poo Ling, where rotational swiddening is practiced by Karen people.

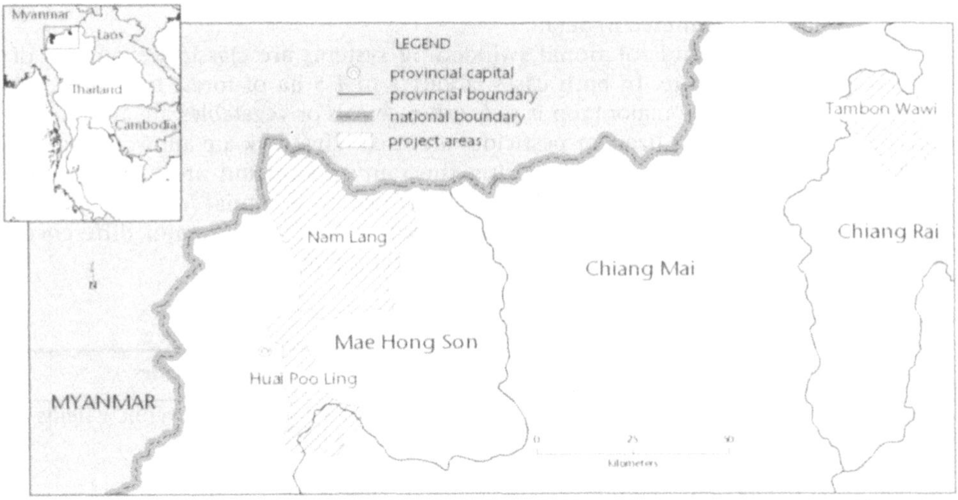

Figure 68-1. TG-HDP Project Areas in Northern Thailand

Traditional Shifting Cultivation

The traditional agricultural systems of the highlanders are based on shifting cultivation, with upland rice and maize on sloping land, paddy rice production in the valleys, and opium poppies and extensive livestock production to meet their cash needs. Agricultural systems and settlement structures differ according to the ethnic origin and cultural background of the various tribes. Nam Lang was the name of the watershed and was renamed Pang Ma Pha sub–district in the course of decentralisation 1996, thus becoming one of four sub–districts in the newly created district Pang Ma Pha of the same name.

In the past, these very extensive farming systems were well adapted to highland conditions, when population density was low and suitable forest land was still abundant and available. The word "suitable" is used with caution, as cutting and burning the forest has considerable effects on the ecology and is only sustainable if the following conditions are observed:

- The area slashed and burned is not too large;
- Soil erosion is not allowed to go on for too long due to cultivation; and
- The fallow period is long enough to allow these areas to regenerate by natural succession to a high level of biological diversity and good ground cover.

These points are mentioned briefly so that they may be kept in mind during discussion of the two types of shifting cultivation. It is difficult to assess the ecological limits of shifting cultivation in terms of the extent of environmental damage inflicted on an area over a period of time. But variations definitely exist, and they need to be considered when assessing the increased demand for land as a result of population growth and migration, mainly from neighboring Myanmar.

Since 1960 the hill tribe population of northern Thailand has rapidly increased from 217,000 (Young 1962) to nearly one million (ADB 2000), but still only accounts for 1.6% of Thailand's population of 62 million. Rapid population growth and a drastic disappearance of forest cover have been blamed exclusively on the hill tribes with their various shifting cultivation systems. Yet one of the few scientific studies carried out to verify that claim concluded that the loss of forest cover correlated more strongly with lowland population increases than with hill tribe growth (Rerkasem 1994). As a result, marginal forest encroachment has accelerated, fallow periods have decreased, and deforestation and forest degradation have occurred. This has resulted in ecological imbalances and has reduced the watershed functions of the highlands.

Given this general trend, shifting cultivation systems and their changing fallow practices need to be examined in detail.

Both the pioneer and rotational swiddening systems are classical examples of slash-and-burn agriculture. In both cases, about 1 to 1.5 ha of forest is burned per household. Although the major crop is rice, other cereals or vegetables are also grown and, traditionally, no fertilizers or pesticides are used. Livestock are allowed to graze freely in the forest or fallow areas during the rainy season but are kept close to villages in the dry season. The annual cycle of cultivation is similar for both systems (see Table 68-1). However, despite the similarities, there are still major differences between the pioneer and rotational swiddening systems (Table 68-2).

Table 68-1. The Annual Cycle of Cultivation

Month	*Swidden Activities*	*Paddy Activities*
January	Select swidden sites, start clearing	No fixed schedule, level new fields, dig new ditches
February	Cut swiddens	Same as January
March	Cut swiddens, burn off	Same as January
April	Same as March, build huts, complete fencing, plant maize	Same as January
May	Complete maize planting, then follow with rice planting	Prepare and sow irrigated nursery; after rains begin, repair dikes and plow fields
June	Start weeding	Plow and harrow, depending on onset of rains; transplant seedlings
July	Continue weeding, some vegetables ripen	Complete soil cultivation and transplanting, weeding
August	Continue weeding, harvest vegetables	Weed and make necessary repairs
September	Final weeding	Weed and general management
October	Rice ripens, some harvesting in late October	Rice ripens, some harvesting
November	Complete harvesting and carry rice to village	Complete harvesting, carry rice to village
December	Finish carrying rice to village	Finish carrying rice to village

Source: van Eckert et al. 1992.

Table 68-2. Characterization of Pioneer and Rotational Swiddening Systems

Pioneer Swiddening	*Rotational Swiddening*
After burning, an area is cultivated for 4 to 5 years till soil fertility declines, then a new area will be chosen. Original area is not cultivated again; there is no cycle.	After burning, an area is cultivated for 1 year only and left to fallow for 8 to 10 years before farmers return to the same area; cyclical cultivation pattern.
Trees are cut and uprooted to allow tillage, so tree regrowth is not possible and fields are later dominated by *Imperata cylindrica* grass.	Trees are cut at breast height but not uprooted. This allows regrowth, mulching, fodder, and seed production.
Rice only is grown in the rainy season, followed by opium. No intercropping.	Mixed cropping on cleared areas, rice with vegetables and tubers. No opium cultivation.
After 1 to 3 years, the rice yields decrease and cash crops are grown until the fields are abandoned.	Cash crops are intercropped with rice on the same area.
Little grouping of households for joint area cultivation, very scattered fields.	Several households grow crops on a joint area; there are a few clusters of large fields.

It can be hypothesized that fallow management, as such, did not exist in traditional swiddening because it was not necessary, and it only came into the picture as development projects attempted to discourage hilltribes first from growing opium and, later, from practicing shifting cultivation, as this was believed to destroy the forest.

TG-HDP Extension Approaches

1984–1987. The first approach employed by TG-HDP in the agricultural sector was similar to that used by other development projects in northern Thailand: crop substitution or replacement of opium poppy cultivation by alternative cash crops, such as coffee or red kidney beans. The new crops were introduced by way of researcher-managed demonstration plots. Seeds or seedlings were provided, and training concentrated on cultivation techniques. Extension was carried out by government-employed field-workers who were rather few in number. The impact was moderate, as farmers' priorities remained fixed on maintaining the yields of subsistence crops. Unfortunately, there was no evaluation of this approach. The TG-HDP felt moved to characterize the traditional farming systems as a vicious circle (Salzer 1987), then went on to introduce a new phase of extension.

1987–1990. Extension of a soil and water conservation (SWC) package began as a means to achieve semipermanent production from sloping land. Vegetative strips, legume rotations, reduced burning, and mulching were the main technical elements demonstrated and extended, and the project provided incentives in cash and kind, including seeds, seedlings, and fertilizer, for both field-workers and farmers (Salzer 1987, 1989; Robert et al. 1989). The incentives also included the provision of Thai identity (ID) cards to hilltribe people who adopted the extension package. There was rapid adoption, but this dropped off markedly once the subsidies were discontinued. A subsequent study found that the perspectives of technicians and farmers, respectively, as to why swiddening and fallowing took place, varied considerably (Bourne 1992). The technicians' view concentrated on the decline in soil fertility, whereas that of the farmers focused on weed buildup. A critical review of the SWC package distinguished between positive incentives in cash and kind up to 1991 and negative incentives following their discontinuation (Enters 1996). These included the threat of resettlement and confiscation of ID cards if the SWC package was not adopted. Farmers soon created "token lines" of grass strips in some areas for demonstration to government officers.

The review by Enters portrayed the TG-HDP as merely going along with the watershed classification by the Royal Forest Department and the subsequent resettlement of hilltribes, which was discussed at parliamentary level. At that time, however, TG-HDP staff had decided to reformulate their extension approaches to provide greater integration of farmers into the general highland development process and to give them the opportunity to demonstrate that their land-use practices were sustainable and did not lead to forest destruction.

After 1990. TG-HDP's assistance programs became based much more on the traditional practices of the different ethnic groups. This was perhaps an acknowledgment of the need to modify both the old SWC package and the hard line taken by the Thai government against the highlanders. Two closely related concepts emerged: community-based land-use planning and local watershed management and sustainable farming systems. All activities were taking place under the abbreviation CLM (Mohns 1989, Patanapongsa 1990, Cheva-Isarakul 1991, Enters 1991, van Eckert 1993, van Eckert and Borsy 1995), and the main aim was the establishment of three types of areas:

- Permanent cultivation areas and permanent settlements;
- Community forest areas; and
- Conservation areas.

The conservation areas were the result of a discussion process that demarcated outer user boundaries, beyond which no activities were permitted. The areas were outlined on land-use maps on a scale of 1:5,000, and were displayed on three-dimensional land-use models made of cardboard or Styrofoam. By measuring the conservation areas in this fashion, they could then be used as a basis for discussions on increasing their size, to demonstrate to government authorities that the villagers could manage and protect forests themselves. In areas set aside for permanent cultivation, the SWC concept was promoted, although without incentives. The community forest areas could be used for wood production as well as collection of nonwood forest products such as mushrooms and bamboo shoots (Chuntanaparb et al. 1995). The whole participatory planning approach was intended to operate by way of land-use planning advisory teams (LUPAT) made up of the various implementing agencies in a holistic, although slightly idealized, process.

The last review of this approach, in 1993, made five recommendations for change. It was admitted, for one thing, that increasing the size of conservation areas by negotiating movements in the outer user boundaries was misunderstood by farmers as taking land away from them. As a consequence, they developed some reluctance to openly display their areas of cultivation (van Eckert 1993). There were also suggestions that adjustments were needed to the LUPAT working approach in order to improve the working relationship between government officers and villagers. As a consequence, the LUPAT system was transformed into a much more informal exchange of information between the responsible implementing agencies. The TG-HDP staff concentrated on enabling farmers to map information at village level, using models, and giving them tools to negotiate land-use and forest issues with the government. These included training in mapping and model building, as well as the aggregation of information at the subdistrict (*tambon*) level, once again using models. It was foreseen that this would become very important with the ongoing process of decentralization in Thailand, which would give tambon administrative offices, representing the lowest level of government in Thailand, the mandate to regulate natural resource issues with respective budget allocations. An additional component in this latter equation was the Community Forestry Act, which was in the stage of public hearings. Once it passed into law, it would provide a legal framework for the resolution of land-use rights and communal forest management. In the final phase (1995-1998), the TG-HDP focused on the aggregation of land use data at Tambon (sub-district) level and on the support of the Pang Ma Pha Hill Tribe Network Organization, which emerged from attempts to resolve conflicts between three neighboring villages over the collection and sale of forest products in 1996 (Jantakad 1998).

Land-use models were built in all target villages and were updated when farmers requested it. The land categories identified on land-use models (see Figure 68-2) were as follows:

- Village and housing area, including home gardens;
- Arable land for annual crops;
- Arable land for perennial crops and agroforestry;
- Pasture areas;
- Social and community forest land; and
- Watershed areas and conservation forest.

As planning approaches became more concerted, it was possible to experiment with new technologies (Puginier 2001). It is against this background that the research project was initiated to develop a method to combine land use planning with remote sensing tools, together with the full participation of the local communities. In order to go beyond land demarcation and to carry the CLM process from the village to higher planning levels, the accompanying PhD research examined possibilities to transfer the data from village maps into a Geographic Information System (GIS), so as to provide visual information understandable by the people who displayed it and by those who would interpret it. Hand-drawn land use maps were collected in villages and digitized using a hand digitizer into the GIS programme Arc Info and then converted into maps using the map-drawing programme Arc View 3 (Puginier 2002). Once the maps had been digitized and printed, they were taken back to villages for modifications or corrections, so as later distribute them in plastified A1 size for longer term use. Maps were also distributed to district forest officials to facilitate their work in land use monitoring. The data and the GIS software were then transferred to the Survey Section of Northern Narcotics Control Office (NNCO) as well as to the International Center for Research in Agroforestry (ICRAF) office in Chiang Mai that collects this data for the whole north of Thailand (Saipothong et al. 1999).

The Future of Shifting Cultivation

When examining the efforts of the Thailand government to establish permanent cultivation in the highlands, and those of the TG-HDP to promote soil and water conservation, one cannot escape the impression that shifting cultivation has no future; everything focuses on limiting forest encroachment and erosion control on agricultural areas. In an attempt to preserve remaining forest areas, any land that can be classified by highlanders as a conservation area in exchange for agricultural extension services is officially recognized as forest land. It immediately falls beyond the outer boundary and is no longer part of the village's fallow areas. This process of agricultural intensification in exchange for a reduction in cultivated area seems to leave little room for fallow management as, ideally, all fallow areas can become secondary forests. This "land deal" only involves swidden intensification as far as it is converted to permanent agriculture through the development of irrigated areas, fruit tree orchards, and vegetative hedgerow management. This leads to a complete modification of indigenous practices up to the point of their abandonment in favor of mainstream permanent farming. However, when looking back at TG-HDP extension approaches since 1984, there has been an evolution from straightforward crop replacement without consideration of indigenous practices, to a more sensitive and less technically oriented approach to the management of natural resources.

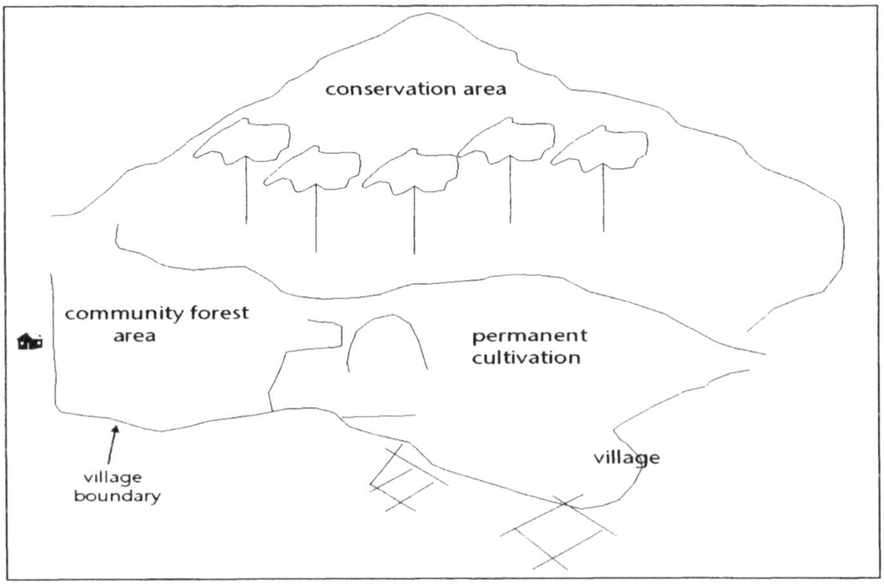

Figure 68-2. Land-Use Mapping Using the CLM Approach.

It is useful, here, to differentiate between the effects of agricultural extension on pioneer swiddening, on one hand, and rotational swiddening, on the other. The main extension effort is quite clear: there is no future for pioneer swiddening. The government wants to stop encroachment into new forest areas that eventually converts them to grassland, with permanent loss of forest. Early efforts concentrated on the establishment of permanent areas with buffer strips of congo grass (*Brachiaria ruziziensis*) and crop rotations involving rice, beans, maize, and fruit trees. However, as the congo grass grew, it spread rapidly and soon became a weed. Either it was to uprooted, or farmers abandoned the land to become pasture in favor of new areas for cultivation. Partly as a result of learning from farmers, TG-HDP and the Department of Land Development modified their extension approach to include more crops and to let farmers choose for hedgerows what best suited their indigenous practices. Hedgerows of pigeon pea (*Cajanus cajan*), *Leucaena* (*Leucaena leucocephala*), vetiver (*Vetiveria zizanioides*), and *Gliricidia* (*Gliricidia sepium*) have lately been promoted. This approach leaves little room for fallow management. On the other hand, fruit trees have become quite popular, and access to the main road between Chiang Mai and Mae Hong Son has increased marketing opportunities for fruit. This has allowed some villages to earn a living from cash crops while slowly giving up their indigenous practices.

Opium cultivation has been reduced dramatically, not so much because villagers are convinced by agricultural alternatives, but more because of the increased efficiency of government aerial surveys spotting opium fields for destruction. Perhaps the most important factor in the transition from pioneer swiddening to permanent agriculture has been an indirect one: the construction of a tarmac road between Chiang Mai and Mae Hong Son.

The situation for rotational swiddening differs in that the population density is much lower and there is no tarmac road, so access to markets is very limited. The extension approach has therefore been modified to suit the traditional fallow system and to make it more productive. Traditionally, fields are only cropped for one year, and pigeon pea planting has been introduced to fallow areas a year after the rice crop (see also Daguitan and Tauli, Chapter 57). There is a dense planting of two to three seeds per hole, roughly every 70 cm, and *Leucaena* is planted every meter, randomly and not in rows. The rapid establishment of pigeon pea and *Leucaena*, with their beneficial side effects of soil maintenance and mulch, leads to a reduction in the fallow cycle of 50%, so that farmers may return to the same area after four years instead of eight. In turn, this enables more land to be reserved as conservation forest. Pigeon pea and *Leucaena* are harvested and sold to the Department of Land Development as an added income and are used to feed poultry and pigs. In parallel with this approach, paddy rice production is expanded through irrigation, reducing reliance on upland rice.

Cattle still graze freely on fallow areas, however, and farmers have not adopted cut-and-carry procedures for feeding them so they can be tethered. As one remarked: "Animals have four legs; we only have two legs, so why should we feed them?"

There is consequently a need to improve livestock management by fencing off fallow areas, as traditional fences protecting areas newly burned for rice cultivation only last for one year. Free-grazing livestock can damage regenerating fallow vegetation, and this is particularly harmful if there is an expectation of reopening the fallow after a reduced period of four years. The TG-HDP offered subsidies on the purchase of barbed wire, but rather than fencing upland areas, farmers fenced mostly areas with permanent irrigation for fruit tree production.

We offer one example of building on traditional practices. Five years ago, in the village of Ban Huai Hee, which had 22 households, some farmers participated in a field trial involving pigeon pea and *Leucaena* as improved fallow species. They were provided with barbed wire but used the wire to protect fruit trees instead of fallow land. Nevertheless, the *Leucaena* and pigeon pea grew well, and in the following slash-and-burn cycle, the rice harvest was perceived to be as good as they might have obtained had they used traditional cultivation practices. The improved fallow was thus meaningful, and the practice has attracted more farmers who are even willing to partly fence their fallow areas with traditional stick fences. Perhaps the success was due to the fact that the village had no paddy area and they were 100% reliant on dryland rice. Results from four other villages were disappointing because grazing cattle destroyed the *Leucaena* and pigeon pea. But, in these cases, there were paddy areas that produced 50% of the rice required for village consumption, and fallow management was not as important. Once again, this example seems to show that with progressing intensification, there will be less room for fallow management. However, there is perhaps more room to accept the Karen rotational swiddening system as sustainable; therefore, this form of shifting cultivation, with fallow management, may have a better future than the pioneer system.

Conclusions

In the highlands of Thailand there is a general trend toward a gradual decrease in fallow areas, and therefore also fallow management, in favor of permanent cropping systems within areas of fixed demarcation. This development can also be observed in other countries and should not be considered unusual in a changing world of agriculture with ever-increasing densities of population. After all, where in highly populated western Europe can one find swiddening systems?

The community-based land-use planning and local watershed management process employed by the TG-HDP has proven to be an entry point for creating land-use groups and community awareness. Villagers themselves, by way of land suitability and capability assessments, can exert control over members who do not follow accepted guidelines, and this increasingly extends to closer monitoring of land use by the government as land in Thailand becomes more and more valuable. This process also involves a gradual reduction of fallow areas in favor of agricultural intensification and forest protection. This may become more effective, with upland communities participating as protectors of the forest, once the Community Forestry Act has passed into law.

Within the geographical scope of this study, a number of practical difficulties emerged once hill tribe farmers were called upon using topographic models and digitized maps without external support. Such tools are useful only if clear goals are set and allow for a certain degree of communal forest management by villagers, which after more than a decade of political debate is still an elusive perspective. The policy framework needs to be reformed to find a compromise between forest protection and agricultural subsistence, and to create a link between national priorities and applications both at the village and Tambon levels. Agreements between villages and government agencies can be made by local Tambon Administrative Organizations (TAO) with their mandate for natural resource management. One potential to deal with these differing priorities at the Tambon level could evolve from the current restructuring project of the Ministry of Agriculture and Cooperatives (MOAC), which is part of the ongoing process of decentralization. Part of this reform has been the introduction of Technology Transfer Centres (TTC) in 1998, with 82 TTCs established nationwide by the Department of Agricultural Extension (DOAE). The initiative aims at covering all Tambons in the next few years (GTZ 2001). There are plans to link the new TTCs with the TAOs, of which all registered villages are members. TAOs are intended to become the major conduit of funds and resources, though the details of responsibilities are still being developed. The application of digitized land use planning now depends on the implementation of policy reform at local level and becomes a testing ground for good governance.

References

ADB. 2000. Addressing the Health and Education Needs of Ethnic Minorities in the Greater Mekong Sub-Region. Asian Development Bank (ADB), Thailand Country Report. Chiang Mai, Thailand: Research Triangle Institute (RTI).

Aguettant, J.L. 1996. Impact of Population Registration on Hilltribe Development Thailand. *Asia-Pacific Population Journal* 11 No. 4, 47-72.

Anonymous. 1998. Review of TG-HDP's Agricultural and Forestry Programs 1984-1998 vol.1 and 2. Internal paper 212. Thai-German Highland Development Program (TG-HDP). Chiang Mai, Thailand: TG-HDP.

Arbhabhirama, A. D. Phantumvanit, J. Elkington, and P. Ingkasuwan. 1987. *Thailand Natural Resources Profile. Is the Resource Base for Thailand's Developmentt Sustainable?* Bangkok, Thailand: Thailand Development Research Institute (TDRI).

Bourne, W. 1992. Nam Lang Impact Survey. Internal paper 165, Thai-German Highland Development Program (TG-HDP). Chiang Mai, Thailand: TG-HDP.

Cheva-Isarakul, B. 1991. Integration of Livestock into SFS. Internal paper 141, Thai-German Highland Development Program (TG-HDP). Chiang Mai, Thailand: TG-HDP.

Chuntanaparb, L., et al. 1995. Study on Non-Wood Forest Product Identification, Availability, Improved Production and Potential for Marketing in Northern Thailand. Internal paper 190, Thai-German Highland Development Program (TG-HDP). Chiang Mai, Thailand: TG-HDP.

Dirksen, H. 1996. Project Progress Report. Internal paper 205, Thai-German Highland Development Program (TG-HDP). Chiang Mai, Thailand: TG-HDP.

Enters, T. 1991. The Transition from Traditional to Sustainable Farming Systems: Opportunities and Limitations, Seven Case Studies of Nam Lang Farming Households. Internal paper 143, Thai-German Highland Development Program (TG-HDP). Chiang Mai, Thailand: TG-HDP.

———. 1996. The Token Line: Adoption and Non-Adoption of Soil Conservation Practices in the Highlands of Northern Thailand. In: *Soil Conservation Extension: From Concepts to Adoption.* Soil and Water Conservation Society of Thailand.

GTZ. 2001. *Community-Based Agricultural Development: Challenges for the Ministry of Agriculture and Cooperatives* (MOAC) 3. Bangkok, Thailand: GTZ.

Jantakad, P. 1998. Natural Resource Management by Network Organization. In: Anonymous. 1998. Review of TG-HDP's Agricultural and Forestry Programs 1984-1998 vol.1 and 2. Internal paper 212. Thai-German Highland Development Program (TG-HDP). Chiang Mai, Thailand: TG-HDP.

Mohns, B. 1989. The Introduction of Land Use Planning and Local Watershed Management in the TG-HDP Project Areas Nam Lang and Huai Poo Ling. Internal paper 113, Thai-German Highland Development Program (TG-HDP). Chiang Mai, Thailand: TG-HDP.

Nelson, M.H. 2000. Local Government Reform in Thailand. Center for the Study of Thai Politics and Democracy, King Prajadhipok's Institute Reports 1. Nonthaburi, Thailand: Center for the Study of Thai Politics.

Patanapongsa, N. 1990. A Study on the Problems and Constraints for Farmers to Continue Participation in the TG-HDP Sustainable Farming Systems Program. Internal paper 135, Thai-German Highland Development Program (TG-HDP). Chiang Mai, Thailand: TG-HDP.

Puginier, O. 2001. Participatory GIS as a Tool for Land Use Planning in Northern Thailand. In Bridges, E.M. et al (eds.): *Responses to Land Degradation.* Enfield, USA: Science Publishers Inc., 288-289.

———. 2002. Hill Tribes Struggling for a Land Deal: Participatory Land Use Planning in Northern Thailand amid Controversial Policies. Ph.D. Thesis at the Humboldt University Berlin, Germany. URL: http//dochost.rz.hu-berlin.de/dissertationen/puginier-oliver-2002-05-16. Digital Thesis.

———. 2003. The Karen in Transition from Shifting Cultivation to Permanent Farming: Testing Tools for Participatory Land Use Planning at Local Level. In Delang, C. (Ed.) *Living at the Edge of Thai Society: The Karen in the Highlands of Northern Thailand.* London, New York: Routledge Curzon, 183-209.

Rerkasem, K. 1994. *Assessment of Sustainable Highland Agricultural Systems.* Bangkok, Thailand: The Thailand Development Research Institute (TDRI).

Robert, G.L., et al. 1989. Soil and Water Conservation Program Impact Measurement Survey. Internal paper 111, Thai-German Highland Development Program (TG-HDP). Chiang Mai, Thailand: TG-HDP.

RTG. 1997. Second Master Plan for Highland Development and Narcotic Crops Control. Bangkok, Thailand: Royal Thai Government, United Nations Fund for Drug Abuse (in Thai language).

Saipothong, P., H. Weyerhäuser, and Thomas, D. 1999. Potential of GIS for Local Land Use Planning: A Case Study in Mae Chaem, Northern Thailand. International Centre for Research in Agroforestry (ICRAF), Chiang Mai University.

Salzer, W. 1987. The TG-HDP Approach toward Sustainable Agriculture and Soil and Water Conservation in the Hills of Northern Thailand. Internal paper 80, Thai-German Highland Development Program (TG-HDP). Chiang Mai, Thailand: TG-HDP.

———. 1989. Monitoring and Evaluation of the TG-HDP with Focus on the Agricultural Sector Program and Specifically to the SWC Cropping Pattern Concept. Internal paper 100, Thai-German Highland Development Program (TG-HDP). Chiang Mai, Thailand: TG-HDP.

van Eckert, M. 1993. Findings of the Review of the Community-based Land Use Planning and Local Watershed Management Process and Recommendations. Internal paper 171, Thai-German Highland Development Program (TG-HDP). Chiang Mai, Thailand: TG-HDP.

van Eckert, M., W. Bourne, K. Buadaeng. 1992. The Karen Farming Systems in Huai Poo Ling. Internal paper 157, Thai-German Highland Development Program (TG-HDP). Chiang Mai, Thailand: TG-HDP.

van Eckert, M., and P. Borsy. 1995. CLM Guidelines in Brief, Community-Based Land Use Planning and Local Watershed Management and Approach to Achieve Sustainable Land Use. Internal paper 189, Thai-German Highland Development Program (TG-HDP). Chiang Mai, Thailand: TG-HDP.

Young, G. 1962. *The Hill Tribes of Northern Thailand.* New York: Ams Pres. (Reprinted from Siam Society Monograph No. 1, 1962.)

PART X

Conclusions

A Batak farmer in Palawan, the Philippines, sets a snare for wild chicken.

Chapter 69

Observations on the Role of Improved Fallow Management in Swidden Agricultural Systems

A. Terry Rambo

Reading this volume from cover to cover, as I necessarily had to do in order to prepare this chapter, was both an enlightening and a sobering experience. It was enlightening because I learned so much about so many different fallow management systems. Clearly, our research community has gathered a great deal of knowledge about fallow management in diverse swidden systems in many different ecological and social settings. Compared to just 30 years ago when, even in the most careful studies of swiddening, the fallow phase was largely ignored, much has been learned. It was a sobering experience because it made me recognize that, despite these advances, our understanding remains patchy and fragmentary. This volume contains many case studies of how fallow management is practiced in specific local contexts. It describes the ways in which a variety of plant species is used to achieve farmers' fallow management objectives, and it examines some of the cultural and economic constraints on improving fallow management in some of these study sites. But the information is incomplete and highly site-specific. Many aspects of fallow management are not yet understood, and explanations rely heavily on commonsense assumptions that have not been empirically verified. We still have a very long way to go before we have an adequate understanding of fallow management.

In this chapter, I will use the case studies in this volume as a base from which to generate questions about the current state of our understanding of fallow management. In particular, I will focus on aspects that remain unclear and questions that still need to be answered.

What Is Shifting Cultivation?

The editor, wisely in my view, does not try to formulate a comprehensive definition of shifting cultivation (or swidden agriculture, as I will usually call it). Many scholars have tried to define shifting cultivation with varying degrees of success. But no generally accepted definition has emerged. In large part, this reflects the extreme diversity of agricultural practices in the tropics. As Harold Brookfield points out, in

A. Terry Rambo, Department of Agronomy, Faculty of Agriculture, Khon Kaen University, Khon Kaen 40002, Thailand.

Chapter 2:[1] "shifting cultivation is a term used to generalize about a huge range of farming systems."

Reading the literature on swidden agriculture brings to mind the story of the blind men attempting to describe the elephant—except that, in this case, it seems like there are several groups of blind men, each touching a different species of animal, not just an elephant but also a tapir and maybe a python. Not surprisingly, swidden researchers have difficulty in reconciling the findings of such disparate investigations. Bitter arguments continue to rage, for example, as to whether swidden plots are abandoned because of declining soil fertility or increasing competition from weeds. Thus, Rattanasorn and Puginier, when detailing the experiences of the Thai-German Highland Development Program in Chapter 68, report that "the perspectives of technicians and farmers, respectively, as to why swiddening and fallowing took place, varied considerably. The technicians' view concentrated on the decline in soil fertility, whereas that of the farmers focused on weed buildup." Roder, Maniphone, Keoboualapha, and Farhney state in Chapter 14 that, in northern Laos, "labor for weed control was the single most important constraint to upland rice production," and was cited by 85% of their farmer respondents, whereas soil fertility was cited by only 21%. Dietrich Schmidt-Vogt, in Chapter 4, also states that weed competition is the main reason given by Lawa farmers for cultivating their swiddens for only one year in northern Thailand. Ole Mertz, in Chapter 7, reports that Iban swiddeners in Sarawak gave a similar explanation for fallowing their fields after only one year. Conversely, Fatima Tangan reports in Chapter 9 that farmers in the Philippine Cordillera complained about declining soil fertility in their swiddens and that this was a major impetus to them planting wild food plants in their fallows, while Balbarino, Bates, de la Rosa, and Itumay state in chapter 18 that "fallowing is a traditional practice among Filipino upland farmers, who understand the necessity of resting the soil under vegetation to restore its fertility." So do farmers fallow their swiddens to rebuild soil fertility, or to control weeds? It may be for both reasons at the same time, as Djogo, Juhan, Aoetpah, and McCallie suggest in Chapter 17 when they say that, in swiddens in Nusa Tenggara Timur, "when the fertility of the soil declines, the land is fallowed to restore its fertility and to control weeds." It may also be the case that declining soil fertility is critical in swidden systems under certain ecological conditions whereas weed build-up is the main problem in other situations.

It seems unlikely that we are ever going to be able to formulate a set of common principles that govern swidden agriculture, because swiddening is not a unitary phenomenon subject to fixed rules of behavior. Rather, different types of swiddens occur in different parts of Asia and the Pacific. As Peter Horne observes in Chapter 10: "shifting cultivation systems are characterized more by their differences than their similarities. Immense variability and complexity in land capability, land-use patterns, and population pressures exist, even across short distances and within individual villages." He draws the conclusion that systems-wide studies are inappropriate because "we will never be able to characterize these systems adequately to satisfy the scientific requirement for certainty before action." I agree with his assessment of the complexity and diversity of swidden systems but do not accept his conclusion that the use of systems approaches is doomed to failure. On the contrary, in my view, the use of a systems perspective offers the most promising way to understand swidden agriculture. The role of fallows would seem to be most effectively studied by looking at the fallow as just one element of a complex agroecosystem that includes a multiplicity of interacting components, not just soil, crops, and forest, but also the farm household, the market, the government, and livestock.

I have left the question *"What is shifting cultivation?"* with which I opened this discussion, unanswered. That is because I cannot formulate a fully satisfactory definition, and I question whether it is possible to do so. We are in a position similar to that of Emile Durkheim who, when he began his study of primitive religion, stated that, initially, he would not attempt to define religion but, instead, would first look

1 Unless otherwise indicated, all references are to chapters in this volume.

at specific examples and then formulate a general definition. But how do we know that the specific cases are examples of the general phenomenon if we can't define it first? Unfortunately, there is no ready solution to this dilemma, so we can only fall back on an assertion that "we know it when we see it." From a practical standpoint, this is probably not a very important weakness. Certainly, most of the cases described in this volume fall rather clearly within the domain that most researchers recognize as that of swidden agriculture.

What Is Fallow Management?

Malcolm Cairns provides a comprehensive typology of fallow management systems in his "roadmap" to this volume, so there is no need to repeat that discussion here. Almost all of the chapters in this volume describe systems that we can readily recognize as involving some type of fallowing. But there are some exceptions. I find it rather a stretch to include continuously cultivated alley cropping systems, such as the *Leucaena leucocephala* system employed in Sulawesi and described in Chapter 25 by Fahmuddin Agus, within this domain. Such systems may have evolved from a swidden precursor, but they are no longer swiddening in any meaningful sense of the term. And to label them as "simultaneous fallows," as Prinz and Ongprasert suggest in Chapter 19, seems to be a bit of linguistic sleight of hand.

Although the Continuum of Indigenous Fallow Management Typologies, proposed by Cairns in Chapter 3, provides a useful roadmap to this volume, I must express some concern that following it too closely may lead our thinking astray. Its necessarily lineal character can all too easily be mistaken for a progressive evolutionary continuum. It starts off with "simple" fallows based on retention or promotion of preferred voluntary species, then moves forward through shrub-based accelerated fallows, herbaceous legume fallows, dispersed tree-based fallows, and perennial-annual crop rotations, until it finally reaches "complex" agroforests. These different types of fallows are not stages on an evolutionary continuum. They are simply descriptive categories for different systems that coexist in time and space in Asia and the Pacific. Each type is an adaptation to specific local ecological and social conditions. Those at the beginning of the sequence are not inferior to those at the end. Depending on local conditions, shrub-based accelerated fallows may, from a farmer's point of view, be superior or inferior to agroforestry systems. It all depends on the context in which they are used.

There is another serious limitation to the typological approach. Although it functions as a convenient framework for organizing the chapters in this volume, it fails to convey any sense of the spatial distribution of fallow management types. The previous 68 chapters assembled within these covers mostly present admirably detailed descriptions of fallow management at specific local sites, be it one barangay in Mindanao or one commune in the mountains of northern Vietnam. It is not possible, however, to fit these diverse local studies together to form a bigger picture of the status of fallow management in the Asia-Pacific region as a whole. This is not a criticism of this volume, which was assembled to meet a different need. But, in future research, it would be extremely useful to study the spatial distribution of different types of fallow management. In particular, it would be useful to see if the existence of specific types of fallow management correlates with variations in topography, soils, climate and hydrology, natural vegetation, population density, ethnicity, access to roads, and so on. Cairns, Keitzar, and Yaden, writing in Chapter 30 on the use of *Alnus nepalensis* in improved fallows by the Nagas, suggest the application of geographic information systems to identify factors determining the sporadic distribution of this system. Such spatial analyses would greatly enhance our understanding of fallow management.

What Is a "Composite Swiddening Agroecosystem"?

In 1992, while doing fieldwork in Vietnam's Hoa Binh Province, I became aware that the Da Bac Tay ethnic minority farmers with whom I was working were practicing a form of swidden agriculture that was quite different from any other system with which I was familiar. Each household farmed both wet rice fields in the valleys and several swidden fields on the surrounding mountain slopes. This mixed system was not a recent response to population pressure but, instead, had been used for as far back as anyone could remember, even when population was much lower and there was abundant unused land to construct additional paddies. Later, I discussed my observations with several knowledgeable colleagues (Michael Dove, Percy Sajise, and Pei Sheng-jie), and they informed me of similar systems existing in other parts of Southeast Asia. I subsequently proposed "composite swiddening agroecosystem" as a label for this specific type of swidden agriculture (Rambo 1996).

Several of the chapters in this volume describe composite systems that appear to be quite similar to that practiced by the Da Bac Tay of Vietnam. Some form of composite swiddening involving both wet rice fields and swiddens is practiced by the Lawa in northern Thailand, described by Schmidt-Vogt in Chapter 4; recently some Kammu in Laos, according to Tayanin in Chapter 6; Thai and Karen in northern Thailand, according to Prinz and Ongprasert in Chapter 19; the Nagas of northeastern India, described by Cairns, Keitzar, and Yaden in Chapter 30; and the Hani of Xishuangbanna in southwestern China, described by Xu in Chapter 56. As Xu reports: "the composite swidden system typically practiced by the Hani includes traditional tea gardens in forests, intensive paddy cultivation, home gardens, shifting cultivation, and livestock grazing."

In Chapter 35, Nicholas Menzies and Nicholas Tapp now propose extending the term "composite swiddening" to other types of swidden systems that incorporate multiple components. In their description of planting of *Cunninghamia lanceolata* in fallows by minorities in southeastern China they observe:

> Commercial production of *Cunninghamia* in mountainous areas, pioneered by upland ethnic minorities, was an integral part of a local economy in which upland dry agriculture, animal husbandry, hunting, and fishing were also important elements. The common practice of intercropping in *Cunninghamia* plantations gives good reason to propose that these farming systems should properly be seen as a form of shifting cultivation. They would appear to fit well with a category of upland farming common in Southeast Asia that Rambo (1996) described as "composite swiddening agroecosystems.

Although I originally proposed this term as a label for a specific type of agricultural system involving simultaneous cultivation of paddies and swiddens, I think there is considerable merit in the suggestion by Menzies and Tapp that it be used for a wider array of swidden systems involving multiple elements. Most, if not all, systems of shifting cultivation practiced in Southeast Asia involve multiple components, so they are indeed composite systems. In particular, livestock play a central role in many swidden systems. I will discuss this later. However, by giving recognition to the complexity of many different forms of shifting cultivation, the use of this label may help to counter the all-too-widely held stereotype of swiddening as a simple, primitive, and backward form of agriculture. In that case, the system I initially described for the Da Bac Tay should be relabeled as the "composite paddy-swidden agroecosystem."

Is Swiddening a "Good" System?

Beginning with Conklin's work on Hanunóo shifting cultivation in 1957, anthropologists have taken on the role of defending swiddening against what they perceive as the wrong-headed and negative assessments of foresters and national governments. Many of the chapters in this volume continue this effort to portray swiddening in a favorable light. Cairns, Keitzar, and Yaden, for example, write in Chapter 30 that "documentation and validation of alder management as a superior fallow system helps debunk the common stereotype of shifting cultivators as wanton destroyers of forest ecosystems and more accurately portrays them as forest planters and managers." In a widely ignored paper (Rambo 1982), I argued a similar case for the Malaysian Orang Asli. Certainly, looking at the chapters in this volume, such as Hoang Xuan Ty's description of indigenous practices employed by swidden farmers to rebuild soil properties during the fallow period in Vietnam (Chapter 55), one cannot fail to be impressed by the ingenuity with which farmers manage their fallow fields to both restore fertility and increase productivity while protecting biodiversity. This is basically a collection of success stories. What this volume largely omits are descriptions of the less admirable manifestations of shifting cultivation. I refer, of course, to the very nonsustainable systems that are lumped together under the label of pioneering swidden systems.

In their extreme form, pioneering shifting cultivators clear primary forest and cultivate their swiddens for as long as 10 years, until soil erosion and loss of fertility become so severe that continued cropping is impossible. Then they abandon the site to move to a new tract of forest. The landscapes they leave behind are virtually lunar, with barren hills or *Imperata* grassland that will not revert to forest for decades, if ever. It is true that many pioneering shifting cultivators are migrants from the lowlands who simply lack the knowledge needed to practice more sustainable forms of swiddening. But many are also members of groups usually classified as "indigenous people" who are now popularly valorized for their presumed abilities to live in harmony with nature. According to Chaleo Kanjunt in Chapter 5, swiddening by Hmong and Lisu farmers in Thailand's Mae Hong Son Province involves "the clearing of steep, erosion prone ridge tops and exposed upper slopes, the removal of most trees, including their stumps, and burning the entire area. Vast areas are now covered with *Imperata cylindrica* and other grasses and herbs."[2] Such destructive pioneering swidden farmers are only a small minority of shifting cultivators. Nevertheless, they have unfortunately provided a basis upon which government officials have formed negative stereotypes covering all shifting cultivation.

In any case, many rotational systems of swiddening, even those that were once admirably productive and sustainable, are now breaking down under the pressures of intensification. This phenomenon is well documented by several chapters in this volume. Yields have fallen to such low levels that they hardly justify the ever-increasing days of labor that farmers must spend in cultivation. The figures are quite shocking: Iban farmers in Sarawak, described by Ole Mertz in Chapter 7, achieve a yield of only 350 kg of rice per ha from their swiddens; Nguyen Tuan Hao, Ha Van Huy, Huynh Duc Nhan, and Nguyen Thi Thanh Thuy report in Chapter 22 that Hmong farmers in Tua Chua in northwestern Vietnam produce only 180 kg of corn per ha; and, according to Michael Robotham in Chapter 58, farmers in Mindoro, the Philippines, obtain maize yields of 1 tonne and rice yields of 600 kg per ha from their swiddens. Such low yields condemn many upland populations to permanent poverty and hunger. Thus, whatever its merits may have been in the good old days, I don't think we should continue to defend swidden agriculture as an intrinsically good system under current conditions. But we must still struggle to get upland development planners to understand that shifting cultivation is not the main cause of deforestation in the uplands of Asia and the Pacific, and that farmers continue to rely on swiddening not because they are ignorant and backward, but because they have no choice, unless and until they are offered viable alternatives.

---

[2] Of course, not all Hmong employ such nonsustainable methods of swiddening. Some groups of Hmong in northern Vietnam have converted large areas of forest to permanent grassland, but other groups employ much more sustainable practices, not unlike those that Menzies and Tapp report of the Hmong that they studied in southwestern China. We should be careful in applying ethnic identifiers to practices that may be used by only some communities belonging to a widespread ethnic group.

What Is the Role of Livestock in Swidden Systems and Fallow Management?

Many of the chapters in this volume make reference to the presence of livestock, but other than Chapter 10 by Peter Horne, on forage management strategies in northern Laos, few provide systematic information about the role of livestock in swidden agroecosystems.[3] It is clear, however, that livestock can exert a strong influence on the manner in which swiddening is practiced and fallows are managed. Cairns, Keitzar, and Yaden suggest in Chapter 30, for example, that the Naga practice of synchronizing the swidden planting cycle is necessary to provide a collective approach to excluding cattle from the crops, thereby reducing the costs. In at least some cases of swidden farming, livestock are far from being an ancillary aspect of the system and are, in fact, one of its core components. In the villages of north central Timor described by Kieft in Chapter 28, livestock provide half of farmers' incomes. There, they grow *Sesbania grandiflora* as a fallow improvement crop as much for the fodder it provides for cattle as for its contribution to restoring soil fertility. As Horne argues, ruminant livestock can play a "substantial role in stabilizing these systems" by providing a ready source of cash that farmers can use to buy rice when local production is insufficient to meet their needs.

Of course, livestock can also have negative impacts on fallow management. In Ban Tat, a Da Bac Tay community I studied in northern Vietnam's mountains, social constraints that prevented farmers from excluding free-ranging livestock from their fallowed fields inhibited efforts to employ improved management techniques (Rambo and Le Trong Cuc 1998). In Chapter 29, Guo, Xia, and Padoch report similar problems of livestock damage to *Alnus* seedlings planted in fallowed swiddens in Yunnan. Under such circumstances, as is noted in Chapter 54, "adequate control of grazing is only possible when an entire village agrees to implement it."

In other communities, construction of fencing to protect fallowed fields represents a major labor cost. As Balbarino, Bates, dela Rosa, and Itumay report in Chapter 18 on short-term fallows in Western Leyte, the Philippines, "the ability of fallow systems to restore soil fertility is hampered by the cultural practice of permitting fallowed farmlands to be used as communal grazing areas." Fencing proved to be too expensive and difficult to maintain and also caused social problems because it conflicted with traditional rights. Farmers resorted to planting *benet* (*Mimosa invisa*) in their fallow fields because it is not eaten by cattle, and its sharp spines, which are elsewhere seen as a negative characteristic, discourage livestock from entering the fields.

Excluding free-grazing livestock from fallows may promote more successful regeneration of these fields, but it may also have deleterious consequences for other components of the agroecosystem. Needing income from the sale of livestock, farmers may resort to grazing them in forest reserves, as happens in Andhra Pradesh and Orissa, in India (Nathan et al., Chapter 54). Alternatively, farmers may reduce the numbers of livestock they keep, thereby lowering supplies of manure needed to intensify production. Such a case is reported in Ta Long's Chapter 53 from northern Vietnam, where San Diu farmers planted litchi orchards on fallow "barren hills" that had previously been used as pastures for grazing cattle.

---

[3] The proceedings of an ACIAR workshop on "Upland Farming Systems in the Lao PDR—Problems and Opportunities for Livestock" contains a number of useful papers dealing with the role of livestock in shifting cultivation systems (ACIAR 1998).

What Are the Benefits and Costs of Adopting Improved Fallow Management Systems and Who Accrues Them?

There are both benefits and costs to adoption of improved fallow management systems. Some of the benefits, and many of the costs, are borne by the farmers, and many chapters in this volume point out that this plays a critical role in the acceptability of such practices. Many of the benefits, and possibly some of the costs, however, accrue to the larger society. Although these are unlikely to influence farmer decision making, they need to be taken into account when formulating policies for upland development.

The most important benefit in the eyes of the farmers seems to be the increase in the productivity of their landholdings. Improved fallows may restore soil fertility, leading to higher yields of annual crops when the fields are cleared again for planting. They may also allow the rotational cycle to be shortened, so that annual crops can be planted more frequently. Furthermore, the species planted in the fallow may, themselves, yield additional products that are valuable to the farmers, such as fruit, fodder, and timber.

Higher yields from annual crops planted on newly cleared fallows, as the result of the soil's fertility having been restored during the fallow, is a seemingly self-evident benefit to farmers from improved management of fallow land. It is not certain, however, that fallowing always results in higher crop yields. Ole Mertz (2002) recently reviewed numerous studies of shifting cultivation and found that empirical evidence for this relationship is weak and often ambiguous. The findings reported in different chapters in this volume are equally ambiguous. Several of them claim that crop yields in the first year of the swidden cycle are significantly higher when improved fallow management practices have been employed. MacDicken, in Chapter 26, on the use of *Leucaena* fallows on Mindoro Island in the Philippines, reports that grain yields were 3.8 t/ha compared to 2.7 t/ha from crops grown on fields cleared from natural fallows. Grist, Menz, and Nelson say in Chapter 32 that profits from maize in a *Gliricidia* fallow system on Mindanao, in the Philippines, were five times higher than those from the *Imperata* fallows previously used. Moreover, maize yields from the *Imperata* system declined over time while they increased in the *Gliricidia* system. On the other hand, Suson, Garrity, and Lasco report in Chapter 33 that farmers employing leguminous hedgerows on Mindanao claimed that soil fertility was not greatly improved by use of this system, and field trials found that the use of prunings as green manure did not always raise crop yields (although other hedgerow trials did result in increased maize production). In the case of some improved fallow management practices, such as the planting of *Sesbania grandiflora* in north central Timor, Indonesia, described by Kieft in Chapter 28, yields of annual crops planted after the fallow were not increased significantly but the duration of the fallow was markedly shortened, from eight years to three or four years, effectively doubling grain production from the same area of land.

A number of the tree-based improved fallow management systems described in this volume generate higher returns, both by increasing the yields of annual crops and by adding harvests of firewood and other products from the trees. However, such systems often involve a considerable delay between the time of planting and the farmer's first direct benefits. A cost-benefit analysis of a *Gmelina* hedgerow system on the island of Mindanao in the Philippines, presented by Magcale-Macandog and Rocamora in Chapter 37, found that it generated negative net income during the first year due to establishment costs and had lower cumulative returns for its first six years than continuous open field maize farming. Only in the seventh year did returns from the improved system begin to exceed those of the conventional system and, by the end of the 14th year when the timber was harvested, returns were double those of the conventional system. Banana and bamboo fallow systems on Mindoro, described by Robotham in Chapter 58, required start-up capital and yielded negative returns in their initial years, but they far outperformed the original system once established. An extreme case of the need for delayed gratification is that of the planting of teak by shifting cultivators in northern Laos, described by Hansen, Sodarak, and Savathvong in Chapter 34. Teak ultimately produces a cash yield of US$13,000/ha, compared to US$1,050 for upland rice if it were planted on the same land, over the same period. But teak cannot be harvested for 20 to 30 years. Few farmers can afford to wait this long, so they sell their newly planted teak plantations to wealthy outsiders for US$700 to US$2,000/ha. As the authors point out, "the rapid expansion of teak in parts of northern Laos could, therefore, result in poor farmers losing much of their best land to richer people."

Some tree-based fallow systems overcome the problem of delayed returns by employing a strategy of successional development, as in the creation of damar agroforests in Sumatra, described by Michon et al. in Chapter 45, and the *Styrax tonkinensis* enriched fallows of northern Laos described by Savathvong et al. in Chapter 46. In another case, described by Paulo Pasicolan in Chapter 63, farmers in Isabela Province of Luzon in the Philippines were encouraged to plant pulpwood trees in their fallows by the provision of start-up capital by a paper company. Now established, the system seems to be self-sustaining because of an assured market for pulpwood. Earlier, the same farmers had immediately abandoned the planting of hedgerows when a government SALT (sloping agricultural land technology) program discontinued the payment of subsidies for planting. A similar fate was encountered by the Thai-German project in northern Thailand, which provided cash incentives to highland farmers to adopt soil conservation measures. As Rattanasorn and Puginier note in Chapter 68: "there was rapid adoption, but this dropped off markedly once the subsidies were discontinued."

Even if improved fallow management does not result in greatly increased annual crop yields, it may still benefit farmers if the species growing in their fallows yield products of value and diversify their sources of income. According to Paul Burgers in Chapter 8, Bidayuh shifting cultivators in Sarawak, Malaysia, who incorporated wild vegetables and fruit, rattan, and bamboo into their fallows not only met some of their own consumption needs, but also gained a steadier income from regular sale of these products in the market. These products also serve as "insurance" in case their annual crops fail. Even the widely despised *Imperata* can be a good source of income to farmers who harvest it from their fallowed fields in Bali and Laos to make thatch, as Lesley Potter and Justin Lee describe in Chapter 11.

Many of the benefits attributed to improved fallow management are of value to the larger society but may not be perceived as being of direct value by the farmers. These include regulation of hydrological cycles, sequestration of carbon, and maintenance of biodiversity.

Stabilization of the hydrological cycle, so as to decrease rainy season flooding and increase dry season stream flow, is often claimed as a benefit of improved fallow management, especially those systems promoting regrowth of forest. Thus, Peter Horne, when discussing forage management strategies in Laos in Chapter 10, reports that villagers noticed that "as the forest started to grow back, the stream flooded less frequently in the wet season and flowed more regularly in the dry season." Maneeratana and Hoare, in their Chapter 13 on northern Thailand, state that, in their analysis, "the assumption is made that regenerating forest has superior watershed characteristics to *Imperata* grassland, which is subject to annual fires." The belief that reforestation provides hydrological benefits is widely accepted and forms the basis for government policies to eliminate shifting cultivation in many Southeast Asian countries (Rambo 1999). But we should not ignore Damrong Tayanin's observation in Chapter 6:

> Contrary to conventional wisdom, the Kammu hold a traditional belief that allowing swiddens to revert back to old forest is dangerous because streams will tend to dry up. If this is true, this would spell calamity and, in the long run, would influence the water levels of larger rivers downstream.

Kammu traditional beliefs are in keeping with the findings of Lawrence S. Hamilton who, after an extensive review of the scientific evidence on the effects of forest on watersheds, concluded that in most cases afforestation "has resulted in lower water tables, less reliable springs, and reduced stream flow, especially in the dry season." (Hamilton 1985, *688*). Hamilton found that well-managed grassland actually appeared to provide greater water yield, with minimal increases in peak flows. He also found little evidence to support the popularly held belief that "large scale forest planting programs have much impact on flooding of lower reaches of rivers" except in those situations where "a major portion of the catchment is degraded, severely gullied, or the soil compacted, such that almost all precipitation is channeled quickly to streams and rivers and sediment from upland erosion is a major contributor to flooding." As he pointed out, such abused watersheds are all too common in the tropics and, in such situations, reforestation "might slow and reduce surface runoff to the point where flooding would be reduced, although not eliminated" (the watershed in northern Laos described by Horne had been severely degraded by wartime bombing and fires) (Hamilton 1985, *688*).

Sequestration of carbon by improved fallows, if it actually occurs, would be a significant benefit to the larger society. However, it is unlikely to be seen as such by farmers.[4] In any case, except for those systems that result in long-term establishment of forests, such as the damar agroforests of Sumatra described by Michon, de Foresta, Kusworo, and Levang in Chapter 45, it does not seem that improved fallow management in systems that involve periodic clearing and burning of swiddens does not seem to achieve long-term sequestration of carbon. Rodel D. Lasco in Chapter 27 on the *Leucaena*-based "Naalad" improved fallow system in Cebu, the Philippines, reports that on average the fallows stored 16 metric tons of carbon per ha, which is far below the 175 to 350 metric tons stored by natural forests in the Philippines. It also appears likely that most of the carbon stored in the *Leucaena* is rather quickly returned to the atmosphere as the trunks decay in the fields, where they are used to construct growing supports for crops.

The contribution that improved fallow systems make to maintenance of biodiversity is highlighted in several chapters in this volume. This would seem to be a major benefit to both farmers and to the larger society, although the extent to which it is valued by the farmers seems to be quite variable.

Wild birds and animals can provide direct benefits to farmers by their dispersal of seeds that stimulate the regrowth of valued trees in fallows. Reed Wadley states in Chapter 41 that one reason the Iban of West Kalimantan preserve some old forests is to provide habitat for the wild animals and birds needed to disperse seed into fallowed swiddens. Farhney, Boonnaphol, Keoboulapha, and Maniphone, in Chapter 40 on management of paper mulberry in swiddens in northern Laos, note that farmers credit birds with playing a major role in spreading the seeds of the mulberry. Decreased populations of wild birds and animals can have negative consequences for swidden farmers. Durno, Deetes, and Rajchaprasit, reporting on Akha regeneration of community forests in northern Thailand in Chapter 12, make the following observation:

> The scarcity of animals and birds, which are important agents of seed dispersal, also threatens the long-term survival of some of the more valuable tree species in the community forest. Currently, only species with wind-dispersed seeds are assured of effective dispersal.

However, wild birds and animals that find a habitat in improved fallow vegetation can also cause losses to the farmers. Reed Wadley, in Chapter 41 on Iban management of complex agroforests in West Kalimantan, Indonesia, notes that forest animals "prey on farm and tree crops, and the recognition of their role in forest

---

4 It is also highly unlikely that farmers will recognize that the release of carbon into the atmosphere by agricultural burning represents a major cost of swiddening to the larger society.

regeneration must be balanced with the reputation of some as crop pests." Malcolm Cairns observes in Chapter 15, on the use of *Austroeupatorium inulaefolium* by the Minangkabau in their fallows in Sumatra, Indonesia, that:

> There is a legitimate concern that fallowed ladang provides an ideal habitat for wild pigs, rats, and insects. During the night, opportunistic wildlife emerges from fallows and feeds on crops in adjacent fields.

Cairns also reports that one Minangkabau farmer claimed that the *Austroeupatorium* served as host for aphids, allowing the insects to spread to neighboring crops.

What Roles Do Social Organization and Cultural Values Play in Fallow Management?

It seems self-evident that social organization and cultural values play major roles in the management of swidden systems. As Nathan et al. observe in Chapter 54: "Management of resources is disparately influenced by politics, legal processes and customary institutions, access to information, and gender definition." Thus, it is hardly surprising that many chapters in this volume refer to the importance of social and cultural variables as factors determining adoption of improved fallow management practices. Unfortunately, however, there is a lack of systematic analyses of how such institutions work. Instead, we are left with what is essentially a collection of anecdotal evidence. Cairns, Keitzar, and Yaden, for example, suggest in Chapter 30 that the custom of village marriage endogamy among the Nagas may constrain the diffusion of the use of alders as a fallow improvement technology to other villages. They also suggest that endemic intervillage warfare in the precolonial period restricted the area available for swidden cultivation, thus serving as a stimulus for intensification and promoting acceptance of interplanting of alders as a means of shortening the fallow cycle. Long Chun-lin reports in Chapter 47 that, among the Lemo of Yunnan, China, planting and tapping of lacquer trees in the fallows is exclusively done by men, so that "girls pay more attention to young men skillful in lacquer cultivation and tapping when considering a suitable marriage partner." Ramakrishnan, in Chapter 21, on the *jhum* system in northeastern India, observes that men from two or three households cooperate in the laborious clearing of swiddens. He concludes that "a well-knit social organization is one of the essential ingredients for such joint efforts, and they help to promote kinship among members of a village, as does the process of allotment of sites for jhum."

Damrong Tayanin, in his first-hand account of Kammu fallow management in Laos (Chapter 6), makes an interesting point about how traditional beliefs can impede adoption of improved fallow management:

> While a family is using a certain area for growing rice and other crops, they are considered its owners and are therefore responsible for it. If something happens in their area—if lightning strikes a tree, for example—that owner is responsible for the situation. They are placed under a taboo until the harvest is finished, and then they have to finish off the old year by driving away the bad luck and the lightning spirit. This is why farmers do not want to retain an area for an extended period, or grow anything after the harvest.

The extent to which local communities are able to control the behavior of their own members and protect their resources against depredations by outsiders are aspects of social organization that can influence the adoption of improved swidden management practices. Da Bac Tay minority farmers I studied in a village in the mountains of northern Vietnam (Rambo and Tran Duc Vien 2001) had eagerly begun to plant ginger in their swiddens because it had a high market price and provided excellent protection to the soil. Although the ginger proved to be highly productive, most farmers soon abandoned planting it. A rapid decline in price due to oversupply was one factor, but crop losses because of theft by other villagers was even more significant. One poor old widow had her entire ginger crop stolen in a single night. Unlike many minority villages, social cohesion in that Da Bac Tay community was extremely low, and some villagers had no compunction against stealing from their neighbors. Outsiders can also threaten local initiatives to intensify fallow production. According to Hoare, Maneeratana, and Songwadhana in Chapter 50, growers of *ma kwaen* (*Zanthoxylum limonella*) in northern Thailand claimed that an outsider middleman had tried to use fire to destroy their trees in an attempt to gain a monopoly over production of this valuable spice.

Of all the social influences affecting the adoption of improved fallow management, land tenure institutions may be the most significant. Many chapters in this volume make the point that allocation of long-term land-use rights to farmers is critical if they are to invest in improving their fallow management practices. According to Paul Burgers' account of the commercialization of fallow species by the Bidayuh in Sarawak, Malaysia, in Chapter 8:

> Village meetings were organized in Pesang, and it was decided that individual ownership of fallow vegetation would provide for a more sustainable and active system of managing fallow or secondary forest. Individual households were permitted to manage their own fallows, and a fine of 25 Malaysian ringits was set for intruders caught harvesting products from other people's fallows.

As Burgers points out, however, such a change to individual ownership is "only feasible if an entire village can be mobilized to revise adat law."

Security of tenure seems to be especially important when farmers are considering shifting to longer-term fallows involving trees. Grist, Menz, and Nelson, in Chapter 32 on adoption of *Gliricidia* in Mindanao, the Philippines, note that "the long term nature of the *Gliricidia* fallow system, requiring four years to be profitable and six years to complete a fallow crop cycle, also requires that land tenure is secure." Adoption of cinnamon-based agroforestry in Kerinci, Sumatra, described in Chapter 65 by Suyanto, Tomich, and Otsuka, in which trees only begin yielding bark after eight years and may continue in production for more than 25 years, is dependent upon giving full property rights to individual farmers, although wet rice lands in the same community continue to be under joint family management.

We should be careful, however, to avoid equating security of tenure with full privatization of land ownership. Assigning individual households legal ownership of plots of formerly communal land can freeze in economic inequalities among households and permanently deprive poor families of natural resources. This is particularly a problem when allocation systems are not transparent and are subject to manipulation by local holders of power. Not surprisingly, some ethnic minority communities in Vietnam's northwestern mountains, like the Black Thai described by Thomas Sikor (2001), have found ways to quietly subvert government-mandated land allocation to households so as to retain communal access to upland areas.

Looking at these examples of the important role played by social institutions and cultural values in fallow management, I can only agree with Brian Belcher's statement in Chapter 64 that, "unless an innovation is socially and economically feasible, its technical merit is only of academic interest." In the future, we need to devote much more effort to understanding the complex interplay between social organization, cultural values, and the ways that swiddens are managed.

How Much Do State Policies Matter?

In many of the chapters, state policies toward swidden agriculture are seen as important influences on the adoption of improved fallow management practices. Some of these influences are positive, but usually they are seen as negative. In some cases, ill-advised policies can have disastrous effects. Such a case is described in Chapter 38 by Sasaki, in which the Benuaq-Dayak of Kalimantan curtailed the planting of rattan in fallows because the Indonesian government attempted to promote value-added local processing by prohibiting exports of unprocessed cane. The failure of the Indonesian state to recognize local ownership of damar agroforests in southern Sumatra and the subsequent allocation of many of these forests to logging concessionaires, discussed in this volume by Michon et al. in Chapter 45, threatened the very survival of a highly sustainable indigenous system. In Laos, according to Savathvong, Fischer, and Pinyopusarerk in Chapter 46, the new policy to allocate land to individual households has caused many to discontinue planting *Styrax tonkinensis* in their fallows since they have not been given sufficient land to allow maintenance of the long fallows required for this tree to reach productive age. On the other hand, as Long Chun-lin points out in Chapter 47 on the Lemo system of lacquer agroforestry in Yunnan, China, government-mandated incorporation of swidden land into production collectives in the 1960s led to a decline in planting of lacquer trees, since they no longer provided any benefit to individual farmers. However, not all government policies have adverse effects. According to Magcale-Macandog and Rocamora in Chapter 37, the Philippine Government's prohibition against logging of natural hardwood trees has caused a scarcity of primary grade wood in the market, raising the price paid for secondary grade wood such as that produced by *Gmelina* trees planted in fallows in the Claveria area of Mindanao.

Discussions of political ecology generally focus on the negative consequences of efforts by colonial governments, or postcolonial national states, to regulate swiddening. However, it is useful to remember that state intervention into local affairs has been occurring for a very long time in many parts of Southeast Asia. As Xu Jianchu reports in Chapter 56 on Hani intensification of swiddening in Xishuangbanna, China, the Hani were subjects of Dai chiefs who lived in the lowlands. All land belonged to these Dai chiefs, who granted the Hani the right to use the land and forest. About 150 years ago, a Dai chief became concerned about the overexploitation of rattan. So, as Xu writes, "drawing on his power as ruler over minorities living in the uplands, [he] designated forest areas with high rattan stocks as a sanctuary...The declaration of the sanctuary forest was made about 150 years ago and it is still respected." The sanctuary serves as a reservoir for rattan seed used to plant fallowed swiddens. Surely, this is one case of a successful "state" intervention in local resource management.

However much state policies can matter in what happens on the ground, it may be prudent to avoid attributing omnipotency to the state. It is only one of many forces involved in determining what farmers actually do, and its policies seem to be subverted by local actors as often as they are implemented. The failure of the Vietnamese state to eradicate shifting cultivation, referred to in Chapter 36 by Tran Duc Vien, is evidence of this. Despite draconian laws and investment of many millions of dollars in "sedentarization" programs (*dinh canh dinh cu*) since the 1960s, there are still millions of households practicing swidden agriculture in Vietnam's uplands (see Hoang Xuan Ty, Chapter 55, and Phan Quoc Sung and Tran Trung Dung, Chapter 61). In one district in the northern mountains where I conducted fieldwork, the farmers, recognizing that the government was hostile to shifting cultivation, began to refer to their swiddens as "mixed gardens" (*vuon tap*) when they reported on their land use to local officials. The latter duly reported to the ministry in Hanoi that shifting cultivation had been eradicated in their district while, in reality, there was no change to the area devoted to swiddens at all. In Chapter 49, Cao and Zhang report a similar disjuncture between official policies to suppress shifting cultivation and the realities of its continued prevalence in Yunnan, China, leading to it being "silently accepted by local officials."

Does the Agroecosystem Contain a Niche for Improved Fallows?

Some chapters in this volume report on fallow management systems that are very successful and have been practiced for many years, even centuries. Others report on systems that are disappearing in the face of changing circumstances, and a number of chapters describe attempts to introduce new systems of fallow management that are rapidly abandoned after a brief trial. This necessarily leads to the question: why do farmers employ improved fallow management in some cases but reject or abandon it in others? An answer may be found by looking at the niche that fallow management occupies, or could potentially occupy, in the total agroecosystem. Does the system have sufficient unused space to permit the adoption of improved fallow management?

In G.E. Hutchinson's classic definition, a niche is "an *n*-dimensional hypervolume" (Hutchinson 1965). The dimensions of the hypervolume are all of the multiplicity of factors that impinge in any way on the survival of the "organism" inhabiting a particular habitat. As originally formulated, the niche was the space occupied by a species in a natural ecosystem. Rerkasem and Rerkasem (1984) extended the niche concept to agroecosystems in their analysis of the coexistence of a multiplicity of different rice varieties within the Chiang Mai Valley agroecosystem. They included not only biophysical factors among their niche dimensions but also market demand, food preferences, place of the variety in the cropping calendar, and labor requirements. They were able to explain why some farmers continued to plant traditional varieties even though improved varieties gave higher yields in some parts of the valley. The niche concept can be further extended to analyze the place of different components within the farmer's total agroecosystem. From this perspective, the fallow can be seen as occupying a specific niche within the swidden system, and the dimensions of this niche will determine whether or not it is possible to employ improved fallow management.

The planting of "jungle rubber" as an improved fallow by Indonesian swidden farmers, described by Penot in Chapter 48, offers an example of how an improved management practice can fit into an existing niche. As he observes, planting of rubber trees in the fallows is acceptable because it achieves the same goals of restoring soil fertility as the natural fallow would within the same period of time, provides an excellent source of cash income and, above all else, does not compete with growing of rice for land or labor. Other systems described in this volume were once successful but are now being abandoned. Examples are the *Styrax tonkinensis* planted fallows used for benzoin production in northern Laos, described by Savathvong, Fischer, and Pinyopusarerk in Chapter 46, and the rattan fallows in Kalimantan described by Sasaki in Chapter 38. Although still ecologically viable, both systems fell victim to changes in the market that undercut their profitability. Other successful systems, such as the *Alnus*-based fallows in Yunnan described by Guo, Xia, and Padock in Chapter 29, are fading because of increasing scarcity of land. Unfavorable government policies on land use can also shrink the niche available for improved fallow management. A case in point is described by Burgers in Chapter 8 on the Bidayuh of Sarawak, where Malaysian government policy seeks to convert shifting cultivation into market-oriented systems based on monocultural production of cash crops. In other such cases, attempts to transform shifting cultivation into permanent cultivation have been abandoned, as in the case in the Philippines, where continuously cropped fields planted with *Senna spectabilis* hedgerows reverted to *Gliricidia sepium*–dominated fallows. In their description in Chapter 33, Suson, Garrity, and Lasco say the change appears to have occurred because "a complex interplay between land, labor, and soil fertility" generated a suitable niche for fallow management.

The agroecological niche concept can be usefully employed to identify swidden systems that can potentially be intensified by incorporating improved fallow management practices. Similarly, it can be used to explain why some improved fallow management systems promoted by government extension agents and nongovernmental organizations' (NGOs) workers have been rejected by farmers. For example, in their Chapter 17 discussion of ways to improve the use of *Tecoma stans* as a fallow improvement plant in East Nusa Tenggara, Djogo, Juhan, Aoetpah, and McCallie rule out its potential use as green manure because the limited labor supply of farm households is already overcommitted during the season when this operation would have to be performed to be effective.

Are We Employing Appropriate Research Methodologies to Study Fallow Management?

There is an American saying that "patriotism is the last refuge of the scoundrel." This refers to the well-known tendency of politicians who have no useful ideas to contribute to the public policy debate to instead invoke national patriotic symbols. By analogy we might say that "discussion of methodology is the last refuge of the scientist with no significant research findings to present." I hope that this is not actually true. But, certainly, discussions of methodology are often arid and dull. Nevertheless, the methodologies we use have such an important influence on what we learn from our research that I don't think this is an issue we should ignore. As Dennis Garrity asserts in his introductory remarks in Chapter 1: "Special care must be taken to develop research methods specifically for the unique conditions of shifting cultivation." Unfortunately, rather than following Garrity's sound advice, all too many researchers have elected to rely on rapid rural appraisal (RRA) and, especially, participatory rural appraisal (PRA) as their methods of choice. Of course, these methods have their place in the swidden researcher's toolkit, but they are not magical means by which to obtain reliable information quickly and cheaply.

Not long ago, in Hanoi, I briefed a group of visiting foreign consultants on a project that the Center for Natural Resources and Environmental Studies and the East-West Center had done for the Swedish International Development Cooperation Agency (SIDA) on development trends in Vietnam's northern mountain region (Le Trong Cuc and Rambo 2001). We had made detailed investigations of five upland communities in order to establish a firm baseline against which to measure changes in the future. We employed a suite of methods, including establishment of ecological transects, vegetation mapping using satellite images, interviewing of randomly selected households using standardized survey questionnaires, and semistructured interviews with community leaders. At the end of my presentation, one of the consultants asked, in quite an aggressive way, why we hadn't used PRA. It was clear from the way she asked her question that she saw us as being somehow morally deficient because we had not involved local people in our research by using participatory rural appraisal as our main method. I responded that we were doing an in-depth scientific study, not working with people to design a community development project. But my explanation really didn't satisfy her. In her eyes, PRA was the only legitimate method to employ in rural research.

This experience got me to thinking about the extent to which rapid appraisal methods had come to dominate upland research in Southeast Asia. The PRA method, in particular, has caught the fancy of NGOs and has been warmly embraced by the World Bank, SIDA, and other development assistance agencies. I don't know of a single development project in the mountains of Vietnam in the past five years that has been launched without first commissioning a PRA.[5] Dozens, if not hundreds, of such studies have been completed, although few have been published so as to be available for public scrutiny. I think this is an unhealthy trend, not because there is anything intrinsically wrong with RRA or PRA, but because excessive reliance on rapid appraisals may be displacing other types of research. Since the methods we use in large part determine the kinds of questions we can answer, I think there is a great risk in allowing RRA or PRA to become the sole method of choice.[6]

---

[5] Some villagers in Vietnam's northern mountains refer to these PRAs as "the four big things:" Big paper and big pens (used to draw the maps and transects), big cars (referring to the researchers' Land Cruisers), and big projects (T. Vien 2002).

[6] In the interests of transparency, I note that I am not an innocent bystander in this matter. I was an early advocate of RRA, incorporating this method in a series of training workshops that I helped organize in the 1980s for SUAN (Southeast Asian Universities Agroecosystem Network). Working

Rapid rural appraisal, as a systematized methodology, was developed in the 1980s by a group of scholars associated with farming and rural systems research projects at Khon Kaen University in Thailand. Terry and Somluckrat Grandstaff provided the initial leadership in this effort, but it was very much a collective enterprise (Khon Kaen University 1987). Researchers at KKU produced a remarkable set of studies using RRA methods (Lovelace et al 1988). These offered new ways of seeing problems of development in the rainfed farming communities of northeastern Thailand and demonstrated the great power of rapid appraisal as a method.[7] But we should look at these studies and try to see why they were so successful. One important and generally unrecognized factor was the quality of the researchers involved. Most held advanced degrees and were recognized experts in their fields. They had long experience in working in northeast Thailand and already had a very good understanding of the local situation. That meant that they could quickly recognize the significance of new information when they encountered it. Most of them also spoke the local language and could communicate easily with the farmers they were interviewing. In their hands, RRA proved to be a productive way to collect information. It was especially fruitful because it created an institutional framework within which specialists from different disciplines could work closely together. Because RRA was perceived as a new method, involvement in the project generated an unusual degree of enthusiasm and commitment on the part of participating researchers. But I suspect that these researchers would have produced useful knowledge using virtually any method that brought them into close contact with the farmers and with each other.

Participatory rural appraisal (PRA) came onto the scene somewhat later,[8] Robert Chambers and Gordon Conway were its early promoters, but NGOs and international development assistance agencies have been its main users and advocates. PRA is seen as a way of empowering poor villagers, "giving them a voice," to use the fashionable jargon. In Conway's words:

> In some ways it has been a revolution: a set of methodologies, an attitude and way of working which has finally challenged the traditional top-down process that has characterized so much development work. Participants from NGOs, government agencies, and the research centers rapidly find themselves, usually unexpectedly, listening as much as talking, experiencing close to first hand the conditions of life in poor households and changing their perceptions about the kinds of intervention and research that are required (Conway 1997, *199*).

As an anthropologist, I must confess to feeling somewhat offended by Conway's claims. Learning from the people through "participation-observation" has been the method of choice in my discipline for almost a century. And we experience the "conditions of life in poor households" by actually living in their houses, often for very extended periods. We don't show up each morning in our air-conditioned Land Cruiser, spend a day listening to people talking in a very artificial situation, and then delude ourselves into thinking that we have had "close to first hand" experience of their lives.

As developed at KKU, RRA was a methodology designed to bring together already knowledgeable specialists to work together in interdisciplinary teams to understand constraints on agricultural development in rural settings. The method was intended

---

together with the late Dr. Terd Charoenwatana I helped organize RRA in Luang Prabang and Savanakhet Provinces in Laos (Gillogly et al 1990; SUAN Secretariat 1992) and with Dr. Le Trong Cuc I organized one of the first applications of this approach in Vietnam's uplands (Le Trong Cuc, Gillogly, and Rambo 1990).

7 See Grandstaff and Messerschmidt (1995) for a useful discussion of RRA methods.

8 PRA is often described as being an outgrowth of RRA. Although PRA employs some of the diagnostic tools that were originally developed for use in RRA, it generally lacks the scientific rigor of the RRA approach, at least as the latter methodology was employed by researchers at Khon Kaen University.

as a way to quickly identify problems for further in-depth research by disciplinary specialists. Thus, doing an RRA was seen as only the first step in a long process. It was never seen as an end in itself, or as a way of providing definitive answers to complex questions. But rapid appraisal has evolved into something very different from what it was when it was first used at KKU. It has become a substitute for conducting long-term, in-depth research. In part this is because the donor agencies need to create the illusion that they have done research to justify their development assistance projects.[9] They love rapid appraisal because it offers a cheap and relatively easy way to create this illusion. Why fund an ethnologist to live in a village for a year when you can claim to understand everything important by sponsoring a three-or-four-day rapid appraisal exercise? It costs a lot less, is ready for publication much sooner, and if its findings are presented in a properly glossy booklet with lots of system diagrams and color photographs, it looks just as impressive. PRA has even greater appeal to the international development assistance agencies because of its populist packaging. How can anyone attack development plans that claim to be designed on the basis of inputs from the local population? A report published by the World Bank in Vietnam, titled "Vietnam: Voices of the Poor," presents a synthesis of several community level "Participatory Poverty Assessments." It claims, rather arrogantly:

> The four PPAs have been accepted by local communities and authorities as sound representations of the reality of poor people's lives. This is an important endorsement. If the people who contributed to the study and the people who have lived in these areas all their lives believe that the studies accurately capture the problems and priorities of the poor, then why should critics living elsewhere remain skeptical? (World Bank and DFID 1999, *4*).

To challenge that claim is to brand oneself as an elitist bent on denying a voice to the poor! But the question of whether or not rapid appraisals accurately capture the complex nature of the development situation in the uplands still needs to be asked. There is not a simple "yes" or "no" answer. Rapid appraisal is useful for generating certain kinds of information but is not suitable for eliciting other kinds of data. Participatory methods can be useful in eliciting valuable new insights from local people about their conditions and needs, but poorly done, they may conceal more than they reveal, especially about power relations and conflicts within a community.[10] Let me stress, however, that both these virtues and these faults are not intrinsic to rapid appraisal methods in principle. Rather, it is the way that these methods are employed, and the questions they are used to answer, that determine their effectiveness. There are some real limitations to the kinds of information that can be generated using rapid appraisal methods, and its accuracy. This applies especially to PRA, as it is commonly done.

RRA and PRA have two major weaknesses. The first is that local knowledge, even if correctly assessed, may be incomplete or even incorrect. For example, Cao Guangxia and Zhang Lianmin report in Chapter 49 that farmers in Yunnan, China, told them that growing rubber involved less labor than cultivating swiddens. However, their studies of labor use showed no difference between the two systems,

---

[9] RRA is also understandably popular among metropolitan development consultants (usually, but not always, Westerners), who are able to "parachute" into the field for a few weeks at most and emerge with what appear to be exciting new understandings of complex rural problems. All too often, those findings have not actually emerged from the RRA exercise. Instead they are obtained by the consultant's skillful extraction of knowledge from local colleagues, who have invested their careers in developing an understanding of the area in which they are working.

[10] In reading many PRA reports I have been struck by how similar they all seem. They all deploy the standard array of transects, maps, and cropping calendars. The local people all sound like they have been to a special school run by Samuel Popkin, to train them to speak as "rational peasants." The descriptions of the communities and their resource management practices are much too neat and orderly to reflect reality. There are no evident contradictions or unanswered questions. And almost never is there any mention of conflicts of interest among members of the community being studied. To the extent that there is any mention of conflict, it is invariably between the community and outside institutions, especially state forestry agencies.

and they ask if these findings suggest "deliberate ignorance" on the part of the farmers. Of course, if rapid assessment generates an accurate picture of how upland people perceive their situation, even if those perceptions are imperfect, it will still be valuable. It is questionable, however, whether most PRAs, as they are actually carried out, are very successful in enlisting the true participation of local people that effectively taps their knowledge and views. Indeed, in some situations, it is very much in the farmers' self-interest to conceal the truth from researchers. Thus, Hoare, Maneeratana, and Songwadhana, in Chapter 50 on planting ma kwaen spice in fallowed swiddens in northern Thailand, report that farmers gave them conflicting information about methods of propagating and maintaining the trees because "existing producers were afraid that their market price would decline if many additional villages started planting the jungle spice."

The second major weakness of participatory methods is that they are an ineffective tool for understanding social organization, particularly for identifying contradictions and conflicts within upland communities. The method itself, in which "local people" are interviewed in a group situation, virtually guarantees that contradictions will be concealed and conflicts hidden beneath the rhetoric of community solidarity. One must ask the questions: how participatory are PRAs? Who participates in PRAs? And under what social constraints do people participate in PRAs? All too often, community participation is organized by the local power elite who stack the sessions with their clients and dependents.[11] Just how open and honest are ordinary villagers likely to be when answering questions about inequities in the land allocation system, when the village head is sitting sipping tea just across the table from them?

In singling out RRA and PRA for criticism, I am not suggesting that these methods should be abandoned entirely; only that they, like all methods, should be employed selectively and without the illusion that they offer a cheap and easy substitute for more intensive methods of data collection. There are no "magic bullets" in research, any more than there are "magic species" for fallow improvement, a topic to which I will now turn.

Why Are We Still Searching for "Magic Bullets" to Improve Fallow Management?

In the mid-1980s, when I was a researcher at the East-West Center in Honolulu, I received an urgent request from an agency in Jakarta asking me to host some Indonesian entomologists who were charged with finding a way to counter the psyllid outbreak that was devastating the *Leucaena* trees in their country. Researchers at the University of Hawaii had promoted *Leucaena* as the "miracle tree" that would solve the problems of poor farmers in the tropics. (Ironically, the lead researcher on that effort had earlier published a paper in which he hypothesized that the collapse of the Maya civilization in central America had been caused by a blight that destroyed their staple crop of corn.) The species had been eagerly embraced by authoritarian leaders like Suharto in Indonesia and Marcos in the Philippines, who actively encouraged its adoption by farmers. Unfortunately, a species of psyllid had found its way from tropical America to Southeast Asia and, free of its natural predators, was enjoying a population explosion with devastating effects on the *Leucaena*. According to Colin Piggin in Chapter 24, productivity of *Leucaena* in Nusa Tenggara Timur quickly fell by 25% to 50%, and sales of livestock, which relied on the trees for fodder, declined by 11%. The vulnerabilities of monocultural plantings of *Leucaena* were being revealed in a way that was both costly to the farmers and politically embarrassing to the dictators. The entomologists arrived in Honolulu and immediately began netting natural insect predators of the psyllids in Hawaii's already

[11] In my first experience with using RRA in a village in Vietnam, my team found itself interviewing a woman who was trembling with malaria chills. It turned out that she was the wife of the hamlet head who had sent us to her so we would not meet with politically less-reliable informants.

badly infested *koa haole* stands and air-shipping them back to Indonesia, where they were immediately released into the wild. I don't know if the predators actually did any good, although, according to Piggin, the productivity of *Leucaena* has gradually recovered as psyllid numbers declined. But I have never forgotten this lesson about the risks of relying on magic bullets in efforts to develop agriculture.

I wonder, however, if we have really learned this lesson. Many of the chapters in this volume are focused on the use of just one species (*Leucaena, Sesbania, Alnus,* etc.) to improve fallow management. Some authors do insert words of caution about excessive reliance on only a single species. Hansen, Sodarak, and Savathvong, for example, in Chapter 34 on teak planting by shifting cultivators in northern Laos, point to the danger that "teak monocropping may lead to a devastating buildup of pests, especially bee-hole borers and caterpillars."

Widespread planting of only a single tree species can also have unanticipated consequences for human health. After World War Two, the Japanese Ministry of Forestry launched an ambitious effort to reforest the country's barren mountains with Japanese cedar (*sugi*), covering 7 million ha, or 12% of Japan's land area. The single-aged stands have now reached sexual maturity and release enormous quantities of pollen every spring, causing untold misery to hay fever victims, of whom there are more than two million in Tokyo alone. The economic costs in lost productivity are in excess of 10 billion yen every year; the human costs, incalculable (*Daily Yomiuri*, 26 March 2001, *9*; *Japan Times*, 25 October 2001, *18*). There are also reports that mass planting of teak in northern Thailand has produced somewhat similar allergy problems, although on a much smaller scale than in Japan (Vityakon 2002.

Many farmers are, themselves, reluctant to adopt monocultures. Menzies and Tapp report in Chapter 35 that upland farmers in southwestern China are concerned about planting *Cunninghamia*. They say that, in times past, the Wu lineage, who engaged in monocultural planting of *shamu* (*Cunninghamia*), encountered soil problems. These problems were explained by a member of another lineage, the She, in the following fashion:

> The Wu people have that problem [the need to rotate shamu off the growing site] because they do not understand the soil. The soil is like a person. Turnips are good to eat and the Wu like to eat a lot of turnips. But if they eat only turnips, then very soon they become sick. Pork is also good to eat, but if we eat only pork, then very soon we also become sick. If we eat turnips and pork and rice and vegetables together we do not become sick, our health is good, and we are happy. The Wu soil is not happy. The Wu give it only shamu.

Eric Penot, in Chapter 48 on the development of jungle rubber in Indonesia, describes how the farmers, acting on their own initiative, diversified their rubber plots by selectively retaining wild species that emerged spontaneously or by deliberately planting desirable trees amongst the rubber, even though "the practice was either forbidden or recognized as undesirable by rubber research and extension services." He reports that, today, the official agencies have finally begun to experiment with such diversified plantings.

If the farmers can understand the dangers of relying on only a single fallow management species, why can't we? I know that there are a number of good reasons for this single-species research focus. For one thing, existing agronomic research methods are almost impossible to apply to multi-species communities. Nevertheless, one must wonder if we are yet on the right track.

Conclusion: Does Improved Fallow Management Have a Future?

Some of the fallow management practices described in this volume, such as the planting of *Alnus* in Yunnan and Nagaland, have histories of many centuries. In the case of the Nagas, endemic warfare caused the concentration of population in defensible sites and this threatened to exceed the production capacity of local swiddens long before the modern population explosion began to have the same effect. Most of the practices, however, represent recent adaptations brought about by the need to intensify swidden productivity. Most of these are indigenous in origin. The causes of their adoption are diverse and include growing population, alienation of land to commercial plantations and forest reserves, and demands to grow more crops to sell for cash. Population growth seems to be an almost universal cause. Although not all places have rates of growth as high as the 4.6% per annum reported in Chapter 30 by Cairns, Keitzar, and Yaden for Nagaland, annual increases of 2% or more are noted by many of the authors for their study sites. Rattanasorn and Puginier in Chapter 68, for example, say that the hilltribe population in northern Thailand nearly quadrupled in the past 30 years, rising from 217,000 in 1960 to 800,000 in 1996. In the Kerinci District of Sumatra, according to Suyanto, Tomich, and Otsuka in Chapter 65, total population increased from 50,000 in 1930 to 286,000 in 1994. Population densities increased from 12 to 68 persons per km^2 in the same period. This increasing density has immense implications for the sustainability of upland farming systems.

In some cases this is an entirely endogenous increase while, in others, it reflects in-migration of outsiders. Alienation of land for commercial plantations, and parks and protected areas that reduce the land available to local communities, lead to the same outcome as population growth: more people must be supported by less land. There is no reason to think that population growth will slow down in most upland communities anytime soon. Given the very large number of people under 15 years of age, demographic momentum alone will ensure that populations double within the next 20 years, no matter what control measures may be introduced. Reduction of population pressure through out-migration of farmers to wage-labor jobs in the cities, such as described in northern Thailand by Maneeratana and Hoare in Chapter 13, is unlikely to be a solution in most other countries.[12] This means that upland farming systems must be able to meet the needs of twice as many people within the next 20 years. The question must be asked: Do any of the improved fallow management systems described in this volume have the potential to achieve that?

Clearly, stabilization of existing swidden systems will not be sufficient, however desirable this goal may be in the short term. For the living standards of shifting cultivators to be maintained at even their current low levels, production must increase by at least 2% per year, every year, for the next 20 years. But this will merely maintain the status quo, leaving many upland farming families suffering poverty and hunger. If their conditions are to be improved significantly, much higher rates of productivity growth must be achieved. The most difficult question we face is how to achieve such growth while maintaining or, in most upland areas of Southeast Asia, restoring sustainability to agricultural systems.

One critical problem that must be addressed is that of food security. At present, few upland systems can produce sufficient grain to meet local needs. Although some improved fallow management systems have shown the ability to boost grain yields to some extent, most systems, especially those involving long rotations with trees, result in lower net outputs of rice and corn. Even those systems that produce higher yields do not have the potential to double productivity in the longer term. So the only alternative seems to be to focus on incorporating income-generating components into upland systems that will provide farmers with sufficient cash to buy grain grown in the lowlands. A number of chapters in this volume illustrate the fact that upland farmers in many places are already following this strategy. For example, Nathan et al. reports in Chapter 54 that farmers in parts of India take advantage of the

[12] It is sometimes suggested that the solution to the problems of excess population in the uplands is out-migration. This is unlikely to be a major factor in most of the mountains of Southeast Asia. There is no way, for example, that Vietnam's already densely populated lowlands and cities can absorb significant numbers of migrants from the mountains. In any case, most people from the uplands lack the education and social connections needed to find employment elsewhere (Jamieson, Le Trong Cuc, and Rambo 1998, *10*).

comparative advantage of the uplands for growing pulses and oilseeds and use their profits to "buy in, rather than produce, their own food grains."

Concentrating on growing high-value cash crops, however, brings high risks as well as potential benefits for upland farmers. There are a limited number of suitable crops and their prices are often highly unstable. Markets have a limited capacity to absorb increased production, and oversupply can lead to steep declines in farm gate prices. For example, the recent precipitous decline in world coffee prices, which was largely triggered by Vietnam's sudden emergence as a major exporter, has had a devastating impact on smallholder producers in Vietnam's central highlands. Farmers raising annual or short-cycle perennial crops can readily switch to other species offering higher returns, provided suitable substitutes are available. But those who plant slow-maturing trees lack such flexibility, a point clearly made in Chapter 67 by Silvia Werner in her discussion of the relative merits of rubber and cassiavera in Sumatra. What is going to happen to the rubber plantations of Yunnan, for example, once China opens its market to lower-cost imported rubber from Thailand and Indonesia, as it may be required to do as a member of the World Trade Organization? The self-evident solution is to diversify planting so as not to put all one's eggs into the same basket. But often, only one, or maybe a few, tree species can readily fit into an existing niche.

There is another problematic aspect to the food security quandary. Upland farmers are reluctant to become totally dependent on imported food because they don't fully trust the market system to always deliver the supplies they need to survive. Many older Southeast Asians still have personal memories of the severe scarcity of food during World War Two, when grain became unavailable for any price and they discovered that they couldn't eat money. In northern Vietnam, for example, floods destroyed most of the rice crop in the Red River Delta in the fall of 1944. Wartime inflation had already made rice unaffordable to most poor people, with the price increasing from 30 piastres per quintal in 1940 to 600 piastres in 1945. Allied bombing of shipping from the Mekong Delta, where there was a surplus, prevented resupply of northern markets, so that even those with cash couldn't find food. The result was a massive famine that claimed between 1.5 million and 2 million victims from a total population of 9.9 million—about 2 people in every 10. (Nguyen The Anh 1998). Not surprisingly, the Vietnamese—government agricultural planners and farmers alike—are unwilling to abandon their pursuit of local food self-sufficiency and put their trust in the market system. It is easy for neoclassical economists to criticize this attitude as irrational, but if we place as much faith in the value of indigenous knowledge as we claim to, then maybe we should have more respect for this hard-earned wisdom of the farmers when thinking about how to improve fallow management systems.

Having discussed some of the constraints that must be confronted by efforts to develop more productive and sustainable swidden systems, I should end this section by trying to answer the question with which I began it: Does improved fallow management have a future? The answer is quite simple: yes, of course it does. It has a future because there are no alternatives to this strategy for the many tens of millions of resource-poor farmers living in the vast upland areas of Asia and the Pacific. Existing systems of swidden cultivation are breaking down under excessive pressures of intensification (what Cairns refers to in Chapter 3 as "the swidden degradation syndrome"), little land is suitable for expanding conventional systems of permanent cultivation, there is no place for increasing populations to move, and there are no miracle technologies on the horizon that are going to transform the situation. So the only option open to us, farmers and researchers alike, is incremental improvement of what we have now. Improved fallow management is not going to solve the long-term problems created by the imbalance of people and resources in the uplands. But, in the short term, it can make a very big contribution to making things better, or, at least, keeping things from getting worse. This is not an unworthy objective, because the long-term future doesn't really matter if we don't survive in the short term. We can only hope that by the time the future becomes the present, upland farmers and fallow management researchers, through hard work and innovative thinking, will have found new solutions to some of these problems.

References

ACIAR. 1998. *Upland Farming Systems in the Lao PDR—Problems and Opportunities for Livestock.* ACIAR Proceedings No. 87. Canberra: Australian Centre for International Agricultural Research.

Conklin, H.C. 1957. *Hanunóo Agriculture in the Philippines.* FAO Forestry Development Paper # 12. Rome: Food and Agriculture Organization of the United Nations.

Conway, G. 1997. *The Doubly Green Revolution: Food for All in the Twenty-first Century.* London: Penguin Books.

Gillogly, K., T. Charoenwatana, K. Fahrney, O. Panya, S. Namwongs, T. Rambo, K. Rerkasem, and S. Smutkupt (eds.). 1990. *Two Upland Agroecosystems in Luang Prabang Province, Lao PDR: A Preliminary Analysis.* A Report on the Southeast Asian Universities Agroecosystem Network (SUAN)-Lao Seminar on Rural Resources Analysis, Vientiane and Luang Prabang, Lao People's Democratic Republic, December 4–14, 1989. Khon Kaen: Khon Kaen Press.

Grandstaff, T. B., and D. A. Messerschmidt. 1995. *A Manager's Guide to the Use of Rapid Rural Appraisal.* Bangkok: UNDP.

Hamilton, L.S. 1985. Overcoming Myths about Soil and Water Impacts of Tropical Forest Land Uses. In: *Soil Erosion and Conservation,* edited by S.A. El-Swaify et al. Soil Conservation Society of America.

Hutchinson, G.E. 1965. *The Ecological Theater and the Evolutionary Play.* New Haven, CT: Yale University Press.

Jamieson, N. Le Trong Cuc, and A.T. Rambo. 1998. *The Development Crisis in Vietnam's Mountains.* East-West Center Special Report No. 6. Honolulu: East-West Center.

Khon Kaen University. 1987 *Proceedings of the 1985 International Conference on Rapid Rural Appraisal.* Khon Kaen: Khon Kaen University Rural Systems Research and Farming Systems Research Projects.

Le Trong Cuc and A. T. Rambo (eds.). 2001. *Bright Peaks, Dark Valleys: A Comparative Analysis of Environmental and Social Conditions and Development Trends in Five Communities in Vietnam's Northern Mountain Region.* Hanoi: National Political Publishing House.

———, K. Gillogly, and A. T. Rambo (eds.). 1990. *Agroecosystems of the Midlands of Northern Vietnam: A Report on a Preliminary Human Ecology Field Study of Three Districts in Vinh Phu Province.* EAPI Occasional Paper No. 12. Honolulu: East-West Center.

Lovelace, G.W., Sukaesinee Subhadira, and Suchint Simaraks (eds.) 1988. *Rapid Appraisal in Northeast Thailand: Case Studies.* Khon Kaen: Khon Kaen University Rural Systems Research Project.

Mertz, O. 2002. The Relationship between Fallow Length and Crop Yields in Shifting Cultivation: A Rethinking. *Agroforestry Systems* 55 (2), 149–159.

Nguyen The Anh. 1998. Japanese Food Policies and the 1945 Great Famine in Indochina. In: *Food Supplies and the Japanese Occupation in South-East Asia,* edited by P.H. Kratoska. New York: St. Martin's Press, 208–226.

Rambo, A.T. 1982. Orang Asli Adaptive Strategies: Implications for Malaysian Natural Resource Development Planning. In: *Too Rapid Rural Development: Perceptions and Perspectives from Southeast Asia,* edited by C. MacAndrews and Chia Lin Sien. Athens, OH: Ohio University Press, 251–299.

———. 1996. The Composite Swiddening Agroecosystem of the Tay Ethnic Minority of the Northwestern Mountains of Vietnam. In: *Montane Mainland Southeast Asia in Transition,* edited by B. Rerkasem. Chiang Mai: Chiang Mai University, 69–89.

———. 1999. The Balance of Nature, the Garden of Eden, and the Power of Policy: Some Observations on Contemporary Environmental Mythology. *The Asian Geographer* 18(1 & 2), 33–46.

———, and Le Trong Cuc 1998. Some Observations on the Role of Livestock in Composite Swidden Systems in Northern Vietnam. In: *Upland Farming Systems in the Lao PDR—Problems and Opportunities for Livestock.* ACIAR Proceedings No. 87. Canberra: Australian Centre for International Agricultural Research, 71–78.

———, and Tran Duc Vien. 2001. Social Organization and the Management of Natural Resources: A Case Study of Tat Hamlet, a Da Bac Tay Ethnic Minority Settlement in Vietnam's Northwestern Mountains. *Southeast Asian Studies (Kyoto)* 39 (3), 299–324.

Rerkasem, B., and K. Rerkasem. 1984. The Agroecological Niche and Farmer Selection of Rice Varieties in the Chiang Mai Valley, Thailand. In: *An Introduction to Human Ecology Research on Agricultural Systems in Southeast Asia,* edited by A.T. Rambo and P. E. Sajise. Los Baños, Laguna: University of the Philippines at Los Baños, 303–311.

Sikor, T. 2001. Agrarian Differentiation in Post-Socialist Societies: Evidence from Three Upland Villages in Northwestern Vietnam. *Development and Change* 32 (5), 923–949.

SUAN Secretariat. 1992. *Swidden Agroecosystems in Sepone District, Savannakhet Province, Lao PDR.* Report of the 1991 SUAN (Southeast Asian Universities Agroecosystem Network)-EAPI-MAF Field Research Workshop. Khon Kaen, Thailand: SUAN Secretariat.

Viên, T. D. 2002. Personal communication between Dr. Tran Duc Vien, Director, Center for Agricultural Research and Environmental Studies, Hanoi Agricultural University, and the author. October 19.

Vityakon, P. 2002. Personal communication between Dr. Patma Vityakon, Associate Professor, Faculty of Agriculture, Khon Kaen University, and the author. October 1.

World Bank and DFID. 1999. *Vietnam: Voices of the Poor*. Hanoi: World Bank.

Botanical Index

Ethnic Group Index

Subject Index

* 9 7 8 1 8 9 1 8 5 3 9 2 0 *
T - #0385 - 101024 - C32 - 254/178/47 - PB - 9781891853920 - Gloss Lamination